W9-AHN-560

MAURITS DEKKER PRESENTS

highlights of organic chemistry

Studies in Organic Chemistry

Executive Editor

Paul G. Gassman

Department of Chemistry
University of Minnesota
Minneapolis, Minnesota

Volume 1 Organoboranes in Organic Synthesis
Gordon M. L. Cragg

Volume 2 Molecular Orbital Methods in Organic Chemistry HMO and PMO
An Introduction *William B. Smith*

Volume 3 Highlights of Organic Chemistry
William J. le Noble

Other Volumes in Preparation

Highlights of Organic Chemistry

an advanced textbook

William J. le Noble
Department of Chemistry
State University of New York
Stony Brook, New York

MARCEL DEKKER, INC. New York

Copyright © 1974 by MARCEL DEKKER, INC.

ALL RIGHTS RESERVED

Neither this book nor any part may be reproduced or transmitted in any form or by any means, electronic or mechanical, including photocopying, microfilming, and recording, or by any information storage and retrieval system, without permission in writing from the publisher.

MARCEL DEKKER, INC.
270 Madison Avenue, New York, New York 10016

LIBRARY OF CONGRESS CATALOG CARD NUMBER: 74-82953
ISBN: 0-8247-6210-X
Current printing last digit:
10 9 8 7 6 5 4 3 2

PRINTED IN THE UNITED STATES OF AMERICA

To Yu-Sen

EDITOR'S FOREWORD

The concept behind the Studies in Organic Chemistry series was to provide, for the chemical community, a group of books which could be used at the advanced undergraduate level and at the graduate level as texts in special topics courses. At the same time, I felt that it was desirable that the members of this series be prepared in a thorough and rigorous fashion, so that they would also prove to be of high value to the practicing organic chemist who might require an authoritative treatise on a given subject. Although I did not originally envision the inclusion of a general advanced organic chemistry text in this series, I am delighted to be able to do so with this publication of Professor le Noble's Highlights of Organic Chemistry.

Highlights of Organic Chemistry provides an advanced conceptual treatment of the basic principles of organic chemistry which is both classically based and impressively modern. The approach is unlike that of any advanced text known to me. The coverage is far broader than that found in many of the "Physical-Organic" texts commonly associated with first-year graduate level courses in organic chemistry. At the same time, basic concepts of modern organic chemistry are dealt with in far more detail and are treated far more rigorously by Professor le Noble's text than by most other "Advanced Organic" texts. In short, I feel that Bill le Noble's Highlights of Organic Chemistry provides a unique work, which should find wide use among advanced students of organic chemistry.

Paul G. Gassman
Columbus, Ohio

PREFACE

This book has grown out of my lecture notes collected over the past ten years for two courses taught in the Chemistry Department at Stony Brook. One of the courses, entitled "Structural Organic Chemistry", is required of those new graduate students interested in Organic Chemistry who haven't had the equivalent elsewhere, but it is also taken by graduate students in the other chemical disciplines and by some of the more ambitious undergraduate students. The other course, called "Reactive Intermediates", is a Special Topics course taught perhaps every fourth year or so; its clientele consists primarily of students and post-doctoral fellows interested in the mechanisms of organic reactions. As these notes grew more voluminous and the reprints on which they were based ran to several thousand, the temptation arose to abstract, combine and edit all this material into some orderly form, and when I was granted a sabbatical leave in 1972-73, I yielded to it.

Most of the writing was done during my stay in Leiden. Many of my colleagues in the United States who have visited that famous, old University commented favorably on the intellectual atmosphere there, and since I speak the native language I thought it would be a good place to do this work. Professor E. Havinga kindly consented to be my host.

Some features of this writing are perhaps somewhat unusual. Thus, I have included a chapter on nomenclature and literature. It has been my experience that only the most elementary nomenclature is introduced in undergraduate work (perhaps appropriately), and that the topic is thereafter neglected to the point that many students are hampered by this hiatus in their knowledge even while screening the titles of papers in organic journals. Consequently good papers that might have aroused their interest are passed over like so much Greek. Similarly, I have been surprised to find on numerous occasions how many students are lost in the library if they are unable to locate a piece of information in Chemical Abstracts. For these reasons I decided to include some remarks on both of these topics.

Organic chemistry has had its share of controversies, and our students should know that even the best minds at work in their field do not always agree on the interpretation of all experiments. Since many of the topics presented are still in a state of active development, not every discussion leads to a clear-cut and inescapable conclusion. By the same token, there are numerous instances in this book in which I have quoted work that was shown to be wrong in later publications. These examples include errors in judgment, errors in experimental work, errors in interpretation, errors sometimes due to haste to get a tantalizing new development in print, errors due to prejudice, and so on. To err is human, it is said, and organic chemistry is a human enterprise. I feel that an element of dullness is introduced if these errors are swept under the rug and only the finished package is presented. Those who come after us should have a chance to learn from our mistakes, and this requires their exposure. I may not have endeared myself this way with the chemists whose misfortune it was to be so human; may it be of solace to them that the author has made his share of missteps, and that perhaps this writing will come to stand in testimony to that fact.

In view of the nature of this book, with an unusually full description of some topics which have attracted much attention in recent years, and perhaps neglect of others, the usual titles such as "Organic Chemistry", "Advanced Organic Chemistry", "Physical-Organic Chemistry", etc. did not seem appropriate. Engaging in one of favorite pastimes one day, the reading of the monthly feature called "Highlights from Current Literature" in Chemistry and Industry, it occurred to me that "Highlights of Organic Chemistry" would be a descriptive title, and I hope it is. This book is primarily meant for graduate students whose interests are strongly research-oriented, but it may also be of use to those who have been unable to keep abreast of the recent literature, and who would like a general overview of what has happened in Organic Chemistry in the past decade.

Unfortunately some very pertinent recent references could not be included in this text. It is hard to describe the sinking feeling one experiences right after putting the finishing touches on a chapter, when the next journal to cross one's desk contains papers describing indisputable evidence pointing to contrary conclusions.

I am deeply indebted to Dr. F. Goudriaan, who taught me the basics of organic chemistry many years ago, and whose encouragement and advice have continued ever since. I owe many thanks to Professor E. Havinga, who gave generously of his time to help in more ways than I can recount. Professor L. Oosterhoff gave me the benefit of his comments on several points bearing on theoretical problems. Mrs. S. Overgaauw was extremely helpful in getting us settled in the initial stages of our visit. Mr. M. Pison provided me with many expert suggestions and comments regarding the drawing of the structures, and to boot, in his own time drew those in the first three chapters as examples. Some of the best students at Leiden critically read parts of the raw manuscript: Mr. S. Eisma read Chapter 7, Mr. H. de Leeuw read Chapter 2, Mr. T. Mulder read Chapter 8, and Mr. F. Wiegerink read Chapter 10. Their comments have a great deal to do with whatever merit this writing may have. Professor P. Gassman read the entire manuscript; he found and corrected innumerable small errors, and a few more serious ones as well. I owe many thanks to Mrs. P. Holmes, who transformed a badly mangled, handwritten manuscript into a faultlessly typed version. After I had once again cut this into pieces, annotated it and taped it back together, Mrs. L. Lawrence transformed it into the camera-ready copy in an incredible eight weeks.

The occasion of the second printing has provided me with an opportunity to remove a number of small errors; I thank Drs. W. R. Drenth, R. C. Kerber, and D. R. McKelvey for compilations of these errors. No attempt has been made to update the text in any other way, however.

In conclusion, I wish to express the hope that readers will continue to call to my attention any further errors they may observe.

W. le Noble
Stony Brook, N.Y.

CONTENTS

Highlights of Organic Chemistry

Chapter 1

ELEMENTARY CONCEPTS IN CHEMISTRY

1-1. THE SCHRÖDINGER EQUATION

The Schrödinger equation [1] is the key to much of our present understanding of organic chemistry. It is true that the organic chemist had by 1925 acquired an impressive amount of physical and chemical information about the numerous compounds of carbon found in nature and many others synthesized in the laboratory. Undeniably, he had succeeded in constructing a comprehensive framework - the structural theory - to account for the number of structural isomers, and he had deduced the spatial relationships of the atoms within a molecule so as to account for stereoisomerism as well. Indeed, each time that new facts came to light that did not fit the theories of the structural theory and the tetravalent, tetrahedral carbon atom, small extensions of these ideas such as the chemical incompatibility of groups, steric hindrance, tautomerism, and so on sufficed to save the ball-and-stick view of his science. However, the development of quantum mechanics after 1925 led to so many new compounds and their interconversions that it is fair to say

---

1. A bibliography of classic papers and books useful to organic chemistry students of quantum chemistry has been given by F. L. Pilar, J. Chem. Educ., 37, 587 (1960).

that organic chemistry owes its decisive new directions to the small group of people, among them Schrödinger, who were at that time rewriting the basic laws of physics.

The Schrödinger equation is a differential equation written as follows:

$$\frac{\partial^2\Psi}{\partial x^2} + \frac{\partial^2\Psi}{\partial y^2} + \frac{\partial^2\Psi}{\partial z^2} + \frac{8\pi^2 m}{h^2}(E_{tot} - E_{pot})\Psi = 0$$

Here Ψ is the square root of the probability of finding a given particle of mass m at a location x, y, z. The equation is the result of a radical departure from classical mechanics, which would attempt to seek definite information about the precise location of the particle. The derivation and background of the equation need not concern us [2]; suffice it here to say that the solution (i.e., integration) of the differential equation in any particular case is a set of one or more ordinary equations giving Ψ as a function of x, y, z, m, and E. The equations then portray a family of surfaces about the origin where Ψ and Ψ^2 have certain values.

1-2. THE PARTICLE IN A BOX

As a simple if hypothetical example of how one might proceed to solve the Schrödinger equation, let us consider the case of an uncharged particle (so that $E_{pot} = 0$ and $E_{tot} = E_{kin}$) of mass m in a one-dimensional box of length a; that is to say, the energy of the particle is set at infinity at any location other than $y = 0$, $z = 0$, and $0 < x < a$ (see Fig. 1-1). The equation then reduces to

$$\frac{\partial^2\Psi}{\partial x^2} + \frac{8\pi^2 m}{h^2} E_{tot}\Psi = 0$$

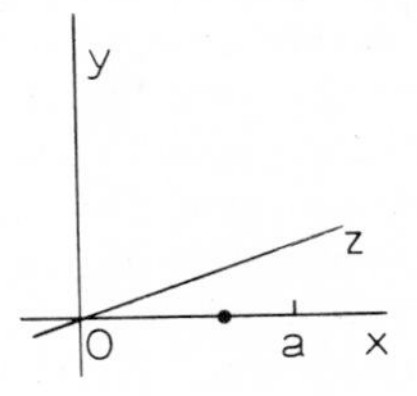

Fig. 1-1. The particle in a box.

---

2. Students interested in this aspect of our discussion are referred to (a) L. Pauling and E. B. Wilson, "Introduction to Quantum Mechanics", McGraw-Hill, New York, 1935; (b) H. Eyring, J. Walter, and G. E. Kimball, "Quantum Chemistry", Wiley, New York, 1944.

The general techniques used in the solution of differential equations cannot be reviewed here [3]; however, it can readily be verified by substitution that the solution of this equation is

$$\Psi = R \sin\left\{\frac{8\pi^2 mE_t}{h^2}\right\}^{1/2} x + S \cos\left\{\frac{8\pi^2 mE_t}{h^2}\right\}^{1/2} x$$

where R and S are integration constants. To evaluate these constants, we apply the boundary conditions, first of all that Ψ and $\Psi^2 = 0$ when $x = 0$:

$$\begin{aligned} 0 &= R \times \sin 0 + S \times \cos 0 \\ &= R \times 0 + S \times 1 \end{aligned}$$

Evidently, S equals zero, and our equation reduces to:

$$\Psi = R \sin\left\{\frac{8\pi^2 mE_t}{h^2}\right\}^{1/2} x$$

Then we apply: $\Psi = 0$ when $x = a$:

$$0 = R \sin\left\{\frac{8\pi^2 mE_t}{h^2}\right\}^{1/2} a$$

R cannot be 0, since then Ψ is 0 everywhere, even within the box. Alternatively then, the sine function must be 0. This is true when the following condition is fulfilled:

$$\left\{\frac{8\pi^2 mE_t}{h^2}\right\}^{1/2} a = n\pi$$

where n is an integer called a quantum number; it may not have the value 0 since then again Ψ becomes 0 everywhere. We can rewrite the result as follows:

$$E_{tot} = \frac{n^2 h^2}{8ma^2}$$

3. An excellent text for self-study of this topic is L. Kells, "Elementary Differential Equations", McGraw-Hill, New York, 1954.

Before we discuss this result, we evaluate R by applying another boundary condition called normalization. Since we know the particle is somewhere in the box, we may write

$$\int_0^a \Psi^2 dx = 1$$

Some rewriting leads to

$$\int_0^a R^2 \sin^2 \frac{\pi n}{a} x \, dx = 1$$

which is readily integrated to yield

$$R = \left(\frac{2}{a}\right)^{1/2}$$

The overall result therefore is the following "eigenfunction":

$$\Psi = \left(\frac{2}{a}\right)^{1/2} \sin \frac{\pi n}{a} x$$

Although this is a purely hypothetical example, it is instructive in that it shows: (a) how a quantum number may make its appearance; (b) that the kinetic energy cannot be zero. This latter result is fundamentally different from any known in classical mechanics, which allows particles to be at rest. We find furthermore that only certain values ("eigenvalues") of E_K are allowed. These values decrease as m increases (to all intents and purposes the two mechanics do not differ in the description of large objects), as a (the size of the box) increases, and as n decreases. These considerations continue to apply as we switch from our uncharged particles to electrons in real molecules, and they are of crucial importance. Figure 1-2 depicts Ψ and Ψ^2 as a function of x and n (the "orbitals" of the particle). When n = 1 and the energy is at a minimum, the particle is said to be in the "ground state"; when n = 2, it is in the first "excited state", and so on. In the excited states there are points ("nodes") within the box where Ψ changes sign; at such points $\Psi^2 = 0$, and thus there is no chance we will find the particle there. We may note also that while Ψ^2 is either zero or positive everywhere in the box, Ψ may have either sign in a given state in a given region of the box.

1-3. THE HYDROGEN ATOM

In the hydrogen atom there is an electrostatic potential energy $E_{pot} = e^2/r$, and hence the Schrödinger equation must be written

$$\frac{\partial^2\Psi}{\partial x^2} + \frac{\partial^2\Psi}{\partial y^2} + \frac{\partial^2\Psi}{\partial z^2} + \frac{8\pi^2 m}{h^2}\left(E_{tot} - \frac{e^2}{(x^2 + y^2 + z^2)^{1/2}}\right)\Psi = 0$$

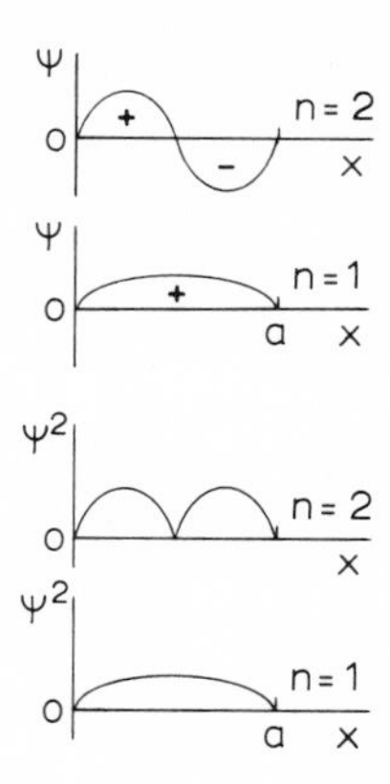

Fig. 1-2. Plots of Ψ and Ψ^2 as functions of a and n.

The complete solution of this equation has been recorded [2a]; there is no need to reproduce it here, but the following features are salient. Three quantum numbers appear:

the principal quantum number n = 1, 2, 3, ---
the orbital quantum number l = 0, 1, 2, --- (n-1)
the magnetic quantum number m = 0, ±1, ±2, --- ±l

Consideration of more complex atoms leads to a fourth quantum number, the spin quantum number s, which may have values +1/2 and -1/2. The energy of the electron is governed first of all by the value of n, with the ground state at n = 1. The number of nodal surfaces will be (n-1). The size of the orbital also depends primarily on n. The shape of the orbital is governed principally by l: the s orbitals (l = 0) are spherically symmetrical; p orbitals (l = 1) are dumbbell shaped; d orbitals (l = 2) appear as a perpendicular pair of dumbbells; and f orbitals appear (l = 3) as a mutually perpendicular trio of dumbbells (see Fig. 1-3). For a given value of n, the energy of an electron is somewhat lower in an s than in a p orbital, which, in turn, is lower than in a d orbital, etc. The magnetic quantum number determines only the orientation of the orbitals. Thus, for l = 1, m may be 1, 0, or -1, corresponding to three mutually perpendicular p-orbitals; for l = 2, m may have five different values each corresponding to a different orientation for the five d orbitals. In the absence of some external feature such as another atom in its proximity, the orbitals corresponding only to different values of m are completely equal in energy (they are "degenerate").

The ground state of a hydrogen atom is therefore described by n = 1, l = 0, m = 0, s = 1/2 (or -1/2). The atom has all the other orbitals derivable from higher values of n, but of course all are unoccupied. If we were to take the electron out of its ground state orbital and "promote" it to one of the higher energy orbitals, we would have an excited hydrogen atom.

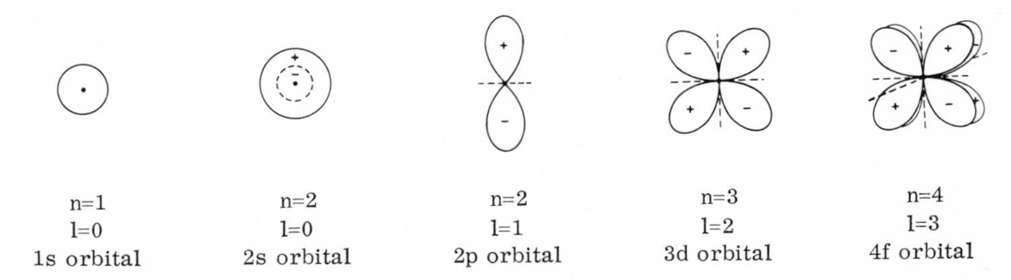

Fig. 1-3. Several orbitals; all but the f-orbital were drawn in cross section. The dotted lines denote nodal planes.

1-4. THE PERIODIC TABLE

If we turn our attention to more complex atoms, we find first of all that the Schrödinger equation immediately becomes much more cumbersome to write, because not only does each electron have an attractive potential vis-à-vis the nucleus, but each also has a repulsive potential toward each of the others. Consequently, exact solutions of the Schrödinger equation can no longer be obtained; however, it has become clear that the solutions can be viewed as simple extensions of the hydrogen atom. Thus, if we began by rewriting the one-electron problem with successively more highly charged nuclei, nothing would change very much except that the various orbitals would shrink somewhat. If we now feed in additional electrons one by one so as to arrive at the complex atoms, these electrons will adopt those quantum numbers that give the system the lowest energy. We find that one more statement is needed to give a good description of the atoms, namely, the Pauli principle. According to this principle, no two electrons within an atom may have all quantum numbers identical: At least one of these numbers must be different. If we review the electronic ground states of the atoms, we therefore arrive at the familiar series:

H : $1s^1$
He: $1s^2$
Li : $1s^2 2s^1$
Be: $1s^2 2s^2$
B : $1s^2 2s^2 2p^1{}_x$
etc.

When one attempts to write out the complete sequence of the periodic table, some further questions arise. First of all, should carbon be described by $1s^2 2s^2 2p^2{}_x$ or by $1s^2 2s^2 2p^1{}_x 2p^1{}_y$? This question, which arises repeatedly as we traverse the sequence of orbitals, has the same answer each time. According to Hund's first rule, whenever a number of electrons must be accommodated in a number of degenerate orbitals, they will singly occupy as many orbitals as possible before beginning double occupancy; this of course makes sense if one considers that this minimizes electron repulsions. Thus, carbon will be $1s^2 2s^2 2p_x{}^1 2p_y{}^1$, nitrogen is $1s^2 2s^2 2p_x{}^1 2p_y{}^1 2p_z{}^1$, etc. The alternative description of carbon is therefore an excited state. A second question deals with the spins of the electrons singly occupying more than one orbital: Will they be parallel or not? The answer to this question is known as Hund's second rule, and again this rule is uniformly valid: In the ground state these spins will be parallel (s will have the same sign for these electrons).

Still another question has to do with the relative energy levels: Since both n and l affect the energies of the orbitals, what is the precise order of the filling of the energy levels? The little mnematic device of Fig. 1-4 shows this order. After the completion of each p subshell, the next electrons must go into the next (higher energy) shell. Thus, there is a natural and large energy gap after each shell has gained two s and six p electrons - since the next electrons must begin the next "higher" shell. As a result, each atom (especially of the first row of the periodic table) appears to have a strong tendency to surround itself with eight outer ("valence") electrons (two in the case of the first few atoms). This trend is known as the octet rule.

For individual atoms, spectroscopists have invented the so-called term symbols $^{A}X_{B}$, which may be derived by means of the following definitions. A is called the multiplicity, and defined by

$$A = 2S + 1$$

where S equals the total spin quantum number (one-half the number of unpaired electrons). The electronic state of the atom is referred to a singlet, doublet, triplet, etc., as A equals 1, 2, 3, etc., respectively. B is the total angular orbital quantum number; it is given by

$$B = L \pm S$$

where L is the resultant orbital quantum number. L vanishes for filled subshells, so that only partially filled subshells need be considered. The negative sign is

Shell	subshells				
K (2 e)	1s				
L (8 e)	2s	2p			
M (18 e)	3s	3p	3d		
N (32 e)	4s	4p	4d	4f	
O (50 e)	5s	5p	5d	5f	5g
P (72 e)	6s	6p	6d		
Q (98 e)	7s	7p			
R (128 e)	8s				
etc.	2 e	6 e	10 e	14 e	18 e

Fig. 1-4. Mnemonic device showing the order of energy levels in the atom.

used when the shell is not yet half filled. Finally, X is described by the symbols S, P, D, F, etc., as L = 0, 1, 2, 3, etc. As an example, we may consider the sulfur atom in its ground state:

$$S = 1s^2 2s^2 2p^6 3s^2 3p^4$$

There will be two unpaired electrons, so that S = 2 x 1/2 = 1 and A = 2 x 1 + 1 = 3. Since for the 3p electrons $\Sigma m = 1 + 0 + (-1) + 1 = 1$, B = 1 + 1 = 2 and the term symbol is $^3P_2$. One of the values of the term symbols is their role in the selection rules, which limit the states to which an atom can be excited by a photon; they state that ΔS must be zero and ΔL must be ±1.

1-5. HYBRID ORBITALS

The set of atomic orbitals described earlier is not unique; the Schrödinger equation has additional solutions, the hybrid orbitals, which are of higher energy and therefore not important in free atoms (we shall see below however that they are very important in molecules).

The simplest way to get a description of these additional solutions is by the use of a mathematical theorem according to which, if a given differential equation has solutions f_1, f_2, etc., then certain linear combinations $f = c_1f_1 + c_2f_2 + \ldots$, etc., may also be solutions. It turns out in fact that the combination of n orbitals (solutions) gives rise to n new orbitals. When this theorem is applied to the 2s and the three 2p orbitals, one finds that there is an alternative solution in which there are four degenerate orbitals called sp^3 hybrid orbitals. They are oriented with respect to one another at tetrahedral angles and are roughly dumbbell shaped, but with one lobe much larger than the other [4]. According to Hund's rules, a free, sp^3 hybridized carbon atom would have four unpaired electrons. If a similar combination of a 2s orbital with two 2p orbitals is studied, one obtains a set of three sp^2 hybrid orbitals, each of a shape similar to that of the sp^3 hybrids; the three are degenerate, at 120° angles with one another, and perpendicular to the remaining p orbital. The use of a 2s orbital and one p orbital similarly gives rise to a pair of sp orbitals, again similar in shape and degenerate, making 180° angles with one another and each perpendicular to the plane of the two remaining and mutually perpendicular p orbitals. Various types of hybrid orbitals involving s, p, and d orbitals are likewise well known, but we need not consider them here.

---

4. L. Pauling, "The Nature of the Chemical Bond", 3d Ed., Cornell University Press, New York, 1960.

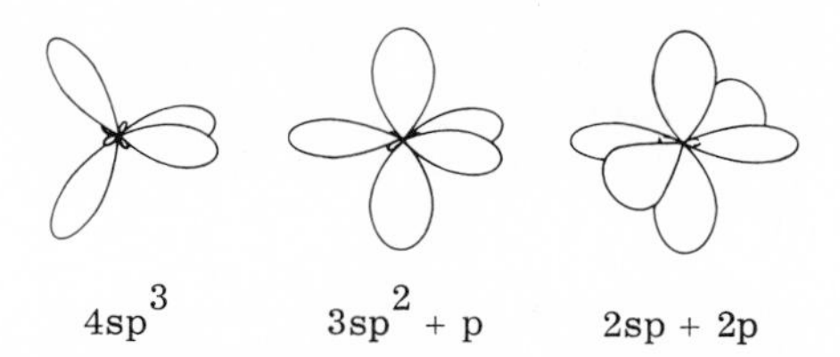

Fig. 1-5. The sp hybrid orbitals and their orientation with respect to each other and any remaining p orbitals. In actual fact, the nodes are not precisely at the nucleus; however this is not important for the discussion here.

1-6. THE VALENCE BOND APPROXIMATION FOR MOLECULES

The Schrödinger equation also moves up a notch in complexity when a second nucleus is present, and rigorous solutions in closed form cannot be obtained. Two approximation methods have been developed that are especially useful in organic chemistry: The valence bond and molecular orbital methods. Both of these methods have pros and cons that have tempted many chemists into believing that they are rival theories; actually both lead to identical results if both are carried out completely rigorously [5]. Since both methods have developed terminologies that are part of the organic chemist's vocabulary, we shall proceed to use words and phrases from both interchangeably wherever it seems convenient.

In the valence bond method the H_2^+ molecule ion would be considered as a hybrid of two structures, in one of which the electron would be considered to be part of the one nucleus and in the other it would belong to the other proton;

$$H_2^{\oplus} = \dot{H}\ \overset{\oplus}{H} \longleftrightarrow \overset{\oplus}{H}\ \dot{H}$$

If Ψ_1 be the wave function describing the first of these two "canonical" or "resonance" structures and if Ψ_2 be the wave function for the other, the hybrid is assumed to be well described by a linear combination of the two:

$$\Psi_{H_2^+} = a_1 \Psi_1 + a_2 \Psi_2$$

In this case there are only two structures that can reasonably be considered, and, furthermore, since the two are completely equivalent, the values of a_1 and a_2 will be equal; however, these statements require modification as soon as more complex situations are encountered. For instance, the hydrogen molecule could be considered as three structures such as (1)-(3); clearly the structure (2) need not have the same "weight" (importance, contribution, or value of a) as structures (1) and (3). Its weight in fact will be far greater, as our intuition tells us.

$$\overset{\oplus}{H}\ \overset{\ominus}{H} \longleftrightarrow \dot{H}\ \dot{H} \longleftrightarrow \overset{\ominus}{H}\ \overset{\oplus}{H}$$

(1) (2) (3)

5. G.W. Wheland, "Resonance in Organic Chemistry", Wiley, New York, 1955.

On the other hand, in the still more complex molecule of lithium fluoride a structure such as (4)

$$\overset{\oplus}{\mathrm{Li}}\ \overset{\ominus}{\mathrm{F}}$$

(4)

may well be the chief contributor to the hybrid. Since the use of a large number of structures obviously leads us into complex mathematics, it is important to ask whether there are any limitations to the structures which must be considered. There are in fact several formal rules to be observed. Thus, the nuclei must be in the same positions in all of them. The number of unpaired electrons must be the same in all; for example, since the electrons are obviously paired in (1) and (3), they also must be in (2).

The following rule is obviously a less formal one: The only structures which are of much importance are those that one would have guessed based on chemical intuition. Thus, structure (2) is certainly more important than (1) and (3); for lithium fluoride (4) is more important than (5), and one would not make a very serious error by not considering (5) at all. Thus, organic chemists rarely consider more than a few structures even when highly complex molecules are discussed.

$$\overset{\bullet}{\mathrm{Li}}\ \overset{\bullet}{\mathrm{F}}$$

(5)

The properties predicted for molecules examined by the valence bond method are in general intermediate between those of the contributing structures, as might be expected; however, there is one very important exception to that. The energy is always lower than that of any of the contributing structures [5]. The difference in energy between the hybrid and the most important contributing structure is called the resonance energy. The resonance energy is small or negligible when the contributing structures differ widely in energy, and it is greatest when two or more contributing structures are equivalent, as is suggested in Fig. 1-6; it also illustrates our earlier statement to the effect that the act of neglecting high energy (intuitively unreasonable) structures would not lead to serious errors.

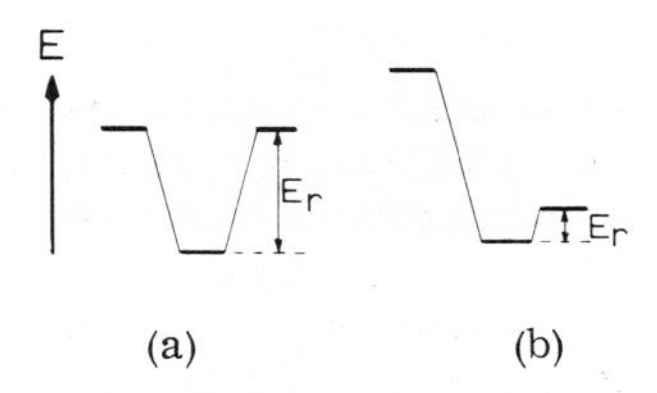

Fig. 1-6. Resonance energies with (a) two equivalent structures; (b) two structures with very different energies.

In the case of the hydrogen molecule ion the resonance energy is essentially the bond energy and in fact we might consider all bond energies resonance energies; once again however, when organic chemists use the term resonance energy they usually refer to the difference in energy between the real molecule (the "hybrid") and the single most important contributing structure. The uses of resonance in organic chemistry are discussed further in Chapter 8.

1-7. THE MOLECULAR ORBITAL APPROXIMATION FOR MOLECULES

In this method we concern ourselves - at least initially - with the orbitals rather than the electrons. For instance, for hydrogen (whether H_2 or H_2^+) we begin by putting the two nuclei at their actual bonding distance and then asking the question: What will be the "molecular orbitals" (MO's) for the electrons that might find themselves in the proximity of both of these nuclei? It is assumed that such orbitals will be a linear combination of the atomic orbitals (LCAO):

$$\Psi = c_1\Psi_1 + c_2\Psi_2$$

In the process of solving this equation, use is made of the proposition that in the molecular orbital the energy must be at a minimum (i.e., $\partial E/\partial c_1$ and $\partial E/\partial c_2$ must be zero). The resulting decrease in energy is now called the "delocalization energy"; it is to most intents and purposes the same as the resonance energy. Once it has been determined what molecular orbitals this combination process gives rise to, the electrons are fed in; the Pauli principle and Hund's rules apply in this process as they do in the case of atomic orbitals. Although no attempt will be made here to reproduce the complete treatment [6], some of the more important conclusions may be stated as follows. First of all, as noted earlier, the combination of any number of atomic orbitals will lead to the same number of molecular orbitals. In the case of hydrogen molecules, the first of these two molecular orbitals (i.e., the lowest in energy) has no nodes and may be thought of as having resulted from the combination of two 1s orbitals in which Ψ had the same sign (whether positive or negative). This molecular orbital has a lower energy than either of the atomic orbitals; this actually holds the two atoms together, and hence it is referred to as the "bonding" orbital. The other has a nodal plane between the atoms and may be considered to have resulted from the combination of two orbitals in which Ψ had opposite signs. Its energy is higher than that of the atomic orbitals and hence it is called the "antibonding" orbital. Of course, when the two electrons are fed in, both will preferably occupy the bonding orbital, and hence they must be paired. Figures 1-7 and 1-8 illustrate the descriptions just given.

---

6. A. Streitwieser, "Molecular Orbital Theory for Organic Chemists", Wiley, New York, 1961.

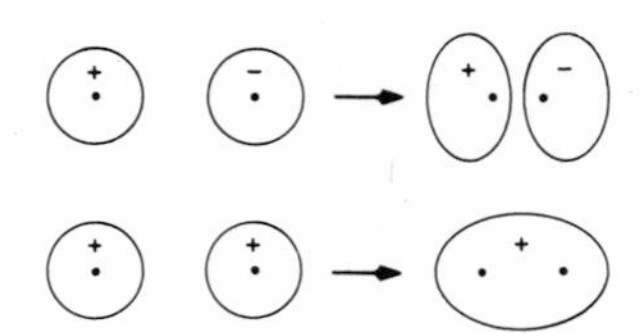

Fig. 1-7. The bonding and antibonding molecular orbitals of hydrogen.

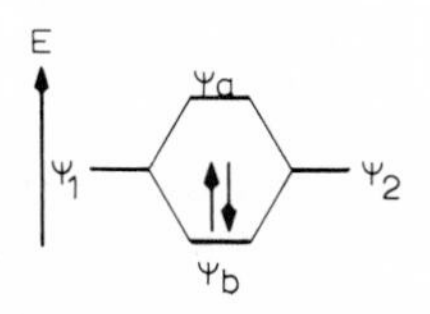

Fig. 1-8. The correlation diagram for H_2 showing the separate and combined levels, and the electron spins.

This method gives a very simple picture of excited states; another advantage is that it often easily accounts for the presence of unpaired electrons. Thus, when the oxygen molecule is considered, the picture that emerges is that of Fig. 1-9. Note that if the $2p_x$ orbitals are the ones in line with the nuclei, their combination will lead to MO's of energy different from those of the other two. If 16 electrons are fed in, the application of Hund's rules leads us to the diradical nature of oxygen.

At this point we may briefly return to hybrid atomic orbitals and to the question why these are preferred to the pure s, p, and l orbitals in most chemical compounds but not in the isolated atoms. Chemical bonding results in a lowering of the overall energy, and it should be expected that this process will be most pronounced when the orbitals "overlap" to the highest degree consistent with the inevitable internuclear repulsions. Inspection of the various types of orbitals immediately reveals that hybrid orbitals can overlap far more efficiently than the pure atomic orbitals in an extended (σ-) approach, though not in a parallel (π-) approach (cf. also next section). Perhaps more important, the hybrid type of atom can sometimes simply form more bonds; thus, while it is easy to understand how an sp^3-hybridized carbon atom can bond four hydrogen atoms, it is hard to see how a nonhybridized atom would do it.

It is also instructive to consider what would happen if we were to force together atoms with certain of their pure orbitals in opposite phase. In the limit as the nuclei occupy the same point (the "united atom"), s orbitals can be seen to give rise to p orbitals, p orbitals to d, etc. (see Fig. 1-10).

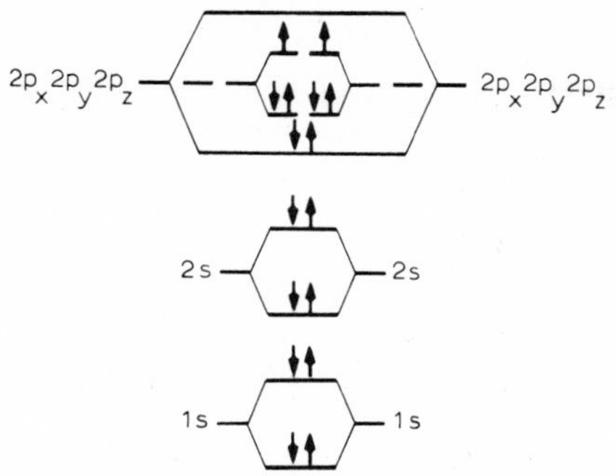

Fig. 1-9. Schematic correlation diagram for the combination of the atomic orbitals of oxygen.

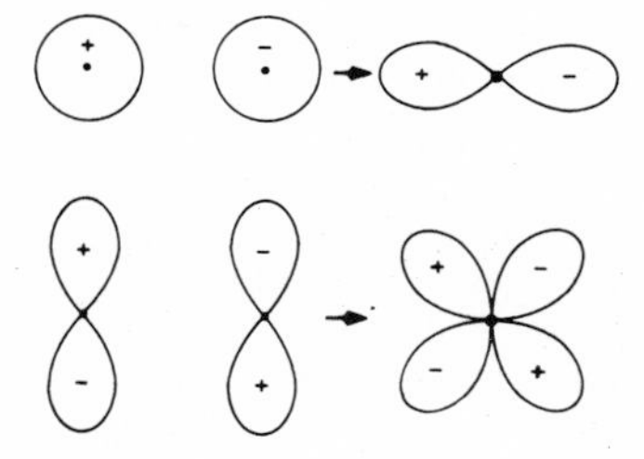

Fig. 1-10. The united atom viewpoint of how s orbitals give rise to p orbitals, etc.

1-8. THE SIMPLE TYPES OF BONDING

Many of the terms and arguments used to describe the bonding of hydrogen are readily extended to higher atoms. We briefly mention here several features that arise when the resulting bonds are considered.

One such feature was already hinted at: In symmetrical ("homopolar") bonds the two ionic contributions will be identical but opposite in direction, and hence the bond will be purely covalent. In nonsymmetrical ("heteropolar") bonds these contributions are not equal, and the result is a polar bond which in limiting cases may be a purely electrostatic or ionic bond. In such bonds, if we compare atoms in the same row, those toward the right side of the periodic table - which have a larger nuclear charge and a smaller radius, and hence will attract electrons more strongly (be "more electronegative") - will be at the negative end of the dipole.

Another familiar aspect is that orbital axes may be either common (σ bonds) or parallel (π bonds), but not perpendicular; as Fig. 1-11 shows, the degree of overlap possible in σ bonds is greater than in π bonds, and hence the latter are weaker. An example of a δ bond (in octachlorodirhenium dianion) has been described by Cotton [7].

With this background we may survey the bonds usually found between atoms in organic compounds. Single bonds involving carbon are always derived from hybrid orbitals; thus, the bond angles of singly bound carbon atoms are usually close to the tetrahedral value of 109.5°, although the atoms may be forced to deviations by being part of a small ring, by being bound to bulky substituents, and so on. Likewise, in doubly bound carbon the bond angles are usually close to 120°, and in triply bound and allenic carbon atoms the bond angles are always 180° or close to it. In tetravalent compounds of silicon, germanium, tin, and lead the bond angles again indicate hybridization.

In compounds of nitrogen and oxygen it is not always clear whether hybridization is involved or not. For example, bond angles of 109° or so are expected for ammonia if the three N-H bonds are based on sp^3 orbitals, and 90° if the nitrogen unshared pair is in a pure 2s orbital. In fact, the bond angle is known to be 104°. In phosphine, the bond angles are 91° so in that case the hydrogen atoms are apparently bound to the phosphorus p orbitals. In ammonium and phosphonium salts, however, the arrangement is tetrahedral. Similar remarks are valid for oxygen; in water, the bond angles are about 105°; in hydrogen sulfide they are 92°. Since as a rule the bond angles at singly bound nitrogen and oxygen are at least not very much less than 109°, we shall hereafter assume that these atoms have hybrid orbitals in all their compounds [7a]. Elements such as boron and magnesium apparently also use hybrid orbitals in simple organic compounds; thus, trimethylboron is planar and dimethylmagnesium is linear.

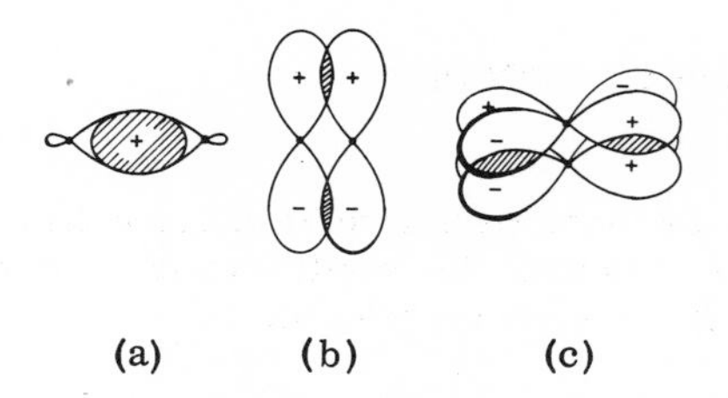

Fig. 1-11. Bond formation: (a) a σ bond; (b) a π bond; (c) a δ bond.

---

7. F.A. Cotton, Accounts Chem. Res., 2, 240 (1969).
7a. Alternatively, bond angles have been very nicely explained as due to unshared electron pair repulsions: See R.J. Gillespie, J. Chem. Educ., 47, 18 (1970).

Double bonds are best viewed as consisting of a σ bond (overlap of two sp^2 hybrid orbitals) and a π bond (overlap of two parallel p orbitals); likewise, triple bonds consist of a σ bond and two π bonds. Allenes are linear, and hence the central carbon atom forms two double bonds, using sp hybrid orbitals and two mutually perpendicular p orbitals (see Fig. 1-12). The σ bonds connecting the carbon atoms result from overlap of sp and sp^2 orbitals. Double bonds are rarely encountered in second row elements; an example is CS_2. In silicon chemistry, not a single example is known. Perhaps the larger size of the atoms prevents efficient overlapping of parallel p orbitals.

In some multiple bonds one of the components is an ionic bond. An example of such a "dative double bond" is provided by amine oxides such as (6); the nitrogen is essentially tetrahedral.

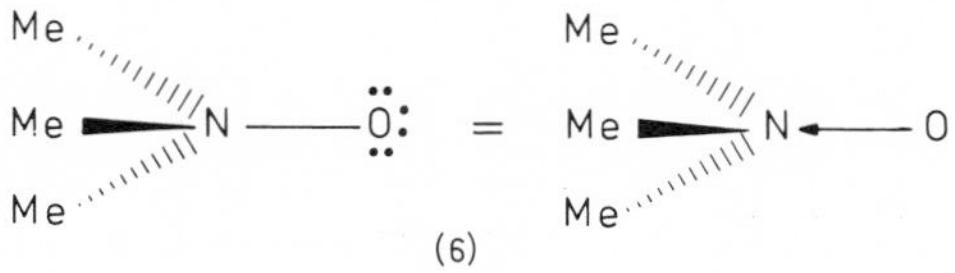

(6)

1-9. MULTICENTER BONDS

The combination of orbitals need not be limited to two, and there are molecular orbitals encompassing more than two atoms. As an example we mention diborane (7). Experiments show that in this symmetrical molecule both boron atoms are tetravalent, and two of the hydrogen atoms are divalent (borane, BH_3, is known but only as a high energy intermediate) [8].

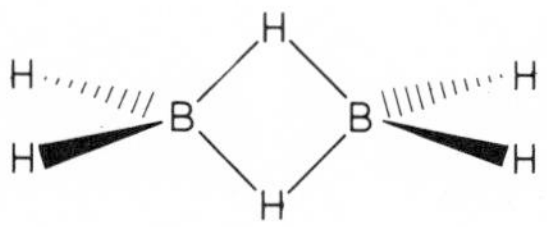

(7)

We can understand this structure if we assume that each boron atom uses sp^3 hybrid orbitals to bind two hydrogen atoms, and that the unused sp^3 orbitals and the hydrogen 1s orbitals are combined as suggested in Fig. 1-13. Of the three pairs of molecular orbitals so formed one is bonding and occupied; the other MO's are nonbonding and antibonding pairs, all of them empty. A similar situation exists in hexamethyldialuminum, in which two methyl groups are bridged between the aluminum atoms [9]. Note that a simple representation such as (7) may be deceptive in that the line segments connecting the atoms do not each represent a pair of electrons.

---

8. G.W. Mappes and T.P. Fehlner, J. Amer. Chem. Soc., 92, 1562 (1970).
9. J.C. Huffman and W.E. Streib, Chem. Commun., 911 (1971).

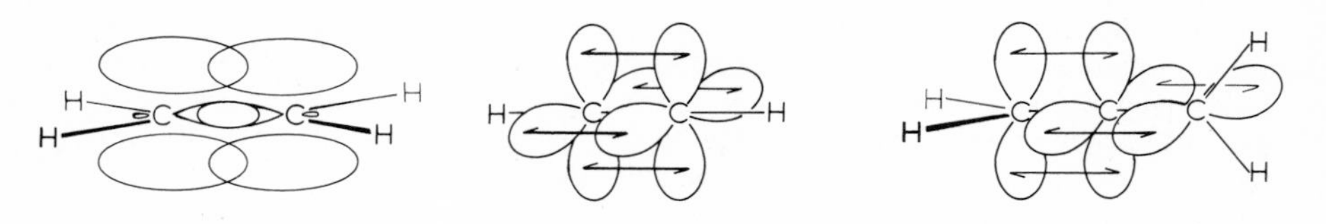

Fig. 1-12. The bonding in ethylene, acetylene, and allene.

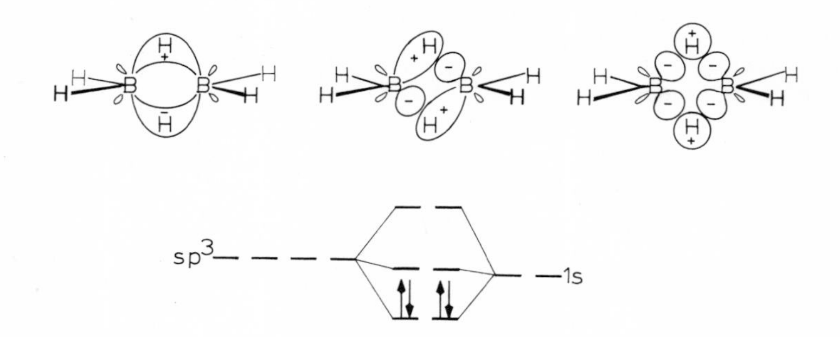

Fig. 1-13. The bonding, nonbonding and antibonding three-center orbitals of diborane and the corresponding correlation diagram.

Still more complex situations may be encountered. Thus, the molecule p-carborane (8) is known, in which two carbon atoms are each bound to five boron atoms and a hydrogen atom, and each boron is bound to a carbon atom, four other boron atoms, and a hydrogen atom [10]. Clearly the combination of many atomic orbitals is required to arrive at so many bonds.

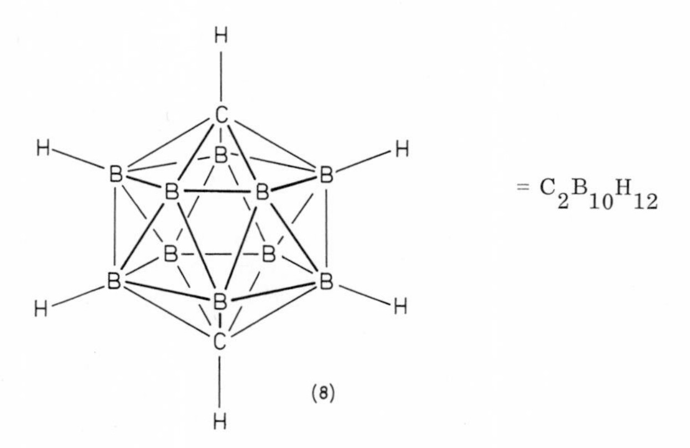

10. (a) W.H. Knoth, J. Amer. Chem. Soc., 89, 1274 (1967); (b) E.L. Muetterties and W.H. Knoth, Chem. Eng. News, 44, 88 (May 9, 1966).

As with hybrid atomic orbitals, the use of multicenter molecular orbitals arises because it allows maximum bonding and hence minimum energies. Such bonds are common in compounds containing boron, since boron cannot satisfy the octet rule by forming only two-electron bonds; in boron-free carbon compounds this kind of process gives rise to coordination numbers greater than four only in special cases such as the nonclassical ions (see Chapter 20).

1-10. THE SIMPLE HÜCKEL MO METHOD

In 1931 Hückel reported a mathematical treatment of conjugated multiple bonds in organic compounds [11]. In such compounds each of several successive carbon atoms in an array has a singly occupied p orbital, and if all these orbitals are parallel, it is possible to combine them so as to give rise to a set of molecular orbitals.

We may begin by learning how the so-called HMO method is applied to the two p orbitals of ethylene. A basic assumption is that the σ framework of the molecule need not be considered in the process, and that we need concern ourselves only with the combination

$$\Psi_{(\pi)} = c_1 \Psi_{(p)1} + c_2 \Psi_{(p)2}$$

We will evaluate the constants c by setting the derivatives $\partial E/\partial c$ equal to zero. We write the Schrödinger equation in the form

$$H\Psi = E\Psi$$

where H is called the Hamiltonian operator; it implies that we mean to carry out an operation on Ψ so as to obtain as a result

$$\frac{h^2}{8\pi^2 m} \left\{ \frac{\partial^2\Psi}{\partial x^2} + \frac{\partial^2\Psi}{\partial y^2} + \frac{\partial^2\Psi}{\partial z^2} \right\}$$

This way of writing the wave equation not only saves paper but, it turns out, this operator has certain convenient properties - which we state here without proof and apply below ($E\Psi$ is simply the product of the energy E and Ψ).

$$\Psi_1 H \Psi_2 d\tau = \int \Psi_2 H \Psi_1 d\tau$$

$$H\,(\Psi_1 + \Psi_2) = H\Psi_1 + H\Psi_2$$

$$H\,(a\Psi_1) = aH\Psi_1$$

Here $\int \Psi d\tau$ means integration of the function Ψ over all space.

11. J.D. Roberts, "Notes on Molecular Orbital Calculations", Benjamin, New York, 1962.

For reasons that will become apparent as we proceed, we multiply both members by Ψ.

$$\Psi H \Psi = E \Psi^2$$

$$\int \Psi H \Psi d\tau = \int E \, \Psi^2 d\tau$$

$$E = \frac{\int \Psi H \Psi d\tau}{\int \Psi^2 d\tau}$$

$$= \frac{\int (c_1 \Psi_1 + c_2 \Psi_2) H (c_1 \Psi_1 + c_2 \Psi_2) d\tau}{\int (c_1 \Psi_1 + c_2 \Psi_2)^2 d\tau}$$

$$= \frac{\int (c_1^2 \Psi_1 H \Psi_1 + 2c_1 c_2 \Psi_1 H \Psi_2 + c_2^2 \Psi_2 H \Psi_2) d\tau}{\int (c_1^2 \Psi_1^2 + 2c_1 c_2 \Psi_1 \Psi_2 + c_2^2 \Psi_2^2) d\tau}$$

At this point further abbreviations are introduced:

$$\int \Psi_a H \Psi_b d\tau = H_{ab}$$

and

$$\int \Psi_a \Psi_b d\tau = S_{ab}$$

These integrals are referred to as follows: H_{aa} = Coulomb integral; H_{ab} = resonance integral; and S_{ab} = overlap integral.

Let us now regard our equation as a ratio

$$E = \epsilon / \Delta$$

in which both ϵ and Δ are functions of our constants c. We recall from calculus that

$$\frac{\partial E}{\partial c} = \frac{\Delta \partial \epsilon / \partial c - \epsilon \partial \Delta / \partial c}{\Delta^2}$$

In our case $\partial E / \partial c = 0$, and since $\Delta \neq 0$,

$$\frac{\partial \epsilon}{\partial c} - \frac{\epsilon}{\Delta} \times \frac{\partial \Delta}{\partial c} = 0$$

or

$$\frac{\partial \epsilon}{\partial c} - E \frac{\partial \Delta}{\partial c} = 0$$

Applying this result to our expression for E, we obtain

$$2c_1H_{11} + 2c_2H_{12} - E(2c_1S_{11} + 2c_2S_{12}) = 0$$

and

$$2c_1H_{12} + 2c_2H_{22} - E(2c_1S_{12} + 2c_2S_{22}) = 0$$

These equations, rewritten in the form

$$c_1(H_{11} - ES_{11}) + c_2(H_{12} - ES_{12}) = 0$$

and

$$c_1(H_{12} - ES_{12}) + c_2(H_{22} - ES_{22}) = 0$$

are called the "secular equations". This set of equations is homogeneous (all right members are equal to zero); this means that the nonzero roots can at this point only be written as indeterminate ratios of determinants which are all equal to zero. The denominator common to all these ratios is called the secular determinant:

$$\begin{vmatrix} H_{11} - ES_{11} & H_{12} - ES_{12} \\ H_{12} - ES_{12} & H_{22} - ES_{22} \end{vmatrix} = 0$$

At this point some further simplifications are applied. First of all, the original functions Ψ_1 and Ψ_2 were presumably normalized, so that

$$S_{11} = S_{22} = 1$$

We further assume that the Coulomb integrals have the same value (α) for any and all doubly bound carbon atoms, that the resonance integrals also have a constant value (β) unless the atoms are not σ bonded, in which case H_j is assumed to be negligible; and finally, that the overlap integrals are negligible. What is left may then be written

$$\begin{vmatrix} \alpha - E & \beta \\ \beta & \alpha - E \end{vmatrix} = 0$$

As a final space-saving device, one usually substitutes $X = (\alpha - E)/\beta$:

$$\begin{vmatrix} X & 1 \\ 1 & X \end{vmatrix} = 0$$

Clearly

$$X = \pm 1$$

and

$$E = \alpha \pm \beta$$

Since

$$c_1(\alpha - E) + c_2 \beta = 0$$

$$c_1/c_2 = \pm 1$$

The coefficients are now obtained by normalization:

$$\int \Psi^2 d\tau = 1$$

$$\int (c_1 \Psi_1 + c_2 \Psi_2)^2 d\tau = 1$$

Since $S_{12} = 0$ and $c_1 = \pm c_2$,

$$c_1 = \pm 1 \sqrt{2}$$

and

$$c_2 = \pm 1 \sqrt{2}$$

Finally

$$\Psi = (1/\sqrt{2})(\Psi_1 \pm \Psi_2)$$

These results are schematically demonstrated in Fig. 1-14.

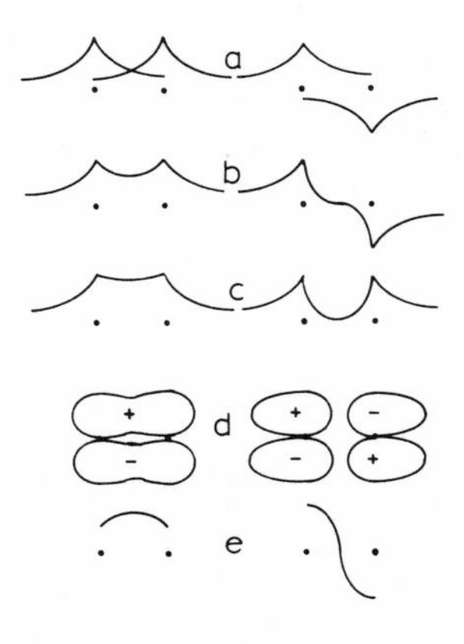

Fig. 1-14. The bonding (π) and antibonding (π *) orbitals of ethylene: (a) individual wave functions when $c_1 = c_2$ and when $c_1 = -c_2$; (b) the combined wave functions; (c) the electron densities; (d) the two molecular orbitals; (e) a frequently used short-hand notation.

1-11. CONJUGATED SYSTEMS

If 1,3-butadiene is considered, there are two reasonable assumptions one can make. One is that the two terminal pairs of carbon atoms are essentially ethylene molecules connected by a single bond; it can indeed be experimentally demonstrated that the central bond is much longer than a normal double bond. If the secular determinant is written on this basis

$$\begin{vmatrix} X & 1 & 0 & 0 \\ 1 & X & 0 & 0 \\ 0 & 0 & X & 1 \\ 0 & 0 & 1 & X \end{vmatrix} = 0$$

one obtains as solutions $E = \alpha \pm \beta$ and $E = \alpha \pm \beta$, a result expected for separate ethylene units.

Alternatively one might argue that the central bond is significantly shorter than normal single bonds and hence that it would make sense to allow an interaction between the p orbitals of atoms 2 and 3. The secular determinant then becomes

$$\begin{vmatrix} X & 1 & 0 & 0 \\ 1 & X & 1 & 0 \\ 0 & 1 & X & 1 \\ 0 & 0 & 1 & X \end{vmatrix} = 0$$

By properly factoring, this determinant is then reduced to

$$X^4 - 3X^2 + 1 = 0$$

The roots of this equation are

$$X = \pm 0.62 \text{ and } X = \pm 1.62$$

so that the four molecular orbitals have energies

$$E = \alpha \pm 0.62\,\beta \text{ and } E = \alpha \pm 1.62\,\beta$$

The coefficients c are found by a process identical to that carried through for ethylene, and one obtains

$$\Psi_a = 0.37\Psi_1 + 0.60\,\Psi_2 + 0.60\,\Psi_3 + 0.37\Psi_4$$
$$\Psi_b = 0.60\,\Psi_1 + 0.37\,\Psi_2 - 0.37\,\Psi_3 - 0.60\,\Psi_4$$
$$\Psi_c = 0.60\,\Psi_1 - 0.37\,\Psi_2 - 0.37\,\Psi_3 + 0.60\,\Psi_4$$
$$\Psi_d = 0.37\Psi_1 - 0.60\,\Psi_2 + 0.60\,\Psi_3 - 0.37\Psi_4$$

The correlation diagram and the MO's are shown in Fig. 1-15.

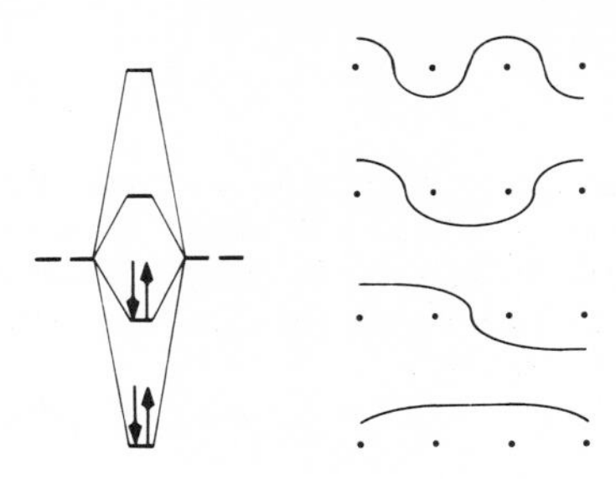

Fig. 1-15. The correlation diagram and the molecular orbitals of conjugated 1,3-butadiene.

The following conclusions are emphasized. The two occupied bonding MO's give the molecule a total energy of

$$E = 2(\alpha + 1.62\beta) + 2(\alpha + 0.62\beta) = 4\alpha + 4.48\beta$$

Since two isolated ethylene units would have an energy of $4\alpha + 4\beta$, the interaction process has led to additional stabilization of 0.48β.

As we shall see (p. 264), the delocalization energy of benzene can be calculated by the Hückel method to be 2β, and since the resonance energy can be experimentally determined to be 36 kcal/mole, the stabilization in the case of 1,3-butadiene as a first guess amounts to 9 kcal/mole. This is of course likely to be somewhat exaggerated: Surely the C_2-C_3 interaction must be smaller than those of C_1-C_2 and C_3-C_4. The calculations can indeed be made somewhat more realistic by assuming, for example, that $H_{23} = 1/2H_{12}$. Further improvements result if one uses $S_{12} = 0.25$ rather than zero, and so on; however, these changes do not affect our main conclusions. If one carries through the same procedure for still longer conjugated systems, the following generalizations can be made. The π and π^* orbitals obtained are to all intents and purposes identical to those we found for the particle in a box. One finds in general that the lowest energy MO has no nodes (other than the plane defined by all the carbon atoms); the next one has one node, the next has two, and so on. So long as the chain has an even number of carbon atoms the nodes always occur between the atoms. When conjugated radicals or ions such as $CH_2 = CH - CH_2$, $CH_2 = CH\text{-}CH\text{=}CH\text{-}CH_2^+$, etc., are considered, some of the nodes occur at the carbon atoms themselves; see, for instance, Fig. 1-16.

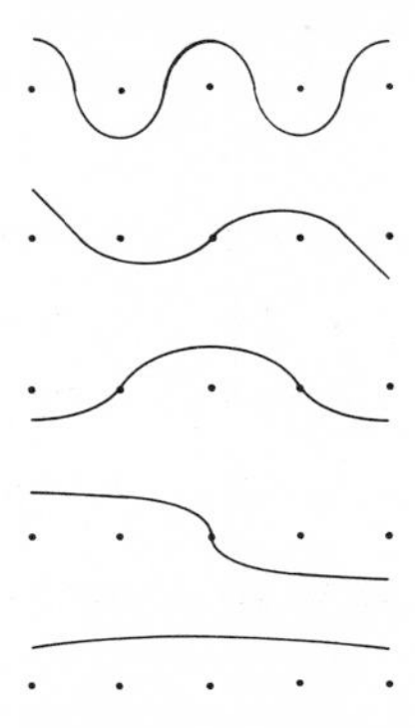

Fig. 1-16. The nodes in the C_5 system.

The delocalization energies and the coefficients of conjugated polyolefins are not the only items that can be extracted by means of these simple Hückel calculations. Heteroatoms may be introduced (with special values for H_{nn}, etc.), charge densities can be calculated, the effect of twisting some of the p orbital axes can be traced, etc.; however, since these calculations are not vital to the further development here, they will not be carried out.

1-12. THE HYDROGEN BOND

The differences in atomic electronegativity cause almost all molecules to be permanently polar; even those that have a net dipole moment of zero may still have polar groups or bonds. In addition, the constant internal and external motions of molecules cause ever shifting and turning induced moments, and as a result molecules are subject to electrostatic attractions (van der Waals forces) to one another.

These attractions are strong enough to have important effects on the physical properties of organic compounds, and one of them in particular is so strong that chemists have adopted the habit of referring to it as a bond: the hydrogen bond. This force comes into play whenever a hydrogen atom bound to an electronegative atom finds itself near the negative end of another dipole (usually also an electronegative atom).

$$\overset{-\delta}{A}—\overset{+\delta}{H} \ldots \overset{-\delta}{B}—\overset{+\delta}{X}$$

The primary reason that hydrogen bonds can be much stronger than other dipole-dipole attractions is that hydrogen is much smaller in size than any other atom, which allows it to approach another dipole more closely. One might expect hydrogen bonding to be especially strong if A is fluorine and if B is a fluoride ion and in fact the F-H ... F^- bond is very strong. Even more interesting however is the fact that this combination is symmetrical, with the hydrogen atom in a position precisely halfway between the fluorine atoms. In other words, a more complete description in this case would be

$$\text{F—H}\ \overset{\ominus}{\text{F}} \longleftrightarrow \overset{\ominus}{\text{F}}\ \overset{\oplus}{\text{H}}\ \overset{\ominus}{\text{F}} \longleftrightarrow \overset{\ominus}{\text{F}}\ \text{H—F}$$

This situation has been encountered only rarely [12], but it serves to raise the question whether the description of a hydrogen bond as a purely electrostatic phenomenon is really adequate. Pimentel has pointed out that boranes also contain symmetrically located hydrogen atoms, and that hydrogen bonds should be considered in terms of molecular orbitals as suggested by Fig. 1-13. This implies that A and B are covalently linked, however weakly, and accordingly that there

12. See for instance (a) D. Hadzi, A. Novak, and J.E. Gordon, J. Phys. Chem., 67, 1118 (1963); (b) D. Hadzi, Chimia, 26, 7 (1972); (c) G.C. Pimentel and A.L. McClellan, "The Hydrogen Bond", Freeman, San Francisco, 1960.

should be an infrared absorption characteristic of this link. This absorption has indeed been observed [13].

Beside fluorine the atoms A and B may be from a first row element such as oxygen or nitrogen, even carbon in certain cases such as HCN and, depending on one's point of view, boron. The electronegativity of second row elements is much less and the hydrogen bonds are correspondingly weaker. It should be emphasized that the primary requirement for B is a high electron density; even such unlikely systems as benzene rings can under certain conditions effectively function as bases in hydrogen bonding. These aspects of the hydrogen bond however are so intimately connected with experimental observation, especially in the infrared and nuclear magnetic resonance spectra, that it seems best at this point to forego further discussion until these techniques have been reviewed (Chapter 2).

1-13. BOND PARAMETERS

There are several descriptive ways of numerically characterizing bonds [4]; they are known as the bond angle, bond length, bond moment, bond polarizibility, bond energy or bond strength, and bond hybridization.

The bond angle is defined by the positions of the nuclei. These positions can experimentally be located by means of electron diffraction in gases and by X-ray diffraction in solids; for reasonably simple gaseous molecules, microwave spectroscopy can also be used. Bond angles depend primarily on the hybridization of the bonds as noted in Section 1-5, but the size of the atoms and any strains that may be present because of cyclic structures may force large deviations. The values shown in Table 1-1 will suffice for most purposes. It may be noted that the term "bond angle" is a little unfortunate in that it implies that the bonds are straight line segments between the nuclei. It is at least probable in many strained molecules such as cyclopropane that the maximum electron densities are not found at the internuclear lines (bent bonds). If we leave excited states out of our considerations for now, it may be said that all bond angles are single-valued; no case is known of isomerism due purely to a difference in bond angles.

Bond lengths likewise are defined by the positions of the nuclei and can be measured by the same techniques as can bond angles. They are determined primarily by the atomic numbers and by the bond multiplicities - multiple bonds being shorter than single ones and heavy atoms being further apart than the lighter ones. Some common values are given in Table 1-2. Bond lengths like bond angles are single-valued, resulting as an energy compromise between the forces giving rise to bonding and internuclear and interelectronic repulsions. They vary somewhat from one molecule to the next; hybridization, strains, and crowding all having their

---

13. (a) T. Miyazawa and K. S. Pitzer, J. Amer. Chem. Soc., **81**, 74 (1959); (b) S. G. W. Ginn and J. L. Wood, Proc. Chem. Soc., (London), 370 (1964); (c) W. J. Hurley, I. D. Kuntz, and G. E. Leroi, J. Amer. Chem. Soc., **88**, 3199 (1966).

effects [14]. An important further effect is that of resonance; thus, the carbon-carbon bonds in benzene are intermediate in length between single and double bonds of that type. In molecules featuring multicenter bonds, adjacent atoms may be usually far apart; thus, the C-C distance in o-carborane is nearly 1.65 Å [15]. In nearly all molecules where such features are absent, however, the bond lengths given in Table 1-2 may be counted on to be correct within a range of about 1-2%.

TABLE 1-1. Approximate Bond Angles Applicable in Most Molecules as a Function of Hybridization

C	(sp^3)		110°
C	(sp^2)		120°
C	(sp)	(acetylenes)	180°
C	(sp)	(allenes)	180°
Si	(sp^3)		110°
N	(sp^3)	(ammonium ions)	110°
N:	(sp^3)		105°
N:	(sp^2)		120°
N	(sp^2)	(immonium ions)	120°
N	(sp)	(nitrilium ions)	180°
N	(sp)	(allenic N)	180°
P:	(p)		90°
O:	(sp^3)	(oxonium salts)	100°
O:	(sp^3)		105°
S:	(sp^3)	(sulfonium salts)	90°
S:	(p)		90°

14. For an interesting discussion of bond shortening due to s character, see R. A. Alden, J. Kraut, and T. G. Traylor, J. Amer. Chem. Soc., 90, 74 (1968).
15. D. Voet and W. N. Lipscomb, Inorg. Chem., 3, 1679 (1964).

TABLE 1-2. Bond Lengths in Angstrom Units

Bond	Length
C–H	1.10
C–C	1.55
C=C	1.35
C≡C	1.20
C–N	1.48
C=N	1.30
C≡N	1.15
C–O	1.43
C=O	1.20
C–F	1.37
C–Cl	1.75
C–Br	1.95
C–I	2.15
N–H	1.00
N–N	1.45
N=N	1.20
O–H	0.95
O–O	1.40

Table 1-3 gives the van der Waals radii of some atoms common to organic chemistry; use of these numbers readily demonstrates the difference in distance between bonded and nonbonded atoms.

Bond moments may be defined by $\mu = q \times l$, where l is the bond length and q is a hypothetical charge one would have to place at the two nuclei in order to produce the "observed" moment. Actually, bond moments cannot be observed directly, but they have to be reproduced from molecular dipole moments by vector analysis. These moments are basically related to the bond lengths and to electronegativities - the more electronegative atoms making more demand on the shared electrons; however, another important consideration is the presence of highly delocalized molecular orbitals. Thus, an electron may be able to achieve a lower energy level by being delocalized in an orbital embracing several electropositive atoms than by being localized at the site of an electronegative atom [5]. Table 1-4 gives some average bond moments in Debye units.

Molecular dipole moments can be determined for molecules in the gas phase by means of microwave spectra; in solution the results are usually based on capacitance measurements.

TABLE 1-3. The van der Waals Radii in Angstroms of Some Common Atoms

Atom	Radius
H	1.2
C[a]	1.7
N	1.5
P	1.9
O	1.4
S	1.9
F	1.4
Cl	1.8
Br	2.0
I	2.2

[a]Half-thickness of benzene.

TABLE 1-4. Bond Moments in Debye Units (10^{-18} esu cm)[a]

C-H	0.4
C-N	1.1
C≡N	4.0
C-O	1.5
C=O	2.7
C-Cl	2.3
C-Br	2.2
C-I	2.0
H-N	1.3
H-O	1.5

[a]Element stated first is the positive end of the dipole.

The polarizability ("softness") of a bond is also of interest. The molecular polarizability can be calculated from the refractive index and the density [16]; inspection of tables of polarizability (see p. 83) leads to the intuitively reasonable conclusions that the larger and less electronegative atoms have the more polarizable bonds and that multiple bonds are much more polarizable than single bonds [17].

Bond energies are defined in terms of the heat necessary to break a bond. It should normally be specified what molecule and what bond is under consideration; thus, the energy necessary to dissociate methane completely into carbon and hydrogen atoms (the "average bond energy") is not necessarily exactly four times the energy necessary to produce hydrogen atoms and methyl radicals (the "bond dissociation energy"). The latter kind is given in Table 1-5.

TABLE 1-5. Bond Energies in kcal/mole for a Number of Bonds Common in Organic Chemistry

Bond	Energy
H-H	105
C-H	100
N-H	100
O-H	115
F-H	135
Cl-H	100
Br-H	85
I-H	70
C-C	85
C=C	145
C≡C	200
C-N	75
C=N	150
C≡N	210
C-O	85
C=O	175
N-N	40
N=N	100
N≡N	225

16. G.W. Castellan, "Physical Chemistry", Chap. 21, Addison-Wesley, Reading, Mass., 1964.
17. J.O. Hirschfelder, C.F. Curtiss, and R.B. Bird, "Molecular Theory of Gases and Liquids", Chap. 13, Wiley, New York, 1954.

TABLE 1-5. (continued)

O-O	33
O=O	120
X...H	0-10

These numbers are derived from many sources. For diatomic species they can be obtained by means of ultraviolet spectroscopy: At energies exceeding the bond energy the spectrum becomes a continuum. For more complex molecules the calorimeter becomes indispensible; it should be realized, however, that bond dissociations can only rarely be carried out directly and hence indirect reactions and the application of Hess' law are usually necessary to derive the results [18].

Like all the other bond parameters, the bond energy varies over a considerable range depending on such factors as local crowding and strain, delocalization both in the molecule itself and in the fragments produced by the bond cleavage, etc. Thus the C-C bond energy in hexa-β-naphthylethane must be close to zero because the molecule is completely dissociated into trinaphthylmethyl radicals; this is due to a combination of substituent crowding in the ethane and to delocalization of the unpaired electron in the radicals. Application of the expression

$$\Delta G = -RT \ln K$$

tells us what fraction of a sample of molecule A-B will be intact at room temperature. If we ignore any difference between the bond energy and ΔG, we find that at 100 kcal/mole

$$K = \exp - (100,000)/(2 \times 300)$$

and hence that very few if any molecules will be dissociated. Even at 5 kcal/mole (approximately the energy of the stronger H bonds) only about 1% would be dissociated.

A related factor is the stiffness of the bond; the weaker bonds would also be expected to be more subject to distortions of various kinds. The resistance to these distortions can often be measured by means of infrared spectroscopy, as will be detailed further in Chapter 2, Section 2-3; suffice it to say at this point that the expected correlation indeed exists.

Finally, hybridization is another important bond parameter. The concept of sp^3, sp^2, and sp orbitals was introduced in Section 1-5, but the remarks made there would not be complete without the comment that hybrid orbitals cannot always be so clearly categorized. In certain molecules, particularly those characterized by crowding and strain, one may find a carbon atom with some of its orbitals having less p character than would be implied by the term sp^3 but more than suggested

---

18. I. M. Klotz and R. M. Rosenberg, "Introduction to Chemical Thermodynamics", 2nd Ed., Chap. 4, New York, 1972.

by sp^2 (a similar remark could of course be made about their s character). Thus, if we consider cyclopropane, the carbon framework - which is necessarily constrained to 60° bond angles - would be strained less if it were held together by p rather than by sp^3 orbitals. That these bonds have indeed more than 75% p character is suggested by several facts. Thus, the HCH angle is much larger (116°) than tetrahedral, as would be expected if these bonds had more than 25% s character. Second, the acidity of cyclopropane is much greater than that of normal alkanes which also points to increased s character (note that acetylene with its sp hybrid orbitals has a pKa some 30 units less than that of methane). Finally, as we shall see in Chapter 2, Section 2-5, the so-called $C^{13}H$ coupling constant which has narrow ranges of well-defined values for sp^3, sp^2, and sp hybridized carbon-hydrogen bonds has values intermediate between those for sp^3 and sp^2 types for cyclopropanes.

1-14. MOLECULAR MOTIONS

The nuclei within a molecule are not at fixed positions except perhaps on an average time scale; if they could be observed over very short time intervals the molecules would be seen to be subject not only to translation, but to vibrations, rotations - both internal and overall - to rocking and bending motions, to inversions, etc. Since these motions are important to our understanding of organic chemistry and to the development of this writing, they are briefly reviewed here.

The rotation of molecules can be interpreted [19] by means of the Schrödinger equation. For a rigid rotator (i.e., a diatomic molecule the length of which is independent of the rotation rate) one finds that

$$E_{rot} = \frac{h^2 J (J + 1)}{8 \pi^2 I}$$

where J is the rotational quantum number (allowed values are 0, 1, 2, 3, ...) and I is the moment of inertia defined by

$$I = \Sigma md^2$$

Here m is the atomic mass and d is the distance of that atom to the axis of rotation (in a diatomic molecule, d is also the distance to the center of mass). The selection rule for rotational levels is $\Delta J = \pm 1$, so that

$$\Delta E = h\nu = \frac{h^2 J}{4 \pi^2 I}$$

19. W. J. Moore, "Physical Chemistry", 5th ed., Chap. 17, Longman, London, 1972.

The absorption of electromagnetic radiation can produce a change in the rotational state only if the molecule has a permanent dipole moment. If that condition is met, the rotational spectrum of the molecule in the gas phase can then be observed in the infrared or microwave region. It allows highly precise measurements of the bond length. In somewhat more complex molecules the geometry can still be very precisely obtained if the same molecule is available with various isotopic substitutions. This is necessary because as the number of atoms increases, so does the number of unknown d terms. One assumes that d is not affected by such substitution. The dipole moment of the molecule in the gas phase can be obtained very precisely by means of the Stark effect; if the rotational spectrum is observed while the sample is in an electric field the absorption lines are found to be split, and the moment can be calculated from the field strength and the splitting.

Vibrational motion of a diatomic molecule can similarly be understood [19] in terms of the wave equation. If we again consider a diatomic molecule and assume that it is a harmonic oscillator (i.e., that it obeys Hooke's law: the restoring force is proportional to the extension of the bond length), the result is

$$E_{vib} = (V + 1/2)\, h\nu$$

where V is the vibrational quantum number. The selection rule is that $\Delta V = \pm 1$; these are the so-called "fundamental" transitions. Transitions with $|\Delta V| \geqq 2$ are called overtones; these are always much weaker. The Hooke's law proportionality constant is given by

$$k = 4\pi^2 \nu^2 \mu$$

where μ is the reduced mass:

$$\mu = \frac{m_1 m_2}{(m_1 + m_2)}$$

The constant k provides a good measure of the stiffness of the bond.

The vibrational transitions are of much higher energy than the rotational ones, and they can as a rule be observed in the infrared. One important aspect of the vibrations in more complex molecules is that they are not strongly coupled, and as a first approximation we may say that the various vibrations displayed in the infrared spectrum are independent of the substituents of both atoms; this of course makes infrared spectroscopy an extremely useful tool to the organic chemist - all the more so because occasional small deviations from the norm are also informative (cf. Sec. 2-3). As with the rotational motion, vibrations cannot be excited unless the bond under consideration has a permanent dipole moment. A difference with rotational motion derives from the (V + 1/2) term: All bonds must be vibrating even in the ground state ("zero point energy" $E_0 = 1/2h\nu_0$).

Not all the absorptions encountered in the infrared are simple along-the-bond back-and-forth vibrations. A complete analysis reveals that a molecule must have (3n - 6) vibrational modes; here n is the number of atoms, 3n is the number of data that must be given to specify their locations, and 6 is the number of translational and rotational motions to which the molecule may be subject (5 for linear molecules). Thus, for a diatomic molecule only a simple vibration is possible; for a nonlinear triatomic molecule three vibrations are possible, one of which would correspond to a bending mode. In still more complex molecules the "normal coordinate analysis" and assignment of all possible modes usually is a difficult task; among the possible modes one then encounters rocking, wagging, twisting, etc.

An extremely important motion within molecules is that of the internal rotations, whereby one group rotates with respect to another around the bond connecting them. These motions, which unlike the vibrations can be frozen out, may be brought about by collisions with and momentum transfer from other molecules. These rotations may be discussed by means of Fig. 1-17, in which a molecule of ethane is pictured with one methyl group rotating around the carbon-carbon single bond. We start with the molecule in the "conformation" shown (the "staggered" one, $\phi = 0$; H_1 and H_2 are said to be in "gauche" positions and H_1 and H_3 are in "trans" positions). If we rotate the front methyl group clockwise, the energy will rise as the C-H bonding electrons are in fact forced into greater proximity; at $\phi = \pi/3$, the energy will reach a maximum, of perhaps 3 kcal/mole. All three pairs of hydrogen atoms are then "eclipsed". The height of the energy barrier is several kcal/mole. This barrier is obviously a function of the substituents at both carbon atoms; it can be made much greater (with $\Delta E > 20$ kcal/mole) if large substituents are used. Another feature that will change when less symmetrical molecules are considered is that the three energy minima need not correspond to degenerate conformations; i.e., these three positions need not portray superimposable and hence indistinguishable molecules.

The three minima would then be isomers, interconvertible by the proper conformational twist ("conformers"). An example would be the 1,2-dibromo-1,2-dichloroethane shown below in Newman projections.

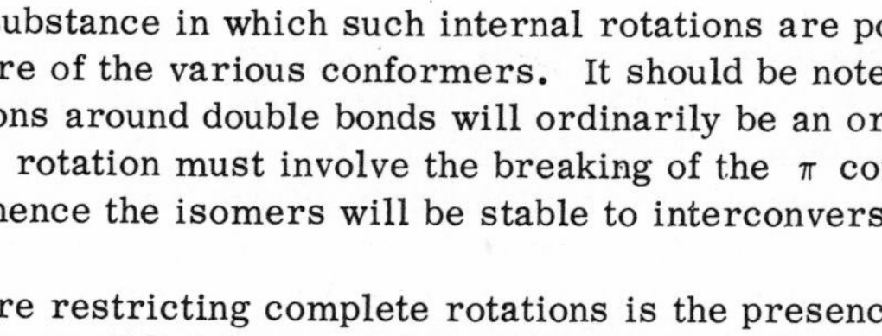

Any organic substance in which such internal rotations are possible will be an equilibrium mixture of the various conformers. It should be noted that the energy barrier for rotations around double bonds will ordinarily be an order of magnitude larger since a 90° rotation must involve the breaking of the π component of the double bond, and hence the isomers will be stable to interconversion in this case.

Another feature restricting complete rotations is the presence of a chain of atoms providing a second link between the two atoms of interest; in other words,

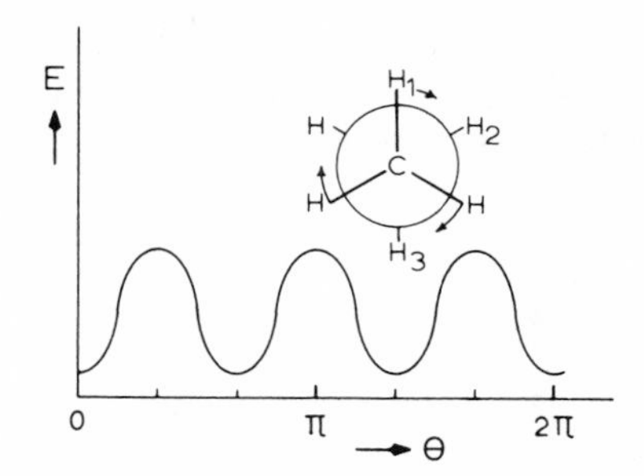

Fig. 1-17. The variation in energy of an ethane molecule as a function of conformation.

a cyclic molecule. Thus in cyclohexane, as the Newman projections below show, only a partial rotation is possible. Such partial rotations are referred to as inversions; they occur in other ring systems as well, and they have become important enough in organic chemistry that we need to consider them in further detail (in Secs. 6-3, 6-4).

Another type of motion referred to as inversion is a process normally associated primarily with nitrogen; it may be described as follows:

The substituent groups or atoms simply move to the right, and simultaneously the nitrogen "flips" to the left through the plane defined by A, B, and C. This process usually has a low activation energy, and it is degenerate unless A, B, and C are all different. It can be prevented if the nitrogen atom is a member of a suitable cyclic system. It is very important to realize that the same operation is not possible with carbon.

While this process may sometimes seem to occur, it will always depend on the ability of the carbon atom to shed - if only temporarily - one of the groups A, B, C, or D, with or without the bonding electrons.

Chapter 2

INSTRUMENTAL TECHNIQUES IN ORGANIC CHEMISTRY [1]

2-1. THE USE OF SIMPLE PHYSICAL CONSTANTS [2]

When the structure of an organic compound of known elemental composition and molecular weight is under investigation, useful initial information can be gained - even as the material is being purified - from the boiling point, melting point, density, refractive index, solubilities, etc. The boiling point for example increases in homologous series with increasing molecular weight, and as the shapes of the molecules vary from highly branched to unbranched; it increases also as the groups present increase in polarity, and especially with the presence and number of hydrogen bonds. A few examples may be helpful.

---

1. The choice of topics for this chapter was governed primarily by the applications described in later chapters. For general reviews, see (a) R.W. Silverstein and G.C. Bassler, J. Chem. Educ., 39, 546 (1962), and bibliography given there; (b) R. Chang, "Basic Principles of Spectroscopy", McGraw-Hill, New York, 1971; (c) P. Laszlo and P.J. Stang, "Organic Spectroscopy; Principles and Applications", Harper and Row, New York, (1971).
2. R.L. Shriner and R.C. Fuson, "The Systematic Identification of Organic Compounds", Wiley, New York, 1968.

The boiling points of methane and methyl fluoride, chloride, bromide, and iodide are -161, -78, -24, 4, and 42°, respectively. The boiling point of formaldehyde is -21°, that of methanol is 65°, that of ethylene glycol is 197°, and that of glycerol is 290°. The boiling points of valeric (pentanoic) acid, 2-methylbutyric acid, 3-methylbutyric acid, and pivalic (2,2-dimethylpropionic) acid are 187, 176, 177, and 164°, respectively; additional examples are readily extracted from any chemical handbook. It should be pointed out that hydrogen bonding raises the boiling point only if it is intermolecular; thus, the boiling point of the ethyl Z-β-hydroxycrotonate (1) is actually lower than that of the isomeric ethyl acetoacetate (2).

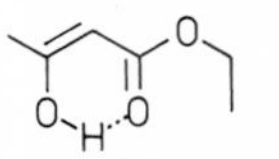

(1) 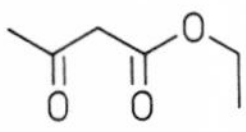 (2)

Melting points are less reliable indicators than boiling points, but in general they tend to be governed by the same considerations except that branching tends to have the opposite effect here. As a consequence highly branched molecules tend to have narrow liquid ranges; thus, 1-butanol is a liquid between -90 and 118°, whereas t-butanol is liquid between 25 and 82°. One might think of unbranched molecules as being more exposed to van der Waals interactions with their neighbors, but also as not fitting in a crystal lattice as neatly. Symmetrical shapes tend similarly to lead to higher melting points; thus, p-disubstituted benzene derivatives invariably have higher melting points than their o- and m-isomers. A spectacular example of this consideration is the hydrocarbon adamantane (3) of tetrahedral symmetry, which has a melting point evidently much higher than its boiling point: When heated at one atmosphere it sublimes at 268°.

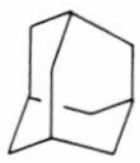

(3)

Densities tend to be in line with melting points, but the most important factor now is the presence of atoms of high atomic numbers. Thus, benzene and fluoro-, chloro-, bromo-, and iodobenzene have densities of 0.88, 1.02, 1.11, 1.50, and 1.83, respectively. One disadvantage of the density is the large amount of material needed to determine it; it is difficult to imagine a way to measure it to the third decimal place with much less than 0.1 g - an enormous amount by present standards. While the material is not lost, the value of the information is usually not worth the tedium involved in recovering the sample.

The index of refraction can be measured more quickly, more accurately, and with somewhat less material, which is not easily recovered, however. The main advantage of the refractive index is that it can be predicted with reasonable accuracy by means of the Lorenz-Lorentz equation

$$n = \left\{ \frac{m + 2dM_D}{m - dM_D} \right\}^{1/2}$$

where m is the molecular weight, d the density, and M_D the so-called molar refractivity. The latter term can be quickly obtained as the sum of the appropriate atomic refractivities and a number of structural correction terms.

2-2. MASS SPECTROMETRY [3]

The mass spectrometer can be used for rather tiny amounts (micrograms) of substance in any state to give information that in many cases suffices to assign the structure even when no other information is available. Basically the idea is that a stream of molecules is bombarded with moderate energy (70 V) electrons; a reaction then occurs and the resulting molecular ion and positively charged fragments

$$M + e^{\ominus} \longrightarrow M^{\oplus} + 2e^{\ominus}$$

of it travel through a combination of magnetic and electric fields which reveals their ratio m/e. The spectrum finally recorded shows a number, usually a large number, of peaks of various intensities; it allows the experienced mass spectrometrist to rule out many structures almost at a glance.

The m/e ratios can be measured to perhaps 10^{-4} amu in the best (highest resolution) machines; with such an instrument one can determine the precise elemental make up of a given ion peak. Thus S^+, O_2^+, $NHOH^+$, CH_3OH^+, and $CH_3NH_3^+$ are isobaric ions (i.e., they all have the same rounded-off mass), but these masses differ considerably in the fourth decimal: 31.9721, 31.9898, 32.0137, 32.0262 and 32.0501, respectively. Even when resolution of only 0.1-0.2 amu is available over a wide range (perhaps to 1000), the elemental composition can often still be quickly ascertained by the use of relative isotopic abundances. Thus, if methyl bromide (molecular weight is 95) is examined, one finds molecular ion ("parent") peaks of almost equal intensity at 94 and 96, corresponding to the two bromine isotropes ^{79}Br and ^{81}Br, which occur in almost equal abundance. Any ion presumably containing a bromine atom should therefore be accompanied

---

3. For brief reviews, see (a) F.W. McLafferty, Science, 151, 641 (1966); (b) K. Biemann, Angew. Chem., Int. Ed. Engl., 1, 98 (1962); (c) E. Schumacher, Helv. Chim. Acta, 46, 1295 (1963); (d) J.L. Franklin, J. Chem. Educ., 40, 284 (1963). Readers interested in the related techniques of ion cyclotron resonance spectroscopy are referred to (e) J.D. Baldeschwieler, Science, 159, 263 (1968), and those interested in the behavior of negative ions in a mass spectrometer to (f) R.T. Aplin, H. Budzikiewicz, and C. Djerassi, J. Amer. Chem. Soc., 87, 3180 (1965).

by another ion with about equal intensity differing from it in m/e by two units. If two bromine atoms are present, there should be three peaks differing in m/e by two units each, and with intensities approximately 1:2:1 (of course, in low resolution machines these ratios may be obscured because of the presence of isobaric ions). Other atoms common in organic chemistry which have stable isotopes are hydrogen (^{2}H, 0.016%), carbon (^{13}C, 1.08%), nitrogen (^{15}N, 0.36%), oxygen (^{17}O, 0.04% and ^{18}O, 0.20%), sulfur (^{33}S, 0.78% and ^{34}S, 4.39%), and chlorine (^{37}Cl, 32.7%) (in each case the abundance is given in percent of the more common isotope).

The interpretation of the mass spectrum basically requires only some rather elementary insights. The parent ion is both a cation and a radical, and it will prefer to break so as to produce a stable cation and a stable radical. For example, allyl methyl ether (4) would be expected to give rise to the resonance stabilized allyl cation very easily.

$$[\text{CH}_2{=}\text{CH}{-}\text{CH}_2{-}\text{O}{-}\text{CH}_3]^{+\cdot} \longrightarrow \text{Me}\dot{\text{O}} + \text{CH}_2{=}\text{CH}{-}\overset{\oplus}{\text{C}}\text{H}_2 \longleftrightarrow \overset{\oplus}{\text{C}}\text{H}_2{-}\text{CH}{=}\text{CH}_2$$

The benzyl ion is (6) likewise a stable one and hence common in mass spectra (actually it is the tropylium ion (5); it has been shown that all isomers C_7H_8 readily produce this ion in the gas phase) [4]. Others are acylium ions (7), alkoxy- or hydroxy-substituted carbonium ions (8), and so on.

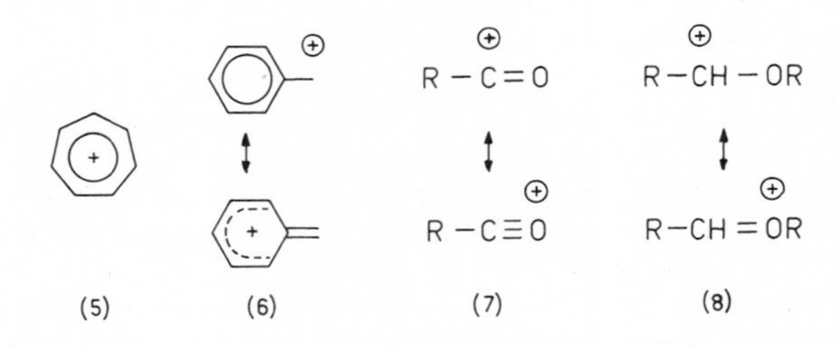

If the likely products of a given bond fission are connected in some additional manner, then the reaction does of course not give rise to "smaller" fragments; hence the absence of very many peaks and an intense parent peak are usually indicative of a highly condensed ring system. If several small groups of peaks are observed at 14 amu intervals, a chain $-(CH_2)_n-$ is likely to be present. If a relatively broad ("metastable") peak not necessarily at an integral value of m/e is observed, a relatively stable ion was present which did not decompose instantly, but did so during the short trip to the detection apparatus. It has been shown that in such a case

$$m_{app} = m_{fin}^2/m_{init}$$

4. S. Meyerson, J. Amer. Chem. Soc., 85, 3340 (1963) and Record Chem. Progr., 26, 257 (1965).

These are several though by no means all the arguments that are used by an experienced mass spectrometrist as he appraises the various possibilities. The necessary calculations are highly repetitive and laborious and lend themselves to computerization; mechanical calculation in fact has become virtually indispensible in this field.

Specific isotopic labeling is a valuable tool in tracking down the general reactions that gaseous organic ions undergo, and many of the insights used have been gained in painstaking investigations of various compounds and their isotopically substituted analogs. Thus, to mention one example, the mass spectra of *trans*-1-decalone (9), *cis*- and *trans*-9-methyl-1-decalone (10) and no fewer than 17 specifically deuterium-labeled analogs of these three compounds have permitted the virtually complete unravelling of the pathways by which the parent ions fragment [5].

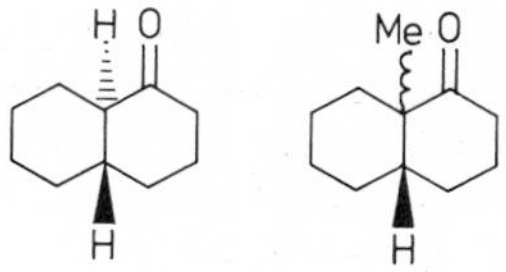

(9) (10)

It should be noted that unlike most of the other techniques discussed below, mass spectrometry can provide only the elementary structural features; the more detailed information such as bond lengths, the dipolar character of bonds, delocalization, strain, and so on must be obtained in other ways.

2-3. INFRARED SPECTROSCOPY [6]

The infrared portion of the electromagnetic spectrum is normally considered to cover the wavelength region from 2 to 15 microns (μ), or as many spectroscopists prefer, from 5000 to 700 wave numbers (cm^{-1}). Since $E = hc/\lambda$, it follows that the energy range is roughly 10 to 1 kcal/mole. As noted in Chapter 1, Section 1-14, the absorptions encountered in this region are primarily molecular distortions of all sorts: Stretching, bending, rocking, etc. Since most of these distortions involve only one bond or a small group of atoms and since they are to a first approximation independent of other atoms present in a molecule, the corresponding

5. E. Lund, H. Budzikiewicz, J. M. Wilson, and C. Djerassi, *J. Amer. Chem. Soc.*, *85*, 941 (1963).
6. For reviews, see (a) K. Whetsel, *Chem. Eng. News*, *46*, 82 (Feb. 5, 1968), and bibliography given there; (b) D. A. Dows, *J. Chem. Educ.*, *35*, 629 (1958). For a review of "near ir," see (c) O. H. Wheeler, *J. Chem. Educ.*, *37*, 234 (1960); for "far ir," see (d) J. W. Brasch, Y. Mikawa, and R. J. Jakobsen, *Appl. Spectr. Rev.*, *1*, 187 (1968).

absorptions are very useful in identifying functional groups. Table 2-1 lists a few of such groups. The assignments are made on the basis of experience (i.e., the same band or bands will generally be encountered if a given group is present) and sometimes on the basis of isotopic substitution (note the mass effect in the equation for the force constant, p. 31; the effect is especially pronounced in deuterium substitution).

Experience has further shown that many groups have several characteristic absorptions; for instance, a methyl group will in general have an absorption due to CH bending at about 1450 cm^{-1} as well as the stretching mode at approximately 2900 cm^{-1}. The deviations caused by neighboring groups are also very useful and may be illustrated by means of the carbon-carbon double bond stretching frequency, which may be found at 1645, 1660, 1675, 1655, or 1670 cm^{-1} depending on whether the olefin is mono-substituted, *cis*-, *trans*-, or gem-disubstituted, or trisubstituted, respectively. Allenes and conjugated or aromatic double bonds occur at still different frequencies. Similarly the carbonyl group position is dependent in a known way on whether the compound under investigation is an aldehyde, ketone, ketene, acid, ester, amide, or anhydride. Benzene rings are subject to skeletal vibrations observable between 1700 and 2000 cm^{-1} that are characteristic of the type of substitution (i.e., *meta*, *sym*-1,3,5-, etc.). Carbon hydrogen stretching frequencies depend on the s character of the bond and on local crowding; Winstein has reported [6e] that in certain highly hindered hydrocarbons such as (11) the stretching frequency of the compressed C-H bond may be shifted to higher energies by as much as 150 cm^{-1}. In few cases are the distinctions so rigid that an

TABLE 2-1. Infrared Absorptions Often Observed in Organic Molecules

Groups	Wave number (cm^{-1})
O-H, N-H	3600
C-H	2850 - 3000
C≡C	2100 - 2250
C≡N	2250
C=O	1700 - 1730
C=C	1650
N=N	1600
C-Cl	700
C-Br	550

6e. D. Kivelson, S. Winstein, P. Bruck, and R. L. Hansen, *J. Amer. Chem. Soc.*, **83**, 2938 (1961).

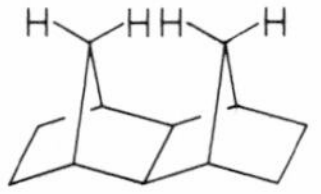

(11)

assignment can be made with absolute certainty; however, the suggestions that are possible on the basis of a careful review of an infrared spectrum are numerous and strong. Even a cursory inspection of the recording can be illuminating; thus, since the absorption of a photon can be used to excite a distortional mode only if this is accompanied by a change in dipole moment, a relatively simple spectrum may well be indicative of a very simple or symmetrical molecule. For this reason carbon tetrachloride and carbon disulfide are the favorite solvents (for volatile substances a vapor spectrum may be taken and for insoluble solids a KBr matrix or hydrocarbon oil (Nujol) suspension is usually employed). The optics of the instrument for the same reason are usually manufactured from fused sodium chloride, potassium bromide, and similar materials.

There are three principal types of use to which the infrared spectrum is put in organic chemistry. One is simply as an aid in structure assignment of new compounds, more or less on the basis of experience and of extensive lists of functional group vs. frequency correlations. A second one is the fingerprinting of a compound; its infrared spectrum characterizes it whether the various absorptions can be assigned or not, and it does so in a way much more unique for instance than can be achieved by means of derivatives and their melting points. A third use is the study of structural details - including such items as bond angles and so on. This latter application is more difficult and it is usually restricted to small, simple molecules; however, certain details of the nature of the bonds may sometimes be revealed even in more complex molecules. The following examples are often encountered.

In conjugated molecules the effect of resonance may be demonstrable. Thus, one finds that in conjugated, unsaturated carbonyl compounds both the C=C and C=O stretching frequencies are lower than in comparable compounds in which this feature is not present; both of these changes may be attributed to a contribution from structure (12). In cyclic ketones one observes a regular increase in the C-O

(12)

stretching frequency as the angle strain is raised; thus, while the absorption appears at 1715 cm^{-1} for cyclohexanone, it has a value of 1813 cm^{-1} for cyclopropanone [7]. Presumably the increased stiffness of the bond can be ascribed to a higher degree of s character (see Sec. 1-3). The hydrogen bond (see Sec. 1-12) is another intriguing example of the kind of detail that can be learned by infrared

7. (a) N. J. Turro, Accounts Chem. Res., 2, 25 (1969); (b) C. S. Foote, J. Amer. Chem. Soc., 86, 1853 (1964); (c) P. v. R. Schleyer, J. Amer. Chem. Soc., 86, 1854, 1856 (1964).

spectroscopy. The formation of such a bond to a basic atom B, in turn, weakens the bond A-H. In compounds containing the OH group there are usually several absorptions: One due to free ROH perhaps at 3650 cm^{-1}, one to dimeric molecules at perhaps 3550, and a very broad and featureless band between 3500 and 3000 cm^{-1} covering presumably more highly polymerized combinations. According to one point of view, the remarkable linewidth of this absorption may be due to exceedingly rapid hydrogen exchange [8], a phenomenon often encountered in nuclear magnetic resonance (see Sec. 2-6 below). In this region Beer's law (optical density is proportional to concentration; see Sec. 2-8) is obviously not obeyed; the free OH band is relatively more intense in dilute solution, whereas the polymer band is most pronounced in concentrated solutions (it may then in fact be the only observable one).

There is much evidence to support the intuitively reasonable proposition known as Badger's rule [9]: The stronger the hydrogen bond is, the larger will be $\Delta\nu$, the difference in ν between free and bonded molecules. Thus, whereas $\Delta\nu$ is 107 cm^{-1} in p-cresol, it is only 53 cm^{-1} in 2,6-diisoprophyl-4-methylphenol [10]. If $\Delta\nu$ for a number of hydrogen-bonding compounds is plotted against alternative measures of the strength of a hydrogen bond such as the difference in their heats of mixing with carbon tetrachloride and chloroform, then a straight line is observed [11]. A similar correlation has been shown to exist between $\Delta\nu$ and the pKa of certain acids - the acid strength is greatly diminished when an internal acceptor atom B is favorably situated nearby [12]. Still another correlation (with nmr observations) will become clear when that topic is discussed below (p. 51).

This simple measure $\Delta\nu$ (and several variations such as the first "overtone" in the near infrared ($\Delta V = 2$) at 7200 cm^{-1} [13], or the bands observed in deuterated molecules) has allowed a number of fascinating observations to be made, and all the following conclusions have depended on it.

By the use of a standard acid HA one can measure the effectiveness of various bases B to accept such a bond and conversely by the use of a standard base one can determine the ability of HA to donate one; one cannot learn very much from the association of a single hydrogen-bonding substance (which functions as both acid and base), since if $\Delta\nu$ is small and little association seems to be occurring, this may be due either to inadequate donor or acceptor properties (or both). Also, the steric requirements of HA and B may be such that the three atoms cannot reach the positions best from the point of view of forming a hydrogen bond [14]. Furthermore, there may be a direct through-the-bonds interaction between A and B

---

8. L. J. Bellamy and P. E. Rogasch, Proc. Roy. Soc. (London), A257, 98 (1960).
9. R. M. Badger and S. H. Bauer, J. Chem. Phys., 5, 839 (1937).
10. N. A. Puttnam, J. Chem. Soc., 486 (1960).
11. (a) M. Tamres and S. Searles, J. Amer. Chem. Soc., 81, 2100 (1959); (b) For a dissenting view on intramolecular hydrogen bonds, see L. P. Kuhn and R. A. Wires, J. Amer. Chem. Soc., 86, 2161 (1964).
12. L. B. Magnusson, C. Postmus, and C. A. Craig, J. Amer. Chem. Soc., 85, 1711 (1963).
13. D. L. Powell and R. West, Spectrochim. Acta, 20, 983 (1964).
14. A. W. Baker and W. W. Kaeding, J. Amer. Chem. Soc., 81, 5904 (1959).

affecting the properties of each in such a way that their normal hydrogen-bonding abilities are changed significantly. Thus, although the nitro group is normally a surprisingly poor base in hydrogen bonds [15], the resonance interaction in o-nitrophenol (13) is such that the intramolecular hydrogen bond is quite strong in that case [16].

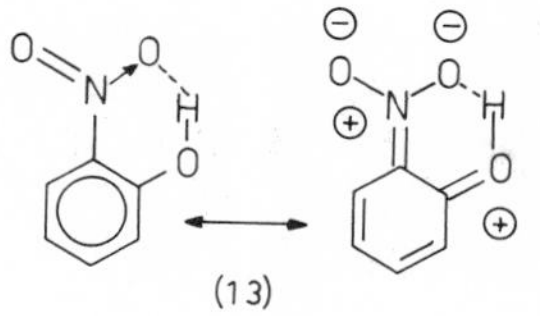

(13)

A study of $\Delta\nu$ of methanol and phenol in dilute solution in $(n\text{-Bu})_rX$ has shown that base strength decreases in the order X = N, P, As > O, S > I > Br > Cl > F [17]. Studies of the substituent effect in phenols and imines have revealed, as might be expected, that electron-withdrawing groups strengthen the hydrogen bonding ability of the group A-H, and that electron donating groups strengthen the ability of B to accept such bonds [18]. In a few cases even the carbon atom may function as either A or B. Thus, mono-substituted acetylenes will associate weakly with dimethylformamide [19], and chloroform will associate with ammonium halides [20]; phenol will associate to some degree with isonitriles [21], and in fact a C-H···C bond can be detected in mixtures of acetylenes with isonitriles [22]. It has been found that π-orbital electrons are capable of functioning as bases; thus, phenols associate weakly with olefins [23] and aromatic hydrocarbons [24]. The increased p character (cf. p. 30) of the carbon-carbon bond in cyclopropane seems to be sufficient to make even that group weakly effective as an acceptor; thus, while the alcohols (14)-(16) show only a single band characteristic of free O-H in very dilute solution in carbon tetrachloride, (17)-(20) show a second weak band at somewhat lower frequency. Interestingly, this hydrogen bond is to the edge rather than to the face of the ring; thus, (21) does not show the presence of a hydrogen bond. The exact orientation of the hydroxy and cyclopropyl groups is of importance; thus, (22) does not show infrared evidence of internal hydrogen bonding. Finally, this phenomenon is not present in larger rings; in the compounds (23) internal hydrogen bonding is observed only when n equals 1 [25].

---

15. H.E. Ungnade, E.M. Roberts, and L.W. Kissinger, J. Phys. Chem., 68, 3225 (1964).
16. W.F. Baitinger, P.v.R. Schleyer, T.S.S.R. Murty, and L. Robinson, Tetrahedron, 20, 1635 (1964).
17. P.v.R. Schleyer and R. West, J. Amer. Chem. Soc., 81, 3164 (1959).
18. A.W. Baker and A.T. Shulgin, J. Amer. Chem. Soc., 81, 1523 (1959).
19. R. West and C.S. Kraihanzel, J. Amer. Chem. Soc., 83, 765 (1961).
20. A. Allerhand and P.v.R. Schleyer, J. Amer. Chem. Soc., 85, 1233, 1715, (1963).
21. A. Allerhand and P.v.R. Schleyer, J. Amer. Chem. Soc., 85, 866 (1963).
22. L.L. Ferstandig, J. Amer. Chem. Soc., 84, 1323 (1962).
23. R. West, J. Amer. Chem. Soc., 81, 1614 (1959).
24. Z. Yoshida and E. Osawa, J. Amer. Chem. Soc., 87, 1467 (1965) and 88, 4019 (1966).
25. L. Joris, P.v.R. Schleyer, and R. Gleiter, J. Amer. Chem. Soc., 90, 327 (1968).

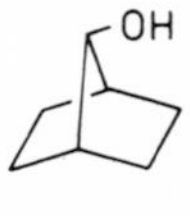

OH OH OH

(14) 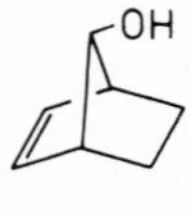 (15)  (16)

OH OH OH OH

(17) 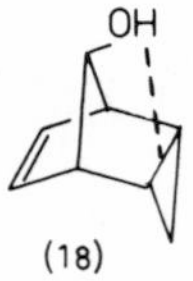 (18) 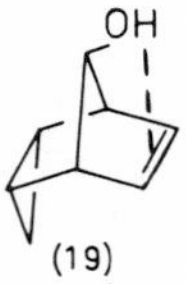 (19) 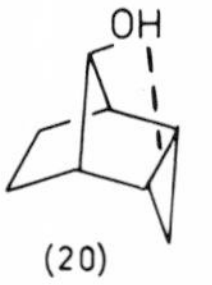 (20)

OH Me CH_2OH $(H_2C)_n$ OH

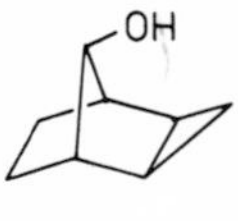

(21) 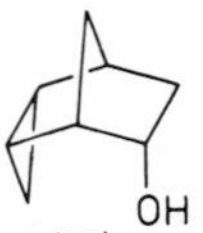 (22) 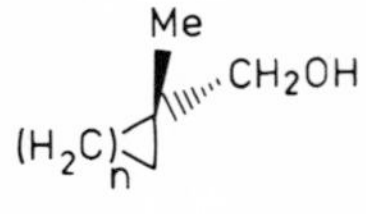 (23)

Another rather amazing development is the finding that certain covalently bound metal atoms may accept a hydrogen bond; for example, ferrocenyl-substituted alcohols (cf. Chapter 10) in dilute carbon tetrachloride solution show, beside free hydroxyl groups, the presence of two weakly hydrogen bonded ones, thought to be metal- and π-bonded (24) and (25). In ether solution, only one band is observed [26].

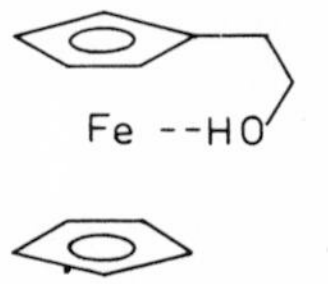

Fe --HO H—O Fe

(24) 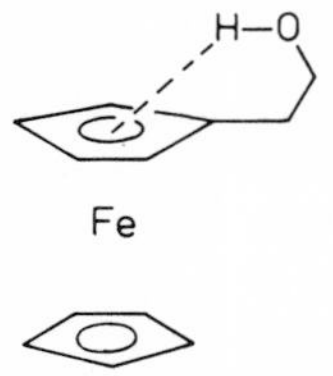 (25)

It is clear from many observations (the symmetrical nature of HF_2^-, the high boiling points of hydroxy compounds, the hydrogen bonding ability of chloroform, etc.) that A must be an electronegative atom or be bound to a number of electronegative atoms to be very effective in the participation of hydrogen bonds. An extreme example is provided by hexafluoroisopropanol ($CF_3CHOHCF_3$), which forms a compound with tetrahydrofuran with a boiling point 30° higher than the base itself [27].

The nature of A and that of B are not the only factors determining the strength of the hydrogen bond. Proximity is obviously another important factor; thus, in the glycols CH_2OHCR_2OH the internal hydrogen bond is stronger the bulkier the group

---

26. D.S. Trifan and R. Bacskai, J. Amer. Chem. Soc., 82, 5010 (1960).
27. W.J. Middleton and R.V. Lindsey, J. Amer. Chem. Soc., 86, 4948 (1964).

R; large alkyl groups would be expected to be further apart and hence to force the two hydroxyl groups more closely together [28].

As noted above, the external hydrogen bond among phenol molecules is weakened somewhat if one o-substituent is present, and severely so if there are two [10]. Orientation is evidently once again an important factor; in the series of phenols (26)-(31), $\Delta\nu$ for the OH-π bond increases from zero up in the order shown, which is exactly the order expected on the degree of nonplanarity anticipated (cf. Sec. 5-11) [29].

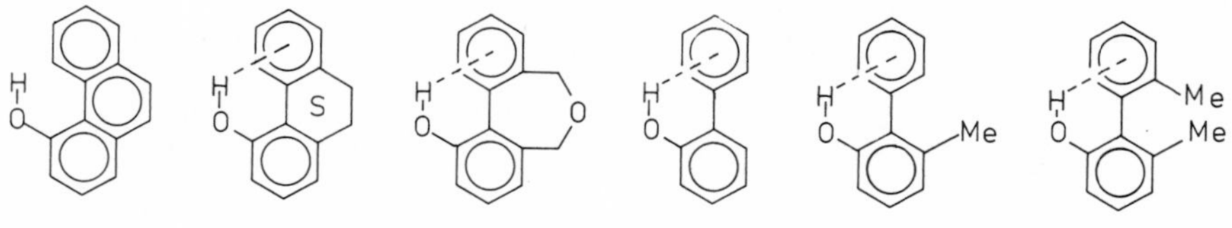

(26) (27) (28) (29) (30) (31)

There are a number of reports according to which there is an inverse relation between $\Delta\nu$ and the dihedral angle θ in vicinal diols (32). Recently however it has been found that this relationship fails badly for certain rigid diols in which θ is precisely known; thus, noradamantane-3,4-diol (33) ($\theta = 0$) has a very weak internal hydrogen bond [30].

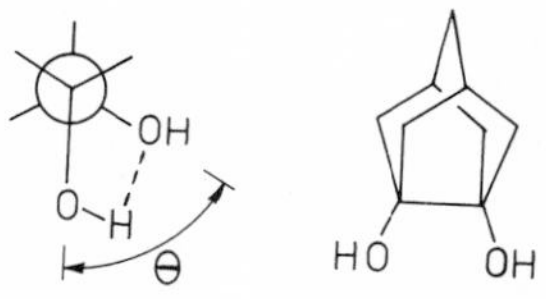

(32) (33)

Infrared spectroscopy can often be used to demonstrate the existence of distinct and different rotational conformations. Even though the activation energies barring the conversion of one rotamer into another are usually far too low to permit their separation (cf. p. 32), if both rotational states have comparable populations they may be recognizable through their infrared spectra. Many examples of this are known. Thus, the two OH bands observed in dilute solutions of 2-phenylethanol have been interpreted as due to a free hydroxyl group in the trans-conformation (34) and a π-bonded one in the gauche rotamer (35) [31]. Another

---

28. L. P. Kuhn, J. Amer. Chem. Soc., 80, 5950 (1958).
29. M. Oki and H. Iwamura, J. Amer. Chem. Soc., 89, 576 (1967).
30. T. M. Gorrie, E. M. Engler, R. C. Bingham, and P. v. R. Schleyer, Tetrahedron Lett., 3039 (1972).
31. P. J. Krueger and H. D. Mettee, Tetrahedron Lett., 1587 (1966).

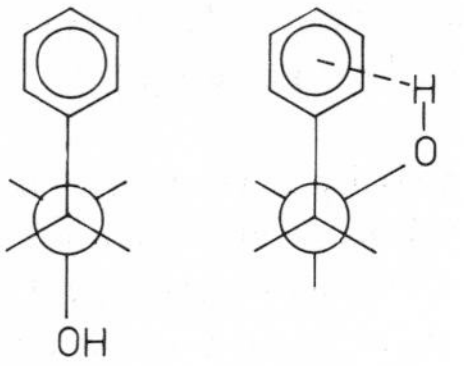

OH

(34) (35)

example occurs in many vinyl ethers (36); resonance involving the unshared oxygen electrons and the π bond leads to a preferred planar structure in which the alkyl and vinyl groups may be either cis or trans (37) or (38). In a number of instances

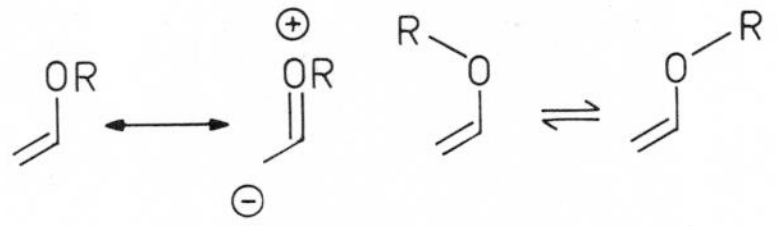

(36) (37) (38)

both rotamers have been observed; a doublet absorption at 1600 cm^{-1} is then seen [32]. Proof that such a doublet has this origin is provided by the temperature effect on the relative intensities; since ΔH for such an equilibrium is likely to be other than zero, temperature changes may be expected to shift the equilibrium.

One additional subject should be mentioned here, namely, the Raman spectrum [33]. It turns out that when a powerful beam of monochromatic visible light is passed through a sample, some of it is found to be scattered; when the scattered light is examined, one finds that beside the original wavelength a number of other frequencies are present (Fig. 2-1). The differences $\Delta\nu$ between the incident and scattered light correspond to various vibrational states; the importance of these lines lies in the fact that some of them may correspond to forbidden and hence "infrared inactive" transitions. Thus, the Raman spectrum complements ir nicely, a fact especially useful when highly symmetrical molecules - with many forbidden transitions - are under study. Another major advantage is that since we are dealing with visible light, glass or quartz cells and optics can be used, and hence aqueous or even sulfuric acid solutions can be employed.

32. N. L. Owen and N. Sheppard, Trans. Faraday Soc., 60, 634 (1964), and references quoted there.
33. R. N. Jones, J. B. DiGiorgio, J. J. Elliott, and G. A. A. Nonnenmacher, J. Org. Chem., 30, 1822 (1965).

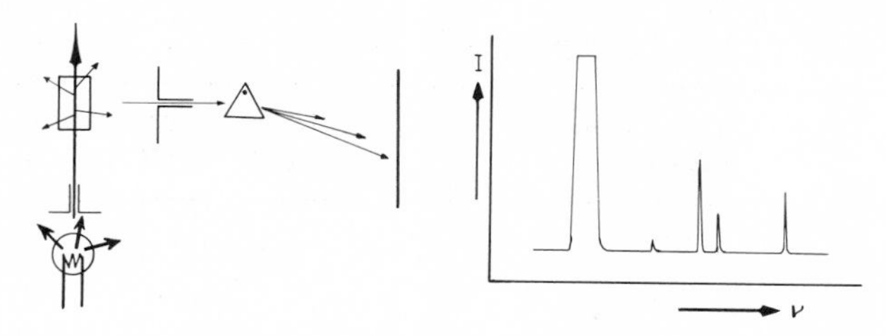

Fig. 2-1. A schematic view of a Raman measurement and a typical Raman spectrum.

2-4. nmr: THE CHEMICAL SHIFT [34]

The phenomenon of nuclear magnetic resonance (nmr) depends on the fact that many nuclei have a net spin, and that the orientation of their spin axes in a magnetic field is restricted to certain angles. One may think of such spinning nuclei as spheres on whose surface a current of positive charge flows. Such a current is associated with a magnetic field, and this field can be oriented with respect to an external field only at certain angles. The spin is said to be quantized (see Fig. 2-2). Nuclei having even numbers of protons and neutrons can spin at only one such orientation; they are said to have a nuclear spin quantum number I = 0. The number of allowed orientations equals (2I + 1). Table 2-2 shows this quantum number for a number of nuclei common in organic chemistry.

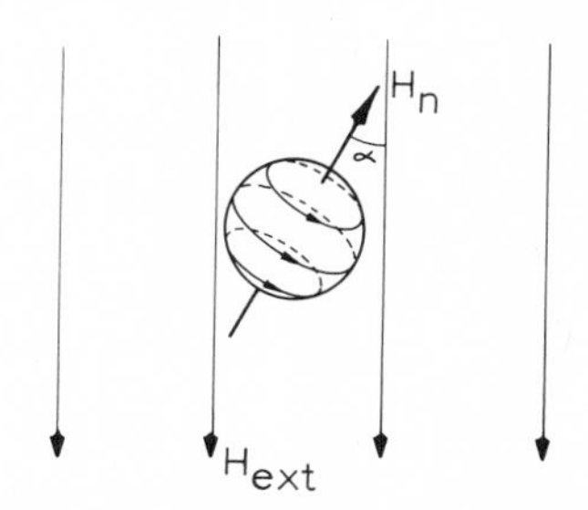

Fig. 2-2. A nucleus spinning in an external magnetic field at an angle α. If the spin quantum number is 1/2, it may also be oriented at angle $(\pi - \alpha)$.

---

34. For reviews, see (a) J.C. Martin, J. Chem. Educ., 38, 286 (1961); (b) J.D. Roberts, "Nuclear Magnetic Resonance; Applications to Organic Chemistry", McGraw-Hill, New York, 1959; (c) J.A. Pople, W.G. Schneider, and H.J. Bernstein, "High-Resolution Nuclear Magnetic Resonance", McGraw-Hill, New York, 1959; (d) D.J. Pasto and C.R. Johnson, "Organic Structure Determination", Prentice-Hall, Englewood Cliffs, New Jersey, 1969, and references given there; (e) J.R. Dyer, "Applications of Absorption Spectroscopy of Organic Compounds", Prentice-Hall, Englewood Cliffs, New Jersey, 1965; (f) F.L.J. Sixma and H. Wynberg, "A Manual of Physical Methods in Organic Chemistry", Wiley, New York, 1964; (g) F.A. Bovey, Chem. Eng. News, 43, 98 (Aug. 30, 1965).

TABLE 2-2. Properties of Some Common Nuclei

Nucleus	Natural abundance (%)	I	Larmor frequency (in MHz at 10^4 gauss)	Relative sensitivity
^{1}H	100	1/2	42.577	1
^{2}H (D)	0.016	1	6.536	0.0096
^{10}B	18.8	3	4.575	0.020
^{11}B	81.2	3/2	13.660	0.17
^{12}C	99	0	-	-
^{13}C	1.11	1/2	10.705	0.0159
^{14}N	100	1	3.076	0.00101
^{15}N	0.37	1/2	4.315	0.00104
^{16}O	100	0	-	-
^{17}O	0.037	5/2	17.235	0.0291
^{19}F	100	1/2	40.055	0.834
^{23}Na	100	3/2	11.262	0.0927
^{24}Mg	90	0	-	-
^{25}Mg	10	5/2	2.606	0.0268
^{28}Si	95	0	-	-
^{29}Si	4.7	1/2	8.460	0.0785
^{31}P	100	1/2	17.235	0.0664
^{32}S	100	0	-	-
^{33}S	0.74 .	3/2	3.266	0.00226
^{35}Cl	75	3/2	4.172	0.00471
^{37}Cl	25	3/2	3.472	0.00272
^{39}K	100	3/2	1.987	-
^{79}Br	50.6	3/2	10.667	0.0786
^{81}Br	49.4	3/2	11.498	0.0984
^{127}I	100	5/2	8.519	0.0935
e^-	-	1/2	27.994	2.85×10^8

A nucleus spinning with its nuclear axis lined up against the field is in a higher energy state than one whose axis is oriented with the external field:

$$E = -\gamma H_0 I h/2\pi$$

where γ is the so-called gyromagnetic ratio. However, the field alone cannot flip it over. When a magnetic nucleus finds itself in a magnetic field, it will precess (see Fig. 2-3); the precession (or Larmor) frequency is proportional to

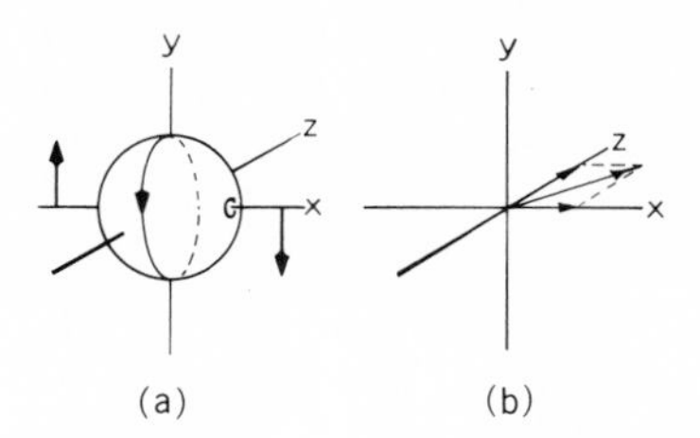

Fig. 2-3. (a) A spinning object subject to forces acting on its axis of rotation (X), and hence forces tending to produce rotation about axis Z. (b) If the rotation is represented by the arrow along the X axis and the rotational forces by the arrow at the Z axis, the resultant is the new axis of rotation. The shift is called precession.

the magnetic field. A flip from one allowed orientation to another can be brought about by photons whose frequency equals the Larmor frequency; the selection rule is $\Delta I = \pm 1$.

$$\nu = \frac{\gamma H_o}{2\pi}$$

When a large number of nuclei are so irradiated and induced to flip, the change induces a current in an appropriately located coil, and this signal is detected. Commercial instruments for protons usually operate at 60, 100, or 200 MHz (14,100, 23,500, and 47,000 gauss, respectively).

Just precisely at what frequency a given proton will precess depends to a minor degree on its chemical environment; the "chemical shift" σ has a range of perhaps 20 ppm (i.e., 20×10^{-6}; hence at 60 MHz, a range of 1200 Hz or cps) with a few exceptional cases lying outside it. Fortunately, the accuracy to which the precession can be routinely measured is extremely high (roughly 1 ppb, or 10^{-9}) and hence proton magnetic resonance has become an extremely useful tool in all phases of organic chemistry.

The frequency of a given chemical type of proton is usually recorded in terms relative to that of a standard substance. The standard is usually tetramethylsilane (TMS), chosen for its volatility and hence easy removability (the boiling point is 26°), chemical unreactivity, and its sharp signal near one extreme of the usual frequency range. The scale used by most chemists is the τ-scale, according to which TMS appears at 10 ppm. Others use the σ-scale, best defined by saying that $\tau + \sigma = 10$; the author prefers the τ-scale because increasingly lower τ values are indeed associated with lower fields (or lower precession frequencies). The use of hertz (Hz) should also be discouraged altogether because the chemical shift is not defined by it unless the external field strength is also given.

6	7	8	9	10	τ
4	3	2	1	0	σ
240	180	120	60	0	Hz (cps, sec^{-1}) at 14,100 gauss
400	300	200	100	0	Hz at 23,500 gauss

The chemical shifts of some common groups are given in Table 2-3.

TABLE 2-3. Chemical Shifts in τ Units for Some Common Proton Types

$-CH_3$	9.1	▷-H	10.5
$-CH_2$	8.7	$-CH_2Cl$	6.3
-CH	8.0	$-CH_2Br$	6.5
$=CH_2$	5.0	$-CH_2I$	6.8
=CH	4.7	$-CHCl_2$	4.2
Ar-H	2-3	$(RO)_2CH_2$	4.7
≡CH	7.5	$=C-CH_3$	8.2
C(=O)H	0.3	$Ar-CH_3$	7.7
OCH_3	6.2	$-COCH_3$	7.7
ROH	5		
COOH	-1		
ArOH	3		

It may be noted that the substituent effects are more or less additive; thus, $C-CH_2-C$ appears at 8.7τ, $C-CH_2-Cl$ appears at 6.3, and $Cl-CH_2-Cl$ appears at 4.2τ. By means of the small list in Table 2-3 one should be able to make reasonable guesses about the chemical shifts of most other groups.

As a rule, the proton appears further downfield as it is bound to a more electronegative atom, or as the carbon atom to which it is bound holds more electronegative groups or atoms. Covalently bound hydrogen atoms tend to be shielded from the external field by the bonding electrons, and hence the presence of electronegative atoms tends to "deshield" them. An extreme example of this effect is the difference in chemical shift between the methyl protons of isobutane (9τ) and those of the tertiary butyl carbonium ion (2τ) (additional examples appear in Chapter 20).

A further effect observable in Table 2-3 is that of magnetic anisotropy; electronic motions in the various types of bonds create small local magnetic fields that may add to or detract from the external field and hence the local protons will be observed at relatively low or high external field, respectively. One of the most dramatic effects of this sort is provided by aromatic rings. The electrons in the cyclic π molecular orbitals are in effect a conducting ring system, and the "ring current" induces a local field effect adding to the external field (see Fig. 2-4); the aromatic protons correspondingly appear at a much lower field: The ring deshields its protons [35]. If a substance under nmr investigation associates weakly

35. For early reference and an example of how this effect may be computed, see S. N. Bhattacharyya, P. N. Sen, and S. K. Bose, Z. Physik. Chem. (Leipzig), 249, 81 (1972).

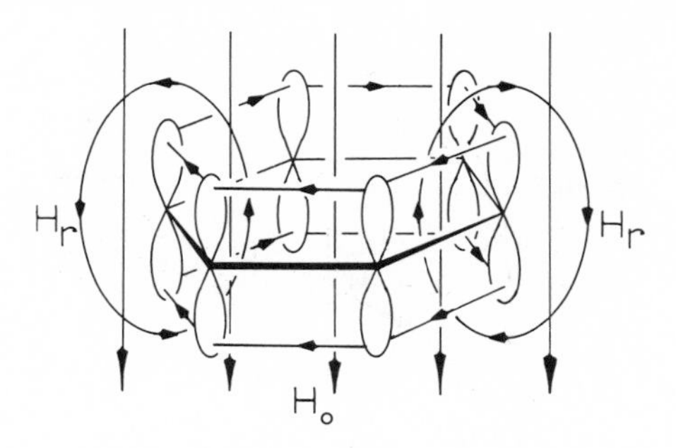

Fig. 2-4. An "aromatic" hydrogen atom and the ring current effect.

with aromatic rings, then the use of benzene as the solvent may in fact influence its chemical shifts somewhat because of this effect.

An effect that is not obvious from Table 2-3 is that of hydrogen bonding, which tends to cause further deshielding. Thus, while hydrogen protons normally appear at about 5 τ, the internally hydrogen bonded proton in ethyl Z-β-hydroxycrotonate (1) appears at -4 τ. One consequence of this is that the chemical shift of acidic protons is often considerably concentration and solvent dependent. Thus, even chloroform which normally forms only weak hydrogen bonds and which normally appears at 2.7 τ, is shifted down to 1.3 τ when mixed with three equivalents of pyridine N-oxide [36] (see further p. 65).

In complex molecules it may of course very well happen that several groups of protons signal in the same chemical shift region, and then little can be concluded from the nmr spectrum. In such a case, the so-called shift reagents are sometimes helpful [37]. The idea is that if the molecule of interest will associate with a reagent that itself has a very intense local field, then those protons that are near to it will undergo a large shift. The spectrum will therefore be more spread out and hence more intelligible. These reagents are of two types; some depend on large aromatic rings, and others on the presence of a paramagnetic ion. The latter are more common. A well-known example is tris-(dipivalomethanato)-europium (39), usually written as $Eu(dpm)_3$.

36. W. J. le Noble, J. LuValle, and A. Leifer, J. Phys. Chem., 66, 1188 (1962).
37. (a) C. C. Hinckley, J. Amer. Chem. Soc., 91, 5160 (1969); (b) R. von Ammon and R. D. Fischer, Angew. Chem., Int. Ed. Engl., 11, 675 (1972); (c) R. E. Sievers, "NMR Shift Reagents", Academic Press, New York, 1973.

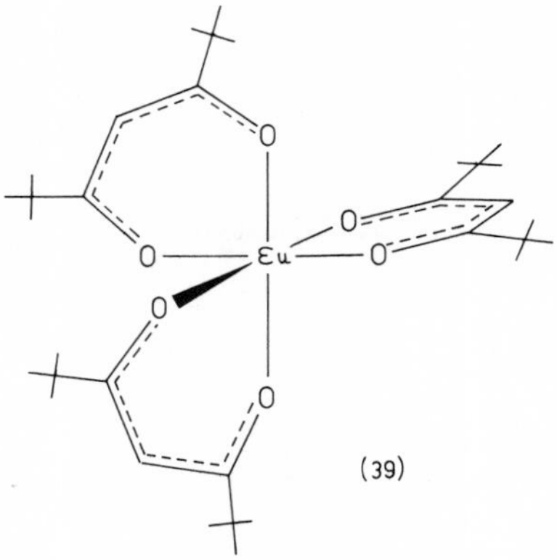

Eu

(39)

Thus, 2-adamantanol (40) normally has only the α- and hydroxy protons signalling separately from all the others which appear at $8.3\,\tau$; but in the presence of (39) the signals due to these protons are scattered from 6 to $-10\,\tau$. Since a molecule such as (40) will associate with the reagent via its oxygen atom, the protons nearest that site are most affected. This is of course helpful in structure elucidations; it can be readily shown this way for example which of the epimeric pair of alcohols is (41) and which is (42) [38].

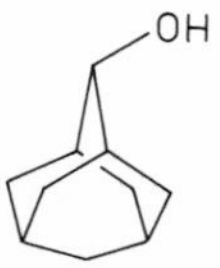

OH

(40)

OH

OH

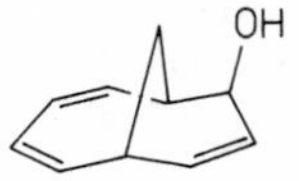

(41) 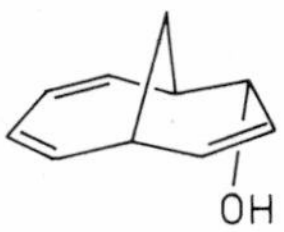 (42)

None of the other atoms in Table 2-2 can presently rival the position of the proton as the most important to organic chemists. ^{12}C, ^{16}O, and ^{32}S have a nuclear spin quantum number of 0. ^{19}F and ^{31}P have spins of 1/2, but these atoms are less important in organic chemistry. ^{13}C is experimentally difficult to use; the combination of low natural abundance and low sensitivity makes it difficult to

38. M.R. Willcott, J.F.M. Oth, J. Thio, G. Plinke, and G. Schröder, Tetrahedron Lett., 1579 (1971).

detect. Various techniques such as the Fourier transform analysis are available now, so that one may avoid the alternative of using specifically enriched compounds, and it may be that ^{13}C will soon rival the proton as the most important nucleus so far as the organic chemist is concerned [39]. Lists of chemical shifts for ^{13}C bound to various groups have been drawn up and can be used in structure assignment. Similar remarks apply to deuterium [40]. Nuclei such as ^{14}N and ^{35}Cl are much more rarely studied; when the spin equals 1 or more the nucleus will have a nuclear quadrupole moment which has the effect of broadening the nmr lines to the point where they are difficult to locate.

Since the sensitivity of any nucleus is independent of its chemical environment (however, see p. 55) one of the important uses of nmr is an analytical one. Thus, in isobutane we may expect two signals in a 9:1 ratio, in *n*-butane two signals in a 3:2 ratio, and so on; these ratios can generally be measured to a few percent precision. An especially useful feature of analysis by nmr is the fact that the mixtures can be analyzed even if the components are not available in pure form for calibration - a statement that cannot be made for any other spectroscopic techniques. Molecular weights can be measured to better accuracy than by melting point depressions and similar means, by comparing the intensities of one of the peaks of the sample compound with one from an added standard [41].

An important question calling for our attention at this point may be illustrated by means of the substance 1-bromo-1-chloro-1-fluoroethane (43), shown here in Newman projection.

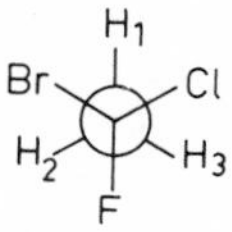

(43)

H_1, H_2 and H_3 are clearly in different environments, and hence one might well expect three signals in a proton nmr spectrum. The situation is quite analogous to an example we encountered earlier (p. 45, 46), in which we could see *cis*- and *trans*-vinyl ethers separately in the infrared. Here, however, only *one* signal is observed. The reason for this is as follows.

The rotation about the carbon-carbon single bond is relatively rapid, and the average lifetime of each conformation is perhaps only 10^{-6} sec or so. Now

39. For reviews, see (a) J. B. Stothers, *Quart. Rev. (London)*, *19*, 144 (1965); (b) *Chem. Eng. News*, *45*, 46 (March 20, 1967); (c) J. B. Stothers, *Appl. Spectr.*, *26*, 1 (1972); (d) E. W. Randall, *Chem. Brit.*, *7*, 371 (1971); (e) J. B. Stothers, "Carbon-13 NMR Spectroscopy", Academic Press, New York, 1972.

40. (a) P. Diehl and T. Leipert, *Helv. Chim. Acta*, *47*, 545 (1964); (b) L. K. Montgomery, A. O. Clouse, A. M. Crelier, and L. E. Applegate, *J. Amer. Chem. Soc.*, *89*, 3453 (1967).

41. S. Barcza, *J. Org. Chem.*, *28*, 1914 (1963).

suppose that H_1 and H_2 have a chemical shift difference of 1 ppm; at 14,000 gauss this would be 60 Hz. According to the Heisenberg uncertainty principle,

$$\Delta E \Delta t \approx h/2\pi$$

Hence

$$h \Delta \nu \Delta t \approx h/2\pi$$

and

$$\Delta t \approx 1/2\pi\Delta\nu$$
$$\approx 0.002 \text{ sec}$$

In other words, with an energy separation as small as 60 Hz, the two conformational states would have to have lifetimes in excess of 0.002 sec or so in order to make H_1 and H_2 separately observable. As a result of the rapid rotation the three protons will therefore appear in the nmr as though they were equivalent. Presumably they could be observed separately at much lower temperatures where the internal rotations are much slower, or in molecules in which the rotation for one reason or another is impossible; we shall encounter examples of this further below (see for example p. 63).

In this connection a subtlety should be pointed out here which may be easily overlooked in the assignment of the peaks and their intensities. This can be understood by considering the acid (44), in which one would be likely at first glance to consider the two methyl groups equivalent in view of the rapid rotation, and hence one might expect a single signal for them six times as intense as each of the other two signals. In fact, one may very well observe two methyl signals because these two groups are actually not equivalent. Thus, one of them is gauche to a hydrogen atom and a carboxylic acid group, and the other is gauche to the hydrogen and chlorine atoms; hence they are in magnetically different environments.

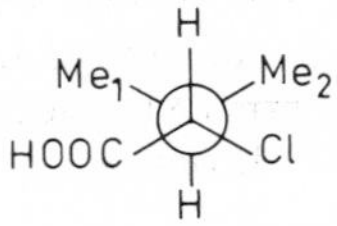

(44)

More important, rotations cannot interchange the methyl groups. Thus, a $\pi/3$ rotation puts the two into new positions in which they are still different, and this remains true no matter how many rotations are carried out or how rapidly this is done. In general this is true whenever we have two protons or groups of protons bound to the same carbon atom in a chiral molecule, although the likelihood of finding two signals will obviously depend on how far the two groups are removed from the center of the asymmetry. The two methyl groups in (44) are sometimes described as diastereotopic; those in β-methylbutyric acid, which are magnetically equivalent, are called enantiotopic.

It should be emphasized that the doublet feature is not due to the presence of enantiomers (R and S forms); if both forms were recorded separately, each would give the same identical spectrum including the doublet [42].

There are two other important features of nmr that can help elucidate structures; one of these is the coupling of nuclei through the bonds (discussed in Sec. 2-5), and the other is via the use of the "nuclear Overhauser effect" [43], which is an effect of nuclei on one another directly through space. Suppose that a given molecule has two protons which have widely different chemical shifts but which are located in close proximity; if one now irradiates the sample at the Larmor frequence of the one proton, an increase in the intensity of the signal of the other may be observed. The increase may reach a theoretical maximum of 50% if the distance between the two is 3 Å or less. For instance, when the two isomeric 3-ethylidene-1-azabicyclo[2.2.2]octanes (45) and (46) are irradiated at the methyl frequency, one shows an increase in intensity at the methyne hydrogen signal whereas the isomer does not respond in that way [44]; the latter is therefore presumably (46).

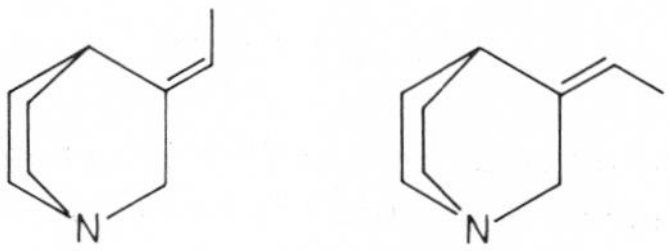

N N

(45) (46)

The theory which explains this effect has been well described elsewhere [43], and we shall not review it further here; it should be obvious however that it provides the organic chemist with a powerful tool in the elucidation of structures of natural and synthetic products.

2-5. nmr: COUPLING

When the proton magnetic resonance spectrum of phosphine (PH_3) is measured, one finds a pair of peaks of equal intensity, even though only one chemical type of protons is present. The explanation is that the phosphorus and hydrogen

---

42. Numerous examples have been reported; see for instance (a) G.M. Whitesides, D. Holtz, and J.D. Roberts, J. Amer. Chem. Soc., 86, 2628 (1964); (b) J.C. Randall, J.J. McLeskey, P. Smith, and M.E. Hobbs, J. Amer. Chem. Soc., 86, 3229 (1964); (c) C. van der Vlies, Rec. Trav. Chim., Pays-Bas, 84, 1289 (1965).
43. See, for example, G.E. Bachers and T. Schaefer, Chem. Rev., 71, 617 (1971).
44. J.C. Nouls, G. Van Binst, and R.H. Martin, Tetrahedron Lett., 4065 (1967).

nuclei are coupled; i.e., the energy of the protons in either spin state is split into two levels, depending on whether the phosphorus nucleus has its axis oriented in the same or opposite direction. This level splitting is ordinarily not the same for the protons whose spin is "up" and those that spin in the opposite direction, and consequently two signals obtain (see Fig. 2-5). Since the sample will contain as many phosphorus nuclei spinning in the one direction as there are spins of the opposite direction, the two peaks will be equal in intensity.

By the same token, the transition of the phosphorus nucleus depends on the spin states of the protons; a simple extension of the argument shows that now a total of four signals may be expected. The intensity ratio will be 1:3:3:1, as may be deduced from Fig. 2-6. In one-eighth of all molecules all three protons will have spins "up"; in three-eighths, only two will be up, etc.

The four lines will be equally spaced (Fig. 2-7), and this spacing will be equal to the spacing between the two signals in the proton spectrum, as can be deduced from Fig. 2-5; i.e.,

$$J_{H-P} = J_{P-H}$$

This spacing or the "coupling constant" (unlike the chemical shift) is independent of the external field and hence is usually expressed in terms of hertz. J may be positive or negative - which, however, is a question that cannot be decided on the basis of a simple nmr spectrum alone.

H↑ P↑
H↑
H↑ P↓
H↓ P↑
H↓
H↓ P↓

Fig. 2-5. A single transition of a proton in the absence of neighboring spins, and a doublet transition because of coupling with a phosphorus atom.

H_1	+	++-	+--	-
H_2	+	+-+	-+-	-
H_3	+	-++	--+	-

Fig. 2-6. A scheme showing all eight possible nuclear spin states of three protons.

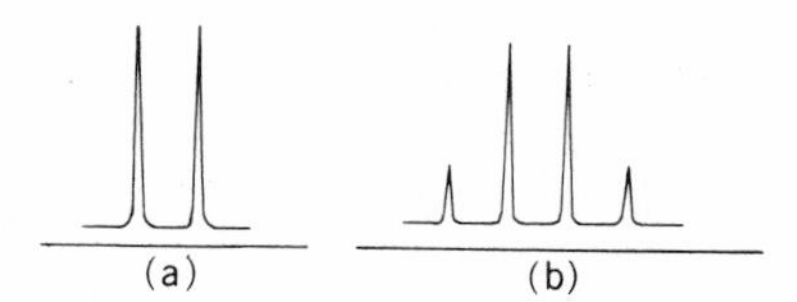

Fig. 2-7. (a) The proton magnetic resonance spectrum for PH_3; (b) the ^{31}P nmr spectrum of PH_3.

The magnitude of the coupling constant depends on the number of bonds intervening between the nuclei. Thus, in trimethylphosphine the proton signal will again be a doublet, and the ^{31}P spectrum will now consist of ten equally spaced lines in the intensity ratio of 1:9:36:84:126:84:36:9:1; J will now be an order of magnitude smaller than in the case of phosphine itself. In hexamethylphosphoramide (47), with three bonds intervening between the protons and the phosphorus nucleus, the coupling constant will be smaller still, perhaps 10 Hz

$P(NMe_2)_3$

(47)

or so, and if the separation is made larger still, J will probably be too small to measure (to avoid confusion about which nuclei are being considered, the number of intervening bonds is sometimes indicated - thus, in the case just mentioned, $^3J_{PH}$ = 9 Hz).

Such couplings arise not only between dissimilar nuclei, but also between the same nuclei in different environments. Thus, the nmr spectrum of acetaldehyde appears as shown in Fig. 2-8. Note that the intensity ratio is not affected by the splitting; it is still one to three (or 1:3:3:1:12:12 overall). Again, on separation by more than three bonds, coupling is, as a rule, no longer observable.

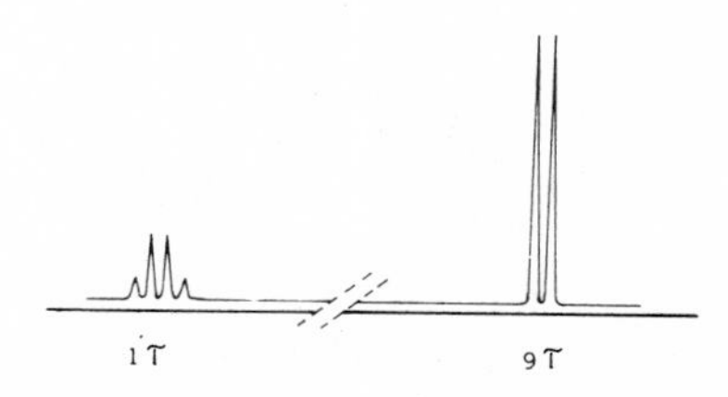

Fig. 2-8. nmr spectrum of acetaldehyde.

At this point, several questions have arisen that must now be considered. First of all, in the case of phosphine one may have wondered why the signal of the three protons was considered only to be split by the phosphorus and not by the protons themselves. In order to explain this, we may refer to Fig. 2-9 which shows what happens to a pair of 1:1 doublets (representing the overall signal of two nearby and dissimilar protons) when the chemical environment of the two is made more and more similar. As $\Delta\sigma$ decreases to only a few times J, the two outer signals become less intense than the inner ones, and this process continues until the outer peaks disappear when the protons become identical. It should be noted that J does not equal zero for chemically identical ("equivalent") nuclei; the transition probabilities for certain signals are zero and the net effect is as though there simply were no splitting. Since the electrons also have a spin ($s = 1/2$), one might furthermore wonder whether there should not be splitting due to them. However, in nearly all organic molecules the electrons are paired, and the effects of the two rapidly moving electrons with opposed spin cancel out. The story is different for the small class of molecules (free radicals) in which one or more electrons are unpaired; that will be taken up in the next section. A third question might be raised about the spectrum of hexamethylphosphoramide, in which the possibility of additional splitting by the nitrogen nuclei was apparently ignored. It is known that all nuclei with I larger than 1/2 have also a nuclear quadrupole, and such quadrupoles provide a mechanism for rapid relaxation between the spin states [45]. As a result the protons "see" only a nitrogen atom in one "average" spin state, and no splitting is observed.

The relaxation rate is dependent on the nature of the compound, and in some of them nitrogen splitting may indeed be observed. In such a case, since ^{14}N has $I = 1$, a 1:1:1 triplet may be expected. By writing out a diagram such as shown in Fig. 2-6, one can readily reduce that if a signal is split by n equivalent nuclei whose nuclear spin quantum number is I, then there will be $(2nI + 1)$ lines. If

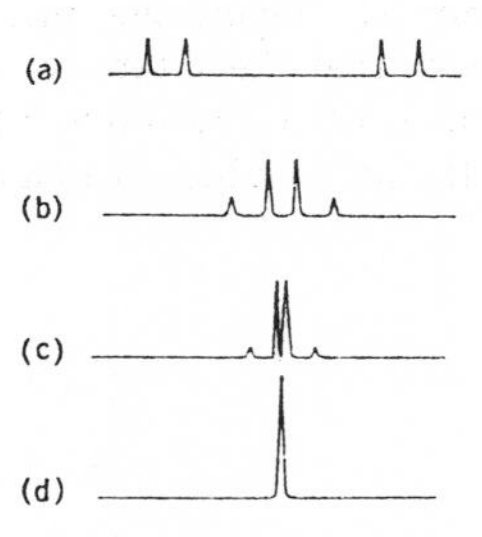

Fig. 2-9. The appearance of an H_A-H_B spectrum as $\Delta\sigma$ is made smaller at constant J.

45. Nuclear quadrupole resonance (nqr) itself can be used to good advantage to learn details of the nature of bonding in solids; see, for example, Chem. Eng. News, 43, 30 (Dec. 20, 1965).

some of the nuclei are nonequivalent, one multiplies the results for all the groups of equivalent nuclei; thus, for 2-methyl-1, 1, 3 triiodopropane (48) the tertiary proton signal would in principle be split into 2 x 3 x 4 = 24 lines (see Fig. 2-10). In practice, since the various J values may not differ as widely as shown many of these peaks may overlap so that the actual patterns are much less straightforward than the one shown in the illustration.

$$CH_3—CH(CH_2I)(CHI_2)$$

(48)

The intensities can be quickly deduced by means of the scheme shown below;

n	I = 1/2	2/2 = 1	3/2
0	1	1	1
1	1 1	1 1 1	1 1 1 1
2	1 2 1	1 2 3 2 1	1 2 3 4 3 2 1
3	1 3 3 1	1 3 6 7 6 3 1	,
4	1 4 6 4 1	,	,
5	,	,	,
,	,	,	,
,	,	,	,

the arrows show how each member results from the addition of the numbers appearing above it.

Although coupling greatly complicates nmr spectra, if one learns the nature of these complications they are in fact exceedingly helpful. While chemical shift differences are useful in assigning the various functional groups present in the molecule, analysis of the coupling patterns will often reveal which of these groups are neighbors. If, as often happens, the spectrum is too complex to yield obvious answers, decoupling can usually be accomplished as follows. If we consider the phosphine molecule again, if we observe the proton nmr spectrum while the sample is being irradiated at the phosphorus frequency, rapid relaxation of the ^{31}P nuclei will then occur, and the doublet collapses to a singlet. In a similar "double resonance" experiment, the ^{31}P quartet can be reduced to a singlet by the irradiation of the sample at the proton frequency. When the nuclei to be decoupled are the same the problem becomes more difficult; since their resonances are very

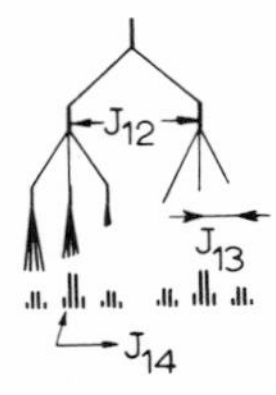

Fig. 2-10. Proton nmr spectrum for (48) if the three coupling constants differ greatly.

similar, highly monochromatic radiation must be used and because of this it cannot very well be highly intense. However, with modern apparatus it is usually possible to "tickle" the sample at each signal and observe the effect on the remainder of the spectrum. This kind of experiment may often be necessary in order to decipher spectra made complex by the presence of several nearly but not quite equivalent neighboring nuclei; for example, the vinyl group gives rise to a rather complex "ABC type" of spectrum in which the significance of the various signals is usually not obvious at first sight [46].

We now turn our attention to some more detailed features of molecular structure that can be readily learned from nmr spectra. One of these is the "dihedral angle", which may be described by saying that if the atoms to which the protons are bound are viewed as lying on a common line with the observer's eye, the two bonds holding the protons define the dihedral angle θ (see p. 45). The coupling constant is related to θ by the Karplus rule:

$$J_{AB} \approx 9 \cos^2\theta - 0.3 \text{ Hz}$$

Thus J will vary from about 0 for hydrogens involving perpendicular bonds to perhaps 10 for those which are either staggered trans or in eclipsed positions (the values of the two constants in the Karplus expression vary somewhat depending on substituents and other features). Several examples may be mentioned here. Thus, in cyclohexanes (49) axial protons such as H_1 are usually coupled much more strongly to neighboring axial protons H_a ($J \approx 10$ Hz) than to equatorial ones H_e ($J \approx 4$ Hz). In the rigid molecule (50) $J_{H1H2} \approx 7$ Hz even though the two are

(49) (50) (51)

separated by four bonds; in (51), on the other hand, $J_{H3H4} = 0$ even though these two protons are closer together. Although the Karplus rule suggests a symmetry of J values about the $\pi/2$ angle, one finds in fact usually that in a planar system

46. For a classification of spin systems, see, for example, A. Ault, J. Chem. Educ., 47, 812 (1970).

where θ is necessarily zero that trans- located protons are coupled more strongly than those that are cis. Thus, in styrene $J_{cis} \approx 7$ Hz whereas $J_{trans} \approx 18$ Hz. Coupling also generally seems more efficient through multiple bonds than through single ones; thus, m- and even p-protons in substituted benzenes are often coupled by 1-2 Hz, and perhaps the record is held by the conjugated triyne (52) in which the methyl and methylene protons are coupled by 0.4 Hz through no fewer than nine bonds!

$$Me-(C\equiv C)_3-CH_2OR$$

(52)

Although the remarks in this section stress the importance of intervening bonds to the process of coupling, there are some instances in which nuclei are apparently coupled directly "through space" rather than via bonds. Thus, in 1-fluoro-8-methylbiphenylene (53) $J_{F-CH_3} = 0$, but in 4-fluoro-5-methylfluorene (54) it is 7 Hz [47].

Me F Me F

(53) (54)

Another interesting feature that can readily be obtained in proton nmr spectra is the $J_{^{13}CH}$ coupling constant. Since ^{13}C occurs in about 1% natural abundance and since $I(^{13}C) = 1/2$, if the spectrum is recorded at high sensitivity one will generally find each proton peak symmetrically flanked by a pair of "^{13}C satellites" (see Fig. 2-11).

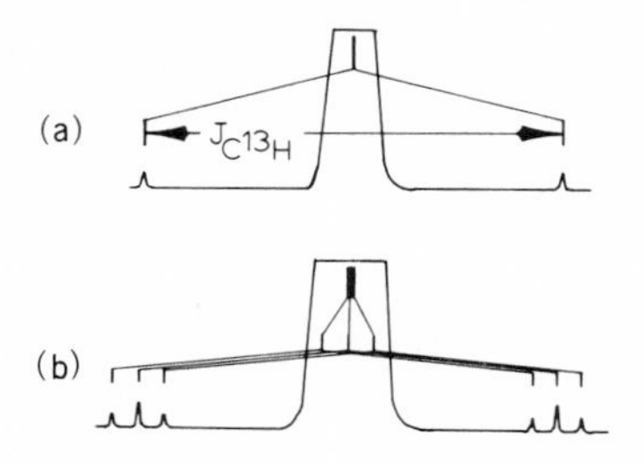

Fig. 2-11. (a) ^{13}C-satellites in methanol (CH_3-signal); (b) similar satellites of the chloromethylene group in 1-chloro-2-methoxyethane.

47. G.W. Gribble and J.R. Douglas, J. Amer. Chem. Soc., 92, 5764 (1970).

It has been discovered that this coupling constant is related to the hybridization of the C-H bond as follows.

$$J_{^{13}CH} \approx 5 \times \% \text{ s character}$$

Thus, in open-chain compounds, $J_{^{13}CH}$ is usually about 125 Hz, in benzenes and in vinyl positions it equals about 160-170 Hz, and in terminal acetylenes it equals about 250 Hz. As was remarked on p. 30, in strained small rings this constant has been indicative of unusually large degrees of s character in the C-H bonds even though formally all bonds may be single. A similar rule relates the percent s character to ^{13}C-^{13}C coupling constants [48].

There is another interesting application of ^{13}C couplings, namely, the measurement of coupling constants involving equivalent protons, which obviously cannot be measured directly as explained in preceding pages. Thus, 1,2-dichloroethane will contain about 2% $Cl^{13}CH_2$-CH_2Cl. In these molecules the protons are no longer equivalent and their mutual coupling will show up in the ^{13}C satellites, each of which will be a 1:2:1 signal. Another method for measuring such constants is by way of deuterium-labeled compounds (see Fig. 2-12); the H-D coupling constants are 6.5 times smaller than the J_{H-H} values.

We may close this section by pointing out a subtlety that is easily overlooked in the anticipated nmr spectrum. Consider for instance o-dichlorobenzene. It is clear that there are two kinds of ring protons present and two of each. However, this equivalence may not be extended to the splitting pattern, as H_2 and H_3 are not equivalent so far as H_1 (or H_4) is concerned; this situation is thus quite different from that, for example, of 1-chloro-2-methoxyethane (Fig. 2-13).

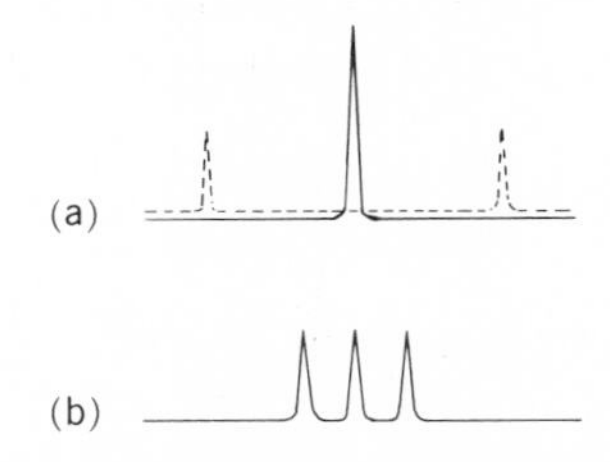

Fig. 2-12. Proton nmr spectrum of a D-labeled compound as a means of obtaining coupling constants for equivalent protons; (a) the real spectrum of trans-stilbene in the vinyl region and the same region with hypothetical coupling; (b) monodeuteriostilbene.

48. (a) F.J. Weigert and J.D. Roberts, J. Amer. Chem. Soc., 94, 6021 (1972); (b) R.D. Bertrand, D.M. Grant, E.L. Allred, J.C. Hinshaw, and A.B. Strong, J. Amer. Chem. Soc., 94, 997 (1972).

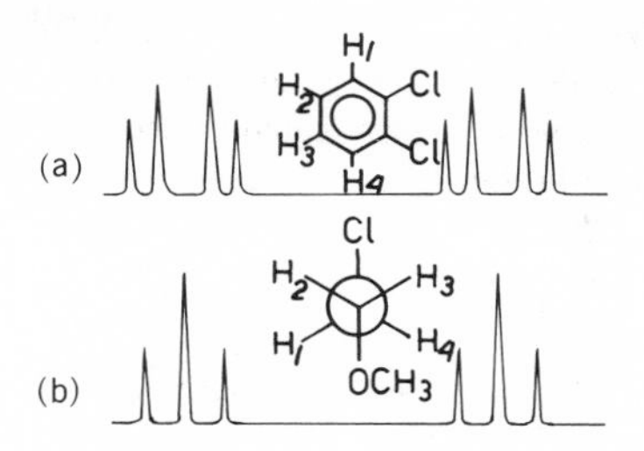

Fig. 2-13. First-order splitting patterns in (a) o-dichlorobenzene and in (b) 1-chloro-2-methoxyethane.

2-6. nmr: EXCHANGE PHENOMENA

In Section 2-4 it was pointed out that the rotation of a methyl group is so rapid that the three hydrogen atoms cannot be observed separately even though at any instant the three might very well occupy different positions. Rotation is not the only process by which hydrogen atoms may become equivalent; others include inversion (such as in (55)), positional exchange such as in (56), and simple, direct proton exchange as in an aqueous solution of acetic acid.

(55) (56)

$$H_2O + MeCOOH' \rightleftharpoons HOH' + MeCOOH$$

Several aspects of this process are worth mentioning here. First, the chemical shift of the averaged protons will be the averaged chemical shift weighted for the number of protons present; thus, in the latter example if the COOH proton for pure acetic acid appears at $-1\,\tau$ and those of pure water at $5\,\tau$, then for an equimolar mixture the signal will appear at $3\,\tau$, and it will have the same intensity as the methyl signal at $8\,\tau$. Exchange is also related to the needs for using nonviscous solvents and for spinning samples. (The spinning often gives rise to a small complication in that "side bands" are recorded for each signal which are separated from it by a number of hertz equal to the spin rate; by varying the spin rate one can easily distinguish them from ^{13}C satellites.) If either of these requirements is not met the peaks may be greatly broadened; rapid tumbling of the molecules is necessary to insure sharp signals.

Exchange may affect both chemical shifts and coupling. Thus, very pure methanol shows the hydroxyl proton as a 1:3:3:1 quartet and the methyl group as a 1:1 doublet; by adding a small amount of hydrochloric acid to catalyze the exchange, both signals change to sharp singlets. One of the most important aspects of the exchange phenomenon is the temperature range in which the spectrum changes ("coalesces") from one typical of slow exchange to one of rapid proton shuffling.

A typical example, shown in Fig. 2-14 is the nmr spectrum of the two methyl groups of dimethylformamide (57). At room temperature the two are evidently chemically different, presumably due to resonance which imparts significant double bond character to the C-N bond (see Sec. 8-5). At 200°, internal rotation is fast

H—C(=O)—N(Me)—Me ⟷ H—C(—O⊖)=N⊕(Me)—Me

(57)

and the two methyl groups give a single sharp signal; at an intermediate coalescence temperature the two peaks merge, and at this temperature the rate constant of exchange (rotational in this case) is given [34c] by

$$k = \pi(\Delta\nu^2 + 6J^2)^{1/2}/\sqrt{2}$$

or, since the protons under discussion here are not coupled (J = 0):

$$k = \pi\Delta\nu/\sqrt{2}$$

One can further calculate the free energy of activation via the absolute rate expression

$$k = \frac{RT}{Nh}\exp(-\Delta G^{\ddagger}/RT)$$

Thus, the exchange phenomenon makes it possible to study the rates of a large number and variety of molecular processes that would be very difficult otherwise, and we shall encounter many examples later on when we further consider rotational motions, inversions, valence isomerizations, and so on.

It should be pointed out that nmr can be used to measure rates in two entirely different ways; this is probably obvious but at the risk of being redundant the author wishes to point to this possible source of confusion. One such way is the traditional one for slow reactions.

$$A \longrightarrow B$$

Since at least some of the nuclei will have different chemical shifts in A and B, the spectrum will gradually (say, over a period of minutes to days) change from that of A to that of B, and by occasionally measuring the spectrum of the mixture, the rate law and the rate constant can be determined in the traditional way. On the other hand, in rapid equilibrations in which K is unity or not far from it, one

$$A \rightleftharpoons B$$

may be able to change the temperature in such a way as to change from the spectrum of a mixture of A and B to one of the average between A and B; at that temperature then, k can be computed from $\Delta\nu$ and J as specified above. The statement

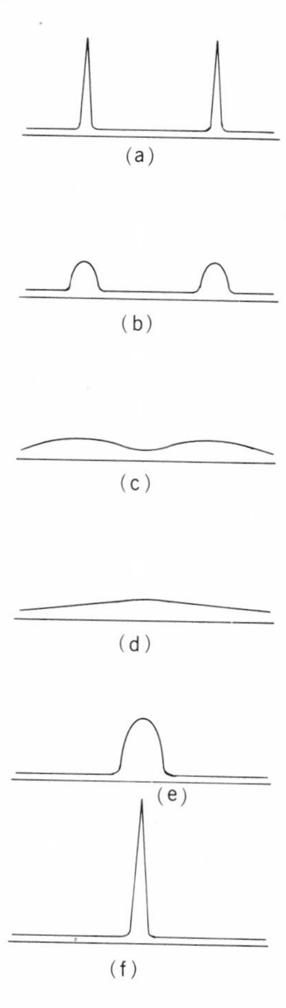

Fig. 2-14. (a) A pair of methyl groups distinguishable by nmr at low but not at high temperatures; (d) is the spectrum at the coalescence temperature.

that a rate constant was measured by nmr does therefore not necessarily mean that the reaction was a rapid and reversible one. In fact, a rapid equilibration in which coalescence occurs at 200°, say, may be a slow reaction that can be followed by nmr in the traditional manner at room temperature.

The hydrogen bond may be mentioned here as an example of how much can be learned by studying the exchange process via nmr. If the chemical shift of the carboxyl proton of a pure carboxylic acid (-1.0τ) is compared with that of the same proton in solution in carbon tetrachloride at various concentrations, little difference is noted until the mole fraction N_{HA} is down to about 0.25 or so; at still greater dilution the signal begins to shift to higher field, and this change continues as far out as it can be measured. This observation [49] is consistent with a

49. L.W. Reeves, Trans. Faraday Soc., 55, 1684 (1959).

picture in which monomers have a higher signal than the hydrogen bonded dimers with which they are in equilibrium (see also p. 51). At very low concentrations the upfield shift becomes very steep and direct extrapolation to infinite dilution is not feasible; however, an indirect method [50] has suggested that the monomeric molecule may have a chemical shift as high as 10 τ. Thus, hydrogen bonding is clearly deshielding the proton. This is possibly an unexpected finding; apparently there is a net repulsion between the unshared electrons of the base and the electrons by which the proton is bound.

Similar findings show that monomeric ethanol in carbon tetrachloride has a σ_{OH} about 5 ppm above the pure substance [51]; for phenol [52], aniline, and thiophenol [53] those shifts are about 2.5, 0.5, and 0.1 ppm, respectively, and for propyl mercaptan [54] it is about 0.3 ppm.

These shifts can also be used as a measure of the strength of various bases as hydrogen bond acceptors; thus the chloroform proton is shifted downfield drastically on the dissolution of pyridine-N-oxide, moderately on admixture of nitrobenzene, and not at all if nitrosobenzene is added [36].

Similar effects are observable in molecules permitting internal hydrogen bonds, which are of course as a rule stronger than those binding separate molecules. Thus, o-carbonyl substituted phenols such as (58) at infinite dilution in carbon tetrachloride have hydroxyl chemical shifts up to 10 ppm below that of infinitely dilute phenol [55].

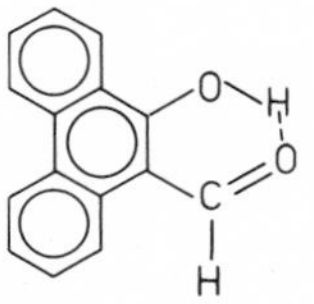

(58)

An even more remarkable case involves the carboxylate anion as a basic group; when racemic α, α'-di-t-butylsuccinate monoanion (59) in 0.13 M concentration in dimethyl sulfoxide is compared with the meso stereoisomer (60), the

---

50. J.C. Davis and K.S. Pitzer, J. Phys. Chem., 64, 886 (1960).
51. E.D. Becker, U. Liddel, and J.N. Shoolery, J. Mol. Spectr., 2, 1 (1958).
52. C.M. Huggins, G.C. Pimentel, and J.N. Shoolery, J. Phys. Chem., 60, 1311 (1956).
53. B.D.N. Rao, P. Venkateswarlu, A.S.N. Murthy, and C.N.R. Rao, Can. J. Chem., 40, 963 (1962).
54. L.D. Colebrook and D.S. Tarbell, Proc. Nat. Acad. Sci. U.S., 47, 993 (1961).
55. A.L. Porte, H.S. Gutowsky, and I.M. Hunsberger, J. Amer. Chem. Soc., 82, 5057 (1960).

former has a carboxyl signal about 15 ppm lower than the latter (the former acid is also about 10^8 times weaker than the latter) [56].

(59) (60)

The downfield shift as a result of hydrogen bonding does not occur when the base is a benzene ring, since, according to our view of the ring current effect, an upfield shift must take place (note that the ring current normally causes a downfield shift for atoms outside the ring, but upfield inside or above it). Thus, in strongly acidic media, protonated ketones occur in *cis* and *trans* mixtures (61) and (62) as do the isoelectronic imines. The positive charge causes the acidic proton

(61) (62)

to occur at very low field. Benzyl methyl ketone under such conditions is observed to be a 1:1 mixture with the acidic proton at $-3.47\ \tau$ in the one case and $-4.60\ \tau$ in the other. In the former isomer this proton is hydrogen π-bonded to the phenyl group, as is shown by the fact that electron-donating substituents increase the intensity of this signal, whereas a *p*-nitro group reduces it to zero [57]. Nuclear magnetic resonance has also permitted a comparison to be made of hydrogen and deuterium bonds. It is based on the fact that fluoroform and fluoroform-d in cyclohexane solution undergo a change in their ^{19}F chemical shifts when tetrahydrofuran is added. From these shifts and their concentration dependence, K could be calculated, the result being that the D bond was a few percent stronger than the H bond [58].

---

56. L. Eberson and S. Forsén, *J. Phys. Chem.*, *64*, 767 (1960).
57. G. C. Levy and S. Winstein, *J. Amer. Chem. Soc.*, *90*, 3574 (1968).
58. C. J. Creswell and A. L. Allred, *J. Amer. Chem. Soc.*, *84*, 3966 (1962).

2-7. ESR AND CIDNP

When the nmr region of the spectrum of a molecule containing an unpaired electron is examined, generally no signals can be found at all; however, the spectrum can in principle still be recorded if the solvent also contains an unpaired electron so that rapid spin exchange occurs. Thus, most of the protons of the free radical (63) can be observed in a di-t-butylnitroxide solvent [59] ((64) is a stable free radical). The enormous sensitivity of the electron (see Table 2-2) makes it

NOCH$_3$

(63) (64)

far simpler, however, to observe the signals of the unpaired electron itself [60]. It is generally possible to observe unpaired electrons down to concentration levels of about 10^{-8} in samples of 0.1 ml (i.e., 10^{11} spins or so). The spectrometers designed for this purpose generally make use of fields of about 3000 gauss; the corresponding frequency is about 9000 MHz. Their resolution is about 0.03 gauss. The operation of the instruments is in principle the same as that of nmr equipment, except that for instrumental reasons it is customary to record the first derivative spectrum rather than the signals themselves (see Fig. 2-15).

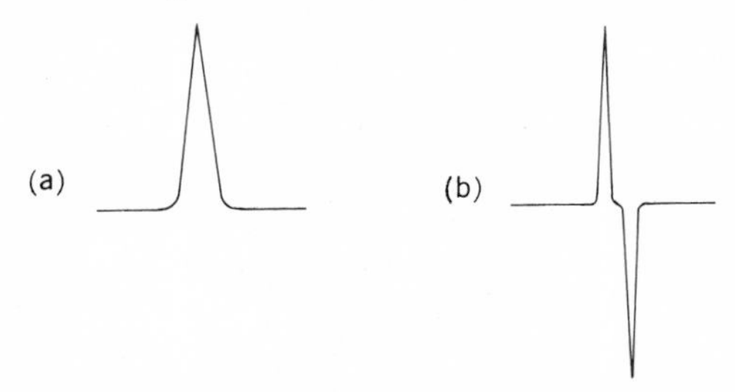

Fig. 2-15. (a) An esr signal; (b) the derivative signal as usually recorded.

---

59. R.W. Kreilick, *J. Amer. Chem. Soc.*, *90*, 2711 (1968).
60. Several excellent reviews are available; among them are (a) J.A. McMillan, *J. Chem. Educ.*, *38*, 438 (1961); (b) E. Müller, A. Rieker, K. Scheffler, and A. Moosmayer, *Angew. Chem., Int. Ed. Engl.*, *5*, 6 (1966); (c) A. Carrington, *Chem. Britain*, *4*, 301 (1968); (d) F. Schneider, K. Möbius, and M. Plato, *Angew. Chem., Int. Ed. Engl.*, *4*, 856 (1965); (e) G.A. Russell, *Science*, *161*, 423 (1968); (f) A.R. Forrester, J.M. Hay, and R.H. Thomson, "Organic Chemistry of Stable Free Radicals", Academic Press, New York, 1968; (g) G.W. Ludwig, *Science*, *135*, 899 (1962); (h) A.D. Baker, C.R. Brundle, and M. Thompson, *Chem. Soc. Rev.*, *1*, 355 (1972). The related Endor technique has been reviewed by (i) K. Möbius and K.P. Dinse, *Chimia*, *26*, 461 (1972).

The considerations governing the general appearance of an esr spectrum are the same as those for the nmr spectrum. The spin (s = 1/2) of an unpaired electron and the nuclear spins interact very strongly with coupling constants (a) often as large as 20 gauss or more. However, it should be remembered that the electron is not restricted to one atom or group and hence, although it may prefer to spend most of its time at or near a given atom, one may observe coupling with nuclei at rather large distances. The following examples are instructive.

When a solution of naphthalene in dry, degassed tetrahydrofuran is treated with sodium, a green color develops which is due to the radical anion (65); at the same time, an esr spectrum can be recorded which shows a total of 25 lines. Clearly the electron is interacting equally with each of the four α-protons and equally with each of the four β-protons. The groups of five lines as well as the individual lines within those groups have intensity ratios of 1:4:6:4:1. The coupling constants are an indication of the spin densities at the carbon atoms; thus, the electron does not spend equal time at the α and β atoms (the resonance structures below do not all have equal weight). It should be noted that the radical anion like

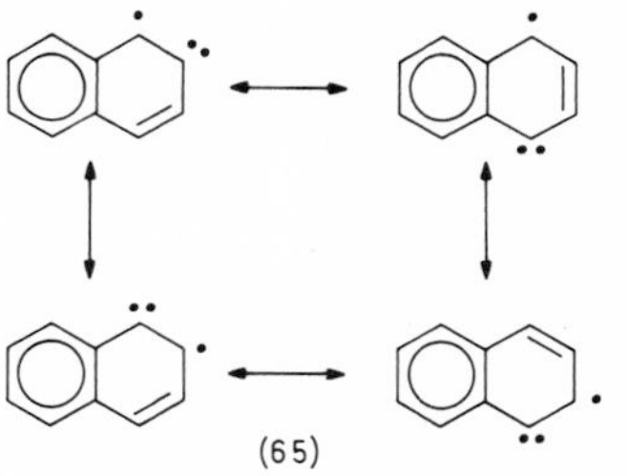

(65)

almost any other radical known to organic chemists is very reactive toward oxygen; this fact makes thorough degassing a necessity in esr experiments.

A second example is provided by bis(trichloromethyl)nitroxide (66) [61].

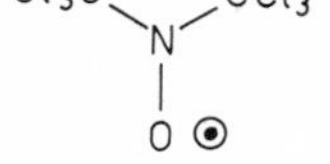

Cl_3C–N(–O⊙)–CCl_3

(66)

Three identical groups of 19 lines each, all equally spaced, can be observed. Since the nitrogen has a nuclear spin quantum number of 1 and both ^{35}Cl and ^{37}Cl have one of 3/2, the spectrum is readily accounted for if one assumes that both chlorine isotopes have almost the same coupling constants (see Table 2-2; a_N = 11.8 gauss and a_{Cl} = 1.25 gauss). It may be noted that quadrupole relaxation of the electron spin is much less efficient than that of nuclear spins, so that splitting of esr signals by nuclei of I>1/2 can be routinely observed.

---

61. H. Sutcliffe and H.W. Wardale, J. Amer. Chem. Soc., 89, 5487 (1967).

The transition, as in nmr, is governed by the selection rule $m = \pm 1$; in special cases (triplet diradicals in low temperature solid matrices) weak $m = \pm 2$ transitions have been observed; cf. Chapters 16, 17, and 18. The energy of the transition is given by the expression

$$h\nu = g\beta H$$

where β is the Bohr magneton and H is the magnetic field. The g factor has a value 2.00232 for a free electron, but in free radicals there will be a small contribution from the orbital angular momentum, and hence a given free radical will be characterized not only by the number of lines, the intensity of the lines, and the coupling constants, but also by g. Thus, g may be as much as 1% larger if heteroatoms are present.

The magnitude of the coupling constants is principally governed by the distance of the atom in question from the radical site, and by the hybridization of the orbital holding the electron. For a free hydrogen atom, a_H is about 500 gauss. It is further determined by factors similar to those encountered in nmr. Thus, the couplings of protons on adjacent carbon atoms vary with their dihedral angles, according to

$$a_H = \beta \cos^2\theta$$

where $\beta = 20-60$ gauss and where θ is defined as by Fig. 2-16. In rigid molecules such as the bicyclic semidione (67) the interaction is much stronger if the hydrogen atom interacts via its back-lobes than if it does so directly; thus, $a_a = 6.5$ gauss whereas $a_s = 0.4$ gauss.

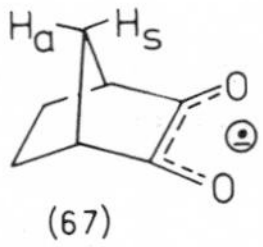

(67)

A very useful concept is the spin density of each of the atoms in the radical; it may be thought of as an indication of which fraction of its time the unpaired electron spends at the site of a given atom. The spin densities ρ are given by

$$\frac{3Ia}{8\pi\mu K}$$

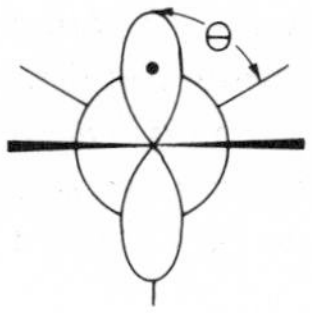

Fig. 2-16. The dihedral angle of a radical site with neighboring protons, shown in Newman projection.

where μ_K is the magnetic moment of the nucleus under consideration; hence they can be calculated from the coupling constants. For radicals holding the unpaired electron in a π-orbital system the McConnell relation holds

$$a_H = Q\rho_C$$

where Q has a value of -22 to -30 gauss. Thus, for the allyl radical (68) Q = -24.7 and the coupling constants a_1, a_2, and a_3 are -13.9, -14.3, and 4.1 gauss, respectively (note that $\Sigma\rho$ must equal unity).

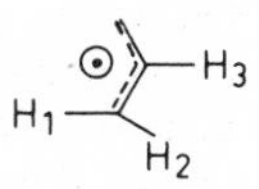

(68)

These various considerations have led to uses of esr that go far beyond the mere assignment of structure in free radical chemistry. Some examples of such uses may be mentioned here (see further Chapter 18). Exchange phenomena can be traced by esr as they can by pmr; since the esr transition is about 1000 times larger in energy, the rate constants so determined are of the order of a microsecond.

For the pentaphenylcyclopentadienyl radical (69) many resonance structures

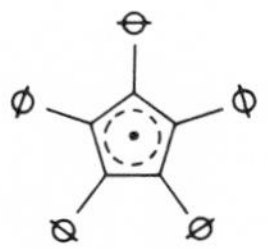

(69)

can be written with the unpaired electron in position at or near every carbon atom in the molecule. One might therefore expect a signal of large overall width; thus, a phenyl radical has a width of about 30 gauss. However, the signal observed for (69) is only 10 gauss wide; this suggests that the unpaired electron is not very able to get delocalized into the phenyl rings, and hence that the latter are far from coplanar with the five-membered ring [62].

The radical anion of p-acetylacetophenone, which theoretically should have a simple spectrum consisting of 35 lines (with groups of four and six equivalent protons), in fact has been shown to have a much more complex spectrum, the explanation of which requires the presence and relatively slow interconversion of the two conformers (70) and (71) [63].

---

62. D.C. Reitz, J. Chem. Phys., 34, 701 (1961).
63. A.H. Maki and D.H. Geske, J. Amer. Chem. Soc., 83, 1852 (1961).

(70) (71)

Chemically induced dynamic nuclear polarization (CIDNP) is a remarkable phenomenon observed in the nmr spectrum during reactions involving free radical intermediates. The observation may be that one or more of the products have an nmr spectrum normal in every way except that some of the signals are much more intense than normal, by a factor of perhaps up to a thousand or so, or in fact emission may occur. Stories abound about the many investigators who observed CIDNP years before it was described in the literature. Any organic chemist who has wrestled with the massive and highly complex early nmr machines should be ready to believe that those who observed emissions or enhanced absorptions by means of them entertained no thoughts other than how to get their obstreperous gadgets to behave. It is indisputable however that while Bargon [64] may not have been the first chemist to witness CIDNP, he was definitely the first one to believe what he saw.

As an example, if the reaction of ethyllithium with isopropyl iodide - which produces ethyl iodide among other products - is monitored by means of nmr, it is observed that the high field lines of both the methyl and methylene signals appear as emissions (see Fig. 2-17) [65].

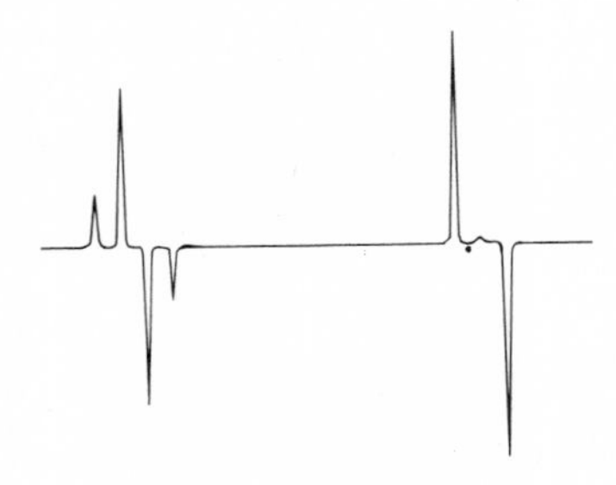

Fig. 2-17. The ethyl iodide nmr spectrum during the reaction of ethyllithium and isopropyl iodide.

---

64. (a) J. Bargon, H. Fischer, and U. Johnsen, *Z. Naturf.*, A22, 1551, (1967); (b) J. Bargon and H. Fischer, *Z. Naturf.*, A22, 1556 (1967).
65. See, for example, (a) *Chem. Eng. News*, *48*, March 9, 1970, p. 36; (b) H. Fischer, *Fortschr. Chem. Forsch.*, *24*, 1 (1971); (c) A. L. Buchachenko and F. M. Zhidomirov, *Russ. Chem. Rev. (English Transl.)*, *40*, 801 (1971); (d) H. R. Ward, *Accounts Chem. Res.*, *5*, 18 (1972); (e) R. G. Lawler, *Accounts Chem. Res.*, *5*, 25 (1972); (f) S. H. Pine, *J. Chem. Educ.*, *49*, 664 (1972); (g) A. R. Lepley and G. L. Closs, "Chemically Induced Magnetic Polarization", Wiley, New York, 1973.

These results are obtained sometimes even when the radical population in the reaction mixture is too low to be observable by means of esr, and hence this phenomenon is a valuable indication of the radical nature of the reaction [66].

These emissions and enhanced absorptions obtain because the protons in the molecules of the compound studied are not in equilibrium so far as their spin states are concerned. This departure from equilibrium, in turn, results from the interactions of two radicals in the so-called cage where the product molecule is born.

These radical pairs with originally parallel spins can collapse to product only after spin inversion. The ease of spin inversion depends on the nuclear spin state, and hence the "cage product" will have mostly that nuclear spin state which was best able to catalyze the inversion. As this spin state reaches equilibrium, enhanced absorption or emission may occur (see Fig. 2-18). Those nuclear spin states that were unfavorable to inversion will be present in radical pairs that diffuse apart, and hence they give rise to "escape products". These clearly may also have nonequilibrium nuclear spin states. Whether any given peak will show enhanced absorption or emission depends on the sign of the product of a number of factors such as the multiplicity of the precursor (triplet vs. singlet), the nature of the product (cage or escape), the coupling constants, and so on [67]. Since the type of spectrum obtained can be compared to those that would obtain for various mechanisms of product formation, CIDNP will be valuable in free radical studies; in fact, it may become possible to recognize "reactions" in which the product and reactant are identical because of return of the radicals to their initial positions.

$$(R-R)_r \rightleftharpoons (\dot{R}\ \dot{R}) \rightleftharpoons (R-R)_p$$

It should be noted that the mere occurrence of weak CIDNP signals should not be unquestioningly accepted as proof for a radical pathway of the main reaction - as has been done in all too many recent instances. It is always possible that such signals arise from insignificant side reactions.

It was recently found that if radicals are produced by a flash of light and the esr spectrum is observed within microseconds afterward, all signals may appear in emission in some cases, although the spectrum reverts to normal absorption very shortly afterwards. This phenomenon is as yet not fully understood [68].

---

66. For other examples, see (a) G. L. Closs and L. E. Closs, J. Amer. Chem. Soc., 91, 4549, 4550 (1969); (b) H. R. Ward and R. G. Lawler, J. Amer. Chem. Soc., 89, 5518 (1967); (c) A. R. Lepley, J. Amer. Chem. Soc., 90, 2710 (1968); (d) R. Kaptein, Chem. Phys. Lett., 2, 261 (1968).
67. (a) G. L. Closs, J. Amer. Chem. Soc., 91, 4552 (1969); (b) G. L. Closs and A. D. Trifunac, J. Amer. Chem. Soc., 91, 4554 (1969); (c). R. Kaptein, Chem. Commun., 732 (1971).
68. P. W. Atkins, I. C. Buchanan, R. C. Gurd, K. A. McLauchlan, and A. F. Simpson, Chem. Commun., 513 (1970).

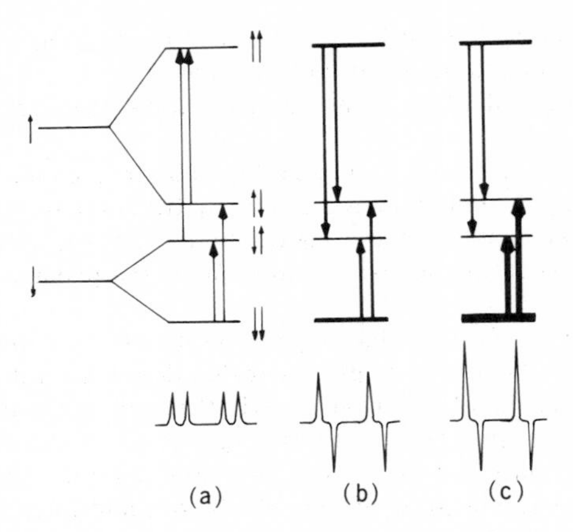

Fig. 2-18. Spin states of two coupled and nonequivalent protons and their nmr spectrum, (a) equilibrium spin state; (b) parallel spins equally overpopulated ("multiplet effect"); (c) parallel spins unequally overpopulated ("net effect").

2-8. ULTRAVIOLET AND VISIBLE SPECTROSCOPY [69]

The ultraviolet and visible regions of the spectrum (200 - 1000 nm or mμ) cover most of the electronic transitions of interest to organic chemists. The two principal features of the various absorptions found there are the wavelength (λ_{max}) and the extinction coefficient (ϵ_{max}) at each of the absorption maxima. The latter term is defined by the Beer-Lambert law for the optical density:

$$\text{O.D.} = \log \frac{I_0}{I} = \epsilon c l$$

where I is the light intensity, c is the concentration of the solute, and l is the path length; ϵ is usually given for a 1 M solution in a 1 cm cell.

As we shall see further below, there are many solvents that are almost perfectly transparent to ultraviolet and visible radiation, and hence it is customary to use highly dilute solutions (down to 10^{-5} M) and a relatively long path (usually 1 cm; the cells and optics are of quartz). Association between solute molecules is usually negligible at such low concentrations, and consequently Beer's law as a rule holds very well for these solutions. As a result ultraviolet-visible spectroscopy is a favorite tool of analytical chemists and kineticists. On the other hand,

69. For reviews and further references, see Ref. 34d-f, and (a) D. H. Whiffen, "Spectroscopy", Wiley, New York, 1966; (b) H. B. Dunford, "Elements of Diatomic Molecular Spectra", Addison-Wesley, Reading, Mass., 1967; (c) S. F. Mason, Quart. Rev., Chem. Soc., 15, 287 (1961); (d) A. I. Scott, "Interpretation of the Ultraviolet Spectra of Natural Products", McMillan, New York, 1964.

deviations from Beer's law can be used to determine the equilibrium constants for association equilibria; thus, such deviations have permitted measurements of the strength of hydrogen bonds between alcohols and various bases [70] and between carboxylic acid molecules [71].

The absorption bands in this region are usually very broad. Each electronic state is associated with its own group of vibrational and rotational levels, and hence a great many transitions are possible. The resolution is usually not good enough (especially for solutions) to observe all these transitions separately, and what is actually observed is the "envelope" of a whole group (see Fig. 2-19). The breadth of these absorptions and the wide range in extinction coefficients makes many of them difficult to observe; often a given transition can only barely be discerned as a "shoulder" on another absorption peak (see Fig. 2-20). "Uv" is therefore not as useful for identification as are some of the techniques previously described; however, there are a number of conclusions that can be drawn, and organic chemists who frequently face this task should know them.

Before these possibilities are identified, we begin by considering the types of electronic transitions that may be encountered. Electrons may be found in bonding, nonbonding, and antibonding levels, and there is a further distinction between σ and π bonds (cf. Sec. 1-8). If we generalize the height of these levels as shown in Fig. 2-21, one may observe that the shortest wavelengths are usually associated with $\sigma \rightarrow \sigma^*$ transitions; these indeed occur at 100 nm or so, and are not observable in the usual uv region. Thus, simple saturated liquid hydrocarbons are usually good solvents for uv experiments.

How tightly nonbonding electrons are held by their atoms can be guessed fairly well on the basis of electronegativity. Thus, $n \rightarrow \sigma^*$ transitions are not observable in the uv for water, alcohols, and ethers and hence these compounds are good solvents; on the other hand, sulfides have absorptions ranging well into the uv region. Halogenated hydrocarbons have uv absorptions whose wavelengths are longer the more halogen atoms there are in the molecule and the higher their atomic numbers are; iodoform is in fact a deep yellow compound.

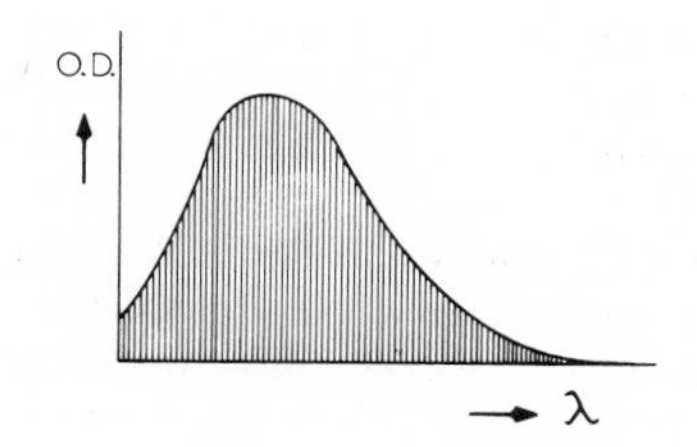

Fig. 2-19. The individual lines and the envelope observed for an electronic absorption.

---

70. A.K. Chandra and S. Basu, Trans. Faraday Soc., 56, 632 (1960).
71. M. Ito, H. Tsukioka, and S. Imanishi, J. Amer. Chem. Soc., 82, 1559 (1960).

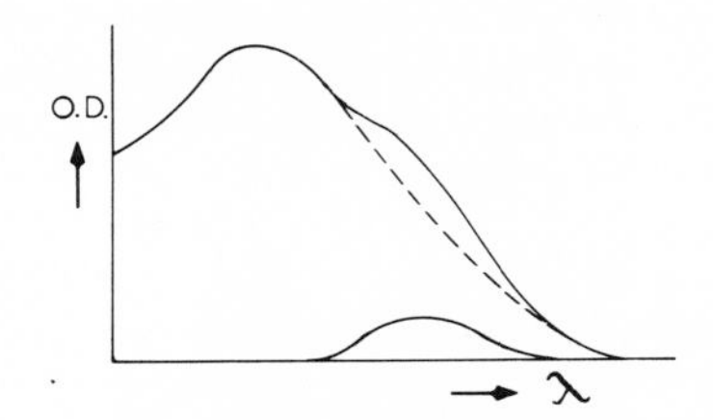

Fig. 2-20. Two individual absorptions of widely different extinction coefficients and the resulting curve.

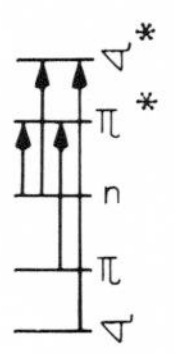

Fig. 2-21. The various electronic levels and the transitions between them.

Perhaps more important is the presence of π electrons. Ethylene has a $\pi \rightarrow \pi^*$ transition not far below 200 nm. In conjugated polyolefins there will be additional π and π^* molecular orbital levels, and the transitions between these involve smaller energy gaps as the number of such levels increases. 1, 3-Butadiene has an absorption at 220 nm, for example, and with further conjugation the absorption continues to shift toward longer wavelength. The same type of absorption is observable in cyclic conjugated (aromatic) compounds, and here also there is a well-known shift toward longer wavelengths as the number of rings increases.

In carbonyl compounds, nitriles, and so on the unshared electrons are subject to $n \rightarrow \pi^*$ transitions, which are generally observable in the uv; however the transition is a forbidden one and its intensity is low. A listing of some of the most important uv transitions are shown in Table 2-4. In each case only the longest wavelength absorption is given.

Because of the broad absorptions and the multitude of "chromophoric groups" that can give rise to them, uv spectroscopy usually does not allow one to conclude very much in the way of a structure assignment (as ir spectra do) if there is no other information about the compound; however, if additional information is available - for example about the source of the compound, ir and nmr spectrum, and so on - the uv spectrum can often be decisive in making a final choice between several possible isomers. This may be done because studies of the uv spectra of large numbers of compounds has led to a number of empirical generalizations, the reasons for which are not always very well understood. Among the structural generalizations there are the following. The longest wavelength absorption for a diene has a λ_{max} of 214 nm; λ_{max} is 253 nm if the diene is part of a carbocyclic ring system. For each additional conjugated double bond there is a shift to longer

TABLE 2-4. Some of the Main uv Absorptions of Organic Compounds

Group	λ_{max}	log ϵ_{max}	Transition
RH	140	4	$\sigma \rightarrow \sigma^*$
ROR	180	3	$n \rightarrow \sigma^*$
RSR	220	3	$n \rightarrow \sigma^*$
RCl	175	3	$n \rightarrow \sigma^*$
RBr	200	3	$n \rightarrow \sigma^*$
RI	250	3	$n \rightarrow \sigma^*$
$R_2C{=}CR_2$	175	4	$\pi \rightarrow \pi^*$
$RC{\equiv}CR$	180	4	$\pi \rightarrow \pi^*$
$R_2C({=}CH{-}CH{=})_2R_2$	220	4	$\pi \rightarrow \pi^*$
$R_2C({=}CH{-}CH{=})_{10}R_2$	350	2	$\pi \rightarrow \pi^*$
Aromatic	260	3	$\pi \rightarrow \pi^*$
$R_2C{=}O$	280	1	$n \rightarrow \pi^*$
$RCX{=}O$	210	2	$n \rightarrow \pi^*$
$R_2C{=}CRCR{=}O$	220	4	$\pi \rightarrow \pi^*$
$RC{\equiv}N$	170	1	$n \rightarrow \pi^*$
$RN{=}NR$	340	1	$n \rightarrow \pi^*$
$RN{=}O$	650	1	$n \rightarrow \pi^*$
$R_2C{=}C{=}CR_2$	210	2	$\pi \rightarrow \pi^*$
$R_2C{=}C{=}O$	330	1	$n \rightarrow \pi^*$
RNO_2	270	1	$n \rightarrow \pi^*$

wavelength ("bathochromic") of 30 nm; for each alkyl or halogen substituent, one adds 5 nm; for a sulfide substituent, 30 nm; for a dialkylamino substituent, 60 nm, and so on [72]. A similar set of rules governs λ_{max} of variously substituted carbonyl compounds [73]. Solvent effects similarly often provide useful information. Thus, when a carbonyl nonbonding electron is promoted to an antibonding level, one may expect that the excited state will be less polar than the ground state since this electron is now shared in the region between the carbon and oxygen atoms. The energy gap between a polar ground state and a less polar excited state will be smaller in a nonpolar solvent than in a polar one, and hence carbonyl compounds are subject to a "hypsochromic" shift of several nanometers if one measures the wavelength of the $n \rightarrow \pi^*$ transition in dioxane rather in hexane. Even longer shifts are observed if water or alcohols are used, since the nonbonding electrons are then involved in hydrogen bonds; in strongly protonating solvents such as sulfuric acid this absorption is absent altogether.

The various electronic states of molecules - at least in the case of simple linear molecules - may be described by term symbols ${}^{\alpha}X$ similar to those used for atoms. Again α represents the multiplicity; $X = \Sigma, \Pi, \Delta$, etc., depending on whether the spin orbital momentum about the molecular axis L = 0, 1, 2, etc. Thus, ground state oxygen is represented by the symbol ${}^{3}\Sigma$. Further designations are often made such as + or -, g or u, etc., depending on the symmetry properties of the molecule [69c]. A fascinating question is that of the physical and chemical properties of the excited molecules; this we will take up in Chapter 13.

2-9. OTHER IMPORTANT TECHNIQUES

Electron spectroscopy [74] is rapidly becoming one of the organic chemist's more important tools. It is based on the photoelectric effect: When a substance is irradiated with an energy exceeding the binding energy of an electron, that electron may get ejected and its excess kinetic energy reveals the binding energy. There are two principal variants of the technique. In one of these called esca (electron spectroscopy for chemical analysis), soft X-rays are used for excitation. These may be of sufficient energy to eject inner electrons. The binding energies of the electrons are subject to chemical shifts; the higher the oxidation state of a given atom, the greater will be the binding energy of its electrons. The carbon 1s esca spectrum of ethyl chloroformate is shown as an example (Fig. 2-22). Spectra of this sort obviously have many uses. Only very tiny samples are needed. One possible problem is that the excitation affects only the material at or near the surface, and contamination of this surface must be scrupulously avoided.

---

72. L. F. Fieser and M. Fieser, "Steroids", Reinhold, New York, 1959.
73. R. B. Woodward, J. Amer. Chem. Soc., 64, 72 (1942).
74. For review, see (a) T. L. James, J. Chem. Educ., 48, 712 (1971); (b) S. D. Worley, Chem. Rev., 71, 295 (1971); (c) S. H. Hercules and D. M. Hercules, Record Chem. Progr., 32, 183 (1971); (d) C. Nordling, Angew. Chem., Int. Ed. Engl., 11, 83 (1972).

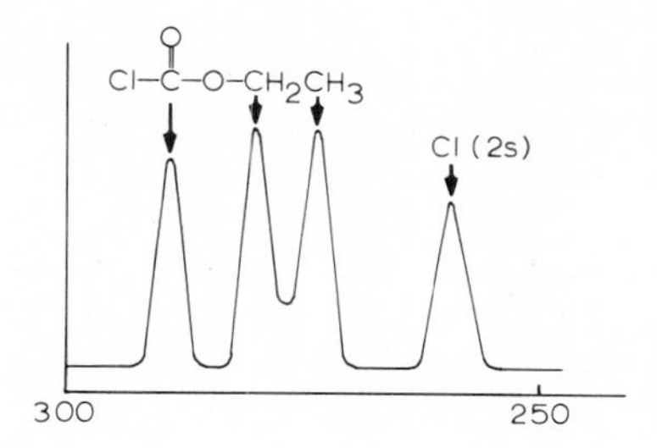

Fig. 2-22. The esca spectrum of ethyl chloroformate; intensity vs. binding energy in electron volts.

In photoelectron spectroscopy, ultraviolet radiation is used to excite, and only valence electrons are then removed. The resolution is better than 1 kcal/mole, so that vibrational fine structure can be observed. Experimental access to various MO levels is of great importance if we wish to understand the structure of organic compounds, and hence rapid development of this technique may be anticipated [75].

X-ray diffraction permits the precise location of all atoms - including the hydrogen atoms in most cases - in the molecule in a solid, as well as the crystal structure. Originally these experiments were so difficult and complex that they were limited to very simple molecules, but improvements in the technique, mechanical computation, and so on have advanced the state of the art to the point where the structures of many complex natural products have been elucidated. The determination of a structure does require the effort of an expert in this field however, and the technique is not yet one routinely applicable by others.

The dipole moment [76] of a molecule is defined by

$$\mu = ql$$

where q is the net charge and l is the length of the dipole; it is usually given in debye units (1 D = 10^{-18} esu cm). The two charges of the dipole are the centers of all positive and negative charge within the molecule. The moment is represented by an arrow, for instance

$$\underset{\longrightarrow}{\text{H–Cl}}$$

75. M. J. S. Dewar and S. D. Worley, J. Chem. Phys., 50, 654 (1969).
76. See H. B. Thompson, J. Chem. Educ., 43, 66 (1966), and the extensive bibliography given at the end of that paper.

There are two main ways of measuring the dipole moment. One of these depends on the Stark effect, i.e., the splitting of rotational lines induced in the microwave spectrum by an electric field; it allows very accurate measurement of the dipole moment of gaseous substances. The other is less accurate but perhaps more widely applicable because it deals with solutions. It is based on the fact that electric dipole tends to orient itself in an electric field, in opposition to the effect of thermal motion. The polarization of the solution is therefore a measure of the dipole moment. The electric field also induces a dipole moment, so that a correction must be made to subtract this contribution.

If the solubility of the compound of interest permits it, a nonpolar solvent such as benzene is used; if not, p-dioxane is often preferred. It can be shown that if the moments are measured in both solvents, there is little difference between the results unless the solute can form a hydrogen bond to the oxygen atoms; in that case the moment is usually larger in dioxane by 0.25-1 D [77].

The moment of a molecule can be fairly successfully guessed by vector addition of the so-called bond moments (see p. 27); thus, the dipole moments of o-, m-, and p-dichlorobenzene (2.50, 1.72, and 0 D, respectively) are about those that would be calculated on the basis of that of chlorobenzene (1.69 D) [78]. This fact allows one to determine the direction of a dipole moment. Thus, if nitrobenzene has a moment μ_A, a substituted benzene C_6H_5X has a moment μ and the p-disubstituted benzene $X\text{-}C_6H_4\text{-}NO_2$ has a moment μ_C such that

$$\mu_C = \mu_A + \mu_B$$

then clearly the group X must be electron donating since the highly electronegative nitro group is electron withdrawing.

One should be careful to distinguish the types of hydrocarbons to which the group X is attached, however, Thus, an amino group would normally be expected to be electron withdrawing; however, in benzenes it is electron donating, presumably because of resonance. In fact, the moment of p-nitroaniline (72) is about 1 D larger than the sum of the moments of nitrobenzene and aniline.

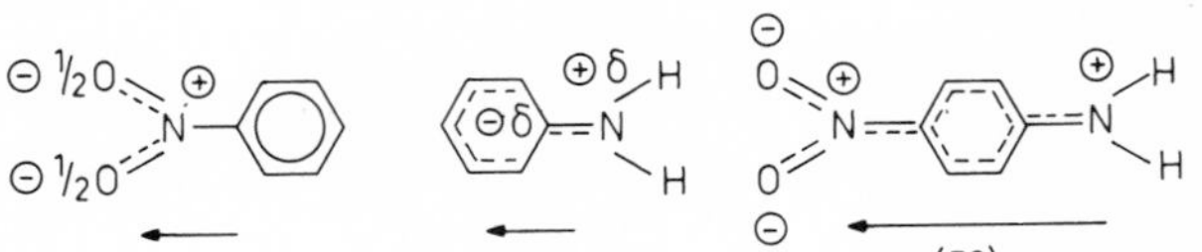

(72)

77. J.H. Richards and S. Walker, Trans. Faraday Soc., 57, 399, 406, 412, 418 (1961).
78. Handbook of Physics and Chemistry, 49th Ed., Chemical Rubber Co., Cleveland, Ohio, 1968-1969.

These resonance interactions can be avoided by the use of a nonconjugated hydrocarbon skeleton such as bicyclo[2.2.2]octane (73).

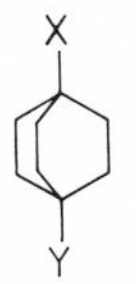

(73)

In the foregoing examples the vector addition is straightforward because all the molecules mentioned in it are rigid and without any conformational freedom. However, the dipole moment of p-dimethoxybenzene (74) is not equal to zero as it would be if the two methoxy groups had s-trans-conformations.

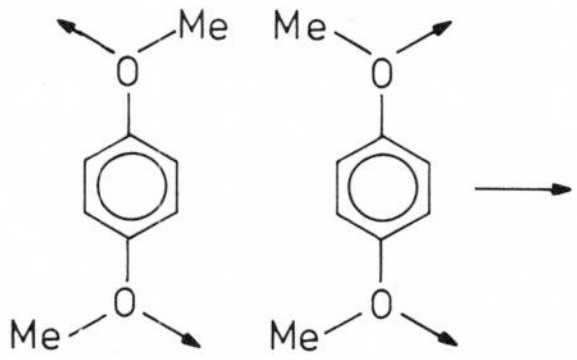

(74)

On the average the two methoxy groups are somewhere between these two extremes and thus the observed moment does not vanish.

Optical rotation [79] has long been one of the organic chemists most important tools. It is well known that asymmetric molecules will rotate the plane of polarized light. This rotation is dependent on a number of experimental factors such as the wavelength (the sodium D line at 589.3 nm is preferred), temperature, and solvent. The phenomenon can be quantitatively expressed in terms of a definition of the "specific rotation"

$$[\alpha]_{\lambda}^{t} = \frac{\alpha}{lc}$$

79. T.M. Lowry, "Optical Rotatory Power", Dover, New York, 1964.

where α is the observed rotation in degrees, l is the polarimeter tube length in dm, and c the concentration in g/cm^3; α is taken to be positive if the rotation is clockwise [80]. The "molecular rotation" is also used; it is defined by

$$M_\lambda^t = [\alpha]_\lambda^t \frac{M}{100}$$

where M is the molecular weight.

The laboratory solution of stereochemical problems very often requires the assessment of the molecular rotations to be expected of new or hitherto unresolved compounds, and hence any method - even if empirical - to estimate these rotations would be of great value to organic chemists. Such a method has been devised by Brewster [81]. The following examples may serve to explain his approach.

The rotation of the plane of polarization results from the interaction of the electric vector of the photons and the electrons in the molecules through which they pass. Let us consider a molecule (75) with an asymmetric carbon atom bound to four different atoms or rigid groups R, S, T, and V.

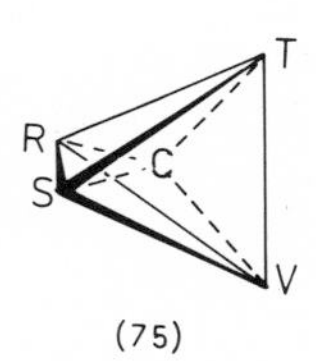

(75)

The molecule may be viewed along the various lines connecting the midpoints of the edges RS, RT, etc., and the carbon atom: a few of these views of the molecule are shown below.

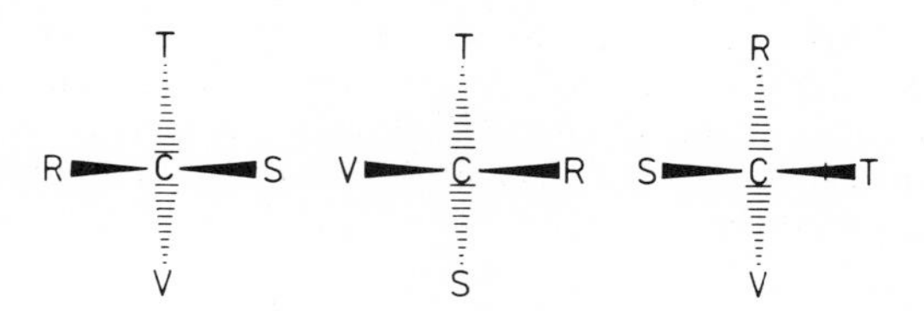

80. Since the observed rotation is not always precisely proportional to the concentration, it has been customary to report the specific rotation by giving the concentration and the solvent. Curiously, the custom is to use g/cm^3 to calculate $[\alpha]_\lambda^t$ but to use g/(100 cm^3) to indicate the concentration used. Thus, if one observed that a pure new substance in chloroform solution at 0.0200g/cm^3 in a 1 dm tube at 25° caused a rotation of polarized light of +0.352° at the D line, one would report

$$[\alpha]_D^{25} = +17.6\ CHCl_3(\underline{c}\ 2.00)$$

81. (a) J.H. Brewster, J. Amer. Chem. Soc., 81, 5475, 5483, 5493 (1959); (b) E.L. Eliel, "Stereochemistry of Carbon Compounds", Chap. 14, McGraw-Hill, New York, 1962.

Of all the ways to view the molecule we select whichever produces a complete right-handed screw turn of increasing or decreasing polarizability, starting with the substituent at the left (in these representations the Fischer convention is used; i.e., the horizontally placed substituents come forward and the vertical ones recede from the plane of the paper). If we find the polarizability increasing in this sequence, the rotation will be positive; otherwise it will be negative. The relative polarizabilities (see p. 28) may be gauged from Table 2-5. Thus, suppose that we are interested in predicting the sign of rotation of the α-iodopropionic acid (76).

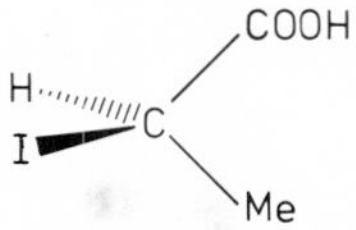

COOH
H
C
I
Me

(76)

TABLE 2-5. Polarizabilities of Various Groups

Substituent	Polarizability (R_D)	$\sqrt{R_D}$
I	13.95	3.73
Br	8.74	2.96
SH	7.73	2.78
Cl	5.84	2.42
CN	3.58	1.89
C_6H_5	3.38	1.82
CO_2H	3.38	1.82
CH_3	2.59	1.61
NH_2	2.38	1.54
OH	1.52	1.12
H	1.03	1.01
D	1.00	1.00
F	0.81	0.90

Since the I atom is the most polarizable group, we write the three representations (77)-(79) as follows

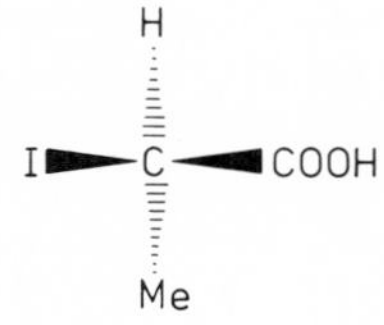

(77) 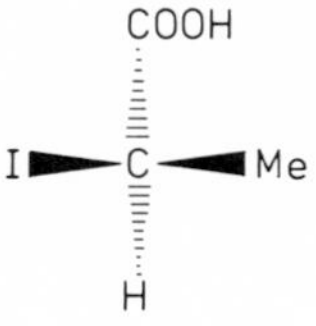 (78) 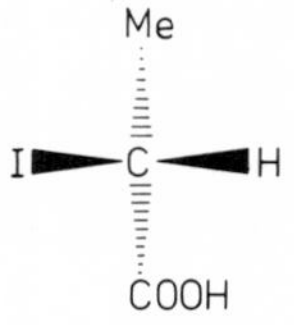 (79)

Since the polarizability order is I>COOH>CH_3>H, (78) is the one of interest, and since the screw pattern is right handed, M_D will be positive. In fact, it has been observed that $M_D = 100^{\circ}$. The corresponding bromo- and chloroacids have M_D values of 47° and 17°, respectively; this decrease in the positive value is also reasonable. It should be pointed out that the method is largely empirical; however, the correlation of the rotations with polarizability is at least intuitively plausible.

The problem becomes somewhat more complex when a flexible chain is considered, since the asymmetry of the molecule is a consequence not only of the four different groups but also because of conformational effects. Brewster's rules allow us to calculate the molecular rotations of such molecules; as an example we consider the 2-chlorobutane molecule (80).

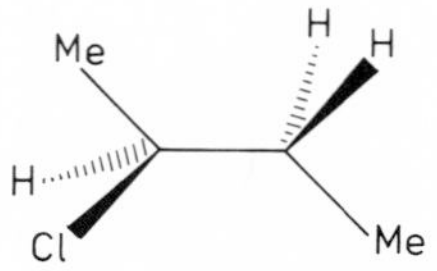

(80)

The conformations corresponding to the energy minima are (81)-(83).

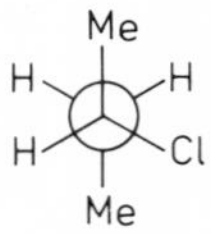

(81) 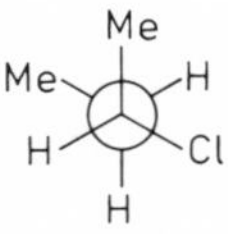 (82) 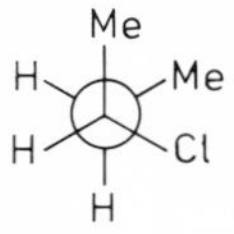 (83)

The third of these is a relatively high minimum and can be neglected; we may assume that half of the molecules will be in each of the two remaining conformations since a methyl group and chlorine atom are similar in size (cf. p. 230). In (81), on circling the central C-C bond in clockwise fashion we note the following succession of pairs of gauche neighboring groups: CH_3-H, Cl-CH_3, and H-H from front to back and H-Cl, CH_3-H, and H-CH_3 from back to front. M_D for this conformation is given by

$$M_D = 160\Sigma(R_D R_D)^{1/2}$$

where the two groups of three pairs have opposite sign. Thus,

$$\begin{aligned} M_D &= 160\ [(1.61 \times 1.01) + (2.42 \times 1.61) + (1.01 \times 1.01) \\ &\quad - (1.01 \times 2.42) - (1.61 \times 1.01) - (1.01 \times 1.61)] \\ &= +135^\circ \end{aligned}$$

conformation (82) is similarly found to have an $M_D = -58^\circ$; hence the substance is predicted to have an average $M_D = 1/2\ (135-58) = +34^\circ$. The observed value is $+36^\circ$. The agreement is not always so impressive; however, there is little doubt about the usefulness of the method. The calculations are relatively simple for cyclic molecules in which few conformations are possible and rather complex for longer chains; however, for such cases the interested reader is referred to Brewster's papers. It may finally be noted that it should be clear that if the four substituents are all different alkyl groups the rotation may be vanishingly small; thus, the 5-ethyl-5-n-propylundecane (85) obtained by the decarboxylation and Raney nickel desulfurization of (+)-3-(5'-ethyl-2'-thienyl)-3-(5'-carboxy-2'-thienyl)hexane (84) has no rotation of polarized light anywhere between 280 and 580 nm [82].

(+) Et–(thienyl)–C(n-Pr)(Et)–(thienyl)–COOH (84) → 1) $-CO_2$; 2) Ni, H_2 → n–Hex–C(n–Pr)(Et)–n–Bu (85)

Although the molecular rotation is wavelength dependent, the variation in the visible region is as a rule not very great for colorless compounds. In fact, this is one of the reasons for Brewster's success in calculating the sign and sometimes the magnitude of this rotation at the sodium D line. However, many optically active organic compounds absorb in the uv, and large effects are often observed when their rotations are measured in that region; they frequently exceed the M_D values by a factor of 100-1000. Thus, many stereochemical studies that might be dismissed as impractical because of small specific rotations may well be feasible if the rotations are measured in the uv. Fig. 2-23 shows some wavelength vs. rotation curves that are often observed. A curve such as B is said to be a manifestation of a positive Cotton effect, whereas C is an example of a negative Cotton effect. These "optical rotatory dispersion" (ord) curves can be directly recorded by means of commercially available instruments [83]. The point at which M_D

---

82. H. Wynberg, G. L. Hekkert, J. P. M. Houbiers, and H. W. Bosch, J. Amer. Chem. Soc., 87, 2635 (1965).
83. For reviews, see Ref. 81b, and (a) Chem. Eng. News, 39, 88 (Aug. 21, 1961); (b) C. Djerassi, "Optical Rotatory Dispersion: Applications to Organic Chemistry", McGraw-Hill, New York, 1960; (c) E. L. Eliel, N. L. Allinger, S. J. Angyal, and G. A. Morrison, "Conformational Analysis", Chap. 3, Wiley, New York, 1965.

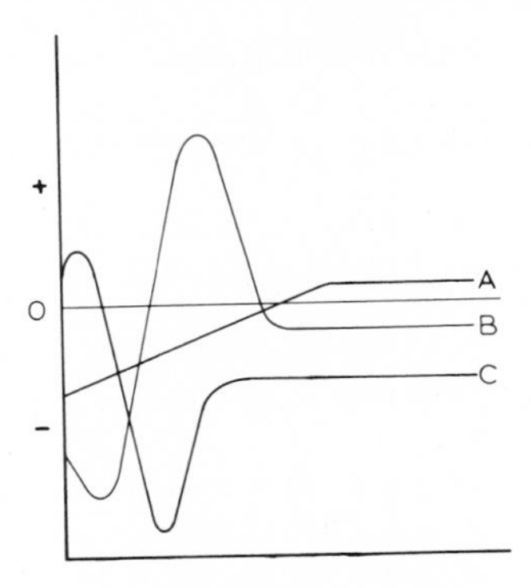

Fig. 2-23. Some ord curves.

becomes zero corresponds to the wavelength λ_{max} at which the substance is absorbing. Obviously, if the absorption is very strong in relation to the rotation the curve becomes hard to measure, and hence ketones with their relatively low ϵ_{max} values (at the $n \rightarrow \pi^*$ transition, see p. 76) are ideal substrates.

The applications of this technique have been primarily in the area of natural product (especially steroid) chemistry. Although these uses are largely based on an empirical understanding of the phenomenon, the following examples may serve to illustrate the applications that have been made.

First, there is a structural application. The ord curves for various oxo-substituted steroids are known, and since distant substituents apparently have small effects on these curves, the curve of a newly found substituted steroid may be helpful in locating the substituent. Configurations and even conformations may likewise be assigned this way in favorable cases. Many of these assignments are based on the so-called octant rule. If cyclohexanone in the conformation shown is viewed along the O=C axis, the various hydrogen atoms may be defined as occupying octants A through D. According to the rule, substituents lying in octants D and B will lead to positive Cotton effects, and those in A and C to negative effects.

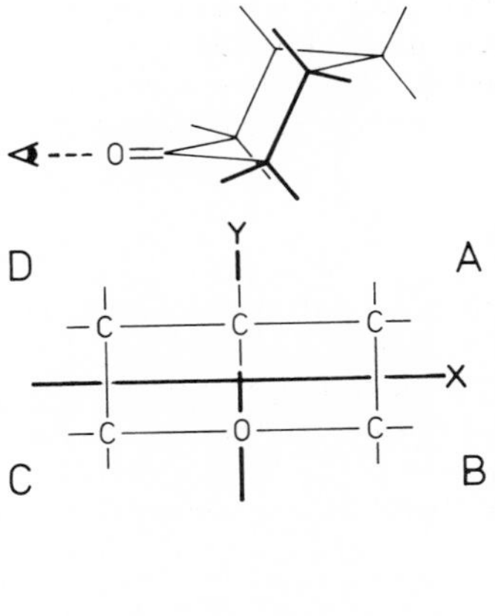

This rule can be used in a variety of ways; even the absolute configuration (see Chapter 5, p.142) can be assigned in favorable cases. Thus, (+)-*trans*-10-methyl-2-decalone (86), which may be either (87) or (88),

Me O H Me O H Me O H Me

(86) (87) (88)

is found to have a positive ord curve and hence is (87).

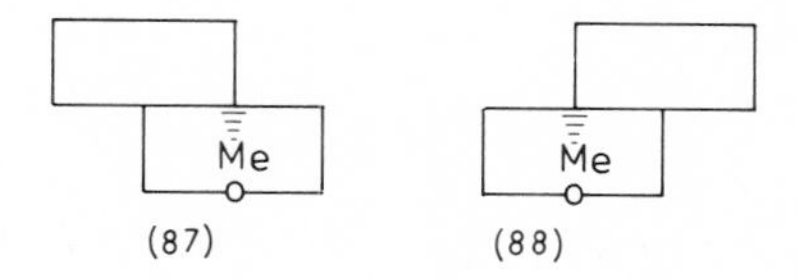

(87) (88)

Other techniques encountered occasionally include circular dichroism, which is related to ord and used especially by chemists interested in stereochemical problems [84]; nuclear quadrupole spectroscopy, useful in the assessment of the charge gradient at nuclei having I > 1/2 and, hence, of electronic structures [85]; Mössbauer spectroscopy, useful in the study of solid compounds containing certain heavy nuclei such as ^{57}Fe, ^{61}Ni, ^{67}Zn, ^{129}I, etc. [86]; and microwave spectroscopy, which is based on the rotational motion of molecules [87], and which is used especially for the precise measurement of bond lengths and angles in small molecules in the gas phase, and of dipole moments of gaseous compounds.

---

84. (a) A. Abu-Shumays and J. J. Duffield, *Anal. Chem.*, *38*, 29A (1966); (b) G. Snatzke, *Angew. Chem., Int. Ed. Engl.*, *7*, 14 (1968).
85. W. J. Orville-Thomas, *Quart. Rev., Chem. Soc.*, *11*, 162 (1957).
86. (a) G. K. Wertheim, "Mössbauer Effect: Principles and Applications", Academic Press, New York, 1964; (b) R. H. Herber, *J. Chem. Educ.*, *42*, 180 (1965).
87. (a) R. H. Schwendeman, H. N. Volltrauer, V. W. Laurie, and E. C. Thomas, *J. Chem. Educ.*, *47*, 526 (1970); (b) W. H. Kirchhoff, *Chem. Eng. News*, *47*, March 24, 1969, p. 88.

Chapter 3

NOTES ON NOMENCLATURE AND LITERATURE

3-1. INTRODUCTION

In the course of reading the chemical literature, the practicing chemist at times finds himself stymied by insufficient knowledge of organic nomenclature, and hence some pages are devoted here to that topic. The objective is not to repeat the rules governing the names of simple noncyclic compounds [1], but to focus on several groups of cyclic molecules on which much attention has centered in recent years for one reason or another, and several of which we shall have reason to mention in various parts of this book. It would require a great deal of space to write and exemplify all the rules that have been written to name even this group of

---

1. (a) R.S. Cahn, "An Introduction to Chemical Nomenclature", 3rd Ed., Butterworths, London, England, 1968; (b) "Chemical Nomenclature", Advances in Chemistry Series, No. 8, 1951; (c) O. Runquist, "Programmed Review of Organic Chemistry Nomenclature", Burgess, Minneapolis, Minn., 1965; (d) J.E. Banks, "Naming Organic Compounds", Saunders, Philadelphia, Pa., 1967.

compounds; the discussion is therefore further limited in such a way that it may be helpful in reconstructing structures from the names and in making educated guesses about the names from given structures; but it will in many cases still be necessary to consult the literature in order to ensure the generation of the correct name.

Many of the names are based on new trivial designations. The novice in organic chemistry often complains that the science has a perfectly logical system for naming the infinite variety of organic compounds in a systematic way, yet its practitioners seem to have a never ending penchant for inventing new trivial names. In principle, the Geneva system could indeed be used for virtually all the classes of compounds described here; however, many of the names so based are so long and cumbersome, requiring so much time and effort in translation from structure to name and back again, that they are more of a hindrance than a help. When the study of a new class of compounds becomes active, the need often arises to replace the Geneva names of basic structural units by simpler terms.

It is furthermore the experience of this author that while every organic chemist knows how to search his library to some degree, many do not use it as effectively as is possible, and hence some remarks on that topic seem in order also.

3-2. NOMENCLATURE: THE ANNULENES

This group of compounds basically comprises the family of completely conjugated cyclic compounds $(\text{-CH=CH-})_n$. The number 2n is written first, in brackets; thus, if benzenes were considered a member of this group, it would be [6]annulene. When the rings become larger than this, it becomes possible for some of the double bonds to assume trans- configurations.

A system of accounting for this can be applied as follows [2]. Consider [12]annulene and its configurations (1), (2), and (3). For each of these, one writes in succession C, T, etc., for cis and trans as one traverses the ring, then assigns 0 for cis and 1 for trans so that a binary number results.

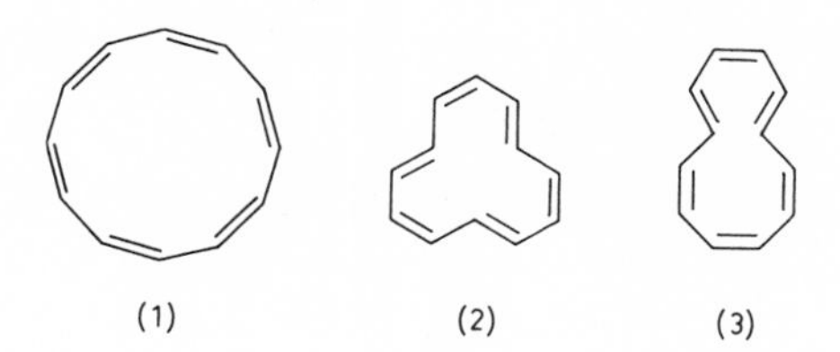

(1) (2) (3)

This is then transposed into a decimal number as shown in the example following.

2. J. F. M. Oth and J. M. Gilles, Tetrahedron Lett., 6259 (1968).

(1)	(2)	(3)
C C C C C C	C T C T C T	C C C T C T
0 0 0 0 0 0	0 1 0 1 0 1	0 0 0 1 0 1
0	16 4 1	4 1
[12]-0-annulene	[12]-21-annulene	[12]-5-annulene

In the case of (2), the bond one elects to start with clearly makes a difference; the choice is then made in such a way as to generate the lowest binary number. The reverse sequence is traversed if one wishes to reconstruct the configuration from the name. The method is unique and unambiguous. However, the names of the two Kekulé structures may be different unless the two can be superimposed in some way. This is the case with (1) and (2), but the Kekulé structure of (3), i.e. (4), leads to the name [12]-9-annulene.

(4)

In practice, since these structures are either resonance structures or valence tautomers, this does not lead to confusion; one simply selects the name involving the lower number. In the literature the number indicating the configuration about the double bond is often omitted - either because it is obvious what the configuration must be, or because the configuration is not known, or, if more than one is possible, because they may be rapidly interconverting.

In some instances transannular bridges are present. The naming is easily extended as shown by the examples below: The name for (5) is 1,6-oxido[10]-5-annulene, that for (6) is anti-1,6:8,13-bismethano[14]-27-annulene, and that for (7) is 1,6:8,13-propanediylidene[14]-27-annulene.

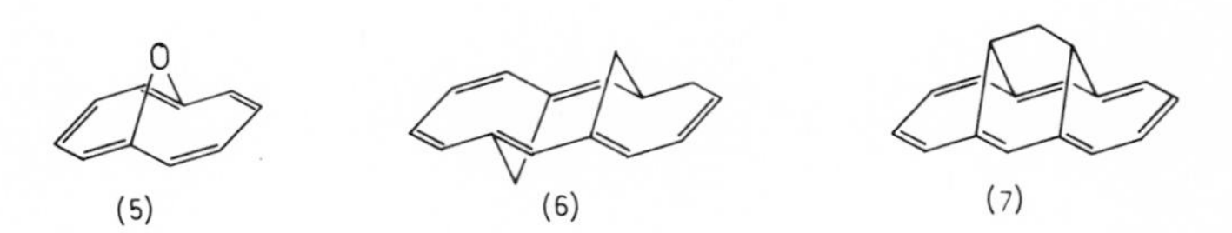

(5) (6) (7)

Triple bonds and cumulative double bonds are often encountered in such molecules also, and their presence is denoted by the prefix dehydro; thus, (8) is 1,6-didehydro[10]annulene.

(8)

3-3. CONDENSED CARBOCYCLIC AROMATIC RING SYSTEMS [3]

A number of these structures received trivial names as they were discovered in the coal tar; some of these are written below (s- end as- mean symmetric and asymmetric, respectively; a single resonance structure is written in each case).

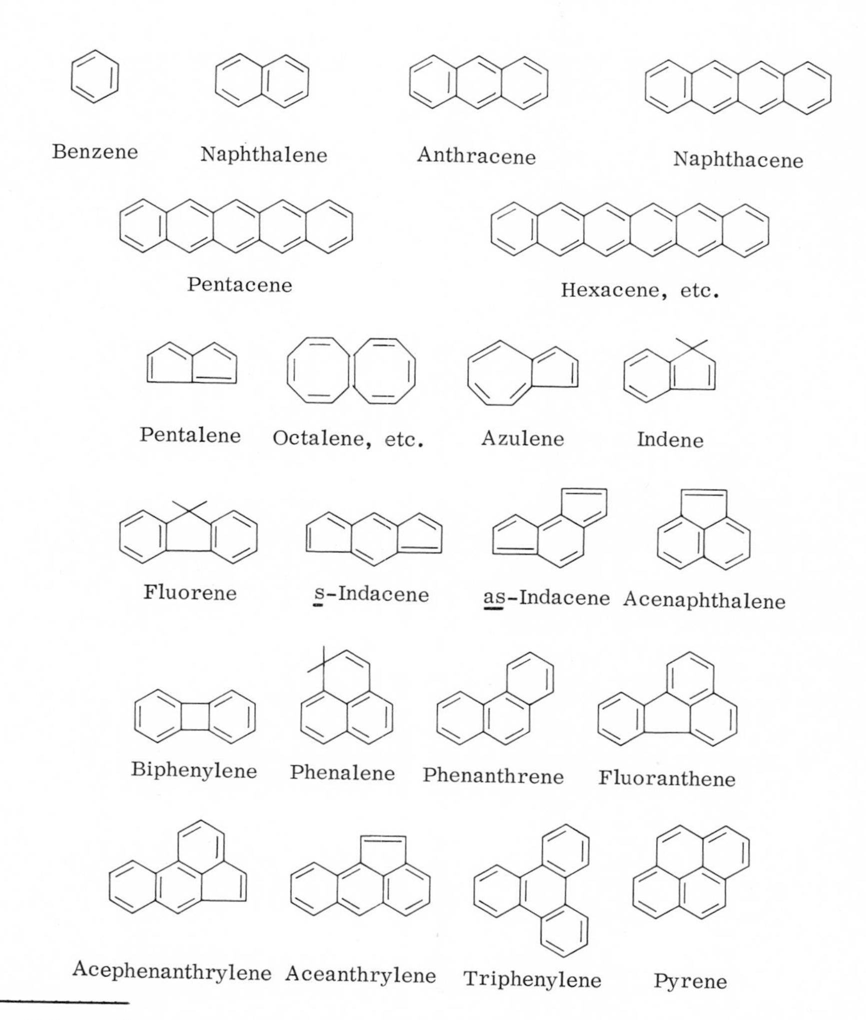

3. This section is meant to touch only on some of the main features of ring nomenclature, so that the compounds encountered later on in this book may be named and referred to by either structure or name. The reader trying out various extensions will find in short order that the rules described below are quite incomplete. For further information, see the Ring Index, Chemical Abstracts, 1959, and subsequent supplements. A wealth of information can also be obtained from the "Handbook of Chemistry and Physics", Chemical Rubber Co., Cleveland, Ohio; see, e.g., the 45th ed., p. C1-C52.

Chrysene

Pleiadene

Picene

Perylene

Tetraphenylene

Pentaphene

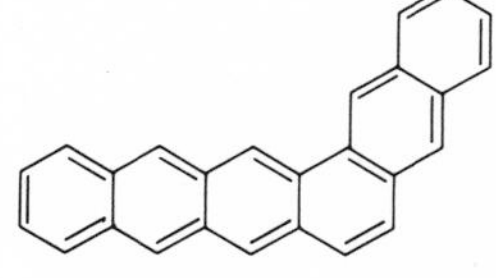

Hexaphene, etc.

Pentahelicene

Hexahelicene, etc.

Rubicene

Coronene

Trinaphthylene

Pyranthrene

Ovalene

When additional rings are fused onto one of these structures a relatively straightforward system is available to name the resulting compound. In order to do so, we need a numbering system, and that means that we should agree on a reproducible way of drawing the structures. Thus, it may be noted that the rings are drawn with two sides vertical, and that as many rings as possible are drawn in a horizontal line. If there are several possible ways to do this, we choose the one that favors the upper right-hand region, and if that still does not decide the issue, we choose the one that avoids rings in the lower left-hand region; all the ring systems with trivial names (p. 91-92) have been written properly. The numbering then starts in the highest ring - if there are more than one, the ring furthest to the right, and at the first nonfused carbon atom while proceeding in a clockwise direction. Fused atoms receive the same numbers at the preceeding carbon atoms plus a letter starting with "a". The outer edges are also labeled by letters, starting with "a" for C_1C_2, and skipping none as we proceed in a clockwise direction. Thus triphenylene is numbered as follows:

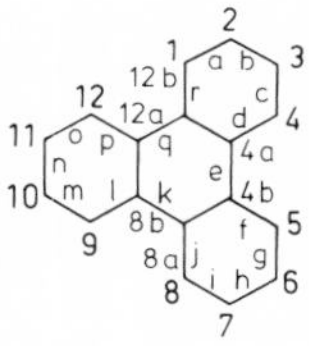

Ambiguity is further prevented by giving atoms common to two or more rings the lowest possible numbers; thus, fluorene is numbered as in (9) rather than in (10).

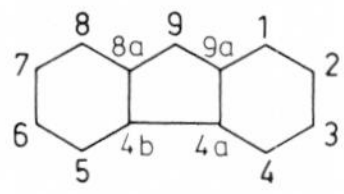

(9) 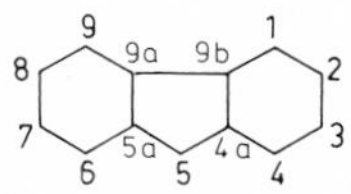 (10)

There are two exceptions, both of a historical nature; anthracene and phenanthrene are numbered as shown below.

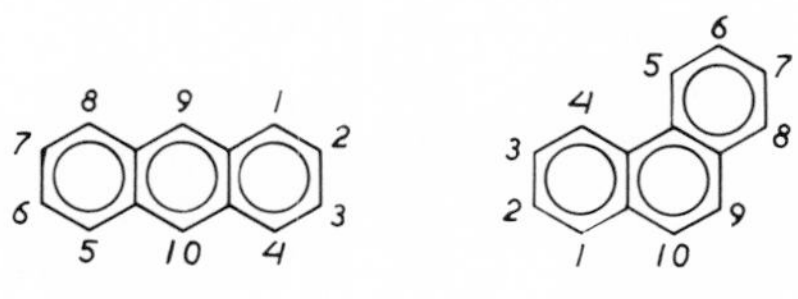

Suppose now that we wish to name the hydrocarbon (11).

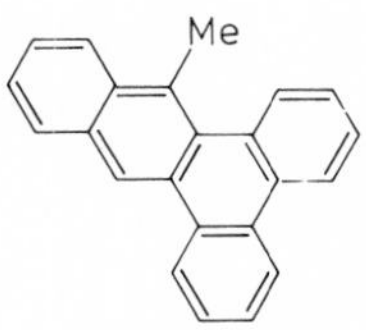

Me

(11)

This is now considered to be built up of a benzene ring and a triphenylene system. After we have written the triphenylene molecule in the required way, there are three ways in which the benzene ring can be fused to it so as to produce (11); we choose the one which involves the lowest numbers. The triphenylene is marked by means of the edges involved and the addend - if other than benzo - by its numbers; the resulting name of the skeleton being benzo[b]triphenylene.

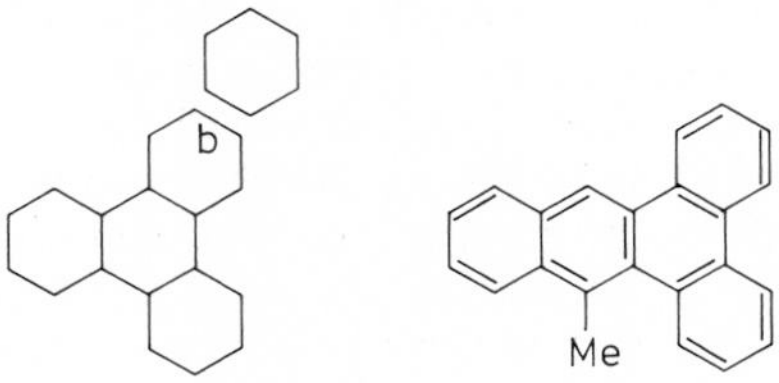

b

Me

The new numbering is then found by observing the same rules as before, and thus the name for (11) is 9-methylbenzo[b]triphenylene. Similarly, (12) is named naphtho[2, 1-a]pyrene;

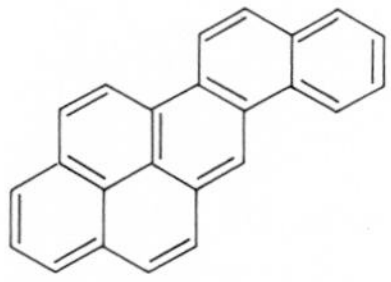

(12)

(13) is named naphtho[8, 1, 2-bcd]perylene, and so on.

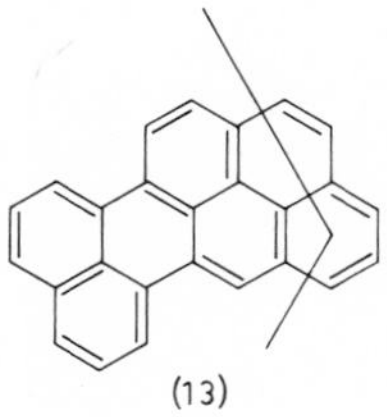

(13)

If a methylene group interrupts the sequence of conjugated double bonds, that group and location are indicated by the symbol H; thus, (14) is 2H-indene,

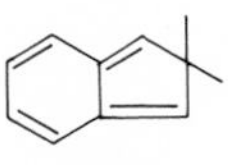

O

(14) 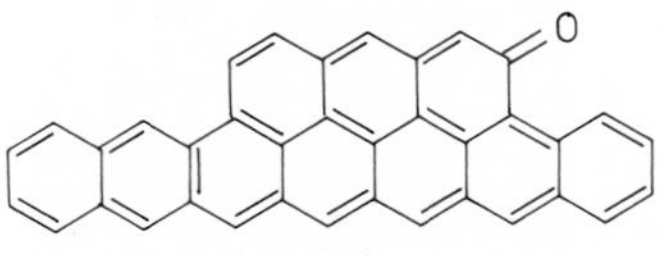 (15)

and (15) is 2H-anthra-[2,1,9,8,7,-fghijk]heptacen-2-one. Finally, if the molecule may be viewed as having been more extensively hydrogenated, the prefix hydro may be used; thus, (16) would be 2,3,9,10-tetrahydroanthracene.

(16)

3-4. HETEROCYCLIC RINGS

Once again we find here the result of ingenious attempts to superimpose order on chaotic beginnings, and therefore one finds many exceptions and apparently arbitrary rules.

The following are some of the more common trivially named hetero ring systems:

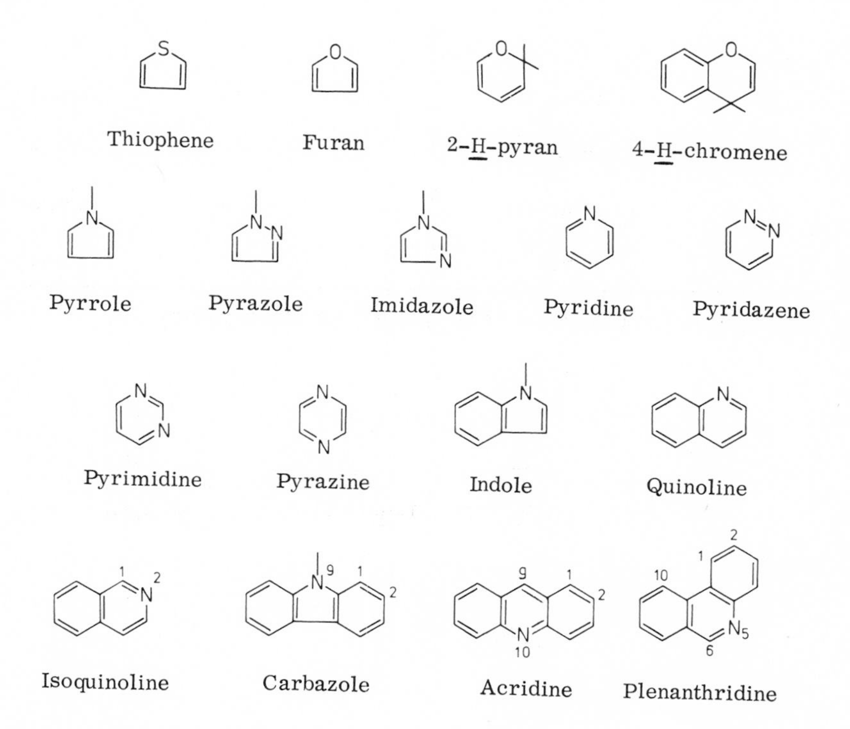

In each case the numbering takes place as in the carbocyclic analogs, starting with a hetero atom if possible and proceeding in clockwise fashion. In a few cases some of the numbers are shown. In a great many rings systematic naming is possible by the use of the prefixes oxa, thia, aza, phospha, sila, bora, etc., for oxygen, sulfur, nitrogen, phosphorus, silicon, and boron, respectively. When more than one hetero element is present, preference is given first to the highest group in the Periodic Table, and then to the lowest atomic weight (hence O, S, N, P, etc.). Use is furthermore made of the suffices in Table 3-1 below; it may be noted that the syllables ri, et, ep, oc, on, and ec stem from the words tri, tetra, hepta, octa, nona, and deca. The examples shown below may be helpful in the process of learning the use of these syllables.

TABLE 3-1. Suffices in Heterocyclic Systems

	N present		N absent	
Ring size	Unsaturated	Saturated	Unsaturated	Saturated
3	irine	iridine	irene	irane
4	ete	etidine	ete	etane
5	ole	olidine	ole	olane
6	ine	a	ine	ane
7	epine	a	epin	epane
8	ocine	a	ocin	ocane
9	onine	a	onin	onane
10	ecine	a	ecin	ecane

[a] Same as the unsaturated compound, and the prefix perhydro.

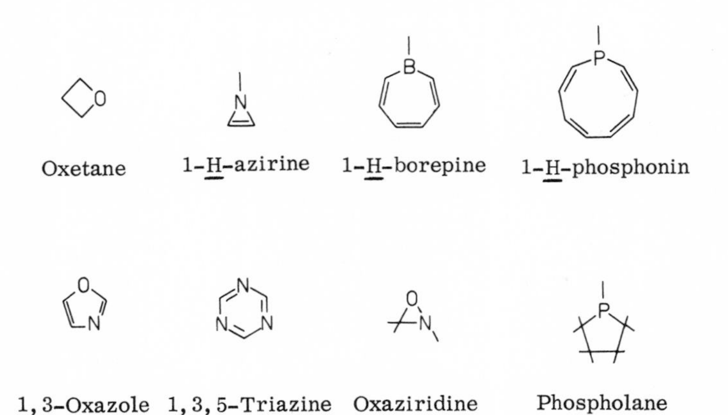

Oxetane 1-H-azirine 1-H-borepine 1-H-phosphonin

1,3-Oxazole 1,3,5-Triazine Oxaziridine Phospholane

When more than one heterocyclic ring occurs in a polycyclic system, the name of the compound is based on rules similar to those quoted in the preceding section. The following examples may be helpful in the design of names from structures and vice versa.

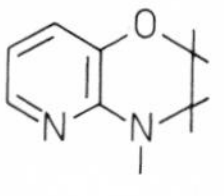

3, 4-dihydro-2*H*-pyrido[3, 2b]-1, 4-oxazine

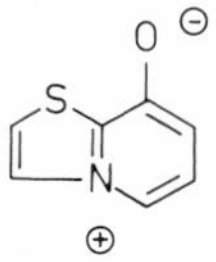

thiazolo[3, 2-a]pyridinium-8-oxide

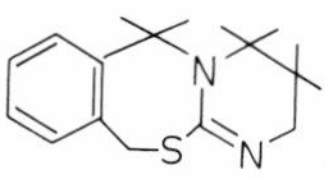

1*H*, 6*H*, 11*H*-2, 3-dihydropyrimido[2, 1-b] benzo[e]-1, 3-thiazepine

3-5. POLYCYCLIC COMPOUNDS

An enormous amount of literature - some of it of crucial importance to the basic concepts of organic chemistry [4] - deals with saturated polycyclic compounds in which the rings have more than two atoms in common. An example is (17).

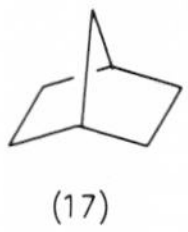

(17)

In such structures we first of all note the presence of bridgehead atoms where the rings are joined together. If there are two, the compound is basically a bicyclo compound; if there are four, a tricyclo compound, etc. Some of these may be single atoms counted twice, as in the tricyclo compound (18).

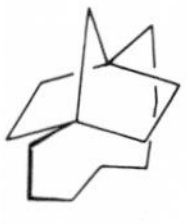

(18)

---

4. For a brief summary of the various concepts that find their main support in polycyclic chemistry, see L. N. Ferguson, *J. Chem. Educ.*, *46*, 404 (1969).

We then count the total number of atoms in each of the bridges as well as in the whole basic system, this information being used as follows: (17), bicyclo[2.2.1] heptane; (18), tricyclo[7.2.2.1]tetradecane, etc. The numbering proceeds, after starting at a bridgehead atom, around the longest bridge first: (19) is 9,9-dimethylbicyclo[3.3.1]nonane,

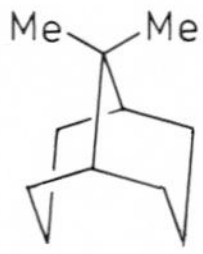

(19)

(20) is 1-azabicyclo[2.2.1]hept-2-ene (sometimes written as 1-azabicyclo[$2^{\Delta 2}$.2.1] heptene), and so on.

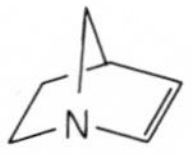

(20)

For substituents, the prefixes *exo*- and *endo*- are used to denote *cis*- and *trans*-, respectively, to the smaller bridge; thus (21) is 2-*exo*-bromobicyclo[3.3.1] nonane

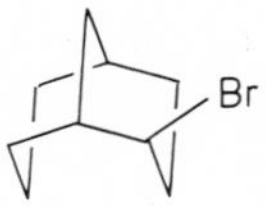

(21)

and (22) is *endo*-3-methylbicyclo[3.2.1]octane.

(22)

The use of the prefixes *syn*- and *anti*- may be illustrated by (23) and (24), which are *syn*-7-methylbicyclo[2.2.1]hept-2-ene and *anti*-7-*exo*-2-dimethylbicyclo[2.2.1] heptane, respectively.

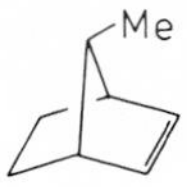

(23) 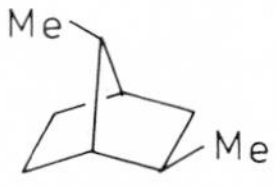 (24)

The problem becomes more difficult when more complex ring systems are considered [5]. As an example, we name (25) (quadricyclane).

(25)

The skeleton is first written in such a way that as many atoms as possible are part of a single ring (the so-called main ring).

≡

We then choose the longest bridge (with the largest number of atoms) spanning this bridge; if two or more are available of equal length, we pick the one that divides the main ring as symmetrically as possible. These three branches are then the basic bridges; the numbering proceeds as in the bicyclic molecules (see below), and the remaining connections are indicated as shown by the name: tetracyclo[3.2.0.$0^{2,7}$.$0^{4,6}$]heptane.

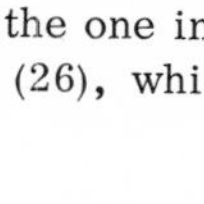

It may be noted that the correct name can usually not be obtained by starting with the designation for a less highly condensed system; thus, (25) should not be named as a $0^{2,6}$.$0^{3,5}$ derivative of (17). As always in organic nomenclature when there are two apparently equivalent ways, the one involving the lowest numbers is chosen. Other examples are cubane (26), which is pentacyclo[4.2.0.$0^{2,5}$.$0^{3,8}$.$0^{4,7}$]octane,

---

5. (a) J. Meinwald and J.K. Crandall, J. Amer. Chem. Soc., 88, 1292 (1966); (b) D.R. Eckroth, J. Org. Chem., 32, 3362 (1967).

(26)

adamantane (27) which is tricyclo$[3.3.1.1^{3,7}]$decane, and

(27)

triamantane (28) which is heptacyclo$[7.7.1.1^{3,15}.0^{1,12}.0^{2,7}.0^{4,13}.0^{6,11}]$octadecane.

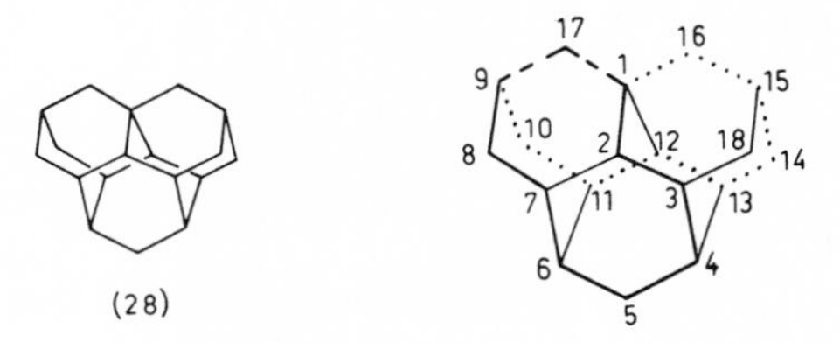

(28)

3-6. TRIVIAL NAMES FOR POLYCYCLIC COMPOUNDS

For the more complex structures encountered among the polycyclic compounds, new simple names have been adopted in recent years for many classes of compounds. Thus, (27) and (28) are examples of the diamondoid molecules, so-called because the carbon atoms are arranged in the same basic manner as in diamond. Diamantane (29) is the second member of this series.

(29)

As in the other groups mentioned below, if substituents are present, the IUPAC name must still be devised so that the proper number can be assigned; thus, (30) is 5-methyldiamantane (or 5-methylpentacyclo$[7.3.1.1^{4,12}.0.^{2,7}0^{6,11}]$tetradecane).

(30)

Another group of compounds are the polyhedranes. There are five regular polyhedra, but in view of the tetravalency of carbon only three polyhedranes are conceivable: Tetrahedrane (31) [6], cubane [7] (at present the only known stable member), and dodecahedrane (32) [8].

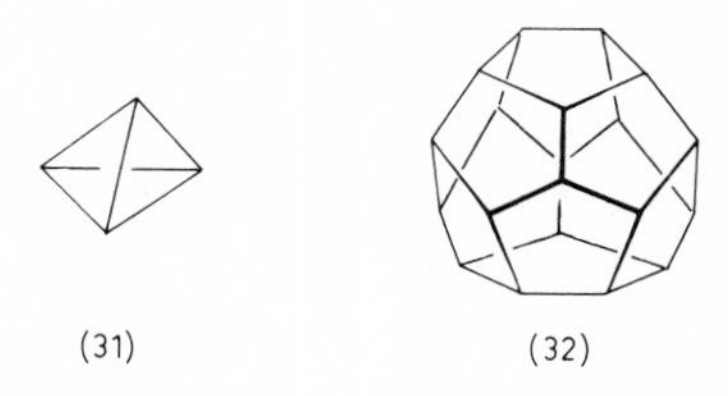

(31) (32)

Further groups of compounds which have earned trivial names in recent years are the asteranes such as triasterane (33) [9], basketanes such as [4-]basketene (34) [10], propellanes such as (35) ([3.2.1]propellane) [11], and radialenes such as [3]radialene (36) [12].

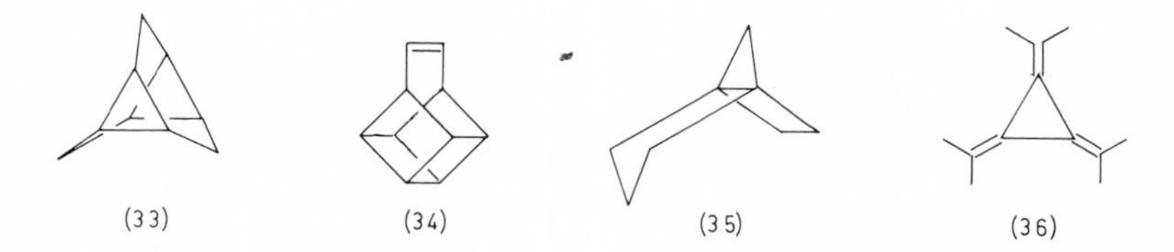

(33) (34) (35) (36)

---

6. P. B. Shevlin and A. P. Wolf, J. Amer. Chem. Soc., 92, 406 (1970).
7. P. E. Eaton and T. W. Cole, J. Amer. Chem. Soc., 86, 3157 (1964).
8. P. E. Eaton and R. H. Mueller, J. Amer. Chem. Soc., 94, 1014 (1972).
9. U. Biethan, U. v. Gizycki, and H. Musso, Tetrahedron Lett., 1477 (1965).
10. S. Masamune, H. Cuts, and M. G. Hogben, Tetrahedron Lett., 1017 (1966).
11. (a) P. G. Gassman, A. Topp, and J. W. Keller, Tetrahedron Lett., 1093 (1969); (b) D. Ginsburg, Accounts Chem. Res., 5, 249 (1972).
12. G. Köbrich and H. Heinemann, Angew. Chem., Int. Ed. Engl., 4, 594 (1965).

Trivial names are invented occasionally for individual compounds that are clearly not members of a class; among them we find prismane (37), brexane (38), brendane (39), twistane (40), barrelene (41), bullvalene (42), calicene (43), and so on.

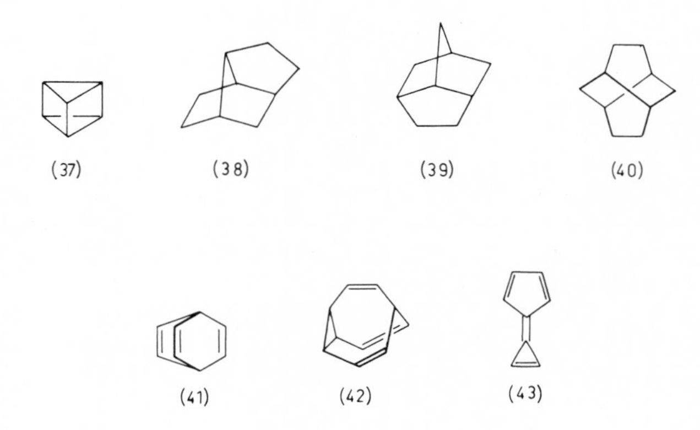

These names would obviously not be very useful if they could not be extended to analogous compounds. The prefizes hydro, dehydro, homo, and nor frequently allow such extensions. Thus, the hydrocarbon (44) is called 1, 3-dehydroadamantane.

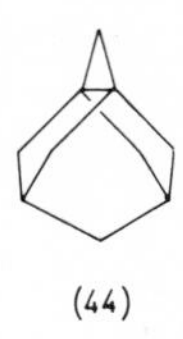

(44)

Homo suggests an extra methylene group; (45) is homocubane. Nor means without the substituents (usually methyl groups) found in the natural product; thus, the name norbornane (17) is derived from the naturally occurring terpene bornane (46).

Me Me

Me

(45) (46)

3-7. SPIROCOMPOUNDS

Such compounds result when two rings are linked together by means of one single atom only. Thus, (47) is called spiro[3.3]heptane, and (48) is dispiro[2.1.3.4]dodecane, with the numbering as indicated.

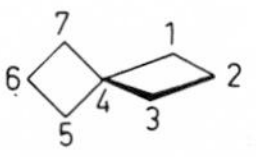

(47) 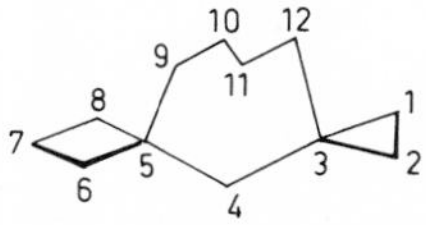 (48)

If one or both of the rings of a monospiro compound are part of fused ring systems, the name consists of the word spiro and then the hydrocarbons so combined in alphabetical order; for example, (49) is spiro[cyclohexane-1,9'-fluorene];

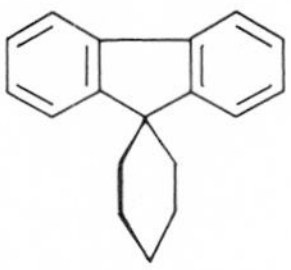

(49)

in a polyspiro case, the beginning of the system is chosen on an alphabetical basis also, as in (50) which is called dispiro[indene-1,1'-cyclopentane-2',1''-phenalene].

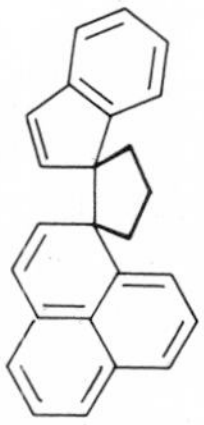

(50)

Recently much interest has centered on the so-called rotanes, in which a central ring is present with all its carbon atoms playing the role of spiro atoms (for example, (51)) [12a].

(51)

The need for well-chosen trivial names is especially clear in the case of spiro compounds; thus, the name spiro[adamantane-2, 2'-adamantane] [12b, c] for (52) is clearly a case in which the full Geneva name is lengthy and unwieldy.

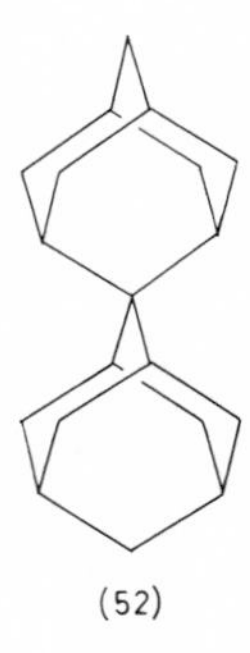

(52)

3-8. FULVENES

Fulvene is the name of compound (53). Compounds of this sort are of considerable theoretical interest, and since one finds a great deal of discussion of their properties in the recent literature, their general nomenclature is briefly discussed here. It should be emphasized that this nomenclature is "unofficial," and that chemical abstracts uses terms such as cyclopentadienylidenealkanes. The preferred names of (53), (54), and (55) are triafulvene, pentafulvene, and heptafulvene; those of (56) and (57) are triafulvalene and pentafulvalene (often fulvalene), respectively; (43) is triapentafulvalene (also calicene), and so on.

12a. A. P. Krapcho and F. J. Waller, J. Org. Chem., 37, 1079 (1972).
12b. E. Boelema, J. Strating, and H. Wynberg, Tetrahedron Lett., 1175 (1972).
12c. W. D. Graham and P. von R. Schleyer, Tetrahedron Lett., 1179 (1972).

(54) (53) (55) (56) (57)

3-9. CYCLOPHANES

Cyclophanes are compounds in which an aromatic ring is bridged by nonadjacent positions. The study of the chemistry of such compounds has taken an enormous flight in recent decades, and as it normally happens in such cases, the ability of the organic chemist to produce an ever widening range of such materials has far outrun his ability to design a reasonable nomenclature. Smith [13] has proposed the system which is used further in this book. If the ring is benzene, the basic compound is a cyclophane; otherwise, a naphthalenophane, pyridinophane, thiophenophane, and so on. The following examples may help illustrate the rules:

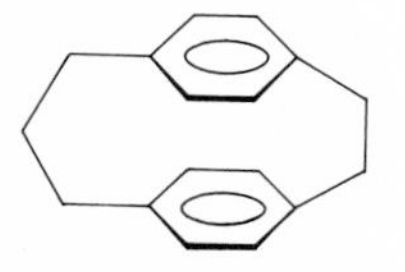

[3.2]paracyclophane

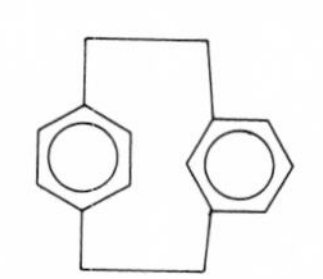

[2.2]metaparacyclophane

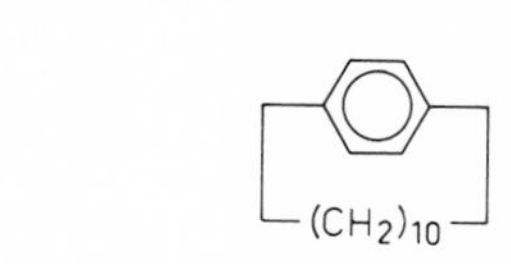

[14]paracyclophane

[8][8]paracyclophane

[12](1, 10)anthracenophane

9, 9-dimethyl-21-nitro [17](1, 5)anthracenophane

13. (a) B. H. Smith, "Bridged Aromatic Compounds", Academic Press, New York, 1964; (b) K. Hirayama, Tetrahedron Lett., 2109 (1972).

3-10. ORGANOMETALLIC COMPOUNDS

The simplest of such compounds are easy to name; thus, phenyllithium, diethylmagnesium, etc., are the prototypes of large numbers of such molecules. In recent decades many chemicals have been discovered in which the metal atom is bound to several carbon atoms of unsaturated ligands, and it has become desirable to indicate this in the name. The system proposed by Cotton [14] now appears to be in general use. In it, the carbon atoms bound to the metal atom are numbered, and the connection is designated by the word hapto. Thus, ferrocene (58) by this nomenclature would be di(pentahaptocyclopentadienyl)iron, which may be written in abbreviated fashion as $(h5\text{-}C_5H_5)_2Fe$.

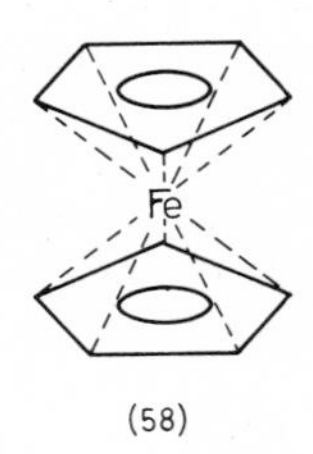

(58)

Other examples may suffice to illustrate this practice; the numbers are indicated only if they are not obvious; the dotted lines indicate bonds that may or may not represent pairs of electrons.

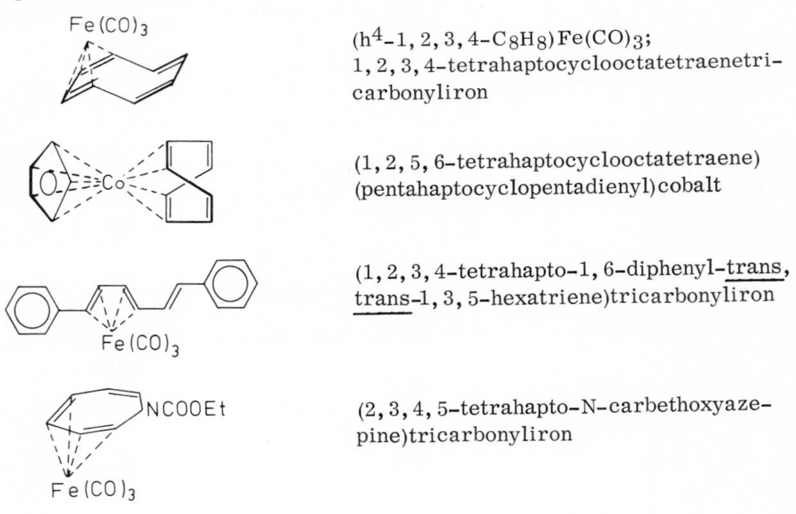

$(h^4\text{-}1,2,3,4\text{-}C_8H_8)Fe(CO)_3$; 1,2,3,4-tetrahaptocyclooctatetraenetricarbonyliron

(1,2,5,6-tetrahaptocyclooctatetraene)(pentahaptocyclopentadienyl)cobalt

(1,2,3,4-tetrahapto-1,6-diphenyl-trans, trans-1,3,5-hexatriene)tricarbonyliron

(2,3,4,5-tetrahapto-N-carbethoxyazepine)tricarbonyliron

---

14. F.A. Cotton, J. Amer. Chem. Soc., 90, 6230 (1968).

(1, 2, 3, 4-tetrahapto: 5, 6, 7, 8-tetrahapto-cyclooctatetraene)<u>trans</u>-bis(tricarbonyliron)

octahaptopentalene-μ-carbonyltetracarbonyldiiron

3-11. CROWN COMPOUNDS

The methyl ethers of ethylene glycol and the corresponding polyglycols are versatile and popular solvents, and hence simple names have been devised for them: Glyme for (60), diglyme for (61), and so on.

(60) (61)

In recent years a group of cyclic polyethers has come into use as chelating agents of alkali metals, and their remarkable properties have insured a great deal of literature for them. Although the standard nomenclature is applicable to them, it is cumbersome and there is a real need for abbreviated names. In view of the crown shape of the molecules, names such as the following have come into common use [15].

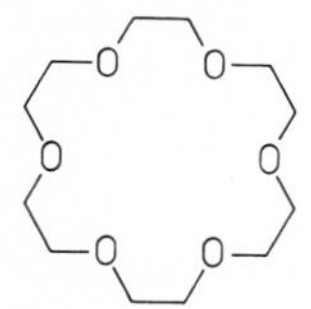

[18]crown-6

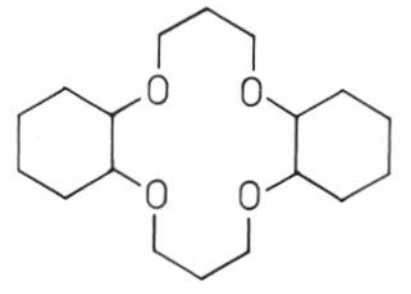

benzo[15]crown-5

perhydrodibenzo[14]crown-4

15. C. J. Pedersen and H. K. Frensdorff, Angew. Chem., Int. Ed. Engl., 11, 16 (1972).

These names are obviously not unambiguous and should not be used unless the structural formula or a reference is given along with it; there is furthermore no clear and self-evident way to extend this nomenclature to include other hetero atoms, the length of the carbon chains, substituents, and so on, without reverting to the IUPAC system.

3-12. ISOTOPICALLY SUBSTITUTED COMPOUNDS

With deuterium or tritium one can devise a name based on that of the parent compound and then indicate the labels as substituents; (62), for example, would be 1,1-dideuterio[2,2]paracyclophane.

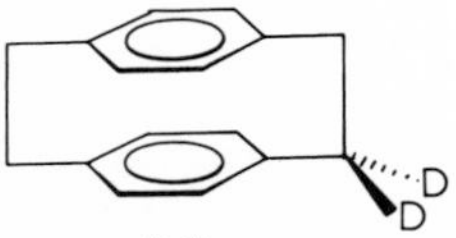

(62)

Further designations in common usage are illustrated below.

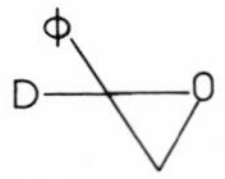

styrene-α-d oxide

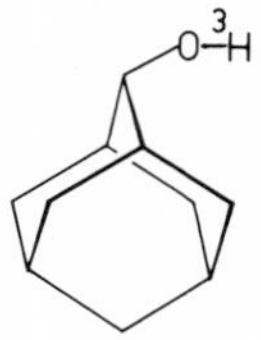

2-adamantanol-O-t

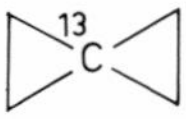

spiro[2.2]pentane-3-^{13}C

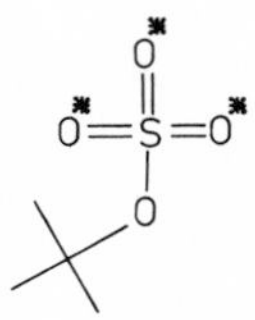

neopentyl tosylate-sulfonyl-^{18}O

3-12. THE CHEMISTRY LIBRARY

The library is by far the most useful tool to the practicing organic chemist, and it would indeed be a basic error for any student ready for his first research problem not to learn thoroughly what sort of information is available there and how to dig it out. One should at that time not merely try to find out what is known about the problem at hand, but study what resources are available and how to use them [16]. Proper use of the library can prevent the loss of months or even years spent in unraveling chemical problems that had already been deciphered. One occasionally hears the remark that the chemical literature is so vast that it is simpler to measure the desired numbers than to find them in the library, and this may be true for relatively trivial pieces of information; but someone who embarks on a major laboratory study without first having thoroughly combed out the library can be certain to waste a great deal of time.

The following resources are available in a well-equipped chemistry library. One finds there an index cabinet containing Library of Congress cards; one set in which the cards are listed alphabetically for authors and another for titles. Very often there will furthermore be an alphabetical list of nouns or key words appearing in the titles. These cards furnish the title, author, and the Library of Congress index number; thus, Bottle's book [16] has the number QD 8.5.B6-1969. It's then a small trick to find this volume on the shelf. Equally important, there will be a small cardex file showing in alphabetical order the journals the library subscribes to, its holdings of back volumes, and their location.

Another important resource is Beilstein's Handbuch der Organischen Chemie - usually referred to simply as "Beilstein", which is a 27-volume compilation of organic compounds and the most basic information about each, with a reference to each bit of information. Beilstein originally completely covered all organic compounds to 1910; a supplement ("erstes Erganzungswerk") covers the period 1910 to 1920, and another covers the period 1920-1930. Still another has been earmarked for 1930-1950. There is furthermore a formula index and a subject index. The former is a listing of gross formulas; thus, ethyl o-nitrosobenzoate must be found through $C_9H_9NO_3$ and - since there are a hundred-odd isomers - the name. Thus, for the o-nitroso ester we are referred to 9, 369, I 150, II 242; in both the original sets and in the supplement we should take out volume 9 and turn to the pages indicated. The subject index is less useful because of the frequent use of trivial names, but if only the name of a given chemical is known the compound may be located that way. Each volume has a list of abbreviations used; basic information about the order of appearance of the various classes or organic compounds has been described in two booklets [17, 18]. Beilstein is useful in that, for a given compound, all the information about a given compound (at least up to 1930) can be found in a single place.

---

16. An excellent way to learn one's way about the library is by means of R. T. Bottle, "The Use of Chemical Literature", 2nd ed., Butterworths, London, 1969.
17. E.H. Huntress, "A Brief Introduction to the Use of Beilstein's Handbuch", 2nd ed., Wiley, New York, 1938.
18. O. Runquist, "Beilstein's Handbuch, A Programmed Guide", Burgess, Minneapolis, Minn., 1966.

Beilstein has been published in German only, so that students who have not mastered that language find themselves handicapped. At this point, it seems worthwhile to digress for a moment to point out that this is a difficulty that will continue to confront and confound any beginning organic chemist who has not been exposed to the use of this language: Most of the early organic chemists were Germans, and hence most of the early literature is in German. Even now, there are several excellent organic chemistry journals printed wholly or partly in German. Similar remarks apply to the use of French in the chemical literature.

The wave of popularity that chemistry has enjoyed as an academic topic since the second World War and the consequent gradual relaxation of a number of standards unfortunately have led to the negligence of foreign languages as chemical tools; hence, it is emphasized here that the acquisition of some modest reading ability in German and French is an absolute must for those who would become professional organic chemists. In consolation it may be said that in many parts of the world English is a foreign tongue also, and that students hopeful of becoming professional organic chemists there will be required not merely to read several foreign languages, but to write in one of them as well.

To return to the topic at hand, it is a common occurrence for an organic chemist to find himself in need of some inorganic compound or reagent not simply available in the stockroom. The inorganic equivalent to Beilstein is Gmelin's "Handbuch der Anorganischen Chemie" ("Gmelin"), the latest edition of which is much more up-to-date than Beilstein. The compilation is based on the elements, and written in German. Mellor's "Comprehensive Treatise on Inorganic and Theoretical Chemistry" is a reasonable alternative source.

For organic chemical information published after 1930, one is heavily dependent on Chemical Abstracts. If we wish to know what else might be known about ethyl o-nitrosobenzoate, we should consult first of all the formula indices: The Collective Formula Index covering the period 1920-1946, and the Decennial Indices for 1907-1916, 1917-1926, and so on, with Quinquennial Indices starting in 1957. There are furthermore semi-annual indices for the most recent period. With each entry there are the names of all known isomers, and the correct name refers the searcher to the page and paragraph number of the appropriate volume of Chemical Abstracts; that paragraph furthermore gives one a condensed version of the paper, the language used, and the full reference (authors, title, and journal reference). Chemical Abstracts is of course further up to date than any handbook, but the collection of the necessary information by this means - especially for a common compound - can be time consuming, since this information is scattered and may be obscured by results available but not of interest for the problem at hand. Subject and author indices are furthermore available and serve in those instances where the primary lead is of that nature.

There are several publications designed to help the organic chemist find his way through the more recent literature not yet indexed by Chemical Abstracts. One of these, Index Chemicus, has been published weekly since 1960 and surveys several hundreds of journals, concentrating especially on new compounds. Each entry receives a number, and structure, preparation, spectral properties, author, and reference are given. There are monthly and annual indices, and a cumulative index 1960-1967 is now available - a valuable addition to the other indices. Another type of service is provided by the Science Citation Index, by means of which one can find papers in which a given earlier article was quoted. A fairly wide range of journals is included in this search. The basic idea is that if someone is publishing an improvement in a certain technique, a refinement in some synthesis, or a correction in a mechanism, that article will surely contain a reference to the earlier work, and this insight can in principle be used to locate related literature.

There are a number of publications that abstract and index more specialized information. Among these are the Sadtler Standard Spectra [19], a collection of several tens of thousands of ir, nmr, and uv spectra indexed by molecular formula, by name, and, to some degree, by the number of peaks and their wavelengths. The "International Critical Tables" [20] and Landolt-Börnstein's Zahlenwerte [21] are further examples of such specialized information.

There are also a number of reference works describing information not specifically related to compounds. Thus, Theilheimer [22], Houben-Weyl [23], and Organic Syntheses [24] are often consulted when synthetic problems arise. Organic Reactions [25] and the Chemistry of Functional Groups [26] may be helpful when the chemical properties of a class of compounds are of interest. Weissberger [27] is the standard reference for anyone interested in all aspects of some standard technique - from simple ones such as distillation to the most complex such as X-ray diffraction.

Furthermore, there are many review sources for almost any one of these topics; among these there are Chemical Reviews, Reviews of Pure and Applied Chemistry, Quarterly Reviews, Russian Chemical Reviews (available in translation), Angewandte Chemie, (available in translation), Dissertation Abstracts, the "Advances" and "Progress" series, and so on. The Accounts of Chemical Research, the Discussions of the Faraday Society, the Chemical Society Special Publications, and others specialize in reviews of currently active areas; Chemistry in Britain and Chemical and Engineering News have occasional feature articles of that sort also.

---

19. "The Sadtler Standard Spectra", Sadtler Research Laboratories, Philadelphia, Pa.
20. "International Critical Tables", E.W. Washburn, ed., 7 Vols and Index, McGraw-Hill, New York, 1930.
21. Landolt-Börnstein, "Zahlenwerte und Functionen aus Physik, Chemie, Astronomie, Geophysik und Technik", New Series, Springer-Verlag, New York.
22. W. Theilheimer, "Synthetic Methods of Organic Chemistry", Kargel, Basel.
23. "Houben-Weyl, Methoden der Organischen Chemie", Georg Thieme Verlag, Stuttgart.
24. Organic Syntheses, Wiley, New York.
25. R. Adams, "Organic Reactions", Wiley, New York.
26. S. Patai, ed., "The Chemistry of Functional Groups", Interscience, New York.
27. A. Weissberger, ed., "Technique of Organic Chemistry", Interscience, New York.

However useful all these various sources may be, they cannot substitute for the type of personalized information stored in the files and memories of the individual practitioners of organic chemistry; information which results from continual intensive reading, browsing, seminars, conversations, symposia, and so on. In organic chemistry especially, many chemists religiously scan the tables of contents and lists of authors of several dozens of journals as the new issues arrive, turning to the abstracts and then to the paper indicated whenever a name or word tickles their curiosity. New book arrival lists, specialized publications such as Chemical Titles and Current Contents, and selections such as Chemistry and Industry's monthly "Highlights of the Chemical Literature" and the Angewandte Chemie's "Rundschau" are further aids to those seeking to stay up-to-date in their science. It may sound like a lot of work to keep up with organic chemistry, and it is; however, those who haven't the time to do it become subject to decay in their ability to teach and contribute to the Science - a sort of first-order process the half-life of which can't be much more than a year or two.

Chapter 4

THE STRUCTURAL THEORY [1]

4-1. THE RULES OF THE STRUCTURAL THEORY

The structural theory was the first comprehensive and lasting theory used by organic chemists. Its primary use was, and is, an accounting for the very large number of organic compounds, a collection which includes not only substances differing in elemental analysis and molecular weight but also compounds (isomers) with identical net formulas, and different only with respect to their physical or chemical properties.

The word "theory" is usually thought of as a stark, single statement or equation that can by means of the appropriate logic be elaborated into the observed facts. The structural theory however is merely a set of verbal rules that must be observed in order to arrive at all the atomic arrangements and geometries of these arrangements that are in principle possible. These rules are the following:

---

1. G.W. Wheland, "Advanced Organic Chemistry", 3rd ed., Wiley, New York, 1960.

1. Each kind of atom has a single valence number
2. Bonds, once formed, are stable at room temperature
3. Bonds do not cross
4. The angles between bonds are single-valued
5. There is free rotation about single bonds, except where this would necessarily lead to bond rupture (such as in cyclic structures)
6. There is no rotation about double bonds
7. Singly bound carbon atoms do not invert
8. Inversion about singly bound nitrogen atoms takes place at a fast rate

The rules are very simple to apply, and it is usually one of the first aspects of organic chemistry that undergraduate students learn. For instance, let us consider C_5H_{10}.

The valence numbers that apply to carbon and hydrogen are four and one, respectively, and consequently we may write the following 13 structures:

(1) 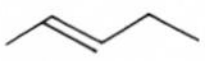 (2) 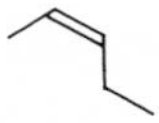 (3)

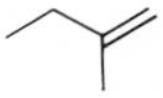

(4) 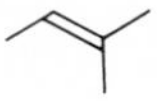 (5) 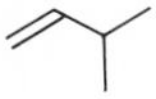 (6)

(7) 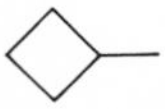 (8) 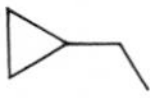 (9)

(10)  (11)  (12)  (13)

We may expect with confidence that there will in fact be 13 substances, neither more nor fewer, with formulas C_5H_{10}; i.e., there will be 13 compounds each containing 85.7% carbon and 14.3% hydrogen and possessing a molecular weight of 70. We may rule out substances with structures such as (14), which violate the first rule.

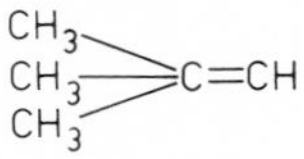

CH_3, CH_3, CH_3 >C=CH

(14)

The second rule tells us that all the structures drawn are kinetically stable and that, to name an example, (8) will not isomerize to (7) at room temperature even though the latter is thermodynamically more stable. The third rule does not apply here; it forbids a structure such as planar (15) for benzene.

(15)

The fourth rule implies that structure (7) is an unique arrangement for cyclopentane and that (16) is not still another isomer.

(16)

Rule 5 means that (1) is not a structure different from (17); on the other hand, (2) and (3) are indeed different in view of rule 6.

(17)

Rule 7 prohibits the interconversion (11) and (12) via a planar state.

(11) (12)

We should note at this point that the first three rules are related to what is traditionally called the structure of the molecule and to structural isomerism, whereas the last five are the basis of its configuration and to stereoisomerism. In the former area we are concerned exclusively with the question of which atoms are bound to which; in the latter we are dealing with more detailed questions of the possible shapes the molecules may have. Thus, (1), (2), and (4)-(11) are the structural isomers of C_5H_{10}; (3) is a stereoisomer of (2), and (12) and (13) are stereoisomers of (11).

In the next few sections we discuss rules 1-3 in further detail; rules 4-8 are reviewed further in Chapters 5-8. It will be clear meanwhile that these rules do not have the stature of physical laws. There are numerous exceptions and qualifications as we shall see below; nevertheless, once these qualifications are known and understood, this basic framework of the structural theory remains useful and in use today in order to deal with the enormous number of organic compounds already known and, hopefully, with the infinitely larger numbers still waiting for their preparation.

4-2. THE "SINGLE" VALENCE NUMBER

The tetravalency of carbon has long been accepted as one of the cornerstones of organic chemistry. Cornerstones do erode, however, as witness what happened to our beliefs concerning the zero valency of the inert gases. Until 1962 every freshman was told the noble gases do not form compounds, and that this fact supports the octet rule [2]. Actually there were many straws in the wind to the effect that there were already so many violations of the octet rule (compounds such as iodine heptafluoride for example) that the finding of an inert gas compound would hardly matter; still it came as something of a shock in 1962 when xenon hexafluoroplatinate was reported [3]. The simple xenon fluorides followed soon after [4], and krypton [5] and radon fluorides [6] as well; these three elements are now the center of a quite respectible body of chemistry [7].

Outside of inert gas chemistry, it never was very difficult to think of elements that are characterized by two or even several valence numbers. For instance, we know both a lead diacetate and a lead tetracetate. Even if we restrict our attention to atoms that are common in organic compounds such as carbon, hydrogen, nitrogen, oxygen, the halogens, etc., exceptions can be cited. Thus, the substance tri-β-naphthylmethyl (18), which has a trivalent carbon atom is well known [8].

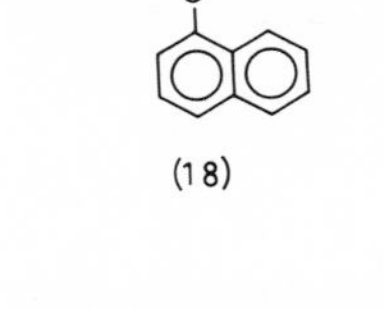

C

(18)

2. See, however, C.L. Chernick, J. Chem. Educ., 41, 185 (1964).
3. N. Bartlett, Proc. Chem. Soc., 218 (1962).
4. H.H. Claassen, H. Selig, and J.G. Malm, J. Amer. Chem. Soc., 84, 3593 (1962).
5. A.V. Grosse, A.D. Kirshenbaum, A.G. Streng, and L.V. Streng, Science, 139, 1047 (1963).
6. P.R. Fields, L. Stein, and M.H. Zirin, J. Amer. Chem. Soc., 84, 4164 (1962).
7. (a) H.H. Hyman, J. Chem. Educ., 41, 174 (1964); (b) J.J. Kaufman, J. Chem. Educ., 41, 183 (1964).
8. C.S. Marvel, J.W. Shackleton, C.M. Himel, and J. Whitson, J. Amer. Chem. Soc., 64, 1824 (1942).

2,2-Diphenyl-picrylhydrazyl (19) [9] has a divalent nitrogen atom, and many nitroxides such as (20) have a univalent oxygen atom (see Chapter 18).

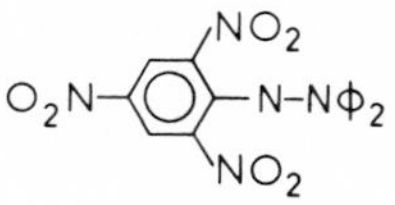

(19) 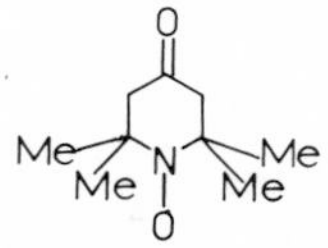 (20)

Such free radicals are stable by virtue of the steric strain and/or the weak bond that would be created if they were to dimerize, and by virtue of the fact that resonance allows the valence deficiency to be shared among several other atoms in the molecule. This and other types of apparent exceptions are well understood now in terms of the electronic structure of atoms and molecules.

Another group of molecules in which the valence number differs from the usual value is that of the onium salts, in which the normally unshared pair of electrons of an atom has become a bonding pair. A well-known example is the ammonium ion.

$$R_3N\!: + R'X \longrightarrow R_3\overset{\oplus}{N}R' + \overset{\ominus}{X}$$

Interesting extensions of such compounds are the ylids, such as (21) in which a trivalent anionic carbon atom leads to an internal type of salt, and the compound (22) in which pentacovalent silicon atoms serve this function [9a].

$$\phi_3\overset{\oplus}{N}-\overset{\ominus}{C}H\phi$$

(21) (22)

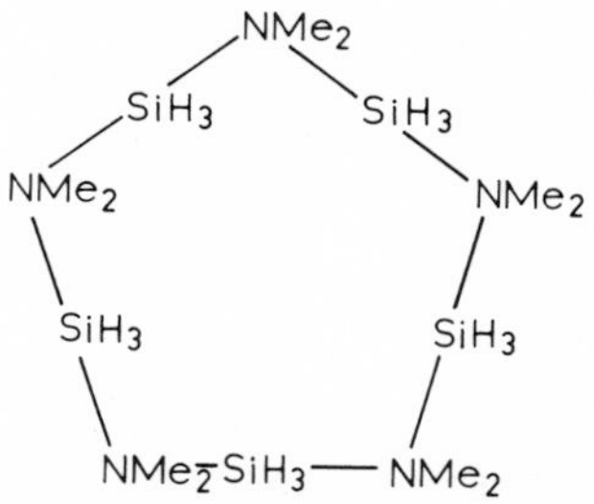

9. E. Müller, I. Müller-Rodloff, and W. Bunge, Ann., 520, 235 (1935).

9a. R. Rudman, W. C. Hamilton, S. Novick, and T. D. Goldfarb, J. Amer. Chem. Soc., 89, 5157 (1967).

In principle, any atom with unshared valence electrons can form such onium salts; thus oxonium salts such as trimethyloxonium tetrafluoroborate (23) have also been characterized, there are sulfonium and phosphonium salts, and we shall encounter others (see Chapter 20).

$$Me_3\overset{\oplus}{O} \quad \overset{\ominus}{B}F_4$$

(23)

In all the examples mentioned thus far the lines drawn in the structures have denoted pairs of electrons binding two nuclei together; however, there are many examples of molecules in which two electrons bind three or more atoms, and this leads to apparent abnormalities in the valence state. For instance, in Chapter 1 we encountered diborane (24). Two hydrogen atoms appear to be bivalent and both boron atoms seem tetravalent; thus, it would seem to add up to 16 electrons although boron has only three outer shell electrons and hydrogen has only one.

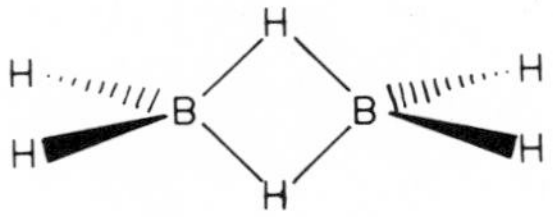

(24)

There is furthermore an extremely large and growing group of molecules which are involved as intermediates in chemical reactions in which one or more atoms deviate from their normal valence states. Thus, it is known [10] that when chloroform hydrolyzes in aqueous base, trichloromethide ion (25) and dichlorocarbene (26) are intermediate stages between the reactant chloroform and the product formate ion.

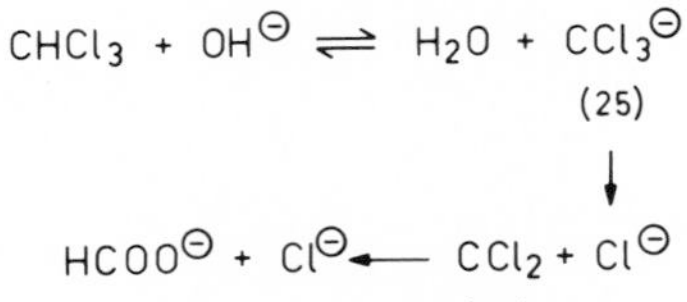

$$CHCl_3 + OH^{\ominus} \rightleftharpoons H_2O + CCl_3^{\ominus} \text{ (25)}$$

$$\downarrow$$

$$HCOO^{\ominus} + Cl^{\ominus} \leftarrow CCl_2 + Cl^{\ominus}$$

(26)

In the very large majorities of such cases the intermediates are extremely reactive and require special methods for their detection and study. While many such species have been found to be stable at very low temperature and in special media such as solid argon, the organic chemist is in most routine applications interested in molecules which are stable in the pure state at room temperature for at least

10. J. Hine, J. Amer. Chem. Soc., 72, 2438 (1950).

an hour or so. Reactive intermediates therefore occupy a special niche in the science of organic chemistry; they need not be considered a hazard to the structural theory. They are considered in some detail in the last chapters of this book.

4-3. KINETIC STABILITY AT ROOM TEMPERATURE

Whereas abnormal valence numbers furnish additional compounds not otherwise expected, kinetic instability leads to "nonexistent compounds" [11]. Kinetic stability may be gauged by means of the Arrhenius expression.

$$k = Ae^{-Ea/RT}$$

The frequency factor A has in most cases a value of roughly 10^{11}. If a given first-order reaction depends on the breaking of a carbon-carbon bond with a bond of energy of 80 kcal/mole, that reaction would at room temperature have a rate constant of about

$$k = 10^{11}/2.72^{80,000/2 \times 300}$$
$$= 10^{-37}\ \text{sec}^{-1}$$

and hence it would be extremely slow. However, the vast majority of chemical reactions do not occur by so brutal a pathway, but rather by cooperative processes in which new, permanent, or temporary bonds compensate for the energy changes attendant on a bond cleavage. At an activation energy of 25 kcal/mole we already have a process with a rate constant of 10^{-4} sec^{-1}, and hence (if the reaction is a first-order process) a half-life of about 2 h; at 5 kcal we have reactions occurring readily even at liquid nitrogen temperatures.

Nature has provided the organic chemist with such a large number of compounds which have such a wide variety of properties that it would be amazing indeed if between the reactive intermediates stable only at a few degrees Kelvin and the compounds decomposing only at several hundred degrees centigrade there were not some compounds that should be included on the basis of the rules of the structural theory but which cannot be isolated because they undergo rapid chemical transformations at room temperature. Examples of this are indeed known, and the reasons for the transformations are manifold. Some structures cannot correspond to stable molecules because they contain incompatible functional groups. Thus, γ-hydroxybutyroyl chloride (27) cannot be prepared because the alcohol group

$$HOH_2C{-}CH_2{-}CH_2{-}C(=O)Cl$$

(27)

11. W.E. Dasent, J. Chem. Educ., 40, 130 (1963).

would immediately react with the acid chloride end of the molecule to eliminate HCl and give butyrolactone (28).

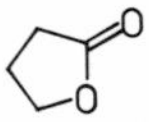

(28)

Other structures such as the unknown hexa-β-naphthylethanes owe their instability to the crowding of atoms that strains the central carbon-carbon bond. Others, such as cyclobutyne, are incapable of existence because the deviations from preferred bond angles are too great. Still others are unknown, in spite of many efforts to prepare them, not because of excessive strains but because just the same, there happens to be a pathway with only a small activation barrier to an isomer which is somewhat more stable; among these are species such as vinyl alcohol (29) which even at room temperature is immediately converted into acetaldehyde (30),

$H_2C{=}CHOH$ (29) $CH_3{-}CHO$ (30)

and norcaradiene (31) which likewise is unstable at room temperature to conversion into tropilidene (32).

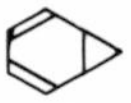

(31)  (32)

Substances of this sort often occur in detectible equilibrium amounts in the more stable isomer. Molecules such as (31) and (32) are known as valence tautomers (see Chapter 11).

Still other molecules are missing with more subtle reasons for their instability. Cyclobutadiene (33) is an example of a molecule which cannot be prepared in pure form because of a combination of strain (which alone would not be enough) and violation of the Hückel rule, according to which in a completely conjugated monocyclic molecule the number of double bonds should preferably be odd.

(33)

Kinetic instability does not necessarily always refer to the pure compound: It happens now and then that compounds presumed to be unstable for preparation are in fact capable of a precarious existence under the right conditions. Thus, carbonic acid (34) has long been thought to exist only in dilute aqueous solution, and in fact it has been argued that such solutions do not contain carbonic acid but only carbon dioxide. However, in 1960 it was reported that sodium carbonate suspensions in dimethyl ether can be treated with an ether solution of hydrogen chloride to give a solution of carbonic acid, which can, in turn, be evaporated to give the highly unstable ether complex (35) [12]. Similarly, it has long been known that the simple conjugated polyacetylenes (36) cannot be prepared unless bulky terminal substituents such as t-butyl or phenyl are present to prevent the triple bonds of neighboring molecules from reacting with one another. However, in 1972 these compounds (up to n = 5) were reported complete with their uv and nmr spectra [13].

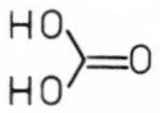

$(HO)_2C{=}O$ (34) $H_2CO_3 \cdot Me_2O$ (35) $H{-}(C{\equiv}C)_n{-}H$ (36)

4-4. CATENANES AND RELATED COMPOUNDS [14]

The rule of noncrossing bonds can be shown to apply not only to permanent structures such as (15), but to high energy transition states, since this leads to the conclusion that looped rings should be capable of a stable existence, a conclusion which is experimentally verifiable. Such catenanes, as they are called, differ from other molecules in one important respect: Their parts are held together by repulsive (nonbonding) forces rather than attractive ones as is the case in nearly all other molecules (see also Chapter 23 for references to the clathrates). One approach to such compounds is a statistical one. A long straight-chain molecule is induced to undergo a cyclization reaction in a medium consisting of large ring molecules, and it is hoped that at least some of the chains are threaded through the rings at the time of cyclization. One catenane has actually been obtained by this route [15]. Thus, when diethyl tetratriacontanedioate (37) is made to undergo the acyloin condensation in the presence of the tetradeuteriotetratriacontane (38), one obtains beside the simple acyloin (39) also a small amount of product (40) associated with the deuterated hydrocarbon in some way. Oxidation of this product to the diacid (41) liberates the deuterated hydrocarbon, and hence it seems reasonable that the source of (38) in the oxidation must have been structure (40).

---

12. G. Gattow and U. Gerwarth, Angew. Chem., Int. Ed. Engl., 4, 149 (1965).
13. E. Kloster-Jensen, Angew. Chem., Int. Ed. Engl., 11, 438 (1972).
14. (a) G. Schill, "Catenanes, Rotaxanes and Knots", Academic Press, New York, 1971; (b) G. Schill and C. Zürcher, Naturwiss., 58, 40 (1971).
15. E. Wasserman, J. Amer. Chem. Soc., 82, 4433 (1960).

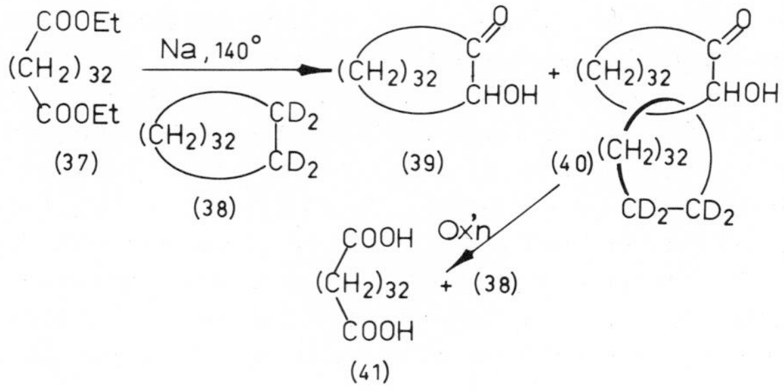

In such experiments it should be kept in mind that the core of the ring must be large enough to permit the facile entry of a polymethylene chain; this consideration leads to a minimum ring size of 20 carbon atoms for each ring [16].

Another approach to catenanes has been based on the methathesis (or dismutation) of olefins, a reaction catalyzed by a mixture of tungsten hexachloride, ethanol, and ethyl aluminum dichloride [17].

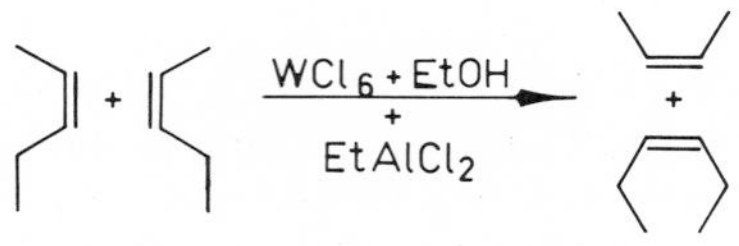

The mechanism is more complex than suggested above, since trans-olefins also form. The reaction can be applied to cyclic diolefins, which, if of sufficient ring size, may react in the folded form drawn below to give catenanes.

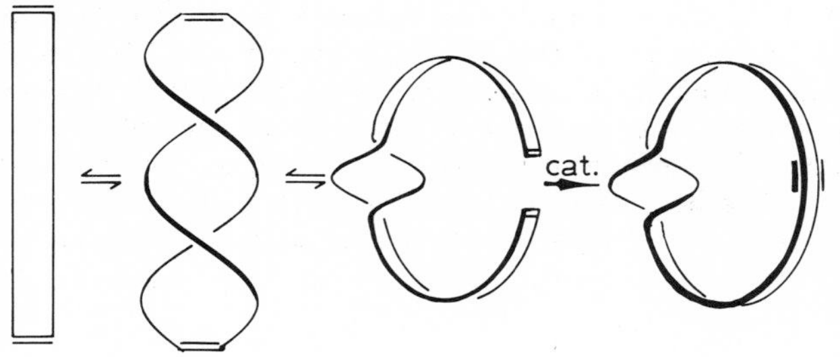

Although complex mixtures are expected and obtained, it has been shown by means of mass spectrometry that catenanes are indeed part of this mixture [18].

16. H. L. Frisch and E. Wasserman, J. Amer. Chem. Soc., 83, 3789 (1961).
17. N. Calderon, H. Y. Chen, and K. W. Scott, Tetrahedron Lett., 3327 (1967).
18. (a) R. Wolovsky, J. Amer. Chem. Soc., 92, 2132 (1970); (b) D. A. Ben-Efraim, C. Batich, and E. Wasserman, J. Amer. Chem. Soc., 92, 2133 (1970); (c) R. Wolovsky and Z. Nir, Synthesis, 134 (1972).

Perhaps the most elegant, if also the most complex, approaches to catenanes have been reported by Schill and Lüttringhaus [19], whose synthesis of (39) is schematically shown below. The structure proof makes use of the observation that in a mass spectrum all those ions which depend on ring opening can also be observed with the rings separately - in contrast to those whose formation is not associated with ring opening [20]. Similar approaches have even led to the synthesis of a [3]-catenane [21].

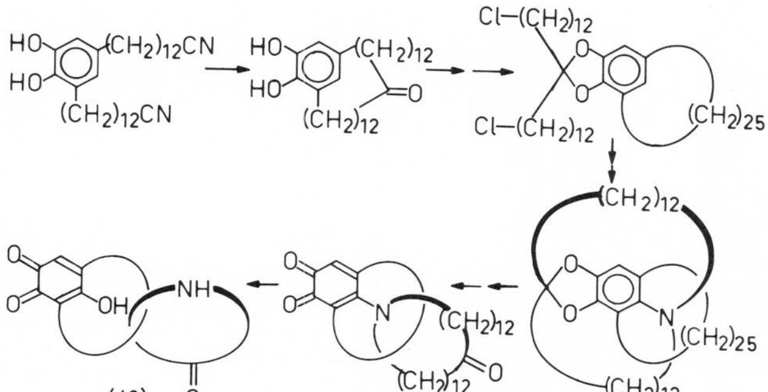

A semistatistical approach has been used by Harrison to produce another type of substance in which two molecules are again literally held together by repulsive forces [22], namely, the rotaxanes. A column packed with a polymer chemically bound to large rings (43) is treated repeatedly with a mixture of pyridine, 1,10-dihydroxydecane, and trityl chloride and finally washed and hydrolyzed to give (44).

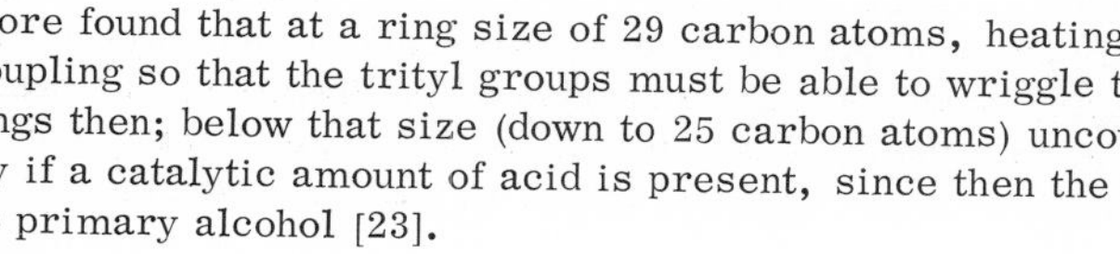

It was furthermore found that at a ring size of 29 carbon atoms, heating to 120° led to slow uncoupling so that the trityl groups must be able to wriggle their way through such rings then; below that size (down to 25 carbon atoms) uncoupling could occur only if a catalytic amount of acid is present, since then the terminal is only a simple primary alcohol [23].

---

19. (a) G. Schill and A. Lüttringhaus, Angew. Chem., 76, 567 (1964); (b) A. Lüttringhaus, F. Cramer, H. Prinzbach, and F.M. Henglein, Ann., 613, 185 (1958); (c) G. Schill, Chem. Ber., 98, 3439 (1965); (d) A. Lüttringhaus and G. Isele, Angew. Chem., Int. Ed. Engl., 6, 956 (1967).
20. W. Vetter and G. Schill, Tetrahedron, 23, 3079 (1967).
21. G. Schill and C. Zürcher, Angew. Chem., Int. Ed. Engl., 8, 988 (1969).
22. I.T. Harrison and S. Harrison, J. Amer. Chem. Soc., 89, 5723 (1967).
23. I.T. Harrison, Chem. Commun., 231 (1972).

Chapter 5

STEREOCHEMISTRY

5-1. INTRODUCTION [1]

At an early stage in the development of their science, the organic chemists began to find cases in which the number of isomers actually found exceeded the number predicted by the theories then in vogue. Starting with Pasteur, Kekulé, LeBel, and van't Hoff, they deduced from these examples that in order to arrive at the correct total number of isomers, further assumptions were necessary, these assumptions being essentially rules 4-8 of the preceding chapter (p. 114). It is with this problem that we concern ourselves in this chapter. Although many of the ideas and concepts accepted now in stereochemistry are presented as principles leading to observed facts, it will do well to point out that most progress in our understanding of this area has been deduced from the facts.

---

1. There are many excellent discussions of this topic. Among the best are (a) G.W. Wheland, "Advanced Organic Chemistry", 3rd ed., Wiley, New York, 1960; Chaps. 6-9; (b) K. Mislow, "Introduction to Stereochemistry", Benjamin, New York, 1965, and references cited in the bibliography of that book.

5-2. CONFIGURATION VS. STRUCTURE

The difference between configuration and structure may be illustrated by means of the following example. If all the structural isomers corresponding to the formula $C_3H_6O_3$ are written, we will encounter among them such examples as dimethyl carbonate (1) and lactic acid (2).

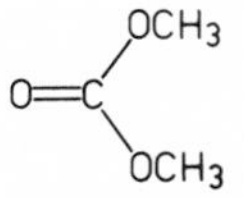

(1) 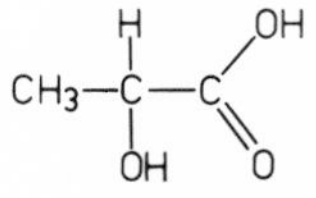 (2)

Clearly (1) and (2) represent different structures: They cannot be interconverted without breaking bonds.

It turns out that there are two compounds, both having the structure of (2). These compounds differ in certain physical properties. Thus, they are said to have optical activity or to be optically active; i.e., they are observed to cause rotation of the plane of polarized light passing through them (cf. p. 81f). They also differ in certain chemical properties; thus, salts of different solubility are often formed with certain alkaloid bases. Furthermore, they differ especially in their physiological properties; for example, only one of the isomers (2a) and (2b) is involved in the chemistry describing the production of useful work from glucose during muscle contraction.

The existence of two isomers of compounds such as (2a) and (2b) can be traced to the tetrahedral carbon atom. If we accept the premise of such atoms, (2) can be written either as (2a) or as (2b); these pictures are attempts to represent three-dimensional molecules in a two-dimensional plane.

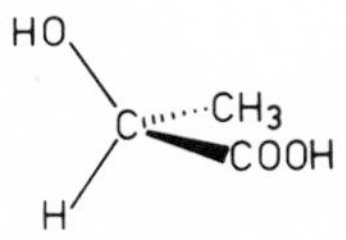

(2a) 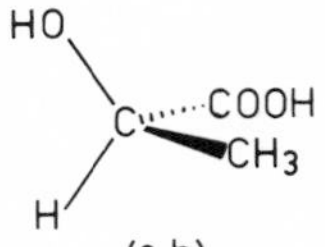 (2 b)

It may be noted that (2a) and (2b) are not two different structures. No bonds need be broken - at least on paper - to convert one into the other; this can be done, for example, by holding the hydrogen and hydroxyl groups where they are, and twisting the other two substituents about the axis bisecting the angle between them.

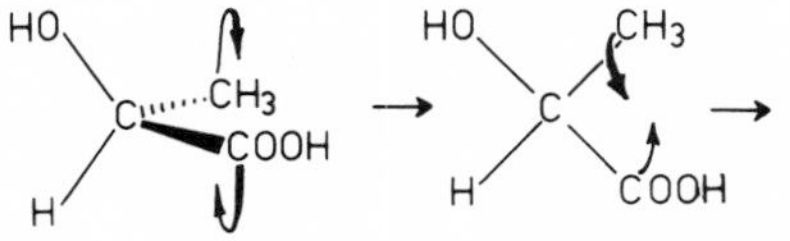

→ → (2b)

Since (2a) and (2b) are not different structures, we need another word relating the fact that structure (2) may represent two different molecules; we say that (2a) and (2b) are two configurations of (2). It will be clear that besides a tetrahedral carbon atom we need a rule which states that the interconversion such as was hypothetically carried out here ("inversion") cannot occur in reality. In fact, although interconversions of (2a) and (2b) are possible, there is no known instance in which the mechanism involves an intermediate state in which the four valences of a carbon atom all lie in a plane; in all cases it is necessary to break one of the four bonds (Section 5-7; for the possible occurrence of square planar carbon atoms, see p. 207).

We shall see below (Section 5-10) that the word configuration is also used in connection with other types of stereoisomerism. However, it remains true in each case that configurations are interconvertible by twisting motions (on paper only), whereas isomeric structures are distinguished by the feature that interconversion without the breaking of bonds is not even conceivable. The distinction may seem obvious and, in fact, it defines the difference between structure and configuration; nevertheless, the two concepts are often confused and hence it seems worthwhile to describe them.

Since (2a) and (2b) appear very similar, we may ask when two molecules are the same and when they are different. Again the author apologizes if the question is obvious and superfluous. Two molecules are the same and identical if they are superimposable; different is they are not. It is not always obvious by looking at two-dimensional representations. Thus, (2a) and (2c) are identical, as may be shown by carrying out a 180° rotation about the X axis, and (2a) and (2d) are as well, as may be shown by a 120° rotation about the C-OH bond.

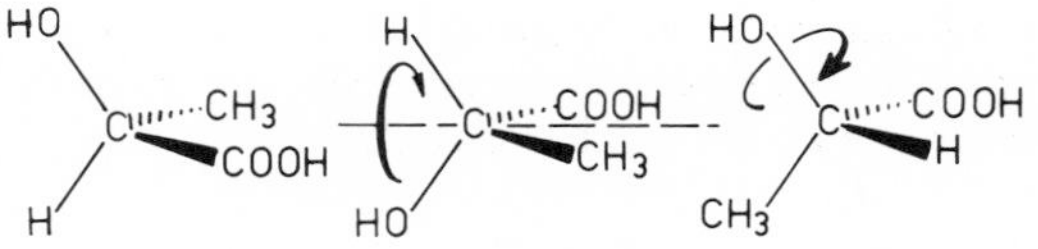

(2a) (2c) (2d)

However, no rotations of the rigid configuration (2b) will make it superimposable on (2a). These observations are somewhat tedious to pursue on paper, but they are immediately obvious and trivial if a set of models is available. Finally, at the risk of redundancy, it is pointed out that single bonds as a rule will freely rotate in a first approximation, so that molecules that are not superimposable but become so by an appropriate rotation about a single bond are considered the same for the purpose of counting isomers.

5-3. ENANTIOMERS, RACEMIC MIXTURES, AND THEIR PROPERTIES

The hypothetical interconversion of (2a) and (2b) by twisting two of the four valences to new positions makes sense only when all four substituents are different; otherwise, we can always select two identical substituents as those which are going to be twisted about, and if we do it (again, on paper), we merely obtain the original molecule.

It may also be observed that the same molecule (2b) is always produced no matter which one of the six pairs of substituents bound to the central carbon atom in (2a) is interchanged; thus (2b) and (2e)-(2i) are identical, as one can readily show by means of models and somewhat less readily without them.

(2-b) (2-e) (2-f)

(2g) (2h) (2 i)

Thus, there are in this case only two configurations, and hence two stereosiomers (2). The relationship between these, (2a) and (2b), is a most interesting one: They are mirror images of one another. It is instructive to look at the mirror image of a model of (2a) and observe that it is identical with a model of (2b) (Fig. 5-1). Molecules which are not superimposable on their mirror image are said to be chiral, or to possess chirality. Stereoisomers which are mirror images of one another are called enantiomers. Another description often used is to say that they are optical isomers, or optical antipodes, or that they constitute an enantiomeric pair. Molecules which have no stereoisomers and hence whose structures have only one possible configuration will be superimposable on their own mirror images; thus, (3a) and (3b) are identical (Fig. 5-2).

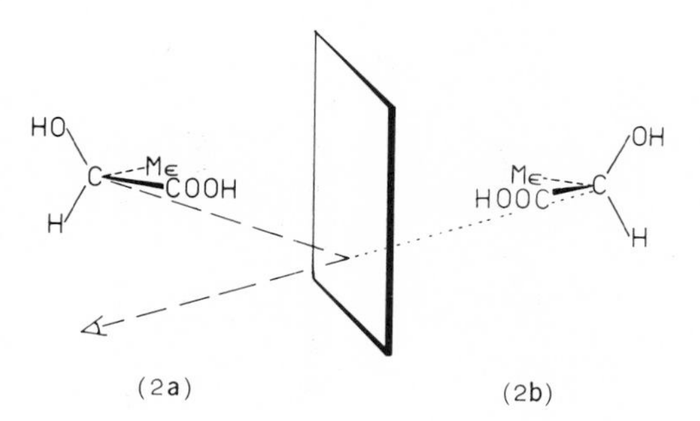

Fig. 5-1. A molecule of (2a) in front of a mirror, and the image (2b).

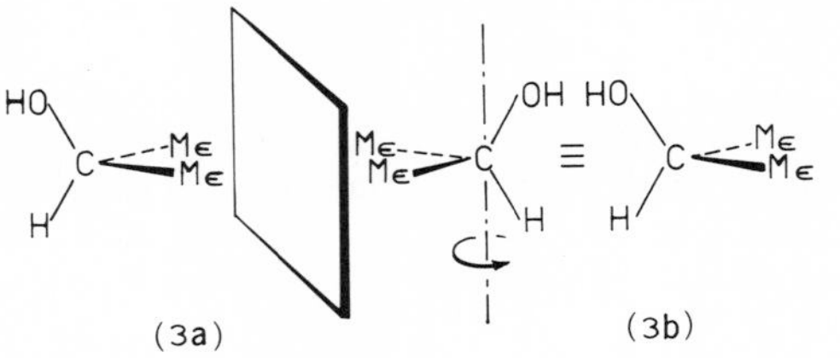

(3a) (3b)

Fig. 5-2. An achiral molecule in front of a mirror, and its mirror image, superimposable, as shown by a 180° rotation.

We may now raise the question, which properties of a pair of enantiomers will reflect the difference? Most of the physical properties of enantiomers are identical; thus, the melting points, vapor pressures, densities, measurements of ir and uv spectra, nmr spectra in achiral solvents (see further below), etc., all lead to identical results. In order to predict whether a given physical property will be different for a pair of enantiomers or not, one imagines that the experiment is done in front of a mirror. In most instances, the result observed directly is obviously identical with that observed in the mirror. Thus, in a vapor pressure measurement, (2a) vapor molecules will obviously compress a spring to the same extent as do those of (2b) (Fig. 5-3). Of course the experiment and its mirror image are not superimposable, if only because (2a) and (2b) are not; however, the question at hand is whether the two springs are compressed to the same extent or not, and clearly they are.

One experiment that does appear differently when viewed in the mirror is the passage of polarized light (see Fig. 5-4). Thus, while for (2a) the plane has been rotated clockwise, the mirror image experiment leads to counterclockwise rotation. The sign of this rotation is indeed equal and opposite for (2a) and (2b). The principal difference between this experiment and the preceding ones is that the polarization experiment is itself inherently chiral.

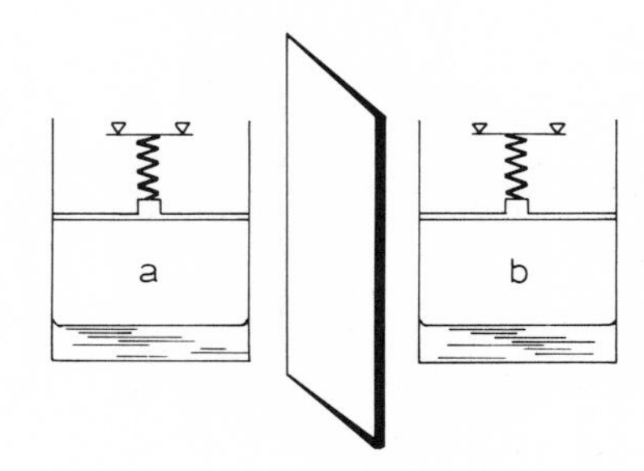

Fig. 5-3. A vapor pressure measurement of a chiral substance, and the mirror image experiment.

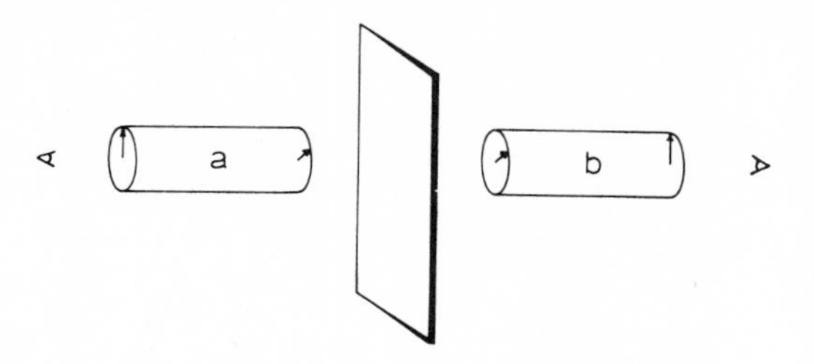

Fig. 5-4. Optical rotation and the mirror image experiment. The arrows suggest the plane of the polarized light.

The same thought experiment can be extended to chemical reactions. Suppose for instance that we approach a protonated molecule (2a) with a methanol molecule as shown in Fig. 5-5, a situation that might apply in an acid catalyzed esterification. It clearly makes no difference whether we approach (2a) or (2b) with methanol or with a mirror image of methanol, since methanol is not chiral. Therefore such factors as the rate of methyl ester formation of (2a) and (2b) will be the same. The story becomes different if our probe possesses chirality. Thus, if the alcohol molecule is so substituted that it is not superimposable on its own mirror image (e.g., the 2-butanol (4a)), then the reactions of (2a) and (2b) with (4a) are different, characterized by different rate constants.

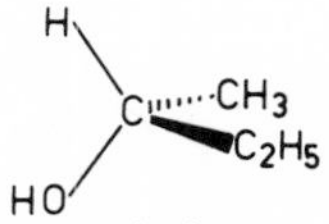

(4a)

To put in another way: If we wish to have the esterifications proceed at the same rate, we must then allow (2a) to react, not with (4a), but with the mirror image molecule (4b), which itself is not superimposable on (4a) (Fig. 5-6).

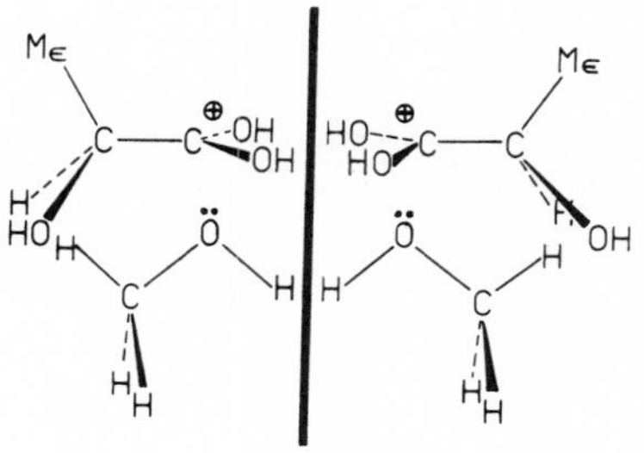

Fig. 5-5. Treatment of a chiral molecule with a achiral one and the mirror image experiment.

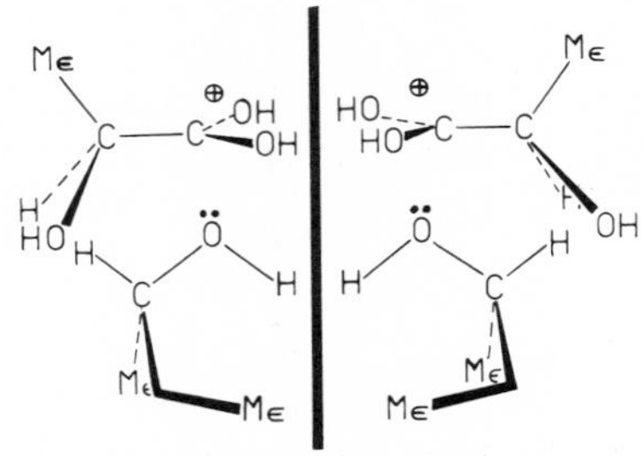

Fig. 5-6. The reaction of (2a) with (4b) in front of the mirror.

The physical properties of a given substance as well as the chemical ones are often affected by the presence of another compound. Now if one determines the effect of a given chiral compound A on the physical properties of another chiral substance B, and also on those of the optical antipode of B, we generally find that these effects are different - again because these two experiments are not related by mirror image.

Thus, an equimolar mixture of the chiral molecules of 2, 2, 2-trifluoro-1-phenylethanol (5a) and their optical antipode (5b) has two F^{19} doublets in solution in optically active α-(1-naphthyl)ethylamine (6).

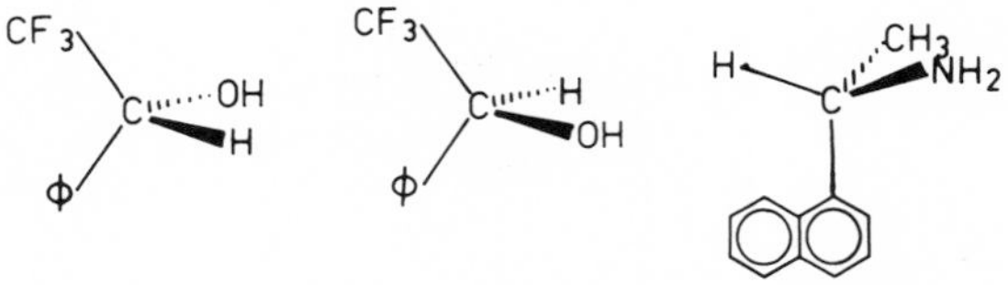

(5a) (5b) (6)

The main splitting is due to coupling with the neighboring proton (J_{HF} = 7 Hz), the other is a chemical shift difference between the enantiomers of (5) in an asymmetric environment [2]. This kind of effect is most likely to be encountered if there is some type of solute-solvent interaction such as hydrogen bonding.

The conclusion therefore is that enantiomers have the same properties when measured by an achiral standard, and that they behave differently when approached with a chiral probe. Since our senses are operated by molecules comprising highly chiral systems, enantiomers differ often most spectacularly in taste and smell.

---

2. W.H. Pirkle and T.G. Burlingame, Tetrahedron Lett., 4039 (1967).

Finally, we should refer to the fact that the chemist very frequently deals with mixtures containing equal numbers of chiral molecules and their optical antipodes. Such mixtures are described as racemic. They obviously have no optical rotation - with the one kind of molecules rotating the plane of polarization in one direction and the mirror image kind rotating it equally far in the opposite sense.

5-4. DIASTEREOMERS AND EPIMERS

If we now turn our attention to the 2-butyl lactates that may form in the reaction of lactic acid with 2-butanol, we realize that four substances may form: (2a)-(4a), (2b)-(4a), (2a)-(4b), and (2b)-(4b).

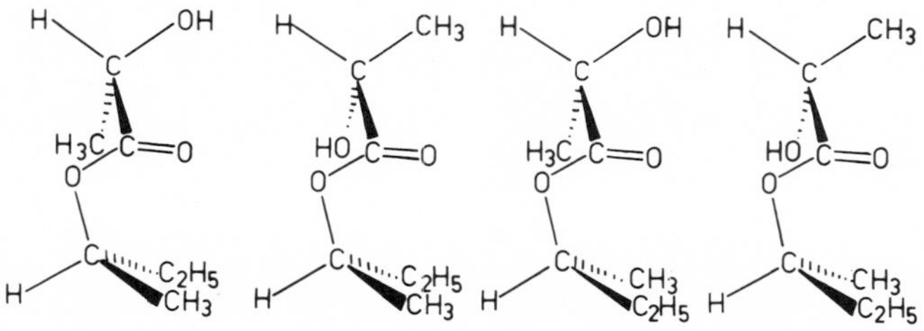

(2a-4a) (2b-4a) (2a-4b) (2b-4b)

The first two (and the last two) of these esters are each an enantiomeric pair; that is, they are mirror images of each other and have identical melting points, nmr spectra, and so on. On the other hand, the first and last of these combinations (or the second and third) are not related in this way. They are clearly different, and they are equally clearly also not mirror images of one another. They are called diastereomers; in general this term is used for any pair of stereoisomers which are not enantiomers. Diastereomers may and generally do differ in all physical, chemical, and physiological properties; thus, solubilities, nmr spectra, taste, and so on may all be different, sometimes substantially so.

If we extend the molecule to contain a third "asymmetric" center (usually a carbon atom having four different substituents), it may be readily verified that a total of eight stereoisomers obtain (four enantiomeric pairs). With n such atoms, a total of 2^n stereoisomers obtain (2^{n-1} pairs). For simple, open-chain molecules, these numbers are maxima; there cannot be more than those numbers of stereoisomers, though there may be fewer. As an example, consider 2,3-dichlorobutane (7).

On the basis of the foregoing discussion, four stereoisomers should exist (7a)-(7d). It can readily be seen, however, that (7b) and (7c) are in fact identical; a 180° rotation of molecule (7c) in the plane of the paper about the point indicated puts it into such a position that its superimposability on (7b) is obvious.

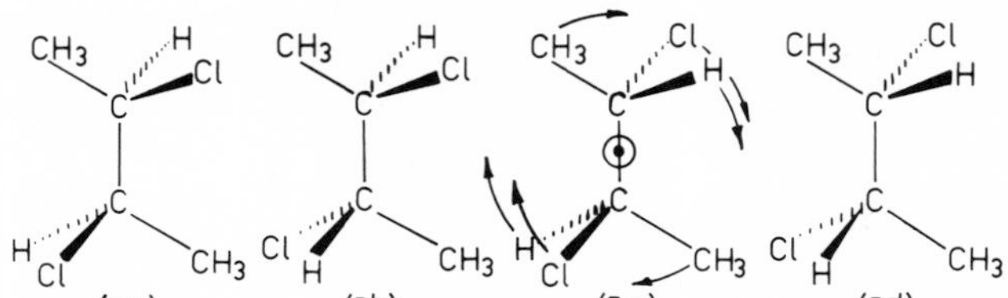

(7a) (7b) (7c) (7d)

A molecule such as (7b) is often called a meso form. It is not optically active; it is often argued that one-half of the molecule precisely counteracts the effect of the other half on the plane of polarized light ("internal compensation"). This term is somewhat out of fashion now that the organic chemist has learned to judge the possibilities for stereoisomerism by means of its symmetry properties (see p. 146), but especially the older literature frequently refers to it.

Organic chemists frequently use molecules containing two asymmetric centers in which two of the three groups at each center are identical, e.g., (8) and (9).

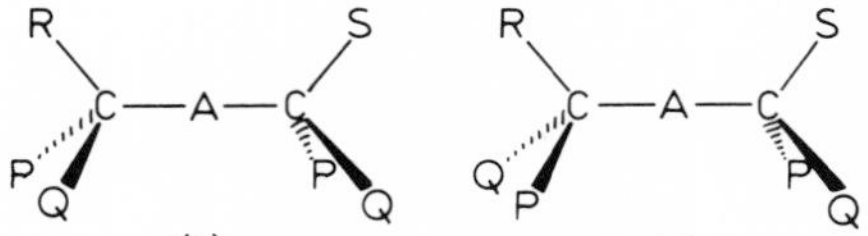

(8) (9)

These two diastereomers (and their mirror images) are referred to, respectively, as the _erythro_ and _threo_ forms of that structure. It may be noted that if the groups R and S were identical, the _erythro_ forms would be a meso form, but the _threo_ forms would still constitute an enantiomeric pair.

Another term often used is the word epimer. When two diastereomeric compounds containing several asymmetric carbon atoms have all but one of these in the same configuration, they are said to be epimers of one another.

5-5. RESOLUTION

Stereochemistry is one of the very most powerful tools available to the organic chemist as he investigates the various intricacies of reacting molecules. What is usually needed in such researches is a batch of some chiral compound in a pure enantiomeric form, and since such pure enantiomers are not available as a rule, one of the first tasks is then either the separation of a racemic compound or the direct synthesis of the enantiomer in question. In this section we take up the first of these problems, which chemists refer to as the resolution of the racemic mixture (or simply as the resolution of that compound).

Since enantiomers always behave in the same way except when approached by means of some chiral probe, the separation must make use of such a probe. Physical separations of enantiomeric molecules can of course only be achieved if they are presented with a choice of chiral environments at the molecular level. To mention an analogy, it is not hard to imagine how one might be able to resolve a racemic mixture of left- and right-handed nuts by means of a right-hand bolt. Gas chromatography has been successfully applied [3]; thus, racemic esters of alanine have been separated by means of gas chromatography in which an optically active substance of high molecular weight was used as substrate. Such separations are not likely to become routine, however, since often a hit-or-miss search is necessary to find a substrate that will work. Furthermore, the difference in physical properties is usually just too small. Thus, it has been found that optically active limonene (10) forms azeotropic mixtures with each of the enantiomers of 2-octanol with boiling points half a degree apart [4], so that a racemic mixture of the alcohol is in principle separable that way; but in practice it is not.

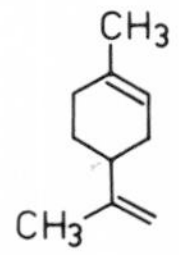

CH_3

CH_3

(10)

Chemical separations are generally the only really satisfactory way. The first step is the conversion of the racemic mixture into a diastereomeric mixture by means of a reaction with an optically active substance. Since the diastereomers will have different physical properties, they might be separated, for example, by recrystallization. Once this is accomplished, the next step is the reverse reaction to give back the original enantiomers. The whole process is then repeated until the optical rotation reaches a maximum.

The optically active material used in the process is very often some naturally occurring alkaloid base such as brucine. If the racemic substance is an acid, a mixture of diastereomeric salts is readily obtained, and once they have been separated, the salts are readily decomposed again [5a]. If the racemic substance is not an acid, it may be possible to synthesize it from an acid which can be prepared in optically active form. Thus, if we wish to separate racemic 2-octanol, we might first convert it into a racemic mixture of half-phthalate esters (11), and these, in turn, into diastereomeric salts, and so on [5b]. Various similar sequences may readily be devised for other types of racemic mixtures.

---

3. E. Gil-Av and B. Feibush, Tetrahedron Lett., 3345 (1967).
4. C. J. McGinn, J. Phys. Chem., 65, 1896 (1961).

5a. An ion exchange method for this has been found to have many advantages; see, e.g., F. W. Bachelor and G. A. Miana, Can. J. Chem., 45, 79 (1967).

5b. B. A. Klyashchitskii and V. I. Shvets, Russ. Chem. Rev. (English Transl.), 41, 592 (1972).

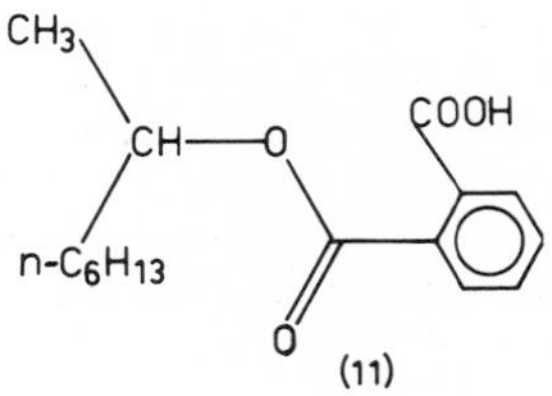

(11)

The main requirement of any intermediate reactions used is that they must be capable of being carried out rapidly, in high yield, and, of the last step, that no racemization occurs. Resolution is often a time consuming job. The solubilities of the diastereomers may not be very different, so that recrystallizations produce only marginal results; furthermore, it is often not even immediately apparent what results were obtained in a crystallization until a sample of the crystals has been decomposed all the way back to the pure starting material, so that the optical activity can be measured.

The optical purity of an optically active substance is obviously an important question. It is defined as follows. If the ratio of the amounts of desired to contaminating enantiomers equal R, then the optical purity of the sample is said to equal $(R-1)/(R+1)$. Alternatively, the optical purity is equal to the ratio of observed to theoretical rotations. It varies therefore from zero for racemic mixtures to unity for a pure enantiomer. One way to measure it is to make use of the isotope dilution technique [6]. A weighed sample of the optically active - though not necessarily optically pure - compound is mixed with a weighed sample of isotopically labeled racemic material. The mixture is treated so as to yield a small amount of racemic mixture, of zero optical rotation; isotopic analysis is carried out, and from the result the optical purity can be calculated. Alternatively, the mixture may be reresolved until one reaches the same degree of optical rotation as was available before; this is then the sample to be examined on its isotope ratio. Either way, the computation is straightforward, though the laboratory operation is of course time consuming.

Another method of measuring optical purity [7] makes use of nmr. It is simpler in that it does not depend on the isolation of pure racemic mixture or reresolution, but instead on the complete conversion into some diastereomeric form. Since nmr spectra can be expected to be somewhat different for diastereomers, the ratio can be obtained by integration. A related innovation is the use of chiral shift reagents such as (12) [8] and of chiral solvents [2].

---

6. J.A. Berson and D.A. Ben-Efraim, *J. Amer. Chem. Soc.*, **81**, 4083 (1959).
7. (a) M. Raban and K. Mislow, *Tetrahedron Lett.*, 4249 (1965) (this paper lists several additional methods and leading references thereto); (b) J.A. Dale, D.L. Dull, and H.S. Mosher, *J. Org. Chem.*, **34**, 2543 (1969).
8. (a) H.L. Goering, J.N. Eikenberry, and G.S. Koermer, *J. Amer. Chem. Soc.*, **93**, 5913 (1971); (b) V. Shurig, *Tetrahedron Lett.*, 3297 (1972).

(12)

A simple chemical method is available if the substance whose optical purity is in question can be converted into one whose specific optical rotation has already been determined. The most important question in this procedure is whether racemization can possibly occur during the conversion [9].

5-6. ASYMMETRIC SYNTHESIS

When an optically inactive mixture of molecules (whether of a racemic or achiral nature) is allowed to undergo a chemical reaction so as to form chiral molecules, a racemic mixture must inevitably result since the energy profile of the reaction is identical with that of the mirror image reaction. Consider for example the addition of bromine to propylene. In this reaction Br^+ is first added to give an intermediate bromonium ion (see p. 709), and finally bromide ion adds on the trans side to give the product. The mirror image reaction, though not superimposable, is identical in its energy profile and hence in rate so that equal amounts of the enantiomers will form (see Fig. 5-7). To put it another way, the initial attack by Br^+ is exactly equally likely on both sides of the molecule.

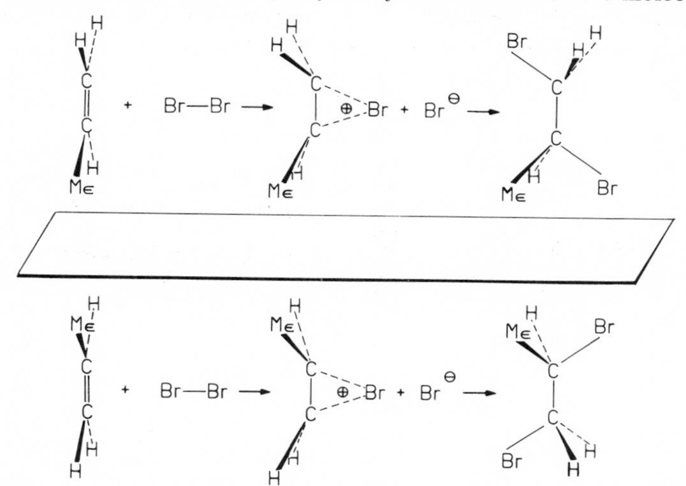

Fig. 5-7. The bromination of propylene and the mirror image reaction.

9. See, for example, H.R. Hudson, Synthesis, 1, 112 (1969).

There are various ways in which the mirror image relation can be made inapplicable. Thus, if an optically active solvent is used, the two reactions going on are no longer mirror images of one another, and hence need no longer have the same rate constant. In practice such devices rarely lead to a pronounced preponderance of one of the enantiomers; however, it has proved possible in many cases to achieve a very great preference for one reaction over the other if another chiral center is already present in the molecule. Thus, if we brominate one of the optical isomers of 3-methylpent-1-ene (13), the two "sides" of the olefinic site are not equally accessible, and hence one diastereomer may well be formed faster than the other.

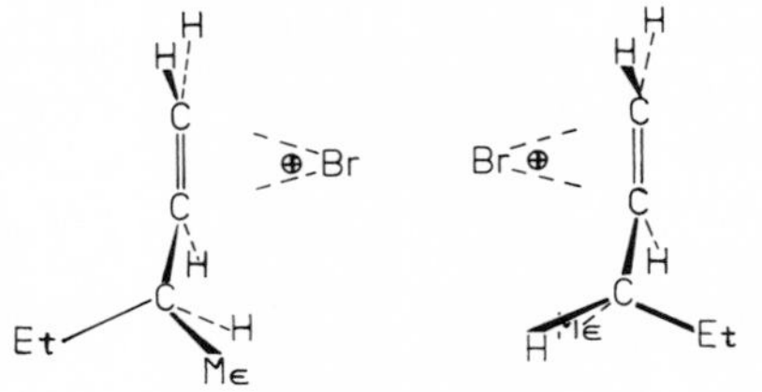

(13)

When such a difference is indeed observed, we refer to the reaction as an example of asymmeteric induction.

In the synthesis of specific stereoisomers of molecules having two or more asymmetric centers, such induction is highly desirable since repeated resolution is simply too wasteful in time and in material. Virtually all biochemical reactions are of this type. Among the most spectacular laboratory examples are the formation of isotactic polypropylene (14) [10] and of syndiotactic polymethyl methacrylate (15) [11]; in each case the asymmetric center just created by the addition of a monomer molecule directs the addition of the next one.

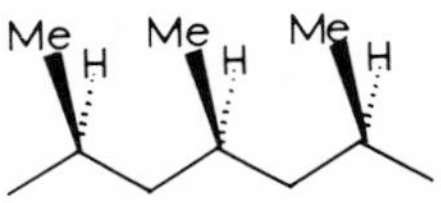

(14) 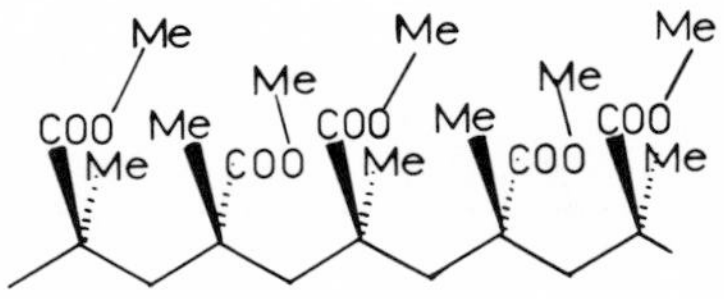 (15)

10. G. Natta, Gazz. Chim. Ital., 89, 52 (1959).
11. T.G. Fox, Gazz. Chim. Ital., 88, 1769 (1958).

A related approach is the use of resolved reagents to alter functional groups; this is called asymmetric synthesis. In view of the tediousness of the resolution process, chiral reagents are used whenever possible to achieve stereospecific synthesis. Some of the most successful of these are the optically active sym-tetrasubstituted diboranes prepared by hydroborating naturally occurring optically active terpenes [12]. Thus, the preparation of optically active alcohols became a simple matter where the resolution of racemic alcohols via half-phthalates, brucine salts, and so on had previously been a difficult and time-consuming operation [13]. The question which enantiomer is likely to be the predominent one formed can usually be guessed quite easily on steric grounds.

Since the synthesis of optically active substances inevitably involves the use of an asymmetric environment, the intriguing question arises, how the extreme selectivity in biochemical systems ever got started. There are actually several possible ways. Thus, the earth's magnetic field provides an asymmetric feature that could slightly favor the primordial synthesis of one optical isomer over that of another; polarized sunlight could play a similar role if such a synthesis were photochemical (see p. 155 for an example); optically active quartz crystals could have this function if the reaction occurred at its surface. Such crystals do occur in nature; they can grow from a optically inactive melt if the nucleation site where the crystal started is an asymmetric one [13a, b]. The requirements for such a process are that the enantiomeric forms must rapidly racemize in solution but not in the solid phase, and that very few (preferably just one) nucleation sites are available. An outstanding recent example is the finding that the melt of racemic 1, 1'-binaphthyl (16) on cooling can be completely converted into a pure enantiomeric form [13c] (see p. 151).

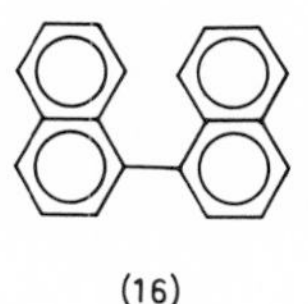

(16)

12. H. C. Brown, N. R. Ayyangar, and G. Zweifel, J. Amer. Chem. Soc., 86, 397 (1964).
13. A number of such reagents have recently been reviewed: T. D. Inch, Synthesis, 2, 466 (1970).
13a. E. Havinga, Chem. Weekbl., 38, 642 (1941).
13b. E. Havinga, Biochim. Biophys. Acta, 13, 171 (1954).
13c. R. E. Pincock and K. R. Wilson, J. Amer. Chem. Soc., 93, 1291 (1971).

5-7. RETENTION, RACEMIZATION, AND INVERSION

When optically active lactic acid is converted into methyl lactate, none of the four bonds of the α-carbon are broken in the process and we have therefore every reason to believe that its configuration remains unchanged. We say then that the reaction has occurred with retention of configuration or, simply, with retention. As we shall see below (e.g., p. 140), retention is not unique to reactions that do not involve the asymmetric carbon atom. If the reaction of an optically active compound results in the formation of a racemic product, then racemization is said to have occurred, and if the configuration of the four substituents in the product is opposite to that of the starting material, then by definition inversion has taken place.

The first question that now arises is how one can tell which of these features characterizes a given reaction involving an optically active compound. Racemization is an obvious phenomenon: If the product in such reactions never has any optical activity, then racemization has occurred (although in an individual case the possibility of a very small specific rotation cannot be dismissed out of hand). Inversion or retention is not so easy to determine. The sign and magnitude are to some degree predictable on the basis of Brewster's rules (cf. p. 82), but these rules are not so reliable that organic chemists would be convinced on that basis alone; they furnish a useful guide in making an initial guess, but not more than that.

A more reliable way is to carry out the analogous reaction with a substrate in which only one kind of stereochemistry is possible. Suppose for instance that we are interested in the stereochemical consequences of the reaction of an alcohol with thionyl chloride to give an organic chloride.

$$ROH + SOCl_2 \longrightarrow SO_2 + RCl + HCl$$

Thus, if we study the reaction of optically active 2-butanol, the signs of rotation of the original alcohol and the final chloride cannot definitely tell us what happened. However, we can compare the rate of this reaction with that of 1-bicyclo[2.2.2] octanol (17).

CH3 C—OH H C2H5 —SOCl2→ CH3 C—Cl H C2H5 or Cl—C CH3 C2H5 H

—OH —SOCl2→ —Cl (no Cl)

(17)

In that case, inversion is obviously not possible; if nevertheless the reaction occurs at a normal rate, it may be assumed that the reaction normally occurs with retention. If the reaction becomes extremely slow or does not occur at all, it is reasonable to conclude that inversion is the norm.

A second method depends on the use of cyclic structures also. If we hydrolyze the cis-epoxide (18) and find that the glycol obtained has the hydroxy group in trans-positions (e.g., by means of an ir study of its hydrogen bond behavior), then it is reasonable to suppose that epoxides always open with inversion.

(18) $\xrightarrow{H_3O^{\oplus}}$ OH, OH

One of the classic experiments in organic chemistry was that by Hughes et al. [14], who found that iodide ion catalyzed the racemization of optically active 2-iodoctane. When radioactive iodide ion was used, the radioactivity became incorporated in the organic molecule; furthermore, the initial rate of exchange is precisely half of that of racemization. The displacement therefore clearly occurs with inversion (an exchange act leads to one radioactive molecule and to a racemic pair of molecules). By such methods the stereochemistry of virtually every important organic reaction has been studied. Although a thorough review of all these studies is beyond the scope of this book, a brief outline of the results is desirable; furthermore frequent reference is made to them in the chapters dealing with reactive intermediates (e.g., p. 551).

It turns out that the simple direct nucleophilic displacement (S_N2) reaction is characterized by inversion.

$Y{:}\rightarrow C(R_1)(R_2)(R_3)-X \longrightarrow Y-C(R_1)(R_2)(R_3) \rightarrow {:}X$

In solvolysis (S_N1) reactions involving a free and long-lived carbonium ion, racemization is the result. This could be explained either by means of a planar cation, superimposable on its mirror image and hence equally likely to react with solvent on either side, or by means of a pyramidal ion which inverts rapidly compared to its capture by solvent, and hence which racemizes before giving the final product (note that the forbiddenness of inversion refers to the tetravalent carbon atom).

---

14. E.D. Hughes, F. Juliusburger, S. Masterman, B. Topley, and J. Weiss, J. Chem. Soc., 1525 (1935).

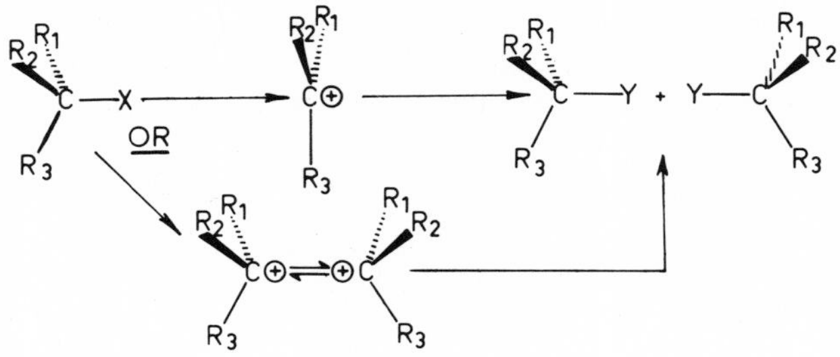

Variations on this theme have been observed. If capture occurs before the leaving group has receded very far, inversion may still result. If an internal group of the molecule temporarily provides the rearside of the electron-deficient carbonium ion center with a pair of electrons, overall retention may result (via a pair of inversions).

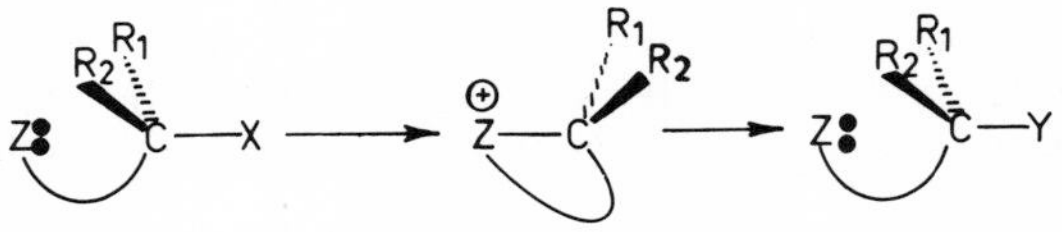

These remarks also apply to reactions in which carbanions or carbon free radicals are formed. In rearrangement reactions the centers to which and from which the moving group shifts usually undergo inversion, while the shifting group itself retains its configuration.

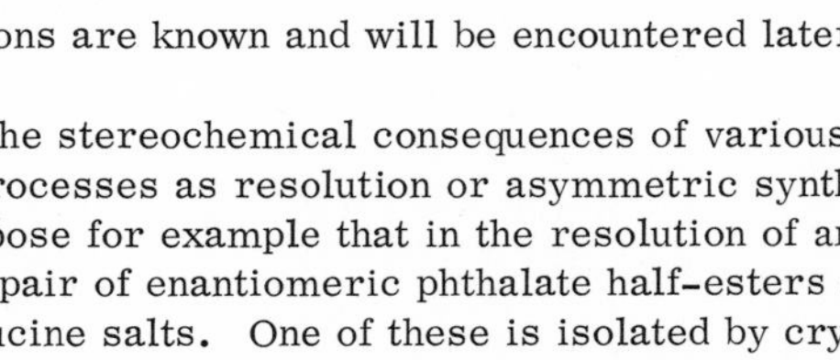

Here again exceptions are known and will be encountered later on (Chapter 14).

Knowledge of the stereochemical consequences of various organic reactions is essential if such processes as resolution or asymmetric synthesis are to be done successfully. Suppose for example that in the resolution of an alcohol we have converted it into a pair of enantiomeric phthalate half-esters and, in turn, into the diastereomeric brucine salts. One of these is isolated by crystallization processes, and the brucine base is removed to yield the optically active phthalate, and only hydrolysis is necessary to give us the optically active alcohol. If at this point the hydrolysis is carried out with strong acid catalysis, we will end up with racemic alcohol, since this reaction proceeds via carbonium ions.

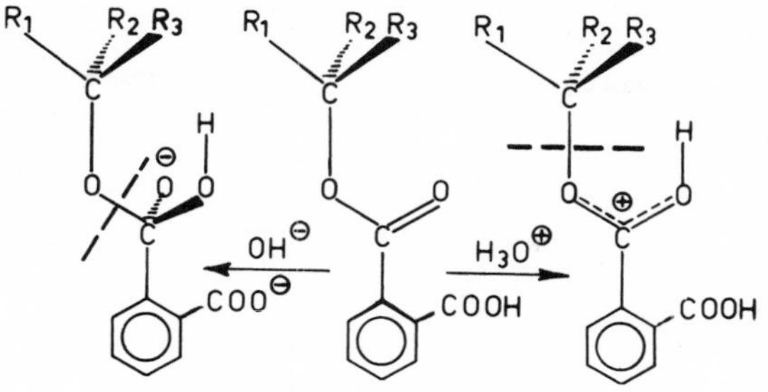

Base catalysis would avoid this problem since none of the bonds of the asymmetric carbon would be broken. On the other hand, if the alcohol is tertiary, base-catalyzed hydrolysis is usually very slow at best or accompanied by side reactions. In that event it might have been best to use oxalate half-esters instead and set the alcohol free by oxidation with lead tetracetate [15].

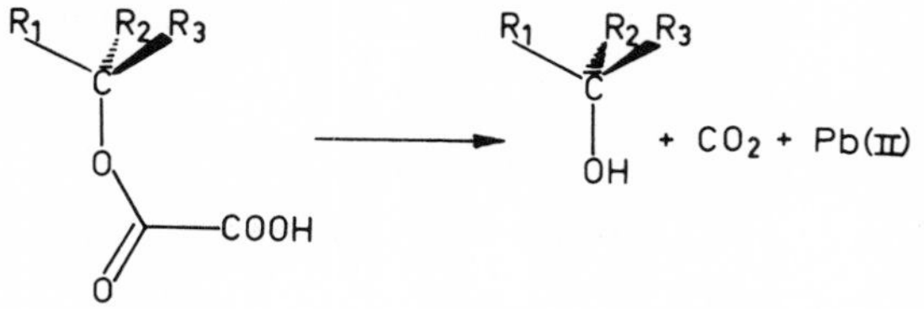

5-8. CONVENTIONS AND DESIGNATIONS IN THE CONFIGURATION OF MOLECULES

There are several conventions that have enjoyed popularity at various times, and all these are likely to be encountered in the literature; hence it is desirable to describe them. The designations (+) and (–) simply refer to the experimentally observed rotations; they are therefore not very useful in the actual description of molecules. Thus, if we compare a large number of molecules $CR_1R_2R_3X$, all with the same groups R_1, R_2, and R_3, all with the same configuration and differing only in X, we will generally find that not all these molecules have the same sign of rotation. This same conclusion also follows from the success of Brewster's rules (cf. p. 82): Given R_1, R_2, and R_3, the sign for a given configuration will be a function of the nature of X.

15. J. G. Molotkovsky and L. D. Bergelson, Tetrahedron Lett., 4791 (1971).

In order to put emphasis on the relation between optically active molecules rather than on the signs of the rotation, the organic chemists adopted a new convention similar to one originally proposed by Fischer. This convention is based on the two glyceraldehydes (19a) and (19b), which are dextro- and levorotatory (and hence (+) and (-)), respectively; these compounds are now considered to be the parents of two optically active series of compounds, all of which will be referred to as D- and L-, respectively.

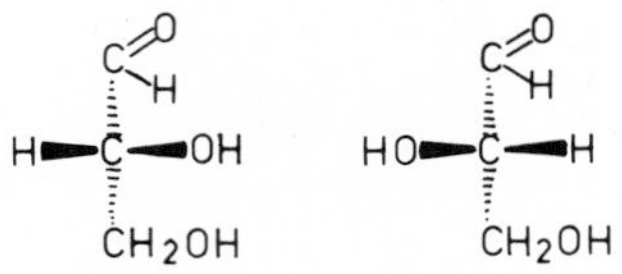

(19a) (19b)

Thus, the glyceric acids (20a) and (20b) that can be obtained by oxidation of (19a) and (19b) are called D- and L-glyceric acid, even though their signs of rotation are (-) and (+), respectively. The designation D- and L- makes use of the experience that inversion will not occur in a reaction in which none of the four bonds to the asymmetric carbon atom break. The complete designations for (20a) and (20b) are D-(-)-glyceric acid and L-(+)-glyceric acid, respectively.

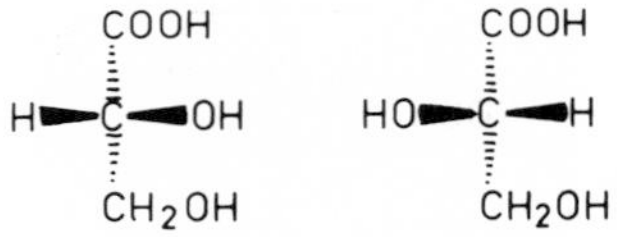

(20a) (20b)

When this convention of relative configurations was first proposed [16], the absolute configurations were not known, and hence there was a 50% chance that the assignments were wrong. It is amusing to speculate what would have happened if that turned out to be the case; however, Fischer's choice was in fact shown to be correct in 1951 by means of an X-ray diffraction study [17].

In spite of this good fortune, much confusion has surrounded the Fischer designation of configuration. Thus, a number of chemists began to use the prefixes d- and l-, and some of these had the optical rotations in mind whereas others sought to convey the relative configuration. Furthermore, it is not always clear how a given chiral carbon compound is related to glyceraldehyde. Even the glyceric acids can presumably be obtained from the aldehydes in either of two ways;

16. E. Fischer, Chem. Ber., 24, 2683 (1891).
17. (a) J. M. Bijvoet, A. F. Peerdeman, and A. J. van Bommel, Nature, 168, 271 (1951); (b) J. L. Abernethy, J. Chem. Educ., 33, 88 (1956).

by the simple oxidation of the aldehyde group or by converting this group into another insensitive to oxidizing agents, followed by oxidation of the primary alcohol grouping and, finally, by reduction of the protected aldehyde end of the molecules. These two pathways would produce - from the same enantiomer of glyceraldehyde - opposite enantiomers of glyceric acid. It would be clear of course which reactions had produced which enantiomer; however, for molecules very different from glyceraldehyde which might require many steps for the conversion it is obviously necessary to state how (i.e., by what sequence of reactions) the compound is related to the aldehyde, rather than merely stating that it is related.

For these various reasons it has become desirable to design a new convention, and the one now being gradually accepted by organic chemists is that of Cahn, Ingold, and Prelog [18]. This system begins by assigning priorities to the various substituents that may be bound to an asymmetric carbon atom. These priorities are decided by the atomic number; thus, iodine has priority over bromine, chlorine over sulfur, and so on. When two substituents are bound via the same kind of atom, then the second highest order atom decides; thus, OCH_3 has priority over OH, C_2H_5 over CH_3, $CBrF_2$ over CCl_3, and so forth. Multiple bonds are regarded as equivalent to n single bonds to the same kind of atom; thus, C≡N has a higher priority than $C(Me)(NMe_2)_2$. The following is a partial list of groups in order of priority:

I
Br
Cl
SR
SH
F
OR
OH
NR_2
NHR
NH_2
COOR
COOH
$CONH_2$
COR
COH
C_6H_5
CMe_3
CHR_2
CH_2R
CD_3
CH_3
D
H

We now view the asymmetric atom in such a way that the lowest priority group points away from us; if the three other groups have a decreasing order of priority in a clockwise direction, the name of the molecule preceded by the prefix R-, and, otherwise, by the prefix S-. In this system, D-glyceraldehyde is glyceraldehyde.

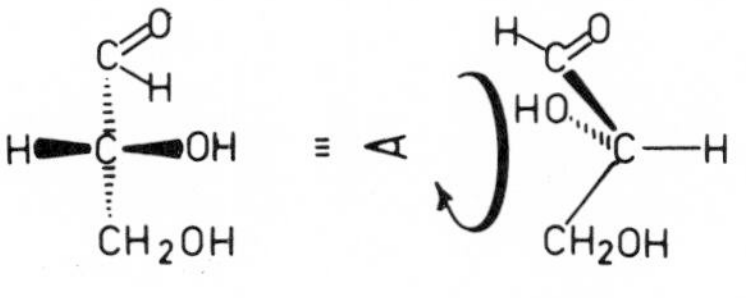

D-(+)-glyceraldehyde R-glyceraldehyde

18. (a) R.S. Cahn and C.K. Ingold, J. Chem. Soc., 612 (1951); (b) R.S. Cahn, C.K. Ingold, and V. Prelog, Experientia, 12, 81 (1956).

The prefixes *D*- and *R*- are obviously not always synonymous. The advantage of the new convention is that the prefix for any chiral compound can be written without knowledge of how the compound might be related to glyceraldehydes or to any other substance. It may be noted that the system is related to the Brewster rules in that the order of priorities used in those rules (relative polarizibilities) happens to be more or less related to atomic number also.

We should at this point also briefly discuss the pictorial representations in use to discuss stereochemical problems. When it first became clear that the carbon atom was tetrahedral in its bonding directions it was understood that (21) and (22) were not designations of different molecules, but merely evidence of the inadequacy of plane paper in an effort to write three-dimensional objects.

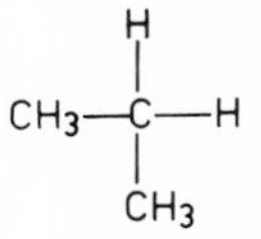

(21) 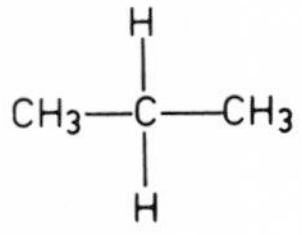 (22)

The story takes on a different slant when the configuration is considered: Does the drawing (23) represent (24) or (25)?

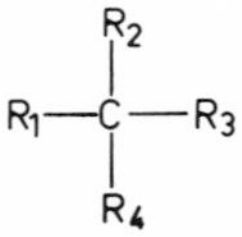

(23)

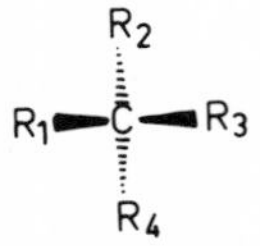

(24)  (25)

Fischer proposed that it represents (24); all horizontally drawn lines incline from the paper, and the vertical ones recede into it. Thus, *R*- (or *D*-) glyceraldehyde would be (19a).

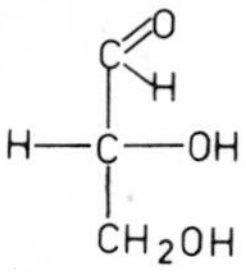

(26) 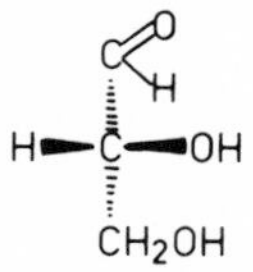 ≡ (19a)

The problem now is that when a picture like (26) is encountered, it is not clear whether the configuration is actually known or not, and in recent years the organic chemist has begun to add a small amount of artistry into his drawings to imply configuration. Thus, drawings such as (27) or (28) are much better than (26) plus a verbal understanding such as that of the Fischer convention.

(27) (28)

Other drawings often used include the Newman projections (29) of the view of a bond along the line of two nuclei, usually with a staggered conformation;

(29)

the indication, by means of heavy dots, which hydrogen atoms are forward and which point to the rear (e.g., (30) and (31))

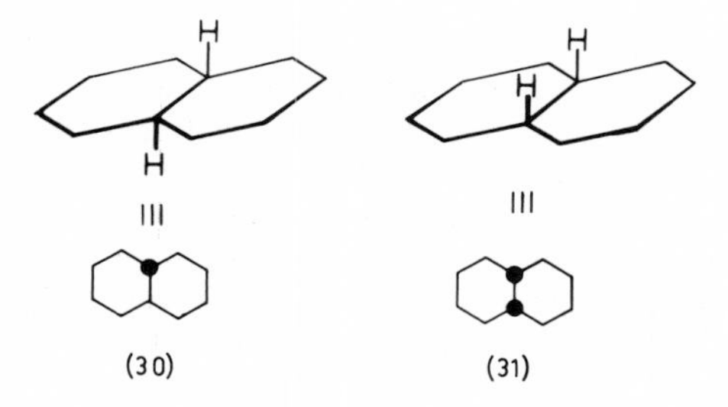

(30) (31)

and, by means of broken lines, which bonds lie behind which (e.g., adamantane (32)).

(32)

Designations such as in (33) are also encountered frequently in the literature; it means configuration at that center is not known or not stable, or that it does not matter, or that both epimers are present.

NC

(33)

5-9. CHIRALITY AND SYMMETRY

All the foregoing discussion has been based on the tentative assumption that chirality depends on the presence of asymmetric carbon atoms, and we should now examine this basis. It is not difficult to show that it is in fact not correct: There are numerous compounds which are chiral (i.e., nonsuperimposable on their mirror image) although they contain no asymmetric carbon atoms in the traditional sense, and there are numerous others which have asymmetric carbon atoms but which are still achiral. What we need first of all is a means of examining which molecules are chiral and which are not. One of these was mentioned previously (cf. p. 127): One makes a ball-and-stick model of the molecule and its mirror image using a mirror if one likes and then tries simply to superimpose the one molecule on the other by freely turning each as a whole and, where necessary, rotating singly bound groups about the bonds connecting them to the rest of the molecule. This method literally goes back to the definition and should be within the cranial capabilities of any chimpanzee. A similar approach can be tried on paper but this is almost as time consuming, since after each rotation another drawing should be made showing the molecule in its new orientation either until it and its mirror image are obviously superimposable or until they are obviously not.

A generalization taught in elementary organic chemistry courses is that one may look for planes or a center of symmetry; if one or more of these symmetry elements are present, the molecule will be "symmetrical" and superimposable on its mirror image. A plane of symmetry requires that for every atom on one side of it there must be another one just like it on the other side such that the plane is perpendicular to the line connecting them at its midpoint. Thus, for the points A and A' in Fig. 5-8, σ is a plane of symmetry, but for the points B and B' or C and C', it is not. A molecule has a center of symmetry if for each atom there is

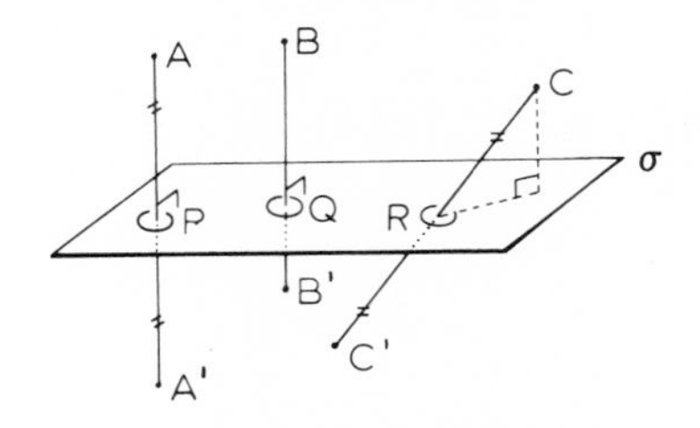

Fig. 5-8. Symmetrical and nonsymmetrical locations above and below a plane.

another like it, the pair being so situated that the center is the midpoint of the line segment connecting them. Thus, P is the center for A and A', and R is that for C and C'.

Thus, among some of the familiar molecules, methane has six planes but no center of symmetry; staggered ethane has a center and three planes; eclipsed ethane has no center but four planes; completely staggered propane has two planes; benzene has seven planes and a center; naphthalene has three planes and a center; cubane has nine planes and a center; and norbornane has two planes and no center.

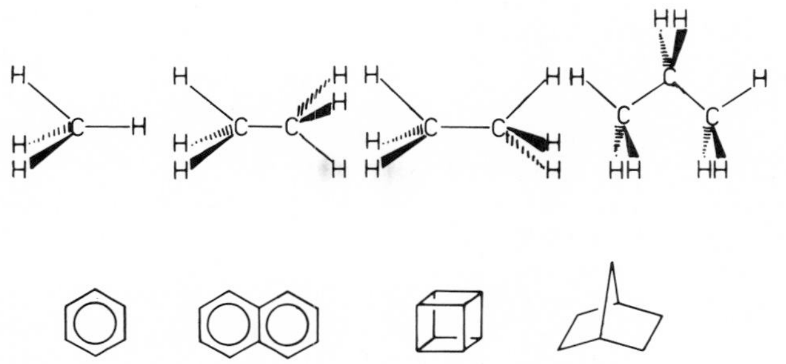

This generalization is relatively easy to apply on paper and it is virtually infallible: If a plane or center of symmetry is correctly identified, the molecule can always be superimposed on its mirror image, and if these elements are absent, it almost never can be. The word almost is used here because a few cases are known of achiral molecules which have no planes or center of symmetry. These cases are discussed in the paragraphs below (p. 148).

Mathematically, the correct statement is that a molecule is superimposable on its mirror image if and only if it has an n-fold alternating axis of symmetry. The presence of such an axis means that the molecule can be rotated about it by 360/n degrees, and if thereafter a reflection is carried out from a mirror perpendicular to the axis, the resulting image should be superimposable on the original object *without any further reorientations or rotations.*

As an example, consider tricyclo[4.2.1.0^{3,7}]nonane, also known as brendane, and the line through carbon atoms 1 and 7 (Fig. 5-9). This line is *not* a one-fold alternating axis, because after a 360° rotation and reflection we do *not* get an image which is superimposable on its mirror image; the fact that we *can* superimpose it after turning it around is not relevant. On the other hand, *the* line passing through A and B is such an axis. Whenever any onefold alternating axis is present, there is also a plane of symmetry; in the present instance, it is defined by atoms 1, 7, and 8. Another example is *trans*-1,3-dibromo-*trans*-2,4-dichlorocyclobutane (Fig. 5-10). The axis through the center of the ring is a twofold alternating axis. Whenever such an axis is found, a center of symmetry is also present.

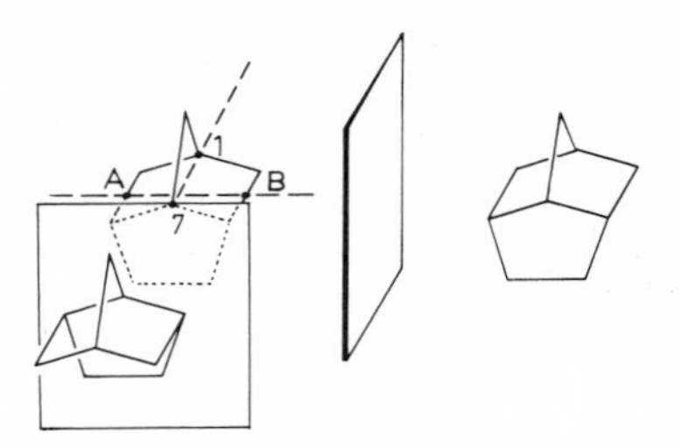

Fig. 5-9. A onefold alternating axis in brendane.

Some compounds are now known whose molecules possess a fourfold alternating axis. Thus, the rather complex quarternary ammonium ion (34) has such an axis, and hence it is achiral even though neither a center nor a plane of symmetry is present [19]; the ion is superimposable on its mirror image, and the salt is optically inactive as it should be.

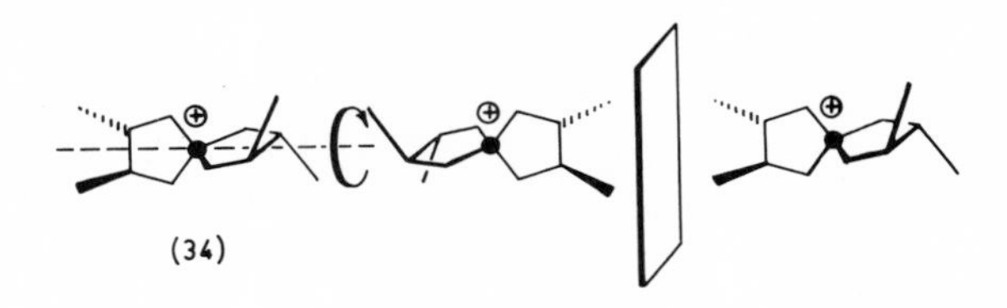

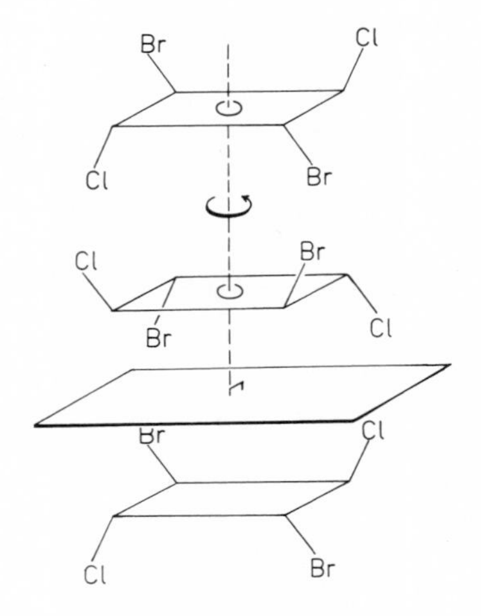

Fig. 5-10. Twofold alternating axis in a cyclobutane.

19. Several examples have been discussed and realized: See G. E. McCasland, R. Horvat, and M. R. Roth, J. Amer. Chem. Soc., 81, 2399 (1959).

This compound is one of a large number whose molecules contain asymmetric carbon atoms, and which are nevertheless achiral. Other examples are (35) which has a plane of symmetry, and (36) which has the fourfold alternating axis indicated.

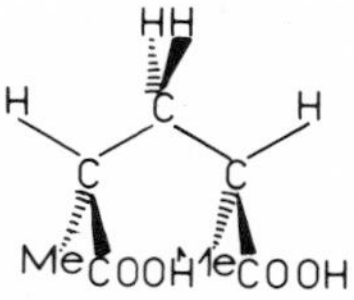

(35)

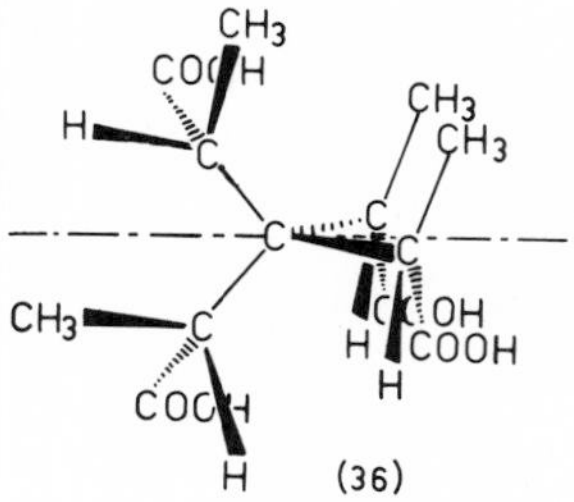

(36)

Such compounds, achiral though containing asymmetric carbon atoms, are often called meso forms whose inactivity was presumed due to "internal compensation" (negation of the rotation caused by one part of the molecule by the other); however, if it is recognized that these are simply molecules having the necessary symmetry element, and hence molecules superimposable on their mirror images, we need not regard them as a special group. The word meso is a handy designation in the nomenclature of stereoisomers, however.

5-10. GEOMETRIC ISOMERISM

One assumption on which our discussion thus far has been based is that all single bonds can freely rotate. Thus, if there were no such freedom, we should have to conclude that a molecule such as (37) must be resolvable; it has no alternating axes of symmetry and is nonsuperimposable on its mirror image.

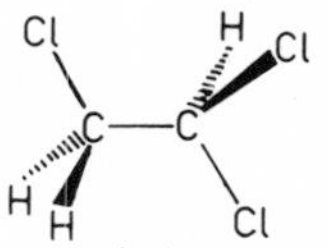

(37)

If all the 1, 1, 2-trichloroethane molecules could be preserved and isolated in this conformation, perhaps at a few degrees Kelvin one would indeed find the substance to be chiral and optically active. However, at only slightly higher temperatures, rotation would set in, and we soon would have significant conversion as follows.

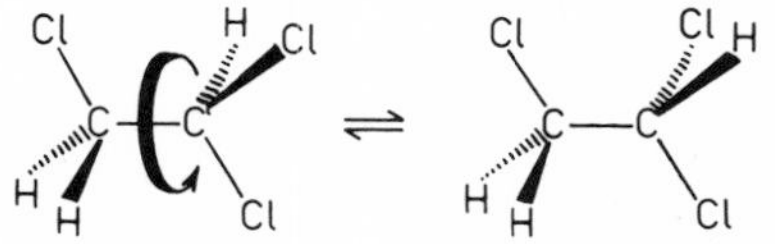

$$\rightleftharpoons$$

It may be noted that the product is the mirror image of (37). Thus, this reaction amounts to racemization, and this is the basic reason why many molecules which on the basis of their structure would seem to be chiral can in fact not be resolved. If the freedom of rotation is removed, first of all a new type of stereoisomerism results which has come to be known as geometric, or *cis*-*trans*-isomerism. Thus, 1, 2-dichlorocyclopropane occurs in the *cis* form (38) and a *trans* form (39); furthermore, the latter is chiral and hence occurs as a pair of enantiomers. It may be noted that while the designations *cis* and *trans* are commonly used, the *R-S* descriptions are applicable to these molecules also.

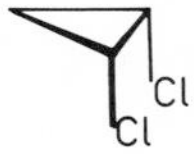

(38) 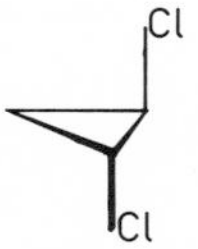 (39a) 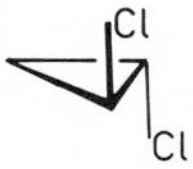 (39b)

Another feature that can be introduced to stop internal rotation is the double bond (it could be considered a two-membered ring). Such molecules are planar so far as the two carbon atoms and the four substituting atoms are concerned, and rotation would necessarily result in the breaking of the π component of the double bond; hence, geometric isomerism may again result, as in *cis*- and *trans*-dichloroethylene, (40) and (41), respectively.

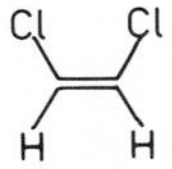

(40) 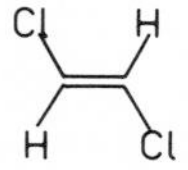 (41)

Because of the planar framework, olefins are achiral unless there are additional groups in the molecules to make them so. It may be noted here that the designations *cis* and *trans* are on the way out, since for olefinic molecules with four different groups these terms are not merely undefined and useless, but give rise to confusion when their use is attempted anyway. In 1968 a new system was proposed that was quickly adopted by organic chemists [19a]. The system makes use of the priorities of substituents also used in the *R, S* convention. One begins by finding the lowest priority substituent; if the substituent *cis* to it is of lower priority than the one trans, then the designation is *Z* (for Zusammen), otherwise *E* (for Entgegen). Thus, (42) is *Z*-1-bromo-2-methyl-1-butene, (43) is *Z*-cyclooctene though (44) is *E*-1-chlorocyclooctene, and so on.

19a. J. E. Blackwood, C. L. Gladys, K. L. Loening, A. E. Petrarca, and J. E. Rush, *J. Amer. Chem. Soc.*, *90*, 509 (1968).

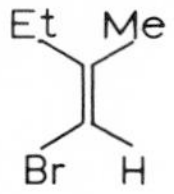

(42) 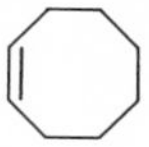 (43) 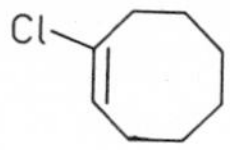 (44)

5-11. STEREOISOMERISM; SPECIAL CATEGORIES

We are now in a position to understand a number of classes of compounds which exhibit stereoisomerism of one sort or another for various special reasons. Among these are the biphenyls. Compounds such as 2,2'-dinitro-6,6'-diphenic acid (45) are known to be resolvable [20], and the reason for this is that the _o_-substituents are too large to permit free rotation of the central carbon-carbon bond and hence racemization. Thus, appropriately substituted biphenyls are compounds in which steric hindrance is large enough to prevent rotation about single bonds;

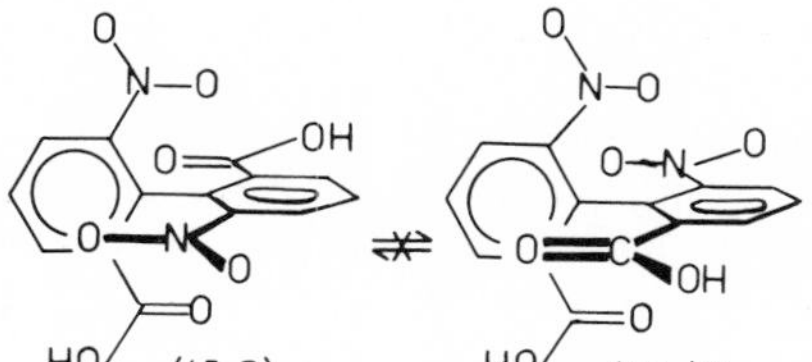

(45-a) $\not\rightleftharpoons$ (45-b)

hence one is equally justified in calling such pairs of molecules conformers or enantiomers. It may be noted that in written presentations the biphenyls are not always shown in the perspective drawn above; thus, (45) may simply be shown as (45c) or (45d). It should be realized by the reader that the molecule is in fact not planar, and hence that (45c) does not really have two planes of symmetry and that (45d) does not actually have a center of symmetry.

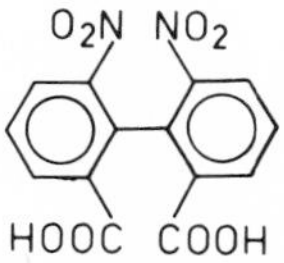

(45-c) 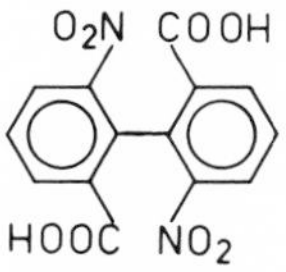 (45-d)

There are two requirements that must be fulfilled if a given biphenyl is to be resolvable. First of all, the 2- and 6-substituents must be different, since otherwise a plane of symmetry exists which bisects this ring and contains the other.

---

20. G.H. Christie and J. Kenner, _J. Chem. Soc._, _121_, 614 (1922).

The 2'- and 6'-substituents must also be different, but neither the 2- and 2'- nor the 6- and 6'-substituents need be. Various obvious modifications are possible; thus, 2,2'6,6'-tetraiodo-3,3'-diphenic acid (46) will be resolvable even though all *o*-substituents are the same.

I I

HOOC I I COOH

(46)

The second requirement is that the groups in the *o*-positions must be of sufficient size to prevent rotation. If we refer to Fig. 5-11 in which the situation involving monatomic substituents is sketched, inspection may suffice to convince the reader that in order to predict the size at which rotation becomes difficult, one needs to know the distance of the *o*-carbon atoms at closest approach, the covalent bond radii of those atoms and the atoms bound to them, and the van der Waals radii of the substituent atoms (on both sides of the molecule; note that there are in each case two possible rotations leading to the enantiomer). If this is done, one finds that even when the substituent atoms interfere fairly seriously, resolution is still not possible, and the conclusion is that some bending of the bonds must be possible. A more sophisticated treatment by Westheimer takes account of this bending (see Chapter 7, Section 7-5); suffice it here to say that it accounts reasonably well for the experimentally observed size requirement. These observations have it that if the 2-position is unsubstituted, the 2'- and 6'-positions must be substituted by groups not smaller than a bromine atom if rotation is to be prevented at room temperature; a methyl group requires at least a methyl group or chlorine atom in opposition, and so forth [21]. It is noteworthy that in optically active biphenyls one cannot point to any carbon atom asymmetric in the traditional sense

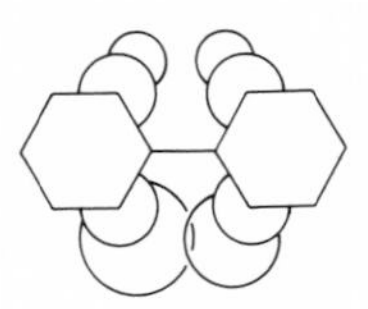

Fig. 5-11. *o*-Substituent interactions in planar biphenyls.

---

21. R. L. Shriner, R. Adams, and C. S. Marvel, "Organic Chemistry", H. Gilman, Ed., 2nd ed., Vol. I, Wiley, New York, 1942; p. 347.

(bound to four different groups). It is interesting to note that nevertheless the methylene protons in a biphenyl such as (47) show up as AB quartets in an nmr spectrum [21a].

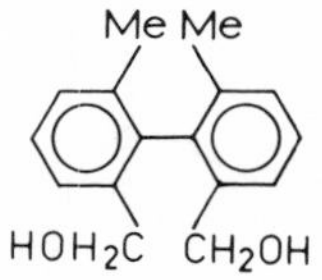

Me Me

HOH₂C CH₂OH

(47)

Certain substituted allenes (48) are a second group of resolvable molecules. Rotation here is impossible because of the double bonds; since we have two cumulative double bonds, the two end groups are in perpendicular planes. The basic shape is the same as that of the biphenyls as is suggested by the end-on views in Fig. 5-12.

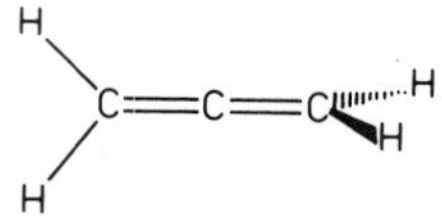

(48)

The requirements are therefore in part the same; R_1 and R_2 must be different, and R_3 and R_4 must also be; it is immaterial whether R_1 or R_2 are the same as or different from R_3 or R_4. The size of the substituents is immaterial in this case.

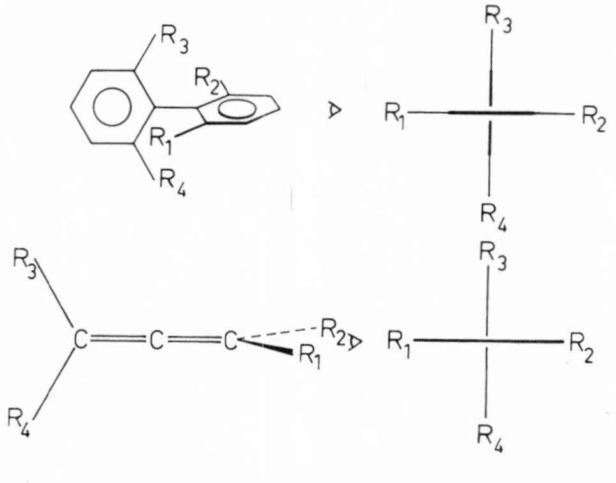

Fig. 5-12. Two views of biphenyls and of allenes.

21a. W. L. Meyer and R. B. Meyer, J. Amer. Chem. Soc., 85, 2170 (1963).

Optically active allenes can be prepared from resolved chiral cyclopropane-carboxylic acids by a sequence involving a carbene reaction at the end [22] (see p. 575).

$$\text{CH}_3,\ \text{C}_2\text{H}_5,\ \text{COOH cyclopropane} \xrightarrow[3-\ \Delta]{1-\ \text{SOCl}_2;\ 2-\ \text{NaN}_3} \text{CH}_3,\ \text{C}_2\text{H}_5,\ \text{NCO cyclopropane} \xrightarrow[6-\text{OCH}_3^{\ominus}]{4-\text{C}_2\text{H}_5\text{OH};\ 5-\text{N}_2\text{O}_4} \text{CH}_3\text{CH}{=}\text{C}{=}\text{CHC}_2\text{H}_5$$

Asymmetric synthesis has been applied also; thus, partial reduction of racemic 1,3-dimethylallene with the hydroboration product derived from (-)-α-pinene (49) in diglyme at 0° leaves the allene enriched in the (-)-enantiomer [23].

$$(49) \xrightarrow[\text{BF}_3]{\text{NaBH}_4} [\ \ldots\ ]_2\text{BH}$$

The absolute configurations of allenes are known, and they are in accord with predictions that can be made for them [24] on the basis of Brewster's rules. It should also be noted that since double bonds may be considered two-membered rings, the stereoisomerism discussed for allenes is in general applicable to spiro compounds.

There is a further large group of aromatic molecules which normally is expected to be planar but which has groups interfering with that geometry. For example, 4,5,8-trimethyl-1-phenanthrylacetic acid (50) is resolvable because the 4- and 5-methyl groups interfere and bend one another out of the plane [25].

---

22. (a) J. M. Walbrick, J. W. Wilson, and W. M. Jones, J. Amer. Chem. Soc., 90, 2895 (1968); (b) W. T. Borden and E. J. Corey, Tetrahedron Lett., 313 (1969).
23. W. L. Waters, W. S. Linn, and M. C. Caserio, J. Amer. Chem. Soc., 90, 6741 (1968).
24. G. Lowe, Chem. Commun., 411 (1965).
25. M. S. Newman and A. S. Hussey, J. Amer. Chem. Soc., 69, 3023 (1947).

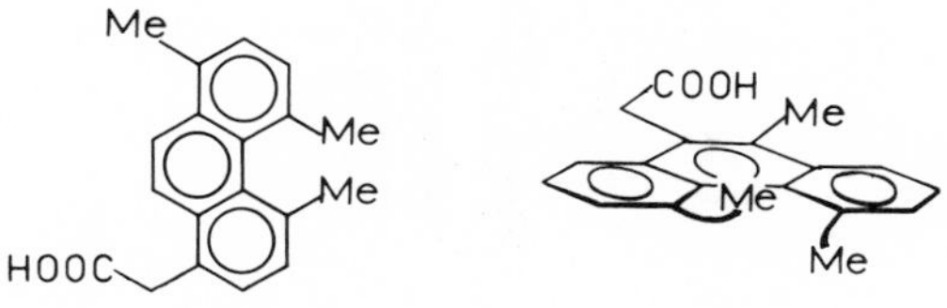

Me Me Me HOOC COOH Me Me Me

(50)

This type of interference gets stronger as additional benzo-rings are added in circular fashion (helicenes); examples of optically active compounds of this type are 1-methyl-4-benzo[c]phenanthrylacetic acid (51) [26], pentahelicene (52) which racemizes rapidly at room temperature [27], and hexahelicene (53) which racemizes slowly at its melting point of 266° [28].

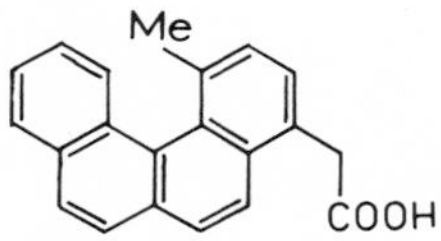

Me COOH

(51) 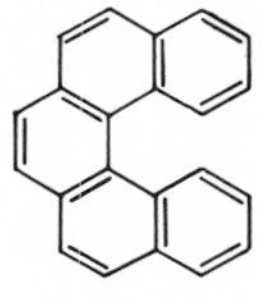 (52) 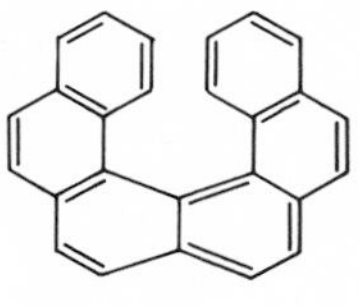 (53)

It was at first taken for granted that the larger helicenes cannot be racemized; however, recently the surprising result was published that the [7]-, [8]-, and [9]-helicenes all readily racemize at their melting points (about 200°) [28a, b]. Since the activation entropies are negative, this process is not likely to involve the breaking of any carbon-carbon bonds (for example, at the inner periphery). Models show that severe bond bending is another possibility. The helicenes also provide us with the only really well-documented case of asymmetric synthesis induced by circularly polarized light; when racemic Z-1, 2-diarylethylenes such as (54) are irradiated with such light in the presence of iodine and oxygen, slightly but measurably optically active helicenes are obtained - the sign of the rotation depending on the sense of the polarization [29].

---

26. M.S. Newman and W.B. Wheatley, J. Amer. Chem. Soc., 70, 1913 (1948).
27. C. Goedicke and H. Stegemeyer, Tetrahedron Lett., 937 (1970).
28. M.S. Newman and D. Lednicer, J. Amer. Chem. Soc., 78, 4765 (1956).
28a. R.M. Martin and M.J. Marchant, Tetrahedron Lett., 3707 (1972).
28b. R.H. Martin, M. Flammang-Barbieux, J.P. Cosyn, and M. Gelbcke, Tetrahedron Lett., 3507 (1968).
29. W.J. Bernstein, M. Calvin, and O. Buchardt, J. Amer. Chem. Soc., 94, 494 (1972).

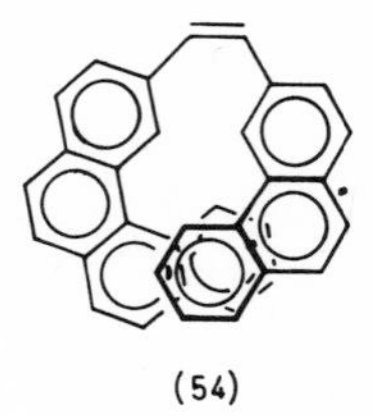

(54)

All other special groups of resolvable compounds depend for their asymmetry on hindrance or inhibited rotation or ring inversion (see Chapter 6, p. 187). An example is trans-cyclooctene (55) [30], in which the vinyl hydrogen atoms prevent the motion that would lead to racemization.

(55)

The cyclophanes are related to such olefins in that internal rotation as indicated is difficult for steric reasons. Thus, the [2.2]paracyclophanecarboxylic acid (56) is resolvable [31].

(56)

The [2.2]metaparacyclophanes (57) and (58) represent a diastereomeric pair which equilibrate at room temperature; the rate and equilibrium constants can readily be measured by means of nmr.

(57) (58)

---

30. A.C. Cope, C.R. Ganellin, H.W. Johnson, T.V. Van Auken, and H.J.S. Winkler, J. Amer. Chem. Soc., 85, 3276 (1963).
31. D.J. Cram and N.L. Allinger, J. Amer. Chem. Soc., 77, 6289 (1955).

At 180°, coalescence occurs, so that the constants can then be evaluated from the coalescence temperature (see p. 64). The resolved compounds do not racemize in the process; this means that only one ring (presumably the m-ring) rotates since racemization would result if both rotated [32].

A different group is the one in which heteroatoms play a special role [33]. When the heteroatom is a tetrahedral, asymmetric atom, the analogy is straightforward, as indeed numerous optically active salts of ammonium, phosphonium, and arsonium cations are known in which the heteroatom is clearly asymmetric. The analogy is also easily extended to neutral silicon and germanium compounds, to amine and phosphine oxides, and to boron anionic compounds such as (59).

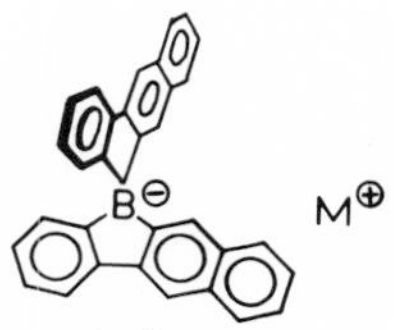

(59)

More interesting perhaps, heteroatoms may lead to molecules with different shapes. Amines are an interesting example of this. Ammonia is a pyramidal molecule, and hence it is reasonable to expect that appropriately substituted amines should be resolvable. However, this has never successfully been done, the reason being that such molecules can and do very rapidly invert and hence racemize.

This process is not observed with the tetrahedral carbon atom. The nitrogen atom in amines has an unshared pair instead of a fourth substituent, and hence inversion is an easier process for it. As we shall see in Chapter 6 (p. 179), the inversion for the nitrogen atom can be slowed by making it part of a small ring.

When the ring is two-membered, i.e., the nitrogen is doubly bound, inversion is generally always slow enough to lead to stereoisomerism. Thus, syn- and anti-oximes such as (60) and (61) are well known, and cis- and trans-azobenzene (62) and (63) can be isolated. On the other hand, carbodiimides invert too rapidly

32. D. T. Hefelfinger and D. J. Cram, J. Amer. Chem. Soc., 93, 4767 (1971).
33. Reviewed by V. I. Sokolov and O. A. Reutov, Russ. Chem. Rev. (English Transl.), 34, 1 (1965).

to permit resolution; thus, (64), which normally has a methyl doublet in the nmr spectrum with J = 6.5 Hz, at -150° begins to show a quartet [34] as expected of an isopropyl group bound to an asymmetric center (see p. 54).

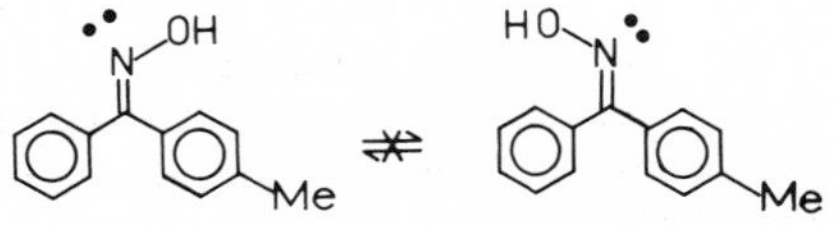

(60) (61)

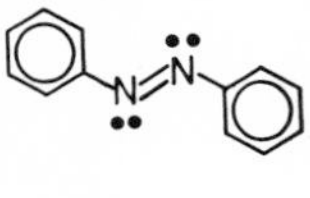

(62) 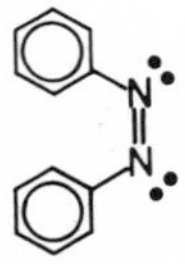 (63) 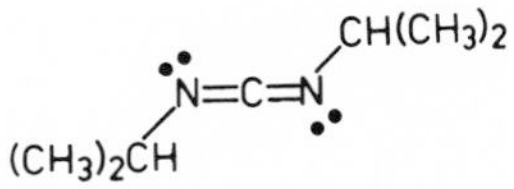 (64)

The bond angles of phosphine, arsine, etc., are smaller than those of ammonia, and hence appropriately substituted phosphines (65) might be expected to be resolvable. This has indeed been accomplished in many cases [35].

(65)

In analogy, many sulfoxides (66) have been resolved [36].

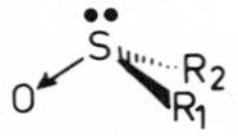

(66)

Oddly enough, while chiral arsines cannot be made to racemize thermally [37], diastereomeric diarsines interconvert rapidly at 200°, as shown by the fact that a

---

34. F.A.L. Anet, J.C. Jochims, and C.H. Bradley, J. Amer. Chem. Soc., 92, 2557 (1970).
35. (a) M.S. Lesslie and E.E. Turner, J. Chem. Soc., 1051 (1935); (b) I.G.M. Campbell and D.J. Morrill, J. Chem. Soc., 1662 (1955).
36. P.W.B. Harrison, J. Kenyon, and H. Phillips, J. Chem. Soc., 129, 2079 (1926).
37. L. Horner and W. Hofer, Tetrahedron Lett., 3281 (1965).

mixture of *meso*- and *dl*-1, 2-dimethyl-1, 2-diphenyldiarsine (67), which at room temperature shows two methyl peaks [75] (As decoupled), at 200° has only a single peak in the nmr spectrum. This phenomenon is independent of the concentration and hence cannot be ascribed to dissociation into divalent arsenic radicals [38].

(67)

Still further shapes can be encountered. Some atoms such as platinum may be coordinated in square planar fashion, leading to *cis*-*trans*-isomerism.

Br—Pt(Cl)(Cl)—Br ⊖   Cl—Pt(Cl)(Br)—Br ⊖

The octahedrally coordinated transition metal atom is well known in inorganic chemistry; with six substituents at the corners there are obviously all sorts of possibilities of stereoisomerism. Several resolvable inorganic compounds of this sort are known which do not contain carbon [39].

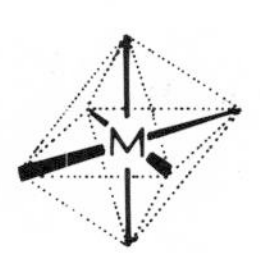

5-12. INTERCONVERSION OF GEOMETRIC ISOMERS

The process of racemization of enantiomers (Section 5-7) has a counterpart in *cis*-*trans*- and *syn*-*anti*-isomerizations. As in the racemization process, when only carbon atoms are involved a bond must be broken. Although the intermediates that are involved are discussed in later chapters, they may be briefly referred to here. *Cis*- and *trans*-disubstituted cyclopropanes can sometimes be interconverted by heating them. A diradical is probably involved as an intermediate (p. 650).

---

38. J. B. Lambert and G. F. Jackson, *J. Amer. Chem. Soc.*, **90**, 1350 (1968).
39. F. Basolo, *Chem. Rev.*, **52**, 459 (1953).

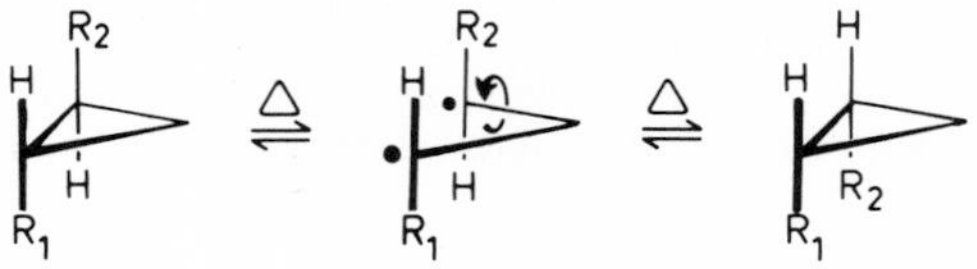

Other ring systems are known to be subject to reversible openings, but many of these reactions are stereospecific and do not lead to isomerization (Chapter 14).

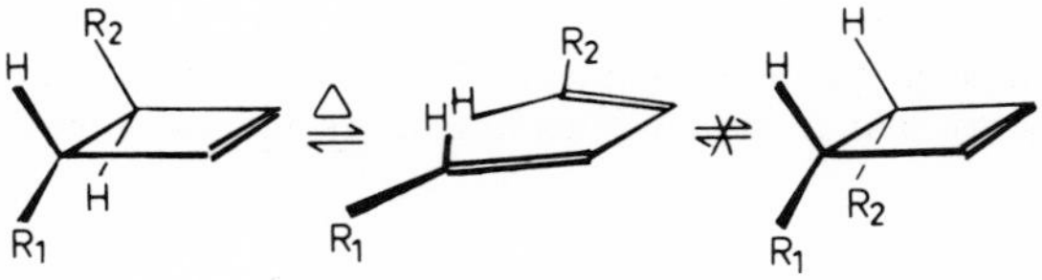

The reversible removal of any of the substituents always leads to epimerization; for example

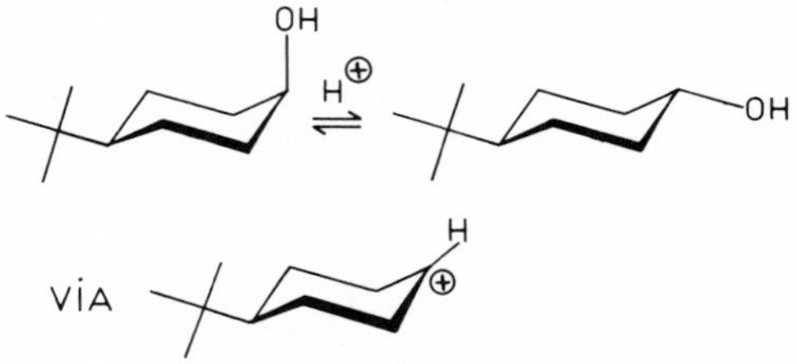

In olefins the π component of the double bond must be reversibly broken. This can often be done photochemically (p. 431).

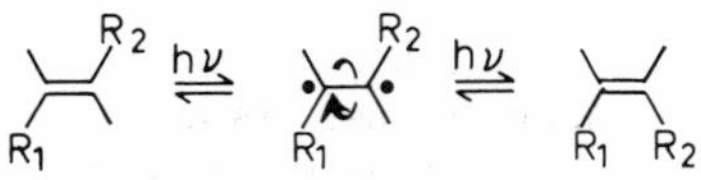

If the π component is not too strong, weakened for example by resonance, isomerization may be possible thermally.

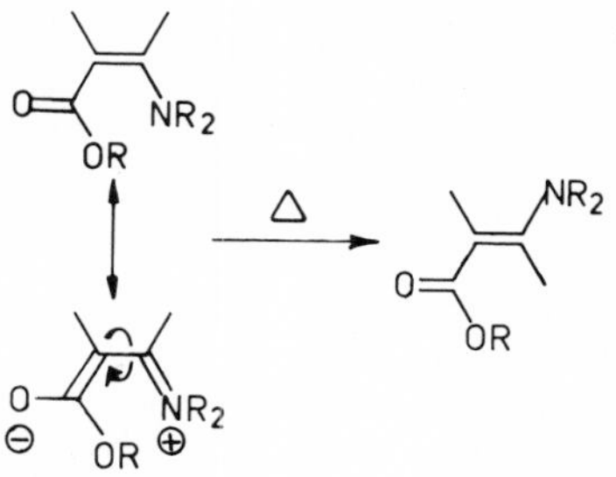

Reversible ion or radical addition also accomplishes this goal; thus, such interconversions are often catalyzed by iodide ion or strong acids.

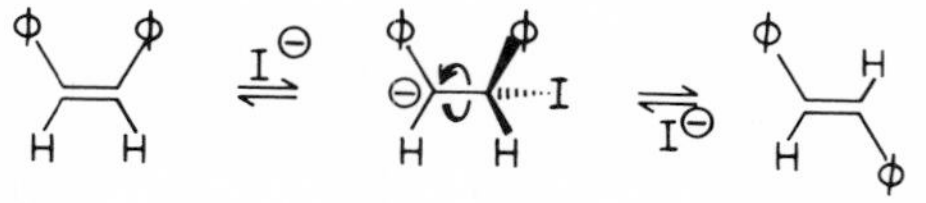

Syn-anti-isomerization of oximes and similar compounds could in principle occur without bond cleavage via inversion; this question is taken up in Chapter 6 (see p. 182).

5-13. SYMMETRY ELEMENTS

It is at this point desirable to introduce the definitions and symbols in use in discussions involving symmetry [40]. First, we recognize that there are various symmetry elements such as planes and centers of symmetry, n-fold alternating axes, and n-fold simple axes (the latter are axes such that rotation of $360/n$ degrees about them produces no observable change). Molecules [41] (or any other objects) are called asymmetric if they have no symmetry elements at all (we ignore the simple onefold axis for this purpose). A molecule containing a single asymmetric carbon atom is an example of an asymmetric molecule; a molecule that has no alternating axis of any kind is called dissymmetric or chiral. Thus, many dissymmetric molecules are not asymmetric; i.e., they may have a simple axis even though they are not superimposable on their mirror images.

---

40. See also M. Zeldin, J. Chem. Educ., 43, 17 (1966).
41. Strictly speaking, we should define the conformation of a molecule when its symmetry elements are under discussion; this aspect is left out except further below in our examples.

Molecules possessing various combinations of symmetry elements are said to belong to one point group or another. The definitions of the various point groups are given with some examples below. Since a onefold alternating axis is equivalent to a plane of symmetry, we refer to such axes simply as planes.

Symbol of point group	Description	Examples
C_1	No elements other than simple onefold axes	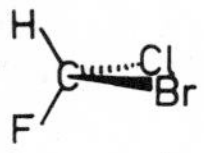 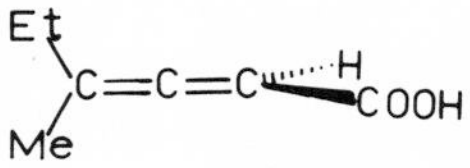
C_2	Only a simple twofold axis	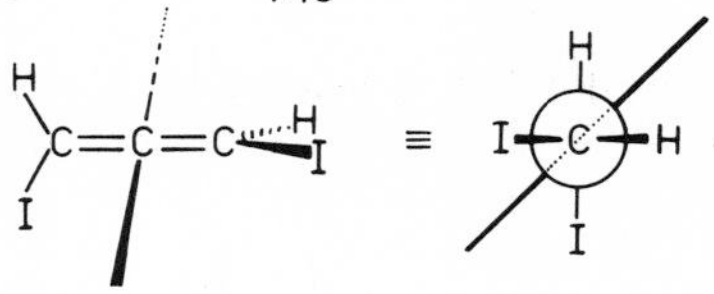
C_3	Only a simple threefold axis	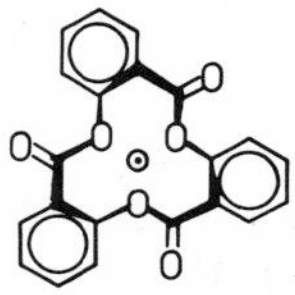
C_n	etc.	
D_2	A C_2 axis and 2 additional C_2 axes in a plane perpendicular to the first	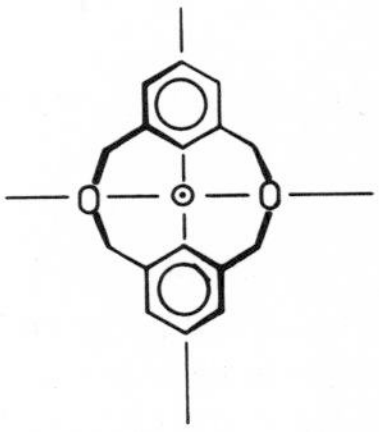
D_3	A C_3 axis and 3 C_2 axes in a plane perpendicular to the C_3 axis	A three bladed propeller; no known examples among molecules
D_n	etc.	
C_s	A plane; no axes	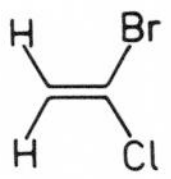

Symbol	Symmetry elements	Example
S_2	A twofold alternating axis (center of symmetry) only	
S_4	A fourfold alternating axis only	
C_{nv}	A C_n simple axis and n planes containing it	(C_{2v}), (C_{3v})
C_{nh}	A C_n simple axis and a plane perpendicular to it	(C_{2h})
D_{nd}	A C_n simple axis, n planes containing it, and n C_2 simple axes	(D_{2d}), (D_{5d})
D_{nh}	A C_n simple axis, n planes containing it, one plane perpendicular to it, and n C_2 simple axes	(D_{3h})
T_d	Tetrahedral symmetry; i.e., 6 planes, 4 C_3 axes, and 3 C_2 axes	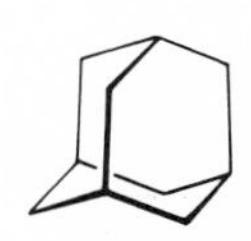

O_h	Octahedral symmetry; i.e., 9 planes, 3 C_4 axes, 4 C_3 axes, and 6 C_2 axes	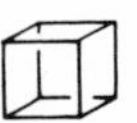

Note once again that molecules will not be resolvable unless all conformations open to them have at most C_n or D_n symmetry (lack reflection symmetry).

5-14. EXAMPLES OF MORE COMPLEX STEREOISOMERISM

All molecules examined thus far in this chapter were simple cases of stereoisomerism or geometric isomerism; not infrequently, however, several features are operative at once. For instance, the nmr spectrum of the allenic diether (68) [42] shows the presence of two diastereomers, one with a C_1H quartet (C_1 is α to the t-butoxy group; coupling with both C_3H and C_4H) centered about 0.05 ppm from that of the other.

(68)

This could be expected since both an asymmetric carbon atom and an appropriately substituted allenic grouping are present.

The examples that follow are for the most part hypothetical - chosen here only for their pedagogic value. While the assurance that all the stereoisomers could indeed be made if anyone were interested may seem glib, there are many examples of the actual existence of large numbers of such isomers in very complicated cases. To name one, the cyclopropanation of biallenyl by means of diazomethane could theoretically provide a total of 20 products (four inactive substances and eight enantiomeric pairs); all have been isolated and identified (no resolution was attempted) [42a]. It should be pointed out that configurations in print usually are not drawn in perspective, but such that the molecule appears to lie in the plane of the paper. The problems below are posed that way; in practice one should realize that one must make his own corrections for this, and that any apparent symmetry may be illusory.

42. M. L. Martin, R. Mantione, and G. J. Martin, Tetrahedron Lett. 4809 (1967).
42a. F. Heinrich and W. Lüttke, Angew. Chem., Int. Ed. Engl., 11, 234 (1972).

For the terphenyl (69), we should realize that the *o*-substituents are of such size that none of the rings can rotate with respect to the others. Thus, the outer rings lie in a common plane, and they are so substituted that *cis*- and *trans*-isomers are possible. The trans- form (70) has a center of *symmetry and* will be optically inactive; the cis-isomer lacks the necessary symmetry and will be an enantiomeric pair. There will be three stereoisomers.

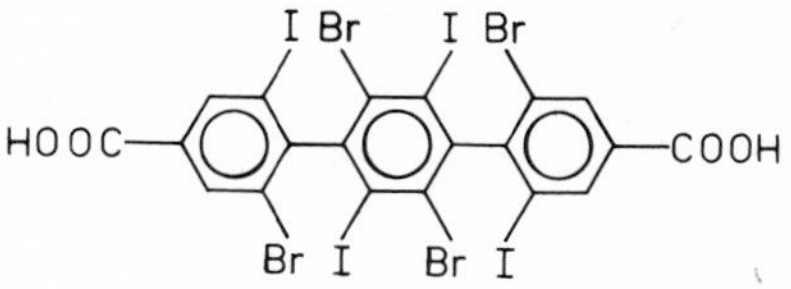

(69)

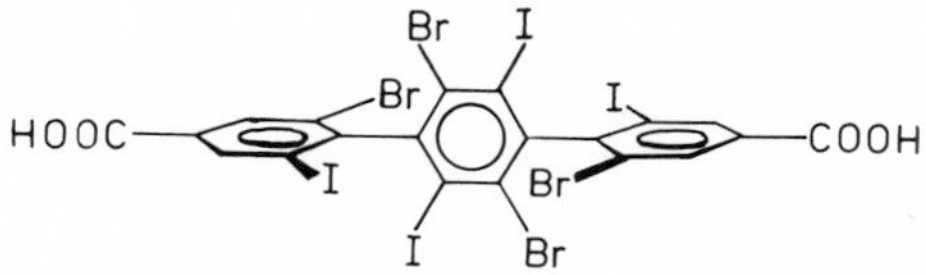

(70)

The diacid (71) has basically a planar carbon framework. *Cis-trans*-isomerism is possible about the double bond; furthermore, *cis-trans*-isomerism is involved in another way since the carboxy groups may be on the same or opposite sides of the basic plane.

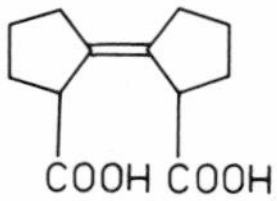

(71)

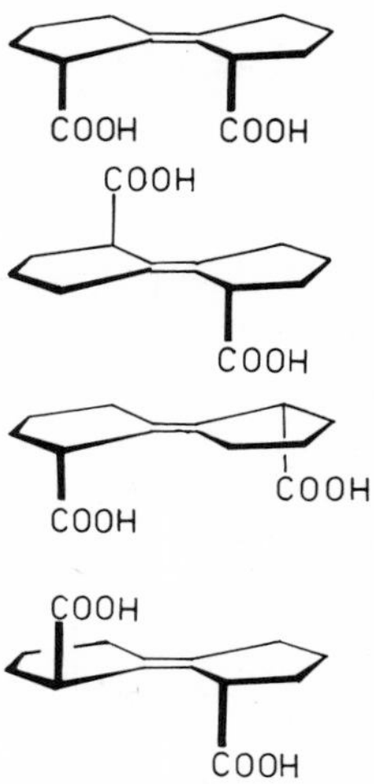

The cis-cis- and trans-trans-isomers are mesoforms; the cis-trans- and trans-cis-isomers are both enantiomeric pairs. Six stereoisomers are therefore capable of existence for this structure.

The diacid (72), if viewed in Newman projection, basically occurs in three forms: syn-syn, anti-anti, and syn-anti. All three lack reflection symmetry and hence can exist as enantiomeric pairs (the anti-syn-isomer is identical with the syn-anti), hence there are six stereoisomers.

HOOC— —COOH

(72)

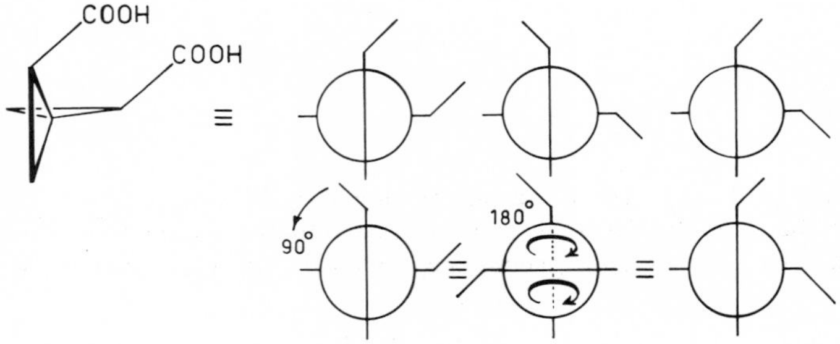

COOH COOH ≡

90° ≡ 180° ≡

The carbon framework of the triene (73) is entirely planar; only cis- and trans-isomers can occur.

$CH_3CH{=}C{=}C{=}CHCH_3$

(73)

As in the helicenes, the pairs of methyl groups in greatest proximity in (74) will be opposite, out of the plane of the aromatic nucleus.

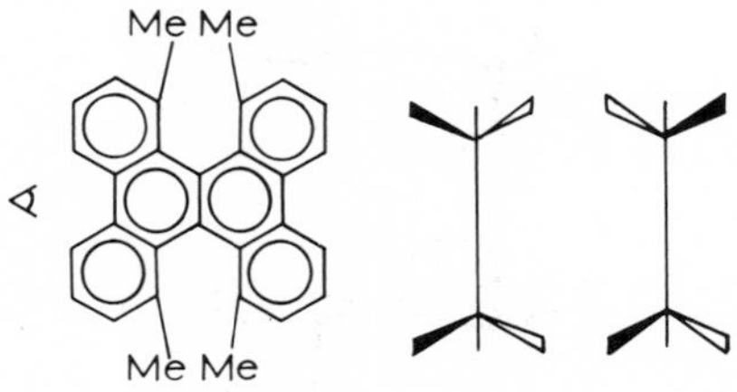

Me Me

Me Me

(74) 1 2

The cis form (1) has both a plane and center of symmetry; the trans form (2) cannot be superimposed on its mirror image. There are three stereoisomers.

We note that in the bicyclic bromide (75) that the six-membered ring is in the endo-position, that the common edge is cis and that the bromine is anti.

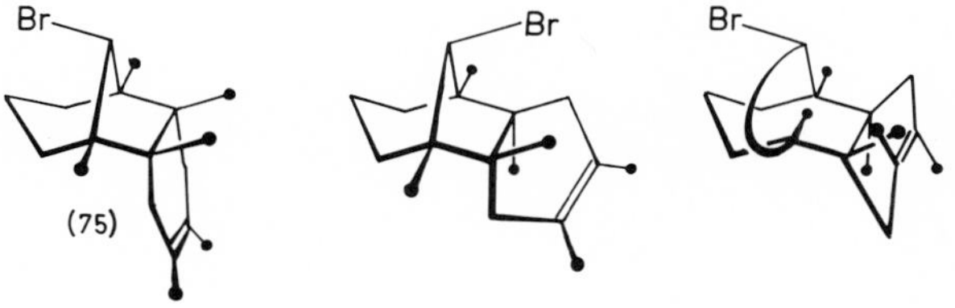

Br Br Br

(75)

There will be an isomer with the bromine syn, and there will be another syn-anti pair with the ring in the exo-position; all of these isomers have a plane of symmetry. There is another syn-anti pair (actually a pair of enantiomeric pairs) with a trans-ring junction; hence here we have eight stereoisomers. Strictly speaking, we might also consider stereoisomers with one or both of the bridgehead hydrogen atoms pointing in instead of out, or with the double bond trans- rather than cis-, but at this point we have undoubtedly exceeded the bounds of strain that real molecules will put up with. In very large rings such strains are not too great, and the in-configuration may be able to compete with out. Thus, Simmons [43] has found by nmr that the bis-protonated 1,10-diazabicyclo[8.8.8]hexacosane (76) occurs in both the out-out and in-in forms, and that these forms slowly equilibrate and that the latter is actually favored by a factor of 100.

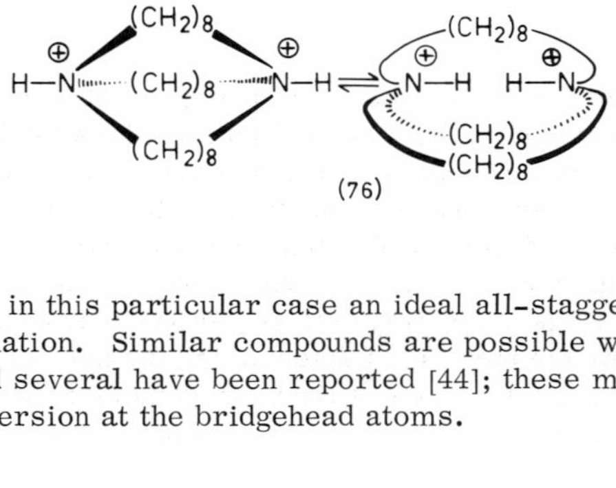

$(CH_2)_8$ ⊕ ⊕ H—N $(CH_2)_8$ N—H ⇌ N—H H—N $(CH_2)_8$ $(CH_2)_8$ $(CH_2)_8$ $(CH_2)_8$

(76)

Models show that in this particular case an ideal all-staggered conformation favors the in-in conformation. Similar compounds are possible with the isoelectronic hydrocarbons and several have been reported [44]; these molecules are of course not subject to inversion at the bridgehead atoms.

43. For review and early references to these species, see H.E. Simmons, C.H. Park, R.T. Uyeda, and M.F. Habibi, Trans. N.Y. Acad. Sci., 32, 521 (1970).
44. (a) P.G. Gassman and R.P. Thummel, J. Amer. Chem. Soc., 94, 7183 (1972); (b) C.H. Park and H.E. Simmons, J. Amer. Chem. Soc., 94, 7184 (1972).

Chapter 6

CONFORMATIONAL ANALYSIS

6-1. INTRODUCTION

In Chapters 1, 4, and 5 we touched several times on the topic of the rotation about bonds. It was convenient on those occasions to make simple assumptions about such conformational processes, namely, that groups connected by means of a single bond may rotate freely at room temperature, that there is no such freedom for groups connected by double bonds, that singly bound carbon atoms never invert unless one of the four bonds is first broken, that nitrogen atoms are able to invert directly, and that rings containing only single bonds around the periphery may be considered flat for the purpose of determining the number of stereoisomers. At this point we need to recount the work done by organic chemists in their efforts to find to what limits these assumptions may be pushed.

6-2. ROTATION ABOUT SINGLE BONDS

We have already run into one instance in which rotation about a single bond is so hindered that it cannot occur at room temperature, viz., the biphenyls bearing large *o*-substituents. As a consequence, the rings in such molecules take up positions in two perpendicular planes, and if the substitution is of such a nature that neither is a plane of symmetry, then stereoisomers may result. In retrospect, it is not surprising that the first example of this type of conformational isomerism to be discovered should have involved biphenyls; in the planar state the o- and o'-substituents are in close proximity and point toward one another for maximum hindrance. As we shall see most, though not all, other known examples of restricted rotation also involve benzene rings.

Ackerman and co-workers [1] discovered that their synthesis of 5-amino-2,4,6-triiodo-N,N,N',N'-tetramethylisophthalamide (1) led to a mixture of two isomeric products which could be separated by means of thin-layer chromatography.

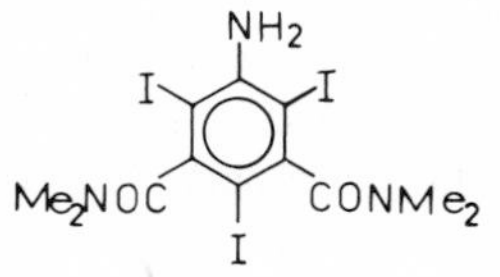

NH_2 I I Me_2NOC $CONMe_2$ I

(1)

These materials had virtually identical ir, uv, and nmr spectra but slightly different melting points. The authors ascribed this observation to the occurrence of rotamers or conformers (2) and (3).

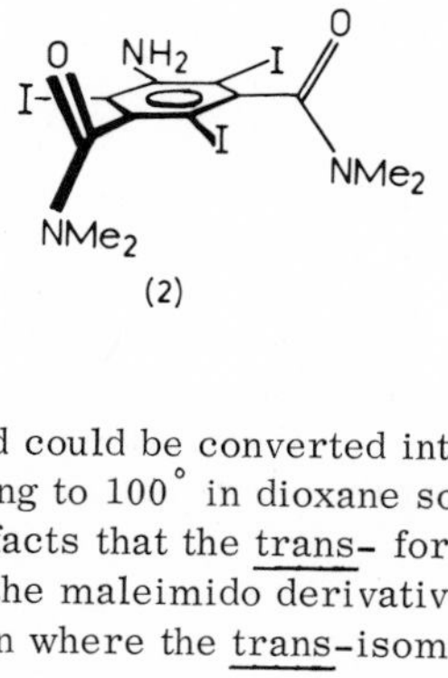

O NH_2 I O I I NMe_2 NMe_2

(2)

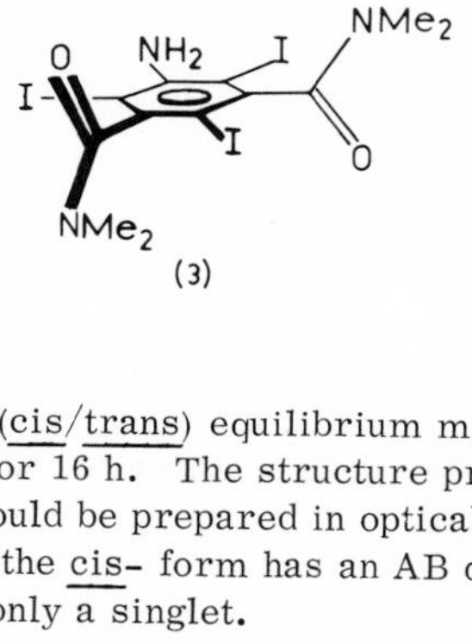

O NH_2 I NMe_2 I I O NMe_2

(3)

Each compound could be converted into a 1:5 (cis/trans) equilibrium mixture by means of heating to 100° in dioxane solution for 16 h. The structure proof depended on the facts that the trans- form (3) could be prepared in optically active form and that the maleimido derivative (4) of the cis- form has an AB quartet in the vinyl region where the trans-isomer has only a singlet.

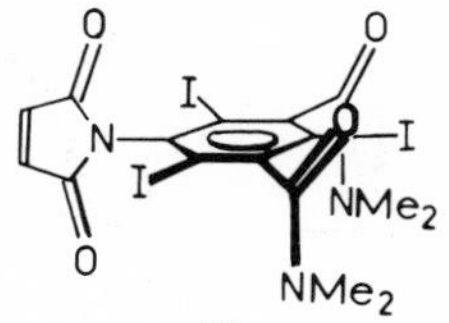

O O I N I I O NMe_2 NMe_2

(4)

1. (a) J. H. Ackerman, G. M. Laidlaw, and G. A. Snyder, Tetrahedron Lett., 3879 (1969); (b) J. H. Ackerman and G. M. Laidlaw, Tetrahedron Lett., 4487 (1969). Several other closely related cases are referred to in those papers.

When the *cis*-isomer was converted into the oxalanilic acid derivative, the product, in turn, was found to consist of a pair of rotational isomers (5) and (6) [2].

(5) (6)

Restricted rotation can usually be found by means of nmr, even when it does not lead to stereoisomerism [3]. Thus, di-*t*-butyl-*p*-methoxyphenylcarbinol (7) has a complex signal in the aromatic region at room temperature, but this part of the spectrum coalesces on heating and finally resolves into a simple A_2B_2 pattern at 175° [4].

(7)

Restricted rotation about the benzylic carbon-carbon bond is the best explanation for this observation. Another interesting example of restricted rotation is *o*-dineopentyltetramethylbenzene (8) [5]. Because of their size, the two *t*-butyl groups must be on opposite sides of the ring at all times. The substance therefore exists as an enantiomeric pair, and accordingly the methylene protons show up as a doublet in the nmr spectrum at room temperature. A concerted switch is evidently possible; coalescence occurs at elevated temperatures, and at 100° the methylene signal is sharp.

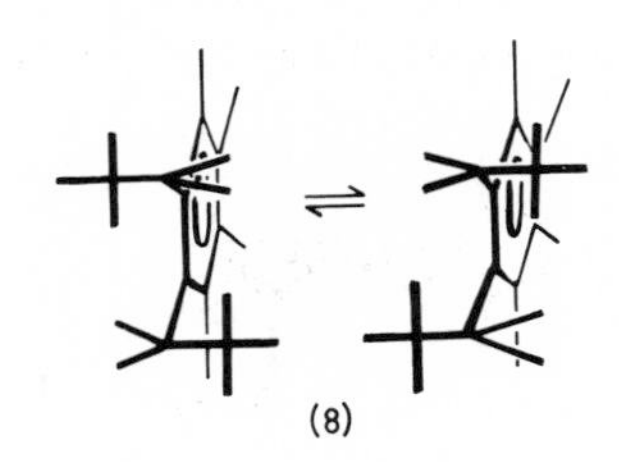

(8)

---

2. J.H. Ackerman and G.M. Laidlaw, *Tetrahedron Lett.*, 2381 (1970).
3. See for review, H. Kessler, *Angew. Chem., Int. Ed. Engl.*, *9*, 219 (1970).
4. G.P. Newsoroff and S. Sternhell, *Tetrahedron Lett.*, 2539 (1967).
5. D.T. Dix, G. Fraenkel, H.A. Karnes, and M.S. Newman, *Tetrahedron Lett.*, 517 (1966).

Even groups smaller than *t*-butyl or the iodine atom may interfere with free rotations. Hexaisopropylbenzene has been prepared [6]; it is clear from models that the isopropyl groups cannot rotate. There are as yet no cases of stereoisomerism because of this; however, the strain in this compound is evident from the fact that the Me_2C-H bond appears in the ir at 3070 cm^{-1} instead of the normal 2900 cm^{-1}. Hexaisopropenylbenzene (9) [7] likewise has unique properties (such as its inability to add bromine) that make it seem likely that rotation is impossible and that various stereoisomers should exist; however, to date no reports have appeared on the configuration of this substance.

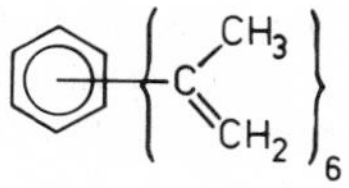

(9)

Rotation is clearly inhibited in the trifluoromethyl group of *o*-trifluoromethylnitrobenzene (10). In the radical anions of the *m*- and *p*-isomers, 1:3:3:1 splitting of the esr signal by the fluorines occurs at room temperature; in the *o*-position however, the trifluoromethyl group causes splitting into a triplet of doublets [8], so that the rotation of this group cannot be free.

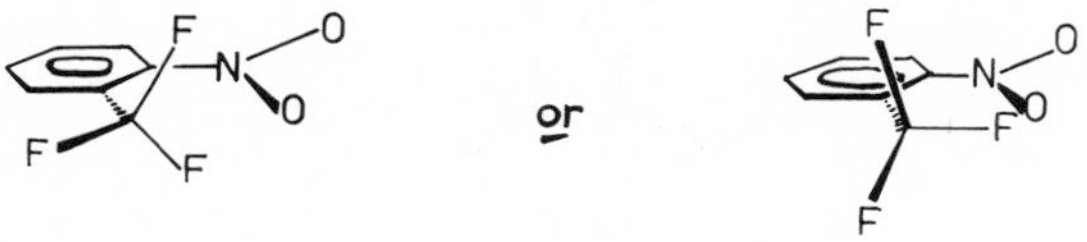

(10)

The sharpness of the signals allows one to estimate that the half-life of the stable conformer must be at least 10^{-6} sec at room temperature. Temperature-dependent nmr has shown that a barrier of 16 kcal/mole inhibits the rotation of the CH_2Cl group in the chloromethyltriptycene (11) [9].

---

6. E.M. Arnett and J.M. Bollinger, *J. Amer. Chem. Soc.*, *86*, 4729 (1964).
7. (a) E.M. Arnett, J.M. Bollinger, and J.C. Sanda, *J. Amer. Chem. Soc.*, *87*, 2050 (1965); (b) J.M. Bollinger, J.J. Burke, and E.M. Arnett, *J. Org. Chem.*, *31*, 1310 (1966).
8. E.G. Janzen and J.L. Gerlock, *J. Amer. Chem. Soc.*, *89*, 4902 (1967).
9. N.M. Sergeyev, K.F. Abdulla, and V.R. Skvarchenko, *Chem. Commun.*, 368 (1972).

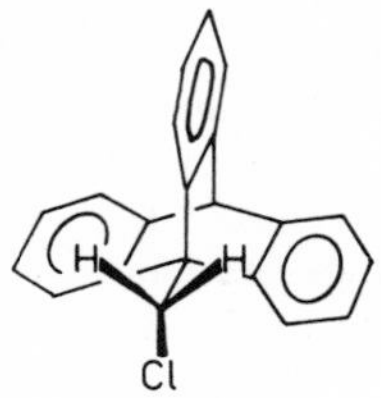

(11)

Even o-methyl groups may interfere sufficiently to inhibit rotation. The heat capacity $\overline{Cp}$ of hexamethylbenzene has been measured from 13 to 340°K, and the discrepancy of the absolute entropy so determined from the value calculated on the basis of an assumption of freely rotating methyl groups has been interpreted on the basis of a barrier to such rotation of 3-8 kcal/mole [10].

There are also examples of restricted rotation in single bonds other than those between carbon atoms. Thus, the nmr spectrum of the borane (12) has a single t-butyl signal at all temperatures, but the signals due to the o-protons and the m-methyl groups begin to be resolved into doublets at about -30°; one can calculate that the free energy of activation for the rotation of this ring is about 12 kcal/mole. The preferred configuration of this molecule must therefore be as shown [11].

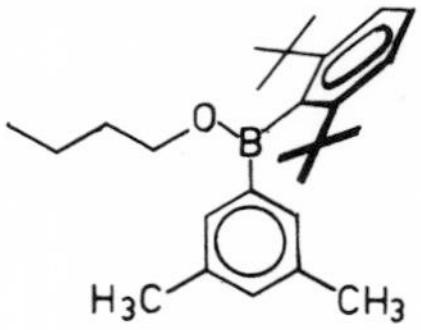

(12)

It is not fundamentally necessary that one of the two groups connected by the nonrotating single bond be a benzene ring. Thus, the t-butyl group in 2-benzyl-2-chloro-3,3-dimethylbutane (13) is at room temperature represented by a singlet signal in the nmr; broadening of this signal sets in at -20°, and at -90° a 1:1:1 triplet is observed [12].

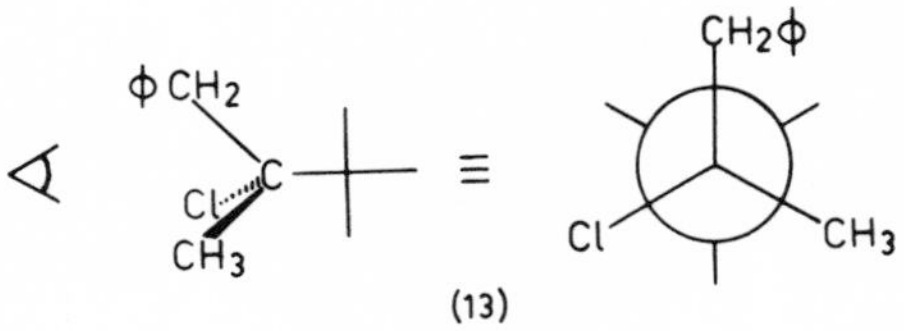

(13)

10. M. Frankosky and J.G. Aston, J. Phys, Chem., 69, 3126 (1965).
11. B. Meissner and H.A. Staab, Ann., 753, 92 (1971).
12. J.E. Anderson and H. Pearson, J. Chem. Soc. (B), 1209 (1971); other examples are referred to in this paper.

Furthermore, 2:1 doublets are usually obtained in the low-temperature nmr spectra of t-butylcycloalkanes (14) [13], which is indicative of restricted rotation. Similarly, t-butyldimethylamine (15) which has two sharp signals in a 3:2 ratio at -100° has three peaks in a 2:2:1 ratio at -166° [14]; $\Delta G^{\ddagger}$ is about 6 kcal/mole in this case.

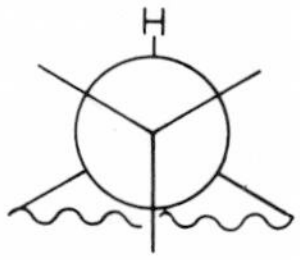

$(CH_3)_2N$—

(14) 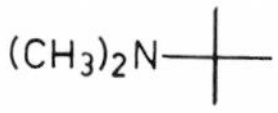 (15)

There is a second major reason why single bonds can sometimes rotate only with difficulty: Resonance may lead to sufficient double bond character to raise the activation energy. Thus, N-benzyl-N-nitroso-2,6-dimethylaniline (16) has been shown to exist as a pair of stereoisomers that can be separated by means of thin-layer chromatography; they interconvert fairly readily at room temperature ($\Delta G^{\ddagger}$ = 23 kcal/mole) [15].

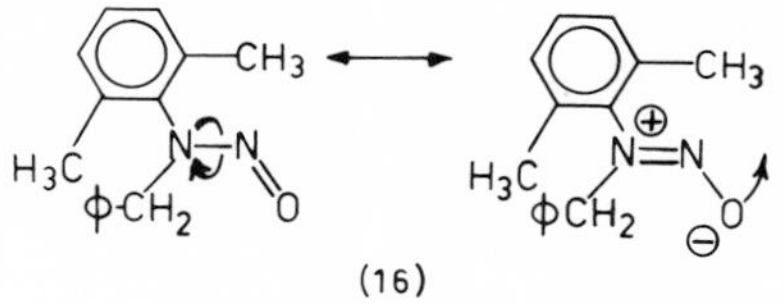

(16)

The authors assume that this is a case of stereoisomerism due to a stiffened single bond. It is possible of course (as in several other instances cited below) that the isomerization occurs via inversion rather than by rotation.

There are numerous cases in which similar resonance hybrids have bonds of sufficient double bond character to permit the freezing out of rotation, although stable stereoisomers can usually not be isolated. One of the better known examples are the amides; thus, dimethylformamide (17) has two methyl nmr signals that do not coalesce until the temperature is raised to 110° or so [16].

---

13. F.A.L. Anet, M.S. Jacques, and G.N. Chmurny, J. Amer. Chem. Soc., 90, 5243 (1968).
14. C.H. Bushweller, J.W. O'Neil, and H.S. Bilofsky, J. Amer. Chem. Soc., 92, 6349 (1970); Tetrahedron, 27, 5761 (1971).
15. (a) A. Mannschreck, H. Muensch, and A. Mattheus, Angew. Chem. Int. Ed. Engl., 5, 728 (1966); (b) A. Mannschreck and H. Muensch, Tetrahedron Lett., 3227 (1968).
16. See for example, J.R. Dyer, "Applications of Absorption Spectroscopy of Organic Compounds", Prentice-Hall, Englewood Cliffs, N.J., 1965, p. 114.

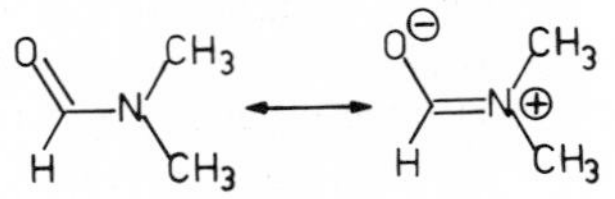

(17)

Even in vinyl ethers evidence for this type of inhibited rotation is easily found. Methyl vinyl ether (18) has two C=C stretching frequencies;

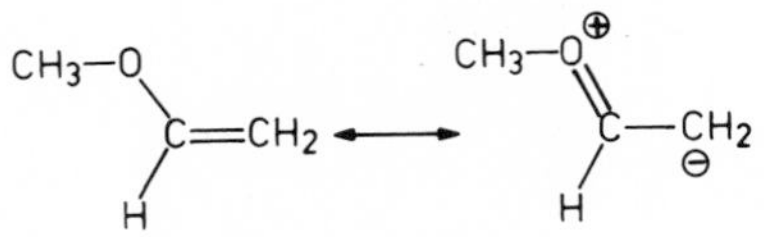

(18)

the ratio of their intensities is temperature dependent [17] and hence we presumably are observing the equilibrium.

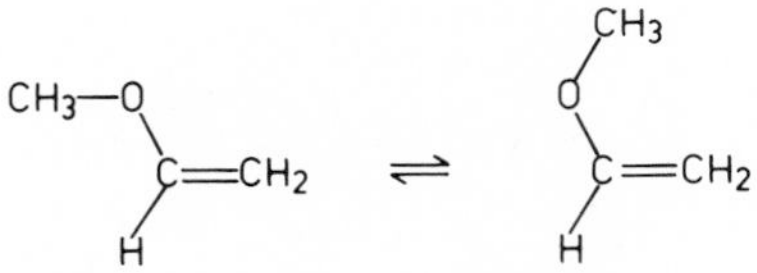

The temperature dependence of the spectrum permits the calculation of ΔH (1.5 kcal/mole); oddly enough, the cis-conformer is the more stable (see also p. 185). Double bond character need not result from adjacent multiple bonds, but may occur in bonds flanked by highly strained small ring systems. As described in Chapter 2, p. 62, there are reasons to believe that the bonds holding substituents to cyclopropyl rings have more than 25% s character. The ring bonds themselves have more p character, and hence bicyclopropyls (19) resemble conjugated olefins to some extent.

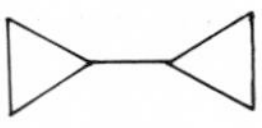

(19)

A spectacular example of this effect was found by Moore and Costin [18], who observed that 2, 2, 3, 2', 2', 3'-hexamethylbis-(1-bicyclo[1.1.0]butyl), which is a

---

17. N. L. Owen and N. Sheppard, Proc. Chem. Soc., 264 (1963).
18. W. R. Moore and C. R. Costin, J. Amer. Chem. Soc., 93, 4910 (1971).

mixture of d, l and meso-isomers (20) and (21), has conjugation between the two bicyclic ring systems as clearly demonstrated by the fact that these substances have a much more intense and much redder (longer wavelength) absorption than 1, 2, 2, 3-tetramethylbicyclo[1. 1. 0]butane (22).

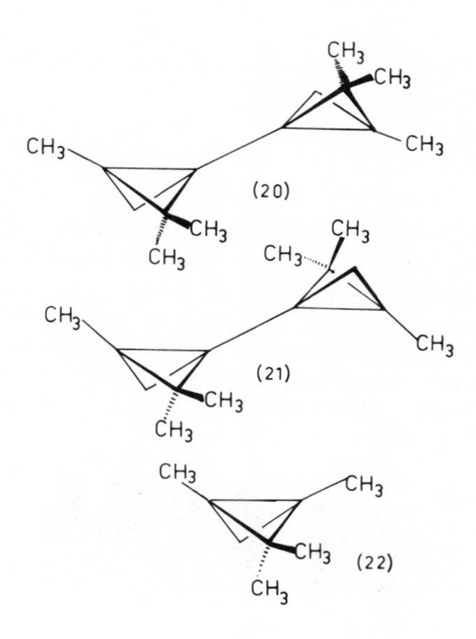

(20)

(21)

(22)

6-3. ROTATION ABOUT DOUBLE BONDS

There are a number of ways in which cis- and trans-olefin isomers can be made to interconvert: iodine catalysis, photochemical treatment, etc. We are concerned here with a different phenomenon. In some instances the double bond is so weak that no external agent is required to bring about rotation. One category of such olefins was encountered when organic chemists attempted to extend the geometric isomerism of olefins and the optical isomerism of allenes to longer cumulative systems. It has been found that 1, 4-diphenyl-1, 4-di-t-butylbutatriene can be separated into the cis and trans forms (23) and (24), respectively [19];

C=C=C=C (23)

C=C=C=C (24)

---

19. R. Kuhn and B. Schulz, Chem. Ber., 98, 3218 (1965).

however, the corresponding pentaenes (25) and (26) evidently rapidly interconvert [20]. Thus, while two methyl signals are observed in perdeuteriodimethylsulfoxide, the signals coalesce at 100°.

C=C=C=C=C=C

(25)

C=C=C=C=C=C

(26)

The free energy of activation is 20 kcal/mole (that for the interconversion of (23) and (24) is about 30 kcal/mole; in both cases the equilibrium constants are about equal to 1). It is actually not surprising that it should become less and less difficult to isomerize such cumulenes as they get longer; whereas the transition state for isomerization in 2-butene requires the complete breakage of a π bond, the same state (27) in a hexapentaene may require only a minor twist so far as each pair of neighboring carbon atoms is concerned.

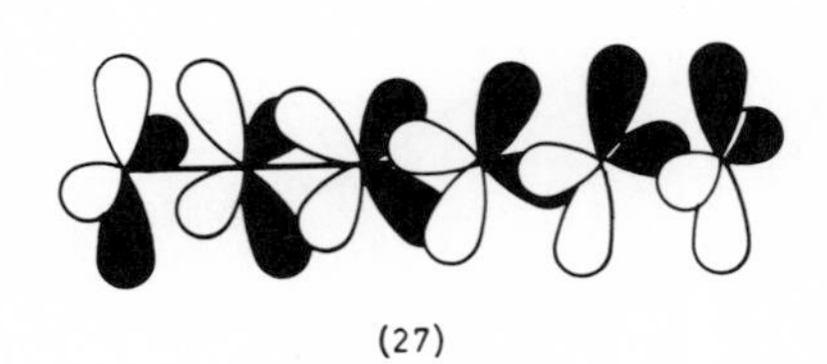

(27)

This phenomenon appears also in long chains of conjugated double bonds; for example, in the nmr spectrum of 3,3',5,5'-tetra-*t*-butyl-2,4'-diphenoquinone (28) which shows two AB quartets in the vinyl region at room temperature, one of these collapses to a singlet at 155° while the other remains unaffected [21]. The activation energy was determined to be 21 kcal/mole; evidently rotation about the central double bond is relatively facile.

(28)

20. R. Kuhn, B. Schulz, and J.C. Jochims, *Angew. Chem., Int. Ed. Engl.*, **5**, 420 (1966).
21. H. Kessler and A. Rieker, *Tetrahedron Lett.*, 5257 (1966).

Just as single bonds sometimes have considerable double bond character because of resonance, double bonds may have single bond character for the same reason. Thus, 1-nitro-2, 2-bis[thiomethyl]ethylene (29) has only one methyl signal in the nmr spectrum at room temperature, and resolution into a pair of such signals only begins at 0° ($\Delta G^{\ddagger} = 15$ kcal/mole) [22].

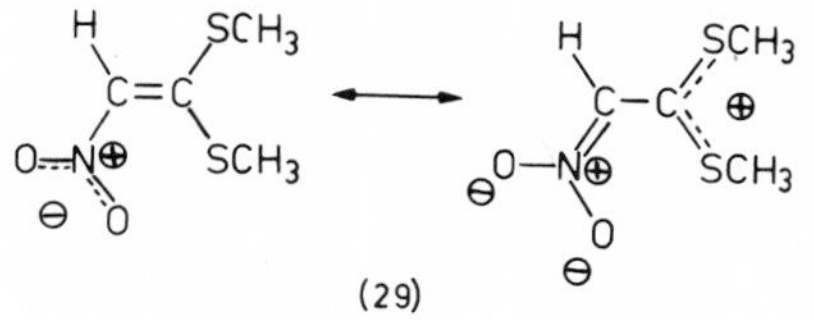

(29)

An even lower barrier was found for the cyclic aminal (30), for which separate methyl or methylene signals could not be found even at -65°.

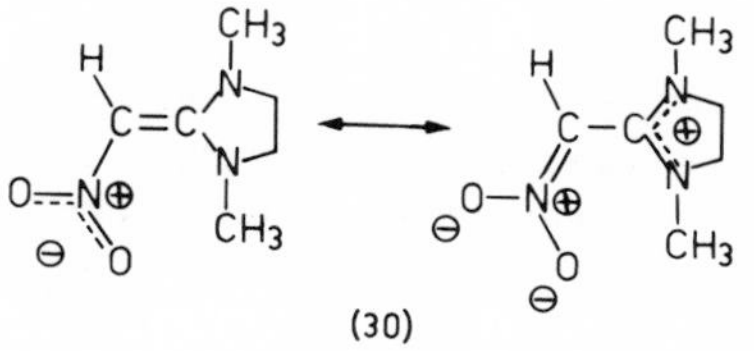

(30)

It may be noted that in each of these examples as a double bond is weakened by resonance, an adjacent single bond is presumably stiffened. Evidence to this effect was found by Taylor [23] in β-aminoacrylate esters of the type (31).

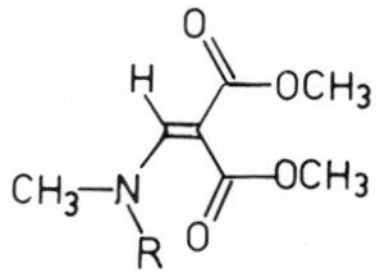

(31)

These compounds have a barrier to rotation of the double bond of 12-22 kcal/mole, as demonstrated by the nmr nonequivalence of the two carbomethoxy groups; $\Delta G^{\ddagger}$ is 12 kcal/mole when R is p-anisyl and 22 kcal/mole when R is p-nitrophenyl. When R is methyl, the two aminomethyl groups are not nmr equivalent at low temperature either; for the C-N rotation, $\Delta G^{\ddagger}$ is 13 kcal/mole.

---

22. G. Isaksson, J. Sandström, and I. Wennerbeck, *Tetrahedron Lett.*, 2233 (1967); several other similar reports are quoted in that paper.
23. Y. Shvo, E.C. Taylor, and J. Bartulin, *Tetrahedron Lett.*, 3259 (1967).

A further extension is possible: The two structures written (one doubly bonded, the other singly) need not be resonance structures, but they may be valence tautomers (see Chapter 11). In other words, the effect is the same whether the real molecule is a hybrid of the two structures or whether we have two rapidly equilibrating molecules. A situation that comes close to the latter type has been described [24]: The silaketene acetals of the type (32) have nmr spectra in which all methoxy groups are equivalent down to at least -60°. The scheme below shows how this is possible.

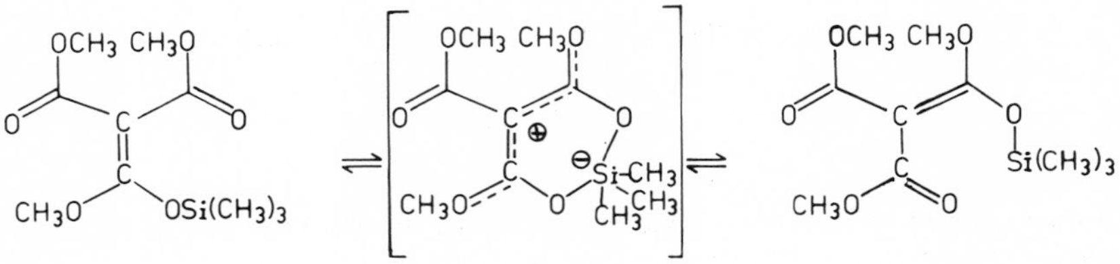

(32)

Two features are necessary in order to explain this result. First of all, since silicon has relatively low-lying d-orbitals and hence can easily expand its valence shell to pentacovalency, the transition state for the 1,5-shift shown is of low energy; however, this renders only two of the three methoxy groups equivalent. The third also loses its identity, because of rotation facilitated by the resonance suggested by structure (33).

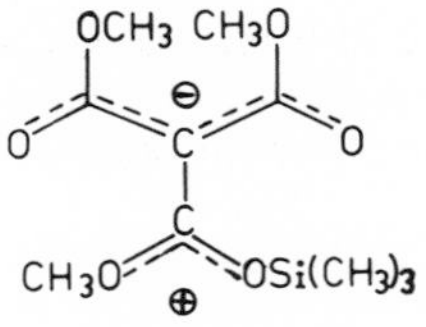

(33)

6-4. INVERSION AT NITROGEN [3]

In Chapter 5, p. 157, mention was made of the fact that the inversion rate of amines is so high that resolution of racemic mixtures of optically active amines can ordinarily not be achieved. We shall now review the evidence bearing on the question of what if anything can be done about this inversion rate.

---

24. Y. N. Kuo, F. Chen, and C. Ainsworth, Chem. Commun., 137 (1971).

The method of choice is nmr in the measurement of inversion rates. Thus, in asymmetric tertiary amines in which one of the substituents is a benzyl group, the AB-methylene quartet will coalesce at temperatures where inversion is sufficiently rapid [25]. If the inversion is still too rapid at low temperatures, it can also be effectively slowed down by the use of solutions sufficiently acidic so that a large fraction is in the form of an ammonium ion which cannot invert at all [26].

By means of such experiments it has been learned that five factors largely determine the inversion rate. First of all, if one or more of the substituents is hydrogen, the rates are too fast to be seen by nmr. In the gas phase, the rapid inversion has been ascribed to tunneling [27]; in solution, it is complicated by proton exchange. Second, in tertiary amines, the size of the substituents is of some importance: They are expected to be crowded together more severely in the pyramidal state than in the planar transition state. Hence amines with larger groups invert somewhat more rapidly then those with smaller substituents - as long as these are of the same electronic type. However, this effect is minor and not well documented.

The third effect was actually suggested long before it was found: Small cyclic tertiary amines should invert more slowly, since the transition state will have all three bond angles at the nitrogen atom deviating greatly from their normal values (the pyramidal state can at least have two normal bond angles) [28]. In support, there are innumerable cases of stable geometric (syn-anti) isomers of oximes in which the nitrogen atom might be considered part of a two-membered ring. The facts seem to bear out this proposal; thus, from the temperature dependence of the nmr spectrum of 1, 2, 2-trimethylaziridine (34) the barrier to inversion has been calculated to be 20 kcal/mole; by comparison, that for N-methylpyrrolidine (35) equals only 7 kcal/mole [30].

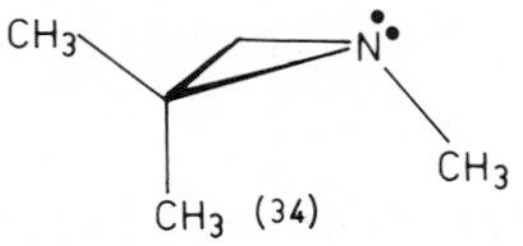

CH_3 CH_3 CH_3 (34) CH_3 (35)

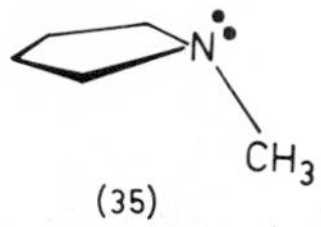

---

25. H.S. Gutowsky, J. Chem. Phys., 37, 2196 (1962).
26. M. Saunders and F. Yamada, J. Amer. Chem. Soc., 85, 1882 (1963).
27. W. Gordy, W.V. Smith, and R.F. Trambarulo, "Microwave Spectroscopy", Wiley, New York, 1953.
28. J.F. Kincaid and F.C. Henriques, J. Amer. Chem. Soc., 62, 1474 (1940).
29. A. Loewenstein, J.F. Neumer, and J.D. Roberts, J. Amer. Chem. Soc., 82, 3599 (1960). Very shortly later 2, 2, 3, 3-tetramethylaziridine was reported to have only a single, sharp methyl signal (G. L. Closs and S. J. Brois, J. Amer. Chem. Soc., 82, 6068 (1960)); however, it was later found that a doublet is indeed present which coalesces at 40°, but at much lower temperatures if a trace of water is present (T. J. Bardos, C. Szantay, and C.K. Navada, J. Amer. Chem. Soc., 87, 5796 (1965)). This points up one of the difficulties in nmr measurements of amine inversion rates. See also the discussion of piperidine inversions, p. 231.
30. J.B. Lambert and W. L. Oliver, J. Amer. Chem. Soc., 91, 7774 (1969).

When one of the substituents bound to the nitrogen atom tends to be doubly bound because of resonance, that obviously reduces the barrier to inversion. In fact, amides are as a rule characterized by a planar nitrogen atom. It is not yet clear whether small bond angles at amidic nitrogen atoms will render it nonplanar again. No temperature dependence has been noted in the nmr spectrum of N-acetylaziridine (36), not even at the lowest temperatures examined [31] (see also p. 220).

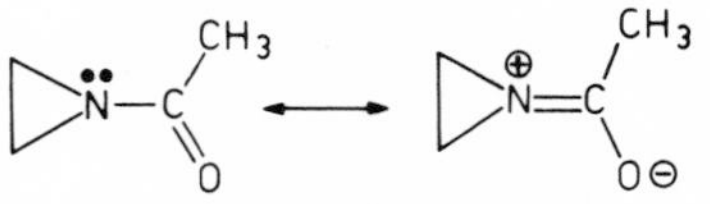

(36)

A peculiar effect as yet not well understood is the so-called "hetero effect"; it is suggested by laboratory observations that substitution by atoms other than hydrogen and carbon tend to lower the inversion rates. Thus, in sharp contradiction to earlier predictions and results [32], Brois discovered that N-chloro- and N-bromo-2,2-dimethylaziridine and -2,2,3,3-tetramethylaziridine had clearly nmr-nonequivalent geminal substituents at room temperature, even at 120°; he proceeded by actually separating cis- and trans-N-chloro-2-methylaziridine, (37) and (38), respectively [34]. This was the first case of a non-bridgehead amine stable to inversion at room temperature.

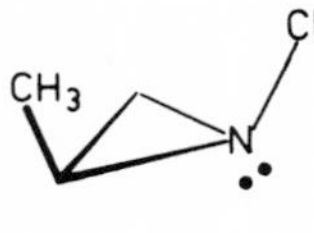

(37) 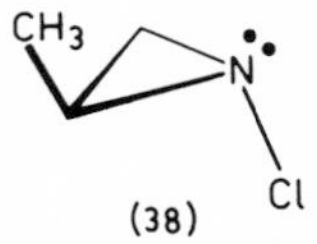 (38)

Almost simultaneously exo- and endo-7-chloro-7-azabicyclo[4.1.0]heptane (39) and (40) were separated [35].

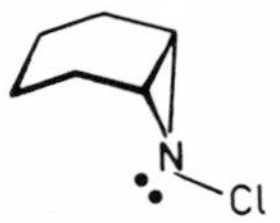

(39) 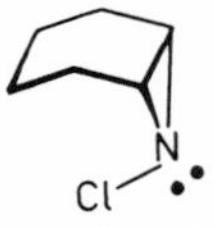 (40)

31. F. A. L. Anet and J. M. Osyany, J. Amer. Chem. Soc., 89, 352 (1967).
32. (a) F. A. L. Anet, R. D. Trepka, and D. J. Cram, J. Amer. Chem. Soc., 89, 357 (1967); (b) D. L. Griffith and J. D. Roberts, J. Amer. Chem. Soc., 87, 4089 (1965).
33. S. J. Brois, J. Amer. Chem. Soc., 90, 506 (1968).
34. S. J. Brois, J. Amer. Chem. Soc., 90, 508 (1968).
35. D. Felix and A. Eschenmoser, Angew. Chem., Int. Ed. Engl., 7, 224 (1968).

N-amino-aziridines are evidently also stable to inversion [36], and further cases are now known [3, 37]. If two heteroatoms are present, even the three-membered ring is no longer necessary: The epimeric N-methoxy-3, 3-dicarbomethoxy-5-cyano-1, 2-oxazolidines (41) and (42) can be separated; the activation energy of interconversion is about 30 kcal/mole [38].

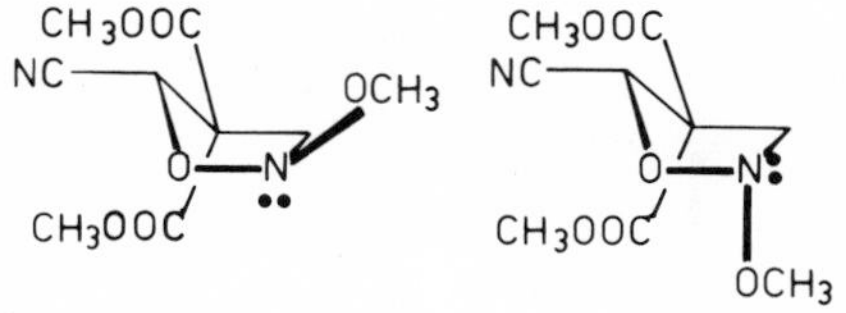

(41) (42)

In some cases, optically active compounds have now been isolated in which a pyramidal nitrogen atom is the chiral center [39]; an example is 2-methyl-3, 3-diphenyloxaziridine (43).

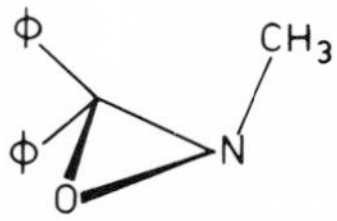

(43)

The nature of the heteroatom effect is not yet understood. It is in fact likely that more than one effect is operating. Thus, N-silylamines and several N-sulfuramines are actually planar [40]. This might be due to the availability of low-lying d-orbitals in these atoms, as in (44); however, if this feature is also important in sulfuramines, it would be difficult to explain the following observation.

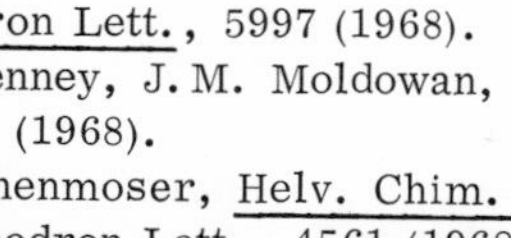

(44)

---

36. S. J. Brois, Tetrahedron Lett., 5997 (1968).
37. M. Raban, G. W. J. Kenney, J. M. Moldowan, and F. B. Jones, J. Amer. Chem. Soc., 90, 2985 (1968).
38. K. Müller and A. Eschenmoser, Helv. Chim. Acta, 52, 1823 (1969).
39. (a) D. R. Boyd, Tetrahedron Lett., 4561 (1968); (b) F. Montanari, I. Moretti, and G. Torre, Chem. Commun., 1694 (1968); (c) R. Annunziata, R. Fornasier, and F. Montanari, Chem. Commun., 1133 (1972).
40. For review, see A. Rauk, L. C. Allen, and K. Mislow, Angew. Chem., Int. Ed. Engl., 9, 400 (1970).

When aziridines (45) are examined in the nmr, coalescence of the methylene group occurs at low temperature as an indication of inversion; but the barrier is completely insensitive to the nature of the group X. Both nitro and methoxy groups lead to a value of 12.5 kcal/mole [41].

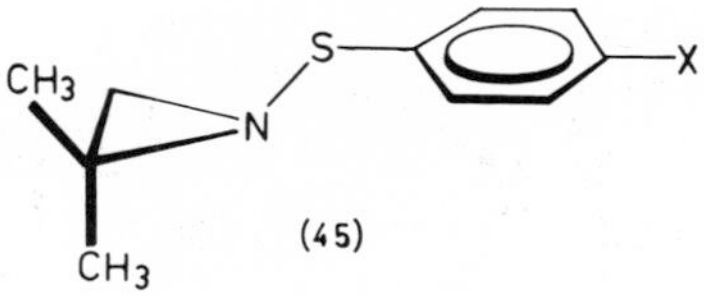

(45)

At any rate, whatever effect makes silyl amines planar, it is evidently opposed by a second factor that increasingly takes hold as the heteroatom considered becomes more electronegative ("up" and "to the right" in the periodic table). This is at present best explained by the statement that the bonds to the nitrogen atom have maximum s character in the planar state, and since the s electrons are most tightly held, an electronegative substituent is least likely to tolerate that state.

In oximes (46) stability to inversion is the rule rather than the exception, as mentioned on p. 179. It is not clear whether the hetero effect is operating in this case.

(46)

Thus, the guanidines (47) have inversion barriers of only 10-12 kcal/mole as determined by means of temperature-dependent nmr spectra [42].

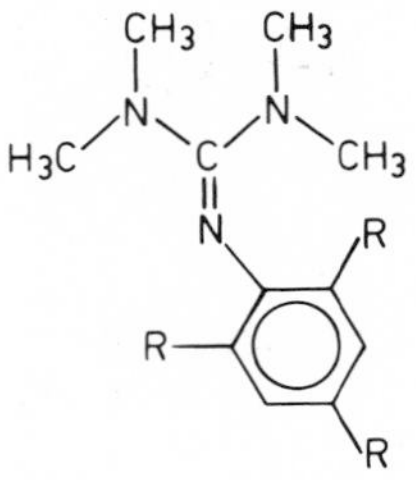

(47)

41. (a) D. Kost, W.A. Stacer, and M. Raban, J. Amer. Chem. Soc., 94, 3233 (1972); (b) M. Raban and D. Kost, J. Amer. Chem. Soc., 94, 3234 (1972).
42. H. Kessler and D. Leibfritz, Tetrahedron, 25, 5127 (1969).

That inversion of the nitrogen atom rather than rotation of the C=N double bond is involved is strongly suggested by the fact that the rates slightly increase as the size of R is increased from hydrogen to *i*-propyl; when the N-methylguanidinium salts (48) are examined [43], the activation energies rise from 15 to 24 kcal/mole between hydrogen and *i*-propyl. In (48) rotation is the only possible mode.

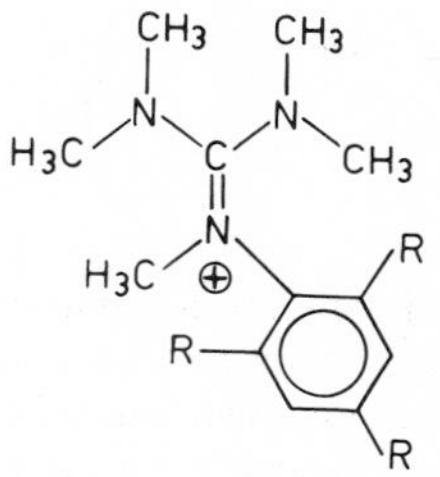

(48)

It does seem likely however that the facile isomerization of the N-phenylimines is in part due to resonance contributions from structures such as (49). N-alkylimines (50) are stable to isomerization to 180°.

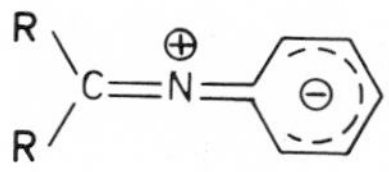

(49) 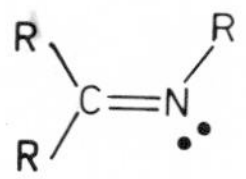 (50)

These studies are further complicated by the fact that tautomerization (see Chapter 12) to enamines (51) is possible if one of the substituents at carbon bears a hydrogen atom and that this enamine is responsible for relatively facile isomerization [44].

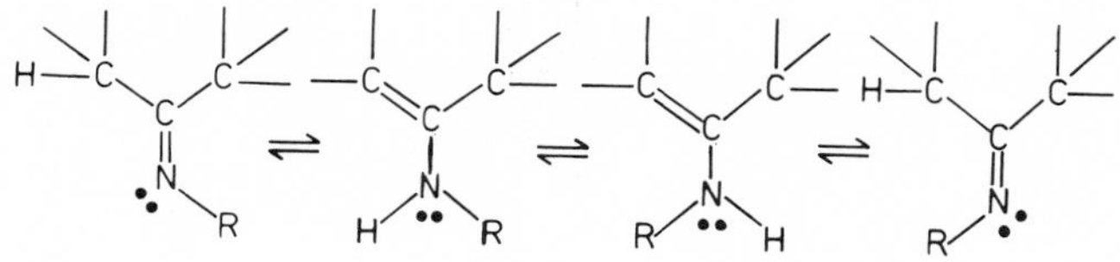

(51)

---

43. H. Kessler and D. Leibfritz, *Tetrahedron Lett.*, 427 (1969).
44. W. B. Jennings and D. R. Boyd, *J. Amer. Chem. Soc.*, *94*, 7187 (1972).

Unlike allenes, carbodiimides (52) cannot be resolved, so that they presumably have rapidly inverting nitrogen atoms or rotating C=N bonds. From the fact that diisopropylcarbodiimide has a pair of 1:2:1 triplets below -150° which coalesce above that temperature, it has been deduced that the barrier to racemization is only 7 kcal/mole [45].

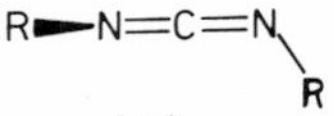

R—N=C=N—R

(52)

In phosphines and arsines, inversion is normally very slow as noted in Chapter 5 (p. 158), and research in this area has been concerned with the questions of whether and of how the rate of this process can be enhanced. Silicon again lowers the energy of the planar state. Thus, while in many phosphines the inversion barrier typically is 35 kcal/mole, this quantity is lowered to about 20 kcal/mole if one of the substituents is silicon [46]. In the trissilylphosphine (53) no resolution of the $SiMe_2$ signal was observed even at -80°, and $\Delta G^{\ddagger}$ must be less than 10 kcal/mole in that case [47].

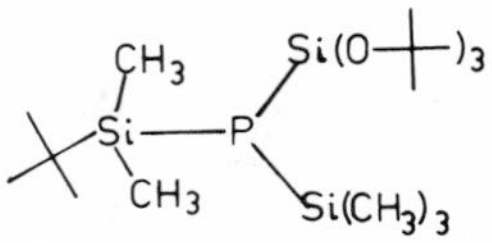

(53)

Another way to lower the inversion barrier is to make the phosphorus or arsenic atom and its unshared pair part of an aromatic system. For example, while in typical acyclic arsines the barrier is as high as 45 kcal/mole, in arsoles (54) it is only 35 kcal/mole.

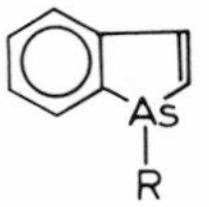

(54)

45. F. A. L. Anet, J. C. Jochims, and C. H. Bradley, *J. Amer. Chem. Soc.*, *92*, 2557 (1970).
46. R. D. Baechler and K. Mislow, *J. Amer. Chem. Soc.*, *93*, 773 (1971).
47. O. J. Scherer and R. Mergner, *J. Organometal. Chem.*, *40*, C64 (1972).

Evidently the planar state is stabilized by 4p–2p conjugation, in spite of the difference in size of these orbitals. Both effects are apparently at work in silicon-substituted arsole (55), in which the barrier is down to 25 kcal/mole [48].

(55)

We may also briefly mention inversion at oxygen here. The methyloxonium salt of oxirane (56) has a methylene singlet nmr signal that resolves at about $-50°$ into a quartet; the inversion barrier is 10 kcal/mole [49]. The analogous salt of tetrahydropyran (57) does not show this phenomenon; evidently a small ring size increases the barrier to inversion at oxygen as well as at nitrogen.

(56) (57)

An interesting suggestion has been made proposing the possible resolution of suitably o-substituted aryl ethers such as (58) [50];

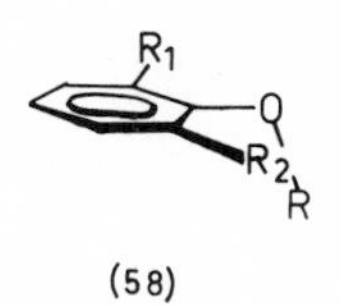

(58)

racemization would require the inversion of divalent oxygen if R_1 and R_2 are large enough. A related case is represented by the diphenyl ethers (59) and (60); here angle strain makes inversion difficult [51] (it may be noted that these molecules are

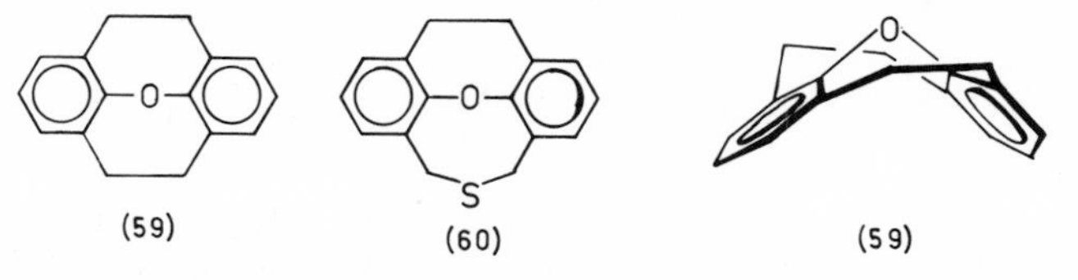

(59) (60) (59)

48. R.H. Bowman and K. Mislow, J. Amer. Chem. Soc., 94, 2861 (1972).
49. J.B. Lambert and D.H. Johnson, J. Amer. Chem. Soc., 90, 1349 (1968).
50. F. von Bruchhausen, H. Oberembt, and A. Feldhaus, Ann., 507, 144 (1933).
51. A.J. Gordon and J.P. Gallagher, Tetrahedron Lett., 2541 (1970).

strained metacyclophanes). It was found [51] that (60), in which this strain is not too severe, has an AA'BB' pattern and an AB quartet in the nmr spectrum at 80°; these signals change to two sharp singlets at 175°. Conversely, (59) has an AA'BB' pattern that remains unchanged to 190°.

6-5. THE INVERSION OF CYCLOHEXANE [52]

Since saturated six-centered rings are extremely common in molecules of biochemical and organic chemical systems, it is not hard to understand why chemists have taken a great interest in the chemistry of cyclohexane and its derivatives. It is surely natural to begin by thinking of cyclohexane as a flat molecule (61).

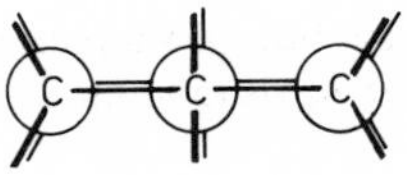

(61)

It is the simplest model; it is for the sake of convenience usually drawn that way, and it may actually be considered to be flat for the purpose of determining the number of stereoisomers of substituted cyclohexanes. However, it is easy to see that the real molecule cannot be flat. Thus, the CCC bond angles would all be 120° as they are in benzene; although this is just right for the sp^2 type of carbon atoms in benzene, it is far too large for the sp^3 type of carbon atoms present in cyclohexane rings. For a second reason, in such a flat ring all of the C-C and all of the C-H bonds would be eclipsed (61), adding intolerably to the energy of such a species. Both of these features would be greatly relieved by allowing the ring to become puckered, i.e., nonplanar in some way.

It is most instructive at this point to make a ball-and-stick model of this molecule; it then immediately becomes obvious that there is one way in which all bond angles can become tetrahedral and all bonds can be perfectly staggered, namely, in the so-called chair form of cyclohexane (62).

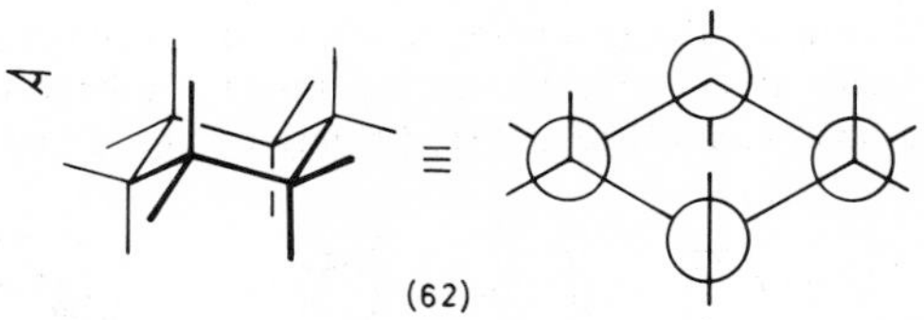

(62)

52. For reviews and references, see (a) E. L. Eliel, J. Chem. Educ., 37, 126 (1960); (b) E. L. Eliel, Angew. Chem., Int. Ed. Engl., 4, 761 (1965); (c) N. C. Franklin and H. Feltkamp, Angew. Chem., Int. Ed. Engl., 4, 774 (1965); (d) J. E. Anderson, Quart. Rev. (London), 19, 426 (1965).

While this is gratifying, it is now incumbent on us to explain why a substance such as cis-1, 2-dichlorocyclohexane (63), which lacks any symmetry altogether, is nevertheless not resolvable.

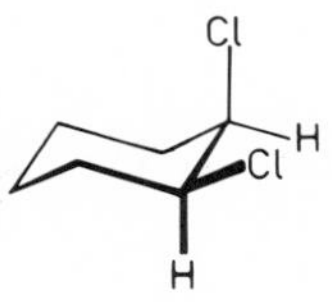

(63)

The reason is, as it is in so many open-chain compounds, amines, and so on, that somehow such compounds must be able to racemize very rapidly. The model again is helpful to show how this can happen. An "inversion" is possible (64) which does not seem to strain the model unduly, and the result of this inversion is the mirror image of what we had before.

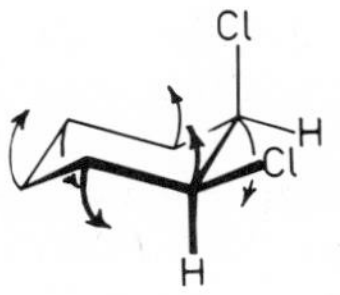

(64)

In other words, inversion amounts to racemization. Inversion in cyclohexane is really only a net twist of 60° in one direction or the other for each of the carbon-carbon bonds, and it does not produce the same result that would have been obtained if all six carbon atoms had been inverted, and hence the word inversion is perhaps not appropriate; however, it is in common usage.

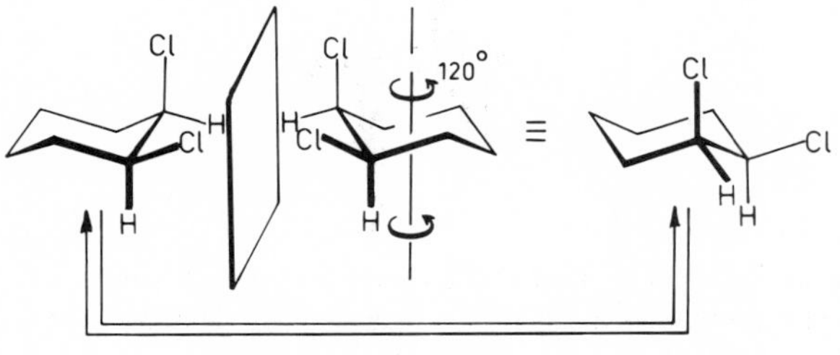

Further inspection of our model shows that six of the twelve hydrogen atoms lie near the plane roughly defined by the carbon atoms; they are referred to as the equatorial hydrogen atoms. The other six point in directions perpendicular to this plane and away from it: Three on the one side and three on the other. These are called the axial hydrogen atoms. The process of inversion interchanges all six equatorial hydrogen atoms with the axial ones.

If a single substituent is present such as in chlorocyclohexane (65), it may occupy either an equatorial or axial position. Neither is a stable stereoisomer, however; the rapid interconversion prevents their isolation and (65) is therefore actually a mixture.

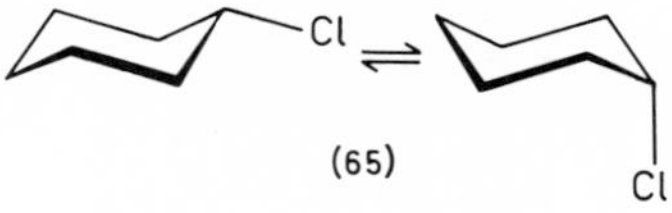

(65)

That this is so can be shown by means of ir; two carbon-chlorine frequencies can be observed [53]. The nmr spectrum at room temperature appears like that of a single substance, but at low temperature the methine hydrogen signal resolves into two; at -115° two clearly separated signals can be seen at 5.5 and 4.2 τ. At that temperature the half-life for interconversion is several minutes. At -150° equatorial cyclohexyl chloride crystallizes from a pentadeuterochloroethane solution; it can be filtered and redissolved in fresh solvent to show only the 4.2 τ signal, while the mother liquor shows a greatly enhanced signal at 5.5 τ. Warming either solution rapidly regenerates the equilibrium mixture [54].

In order to understand the properties of cyclohexane derivatives under less drastic conditions we should know the equilibrium constants that characterize the inversion process. These constants obviously depend on the nature of the substituents; for the unsubstituted compound (if it still makes sense to speak of an equilibrium then), K = 1.

Inspection of the model shows that the axial atoms in positions 1, 3, and 5 (and 2, 4, and 6) are not far apart, and that they will crowd and hinder one another in a way quite unlike the equatorial ones. Consequently, large groups tend to occupy the equatorial positions.

The methods whereby this fact can be ascertained are described below, and the magnitude of this hindrance is considered further in Chapter 7; suffice it to say at this point that an axial t-butyl group interferes with the 3- and 5-hydrogen atoms so strongly that t-butylcyclohexane is essentially pure equatorial t-butylcyclohexane.

---

53. K. Kozima and K. Sakashita, Bull. Chem. Soc. Japan, 31, 796 (1958).
54. F.R. Jensen and C.H. Bushweller, J. Amer. Chem. Soc., 88, 4279 (1966) and 91, 3223 (1969).

The intensity ratios of the methine hydrogen signals suggest that in the case of chlorocyclohexane, the equatorial isomer is favored by roughly a factor of 10 at -115°, and this direct approach is one of the methods by means of which K can be measured. This method is obviously not always applicable; thus, signal resolution may not occur in an accessible temperature range. Also, if information concerning the magnitude of K at room temperature is desired, extrapolation is necessary. In a related method the chemical shift of the methine hydrogen atom is used by means of the following expressions,

$$\sigma = N_e\sigma_e + N_a\sigma_a$$

and

$$K = \frac{N_e}{N_a} = \frac{\sigma_a - \sigma}{\sigma - \sigma_e}$$

where σ is the chemical shift, N is a mole fraction, and e and a stand for equatorial and axial, respectively; they essentially restate the observation (described in Chapter 2, Section 2-6, p. 63) that the chemical shift of a rapidly equilibrating proton is the weighted average of the extreme positions. To determine σ_a and σ_e, use can be made of the fact that the t-butyl group must occupy the equatorial position; thus, if cis- and trans-4-chloro-t-butylcyclohexane (66) and (67) can be synthesized, we may be confident that we will be observing substances containing exclusively axial and equatorial chlorine atoms (and, hence, equatorial and axial methine hydrogens), respectively.

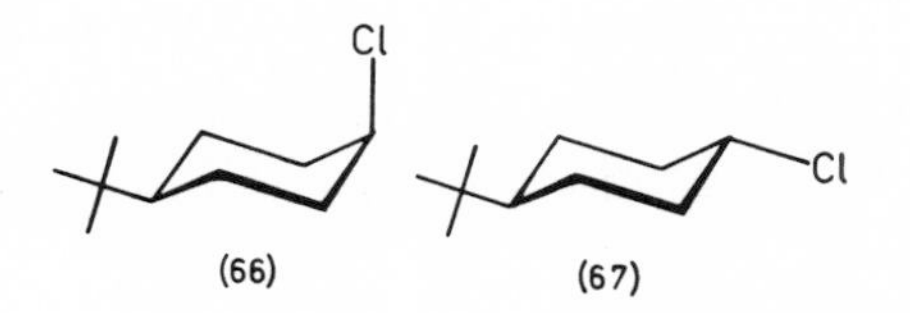

The basic assumption is that there will be no direct interaction across the ring and that there will be no distortions by the t-butyl group of the ring itself, so that this group will not affect the magnitude of σ_e and σ_a. It may be pointed out in this connection that cis-1,2- and 1,4- and trans-1,3-disubstituted cyclohexanes, and even perfluorocyclohexane, all tend to invert at about the same rate, with coalescence occurring at about -60° [52a]; evidently these substituents do not produce extreme ring distortions.

A kinetic method has also been used. If we consider some reaction of a substituted cyclohexane, the observed rate constant k_0 will actually be the result of two simultaneous reactions.

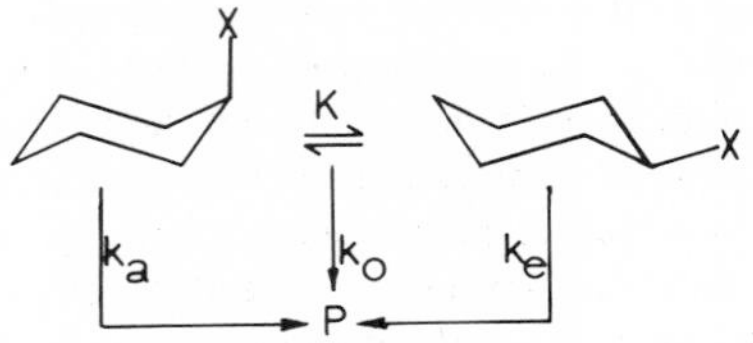

$$k_o = N_e k_e + N_a k_a$$

Since $K = N_e/N_a$ and $N_e + N_a = 1$, we find that

$$N_a = 1/(K + 1) \text{ and } N_e = K/(K + 1)$$

Substitution of these values into the expression for k_o and rearrangement gives

$$K = (k_o - k_a)/(k_e - k_o)$$

The constants k_a and k_e can be obtained from the cis- and trans-4-t-butyl-substituted cyclohexanes, but again one should consider the possibility that the t-butyl groups may affect these rate constants in other ways; in fact, the study of more than one reaction is desirable [55]. In some cases direct equilibration can be studied. Thus, cis- and trans-4-t-butylcyclohexanol equilibrate slowly in the presence of aluminum i-propoxide, and K can be measured directly by conventional means. Still other methods may be found in the review references cited earlier. The results (the equilibrium constants or free energies differences as a function of substituent) have an important bearing on steric effects and hence will be discussed in some detail in Chapter 7.

There are two other basic shapes that are possible: The twist and the boat conformation, (68) and (69), respectively. The former is commonly assumed to be a high energy intermediate in chair interconversions.

(68)

(69)

55. For a reservation about this method, see J. L. Mateos, C. Perez, and H. Kwart, Chem. Commun., 125 (1967).

The twist form can be achieved as a relatively stable form by forcing a large substituent into an axial position. Thus, *cis*-1,2-di-*t*-butylcyclohexane (70) shows the following nmr behavior. The single sharp methyl peak observed at elevated temperatures resolves into two of equal intensity on cooling to 35° (E_a = 16 kcal/mole), and the upfield signal resolves into two signals of roughly 2:1 ratio at -80° (E_a = 10 kcal/mole). The interpretation is that the twist form is actually the more stable one compared to the chair form by a fraction of 1 kcal/mole, that at -80° the interconversion of twist into chair form becomes rapid on the nmr time scale, and that the twist-form interconversion becomes rapid at 35° [56].

(70)

The boat form is probably the transition state between twist forms; its high energy is due in part to the eclipsing of two pairs of hydrogen atoms and to the interaction of two *cis*-hydrogen atoms in 1,4-positions (71).

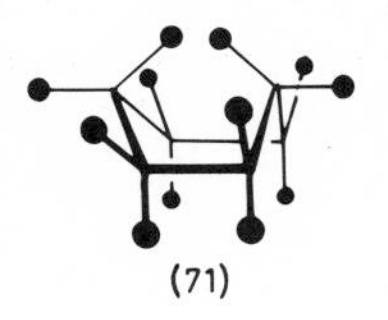

(71)

It is generally not possible to stabilize a boat form except by tying up the ring in such a way that escape is impossible; norbornane (72) is the best-known example.

≡

(72)

---

56. H. Kessler, V. Gusowski, and M. Hanack, Tetrahedron Lett., 4665 (1968); see also the references given there to the similar results obtained with *trans*-1,3-di-*t*-butylcyclohexane.

Another approach is the presence of a 1, 3-methylene bridge; of the two six-membered rings one is then necessarily a boat form. Since inversion of the free end of the ring can only interconvert two rather high energy forms, the barrier is relatively low; that in the bicycloheptane derivative (73) is less than 4 kcal/mole [57], with the ^{19}F nmr signals still sharp at -175°.

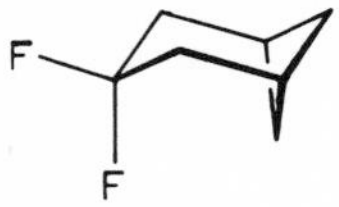

F
F
(73)

Bridging is often the method of choice to stabilize or "freeze" the other forms of cyclohexane. Thus, trans-decalin (74) is incapable of complete inversions; the two tertiary hydrogen atoms are both axial (in the cis-isomer (75) one is axial, the other equatorial).

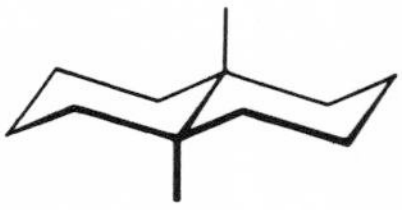

(74)  (75)

In adamantane (76), four chair forms of cyclohexane are present and all bridgehead hydrogen atoms are equatorial with respect to each of the rings.

(76)

57. K. Grychtol, H. Musso, and J. F. M. Oth, Chem. Ber., 105, 1798 (1972).

Extensions of such molecules such as diamantane (77) [58] and triamantane (78) [59] are known; the ultimate extrapolation is diamond [60].

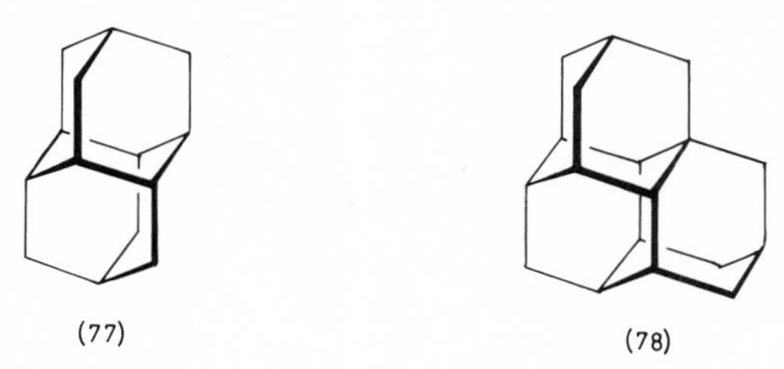

(77) (78)

The ultimate example of the twist form is twistane, or tricyclo[4.4.0.0^{3,8}]decane (76) [61].

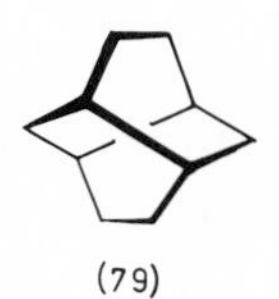

(79)

Boat-form cyclohexanes are present in the asteranes, of which pentaasterane (80) is shown as an example [62].

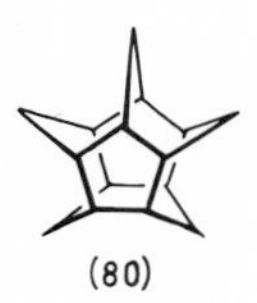

(80)

---

58. (a) C. Cupas, P.v.R. Schleyer, and D.J. Trecker, J. Amer. Chem. Soc., 87, 917 (1965); (b) T.M. Gund, V.Z. Williams, E. Osawa, and P.v.R. Schleyer, Tetrahedron Lett., 3877 (1970); (c) T.M. Gund, M. Nomura, V.Z. Williams, and P.v.R. Schleyer, Tetrahedron Lett., 4875 (1970); (d) T.M. Gund, P.v.R. Schleyer, and C. Hoogzand, Tetrahedron Lett., 1583 (1971).
59. V.Z. Williams, P.v.R. Schleyer, G.J. Gleicher, and L.B. Rodewald, J. Amer. Chem. Soc., 88, 3862 (1966).
60. Reviews: (a) H. Stetter, Angew. Chem., Int. Ed. Engl., 1, 286 (1962); (b) R.C. Fort and P.v.R. Schleyer, Chem. Rev., 64, 277 (1964); (c) R.C. Bingham and P.v.R. Schleyer, Fortschr. Chem. Forsch., 18, 1 (1971); (d) V.V. Sevost'yanova, M.M. Krayushkin, and A.G. Yurchenko, Russ. Chem. Rev. (English Transl.), 39, 817 (1970).
61. K. Adachi, K. Naemura, and M. Nakazaki, Tetrahedron Lett., 5467 (1968).
62. U. Biethan, U.v. Gizycki, and H. Musso, Tetrahedron Lett., 1477 (1965).

Other polycyclic systems such as the propellanes (for instance (81)) will probably also be scrutinized by organic chemists interested in conformational problems [63].

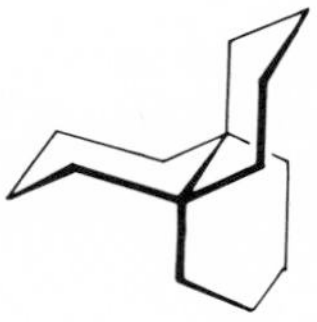

(81)

6-6. OTHER RING SYSTEMS [64]

Cyclohexane is of course not unique in its inversion properties, and in this section we shall review some of the other systems that have been examined to date.

The introduction of a double bond tends to flatten the ring and consequently to lower the inversion barrier. Cyclohexene has been examined by nmr; thus cyclohexene-3, 3, 4, 5, 6, 6-d_6 (82) has a sharp signal at 8.42 τ which at -160° begins to resolve into a doublet.

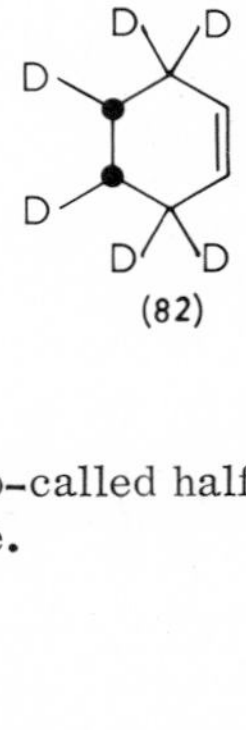

(82)

The interpretation involves the so-called half-chair forms (83), which interconvert very rapidly at room temperature.

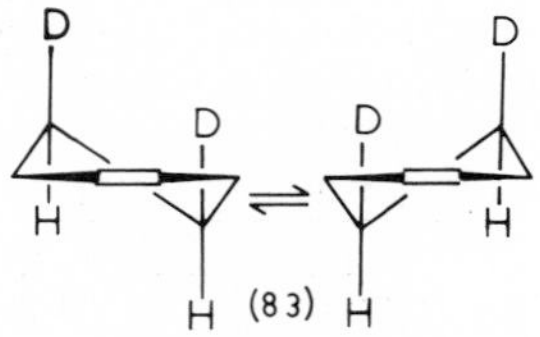

(83)

63. F. Nerdel, K. Janowsky, and D. Frank, Tetrahedron Lett., 2979 (1965).
64. For review, see J.E. Anderson, Quart. Rev. (London), 19, 426 (1965).

Calculation shows that $\Delta G^{\ddagger} = 5.3$ kcal/mole, and k = 50 sec^{-1} at -160°; at 25°; k is estimated to be 10^9 sec, about 100,000 times faster than cyclohexane [65]! Other cyclohexanes containing sp^2 carbon atoms, such as exomethylenecyclohexane, cyclohexanone, etc., likewise have very low inversion barriers of 5.7 kcal/mole or less [66].

With hetero atoms such as nitrogen the possibility of inversion of this atom complicates the study of the inversion of the ring.

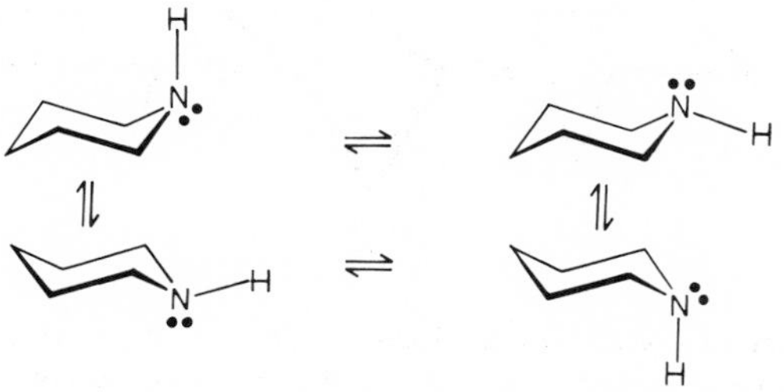

The nmr spectrum of piperidine-3,3,5,5-d_4 (84) has two AB patterns below -10° that coalesce at that temperature [67]; since the hydrogen atoms at C_4 should not be affected by the nitrogen configuration, ring inversion must be responsible.

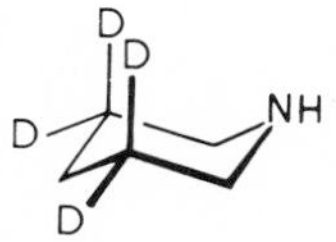

(84)

Evidently this ring inverts with greater difficulty than cyclohexane. This conclusion is somewhat uncertain since the piperidine was in methanol-d_4 solution; hydrogen bonding may have hampered inversion in this case (see further p. 231).

The opposite conclusion indeed applies to 1,4-dioxane. It was at first thought that inversion is still extremely fast even at -160°, since no signal resolution of the singlet occurs down to that temperature [68]. However, Jensen argued that this might be due to an accidental equivalence of axial and equatorial hydrogen atoms, and this was found to be the case [69]. Thus, when 1,4-dioxane-d_7 is

---

65. F.A.L. Anet and M.Z. Haq, J. Amer. Chem. Soc., 87, 3147 (1965).
66. (a) J.T. Gerig, J. Amer. Chem. Soc., 90, 1065 (1968); (b) F.R. Jensen and B.H. Beck, J. Amer. Chem. Soc., 90, 1066 (1968).
67. J.B. Lambert and R.G. Keske, J. Amer. Chem. Soc., 88, 620 (1966).
68. B. Petersen and J. Schaug, Acta Chem. Scand., 22, 1705 (1968).
69. F.R. Jensen and R.A. Neese, J. Amer. Chem. Soc., 93, 6329 (1971).

employed, the isotopic substitution affects the chemical shift of the remaining proton somewhat, and, in fact, σH_a is now slightly different from σH_e. With deuterium decoupling, a doublet results on cooling to -94°.

With five-membered rings the angles in the flat, regular pentagon are 108°, almost exactly equal to the tetrahedral angle. Nevertheless, such rings are not planar because this would require complete eclipsing of all substituents, and hence they assume the shape of envelopes (85).

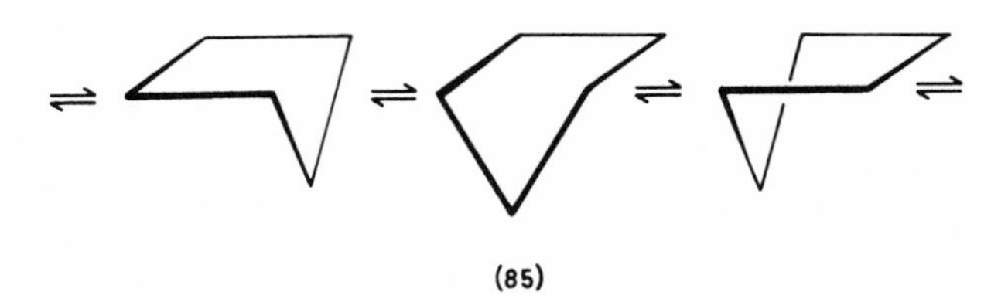

(85)

Five-membered rings have their analogs to adamantane in tricyclo[3.3.0.0^{3,7}] octane (86) [70] and dodecahedrane (87), a molecule not as yet synthesized [71].

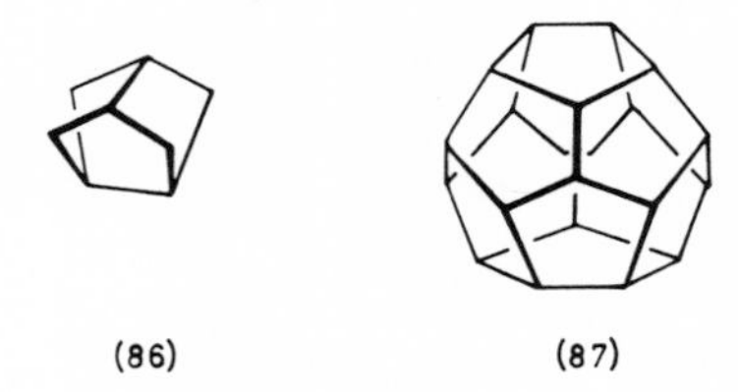

(86) (87)

Even in four-membered rings, the four carbons often accept additional angle strain ($\alpha < 90°$) in order to avoid eclipsing to some degree [72].

Thus, it has been found that methyl trans-3-methylcyclobutanecarboxylate (88) is more stable than the cis-isomer (89), which makes sense only if both substituents are in equatorial positions [73].

70. P.K. Freeman, V.N.M. Rao, and G.E. Bigam, Chem. Commun., 511 (1965).
71. (a) R.B. Woodward, T. Fukunaga, and R.C. Kelly, J. Amer. Chem. Soc., 86, 3162 (1964); (b) P.E. Eaton and R.H. Mueller, J. Amer. Chem. Soc., 94, 1014 (1972).
72. For review, see A. Wilson and D. Goldhamer, J. Chem. Educ., 40, 504 (1963).
73. J.M. Conia, J.L. Ripoll, L.A. Tushaus, C.L. Neumann, and N.L. Allinger, J. Amer. Chem. Soc., 84, 4982 (1962).

(88) (89)

Other evidence includes the fact that cis- and trans-3-i-propylcyclobutylamine ((90) and (91), respectively), give the same five rearranged products when treated with nitrous acid, but in very different ratios [74]; this suggests that the i-propyl group exerts some conformational control on the amino group as the t-butyl group does in cyclohexane.

(90) (91)

Not all four-membered rings are puckered, however. Thus, while the trans-1,3-cyclobutanedicarboxylic acid (92) is puckered as shown by X-ray diffraction [75], the disodium salt is planar. Packing forces in the crystals may be responsible for these apparently capricious variations.

(92)

In fact, the dihedral angle may vary from 120° as it does in bicyclo[1.1.1]pentane (93) [76] to 180° as in cubane (94) [77].

(93) (94)

74. I. Lillien and R.A. Doughty, Tetrahedron Lett., 3953 (1967).
75. E. Adman and T.N. Margulis, J. Amer. Chem. Soc., 90, 4517 (1968).
76. J.F. Chiang and S.H. Bauer, J. Amer. Chem. Soc., 92, 1614 (1970).
77. E.B. Fleischer, J. Amer. Chem. Soc., 86, 3889 (1964).

In the basketene derivative, *trans*-9,10-pentacyclo[4.4.0.$0^{2,5}$.$0^{3,8}$.$0^{4,7}$]decanedioic acid (95), two of the four cyclobutane rings are flat, the other two are puckered [78].

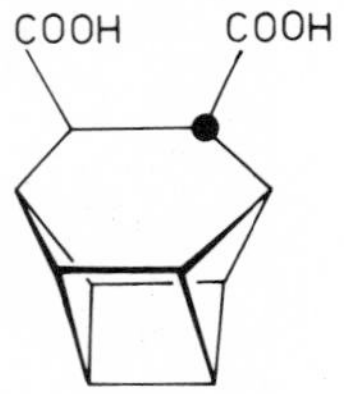

(95)

It is clear from all these examples that angle strain and eclipsing problems are in a delicate balance in cyclobutane derivatives.

In rings larger than the six-membered ones, puckering is obviously expected [52d], but only a few examples will be mentioned. Aluminum *i*-propoxide equilibration of *cis*- and *trans*-4-*t*-butylcycloheptanol (96) shows that the former is only slightly lower in energy (by less than 0.5 kcal/mole); the hydroxy group evidently favors neither position.

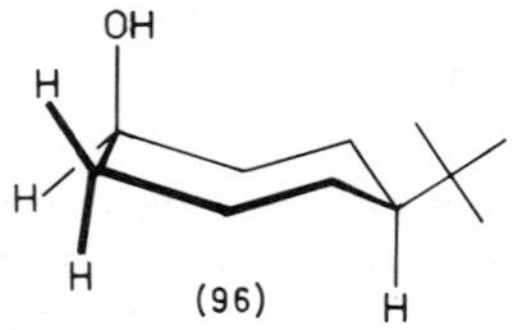

(96)

Thus, in this larger and less rigid ring the *t*-butyl group does not effectively function as a conformational anchor [79]. Double bonds again seem to promote low inversion barriers; thus, tropilidene (97) shows nmr methylene signal coalescence at -145° [80].

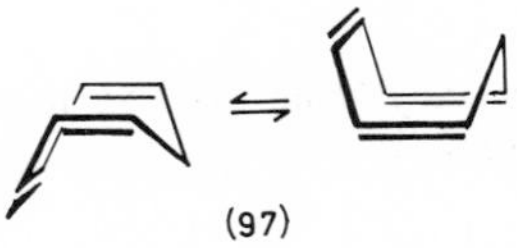

(97)

78. J.P. Schaefer and K.K. Walthers, *Tetrahedron*, **27**, 5281 (1971). This reference contains a complete and up-to-date list of all the cyclobutane derivatives whose conformations have been determined.
79. M. Mühlstädt, R. Borsdorf, and F.J. Strüber, *Tetrahedron Lett.*, 1879 (1966).
80. F.A.L. Anet, *J. Amer. Chem. Soc.*, **86**, 458 (1964).

This is not true in all medium size rings, however. Models suggest for example that trans, trans-1, 5-cyclooctadiene should be able to occur in two conformations (98) and (99), called perpendicular and parallel, respectively, and that interconversion might be difficult enough to allow separation [81].

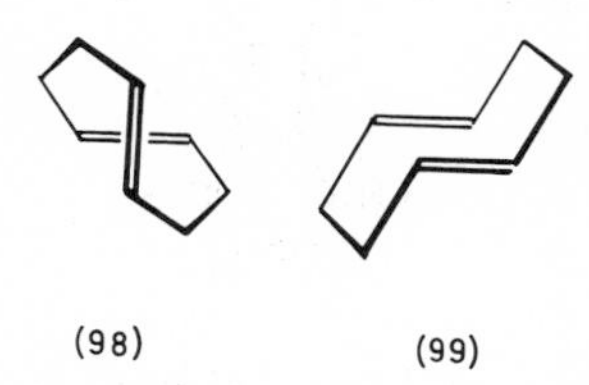

(98) (99)

This fact was recently accomplished, not with the cyclooctadienes themselves, but with the cyclopropane derivatives (100) and (101) as well as with several other derivatives [82].

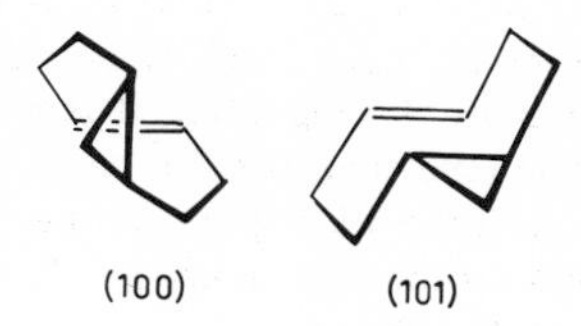

(100) (101)

Eight-membered rings have been studied especially by Roberts [83]. A twelve-membered ring (cyclotetraveratrylene (102)) was recently reported [84] to undergo nmr spectral changes a little below 0°; the F^{19} nmr spectrum of 1, 1-difluorocyclododecane [85] shows the presence of some conformational equilibrium at about -150°, and so on.

81. G. M. Whitesides, G. L. Goe, and A. C. Cope, J. Amer. Chem. Soc., 91, 2608 (1969).
82. A. Deyrup, M. Betkouski, W. Szabo, M. Mathew, and G. J. Palenik, J. Amer. Chem. Soc., 94, 2147 (1972).
83. See for example, C. Ganter, S. M. Pokras, and J. D. Roberts, J. Amer. Chem. Soc., 88, 4235 (1966).
84. J. D. White and B. D. Gesner, Tetrahedron Lett., 1591 (1968).
85. E. A. Noe and J. D. Roberts, J. Amer. Chem. Soc., 94, 2020 (1972).

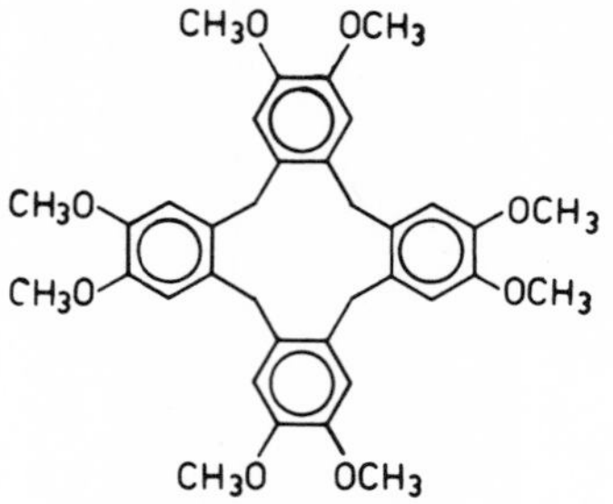

(102)

Inversions of even larger rings will be considered in Chapter 9 (p. 273 and p. 284). At this point, our general conclusions are, first, that virtually all cyclic compounds other than the aromatic ones are nonplanar, and, second, that nmr is indispensible in the search for the various motions that are possible, for the puckered states that occur and the energies that separate them. This is all the more important because in open-chain compounds these motions are usually too rapid and the conformations too many to be accessible to study. Finally, it may be obvious now that substituents separated by many bonds may actually be near one another, which gives rise to many chemical and physical properties that would otherwise be unexpected. Thus, 1-thiacyclooctan-5-one (103) has two carbonyl stretching frequencies at 1703 and 1684 cm^{-1}, unlike 1-thiacycloheptan-4-one (104) which has only one (at 1711 cm^{-1}); similarly, it has a dipole moment $\mu = 3.81$ D substantially higher than that of latter compound ($\mu = 3.04$ D).

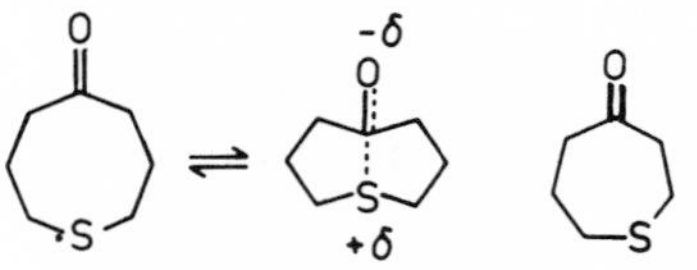

(103) (104)

These facts are attributed to the presence of a form of (103) in which the keto- and sulfide groupings are close enough for a transannular interaction [86]; furthermore, the corresponding alcohol (105) is readily dehydroxylated to give the bicyclic sulfonium salt (106) [87].

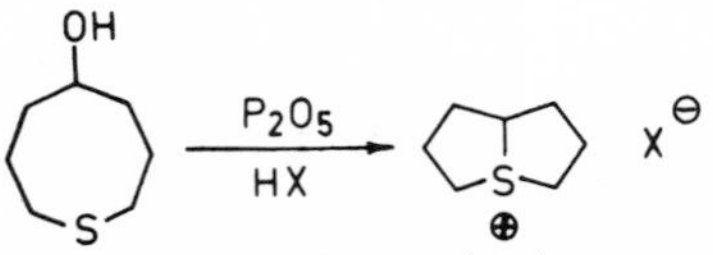

(105) $\xrightarrow[HX]{P_2O_5}$ (106) $X^{\ominus}$

86. N. J. Leonard, T. L. Brown, and T. W. Milligan, J. Amer. Chem. Soc., 81, 504 (1959).
87. C. G. Overberger and A. Iusi, J. Amer. Chem. Soc., 81, 506 (1959).

Chapter 7

STRAIN AND STERIC HINDRANCE [1]

7-1. ANGLE STRAIN AT SINGLY BOUND ATOMS.

The CCC bond angles at singly bound carbon atoms may differ greatly from the preferred tetrahedral value. This is clear from the fact that cyclopropane and large numbers of its derivatives are well-known and well-behaved substances. They are thermally less stable than such compounds as cyclohexane and have a higher combustion enthalpy by about 10 kcal per methylene group; evidently such strains (often termed internal or I-strains [1b]) are not great enough to prohibit the existence of such compounds. We shall see, in fact, that many organic molecules are able to absorb even more strain; we begin here, however, with some remarks about the term bond angle, about cyclopropane, and about ways to measure strain.

---

1. (a) For a general review, see M.S. Newman, "Steric Effects in Organic Chemistry", Wiley, New York, 1956; (b) H.C. Brown, Record Chem. Progr., **14**, 83 (1953).

The definition of the bond angle makes use of the nuclei as reference points; thus, the bond angles in cyclopropane are simply 60°. In a way, this does not present a true picture of the angle between the bonds, in that the bonds may be bent. Another way to express this is to say that the locus of maximum electron density between the nuclei need not be simply the straight line between the nuclei (Fig. 7-1).

There is some support for the notion of bent bonds in cyclopropane; thus, the carbon-carbon bond lengths (i.e., the straight line distances between the carbon atoms) in this hydrocarbon and its derivatives are generally a few hundredths of 1 Å shorter than in ethane or cyclohexane. Further support derives from the success of Walsh's theoretical model of cyclopropane [2] in predicting a number of properties of cyclopropanes. The basic idea in Walsh's model is that the carbon-carbon bonds in cyclopropane have more than a normal share (75%) of p character in order to reduce angle strain to a minimum, and as a result that the ring-substituent bonds have a more than normal amount of s character. Suppose we assume that the two CH bonds utilize sp^2 orbitals; this leaves for each carbon atom one sp^2 orbital (directed to the center of the ring) and one p orbital (in the plane of the ring and perpendicular to the plane of the methylene group). The three sp^2 orbitals are combined in the usual way, as are the three p orbitals; the six electrons are then distributed over the resulting orbitals in accordance with the energies, which may in turn be gauged from the number of nodes in each (see Fig. 7-2).

As a result, cyclopropyl bound hydrogen is more acidic than hydrogen atoms in unstrained saturated hydrocarbons; proton coupling with ring ^{13}C is larger than the normal 125 Hz, the HCH bond angle is 116°, and so on. Finally, it has been learned in recent years that conjugation of cyclopropyl rings with unsaturated systems is often characterized by effects as large as those attending the conjugation of multiple bonds; for example, the conformation in cyclopropylcarboxaldehyde (1) is such that the ring prefers to be parallel to the carbonyl π orbital (bisected conformation), and the barrier to interconversion of the two energy minima is even

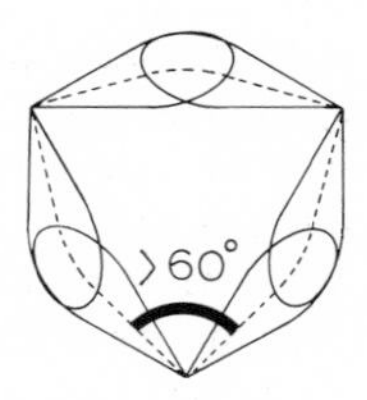

Fig. 7-1. Bent bonds in cyclopropane.

2. (a) A. D. Walsh, Trans. Faraday Soc., 45, 179 (1949); (b) M. Y. Lukina, Russ. Chem. Rev. (English Transl.), 31, 419 (1962); (c) W. A. Bernett, J. Chem. Educ., 44, 17 (1967); (d) R. Jesaitis, private communication.

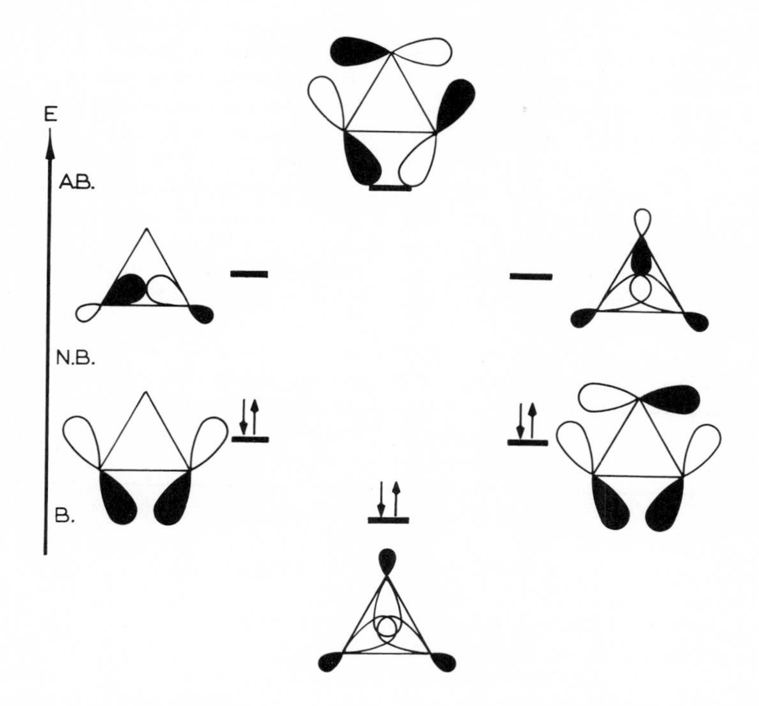

Fig. 7-2. The Walsh molecular orbitals of the ring in cyclopropane.

larger than in acrolein (2) and benzaldehyde [3]. Still another significant observation is that the cyclopropyl group has a strong tendency to donate electrons to nearby electron deficient centers - with its edges, but not with its face (see Chapter 20)!

(1) (2)

It may be noted in this connection that short bonds are not an infallible indication of bond bending. Other factors are also important in determining bond lengths. Thus, cyclobutane and its derivatives tend to have abnormally long bonds; this has been ascribed to cross-ring repulsion between the two diagonally placed and relatively unshielded sets of carbon nuclei [4]. The central CC bonds in tri-*t*-butylmethane are as long as 1.61 Å - obviously because of steric crowding [5].

---

3. For discussion and references, see R. Hoffmann, Tetrahedron Lett., 3819 (1965).
4. For a discussion of the bond lengths in small rings, see E. Goldish, J. Chem. Educ., 36, 408 (1959).
5. H. B. Burgi and L. S. Bartell, J. Amer. Chem. Soc., 94, 5236 (1972).

The angle strain at a given carbon atom cannot always be accurately assessed by means of calorimetry: In combustion experiments one must rely on small differences between large numbers. A very convenient though empirical correlation known as the Foote-Schleyer relationship relates the angle strain to the infrared frequency of the carbonyl group in the corresponding ketone [6]. In Table 7-1 this change in frequency clearly parallels the change in bond angle in the hydrocarbon. A second indication of strain is the coupling constant $J_{C^{13}H}$ in the hydrocarbons themselves; this is related to the % s character in the C-H bonds (see p. 62). Relatively high field chemical shifts for protons bound to strained small rings are also characteristic. Thus, the cyclopropane H usually appears at 10 τ or more.

TABLE 7-1. The C=O Stretching Frequency of Several Cyclic Ketones and the ^{13}CH Coupling Constants of the Corresponding Hydrocarbons

ΔH_{Strain}	Ketone	$\nu_{C=O}$, cm^{-1}	$J_{C^{13}H}$, Hz
28	Cyclopropanone	1815	161
26	Cyclobutanone	1791	134
7	Cyclopentanone	1748	-
0	Cyclohexanone	1716	125
6	Cycloheptanone	1705	-

Another consequence of angle strain is the fact that the angle between the other two groups bound to the carbon atom tends to deviate from the normal value in the opposite direction; thus, the angle HCH in most cyclopropane derivatives is approximately 116° (this fact of course also supports Walsh's model). As a result strained small rings can be stabilized by large groups which of steric necessity require a rather large bond angle between them; many exotic small ring systems synthesized for the first time in recent years carry t-butyl or adamantyl groups at the corners (several instances are scattered through this chapter). A related effect is the fact that the internal hydrogen bond in 2,2-dialkyl-1,3-propanediols (3) becomes stronger (as judged by ir; see p. 42) as the alkyl groups become larger [7].

(3)

6. (a) C.S. Foote, J. Amer. Chem. Soc., 86, 1853 (1964); (b) P.v.R. Schleyer, J. Amer. Chem. Soc., 86, 1854, 1856 (1964).
7. P.v.R. Schleyer, J. Amer. Chem. Soc., 83, 1368 (1961); many earlier references to this effect are given in this paper.

Conversely, if the external bond angle is also constrained to a small value the energy of the molecule goes up further. An example is spiropentane (4); evidence of its strain may be found in its reactivity with many reagents such as chlorine under conditions in which cyclopropane is relatively inert [8], and in the preponderance of ring-opened products.

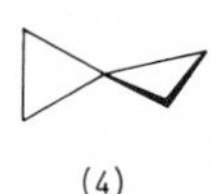

(4)

Several members of the asteranes such as pentaspiro[2.0.2.0.2.0.2.0.2.0] pentadecane (5) and trispiro[4.0.4.0.4.0]pentadecane (6) are known. In this series the bond angle combinations can be varied almost at will, but the resulting strains have not yet been assessed [9].

(5) (6)

By extrapolation it would seem likely that exo double bonds increase strain still further, and apparently they do, as is detailed in the following section.

Small rings with a common edge are highly strained, as revealed in the bicyclo[n.1.0]hydrocarbons; bicyclo[1.1.0]butane (7) is the extreme example [10]. This hydrocarbon and several derivatives have been isolated [11]; its strain (65 kcal/mole) is evident from the facts [12] that it polymerizes rapidly at 150°, that opening of at least one ring occurs in acid media [13], and that the common edge

(7)

8. D.E. Applequist, G.F. Fanta, and B.W. Henrikson, J. Amer. Chem. Soc., 82, 2368 (1960).
9. See for instance, A.P. Krapcho and F.J. Waller, J. Org. Chem., 37, 1079 (1972).
10. For reviews, see (a) D. Seebach, Angew. Chem., Int. Ed. Engl., 4, 121 (1965); (b) K.B. Wiberg, Record Chem. Progr., 26, 143 (1965).
11. D.M. Lemal, F. Menger, and G.W. Clark, J. Amer. Chem. Soc., 85, 2529 (1963).
12. R. Srinivasan, J. Amer. Chem. Soc., 85, 4045 (1963).
13. J. Bayless, L. Friedman, J.A. Smith, F.B. Cook, and H. Shechter, J. Amer. Chem. Soc., 87, 661 (1965).

bond has a length of 1.64 Å - at the moment one of the longest known single C-C bonds [14]. Still further strain can be incorporated by tying together the two remaining methylene groups. Thus, the tricyclo[1.1.1.$0^{4,5}$]pentane system (8) is known [15]. The ultimate example of a compound with multiple angle strains is tetrahedrane (9).

(8) (9)

According to one theoretical estimate, this compound is incapable of existence and should spontaneously decompose to acetylene [16]. Several claims to tetrahedron derivatives have been shown to result from errors of one sort or another [17]; thus cyclobuteno[l]phenanthrene (10) has been mistaken for diphenyl tetrahedrane (11).

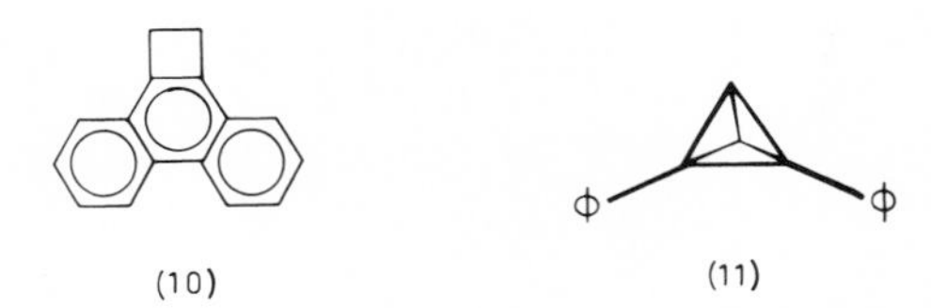

(10) (11)

Tetrahedrane has been recognized as a transient intermediate; here, however we are outside the domain of isolable substances and hence refer the interested reader to Chapter 24 (p. 861) for further discussion of this compound. It may be pointed out, however, that tetrahedrane does have known analogs in inorganic chemistry. Thus, white phosphorus (12) and tetrachlorotetraborane (13) have the tricyclo [1.1.0.$0^{2,4}$] structure.

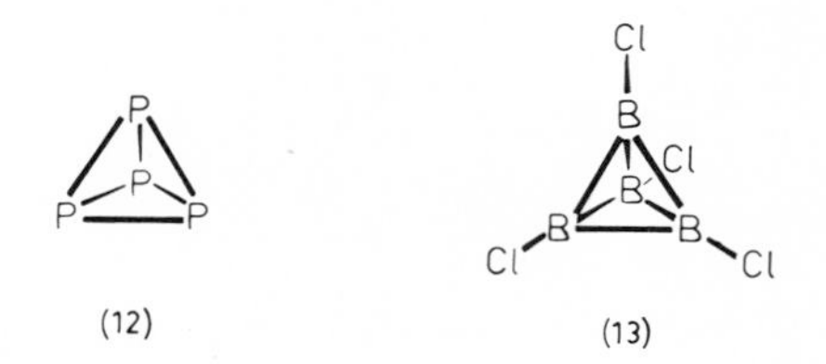

(12) (13)

---

14. Only that in o-carborane (see p. 16) is longer, but in that case of course fewer than 2 electrons are available for the carbon carbon bond.
15. (a) S. Masamune, J. Amer. Chem. Soc., 86, 735 (1964); (b) G. L. Closs and R. B. Larrabee, Tetrahedron Lett., 287 (1965).
16. R. J. Buenker and S. D. Peyerimhoff, J. Amer. Chem. Soc., 91, 4342 (1969).
17. See for example, (a) S. Masamune and M. Kato, J. Amer. Chem. Soc., 87, 4190 (1965) and 88, 610 (1966); (b) E. H. White, G. E. Maier, R. Graeve, U. Zirngibl, and E. W. Friend, J. Amer. Chem. Soc., 88, 611 (1966); (c) H. O. Larson and R. B. Woodward, Chem. Ind. (London), 193 (1959).

Angle strain can be introduced in organic compounds not only by squeezing together one pair of bonds, but also by twisting two pairs relative to each other. A combination of these motions could even lead - at least on paper - to square planar carbon.

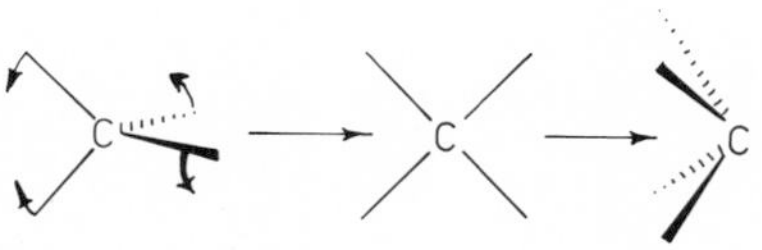

Beyond that, we may even imagine a carbon atom with all four of its valences lying in one hemisphere. A rational experimental approach in those directions is the synthesis of trans-fused bicyclo[n.1.0]alkanes; the most highly strained derivative known to date is with n = 4, (14) [18].

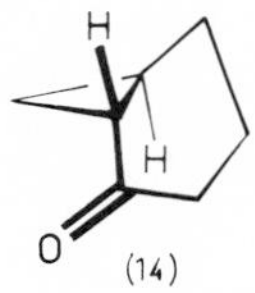

(14)

Even a model with n = 3 can be constructed, but it is obviously strained and whether it can be synthesized remains to be seen. Trans-bicyclo[3.2.0]heptan-1-ol (15) has been prepared by Wiberg [19].

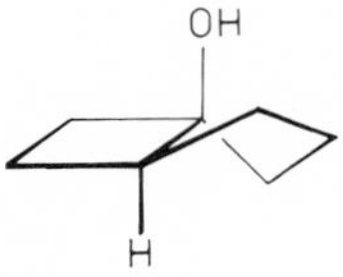

(15)

Examples of square planar carbon are not yet known, although a theoretical justification for this geometry has appeared. Hoffmann has reasoned [20] that a carbon atom may adopt a planar geometry in order to allow a large, conjugated string of

---

18. J. V. Paukstelis and J. Kao, J. Amer. Chem. Soc., 94, 4783 (1972).
19. K. B. Wiberg, J. E. Hiatt, and G. Burgmaier, Tetrahedron Lett., 5855 (1968).
20. R. Hoffmann, Pure Appl. Chem., 28, 181 (1971).

doubly bound carbon atoms to adopt a low-energy planar form (see for instance p. 301); the molecule (16) could be an example. The central carbon atom in fenestrane (17) might be constrained to planarity by torsional forces imposed by the cyclooctane periphery [21].

(16) (17)

Still other molecules have been postulated to possess a square planar carbon atom in an excited state (see p. 430).

Incredible as it may seem, several compounds are now known in which all four groups bound to a carbon atom lie to one side of the nucleus. So far, all known examples are propellanes [22]. 1,3-Dehydroadamantane (18) is an example of this type of compound [23]; it is highly reactive, and the 1,3-bond is broken at room temperature by oxygen, halogens, weak acids, and so on.

(18)

Its chemical behavior contrasts sharply with that of 2,4-dehydroadamantane (19) [24] and 2,4,6,9-bisdehydroadamantane (20); thus, the latter compound survives brief heating to 400° [25].

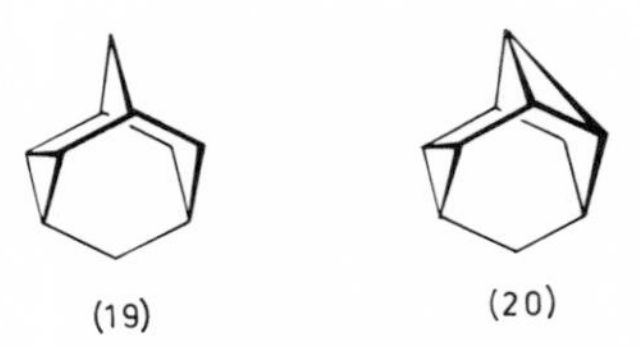

(19) (20)

---

21. V. Georgian and M. Saltzman, Tetrahedron Lett., 4315 (1972).
22. For review, see D. Ginsburg, Accounts Chem. Res., 5, 249 (1972).
23. R.E. Pincock and E.J. Torupka, J. Amer. Chem. Soc., 91, 4593 (1969).
24. A.C. Udding, J. Strating, H. Wynberg, and J.L.M.A. Schlatmann, Chem. Commun., 657 (1966).
25. H.W. Geluk and T.J. de Boer, Chem. Commun., 3 (1972), and Tetrahedron, 28, 3351 (1972).

Tricyclo[3.2.1.0^{1,5}] octane or [3.2.1]propellane (21) is an even more highly strained molecule of this sort; combustion experiments suggest that this skeleton has a strain energy of about 60 kcal/mole, and it reacts rapidly with a variety of reagents [26].

(21)

Remarkably enough, the compound is thermally stable to nearly 200°; the reason for this stability is probably simply that there is no pathway to any reasonable product. Another remarkable example is (22), a member of a class of compounds mentioned again on p. 295 [27]. [2.2.2]Propellane has not yet been synthesized, and may well be incapable of isolation [28]; tricyclo[2.2.1.0^{1,4}]heptane, still more highly strained, has been encountered as a possible intermediate (see p. 863).

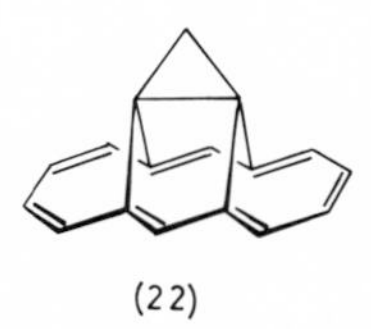

(22)

A variety of additional polycyclic strained small ring compounds are known [10a], [29]; they include prismane (23), cubane (24) and quadricyclane (25).

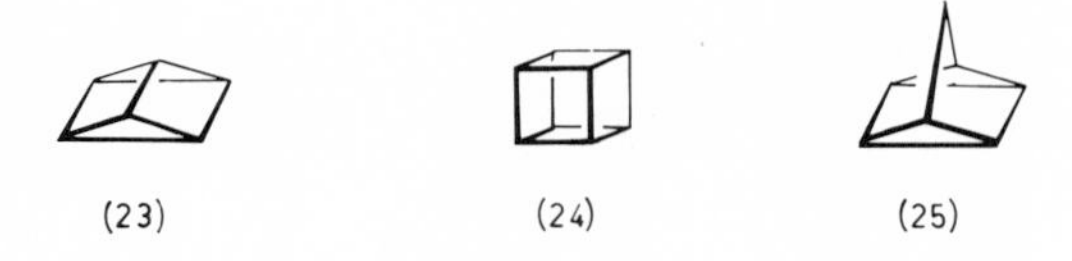

(23) (24) (25)

---

26. (a) K. B. Wiberg and G. J. Burgmaier, Tetrahedron Lett., 317 (1969) and J. Amer. Chem. Soc., 94, 7396 (1972); (b) P. G. Gassman, A. Topp, and J.W. Keller, Tetrahedron Lett., 1093 (1969); (c) K. B. Wiberg, E. C. Lupton, and G. J. Burgmaier, J. Amer. Chem. Soc., 91, 3372 (1969); (d) K. B. Wiberg, G. J. Burgmaier, K.W. Shen, S. J. La Placa, W. C. Hamilton, and M. D. Newton, J. Amer. Chem. Soc., 94, 7402 (1972).
27. F. Gerson, K. Müllen, and E. Vogel, J. Amer. Chem. Soc., 94, 2924 (1972).
28. M. D. Newton and J. M. Schulman, J. Amer. Chem. Soc., 94, 4391 (1972).
29. Some of these are reviewed by L. N. Ferguson, J. Chem. Educ., 47, 46 (1970) and 46, 404 (1969).

7-2. ANGLE STRAIN AT DOUBLY BOUND ATOMS.

The ideal bond angle at olefinic carbon is thought to be 120°. As with single bonds, this angle may be reduced or expanded to some degree, and the bonds themselves may be twisted and bent in various ways:

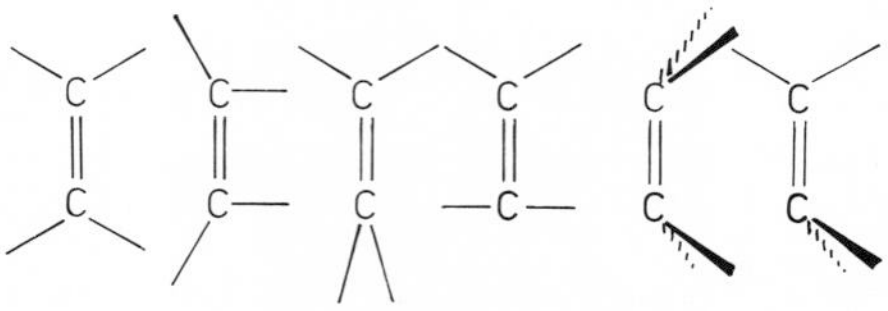

How severely the olefinic bond may be strained is obvious at once from the fact that cyclopropene (26) does exist [30].

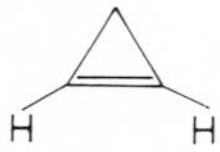

(26)

The angle strain in this and similar compounds leads again to a high degree of s character in the C-H bond, as revealed by the $JC^{13}H$ constants and high acidity; thus, lithium salts such as (27) can be readily prepared [31].

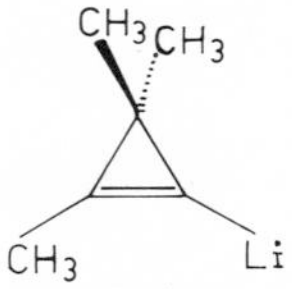

(27)

---

30. See for the synthesis and properties of various cyclopropenes: (a) G. L. Closs and K. D. Krantz, J. Org. Chem., 31, 638 (1966); (b) F. Fisher and D. E. Applequist, J. Org. Chem., 30, 2089 (1965); (c) G. L. Closs and L. E. Closs, J. Amer. Chem. Soc., 83, 1003 (1961); (d) G. L. Closs, L. E. Closs, and W. A. Böll, J. Amer. Chem. Soc., 85, 3796 (1963); (e) K. B. Wiberg and W. J. Bartley, J. Amer. Chem. Soc., 82, 6375 (1960).
31. G. L. Closs and L. E. Closs, J. Amer. Chem. Soc., 85, 99 (1963).

The cyclopropenes are not very stable and most of them polymerize rapidly even below 0°; nevertheless at least some of them, derivatives of sterculic acid (28), occur as natural products [32].

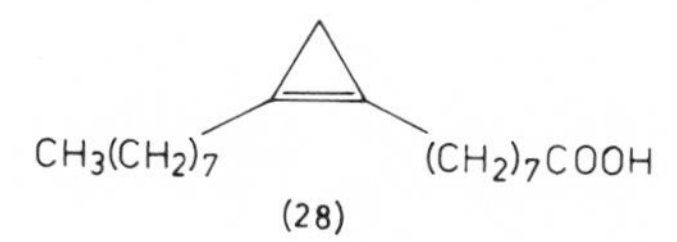

(28)

The single bonds at an olefinic site may also be pressed together to a significant extent. Thus, methylenecyclopropane (29) is a well-known compound [33], and in fact bis- [34] and trismethylenecyclopropane [35, 36] (30) and (31) have been reported.

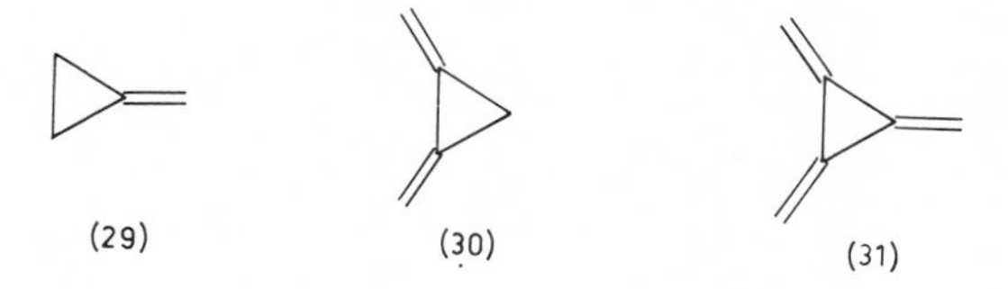
(29) (30) (31)

The latter compound is not very stable in spite of the conjugation; it polymerizes rapidly at room tempersture [36]. It is part of a larger group of compounds called the radialenes of which several members are now known; this will permit a systematic study of the properties of strained olefins [37].

---

32. (a) J.R. Nunn, J. Chem. Soc., 313 (1952); (b) K. Hofmann and C.W. Yoho, J. Amer. Chem. Soc., 81, 3356 (1959); (c) K.L. Rinehart, C.L. Tarimu, and T.P. Culbertson, J. Amer. Chem. Soc., 81, 5007 (1959); (d) N.T. Costellucci and C.E. Griffin, J. Amer. Chem. Soc., 82, 4107 (1960).
33. (a) E.F. Ullman and W.J. Fanshawe, J. Amer. Chem. Soc., 83, 2379 (1961); (b) T.C. Shields, B.A. Shoulders, J.F. Krause, C.L. Osborn, and P.D. Gardner, J. Amer. Chem. Soc., 87, 3026 (1965).
34. (a) R.F. Bleiholder and H. Shechter, J. Amer. Chem. Soc., 86, 5032 (1964); (b) J.K. Crandall and D.R. Paulson, J. Amer. Chem. Soc., 88, 4302 (1966).
35. (a) G. Köbrich and H. Heinemann, Angew. Chem., Int. Ed. Engl., 4, 594 (1965); (b) P.A. Waitkus, L.I. Peterson, and G.W. Griffin, J. Amer. Chem. Soc., 88, 181 (1966).
36. E.A. Dorko, J. Amer. Chem. Soc., 87, 5518 (1965).
37. P.A. Waitkus, E.B. Sanders, L.I. Peterson, and G.W. Griffin, J. Amer. Chem. Soc., 89, 6318 (1967).

Exo-methylenecyclopropenes incorporate both kinds of strain, and some of these such as (32) have been reported [38].

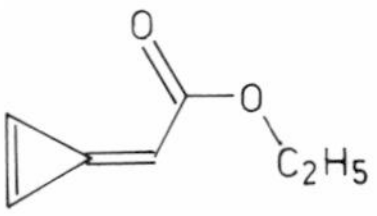

(32)

Surprisingly enough, these compounds are quite stable; the reason for this observation is the aromaticity in one of the resonance structures (33) of the three-membered ring (see p. 302).

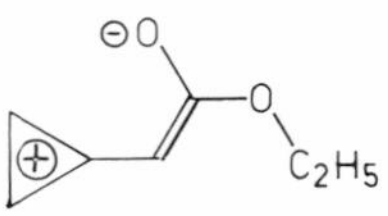

(33)

Trans double bonds in cyclic compounds are subject to strains bending two substituents out of the plane of the bond. The smallest known stable *trans*-cycloalkene is the eight-membered ring (34). Models suggest that *trans*-cyclooctene is a distorted chair form unable to invert. This compound should therefore be resolvable, and this feat has indeed been accomplished (p. 389).

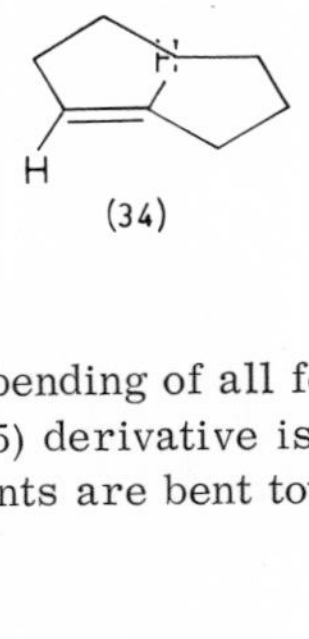

(34)

Out-of-plane bending of all four substituents occurs in various compounds. The anthracene (35) derivative is known [39]; models show that in that compound all four substituents are bent toward one side of the ethylene skeleton.

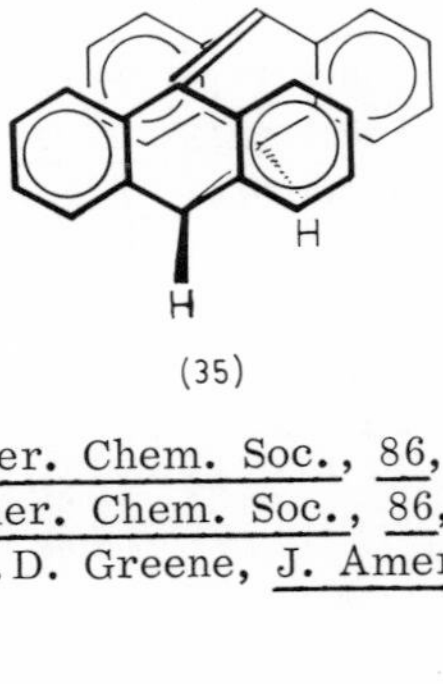

(35)

---

38. (a) M.A. Battiste, *J. Amer. Chem. Soc.*, *86*, 942 (1964); (b) W.M. Jones and J.M. Denham, *J. Amer. Chem. Soc.*, *86*, 944 (1964).
39. N.M. Weinshenker and F.D. Greene, *J. Amer. Chem. Soc.*, *90*, 506 (1968).

Twisting of double bonds can be accomplished to various degrees. In twistene (36) the double bonds have been twisted by about 20° [40].

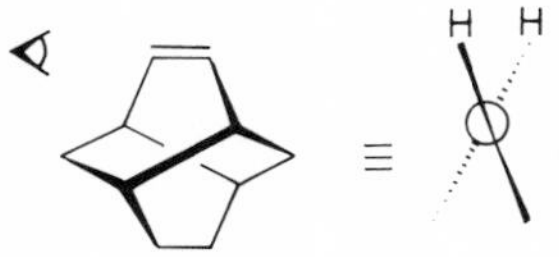

(36)

This rigid molecule is asymmetric and hence should be resolvable; it has indeed been obtained in optically active form. An even more severe twist occurs in the unsaturated steroid derivative (37); X-ray diffraction experiments have shown a deviation from planarity of as much as 30° [41].

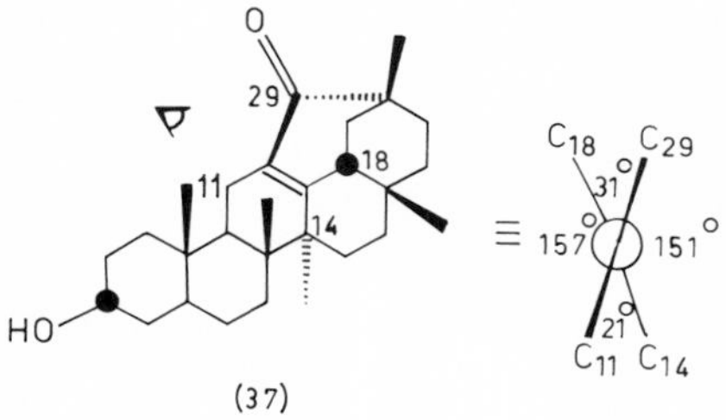

(37)

This compound may be considered as a rather complex example of Bredt's rule; according to this rule, in bicyclic compounds none of the bridgehead carbon atoms may be doubly bound [42]. Thus, bicyclo[2.2.1]hex-1-ene (38) should be highly strained and incapable of existence (but see p. 863).

(38)

40. M. Tichy and J. Sicher, Tetrahedron Lett., 4609 (1969).
41. W.E. Thiessen, H.A. Levy, W.G. Dauben, G.H. Beasly, and D.A. Cox, J. Amer. Chem. Soc., 93, 4312 (1971).
42. J. Bredt, Ann., 395, 26 (1913).

The rule can of course be violated by making one of the bridges long enough to remove the strain, and Prelog has reported the bicyclic "anti-Bredt" ketone (39) [43].

(39)

With the great advances that have been made in synthesis in recent years it has been possible to subject Bredt's rule to further scrutiny. Thus, bicyclo[3.3.1] non-1-ene (40) can be prepared, and it's found to be stable in the absence of air [44], acids [45] and other reagents [46].

(40)

The isomers (41) and (42) have also been prepared [47].

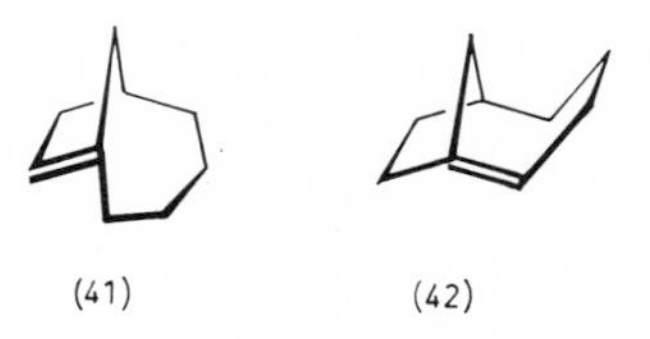

(41) (42)

All these are essentially examples of bridged *trans*-cyclooctenes [46, 48]. In the isomers (43) and (44) the *trans*-double bond is in a bridged cycloheptene ring, and accordingly these would be expected to be even more unstable.

---

43. V. Prelog, P. Barman and M. Zimmermann, *Helv. Chim. Acta*, **32**, 1284 (1949).
44. (a) J. A. Marshall and H. Faubl, *J. Amer. Chem. Soc.*, **89**, 5965 (1967); (b) J. R. Wiseman, *J. Amer. Chem. Soc.*, **89**, 5966 (1967).
45. P. M. Lesko and R. B. Turner, *J. Amer. Chem. Soc.*, **90**, 6888 (1968).
46. J. R. Wiseman and W. A. Pletcher, *J. Amer. Chem. Soc.*, **92**, 956 (1970).
47. J. R. Wiseman, H. F. Chan and C. J. Ahola, *J. Amer. Chem. Soc.*, **91**, 2812 (1969).
48. G. L. Buchanan and G. Jamieson, *Tetrahedron*, **28**, 1123, 1129 (1972).

(43)  (44)

These compounds undoubtedly are at or near the limit of the definition of stability; they were prepared [49] via pyrolysis of the quaternary ammonium hydroxide (45) followed by the freezing of the pyrolysate vapor on a cold finger.

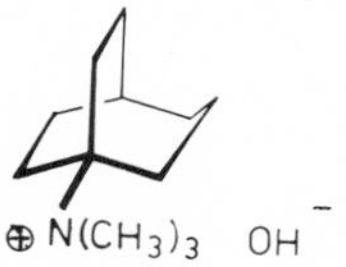

(45)

Although the nmr spectra could be measured [50] - separation or purification proved impossible since these olefins dimerize rapidly even at 0° to give various isomeric hydrocarbons such as (46). Similar remarks apply to the bridged *trans*-cycloheptene (47) [51];

(46) 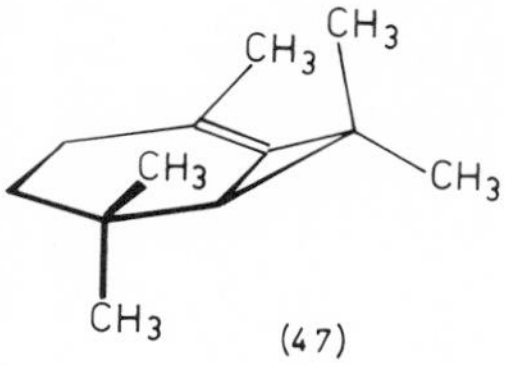 (47)

however, the [3.2.0] analog (48) has been isolated in pure form [52].

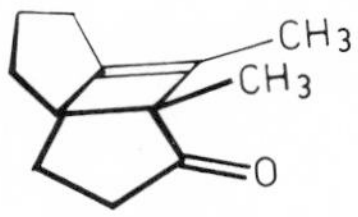

(48)

---

49. J. R. Wiseman and J. A. Chong, *J. Amer. Chem. Soc.*, *91*, 7775 (1969).
50. Initial identification was made by means of the odor, said to be powerful and characteristic of such compounds.
51. G. Köbrich and M. Baumann, *Angew. Chem., Int. Ed. Engl.*, *11*, 52 (1972).
52. R. L. Cargill and A. B. Sears, *Tetrahedron Lett.*, 3555 (1972).

More highly strained "anti-Bredt" compounds are not likely to be isolated, although their transient intermediacy can of course be pursued (cf. p. 713 and p. 863).

The reactivity of strained ring systems [29] can often be predicted on the basis of changes in strain. Small rings that include an sp^2 hybridized carbon atom enjoy strain relief if it can be converted into an sp^3 center. Thus, whereas ketones are usually not hydrated in aqueous solution, many small ring ketones are actually isolated as 1, 1-dihydroxy derivatives.

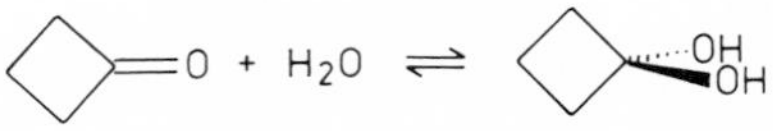

=O + H_2O ⇌ OH OH

Conversely, solvolysis reactions of such compounds as cyclopropyl chloride are very slow because of a hybridization change in the opposite direction.

7-3. OTHER TYPES OF STRAINED MULTIPLE BONDS.

Among the modifications of strained doubly bound molecules that can be made, the substitution by a benzo group is one of the more severe. Nevertheless, even benzocyclopropene derivatives are known. Thus, the ester (49) has been prepared. It is unstable to acids and to elevated temperatures [53]; however, the uv and nmr spectra were measured, and found to be similar to those of methyl benzoate.

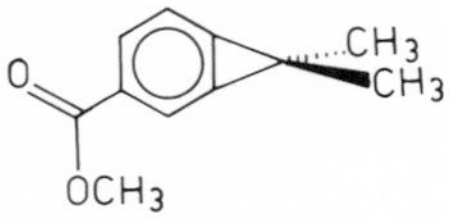

O CH_3 CH_3 OCH_3

(49)

Benzocyclopropene itself was reported shortly thereafter. It was synthesized by means of a technique that has produced many unstable molecules at moderate temperatures, namely, the elimination of a benzene derivative, in this case from the radialene (50) [54].

53. R. Anet and F. A. L. Anet, J. Amer. Chem. Soc., 86, 525 (1964).
54. (a) E. Vogel, W. Grimme, and S. Korte, Tetrahedron Lett., 3625 (1965); (b) B. Halton and P. J. Milsom, Chem. Commun., 814 (1971); (c) W. E. Billups, A. J. Blakeney, and W. Y. Chow, Chem. Commun., 1461 (1971).

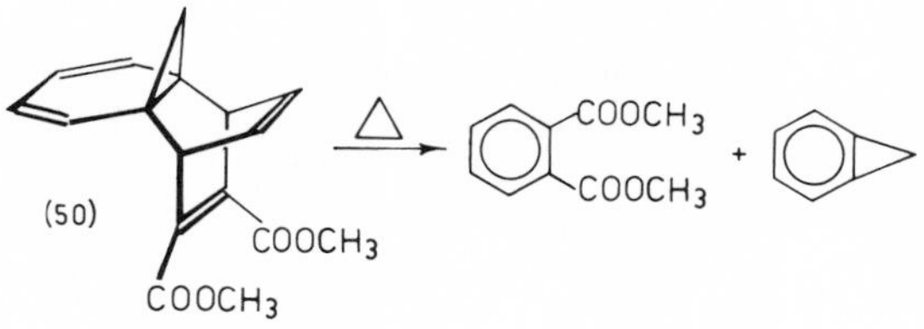

(50)

The methylene protons have a coupling constant $J_{C^{13}H} = 180$ cps. Benzene rings may be distorted in other ways. Very large strains are obviously built into the cyclophanes such as (51); these systems will be discussed in Chapter 10, Section 10-2.

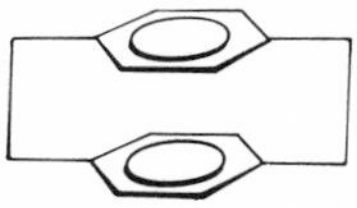

(51)

Turning now to small rings with multiple bonds and heteroatoms, we find that such atoms generally decrease the stability as compared with the carbon analogs because they provide an additional site of attack to various reagents; furthermore, there is the possibility that stabilization may result from ring opening. Thus, cyclopropanone (52) might conceivably open to the oxyallyl zwitterion (53).

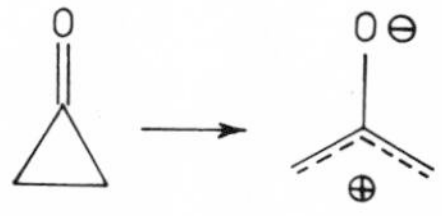

(52) (53)

Cyclopropanone has been obtained in dilute solution by the photodecarbonylation of 1,3-cyclobutanedione (54) (cf. p. 437) and by the reaction of diazomethane with ketene [55]; however, it is a highly unstable compound that rapidly decarbonylates further to ethylene.

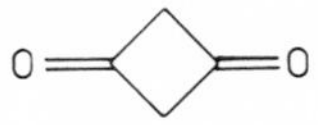

(54)

55. (a) W. B. Hammond and N. J. Turro, J. Amer. Chem. Soc., 88, 2880 (1966); (b) N. J. Turro and W. B. Hammond, J. Amer. Chem. Soc., 88, 3672 (1966); (c) H. H. Wasserman and D. C. Clagett, J. Amer. Chem. Soc., 88, 5368 (1966). For review, see (d) N. J. Turro, Accounts Chem. Res., 2, 25 (1969).

Taking advantage of the stabilizing effect of large alkyl substituents, these workers were able to obtain a fairly concentrated solution of tetramethylcyclopropanone (55) by the same path [56].

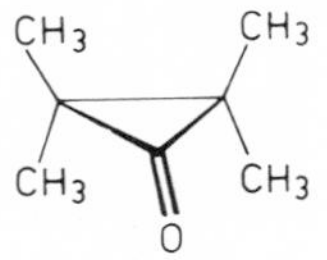

(55)

The strain in the ring is evident from the infrared spectrum, which shows ν C=O to be 1840 cm^{-1}. This number also clearly suggests that the classical structure describes this compound better than the oxyallyl alternative (57). *Trans*-1,2-di-*t*-butylcyclopropanone (56) can actually be isolated [57];

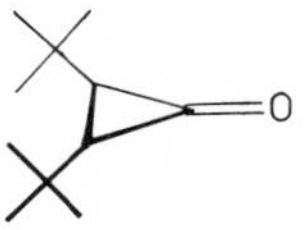

(56)  (57)

it has a melting point of 25°, and it has a half life of six hours at 150° in carbon tetrachloride. The *trans*-configuration follows from the fact that the benzyl hemiacetal (58) has a benzylic proton quartet rather than a singlet.

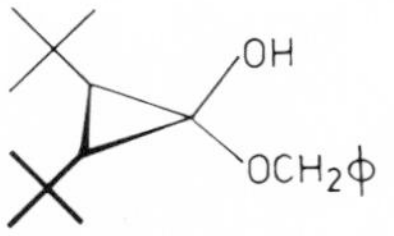

(58)

Later (56) was isolated in optically active form, which rules out the oxyallyl structure [58]. However, this compound readily racemizes at 80°, and the rate constant is ten times greater in acetonitrile than in isooctane, so that (57) may be an intermediate or low energy transition state.

---

56. N. J. Turro, W. B. Hammond, and P. A. Leermakers, *J. Amer. Chem. Soc.*, *87*, 2774 (1965).
57. J. F. Pazos and F. D. Greene, *J. Amer. Chem. Soc.*, *89*, 1030 (1967).
58. D. B. Sclove, J. F. Pazos, R. L. Camp, and F. D. Greene, *J. Amer. Chem. Soc.*, *92*, 7488 (1970).

Still another type of isolable cyclopropanone is the diquinocyclopropanone (59) [59]. Cyclopropanones are also intermediates in the Favorsky reaction (see p. 864); for a discussion of the remarkable cyclopropenones, the interested reader is referred to p. 304.

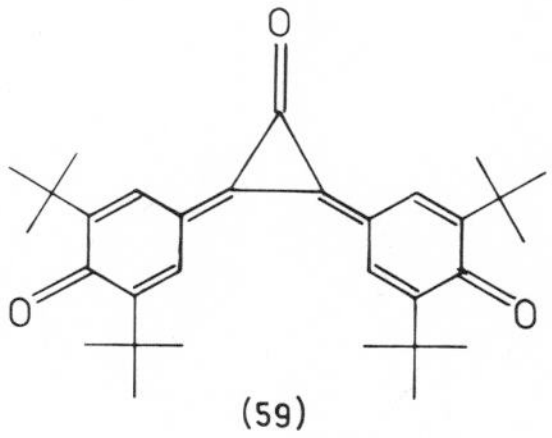

(59)

Both azirines such as (60) [60] and diazirines such as (61) [61] are known.

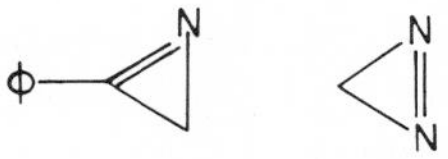

(60) (61)

Diazirine has a methylene singlet at 8.7 τ - a remarkably high value for a methylene proton having two unsaturated α-nitrogen atoms; oxirenes are not known except possibly as intermediates in the oxidation of acetylenes and in Wolff rearrangements (see p. 588). There are a number of compounds which combine some of the various features discussed above. Several α-lactams such as 1,3-di-t-butyldiaziridinone (62) are known [62].

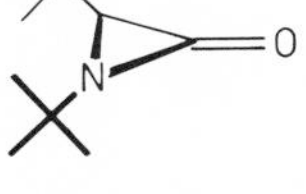

(62)

59. (a) R. West and D.C. Zecher, J. Amer. Chem. Soc., 89, 152 (1967); (b) D.C. Zecher and R. West, J. Amer. Chem. Soc., 89, 153 (1967).
60. G. Smolinsky, J. Amer. Chem. Soc., 83, 4483 (1961).
61. W.H. Graham, J. Amer. Chem. Soc., 84, 1063 (1962).
62. Reviewed by I. Lengyel and J.C. Sheehan, Angew. Chem., Int. Ed. Engl., 7, 25 (1968).

This compound can be heated to 140°, and has a $J_{C^{13}H}$ of about 170 Hz (see further p. 877). The bis-(1-adamantyl) analog (63) has been examined by means of X-ray diffraction and found to have a pyramidal nitrogen atom; the two hydrocarbon groups are in trans-positions [63].

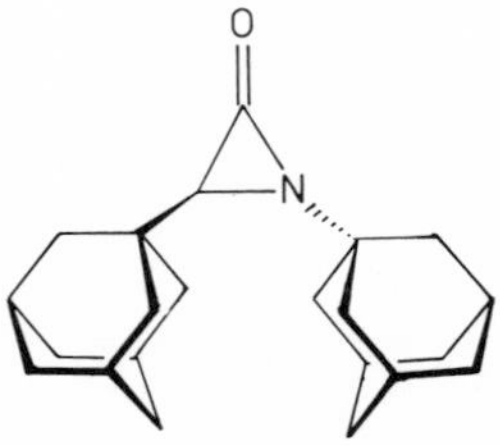

(63)

α-Lactones are again mostly the story of unstable intermediates (see p. 866). It may be mentioned here however, that the careful photolysis of 1,2-dioxolanediones (64) at 77°K leads to compounds (65), which polymerize to polyesters (66) even at -100°C. The carbonyl frequency of 1895 cm^{-1} suggests that the ring is closed as with the aziridinones [64].

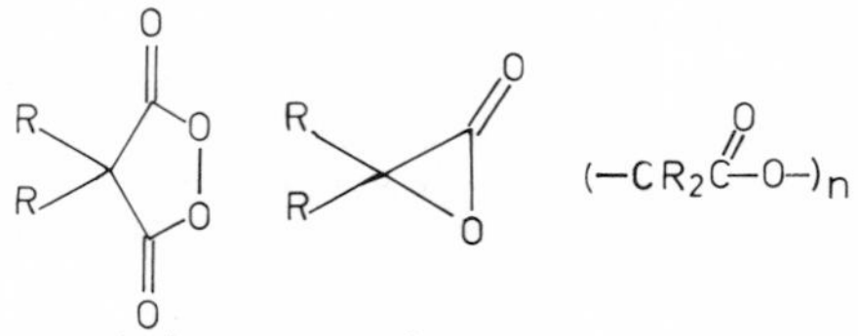

$(-CR_2C(=O)-O-)_n$

(64) (65) (66)

2,3-Diazocyclopropanone (67) has been isolated; it survives brief heating to 175° [65].

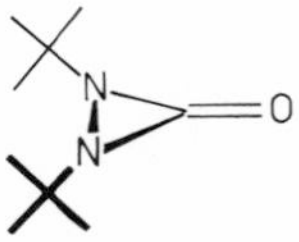

(67)

63. A.H.J. Wang, I.C. Paul, E.R. Talaty, and A.E. Dupuy, Chem. Commun., 43 (1972).

64. O.L. Chapman, P.W. Wojtkowski, W. Adam, O. Rodriquez, and R. Rucktäschel, J. Amer. Chem. Soc., 94, 1365 (1972).

65. F.D. Greene and J.C. Stowell, J. Amer. Chem. Soc., 86, 3569 (1964).

Bending at the site of acetylenic carbon can be tolerated to some degree. Cyclooctyne (68) has been described [66], but cycloheptyne (69) must be counted among the unstable intermediates (cf. Chapter 21).

(68) (69)

The introduction of α-gem-dimethyl groups and a sulfur atom in the ring leads to enhanced stability, and 3, 3, 6, 6-tetramethyl-1-thia-4-cycloheptyne (70) can be isolated [67]. The compound can be heated without decomposition to 140°; nevertheless, the strain is evident from the rapid reactions with oxygen and with dienes.

O_2

(70)

The role of the sulfur atom is evidently that its size permits some relaxation of the bond angles in the ring. The carbon analog (71) has also been prepared [68], but it dimerizes at room temperature within an hour. Models suggest that the deviation from linearity at each acetylenic carbon must be close to 30°.

(71)

---

66. W. Tochtermann, G. Schnabel, and A. Mannschreck, Ann., 705, 169 (1967), and earlier papers quoted there.
67. A. Krebs and H. Kimling, Tetrahedron Lett., 761 (1970).
68. A. Krebs and H. Kimling, Angew. Chem., Int. Ed. Engl., 10, 509 (1971).

1,2-Cyclononadiene (72) is the smallest cyclic allene capable of isolation; it does not appear to be unduly strained [69]. The next smaller members are mentioned in Chapter 10 (p. 384 and 389).

(72)

Still longer cumulenic chains obviously require larger rings to be accommodated; thus, 1,2,3-cyclodecatriene (73) has been isolated and it behaves like a normal 1,2,3-triene [70].

(73)

It might be pointed out that multiple features of unsaturation in a ring system do not necessarily build up the strain in proportion; thus, 1,7-cyclododecadiyne (74) is a stable compound [71], as a model would suggest.

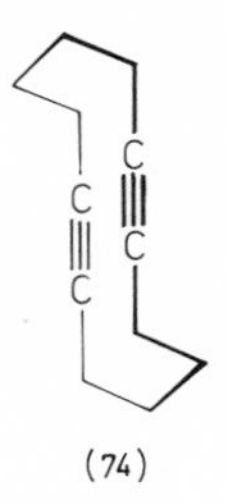

(74)

69. L. Skattebol, Acta Chem. Scand., 17, 1683 (1963).
70. W.R. Moore and T.M. Ozretich, Tetrahedron Lett., 3205 (1967).
71. D.J. Cram, Record Chem. Progr., 20, 71 (1959).

7-4. STERIC HINDRANCE

Historically speaking, steric hindrance became known as a consequence of a number of reactions that are either retarded or prevented altogether when the reaction sites are flanked by methyl groups [72], and the best known examples of the effect (often termed front or F-strain [1b]) continue to be in that area [1]. It should be pointed out that steric effects do not always cause rate retardation, but relief of steric crowding such as in the ionization of crowded tertiary halides (back or B-strain) may in fact lead to rate accelerations. In this section we shall concern ourselves with the limits to which large substituents can be crowded together, and measures by means of which the steric requirements of various groups may be measured.

The largest common substituents that organic chemists frequently have to contend with are the *t*-butyl and the phenyl groups (increasingly also, the adamantyl group). The former is more or less spherical, and little relief can be provided by rotating it, but interference by phenyl groups depends strongly on their orientation.

Two *tert*-butyl groups can be crowded quite closely together; thus, many *o*-di-*t*-butylbenzenes (75) are known.

(75)

Their effect on the structure of benzene has been measured by X-ray diffraction [73]; the effect is considerable, as is shown in (76). The ring is still close to exactly planar, but the angles show the stretching.

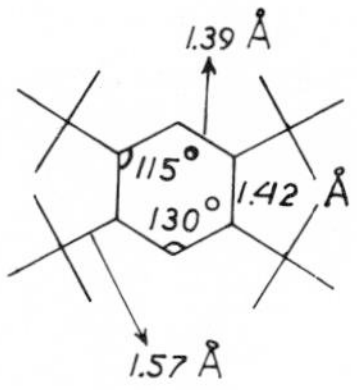

1.39 Å
115°
130°
1.42 Å
1.57 Å

(76)

72. A discussion of the historical aspects of the phenomenon is given by G.W. Wheland, "Advanced Organic Chemistry", Wiley, New York, 1960.

73. A. van Bruijnsvoort, L. Eilermann, H. van der Meer, and C.H. Stam, *Tetrahedron Lett.*, 2527 (1968).

The introduction of a third *t*-butyl group in the 3-position rarely succeeds. At present only one example is known [74]; curiously, when *t*-butylfluoroacetylene is allowed to stand unperturbed, 1,2,3-tri-*t*-butyltrifluorobenzene (77) is one of the substances formed.

F F F

(77)

Calorimetric experiments [75] have shown that the isomerization of 1,2- to 1,3-di-*t*-butylbenzene is exothermic by no less than 17-22 kcal/mole. An *i*-propyl group has of course much less severe steric requirements, and the existence of hexa-*i*-propylbenzene [76] has already been commented on (see p. 171). That these groups have little maneuvering room is clear from the fact that the double bonds in hexa-*i*-propenylbenzene do not react with bromine and permanganate ion [77a].

The interference of *o*-neopentyl groups has already been noted (see p. 170). The highly hindered hexakistrimethylsiloxybenzene (78) has been synthesized.

$(OSi(CH_3)_3)_6$

(78)

No evidence for restricted rotation was found; only one sharp singlet at 9.86 τ can be observed even at -100° [77b].

Peri-di-*t*-butylnaphthalenes such as (79) have been prepared [78]. In the parent compound the two *t*-butyl groups are bound to carbon atoms that are more widely separated, but they are as a first approximation oriented in a parallel direction and therefore interfere even more than *o*-*t*-butyl groups in benzene;

---

74. (a) H.G. Viehe, R. Merényi, J.F.M. Oth, and P. Valange, *Angew. Chem., Int. Ed. Engl.*, 3, 746 (1964); (b) H.G. Viehe, *Angew. Chem., Int. Ed. Engl.*, 4, 746 (1965).
75. Reviewed by H.H.J. Oosterwijk, *Chem. Weekblad*, 62, 325 (1966).
76. E.M. Arnett and J.M. Bollinger, *J. Amer. Chem. Soc.*, 86, 4729 (1964).
77a. E.M. Arnett, J.M. Bollinger, and J.C. Sanda, *J. Amer. Chem. Soc.*, 87, 2050 (1965).
77b. S. Murai, T. Murakawa, and S. Tsutsumi, *Chem. Commun.*, 1329 (1970).
78. R.W. Franck and E.G. Leser, *J. Amer. Chem. Soc.*, 91, 1577 (1969) and *J. Org. Chem.*, 35, 3932 (1970).

(79)

the strain energy has been estimated to be nearly 40 kcal/mole. Low temperature nmr studies [79] show that the *t*-butyl groups have a barrier to free rotation of about 6-7 kcal/mole. Nmr also shows that they are bent out of the plane in opposite directions, and that the barrier to flipping is more than 24 kcal/mole; thus, the quartet of the benzyl group in (80) does not broaden even at 195°.

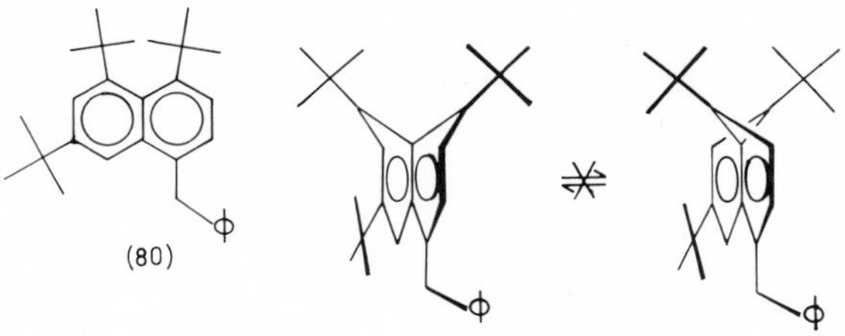

(80)

Cis-1,2-di-*t*-butylethylene (81) is likewise a known substance; again the two substituents are crowded together somewhat more than in the *o*-positions of benzene because of the shorter double bond.

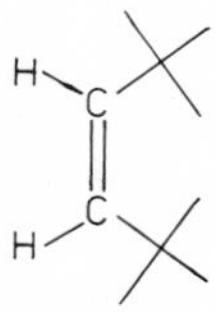

(81)

It may be noted at this time that neighboring groups do not have to interact in a repulsive manner; there are attractive van der Waals forces between them as well, and which are stronger depends on the individual case. Thus, it may seem surprising that in each of the ten 1,2-dihaloethylenes (except diiodo-), the *cis*-isomer is the more stable [80].

---

79. J.E. Anderson, R.W. Franck, and W.L. Mandella, *J. Amer. Chem. Soc.*, *94*, 4608 (1972).
80. (a) A.R. Olson and W. Maroney, *J. Amer. Chem. Soc.*, *56*, 1320 (1934); (b) H.G. Viehe, *Angew. Chem., Int. Ed. Engl.*, *3*, 152 (1964); (c) cf. also, *Chem. Eng. News*, *41*, 38 (Oct. 7, 1963).

At the other extreme, very tiny steric effects have been observed. Thus, optically active 9, 10-dihydro-4, 5-bis(trideuteriomethyl)phenanthrene (82) racemizes 13% faster than the protio analog. C-D bonds have a lower zero-point vibrational energy than C-H bonds, and hence presumably a smaller amplitude; in effect, this would make the CD_3-group smaller [81].

D_3C CD_3

(82)

Similarly, the enthalpy of association of boron trifluoride with hexadeuterio-2,6-lutidine (83) is slightly larger than with 2, 6-lutidine itself; this difference is not observed with the picolines (84), or in the formation of the lutidine-borane adducts (85) [82].

D_3C N CD_3 — N CD_3 — D_3C $\oplus$N CD_3 $\ominus BH_3$

(83) (84) (85)

An interesting recent example involves 8-deuterio[2. 2]metaparacyclophane (86). The AA'XX' nmr pattern of the para-ring coalesces at a temperature 2-5° lower than that of the protio analog; $k_D/k_H = 1.20$, and hence $\Delta\Delta G^{\ddagger} \approx 110$ cal/mole [83].

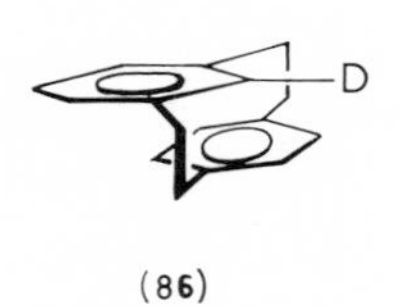

(86)

There are many ways in which steric hindrance can be demonstrated and measured. Highly crowded protons tend to be deshielded by the crowding atoms (compare the effect of H-bonding, p. 66); thus, in hexa-i-propylbenzene the methine septet is shifted downfield by 0.7 ppm compared to cumene, (87) [76].

---

81. K. Mislow, R. Graeve, A.J. Gordon, and G.H. Wahl, J. Amer. Chem. Soc., 86, 1733 (1964).
82. H.C. Brown, M.E. Azzaro, J.G. Koelling, and G.J. McDonald, J. Amer. Chem. Soc., 88, 2520 (1966).
83. S.A. Sherrod and V. Boekelheide, J. Amer. Chem. Soc., 94, 5513 (1972).

(87)

The infrared frequency of a C–H bond with a crowded hydrogen atom tends to shift to greater energies; thus, in the polycyclic hydrocarbons such as (88) the C–H bonds indicated have a fundamental stretching frequency about 100 cm^{-1} higher than those in cyclohexane (see p. 40) [84].

(88)

Ultraviolet spectra are likewise often informative. Thus, while alkyl substitution usually has only minor effects on the uv spectrum of benzene, 1,2-di-*t*-butyl substitution leads to large bathochromic shifts [75]. Evidently the substituents cause distortions that have less effect on the energy of the excited state than on that of the ground state.

There are several widely used chemical methods of measuring and expressing the steric requirements of groups. In one of these, the rate of racemization of optically active biphenyls (89) can be measured (cf. Section 5-11).

(89)

The rate constants are obviously related to the sizes of substituents A and B. Although a correlation of that sort does exist, it is not as useful as some others described below for the following reasons. First, the information so obtained

---

84. D. Kivelson, S. Winstein, P. Bruck, and R. L. Hansen, *J. Amer. Chem. Soc.*, **83**, 2938 (1961).

does not really pertain to normal ground state molecules, but to transition states in which undoubtedly several bonds are severely stretched or bent and hence comparisons with normal molecules may not be valid. Second, the configuration of the groups A and B is a rather special one that does not occur in many molecules. As we shall see below, groups that interfere especially severely in biphenyls need not hinder one another as severely in other molecules.

A second method depends on the determination of the equilibrium constant (or heat of reaction) for reversible associations. For example, when pyridine is compared with 2, 6-dialkylpyridines in its reaction with various Lewis acids [85], it becomes clear from the values of K and ΔH that the alkyl groups diminish the stability of the couples as the number of carbon atoms increases, and also as the Lewis acid itself is made more voluminous.

For situations in which the steric problem to be evaluated is similar to that which obtains in these acid-base complexes, the large amount of information that has been gathered in these reactions is very helpful, the only problem being the fact that the complexes are either charged (e.g., (90)) or zwitterionic (e.g., (91)), and the nature of the solvent interaction with such a species may be quite different than in the molecule being compared with it.

⊕N–H (90)   ⊕N–⊖BF_3 (91)

By far the most popular measure of the steric requirements is the so-called A value, which is simply the free energy difference ΔG (= -RT lnK) involved in the axial-equatorial equilibrium of substituted cyclohexanes [86].

⇌

85. For review, see (a) H. C. Brown, J. Chem. Educ., 36, 424 (1959); (b) H.C. Brown, Record Chem. Progr., 14, 83 (1953); (c) H. C. Brown, J. M. Brewster, and H. Shechter, J. Amer. Chem. Soc., 76, 467 (1954).
86. See E. L. Eliel, J. Chem. Educ., 37, 126 (1960) and Angew. Chem., Int. Ed. Engl., 4, 761 (1965).

The methods by means of which K may be measured were reviewed in Chapter 6 (p. 188); the interference in the axial form is primarily by the 3- and 5-hydrogen atoms.

The reasons for this popularity is first of all that the several methods by means of which A can be measured give reasonably similar answers [86]. Thus, when the A value of the hydroxy group is measured by means of nmr, it was found to be more or less independent of the presence of various groups in the 4-position [87]. Similar conclusions have been drawn for the A value of a methyl group [88].

Even more remarkable support of this sort was recently described [89] by Perrin. It is based on the Cope-Claisen rearrangement, which may be written in the general form

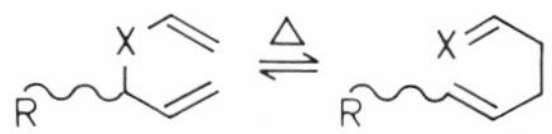

It has been shown (see p. 527) that the transition state in these reactions has the chair form, and the reactant may give either chair by means of simple internal rotations.

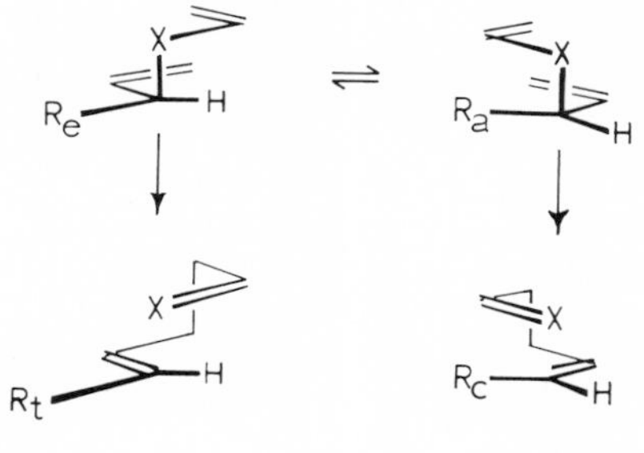

From the product ratios one can calculate $\Delta\Delta G^{\ddagger}$ for the two transition states, and this turns out to be almost identical to the known A value of R.

A second reason for the popularity of A values is that the type of interactions giving rise to it (with 3- and 5-hydrogen atoms) are very similar to those that occur in simple aliphatic derivatives.

---

87. R. D. Stolow, T. Groom, and P. D. McMaster, Tetrahedron Lett., 5781 (1968).
88. J. J. Uebel and J. C. Martin, J. Amer. Chem. Soc., 86, 4618 (1964).
89. C. L. Perrin and D. J. Faulkner, Tetrahedron Lett., 2783 (1969).

Table 7-2 shows a partial list of A values as recommended by Eliel [86].

TABLE 7-2. A Values for Common Substituents

H	0
F	0.2
Cl	0.4
Br	0.4
I	0.4
OH	0.6
OMe	0.7
NO_2	1.0
NH_2	1.2
NH_3	1.9
NMe_2	2.1
Me	1.7
C CH	0.2
Et	1.8
i-Pr	2.1
t-Bu	>4.4
C_6H_5	3.1
CN	0.2
COOH	1.2
2 e^-	?

Several features can be pointed out now that might at first sight seem surprising, but which should in fact logically be expected. First of all, there is the fact that all the simple spherical atoms have nearly equal A values although there is obviously a very large difference in size between hydrogen and iodine atoms. This is due to the simple fact that as the atom considered is increased in size it will also be further away (see Fig. 7-3).

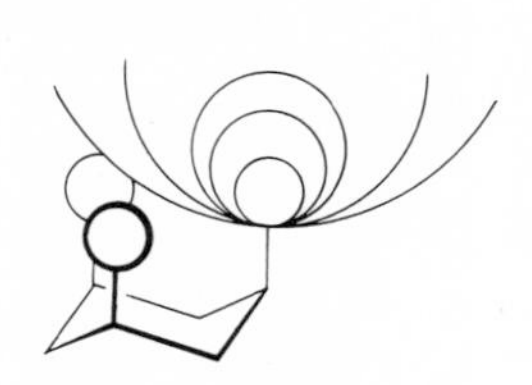

Fig. 7-3. A drawing suggesting that large axial atoms in cyclohexane interfere with a 3-hydrogen atom but less than in direct proportion to their radii. The vertical lines are the covalent bond radii of the carbon atoms.

The important contribution to steric hindrance comes obviously from branched or angular groups - particularly if they cannot be rotated out of the way. Thus, methyl, ethyl, and i-propyl groups have similar A values, but that of t-butyl is very much larger. In turn, the A value of methyl is larger than those of ethynyl and hydroxy (somewhat erratic A values are observed for the hydroxy group - probably due to strong interaction with the solvent by hydrogen bonding and the like).

One of the more difficult puzzles to sort out is the A value of the unshared pair. Thus, piperidine (92) may exist in either of two conformations, and an A value might be assigned to the unshared pair.

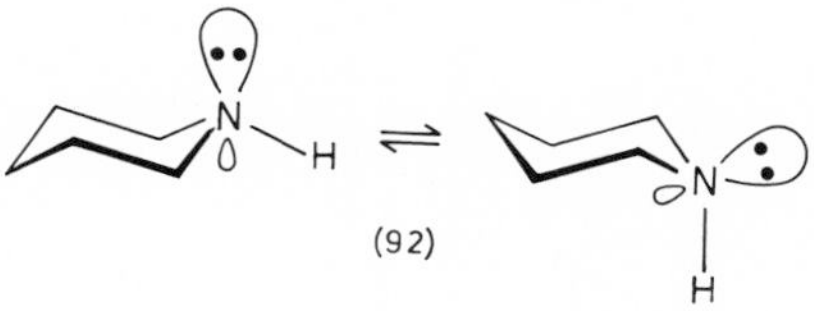

(92)

The problem is made difficult by hydrogen bonding and by the fact that the ring and the nitrogen atom can both invert and may do so independently. According to an nmr study of the neighboring methylene groups in N-substituted 3, 3, 5, 5-tetradeuteriopiperidines (93), the α-methylenes give rise to an AB-quartet in the proton nmr spectrum below $-10^\circ C$; $\Delta\sigma$ is 0.44 ppm when R is hydrogen, 0.94 ppm when R is methyl, and 1.00 ppm when R is t-butyl.

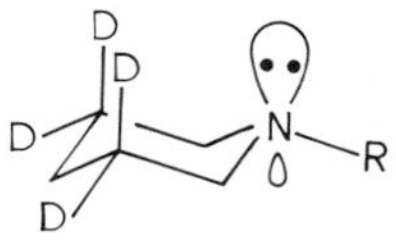

(93)

With all the hydrochlorides and with cyclohexane itself this difference equals 0.45 ppm. On this basis it was concluded that in piperidine the N-bound hydrogen must be axial and that the lone pair must be "bigger than" the proton [90]. This conclusion has been contested on the basis of an ir study in which the temperature dependence of the stretching frequencies of NH in piperidine was measured; equatorial hydrogen was concluded to be preferred by about 0.5 kcal. Both bands disappear when deuterium is substituted for the N-bound hydrogen atom; the band shapes were used to assign ν_e and ν_a [91].

---

90. (a) J. B. Lambert and R. G. Keske, J. Amer. Chem. Soc., 88, 620 (1966) and references given there. The assumptions on which this interpretation is based have been criticized, however; see (b) M. J. T. Robinson, Tetrahedron Lett., 1153 (1968).
91. R. W. Baldock and A. R. Katritzky, Tetrahedron Lett., 1159 (1968).

Another contradictory observation has been reported with the 1, 3-dioxanes (94), which undergo a carbonium ion type of epimerization in the presence of boron trifluoride as shown below.

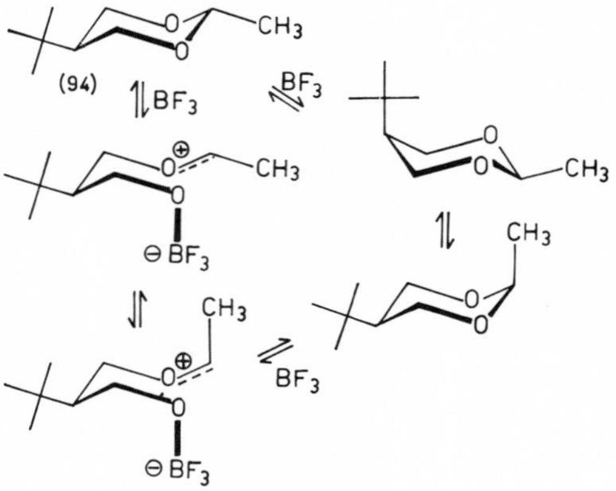

The reason for believing that in the *cis*-epimer the *t*-butyl rather than the methyl group is axial is the magnitude of the coupling constants J_{ab} and J_{ac}, both of which are less than 4 Hz; the Karplus rules would predict that J_{ab} should be at least 12 Hz or so for the conformer with equatorial *t*-butyl.

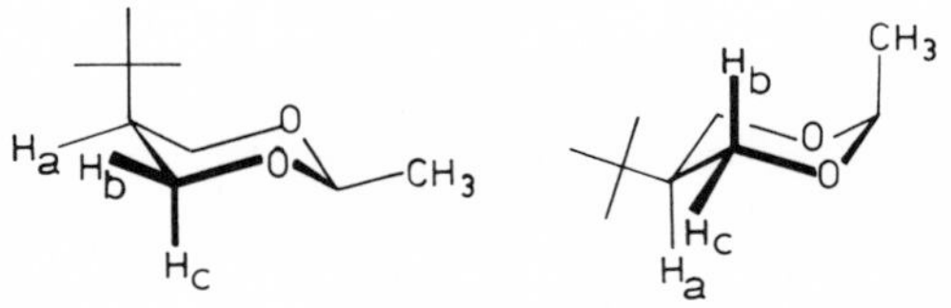

The equilibrium constant for the BF_3-catalyzed epimerization favors the *trans*-epimer by only 1.5 kcal/mole; evidently the interaction between the *t*-butyl group and the 1- and 3-unshared pairs is not severe [92].

Katritzky [93] calculated the theoretical dipole moments for the axial and equatorial conformations of N-*t*-butyldihydro-1, 3, 5-dithiazine (95) (1.59 and 2.75 D, respectively) and experimentally observed a value of 1.85 D. This also justifies the belief that axial *t*-butyl groups are possible if the 3- and 5-positions present only unshared pairs.

---

92. E. L. Eliel and S. M. C. Knoeber, *J. Amer. Chem. Soc.*, *88*, 5347 (1966).
93. L. Angiolini, R. P. Duke, R. A. Y. Jones, and A. R. Katritzky, *Chem. Commun.*, 1308 (1971).

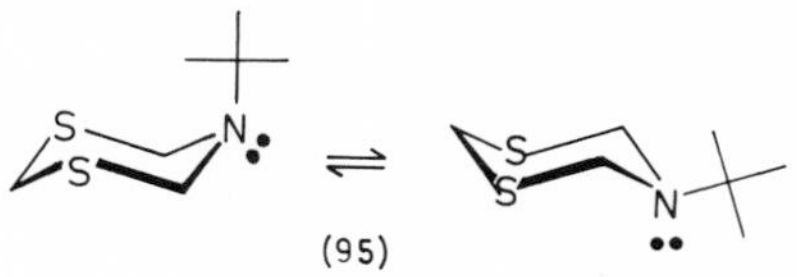

(95)

Thus, the weight of the evidence seems to say that the "size" of the unshared pairs is inconsequential; in fact, Allinger has suggested that many of the reported results can be explained by assuming that the lone pair has no size at all, and that the question whether the N-H group becomes axial or equatorial depends on whether attractive or repulsive interactions between the nitrogen-bound and 1- and 3-hydrogen atoms win out [94].

Perhaps the most clear-cut evidence for the proposition that an unshared pair is smaller than a hydrogen atom is provided by a recent nmr study of the metaphanes (96)-(98). The multiplet of the central methylene group collapses in the first three of these compounds at -75°, -28°, and -50°, respectively; in the case of (99), no coalescence was observed below room temperature [95].

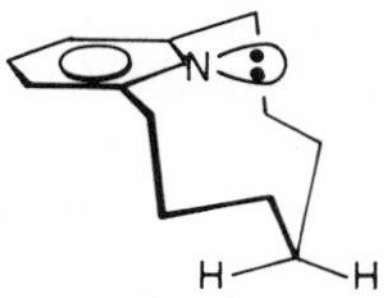

(96) 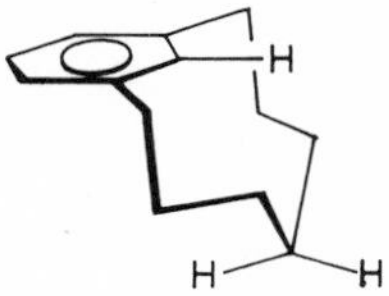 (97)

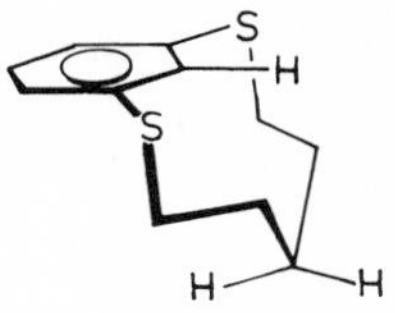

(98) 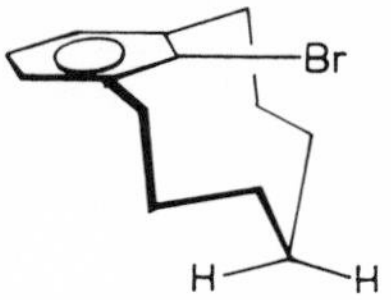 (99)

A final comment concerns the problem of polysubstituted cyclohexanes. It has been observed in many instances that A values are roughly additive. Thus, equilibrium constants can be computed on the basis of a list of values for the inversion processes of molecules containing two or more substituents, providing that in both conformations all 1,3-diaxial interactions involve at least one hydrogen atom.

---

94. N. L. Allinger, J. A. Hirsch, and M. A. Miller, Tetrahedron Lett., 3729 (1967).
95. S. Fujita, S. Hirano, and H. Nozaki, Tetrahedron Lett., 403 (1972).

7-5. THE WESTHEIMER MODEL

Since the structure of molecules appears to be a compromise between the best bond angles, the best bond lengths, and the optimum proximity of nonbonded atoms, it can in principle be calculated on a quantum mechanical basis, in a way that discards such crutches as steric, polar, and resonance effects. In practice, the job is much too complex, and even with the fastest and biggest computers available now or in the foreseeable future it will stay that way. However, if we are willing to compromise a little and accept a number of pieces of experimental information to guide us, remarkably good facsimiles of molecules can be obtained purely on the basis of steric strains by means of the Westheimer model [96].

This model is similar to that by means of which one might calculate the distortions of a mechanical object - such as a bridge - under various stresses. The input data then are the dimensions of the various components, the equations governing their behavior under stretching, bending, and twisting stresses and Newton's laws; the requirement is that all forces must be in equilibrium. By the same token, the calculations as applied to molecules require as input the masses of the atoms, the force constants of the various bonds against stretching, bending, and twisting, and the potential functions governing the forces operating between nonbonded atoms. At a given set of positions of the atoms the overall energy will be at a minimum - this minimum will lie some calculable distance above the level that would obtain if all distances and angles in the molecule could be ideal ones.

The force constants are not a serious problem in these calculations. It is true that the ir spectra of the more complex molecules are not very readily interpreted so as to yield all the fundamental frequencies, but this has been done in enough cases that reasonably accurate assumptions can almost always be made. The potential functions for interactions of nonbonded atoms present a more serious uncertainty, especially for the steep, repulsive part of the upper curve in Fig. 7-4; clearly a small error in r can make a large difference in energy.

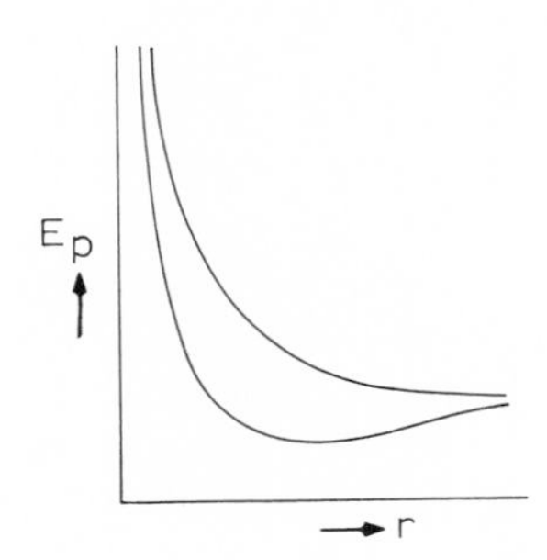

Fig. 7-4. The potential energies of bonded and nonbonded atoms as a function of distance.

96. F.H. Westheimer, "Steric Effects in Organic Chemistry", (M.S. Newman, ed.), Chapter 12, Wiley, New York, 1956, and references quoted there.

The detailed operations to be carried out in order to calculate a given case are beyond the scope of this book; however, an outline in principle is given here for the racemization of the optically active biphenyl (100).

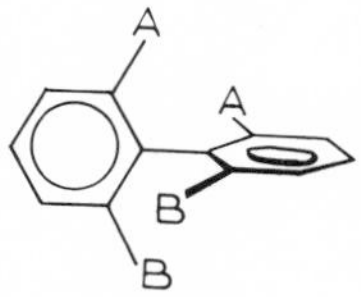

(100)

In the perpendicular conformation, the molecule is generally unhindered and we may set its potential energy equal to zero.

Racemization presumably requires the planar transition state (101), in which the atoms A and B have been bent back by angles of α and β such that the distance between them is d_1, and the inter-ring bond has been stretched by an amount d_2; we ignore all the other deformations that surely occur.

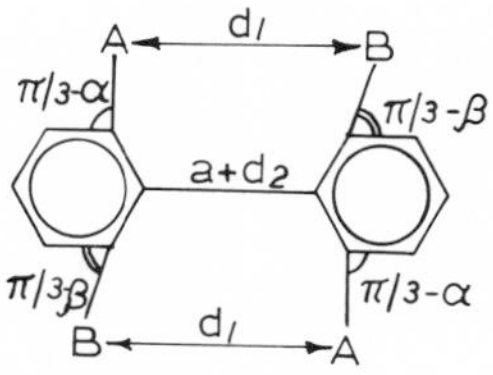

(101)

The distortions of the bonded atoms are assumed to be governed by Hooke's law

$$F = -kq$$

so that

$$E = 1/2kq^2$$

Here k is the force constant and q the displacement. The total energy (the activation energy for racemization) will be

$$E_a = 2k_1(\alpha)^2 + 2k_2(\beta)^2 + k_3(d_2)^2 + 2E(d_1).$$

The force constants of the bending modes CCA and CCB, k_1 and k_2, respectively, and that of the φ–φ stretching mode may be obtained from ir spectra; most likely they can be found listed in the appropriate literature. The potential function $E(d_1)$ can perhaps be estimated from the van der Waals constants a and b of gaseous substances AB, A_2, and B_2. The distance d_1 can be expressed in terms of α, β, and d_2, the three variables in our equation. Finally, we have

$$\frac{\partial E_a}{\partial \alpha} = \frac{\partial E_a}{\partial \beta} = \frac{\partial E_a}{\partial d_2} = 0$$

and it then remains only to solve for these variables and E_a.

These calculations lend themselves to iteration and hence to computer application; a highly sophisticated version is now available, its further refinement is an active area of research [97].

---

97. (a) K. B. Wiberg, J. Amer. Chem. Soc., 87, 1070 (1965); (b) N. L. Allinger, M. T. Tribble, M. A. Miller, and D. H. Wertz, J. Amer. Chem. Soc., 93, 1637 (1971).

Chapter 8

RESONANCE

8-1. INTRODUCTION.

The theory of resonance is a direct off-shoot from one of the main developments in physics in the first part of the twentieth century. In the hunt for approximate solutions of the Schrödinger equation for complicated chemical systems, initially such as the helium atom and the hydrogen molecule, but later for molecules as complex as benzene, the so-called Valence Bond Method was developed, especially by Pauling and Wheland [1].

In spite of its mathematical origins, the theory eventually developed a highly useful descriptive language, and it is actually this language that survives to this day. The theory itself is somewhat passé now; it cannot be said that it is incorrect in any way, but it has been found to lend itself less readily as a basis for new predictions than the rival Molecular Orbital Theory. In this chapter we describe

---

1. The outstanding reference on the history, the concept and the uses of resonance is G.W. Wheland, "Resonance in Organic Chemistry", Wiley, New York, 1955.

the resonance language as well as a variety of examples of how it applies. We begin with a description of the benzene problem.

8-2. THE LANGUAGE OF RESONANCE.

For a long time, benzene presented one of the more difficult structural problems to organic chemists. Making use of the rules of the structural theory, one can easily write all the possible structures for a compound C_6H_6, but then two pieces of the puzzle do not fall into place.

The first of these is that with substituted benzenes such as C_6H_5R, $C_6H_4R_2$, C_6H_4RS, etc., the isomer number cannot always be made to fit the facts. The experience is that the three formulas just listed correspond to 1, 3 and 3 isomers, respectively, and one can easily convince himself that structures such as (1)-(5) do not suffice.

$CH_2{=}C{=}CH{-}CH{=}C{=}CH_2$ (1)

$CH_3{-}C{\equiv}C{-}C{\equiv}C{-}CH_3$ (2)

(3) (4) (5)

The prismatic structure (6) meets the requirement; however, several of the disubstituted members of these are asymmetric and should be resolvable. Since this was never accomplished, one concludes that structure (6) predicts the wrong number of stereoisomers.

(6)

Kekulé contributed the idea of two rapidly equilibrating cyclohexatriene structures (5).

(5)

This would prevent the separation of isomers o-$C_6H_4R_2$ (7), and it gets us successfully around the isomer number problem.

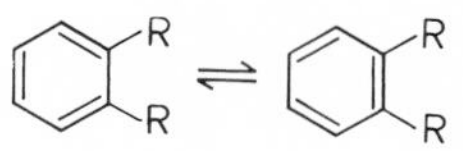

(7)

However, there is another difficulty, namely, that benzene does not have the properties expected for such Kekulé structures on the basis of experience with other olefinic structures. Thus, while olefins with some rare exceptions rapidly add bromine and discolor permanganate ion, benzene is stable to these reagents. Bromine can be added, but the reaction is slow; on the other hand, when Lewis acids are present, a rapid reaction of bromine takes place but it is substitution rather than addition.

In the resonance picture benzene is considered not to be an equilibrium mixture of the two Kekulé structures, but a hybrid of the two. That is to say, it is neither one, but it has a new structure intermediate between the two structures suggested by drawings (5). Structures (5a) and (5b) are really hypothetical; they are called contributing or canonical structures (Kekulé structures in the case of benzene and its derivatives); (8) is called the resonance hydrid structure. In (5a) the carbon-carbon bonds are alternately long and short, and, in (5b), short and long; in the hybrid, all are equally long, and of intermediate length.

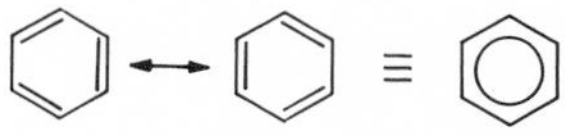

(5a) (5b) (8)

The properties of the hybrid in general will be intermediate between those of the contributing structures; however, an important feature of the theory is that the energy of the hybrid will be lower. If a good estimate can be made of the energy of the contributing structures, the amount by which resonance lowers it is called the resonance energy. A reasonable estimate of the resonance energy can often be made by means of calorimetric experiments. Thus, the heat of hydrogenation of cyclohexene can be determined, and one assumes that the heat of hydrogenation of benzene therefore should be three times as great. The difference between this expected value and the observed one is the resonance energy - which amounts to 36 kcal/mole in the case of benzene. This resonance energy is responsible for the anomalous indifference of benzene to bromine; once a bromine molecule has been added, the resulting dibromide (9) can no longer be written as a resonance hybrid of Kekulé structures. This loss of resonance energy - or at least some part of it - must therefore be added to the activation energy, and this in turn makes for a slower reaction.

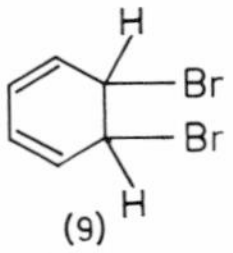

H
Br
Br
H
(9)

The number of contributing structures need not be two, nor need all structures be equivalent and contribute equally. Thus, naphthalene (10) has three canonical structures of which only two contribute equally; the third may contribute more or less ("be more or less important") than these two.

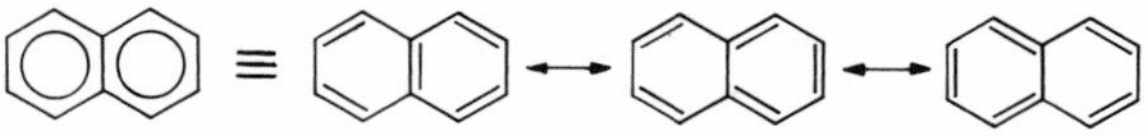

(10)

The three principal rules by which possible contributing structures are judged are the following. First of all, the structures to be considered must have the same nuclear positions. This is obviously not the case, to give an example, in the structures of butane (11) and isobutane (12) and hence these two structures correspond to different substances.

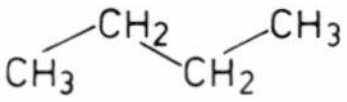

CH_3 CH_2 CH_2 CH_3

(11)

CH_3 CH CH_3 CH_3

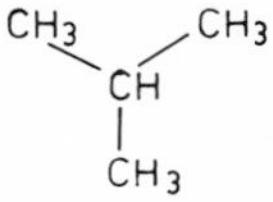

(12)

This rule is actually not truly unambiguous. The classical structures said to contribute to the hybrid in a case such as benzene do differ slightly in the positions of the nuclei, since double bonds are shorter than single ones. It is considered that if the gain in resonance energy warrants it, the nuclei will be displaced from their classical positions, and in extreme cases such as benzene may be at intermediate positions.

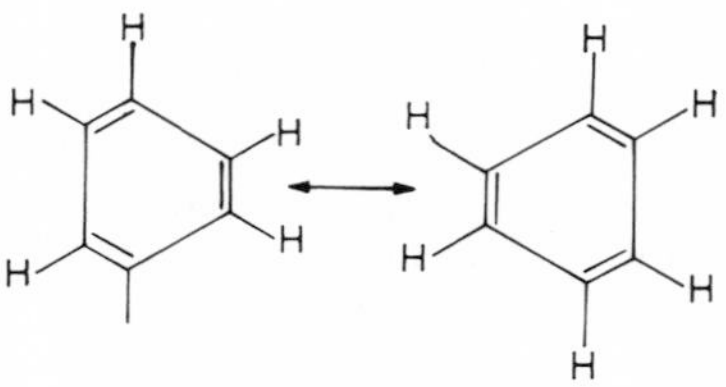

If only small displacements are necessary in order to arrive at an intermediate structure, this does not necessarily mean that a resonance hybrid will result. Thus, we could write two structures for semibullvalene, (13a) and (13b), in which the nuclear positions are nearly the same; however, the actual molecule is only one - or better: A degenerate mixture - of these, with distinctly different single and double bonds.

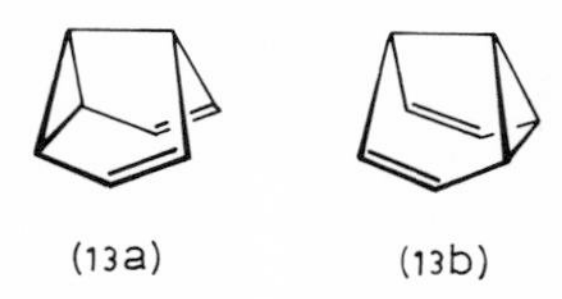

In the diazasemibullvalene (14), a theoretical prediction has it that there is resonance, and a single structure; but this molecule is not yet known (see p. 407).

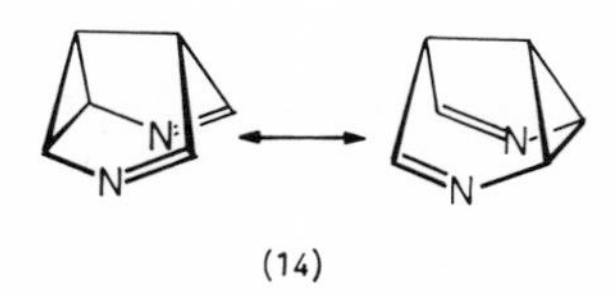

In semibullvalene, a very rapid degenerate interconversion is known to take place between the two structures; such reactions are referred to as valence tautomerizations, and these are discussed further in Chapter 11. Figure 8-1 shows the various possible differences between pairs of identical structures.

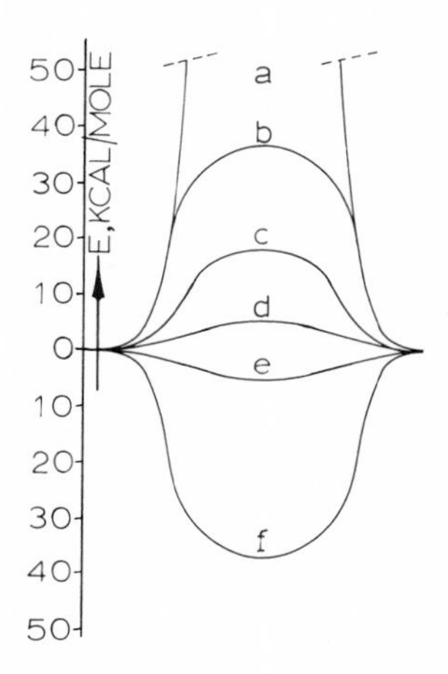

Fig. 8-1. Degenerate rearrangements and resonance. See the text for explanation.

Curve a represents the degenerate isomerization of propylene. This reaction cannot be realized in the laboratory and hence the height of the barrier is not known; it probably lies somewhere between 50 and 100 kcal/mole.

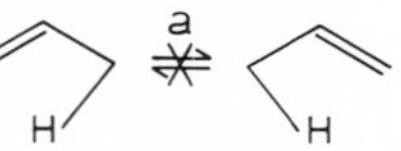

Curve b is for a similar isomerization of 1, 3-pentadiene - a reaction which can occur if the condition of a high temperature is met.

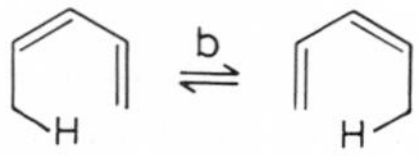

In curve c, with bullvalene as an example, the barrier is so low that the isolation of a specifically labelled species at room temperature would fail.

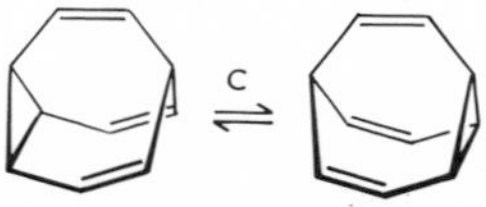

Curve d is an example (semibullvalene) of a barrier so low that sophisticated physical techniques are necessary to show that each of the two structures do have a finite existence (i.e., that some vibrational levels exist in each of the two energy minima).

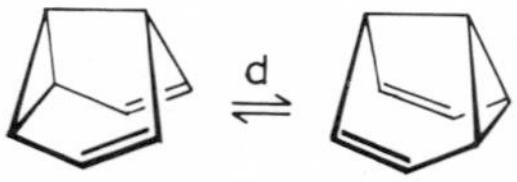

In curve e (the thiothiophthenes are an example), the energy barrier has vanished altogether, and there is now only one molecule which is a resonance hybrid; however, the well is not deep, and the resonance energy is negligible.

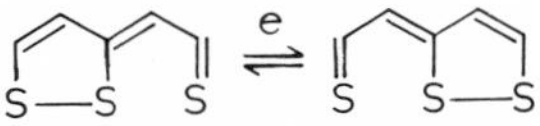

Perhaps only a minor perturbation, such as the presence of otherwise harmless substituents such as phenyl groups may restore the situation to one such as is represented by curve d. Finally, f (benzene) is an example of a hybrid characterized by a large resonance energy.

f

In every case the nuclei are in nearly the same positions, yet one cannot be sure without experiment in any given instance what type of situation one is dealing with. This uncertainty is at the heart of the non-classical ion controversy (Chapter 20).

A second rule is that all the structures considered must have the same number of unpaired electrons. Thus, a structure such as (15) cannot be considered to contribute to the character or properties of benzene.

(15)

A third rule is concerned with the "importance" or "weight" of the various contributing structures (those of the lowest energy contribute the most). Let us consider the isoelectronic molecules ketene and diazomethane as examples. The principal structure of ketene is (16a):

$H_2C{=}C{=}O$

(16a)

(16b)-(16e) might all be written as possible canonical structures.

$H_2C{=}\overset{\oplus}{C}{-}\overset{\ominus}{O}$ (16b) $H_2\overset{\ominus}{C}{-}\overset{\oplus}{C}{=}O$ (16c) $H_2\overset{\ominus}{C}{-}C{\equiv}\overset{\oplus}{O}$ (16d) $H_2\overset{\oplus}{C}{-}C{\equiv}\overset{\ominus}{O}$ (16e)

The classical structure is the "best" of these; i.e., the real molecule is best represented by it, and resembles it most closely. This classical structure obeys the octet rule: Each of the atoms has an octet of valence electrons.

```
H
 ·
  :C::C::O:
 ·
H
```

There is also no charge separation - which would increase the energy of the molecule.

Structures (16b) through (16e) all have this drawback of charge separation; hence all are of relatively high energy and contribute little to the nature of ketene. Among them, (16b) and (16e) have the virtue that at least the negative charge is at the most electronegative atom. Structures (16b), (16c), and (16e) all violate the octet rule; (16e) has ten electrons on oxygen and hence is of very high energy. In general, when two structures of very different energy contribute, the real molecule strongly resembles the structure of lowest energy, and little resonance energy results as is suggested by Fig. 8-2 (it may be noted here in passing that a very similar argument applies, with the curves drawn upside down, when activation barriers are considered; the principle is then known as the Hammond postulate (see p. 545)). For diazomethane no structure can be written in which there is no charge separation and in which the octet rule is satisfied; there is in this case not a single "good" structure.

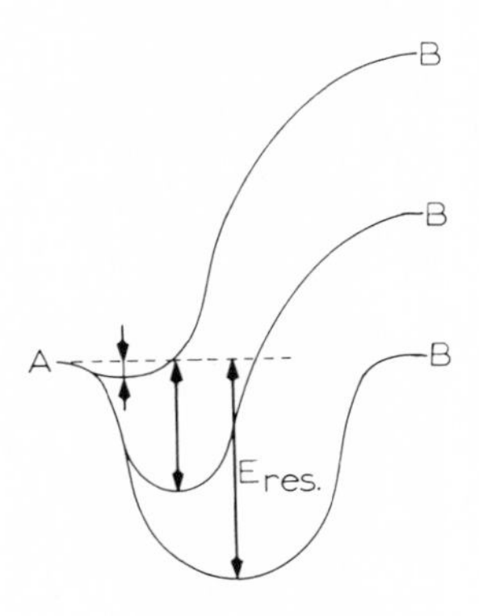

Fig. 8-2. Positions of the nuclei. A diagram suggesting how the resonance hybrid resembles structure A more and more strongly as the energy of structure B increases. Simultaneously, the resonance energy becomes smaller.

Structure (17a) is probably the best. In such a case much can obviously be learned from the experimental bond lengths in the molecule.

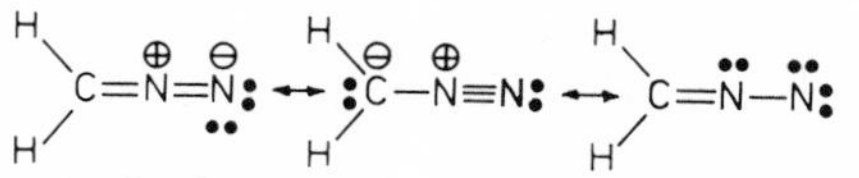

$H_2C{=}\overset{\oplus}{N}{=}\overset{\ominus}{N}{:} \longleftrightarrow H_2\overset{\ominus}{C}{-}\overset{\oplus}{N}{\equiv}N{:} \longleftrightarrow H_2C{=}\ddot{N}{-}\ddot{N}{:}$

(17a) (17b) (17c)

Attempts are often made to circumvent the somewhat cumbersome writing of resonance structures. Aromatic rings are often written with a circle in the center, and a molecule such as diazomethane might be represented by (17d).

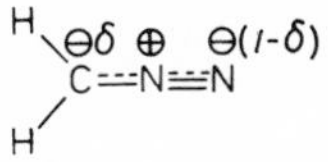

$H_2\overset{\ominus\delta}{C}{\cdots}\overset{\oplus}{N}{\equiv}\overset{\ominus(1-\delta)}{N}$

(17d)

While these symbols are usually clear, in more extended systems the uses of circles and of dotted lines to indicate fractions of pairs of electrons are not recommended. As an example we may consider the chemistry of compounds dihydroheptazethrene (18) and dibenzo[op, uv]-5H, 7H-pentacene (19).

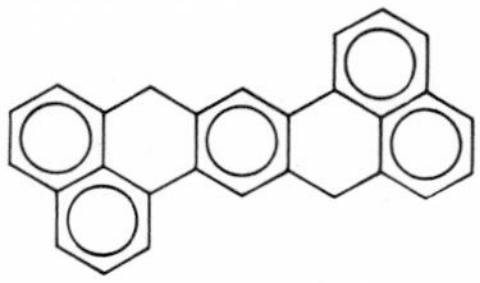

(18) 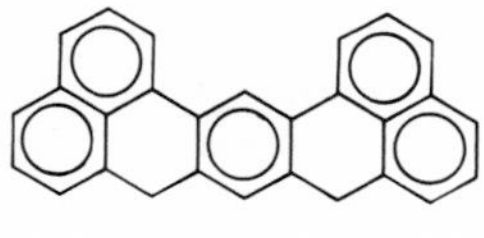 (19)

It might be supposed that upon dehydrogenation (18) would give (20) and (19) would give (21). The former of these products actually exists (heptazethrene, (20)), and can be produced in this manner, but dehydrogenation of (19) leads only to a polymer. Kekulé structures cannot be written for (21).

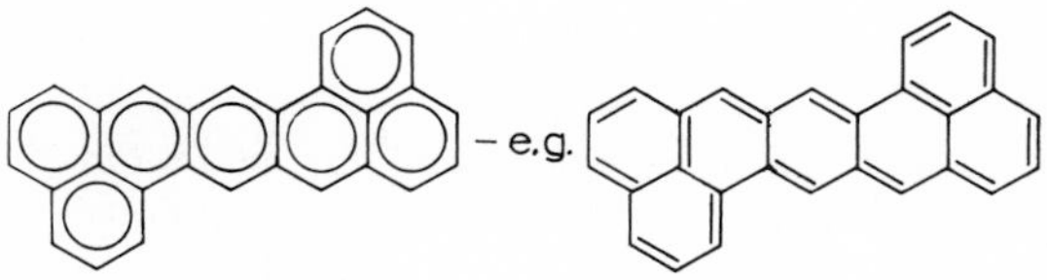

– e.g.

(20) 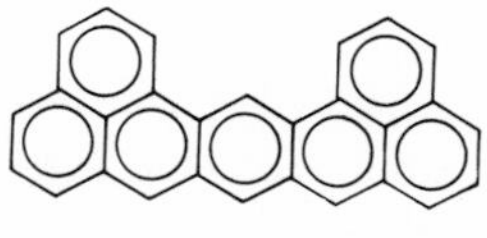 (21)

As far as the mathematical formulation of the theory is concerned, it is superficially similar to that of the molecular orbital theory. It is assumed that the wave function describing the real molecule is a linear combination of the functions for the contributing structures. In the MO theory the number of coefficients is equal to the number of atomic orbitals combined and is therefore limited; by contrast, in the valence bond method the number of contributing structures is ideally infinite, since each individual structure, no matter how "bad", will still lower the energy a little bit if it is included. In practice of course the number of structures considered in a mathematical treatment is limited by the complexity of the operation.

8-3. EFFECT OF RESONANCE ON ACID STRENGTH

Benzene is surely the best known and classic example of resonance but there are many others. One of the most easily observed and dramatic effects is that on acid and base strengths. Thus, the fact that carboxylic acids are some 10 orders of magnitude more acidic than alcohols is due in large part to the resonance of the anion (the inductive effect of the α-carboxyl group acts in the same direction).

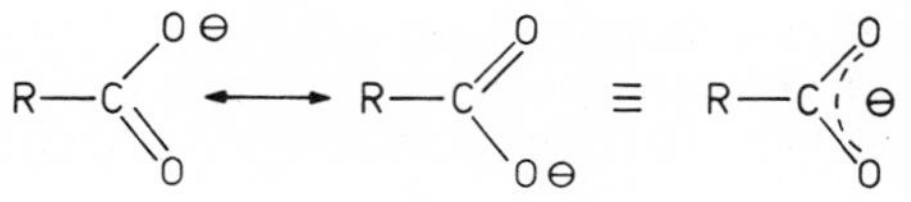

R—C(=O)O⊖ ⟷ R—C(=O)O⊖ ≡ R—C(O)(O) ⊖

X-ray diffraction measurements of carboxylate salts show the two C-O bond lengths to be equal. Many of the simple inorganic oxyacids similarly owe at least part of their strength to resonance. Base strength may likewise be affected. Thus, while amines are generally only weakly basic, guanidine (22) is comparable to the inorganic hydroxides because of the resonance of the protonated structure (23) (this particular type of resonance is sometimes referred to as Y-delocalization) [1a].

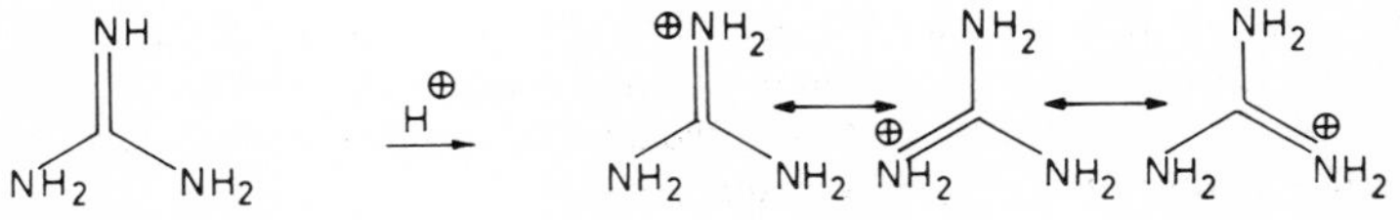

(22) $\xrightarrow{H^\oplus}$ (23)

Even when only non-equivalent structures can be considered, the effects may still be noticeable. Thus, phenol (24) is very much stronger than aliphatic alcohols because several resonance structures may be written:

1a. P. Gund, J. Chem. Educ., 49, 100 (1972).

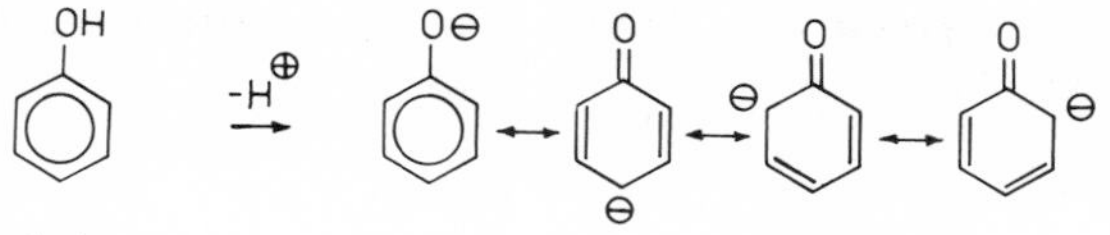

(24)

(or, more briefly but also less clearly as (25)).

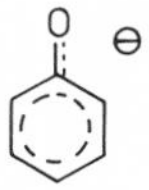

(25)

Of these structures the most important is of course the one in which oxygen carries the negative charge, but the others have enough weight to account for the increase in acidity; the pKa≈10, in contrast to that for most simple alcohols for which pKa ≈ 15. If the negative charge can be delocalized to other electronegative sites, further increases of acid strength can be expected; thus p-nitrophenol (26) has a pKa of 7.2,

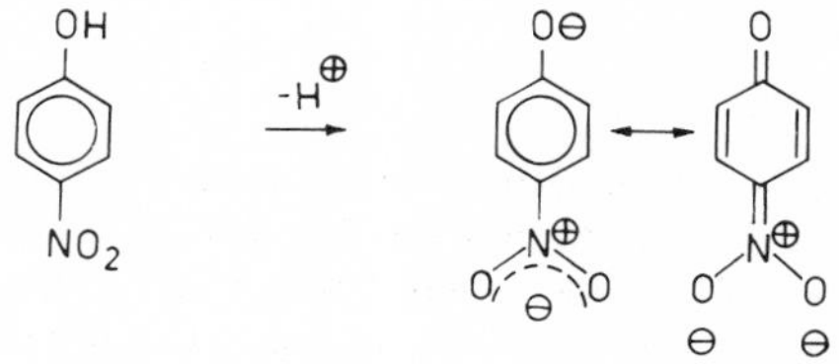

(26)

and picric acid (27) is comparable in strength to the inorganic acids.

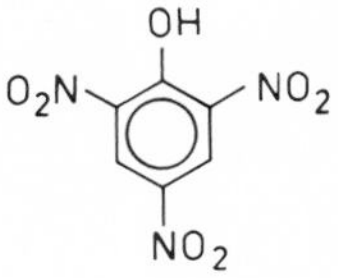

(27)

Another interesting group of examples is the series squaric acid (28), croconic acid (29) and rhodizonic acid (30); these are among the relatively few strong dibasic organic acids [2].

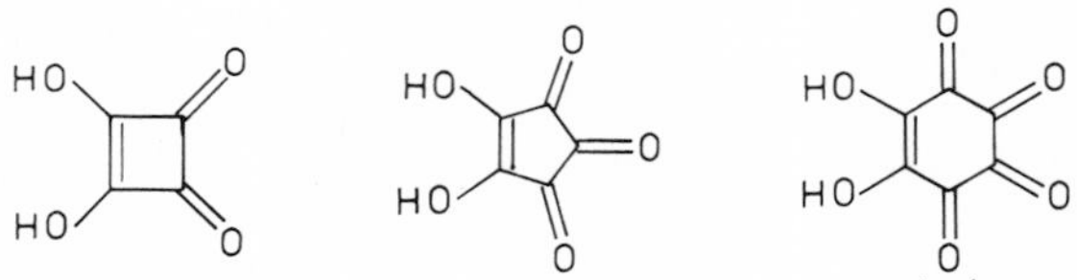

(28) (29) (30)

Thus, squaric acid has a $pK_1 \approx 1$ and $pK_2 \approx 3$, which makes it comparable to sulfuric acid in strength. Unlike the free acid, the salts have no ir absorption in the normal C=O and C=C ranges [3], further suggesting that the dianions are characterized by resonance such as (31).

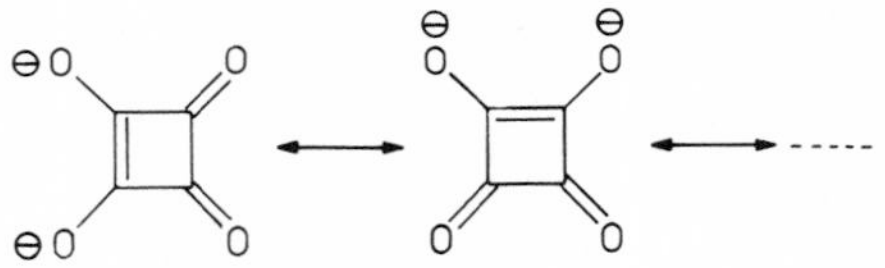

(31)

In fact, a complete coordinate analysis of rhodizonate anion for example, based on ir and Raman data, has shown that this species has D_{6H} symmetry [4]. Methylhydroxycyclobutenedione (32) likewise has a pK_1 of 1 and the anion is subject to rapid base catalyzed deuterium exchange, so that even in that case there is a strong tendency to form dianion (33) [5].

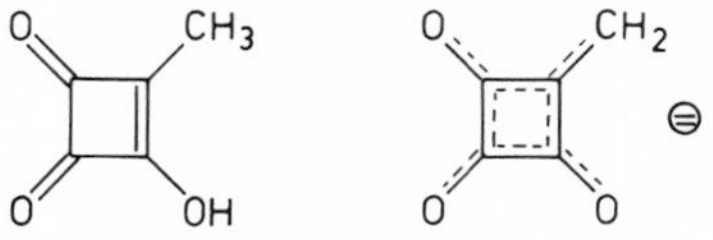

(32) (33)

---

2. (a) R. West and D.L. Powell, J. Amer. Chem. Soc., 85, 2577 (1963), and the several papers immediately following that reference; (b) Chem. Eng. News, 40, 40 (April 23, 1962); (c) D. Alexandersson and N.G. Vannerberg, Acta Chem. Scand., 26, 1909 (1972).
3. S. Cohen, J.R. Lacher, and J.D. Park, J. Amer. Chem. Soc., 81, 3480 (1959).
4. R.T. Bailey, J. Chem. Soc. (B), 627 (1971).
5. J.S. Chickos, J. Amer. Chem. Soc., 92, 5749 (1970).

Even relatively simple hydrocarbons may become quite acidic due to resonance; cyclopentadienes are among the best known examples. Thus, fluorene (34) has a pKa of 23;

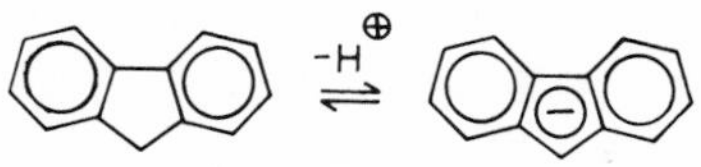

(34)

that of diphenylmethane (35) is 35.

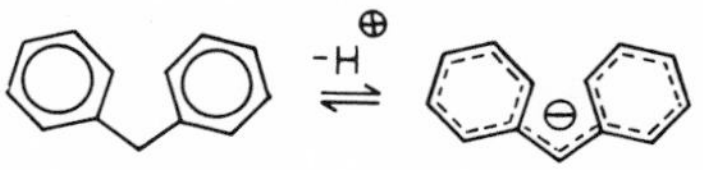

(35)

The effect of resonance need not necessarily be an increase in acid or base strength. The anilines become progressively weaker bases as nitro groups are placed in *o*- and *p*-positions in the ring.

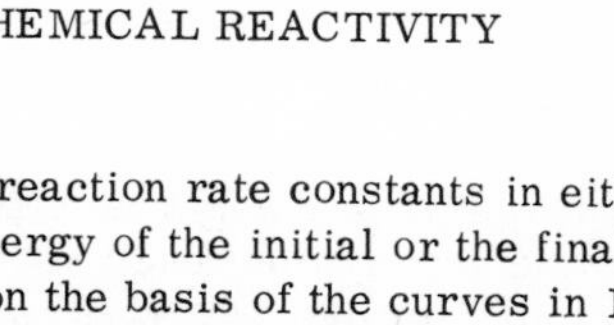

8-4. RESONANCE AND CHEMICAL REACTIVITY

Resonance may affect reaction rate constants in either direction, depending on whether it lowers the energy of the initial or the final state. The situation may be conveniently discussed on the basis of the curves in Fig. 8-3. If we start the reaction with similar starting materials and obtain a resonance stabilized product in one case, then this lower energy will not become evident only as the nuclei reach their final positions but even as they approach them, and a significant lowering of the activation energy results (see also p. 545, p. 701).

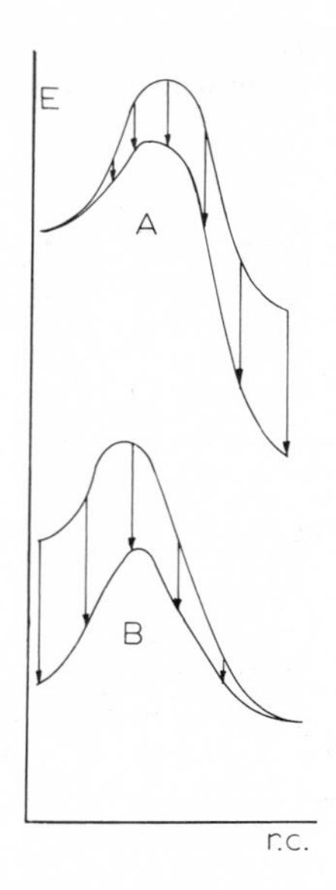

Fig. 8-3. Comparison of two similar reactions, one of them (a) giving a resonance stabilized product, the other (b) starting with a resonance stabilized reactant.

As an example, we may consider the ionization of n-propyl chloride (36) and of allyl chloride (37). Throughout the traversal of the reaction coordinate we may consider the allyl group to be a resonance hybrid, with the structure (38a) contributing virtually 100% at the beginning, and (38a) and (38b) being of equal importance at the end:

Cl Cl ⊕ ⊕

(36) (37) (38a) (38b)

Resonance plays an important role in the directional effects of substituents in the benzene nucleus during aromatic substitution reactions. We may consider the nitration of aniline (39) and of nitrobenzene (40) as examples. The former may be considered a resonance hybrid of structures (39a)-(39d), and the latter of structures (40a)-(40d).

NH_2 ⊕NH_2 ⊕NH_2 ⊕NH_2

(39a) (39b) (39c) (39d)

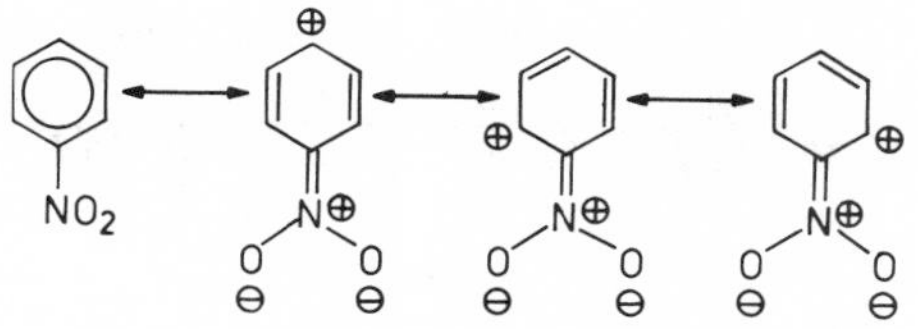

(40a) (40b) (40c) (40d)

The reactive intermediate in nitration reactions is the nitronium ion NO_2^+ (in many solvents, this ion may be transferred by a carrier molecule). This ion clearly will be directed to the o- and p-positions of the aniline molecule, and in fact, the approach of a positively charged ion is likely to reinforce the charge separation (polarize the aniline molecule; increase the weight of such structures as (39b)-(39d). Another way to explain the directional effect is to say that an intermediate such as (41) is of lower energy than (42) because the former has the advantage of resonance structure (41a).

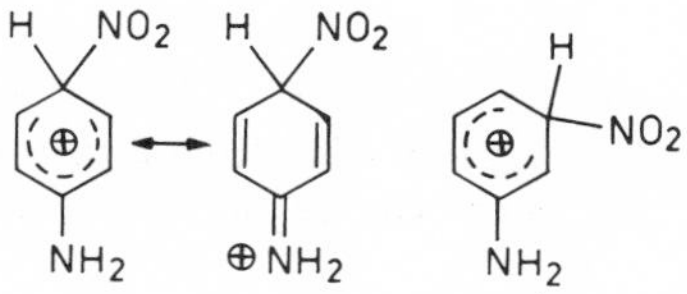

(41) (41a) (42)

In nitrobenzene, the o- and p-positions carry a partial positive charge, and the directional effect is now to the m-position. It is furthermore easily explained on the same basis why the nitration of aniline is faster, and that of nitrobenzene slower, than that of benzene.

8-5. RESONANCE AND PHYSICAL PROPERTIES

There are numerous examples of effects of resonance on physical properties. Quite often, these come about because one of the contributing structures is a zwitterionic one; this may affect even so elementary a property as the boiling point. Thus, it has been reported that cyclopropenone (43) has a boiling point of 30° at 0.5 torr [6]; we may compare that to a value of about -50° for acetone at this pressure. This great difference is surely due to the polar nature of cyclopropenone, which can only be attributed to resonance.

---

6. R. Breslow and M. Oda, J. Amer. Chem. Soc., 94, 4787 (1972).

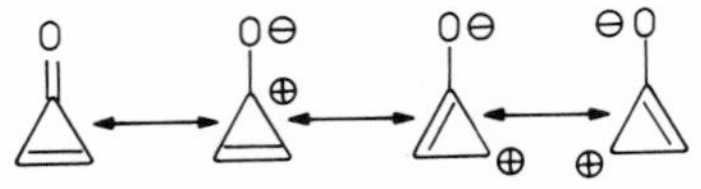

(43)

There are many known cases of resonance effects on the dipole moment. The moments of aniline (39) and of nitrobenzene (40) are 1.53 and 3.95 D, respectively; that of p-nitroaniline is 6.35 D - more than the sum of the individual components [7].

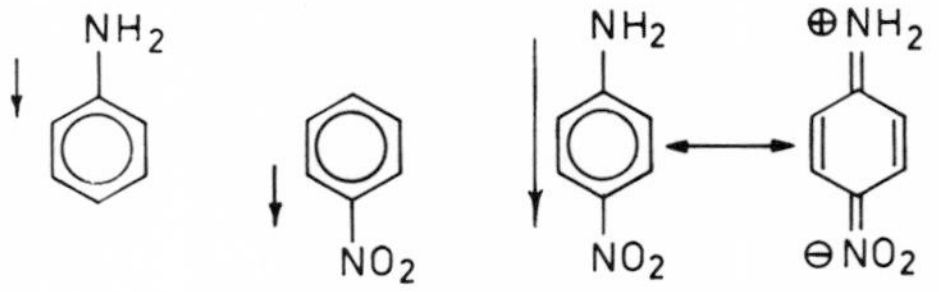

(39) (40) (44) (44a)

The fact that the ring is the negative end of the dipole in aniline itself in spite of the normally greater electronegativity of nitrogen is in itself an indication of resonance.

Resonance can frequently be very easily demonstrated by means of ir spectra. The stretching frequencies of vinyl and carbonyl groups are both depressed by easily measurable amounts when these two bonds are conjugated. This is easily understood in terms of resonance: There is a small but discernible contribution by a structure in which both of these bonds are formally single.

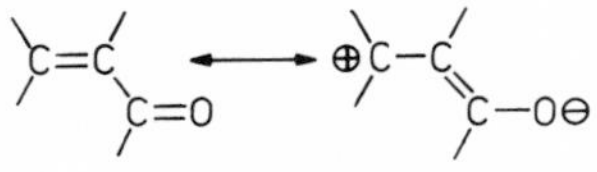

The carbonyl stretching frequency of carboxylic acids is generally about 1700 cm^{-1}; the carboxylate salts instead have two broad absorptions about 1550 and 1350 cm^{-1}. This shows that single bond character is pronounced in the carbonyl bond of the salt [8].

---

7. Ref. 1, p. 233.
8. The simplicity of this argument is perhaps deceptive: It is based on the assumption that when a bond becomes less stiff, less energy will be required to raise it to the first excited vibrational level - which is probably not obvious. See p. 29.

Resonance can often by shown to be characterized by a bathochromic shift (red shift, towards longer wavelength) in the electronic spectrum; thus, λ_{max} is found to be shifted that way when the spectrum of aniline is compared with that of the anilinium ion. While pure aniline and nitrobenzene are colorless or nearly so, p-nitroaniline is distinctly yellow. Furthermore, λ_{max} increases as more and more polar solvents are used; between hexane and dimethylformamide a shift of 60 nm is observed. The use of a polar solvent clearly increases the weight of structure (44a) [9]. Bond lengths and angles can be demonstrated to be changed by the operation of resonance. One interesting example is that of biphenyl: As shown by X-ray diffraction experiments, the molecule is coplanar in the crystalline state and the central carbon-carbon bond is shorter by nearly 0.1 Å than the normal single bond in spite of the repulsive interaction of the o-hydrogen atoms [10].

8-6. FORMAL BONDS AND NO-BOND RESONANCE

The conjugation of simple double bonds in the theory of resonance is described by structures such as those drawn above and below [10a].

No dipolar character can be demonstrated in 1,3-butadiene, and hence structures (45a) and (45b) must have equal weight.

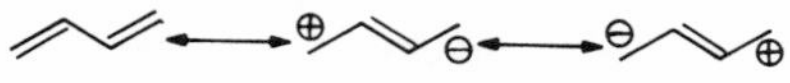

(45) (45a) (45b)

It would seem that the use of ionic structures to describe the properties of a nonpolar hydrocarbon is somewhat arbitrary and unrealistic, and hence some authors prefer structures such as (45c) and (45d).

(45c) (45d)

9. B. D. Pearson, Proc. Chem. Soc., 78 (1962).
10. J. Dhar, Indian J. Phys., 7, 43 (1932).
10a. N. F. Phelan and M. Orchin, J. Chem. Educ., 45, 633 (1968).

The presence of the extremely long and weak "formal bond" in (45c) cannot be directly demonstrated, and this concept would therefore likewise seem artificial and arbitrary. The structure (45d) might be described as a singlet diradical (the arrows are often omitted and written here only to distinguish it from the triplet). While the 1,4-interaction is difficult to prove, the 2,3-bond is indeed considerably shorter than normal single bonds, and the barrier to rotation of this bond is a bit greater than that in ethane.

Similar structures are used to account for the properties of a number of compounds for which classical structures cannot be drawn. Thus, the substance diborane (46) may be considered the hybrid of two structures in which the bridging hydrogen atoms are alternately singly bound or not bound at all; this description of the three center bond (p. 15) is sometimes referred to as no-bond resonance.

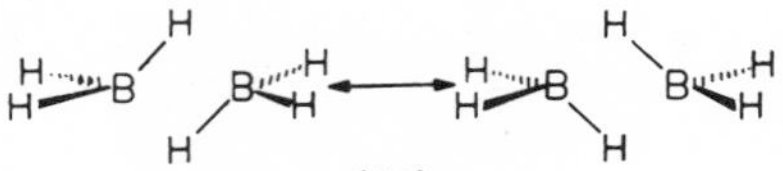

(46)

A related case is the symmetrical hydrogen bond, for instance in HF_2^-.

$$F{-}H \quad \overset{\ominus}{F} \longleftrightarrow \overset{\ominus}{F} \quad H{-}F$$

There are a few cases of organic molecules with a symmetrical hydrogen bond. The nmr spectrum of 2-benzylamino-4-benzylimino-2-pentene (47) shows single sharp methyl and methylene signals, and although this alone does not prove that the compound is a single resonance hybrid rather than a rapidly equilibrating pair, the fact that upon protonation the methylene peak moves upfield rather than down is indicative of resonance, the loss of the ring current offsetting the presence of the positive charge [11]. It can of course be debated whether the free base indeed has a ring current to lose (cf. p. 346).

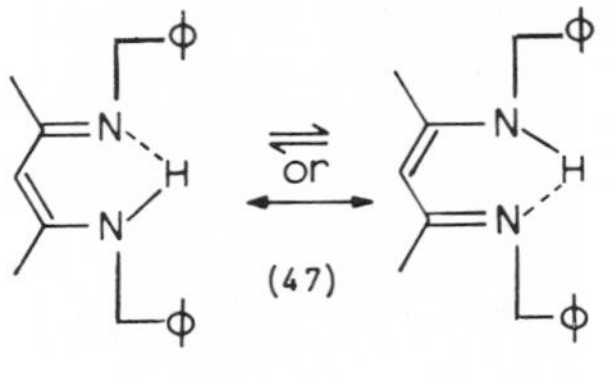

(47)

The thiothiophthenes (48) and their selenium analogs comprise an interesting group of compounds. Attempts to prepare unsymmetrically substituted isomers (49a) and (49b) invariably led to a single substance [12].

---

11. L. C. Dorman, Tetrahedron Lett., 459 (1966).
12. E. Klingsberg, J. Amer. Chem. Soc., 85, 3244 (1963) and Quart. Rev. (London), 23, 537 (1969).

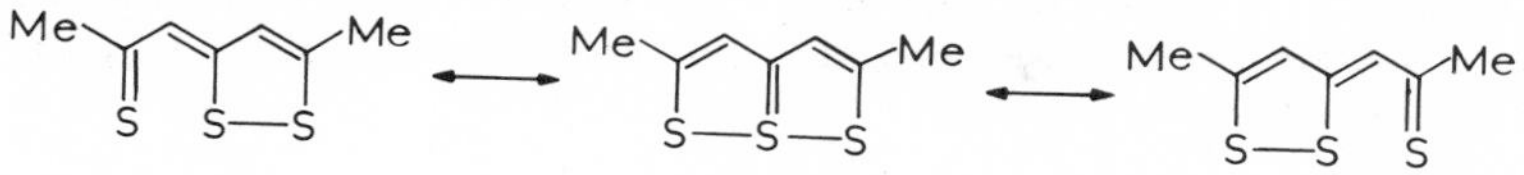

(48)

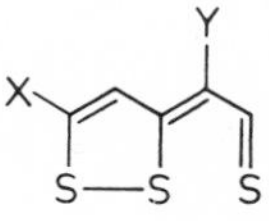

(49a) 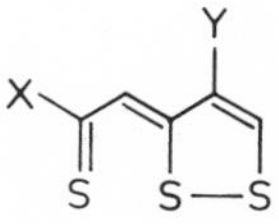 (49b)

In a number of instances it has been shown by X-ray diffraction measurements that the distances between the sulfur atoms are equal, and intermediate between those that are singly bound as in disulfides, and those that are non-bonded (at van der Waals distances) [13]; in these cases the molecules are best described as hybrids. The resonance energy is not very great in these molecules, and the curves relating the Kekulé structures are shallow and flat. Thus, the diphenyl derivative (50) was found by means of X-ray diffraction to have unequal S-S distances. One of the phenyl groups is twisted out of the plane by 45°; thus, a mere conformational effect seems sufficient to cause serious distortion away from symmetry in the thiothiophthene skeleton [13a, b]. In the 2,4-isomer (51) the difference in length of the two SS bonds amounts to several tenths of one Å [13c]; in fact, a recent report has it that all the evidence for resonance in these compounds (including X-ray diffraction and esr) [13d] may have been misinterpreted [13e].

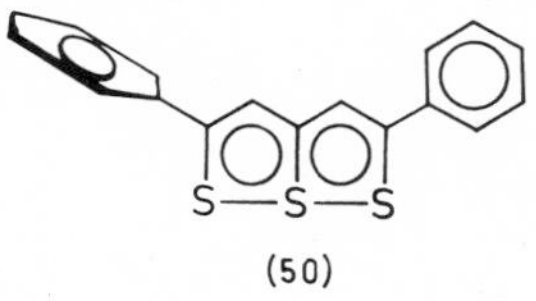

(50) 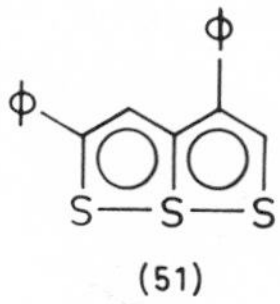 (51)

13. S. Bezzi, M. Mammi, and C. Garbuglio, Nature, 182, 247 (1958).
13a. A. Hordvik, T. S. Rimala and L. J. Saethre, Acta Chem. Scand., 26, 2139 (1972); (b) B. Birknes, A. Hordvik and L. J. Saethre, Acta Chem. Scand., 26, 2140 (1972).
13c. A. Hordvik and K. Julsham, Acta Chem. Scand., 23, 3611 (1969).
13d. F. Gerson, R. Gleiter, J. Heinzer and H. Behringer, Angew. Chem., Int. Ed. Engl., 9, 306 (1970).
13e. R. Gleiter, D. Schmidt and H. Behringer, Chem. Commun., 525 (1971).

A related phenomenon is that of hyperconjugation. In toluene, the p-position is activated toward electrophilic substitution much more than can be accounted for on the basis of an inductive effect (such as would be transmitted by a m-methyl group). It is said that no-bond resonance (in this case referred to as hyperconjugation) is responsible.

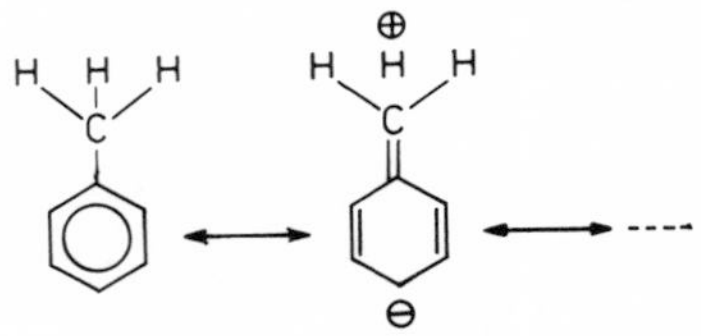

No-bond resonance is also considered by many chemists to account for the peculiar fact that the carbon-fluoride bond in monofluorinated hydrocarbons is much weaker than in polyfluorinated hydrocarbon derivatives - although it is by no means obvious why a bond involved in such resonance should become stronger rather than weaker because of it.

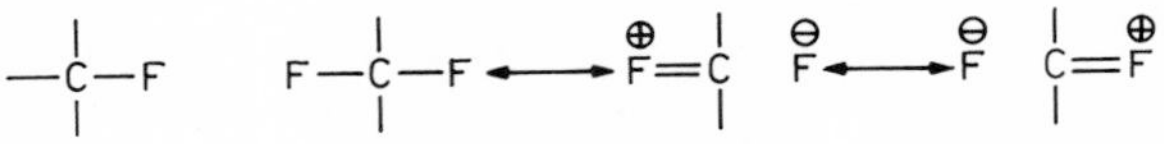

A difficult problem is that of conjugation through three-membered rings. It might be reasonable to ask whether such conjugation is possible since it clearly is effective in the next smaller ring size, the olefinic linkage. Examples of such conjugation can be found in many sections of this book; we may mention one here. Ethyl trans-2-p-nitrophenylcyclopropylcarboxylate (52) undergoes alkaline hydrolysis five times faster than the unsubstituted phenyl derivative (53) [14].

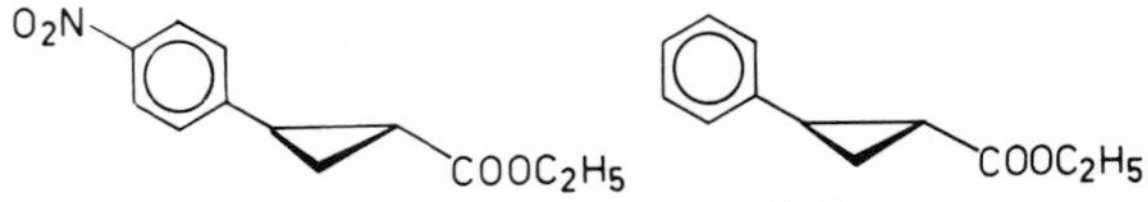

(52) (53)

For the β-phenylproprionate esters (54) this ratio is 2, and for the trans-cinnacinnamates (55) it is 10 [15]. Thus, the transmission of the electron withdrawing effect of the nitro group, while not as good as that through a double bond is better than that through a single bond (the purely inductive effect).

---

14. R. Fuchs and J. J. Bloomfield, J. Amer. Chem. Soc., 81, 3158 (1959).
15. K. Kindler, Ann., 452, 90 (1927), as quoted in Ref. 1.

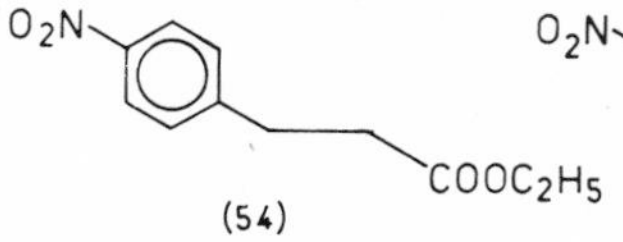

(54) 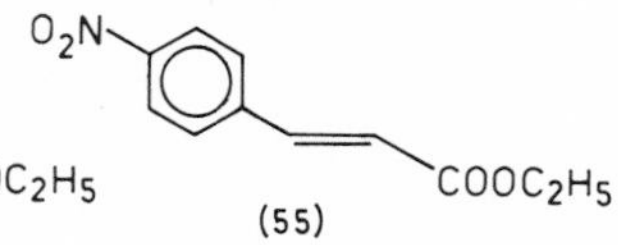 (55)

To account for these facts by means of resonance, we would have to draw structures such as (55a) and (52a). While structure (55a) is a reasonable one so far as bond angles and the planarity of the carbon framework are concerned, structure (52a) is not. This difficulty comes up each time when one tries to use resonance to account for the transmission of electronic effects through a three-membered ring.

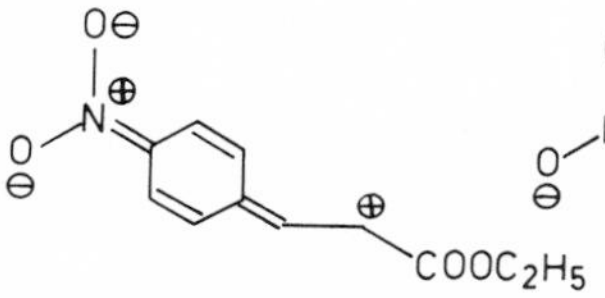

(55a) 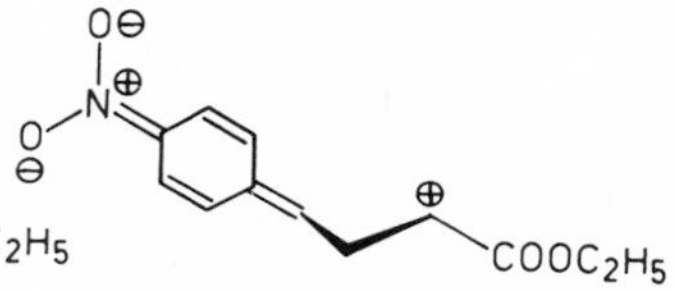 (52a)

8-7. INHIBITION OF RESONANCE

In order to be effective, resonance often requires that the interacting groups adopt one of a number of possible conformations. Thus, in p-nitroaniline both the nitro and the amino group must obviously be coplanar with the ring if the dipolar structure (44a) is to have a low energy. Ortho substituents will interfere with these two groups in this position to a maximum degree. If the interference is so strong as to render coplanarity impossible, than resonance is likewise impeded, and we say that resonance has been sterically inhibited. Thus, while the dipole moment of p-nitroaniline is 6.35 D, in the durene derivative (56) it is only 4.98 D [16].

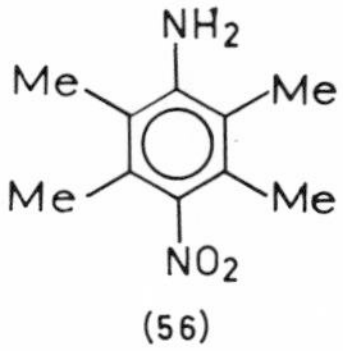

(56)

For similar reasons, 3,5-dimethyl-4-nitrophenol is a weaker acid than p-nitrophenol by about 1 pKa unit [17], the nucleophilic substitution reaction of 3,5-dimethyl-4-nitrobromobenzene with piperidine is about thirty five times slower

16. Ref. 1, p. 234.
17. Ref. 1, p. 368.

than that of p-nitrobromobenzene [18], and so on. It goes without saying that the effects of the inhibiting groups themselves alone must also be considered in such studies.

Steric inhibition of resonance may also be brought about by angle strain. 2, 2a, 3, 3a, 4, 5-Hexahydro-1H-cyclopent[jkl]-as-indacene (57) is an extreme example of a benzene ring so distorted by strain (imposed by the three saturated five-membered rings) that most or all of the normal resonance energy of the benzene is gone [19]; thus, it is rapidly oxidized in air to the diperoxide (58) (of undetermined configuration), and it can be completely reduced by hydrogen at atmospheric pressure.

(57) (58)

8-8. THE MILLS-NIXON EFFECT

It was found by Mills and Nixon in 1930 that 5-hydroxyhydrindene (59) couples with diazonium salts more rapidly at position 6 than at 4.

(59)

Since such coupling also readily occurs with aliphatic enols, they took this to mean that (59) had structure (59a), and that (59b) represented at the most a minor equilibrium component [20].

(59a) (59b)

18. W. C. Spitzer and G. W. Wheland, J. Amer. Chem. Soc., 62, 2995 (1940).
19. H. Rapoport and G. Smolinsky, J. Amer. Chem. Soc., 82, 1171 (1960).
20. W. H. Mills and I. G. Nixon, J. Chem. Soc., 132, 2510 (1930).

While this explanation is hardly taken seriously any more, it is not totally inconceivable that appropriate substituents might distort a benzene ring to the point where it becomes a classical Kekulé pair rather than a regular hexagon (compare the thiothiophthenes, p. 255). Large distortions can indeed be forced; thus, in benzocyclopropenes there is a difference of 0.1 A in length between the shortest and the longest carbon-carbon bond in the six-membered ring [21]. At least one instance is known in which these bonds actually alternate in length on the single bond-double bond pattern (p. 388); however, no case is known in which two Kekulé isomers actually exist - whether equilibrating or not.

8-9. WEAKNESSES OF THE CONCEPT OF RESONANCE

One of the real strengths of the concept of resonance is its simplicity; however, this is also one of its weaknesses, since the simplicity is at times only apparent. Thus, the stabilization of benzene does not obtain in cyclobutadiene, even though the argument is the same: One can write two equivalent structures that could contribute to a square planar molecule.

This molecule is known (see p. 281), but it is extremely unstable. This is not due to the strained bond angles alone; there are even more highly strained molecules such as cyclopropenone which are stabilized by resonance. There are similar difficulties with cyclooctatetraene, which likewise is not stabilized by resonance [22].

It should be pointed out that these difficulties do not really invalidate the theory or concept of resonance; these problems have been known for a long time and they can be accommodated in the more precise, mathematical formulations of the theory; it is true, however, that the simple use of resonance structures breaks down here.

Another weakness is that at least some of its applications seem somewhat artificial, as the reader doubtless felt when the concepts of no-bond resonance were applied to gem-difluorides and cyclopropanes. Another objection one might raise is that it is somewhat cumbersome at times. Thus, while the two Kekulé structures suffice for benzene for most purposes, for a complete description of the tropylium cation one must write seven, for ferrocene twenty five, and so on.

---

21. E. Carstensen-Oeser, B. Müller, and H. Dürr, Angew. Chem., Int. Ed. Engl., 11, 422 (1972).
22. H. Meislich, J. Chem. Educ., 40, 401 (1963).

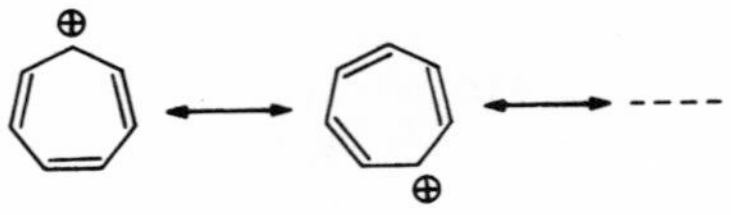

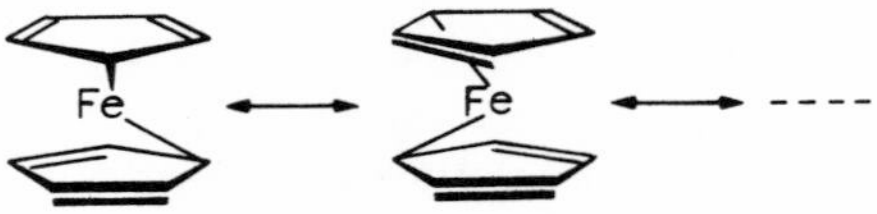

While it is usually not necessary to write all possible structures, the appearance of the dotted lines diminishes the elegance of the argument that the concept of resonance is a simple extension of the structural theory.

Most important however is the fact that the theory of resonance has not been able to maintain its early lead over the molecular orbital theory as a basis for new predictions. One might say that very little if anything is wrong with it; but to survive a theory must be of predictive value, and resonance simply has not been able to match the incredible string of successes attributable to the MO theory - successes some of which are described elsewhere in this book.

Chapter 9

AROMATICITY [1]

9-1. INTRODUCTION

Aromaticity is one of the finest examples of a scientific development organic chemistry has to offer. It is complete with the odd facts that initially did not fit well in the general framework, the various ingenious proposals to make them fit that were in due time shown to lead to wrong predictions, the victorious idea that became gradually recognized as the "truth". Painfully slowly, as clever men ask searching questions, more and more of the hidden story yields to daylight. But there are also stumbling blocks. Then a new development in another science,

---

1. For other discussions, see (a) G. M. Badger, "Aromatic Character and Aromaticity", Cambridge University Press, 1969; (b) A. Courtin and H. Sigel, Chimia, 19, 407 (1965); (c) W. Baker, Chem. Brit., 1, 191, 250 (1965); (d) R. Breslow, Chem. Eng. News, 43, June 28, 1965, p. 90. Further reviews are quoted in several of the sections below. An interesting discussion of cyclic, conjugated chlorocarbons is available; (e) R. West, Pure Appl. Chem., 28, 379 (1971); (f) "Aromaticity", Special Publication No. 21, Chem. Soc. (London), 1967.

leading to a view of benzene and its derivatives entirely new, offered by a scientist not even intimately involved in organic chemistry at all. This viewpoint is not immediately accepted, or even widely known: The chemicals that would tell the story are too far out and cannot be synthesized. Then, with the help of new instrumentation and methods of analysis, they begin to appear to testify to the predictive value of the Hückel rule. How Kekulé, Ladenburg and Thiele would have been fascinated by cyclobutadiene, the annulenes, calicene, by ring currents and esr. It whets our desire to look ahead, and see what is still in store: Surely the present wave of success if not the end of it.

But at this stage, it would not make sense to review the topic in an historical vein [2]. It is so large now, that to fully appreciate it we must make an arbitrary division into digestible pieces. We discuss therefore the concept of aromaticity itself and the criteria that have been established for it; the simple aromatic compounds, the non-aromatic analogs, the bridged aromatic rings, aromatic ions and dipoles, the introduction into the ring of atoms other than carbon, fused rings, interrupted aromatic rings, new types of aromaticity, and finally, the interference with and loss of aromaticity due to strains.

9-2. THE CONCEPT AND CRITERIA [3]

In Chapter 1 (Section 1-10), we briefly discussed how the interaction of neighboring p orbitals could be traced by means of the simple Hückel method to the concept of a delocalization energy, and further how that method predicts certain observations such as the shortened single bonds between conjugated double bonds. The same type of calculation can be done for cyclic conjugated molecules. Thus, for benzene the secular determinant would be written:

$$\begin{vmatrix} x & 1 & 0 & 0 & 0 & 1 \\ 1 & x & 1 & 0 & 0 & 0 \\ 0 & 1 & x & 1 & 0 & 0 \\ 0 & 0 & 1 & x & 1 & 0 \\ 0 & 0 & 0 & 1 & x & 1 \\ 1 & 0 & 0 & 0 & 1 & x \end{vmatrix} = 0$$

---

2. (a) For the references to the history of aromaticity, see G.W. Wheland, "Advanced Organic Chemistry", 3rd Ed., Wiley, New York, 1960, Chr. 10; (b) H. Hartmann, Angew. Chem., Int. Ed. Engl., 4, 729 (1965).
3. (a) J. F. Labarre and F. Crasnier, Fortschr. Chem. Forsch., 21, 33 (1971); (b) A. J. Jones, Rev. Pure Appl. Chem., 18, 253 (1968); (c) A. Streitwieser, "Molecular Orbital Theory for Organic Chemists", Wiley, New York, 1961.

Factoring [4] leads to the following equation of x:

$$x^6 - 6x^4 + 9x^2 - 4 = 0.$$

The solutions are:

$$x = \pm 1, \ \pm 1, \ \pm 2.$$

Hence, the orbitals have wave functions

$$\Phi_1 = (\Psi_1 + \Psi_2 + \Psi_3 + \Psi_4 + \Psi_5 + \Psi_6)/\sqrt{6}$$

$$\Phi_2 = (2\Psi_1 + \Psi_2 - \Psi_3 - 2\Psi_4 - \Psi_5 + \Psi_6)/\sqrt{12}$$

$$\Phi_3 = (\Psi_2 + \Psi_3 - \Psi_5 - \Psi_6)/\sqrt{4}$$

$$\Phi_4 = (\Psi_2 - \Psi_3 + \Psi_5 - \Psi_6)/\sqrt{4}$$

$$\Phi_5 = (2\Psi_1 - \Psi_2 - \Psi_3 + 2\Psi_4 - \Psi_5 - \Psi_6)/\sqrt{12}$$

$$\Phi_6 = (\Psi_1 - \Psi_2 + \Psi_3 - \Psi_4 + \Psi_5 - \Psi_6)/\sqrt{6}$$

and the energies are:

$$E_1 = \alpha + 2\beta$$

$$E_2 = E_3 = \alpha + \beta$$

$$E_4 = E_5 = \alpha - \beta$$

$$E_6 = \alpha - 2\beta$$

Aside from the ever present node through the plane of all the carbon atoms, the lowest energy molecular orbital has no nodes, the next degenerate pair has one, the next pair has two and the upper level has three. This conclusion may be visualized as follows:

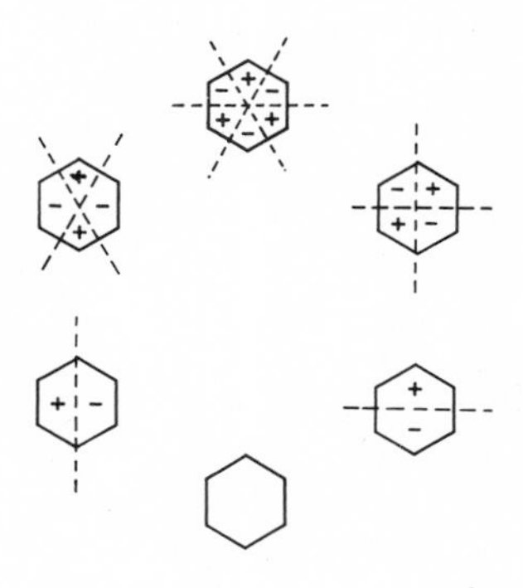

4. Many devices (such as character tables) are known by means of which large determinants can be simplified and solved; see e.g., Ref. 3c for such methods pertinent to the secular determinants.

Pictures such as

are also much in use.

The delocalization energy of benzene equals 2β, as follows from the energies of the three lowest (and only occupied) orbitals (Fig. 9-1). For benzene-like molecules the experimental resonance energy is commonly believed to be about 36 kcal/mole (see p. 239), so that $\beta \approx 18$ kcal.

The calculation can be carried out in a general way, and the conclusion may be expressed in terms of the Hückel polygon. To draw this polygon as is shown in Fig. 9-2, a circle is drawn of radius 2β, and the polygon is inscribed in such a way that one corner points down. The corners then represent the levels on the energy scale.

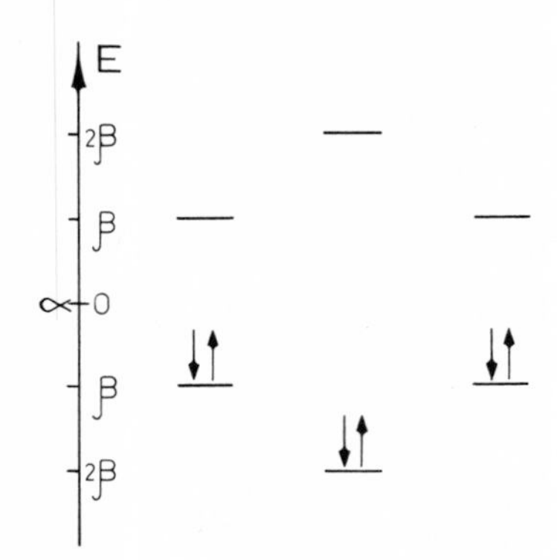

Fig. 9-1. The π-levels of benzene.

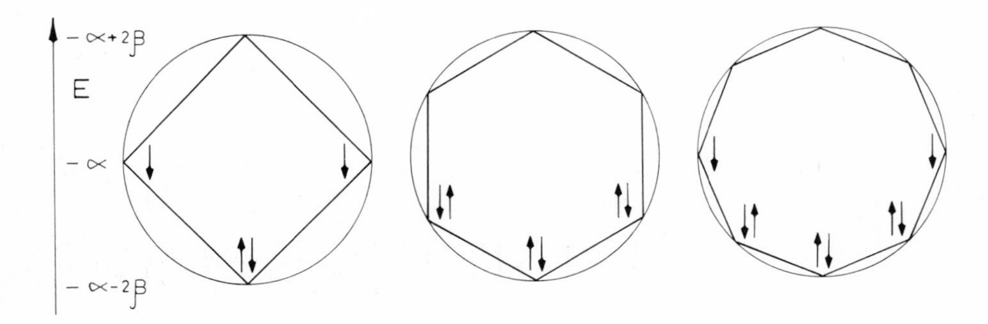

Fig. 9-2. The Hückel polygons for cyclobutadiene, benzene, and cyclooctatetraene.

Thus, in each case the lowest level has an energy ($\alpha + 2\beta$) and the next two levels are degenerate. In the case of cyclobutadiene, they have an energy α. Since there are four electrons, the resonance energy will be $2(\alpha + 2\beta) + 2\alpha - 4(\alpha + \beta) = 0$. In other words, Hückel's method predicts that cyclobutadiene will not be resonance stabilized. Off-hand, one cannot say whether the molecule would still prefer to be square planar, or whether it will have a lower total energy in the rectangular shape. In the former case, it would necessarily have two unpaired electrons as demanded by Hund's rules; in the latter case it would be diamagnetic, so that in principle we have an experimental approach to this question.

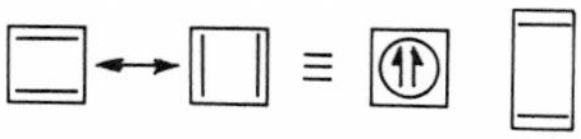

It should be noted that these calculations have some validity only if all carbon atoms are equidistant; if they are not, it is probably more nearly correct to assume that the resonance integrals are alternately zero, and that the molecule is simply a diene.

A little knowledge of trigonometry and geometry suffices at this point to calculate the resonance energies of each of the monocyclic systems (see Fig. 9-2). Although the Hückel calculations suggest that all polyenes beyond cyclobutadiene will be stabilized by delocalization, the more elaborate calculations by Dewar and Gleicher show that the delocalization energies alternate in sign as the rings get larger, until they all become negative at 26 carbon atoms [5].

In general we may conclude therefore that systems in which the number of p electrons is a multiple of 4 (4n systems) will not be able to lower their energies by delocalization; if they are nevertheless delocalized, they will be diradicals and hence reactive. Aromaticity is on the other hand expected for (4n + 2) systems, with a resonance energy 2β for benzene, gradually diminishing for larger rings, and more or less vanishing in the neighborhood of 30 carbon atoms.

Before we begin to look into the experimental picture, we should discuss by what criteria we will distinguish at least the stabilized, aromatic (4n + 2) molecules [3b]. At this point organic chemists differ somewhat in their point of view. Aromatic originally meant that these compounds have a strong and usually pleasant odor; but later it assumed the meaning that they had unusual stability, comparable to benzene itself. Although this criterion is still used and will at times be used below, it is obviously rather arbitrary, and will apply only if the compounds being compared are not too dissimilar. Stability and instability are relative terms; a given compound may be thermally much more stable than another, but it may be more rapidly attacked by oxygen, and so on. Even the important modes of

5. (a) M. J. S. Dewar and G. J. Gleicher, J. Amer. Chem. Soc., 87, 685 (1965); (b) J. F. M. Oth, Pure Appl. Chem., 25, 573 (1971); (c) R. Waack, J. Chem. Educ., 39, 469 (1962).

attack of electrophilic and radical reagents - addition to olefins, substitution in aromatic molecules - does not always serve us to distinguish aromatic from non-aromatic molecules. Complexation with highly electron-deficient species (see Chapter 23) is used now and then.

Further information may be gleaned from the uv spectrum; planar aromatic molecules will generally have absorptions at longer wavelengths than molecules with non-interacting multiple bonds. The molecules shaped as regular polygons will have simpler ir spectra because of their symmetry, with the carbon-carbon stretching bands appearing at lower energies than those of their olefinic cousins. X-ray diffraction is a very powerful tool; thus, while C-C singly bound atoms usually are 1.54 Å apart and C-C double bonds normally have a length of about 1.30 Å, the delocalized aromatic carbon-carbon bonds are intermediate in length (1.40 Å), easily distinguishable from the alternate pattern of polyolefins. A certain degree of planarity is essential for aromaticity, and this is also recognizable by X-ray diffraction. One of the most simple criteria in application is the nmr spectrum; the protons in aromatic molecules will, because of the ring current, have chemical shifts at much lower field than vinylic hydrogen atoms [6]. Important information can also be gleaned from the esr spectra of the radical anions and cations of aromatic hydrocarbons, as discussed in Chapter 18, Section 18-7. It must be said however (as will also be obvious from the further discussion below), that not a single one of these criteria can <u>always</u> be relied upon; all have their exceptional cases, and none, when violated, are good enough by themselves to rule out aromaticity. While this situation may be regretted, it is a fact of life, and perhaps it would be wise to regard it as a spur to a further search for more adequate definitions and criteria.

9-3. NEUTRAL, MONOCYCLIC, CARBOCYCLIC AROMATIC MOLECULES

If we focus our attention on the next higher benzene analog, [10]annulene (1) [7], we note at once what the difficulties may be. If the molecule is shaped like a regular decagon, all the CCC bond angles would be stretched quite far from their optimum value of 120°.

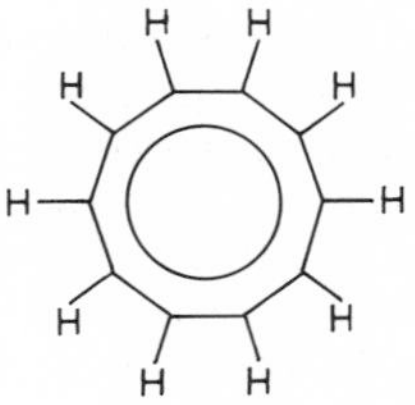

(1)

---

6. See for example, J.A. Elvidge and L.M. Jackman, J. Chem. Soc., 859 (1961).
7. (a) For review, see S. Masamune and N. Darby, Accounts Chem. Res., 5, 272 (1972); (b) For a review of the annulenes in general, see F. Sondheimer, Accounts Chem. Res., 5, 81 (1972).

The necessary strain energy must be subtracted from the resonance energy, and it is hard to say whether the result will be positive or negative. Alternatively, not all the double bonds need be in the cis-conformations. Thus, (2) is another possibility in which the bond angles are no longer strained; now, however, we have to consider the fact that two of the hydrogen atoms lie within the cavity formed by the carbon atoms.

(2)

This is not necessarily bad, but the problem with (2) is that the two hydrogen atoms would clearly interfere with one another (for a method of naming these conformational isomers, see p. 89). Still another way to approach the problem is to introduce one or more triple bonds; thus, one could consider such structures as (3) or (4).

(3)

(4)

These molecules are more or less strainless so far as angles and nuclei are concerned, but now there is the problem that the electrons in the π-orbitals that lie in the plane of the ring will repel each other. These electrons are of course not counted in the aromatic system; both (3) and (4) are 10 π-electron systems for the purpose of our present discussion, though each actually has a total of 14.

It is one of the marvels of this time in organic chemistry that the advances in synthesis have kept pace with those in theory, and some of these highly strained molecules have actually been prepared. Thus, photolysis of cis-9,10-dihydronaphthalene (5) at -60° gave a mixture of (1), (6) and (7), which could be separated by column chromatography at -80°.

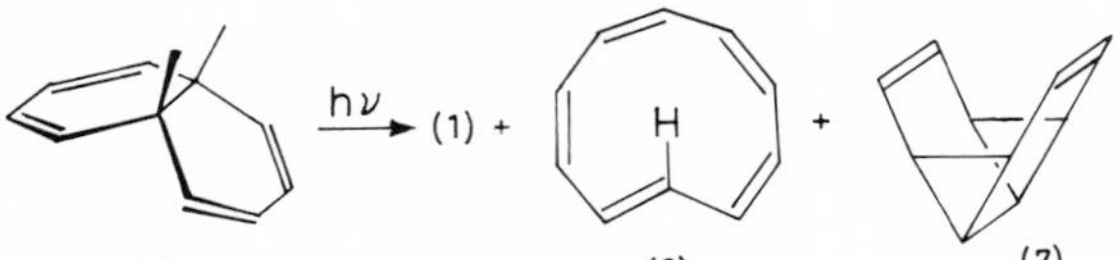

hν → (1) + (6) + (7)

(5) (6) (7)

Both (1) and (6) rapidly isomerize to cis- and trans-9,10-dihydronaphthalene, respectively; hence, they are not especially stable. Their uv spectra resemble that of cyclononatetraene (8), a simple conjugated polyolefin; furthermore, the nmr spectra show signals for (1) and (6) at chemical shifts similar to those that apply to (8), indicating the absence of ring currents [8].

(8)

Benzo annelation often stabilizes sensitive compounds of this sort, and actually the first successful synthesis of a [10]annulene appeared in 1967 [9], when the tribenzo[10]annulene (9) was reported.

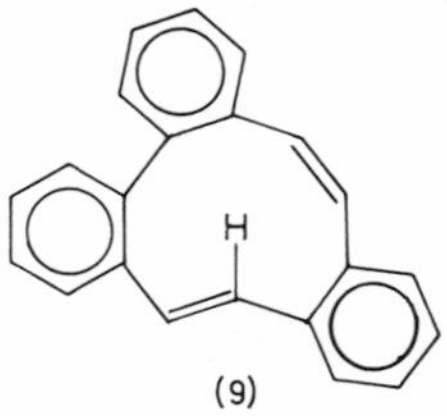

(9)

---

8. (a) J. J. Bloomfield and W. T. Quinlin, J. Amer. Chem. Soc., 86, 2738 (1964); (b) E. E. van Tamelen and T. L. Burkoth, J. Amer. Chem. Soc., 89, 151 (1967); (c) S. Masamune and R. T. Seidner, Chem. Commun., 542 (1969); (d) E. E. van Tamelen, T. L. Burkoth, and R. H. Greeley, J. Amer. Chem. Soc., 93, 6120 (1971); (e) S. Masamune, K. Hojo, K. Hojo, G. Bigam, and D. L. Rabenstein, J. Amer. Chem. Soc., 93, 4966 (1971); (f) E. E. van Tamelen and R. H. Greeley, Chem. Commun., 601 (1971).
9. P. J. Mulligan and F. Sondheimer, J. Amer. Chem. Soc., 89, 7119 (1967).

This material is thermally stable, having a melting point of 122°; however, this stabilization is due to the benzene rings, and not to aromaticity of the ten-membered ring. The nmr spectrum has been analyzed and shows that the four olefinic protons appear at 3.25-4.6τ, the other twelve at 2.5-3.2τ. Further evidence for the lack of any special stability of these compounds is the fact that trienediones such as (10) fail to enolize [10]; this is in marked contrast to behavior of the enedione (11) [11] (see p. 422).

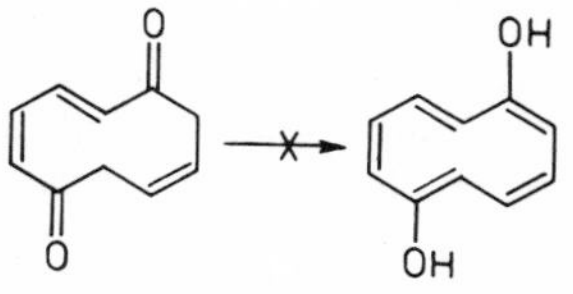

O OH

OH O

(10)

O OH

O OH

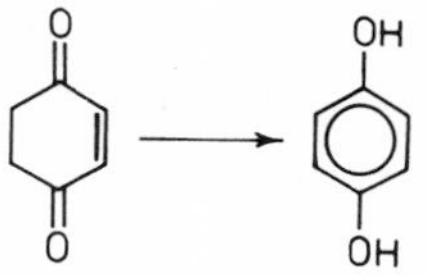

(11)

The synthesis of dehydro compounds such as (3) or (4) has not as yet succeeded. All attempts to generate structure (3) by various means lead only to azulenes and naphthalenes instead; if (3) is formed at all, it immediately stabilizes itself by oxidizing the solvent [12].

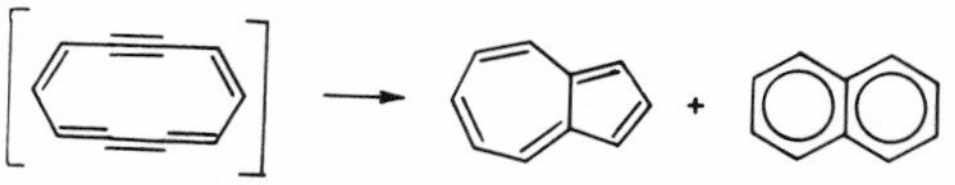

(3)

The solvolysis of the dibrosylate (12), which might be expected to give the annelated derivative (13) [13], produces the isomer (14) instead. Thus, in this reaction (13) may at best be an intermediate that rapidly and completely isomerizes to the far more stable benzene compound (14).

---

10. P. J. Mulligan and F. Sondheimer, J. Amer. Chem. Soc., 89, 7118 (1967).
11. E.W. Garbisch, J. Amer. Chem. Soc., 87, 4971 (1965).
12. See for example, R.W. Alder and D.T. Edley, J. Chem. Soc. (C), 3485 (1971).
13. N. Darby, C.U. Kim, J.A. Salaün, K.W. Shelton, S. Takada, and S. Masamune, Chem. Commun., 1516 (1971).

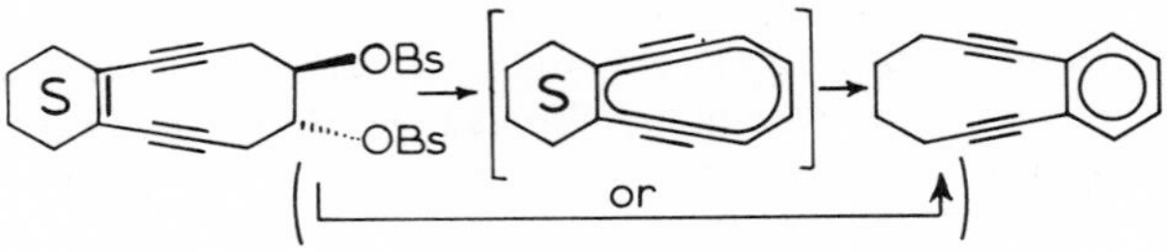

(12) (13) (14)

A related question of interest is concerned with poly-unsaturated hydrocarbons in which the prerequisite (4n + 2) number of p electrons is present, but in which the p orbitals cannot all be parallel. [4.4.2]Propella-2,4,7,9,11-pentaene (15) is an example of such a compound. Its nmr signals are all centered at 4.1 and 4.3 τ, and its uv spectrum likewise resembles those of conjugated open chain dienes; it readily reacts with dienophiles to give Diels-Alder products such as (16) - behavior not normally associated with aromatic species [14].

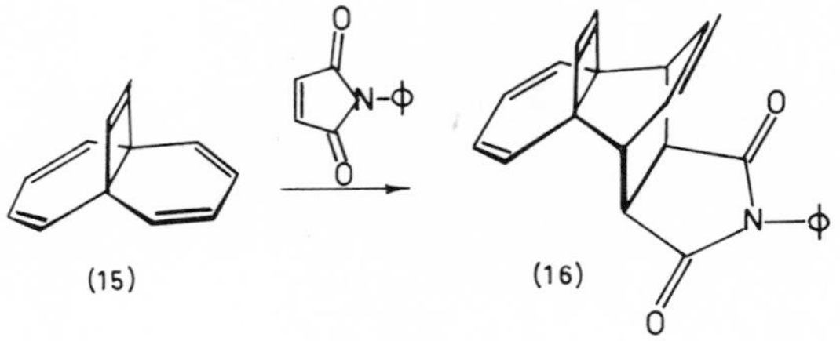

(15) (16)

Similar behavior had earlier been noted with the six π-electron species barrelene (17); the heat of hydrogenation to bicyclooctadiene (18) is 38 kcal/mole, comparable to that for cis-1,2-di-t-butyl-ethylene (36 kcal/mole) and much larger than that of cyclohexene (28 kcal/mole) [15].

(17) (18)

As we extrapolate to the larger-membered rings, the synthetic difficulties fortunately do not keep pace. Angle strain can now be avoided by the incorporation of trans- double bonds, and as the rings considered are made larger and larger, the inner hydrogen atoms and/or the in-plane p orbitals are more distant and less prone to interfere. A further fortunate circumstance is that ingenious but simple

14. L.A. Paquette and J.C. Philips, J. Amer. Chem. Soc., 91, 3973 (1969).
15. R.B. Turner, J. Amer. Chem. Soc., 86, 3586 (1964).

synthetic approaches to the larger rings have been invented, principally by Sondheimer and his students [16]. In principle, what is involved is first of all oxidation of partially saturated, terminal diacetylenes to a complex mixture of various polymeric materials, from which one or more cyclic species can usually be isolated; these are then in turn isomerized to conjugated poly-unsaturated isomers by treatment with base. Partial hydrogenation or dehydrogenation finally converts some or all of the triple bonds remaining into double bonds or vice versa.

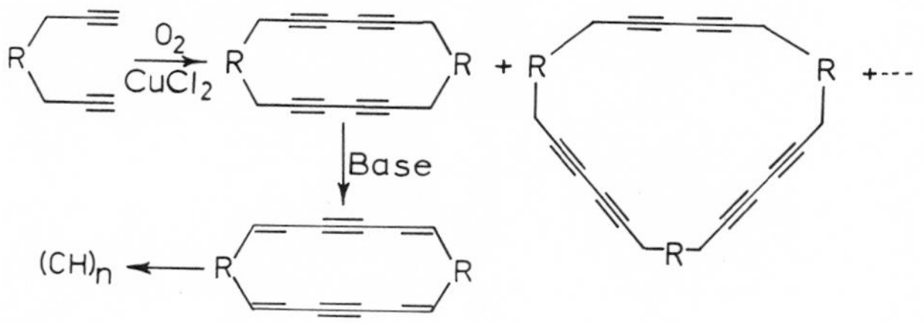

O_2, $CuCl_2$; Base; $(CH)_n$

For example, oxidative coupling of trans-, trans-4, 10-tetradecadiene-1, 7, 13-triyne (19) and subsequent treatment with t-butoxide gives low yields of 1, 8-bisdehydro[14]annulene (20) and of the dehydro[14]annulene (21) [17].

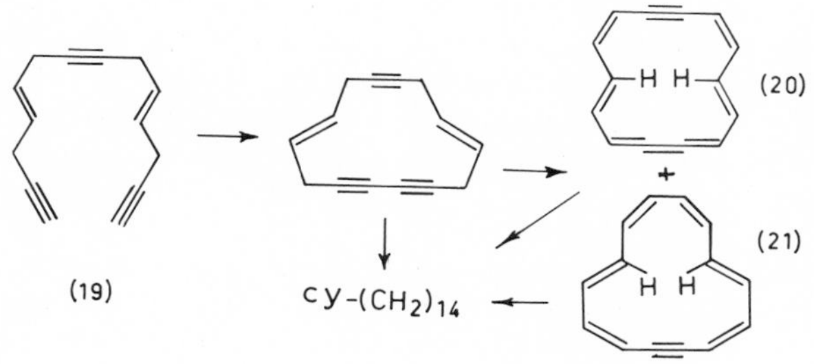

(19) (20) (21) cy-$(CH_2)_{14}$

Of these, especially the former compound is interesting. Models suggest that both the inner hydrogen atoms and the in-plane π orbitals are far enough apart that no significant steric interactions are likely to occur. The nmr spectrum shows a double doublet at 0.45 τ, a doublet at 1.57 τ, and a triplet at 15.54 τ, with the intensity ratio of 2:2:1. The fully symmetric nature of the molecule is evident

---

16. (a) F. Sondheimer and R. Wolovsky, J. Amer. Chem. Soc., 81, 1771 (1959); (b) F. Sondheimer, Y. Amiel, and R. Wolovsky, J. Amer. Chem. Soc., 81, 4600 (1959); (c) F. Sondheimer and Y. Gaoni, J. Amer. Chem. Soc., 81, 6301 (1959); (d) A. S. Hay, J. Org. Chem., 25, 1275 (1960).
17. F. Sondheimer, Y. Gaoni, L. M. Jackman, N. A. Bailey, and R. Mason, J. Amer. Chem. Soc., 84, 4595 (1962).

from the fact that the inner-proton signal is split in such a way as to show that the two neighboring protons are completely equivalent; the "acetylenic" and "allenic" proton neighbors are not different. Thus, the molecule should be written as a resonance hybrid of two Kekulé forms such as (20) and (22), or as (23).

(20) (22) (23)

The values of J_{ab} and J_{bc}, 8.0 and 13.3 Hz, respectively, confirm the cis- and trans- conformations of these CH bonds. Even more remarkable is the enormously high field at which the inner protons appear; this is fully in accordance with the proposition that a ring current is responsible for the low field at which outer protons normally signal their presence in the more conventional aromatic molecules such as benzene (see p. 50). The completely symmetrical and planar nature of the molecule is furthermore confirmed by X-ray diffraction. The compound is stable under ordinary laboratory conditions, and is colored deep red with a λ_{max} at 586 nm. It is clearly an example of an aromatic compound, fulfilling the prophesy that an unstrained, planar, $(4n + 2)$ system (here with $n = 3$) will be unusually stable and delocalized. The nmr spectrum of (21) shows that that compound is aromatic also; two signals appear in 5:1 ratio at 1.7–2.7 and 10.7 τ, respectively. Both compounds react with electrophilic reagents such as copper(II) nitrate in acetic anhydride to give nitro derivatives; sulfonic acids and Friedel-Crafts products have likewise been prepared, and complexes form upon treatment with 1,3,5-trinitrobenzene [18].

Partial reduction of either compound produces [14]annulene (24) itself. Aromaticity was initially not obviously indicated. Electrophilic reagents destroyed it; no complexes could be isolated, and the thermal stability was far less than that of its precursors. Finally, the nmr spectrum shows a single, sharp peak at 4.42 τ, similar to that of cyclooctatriene (25) [19]. In other words, it appeared as though the four inner hydrogen atoms interfered enough to render the molecule essentially nonaromatic.

(24) (25)

18. Y. Gaoni and F. Sondheimer, J. Amer. Chem. Soc., 86, 521 (1964).
19. L.M. Jackman, F. Sondheimer, Y. Amiel, D.A. Ben-Efraim, Y. Gaoni, R. Wolovski, and A.A. Bothner-By, J. Amer. Chem. Soc., 84, 4307 (1962).

It seemed an odd coincidence, however, that all fourteen of the protons should absorb at a single chemical shift, and it was suspected that perhaps some sort of averaging process was giving rise to that result. A subsequent reinvestigation showed this to be the case [20]: Upon cooling a solution of the substance, broadening occurs, and at -60° resolution occurs into two bands at 2.4 and 10.0 τ, in an area ratio of 5:2. Thus, the room temperature signal is merely the weighted average of the high and low field signals observable at -60°, and the substance is aromatic after all - even though it is not especially stable. The coalescence of the spectrum is indicative of the remarkable fact that the molecule is able to turn inside out with great ease and scramble all its proton positions in the process. This phenomenon is observed with many of the large, fully conjugated rings [21]. A complete analysis of the nmr spectrum of [14]annulene at -135° shows it to be an equilibrium mixture of (24) and (26), in a 92:8 ratio [19]; thus, the strains of interfering hydrogen atoms and of distorted bond angles are delicately balanced.

(24) ⇌

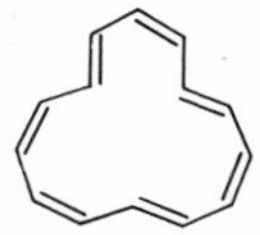

(26)

A further complication in the [14]annulene story is that these two isomers were originally thought to be stereoisomers such as (27) and (28) [20].

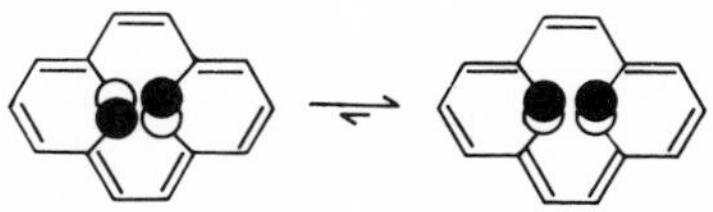

(27) (28)

The X-ray diffraction further shows that there is no significant bond alternation in (24) and hence that the molecule is indeed aromatic; however, the ring is distorted quite a bit from planarity and the resonance energy must be low. The inside protons are located neither like in (27) nor like in (28), but in such a way that the molecule - like (28) - has a center of symmetry [22].

---

20. Y. Gaoni, A. Melera, F. Sondheimer, and R. Wolovsky, Proc. Chem. Soc., 397 (1964).
21. For review, see J. F. M. Oth, Pure Appl. Chem., 25, 573 (1971).
22. C.C. Chiang and I.C. Paul, J. Amer. Chem. Soc., 94, 4741 (1972).

[18]Annulene (29) is the first in the series of (4n + 2) polyenes virtually free from interference of the internal hydrogen atoms.

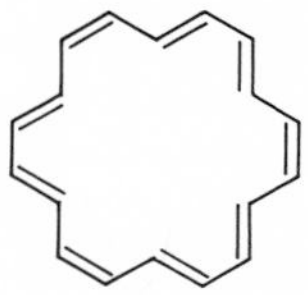

(29)

Its synthesis was reported in 1959 [23]. The compound is not especially stable; bromine adds to it, and it readily undergoes Diels-Alder reactions with dienophiles. The uv spectrum was interpreted in terms of bond alternation and a polyolefinic nature [24]. The nmr spectrum tells a different story, however. Two broad bands were observed, at 1.1 and 11.8 τ, and in a 2:1 ratio. A strong ring current is clearly indicated [19]. At elevated temperatures, coalescence occurs [20] and at 110° only a single sharp bond is observed at 4.42 τ; the inside-out process is evidently occurring in this case also. The heat of combustion also supports the conclusion that the molecule is aromatic; the data suggested that stabilization amounts to no less than 100 kcal/mole [25]. X-ray diffraction experiments [26] have shown the entire molecule to be very nearly planar. Finally, it has been reported that substitution products such as nitro- and acetyl[18]annulene can be prepared under sufficiently gentle conditions [27]. A number of dehydro[18]annulenes are also known; among them are compounds (30) [28] and (31) [29]. All these are typically aromatic substances; thus (30) forms complexes with 2,4,7-trinitrofluorenone (32) and has an inside-hydrogen triplet at 15.24 τ in the nmr.

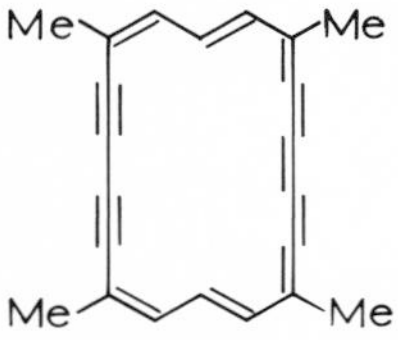

(30) 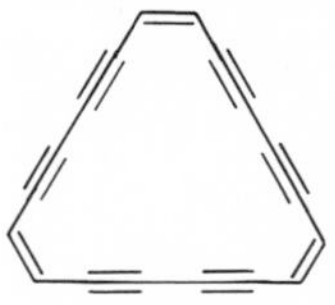 (31) 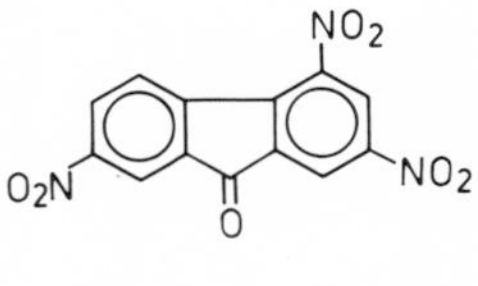 (32)

23. (a) F. Sondheimer and R. Wolovsky, Tetrahedron Lett., 3 (1959); (b) F. Sondheimer, R. Wolovsky, and Y. Amiel, J. Amer. Chem. Soc., 84, 274 (1962).
24. H.C. Longuet-Higgins and L. Salem, Proc. Roy. Soc. (A), 257, 445 (1960).
25. A.E. Beezer, C.T. Mortimer, H.D. Springall, F. Sondheimer, and R. Wolovsky, J. Chem. Soc., 216 (1965).
26. See footnote 14 of Ref. 25.
27. I.C. Calder, P.J. Garratt, H.C. Longuet-Higgins, F. Sondheimer, and R. Wolovsky, J. Chem. Soc. (C), 1041 (1967).
28. (a) K. Fukui, T. Okamoto, and M. Nakagawa, Tetrahedron Lett., 3121 (1971); (b) J. Ojima, T. Katakami, G. Nakaminami, and M. Nakagawa, Tetrahedron Lett., 1115 (1968); see also (c) M. Iyoda and M. Nakagawa, Tetrahedron Lett., 3161 (1972).
29. W.H. Okamura and F. Sondheimer, J. Amer. Chem. Soc., 89, 5991 (1967).

The availability of substituted [18]annulens gives a further insight into the conformational mobility of these molecules [30]. Since in the parent compound all protons become equivalent on the nmr time scale at elevated temperatures and since only six protons can be inside at any time, three degenerate conformations must be achievable. With nitro[18]annulene (33) the signal ratio at low temperature is 11:6. Thus, there is no room for the nitro group inside, not even temporarily: On heating coalescence again occurs but now five protons remain at low field [27]. 1-Chloro-2-fluoro[18]annulene (34) has a 10:6 signal ratio, and the spectrum is temperature independent, at least up to 100° [31].

NO_2 Cl F

(33) (34)

The number or type of conformations do not tell us very much about the mechanism of the process. The activation energies are so low that the carbon atoms obviously do not all roll their p orbitals upside down at the same time; somehow only a single pair of neighboring carbons must do it at a time, or if the whole process is concerted, it must be done in such a snake-like manner that little overlap is lost in the process [21, 32].

As we look at still larger (4n + 2) rings the question of interest is whether and where aromaticity stops. The simple Hückel polygon suggests even without detailed computation that as n is made ever larger, there are more levels at energies between $(\alpha + \beta)$ and $(\alpha + 2\beta)$ than between $(\alpha + \beta)$ and α (Fig. 9-3); but as noted (p. 265), the refinements introduced by Dewar and Gleicher [5a] suggest that aromaticity will end between [22]- and [26]annulene.

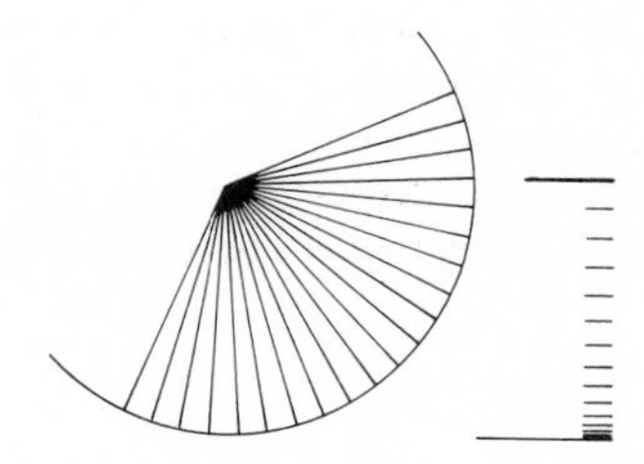

Fig. 9-3. Energy levels of very large annulenes according to the simple Hückel approach.

30. J. M. Gilles, J. F. M. Oth, F. Sondheimer, and E. P. Woo, J. Chem. Soc. (B), 2177 (1971).
31. G. Schröder, R. Neuberg, and J. F. M. Oth, Angew. Chem., Int. Ed. Engl., 11, 51 (1972).
32. For review of this topic, see R. C. Haddon, V. R. Haddon, and L. M. Jackman, Fortschr. Chem. Forsch., 16, 103 (1971).

Both [22]annulene [33] and dehydro[22]annulene [34] have recently been prepared; they have low and high field groups of signals in 14:8 and 13:7 ratios, and hence structures (35) and (36), respectively.

(35) (36)

Two dehydro[26]annulenes are known: (37) and (38). Of these, the former shows all its proton signals between 2.0 and 4.5 τ even at -60°, and hence one concludes that this compound is not aromatic, unless it happens to have an unprecedentedly low activation barrier to conformational inversions [35].

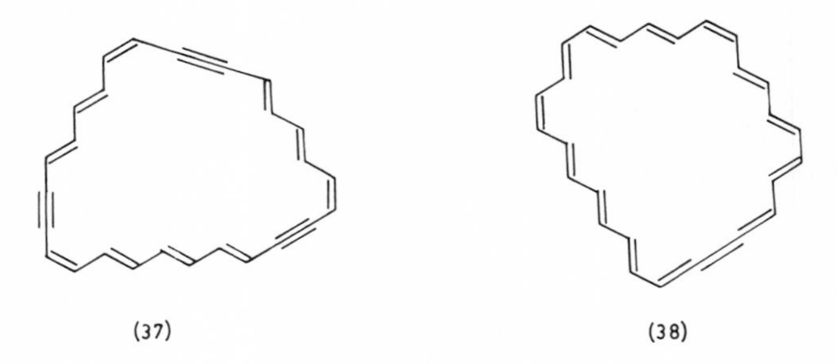

(37) (38)

There is of course the possibility that since the Kekulé structures are not degenerate (the other has three allenic multiple bonds), less fully dehydrogenated [26]annulene might still be aromatic, and this was found to be the case: At -90°, (38) has signals at 2.1-3.8 and at 5.2-6.0 τ in the area ratio 5:3. In this case a weak ring current is therefore still present [36]. In neither case ((37) nor (38)) is the exact conformation or shape known as yet; the formula's drawn here merely represent a possibility. The bisdehydro[26]annulene (39) has been fully characterized by nmr; it has a ring current indicating aromaticity [37].

---

33. R.M. McQuilkin, B.W. Metcalf, and F. Sondheimer, Chem. Commun., 338 (1971).
34. R.M. McQuilkin and F. Sondheimer, J. Amer. Chem. Soc., 92, 6341 (1970).
35. C.C. Leznoff and F. Sondheimer, J. Amer. Chem. Soc., 89, 4247 (1967).
36. B.W. Metcalf and F. Sondheimer, J. Amer. Chem. Soc., 93, 5271 (1971).
37. M. Iyoda and M. Nakagawa, Tetrahedron Lett., 4253 (1972).

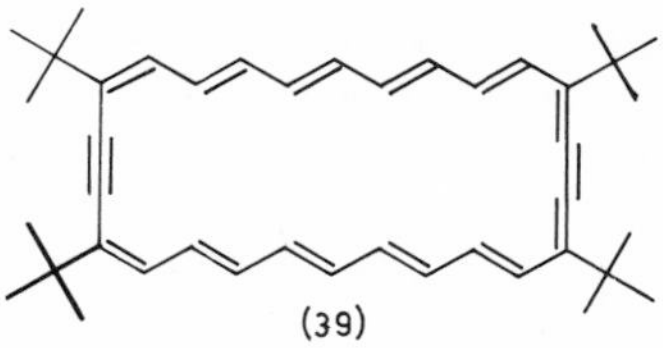

(39)

[30]Annulene (40) has already been fully described in 1962 [23b]. The compound decomposes rapidly at room temperature and although detailed X-ray diffraction and nmr patterns have not as yet been published, its general behavior shows that the compound differs in its properties from its predecessors, and that it is a polyolefin.

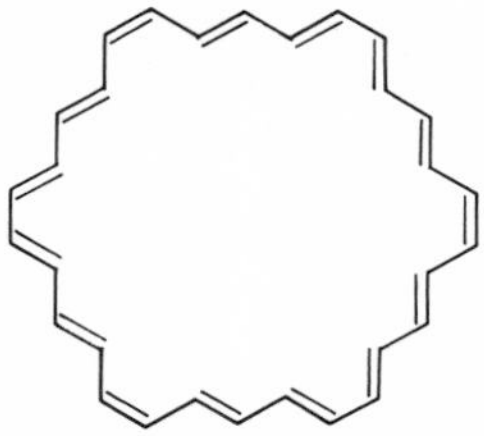

(40)

It is possible of course that by the various devices discussed below, it may prove possible to generate aromatic systems involving 30 or more π electrons; however, so far as the simple annulenes are concerned, the Dewar-Gleicher picture seems essentially correct.

9-4. THE 4n SYSTEMS. NON-AROMATICITY VS. ANTI-AROMATICITY

In 4n systems, it is not clear from the Hückel treatment or its later refinements what is likely to happen other than that resonance stabilization will be absent. Such molecules could simply be polyolefins, with alternating single and double bonds as would be obvious from X-ray diffraction, with typically olefinic nmr absorptions, no pronounced tendency to form a planar structure, and sensitivity to reagents that attack simple double bonds. Such molecules would be considered non-aromatic.

On the other hand, these molecules might still be planar, with approximately equally long bonds. The π system would again not be stabilized, and if the upper two electrons had a degenerate pair of orbitals available to them as is suggested by the Hückel polygon, such molecules would be diradicals. In recent years the realization has furthermore taken hold that completely conjugated cyclic systems - whether with equally long bonds or not and whether planar or not - are destabilized as compared to an open chain polyolefin of the same number of π electrons if the latter can assume a planar conformation. The reason for this is simply that an open chain conjugated polyolefin does have a small but discernable delocalization energy (see p. 22), and that a closed 4n system as a first approximation does not. The planar cyclic 4n systems are referred to as anti-aromatic, and as we shall see below (p. 290), there is a straightforward criterion for recognizing that quality. It must be admitted that in fact no one had given much thought to this possibility - until the experimental facts unceremoniously forced the organic chemist to do so. An extenuating circumstance is that the conformational mobility was not expected either, and it is clear in retrospect that such mobility can make either aromaticity or anti-aromaticity look like non-aromaticity. Since the nmr spectra of the 4n type of molecules thus appeared to indicate polyolefinic structures, these results were accepted without further ado. Such is human nature, and the most experienced, cleverest chemist may fall prey to it as easily as does the average freshman. The nmr spectra play a vital role in experiments directed at these questions. As is pointed out elsewhere in this book (p. 241), when resonance between two Kekulé structures does not occur, bond shifts are still possible, and if they are rapid, the nmr spectrum may be the average one.

The simplest 4n system is cyclobutadiene (41), in which n = 1. Ever since organic chemists first dimly perceived what the aromatic nature of benzene was all about, this elusive compound has been the object of synthetic efforts without parallel.

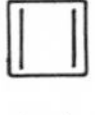

(41)

The ingenuity that went into them as well as the results are one of the most fascinating stories of organic chemistry; however, only the final conclusions can be discussed here [38]. Cyclobutadiene is an extremely unstable compound, completely incapable of isolation at room temperature, or even at the lowest temperatures routinely accessible in the organic laboratory. It can be stabilized by metal complexation, and these complexes are discussed in Chapter 10 (p. 379); free cyclobutadienes occur almost surely in some reactions as intermediates, and those experiments are discussed in Chapter 24 (p. 860); here, we shall discuss only those indirect experiments that give some information on the structure of this compound.

One method of stabilizing cyclobutadiene is by means of well-chosen substituents, and although a measure of success has been achieved, the substituents perturb the carbon skeleton to such a degree that the conclusions have no general

38. For summaries of and references to early and recent efforts, see: (a) R. Criegee, Angew. Chem., Int. Ed. Engl., 1, 519 (1962); (b) R. Pettit, Pure Appl. Chem., 17, 253 (1968).

validity. As an example [39] we may mention diethyl 2,4-bis(diethylamino)cyclobutadiene-1,3-dicarboxylate (42).

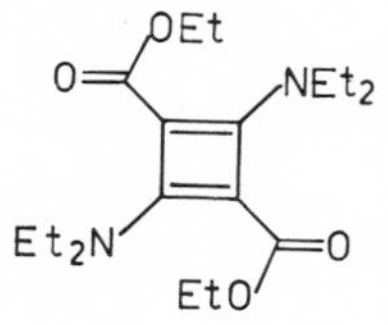

(42)

This compound is reasonably stable, melting at 52°, and the ring is square planar as revealed by X-ray diffraction measurements; however, the C-C bonds of the ring are rather long (1.46 Å), especially if it is considered that they may be somewhat bent. The exocyclic C-C and C-N bonds on the other hand are quite short (1.45 and 1.32 Å, respectively). It is hard to avoid the conclusion that there is a major, if not overwhelming contribution from structure (43).

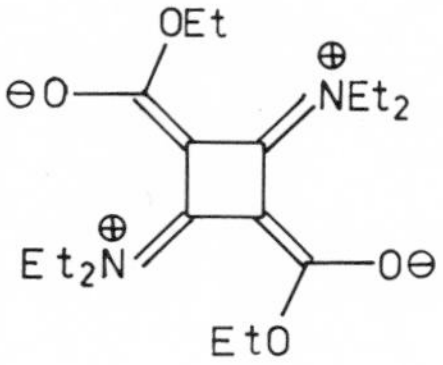

(43)

The stabilization by bulky substituents - a trick that has worked in the synthesis of so many other small rings (see p. 204) - does apparently not help. Attempts to prepare (44) lead to the isomer (45) instead; it is possible that the cyclobutadiene is a precursor of the final product [40].

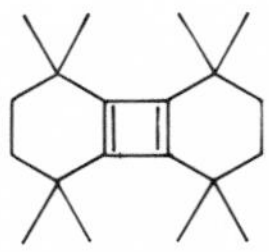

(44) 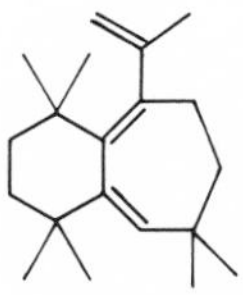 (45)

---

39. This and several related cases are discussed in: (a) H. J. Lindner and B. von Gross, Angew. Chem., Int. Ed. Engl., 10, 490 (1971); (b) R. Gompper and G. Seybold, Angew. Chem., Int. Ed. Engl., 7, 824 (1968); (c) M. Neuenschwander and A. Niederhauser, Chimia, 22, 491 (1968) and Helv. Chim. Acta., 53, 519 (1970).
40. D.E. Applequist, P.A. Gebauer, D.E. Gwynn, and L.H. O'Connor, J. Amer. Chem. Soc., 94, 4272 (1972).

Benzo annelation has also been used as a method of stabilization. Biphenylene (46) has long been known, and 1,2-diphenylnaphtho(b)cyclobutadiene (47) is now known [41].

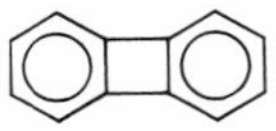

(46) 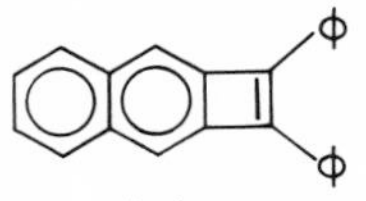 (47)

X-ray diffraction studies [42] indicate that the two bonds connecting the benzo rings in biphenylene are of approximately single bond length, and hence we might expect that the four-membered ring is an anti-aromatic singlet. The nmr spectrum has indeed recently been interpreted as lending some support to an anti-aromatic structure for the four-membered ring [43].

There is one technique that can be used in some cases when the molecule of interest is extremely unstable, namely the matrix isolation technique (see further p. 553). The main requirements are that the molecule must have a precursor from which it can be set free by irradiation, and that the other fragments so produced must not react with it so as to return to the initial state or other products. The precursor is vaporized and mixed with a very large excess of some inert gas such as argon and then frozen on a cold surface, e.g., to 4°K. This so-called matrix (in which the precursor molecules are presumably widely separated) is then irradiated and the products can be studied by ir or uv-visible spectroscopy; if desired, the spectrum can be measured after warming the matrix to various temperatures, so that products and reactivities can also be studied. By means of this technique, it was found [44] that the anhydride (48) apparently gives rise to tetramethylcyclobutadiene upon photolysis at -196°. A yellow color is observable, which irreversibly disappears upon warming; among the products one finds the Diels-Alder product (49).

41. (a) M. P. Cava, B. Hwang, and J. P. Van Meter, J. Amer. Chem. Soc., 85, 4032 (1963); see also (b) B. W. Roberts and A. Wissner, J. Amer. Chem. Soc., 94, 7168 (1972).
42. G. W. Wheland, "Resonance in Organic Chemistry", Wiley, New York, 1955, p. 172.
43. H. P. Figeys, Angew. Chem., Int. Ed. Engl., 7, 642 (1968).
44. (a) G. Maier and U. Mende, Tetrahedron Lett., 3155 (1969). See also (b) G. Maier and M. Schneider, Angew. Chem., Int. Ed. Engl., 10, 800 (1971); (c) G. Maier and F. Boszlet, Tetrahedron Lett., 1025 (1972).

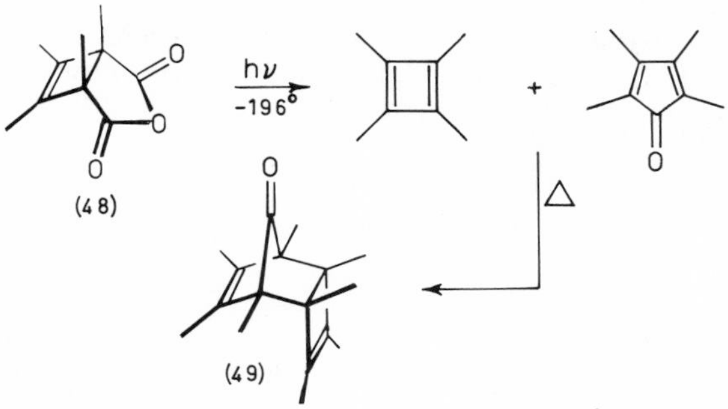

hν
−196°
(48)
(49)

The dimethyl analog (50) similarly gives rise to both adducts (51) and (52) (among other products), so that 1, 2-dimethylcyclobutadiene either is square planar (and presumably diradical) or it manages to equilibrate rapidly even at these low temperatures [44].

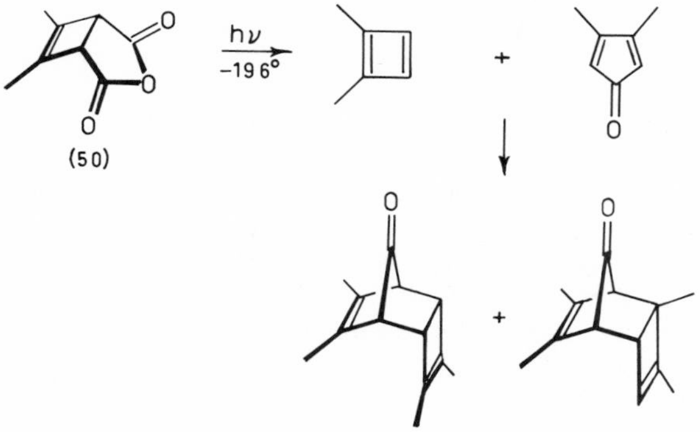

hν
−196°
(50)
(51)
(52)
or

Cyclobutadiene itself has been generated in a matrix by photolysis of α-pyrone (53) and observed by ir; however, no conclusion concerning the question of a planar vs. rectangular cyclobutadiene has as yet been reached [45].

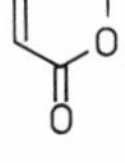

(53)

---

45. (a) C. Y. Lin and A. Krantz, Chem. Commun., 1111 (1972); (b) O. L. Chapman, C. L. McIntosh, and J. Pacansky, J. Amer. Chem. Soc., 95, 244 (1973).

Another aspect of the problem is shown in reports describing the trapping of cyclobutadienes in solution at higher temperatures - at which they are not stable enough to be observed directly. Freedman and Pettit have observed that the trapping of cyclobutadiene itself by means of dimethyl maleate (54) and fumarate (55) is stereospecific [46].

If this hydrocarbon were square planar and of a diradical nature, the intermediate diradical likely to be present would be able to equilibrate via rotational motions, and at least some cross-over of products should have resulted (see p. 491). Again this suggests a rectangular (though possibly equilibrating) cyclobutadiene. Finally, when o-diphenylcyclobutadiene (56) is generated, the location of the phenyl groups in the product is a strong function of the reactivity of the dienophile; with the moderately reactive N-phenylmaleimide (57) only the adduct (58) is obtained, but with the very reactive tetracyanoethylene (59) a 1:7 mixture of (60) and (61) is formed. This suggests that (62) initially is released as the 1,4 isomer, that the 1,2-isomer (63) is more stable, and that the degree of isomerization depends on the time available before capture [47].

---

46. (a) L. Watts, J. D. Fitzpatrick, and R. Pettit, J. Amer. Chem. Soc., 88, 623 (1966); (b) P. Reeves, J. Henery, and R. Pettit, J. Amer. Chem. Soc., 91, 5888 (1969); (c) H.H. Freedman, J. Amer. Chem. Soc., 83, 2194, 2195 (1961).
47. P. Reeves, T. Devon, and R. Pettit, J. Amer. Chem. Soc., 91, 5890 (1969).

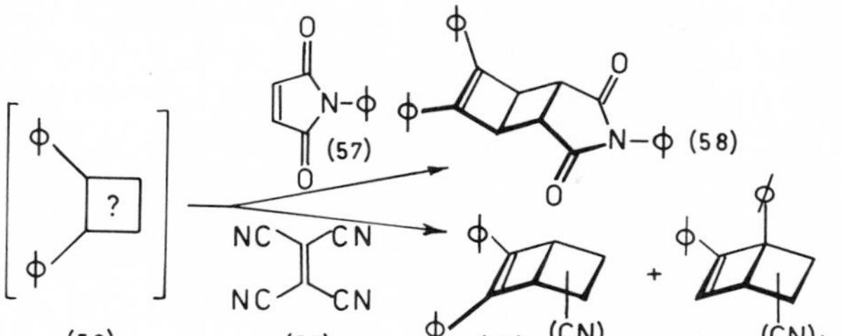

This conclusion is a bit tentative, of course; it could be, for example, that the more reactive dienophile is less finicky about which side of the substituted square planar molecule it attacks. It would be more convincing if the product ratio in the capture by dienophiles depended not only on their reactivity, but also on their concentration.

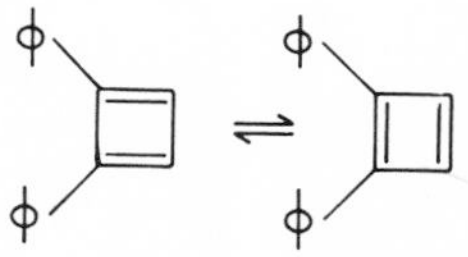

Further insight into the problem is emerging on the basis of electrochemical experiments [48]. It has been found that the polarographic oxidation of dienes such as (64) (which eventually produces a dimer) requires a significantly higher voltage than the saturated analog (65). This supports the opinion that this difference - which amounts to at least 10 kcal/mole or so - is indicative of destabilization of the π-system, and hence of antiaromaticity.

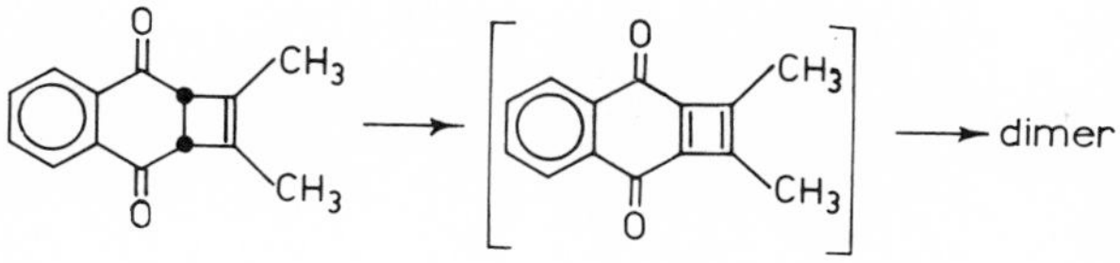

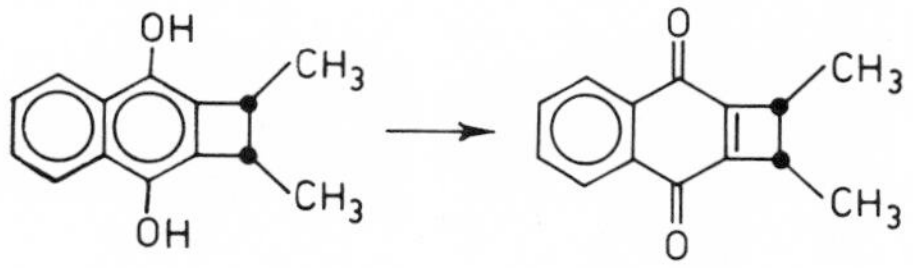

48. (a) R. Breslow, R. Grubbs, and S.I. Murahashi, J. Amer. Chem. Soc., 92, 4139 (1970); (b) A. Pullman, J. Amer. Chem. Soc., 93, 1825 (1971).

Cyclooctatetraene (66) has likewise been the objective of an intensive synthetic effort, but this one was crowned with success relatively early: In 1911, after eight years of work, Willstätter succeeded in converting the alkaloid pseudopelletierine (67) into this hydrocarbon via twelve steps; only a gram or so was obtained [49].

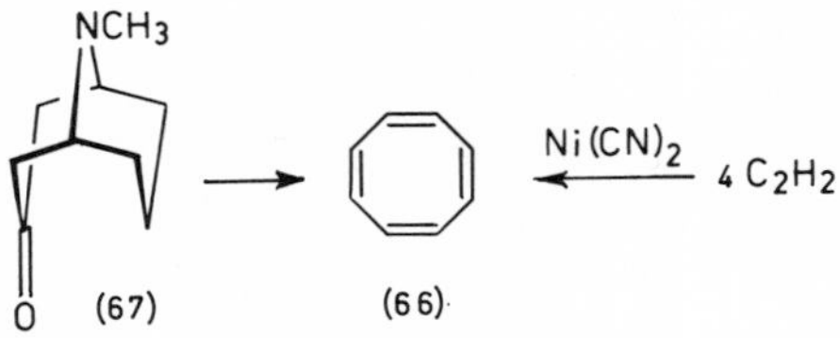

NCH_3 (67) → (66) $\xleftarrow{Ni(CN)_2}$ $4\ C_2H_2$

By grim irony, Willstätter complained in his biography in 1942 - the last year of his life - that the synthesis had never been repeated - unaware that at that time Reppe had accidentally stumbled on a single step catalytic conversion of acetylene into the tetramer, and that this material was then being produced in tonnage quantities. The lack of interest was perhaps induced by Willstätter's disappointing conclusion that the yellow material behaved like a typical olefin with various reagents, and hence it was not a close cousin of benzene. Because of its availability in industrial quantities the properties of cyclooctatetrane are now virtually as well known as those of benzene, and little need be said here except that all subsequent experiments have borne out Willstätter's conclusion. The molecule is sometimes referred to as bathtub or saddle shaped; electron diffraction measurements have shown that the molecule has D_{2d} symmetry, with all double bonds 1.34 Å long and the single bonds of 1.48 Å length [50]. The non-planarity removes any possible question about its electronic nature: There is no possibility either of any delocalization or of a diradical species in this case. As we shall see in the following paragraphs however, it is possible to make an estimate of the energy separating the anti-aromatic state from the non-aromatic real molecule.

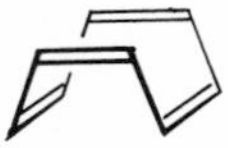

If the conformational mobility encountered in the larger aromatic rings may be extended to cyclooctatetraene, one may except that inversions might occur, as follows.

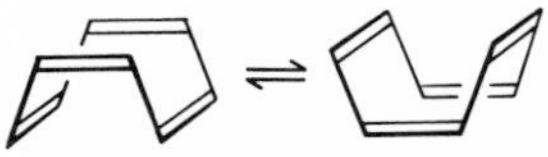

⇌

49. (a) R. Willstätter and E. Waser, Chem. Ber., 44, 3423 (1911); (b) R. Huisgen, J. Chem. Educ., 38, 10 (1961).
50. M. Traetteberg, Acta Chem. Scand., 20, 1724 (1966).

In the parent compound, such a motion would merely regenerate the molecule, but in monosubstituted derivatives such as fluorocyclooctatetraene (68), in effect this equilibration would amount to racemization.

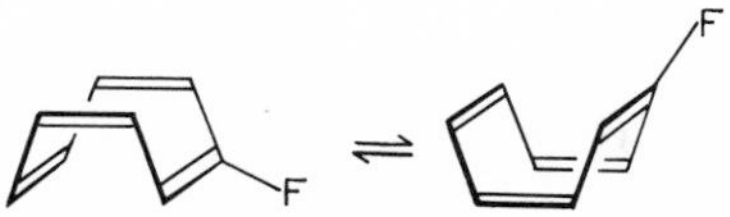

(68)

If furthermore the substituent itself had any asymmetric center, inversion would produce a new diastereomer; there are therefore potentially possibilities for stereoisomerism such as there are in the amines.

The simple olefinic nature of cyclooctatetraene presumably also gives rise to the so-called bond-shift isomerism, exemplified by molecules such as (69)-(72). This is the very kind of isomerism that the pre-Kekulé organic chemist had vainly looked for in o-disubstituted benzenes.

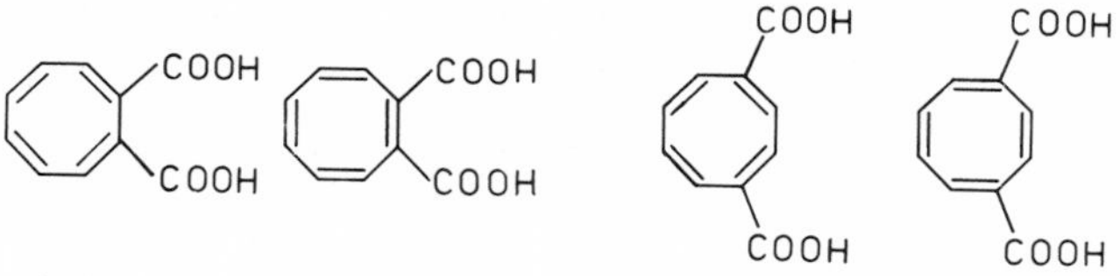

(69) (70) (71) (72)

Such isomers should in more general terms be referred to as valence tautomers; by this term one implies isomers which are interconvertible without disconnecting any substituents from the carbon skeleton (see p. 393). Although there are as yet no examples in which two such isomers have been isolated, the experiments below suggest that such examples may well be forthcoming.

The problem one faces in a search for evidence for the processes of inversion and bond-shift isomerization is the relation between the two. Thus, the most complicated situation one could imagine is envisioned in the scheme below; two corners have been labeled so as to suggest what might be happening.

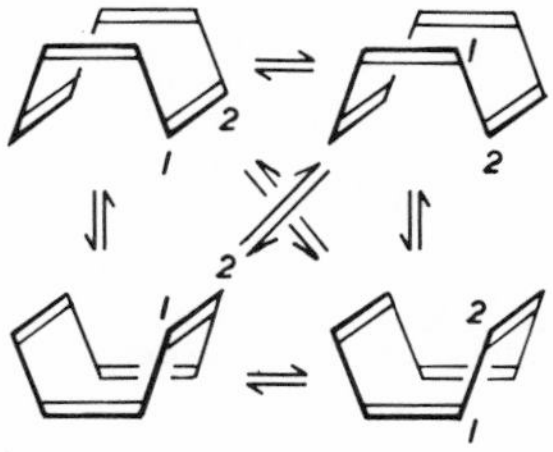

Keeping these complications in mind, we are now in a position to review some recent experiments. The F^{19}-nmr spectrum of fluorocyclooctatetraene (68), which shows a broad band at room temperature, sharpens when the temperature is lowered and finally a doublet is obtained indicating splitting by a single proton (H_1).

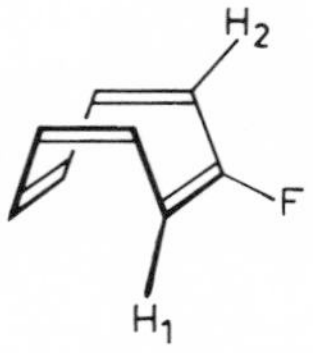

(68)

Evidently, the fluorine atom at this temperature does not "see" its second proton neighbor, since the dihedral angle between them is about 90°. The activation energy is about 12 kcal/mole [51]; a similar experiment has been based on the temperature dependence of the C^{13} satellites in cyclooctatetraene itself [52].

It should be pointed out that such experiments <u>may</u> involve inversion, but the results <u>require</u> a bond shift. If inversion alone occurred, the fluorine signal would still be split by the same proton and there would be no loss of sharpness; broadening can occur only if a rapid bond shift isomerization is occurring.

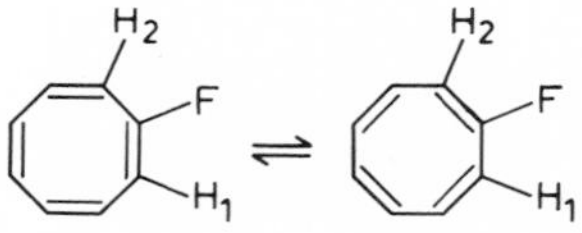

The *ortho*-dicarboxylic acid in fact apparently has the 1,8-structure (69) rather than the 1,2-form (70). This is concluded from the facts that it forms an anhydride (73) only with great difficulty and in low yield, and that this product is unstable to polymerization. By contrast, the benzo analog (74) - which surely must have the 1,2-structure - forms a stable anhydride (75) very easily [53]. The implication is that the acid occurs only or predominantly in form (69), and that furthermore anhydride (73) does not readily isomerize to its bond shift isomer; this latter conclusion, if correct, raises questions about the assumption often made that such isomerizations occur via a planar transition state.

---

51. D.E. Gwynn, G.M. Whitesides, and J.D. Roberts, *J. Amer. Chem. Soc.*, *87*, 2862 (1965).
52. F.A.L. Anet, *J. Amer. Chem. Soc.*, *84*, 671 (1962).
53. D. Bryce-Smith, A. Gilbert, and J. Grzonka, *Angew. Chem., Int. Ed. Engl.*, *10*, 746 (1971).

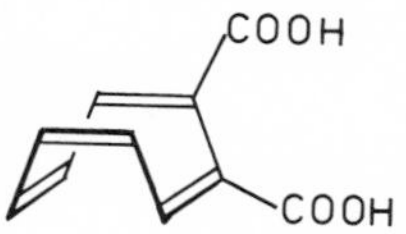

(69) 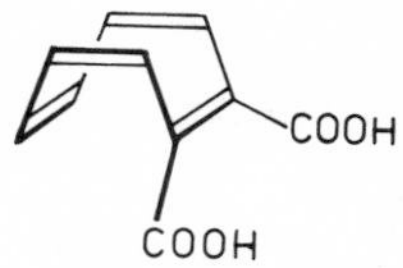 (70) 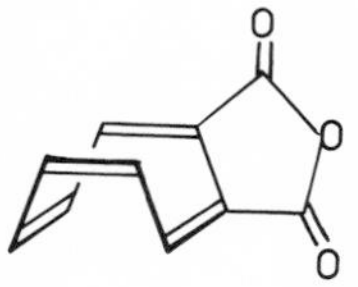 (73)

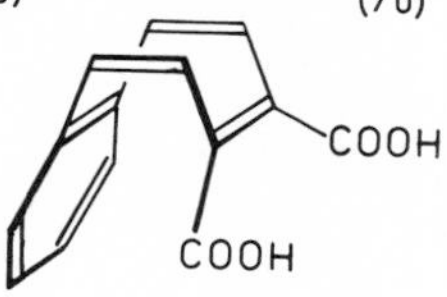

(74) 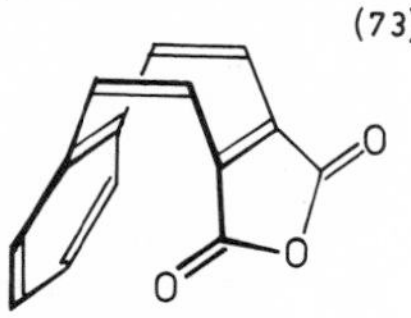 (75)

Two most illuminating experiments have recently been reported by Anet [54]. It was found, first of all, that the ester (76) has two pairs of methyl pmr signals, both in the ratio of about 17:1; thus, two bond-shift isomers (77) and (78) are present, and in a 17:1 ratio. If a solution of (76) is photolyzed, these nmr signals change in relative intensity until the minor isomer becomes predominant. Warming to -30° rapidly restores the original ratio; warming to 90° raises the rate of isomerization to such a degree that both signal pairs coalescence. These experiments show that isomerization occurs, and that ΔE_a for this process is about 19 kcal/mole (the major isomer was assigned the 1,8-structure (77), but this is relatively unimportant).

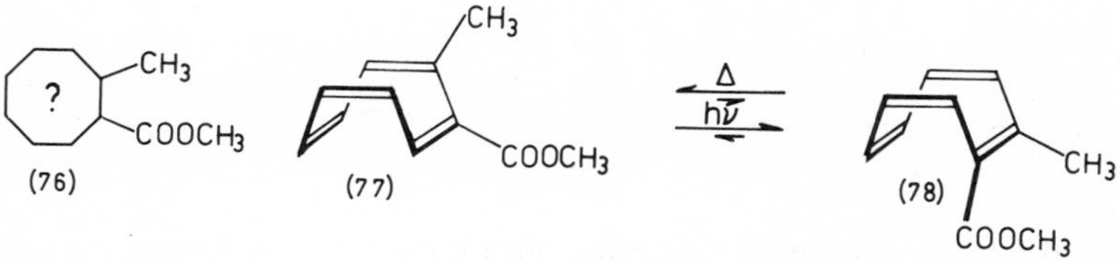

(76) (77) (78)

Equally interesting is the nmr spectrum of ester (79). In this case, inversion without bond shift would interconvert a diastereomeric pair (note that two asymmetric "centers" are present). Altogether four ring-methyl signals can be seen in the nmr spectrum. Heating to 70° leads to the coalescence of this group to two peaks: Evidently ring inversion is already rapid when bond shift isomerization is still slow in the nmr time scale.

54. F.A.L. Anet and L.A. Bock, J. Amer. Chem. Soc., 90, 7130 (1968).

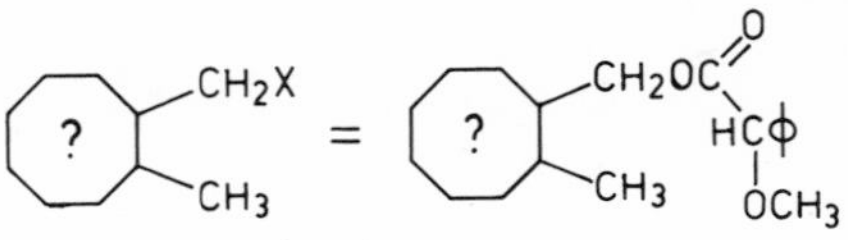

(79)

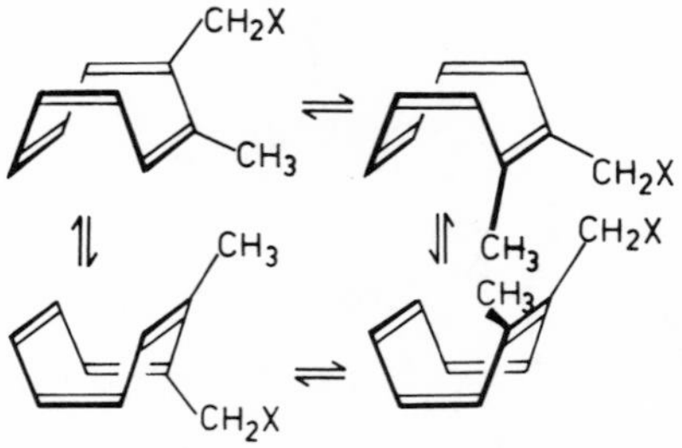

We may now consider the question whether the hypothetical planar cyclooctatetraenes would be antiaromatic or not. It is generally assumed that the transition states in the inversion and or isomerization processes are at least close to planar, but this is only an assumption. In molecules such as (80) and (81), the methyl groups give rise to separate nmr signals as the gem-dimethyl group is bound to an asymmetric structure.

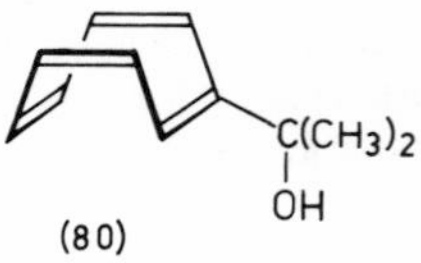

(80) 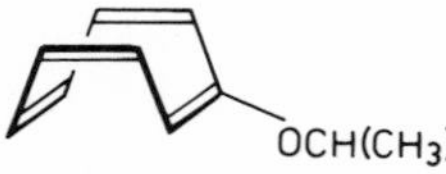 (81)

Inversion rapid on the nmr time scale renders them equivalent, and at some appropriate temperature coalescence occurs. From the coalescence temperatures the activation free energies can be calculated (see p. 64); for (80) and (81) the activation energies are 15 and 13 kcal/mole, respectively.

An instructive story can be told concerning the tetrabenzo analog (82). The benzo annelation can be expected to minimize delocalization, and hence the temperature at which coalescence of the methyl groups occurs may be informative about the nature of planar cyclooctatetraene.

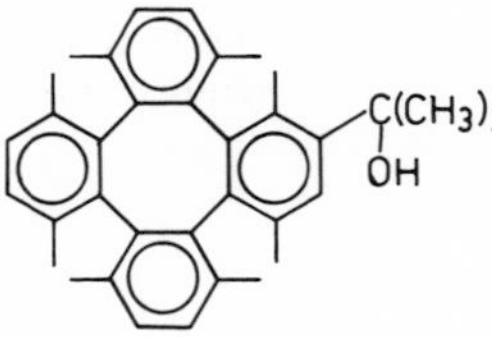

(82)

As it turned out, only a single methyl signal was observed, even at the lowest temperatures accessible. The authors concluded that E_a could not be more than 5.7 kcal/mole, and hence that the annelation lowered the energy of planar cyclooctatetraene [55].

This argument is based on the assumption that the two methyl signals are not coincidentally isochronous, and the report was almost immediately followed by an avalanche of papers showing that the assumption was inapplicable [56]. Thus, a Westheimer calculation shows that the interference of the o-hydrogen atoms is so severe that $\Delta G^{\ddagger}$ must be over 200 kcal/mole. With a compound similar to (82) but with an isopropylketone grouping rather than i-propyl, two methyl doublets are observed and no coalescence occurs up to 190°, so that $\Delta G^{\ddagger}$ then must be at least 26 kcal/mole.

The result of these various experiments is that planar cyclooctatetraene is at least 10-20 kcal less stable than the saddle shaped tetraolefin - presumably at least in part for reasons of angle strain and steric hindrance. Since we do not really know whether the transition states are planar, the energy difference may be much greater.

Research leading to the [12]annulenes provided organic chemists with another of those tantalizing surprises so frequently encountered in this field. The trisdehydro derivative (83) was prepared in 1966 [57]; even though this molecule is free of angle strain, it was found to be sensitive to air.

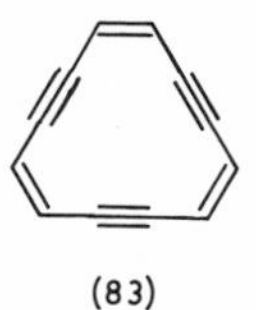

(83)

The nmr spectrum was found to consist of a singlet at 5.5τ, and hence the discoverers were satisfied that the compound was not aromatic [58].

---

55. H. P. Figeys and A. Dralants, Tetrahedron Lett., 3901 (1971).
56. (a) A. Rosdahl and J. Sandström, Tetrahedron Lett., 4187 (1972); (b) C. J. Finder, D. Chung, and N. L. Allinger, Tetrahedron Lett., 4677 (1972); (c) G. H. Senkler, D. Gust, P. X. Riccobono, and K. Mislow, J. Amer. Chem. Soc., 94, 8626 (1972); (d) D. Gust, G. H. Senkler, and K. Mislow, Chem. Commun., 1345 (1972).
57. In retrospect it became clear that (83) was already available in 1962 but it was then mistakenly identified as a bisdehydro[12]annulene.
58. (a) K. G. Untch and D. C. Wysocki, J. Amer. Chem. Soc., 88, 2608 (1966); (b) F. Sondheimer, R. Wolovsky, P. J. Garratt, and I. C. Calder, J. Amer. Chem. Soc., 88, 2610 (1966).

The nmr spectrum of the bisdehydro derivative (84) was similarly believed simply to indicate the absence of aromaticity; all hydrogens signalled their presence at 5-6 τ except for two (H_1 and H_2) which showed up at -1τ. This very odd fact was explained as possibly due to deshielding by the triple bonds [59].

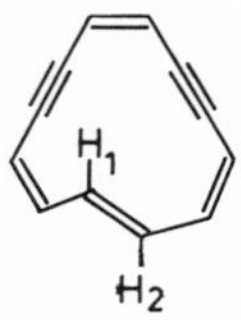

(84)

However, at this time the notion of anti-aromaticity was beginning to take hold, and it was suspected that in such compounds a paramagnetic ring current would shift normal outside protons to higher field (as in the two examples just mentioned), and inside protons to extremely low field, exactly opposite to aromatic compounds. The signal at -1τ for H_1 and H_2 in (77) could be due to averaging caused by conformational mobility. Untch [60] succeeded in preparing the bromo derivative (85), in which the large size of the bromine atom prevents this mobility, and indeed a single proton, obviously the inner one, is now observed at extremely low field: -6.4τ.

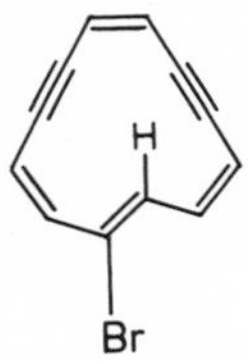

(85)

Anti-aromaticity has since also been indicated for the remarkable tetrakisdehydro derivative (86) by comparison with the open chain analog (87): The former has a pair of methylene signals, the lower one of which occurs at 8.15-8.55 τ; in (87) the corresponding signal is at 7.65-7.95 τ [61].

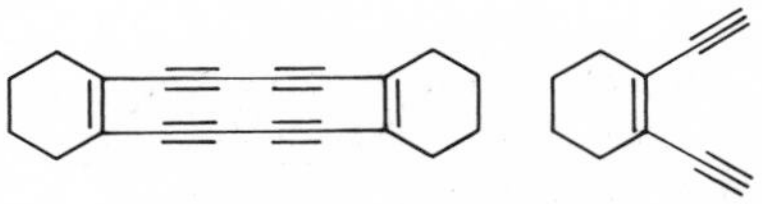

(86) (87)

59. R. Wolovsky and F. Sondheimer, J. Amer. Chem. Soc., 87, 5720 (1965).
60. K.G. Untch and D.C. Wysocki, J. Amer. Chem. Soc., 89, 6387 (1967).
61. G.M. Pilling and F. Sondheimer, J. Amer. Chem. Soc., 93, 1970 (1971).

[12]Annulene itself (88) is a difficult compound to obtain; the three internal hydrogen atoms render the molecule quite unstable. It can be made by photolysis of the isomer (89) at -100°, and isomerizes at -40° to (90).

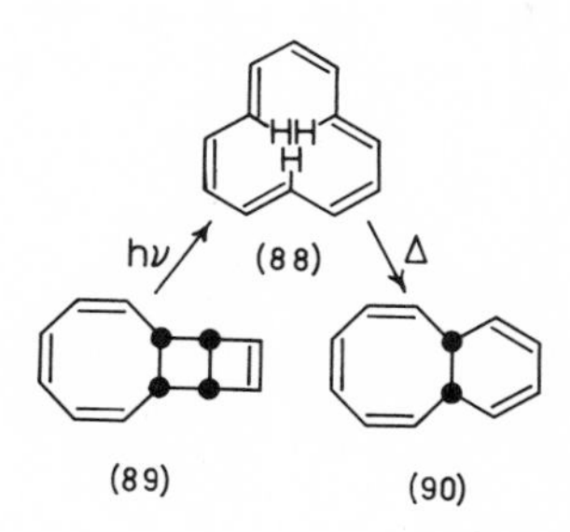

At -170°, nmr signals can be observed above and below the benzene reference signal (by 1.39 and -0.56 ppm, respectively), in a 3:1 ratio. The steric compression leaves the molecule subject to very great conformational mobility: At only slightly higher temperature coalescence occurs to give rise to a 1:1 signal. Evidently only two conformations are possible, and in spite of the mobility, half of the protons are doomed forever to stay outside [62, 21]. It may be noted here that since the terms aromatic and anti-aromatic are subject to confusion and the nature of the ring current is clear, the terms diatropic and paratropic have come into use recently.

The chemistry of [16]annulene (91) is now very accessible to study since it can be simply obtained by photolysis of this *cis*-cyclooctatetraene dimer (92).

hν

HH
HH

(92) (91)

The main information is that at room temperature all protons signal at 3.3 τ, but at -110° a 1:2:1 triplet appears at -0.4 τ and a multiplet centered at 4.6, in a 1:3 overall ratio. Thus, this compound is anti-aromatic [63]. The X-ray diffraction pattern reveals that the molecule is approximately planar, and that the carbon-carbon bonds alternate in length between single and double bond values [64].

---

62. (a) H. Röttele, W. Martin, J. F. M. Oth, and G. Schröder, *Chem. Ber.*, *102*, 3985 (1969); (b) J. F. M. Oth, H. Röttele, and G. Schröder, *Tetrahedron Lett.*, 61 (1970); (c) J. F. M. Oth, J. M. Gilles, and G. Schröder, *Tetrahedron Lett.*, 67 (1970).
63. G. Schröder and J. F. M. Oth, *Tetrahedron Lett.*, 4083 (1966).
64. S. M. Johnson and I. C. Paul, *J. Amer. Chem. Soc.*, *90*, 6555 (1968).

Similar conclusions have been drawn concerning several dehydro[16]annulenes [65]. The [20]- and [24]annulenes [66, 67] and some of their dehydroannulenes [68] also show this behavior, and we may conclude this section with the observation that for n = 1-6, all (4n + 2) conjugated carbocycles are observed to be aromatic except [10]annulene, and all the (4n) analogs are anti-aromatic except cyclooctatetraene; hence these medium size rings are governed more by factors of strain than by electronic considerations.

9-5. BRIDGED AROMATIC SPECIES

In those of the annulenes in which the internal hydrogen atoms impose such strain that aromaticity does not occur or leads to only a marginal stability increase, one might be able to avoid some of the strain by replacing those atoms by a single bridging atom. Thus, it would be interesting to compare [10]annulene with 1,6-methanocyclodecapentaene (92) if it were possible to synthesize it; models suggest that the ten-membered ring should be flatter.

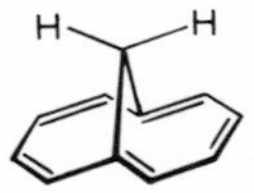

H H

(92)

Vogel and his co-workers succeeded in preparing the compound, and its aromatic nature was soon evident [69]. It is stable to oxygen and to heating to at least 220°; it will undergo substitution reactions with bromine, copper nitrate and acetyl chloride (primarily in the α-positions as in naphthalenes). The nmr spectrum is most informative; an A_2B_2 pattern is observed at 2.8 τ and a singlet at 10.5 τ in a 4:1 ratio; a ring current is clearly indicated. All this evidence mitigates against an alternative structure such as the propellatetraene (93). Especially convincing is the fact $J_{C^{13}H}$ for the methylene bridge is 142 Hz; the %

---

65. I.C. Calder, Y. Gaoni, P.J. Garratt, and F. Sondheimer, J. Amer. Chem. Soc., 90, 4954 (1968).
66. B.W. Metcalf and F. Sondheimer, J. Amer. Chem. Soc., 93, 6675 (1971).
67. I.C. Calder and F. Sondheimer, Chem. Commun., 904 (1966).
68. (a) K. Stöckel and F. Sondheimer, J. Chem. Soc., Perkin I, 355 (1972); (b) F. Sondheimer and R. Wolovsky, J. Amer. Chem. Soc., 81, 4755 (1959); (c) I.C. Calder and P.J. Garratt, J. Chem. Soc. (B), 660 (1967).
69. (a) E. Vogel and H.D. Roth, Angew. Chem., Int. Ed. Engl., 3, 228 (1964); (b) E. Vogel and W.A. Böll, Angew. Chem., Int. Ed. Engl., 3, 642 (1964); (c) Chem. Eng. News, 42, 40, Oct. 5, 1964; (d) E. Vogel, W.A. Böll, and M. Biskup, Tetrahedron Lett., 1569 (1966).

(93)

s character of these CH bonds is clearly not very high. The X-ray diffraction pattern of the α-carboxylic acid (94) tells the full story [70]; the C_1 - C_6 distance is 2.5 Å, and the peripheral C-C bonds are all nearly equal in length at (1.40 ± 0.02) Å.

(94)

The bridging bonds (C_1 - C_{11} and C_6 - C_{11}) are shorter than normal (1.48 Å), and the methylene abond angle CCC is 99°. The C_{10}-ring atoms very nearly lie in one plane, with the atoms C_1 and C_6 somewhat to one side of it.

We therefore conclude that aromatic 10 π-electron systems can exist after all; however, the difference in energy between species (92) and (93) is not very great, and the imposition of additional strains changes the picture. The presence of an ethano bridge instead of a methano bridge for example (compound (95)) causes decomposition at 150° to ethylene and naphthalene; uv absorption now occurs at much higher energies, and the nmr spectrum has a quartet at 4.5 τ and a singlet at 7.5 τ in equal intensity [71]. Evidently (95) is indeed a propellatetraene; the extra methylene group is just too much for the 10-membered ring to accommodate.

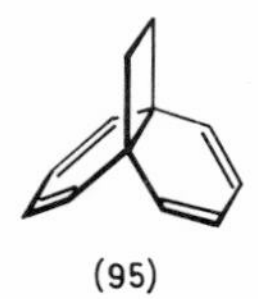

(95)

70. M. Dobler and J.D. Dunitz, Helv. Chim. Acta, 48, 1429 (1965).
71. E. Vogel, W. Mayer, and J. Eimer, Tetrahedron Lett., 655 (1966).

The introduction of certain groups at the bridge position likewise interferes with aromaticity. The methylene, methyl- and difluoro-derivatives (96) [72], (97) [73], and (98) [74] are all stable aromatic species, but (99) [75] is another propellane. X-ray diffraction of this species has revealed an angle of 136° between the six-membered rings and that the carbon-carbon distances in the ten-membered ring are alternately 1.34 and 1.45 Å long; the C_1C_6 bond length is 1.80 Å, probably a record for single bonds [76].

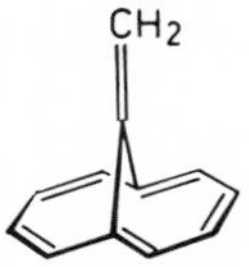

CH_2 CH_3 F F H_3C CH_3

(96) 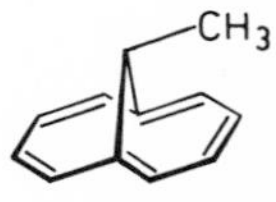 (97) 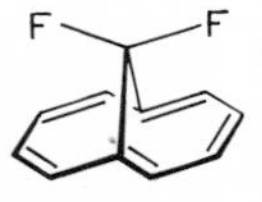 (98) 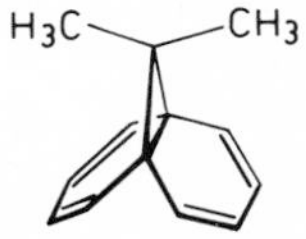 (99)

This general principle of inducing or accentuating aromaticity can be extended to [14]annulenes. Syn-1,6:8,13-bismethano[14]annulene (100) is aromatic as judged by all the criteria described above for the lower homolog; however, in the anti-isomer (101) the strains introduced by the two methylene bridges collaborate in such a way as to generate a non-aromatic species, which is not simply of the propellane type, but which has alternating single and double bonds (about 1.47 and 1.35 Å in length, respectively) along the 14-membered periphery as shown by X-ray diffraction [77].

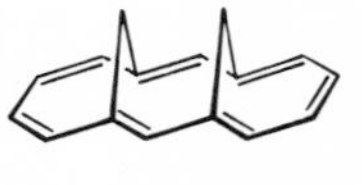

(100) 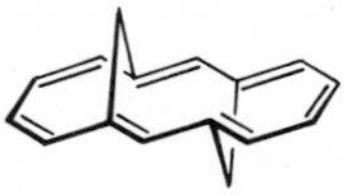 (101)

The nmr spectrum of (101) shows that a very rapid bond-shift isomerization is occurring (note that in (101) all four bridge protons should be different). Even at -100° only two doublets can be seen; however, coalescence is setting in and at -140° the species (101) is frozen out [78].

---

72. E. Vogel, F. Weyres, H. Lepper, and V. Rautenstrauch, Angew. Chem., Int. Ed. Engl., 5, 732 (1966).
73. E. Vogel, Special Publication Nr. 21, Chem. Soc. (London), 113 (1967).
74. C.M. Gramaccioli and M. Simonetta, Tetrahedron Lett., 173 (1971).
75. V. Rautenstrauch, H.J. Scholl and E. Vogel, Angew. Chem., Int. Ed. Engl., 7, 288 (1968).
76. R. Bianchi, A. Mugnoli, and M. Simonetta, Chem. Commun., 1073 (1972).
77. C.M. Gramaccioli, A. Mimun, A. Mugnoli, and M. Simonetta, Chem. Commun., 796 (1971).
78. E. Vogel, U. Haberland, and H. Günther, Angew. Chem., Int. Ed. Engl., 9, 513 (1970).

In the syn-isomer (100) the two inner bridge hydrogen atoms must interfere with one another - and therefore with the aromatic stability - to some degree; this is nicely demonstrated by replacing the two with still another methylene bridge such as in 1,6:8,13-propanediylidene[14]annulene, (102) [79].

(102)

A very stable and typically aromatic system results, with the ring protons flipping at 2.1 - 2.4 τ, and the bridge hydrogen atoms at 10.6 - 11.2 τ ; various aromatic substitutions can be done with it, and the bonds in the aromatic group of carbon atoms are all of nearly equal and typically aromatic length [80]. Interestingly, even the insertion of a second methylene as in (103) does not significantly alter this picture; the strain so introduced is evidently accommodated by bending the bridge rather than the 14-membered ring.

(103)

On the other hand, (104) and even (105) are also stable aromatic species [82].

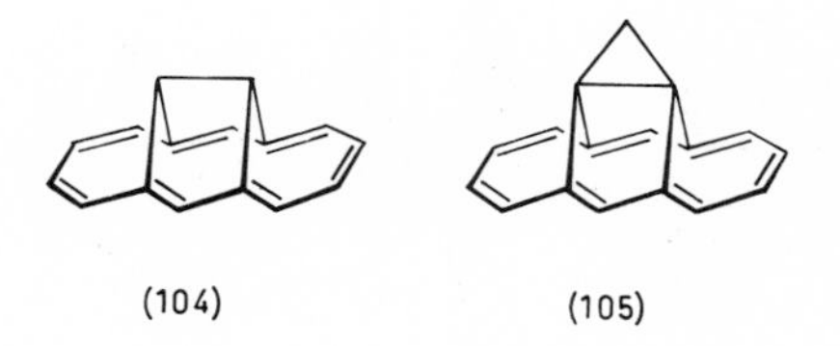

(104) (105)

79. (a) E. Vogel, A. Vogel, H.K. Kübbeler, and W. Sturm, Angew. Chem., Int. Ed. Engl., 9, 514 (1970); (b) W. Bremser, J.D. Roberts, and E. Vogel, Tetrahedron Lett., 4307 (1969).
80. A. Gavezzotti, A. Mugnoli, M. Raimondi, and M. Simonetta, J. Chem. Soc., Perkin II, 425 (1972).
81. (a) E. Vogel, W. Sturm, and H.D. Cremer, Angew. Chem., Int. Ed. Engl., 9, 516 (1970); (b) C.M. Gramaccioli, A. Mugnoli, T. Pilati, M. Raimondi, and M. Simonetta, Acta Cryst., B28, 2365 (1972).
82. E. Vogel and H. Reel, J. Amer. Chem. Soc., 94, 4388 (1972).

The bridging is in principle equally possible with atoms other than carbon and several examples are known: The epoxide (106) and the amine (107) are well-behaved aromatic compounds similar in their behavior to the methano-bridged species [83].

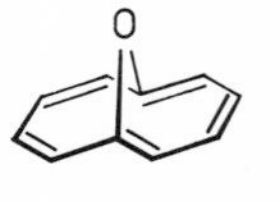

O NH

(106) 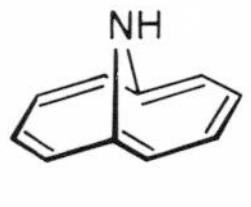 (107)

In the bridged [14]annulenes, the syn-diepoxide (108) is likewise a stable aromatic compound [84]. The anti-isomer (109) to date has not been made; all attempts so far have only succeeded in producing the syn-isomer. Evidently the anti-epoxide is not aromatic and stabilizes itself by flipping one of the oxygen atoms right through the ring system to the other side [85].

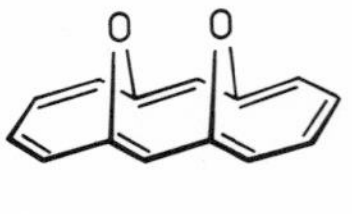

O O O O

(108) 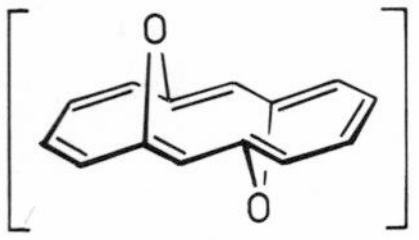 (109)

The bridged [10]annulenes have already been put to certain uses. Thus, the synthesis of the highly strained benzocyclopropenes may conveniently be carried out as follows [86].

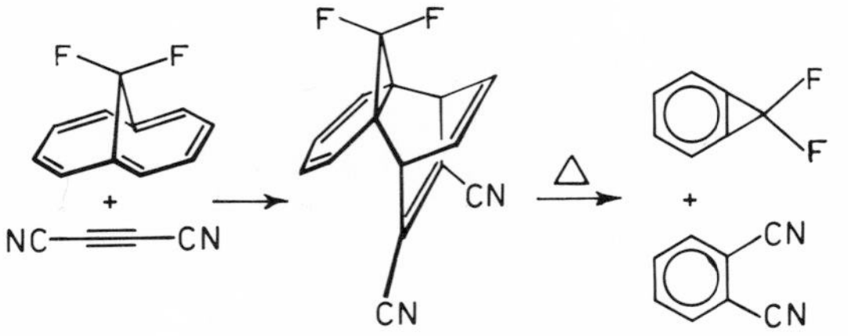

F F + NC—≡—CN → F F CN CN Δ → F F + CN CN

83. E. Vogel, M. Biskup, W. Pretzer, and W.A. Böll, Angew. Chem., Int. Ed. Engl., 3, 642 (1964).
84. E. Vogel, M. Biskup, A. Vogel, and H. Günther, Angew. Chem., Int. Ed. Engl., 5, 734 (1966).
85. E. Vogel, Pure Appl. Chem., 28, 355 (1971).
86. E. Vogel, S. Korte, W. Grimme and H. Günther, Angew. Chem., Int. Ed. Engl., 7, 289 (1968).

The [14]annulene analogs considered here do not have the same conformation as does [14]annulene itself. In the latter species, the inside hydrogen atoms can be avoided in an alternative manner, namely by a pair of carbon atoms in such a way that a virtually strainless and practically planar ring results. More important, Boekelheide and his students [87] succeeded in a 17-step synthesis of such compounds as trans-15, 16-dimethyldihydropyrene, (110).

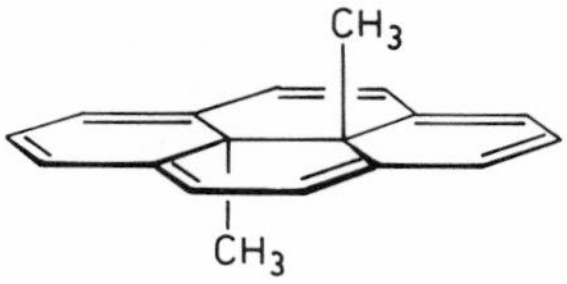

CH_3

CH_3

(110)

The methyl groups show up at no less than 14.25 τ in the nmr spectrum; the ring protons are found at 1.3-2.0 τ, and clearly a strong diamagnetic ring current is induced. Electrophilic substitution occurs readily at the 2- and 7-positions. X-ray diffraction shows the whole system to be planar with only the two inner carbon atoms somewhat out of this plane; all peripheral carbon atoms are nearly equidistant from their neighbors [88]. The compound is quite stable; while photolysis converts it into the metacyclophane (111), the latter reverts to (110) on warming.

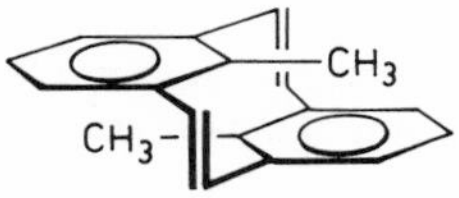

CH_3

CH_3

(111)

In the diethyl homolog the methylene hydrogen atoms appear at 14.0 τ and the methyl protons at 11.9; in this case the cyclophane is thermally more stable. Eventually, the parent compound (112) was reported; the inner protons appear at 15.5 τ in this case [89].

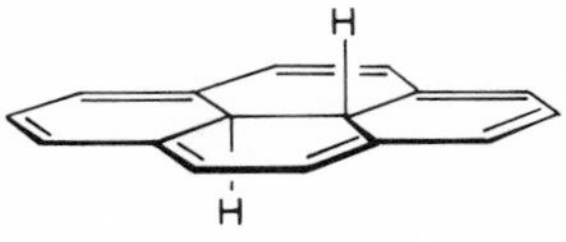

H

H

(112)

---

87. (a) V. Boekelheide and J.B. Phillips, J. Amer. Chem. Soc., 89, 1695 (1967); (b) J.B. Phillips, R.J. Molyneux, E. Sturm, and V. Boekelheide, J. Amer. Chem. Soc., 89, 1704 (1967); (c) V. Boekelheide and T. Miyasaka, J. Amer. Chem. Soc., 89, 1709 (1967); (d) Chem. Eng. News, 41, 42, May 27, 1963; (e) H.B. Renfroe, L.A.R. Hall and J. Gurney, Abstracts, 153d National Meeting of the A.C.S., Miami Beach, Fla., April 1967, paper O-184.
88. A.W. Hanson, Acta Cryst., 23, 476 (1967).
89. R.H. Mitchell and V. Boekelheide, J. Amer. Chem. Soc., 92, 3510 (1970).

The substituents bound to the inner carbon atoms may be replaced by a single atom in certain cis-15, 16-dihydropyrene derivatives; now however, the tendency for isomerization to the cyclophanes becomes strong, and in fact only these molecules result. Examples are (113) [90], (114) [91] and (115) [92] and their derivatives.

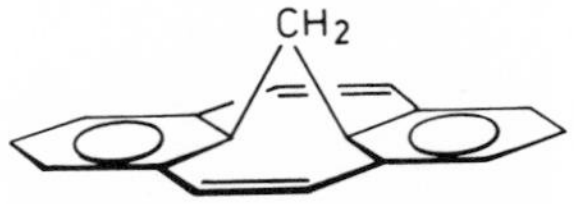

(113)

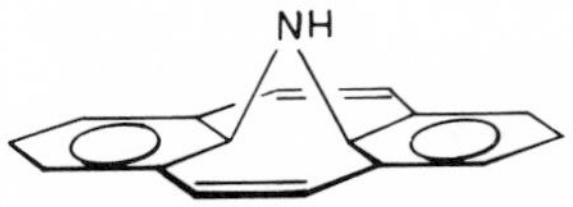

(114)  (115)

In (113), the bridge hydrogens are found at 7.5 τ in the nmr spectrum, and unlike the green, aromatic 15, 16-dihydropyridine derivatives, (113)-(115) are colorless; it was noted however that when (115) is dissolved in strong acid, a green color does develop briefly before decomposition sets in.

Still another conceivable way to span the [14]annulene periphery is by means of a single carbon atom, as hinted at in Chapter 7 (p. 208); this atom would have to be square planar. No such species has as yet been reported, however.

One bridged [16]annulene is known [93], namely (116). The compound is green also. The nmr spectrum shows a methyl signal at 5.2 τ and all other protons at higher field: Thus, the compound is very likely anti-aromatic.

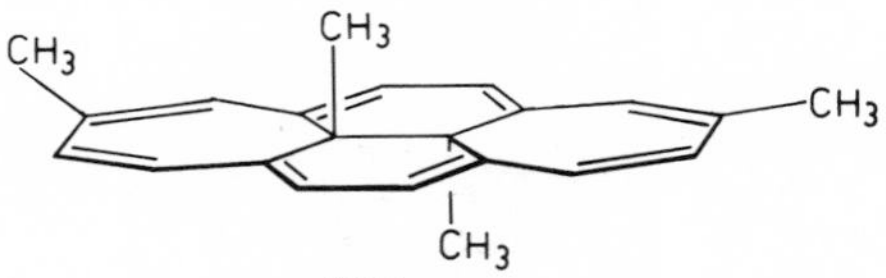

(116)

---

90. H.B. Renfroe, J. Amer. Chem. Soc., 90, 2195 (1968).
91. (a) B.A. Hess, A.S. Bailey, and V. Boekelheide, J. Amer. Chem. Soc., 89, 2746 (1967); (b) B.A. Hess, A.S. Bailey, B. Bartusek, and V. Boekelheide, J. Amer. Chem. Soc., 91, 1665 (1969).
92. B.A. Hess and V. Boekelheide, J. Amer. Chem. Soc., 91, 1672 (1969).
93. R.H. Mitchell and V. Boekelheide, Chem. Commun., 1557 (1970).

It is possible not only to induce or facilitate aromaticity by bridging, one can try to interfere or compete with it in the same way in appropriate cases. Thus, if the internal hydrogen atoms of [18]annulene are replaced by three oxygen atoms, a trioxide (117) results which might be a molecule basically consisting of three aromatic furan rings held together by vinyl bridges rather than an [18]annulene with internal oxygen bridges.

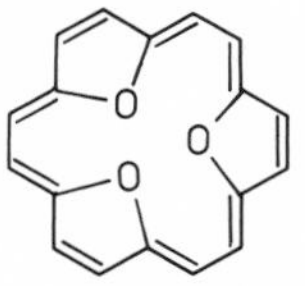

(117)

In fact, this red-colored compound is stable to heating to nearly 200° and shows nmr signals at 1.32 and 1.34 τ; those in [18]annulene appear at 1.1 τ, and hence (117) is still an annulene [94]. On the other hand, trisulfide (118) has signals at 3.27 and 3.23 τ; evidently, there is not enough room for three sulfur atoms, and the result is a trio of vinyl-linked thiophene rings [95].

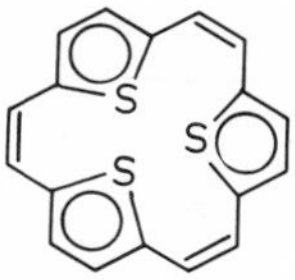

(118)

This is especially obvious when (118) is allowed to react with potassium metal and an esr spectrum is measured of the resulting radical anion; the electron signal is then found to be split by six pairs of equivalent hydrogens, rather than by two groups of six [96]. Two oxygen atoms and one sulfur still results in an overall ring current [97], and one oxygen and two sulfur atoms again leads to six π-electron systems [98].

---

94. G. M. Badger, J. A. Elix, and G. E. Lewis, Austr. J. Chem., 19, 1221 (1966).
95. G. M. Badger, J. A. Elix, and G. E. Lewis, Austr. J. Chem., 18, 70 (1965).
96. F. Gerson and J. Heinzer, Helv. Chim. Acta, 51, 366 (1968).
97. G. M. Badger, G. E. Lewis, and U. P. Singh, Austr. J. Chem., 19, 1461 (1966).
98. G. M. Badger, G. E. Lewis, U. P. Singh and T. M. Spotswood, Chem. Commun., 492 (1965).

Even larger rings have been spanned in this way [99]; (119) and its geometric isomer (120) are both clearly anti-aromatic, with the inside protons at 1.3 τ, the outside vinyl protons at 5.2 and the furan ring protons at 4.3 τ (3.0 τ is normal for furan β-protons).

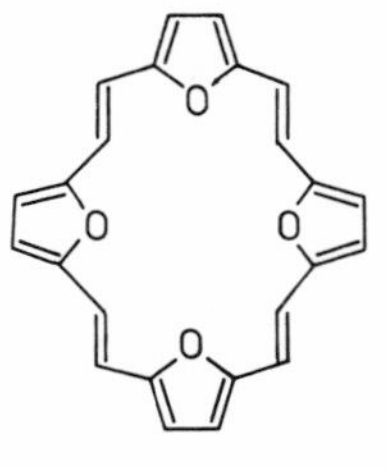

(119) 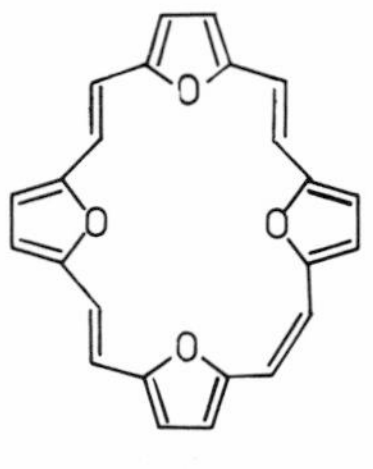 (120)

This, although in agreement with the possible [20]annulenic nature of the compound, raises the question why the material apparently does not take advantage of the possibility of at least the furan resonance. The two still larger ring systems with potential 30 and 36 π electron rings (121) and (122) both have nmr absorptions typical of furans and vinyl hydrogen atoms.

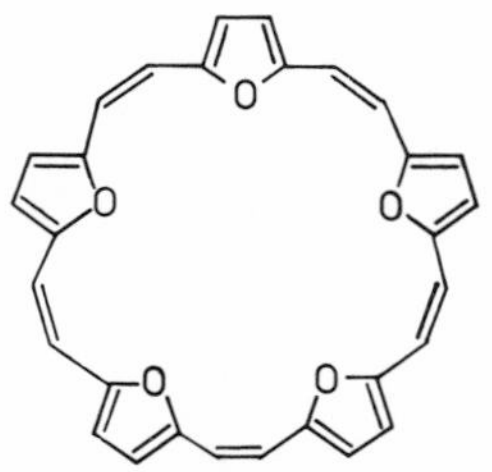

(121) 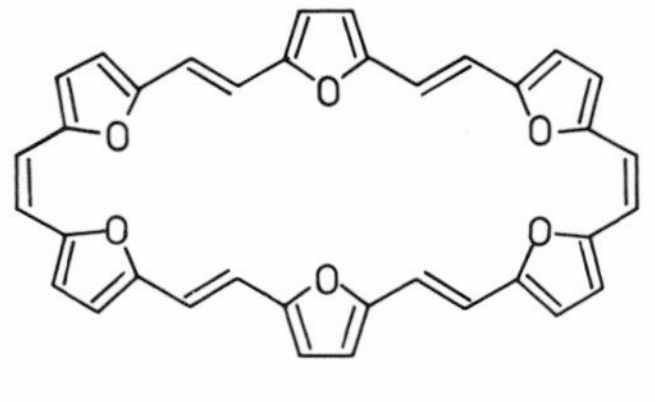 (122)

The [22]annulene (123) is clearly aromatic [100]; the four vinyl protons appear at -2.28 τ. Addition of an acid to give the monoprotonated base yields NH-peaks at 15.5 and 16.8 τ.

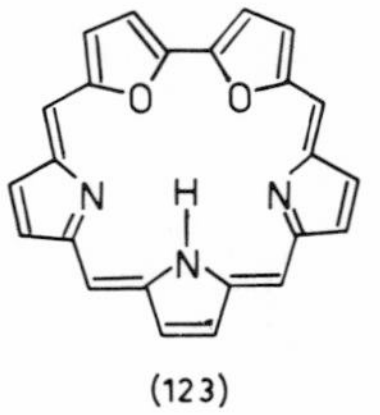

(123)

---

99. J.A. Elix, Austr. J. Chem., 22, 1951 (1969).
100. M.J. Broadhurst, R. Grigg, and A.W. Johnson, Chem. Commun., 23 (1969).

Porphin (124) - the parent compound of the porphyrins is another among the large, bridged aromatic ring systems (it is an [18]annulene); the internal protons appear at 14.4 τ and the external ones at 0.08 and -1.22 τ [101]. The stability of the phtalocyanines (125) and their ability to coordinate almost any metal (even beryllium) in a square planar configuration [102] strongly suggests that these molecules are essentially large aromatic ring systems.

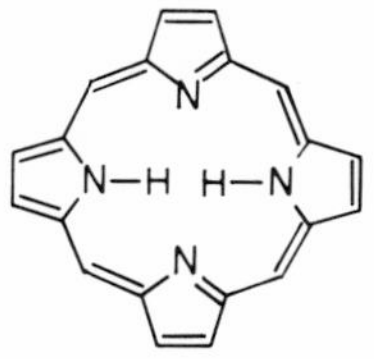

(124) 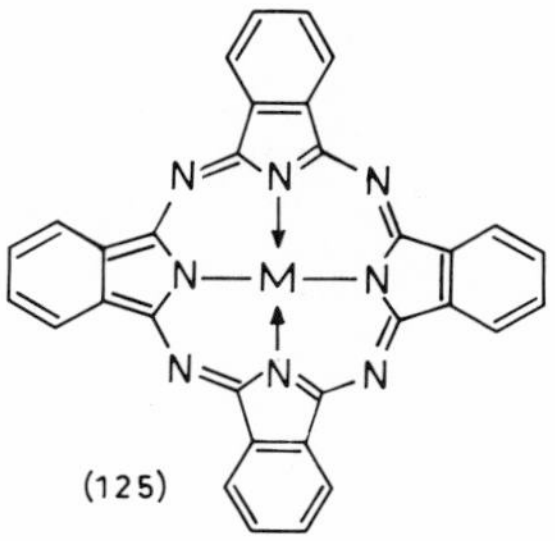 (125)

9-6. AROMATIC IONS AND DIPOLES

The present section is largely one of discussion of ionic derivatives of odd-membered olefinic rings C_nH_{n+1}. The removal of a proton or hydride ion from the methylene group leaves behind either an empty or filled p orbital that should be capable of interacting with its neighbors, and generate a completely conjugated ion that should in principle follow the Hückel development or its successors.

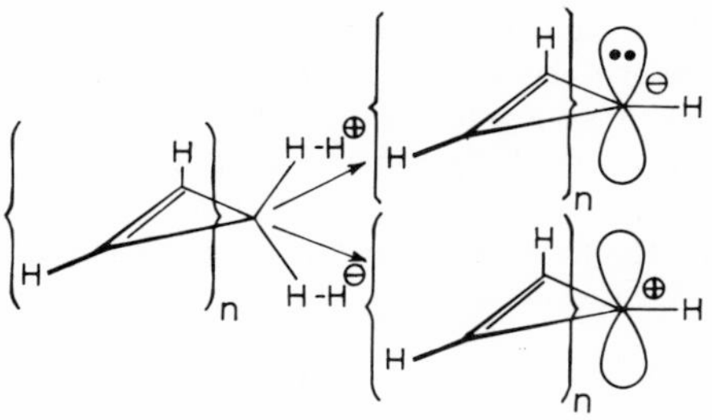

On paper the simplest one of such is the cyclopropenium cation, (126), which has 2 π electrons and hence obeys the Hückel rule for aromaticity [103]. Whether such a species might actually be observable or even isolable seemed very doubtful at

101. Ref. 1a, p. 105-106.
102. F.H. Moser and A.L. Thomas, J. Chem. Educ., 41, 245 (1964).
103. For review, see A.W. Krebs, Angew. Chem., Int. Ed. Engl., 4, 10 (1965).

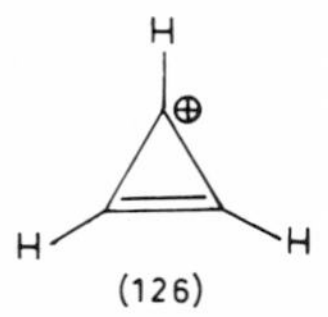

(126)

best, and it was not seriously discussed before 1957, when Breslow [104] reported the synthesis of triphenylcyclopropenium salt (127). To appreciate the impact of this paper it should be realized that at that time carbonium ions were regarded as hypothetical and unproven intermediates, and that few organic chemists seriously believed even in the existence of cyclopropenes.

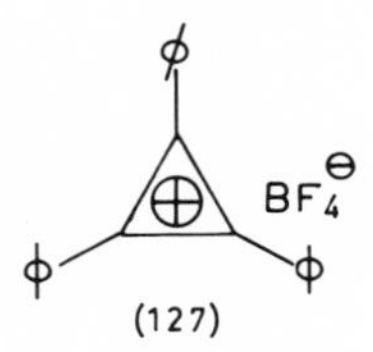

(127)

The author vividly recalls a seminar from his graduate student days at the University of Chicago in the mid nineteen fifties at which the speaker discussed an unusual cyclopropane rearrangement. During the question period that followed one of the graduate students suggested that the reaction could have occurred via a cyclopropene intermediate. That brought about an immediate and generous round of laughter; but the student - not anxious to accept sole responsibility for all this mirth - heatedly maintained that one of the professors in the Department had suggested a cyclopropene intermediate as a possibility in his research. "He's right here in this room," he added, "sitting in the front row." The several professors seated there looked at one another, with what feelings I cannot say; but none rose to acknowledge having said anything of the sort, and the group returned to its business. In fact, cyclopropene had been reported [105] well before this event, but it was evidently either little known or not believed [106]; at any rate, the episode typifies the atmosphere in which Breslow's paper was received.

Breslow showed that in alcohols, (127) was in reversible equilibrium with ethers such as (128), and that it was unbelievably stable, surviving being heated to 300° [107].

---

104. (a) R. Breslow, J. Amer. Chem. Soc., 79, 5318 (1957); (b) R. Breslow and C. Yuan, J. Amer. Chem. Soc., 80, 5991 (1958).
105. N. J. Demjanow and M. Dojarenko, Chem. Ber., 56, 2200 (1923).
106. There are of course a great many unlikely claims in the early literature written before the limitations of the structural theory were well established.
107. R. Breslow and H. Höver, J. Amer. Chem. Soc., 82, 2644 (1960).

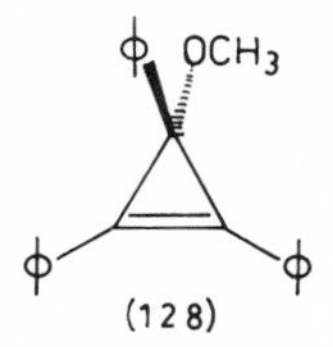

(128)

Even the carbon-azide linkage - normally a model of covalency - takes on ionic character in a compound such as (129). Thus, the nmr spectrum of this compound which shows two types of t-butyl group at low temperature, shows only a single one at or near room temperature, and the coalescence temperature is inversely related to the polarity of the solvent [108].

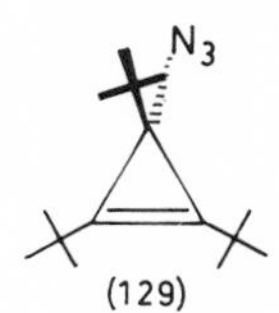

(129)

A number of dialkyl and diaryl analogs have now been reported. This permitted a chemical shift measurement of the ring bound proton [109]; for the dipropyl ion (130), this proton is shifted down by 4.15 ppm due to the combined effects of ring current and positive charge.

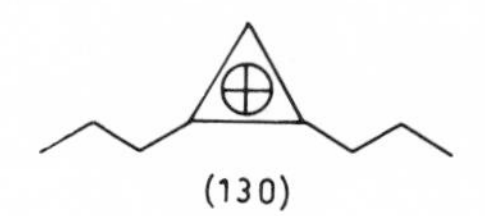
(130)

Substituents of all sorts stabilize the molecule, but eventually even the parent compound (131) was isolated as the hexachloroantimonate [109]. The nmr spectrum shows a singlet at -1.0 τ.

(131)

---

108. R. Curci, V. Lucchini, P. J. Kocienski, G. T. Evans, and J. Ciabattoni, Tetrahedron Lett., 3293 (1972).
109. (a) R. Breslow, J. T. Groves, and G. Ryan, J. Amer. Chem. Soc., 89, 5048 (1967); (b) R. Breslow and J. T. Groves, J. Amer. Chem. Soc., 92, 984 (1970).

An especially stable case is that of the tris-(dimethylamino)cyclopropenium salts; the perchlorate can be recovered unchanged from boiling water [110]. The three carbon-carbon bonds in this case have equal length at 1.36 Å as shown by X-ray crystallography [111].

Perhaps equally startling is the fact that cyclopropenones are isolable and, depending on the substituents, marvels of stability. This stability may be appreciated perhaps best by comparing these materials with cyclopropanones. The latter are intermediates in the Favorsky rearrangement of α-haloketones to acids [112]. They cannot be isolated in this reaction. However, if α,α'-dihaloketones are used, the reaction gives cyclopropenones as the final product [113].

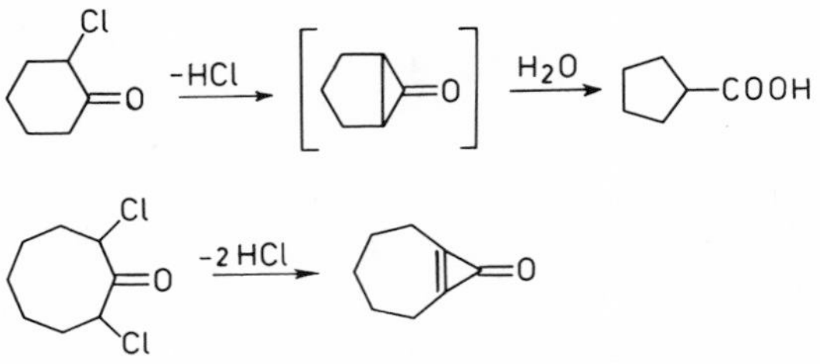

Cl
-HCl
O
O
H_2O
COOH

Cl
-2 HCl
O
O
Cl

The stability obviously derives from resonance with the aromatic dipole;

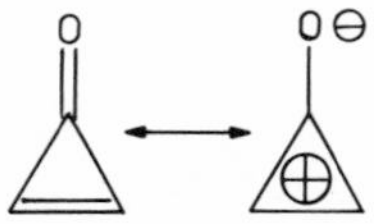

O
O⊖
⊕

thus, in diphenylcyclopropenone (132), the stretching frequency of the carbonyl group is only 1650 cm^{-1}, and the compound has a dipole moment of more than 5 D [114].

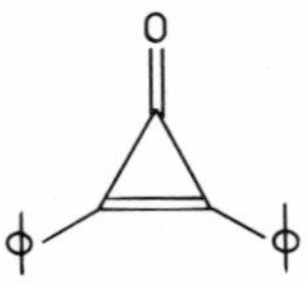

O
ϕ
ϕ

(132)

110. Z. Yoshida and Y. Tawara, J. Amer. Chem. Soc., 93, 2573 (1971).
111. A.T. Ku and M. Sundaralingam, J. Amer. Chem. Soc., 94, 1688 (1972).
112. A. Kende, Organic Reactions, 11, 261 (1960).
113. R. Breslow, J. Posner, and A. Krebs, J. Amer. Chem. Soc., 85, 234 (1963).
114. R. Breslow, T. Eicher, A. Krebs, R.A. Peterson, and J. Posner, J. Amer. Chem. Soc., 87, 1320 (1965).

A large number of cyclopropenones are now known; among them the parent compound [116], which has an nmr singlet at 1.0 τ. Its $J_{C^{13}H}$ value is 217 Hz in chloroform, 250 Hz in sulfuric acid.

There are furthermore a number of so-called triafulvenes that are of interest in this connection. They are not very stable and the parent species (133) is unknown; however, a number of substituted examples have been reported that surely owe their ability to resonance similar to that of cyclopropenone.

(133)

As examples we may mention the compounds (134) [117] and (135) [118];

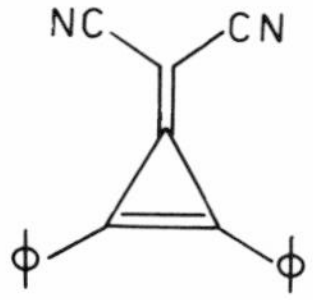

NC CN

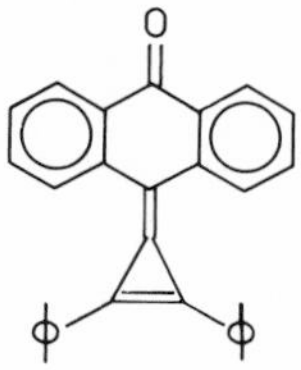

O

ϕ ϕ ϕ ϕ

(134) (135)

the dipole moment of (134) is several D larger than that of (136), and that of (135) was given [118] as 9.4 D. Still further examples are mentioned later in this section.

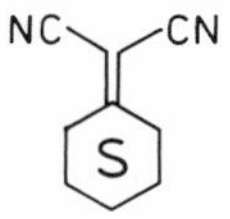

NC CN

S

(136)

---

115. R. Breslow, L.J. Altman, A. Krebs, E. Mohacsi, I. Murata, R.A. Peterson, and J. Posner, J. Amer. Chem. Soc., 87, 1326 (1965).
116. (a) R. Breslow and G. Ryan, J. Amer. Chem. Soc., 89, 3073 (1967); (b) R. Breslow and M. Oda, J. Amer. Chem. Soc., 94, 4787 (1972); (c) R. Breslow, M. Oda, and J. Pecoraro, Tetrahedron Lett., 4415 (1972); (d) M. Oda, R. Breslow, and J. Pecoraro, Tetrahedron Lett., 4419 (1972).
117. (a) E.D. Bergmann and I. Agranat, J. Amer. Chem. Soc., 86, 3587 (1964); (b) A.S. Kende and P.T. Izzo, J. Amer. Chem. Soc., 86, 3587 (1964).
118. B. Föhlisch and P. Bürgle, Tetrahedron Lett., 2661 (1965).

After all this success it was only natural that a search was made for another unlikely 2 π electron system, namely that of the cyclobutenium dication (137) (see also p. 807).

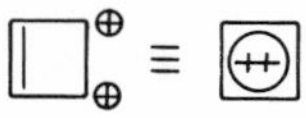

(137)

In quinones such as (138) a resonance contribution from a zwitterionic dication is possible [119], but it is obviously difficult to judge how important it is. This purple compound is very stable with a melting point of over 200°; the nmr spectrum shows that it is a symmetrical molecule, and the only ir absorption attributable to carbonyl stretching was a broad bond between 1590 and 1650 cm^{-1}.

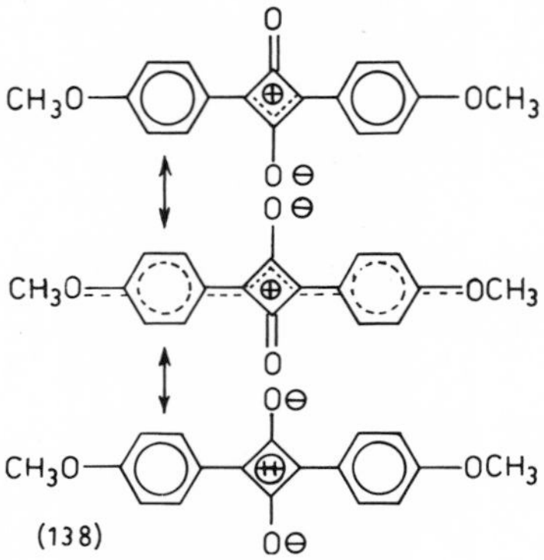

CH_3O OCH_3

(138)

1,2,3,4-Tetramethyl-3,4-dichlorocyclobutene (139) reacts with silver hexafluoroantimonate in liquid sulfur dioxide to give only one mole of silver chloride and the remaining solution showed three methyl absorptions in the nmr in a 2:1:1 ratio at 7.0, 7.5 and 7.6 τ, respectively; in this case the cation (140) was evidently generated [120].

---

119. (a) H.E. Sprenger and W. Ziegenbein, Angew. Chem., Int. Ed. Engl., 6, 553 (1967); (b) D.G. Farnum, B. Webster, and A.D. Wolf, Tetrahedron Lett., 5003 (1968).
120. T.J. Katz, J.R. Hall, and W.C. Neikam, J. Amer. Chem. Soc., 84, 3199 (1962).

(139) (140)

Under similar conditions the tetraphenyl analog (141) gave a solid salt at first believed to be the dicationic salt (142), since a single type of phenyl was observed in the nmr spectrum [121]; however, X-ray diffraction [122] showed it to be the isomeric salt (143), which in sulfur dioxide solution evidently equilibrates very rapidly.

(141) (142) (143)

Later it was learned that with a little more patience, the tetramethyl dication might have been found; Olah observed that the nmr spectrum of Katz's solution upon standing for several minutes changed to a much simpler spectrum showing only a single peak at 6.32 τ [123]. Similarly the tetraphenyl dication could be observed if a more acidic solvent such as fluorosulfonic acid was used [124].

The stability of the cyclopropenium cation contrasts very sharply with that of the anion. 1,2,3-Triphenylcyclopropene (144) does not undergo exchange with ammonia-d_3 [125].

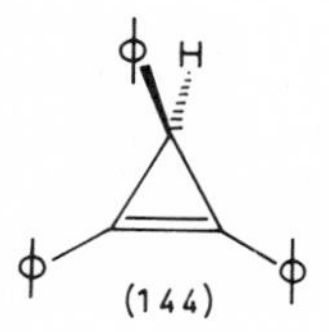

(144)

---

121. H.H. Freedman and A.M. Frantz, J. Amer. Chem. Soc., 84, 4165 (1962).
122. (a) R.F. Bryan, J. Amer. Chem. Soc., 86, 733 (1964); (b) H.H. Freedman and A.E. Young, J. Amer. Chem. Soc., 86, 734 (1964).
123. G.A. Olah, J.M. Bollinger, and A.M. White, J. Amer. Chem. Soc., 91, 3667 (1969).
124. G.A. Olah and G.D. Mateescu, J. Amer. Chem. Soc., 92, 1430 (1970).
125. R. Breslow and P. Dowd, J. Amer. Chem. Soc., 85, 2729 (1963).

By an indirect method based on the polarographic reduction of the cation to the anion, Breslow [126] was able to calculate that the pKa of (144) is 51 - a value substantially higher than even that for methane, and undoubtedly due to the fact that now four π electrons are present in the system. Tripropylcyclopropene has a pKa estimated at 65 [127].

The nature of the π electron system of cyclopropenium anion is nevertheless known now. If one electron withdrawing group is present such as a cyano group (145), base promoted deuterium exchange does occur, though some 10,000 times more slowly than in (146) [128]. When a resolved (p-substituted) analog of (145) is treated with base, the exchange and racemization rate constants can be measured independently. Their ratio is found to be 4; the same ratio for resolved (147) is 80.

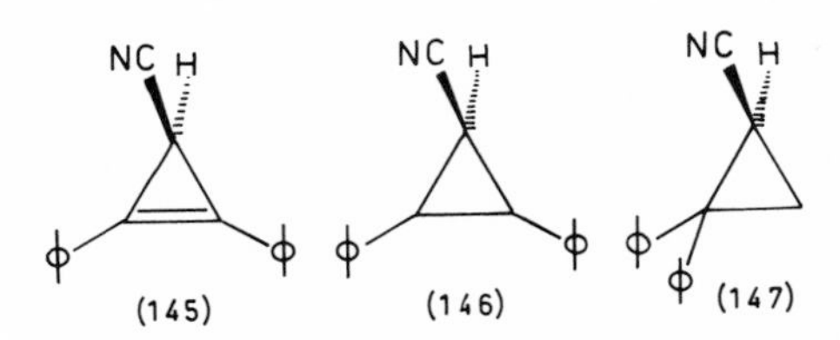

Thus, while the anion (148) forms much more slowly than (149), it becomes planar much more easily - which hints at an antiaromatic species [129].

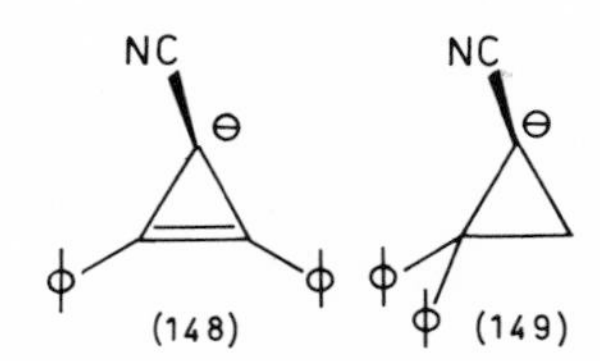

Cyclopropene itself (150) is about as acidic as acetylene, and its anion can readily be studied in liquid ammonia. The nmr spectrum shows a doublet at 10.45 τ and a triplet at 3.27 τ; thus, the anion is (151) rather than (152). The addition of an excess of ethyl iodide then produces the derivatives (153) [130].

---

126. R. Breslow and K. Balasubramanian, J. Amer. Chem. Soc., 91, 5182 (1969).
127. R. Breslow, Pure Appl. Chem., 28, 111 (1971).
128. R. Breslow, J. Brown, and J.J. Gajewski, J. Amer. Chem. Soc., 89, 4383 (1967).
129. R. Breslow and M. Douek, J. Amer. Chem. Soc., 90, 2698 (1968).
130. A.J. Schipperijn, Recl. Trav. Chim. Pays-Bas, 90, 1110 (1971).

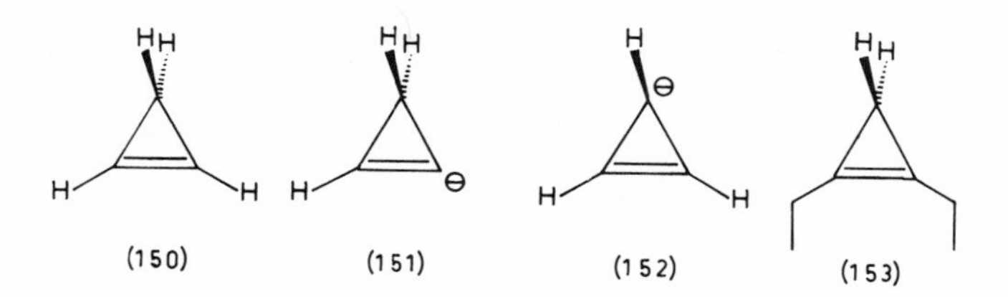

(150) (151) (152) (153)

A striking contrast exists between cyclopropenone and cyclopentadienone (154). All attempts to obtain (154) have only resulted in the Diels-Alder dimer (155), and the spectrum can be determined only by the matrix isolation technique [131].

(154) (155)

Some substituted cyclopentadienones such as (156) can be isolated, however, but even that molecule slowly dimerizes at room temperature [132].

(156)

Perhaps the best studies concerning the 4 π-electron system to date have been those with the cyclopentadienylium cation. When boron trifluoride is passed into a solution of pentaphenylcyclopentadienol (157) at -60°, a blue solution results which with hydride sources gives the cyclopentadiene (158) and with methyl alcohol the ether (159). The nmr spectrum shows no trace of phenyl absorption; however, an esr signal could be found unambiguously attributable to a triplet state, showing the ion (160) has a diradical state which is either the ground state or only very slightly higher in energy than the singlet state [133].

131. O. L. Chapman and C. L. McIntosh, Chem. Commun., 770 (1971).
132. (a) E. W. Garbisch and R. F. Sprecher, J. Amer. Chem. Soc., 88, 3433, 3434 (1966); (b) Chem. Eng. News, 44, 40, April 11, 1966.
133. (a) R. Breslow, H. W. Chang, and W. A. Yager, J. Amer. Chem. Soc., 85, 2033 (1963); (b) H. Volz, Tetrahedron Lett., 1899 (1964); (c) R. Breslow, H. W. Chang, R. Hill, and E. Wasserman, J. Amer. Chem. Soc., 89, 1112 (1967).

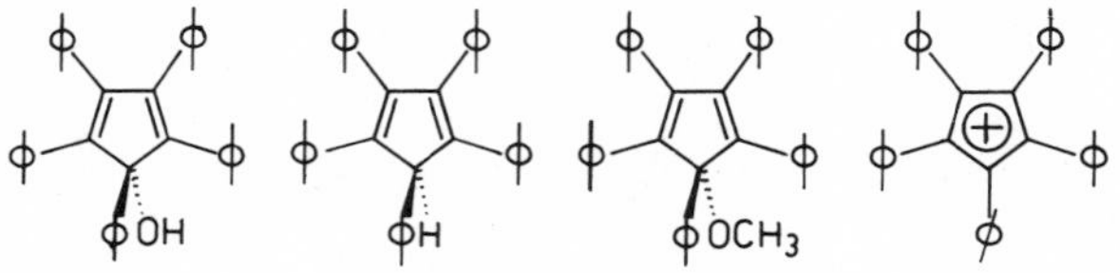

(157) (158) (159) (160)

In the triplet, the second and third π levels are clearly degenerate, and Hund's rules are obeyed. Similar work has been done with the pentachloropentadienylium cation (161) and with the parent cyclopentadienyl cation itself [134].

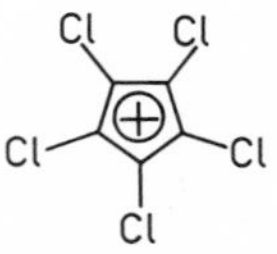

(161)

The anti-aromaticity of the parent ion is suggested by the great difficulty of its formation. 5-Iodocyclopentadiene (162) solvolyzes at least 10^5 times more slowly than iodocyclopentane (163), quite unlike what would be expected on the basis of allylic resonance (see Chapter 8) [135].

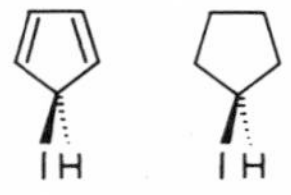

(162) (163)

In one case isolation of a salt has succeeded: Under rather special conditions, the 9-*t*-butylfluorenylium salt (164) can be obtained in pure form, but owing to its instability to Wagner-Meerwein rearrangements (Chapter 20) its physical properties are still unknown [136].

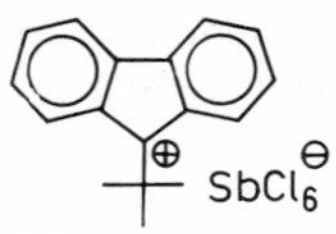

(164)

---

134. (a) R. Breslow, R. Hill, and E. Wasserman, J. Amer. Chem. Soc., 86, 5349 (1964); (b) M. Saunders, R. Berger, A. Jaffe, J.M. McBride, J. O'Neill, R. Breslow, J.M. Hoffman, C. Perchonock, E. Wasserman, R.S. Hutton, and V.J. Kuck, J. Amer. Chem. Soc., 95, 3017 (1973).
135. R. Breslow and J.M. Hoffman, J. Amer. Chem. Soc., 94, 2110 (1972).
136. H. Volz, G. Zimmermann, and B. Schelberger, Tetrahedron Lett., 2429 (1970).

In analogy to the 2- and 4 π-electron systems, benzene is not the only 6 π-electron ring. The so-called active hydrogen atom of cyclopentadiene is obviously a direct consequence of the aromaticity of the anion. The acidity can be enhanced by groups which are able to delocalize the unshared pair further. Thus, pentacyanocyclopentadiene (165) has a pKa estimated to be less than -11, and hence it is an acid much stronger even than perchloric acid [137].

CN CN
NC CN
NCH

(165)

As in the case of the cyclopropenium salts, the cyclopentadiene anions can be obtained in a state bound to the counterion. An interesting example is diazocyclopentadiene (166), an isolable and thermally quite stable compound that should be contrasted with diazoalkanes to appreciate the effect of the aromatic ring [138].

(166)

Other examples are (167) [139], (168) [140], (169) [141], (170) [142], (171) [143], and (172) [144].

(167) (168) (169) (170) (171) (172)

137. O.W. Webster, J. Amer. Chem. Soc., 88, 3046 (1966).
138. W. von E. Doering and C.H. DePuy, J. Amer. Chem. Soc., 75, 5955 (1953).
139. I.B.M. Band, D. Lloyd, M.I.C. Singer, and F.I. Wasson, Chem. Commun., 544 (1966).
140. (a) F. Ramirez and S. Levy, J. Amer. Chem. Soc., 79, 67, 6167 (1957); (b) D. Lloyd, M.I.C. Singer, M. Regitz, and A. Liedhegener, Chem. Ind. (London), 324 (1967).
141. D. Lloyd and M.I.C. Singer, Chem. Ind. (London), 118 (1967).
142. (a) D. Lloyd and M.I.C. Singer, Chem. Ind. (London), 510 (1967); (b) B.H. Freeman and D. Lloyd, J. Chem. Soc. (C), 3164 (1971).
143. D. Lloyd and M.I.C. Singer, Chem. Ind. (London), 787 (1967).
144. D. Lloyd and M.I.C. Singer, Chem. Commun., 1042 (1967).

Fulvene (173) is not very stable [145], but its life time and dipole moment can be enhanced by cyclopropyl groups (174) [146] (which are known to stabilize carbonium ions; see p. 751) and amino groups (175) [147]. These compounds have dipole moments of 1.1, 1.7 and 5.4 D, respectively.

NMe_2 NMe_2

(173) (174) (175)

An interesting approach is the linking of potentially negatively and positively charged aromatic systems [148]; the best known example at this time is probably the molecule triapentafulvene or calicene, (176) (calix means cup).

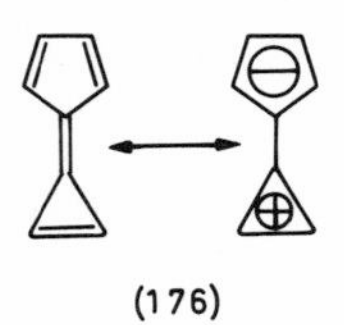

(176)

A number of substituted calicenes are now known [149]; a degree of aromaticity is indicated by the fact that some undergo electrophilic substitution reactions such as bromination and nitration [150], some have high dipole moments [151] (that of hexaphenylcalicene is 6.3 D, a record for a hydrocarbon [152]), and some have uv

---

145. (a) D. Meuche, M. Neuenschwander, H. Schaltegger, and H.U. Schlunegger, Helv. Chim. Acta, 47, 1211 (1964); (b) K. Hafner, K.H. Hafner, C. König, M. Krender, G. Ploss, G. Schulz, E. Sturm, and K.H. Vöpel, Angew. Chem., Int. Ed. Engl., 2, 123 (1963).
146. R.C. Kerber and H.G. Linde, J. Org. Chem., 31, 4321 (1966).
147. A.P. Downing, W.D. Ollis, and I.O. Sutherland, J. Chem. Soc. (B), 111 (1969).
148. For review, see H. Prinzbach, Pure Appl. Chem., 28, 281 (1971).
149. (a) Chem. Eng. News, 43, 42, April 19, 1965; (b) H. Prinzbach, D. Seip, and U. Fischer, Angew. Chem., Int. Ed. Engl., 4, 242 (1965); (c) W.M. Jones and R.S. Pyron, J. Amer. Chem. Soc., 87, 1608 (1965); (d) A.S. Kende and P.T. Izzo, J. Amer. Chem. Soc., 87, 1609 (1965); (e) A.S. Kende, P.T. Izzo, and P.T. MacGregor, J. Amer. Chem. Soc., 88, 3359 (1966).
150. A.S. Kende, P.T. Izzo, and W. Fulmor, Tetrahedron Lett., 3697 (1966).
151. E.D. Bergmann and I. Agranat, Tetrahedron, 22, 1275 (1966).
152. E.D. Bergmann and I. Agranat, Chem. Commun., 512 (1965).

spectra strongly dependent on the solvent as might be expected for a dipolar species [153]. The large contribution from the dipolar structure may be gauged from the observation that the two α-methylene triplets in (177) coalesce near room temperature [150].

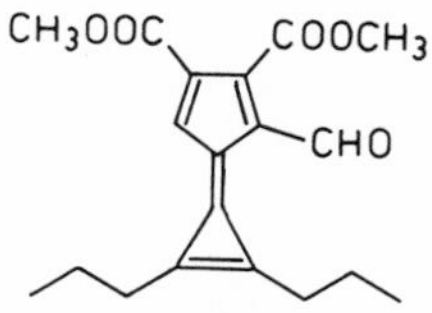

CH_3OOC $COOCH_3$ CHO

(177)

It does not exclusively determine the properties of the molecule, however; the X-ray diffraction pattern of (178) clearly shows that the double bonds are 1.35 Å in length, and the single ones are 1.41 - 1.44 Å long [154].

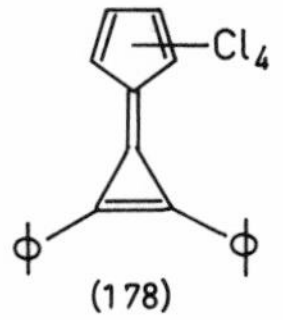

Cl_4 ϕ ϕ

(178)

Finally, it may be noted that very few symmetrical fulvalenes are known; (179) is one of them [155].

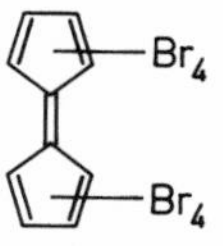

Br_4 Br_4

(179)

Tropylium salts such as the bromide (180) occupy an important position in the historical development of simple aromatic species.

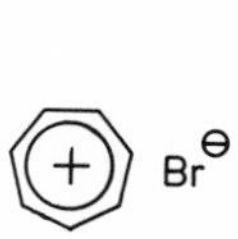

$Br^{\ominus}$

(180)

---

153. A.S. Kende and P.T. Izzo, J. Amer. Chem. Soc., 87, 4162 (1965).
154. O. Kennard, K.A. Kerr, D.G. Watson, and J.K. Fawcett, Proc. Roy. Soc. (A), 316, 551 (1970).
155. P.T. Kwitowski and R. West, J. Amer. Chem. Soc., 88, 4541 (1966).

The compound was first obtained in 1891, but it was found to be deliquescent in air and could not be adequately analyzed [156]; its first opportunity for recognition therefore was lost, and it was not until 1954 that its aromaticity was realized [157]. The ease of its formation for example by oxidation of tropilidene (181) [158], its water solubility and simple ir spectrum all attest to its stability and symmetrical shape [159].

(181)

The tropylium ion is in many ways similar to the cyclopropenium ion. Thus, in analogy to cyclopropenone, tropone (182) [160] is a polar ketone (the hydrogen bonded tropolone (183) [161] is also a well-known species); 3-azidotropone (184) was recently shown by means of X-ray diffraction to have alternately single and double bonds but a planar ring [162].

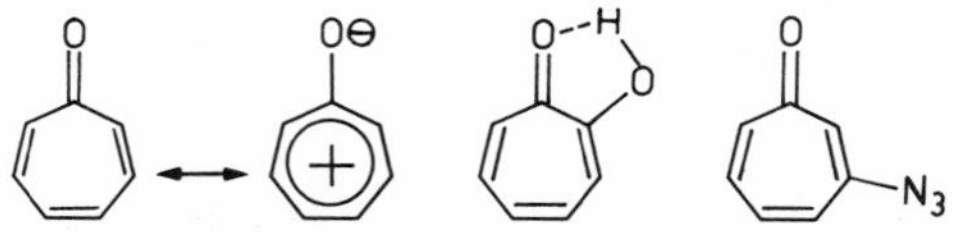

(182) (183) (184)

---

156. G. Merling, Chem. Ber., 24, 3108 (1891).
157. W. von E. Doering and L.H. Knox, J. Amer. Chem. Soc., 76, 3203 (1954).
158. (a) D.H. Geske, J. Amer. Chem. Soc., 81, 4145 (1959); (b) K. Conrow, J. Amer. Chem. Soc., 81, 5461 (1959); (c) A.P. ter Borg, R. van Helden, and A.F. Bickel, Recl. Trav. Chim. Pays-Bas, 81, 164 (1962).
159. For review, see W. von E. Doering and H. Kranch, Angew. Chem., 68, 661 (1956).
160. (a) H.J. Dauben and H.J. Ringold, J. Amer. Chem. Soc., 73, 876 (1951); (b) W. von E. Doering and F.L. Detert, J. Amer. Chem. Soc., 73, 876 (1951).
161. (a) W. von E. Doering and L.H. Knox, J. Amer. Chem. Soc., 72, 2305 (1950); (b) R.D. Haworth and J.D. Hobson, Chem. Ind. (London), 441 (1950); (c) J.W. Cook, A.R. Gibb, R.A. Raphael, and A.R. Somerville, Chem. Ind. (London), 427 (1950); (d) T. Nozoe, S. Seto, Y. Kitahara, M. Kunari, and Y. Nakayama, Proc. Japan Acad., 26, 38 (1950); cf. Chem. Abstr., 45, 7098e (1951).
162. D.W.J. Cruickshank, G. Filippini, and O.S. Mills, Chem. Commun., 101 (1972).

Immonium salts such as (185) have a very low barrier to rotation of the C-N bond (15 kcal/mole) [163].

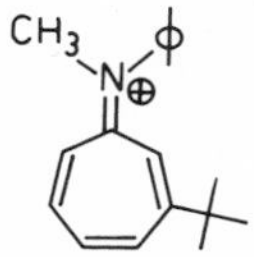

(185)

Finally, heptafulvene (186) [164] and heptafulvalene (187) are known though unstable compounds;

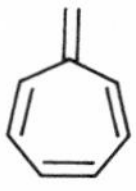

(186)

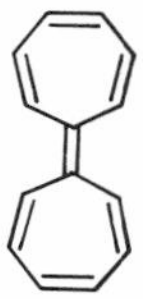

(187)

"sesquifulvalene" (188) [165] is apparently stabilized by the possibility of zwitterionic resonance structures: It has nmr absorptions at about 1.55 τ, and melts at 210°.

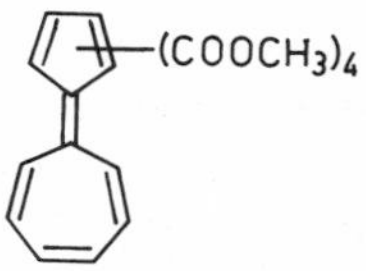

$-(COOCH_3)_4$

(188)

The parent compound is also known but it is not very stable, polymerizing even at 30° [166].

163. A. Krebs, Tetrahedron Lett., 1901 (1971).
164. W. von E. Doering and D.W. Wiley, Tetrahedron, 11, 183 (1960).
165. G. Seitz, Angew. Chem., Int. Ed. Engl., 5, 82 (1967).
166. M. Neuenschwander and W.K. Schenk, Chimia, 26, 194 (1972).

It is not clear whether the three- or seven-membered ring is better able to accommodate the positive charge. Of the two ketones (189) and (190), the former has a much higher dipole moment (9.4 vs. 3.8 D) [167]; but in (191), where the two rings compete directly for the charge, 5:3 signals are observed in the nmr centered at 1.9 and 2.3 τ, respectively [168]. X-ray diffraction of the perchlorate salt (192) shows that the two exo-bonds are roughly equally long [169].

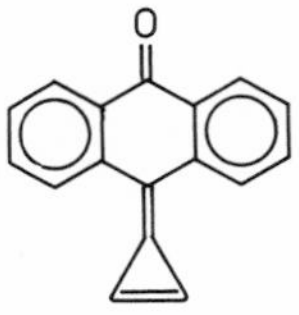

(189) 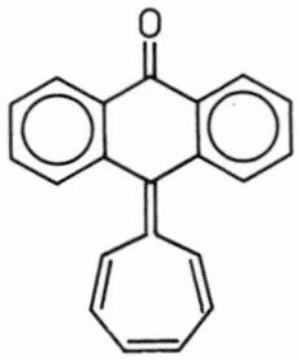 (190) 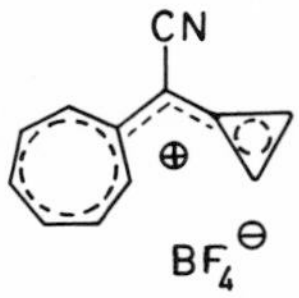 (191) 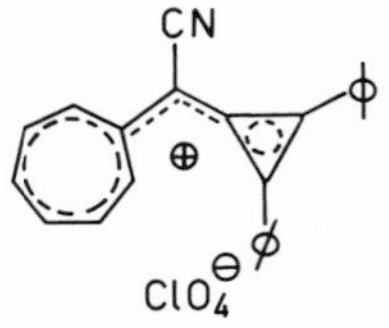 (192)

The two rings in (193) collaborate to give a dipolar species; $\nu_{C=O}$ is only 1600 cm^{-1} in this case [170].

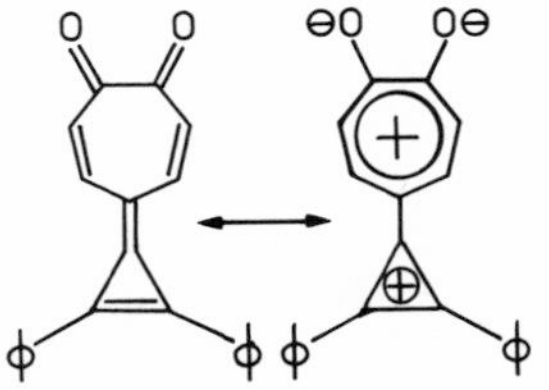

(193)

It seems to be as difficult to put six π electrons into the cyclobutadiene ring as it is for two. Thus, if cyclobutene is treated with butyllithium and subsequently quenched with D_2O, no deuterium incorporation in the recovered cyclobutene is encountered [171]. As with the dication, patience would seem to be a virtue in these experiments: If such solutions are quenched after some time has elapsed, some of the dideuterated compound (194) is indeed present [172].

---

167. (a) B. Föhlisch and D. Krockenberger, Chem. Ber., 101, 3990 (1968); (b) B. Föhlisch, P. Bürgle, and D. Krockenberger, Chem. Ber., 101, 4004 (1972).
168. M. Oda, K. Tamate, and Y. Kitahara, Chem. Commun., 347 (1971).
169. C. Kabuto, M. Oda, and Y. Kitahara, Tetrahedron Lett., 4851 (1972).
170. K. Takashashi and K. Takase, Tetrahedron Lett., 2227 (1972).
171. W. Adam, Tetrahedron Lett., 1387 (1963).
172. J.S. McKennis, L. Brener, J.R. Schweiger, and R. Pettit, Chem. Commun., 365 (1972).

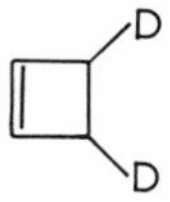

(194)

The enhanced stability of these dianions may perhaps also be gauged by the fact that the biphenylene radical anion has a considerable tendency to disproportionate [173]. It may be mentioned in this connection that the cyclopropenolone (195) is strongly acidic with a pKa of 1, and that this acidity is sometimes attributed to the presence of a six π-electron system (compare tropolone) [174].

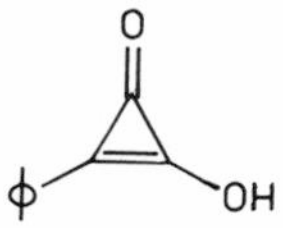

(195)

Several studies have been reported concerning charged or dipolar eight π-electron systems. When heptaphenyltropilidene is treated with potassium metal, the anion (196) is indeed formed; it is a singlet, as was concluded from magnetic studies [175].

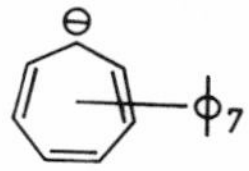

(196)

It is clear from exchange studies that the tropenide anion itself can be generated by strong bases, and stable solutions can be obtained by reaction of tropyl methyl ether with sodium. These solutions are diamagnetic [176]. The eight π-electron cyclononatetraenyl cation (197) can also be generated, by allowing the chloride

---

173. (a) N. L. Bauld and D. Banks, J. Amer. Chem. Soc., 87, 128 (1965); however, see also (b) R. Waack, M. A. Doran, and P. West, J. Amer. Chem. Soc., 87, 5508 (1965).
174. (a) D. G. Farnum and P. E. Thurston, J. Amer. Chem. Soc., 86, 4206 (1964); (b) D. G. Farnum, J. Chickos, and P. E. Thurston, J. Amer. Chem. Soc., 88, 3075 (1966).
175. R. Breslow and H. W. Chang, J. Amer. Chem. Soc., 87, 2200 (1965) and 84, 1484 (1962).
176. (a) H. J. Dauben and M. R. Rifi, J. Amer. Chem. Soc., 85, 3041 (1963); (b) W. von E. Doering and P. P. Gaspar, J. Amer. Chem. Soc., 85, 3043 (1963).

(198) to ionize in liquid sulfur dioxide. If the 1-deuteriochloride is used the products that eventually form have the deuterium completely scrambled, so that the completely symmetrical ion is probably present; however, its magnetic properties are not known [177].

(197) (198)

Finally, the annulenone (199) has been prepared; the two ring protons signal at 3.37 τ, and hence no paramagnetic ring current is indicated [178].

(199)

Many successes have been scored in attempts to incorporate 10 π-electrons in various rings which is remarkable in view of the fact that the non-bridged [10] annulenes themselves are purely olefinic in nature. One of the more spectacular early results was that cyclooctatetraene readily reacts with potassium metal in tetrahydrofuran to give a solution of the dianion (200) [179].

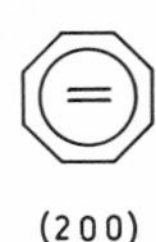

(200)

The planar aromatic nature of the anion is clear from the nmr spectrum: A single peak is observable at a chemical shift only barely above that of cyclooctatetraene itself; the ring current evidently offsets the shielding effect normally expected from two extra electrons. Equally interesting is the fact that the radical mono-anion (201) is not stable to disproportionation [180].

---

177. A.G. Anastassiou and E. Yakali, Chem. Commun., 92 (1972).
178. M. Rabinovitz, F.D. Bergmann, and A. Gazit, Tetrahedron Lett., 2671 (1971).
179. T.J. Katz, J. Amer. Chem. Soc., 82, 3784 (1960).
180. T.J. Katz, J. Amer. Chem. Soc., 82, 3785 (1960).

(201)

It may be however, that these conclusions are dependent on the conditions; it has been reported that the electrochemical reduction of cycloctatetraene does not go beyond the radical anion stage [181].

Cyclononatetraenes are weakly acidic and will form the corresponding aromatic anions; both the all *cis*- and *cis*, *cis*, *cis*, *trans*-anions (202) and (203) are known. The nmr of the former consists of a singlet at 3.3 τ [182]; the latter has six protons at 3-3.6 τ, two at 2.7 τ and one at 13.5 τ [183].

(202) (203)

The dibenzo derivative (204) is especially interesting in that both the nonaromatic and aromatic anions (205) and (206) are known; with (204), butyllithium gives (205), which then at room temperature slowly isomerizes to the planar (206).

(204) (205) (206)

In (205) and (206), H_a, H_b and H_c appear at 5.67, 4.49 and 7.04, and at 3.42, 2.65-3.1 and 3.65 τ, respectively; thus the effect of the ring current on the chemical shifts is clearly demonstrated in this case [184]. Some nonafulvenes [185)

181. D.R. Thielen and L.B. Anderson, J. Amer. Chem. Soc., 94, 2521 (1972).
182. E.A. LaLancette and R.E. Benson, J. Amer. Chem. Soc., 87, 1941 (1965).
183. G. Boche, D. Martens, and W. Danzer, Angew. Chem., Int. Ed. Engl., 8, 984 (1969).
184. P.J. Garratt and K.A. Knapp, Chem. Commun., 1215 (1970).
185. (a) K. Hafner and H. Tappe, Angew. Chem., Int. Ed. Engl., 8, 593 (1969); (b) M. Rabinovitz and A. Gazit, Tetrahedron Lett., 721 (1972).

such as (207), the diazocyclononatetraene (208) [186] and the phosphorane (209) [187] are known, and exhibit properties similar to the six π-electron analogs; the nona-heptafulvalene (210) has not much dipolar character, however [188].

(207) (208) (209) (210)

The bridged anion (211) has been reported [189], as have the cation (212) [190] and annulenones such as (213) [191] and (214) [192]; all have the properties expected on the basis of simple extensions of the six π-electron analogs and of 1,6-methano [10]annulene.

(211) (212) (213) (214)

A bridged pentaundecafulvalene (215) ("Fidecene", for violin) has also been synthesized [193]; it is thermally very stable and the methylene protons are found at 6.4 and 9.8 τ in the nmr, so that a strong ring current is indicated.

---

186. D. Lloyd and N.W. Preston, Chem. Ind. (London), 1039 (1966).
187. M. Rabinovitz and A. Gazit, Tetrahedron Lett., 3361 (1972).
188. M. Rabinovitz and A. Gazit, Tetrahedron Lett., 3523 (1972).
189. (a) W. Grimme, M. Kaufhold, U. Dettmeier, and E. Vogel, Angew. Chem., Int. Ed. Engl., 5, 604 (1966); (b) P. Radlick and W. Rosen, J. Amer. Chem. Soc., 88, 3461 (1966).
190. W. Grimme, H. Hoffmann, and E. Vogel, Angew. Chem., Int. Ed. Engl., 4, 354 (1965).
191. (a) W. Grimme, J. Reisdorff, W. Jünemann, and E. Vogel, J. Amer. Chem. Soc., 92, 6335 (1970); (b) H. Ogawa, H. Kato, and M. Yoshida, Tetrahedron Lett., 1793 (1971).
192. J. Reisdorff and E. Vogel, Angew. Chem., Int. Ed. Engl., 11, 218 (1972).
193. (a) H. Prinzbach and L. Knothe, Angew. Chem., Int. Ed. Engl., 6, 632 (1967); (b) H. Prinzbach and L. Knothe, Angew. Chem., Int. Ed. Engl., 7, 729 (1968); (c) H. Prinzbach, L. Knothe, and A. Dieffenbacher, Tetrahedron Lett., 2093 (1969).

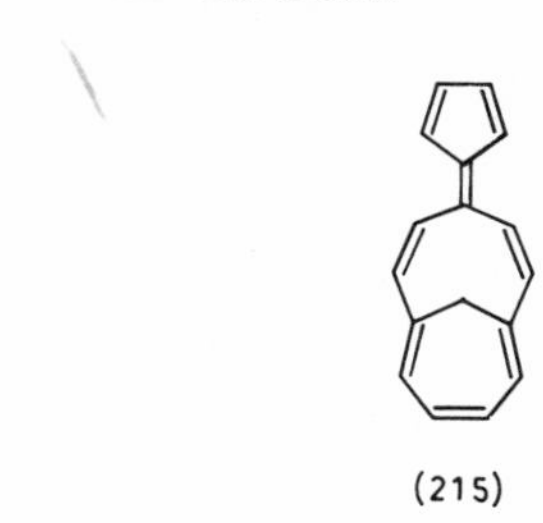

(215)

Aromatic ions and dipoles promise to play an important role in the chemistry of the larger annulenes. The nmr signals of the α-methylene protons of the ketone (216) and the fulvalene (217) are at somewhat higher field than in open chain precursors [194], which suggests that they may be anti-aromatic; an X-ray diffraction of the former study has revealed only bond alternation and no delocalization [195]. (218) is likewise anti-aromatic: The internal protons appear at -0.3 τ in the nmr [196].

(216) (217) (218)

The bridged anhydride (219) is aromatic in nature, but the polyenones (220) and (221) become so only upon protonation in strong acid with the outside protons generally signalling at -0.5 to 1.0 τ and the inside proton at 14 τ [197].

(219) (220) (221)

---

194. G.M. Pilling and F. Sondheimer, J. Amer. Chem. Soc., 90, 5610 (1968) and 93, 1977 (1971).
195. V.F. Duckworth, P.B. Hitchcock, and R. Mason, Chem. Commun., 963 (1971).
196. (a) G.W. Brown and F. Sondheimer, J. Amer. Chem. Soc., 91, 760 (1969); for further examples, see (b) P.D. Howes, E. LeGoff, and F. Sondheimer, Tetrahedron Lett., 3691, 3695 (1972).
197. (a) H. Ogawa, N. Shimojo, and M. Yoshida, Tetrahedron Lett., 2013 (1971); (b) H. Ogawa, M. Yoshida, and H. Saikachi, Tetrahedron Lett., 153 (1972); see also (c) T.M. Cresp and M.V. Sargent, Chem. Commun., 807 (1972).

In these large rings the presence of two extra electrons frequently produces enormous shifts in the nmr spectra. Thus, when the annulenone (222) is converted into the anion (223), the inner protons move from -0.5 to about 18.5 τ [198].

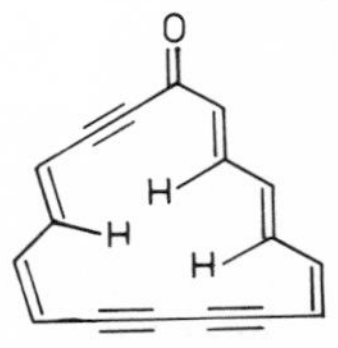

(222) 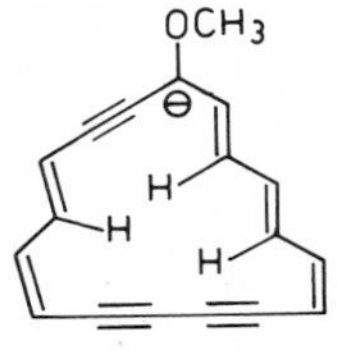 (223)

Conversely, when 15,16-trans-dimethyldihydropyrene is reduced with two electrons to (224), the methyl groups move from +14.3 to -11.0 τ, a downfield shift of 25 ppm in spite of the two extra electrons; the outer protons move up from 1.5 to 13.6 τ [199].

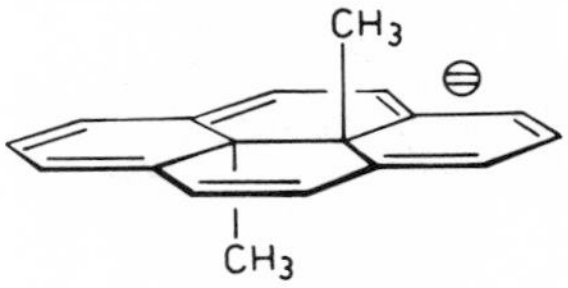

(224)

This result makes clear the importance of planarity to achieve anti-aromaticity; in the much more flexible dianion (225), the bridge protons are shifted down by only three ppm, and the peripheral protons up by two [199a].

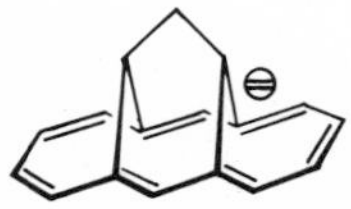

(225)

---

198. J. Griffiths and F. Sondheimer, J. Amer. Chem. Soc., 91, 7518 (1969).
199. R.H. Mitchell, C.E. Klopfenstein, and V. Boekelheide, J. Amer. Chem. Soc., 91, 4931 (1969).
199a. F. Gerson, K. Müllen, and E. Vogel, Angew. Chem., Int. Ed. Engl., 10, 920 (1971).

The extra electrons also increase the stability in many cases. Thus, in the case of [16]annulene, which has four inner protons at -0.6 τ and which readily isomerizes at room temperature to a tricyclic hydrocarbon, the dianion (226) has its inner protons absorbing at +18.2 τ, and the solution can be heated to 100° for hours without decomposition [200].

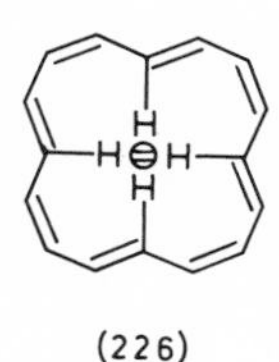

(226)

Similar observations have been described for the [12]annulene dianions [201]. It may in fact be that aromaticity can be extended beyond 26 π electrons by means of this effect. Already one aromatic dianion of such dimensions is known; when the octadehydro[24]annulene (227) is treated with potassium to give (228), the nmr does not noticeably shift upfield as it should when a simple unsaturated compound gains electrons. Surely this must be due to a diamagnetic ring current [202].

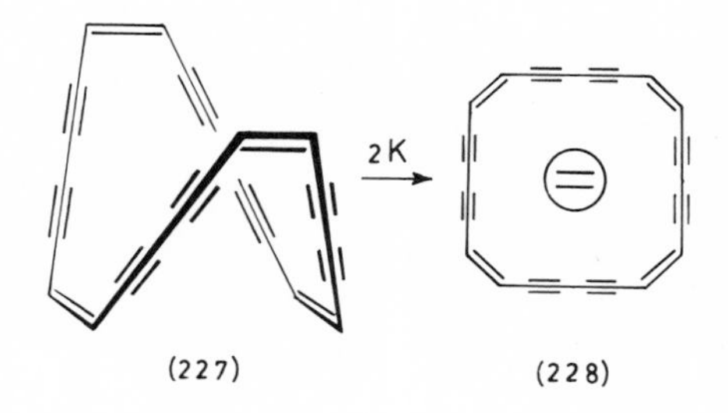

(227) (228)

We close this section with a brief discussion of the effect of charge and of ring current on the chemical shift. Cyclopentadienide has its protons at 1.9 ppm upfield from benzene; those of tropylium ion are equally far downfield from it [203]. That then must be about the magnitude of the field shift due to one electron.

---

200. (a) J. F. M. Oth, G. Anthoine, and J. M. Gilles, Tetrahedron Lett., 6265 (1968); (b) J. F. M. Oth, H. Baumann, J. M. Gilles, and G. Schröder, J. Amer. Chem. Soc., 94, 3498 (1972).
201. (a) J. F. M. Oth and G. Schröder, J. Chem. Soc. (B), 904 (1971); (b) P. J. Garratt, N. E. Rowland, and F. Sondheimer, Tetrahedron, 27, 3157 (1971).
202. R. M. McQuilkin, P. J. Garratt, and F. Sondheimer, J. Amer. Chem. Soc., 92, 6682 (1970).
203. G. Fraenkel, R. E. Carter, A. Mclachlan, and J. H. Richards, J. Amer. Chem. Soc., 82, 5846 (1960).

That a ring current is really present in cyclopentadienide has been demonstrated by means of the anion of [9]-1,3-cyclopentadienophane (229) [204]: One of the nine methylene groups then appears at 9.45 τ, unlike the eight others which show up at 8.5 - 9.0 τ. This must be the central methylene group which presumably lies above the aromatic ring.

(229)

Upfield shifts upon electron addition have also been noted for the cmr chemical shifts; a regular increase from -15 ppm (down from carbon disulfide) to +110 ppm has been noted between tetramethyl cyclobutadiene dication and cyclooctatetraene dianion [124].

9-7. THE EFFECT OF HETEROATOMS

The presence of heteroatoms leads to a host of new possibilities. In this section we shall consider one by one what these possibilities are, and what has been learned about them.

A boron atom is isoelectronic with a carbonium ion site, and hence one interesting question is whether the neutrality possible when a boron is substituted for a carbonium ion site helps or hinders the stability of the system.

Thus, boririne (230) might well be an aromatic compound; however, it is at present not known.

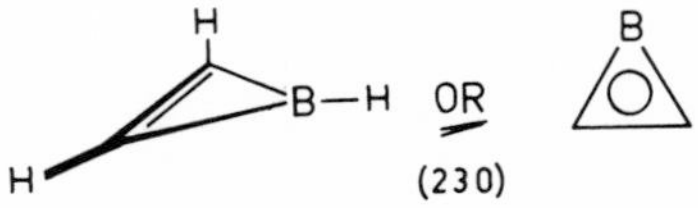

(230)

204. S. Bradamante, A. Marchesini, and G. Pagani, Tetrahedron Lett., 4621 (1971).

By the same token, boroles, if isolable, might be expected to be anti-aromatic, or ground state triplets. Pentaphenylborole (231) is known [204a] as a thermally reasonably stable green compound sensitive to oxygen.

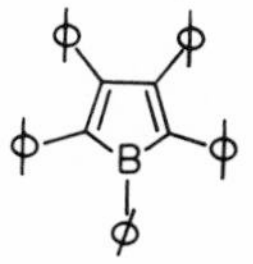

(231)

An esr signal was observed, but it may be due to the radicals formed by adventitious oxidation. Some aluminoles such as (232) are also known, but again the only pertinent information is that these compounds are labile to various reagents but thermally stable [205].

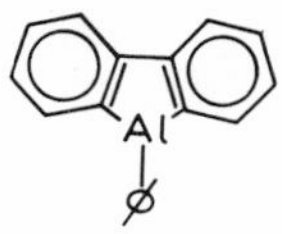

(232)

Several borepins have been synthesized, and at least some of them definitely display aromatic character. The benzoborepin (233), its complex with trimethylamine (234) and the partially saturated analog (235) yield solid information on this point [206].

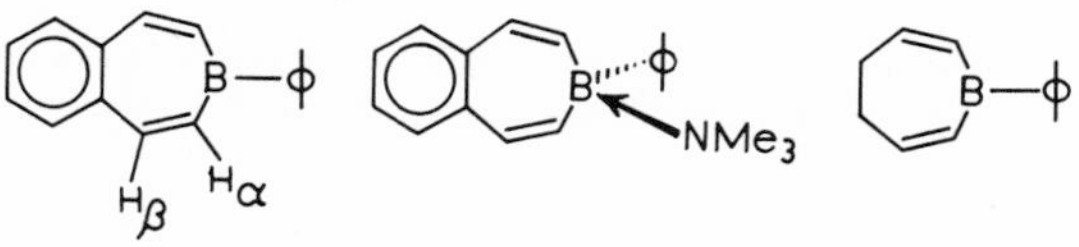

(233) (234) (235)

The α-protons in these compounds precess at 1.78, 2.82 and 2.95 τ, and the β-protons at 2.28, 3.25 and 3.64 τ, respectively; thus, the effect of interrupting conjugation by either saturation or complex formation is clearly revealed.

204a. J. J. Eisch, N. K. Hota, and S. Kozima, J. Amer. Chem. Soc., 91, 4575 (1969).

205. J. J. Eisch and W. C. Kaska, J. Amer. Chem. Soc., 88, 2976 (1966).

206. A. J. Leusink, W. Drenth, J. G. Noltes, and G. J. M. van der Kerk, Tetrahedron Lett., 1263 (1967).

Similarly, the benzoborapinol (236) is said to have uv and nmr spectra that strongly resemble those of the benzotropone (237) and its conjugate acid (238) [207].

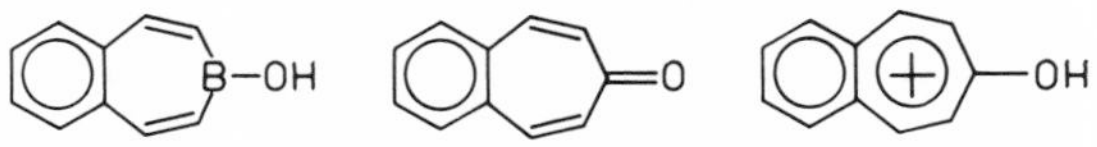

(236) (237) (238)

If a boron atom takes the place of a carbon atom in a neutral aromatic species, an anion of course results, and some such species have been reported; among them are the lithium salt of 1-phenylborabenzene (239) and that of the boraanthracene (240), as well as the zwitterionic species (241) [208].

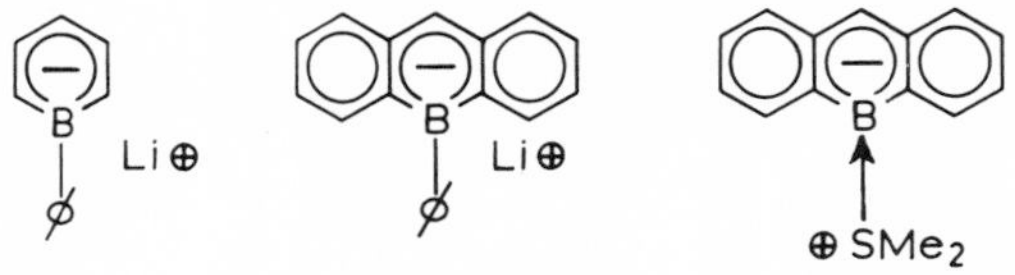

(239) (240) (241)

In all, the aromatic protons appear slightly upfield compared to the carbocyclic species; the B^{11}-nuclei precess far upfield compared to neutral boron compounds. There is no doubt that all these compounds display benzene-like aromaticity.

Nitrogen is by far the most common non-carbon atom in heterocyclic chemistry. Just as with boron, basically two kinds of incorporation are possible. In one of these the unshared pair is directed outward, and in the other it is parallel with the p orbitals in the system. To the extent that these two electrons then get delocalized, the nitrogen will formally be positively charged - unless the substituent (for example, a proton) is removed by ionization; then the second case reverts to the first. Pyridine and pyrrole are the best known examples of these types.

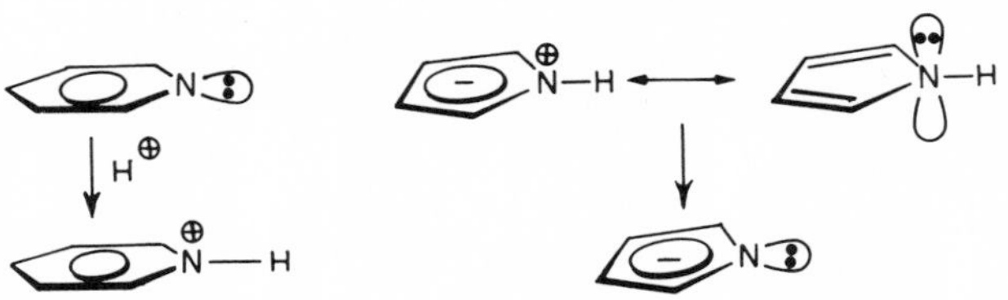

---

207. G. Axelrad and D. Halpern, Chem. Commun., 291 (1971).
208. P. Jutzi, Angew. Chem., Int. Ed. Engl., 11, 53 (1972).

It may be noted at this point that many of the pyrrole analogs so far reported have nitrogen atoms that - for reasons of synthetic convenience - bear an α-carbonyl group. In all these cases the interaction of this group with the unshared pair has turned out to be so great as to leave the basic question unanswered.

Nitrogen is easily accommodated in simple Hückel calculations, for instance by the use of $(\alpha+\beta)$ for the Coulomb integral. We should also consider what is a likely value for a ring-bound α-methyl nmr signal if the nitrogen is part of the ring. Elvidge has argued on the basis of the methyl signals from poly-unsaturated hydrocarbons that a value of 8.03 τ can be expected for non-aromatic N-CMe, and 7.82 if a diamagnetic ring current is induced [209].

Substitution of nitrogen for carbon in the cyclopropenium ion does not appear to destroy the stability of the ion; this can be concluded from the facts that the two chlorides (242) and (243) rapidly equilibrate even at -10°, and that both produce immediate precipitates with silver nitrate [210].

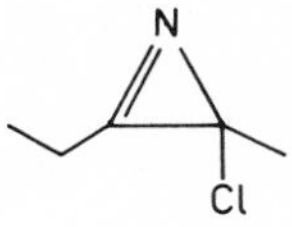

(242) 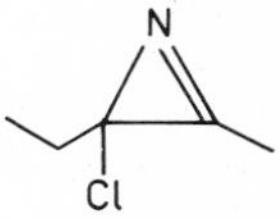 (243)

Several 8 π-electron azocines such as (244) have been synthesized by Paquette [211].

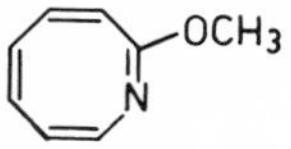

(244)

They can be electrochemically reduced to the dianions; the radical anions are in fact more easily reduced than the parent molecule, and do not reach a high concentration, which indicates special stability for the 10-electron heterocyclic dianion. The compound (245) is not stabilized by a pyrrole type of aromaticity [212]; it decomposes even at room temperature to a bisenamine, and the vinyl protons signal at 3.4 τ.

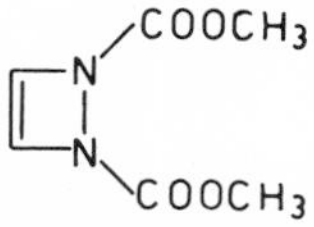

(245)

209. J.A. Elvidge, Chem. Commun., 160 (1965).
210. J. Ciabattoni and M. Cabell, J. Amer. Chem. Soc., 93, 1482 (1971).
211. L.B. Anderson, J.F. Hansen, T. Kakihana, and L.A. Paquette, J. Amer. Chem. Soc., 93, 161 (1971).
212. E.E. Nunn and R.N. Warrener, Chem. Commun., 818 (1972).

Anti-aromaticity has been sought but not encountered in several eight π-electron systems containing nitrogen: The dihydropyrazine (246) [213] and various azepines such as (247) all behave like typical enamines [214]. Possibly the only indication of destabilization in such systems has been the remarkable nitrogen atom extrusion from the anion (248) by sodium, to give 2,6-diphenylpyridine and sodium azide [215].

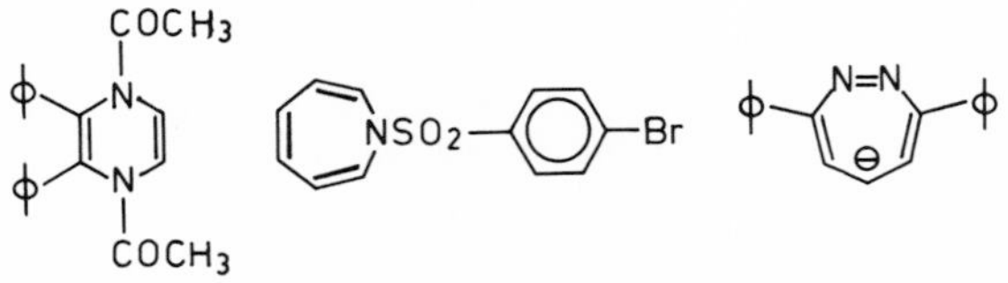

(246) (247) (248)

Aromaticity can in certain cases be achieved in the nitrogen containing anions derived from certain dihydroazocines such as (249); thus, this compound is about 100 times more acidic than the analog (250), presumably because the anion contains a ten π-electron ring [216]; on the other hand, (251) [217] and (252) [218] are non-aromatic as judged by nmr and uv spectra.

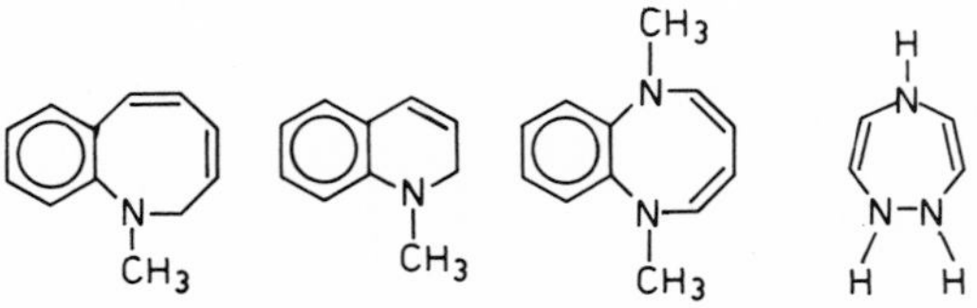

(249) (250) (251) (252)

1-Azonine (253) and its N-methyl derivative have been characterized; both are aromatic as revealed by their nmr spectra.

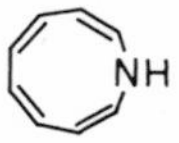

(253)

213. S.J. Chen and F.W. Fowler, J. Org. Chem., 36, 4025 (1971).
214. I.C. Paul, S.M. Johnson, L.A. Paquette, J.H. Barrett, and R.J. Haluska, J. Amer. Chem. Soc., 90, 5023 (1968).
215. R.R. Schmidt and H. Vatter, Tetrahedron Lett., 4891 (1972).
216. R.M. Coates and E.F. Johnson, J. Amer. Chem. Soc., 93, 4016 (1971).
217. H.J. Shue and F.W. Fowler, Tetrahedron Lett., 2437 (1971).
218. N.L. Allinger and G.A. Youngdale, J. Amer. Chem. Soc., 84, 1020 (1962).

The parent compound is furthermore fairly acidic (about 2 pKa units stronger than pyrrole). The salt has two protons precessing at 1.4 τ and the others at 3.4 τ in DMSO; in acetone the α-proton chemical shift depends on the counterion. The uv spectrum is similar to that of the carbocyclic anion [219].

Interestingly, if the unshared pair on the nitrogen atom is made less readily available such as is the case in the N-carbethoxy derivative (254), then aromaticity ends; the uv absorption moves to shorter wavelength, the chemical shift of the ring protons moves up by 1/2-1 τ, and the compound is much more prone to rearrangement [220].

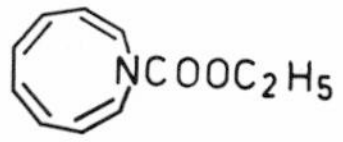

$NCOOC_2H_5$

(254)

The 1,2,3-triazonin (255) is strongly aromatic; its nmr spectrum shows the ring hydrogen atom at chemical shifts typical of pyrroles and below those of the azepins; the compound is stable thermally and to oxygen and catalytic hydrogenation, and does not complex with $(DPM)_3Eu$ (see p. 51) [221].

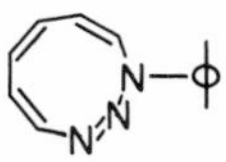

(255)

The largest analog of pyrrole yet known is the bisdehydroaza[17]annulene (256); when the nmr spectrum of this material is compared with that of the open chain analog (257), one finds that the inside protons have been shifted upfield almost 4 ppm, and the outside protons downfield by 1.

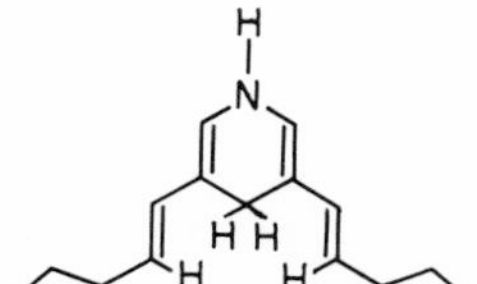

(256)

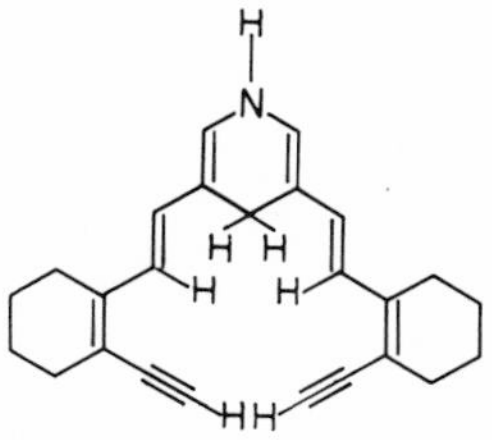

(257)

219. (a) A.G. Anastassiou and S.W. Eachus, J. Amer. Chem. Soc., 94, 2537 (1972); (b) A.G. Anastassiou and H. Yamamoto, Chem. Commun., 286 (1972); (c) R.T. Seidner and S. Masamune, Chem. Commun., 149 (1972); (d) A.G. Anastassiou, Accounts Chem. Res., 5, 281 (1972).

220. (a) A.G. Anastassiou, S.W. Eachus, R.P. Cellura, and J.H. Gebrian, Chem. Commun., 1133 (1970); (b) A.G. Anastassiou and J.H. Gebrian, J. Amer. Chem. Soc., 91, 4011 (1969) and Tetrahedron Lett., 825 (1970).

221. L.A. Paquette and R.J. Haluska, J. Amer. Chem. Soc., 94, 534 (1972).

These numbers are only half as large if the N-carbethoxy derivatives are compared [222]. Finally, the porphyrins were already mentioned.

Oxygen is a much less important atom in heterocyclic chemistry; its electronegativity is such that its unshared electrons cannot be easily induced to participate in extended π systems. The aromaticity of furan and pyrylium salts is well-known, but there are few other examples. Oxonin (258) for example is simply an olefinic ether devoid of aromatic properties [219d]; the same is true of oxa[17] annulene [223].

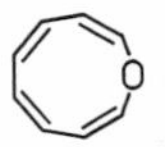

(258)

The situation might be better with anionic oxygen if it could be incorporated in a cyclic system. For a number of years it appeared as though the acetylacetonate salts (259), and in general the salts of β-dicarbonyl compounds were aromatic, 6 π-electron systems.

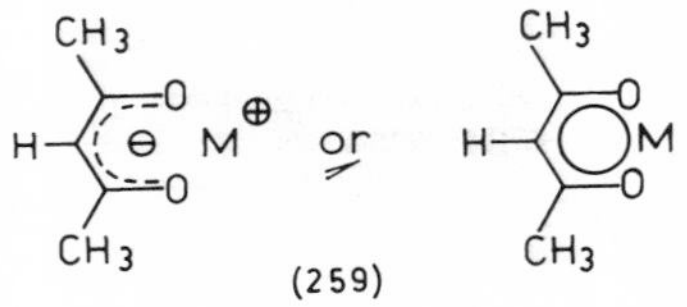

(259)

Thus, ir analysis showed these salts to be completely symmetrical, and it was found that the central carbon atom is subject to electrophilic substitution such as nitration, acetylation and so on [224].

On the other hand, the chemical shifts of the ring proton and methyl groups do not seem low enough to support a ring current [225]. A further test had the following devastating result. In 9,9-bianthryl (260) the chemical shift of the 1- and 8-protons is more than 1 ppm higher than those at positions 4 and 5 because of their locations over the ring. However, in the C-anthrylacetylacetonate salts

---

222. (a) P. J. Beeby and F. Sondheimer, J. Amer. Chem. Soc., 94, 2128 (1972); (b) G. Schröder, G. Heil, H. Röttele, and J. F. M. Oth, Angew. Chem., Int. Ed. Engl., 11, 426 (1972).
223. G. Schröder, G. Plinke, and J. F. M. Oth, Angew. Chem., Int. Ed. Engl., 11, 424 (1972).
224. J. P. Collman, R. A. Moss, S. D. Goldby, and W. S. Trahanovsky, Chem. Ind. (London), 1213 (1960).
225. R. C. Fay and N. Serpone, J. Amer. Chem. Soc., 90, 5701 (1968).

(261) this difference is much smaller (possibly now due to the negatively charged carbon moiety), but more important, it is the same in the salts as it is in the free ions and enols. Thus, there is no ring current, and no aromaticity [226].

(260) (261)

Sulfur atoms, though much less electronegative, have the larger 3p orbitals that may not interact effectively with the 2p variety of the neighboring carbon atoms. There are certain examples of such interaction, such as the stable sulfonium ylids (R-S-CH_2^+) and compounds such as carbon disulfide and thiophene; however, most extensions into rings of other sizes have not been successful. Among the substances investigated one finds the thionins (262) and (263) [227], the thiaannulenes (264)-(266) [228], and the bridged species (267) and (268) [229].

(262) (263) (264) (265)

(266) (267) (268)

In all cases, although the compounds were thermally reasonably stable, the nmr spectra were invariably found to be in the olefinic region, no differences were observed between inside and outside protons, and no shifts were found upon oxidation to the sulfoxides. There are two possible exceptions. One is the dithiocin (269) [230], which does not react with bromine and cannot be oxidized to a sulfoxide;

226. M. Kuhr and H. Musso, Angew. Chem., Int. Ed. Engl., 8, 147 (1969).
227. P.J. Garratt, A.B. Holmes, F. Sondheimer, and K.P.C. Vollhardt, J. Amer. Chem. Soc., 92, 4492 (1970).
228. (a) A.B. Holmes and F. Sondheimer, J. Amer. Chem. Soc., 92, 5284 (1970) and Chem. Commun., 1434 (1971); (b) P.J. Garratt, A.B. Holmes, F. Sondheimer, and K.P.C. Vollhardt, Chem. Commun., 947 (1971).
229. T.M. Cresp and M.V. Sargent, Chem. Commun., 1458 (1971).
230. M.O. Riley and J.D. Park, Tetrahedron Lett., 2871 (1971).

more important, in the fmr spectrum the fluorines of (269) are shifted to considerably lower field as compared to (270), and to the same region where one finds those of (271). It appears as though the fluorine atoms need the ten π-electron system to satisfy their electron demand.

(269) (270) (271)

The other possible exception is the rather complex thiepin (272). In this case when the sulfur is removed (with triphenylphosphine), the o-protons of the N-phenyl ring shift upfield by 0.5 ppm; this was interpreted as evidence for a paramagnetic ring current in the disappearing thiepin ring [231]. However, it would seem equally possible that this shift is produced by the newly forming benzene ring.

(272)

There is another interesting aspect of the participation of sulfur atoms (and others in the second row) in aromatic systems: They have relatively low-lying empty d orbitals that might be used to construct delocalized π systems [232]. The evidence for this proposition is most convincing in the case of certain cyclic sulfones.

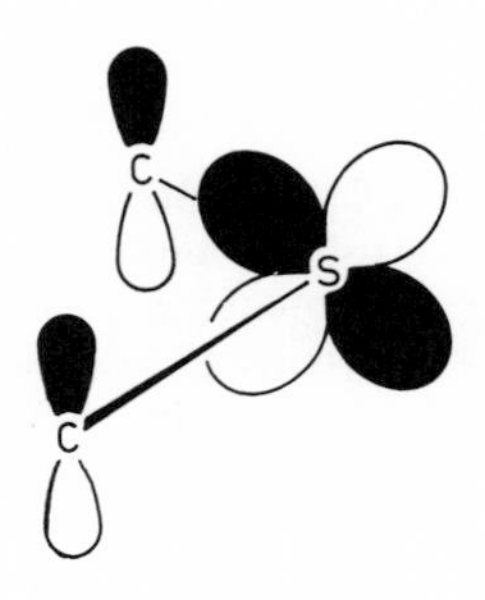

231. J.M. Hoffman and R.H. Schlessinger, J. Amer. Chem. Soc., 92, 5263 (1970).

232. (a) C.C. Price, Chem. Eng. News, 42, 58, Nov. 30, 1964; (b) M.C. Caserio, R.E. Pratt, and R.J. Holland, J. Amer. Chem. Soc., 88, 5747 (1966) and references given there; (c) D.L. Coffen, Record Chem. Progr., 30, 275 (1969).

Thus, there is reason to believe that the reason for the facile Ramberg-Bäcklund reaction of α,α-dihalosulfones is the relatively low energy content of the intermediate thiirene dioxides [233].

$(RCH_2)_2SO_2 \rightarrow\rightarrow\rightarrow$ [thiirene dioxide (R, R, SO_2) $\leftrightarrow$ ($\oplus$S$\cdots$O$\ominus$)] $\xrightarrow[-SO_2]{\Delta}$ $RC\equiv CR$

This was confirmed by the finding that these intermediates can be isolated. It was found that the unsaturated species (273) is considerably more stable than the saturated analog (274) [234].

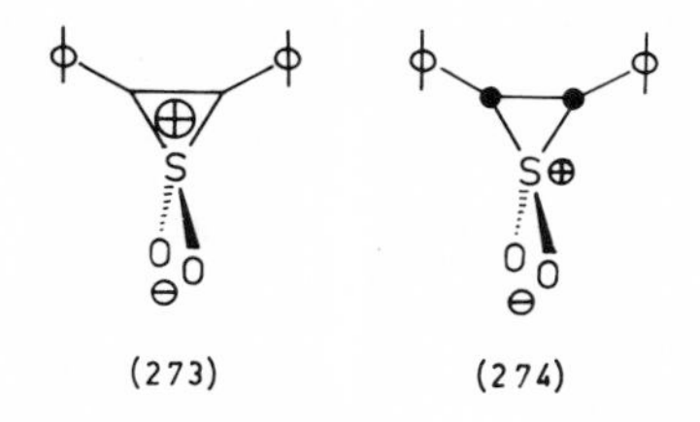

(273) (274)

Even the methyl derivative (275) has now been synthesized, and an nmr comparison with 1-methylcyclopropenone (276) is most instructive: The methyl groups appear at 7.5 and 7.6 τ, and the ring protons at 1.0 and 1.3 τ, respectively [235].

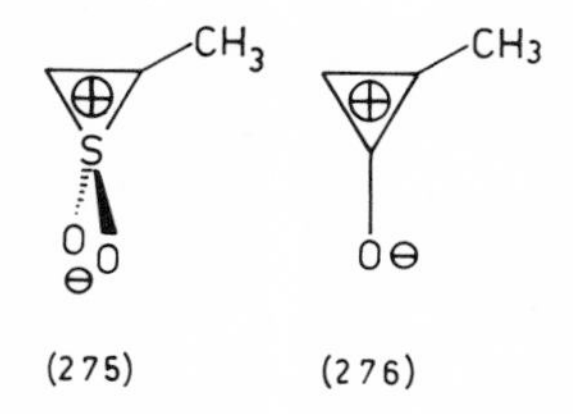

(275) (276)

Sulfoxides of this type are thermally even more stable than the sulfones. This may be due in part to the lower energy of the elimination fragment SO_2 as compared to SO; however, it has been noted that the uv spectrum of sulfoxide (277) resembles that of diphenylcyclopropenone (278) even more closely than that of the sulfone [236].

---

233. L.A. Paquette and L.S. Wittenbrook, J. Amer. Chem. Soc., 89, 4483 (1967).
234. L.A. Carpino and L.V. McAdams, J. Amer. Chem. Soc., 87, 5804 (1965).
235. L.A. Carpino and R.H. Rynbrandt, J. Amer. Chem. Soc., 88, 5682 (1966).
236. L.A. Carpino and H.W. Chen, J. Amer. Chem. Soc., 93, 785 (1971).

(277) (278)

Six π-electron systems of this sort are also known; thus, it has been found that cyclic sulfoxides such as (279) are acids 3-5 pKa units stronger than open chain analogs such as (280) [237].

(279) (280)

There are also exceptions; thus, thiepin-1,1-dioxide (281) is clearly non-planar as found by means of X-ray diffraction although some delocalization is still possible [238].

(281)

Furthermore, X-ray experiments have shown that the six π-electron six-membered ring (282) is nearly planar but has alternately single and double carbon-carbon bonds [239].

(282)

237. S. Bradamante, S. Maiorana, A. Mangia, and G. Pagani, J. Chem. Soc. (B), 74 (1971).
238. H. L. Ammon, P. H. Watts, J. M. Stewart, and W. L. Mock, J. Amer. Chem. Soc., 90, 4501 (1968).
239. W. E. Barnett, M. G. Newton, and J. A. McCormack, Chem. Commun., 264 (1972).

Finally, 1-thiabenzene (283) [240] and the oxide (284) [241] have been reported, but the information available does not yet permit an evaluation of their electronic make-up. The thiothiophthenes were mentioned earlier (p. 254); still another type of aromatic sulfur compound is described in the next section.

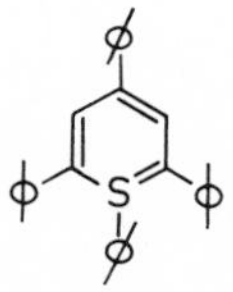

(283) 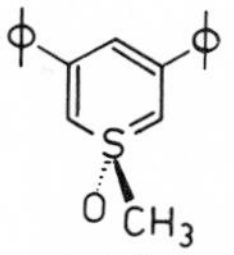 (284)

Similar developments have occurred in the chemistry of phosphorus. As noted earlier (Chapter 1), phosphorus, using its p orbitals, forms single bonds at angles of about 90°; phospholes (285) are therefore not aromatic. It was found [242] however that with compound (286) the methyl octet (due to H-splitting, P-splitting and a neighboring asymmetric center) collapses to a quartet at 42°C; $\Delta G^{\ddagger}$ for this inversion is therefore 16 kcal/mole, about 20 kcal/mole lower than is generally observed for phosphines.

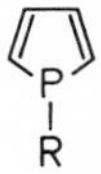

(285) 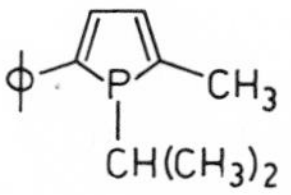 (286)

This lowering of the activation barrier must be due to aromaticity of the transition state. Phospha- and arsabenzene have been prepared and found to be somewhat labile compounds with ir, uv and nmr spectra similar to those of pyridine; thus, these molecules may be assumed to be aromatic; bismabenzene could not be isolated in pure form [243]. A third example is that of the phosphine oxide (287), in which the C-C and C-N bonds have found to be about equally long at 1.42 Å; the ring is planar, and clearly aromatic.

---

240. C.C. Price, M. Hori, T. Parasaran, and M. Polk, J. Amer. Chem. Soc., 85, 2278 (1963).
241. A.G. Hortmann, J. Amer. Chem. Soc., 87, 4972 (1965).
242. W. Egan, R. Tang, G. Zon, and K. Mislow, J. Amer. Chem. Soc., 92, 1442 (1970).
243. (a) A.J. Ashe, J. Amer. Chem. Soc., 93, 3293 (1971); (b) A.J. Ashe and M.D. Gordon, J. Amer. Chem. Soc., 94, 7596 (1972).

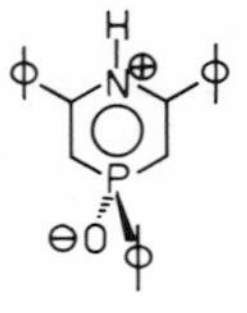

(287)

In this case the conjugation depends on an empty 3d orbital of phosphorus [244]. The same is probably true of a large number of remarkably stable phosphonitriles, compounds of the general type $(NPX_2)_n$ [245]. These compounds, exemplified by (288)-(290) have equally long P-N or S-N bonds. It should be added that the Hückel rule may not apply in these instances, and in fact, planarity may not be vital.

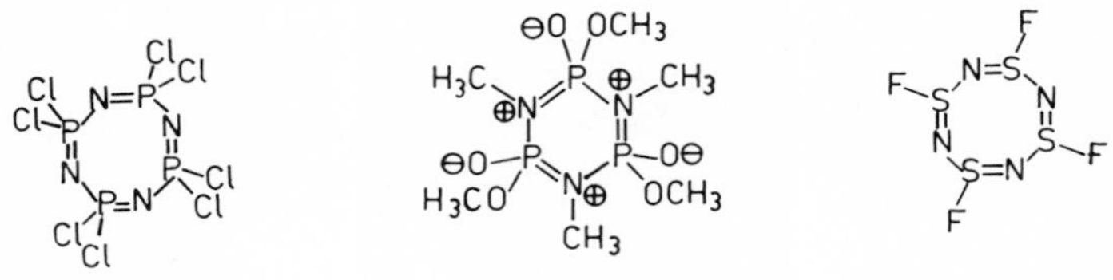

(288) (289) (290)

The reluctance of elements of the second row or below to form double bonds with atoms of the first row is especially great for silicon and there are no instances of stable compounds having such bonds (see however p. 859). Since furthermore the d orbitals are relatively higher in energy (because of the lesser nuclear charge), the prospects for incorporating silicon in aromatic systems are not good.

Early claims of the acidity of silole (291) [246] could not be substantiated, and were later withdrawn [247].

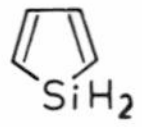

(291)

---

244. J.C. Williams, J.A. Kuczkowski, N.A. Portnoy, K.S. Yong, J.D. Wander, and A.M. Aguiar, Tetrahedron Lett., 4749 (1971).
245. (a) K. Harada, S.W. Fox, and A. Vegotsky, Chem. Eng. News, 37, 57, April 20, 1959; (b) G.B. Ansell and G.J. Bullen, Chem. Commun., 493 (1965); (c) H.R. Allcock, Chem. Eng. News, 46, 68, April 22, 1968, and Chem. Rev., 72, 315 (1972).
246. R.A. Benkeser, R.F. Grossman, and G.M. Stanton, J. Amer. Chem. Soc., 83, 3716, 5029 (1961) and 84, 4723, 4727 (1962).
247. R.A. Benkeser and G.M. Stanton, J. Amer. Chem. Soc., 85, 834 (1963).

Some siloles such as (292) were later characterized, but although they react with strong bases such as phenyllithium, the reaction is irreversible, and protonation does not regenerate them [248].

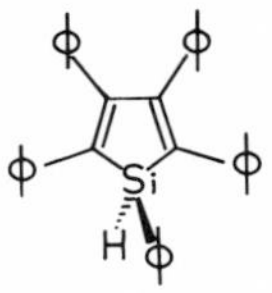

(292)

The silole (293) can be converted into a dianion via exposure to alkali metals; whether aromaticity characterizes this species is not yet clear, but the radical anion which precedes it in the reaction does not disproportionate [249].

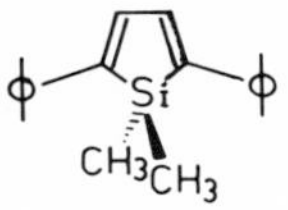

(293)

Oddly enough, the pentaphenyl germole (294) has been found to be at least 10^6 times more acidic than thiphenylgermane (295); this is similar to the difference in pKa (about 10 units) between the pentaphenylcyclopentadiene (296) and triphenylmethane (297) [250].

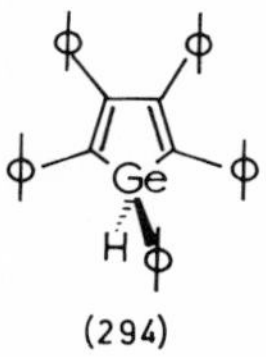

$GeH\phi_3$ $CH\phi_3$

(294) (295) (296) (297)

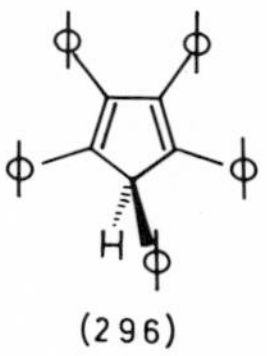

The silepin (298) and the stannepin (299) have ir spectra clearly showing the presence of simple double bonds similar to those of simple vinylsilanes, and the nmr spectra of neither compound give any evidence of a ring current [251].

---

248. M.D. Curtis, J. Amer. Chem. Soc., 89, 4241 (1967).
249. E.G. Janzen, J.B. Pickett, and W.H. Atwell, J. Organometal. Chem., 10, P 6 (1967).
250. M.D. Curtis, J. Amer. Chem. Soc., 91, 6011 (1969).
251. (a) L. Birkofer and H. Haddad, Chem. Ber., 102, 432 (1969); (b) A.J. Leusink, J.G. Noltes, H.A. Budding, and G.J.M. van der Kerk, Recl. Trav. Chim. Pays-Bas, 83, 1036 (1964).

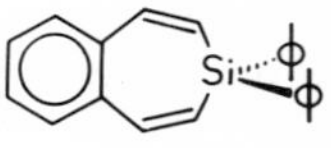

(298) 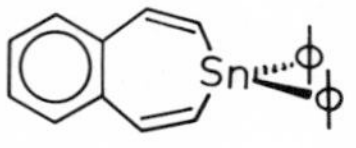 (299)

Considerable excitement followed the announcement that silirenes and germirenes such as (300) and (301) had been synthesized and found to be incredibly stable compounds [252]; alas, it turned out that these materials were in fact dimers of these structures and should be described as (302) and (303), respectively [253].

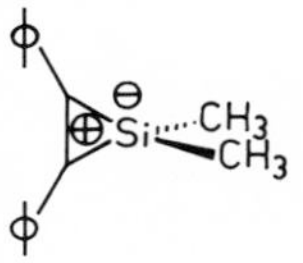

(300) 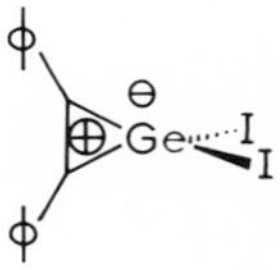 (301) 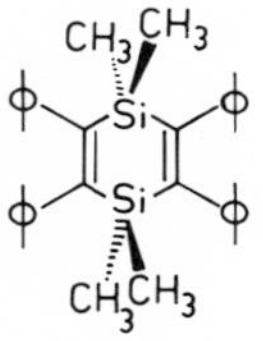 (302) 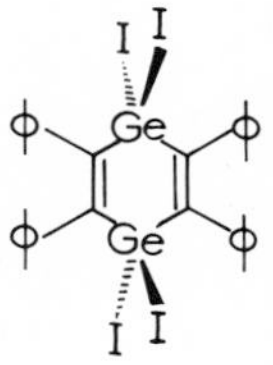 (303)

9-8. POLYCYCLIC AROMATIC SPECIES

There is a staggering amount of literature on this general topic. We shall not attempt a general review here, but certain aspects of this area are both interesting and timely, and a brief discussion of them follows.

One of the first questions that might be raised is whether the simple Hückel rule is applicable. HMO calculations for various polynuclear systems do not show that the $(4n + 2)$ π-electron rule remains valid, and although it does seem to be obeyed in many cases, to use it indiscriminately in polynuclear systems is to venture onto thin ice. Much depends actually on the length of each common edge. If it is short, the resonance integral will be large and a significant effect on the overall energy may be expected; if it is long and essentially single, one may essentially regard the molecule as a bridged monocyclic molecule. In polynuclear hydrocarbons especially it is tempting to regard the internal bridges as irrelevant except in that they replace hydrogen atoms that would probably be subject to severe

252. M.E. Volpin, Y.D. Koreshkov, V.G. Dulova, and D.N. Kursanov, Tetrahedron, 18, 107 (1962).
253. (a) R. West and R.E. Bailey, J. Amer. Chem. Soc., 85, 2871 (1963); (b) F. Johnson, R.S. Gohlke, and W.A. Nasutavicus, J. Organometal. Chem., 3, 233 (1965); (c) L.V. Vilkov, L.N. Gorokhov, V.S. Mastryukov, and A.D. Rusin, Russ. J. Phys. Chem. (English Transl.), 38, 1451 (1964).

non-bonded interactions. The general idea that polycyclic conjugated systems may be regarded simply on the basis of their peripheries with disregard to internal cross linkages is due to Platt [254].

Starting with naphthalene, one extension leads us to consider heptalene (304), octalene (305), etc. When the rings are even, the fused edge can be written as a double bond; if they are odd, it can not, unless zwitterionic structures are considered.

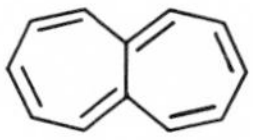

(304) 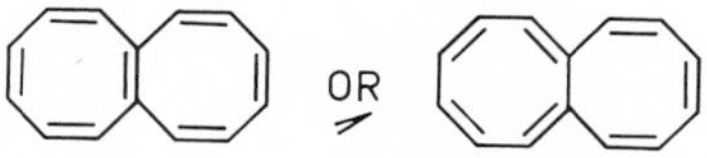 (305) OR

In the even types furthermore, the total number of π-electrons is (4n + 2), and in the odd ones it is 4n. One of the interesting questions that arises here is whether (305) is one of a class of compounds in which the magic number of (4n + 2) electrons is achieved by the fusion of two rings, each having 4n electrons. In any event, the results so far known have it that none of these compounds other than naphthalene have any significant delocalization. Heptalene [255] is a rather unstable compound with complex nmr signals centered at 4.2 and 5.0 τ ; the carbon-carbon bonds are alternately long and short [256]. The benzooctalene (306) has been found [257] to have terminal cyclooctatetraene proton signals in the same position as cyclooctatetraene itself, to have a uv spectrum similar to benzocyclooctatetraene, and to possess no noteworthy stability; hence it is a simple benzo-annelated polyene.

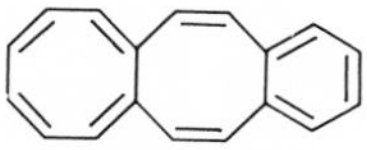

(306)

Several substituted pentalenes are known, such as hexaphenylpentalene, (307) [258], 1,3-bis(dimethylamino)pentalene (308) [259] and the derivative (309) [260].

---

254. J.R. Platt, J. Chem. Phys., 22, 1448 (1954).
255. H.J. Dauben and D.J. Bertelli, J. Amer. Chem. Soc., 83, 4659 (1961).
256. R. Zahradnik, Angew. Chem., Int. Ed. Engl., 4, 1039 (1965).
257. R. Breslow, W. Horspool, H. Sugiyama, and W. Vitale, J. Amer. Chem. Soc., 88, 3677 (1966).
258. E. LeGoff, J. Amer. Chem. Soc., 84, 3975 (1962).
259. K. Hafner, K.F. Bangert, and V. Orfanos, Angew. Chem., Int. Ed. Engl., 6, 451 (1967).
260. K. Hartke and R. Matusch, Angew. Chem., Int. Ed. Engl., 11, 50 (1972).

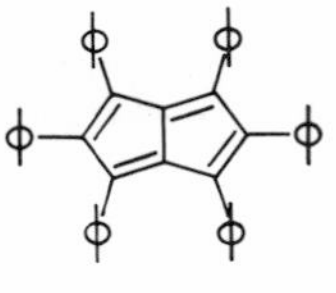

(307) 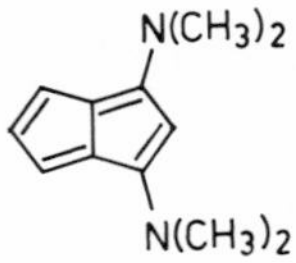 (308) 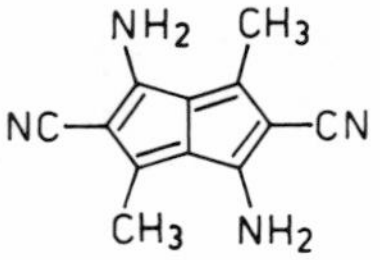 (309)

The stability greatly increases in that order, no doubt due to resonance.

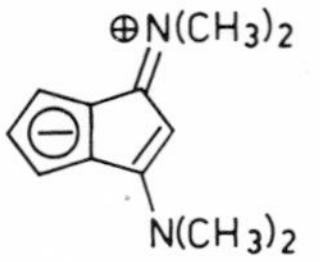

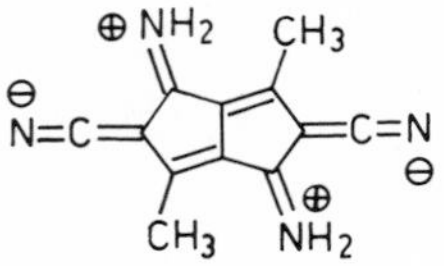

1-Methylpentalene (310) has been deposited on a cold finger after the flash thermolysis (and retro Diels-Alder reaction) of (311), but this compound starts dimerizing even at -140° [261], and pentalene itself has as yet not been observed at all.

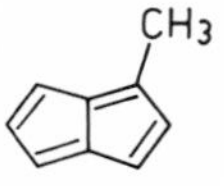

(310) 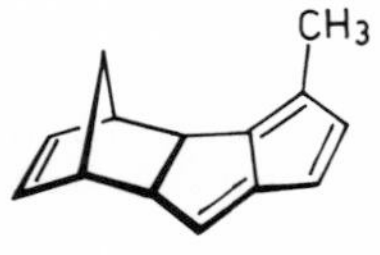 (311)

In view of these discouraging results, it seems surprising that the dianion (312) of pentalene is well known, and easily obtained (by treatment of dihydropentalene with butyllithium) [262].

(312)

The salt can in fact be crystallized from THF; in this solvent, the nmr signals occur at 4.27 (triplet) and at 5.02 (doublet) τ. This result shows that at least in the case of pentalene the ten π-electron system is greatly preferred over one with

---

261. R. Bloch, R.A. Marty, and P. de Mayo, J. Amer. Chem. Soc., 93, 3071 (1971).
262. (a) T.J. Katz and M. Rosenberger, J. Amer. Chem. Soc., 84, 865 (1962); (b) T.J. Katz, M. Rosenberger, and R.K. O'Hara, J. Amer. Chem. Soc., 86, 249 (1964).

eight, and it has sparked a great deal of further fruitful work. Thus, methyl derivatives of the 2-oxa-, aza- and thiapentalenyl anions (313) [263], (314) [264] and (315) [265] are known and reasonably stable ten π-electron systems, with nmr spectra testifying to the aromatic nature of their stability.

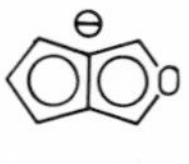

(313) 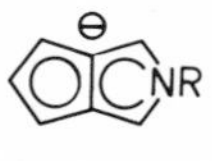 (314) 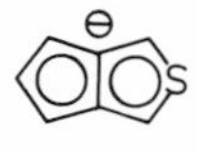 (315)

But even more promising are the neutral diaza- and tetraazapentalenes. 3a, 6a-Diazapentalene (or pyrazolo[1, 2-a]pyrazole) (316) and many of its derivatives have been reported in recent years [266].

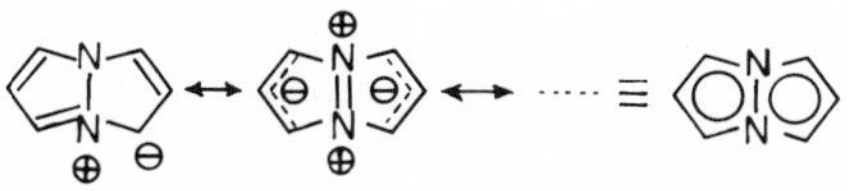

(316)

The simple resonance structures all have negative charge at the carbon atoms, and the present system is therefore subject to oxidation; electron withdrawing substituents stabilize it. The parent compound has a simple nmr spectrum: A doublet at 3.0 τ and a triplet at 3.4 τ. Its easy synthesis, neutrality and stability foreshadow an important role for this aromatic molecule.

The substitution of some of the CH groups by nitrogen atoms stabilizes the molecule further. The dibenzotetraazapentalenes (317) and (318) both are stable

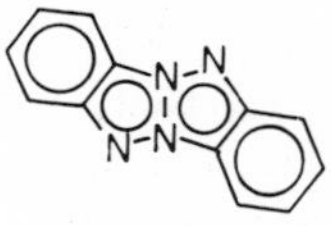

(317) 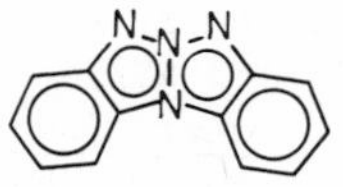 (318)

263. T. S. Cantrell and B. L. Harrison, Tetrahedron Lett., 1299 (1969).
264. (a) T. S. Cantrell and B. L. Harrison, Tetrahedron Lett., 4477, (1967); (b) W. H. Okamura and T. J. Katz have described the 4-isomer: Tetrahedron, 23, 2941 (1967).
265. H. Volz and B. Meszner, Tetrahedron Lett., 4111 (1969).
266. (a) S. Trofimenko, J. Amer. Chem. Soc., 87, 4393 (1965) and 88, 5588 (1966); (b) T. W. G. Solomons and C. F. Voigt, J. Amer. Chem. Soc., 87, 5256 (1965) and 88, 1992 (1966).

to heating to 300°, and to hot aqueous base and acid [267]. A structure proof for these compounds by means of X-ray diffraction has been reported [268].

Another amazing development has been the synthesis and thermal stability of the thieno[3,4-c]thiophenes such as (319). The only sensible resonance structures one can write for this type of compound clearly requires the sulfur d orbitals, hence these molecules have been referred to "non-classical".

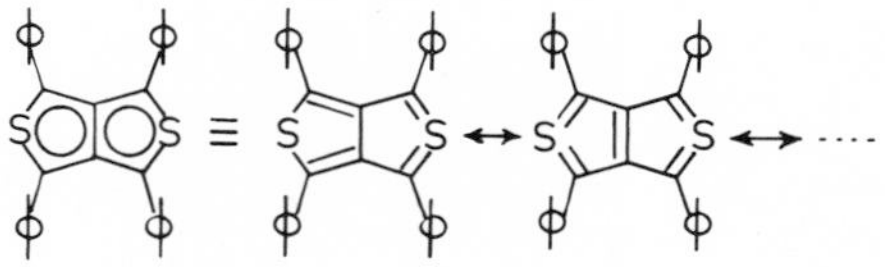

(319)

The melting point of (319) is given as 250°, which leaves little doubt about the stability of the basic framework [269]. The analog (320) is of interest in this connection.

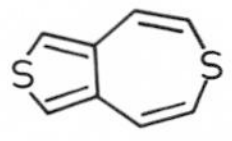

(320)

Although all other known thiepins are unstable and rapidly lose sulfur, (320) is thermally stable. Its nmr spectrum shows a singlet at 3.4 τ and doublets at 3.9 and 4.8 τ [270]. The thienopyrrole (321) and thienofuran (322) are likewise

---

267. (a) R.A. Carboni and J.E. Castle, J. Amer. Chem. Soc., 84, 2453 (1962); (b) R. Pfleger, E. Garthe, and K. Rauer, Chem., Ber., 96, 1827 (1963); (c) R.A. Carboni, J.C. Kauer, J.E. Castle, and H.E. Simmons, J. Amer. Chem. Soc., 89, 2618 (1967); (d) R.A. Carboni, J.C. Kauer, W.R. Hatchard, and R.J. Harder, J. Amer. Chem. Soc., 89, 2626 (1967); (e) J.C. Kauer and R.A. Carboni, J. Amer. Chem. Soc., 89, 2633 (1967); (f) Y.T. Chia and H.E. Simmons, J. Amer. Chem. Soc., 89, 2638 (1967); (g) R.J. Harder, R.A. Carboni, and J.E. Castle, J. Amer. Chem. Soc., 89, 2643 (1967).
268. M. Brufani, W. Fedeli, G. Giacomello, and A. Vaciago, Chem. Ber., 96, 1840 (1963).
269. (a) R.H. Schlessinger and I.S. Ponticello, J. Amer. Chem. Soc., 89, 3641 (1967); (b) M.P. Cava and G.E.M. Husbands, J. Amer. Chem. Soc., 91, 3952 (1969); (c) J.M. Hoffman and R.H. Schlessinger, J. Amer. Chem. Soc., 91, 3953 (1969); (d) J.D. Bower and R.H. Schlessinger, J. Amer. Chem. Soc., 91, 6891 (1969).
270. R.H. Schlessinger and G.S. Ponticello, J. Amer. Chem. Soc., 89, 7138 (1967).

(321) (322)

remarkably stable [271]. There are at present no other analogs of naphthalene; butalene (323) may play a role as a transient intermediate [272], and the only news about propalene (324) is Dewar's prediction that "it should be .... feasible to make it" [273].

(323) (324)

Another interesting series to consider is (325)-(328).

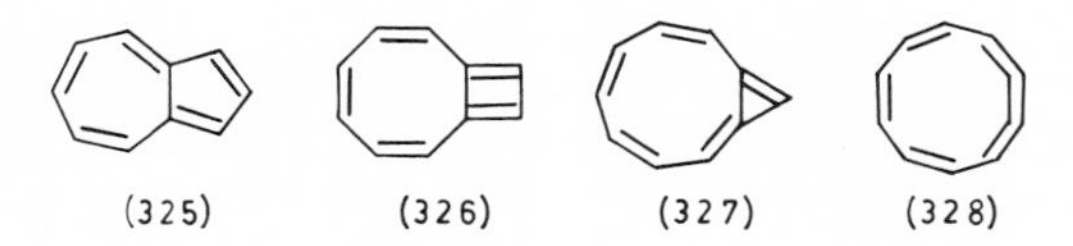

(325) (326) (327) (328)

The first of these is commonly known as azulene, a beautifully colored and stable aromatic compound the structure of which evidently has a fairly important contribution from the dipolar form (329); this can be concluded from a study of the variation of dipole moment as a function of substituents [274].

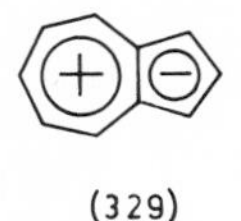

(329)

Thus, the molecule appears to derive some of its stability from fusion of a tropylium and a cyclopentadienide ion, just as naphthalene can be regarded as two fused benzene rings as well as a bridged [10]annulene. A crucial point is

---

271. (a) M.P. Cava and M.A. Sprecker, J. Amer. Chem. Soc., 94, 6214 (1972); see also (b) K.T. Potts and D. McKeough, J. Amer. Chem. Soc., 94, 6215 (1972).
272. R.R. Jones and R.G. Bergman, J. Amer. Chem. Soc., 94, 660 (1972).
273. N.C. Baird and M.J.S. Dewar, J. Amer. Chem. Soc., 89, 3966 (1967).
274. G.W. Wheland and D.E. Mann, J. Chem. Phys., 17, 264 (1949).

reached with bicyclo[6.2.0]decapenta-1,3,5,7,9-ene, (326). If this is a bridged and hence relatively unstrained [10]annulene, it is expected to be stable; if it is essentially a fused cyclobutadienocyclooctatetraene, the synthetic chemist faces a grim task. Nature has yielded on this point, but not without malice; although several derivatives have been prepared (as red, thermally stable, but air-sensitive solids), the nmr spectral information is tantalizingly inconsistent. Thus, the methyl derivative (330) basically has three signals: A singlet at 3.88 τ, a doublet at 7.52 τ and a quartet at 2.34 τ, in a 6:3:1 ratio.

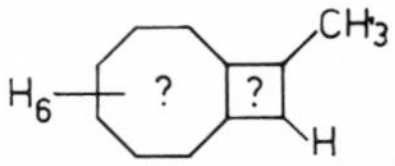

H_6 ? ? CH_3 H

(330)

Cyclooctatetraene itself has a singlet at 3.88 τ; on the other hand, the triene (331) has a methyl signal at 8.12 τ and the ring proton signal at 3.40 τ, and the triene (332) has its four-membered ring protons absorbing at 3.30 τ [275]. Thus, it seems that one end of the system is subject to a ring current, and the other end is not. X-ray diffraction will probably be more informative in this case.

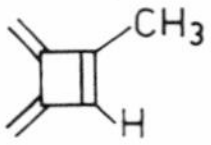

CH_3 H

(331) 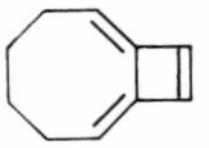 (332)

Other substances of this type have not yet been reported, but there is a large and rapidly growing assortment of other facts bearing on the general question. For instance, the triene (333) can be prepared, albeit only in dilute solutions in which it rapidly dimerizes at room temperature [276].

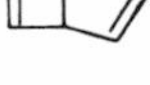

(333)

By comparing the rate of dimerization with that of deuterium exchange with various weak acids, Breslow managed to make an estimate of its pKa. This amounted to about 29, which should be compared with a value of 19 for cyclopentadiene.

---

275. (a) G. Schröder and H. Röttele, Angew. Chem., Int. Ed. Engl., 7, 635 (1968); (b) G. Schröder, S.R. Ramadas, and P. Nikoloff, Chem. Ber., 105, 1072 (1972).

276. R. Breslow, W. Washburn, and R.G. Bergman, J. Amer. Chem. Soc., 91, 196 (1969).

This large destabilization must mean that the anion (334) is anti-aromatic [277].

(334)

Isoindole is said to be an equilibrium mixture of (335) and (336), with the latter predominating; the nmr spectrum shows signals only from 2.5 to 3.2 τ [278].

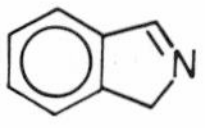

(335) 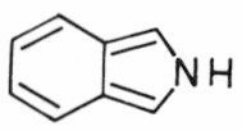 (336)

Isobenzofuran (337) has also been isolated; like isoindole, it is unstable to dienophiles and is rapidly destroyed at room temperature by dimerization. The nmr spectrum shows the furan protons at 1.60 τ; this compares with 2.80 τ for cyclooctatetraeno[c]furan (338), and hence is strongly indicative of aromatic character [279].

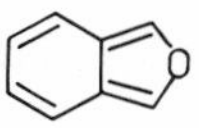

(337) 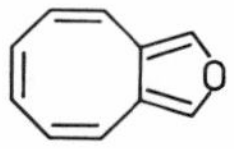 (338)

The eight-membered ring protons of (338) appear to be shifted upfield by about 0.5 ppm compared to cyclooctatetraene itself [280], and hence in that case there is some evidence for anti-aromaticity. The cation (339) has been isolated as the stable tetrafluoroborate; it is stable to triphenylmethane (340) [281].

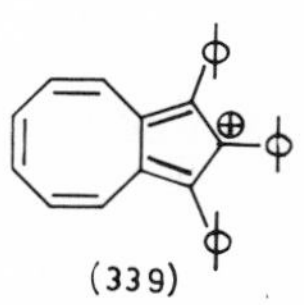

$HC\phi_3$

(339) (340)

277. R. Breslow and W. Washburn, J. Amer. Chem. Soc., 92, 427 (1970).
278. R. Bonnett and R. F. C. Brown, Chem. Commun., 393 (1972).
279. (a) R. N. Warrener, J. Amer. Chem. Soc., 93, 2346 (1971); (b) D. Wege, Tetrahedron Lett., 2337 (1971).
280. E. Le Goff and R. B. La Count, Tetrahedron Lett., 2787 (1965).
281. H. Dürr and G. Scheppers, Tetrahedron Lett., 6059 (1968).

In (341), which is not very stable to polymerization but which can be obtained in solution from (342) by isomerization, the methyl group appears at 7.80 τ in the nmr, and the compound is about as basic as diphenylcyclopropenone; thus, there does seem to be some resonance stabilization in this case [282].

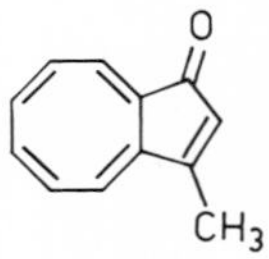

(341) 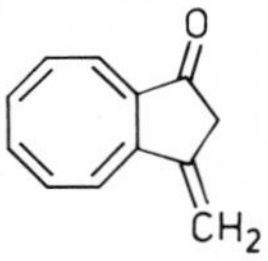 (342)

Finally, the aromaticity of the green bisdehydro ring system (343) - which has nmr signals in the same range as azulene - shows that these questions can be pursued in the large ring annulene series as well [283].

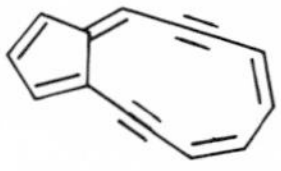

(343)

Among the various interesting bicyclic conjugated species, the 1-amino-7-iminocycloheptatrienes (344) deserve to be mentioned.

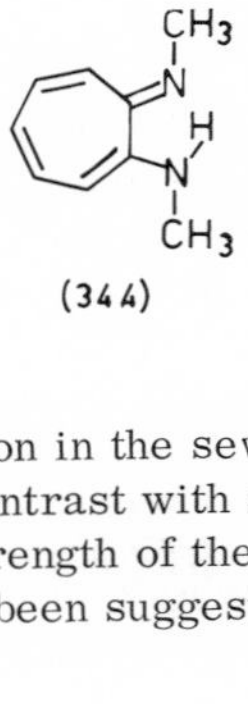

(344)

It was found that bromine substitution in the seven-membered ring led to an increase in the dipole moment - in contrast with such molecules as tropolone and azulene; on that basis and on the strength of the fact that the two methyl groups are equivalent even at -80°, it has been suggested [284] that (344) should not be

---

282. R. Breslow, W. Vitale, and K. Wendel, Tetrahedron Lett., 365 (1965).
283. J. Mayer and F. Sondheimer, J. Amer. Chem. Soc., 88, 602 (1966).
284. (a) R.E. Benson, J. Amer. Chem. Soc., 82, 5948 (1960); (b) W.R. Brasen, H.E. Holmquist, and R.E. Benson, J. Amer. Chem. Soc., 83, 3125 (1961); (c) L.C. Dorman, Tetrahedron Lett., 459 (1966).

considered as a tropylium derivative (345), but as the ten π-electron system (346), incorporating a hydrogen atom as part of the ring.

(345) (346)

It seems surprising that a hydrogen atom - whose p orbitals must be of rather high energy - might become part of a π system; at any rate, support for this proposition came from an X-ray diffraction pattern [285], which seemed to suggest that the hydrogen atom is symmetrically located between the nitrogen nuclei. However, the location of hydrogen atoms is not easily judged that way. In a related case, nmr work decisively ruled out a similar symmetrical structure: In (347) the N-H proton, at room temperature split into a 1:2:1 triplet by coupling with two equivalent neighboring protons, at -80° appears to be coupled to only one of them [286].

(347)

The construction of linear doubly-bridged, tricyclic systems so far has led mostly to simple polyenes or molecules in which at least one ring is non-aromatic. Among these attempts [287], one might mention the species (348)-(352);

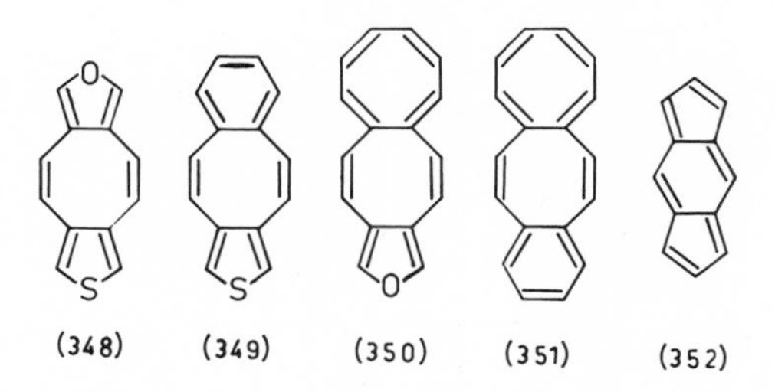

(348) (349) (350) (351) (352)

285. P. Goldstein and K.N. Trueblood, Acta Cryst., 23, 148 (1967).
286. U. Müller-Westerhoff, J. Amer. Chem. Soc., 92, 4849 (1970).
287. (a) A.P. Bindra, J.A. Elix, and M.V. Sargent, Tetrahedron Lett., 5573 (1968); (b) J.A. Elix, M.V. Sargent, and F. Sondheimer, J. Amer. Chem. Soc., 92, 973 (1970); (c) J.A. Elix, M.V. Sargent, and F. Sondheimer, J. Amer. Chem. Soc., 89, 5080 (1967); (d) J.A. Elix, M.V. Sargent, and F. Sondheimer, Chem. Commun., 509 (1966); (e) K. Hafner, Angew. Chem., Int. Ed. Engl., 3, 165 (1964).

among these, especially s-indacene is of interest in that a variety of reagents quickly aromatize it in one fashion or another.

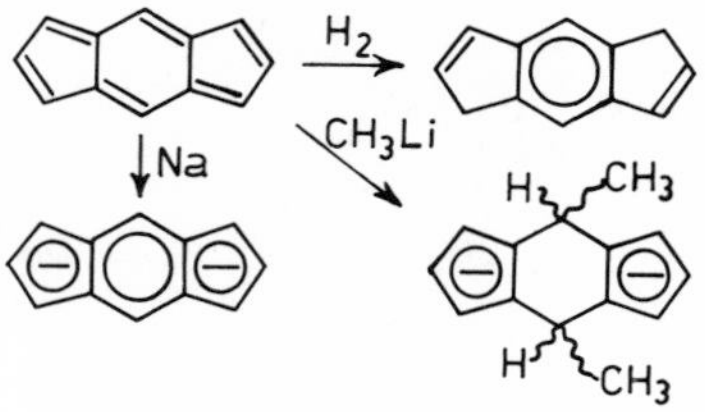

H_2 Na CH_3Li H CH_3 H CH_3

If the three rings are so fused as to produce a common corner, a new problem arises in that cross-conjugation (by which is meant conjugation such as (353)) must now necessarily be one of the features describing the system.

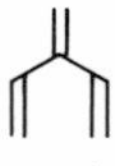

(353)

This means, for example that one cannot ignore the common-corner atom, and consider it part of an otherwise irrelevant bridge. An interesting and basic example is phenalene (354) or rather what is left of this 1,8-substituted naphthalene after one hydrogen atom is removed.

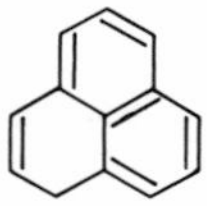

(354)

Simple Hückel calculations now show the presence of six bonding levels and a single non-bonding level (see Fig. 9-4) [288].

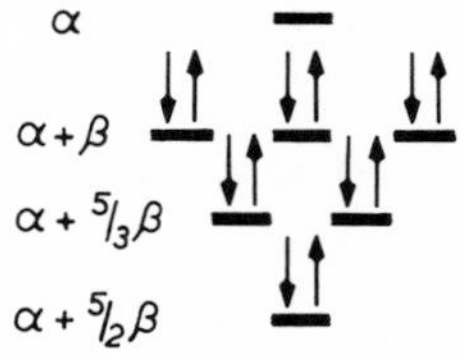

α

$\alpha+\beta$

$\alpha+\frac{5}{3}\beta$

$\alpha+\frac{5}{2}\beta$

Fig. 9-4. The M.O. levels of the phenalenium system.

288. D.H. Reid, Quart. Rev. (London), 19, 274 (1965).

Accordingly, the 12 π-electron cation and the 14 π-electron anion should both be singlets of comparable stability, and indeed, the perchlorate, the lithium salt and the free radical have all been prepared [289]. Furthermore, the molecule should be equally capable of accommodating electron-withdrawing systems such as cyclopentadienylidene and donating ones such as cyclopropenylidene. To date several examples have appeared; among them there are derivatives of the pentaphenafulvalene system (355) and of the pyranylidenephenalene nucleus (356).

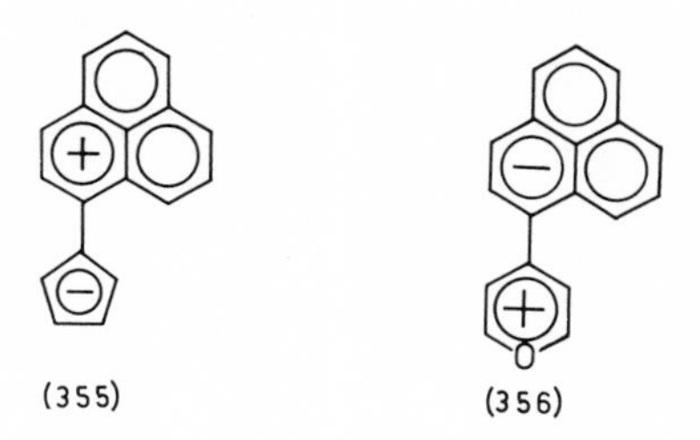

All these compounds are thermally quite stable, and measurements of the magnitude of the electric moments and their dependence on substituents leave no doubt about the direction of the dipoles [290].

Of similar interest are the ketone (357) and the diazo derivative (358). The latter unfortunately is not known; the former has a dipole moment of 3.9 D, a stretching frequency of 1637 cm^{-1} and a pKb value of 0.4.

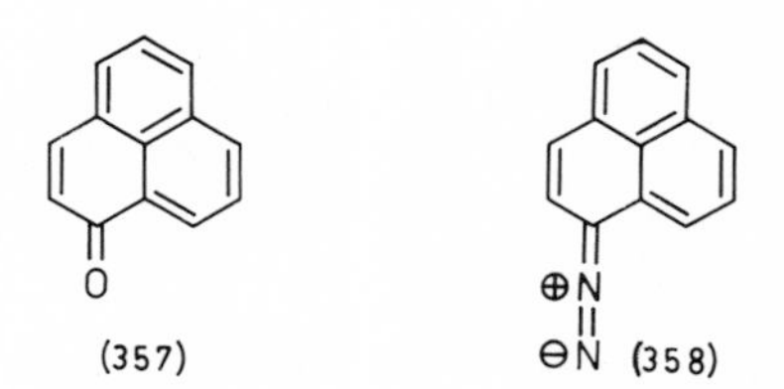

The bridged phenalenone (359) behaves similarly; the corresponding values here are 4.36 D, 1605 cm^{-1} and -2.6 [291].

---

289. (a) R. Pettit, J. Amer. Chem. Soc., 82, 1972 (1960) and papers quoted there; (b) H. Prinzbach, V. Freudenberger, and U. Scheidegger, Helv. Chim. Acta, 50, 1087 (1967).
290. (a) I. Murata, T. Nakazawa, and M. Okazaki, Tetrahedron Lett., 1921 (1969); (b) I. Murata, T. Nakazawa, H. Shimanouchi, and Y. Sasada, Tetrahedron, 26, 5683 (1970); (c) I. Murata, T. Nakazawa, and S. Tada, Tetrahedron Lett., 4799 (1971).
291. I. Murata, T. Nakazawa, and T. Tatsuoka, Tetrahedron Lett., 1789 (1971).

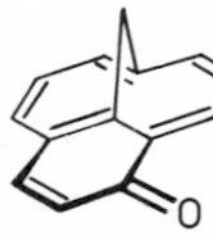

(359)

The isomer of phenalene, 2H-benz[cd]azulene (360), is completely analogous in that it can easily be converted into either the carbonium ion, free radical or carbanion [292].

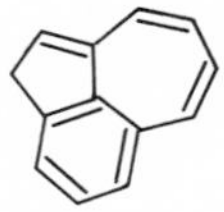

(360)

Fascinating questions arise when slight extensions of these molecules are considered, such as leaving out a methylene group so to leave the 12 π-electron system cyclopent[cd]azulene (361): Will this simply be an unstable pentalene, annelated with a butadiene moiety, will it be an annelated azulene (as acenaphthalene is simply a vinyl-substituted naphthalene), or will it be a new and stable species involving all carbon atoms?

OR OR

(361)

The theoretical prediction was that the latter alternative would prevail. One finds on reading the literature in this general area that the theoreticians are not always agreed in their calculations until after the facts are in, but in this case they were and the molecule is indeed completely delocalized and very stable [293].

The electronically similar species (362) is likewise aromatic, as shown by nmr [294] and so is cycl[3.2.2]azine (363) [295]; the latter compound is non-basic;

---

292. (a) V. Boekelheide and C.D. Smith, J. Amer. Chem. Soc., 88, 3950 (1966); (b) E. Galantay, H. Agahigian, and N. Paolella, J. Amer. Chem. Soc., 88, 3875 (1966).
293. K. Hafner, Pure Appl. Chem., 28, 153 (1971).
294. K. Valentin and A. Taurins, Tetrahedron Lett., 3621 (1966).
295. R.J. Windgassen, W.H. Saunders, and V. Boekelheide, J. Amer. Chem. Soc., 81, 1459 (1959).

and undergoes the gamut of electrophilic substituted reactions. Furthermore, the hydrocarbon (364) is quite acidic and easily converted into the symmetrical lithium salt [296].

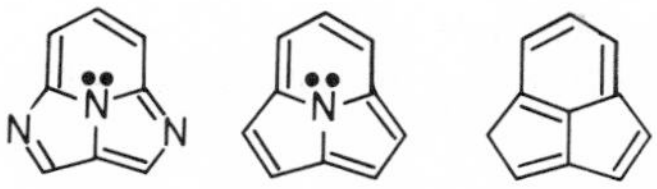

(362) (363) (364)

Both acepleiadene (365) and aceheptalene (366) have been found to be aromatic in the sense that not only are they very stable, but that the carbon-carbon bonds are all about equally long at an average of about 1.40 Å [297].

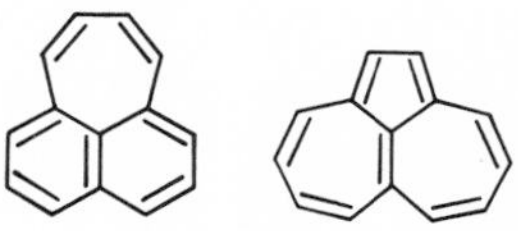

(365) (366)

The tetranuclear ring systems provide us with a fine tool to test Platt's notion of the exclusive importance of the periphery, since now an entire ethylene unit can potentially be contained inside. We shall review both the [12] and [14] bridged annulenes - if it is proper to call them that. The main [12] membered ring system known is that of pyracylene and its derivatives - an area in which the principal contributions have been made by Trost. The general conclusion is that - as suggested by the number of peripheral but not by the total number of π electrons, there is a paramagnetic ring current and a certain degree of destabilization. Thus, pyracyloquinone (367) is exceedingly easily reduced to the diketone (368), which in turn does not under any circumstances enolize to the pyracylenediol (369).

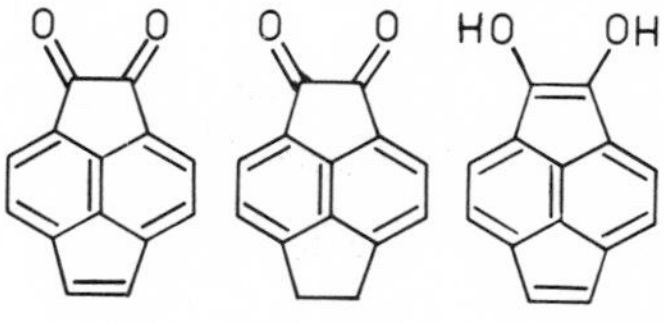

(367) (368) (369)

---

296. P. Eilbracht and K. Hafner, Angew. Chem., Int. Ed. Engl., 10, 751 (1971).
297. (a) V. Boekelheide and G.K. Vick, J. Amer. Chem. Soc., 78, 653 (1956); (b) P.D. Gardner and R.J. Thompson, J. Org. Chem., 22, 36 (1957); (c) E. Carstensen-Oeser and G. Habermehl, Angew. Chem., Int. Ed. Engl., 7, 543 (1968).

Even when (367) is converted into a dianion by means of potassium metal and this species is protonated, (368) is the exclusive result [298]. Pyracylene itself (370) and several simple derivatives [299] are now known (the parent hydrocarbon only in solution), and they seem to leave little doubt about their anti-aromatic nature. Thus, (370) has singlets of equal intensity at 3.5 and 4.0 τ [299]; about 1 ppm higher than the dihydro- and tetrahydro derivatives (371) and (372);

(370) (371) (372)

furthermore, (372) is easily oxidized to (371) by means of chloranil (373) but not further.

(373)

It was found also that the dianion of the isomeric dibenzo[cd, gh]pentalene (374) (the hydrocarbon itself is unknown) has chemical shifts such as to require the assumption that the two electrons are exclusively on the periphery; the argument is based on a comparison with the dianions of *s*- and *as*-indacene (375) and (376) [300]. Unfortunately, in none of these cases have X-ray or C^{13} nmr data appeared in print as yet.

(374) (375) (376)

---

298. B.M. Trost, *J. Amer. Chem. Soc.*, 88, 853 (1966) and 91, 918 (1969).
299. (a) B.M. Trost and D.R. Brittelli, *Tetrahedron Lett.*, 119 (1967); (b) B.M. Trost and G.M. Bright, *J. Amer. Chem. Soc.*, 89, 4244 (1967); (c) B.M. Trost, G.M. Bright, C. Frihart, and D. Brittelli, *J. Amer. Chem. Soc.*, 93, 737 (1971).
300. B.M. Trost and P.L. Kinson, *J. Amer. Chem. Soc.*, 92, 2591 (1970).

These conclusions are also supported by a study of 8b, 8c-diazacyclopent[fg]acenaphthylene (377) and its di- and tetrahydro derivatives (378) and (379).

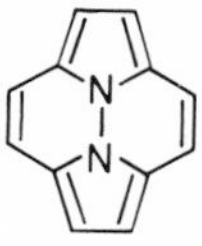

(377) 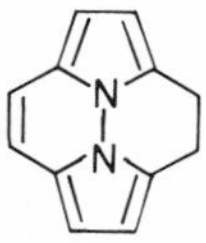 (378) 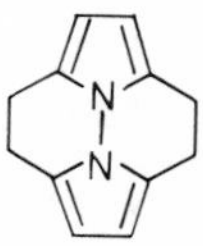 (379)

The five-membered ring protons in these compounds appear in the nmr spectrum at 4.8, 3.2 and 4.2 τ, respectively, and it was furthermore found that (377) is very easily reduced to (379) [301]. In the related cycl[3.3.3]azine (380) the pmr signals were quoted [265] as occurring at 6.35 and 7.93 τ in a 1:2 ratio.

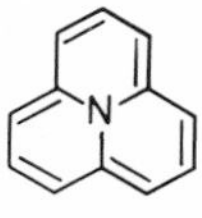

(380)

Finally, in 5H-cyclopent[cd]phenalen-5-one (381) and its dehydro analog (382), all CH-signals of the former occur at 0.5-1 ppm upfield of those of the latter; in sulfuric acid these differences increase even further, to 1-1.5 ppm [302].

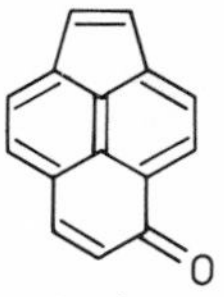

(381) 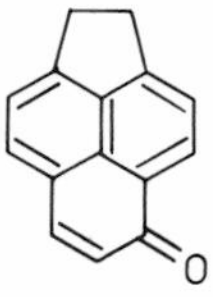 (382)

There are now eight known examples of [14]annulenes (or their derivatives) with an internal ethylene unit [303]. They are: Cycloocta[def]biphenylene (383) [304], dicyclopenta[ef, kl]heptalene (384) [305], naphth[2,1,8-cde]azulene (385) [306],

---

301. W.W. Paudler and E.A. Stephan, J. Amer. Chem. Soc., 92, 4468 (1970).
302. I. Murata, K. Yamamoto, T. Hirotsu, and M. Morioka, Tetrahedron Lett., 331 (1972).
303. P. Baumgartner, E. Weltin, G. Wagnière, and E. Heilbronner, Helv. Chim., Acta, 48, 751 (1965).
304. C.F. Wilcox, J.P. Uetrecht, and K.K. Grohman, J. Amer. Chem. Soc., 94, 2532 (1972).
305. (a) A.G. Anderson, A.A. MacDonald, and A.F. Montana, J. Amer. Chem. Soc., 90, 2993 (1968); (b) C. Jutz and E. Schweiger, Angew. Chem., Int. Ed. Engl., 10, 799 (1971).
306. P.D. Gardner, C.E. Wulfman, and C.L. Osborn, J. Amer. Chem. Soc., 80, 143 (1958).

pentaleno[2, 1, 6a, 6-def]heptalene (386) [307], cyclohepta[def]fluorene (387) [308], acepleiadylene (388) [309], pyrene (389) and cyclohept[bc]acenaphthylene (390) [310].

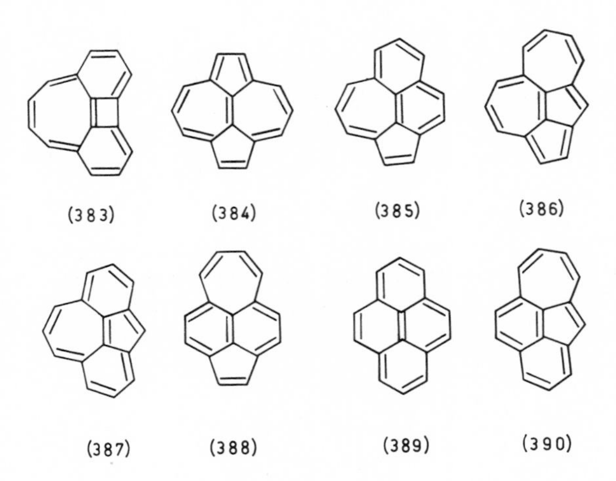

The type and amount of information unfortunately varies greatly from case to case, and the only possible general conclusion is that certainly most but not all of these compounds are characterized by chemical stability and low field nmr signals (3.0 τ or below). The most noteworthy exception in this series is (383), whose nmr signals lie between 3.3 and 4.7 τ; it seems likely that the potential cyclobutadiene ring is responsible.

In an interesting comparison with (352) and (353), Murata has found that the peripheral protons of 7H-cycloehpta[cd]phenalen-7-one (391) are a bit downfield as compared to those of (392); the difference becomes greater in acid solution [311].

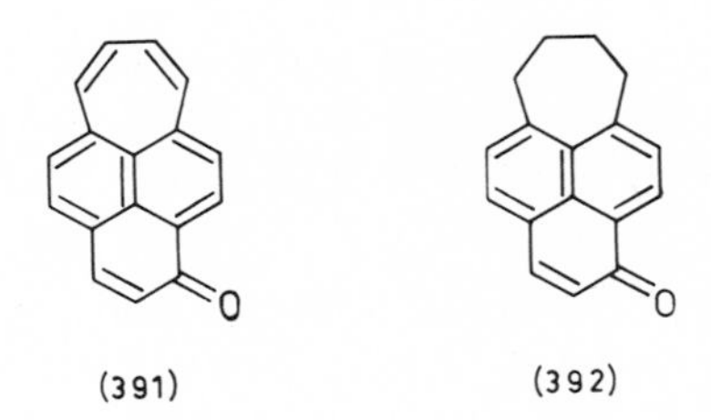

307. K. Hafner, R. Fleischer, and K. Fritz, Angew. Chem., Int. Ed. Engl., 4, 69 (1965).
308. R. Munday and I. O. Sutherland, Chem. Commun., 569 (1967).
309. V. Boekelheide and G. K. Vick, J. Amer. Chem. Soc., 78, 653 (1956).
310. D. H. Reid, W. H. Stafford, and J. P. Ward, J. Chem. Soc., 1193 (1955).
311. I. Murata, K. Yamamoto, and T. Hirotsu, Tetrahedron Lett., 3389 (1972).

In connection with these aromatic cavities, a group of compounds may be pointed out which are characterized by one completely conjugated ring within another ("circulenes", according to a suggestion by Wynberg [312]). Thus, coronene (393) is an example of such a compound, with a [6]annulene within an [18]annulene [313].

(393)

It is of course difficult to assess the merit of such a proposal, although C^{13}-nmr may be of help. Known examples include corannulene (394) [314] and the heterocyclic species (395) [312].

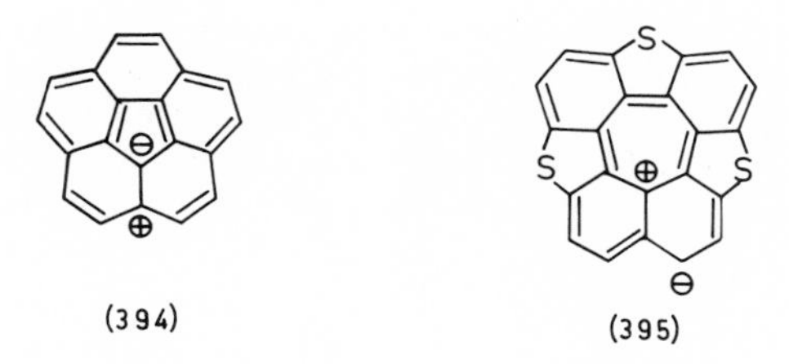

(394) (395)

9-9. HOMOAROMATICITY

In 1956, Doering suggested that the termini of the conjugated chain in tropilidene (396) were close enough together to interact, and that in fact the molecule should be written as (397) [315].

---

312. J. H. Dopper and H. Wynberg, Tetrahedron Lett., 763 (1972).
313. See also, (a) F. Vögtle and H. A. Staab, Chem. Ber., 101, 2709 (1968); (b) D. Hellwinkel and G. Reiff, Angew. Chem., Int. Ed. Engl., 9, 527 (1970); (c) G. Ege and H. Vogler, Z. Naturforsch. (B), 27, 918 (1972).
314. (a) W. E. Barth and R. G. Lawton, J. Amer. Chem. Soc., 93, 1730 (1971) and 88, 380 (1966); (b) J. Janata, J. Gendell, C. Y. Ling, W. Barth, L. Backes, H. B. Mark, and R. G. Lawton, J. Amer. Chem. Soc., 89, 3056 (1967).
315. W. von E. Doering, G. Laber, R. Vonderwahl, N. F. Chamberlain, and R. B. Williams, J. Amer. Chem. Soc., 78A, 5448 (1956).

(396) (397)

In other words, the π system is the same as in benzene, but instead of six σ bonds between the carbon atoms there are only five and one is a methylene bridge; this amounts to no-bond resonance structures.

When such structures are drawn one should be careful to remember that the single line connecting these atoms is a π bond, and does not signify a σ bond as is the case any other time.

It turned out that definite evidence (such as, for example, an unusually low heat of hydrogenation) was not found, and this kind of homoconjugation is no longer believed to be significant in tropilidene [316]. Nevertheless, in 1959 Winstein generalized the concept and called it homoaromaticity [317]. Accordingly, tropilidene is homobenzene, (398) is hexahomobenzene, (399) is a bishomothiophene, and so on.

(398)

(399)

It is obvious that in order to be effective, homoaromaticity requires the σ-non-bonded, π-bonded atoms to be near one another. This means that the methylene carbon atoms will be subject to an angle strain; the molecule will be homoaromatic if the potential resonance energy is greater than the additional strain. Ideally, the atoms are connected in such a way that the π termini are rigidly forced close

316. K. Conrow, J. Amer. Chem. Soc., 83, 2958 (1961).
317. S. Winstein, J. Amer. Chem. Soc., 81, 6524 (1959).

together and there is no escape from the strain. We shall now review the literature reporting the pursuit of evidence, beginning with some non-aromatic cases.

The π-interaction between non-conjugated groups can in many instances be recognized by means of uv spectroscopy [318]. Thus, simple ketones usually have a weak $\pi^* \leftarrow n$ transition at about 280 nm, with an extinction coefficient well below 50. When the carbonyl group is conjugated with an olefinic group, the wavelength moves up to 325 nm and ϵ to about 50. In 6-norbornenone (400) one finds that λ_{max} equals 290 nm and that ϵ there equals 110.

(400)

Similarly, it is observed that 1, 3, 6, 8-nonatetraene (401) has a λ_{max} 10 nm greater than 1, 3-nonadiene (402), and norbornadiene (403) has a wavelength 30 nm above that of norbornene (404).

(401) (402) (403) (404)

Cis-1, 4, 7-cyclononatriene (405) could be an example of a trishomobenzene, and it has indeed been shown by means of ir that the molecule prefers the necessary crown conformation [319].

(405)

318. (a) Reviewed by L. N. Ferguson and J. C. Nnadi, *J. Chem. Educ.*, *42*, 529 (1965); (b) For an example and recent references, see H. D. Martin, and D. Forster, *Angew. Chem., Int. Ed. Engl.*, *11*, 54 (1972).
319. S. J. Wilt and M. A. El-Sayed, *J. Amer. Chem. Soc.*, *88*, 2911 (1966).

The heat of combustion revealed no significant resonance energy, but this is of course difficult to say because of the various types of strain involved [320]; there is photoelectron spectroscopic evidence for a resonance energy of 5 kcal/mole [321].

In order to reduce the conformational flexibility, Woodward synthesized triquinacene (406) [322].

(406)

The nmr and uv spectra were very similar to those of cyclopentene, and although a long wavelength shoulder was observable in the uv, the authors did not consider it sufficient evidence for homoconjugation. Recently the isomer (407) (diademane) has also been prepared; it reverts to (406) upon heating to 90° [323].

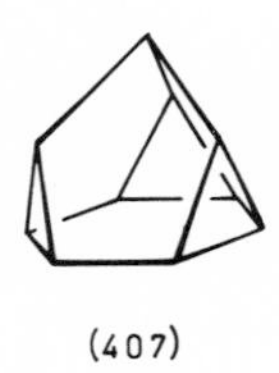

(407)

It should be noted that the existence of this isomer does not rule out homoconjugation in triquinacene; it differs from the latter in that σ bonds connect the six equivalent carbons rather than π bonds. Another approach was tried by Untch: In order to reduce angle strains he studied Z, Z, Z-1, 5, 9-cyclododecatriene (408).

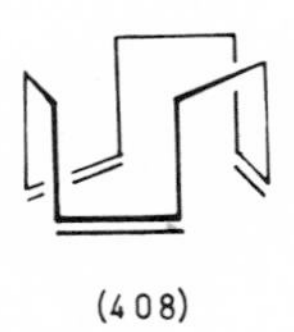

(408)

---

320. W.R. Roth, W.B. Bang, P. Goebel, R.L. Sass, R.B. Turner, and A.P. Yü, J. Amer. Chem. Soc., 86, 3178 (1964).
321. P. Bischof, R. Gleiter, and E. Heilbronner, Helv. Chim. Acta, 53, 1425 (1970).
322. R.B. Woodward, T. Fukunaga, and R.C. Kelly, J. Amer. Chem. Soc., 86, 3162 (1964).
323. A. deMeijere, D. Kaufmann, and O. Schallner, Angew. Chem., Int. Ed. Engl., 10, 417 (1971).

No evidence of any sort could be found that would point to conjugation. It was concluded that the lack of π-interactions in this case was due to another type of strain, namely the eclipsing of neighboring methylene units that would be required [324].

A very good possibility has been based on an analogy with the unusually low barrier to isomerization in the semibullvalenes (see p. 242 and p. 407). When the nmr spectrum of (409) is compared with that of the equilibrating pair (410), a ring current seems clearly indicated since the vinyl protons of (409) appear more than one ppm lower and its bridge protons are on either side of those in (410). The authors were cautious at first not to claim complete success however, since the overall appearance of the proton nmr spectrum is similar to that of (411) - in which case homoaromaticity seems much less likely; however, the ^{13}C nmr spectra of (409) and of 1,6-methanocyclodecapentaene appeared so strikingly similar that they were led to conclude that the former compound is truly homoaromatic [325a,b]. But hardly had this paper made its appearance when Vogel showed decisively that the conclusion was wrong, and that it rested on a misassignment of one of the cmr signals [325c,d]. This is perhaps one of the more remarkable examples of the fact that the step between a recorder blip and the final conclusion is not a trivial one.

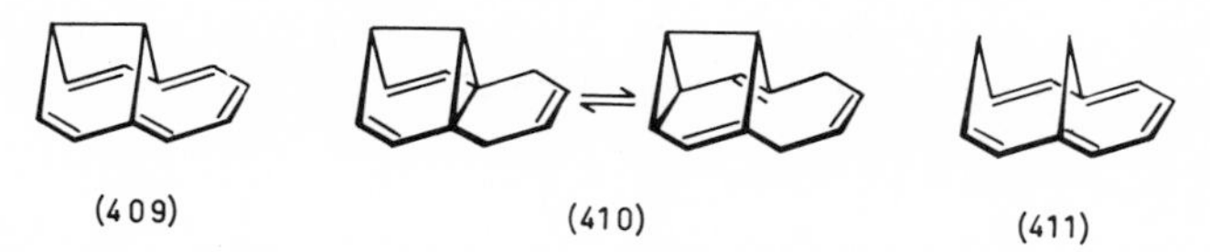

(409) (410) (411)

Still another type of approach has been reported [326]. It was found that the tetracyclic azo compound (412) eliminates nitrogen to give a mixture of 2,6-diphenyldihydronaphthalenes, but this reaction is not significantly faster (if anything, it is slower) than that of the saturated analog (413), so that the transition state or intermediate involved is not stabilized by homoconjugation as a tetrakishomobenzene (414).

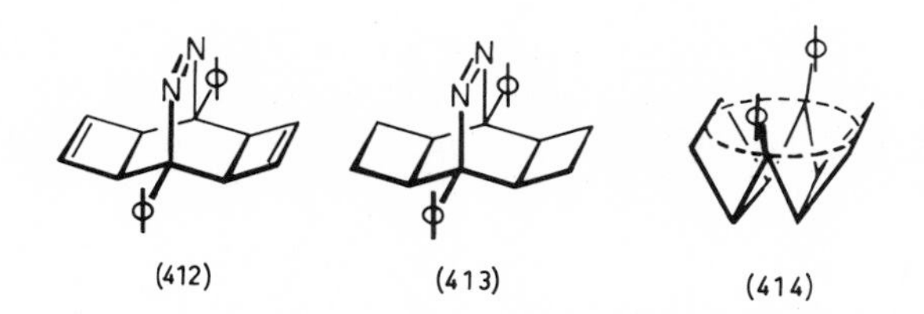

(412) (413) (414)

---

324. K.G. Untch and D.J. Martin, J. Amer. Chem. Soc., 87, 3518 (1965).
325. (a) L.A. Paquette, R.E. Wingard, and R.K. Russell, J. Amer. Chem. Soc., 94, 4739 (1972); (b) E. Wenkert, E.W. Hagaman, L.A. Paquette, R.E. Wingard, and R.K. Russell, Chem. Commun., 135 (1973); (c) E. Vogel, U.H. Brinker, K. Nachtkamp, J. Wassen, and K. Müllen, Angew. Chem., Int. Ed. Engl., 12, 758 (1973); (d) H. Günther, H. Schmickler, U.H. Brinker, K. Nachtkamp, J. Wassen, and E. Vogel, Angew. Chem., Int. Ed. Engl., 12, 760 (1973).
326. L.A. Paquette, M.R. Short, and J.F. Kelly, J. Amer. Chem. Soc., 93, 7179 (1971).

Success has apparently been encountered among the macrocyclic annulenes. The first claim came from Zollinger, who found that in (415), H_1 and H_2 occur as a quartet at about 1.35 τ [327].

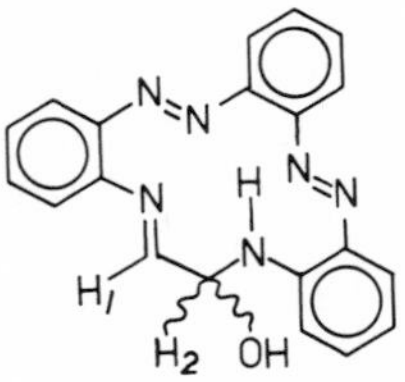

(415)

This unusual chemical shift was considered evidence for a ring current. The lone inside proton was found at 0.24 τ. This low value was attributed to H-bonding; whether this interpretation is right or not remains to be seen. Oddly enough, the most unambiguous evidence so far seems to apply to anti-homoaromatic compounds, namely the homo[16]annulenes (416)-(418) [328]. In each case, the inside protons are found at 1-2 τ and the outside protons at 3-5 τ.

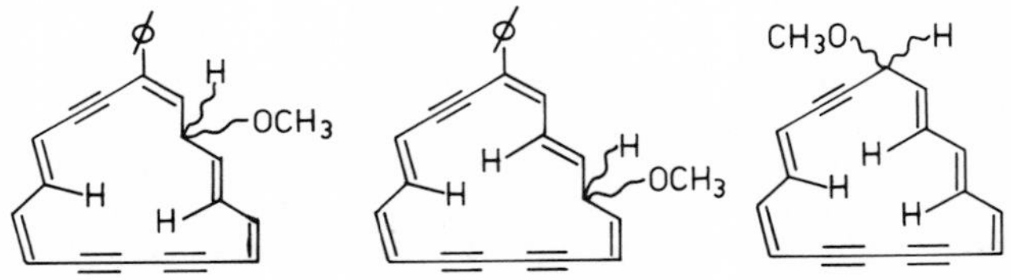

(416) (417) (418)

In view of the fleeting existence of many carbonium ions and carbanions, one might expect that homoaromaticity would not be found among the charged aromatic species but in fact the most clear-cut examples are found there. Perhaps the reason is that aromaticity in such species permits a more extensive delocalization of charge.

When cyclooctatetraene is dissolved in fluorosulfonic acid, the nmr spectrum shows 1:2:1 triplets at 1.43, 1.61 and 1.73 τ, respectively, in a 2:2:1 ratio, and further signals occur at 3.52, 4.87 and 10.73 τ in a 2:1:1 ratio. These last two signals are due to the methylene protons, as is clear from a measurement of their coupling constants. The remarkable contrast between these chemical shifts is very difficult to explain in terms other than of the homotropylium ion, (419) [329].

327. P. Skrabal and H. Zollinger, Helv. Chim. Acta, 54, 1069 (1971).
328. F. Sondheimer, Pure Appl. Chem., 28, 331 (1971).
329. J. L. von Rosenberg, J. E. Mahler, and R. Pettit, J. Amer. Chem. Soc., 84, 2842 (1962).

Thus, structure (420) [330] is not satisfactory, if only because of the τ-values observed when cyclooctatriene is protonated to give (421) (see also p. 389) [331]. The ion is isolable as the hexachloroantimonate salt.

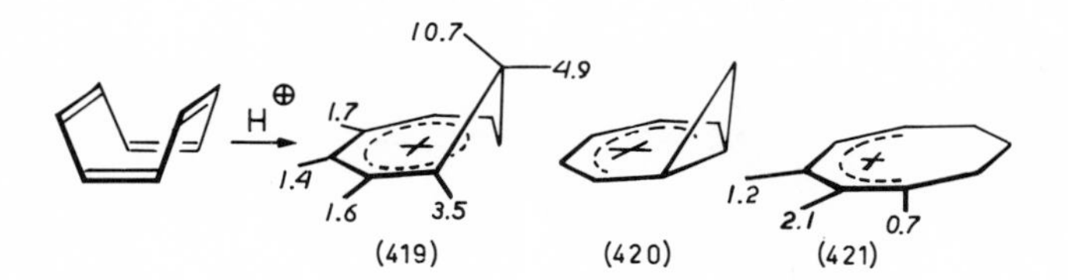

(419) (420) (421)

If D_2SO_4 is used as the acid at -10°, it is found that the deuteron is originally deposited at the inside position. Equilibration slowly occurs at that temperature, and ΔE_a is found to be 22 kcal/mole; this then must be a minimum measure of the energy separating the simple flat ion from the homotropylium species [332].

The evidence for homoaromaticity in homotropones is weak [333]. Compounds such as (422) and (423) are more basic than dienones such as (424), but the difference in chemical shift between the inside and outside methylene protons is insignificant.

O O CH_3 O CH_3

(422) (423) (424)

This changed drastically when such ketones are dissolved in strong acids; in the hydroxyhomotropylium ions which then form, all protons have been shifted downfield except the inside proton which is shifted in the opposite direction. In the ion derived from (422) no inside-outside isomerization can be observed by nmr even at +80° [334].

330. N.C. Deno, Progr. Phys. Org. Chem., 2, 129 (1964).
331. (a) P. Warner, D.L. Harris, C.H. Bradley, and S. Winstein, Tetrahedron Lett., 4013 (1970); (b) C.E. Keller and R. Pettit, J. Amer. Chem. Soc., 88, 604, 606 (1966).
332. S. Winstein, C.G. Kreiter, and J.I. Brauman, J. Amer. Chem. Soc., 88, 2047 (1966).
333. (a) J.D. Holmes and R. Pettit, J. Amer. Chem. Soc., 85, 2531 (1963); (b) Y. Sugimura, N. Soma, and Y. Kishida, Tetrahedron Lett., 91 (1971).
334. M. Brookhart, M. Ogliaruso, and S. Winstein, J. Amer. Chem. Soc., 89, 1965 (1967).

Both the 8-chlorohomotropylium ions (425) and (426) can be obtained as salts in completely pure form [335].

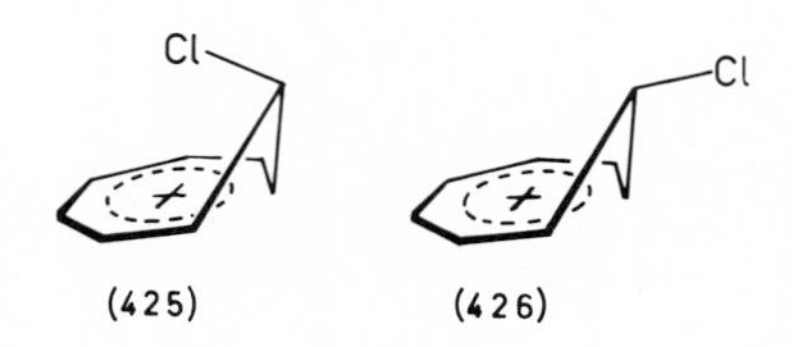

Still another route to these ions is exemplified by the alcohol (427), which is dehydrated upon dissolution in strong acid to give (428) [336]. The isomeric ion (429) was similarly made [337].

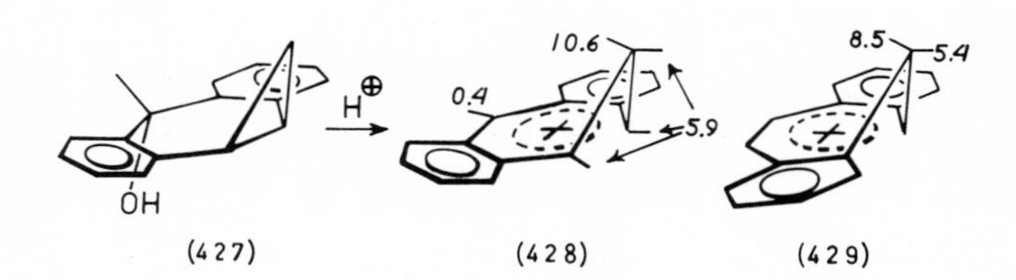

Several bishomotropylium ions are known [338]. They include (431) (inside protons at 8.1 τ, outside bridge protons at 6.2 τ), obtained by the protonation of homocyclooctatetraene (430) and stable only at -130° [339];

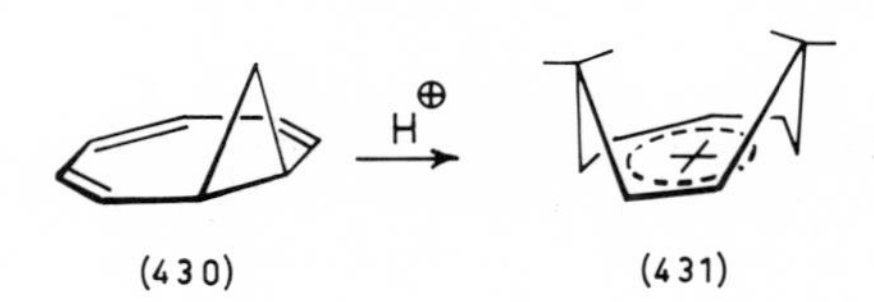

---

335. (a) G. Boche, W. Hechtl, H. Huber, and R. Huisgen, J. Amer. Chem. Soc., 89, 3344 (1967); (b) R. Huisgen, G. Boche, and H. Huber, J. Amer. Chem. Soc., 89, 3345 (1967); (c) R. Huisgen and J. Gasteiger, Tetrahedron Lett., 3661 (1972); (d) J. Gasteiger and R. Huisgen, Tetrahedron Lett., 3665 (1972); (e) J. Gasteiger and R. Huisgen, J. Amer. Chem. Soc., 94, 6541 (1972).
336. (a) R. F. Childs and S. Winstein, J. Amer. Chem. Soc., 89, 6348 (1967); (b) R. F. Childs, M. A. Brown, F. A. L. Anet, and S. Winstein, J. Amer. Chem. Soc., 94, 2175 (1972).
337. G. D. Mateescu, C. D. Nenitzescu, and G. A. Olah, J. Amer. Chem. Soc., 90, 6235 (1968).
338. P. Ahlberg, D. L. Harris, M. Roberts, P. Warner, P. Seidl, M. Sakai, D. Cook, A. Diaz, J. P. Dirlam, H. Hamberger, and S. Winstein, J. Amer. Chem. Soc., 94, 7063 (1972).
339. P. Warner and S. Winstein, J. Amer. Chem. Soc., 93, 1284 (1971).

the bridged species (432), obtainable from several precursors and stable to -80° [340], and the still more complex ion (433) of which the methylene protons signal at 9.0 and 10.0 τ [341].

(432) (433)

Interestingly, while the cation derived from the cis-conformation (434) is homoaromatic as judged by nmr, that obtained from the trans-isomer (435) is not [342].

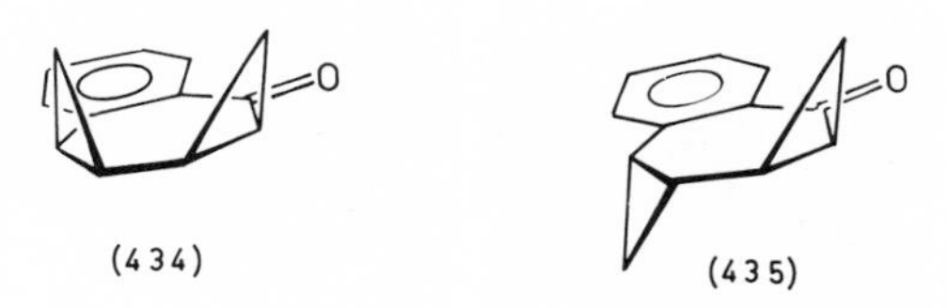

(434) (435)

No evidence has as yet been found for the ultimate extension of this series, the heptahomotropylium ion (436). This has not been for lack of effort; when the deuterium labeled precursor (437) is solvolyzed, the alcohol and olefin products show no evidence of scrambling [343].

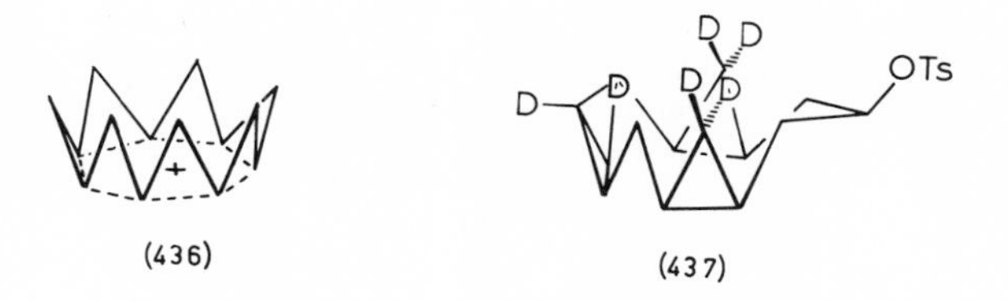

(436) (437)

Homocyclopentadienyl cations would be antihomoaromatic. Evidence for such species has been found: The solvolysis of the diene esters (438) is slower than

---

340. P. Seidl, M. Roberts, and S. Winstein, J. Amer. Chem. Soc., 93, 4089 (1971).
341. (a) G. Schröder, U. Prange, N.S. Bowman, and J.M.F. Oth, Tetrahedron Lett., 3251 (1970); (b) M. Roberts, H. Hamberger, and S. Winstein, J. Amer. Chem. Soc., 92, 6346 (1970).
342. H.A. Corver and R.F. Childs, J. Amer. Chem. Soc., 94, 6201 (1972).
343. R.W. Thies, M. Sakai, D. Whalen, and S. Winstein, J. Amer. Chem. Soc., 94, 2270 (1972).

that of the allylic analogs (439) by about 100, signifying a possible destabilization of the intermediate carbonium ion because of antihomoaromaticity (440) [344].

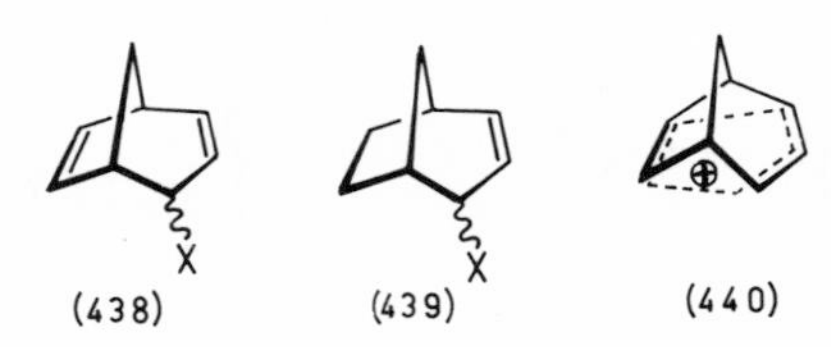

(438) (439) (440)

Pentahomocyclopentadienyl cation has been hunted by means of labelling studies, and the results may be summarized by the statement that the twofold symmetrical trishomocyclopropenyl (441) rather than the fivefold symmetric antihomoaromatic ion (442) was indicated [343].

OTs

(441)

(442)

The threefold symmetric trishomocyclopropenyl ion (443) was the first homoaromatic species encountered [345] and it led directly to Winstein's general proposal. Its existence was based on the observations that cis-bicyclo[3.1.0]hexyl esters (444) solvolyzed faster than the trans-, (445), and that complete scrambling of the deuterium among the three possible positions occurred during the solvolysis of (444), but not in that of (445).

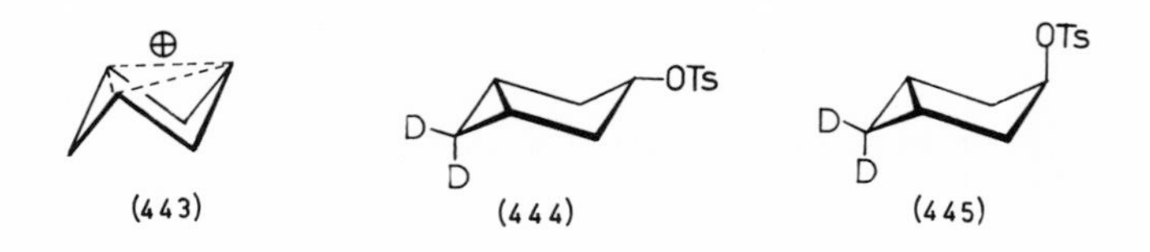

(443) (444) (445)

---

344. (a) A. F. Diaz, M. Sakai, and S. Winstein, J. Amer. Chem. Soc., 92, 7477 (1970); (b) R. M. Moriarty and C. L. Yeh, Tetrahedron Lett., 383 (1972).
345. (a) S. Winstein, J. Sonnenberg, and L. de Vries, J. Amer. Chem. Soc., 81, 6523 (1959); (b) S. Winstein and J. Sonnenberg, J. Amer. Chem. Soc., 83, 3235, 3244 (1961); (c) for a discussion of the phenyl substituted ions, see E. J. Corey and H. Uda, J. Amer. Chem. Soc., 85, 1788 (1963); (d) W. Broser and D. Rahn, Chem. Ber., 100, 3472 (1967).

At the other end of the spectrum, protonation of the higher annulenes may well provide results of great interest. To date this has been done only with 1,6-methanocyclodecapentaene, but the nmr positions of the bridge protons revealed no paramagnetic ring current, and hence no antihomoaromatic species (446) [346].

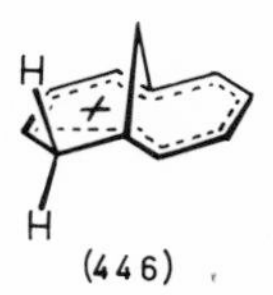

(446)

The bishomocyclobutadienylium dication (447) was, at one point, believed to be an intermediate in the solvolysis of the ditosylate (448). This is no longer thought to be correct (discussed further on p. 808).

(447) (448)

Anions provide us with an equally good opportunity to synthesize homoaromatic species. The diene (449) exchanges the allylic protons with deuterated base some 100,000 times faster than the monoene (450), clearly via the ion (451) [347] (the contrast of this finding with that of the chlorides (438) and (439) should be noted).

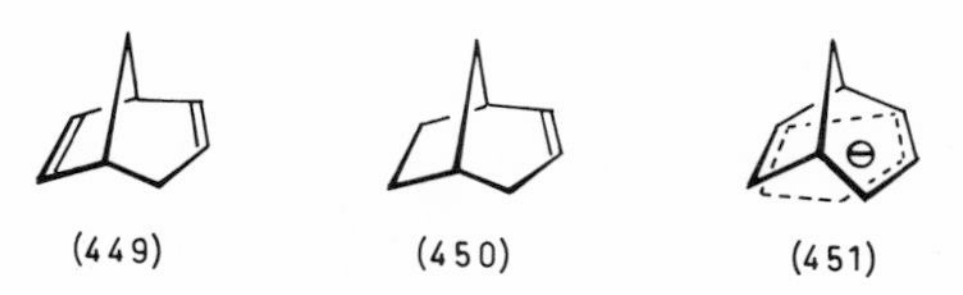
(449) (450) (451)

This ion can be prepared as a stable solution in a non-protic medium; the nmr spectrum shows large upfield shifts for the methylene protons [348].

---

346. P. Warner and S. Winstein, J. Amer. Chem. Soc., 91, 7785 (1969).
347. J. M. Brown and J. L. Occolowitz, Chem. Commun., 376 (1965).
348. (a) J. M. Brown, Chem. Commun., 638 (1967); (b) S. Winstein, M. Ogliaruso, M. Sakai, and J. M. Nicholson, J. Amer. Chem. Soc., 89, 3656 (1967).

A very interesting case is that of the stereoisomeric hydrocarbons (452) and (453).

(452) (453)

Exposure of (452) to potassium metal in THF at -80° gives a radical anion of which the bridge proton coupling constants in the esr spectrum are 17 Gauss - a large value suggesting that the electron interacts fully with the bridgehead positions as in (454) [349]. By the same token, the *trans*-isomer which cannot flatten and which can therefore only form anion radical (455) has corresponding coupling constants of only 0.5 Gauss [350]. Further exposure of (454) to the metal generates the homocyclooctatetraene dianion (456), in which perhaps the most remarkable feature is the lowering of the chemical shift of the bridgehead protons by nearly 3 ppm in spite of the two electrons [351].

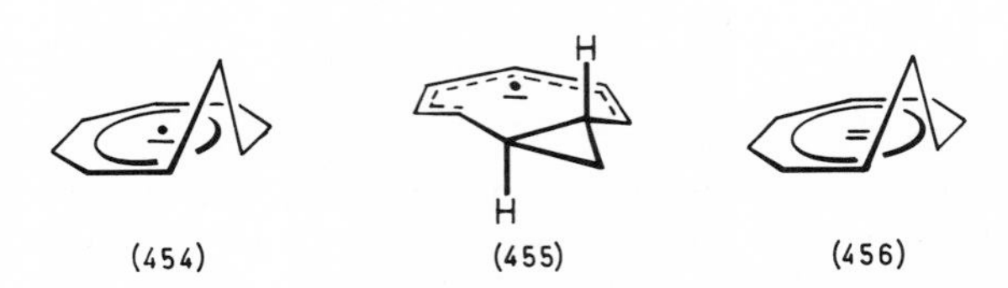

(454) (455) (456)

Another example was described by Böll, who found that 1, 6-methanocyclodecapentaene reacts with strong base in DMSO to give the homoaromatic anion (457) [352].

(457)

349. (a) R. Rieke, M. Ogliaruso, R. McClung, and S. Winstein, *J. Amer. Chem. Soc.*, *88*, 4729 (1966); (b) T. J. Katz and C. Talcott, *J. Amer. Chem. Soc.*, *88*, 4732 (1966).
350. G. Moshuk, G. Petrowski, and S. Winstein, *J. Amer. Chem. Soc.*, *90*, 2179 (1968).
351. M. Ogliaruso, R. Rieke, and S. Winstein, *J. Amer. Chem. Soc.*, *88*, 4731 (1966).
352. W.A. Böll, *Tetrahedron Lett.*, 5531 (1968).

In all this it should be remembered that most of the evidence concerning the homoaromatic ions stems from nmr spectra, which can sometimes be simulated by rapidly equilibrating ions, just as the spectra of benzene derivatives could be generated by pairs of rapidly equilibrating Kekulé structures. What these spectra show is that the activation energies for degenerate rearrangement cannot be high; they cannot prove that they are negative. There are in principle of course means by which these questions can be settled; X-ray diffraction and ir spectroscopy are among them. Unfortunately, they are difficult or impossible to apply in most of the cases discussed here. A further complication may be illustrated by means of the homoaromatic anion (458) [353], prepared from the corresponding hydrocarbon and a strong base.

(458)

The ion does not achieve full symmetry, as indicated by exchange experiments, the reason being that the counterion maintains a prejudice toward the original methylene site, and exchanges positions only slowly. While certain experiments such as the esr experiment cited above for the homocyclooctatetraenes are difficult to refute, in several other cases the proofs are less than absolute. A related problem involving cations is discussed in Chapter 20.

9-10. FURTHER TYPES OF AROMATICITY

It is useful to look at homoaromatic ions in another way, which allows one to generalize the concept further. If the anion (459) is considered as having non-interacting bridges, the π orbitals of these may be written as shown in Fig. 9-5.

(459)

It can be seen that the allylic non-bonding level and the ethylenic anti-bonding level have the same symmetry with respect to the plane bisecting the anion, so that interaction is therefore possible which lowers the energy of the two electrons in the non-bonding level, and (homoaromatic) stabilization occurs. By the same

353. J.W. Rosenthal and S. Winstein, Tetrahedron Lett., 2683 (1970).

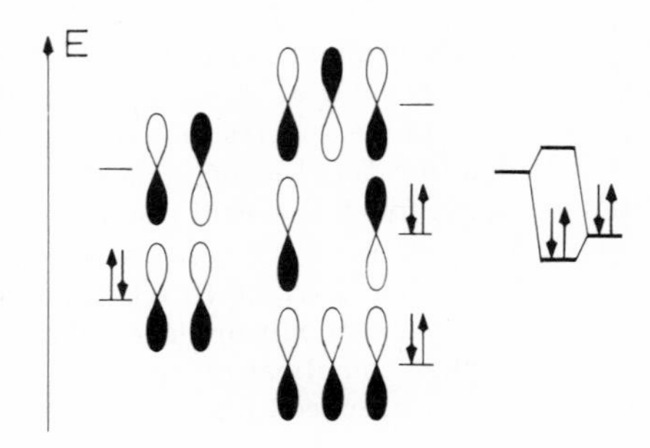

Fig. 9-5. The MO levels of a olefin and an allyl group, and their interaction.

argument (Fig. 9-6), in the anion (460) no such stabilization occurs; but in the corresponding cation it does (then the energy of the electrons in the highest filled butadiene orbital is lowered by interaction with the non-bonding level).

(460)

Goldstein [354] has extended this argument to bicyclic ions in which all three bridges are unsaturated; the resulting stabilized species are referred to as bicycloaromatic.

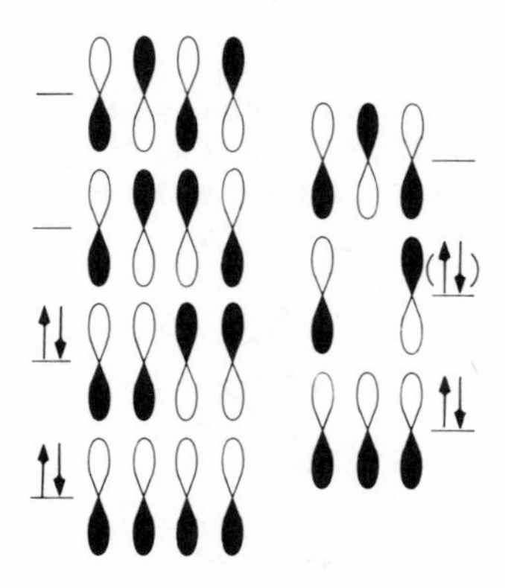

Fig. 9-6. The MO levels of a 1,3-butadiene and an allyl group.

354. M.J. Goldstein, J. Amer. Chem. Soc., 89, 6357 (1967).

Thus, in an [p.q.r] system where r is the odd bridge and where p and q are equal (and even), either the p or q bridge would lead to stabilization or neither would; if p and q are not equal, the longest one will dominate - on the reasonable assumption that it will have the lowest empty anti-bonding level and the highest filled bonding orbital. The total number of π electrons should be $(4n + 2)$ for the odd and longest even bridge; furthermore, since we wish to compare the system with a destabilized isoelectronic reference ion, it should be $4n$ π electrons overall [355].

Several examples are known. The 7-norbornadienyl cation (461) is one of them; thus (462) solvolyzes much more rapidly than any of the esters (463)-(465) (for an extensive discussion of this species, see Chapter 20).

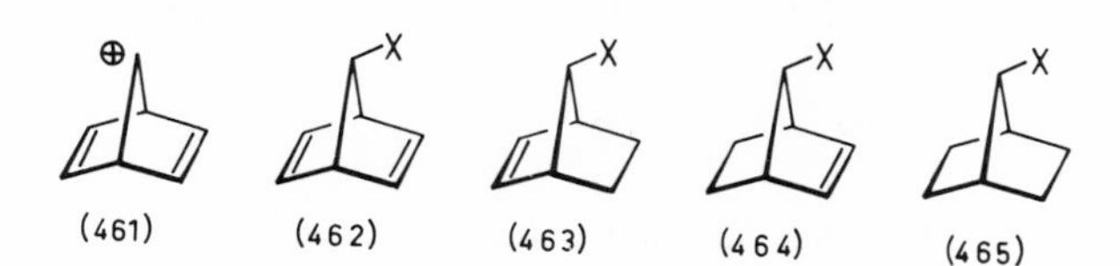

Similarly, the triene (466) undergoes base promoted D-exchange roughly 1000 times faster than related olefins, presumably via the bicycloaromatic anion; on the other hand, the triene (467) does not seem to solvolyze significantly faster than simple allylic analogs, and the cation is therefore not bicycloaromatic [355, 356].

From these arguments it will be obvious that barrelenes (468) [357] are not stabilized, and indeed, the heat of hydrogenation of such compounds is extremely high as noted earlier (p. 270).

---

355. See J.B. Grutzner and S. Winstein, J. Amer. Chem. Soc., 94, 2200 (1972), for a detailed discussion of this point.
356. (a) M.J. Goldstein and B.G. Odell, J. Amer. Chem. Soc., 89, 6356 (1967); (b) J.C. Barborak and P. v. R. Schleyer, J. Amer. Chem. Soc., 92, 3184 (1970); (c) J.B. Grutzner and S. Winstein, J. Amer. Chem. Soc., 92, 3186 (1970); (d) P. Ahlberg, J.B. Grutzner, D.L. Harris, and S. Winstein, J. Amer. Chem. Soc., 92, 3478 (1970).
357. (a) H.E. Zimmerman and R.M. Paufler, J. Amer. Chem. Soc., 82, 1514 (1960); (b) H.E. Zimmerman and G.L. Grunewald, J. Amer. Chem. Soc., 86, 1434 (1964); (c) C.G. Krespan, B.C. McKusick, and T.L. Cairns, J. Amer. Chem. Soc., 82, 1515 (1960) and 83, 3428 (1961); (d) C.G. Krespan, J. Amer. Chem. Soc., 83, 3432 (1961).

(468)

Another extension is to spiroaromaticity. When two conjugated systems are insulated by a tetrahedral atom, interaction between the termini need not be zero, and it has been predicted that stabilizing interactions should occur in such species as (469)-(471) [358].

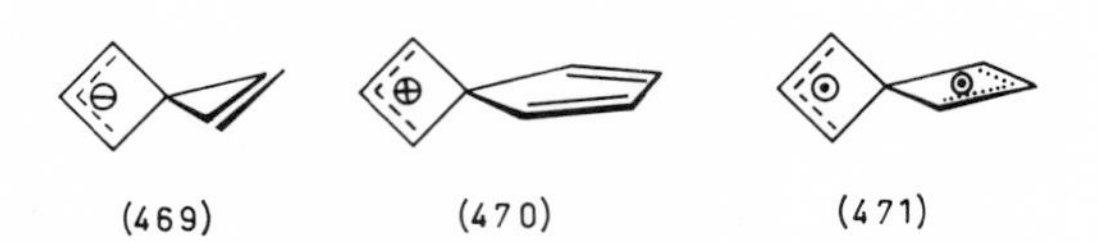

(469) (470) (471)

As yet, only spectral evidence has been quoted in support of this phenomenon [359]; the only kinetic experiment so far has shown that base promoted tritium exchange of (472) is not significantly faster than that of such analogs as (473) [360]. Finally, it may be mentioned that these and still other types of aromaticity have been generalized in a recent discussion by Goldstein and Hoffmann [361].

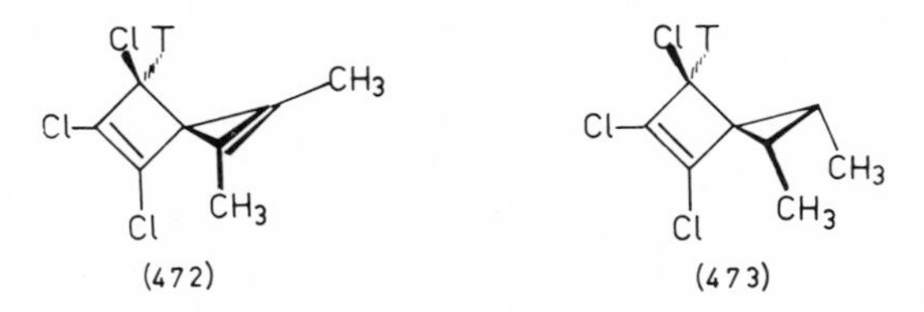

(472) (473)

358. (a) H. E. Simmons and T. Fukunaga, J. Amer. Chem. Soc., 89, 5208 (1967); (b) R. Hoffmann, A. Imamura, and G. D. Zeiss, J. Amer. Chem. Soc., 89, 5215 (1967).

359. (a) R. A. Clark and R. A. Fiato, J. Amer. Chem. Soc., 92, 4736 (1970); (b) H. Dürr and L. Schrader, Z. Naturforsch. (B), 24, 536 (1969); (c) R. Boschi, A. S. Dreiding, and E. Heilbronner, J. Amer. Chem. Soc., 92, 123 (1970); (d) F. Gerson, R. Gleiter, G. Moshuk, and A. S. Dreiding, J. Amer. Chem. Soc., 94, 2919 (1972); (e) H. Dürr, R. Sergio, and W. Gombler, Angew. Chem., Int. Ed. Engl., 11, 224 (1972); (f) H. Dürr and B. Ruge, Angew. Chem., Int. Ed. Engl., 11, 225 (1972).

360. M. F. Semmelhack, R. J. DeFranco, Z. Margolin, and J. Stock, J. Amer. Chem. Soc., 94, 2115 (1972).

361. M. J. Goldstein and R. Hoffmann, J. Amer. Chem. Soc., 93, 6193 (1971).

9-11. STRAINED AROMATIC SYSTEMS

Several types have been discussed elsewhere: The benzocyclopropenes (p. 216), the t-butyl substituted benzenes (p. 223), corannulene (p. 355), and enforced non-planarity by annelation with saturated rings (p. 257, 258). Little is known as yet about the result of still a different type of distortion, namely that of fusing an o-carborane cage to benzene. Benzocarborane (474) is known [362]; the common edge is likely to be very long (1.67 Å in o-carborane itself [363]), and possibly as a consequence the opposite edge may be quite short.

The uv spectrum is similar to that of benzene, the nmr signals are half-way between those of benzene and dienes. The compound is unusually stable, and endures bromine and hot sulfuric acid.

Some of the most interesting information concerning distorted aromatic rings comes from the field of the cyclophanes of which a large variety is now known; these include para- [364] and metaparacyclophanes [365] such as (475) and (476), heterocyclic analogs such as (477) [366] and simple phanes such as (478) [367], multiply bridged rings such as (479) [368], and multiply layered ones such as (480) [369].

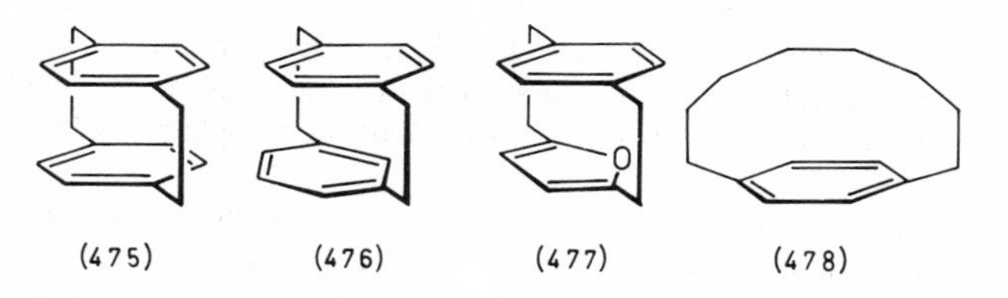

---

362. (a) N.K. Hota and D.S. Matteson, J. Amer. Chem. Soc., 90, 3570 (1968); (b) D.S. Matteson and N.K. Hota, J. Amer. Chem. Soc., 93, 2893 (1971).
363. J.A. Potenza and W.N. Lipscomb, Inorg. Chem., 5, 1471, 1478, 1483 (1966).
364. D.J. Cram and J.M. Cram, Accounts Chem. Res., 4, 204 (1971).
365. F. Vögtle and P. Neumann, Chimia, 26, 64 (1972), Angew. Chem., Int. Ed. Engl., 11, 73 (1972), and Synthesis, 85, 103 (1973).
366. D.J. Cram and G.R. Knox, J. Amer. Chem. Soc., 83, 2204 (1961).
367. V. Boekelheide and R.A. Hollins, J. Amer. Chem. Soc., 92, 3512 (1970).
368. (a) T. Otsubo, Z. Tozuka, S. Mizogami, Y. Sakata, and S. Misumi, Tetrahedron Lett., 2927 (1972); (b) see also, A.J. Hubert, J. Chem. Soc. (C), 11 (1967); (c) A.J. Hubert and M. Hubert, Tetrahedron Lett., 5779 (1966).
369. T. Otsubo, Z. Tozuka, S. Mizogami, Y. Sakata, and S. Misumi, Tetrahedron Lett., 2927 (1972).

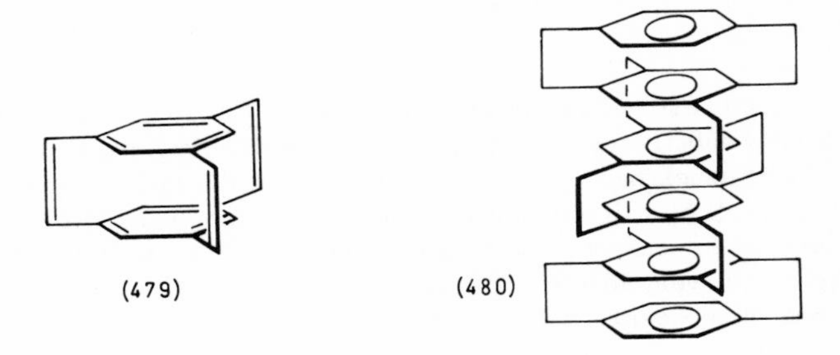
(479) (480)

Strain energy in (475) was shown by combustion to be 31 kcal/mole [370]. X-ray diffraction experiments [371] have shown the distortion in this case: $\alpha = 155^\circ$, $\beta = 11^\circ$, and $d(CH_2\text{-}CH_2) = 1.56$ Å.

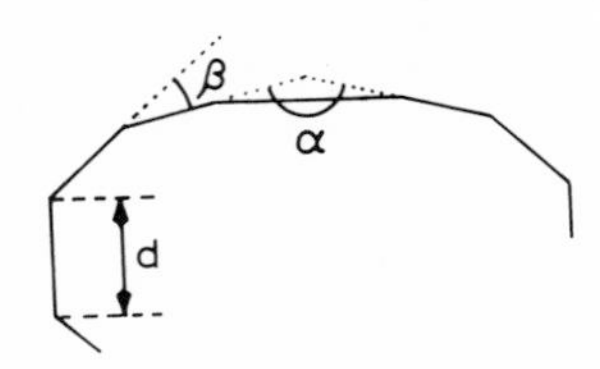

This case furthermore presents an interesting dilemma of opposed strains; the benzene bending vs the methylene eclipsing. The X-ray results show that the two rings are slightly twisted with respect to one another to avoid complete eclipsing. The p-ring is bent even more severely in (476) [372]. The uv spectra are generally subject to red shifts, and the nmr spectra show that the protons are shifted upfield, the more so the greater the strain [373]; thus, in [2.2]metacyclophane the central protons appear at 5.75 τ [365]. In (479), in fact, the aromatic protons at 3.76 τ appear well upfield relative to the vinyl hydrogen atoms (2.63 τ) [367]. Similar evidence for strain has been found in [7]paracyclophane [373a, b].

Because of their unique geometry, these molecules have been used for many special purposes. Thus, the operation of a direct field effect in chemical reactivity has been demonstrated this way; the phane dication (481) is nitrated 200 times more slowly than the open analog (482) [374], and in mono-substituted [2.2]paracyclophanes such as (483) the group already present has definite orienting ability as regards the unsubstituted ring [375].

---

370. R.H. Boyd, Tetrahedron, 22, 119 (1966).
371. C.L. Coulter and K.N. Trueblood, Acta Cryst., 16, 667 (1963).
372. K.N. Trueblood and M.J. Crisp, cited in Ref. 364 (Ref. 17).
373. D.J. Cram and R.C. Helgeson, J. Amer. Chem. Soc., 88, 3515 (1966).
373a. N.L. Allinger and T.J. Walter, J. Amer. Chem. Soc., 94, 9267 (1972); (b) A.D. Wolf, V.V. Kane, R.H. Levin and M. Jones, J. Amer. Chem. Soc., 95, 1680 (1973).
374. G. Mossa, A. Ricci, and J.H. Ridd, Chem. Commun., 332 (1971).
375. H.J. Reich and D.J. Cram, J. Amer. Chem. Soc., 90, 1365 (1968).

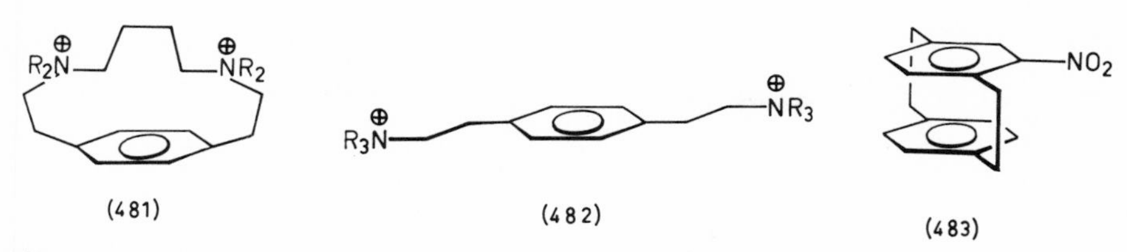

(481) (482) (483)

There are furthermore some interesting features about the mechanism of racemization of optically active cyclophanes; these are discussed in Chapter 18 (p. 648).

The helicenes were mentioned earlier (p. 155). The fact that [5] and [6]helicene could be resolved shows that a twist can be introduced to separate one layer from the next. Larger systems such as [7]- to [13]helicene ((484) [376]-(485) [377]) are known, as well as such variants as dl- (486) and the meso-isomer [378]. The benzene ring is relatively soft against twist; no great strain seems to be present in these compounds.

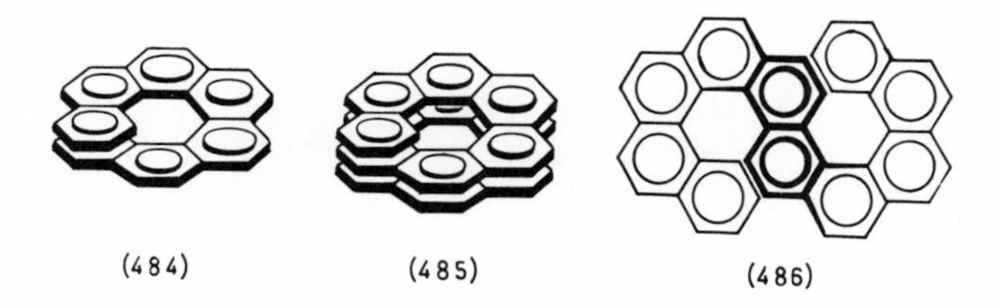

(484) (485) (486)

376. M. Flammang-Barbieux, J. Nasielski, and R.H. Martin, Tetrahedron Lett., 743 (1967).

377. R.H. Martin, G. Morren, and J.J. Schurter, Tetrahedron Lett., 3683 (1969).

378. W.H. Laarhoven and T.J.H.M. Cuppen, Tetrahedron Lett., 163 (1971).

Chapter 10

ORGANOMETALLIC COMPOUNDS

10-1. INTRODUCTION

In 1827, the Danish chemist Zeise prepared the salt that still bears his name, and which has the composition $KPtCl_3(C_2H_4) \cdot H_2O$, as well as the anhydrous compound. He described it as a derivative of ethylene, and because of that got into a controversy with Liebig, who argued that the compound must be an ethyl derivative. It did not help matters that Zeise made his arguments for the anhydrous compound, and that Liebig based his case on the hydrate [1]. While it gradually became clear that Zeise was right, the compound did not fit well into the structural theory of organic chemistry, and its structure was not really clear until ferrocene had been encountered and characterized in the early nineteen fifties. But then the stage was set for the enormous development of what is now called "organometallic" chemistry. This growth is still under way, and it is clear that it will have a tremendous impact in both organic and inorganic chemistry.

---

1. J.S. Thayer, J. Chem. Educ., 46, 442 (1969).

In this chapter we survey the several types of organometallic compounds, consider the nature of the metal-carbon bonds, and mention some of the uses to which these compounds have been put so far.

10-2. SIMPLE OLEFIN - METAL COMPLEXES

It should be mentioned at the beginning that many organometallic compounds have perfectly normal, classical bonds. Some of these may be primarily ionic in nature (such as in methylsodium), others are covalent (such as tetramethyltin), and still others have components of both types (for instance, Grignard derivatives). Furthermore, a number have three center bonds basically constructed of s and p orbitals; thus trimethylaluminum (1) is a dimer in which two carbon atoms are in bridging positions (p. 15).

(1)

Another type of bonding prevalent in organometallic compounds may be illustrated by means of the silver complex of ethylene [2]. The π electrons of the ethylene are donated to the empty orbital of silver [5s] (2); a second bond results from the "back-donation" of a pair of silver 4d electrons to the empty ethylene π^* orbital (3).

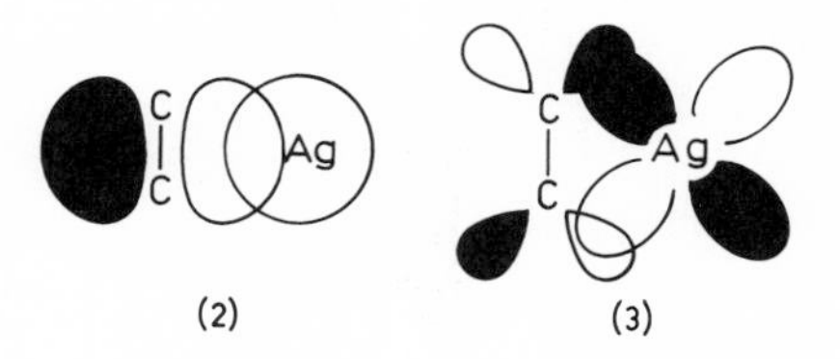

(2) (3)

This type of bonding clearly involves the face of the ethylene molecule [3]. The resultant compound is consequently often referred to as a complex (it is difficult to describe a fundamental difference between a compound and a complex; see Chapter 23).

2. (a) For review, see H.D. Kaesz, J. Chem. Educ., 40, 159 (1963); (b) G.F. Emerson, K. Ehrlich, W.P. Giering, and D. Ehntholt, Trans. N.Y. Acad. Sci., 32, 1001 (1968).
3. J. Chatt and L.A. Duncanson, J Chem. Soc., 2939 (1953).

In Zeise's salt, the donating orbital of the metal is a dsp^2 hybrid, since the configuration of the platinum is square planar (4) [4].

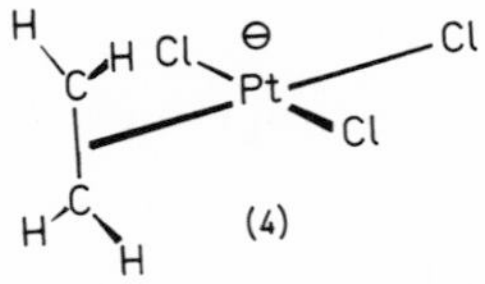

(4)

A functional group often encountered in organometallic chemistry is a metal atom bound to several carbon monoxide or trifluorophosphine molecules. The bonding to each of these is similar to that just described. An unshared sp pair of the carbon atom is delocalized to an empty metal orbital so as to form a σ bond (5); an unshared d pair of the metal is back-donated to the carbon monoxide molecule via its π^* orbital (6) (or by way of a d orbital on phosphorus).

(5) (6)

Examples of completely carbonylated metals are $Ni(CO)_4$, $Co_2(CO)_8$, $Fe(CO)_5$, $Mn_2(CO)_{10}$ and $Cr(CO)_6$. These are all very stable compounds, because in each of them the metal atom has reached inert gas configuration [5]; all of them are diamagnetic.

The olefins that can form metal complexes are not limited to simple alkenes. Conjugated dienes form many strongly bound complexes. Thus, consider tetrahapto-1,3-butadienetricarbonyliron (7).

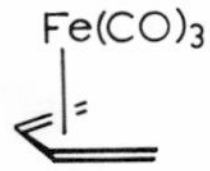

$Fe(CO)_3$

(7)

Iron itself has the following electron configuration.

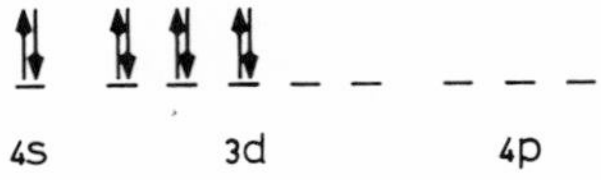

4s 3d 4p

4. J.A. Wunderlich and D.P. Mellor, Acta Cryst., 8, 57 (1955).
5. C.A. Tolman, Chem. Soc. Rev., 1, 337 (1972).

Again we note that maximum stability is achieved if the iron atom can reach the filled inert gas configuration $4s^2 3d^{10} 4p^6$, and if as many of these pairs as possible can be used as bonding pairs [5]. We begin by transplanting the 4s pair to an empty d orbital and by hybridizing the 4s and 4p orbitals; three of the filled d orbitals and three of the empty sp^3 orbitals are then used for binding the carbon monoxide ligends in the manner just described. This leaves us with one filled and one empty d orbital, and with the remaining empty hybrid orbital. It must at this point be kept in mind that the butadiene orbitals have the symmetry shown.

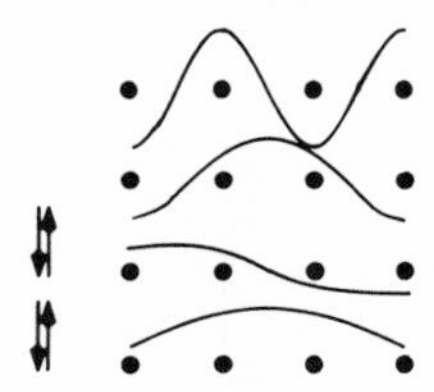

The simplest possibility now would be the donation of the π bonding pairs to the sp^3 and d orbitals, and back-donation of a d pair to the highest π^* orbital.

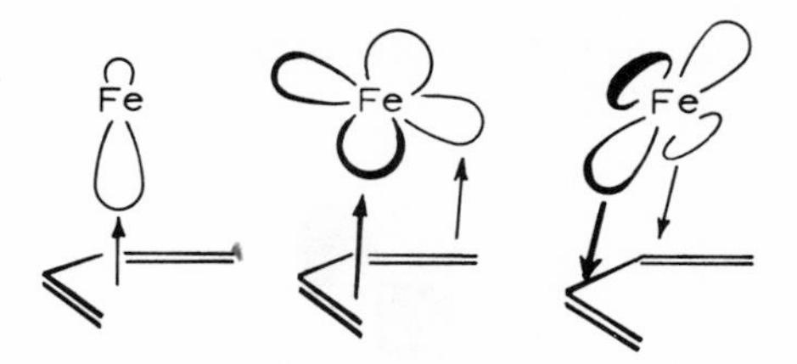

As a result all of the metal orbitals are filled with bonding pairs.

A simple extension of these ideas leads to the hexahaptohexa-1,3,5-triene-tricarbonylmolybdenum (8). A little practice and a glance at the periodic table will suffice to design reasonable bonding schemes for the many new molecules recently described in the literature (cf. Section 3-10, and the review articles quoted above).

$Mo(CO)_3$

(8)

10-3. COMPLEXES WITH CYCLIC MOLECULES

It was discovered in 1951 that the reaction of cyclopentadienylmagnesium bromide with ferric chloride [6], and in 1952 that of cyclopentadiene with iron [7] gave a beautiful, orange compound $C_{10}H_{10}Fe$ of great thermal and chemical stability: It was found that it can be heated without decomposition to 500°. The substance, soon afterwards named ferrocene, undergoes typically aromatic substitution reactions [8]; its structure [9] was shown to be symmetrical, with the iron sandwiched between two cyclopentadienyl rings (9) [10].

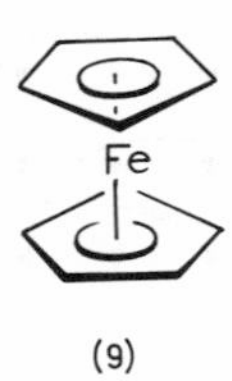

(9)

The bonding can be visualized in terms of the orbitals of the ferrous ion and the cyclopentadienide orbitals, e.g., as shown below; the 12 π electrons of the cyclopentadienide rings are enough to achieve the inert gas configuration with the ferrous ion (naturally back-donation to π^* orbitals helps achieve strong bonding and a nonpolar nature within the molecule).

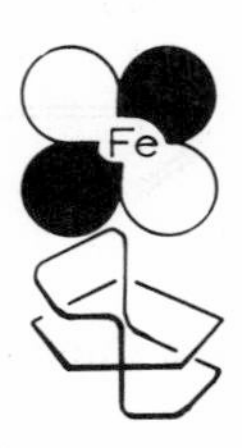

The stabilization of the ring that is thereby realized may be gauged from the fact that the protonation of ferrocene does not lead to coordinated or free cyclopentadiene, but to the complex (10), in which the lone proton appears at about 12 τ in the nmr spectrum and the other ten as a doublet at 5.0 τ [11].

---

6. T. J. Kealy and P. L. Pauson, Nature, 168, 1039 (1951).
7. S.A. Miller, J.A. Tebboth, and J.F. Tremaine, J. Chem. Soc., 632 (1952).
8. R.B. Woodward, M. Rosenblum, and M.C. Whiting, J. Amer. Chem. Soc., 74, 3458 (1952).
9. For review, see M.D. Rausch, J. Chem. Educ., 37, 568 (1960).
10. J.D. Dunitz, L.E. Orgel, and A. Rich, Acta Cryst., 9, 373 (1956).
11. T.J. Curphey, J.O. Santer, M. Rosenblum, and J.H. Richards, J. Amer. Chem. Soc., 82, 5249 (1960).

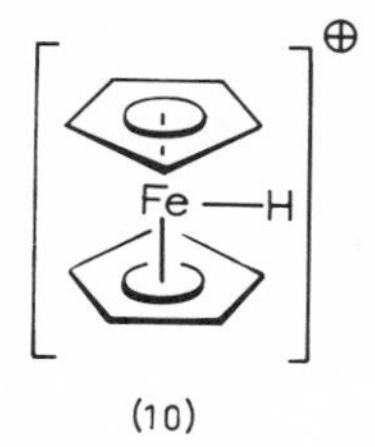

(10)

A related phenomenon is that the iron atom may function as the base in hydrogen bonds. When alcohols such as (11) in dilute carbon tetrachloride solution are examined by ir, three OH bonds can be observed: Free OH, OH hydrogen-bonded to the π system of the ring, and OH hydrogen-bonded to the metal. The assignments were made on the basis of the fact that one of the latter two bonds is absent in the *endo*-alcohol (12), and the other in the epimer (13) [12].

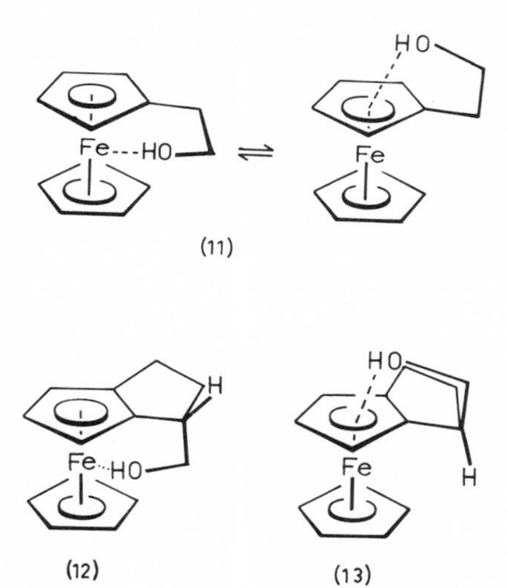

(11)

(12) (13)

As noted however, electrophilic species do attack the five-membered rings and ferrocene is subject to the gamut of aromatic substitution reactions [8]. Furthermore, many other metal atoms may occupy the position between the rings, and a rich and varied chemistry has grown up around these metallocenes [9, 13].

Shortly after the initial discoveries concerning ferrocene, it was predicted on the basis of analogy by Longuet-Higgins and Orgel [14] that cyclobutadiene should likewise be able to interact with and be stabilized by certain transition metal atoms

---

12. D. S. Trifan and R. Bacskai, *J. Amer. Chem. Soc.*, *82*, 5010 (1960).
13. K. Plesske, *Angew. Chem., Int. Ed. Engl.*, *1*, 312, 394 (1962).
14. H. C. Longuet-Higgins and L. E. Orgel, *J. Chem. Soc.*, 1969 (1956).

to form such "sandwich compounds". Shortly thereafter several examples were found [15]. It was also found by means of X-ray diffraction that complexes such as (14) had a square planar carbon skeleton [16]; furthermore they are subject to a variety of simple aromatic substitution reactions [17], and in fact, some substituent effects have been described [18].

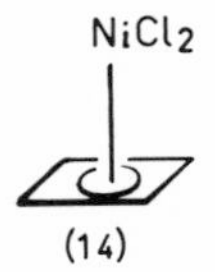

(14)

The iron tricarbonyl complex is of special interest in that it can be used to set free cyclobutadiene itself, either by oxidation with ceric ion [19] or by flash thermolysis [20] (p. 860). The cyclopropenium complex of nickel (15) (an "open-faced sandwich") was recently described as well [21]. Again, the stability found in most of these compounds derives from the filling of the metal p and d orbitals with electrons that are also used for bonding.

Br

Ni

CO

CO

(15)

Benzene and many of its derivatives are especially apt to form chromium complexes such as (16) and (17) [22].

15. (a) R. Criegee and G. Schröder, Ann., 623, 1 (1959); (b) W. Hübel and E.H. Braye, J. Inorg. Nucl. Chem., 10, 250 (1959).
16. J.D. Dunitz, H.C. Mez, O.S. Mills, and H.M.M. Shearer, Helv. Chim. Acta, 45, 647 (1962).
17. (a) Chem. Eng. News, 43, 38, Aug. 23, 1965; (b) J.D. Fitzpatrick, L. Watts, G.F. Emerson, and R. Pettit, J. Amer. Chem. Soc., 87, 3254 (1965).
18. For review, see R. Pettit, Pure Appl. Chem., 17, 253 (1968).
19. L. Watts, J.D. Fitzpatrick, and R. Pettit, J. Amer. Chem. Soc., 87, 3253 (1965).
20. E. Hedaya, I.S. Krull, R.D. Miller, M.E. Kent, P.F. D'Angelo, and P. Schissel, J. Amer. Chem. Soc., 91, 6880 (1969).
21. W.K. Olander and T.L. Brown, J. Amer. Chem. Soc., 94, 2139 (1972).
22. E.O. Fischer and W.Z. Hafner, Z. Naturforsch. (B), 10, 665 (1955).

The profound effect that metal complexation may have on an aromatic ring is nicely demonstrated by the paracyclophanes. An interannular effect was found; bis-complexes (18) cannot be formed unless n is at least six [23].

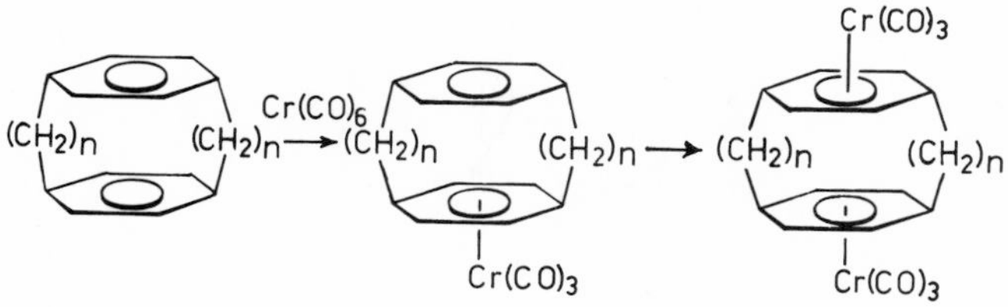

$(CH_2)_n$ $(CH_2)_n$ $\xrightarrow{Cr(CO)_6}$ $(CH_2)_n$ $(CH_2)_n$ (with $Cr(CO)_3$) $\longrightarrow$ $(CH_2)_n$ $(CH_2)_n$ (with two $Cr(CO)_3$)

(18)

Still further variations are the "mixed sandwich compounds" such as (19) [24]; again the total number of π electrons in all of these equals the number of vacancies in the metal p and d orbitals. Something of a novelty is the recently reported "club sandwich" (20) [25].

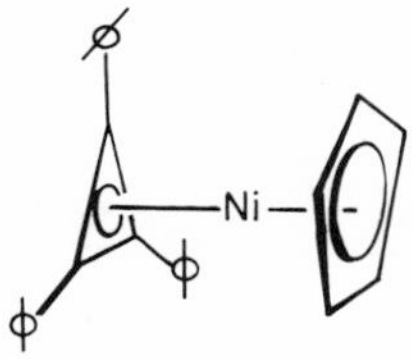

(19) 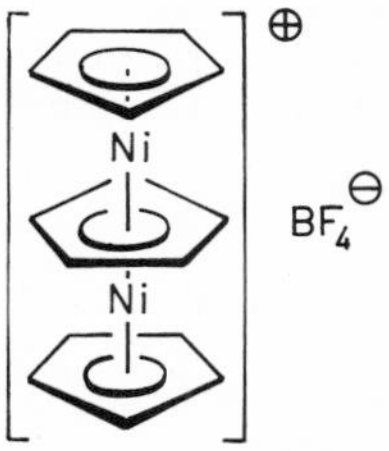 (20) $BF_4^{\ominus}$

One aspect of the chemistry of these mixed sandwich compounds is that one can study the competition of the rings for various reagents. Thus, it was learned that the ring stability to nucleophilic attack decreases in the order of ring size 5>4>6>7 [26]; for instance

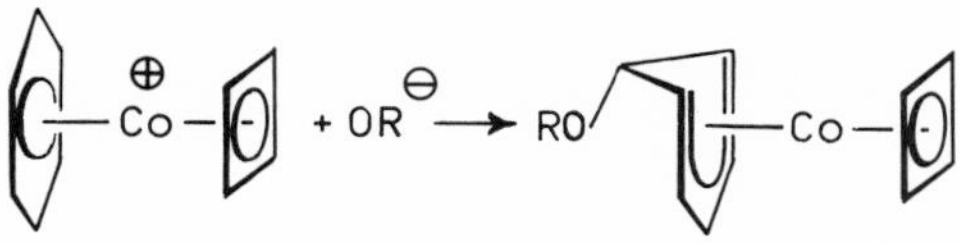

$$\text{Co}^{\oplus} + OR^{\ominus} \longrightarrow RO\text{–}\ldots\text{–Co}$$

23. D. J. Cram and D. I. Wilkinson, J. Amer. Chem. Soc., 82, 5721 (1960).
24. M. D. Rausch, R. M. Tuggle, and D. L. Weaver, J. Amer. Chem. Soc., 92, 4981 (1970).
25. H. Werner and A. Salzer, Inorg. Metalorg. Chem., 2, 239 (1972).
26. A. Efraty and P. M. Maitlis, Tetrahedron Lett., 4025 (1966).

An additional dimension has recently been added to the field with the discovery that some of the complex boranes can participate as a ligand; thus, compounds such as (21) were reported [27].

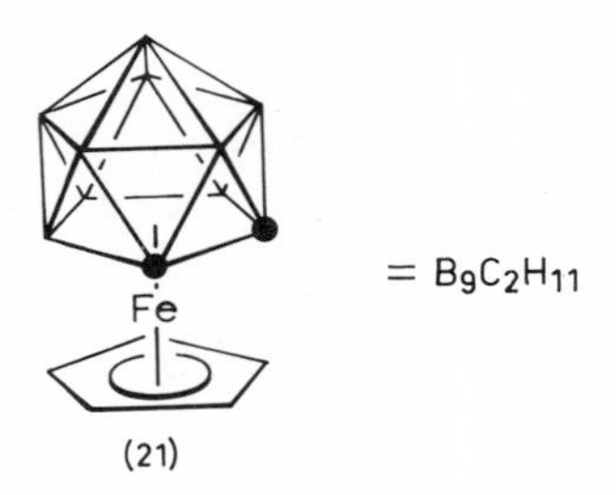

In 1968, some seventeen years after the first reports of ferrocene, Streitwieser had "one of those rare but delightful flashes of insight" [28] that it should be possible to prepare similar compounds from cyclooctatetraene dianion and uranium (IV): The twenty π electrons would be enough to fill the 5f and 7p subshells. One of the possible orbital interactions (between one of the two highest occupied orbitals of the hydrocarbon with an empty f orbital of the uranium) is shown below.

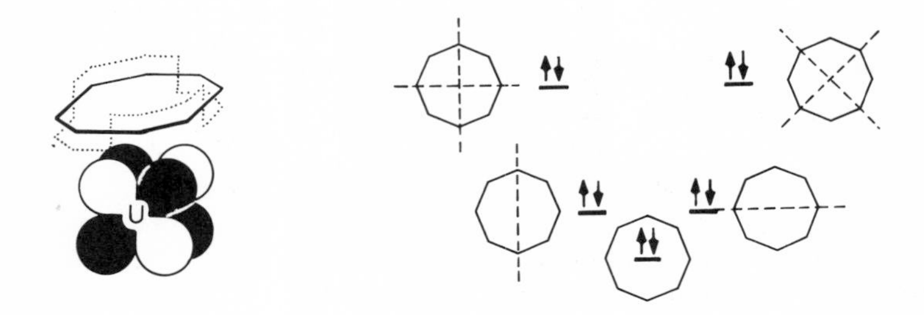

The complex uranocene, $U(C_8H_8)_2$ was indeed obtained [29] as a thermally stable green solid, shown later to be of the sandwich type by means of nmr [30] and X-ray diffraction [31]. Several other such compounds are known now; they include neptunocene (22) [32] and thoracene (23) [33].

---

27. M. F. Hawthorne, Pure Appl. Chem., 17, 195 (1968).
28. Chem. Eng. News, 47, 36, Jan. 6, 1969.
29. A. Streitwieser and U. Müller-Westerhoff, J. Amer. Chem. Soc., 90, 7364 (1968).
30. A. Streitwieser, D. Dempf, G. N. La Mar, D. G. Karraker, and N. Edelstein, J. Amer. Chem. Soc., 93, 7343 (1971).
31. K. O. Hodgson, D. Dempf, and K. N. Raymond, Chem. Commun., 1592 (1971).
32. D. G. Karraker, J. A. Stone, E. R. Jones, and N. Edelstein, J. Amer. Chem. Soc., 92, 4841 (1970).
33. A. Streitwieser and N. Yoshida, J. Amer. Chem. Soc., 91, 7528 (1969).

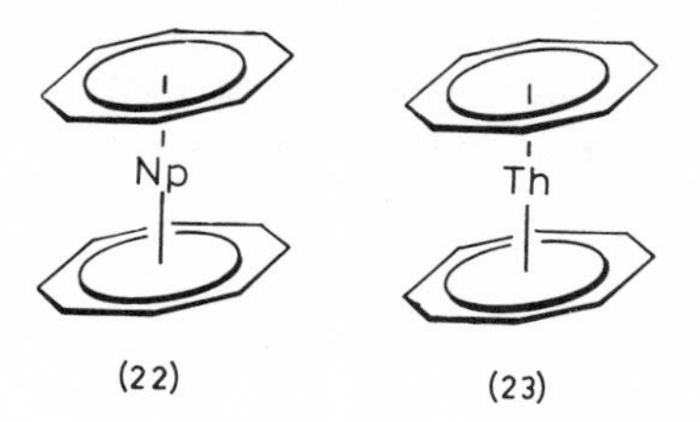

(22) (23)

10-4. STABILIZATION BY METAL COMPLEXATION

The combination of having potentially electrophilic empty orbitals filled and potentially nucleophilic electrons shared, stabilizes molecules complexed with transition metals. Actually, a large number of molecules too unstable to isolate when alone are unreactive in combination with a suitable metal atom. Of course, it will be realized that this stabilization alters the properties of such molecules a great deal and a study of these properties may not mean very much in terms of the non-complexed molecule.

A number of examples will be encountered in the last several chapters which deal with reactive intermediates, but a few will be mentioned here. Surely the most spectacular one is that of cyclobutadiene, as noted in the preceding section. Pentalene and its various derivatives and analogs can be so stabilized; thus, (octahapto-1-dimethylaminopentalene)-μ-carbonyltetracarbonyldiiron (24) has been described as a stable solid [34].

$(OC)_2Fe$ — $Fe(CO)_2$
C=O

(24)

As-indacene can be isolated as the "hero-type" of complex (25) [35] or its geometric isomer.

Fe Fe

(25)

34. D. F. Hunt and J.W. Russell, *J. Amer. Chem. Soc.*, **94**, 7198 (1972).
35. T. J. Katz and J. Schulman, *J. Amer. Chem. Soc.*, **86**, 3169 (1964).

An interesting example is the complexation of allenes with platinum. The smallest known and isolable cyclic allene is 1,2-cyclononadiene; in attempts to prepare the lower homologs only dimers are obtained. In the presence of bis(triphenylphosphine)ethyleneplatinum those reactions however give rise to the complexes (26)-(28), all of which are thermally quite stable.

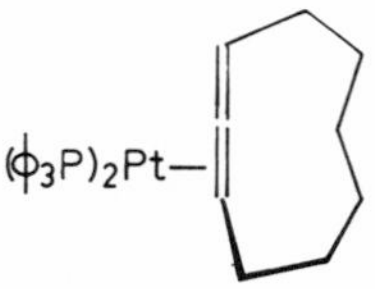

(26) 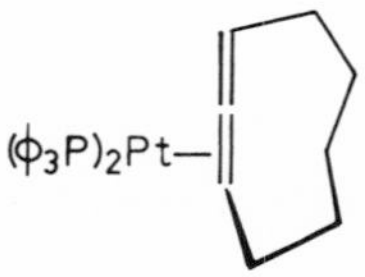 (27) 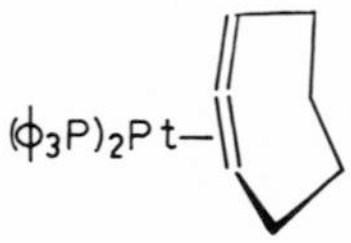 (28)

The allenes can normally be displaced from the complexes by means of carbon disulfide. The cyclononadiene can thus be obtained. Free 1,2-cyclooctadiene can be obtained in solution at -60°; it is sufficiently long lived under these conditions to permit the measurement of its spectra; at higher temperatures it dimerizes rapidly. The 1,2-cycloheptadiene complex can be decomposed only at higher temperatures, and only the dimer is then formed [36].

An example of another sort was found by Emerson [2b, 37], who reported the isolation of the stable iron tricarbonyl complex (29) of trimethylenemethane.

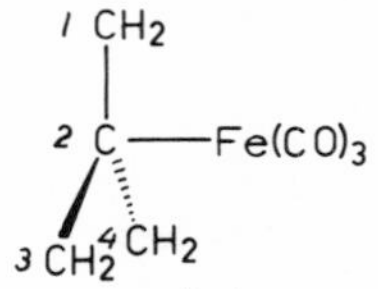

(29)

The secular determinant for this hydrocarbon is written as follows:

$$\begin{vmatrix} X & 1 & 0 & 0 \\ 1 & X & 1 & 1 \\ 0 & 1 & X & 0 \\ 0 & 1 & 0 & X \end{vmatrix} = 0$$

36. J.P. Visser and J.E. Ramakers, Chem. Commun., 178 (1972).
37. (a) G.F. Emerson, K. Ehrlich, W.P. Giering, and P.C. Lauterbur, J. Amer. Chem. Soc., 88, 3172 (1966); (b) K. Ehrlich and G.F. Emerson, J. Amer. Chem. Soc., 94, 2464 (1972).

It is readily solved and shown to lead to a degenerate pair of non-bonding orbitals; this molecule should therefore by an excellent candidate for stabilization by means of metal complexation. It had previously been recognized as an unstable triplet state in a low temperature matrix (obtained by depositing the thermolysis product of 4-methylene-1-pyrazoline (30); or by the photolysis of 3-methylenecyclobutanone (31) (see also p. 651) [38].

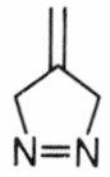

(30)  (31)

Among the further examples of such stabilization one may mention complexes of tetramethyleneethane (32) [39], cyclopentadienone (33) [40], and *o*-xylylene (34) [41].

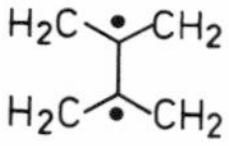

(32)  (33) 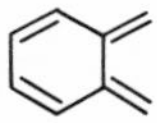 (34)

10-5. FLUXIONAL COMPLEXES AND BOND LOCALIZATION [42]

When the conjugated system contains more electrons than the metal can accommodate, the excess will be ignored. In 1,6-methanocyclodecapentaene for instance, tricarbonylchromium complexation leads to an adduct in which the methylene protons appear as a quartet in the nmr, showing that the complex is not symmetrical (35) [43].

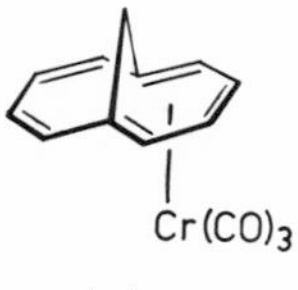

(35)

38. P. Dowd, J. Amer. Chem. Soc., 88, 2587 (1966).
39. A. Nakamura, Bull. Chem. Soc. Japan, 39, 543 (1966).
40. M.L.H. Green, L. Pratt and G. Wilkinson, J. Chem. Soc., 989 (1960).
41. W.R. Roth and J.D. Meier, Tetrahedron Lett., 2053 (1967).
42. For review, see F.A. Cotton, Chem. Brit. 4, 345 (1968) and Accounts Chem. Res., 1, 257 (1968).
43. E.O. Fischer, H. Rühle, E. Vogel, and W. Grimme, Angew. Chem., Int. Ed. Engl., 5, 518 (1966).

In some cases the metal is able to shift its position alternately to interact with different portions of the molecule. The two trieneirontricarbonyl complexes (36) and (37) can be prepared, and slow interconversion occurs ($K \approx 1$) at 125° [44].

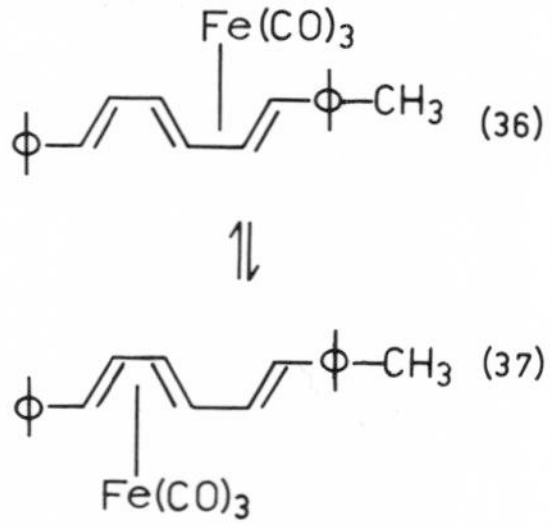

(36)

(37)

When the double bonds are of the cis-configuration, this process appears to be faster; thus, the two tropone complexes (38) and (39) interconvert measurably rapidly at 80° [45].

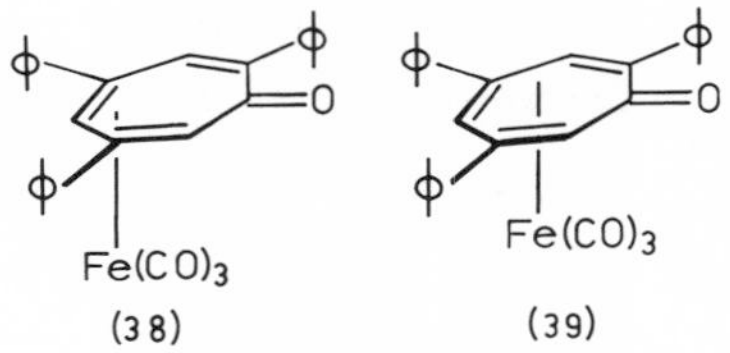

(38) (39)

The isomerization becomes still easier if the system is cyclic and completely conjugated. Thus, the cyclooctatetraene irontricarbonyl complex (40) shows a single peak in the nmr spectrum, and no broadening is observed upon cooling to -60° [46]; however, X-ray diffraction shows that the metal does interact with only a diene portion of the hydrocarbon [47], so that extremely rapid shifts must be occurring.

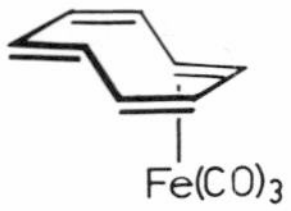

(40)

---

44. H.W. Whitlock and Y.N. Chuah, J. Amer. Chem. Soc., 87, 3605 (1965).
45. E.H. Braye and W. Hübel, J. Organometal. Chem., 3, 25 (1965).
46. A. Davison, W. McFarlane, L. Pratt, and G. Wilkinson, Chem. Ind. (London), 553 (1961).
47. B. Dickens and W.N. Lipscomb, J. Chem. Phys., 37, 2084 (1962).

Organometallic molecules of this sort are now referred to as fluxional molecules; they represent a special group of valence tautomers, a class of compounds further discussed in Chapter 11. As a further example we may discuss the compound $Fe(CO)_2(C_5H_5)_2$. At -100°, the nmr spectrum shows a singlet at 6.5 τ, a singlet at 5.6 τ and an A_2B_2 quartet centered at 3.7 τ, in an overall ratio of 1:5:4; the structure is therefore (41), with one π-bonded and one σ-bonded ring; hence it should be formally called (h^5-cyclopentadienyl) (h^1-cyclopentadieyl)dicarbonyliron.

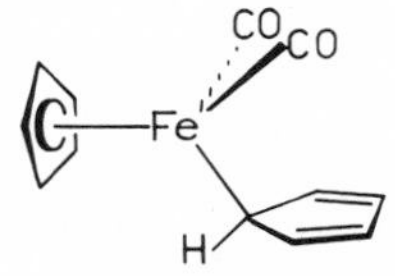

CO CO Fe H

(41)

Coalescence occurs on warming, and at room temperature only two signals remain in a 1:1 ratio at 4.3 and 5.6 τ. Evidently the σ-bonded ring is then undergoing very rapid 1,2 shifts [42].

Very complex behavior is sometimes encountered this way. 1,3,5,7-Tetramethylcyclooctatetraenemolybdenumtricarbonyl at -10 τ shows an eight line (unsplit) spectrum, as expected for structure (42).

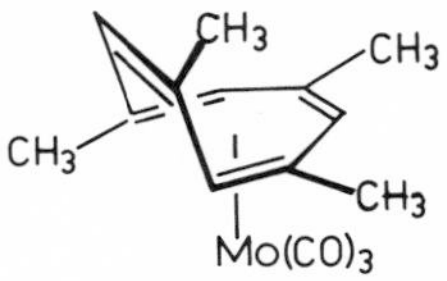

CH_3 CH_3 CH_3 CH_3 $Mo(CO)_3$

(42)

At somewhat higher temperatures spectral changes occur indicative of 1,2-shifts; as the temperature is increased further, transannular jumps begin to be observable, and when the temperature reaches +80°, still another motion - as yet unidentified - becomes important [48].

If the excess π-electrons are part of a completely delocalized system, complexation may localize them (cf. also the discussion of the Mills-Nixon effect, p. 258). The evidence on this point is primarily based on X-ray diffraction data. Thus, in the benzocyclobutadieneirontricarbonyl, (43) the four membered ring is nearly exactly square, with all four sides 1.47 ± 0.02 Å long.

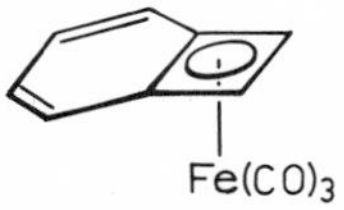

$Fe(CO)_3$

(43)

---

48. Chem. Eng. News, 44, 58, Oct. 10, 1966, and references given there.

The benzene ring now resembles cyclohexatriene more than benzene: The bonds drawn as double in (43) have a length of about 1.34 ± 0.02 Å, and the singly drawn bond between them has a length of 1.48 Å [49]. This shows how greatly the cyclobutadiene ring was stabilized by complexation; the X-ray data stand in stark contrast with those of free biphenylene, in which the aromatic rings are nearly hexagonal and the connecting bonds have essentially single-bond length [50].

Similar observations have been made for the complexed vinyl-substituted benzenes (44)-(46); their structures and the most interesting bond length values (in Å) are shown below [49].

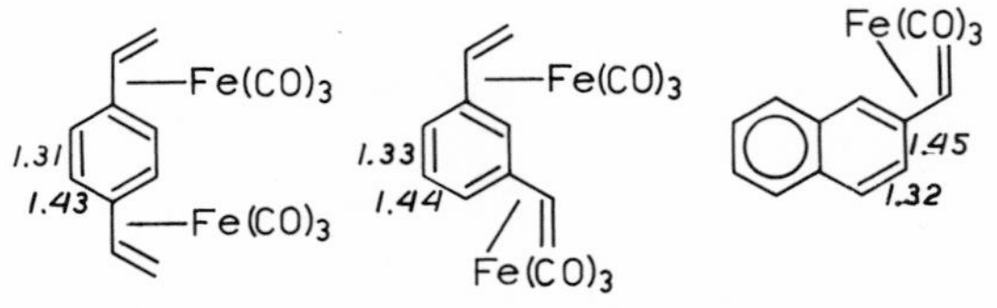

$Fe(CO)_3$ $Fe(CO)_3$ 1.31 1.43 (44)

$Fe(CO)_3$ $Fe(CO)_3$ 1.33 1.44 (45)

$Fe(CO)_3$ 1.45 1.32 (46)

In a related development, it was shown that the photolysis of a mixture of iron-pentacarbonyl and m-isopropenyltoluene produces the two isomeric and separable complexes (47) and (48), in which - as the authors put it [51] - the Kekulé structures had been trapped.

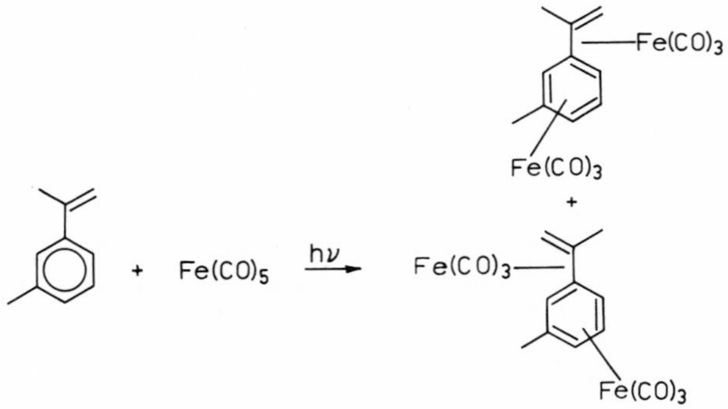

$+ Fe(CO)_5 \xrightarrow{h\nu}$ (47) [$Fe(CO)_3$, $Fe(CO)_3$] + (48) [$Fe(CO)_3$, $Fe(CO)_3$]

49. R.E. Davis and R. Pettit, J. Amer. Chem. Soc., 92, 716 (1970).
50. J.K. Fawcett and J. Trotter, Acta Cryst., 20, 87 (1966).
51. R. Victor, R. Ben-Shoshan, and S. Sarel, Chem. Commun., 1680 (1970).

10-6. APPLICATIONS IN ORGANIC CHEMISTRY

The applications are of course far too numerous for a complete listing, and only a few examples can be given here.

An obvious use is that complexation can be used in separation processes. Thus, olefins can often be successfully separated by gas chromatography if the stationary phase has been treated with silver nitrate. A variety of bond strengths has been found for the olefin-silver ion complex, with low molecular weight olefins more strongly bound than the higher homologs, terminal olefins more strongly than the branched isomers, the *cis*-olefins more strongly than the *trans*, and so on [52].

Resolution of optically active olefins and allenes has been achieved by means of platinum complexes. For instance, when racemic 1,2-cyclononadiene is treated with platinum chloride complexed with optically active α-phenylethylamine, a mixture of diastereomeric complexes is formed which can be separated by crystallization. The allenes are then liberated by means of displacement by cyanide ion [53].

$$\underset{d,l}{\text{(1,2-cyclononadiene)}} + \underset{d'}{Cl_2Pt{-}NH_2CH(\phi)CH_3} \longrightarrow \underset{d'd\,+\,d'l}{Cl_2Pt(NH_2CH(CH_3)\phi)(C_9H_{14})}$$

This technique was first devised by Cope for the resolution of *trans*-cyclooctene [54]. A recent refinement has been use of ethylene at 20 atm as a displacing agent. Since at low pressure ethylene is in turn displaced by other olefins, the reaction sequence is reversible and the platinum complex can be used repeatedly [55].

A truly elegant application in structure determination was devised by Winstein to help solve the problem of the structure of homotropylium ion. As may be recalled from the discussion of p. 361, the difficulty is to distinguish structures (49)

---

52. M.A. Muhs and F.T. Weiss, *J. Amer. Chem. Soc.*, *84*, 4697 (1962).
53. A.C. Cope, W.R. Moore, R.D. Bach, and H.J.S. Winkler, *J. Amer. Chem. Soc.*, *92*, 1243 (1970).
54. (a) A.C. Cope, K. Banholzer, H. Keller, B.A. Pawson, J.J. Whang, and H.J.S. Winkler, *J. Amer. Chem. Soc.*, *87*, 3644 (1965); (b) A.C. Cope, C.R. Ganellin, H.W. Johnson, T.V. Van Auken, and H.J.S. Winkler, *J. Amer. Chem. Soc.*, *85*, 3276 (1963).
55. R. Lazzaroni, P. Salvadori, and P. Pino, *Tetrahedron Lett.*, 2507 (1968).

and (50). While a very large difference was found between the "inside" and "outside" methylene protons, there is no apparent way of knowing how large this difference should be for structure (50).

(49) (50)

Complexation with molybdenum- and irontricarbonyl provides the answer [56], since the former metal can accommodate 12 extra electrons and the latter only 10. Dissolution of the complexes (51) and (52) in acid gives solutions of the ions (53) and (54); in the former the methylene protons are found at 6.6 and 10.2 τ, in the latter at 8.5 and 8.7 τ. In the chromiumtricarbonyl complex (55), they again appear at 6.5 and 11 τ, and the conclusion should be clear [57].

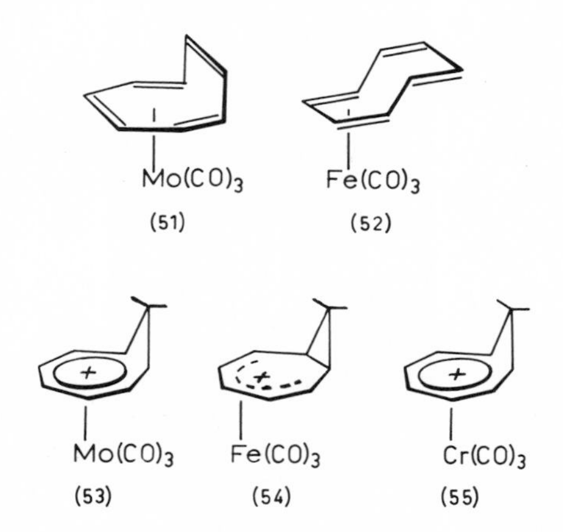

(51) (52)

(53) (54) (55)

Until recently, the reaction

surely would have seemed a most improbable event, to be encountered only on sophomore exam papers. Yet this reaction - called olefin metathesis - can be brought about by means of a complex catalyst of which one of the vital components

---

56. S. Winstein, H.D. Kaesz, C.G. Kreiter, and E.C. Friedrich, J. Amer. Chem. Soc., 87, 3267 (1965).
57. R. Aumann and S. Winstein, Tetrahedron Lett., 903 (1970).

is tungsten hexachloride. While the details of the reaction are likely to remain hidden for some time [58], it seems clear that the reaction must involve reversible complexation steps and a loss of memory on the part of the complex as to which carbon atoms were joined to which.

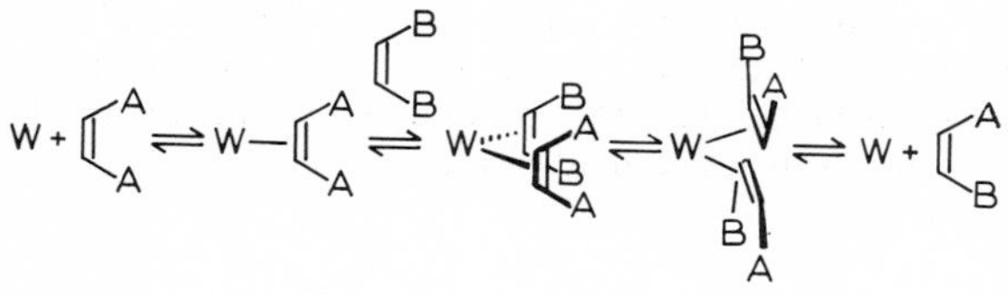

An immediate application of this reaction has been the one step conversion of large cyclic olefins into catenanes (Section 4-4) - another development the prediction of which would have produced benign smiles at best not long ago. The catalyzed trimerization of acetylenes to benzenes and the tetramerization to cyclooctatetraenes are likewise reactions that must surely involve metal complexes [59]. The intervention of acetylene-chromium complexes has been proved (by isolation) [60]. The further question might be raised whether cyclobutadiene complexes are also intermediates.

With chromium catalysis this does not seem to be the case; thus, when the labeled 2-butyne (56) is allowed to trimerize, no (57) is obtained [61].

$CH_3C{\equiv}CCD_3$

(56) (57)

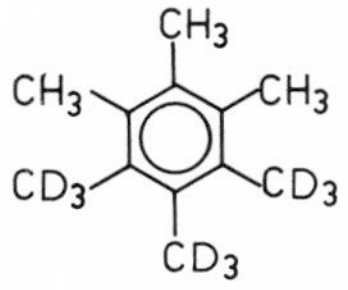

On the other hand, in the presence of aluminum chloride (57) is obtained and a related intermediate - the Diels-Alder adduct (58) - can be isolated.

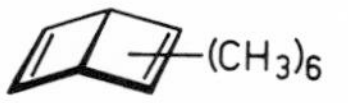

$(CH_3)_6$

(58)

---

58. (a) N. Calderon, Accounts Chem. Res., 5, 127 (1972); (b) Chem. Eng. News, 45, 51, (Sept. 25, 1967).
59. (a) H. Zeiss in "Organometallic Chemistry", H. Zeiss, Ed., American Chemical Society Monograph No. 147, Reinhold, New York, 1960, p. 380; (b) P.M. Maitlis, Advan. Organometal. Chem., 4, 95 (1966).
60. W. Herwig, W. Metlesics, and H. Zeiss, J. Amer. Chem. Soc., 81, 6203 (1959).
61. (a) G.M. Whitesides and W.J. Ehmann, J. Amer. Chem. Soc., 90, 804 (1968); (b) H. Dietl, H. Reinheimer, J. Moffat, and P.M. Maitlis, J. Amer. Chem. Soc., 92, 2276 (1970).

Still another consequence of the developments in organometallic chemistry is that some of the mystery may be disappearing from the host of heterogeneous catalysts used in various processes such as hydrogenation, olefin dimerization, nitrogen fixation, and so on [62]. It is gradually becoming clear that many of these reactions can also be promoted by homogeneous catalysis, and in those cases the mechanism is more amenable to study. The key to the operation of these catalysts is a vacant coordination site which may be used to bind one of the reagents which can then be transferred to the substrate. One of the first such compounds to be used was Vaska's compound (59) [63], an iridium complex that readily adds hydrogen, acetyl chloride, sulfonyl chlorides and so on.

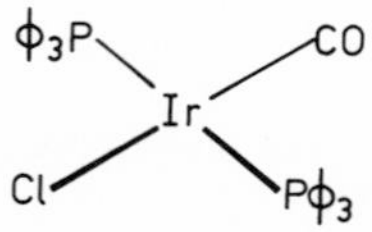

ϕ_3P CO
Ir
Cl Pϕ_3

(59)

Minor differences in metal and ligands produce vital differences in reactivity, and one needs no great imagination to foresee a bright future in this area.

62. (a) J.P. Collman, *Accounts Chem. Res.*, *1*, 136 (1968); (b) R.E. Harmon, S.K. Gupta, and D.J. Brown, *Chem. Rev.*, *73*, 21 (1973).
63. L. Vaska and J.W. DiLuzio, *J. Amer. Chem. Soc.*, *84*, 679 (1962).

Chapter 11

VALENCE ISOMERIZATION

11-1. INTRODUCTION

It is somewhat humiliating for an author to begin the discussion of a new topic with the admission that he is unable to define it. All attempts at a precise definition of valence isomerism either lead to the inclusion of isomers which are generally not so regarded, or to the exclusion of examples (such as the bond-shift isomers, cf. p. 283, 285) that belong in. This somewhat vague area has nevertheless attracted the energies and ingenuities of many organic chemists, and it is included here because their results are interesting.

A general description might be arrived at by saying that compounds are valence isomers if they can in principle be interconverted via Cope rearrangements. This rearrangement is in general terms described by:

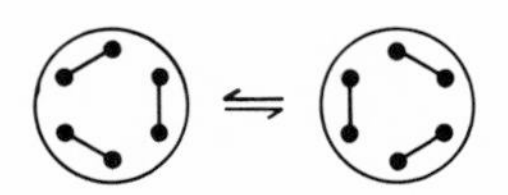

The surrounding circles imply that we are talking about single molecules. It is not necessary that all the centers involved be carbon atoms. The three lines drawn may represent pairs of either σ- or π electrons (in practice however, at least one should be a π pair; labeled hexahomobenzene (1) [1] does not isomerize under any known conditions, for example [2]).

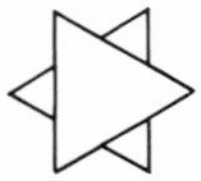

(1)

The centers need not be arranged in a regular hexagon; in fact, they need not all lie in one plane. Their number need not be six; thus, one or more of the pairs could be unshared. These rearrangements need not lead to chemically new species (they may be "degenerate"). An important characteristic is that the hydrogen atoms do not become separated from the carbon atoms holding them in the course of these rearrangements.

The phrase, "in principle" implies that the interconversion need not be demonstrable: The activation energy might be too high. If the activation energy is low enough so that an appreciable rate results at room temperature, the words isomerization and isomers are replaced by tautomerization and tautomers. If the activation energy is negative, as is the case in benzene and the homoaromatic species, a single intermediate structure results because of resonance (p. 241).

The difficulty with this description is that it leaves more room than intended. Thus, while prismane is generally considered a valence isomer of benzene because of the Cope rearrangements,

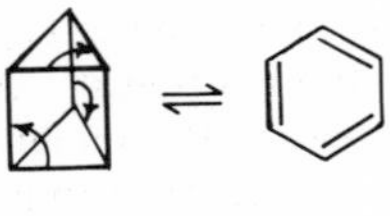

m- and *p*-substituted benzene isomers would also have to be included because they are in principle interconvertible via such rearrangements:

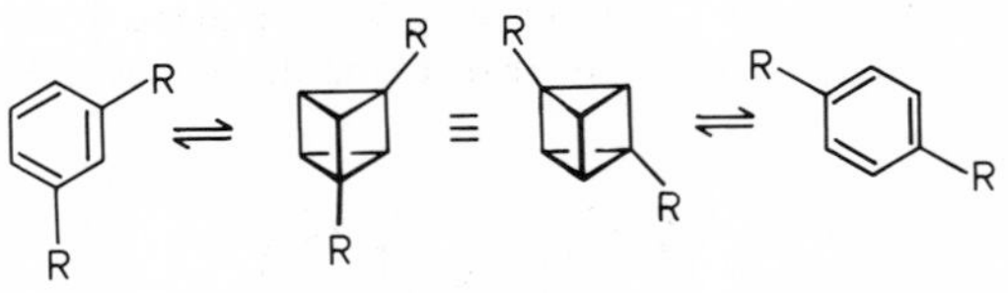

---

1. R. S. Boikess and S. Winstein, *J. Amer. Chem. Soc.*, **85**, 343 (1963).
2. R. S. Boikess, private communication.

If it did in fact turn out that in a given example *m*- and *p*-disubstituted benzenes were interconvertible *and* that the groups which were apparently moving, in fact, remained bound to the atoms that originally held them, these substances would indeed be valence isomers. If both Cope and Lewis acid catalyzed pathways were open to the molecule, the history of the samples would have to be taken into account. This is clearly not a tidy situation. We shall therefore leave it to the debaters, and turn to the literature to see what the laboratory folks have to say about it.

11-2. THE VALENCE ISOMERS OF KEKULÉ BENZENE [3]

Six CH groups can be put together in five different ways, (2)-(6).

3,3'-Bicyclopropenyls (3) are usually not considered as valence isomers because they were never seriously proposed as structures for benzene, and also because they are apparently absent under conditions that generate the others. They were first characterized by Breslow [4]. They can be converted into Kekulé benzene via base catalysis; it was shown that the 3- and 3'-atoms wind up in *o*-positions, and that the process involves anion addition rather than proton abstraction to give cyclopropenide anions:

Structure (4) (bicyclo[2.2.0]hexa-2,5-diene) [5] was first proposed by Dewar [6] in the planar form (7) as a possible structure of benzene.

---

3. For review, see (a) H.G. Viehe, *Angew. Chem., Int. Ed. Engl.*, **4**, 746 (1965); (b) E.E. van Tamelen, *Angew. Chem., Int. Ed. Engl.*, **4**, 738 (1965) and *Accounts Chem. Res.*, **5**, 186 (1972); (c) I.G. Bolesov, *Russ. Chem. Rev. (English Transl.)*, **37**, 666 (1968).
4. (a) R. Breslow and P. Gal, *J. Amer. Chem. Soc.*, **81**, 4747 (1959); (b) R. Breslow and P. Dowd, *J. Amer. Chem. Soc.*, **85**, 2729 (1963).
5. Reviewed by W. Schäfer and H. Hellmann, *Angew. Chem., Int. Ed. Engl.*, **6**, 518 (1967).
6. J. Dewar, *Proc. Roy. Soc. Edinburgh*, 84 (1866/67).

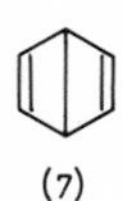

(7)

Later this "Dewar-structure" was considered to make a significant contribution to benzene [7]. As progress was made in the synthesis of small ring polycyclic compounds, many chemists became aware that there might also be a non-planar and more real structure (7) such as (4). Acting on the belief that a sterically hindered and possibly pre-bent benzene such as (8) would probably be most apt to isomerize and least inclined to revert, van Tamelen and Pappas in 1962 succeeded in the first "Dewar benzene" synthesis of (9) [8].

hν

(8) (9)

The stability of this isomer was even greater than expected; it reverts to (8), but with a half-life of fifteen minutes even at 200°. New ways of preparing such compounds were soon discovered [9]. Pettit had learned to generate cyclobutadienes as transient intermediates (see p. 282, p. 380), and their Diels-Alder addition to acetylenes leads to such exotic molecules as (10), called hemi-Dewar biphenyl [10].

(10)

The direct trimerization of 2-butyne in the presence of aluminum chloride to hexamethyl-Dewar benzene (11) was being carried out on a pilot plant scale by 1966 [11].

CH_3

3 ||| $\xrightarrow{AlCl_3}$ $-(CH_3)_6$

CH_3 (11)

7. G.W. Wheland, "Advanced Organic Chemistry", Wiley, New York, 1960; p. 117f.
8. E.E. van Tamelen and S.P. Pappas, J. Amer. Chem. Soc., 84, 3789 (1962).
9. W. Schäfer and H. Hellmann, Angew. Chem., Int. Ed. Engl., 6, 518 (1967).
10. G.D. Burt and R. Pettit, Chem. Commun., 517 (1965).
11. Chem. Eng. News, 44, 20, July 11, 1966.

The parent compound (4) was obtained in dilute solution in pyridine in 1963 by the photolysis of the anhydride (12) followed by oxidation [12].

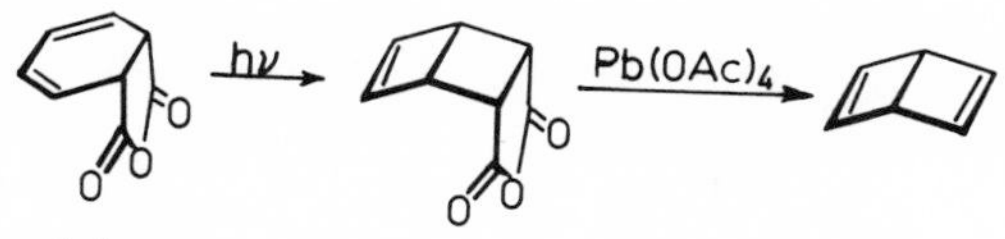
hν
Pb(OAc)$_4$
O
O
O

(12) (4)

A relation of these compounds with the bicyclopropenyls became evident when it was found that 3,3'-dimethylbicyclopropenyl (13) upon treatment with silver ion gives roughly equal amounts of (14) and (15) [13].

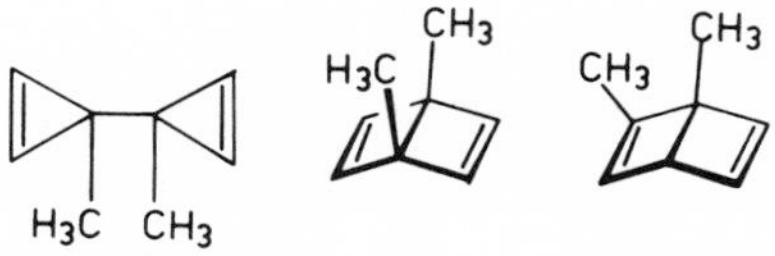
H$_3$C CH$_3$
CH$_3$
H$_3$C
CH$_3$
CH$_3$

(13) (14) (15)

Dewar benzenes have an extremely long central C–C bond (about 1.63 Å) [14] and the heat of isomerization to benzene (62 kcal/mole) [11] likewise testifies of great strains. In view of this fact one might expect a terrific instability to isomerization and a general tendency to produce Kekulé benzene derivatives with various reagents. But this is not the case. The parent compound isomerizes only slowly at room temperature [12], and any substituent seems to increase this stability greatly. When Dewar benzene is treated with the various reagents shown below, no Kekulé benzene derivatives form in any case [15].

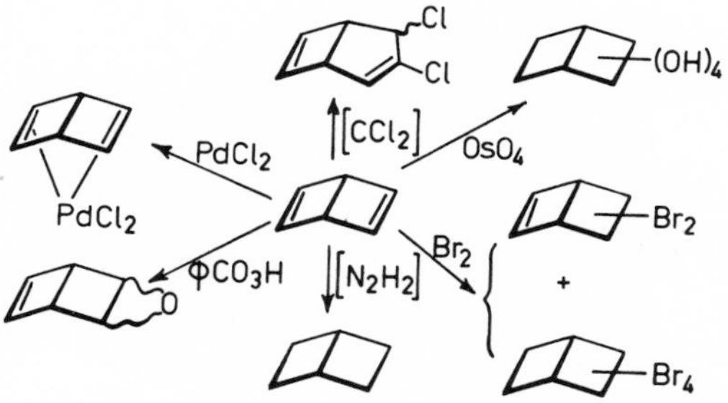
Cl
Cl
(OH)$_4$
[CCl$_2$]
OsO$_4$
PdCl$_2$
PdCl$_2$
Br$_2$
Br$_2$
ΦCO$_3$H
[N$_2$H$_2$]
O
+
Br$_4$

12. E.E. van Tamelen and S.P. Pappas, J. Amer. Chem. Soc., 85, 3297 (1963).
13. W.H. de Wolf, J.W. van Straten, and F. Bickelhaupt, Tetrahedron Lett., 3509 (1972).
14. M.J. Cardillo and S.H. Bauer, J. Amer. Chem. Soc., 92, 2399 (1970).
15. (a) E.E. van Tamelen, S.P. Pappas, and K.L. Kirk, J. Amer. Chem. Soc., 93, 6092 (1971); (b) E.E. van Tamelen and D. Carty, J. Amer. Chem. Soc., 93, 6102 (1971).

Prismane (6) (Ladenburg benzene, or tetracyclo[2.2.0.0^{2,6}.0^{3,5}]hexane) was first proposed as a possible structure for benzene [16]. Several substituted examples of this hydrocarbon are now prepared especially by photolysis of the corresponding Dewar benzenes [17]. They are even more strained than these precursors, to which they can be reisomerized by treatment with acid [18]; but like the Dewar benzenes, they are amazingly stable thermally in solution though not in the pure state.

Finally, benzvalene (5) (Hückel benzene, or tricyclo[3.1.0.0^{2,6}]hex-3-ene) which in planar form was considered as a possible contribution to benzene [19], is known as the tri-t-butyl derivative (16) [17a]; even the parent compound has been obtained by the irradiation of benzene (see p. 445) [20].

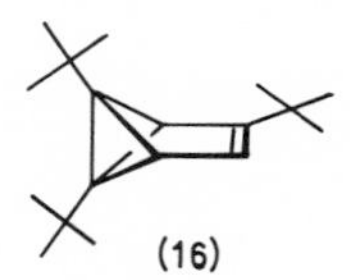

(16)

Serendipity played an important role in the development of this chemistry. It is at the same time sobering and titillating to realize how often it happens, even now, that problems in organic chemistry are solved not by clever design, but by accident. Compounds that defy any rational syntheses unceremoniously crystallize in the flasks of those who are trying to make something else. Limitations pronounced after years of painstaking work are surpassed by individuals who were not even aware of them. Whole new fields spring up because of chance conversations between people active in entirely unrelated areas.

One such event occurred in 1964, not long after the first reports of Dewar benzenes, when Viehe and co-workers found that t-butylfluoroacetylene (17) upon standing in a refrigerator rapidly trimerized to give a mixture of (18), (19), (20) and (21). It was also learned that (21) upon warming is converted into (18) [21].

---

16. A. Ladenburg, Chem. Ber., 2, 140 (1869) and 5, 322 (1872).
17. (a) K.E. Wilzbach and L. Kaplan, J. Amer. Chem. Soc., 87, 4004 (1965); (b) R. Criegee and R. Askani, Angew. Chem., Int. Ed. Engl., 5, 519 (1966); (c) D.M. Lemal and J.P. Lokensgard, J. Amer. Chem. Soc., 88, 5934 (1966).
18. R. Criegee, R. Askani, and H. Grüner, Chem. Ber., 100, 3916 (1967).
19. E. Hückel, Z. Elektrochem., 43, 752 (1937).
20. K.E. Wilzbach, J.S. Ritscher, and L. Kaplan, J. Amer. Chem. Soc., 89, 1031 (1967).
21. (a) Chem. Eng. News, 42, 38, Dec. 7, 1964; (b) H.G. Viehe, R. Merényi, J.F.M. Oth, J.R. Senders, and P. Valange, Angew. Chem., Int. Ed. Engl., 76, 922 (1964); (c) H.G. Viehe, R. Merényi, J.F.M. Oth, and P. Valange, Angew. Chem., Int. Ed. Engl., 3, 746 (1964); (d) Ref. 3a.

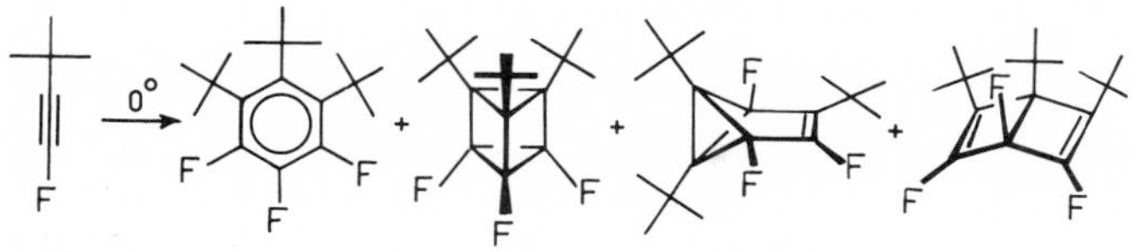

(17) (18) (19) (20) (21)

Thus, not only did all three valence isomers fall out of the blue at one shot, but all were the most crowded possible isomers; to date no other 1, 2, 3-tri-t-butylbenzene has been reported. While the story loses a bit of its punch as (19) was actually found to be a tetrameric substance [3a], it served at the time to convince the organic chemist that the valence isomers of benzene were capable of existence and that time spent in designing synthetic pathways need not be in vain.

It appears now that fluorine substitution is often helpful in the synthesis and preservation of the benzene valence isomers. The electronegativity of the fluorine atoms is probably best satisfied by bonding to carbon atoms via bonds of low s character; the unusually ready oligomerization of the fluoroacetylenes suggests this also. Perfluorohexamethylbenzene (22) has by photochemical means been converted into all three of its isomers (23)-(25), and to date it represents the only case in which all three were isolated in pure form [22].

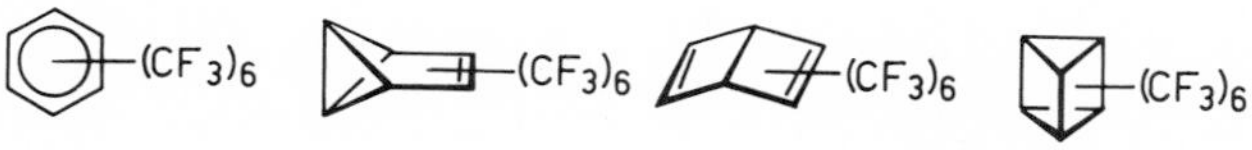

$(CF_3)_6$ $(CF_3)_6$ $(CF_3)_6$ $(CF_3)_6$

(22) (23) (24) (25)

All revert to the benzene upon heating, but they do this remarkably slowly, with half-lives of many hours at 170°. Another remarkable fact is that all three have boiling points near 100°; perhaps intermolecular fluorine-fluorine repulsions are responsible. Only the benzvalene was missing when otherwise similar experiments were carried out with perfluorohexaethylbenzene. The strain of these compounds in general is very high, and the thermal stability observed is not a good measure of the thermodynamic energy contents of these systems as compared to Kekulé benzene. The reason for this phenomenon is now known to be described by the Woodward-Hoffman rules, which are reviewed in Chapter 14. The mechanism of the photochemical preparations will be discussed in Section 13-5.

---

22. (a) M. G. Barlow, R. N. Haszeldine, and R. Hubbard, Chem. Commun., 202 (1969), and J. Chem. Soc. (C), 1232 (1970); (b) D. M. Lemal, J. V. Staros, and V. Austel, J. Amer. Chem. Soc., 91, 3373 (1969); (c) D. M. Lemal and L. H. Dunlap, J. Amer. Chem. Soc., 94, 6562 (1972).

11-3. VALENCE ISOMERS OF RELATED SPECIES

Cyclopentadiene can be converted by photolysis into bicyclo[2.1.0]pent-2-ene (26) [23]. The strain in this compound is even greater than in Dewar benzene; at 25° it has a half-life of a few hours in dilute solution in carbon tetrachloride.

(26)

The first case of valence isomerism in a six-membered ring was that of the epoxyindenone-pyrylium oxide pair, (27) and (28) [24].

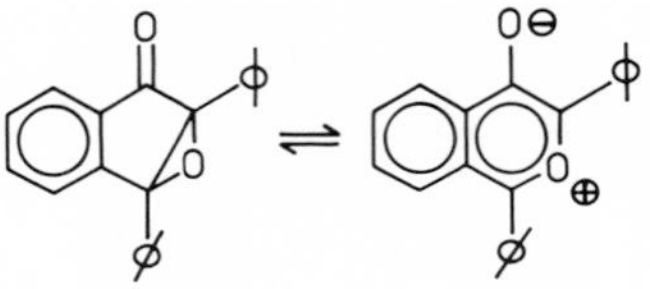

(27) (28)

It had been known since 1921 [25] that the colorless compound (27) was thermochromic; upon warming it becomes red, and this color change is reversible. Ulmann showed by chemical means that the color changes were due to the valence isomerism indicated in the equation.

The observation that the hydrochloride of 2-amino-5-chloropyridine (29) upon irradiation by sunlight gave a material of the same elemental composition that furthermore reverted to the starting material upon heating led to an early claim [26] that the product was a Dewar pyridinium salt (30); this was withdrawn when the product turned out to be the dimer (31) [27].

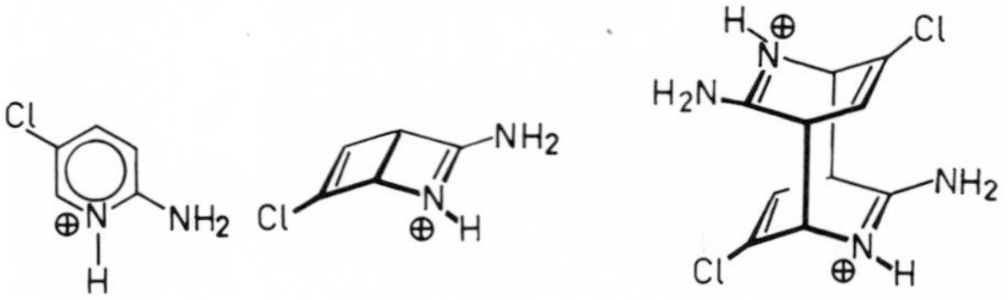

(29) (30) (31)

---

23. J.I. Brauman, L.E. Ellis, and E.E. van Tamelen, *J. Amer. Chem. Soc.*, *88*, 846 (1966).
24. E.F. Ullman and J.E. Milks, *J. Amer. Chem. Soc.*, *84*, 1315 (1962).
25. E. Weitz and A. Scheffer, *Chem. Ber.*, *54*, 2327 (1921).
26. E.C. Taylor, W.W. Paudler, and I. Kuntz, *J. Amer. Chem. Soc.*, *83*, 2967 (1961).
27. E.C. Taylor, R.O. Kan, and W.W. Paudler, *J. Amer. Chem. Soc.*, *83*, 4484 (1961).

Irradiation of pyridine was later shown indeed to give a Dewar pyridine (32), which in water hydrolyzes to (33); both (32) and (33) eventually revert to pyridine, the former within minutes at room temperature [28].

(32) (33)

A different type of Dewar pyridine (34) as well as the azaprismane (35) result from the irradiation of pentakis(perfluoroethyl)pyridine (36), and remarkable thermal stability was again encountered as in the case of the perfluoroalkylbenzenes [29].

(34) (35) (36)

Another interesting example is 2-pyrone, (37), which upon irradiation yields the Dewar pyrone (38); this product in turn undergoes thermal isomerization to (39) [30]. In the same way, α-pyridones (40) can photochemically be converted into Dewar analogs; thus, (41) is a reasonably stable solid melting at 65° [31].

(37) (38) (39) (40) (41)

As some of the strain is removed from such valence isomers, their thermal stability increases still further. Thus, the isomers (42) and (43) of tropone (44) are known; the former must be heated to 350° in order to be converted back to (44) [32].

(42) (43) (44)

28. K. E. Wilzbach and D. J. Rausch, J. Amer. Chem. Soc., 92, 2178 (1970).
29. M. G. Barlow, J. G. Dingwall, and R. N. Haszeldine, Chem. Commun., 1580 (1970).
30. E. J. Corey and W. H. Pirkle, Tetrahedron Lett., 5255 (1967).
31. R. C. De Selms and W. R. Schleigh, Tetrahedron Lett., 3563 (1972).
32. P. R. Story and S. R. Fahrenholtz, J. Amer. Chem. Soc., 87, 1623 (1965).

11-4. THE CYCLOHEPTATRIENE - NORCARADIENE PROBLEM [33a]

In an earlier section (p. 355) tropilidene was mentioned as a forerunner (homobenzene) in the development of the concept of homoconjugation. While such conjugation turned out not to be important and while no significant resonance between the cycloheptatriene and norcaradiene structures (45) and (46) could be shown to occur, there is still the problem whether these two structures can equilibrate - or whether they are stable compounds each capable of an independent existence [34].

(45) (46)

In all cases where both isomers are known, they are in a very rapid equilibration. Complicating the study of such equilibria is the fact that cycloheptatriene is subject to an even more rapid inversion process:

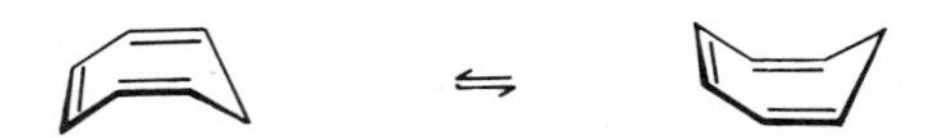

The 1,6-protons will be subject to a large difference in chemical shift in the two isomers, and hence the nmr studies have concentrated on them. The position of the equilibrium between (45) and (46) is strongly dependent on the 7-substituents, and in a way that is not really understood at present. Thus, it has been found that in the liquid phase, two hydrogen atoms [35], a cyano group [36] or bistrifluoromethyl substitution give rise the cycloheptatriene structures (47), (48) and (49) [37]; on the other hand, the 7,7-dicyano- and 7-cyano-7-carbomethoxy analogs are norcaradienes (50) and (51) [38].

---

33. (a) For review, see G. Maier, Angew. Chem., Int. Ed. Engl., 6, 402 (1967); (b) W.D. Stohrer, Chem. Ber., 73, 21 (1973).
34. (a) E.J. Corey, H.J. Burke, and W.A. Remers, J. Amer. Chem. Soc., 77, 4941 (1955); (b) W. von E. Doering, G. Laber, R. Vonderwahl, N.F. Chamberlain, and R.B. Williams, J. Amer. Chem. Soc., 78, 5448 (1956). The problem was first recognized by (c) R. Willstätter, Ann., 317, 204 (1901).
35. W. von E. Doering and M.R. Willcott, as quoted (Ref. 9) by J.A. Berson and M.R. Willcott, J. Amer. Chem. Soc., 88, 2494 (1966).
36. C.H. Bushweller, M. Sharpe, and S.J. Weininger, Tetrahedron Lett., 453 (1970).
37. J.B. Lambert, L.J. Durham, P. Lepoutere, and J.D. Roberts, J. Amer. Chem. Soc., 87, 3896 (1965).
38. (a) E. Ciganek, J. Amer. Chem. Soc., 87, 652 (1965); (b) C. Ganter and J.D. Roberts, J. Amer. Chem. Soc., 88, 741 (1966); (c) C.J. Fritchie, Acta Cryst., 20, 27 (1966).

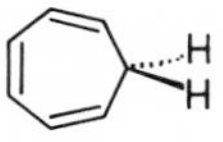

(47) 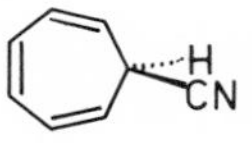 (48) 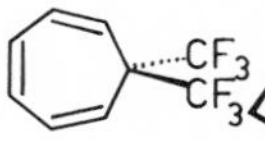 (49) 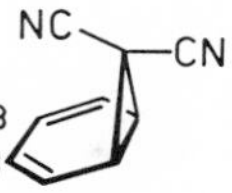 (50) 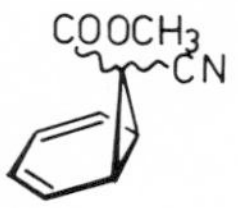 (51)

The presence of both isomers can be demonstrated when the methylene group is substituted as in (52)-(57) [39]. To complicate matters still further, several compounds in this group are exclusively noncaradienes in the solid phase; furthermore, the position of the equilibrium for (55), (56) and (57) does not appear to be related to the nature of the p-substituent in any sensible way (with (57) intermediate between (55) and (56)) [39a].

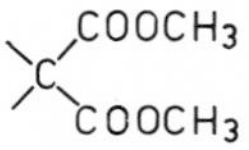

(52) 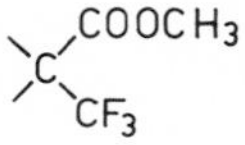 (53) 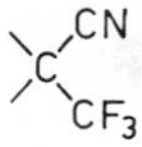 (54) 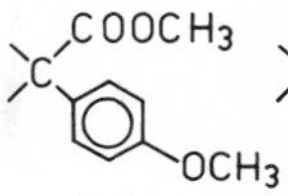 (55) 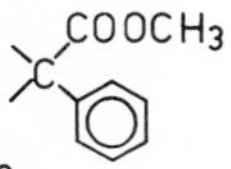 (56) 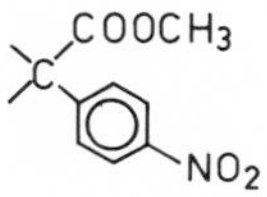 (57)

It is not difficult to list ways in which substituents could affect the equilibrium. They may sterically repel each other, and thus force the ring to adopt a small bond angle at C_7; this would favor the norcaradiene structure. The substituent interaction may be of a dipolar nature, and if this is attractive it might widen the bond angle within the ring. Inductive electron withdrawal may affect whatever small stabilization is contributed by homoaromaticity in the seven-membered ring. It has not proved possible however, to give a single convincing explanation of all the observed substituent effects [40a-c]. Benzo annelation has an important effect, quite predictable since the benzene resonance energy is lost in one of the forms [41].

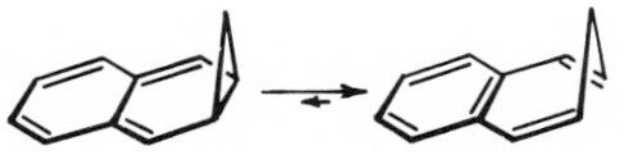

Spiro substitution may give some indication how important the bond angle may be. Surely, a small ring will force the C_7 bond angle to be large, and hence the cycloheptatriene form will be favored then:

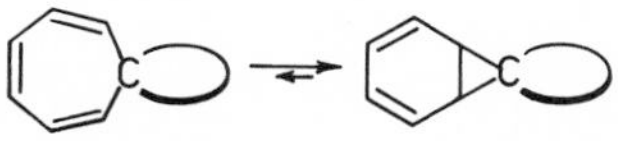

39. E. Ciganek, J. Amer. Chem. Soc., 93, 2207 (1971).
39a. G.E. Hall and J.D. Roberts, J. Amer. Chem. Soc., 93, 2203 (1971).
40. See further (a) R. Hoffmann, Tetrahedron Lett., 2907 (1970); (b) H. Günther, Tetrahedron Lett., 5173 (1970); (c) M.J.S. Dewar and D.H. Lo, J. Amer. Chem. Soc., 93, 7201 (1971).
41. (a) W. von E. Doering and M.J. Goldstein, Tetrahedron, 5, 53 (1959); (b) E. Vogel, D. Wendisch, and W.R. Roth, Angew. Chem., Int. Ed. Engl., 3, 443 (1964).

The information available to date is that with a six-membered ring such as in (58), and some five-membered ones such as in (59) the norcaradiene structure prevails [42]; with some five-membered rings such as with (60) [43] and (61) [44] the cycloheptatriene wins and in still others such as (62) equilibration between the two forms can be demonstrated by low temperature nmr [45]. It may be noted that in examples such as (59) and (61), spiro conjugation may complicate matters further (p. 370).

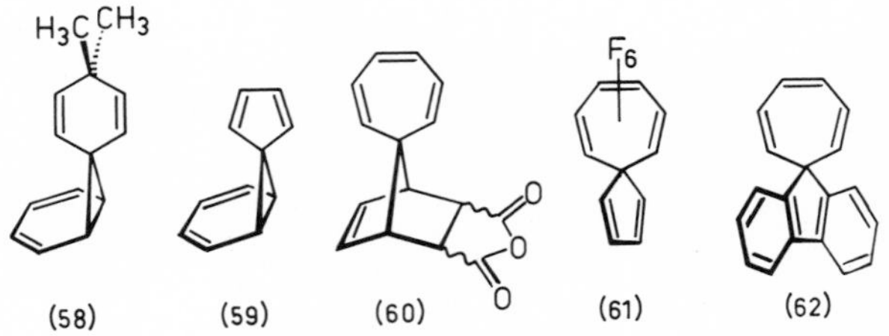

H_3C CH_3 F_6

(58) (59) (60) (61) (62)

Another approach is to bridge the 1- and 6-positions independently; a short bridge will force these atoms close enough together that they may form a σ-bond. Here some success has been achieved. A trimethylene bridge such as in (63) leads to an nmr spectrum typical of a cyclopropane, with one signal at more than 10 τ, whereas the pentamethylene bridged analog (64) has an nmr spectrum typical of cycloheptatrienes [46]. The tetramethylene bridge likewise produces this form (65). There is no change if 7-anti-cyano group (66) is now introduced, but the 7-syn-cyano group has the opposite effect (67) [47].

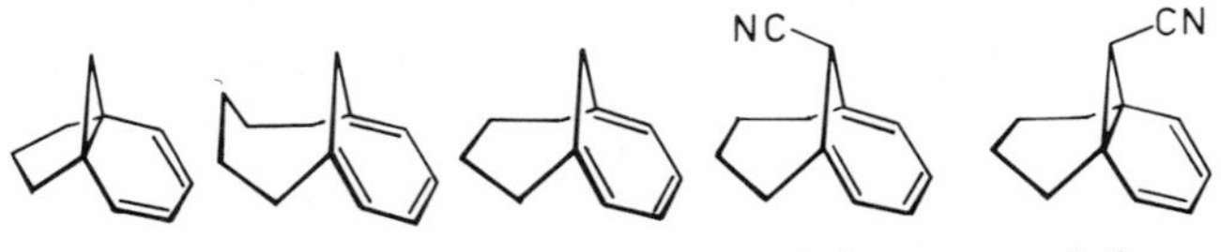

NC CN

(63) (64) (65) (66) (67)

It may also be recalled from Chapter 10 that this structural problem also occurs in the 1,6-methanocyclodecapentaenes and that substituents likewise have a large and seemingly capricious effect there. Non-connecting 1,6-substituents such as methyl generally force these atoms apart, which results in cycloheptatriene structures [48].

---

42. (a) D. Schönleber, Chem. Ber., 102, 1789 (1969); (b) M. Jones, A.M. Harrison, and K.R. Rettig, J. Amer. Chem. Soc., 91, 7462 (1969).
43. D. Schönleber, Angew. Chem., Int. Ed. Engl., 8, 76 (1969).
44. M. Jones, Angew. Chem., Int. Ed. Engl., 8, 76 (1969).
45. H. Dürr and H. Kober, Angew. Chem., Int. Ed. Engl., 10, 342 (1971).
46. E. Vogel, W. Wiedemann, H.D. Roth, J. Eimer, and H. Gunther, Ann., 759, 1 (1972).
47. E. Vogel, Pure Appl. Chem., 20, 237 (1969).
48. H. Günther and H.H. Hinrichs, Tetrahedron Lett., 787 (1966).

The introduction of a transition metal atom capable of interacting with three double bonds such as chromium (68) has a decisive effect. Even the trimethylene bridged system now shows the cycloheptatriene structure, [49] and in fact, a very high field methano signal and an X-ray diffraction pattern showing C_1C_6 to have a length of 1.65 Å [50] induced the authors to claim a homoaromatic structure (69) [49]. Irontricarbonyl complexation (70) of course does not have this effect; one is reminded here of Winstein's solution to the problem of the homotropylium ion structure (p. 389).

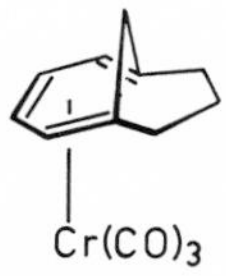

Cr(CO)3 Cr(CO)3 Fe(CO)3

(68) 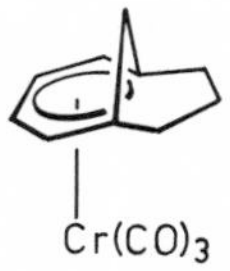 (69) 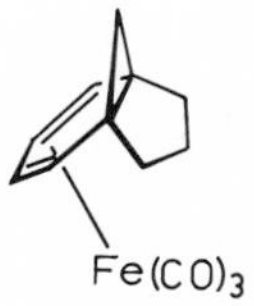 (70)

A similar situation is encountered when the methylene bridge is replaced by a hetero atom. Thus, a rapid equilibration has been observed between benzene oxide (71) and oxepin (72) [51]. It was furthermore noted that the uv spectrum is strongly solvent dependent [52].

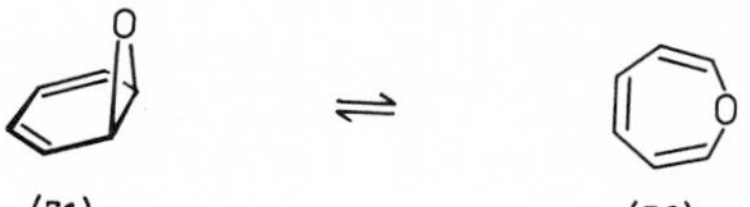

(71) ⇌ (72)

The unshared electrons of the oxygen give the seven-membered ring an undesirable 4n total and hence the oxide competes more favorably than noncaradiene did: Both forms are present in roughly equal amounts, as has been learned by means of low temperature nmr. The reactions of this substance reflect the nature of the parent molecules: Reduction leads to hexamethylene oxide (73), but maleic anhydride leads largely to the oxide adduct (74) [52].

(73) 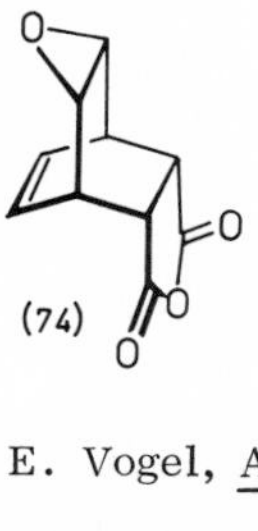 (74)

---

49. W.E. Bleck, W. Grimme, H. Günther, and E. Vogel, Angew. Chem., Int. Ed. Engl., 9, 303 (1970).
50. R.L. Beddoes, P.F. Lindley, and O.S. Mills, Angew. Chem., Int. Ed. Engl., 9, 304 (1970).
51. H. Günther, Tetrahedron Lett., 4085 (1965).
52. E. Vogel, W.A. Böll, and H. Günther, Tetrahedron Lett., 609 (1965).

In this instance as in that of tropilidene, it has proved possible to introduce substituents into the molecule in such a way as to force the equilibrium all the way in one direction or the other. Thus, 8,9-indanoxide (75) [53], 2,7-dimethyloxepin (76) [53], 1,2-naphthalene oxide (77) [54] and trans-benzene trioxide (78) [55] all have the structures shown below, and all have solvent independent uv spectra.

(75) (76) (77) (78)

Little is known as yet about valence isomerism when nitrogen is the heteroatom; the few azepins and thiepins known are all trienes (see p. 328 and p. 334).

11-5. BULLVALENE AND ITS ANALOGS

The Cope rearrangement is usually not particularly facile, often requiring the heating to almost 200° for some hours for its completion [56]. As is usually the case, however, the rate is found to be faster when the reacting groups are rigidly held in the positions necessary for reaction. With medium sized rings the reaction goes in the opposite direction [57], but when a strained ring such as (79) is considered, the Cope rearrangement occurs rapidly at 120° [56].

(79)

Cis-1,2-divinylcyclopropane (80) cannot be isolated even at -40°; any attempt to prepare it at that temperature by the reaction of diazomethane with cis-1,3,5-hexatriene leads only to cycloheptadiene (81) [58].

(80) (81)

53. E. Vogel and H. Günther, Angew. Chem., Int. Ed. Engl., 6, 385 (1967).
54. E. Vogel and F.G. Klärner, Angew. Chem., Int. Ed. Engl., 7, 374 (1968).
55. C.H. Foster and G.A. Berchtold, J. Amer. Chem. Soc., 94, 7939 (1972).
56. E. Vogel, Angew. Chem., Int. Ed. Engl., 2, 1 (1962).
57. C.A. Grob and P. Schiess, Angew. Chem., 70, 502 (1958).
58. W. von E. Doering and W.R. Roth, Tetrahedron, 19, 715 (1963).

An interesting extension of this is the homotropilidene (82) [59]. When this species undergoes the Cope rearrangement it only reproduces itself; in other words, the Cope rearrangement is degenerate in this case; but it can be shown by low temperature nmr that the equilibration is indeed occurring [60].

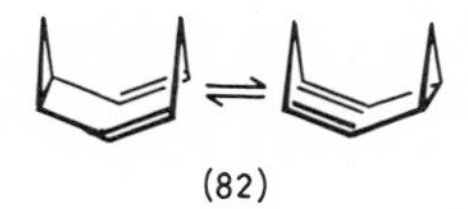

(82)

If the two most distant carbon atoms are connected, the extra rigidity increases the rate still further: Semibullvalene, (83), obtained by the photolysis of barrelene (84), looks symmetrical in the nmr even at the lowest temperatures available for that approach.

(83) (84)

A symmetrical bishomobenzene structure (85) is ruled out however by its uv spectrum, which is similar to that of dihydrobullvalene (86) - a species whose degenerate Cope rearrangements are not so rapid [61]. The symmetrical structure (85) is presumably the transition state for rearrangement of (83); these two states cannot be separated by more than a few kcal/mole.

(85) (86)

In fact, Dewar has predicted that the introduction of electronegative substituents will lower the activation energy still more, and that molecules such as (87)-(89) will be non-classical bishomoaromatic species [62] (see also p. 359).

---

59. G. Schröder, J. F. M. Oth, and R. Merényi, Angew. Chem., Int. Ed. Engl., 4, 752 (1965).
60. W. von E. Doering and W. R. Roth, Angew. Chem., Int. Ed. Engl., 2, 115 (1963).
61. H. E. Zimmerman and G. L. Grunewald, J. Amer. Chem. Soc., 88, 183 (1966).
62. M. J. S. Dewar, Z. Náhlovská, and B. D. Náhlovský, Chem. Commun., 1377 (1971).

(87) (88) (89)

None of the other bridged species known so far - such as barbaralane (90) [63] - equilibrate quite as rapidly; but they include one of special interest, namely bullvalene (91).

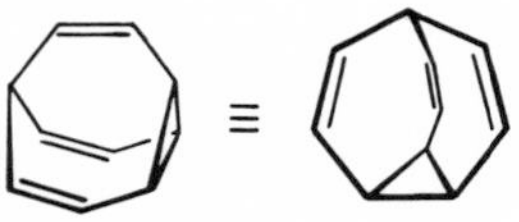

(90) (91)

The interest in this molecule derives from its threefold symmetry. Three Cope rearrangements are possible, and each regenerates the molecule. Via a succession of rearrangements such as shown below, the ten CH- groups of the molecule can come to occupy any conceivable combination of positions; no fewer than $10!/3 \approx 1.2 \times 10^6$ combinations are possible.

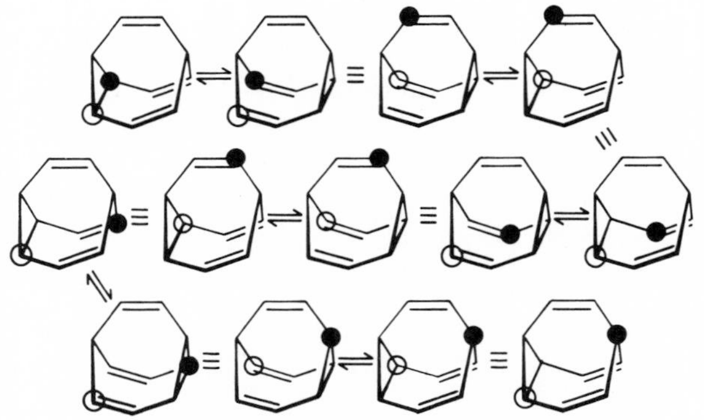

If these rearrangements are rapid on the nmr time scale, a single unsplit signal would be observed rather than the four split signals to be expected for slowly scrambling species. These properties were foreseen by Doering, who first proposed this structure [60]. A remarkable simple synthesis is now available [64],

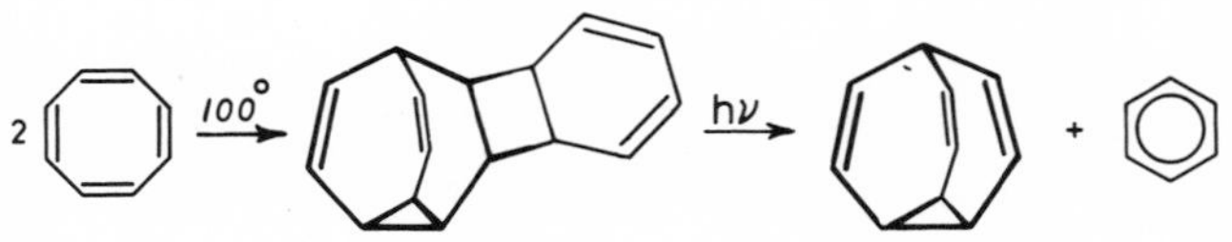

2 → 100° → hν → +

63. J.B. Lambert, Tetrahedron Lett., 1901 (1963).
64. G. Schröder, Chem. Ber., 97, 3140 (1964).

and the degenerate rearrangement does indeed occur: At 100°, a single peak is observed in the nmr at 5.78 τ, and below that, resolution occurs finally yielding two complex signals at 4.35 and 7.92 τ in a 6:4 ratio [65].

If substituents are present, rearrangements still occur, although the substituent is not necessarily indifferent to which of the four positions it occupies. Fluorobullvalene has only a doublet in the nmr at 140°, with J_{HF} = 6 Hz; bibullvalenyl (92) shows only a singlet at 80° though in principle 10 isomers could be present, and some of these have in principle 12 non-equivalent protons [66]. In homobullvalenone (93), the presence of a small amount of isomer (94) could be inferred from the ir spectrum.

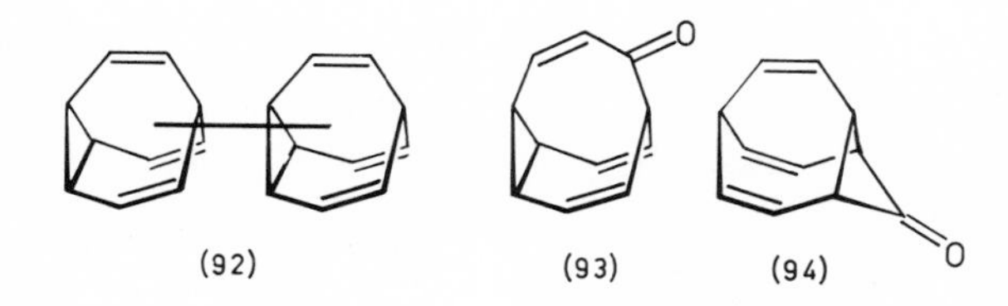

(92) (93) (94)

A deuterium labelling experiment was done in which a D-atom was found not to wander away from the positions α to the carbonyl, so that only (93) and (94) participate in the equilibration process [67].

11-6. OTHER TYPES

Homobullvalenone is one of many examples in which a Cope rearrangement involves an eight membered ring. One of the earliest of these was noted by Cope himself [68], who found that 1,3,5-cyclooctatriene, (95), contains about 15% of the isomeric bicyclic hydrocarbon (96) in equilibrium with it.

(95) ⇌ (96)

65. R. Merényi, J.F.M. Oth, and G. Schröder, Chem. Ber., 97, 3150 (1964).
66. G. Schröder and J.F.M. Oth, Angew. Chem., Int. Ed. Engl., 6, 414 (1967).
67. M.J. Goldstein, R.C. Krauss, and S.H. Dai, J. Amer. Chem. Soc., 94, 680 (1972).
68. A.C. Cope, A.C. Haven, F.L. Ramp, and E.R. Trumbull, J. Amer. Chem. Soc., 74, 4867 (1952).

Cyclooctatetraene had long been suspected of being contaminated with the isomer (97), since reaction with dienophiles such as maleic anhydride yields adducts such as (98) [69].

(97) (98)

Huisgen reasoned that such adducts form exclusively from valence isomer (97), since cyclooctatetraene itself does not have the preferred planar diene feature. Normally, if the isomerization rates are high, the rate of formation of adducts such as (98) should be a function of the concentration of the dienophile; however, if the latter is extremely reactive the rate should become independent of the latter since then the tautomerization becomes rate limiting. Indeed, Huisgen found dicyanomaleic anhydride to be such a reagent, and from the kinetic data he could calculate that the equilibrium percentage of (97) is about 0.01% [70].

In general, degenerate Cope rearrangements in bicyclic systems may be expected for structures (99). Several cases have now been realized in which A is a group more complex than methylene. Even in the trans-connected bicyclic system (100) such isomerization was observed: The optically active material racemizes as shown [71].

(99) (100)

Similarly, a D-labelling study [72] has revealed a degenerate rearrangement of the bicyclic olefin (101) at 185°.

(101)

---

69. W. Reppe, O. Schlichting, K. Klager, and T. Toepel, Ann., 560, 1 (1948).
70. R. Huisgen and F. Mietzsch, Angew. Chem., Int. Ed. Engl., 3, 36 (1964).
71. P.S. Wharton and R.A. Kretchmer, J. Org. Chem., 33, 4258 (1968).
72. W. Grimme, J. Amer. Chem. Soc., 94, 2525 (1972).

As in the previous case, this rearrangement is too slow to cause nmr coalescence, and chemical means were required to reveal it. This is also true of hypostrophene (102), which is capable of five-fold degeneracy [73],

(102)

and of lumibullvalene (103), which scatters a deuterium atom to all ten possible positions - though probably via diradical (104) rather than via Cope rearrangements [74]. Like bullvalene itself, these two compounds are among a large number of $(CH)_{10}$ valence isomers; however, further review of this fascinating topic is outside the scope of this writing [75].

(103) (104)

---

73. J.S. McKennis, L. Brener, J.S. Ward, and R. Pettit, *J. Amer. Chem. Soc.*, *93*, 4957 (1971).
74. L.A. Paquette and M.J. Kukla, *J. Amer. Chem. Soc.*, *94*, 6874 (1972).
75. See for example: (a) L.A. Paquette, *J. Amer. Chem. Soc.*, *93*, 7110 (1971); (b) E.E. van Tamelen and B.C.T. Pappas, *J. Amer. Chem. Soc.*, *93*, 6111 (1971); (c) E.E. van Tamelen, T.L. Burkoth, and R.H. Greeley, *J. Amer. Chem. Soc.*, *93*, 6120 (1971); (d) S. Masamune and N. Darby, *Accounts Chem. Res.*, *5*, 272 (1972).

Chapter 12

TAUTOMERISM

12-1. INTRODUCTION

Tautomers are isomers so easily interconvertible that they cannot be individually isolated under ordinary conditions; the valence tautomers described in the preceding chapter are an example. The phenomenon of tautomerism presented one of the earliest difficulties the practitioners of organic chemistry had to cope with in their structure assignments. It is probable that several aspects of this phenomenon were seen as separate and unrelated problems at first. Thus, there was the curious observation that all reactions that should produce simply vinyl alcohols gave the isomeric ketones and aldehydes instead. Another problem was that many carbonyl compounds in some of their chemical reactions behaved as though they were vinyl alcohols. It was eventually realized that the basic cause was the acidity - however weak - of carbon-bound hydrogen atoms in general and those in position α to a carbonyl group in particular. This gave rise to proton exchange and equilibration, and when these were rapid or easily demonstrated, the phenomenon was referred to as tautomerism. Since acidity plays an important role in tautomerism, a section is devoted to it in this chapter. In it, only the strengths of the various organic acids are discussed; the broader problems of the nature of acids (e.g. proton-donors *vs.* Lewis acidity) and of the Hammett functions are not reviewed here, and they are only briefly referred to elsewhere in this book (e.g., p. 549). The various methods whereby the pKa of an acid is determined is also considered to be outside the scope of this writing [1].

12-2. ACIDS AND ACID STRENGTHS

Since 1955, when Brown, McDaniel and Häfliger wrote their classic review [1b], there has been little change that might be described as fundamental, although the information now available is of astronomical proportions. The most important effects are the same as those operating in inorganic acids. Thus, as every freshman knows, acid strength is a very important function of the nature of the atom holding the proton. First of all, when the hydrides of atoms in the same row of the Periodic Table are compared, acid strengths rapidly increase with atomic number; hydrofluoric acid is much stronger than water, which is much stronger than ammonia, which is in turn much stronger than methane, and so on. This same sequence holds in every row. The reason for this is that the increase in electronegativity brings about an increase in the polarity of the A-H bond; the increase in electronegativity is in turn caused by the increase in nuclear charge. This relation is offset in only a minor way by the small decrease in A-H bond length which the increase in electronegativity also produces. A second effect occurs in every column (or "group"): Again there is an increase in the acid strength of the hydrides as increasing atomic numbers are considered. Thus, hydriodic acid is a stronger acid than hydrofluoric, hydrogen sulfide is stronger than water, and so on. This is so in spite of the fact that electronegativity decreases as we compare the upper atoms within a group with the lower ones; the change in bond length is now more drastic than within a row, and since the coulombic attraction of the ions in inversely proportional to the square of the separation of the two, an increase in acid strength results. The third trend easily discernable among inorganic acids is that increases in oxidation state in the central atom produce parallel increases in acid strength: Perchloric acid is stronger than chloric, which is stronger than chlorous, and so on (but note that one should not extrapolate from HClO to HCl). This may be related to the increased polarity of the bond by which the proton is bound, or the increased resonance of the anion, or both.

1. (a) D.J. Cram, "Fundamentals of Carbanion Chemistry", Academic Press, New York, 1965; (b) H.C. Brown, D.H. McDaniel, and O. Häfliger in "Determination of Organic Structures by Physical Methods", E.A. Braude and F.C. Nachod, Ed., Academic Press, New York, 1955, Vol. 1, Chr. 14. The data quoted in this section were all taken from this reference except where indicated.

These three effects also operate in organic acids: Amines are weaker than alcohols, which in turn are weaker than mercaptans; sulfonic acids are stronger than sulfinic, and so on. However, the endless variations that can be introduced in organic molecules both by way of substitution and structure changes have given us an opportunity to look beyond these gross effects, and look for more subtle nuances; in fact, a thorough understanding of these subtleties now allows the organic chemist to claim the records in both acid strength - with pentacyanocyclopentadiene (1) - and in acid weakness - with 1, 2, 3-trisubstituted cyclopropenes (2). These two differ in acid strength by some eighty powers of ten (see p. 308, 311). It would be difficult indeed to think of another physical property to which so wide a range applies.

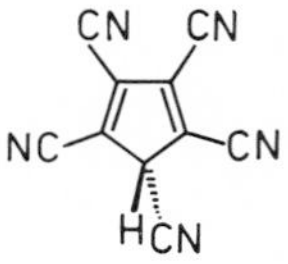

(1) 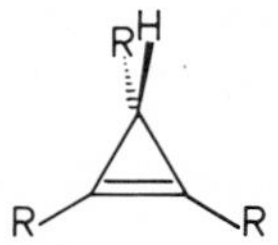 (2)

For the sake of discussion it is convenient - if not always realistic - to separate the effects operating in organic acids into five categories: The polar, resonance, steric, H-bond and hybridization effects. The polar or inductive effect is simply an extension of one already mentioned, namely that of the effect of the oxidation state. Polar or electron withdrawing groups tend to make an acid stronger. Thus, if we compare chloroacetic acid (3) with acetic acid (4), the pKa values of these two are 2.7 and 4.8, respectively.

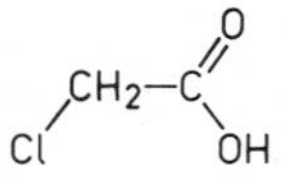

(3) 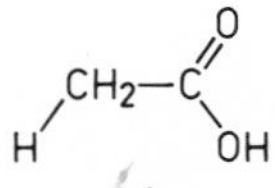 (4)

Reasonably good correlations can be found with the nature of the substituent, its distance from the ionizing group, and the number of substituents. The α-NH_3^+ substituent lowers the pKa of acetic acid to 2.3, and the α-carboxylate group raises it to 5.7; in valeric acid, the ω-NH_3^+ group lowers the pKa by only 0.5, and an ω-CO_2^- group raises it by a like amount. These inductive effects can become very large; thus, in the series methane, fluoroform and tris-trifluoromethylmethane [2] the pKa drops from 40 to 28 to 11. This effect can be treated in a semiquantitative manner on the basis of simple electrostatics; the difficulty in the treatment is our ignorance of the effective dielectric constant for the medium through which the polar effect is transmitted [3].

---

2. S. Andreades, J. Amer. Chem. Soc., 86, 2003 (1964).
3. G.W. Wheland, "Advanced Organic Chemistry", 3rd Ed., Wiley, New York, 1960; Chr. 14.

The resonance effect is of major importance, as might be gauged not only from the extreme strength of the carbon acid (1), but also from much more mundane examples. The acidity of carboxylic acids and to a lesser degree that of phenols as compared to that of simple alifatic alcohols surely derives its origin partly from resonance (the inductive effect operates in the same direction in these cases). Other remarkable examples include picric acid with a pKa of about 1 - comparable in strength to several of the stronger inorganic acids; indeno[1,2,3-jk]fluorene or fluoradene (5), which has a pKa of 11 [4], and tris-(7H-dibenzo[c,g]fluorenylidene-methyl)methane (6), which with its pKa of 5.9 is the most acidic hydrocarbon yet known [5].

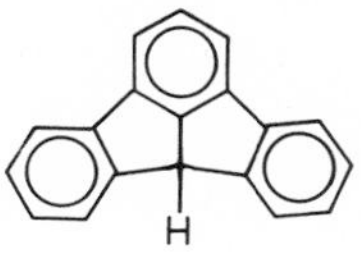

(5) 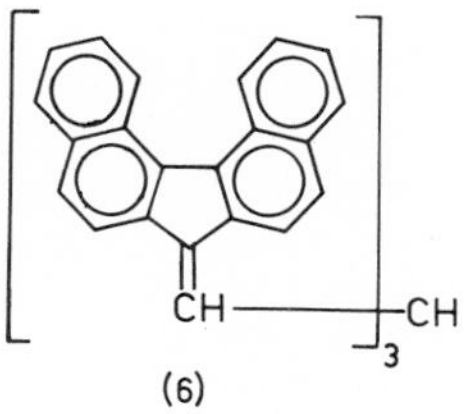 (6)

Steric effects are generally not large, but they can be used in interesting ways to judge the importance of other effects. For instance, Wheland has reported the pKa's of the phenols (7)-(10) as 10.60, 7.16, 10.17 and 8.25, respectively [6].

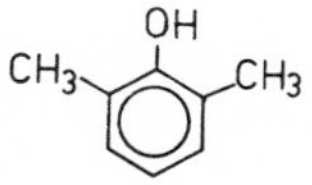

(7) 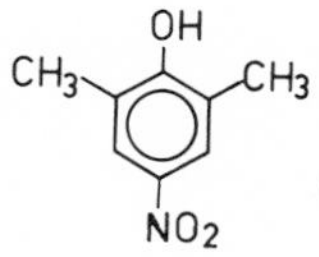 (8) 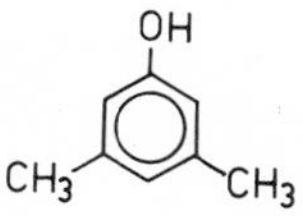 (9) 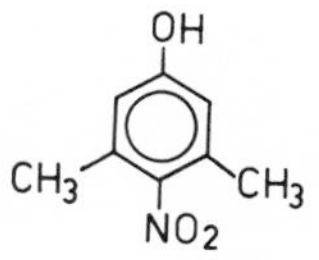 (10)

The acid strength of phenol is a little greater when 3,5-methyl groups are present than when they are in 2,6-positions; the opposite is true with p-nitrophenol. In the latter case they sterically inhibit the resonance interaction of the nitro- and hydroxy groups.

Hydrogen bonds may have fairly large effects, and they can be in either direction depending on whether the free acid or the anion is stabilized by such a bond. In the former category one might mention the diacid monoanions; those in which the two groups are rigidly held in *cis*-positions have a K_1/K_2 ratio much larger than

---

4. H. Rapoport and G. Smolinsky, J. Amer. Chem. Soc., 80, 2910 (1958).
5. R. Kuhn and D. Rewicki, Angew. Chem., Int. Ed. Engl., 6, 635 (1967).
6. G.W. Wheland, R.M. Brownell, and E.C. Mayo, J. Amer. Chem. Soc., 70, 2492 (1948).

the geometric *trans*-isomers. One extreme example is *cis*-caronic acid, (11), for which K_1/K_2 is about 9.3×10^5, whereas for the *trans*- isomer (12) the ratio equals only 32 [7].

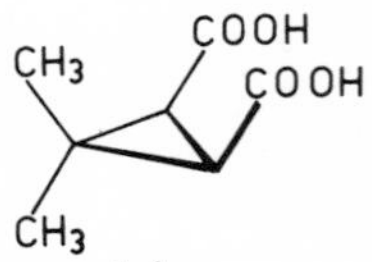

(11) 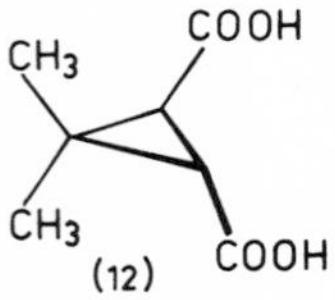 (12)

As an example of an ionization product stabilized by the hydrogen bond one might mention the 2,6-dihydroxybenzoate anion (13); the corresponding acid has a pKa of 2.30 compared to 4.20, 4.08 and 4.58 for benzoic acid, and *m*- and *p*-hydroxybenzoic acids (14)-(16), respectively.

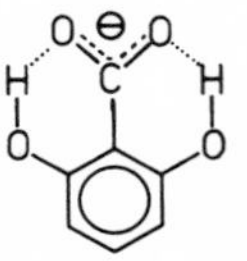

(13) 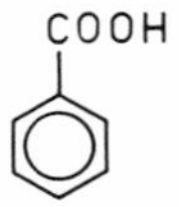 (14) 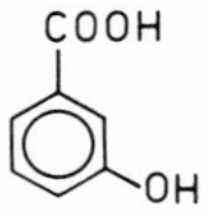 (15) 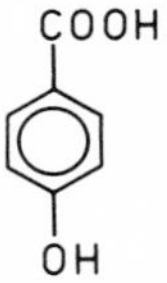 (16)

Steric effects may hinder or promote hydrogen bonding and its effect on acid strength. A spectacular example of this was found by Eberson [8]: Whereas the K_1/K_2 ratio for succinic acid (17) is 81 and that for *meso*-α,α'-di-*t*-butylsuccinic acid (18) is 73, for the corresponding racemic mixture (19) it equals 3.5×10^9. Both diastereomers of α,α'-di-*t*-butylglutaric acid (20) have a ratio of 50.

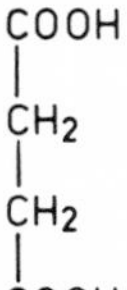

(17) 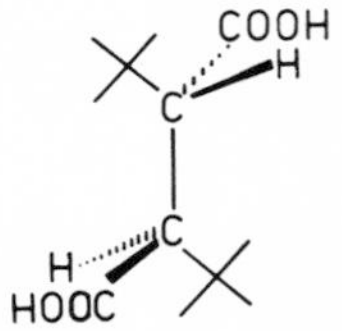 (18) 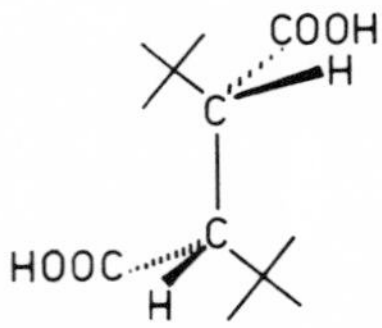 (19) 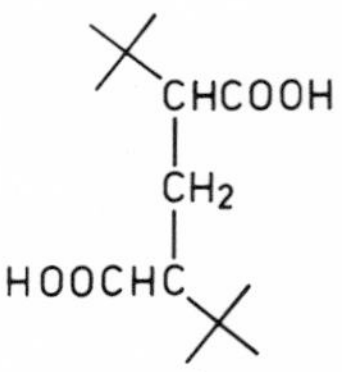 (20)

There is a definite correlation between the acidity of a given CH group and the degree of s character of the bond connecting these atoms (see p. 30). The normal alkanes with 25% s character generally have a pKa estimated at somewhat over 40; ethylene has a pKa of 36, and acetylene, with 50% s character has a pKa of 25.

---

7. I. Jones and F. G. Soper, *J. Chem. Soc.*, 133 (1936).
8. L. Eberson, *Acta Chem. Scand.*, *13*, 211 (1959).

Cyclopropane has a pKa between those of ethylene and ethane; the acidity of cyclopropenes (at the site of the double bond) is comparable to that of acetylenes [9]. Use can be made of this fact to deuterate selected positions in fused small ring compounds such as (21) [10].

H H

(21)

12-3. KETO-ENOL EQUILIBRIA

In carboxylic acids, the α-carbonyl group stabilizes the anionic site inductively as well as by resonance. This is also true of a methylene with an α-carbonyl group. The resulting anions are known as ambident, since they may displace the leaving group from a suitable alkylating agent with either site.

$$R-C(=O)OH \underset{}{\overset{-H^{\oplus}}{\rightleftharpoons}} R-C(\cdots O)_2^{\ominus}$$

$$R_1-C(=O)-CHR_2R_3 \overset{-H^{\oplus}}{\rightleftharpoons} R-C(\cdots O)(\cdots CR_2R_3)^{\ominus}$$

This behavior is further discussed in Chapter 22, Section 2; the question of interest here is what happens on reprotonation. In principle either site can be protonated, and since the energy of the parent carbonyl compound is generally substantially lower than that of the vinyl alcohol, at equilibrium the former will be greatly favored. Since both bases or acids catalyze these reactions, the equilibrations are furthermore almost always very rapid, and simple enols can as a rule not be isolated or even directly observed.

$$R-C(=O)-HCR_1R_2 \overset{H^{\oplus}}{\rightleftharpoons} R-C^{\oplus}(OH)-HCR_1R_2 \overset{-H^{\oplus}}{\rightleftharpoons} R(OH)C=CR_1R_2$$

---

9. A. J. Schipperijn, Rec. Trav. Chim. Pays Bas, 90, 1110 (1971).
10. G. L. Closs and L. E. Closs, J. Amer. Chem. Soc., 85, 2022 (1963).

An exception was reported [11] only recently; it resulted from the benzoic acid catalyzed cleavage of ketal (22) to dimethylformamide (23) and enol (24). The enol isomerizes further to the ketone (25), but if the catalyst concentration is low (one thousandth of that of (22)), the enol (24) can briefly be observed by nmr; thus an AB quartet is visible in the vinyl region at 5 τ, and the hydroxyl proton appears at 4 τ. The OH stretching frequency is 3390 cm^{-1}.

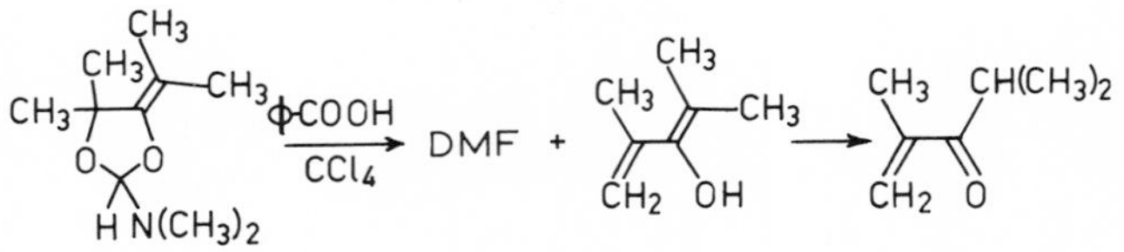

$\phi COOH$, CCl_4 → DMF + ... → ...

(22) (23) (24) (25)

When the enols under consideration are conjugated dienols, they can be stabilized by complexation with the $Fe(CO)_3$ group (see p. 383). Thus, each of the complexes (26)-(28) has been synthesized and found to be stable. With methyllithium they react to form the corresponding methyl ethers [12].

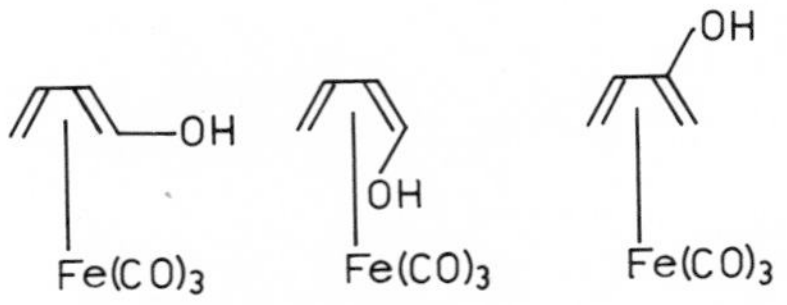

(26) (27) (28)

It has become clear from many combustion and hydrogenation experiments involving carbonyl compounds, alkenes, vinyl ethers and so on, that the energy content of the combined groups OH, C=C and C-O is somewhat higher than that of the groups C=O, C-C and C-H, and although the reasoning is of course a bit circular, this difference may be seen as the cause of the fact that vinyl alcohols are thermodynamically not stable to isomerization to the isomeric carbonyl compounds.

One can of course design enolic molecules which are stabilized in some way such that interconversion to the carbonyl isomers is not so favored. One trick is to introduce additional double bonds in the molecule such that conjugation is achieved by means of enolization; another is to make it possible for the enolic proton to engage in a hydrogen bond. Thus, if we consider the acetoacetic ester

---

11. (a) E.A. Schmidt and H.M.R. Hoffmann, J. Amer. Chem. Soc., 94, 7832 (1972); (b) H.M.R. Hoffmann and E.A. Schmidt, J. Amer. Chem. Soc., 94, 1373 (1972); see also Ref. 4 of that paper.
12. C.H. DePuy, R.N. Greene, and T.E. Schroer, Chem. Commun., 1225 (1968).

(29), the enol is stabilized by both conjugation and by an internal hydrogen bond; this changes the picture to such a degree that the pure substance contains about 7-8% enol, easily recognizable by nmr. The use of quartz apparatus (glass contains many acidic and basic sites) allows both tautomers to be separated by distillation and to be isolated in pure form. The ester (29) was indeed one of the first substances for which the phenomenon of tautomerism seemed to be clearly indicated [13].

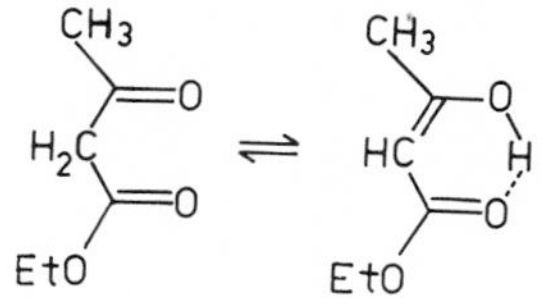

(29)

With the equilibrium constant for (29) so near unity, it is easy to understand that further, relatively minor changes in the molecule may suffice to make either tautomer predominant. Much has been written about the percentage enol in β-dicarbonyl compounds as a function of many variables such as the solvent, temperature, pressure, substituents, the presence of a cyclic feature in the ketone and so on. Seemingly large and sometimes capricious changes in the equilibrium composition can often be observed [3], but it should be pointed out that a change of 10% enol to 90% corresponds to a change of only three kcal/mole in free energy (RT ln 81).

Some of the experiments give an interesting clue to the importance of the hydrogen bond in enol formation. In those rare instances in which both the cis- and trans-enols (e.g. (30)) have been isolated, the latter quickly isomerize into the former [14].

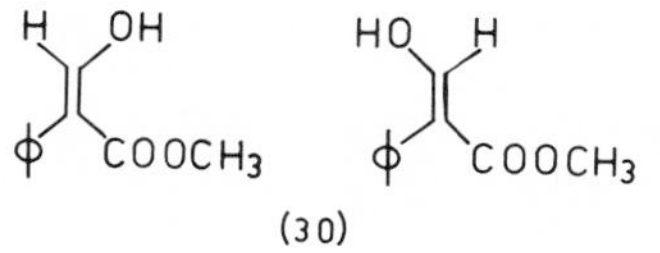

(30)

In the series of "cis-fixed" bicyclic diketones (31), (32) and (33) the % enol varies from 100, 80, to 1.4, respectively; in this series the hydrogen bond becomes obviously longer and weaker [15].

13. See Ref. 3 for a discussion of the historical development (Chr. 14).
14. S.T. Yoffe, P.V. Petrovskii, E.I. Fedin, K.V. Vatsuro, P.S. Burenko, and M.I. Kabachnik, Tetrahedron Lett., 4525 (1967).
15. H. Stetter, I. Krüger-Hansen, and M. Rizk, Chem. Ber., 94, 2702 (1961).

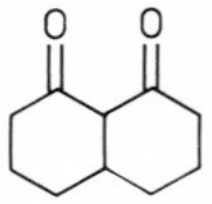

(31) 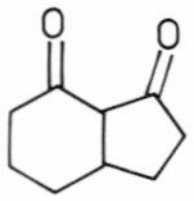 (32) 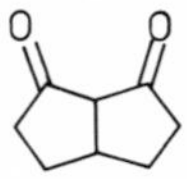 (33)

For disubstituted 1,3-diketones (34), the enol content increases rapidly as the size of the alkyl substituent is increased [16]; when R is t-butyl, the ketoform is undetectable.

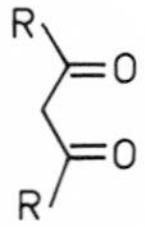

(34)

The analysis of these mixtures present in β-dicarbonyl compounds has been a difficult problem, particularly before the advent of modern instrumentation. The first apparently reliable technique was the so-called Kurt Meyer method. In this procedure, an excess of bromine is added to the sample; this bromine reacts with the enol to give an α-bromoketone and hydrobromic acid. The excess bromine is then destroyed with β-naphthol; the α-bromoketone is reduced with iodine, and the iodine so formed is finally titrated [3], the equations being:

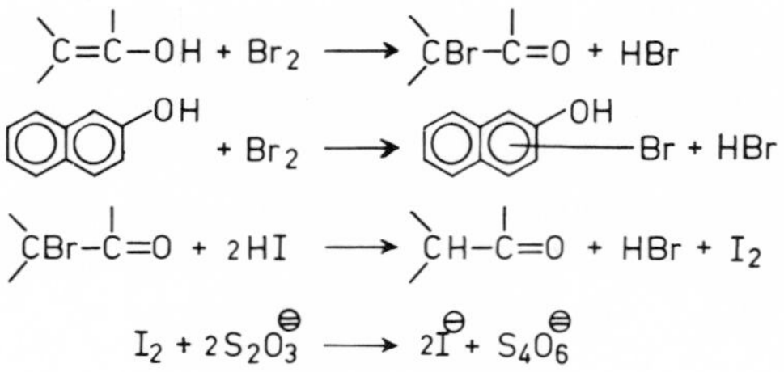

$$>C=C(-)-OH + Br_2 \longrightarrow >CBr-C(-)=O + HBr$$

$$\text{β-naphthol (OH)} + Br_2 \longrightarrow \text{1-bromo-2-naphthol} + HBr$$

$$>CBr-C(-)=O + 2HI \longrightarrow >CH-C(-)=O + HBr + I_2$$

$$I_2 + 2S_2O_3^{\ominus} \longrightarrow 2I^{\ominus} + S_4O_6^{\ominus}$$

One problem with this procedure is that HBr catalyzes the keto-enol interconversion, and another that the reduction by HI is often slow and possibly incomplete. Various modifications have been proposed and used, but now all seem supplanted by nmr [17]. The change-over is likely to produce some revised data; thus, cyclooctanone, which by a modified Meyer method [18] contains about 10% enol, does not have any as judged by its nmr spectrum [19].

---

16. G.S. Hammond, W.G. Borduin, and G.A. Guter, J. Amer. Chem. Soc., 81, 4682 (1959).
17. (a) J.L. Burdett and M.T. Rogers, J. Phys. Chem., 70, 939 (1966); (b) For review, see A.I. Kol'tsov and G.M. Kheifets, Russ. Chem. Rev. (English Transl.), 40, 773 (1971).
18. A. Gero, J. Org. Chem., 19, 469, 1960 (1954) and 26, 3156 (1961).
19. W.J. le Noble, unpublished work.

If aromatization is part of the process of enolization, the picture changes completely. Thus, phenols are enols, and no physical technique gives any hint of the presence of the tautomeric cyclohexadienones (35). When the resonance energies of benzene and a conjugated dienone chain are added in, the same total comes out greatly in favor of the enol. In fact, the keto-enol equilibrium has often been used as a measure of the resonance stabilization; for instance, while no enolization occurs in 3-cyclobutenones (37), no ketone can be detected in phenol (36).

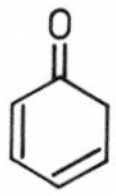

O OH O

(35) 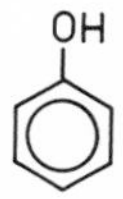 (36)  (37)

A mixture can be demonstrated to be present in 9-anthranol (38) [20]; the resonance energy of anthracene is less than three times as great as that of benzene.

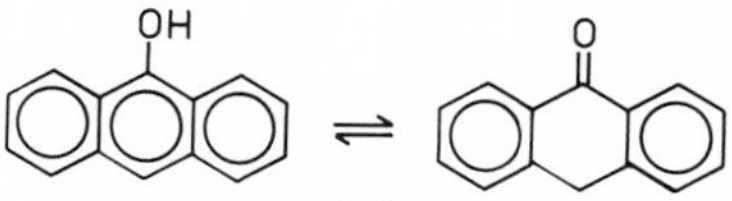

OH O

(38)

There are a number of simple phenols in which the keto form is of some importance. If more than one hydroxy group is present, the energy balance may shift the molecule in the direction of the keto form once again. Phloroglucinol (39) chemically behaves more like a triketone than a triol; thus, it readily forms the trioxime (40) with hydroxylamine.

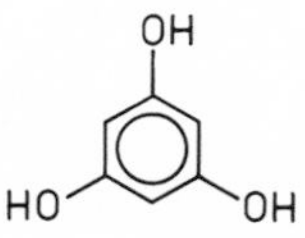

OH NOH

HO OH HON NOH

(39) 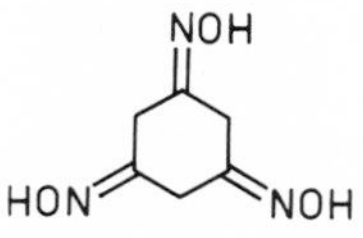 (40)

The tautomer of hydroxyquinone (41), cyclohex-2-ene-1,4-dione (42) can be isolated and it is stable to 10°, but at higher temperatures it quickly reverts to (41) [21].

20. For references, see Z. Majerski and N. Trinajstić, Bull. Chem. Soc. Japan, 43, 2648 (1970).
21. E.W. Garbisch, J. Amer. Chem. Soc., 87, 4971 (1965).

(41) 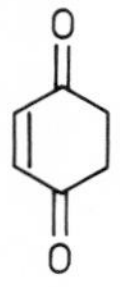 (42)

The keto form of phenol itself cannot be isolated, although transition metal stabilization has made it possible to prepare (43) [22].

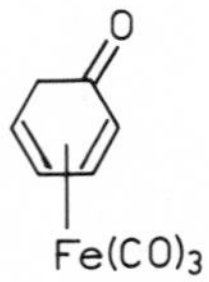

$Fe(CO)_3$

(43)

Some highly substituted cyclohexadienones such as (44) are known [23], but all tautomerize quickly in the presence of proton sources.

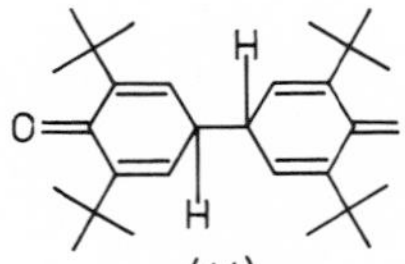

(44)

As noted above, the keto-enol equilibrium has often been used as a device to gauge the stability of a double bond in a given position. An interesting example is provided by (45). This molecule is a stable mono-enol: Its total bond angle strains are a little less severe than those of the diketone (46). The nmr spectrum shows two methyl signals that coalesce upon heating. Line shape analysis of the spectra showed that the species responsible was this diketone rather than the cyclobutadiene (47) [24].

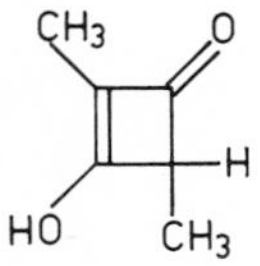

(45) 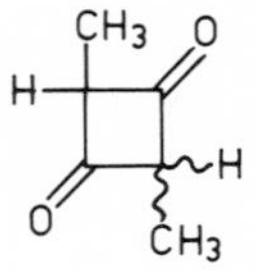 (46) 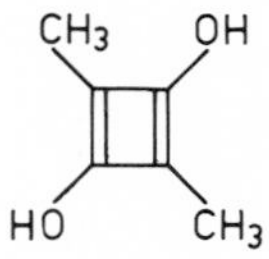 (47)

---

22. A. J. Birch, P. E. Cross, J. Lewis, D. A. White, and S. B. Wild, J. Chem. Soc. (A), 332 (1968).
23. M. S. Kharasch and B. S. Joshi, J. Org. Chem., 22, 1439 (1957).
24. J. S. Chickos, D. W. Larsen, and L. E. Legler, J. Amer. Chem. Soc., 94, 4266 (1972).

12-4. OTHER TYPES OF TAUTOMERISM

Cyclopropyl alcohols (48) may be considered condidates for tautomerism, since in so many ways cyclopropane rings mimick double bonds in their behavior.

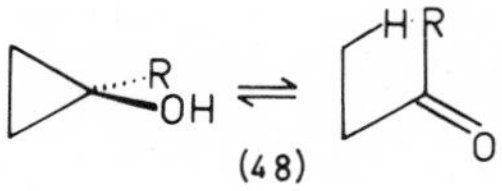

(48)

Although cyclopropanols are well known compounds that do rearrange to carbonyl compounds in the presence of acids or bases [25], the reverse of this process was not known until Nickon in 1962 demonstrated its occurrence by means of optically active camphenilone, (49) [26].

(49)

It was found that it racemizes at 185° in the presence of a strong base. This was shown to be due to reversible homoenolization.

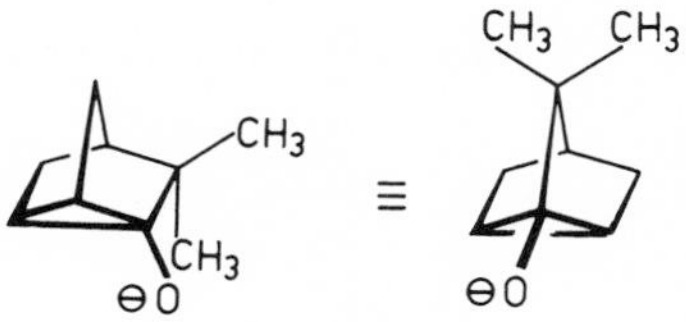

As an alternative explanation for the exchange observed, the enol (50) is ruled out because it violates Bredt's rule (see p. 213);

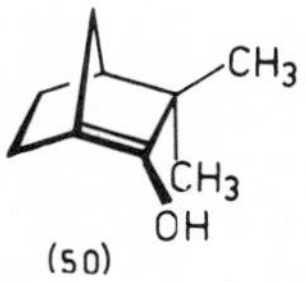

(50)

---

25. See for example C. H. DePuy and F. W. Breitbeil, J. Amer. Chem. Soc., 85, 2176 (1963).
26. A. Nickon and J. L. Lambert, J. Amer. Chem. Soc., 84, 4604 (1962) and 88, 1905 (1966).

reversible addition and ring opening such as the process producing carbanion (51) is ruled out because of the fact that esters such as (52) do not give camphenilone under these conditions.

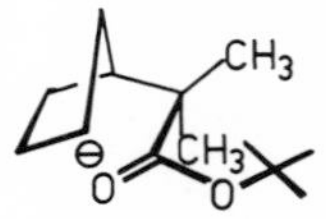

CH_3 CH_3 O O CH_3 O CH_3 OC_2H_5

(51) 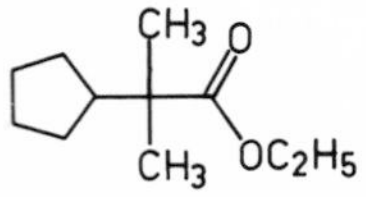 (52)

A related type of tautomerism is encountered in the carbohydrate field. Thus, the well-known phenomenon of mutarotation is in fact an example of tautomerism, in this area often referred to ring-chain tautomerism [27].

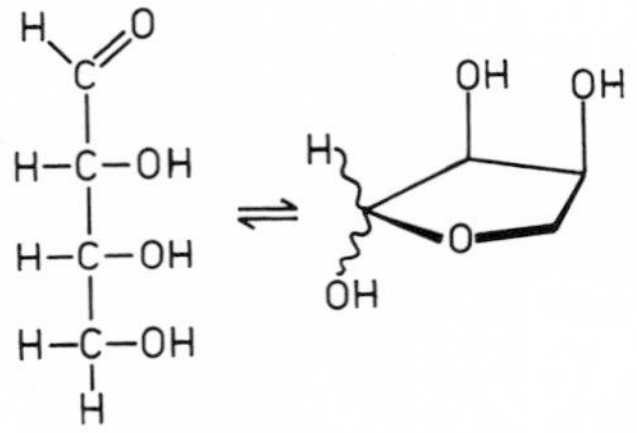

H C O
H–C–OH
H–C–OH
H–C–OH
H
⇌
H OH O OH OH

The hemiacetals or ketals in this reaction often predominate at equilibrium - an indication of the stability of five- or six-membered rings.

The tautomeric equilibrium in more general terms may be written:

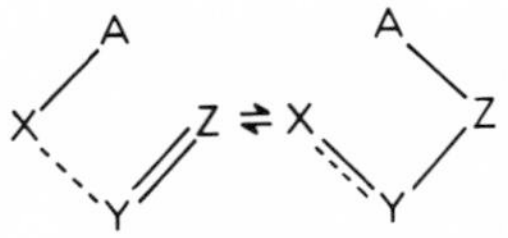

A X Y Z ⇌ A X Y Z

where the dotted line represents a connection which may but does not have to be a bond (it could be a $-CH_2-$ group as in homoenolization, or a chain of atoms such as in the carbohydrates). Z does not have to be oxygen. Thus, Schiff bases of β-ketoesters such as (53) are in equilibrium with β-aminocrotonates (54), and in some cases both are isolable [28].

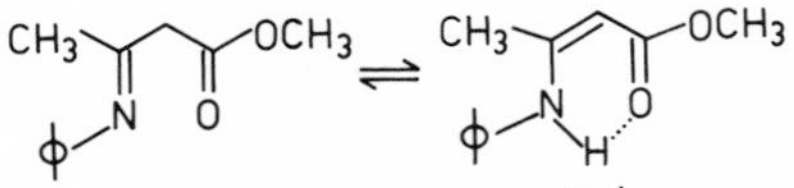

CH_3 OCH_3 ϕ N O ⇌ CH_3 OCH_3 ϕ N H O

(53) (54)

27. Reviewed by P.R. Jones, Chem. Rev., 63, 461 (1963).
28. G.O. Dudek and G.P. Volpp, J. Amer. Chem. Soc., 85, 2697 (1963).

Neither does Y need to be carbon; the so-called aci-nitro tautomerism is an example with Y is nitrogen.

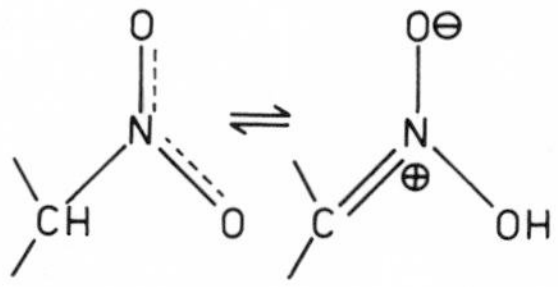

Finally, A does not have to be hydrogen; thus, while (55) and (56) can both be isolated, they are very easily interconvertible [29].

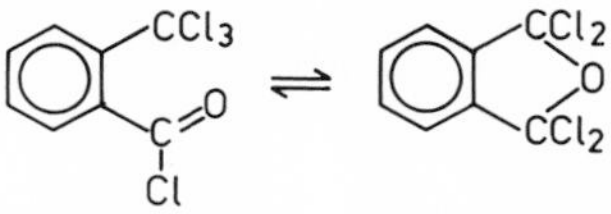

(55) (56)

In hydrocyanic acid (57), X---Y is a single atom.

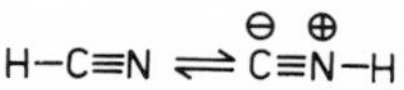

$$H-C\equiv N \rightleftharpoons \overset{\ominus}{C}\equiv\overset{\oplus}{N}-H$$

(57)

Among the further variations often encountered are the α-pyridones (58) which are favored in equilibrium with α-hydroxypyridines (59),

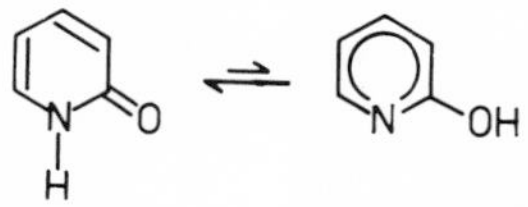

(58) (59)

and the lactam-lactim tautomerism which is described by the following equation.

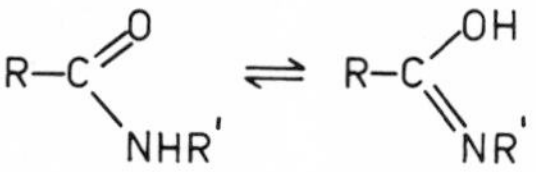

$$R-C(=O)-NHR' \rightleftharpoons R-C(-OH)=NR'$$

Finally, it should be mentioned that many of the innumerable rearrangements encountered in organic chemistry can be described in the same general terms as tautomerism, and in fact, many are just as facile if not more so. Whether these chemical changes should be described as tautomerism or as rearrangements depends on their historical background only.

---

29. E. Ott, Chem. Ber., 55, 2108 (1922).

Chapter 13

PHOTOCHEMISTRY

13-1. INTRODUCTION

Organic photochemistry is an area in a state of rapid development. Practically every new issue of the important journals contains articles about a new reaction, a new mechanistic insight, or a new photochemical way to prepare organic compounds. This feeling becomes even stronger when it is realized what is as yet unknown: The geometry of excited molecules is in most cases not known, and in many instances the electron distribution is not either. But the facts that photochemical reactions are largely those of π-electron systems and that such systems are amenable to a degree to quantum mechanical understanding have tempted the organic chemist to get into the field and survey the possibilities. Already so much has been uncovered that it is not possible to do thorough justice to it in a chapter, and hence only some of the most important aspects are discussed here [1]; these

---

1. For reviews see: (a) G. S. Hammond and N. J. Turro, Science, 142, 1541 (1963); (b) N. J. Turro, Chem. Eng. News, 44, 84, May 8, 1967; (c) "Organic Photochemistry", Vol. 1 and 2, O. L. Chapman, Ed., Marcell Dekker, New York, 1969.

include some of the basic language of the photochemist, some of the more interesting photoreactions of olefins, carbonyl compounds and aromatic species, chemiluminescence and "photochemistry without light". A very important recent development, the operation of symmetry-based selection rules, is discussed in the following chapter; for another topic important in photochemistry, that of photooxidations, the reader is referred to p. 878.

13-2. THE LANGUAGE OF THE PHOTOCHEMIST

Let us see what happens to a molecule about to absorb a quantum of electromagnetic radiation. There will probably not be very many of such molecules in the photochemical reaction: The photochemist usually works with small quantities of matter, and often in dilute solution. The reason for this is that it is difficult to generate light with a very high flux. The human eye being as sensitive to light as it is, what would seem to be a very powerful light source may actually be capable of only a very meager output of light in terms of moles of quanta (einsteins): Not much more than 10^{19}–10^{20} photons per sec can be achieved. If the number of quanta passing through a square cm per sec is considered, one realizes that photochemistry is not for the production of large amounts of material - even if the so-called quantum yield (ϕ) is close to 1 (this may not be true of photochemical chain reactions, such as photoinitiated polymerization, in which ϕ may be much larger than 1). The reactor is usually so constructed that simple mechanical loss is minimized, with a central light source concentrically surrounded by the solution to be irradiated and beyond that by a reflecting surface.

It is of course often necessary in quantitative experiments to measure the light flux. This is done by means of the so-called actinometers such as the uranyl oxalate actinometer; this is simply a reaction vessel of the same dimensions as the one used in the study of interest, and filled with a solution capable of a photochemical reaction of accurately known quantum yield.

While the number of quanta may be small, the energy per photon is very high. Thus, light of 300 nm has an energy

$$E = h\nu = hc/\lambda,$$

$$E = \frac{6.67 \times 10^{-27} \times 3.00 \times 10^{10}}{300 \times 10^{-7}} \times 6.02 \times 10^{23} \times \frac{10^{10}}{4.20}$$

$$= 95 \text{ kcal/mole}.$$

This is about the energy of most carbon-carbon single bonds. It compares with a thermal energy of RT ≈ 0.6 kcal/mole (at room temperature) available to the average molecule; at one end of the Boltzman distribution perhaps 20 kcal/mole may be available to a small number of molecules, and perhaps 30 kcal at the highest temperatures possible before a general chemical decomposition process sets in. The really interesting part of photochemical processes arises as this extremely large amount of energy is available only to certain molecules - in fact, often to only a selected part of these molecules. This selective administration of large doses of energy leads to a similarity between the processes occurring in the mass spectrometer and in the photochemical reactor; the same fragments are often produced in these two operations with the same molecule [2].

Before one sets out to irradiate a given substance it should of course be ascertained from spectroscopic observation that it has an absorption band in the region of the wavelength to be used, and equally obviously, the solvent and the reacton itself should be materials transparent to this radiation; quartz - which is transparent down to 190 nm - is often used. Preferably, the molecule to be photolyzed should also have a rather large extinction coefficient at the wavelength used. It is not sufficient that the molecule merely have an energy level at the appropriate energy difference above the ground level; several considerations go into the question of whether it can absorb the photon or not. One of these is the Franck-Condon principle. In Fig. 13-1, the three curves represent electronic energy levels, and the horizontal lines are vibrational levels. The ground state (curve 0) will be near the extreme points A and B most of the time, since at these points the vibrating parts of the molecule will be moving most slowly; thus, absorption is most likely at that time. The absorption process is essentially instantaneous, and the new level must correspond to these positions. Positions near A and B do not correspond to favorable positions in level 1; however, there are some favorable vibrationally excited

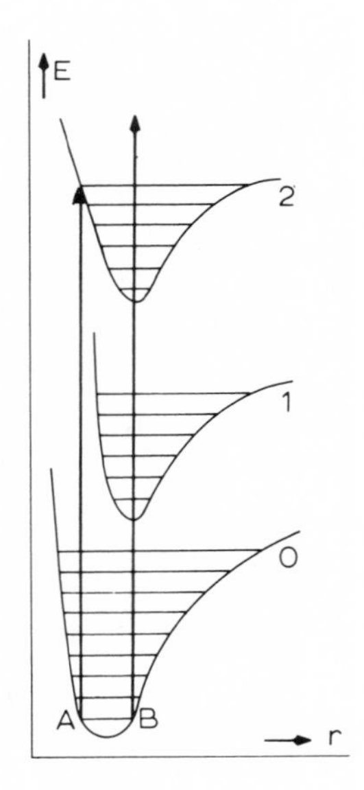

Fig. 13-1. Some of the electronic and vibrational levels of a hypothetical molecule.

2. N. J. Turro, D. C. Neckers, P. A. Leermakers, D. Seldner, and P. D'Angelo, J. Amer. Chem. Soc., 87, 4097 (1965).

levels in electronic level 2, and so if the energy is available, these are likely transitions to occur. An electronic excitation which does not at the same time involve vibrational excitation is described as vertical; on the other hand, an excitation of the type suggested by Fig. 13-1 is called adiabatic. A second important consideration is that selection rules must be obeyed. These rules find their origin in the conservation principles of spin, angular momentum, etc.; it is not necessary to discuss them here, however.

What happens after the photon has been absorbed is something that may be conveniently described by means of the so-called Jablonski diagram (Fig. 13-2).

Absorption amounts to the raising of an electron from a molecular orbital occupied in the ground state to another orbital normally vacant. Because of the requirement of spin conservation, if - as is almost always the case - the ground state molecule is a singlet (S_0) with all the electrons paired, the absorption process will lead to another singlet state, for instance S_1. The molecule may very well be in an excited vibrational state now; i.e., it is extremely hot. The first thing it will do is relax (cool) during the next few collisions with its nearest neighbors. This cooling-off period is very short, of the order of the time length of most molecular vibrations; the molecule is said to "cascade down" to the vibrational ground state. At that point, it may do one of several things. It may get back down to the electronic ground level by emission of a photon; this process is called fluorescence. Fluorescence generally is observed to stop 10^{-7}-10^{-8} sec after irradiation. A second choice open to the molecule is that it may manage to match its energy with a vibrationally excited state of the S_0 level and be converted to it; this process is called internal conversion. Like the thermal relaxation, it amounts also to a "radiationless transition". Still another process is that it may transfer its energy to another molecule and excite it according to:

$$A^*(S_1) + B(S_0) \longrightarrow A(S_0) + B^*(S_1)$$

This process is called sensitization, and since only singlets are involved, it should be specifically termed singlet sensitization. Because of the short life time of the excited state, this process is not observed very often.

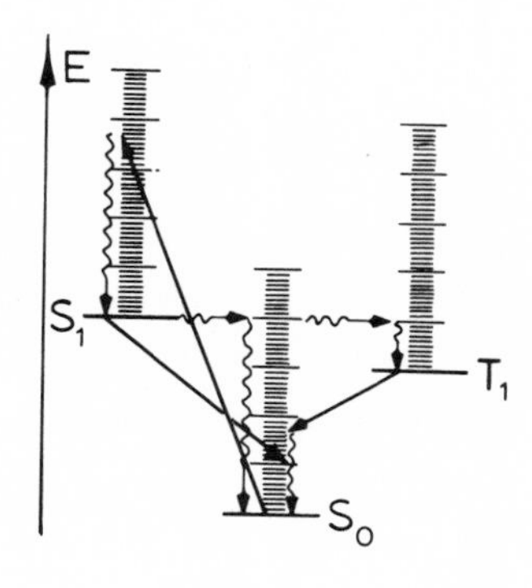

Fig. 13-2. A Jablonski diagram.

The excited state may also react chemically, for instance by isomerization, or by attacking another molecule. The chemical properties of excited molecules are often very different from those of the ground state precursors, which is the basic reason for the interest of so many organic chemists. Finally, the excited molecule may match its energy with that of a vibrationally excited triplet level and be converted to it; in this state the two upper electrons are unpaired. This is called "intersystem crossing." It would seem to be a forbidden process because of spin inversion; however, the process may become allowed provided that the electronic spins and orbits are properly matched so as to produce overall conservation of spin and angular momentum.

The excited triplet now quickly cascades down to the T_1 level, which, because of the forbiddenness of any move back to the S_0 level has a relatively long life time. If emission occurs, this is referred to as phosphorescence. Intersystem crossing back to S_0 may also occur; perhaps more interesting is the sensitization of another molecule to its triplet state. Molecules which undergo intersystem crossing with high efficiency - such as benzophenone - are often used as triplet sensitizers, or simply as "sensitizers." Besides this efficiency, the requirements of such molecules are a triplet state of a relatively high energy (i.e., above that of most other molecules), and a relatively long life time (generally at least 10^{-4} sec or so is available) [3]. Materials such as oxygen and piperylene (1,3-pentadiene) which are especially easily sensitized and hence which readily deactivate excited states are called quenchers. Chemical reactions of the triplet state are of course also encountered.

The measurement of fluorescence and phosphorescence spectra is vital to the use of sensitizers and quenchers; they provide the photochemist with information about the energies of the excited states involved and often allow the identification of such phenomena if they are part of a given photochemical experiment. Several other concepts unique to photochemistry remain to be introduced; this is done below in the course of the discussion of several photochemical conversions. It should be remembered in these discussions that the shape of the excited molecule may be quite different from that of the ground state. Thus, it is thought [4] that the σ^* state of spiropentane (1) has a planar carbon skeleton and that the T_1 state of methane is planar [5].

(1)

These shapes are in most cases not known, but where they are, it will be noted.

3. N.J. Turro, J. Chem. Educ., 43, 13 (1966).
4. R. Hoffmann, Tetrahedron Lett., 3819 (1965).
5. S. Durmaz, J.N. Murrell, and J.B. Pedley, Chem. Commun., 933 (1972).

13-3. REACTIONS OF OLEFINS

If we raise one of the π electrons of an olefin to the π^* antibonding level, there is no longer any reason for the molecule to be planar, and the excited state probably assumes a shape in which the two "halves" lie in perpendicular planes. The molecule may therefore return to the original ground state or that of its geometric isomer. Thereafter, both isomers have some probability of being excited again, and so on; finally, a photostationary state is reached. This mixture may have a composition very different from that which would result from thermodynamic equilibration, since it is determined not by free energy content but, in part, by the extinction coefficients. If the *cis*- isomer has a much lower value of ϵ at the exciting wavelength, it will soon predominate in the mixture, and this is often the case. Thus, *cis*-stilbenes (2) can often be conveniently prepared from the *trans*-isomers (3) by irradiation [6].

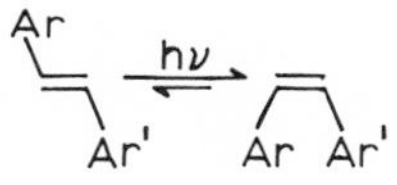

This reaction need not be sensitized, but it *can* be, and one of the more interesting facts to emerge from these isomerization studies is the close approach evidently necessary for sensitization. Thus, while benzophenone (4) and 2,6-dimethylbenzophenone (5) sensitizers do not effect the composition of the photostationary state of stilbene, the diisopropyl analog (6) generates a different mixture, suggesting that the *i*-propyl groups are large enough to interfere with sensitization [7].

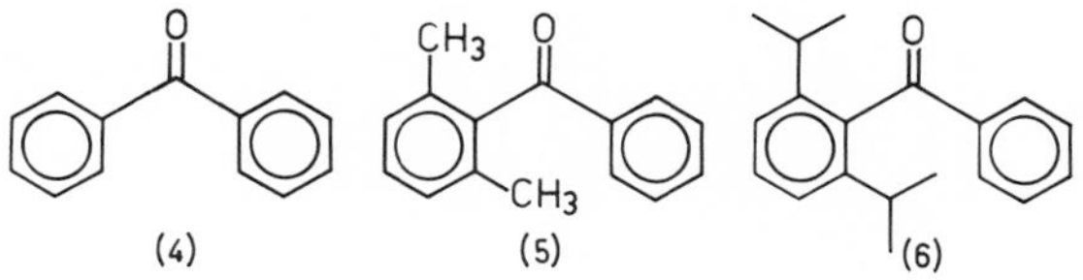

The excited molecule has two of its upper electrons in different orbitals, and the radical nature of photochemical intermediates is often evident. Thus, benzonorbornadiene (7) upon irradiation in the presence of acetophenone rearranges to (8) - surely via the pathway indicated [8].

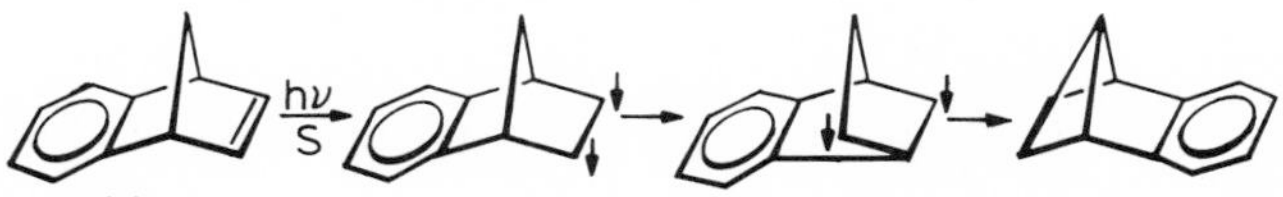

---

6. D. Schulte-Frohlinde, *Ann.*, *615*, 114 (1958).
7. W.G. Herkstroeter, L.B. Jones, and G.S. Hammond, *J. Amer. Chem. Soc.*, *88*, 4777 (1966).
8. J.R. Edman, *J. Amer. Chem. Soc.*, *88*, 3454 (1966).

This reaction is also an example of a second likely process, namely the conversion of double bonds to single ones: The latter are not able to absorb in the same wavelength region. This can happen by the cycloaddition route, in which two olefin molecules form one of cyclobutane. If the two double bonds are part of the same molecule, rather strained products may be formed; indeed, of the highly strained compounds that have been prepared in recent years, many have been the result of photochemical synthesis [9].

The mercury sensitized photoreaction of 1,5-hexadiene leads to bicyclo[2.1.1] hexane (9) rather than the [2.2.0]isomer [10]; this suggests the pathway:

Hg → Hg*(1) → Hg*(3) → ... (9)

OR

The intermediate ring size may be related to the direction of this reaction; thus, the mercury sensitized reaction of 1,6-heptadiene gives a mixture of (10) and (11) [11].

hν / Hg → (10) + (11)

The difficulty of deciphering the behavior of such dienes may be gauged from the fact that myrcene (12) in the presence of ketone sensitizers yields only the bicyclic species (13) [12], whereas in the absence of sensitization (14) is obtained as the principal product [13].

(12) S → (13); hν → (14)

---

9. For review, see (a) W. L. Dilling, Chem. Rev., 66, 373 (1966); (b) R. N. Warrener and J. B. Bremner, Rev. Pure Appl. Chem., 16, 117 (1966).
10. P. de Mayo and S. T. Reid, Quart. Rev. (London), 15, 393 (1961).
11. R. Srinivasan and K. A. Hill, J. Amer. Chem. Soc., 87, 4988 (1965).
12. K. J. Crowly, Tetrahedron, 21, 1001 (1965).
13. R. S. H. Liu and G. S. Hammond, J. Amer. Chem. Soc., 86, 1892 (1964).

As noted above, many of the most exotic molecules were first synthesized photochemically. The following examples may be recorded here. Tricyclo[3.3.0.0^{2,6}] octane (15) has been prepared [14] from 1,5-cyclooctadiene, (16);

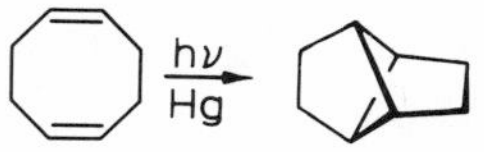

hν Hg

(16) (15)

quadricyclane (17) can be obtained by the irradiation of norbornadiene, (18) [15];

hν

(18) (17)

Dewar benzenes (19) can be converted into prismanes (20) [16];

hν

(19) (20)

cyclopentadiene Diels-Alder adducts such as (21) can be tied up into bishomocubane derivatives such as (22) [17];

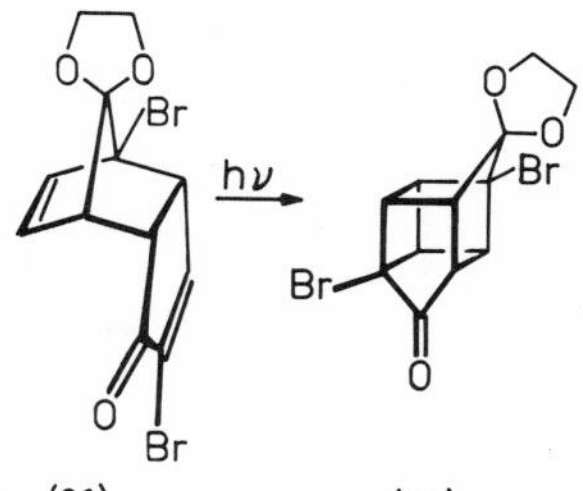

O O Br hν O O Br Br O Br O

(21) (22)

14. R. Srinivasan, J. Amer. Chem. Soc., 86, 3318 (1964).
15. G.S. Hammond, P. Wyatt, C.D. DeBoer, and N.J. Turro, J. Amer. Chem. Soc., 86, 2532 (1964).
16. K.E. Wilzbach and L. Kaplan, J. Amer. Chem. Soc., 87, 4004 (1965).
17. P.E. Eaton and T.W. Cole, J. Amer. Chem. Soc., 86, 3157 (1964).

the cycloaddition of tolane (23) to naphthalene, which presumably involves the intermediate (24), gives a high yield of (25) [18].

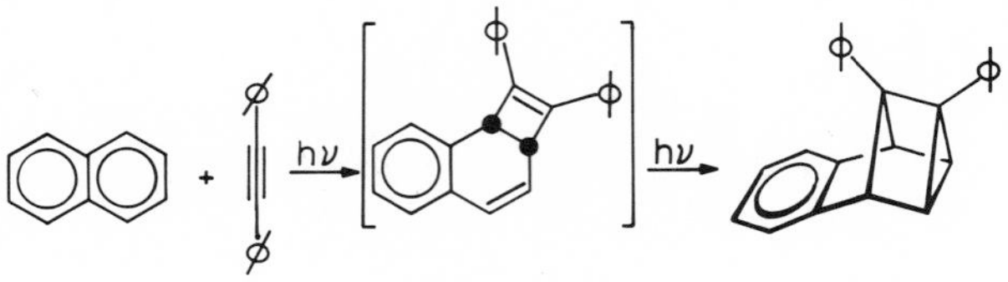

hν hν

(23) (24) (25)

With conjugated polyenes, ring closures and openings may be observed [19].

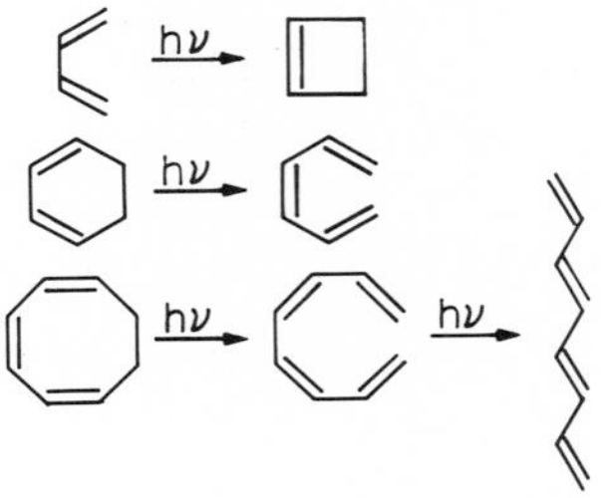

hν
hν
hν hν

These appear to be concerted reactions, unaffected, for example, by quenchers such as oxygen or piperylene [19a]. Triplet sensitizers do not promote them, and in fact may hinder them by chemically diverting the excited singlets [20]. The stereochemistry of these "electrocyclic reactions" is discussed in Section 14-2.

In spite of the simplicity of these reactions, surprises still seem the order of the day in photochemistry; thus, 9,10-cis-dihydronaphthalene (26) gives bullvalene (27) on irradiation at 254 nm [21].

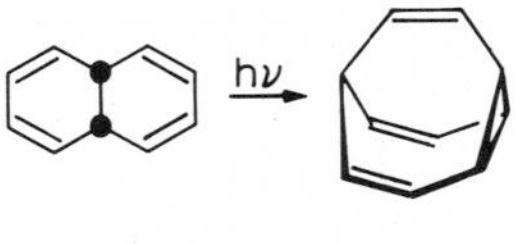

hν

(26) (27)

---

18. W.H.F. Sasse, P.J. Collin, and G. Sugowdz, Tetrahedron Lett., 3373 (1965).
19. (a) W.G. Dauben, Chem. Weekblad, 60, 381 (1964); (b) T.D. Goldfarb and L. Lindqvist, J. Amer. Chem. Soc., 89, 4588 (1967).
20. J. Saltiel, R.M. Coates, and W.G. Dauben, J. Amer. Chem. Soc., 88, 2745 (1966).
21. W. von E. Doering and J.W. Rosenthal, J. Amer. Chem. Soc., 88, 2078 (1966).

Another interesting photoreaction is the conversion of o-divinylbenzene to benzobicyclo[3.1.0]hexene (28); labeling experiments have shown that no hydrogen atoms are shifted in the process [22].

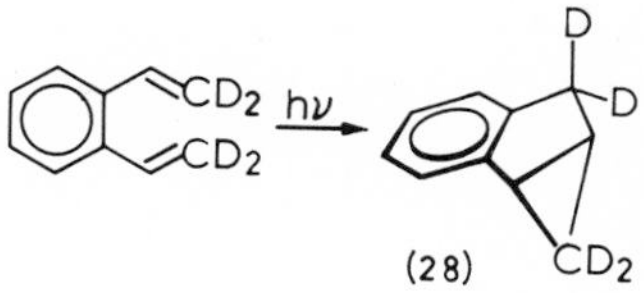

Also of interest is the closure reaction of the vinylcyclopropane derivative (29), which yields product (30) in addition to minor amounts of several other compounds [23].

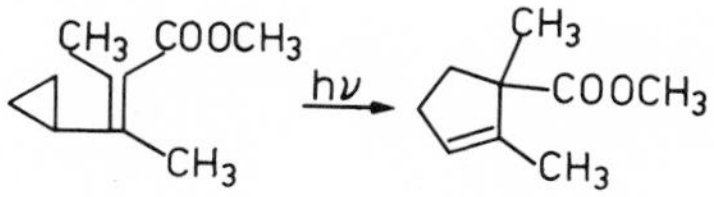

13-4. REACTIONS OF CARBONYL COMPOUNDS

In contrast to the simple olefins in which only $\pi^* \leftarrow \pi$ transitions are possible, in carbonyl compounds a non-bonding electron is available for promotion into a π^* orbital, and much of the photochemistry of carbonyl compounds is that of $\pi^* \leftarrow n$ states. This has a great effect on the properties of this group as may be demonstrated by means of formaldehyde - one of the few molecules for which the physical properties of the excited state are known. The dipole moment of the triplet is 1.3 D as compared to 2.3 D for the ground state, the C-O length is 1.31 Å as compared to 1.22 for the ground state, and $\nu_{C=O}$ is 1180 cm^{-1} as compared to 1745 cm^{-1} for the unexcited molecules; the excited molecule is pyramidal and inverts at a very fast rate [24]. The oxygen atom, as expected, is strongly electrophilic, and the molecule has much radical character. Intersystem crossing is relatively easy in these states, and the question whether the singlet or triplet state is responsible for a given product often arises; by the proper use of quenchers and sensitizers this can frequently be determined. The main chemical

---

22. (a) M. Pomerantz, J. Amer. Chem. Soc., 89, 694 (1967); (b) J. Meinwald and P.H. Mazzocchi, J. Amer. Chem. Soc., 89, 696 (1967).
23. (a) M.J. Jorgenson, J. Amer. Chem. Soc., 88, 3463 (1966); (b) M.J. Jorgenson and C.H. Heathcock, J. Amer. Chem. Soc., 87, 5264 (1965).
24. D.E. Freeman and W. Klemperer, J. Chem. Phys., 45, 52 (1966).

properties of excited carbonyl compounds are abstraction reactions, α-cleavage and decarbonylations, and addition to carbon-carbon multiple bonds [25].

Hydrogen abstraction is a common reaction, and something that should be kept in mind when ketones such as benzophenone are going to be used as sensitizers: If a primary or secondary alcohol solvent is used, much benzhydrol and benzpinacol (31) will be obtained.

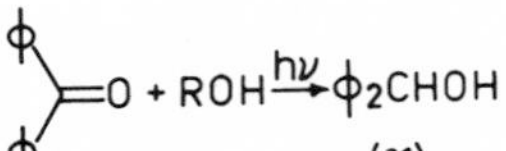

$$\phi_2C{=}O + ROH \xrightarrow{h\nu} \phi_2CHOH$$

(31)

Examples of internal hydrogen abstractions are also common. 1-Adamantylacetone (32) rearranges to the epimeric mixture of alcohols (33) upon irradiation [26]; this type of a γ-hydrogen atom abstraction to produce a diradical is known as a Norrish type II reaction.

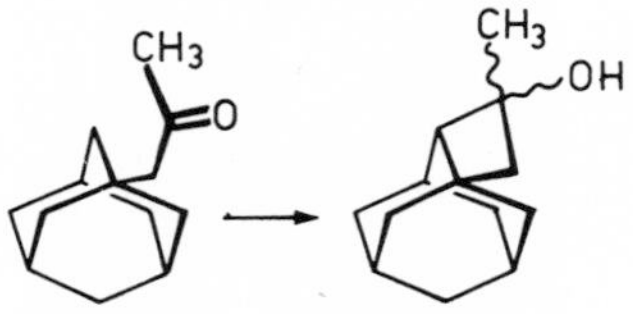

CH_3 … $=O$ → CH_3 … OH

(32) (33)

It is not always followed by recombination; sometimes fission occurs as shown [27].

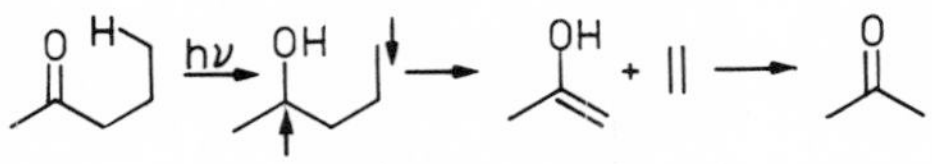

O H $\xrightarrow{h\nu}$ OH → OH + || → O

It appears that the competition between these various reactions depends on the multiplicity of the excited state. It has been found in one example, namely that of *cis*-dibenzoylethylene (34), that irradiation in the presence of benzophenone in alcohol solvents leads to simple hydrogenation whereas without the sensitizer a phenyl 1,5-shift occurred followed by reaction of the intermediate ketene with the solvent [28].

---

25. For review, see N.J. Turro, J.C. Dalton, K. Dawes, G. Farrington, R. Hautala, D. Morton, M. Niemczyk, and N. Schore, *Accounts Chem. Res.*, *5*, 92 (1972).
26. R.B. Gagosian, J.C. Dalton, and N.J. Turro, *J. Amer. Chem. Soc.*, *92*, 4752 (1970).
27. G. Adam, *Z. Chem.*, *8*, 441 (1968).
28. (a) G.W. Griffin and E.J. O'Connell, *J. Amer. Chem. Soc.*, *84*, 4148 (1962); (b) H.E. Zimmerman, H.G.C. Dürr, R.G. Lewis, and S. Bram, *J. Amer. Chem. Soc.*, *84*, 4149 (1962).

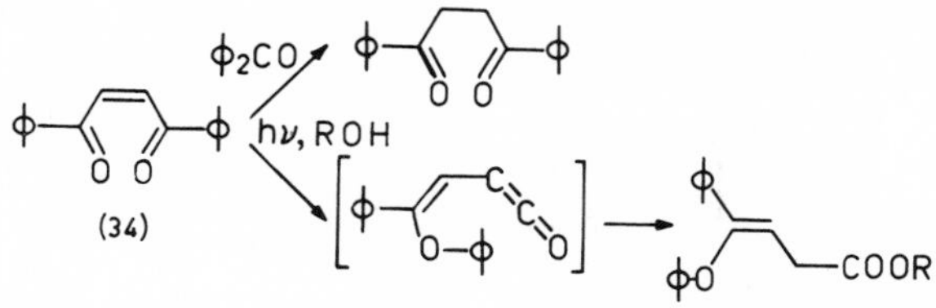

α-Cleavage produces a radical pair which may disproportionate or recombine; the former reaction leads to ketene derivatives, and the latter to epimerization [27].

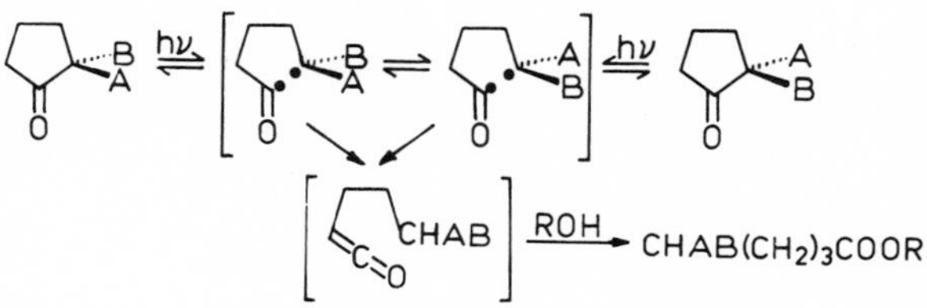

Decarbonylation - a double α-cleavage - is referred to as a Norrish type I reaction. It has been used for the synthesis of many strained ring compounds such as the cyclopropanone (35) [29] and bicyclo[1.1.1]pentane (36) [30].

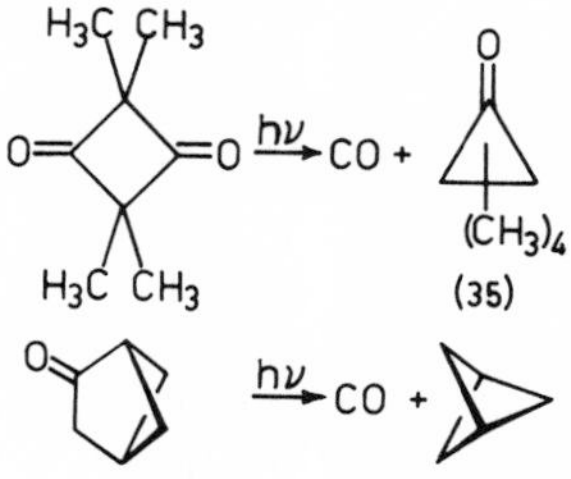

These reactions are apparently concerted, and not simply a succession of two α-cleavages to give a radical pair; thus, if a mixture of (37) and (38) is irradiated, no cross-over products result, and radical traps such as mercaptans do not interfere even at high concentration [31].

29. N. J. Turro, W. B. Hammond, and P. A. Leermakers, J. Amer. Chem. Soc., 87, 2774 (1965).
30. J. Meinwald, W. Szkrybalo, and D. R. Dimmel, Tetrahedron Lett., 731 (1967).
31. M. Akhtar, Tetrahedron Lett., 4727 (1965).

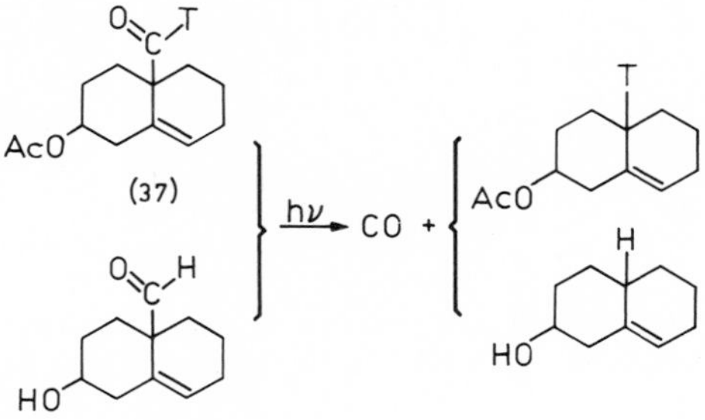

(37)

(38)

Photoaddition of carbonyl groups to double bonds to give oxetanes have been observed in many instances; thus, 2-methylhepten-6-one (39) gives oxetanes (40) and (41) [32], and benzophenone adds photochemically to furan to give (42) [33].

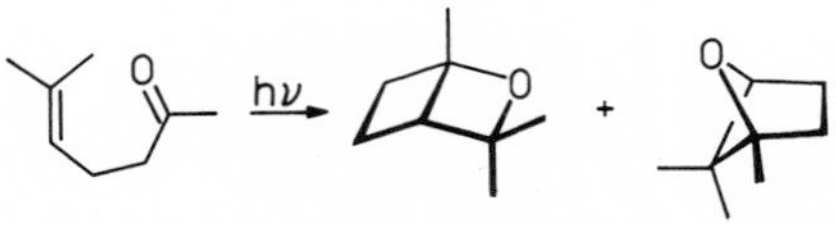

(39) (40) (41)

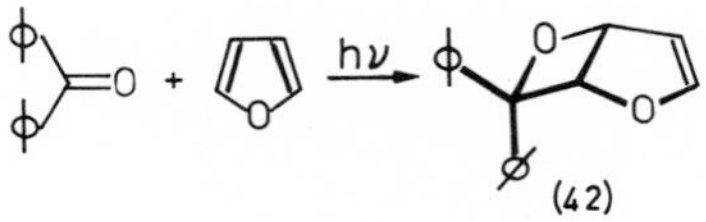

(42)

The mechanism of a photochemical reaction can often be traced in quite a bit of detail and we discuss one example here. Irradiation of cyclobutene in solution in acetone leads to three products [34]. Although the quantum yields for all three processes are high (e.g., up to 30% for (45)), a trace of biacetyl stops the reaction completely.

(43) (44) (45)

Since this compound is a well-known triplet quencher (able to stop benzophenone sensitization, for example), the reaction obviously involves the acetone triplet.

32. N.C. Yang, M. Nussim, and D.R. Coulson, Tetrahedron Lett., 1525 (1965).
33. R. Srinivasan, J. Amer. Chem. Soc., 82, 775 (1960).
34. R. Srinivasan and K.A. Hill, J. Amer. Chem. Soc., 88, 3765 (1966).

One could therefore imagine the reaction to involve the following pathway (M means any molecule, A means acetone and C stands for cyclobutene).

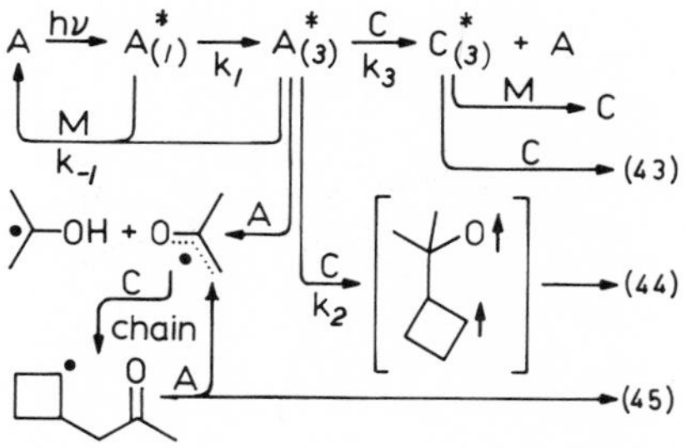

If the scheme is correct, the rate law for the formation of (44) must meet the criterion suggested by the following analysis. If n represent the absorbed light flux (the number of quanta absorbed per liter per sec) and N is Avagadro's number, the steady state approximation (see p. 541) for $A^*_{(3)}$ is:

$$d[A^*_{(3)}]/dt = 0 = n/N - k_{-1}[A^*_{(3)}]\,[M] - k_3[A^*_{(3)}]\,[C] - k_2[A^*_{(3)}]\,[C] - k_4[A]\,[A^*_{(3)}]$$

so that

$$[A^*_{(3)}] = (n/N)/\{k_{-1}M + (k_3 + k_2)\,[C] + k_4[A]\}$$

Now:

$$d[(44)]/dt = k_2[C]\,[A^*_{(3)}] = k_2[C]\,(n/N)/\{k_{-1}[M] + (k_2 + k_3)\,[C] + k_4[A]\}$$

and since we may define the quantum yield of formation of (44)

$$\phi_{(44)} = (d[(44)]/dt)/(n/N)$$

we have

$$\phi_{(44)} = k_2[C]/\{k_{-1}[M] + (k_2 + k_3)\,[C] + k_4\,[A]\}$$

and

$$1/\phi_{(44)} = \frac{k_2 + k_3}{k_2} + \frac{k_{-1}[M] + k_4[A]}{k_2[C]}$$

A plot of $1/\phi_{(44)}$ versus $1/[C]$ should therefore be a straight line, and if this is found to be so then the proposed mechanism has met one criterion. The other quantum yields can be similarly examined. These relations of ϕ are called Stern-Volmer plots; both slope and intercept often give information about the individual rate constants.

The reactions of α,β-unsaturated carbonyl compounds by and large are similar to those of the simple olefins, although $\pi^* \leftarrow n$ transitions are then involved and longer wavelengths can be used. Thus, the simple cycloaddition (46) → (47) can be carried out at a wavelength of 415 nm [35].

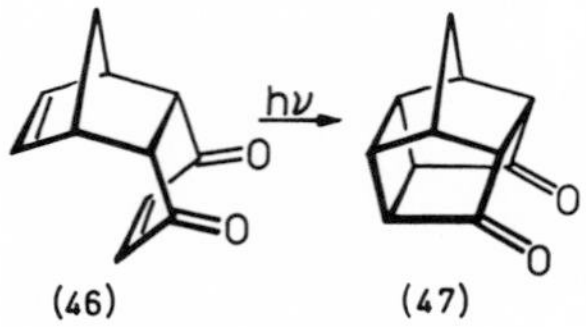

hν

(46) (47)

The conversion of carvone (48) into (49) is another example [36].

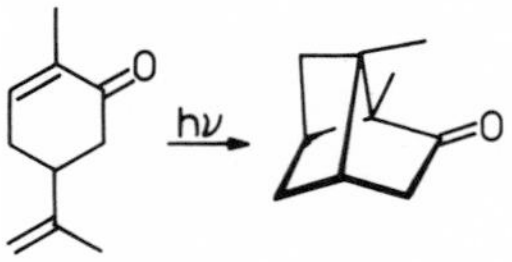

hν

(48) (49)

A double cycloaddition occurs upon the irradiation of p-quinone (50) [37]; unfortunately, the yield of the conversion is very low (less than 1%). The reason for this may be that the reaction may revert thermally; thus, upon irradiation of 2,3-dimethylquinone (51), one soon finds significant quantities of the 2,5-isomer (52) [38].

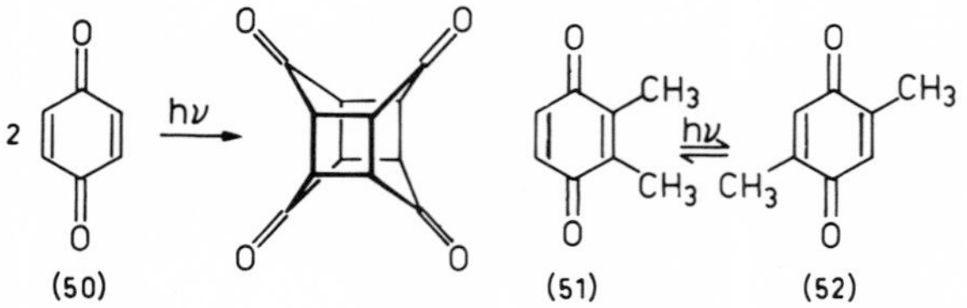

hν

(50) (51) (52)

Benzoquinone is also involved in one of the somewhat rare [4+2]cyclo additions to a conjugated diene: Thus, butadiene yields the spiro compound (53) [39].

---

35. R.C. Cookson, E. Crundwell, R.R. Hill, and J. Hudec, J. Chem. Soc., 3062 (1964).
36. J. Meinwald and R.A. Schneider, J. Amer. Chem. Soc., 87, 5218 (1965).
37. D. Bryce-Smith and A. Gilbert, J. Chem. Soc., 2428 (1964).
38. W. Flaig, J.C. Salfeld, and A. Llanos, Angew. Chem., 72, 110 (1960).
39. J.A. Barltrop and B. Hesp, J. Chem. Soc., 5182 (1965).

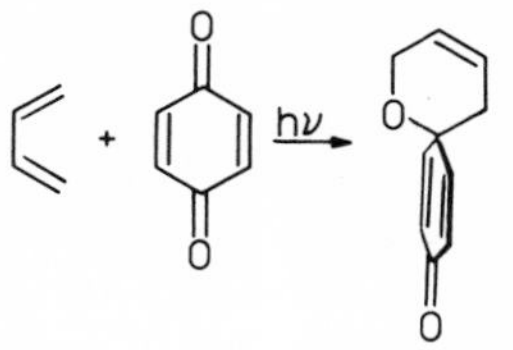

(53)

Tolane can also be added to benzoquinone; the product is (54) [40].

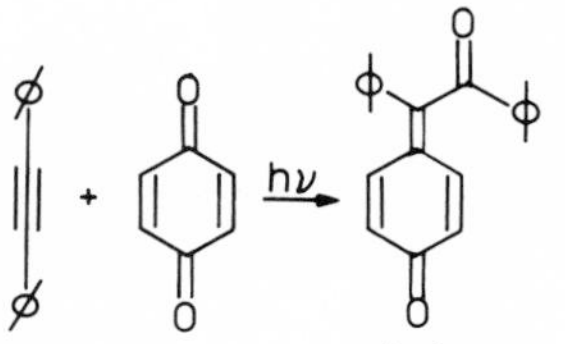

(54)

The cycloaddition of 3-cyclopentenone to cyclohexene affords the routine product (55), but this reaction has an unusual feature [41].

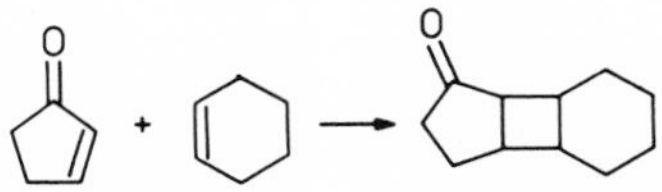

(55)

Benzophenone, which has a triplet energy of 69 kcal/mole, can sensitize the ketone; the latter compound quenches benzophenone phosphorescence and stops its photoreduction by isopropyl alcohol. However, benzophenone is not able to promote the formation of (55); this can only be done by sensitizers (such as cyclopropyl phenyl ketone) which have triplet energies of at least 73 kcal/mole. The conclusion is that the reactive species is the second excited triplet state - a situation which has few analogs. This study was made difficult by the further circumstances that there are a few sensitizers with triplet energies as low as 59 kcal/mole (e.g. acenaphthene (56)) that do promote the cycloaddition; however in each case it was shown that cyclopentenone quenched the fluorescence of these materials, and hence that these were acting as singlet sensitizers.

(56)

40. H.E. Zimmerman and L. Craft, Tetrahedron Lett., 2131 (1964).
41. P. de Mayo, J.P. Pete, and M. Tchir, J. Amer. Chem. Soc., 89, 5712 (1967).

Cyclopentenone then crosses over to the second excited triplet state and goes on to react.

Much work has been done to elucidate the mechanism of the photochemical rearrangement of 4,4-disubstituted cyclohexadienones (57) to form 6,6-disubstituted bicyclo[3.1.0]hex-3-en-2-ones (58) - products which themselves subsequently suffer further rearrangements, ring opening, and so on [42].

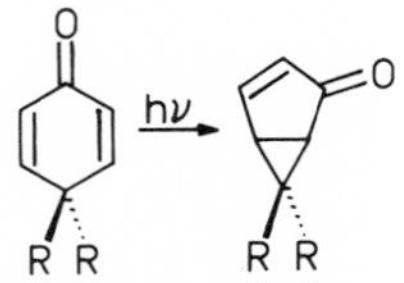

hν

(57) (58)

It was proposed at an early stage (e.g. when R is phenyl [43]) that the initial cold product was a zwitterion (59) which subsequently rearranged.

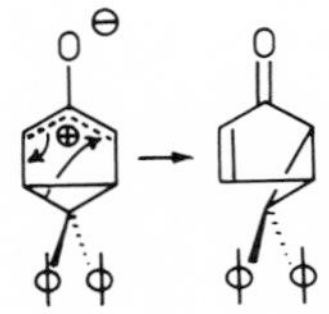

(59) (58)

The best evidence for such an intermediate is the fact that both the dehydrobromination of (60) with base and the debromination of (61) with sodium produce (58) [44].

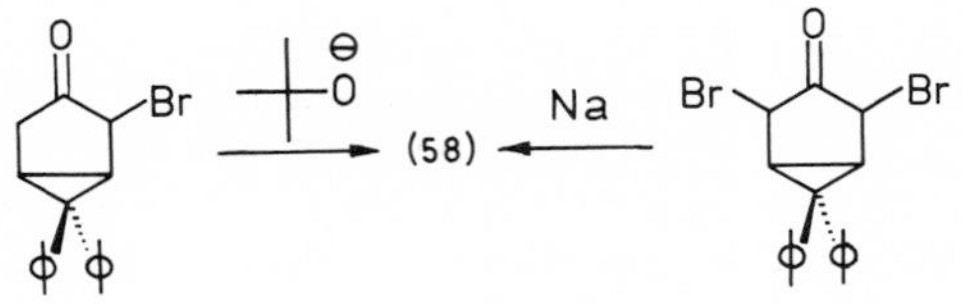

(60) (58) Na (61)

Later however, difficulties developed such as the finding that cyclohexenones such as (62) in like fashion gave rise to products (63) [45].

42. H.E. Zimmerman, Science, 153, 837 (1966).
43. H.E. Zimmerman and D.I. Schuster, J. Amer. Chem. Soc., 84, 4527 (1962).
44. (a) H.E. Zimmerman, D. Döpp, and P.S. Huyffer, J. Amer. Chem. Soc., 88, 5352 (1966); (b) H.E. Zimmerman and G.A. Epling, J. Amer. Chem. Soc., 94, 7806 (1972).
45. For references and discussion, see H.E. Zimmerman, R.G. Lewis, J.J. McCullough, A. Padwa, S.W. Staley, and M. Semmelhack, J. Amer. Chem. Soc., 88, 1965 (1966).

(62) (63)

Furthermore, it was shown that the excited state had more diradical than zwitterionic character as originally postulated; thus, the reaction can be sensitized by benzophenone and quenched by piperylene, and compound (64) in certain solvents gives rise to p-cresol (65), presumably by ejecting the relatively stable CCl_3 radical [46].

(64) (65)

Finally, even in the unsensitized reaction there is no incontrovertible evidence that there is an intermediate at all, and a concerted reaction cannot be ruled out.

(57) (58)

α-Dicarbonyl compounds are a group of chemicals of great interest to photochemists. They tend to have readily accessible triplet states and be excellent quenchers. Hydrogen abstraction is likewise common, and this is presumably involved in the photo rearrangement of 1,2-alkanediones such as (66) into cyclobutanolones (67) [47].

(66) (67)

---

46. (a) D. J. Patel and D. I. Schuster, *J. Amer. Chem. Soc.*, **89**, 184 (1967); (b) D. I. Schuster and C. J. Polowczyk, *J. Amer. Chem. Soc.*, **88**, 1722 (1966); (c) however, cf. also J. King and D. Leaver, *Chem. Commun.*, 539 (1965).
47. N. C. Yang and D. D. H. Yang, *Tetrahedron Lett.*, No. 4, 10 (1960).

An indanolone (68) has been similarly obtained from the α-dione (69) [48].

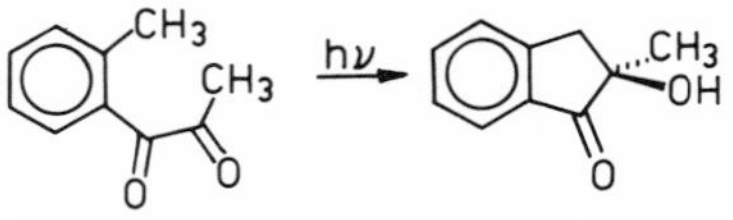

(69) (68)

A reasonably flexible chain seems to be important in these reactions; thus, camphorquinone (70) does not react (a totally different type of reactión sets in if oxygen is admitted) [49].

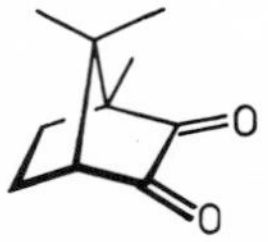

(70)

A [4+2] cyclo addition has been observed in the case of phenanthraquinone (71), which gives the benzodioxane adduct (72) [50].

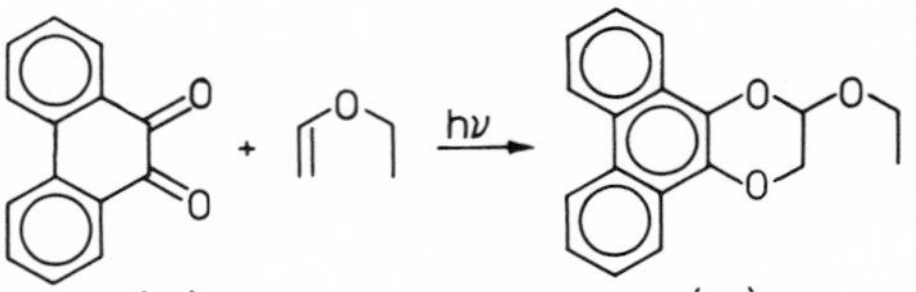

(71) (72)

13-5. REARRANGEMENTS IN AROMATIC RINGS

Some of the earliest successes in the benzene photochemistry resulted from the use of highly substituted and in some cases distorted rings. Thus, either 1,3,5- or 1,2,4-tri-t-butylbenzene (73) and (74) upon irradiation at 245 nm led to

48. R. Bishop and N.K. Hamer, Chem. Commun., 804 (1969).
49. J. Meinwald and H.O. Klingele, J. Amer. Chem. Soc., 88, 2071 (1966).
50. C.H. Krauch, S. Farid, and G.O. Schenck, Chem. Ber., 98, 3102 (1965).

a photostationary mixture containing the several valence isomers (73)-(77) in the ratio of 7:21:1:7:65, respectively [51].

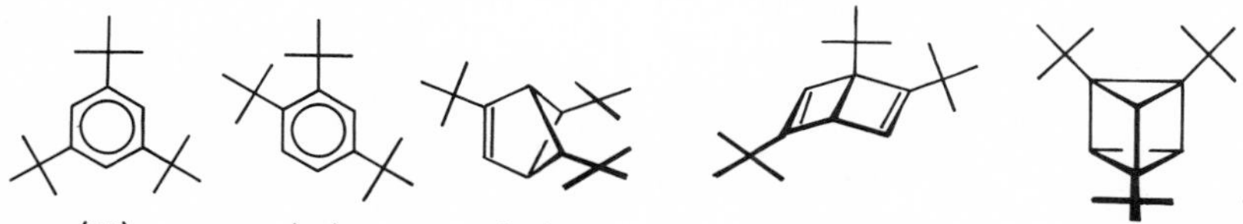

(73) (74) (75) (76) (77)

Hexafluorobenzene vapor (78) upon photolysis gives rise to the Dewar benzene valence isomer (79) [52]; neither of these reactions is subject to sensitization.

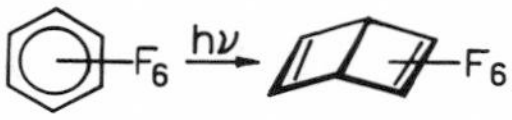

F_6 $\xrightarrow{h\nu}$ F_6

(78) (79)

It had been noted earlier that the various methylbenzenes could be photoisomerized [53]; however, it was not known whether valence isomers were involved as intermediates until it was shown that the methyl groups migrate with the carbons to which they were originally bound. Thus, mesitylene-1,3,5-C_3^{14} (80), prepared from acetone-2-C^{14}, upon photolysis is partly transformed into 1,2,4-trimethylbenzene (81). This in turn can be ozonized and subsequently converted into a mixture of glyoxime (82), methylglyoxime (83) and dimethylglyoxime (84); the radioactivity of the five compounds (80)-(84) is then found to be precisely in the ratio 3:3:0:1:2 [54].

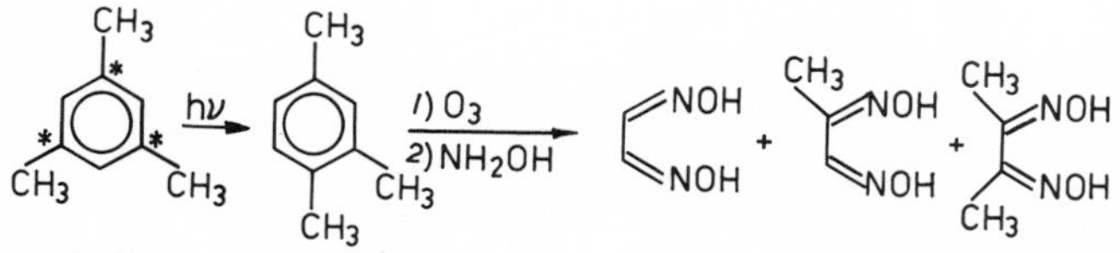

CH_3 $\xrightarrow{h\nu}$ CH_3 $\xrightarrow{1) O_3 \; 2) NH_2OH}$ NOH NOH + CH_3 NOH NOH + CH_3 NOH NOH CH_3

(80) (81) (82) (83) (84)

The photochemistry of benzene itself is very complex, and not readily understood in terms of the highest unoccupied and lowest unfilled molecular orbitals (Fig. 13-3). One might expect that the excited singlet would have some of the

51. K.E. Wilzbach and L. Kaplan, J. Amer. Chem. Soc., 87, 4004 (1965).
52. I. Haller, J. Amer. Chem. Soc., 88, 2070 (1966).
53. (a) K.E. Wilzbach and L. Kaplan, J. Amer. Chem. Soc., 86, 2307 (1964); (b) A.W. Burgstahler and P.L. Chien, J. Amer. Chem. Soc., 86, 2940 (1964).
54. L. Kaplan, K.E. Wilzbach, W.G. Brown, and S.S. Yang, J. Amer. Chem. Soc., 87, 675 (1965).

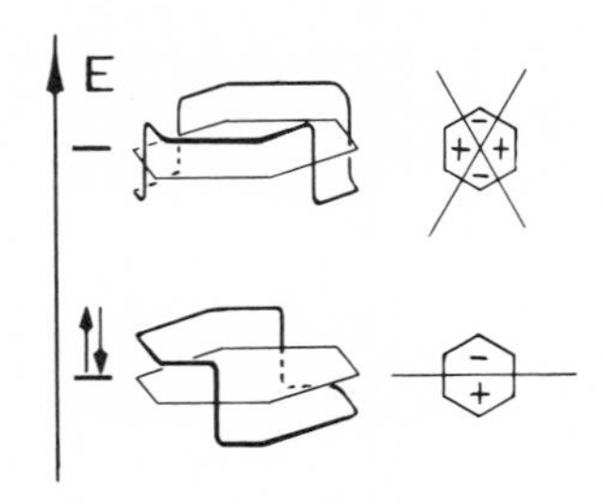

Fig. 13-3. The highest filled and lowest empty orbitals of benzene.

diradical character suggested by structure (85); a distortion bringing two *para*-carbon atoms more closely together has indeed been described [55].

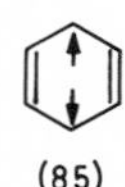

(85)

One might therefore furthermore suppose that this might lead to Dewar benzene and perhaps prismane; however, in the early stages of the work only a small quantity of benzvalene was found in the irradiation of liquid benzene at 254 nm [56]. Both in this experiment and in the photolysis of benzene vapor at 185 nm fulvene (86) is formed [57].

(86)

A rationalization by Bryce-Smith and Longuet-Higgins [58] has proved capable of accommodating a large number of existing and subsequent observations. They assume that (85) stabilizes itself, partly by rearrangement to the singlet diradical (87) and partly by intersystem crossing to (88).

(87) ← (85) → (88)

55. M.S. de Groot and J.H. van der Waals, *Molecular Physics*, *6*, 545 (1963).
56. K.E. Wilzbach, J.S. Ritscher, and L. Kaplan, *J. Amer. Chem. Soc.*, *89*, 1031 (1967).
57. H.R. Ward, J.S. Wishnok, and P.D. Sherman, *J. Amer. Chem. Soc.*, *89*, 162 (1967).
58. D. Bryce-Smith and H.C. Longuet-Higgins, *Chem. Commun.*, 593 (1966).

Intermediate (87) would account for the formation of fulvene and benzvalene, and for the observation that monoolefins give rise to adducts such as (89) and (90); triplet (88) presumably gives rise to (91) [59].

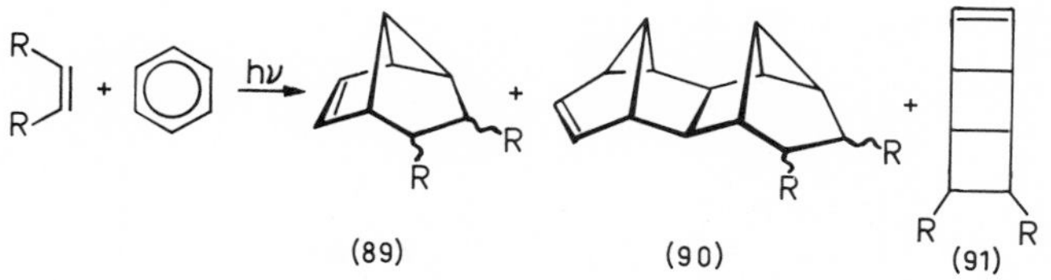

(89) (90) (91)

An example of (91) is the cyclobutene adduct (92) recently obtained by Srinivasan [60].

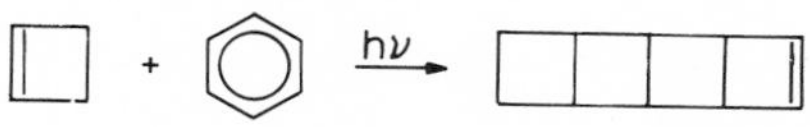

(92)

Dewar benzene itself, which is attributed to this triplet also, was eventually found under conditions somewhat different from those applied by earlier workers [61]. The same intermediate also accounts for the 1,4-addition products such as (93) with dienes [62]; on the other hand alcohols have recently been found [63] to capture (87) to give bicyclic ethers (94) and (95) (the configurations of compounds (89)-(94) are not precisely known at present).

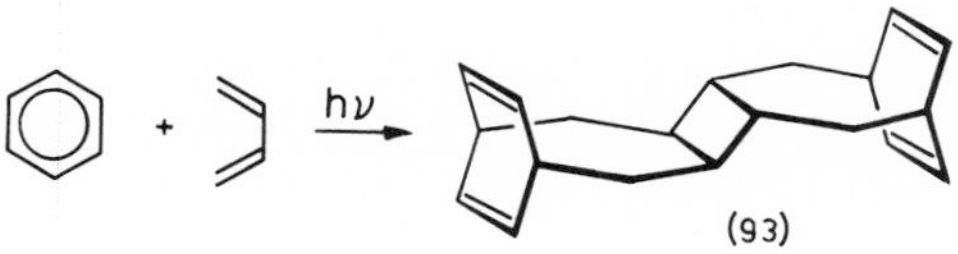

(93)

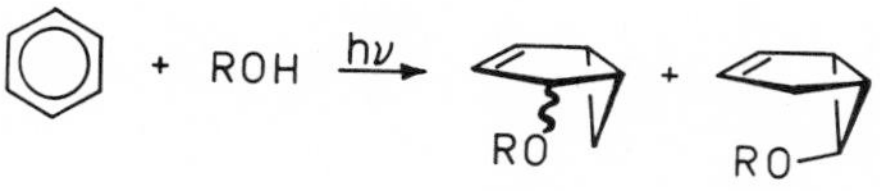

(94) (95)

---

59. (a) K.E. Wilzbach and L. Kaplan, J. Amer. Chem. Soc., 88, 2066 (1966); (b) D. Bryce-Smith, A. Gilbert, and B.H. Orger, Chem. Commun., 512 (1966).
60. R. Srinivasan and K.A. Hill, J. Amer. Chem. Soc., 87, 4653 (1965).
61. H.R. Ward and J.S. Wishnok, J. Amer. Chem. Soc., 90, 1085 (1968).
62. G. Koltzenburg and K. Kraft, Tetrahedron Lett., 389 (1966).
63. L. Kaplan, J.S. Ritscher, and K.E. Wilzbach, J. Amer. Chem. Soc., 88, 2881 (1966).

As mentioned above, strained benzene rings seem more prone to photo-shuffling than others. An extraordinary result has been reported by Wasserman, who found that [2.2]paracyclonaphthane (96) upon irradiation is transformed into dibenzequinine (97); the symmetrical nature of this hydrocarbon was clearly evident from its simple ir and nmr spectrum (three sharp singlets at 2.65, 7.60 and 8.36 τ in a 2:2:1 ratio, respectively) [64].

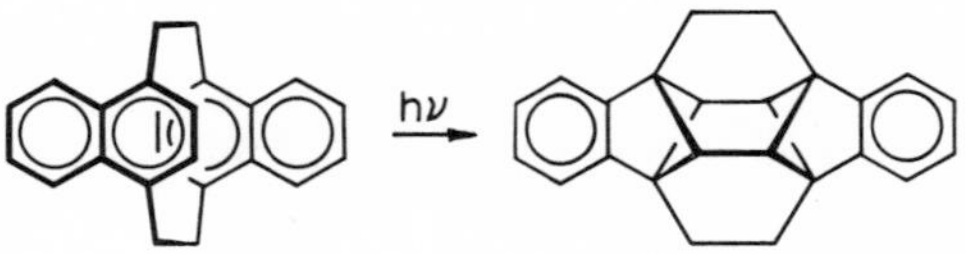

(96) (97)

It seems likely that the tools of photochemistry will increasingly be applied to the infinite variety of non-benzenoid aromatic hydrocarbons. Already it has been found that pyrazine (98) is photochemically transformed into pyrimidine (99) to some degree [65], and that 3,5-diphenylisoxazole (100) upon irradiation leads to the 2,5-isomer (101) [66]; in this latter reaction the azirinyl ketone (102) could be fished out of the reaction mixture and separately converted into (101).

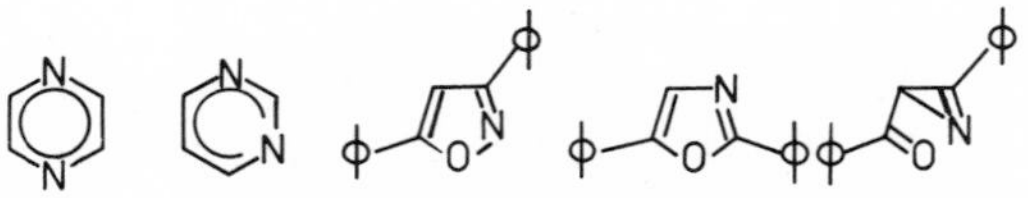

(98) (99) (100) (101) (102)

Similar rearrangements are encountered in thiophene; thus, 2-phenylthiophene (103) upon irradiation gives 3-phenylthiophene (104);

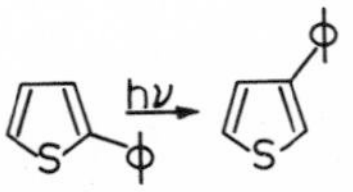

(103) (104)

2,2'-dithienyl (105) similarly gives the 2,3'- and ultimately the 3,3'-isomer ((106) and (107), respectively) [67].

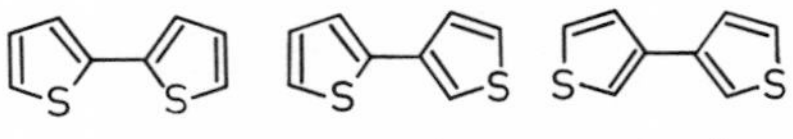

(105) (106) (107)

64. H.H. Wasserman and P.M. Keehn, J. Amer. Chem. Soc., 89, 2770 (1967).
65. F. Lahmani, N. Ivanoff, and M. Magat, Compt. Rend., 263, 1005 (1966).
66. B. Singh and E.F. Ullman, J. Amer. Chem. Soc., 89, 6911 (1967).
67. H. Wynberg and H. van Driel, J. Amer. Chem. Soc., 87, 3998 (1965).

Since the substituent remains bound to the same carbon atom, intermediate valence isomers such as (108)-(110) seemed a possibility [68];

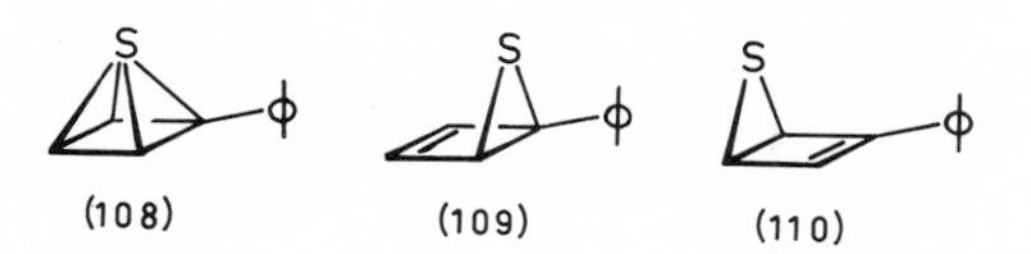

(108) (109) (110)

recently however, it has been learned that D-labeled thiophene (111) gives rise to (112), and hence an intermediate such as (113) seems indicated [69].

(111) (112) (113)

Many of the photo reactions of aromatic species are now known to involve excited state complexes, or excimers (see p. 850).

13-6. EXAMPLES OF PHOTOLYSIS

In our previous discussion, the word photolysis has been used now and then to describe any chemical reaction resulting from irradiation; here we discuss it in terms of its narrower significance as a photo process resulting simply in bond fission. This phenomenon is of course involved in virtually all of the processes referred to earlier, but a number of interesting photoreactions remain that cannot conveniently be classified in any other way.

The photolysis of azo compounds to produce nitrogen and hydrocarbons is a process which is sensitized by excited singlets [70]. Thus, azo-2-methyl-2-propane (114) and 2,3-diazabicyclo[2.2.1]hept-2-ene (115) efficiently quench the fluorescence of anthracene, phenanthrene, and so on, as they decompose; triplet quenchers such as piperylene do not affect the process.

(114) (115)

---

68. H. Wynberg and H. van Driel, Chem. Commun., 203 (1966).
69. H. Wynberg, R.M. Kellogg, H. van Driel, and G.E. Beekhuis, J. Amer. Chem. Soc., 88, 5047 (1966).
70. P.D. Bartlett and P.S. Engel, J. Amer. Chem. Soc., 90, 2960 (1968).

The paracyclophanes are of further interest in that they can be photoracemized. Thus (-)-4-methyl[2.2]paracyclophane (116) racemizes when irradiated, and the same is true of the (-)4,7-dimethyl-analog (117). Mechanisms involving symmetrical intermediates such as (118)-(121) can be ruled out, and only diradicals (122) can reasonably accommodate the evidence; thus, in some cases oxygen was found to inhibit the racemization [71].

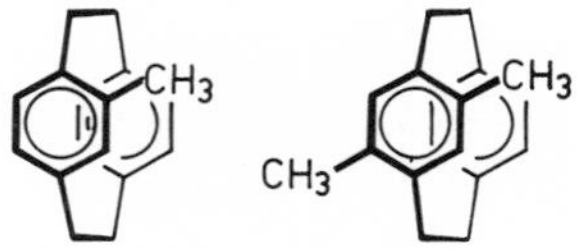

(116) (117)

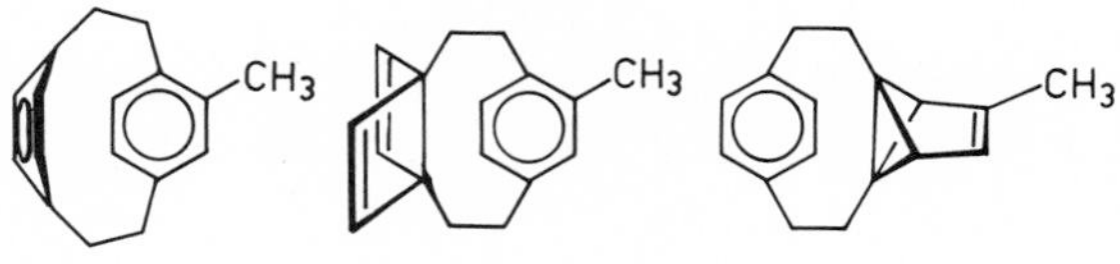

(118) (119) (120)

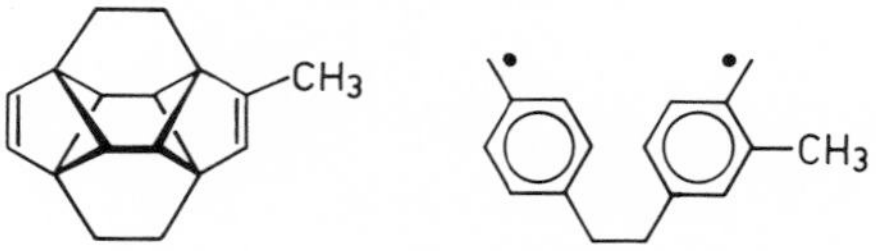

(121) (122)

Reversible bond fission is also involved in certain examples of photochromism (cf. p. 400). The spiropyrans (123) may serve as an example. When these colorless compounds are irradiated, a highly colored isomer is formed which upon standing in the dark slowly reverts to (123) [72]. The colored form has been shown to be (124). Note that (124) is not an excited state but simply an unstable isomer formed by irradiation.

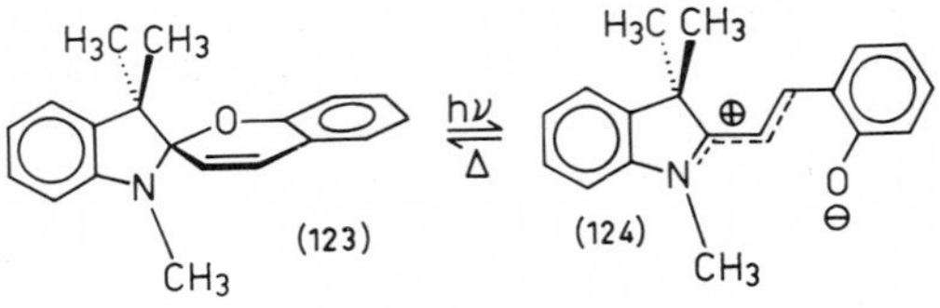

(123) $\underset{\Delta}{\overset{h\nu}{\rightleftharpoons}}$ (124)

It is believed that a simple S-S bond cleavage in lipoic acid (125) is involved in the photosynthetic process [73]; esr spectra suggesting the presence of such radicals have been recorded [74].

---

71. M.H. Delton and D.J. Cram, J. Amer. Chem. Soc., 94, 2471 (1972).
72. E. Berman, R.E. Fox, and F.D. Thomson, J. Amer. Chem. Soc., 81, 5605 (1959).
73. R.B. Whitney and M. Calvin, J. Chem. Phys., 23, 1750 (1955).
74. E.E. Smissman and J.R.J. Sorenson, J. Org. Chem., 30, 4008 (1965).

(125)

Finally, a number of photo-Claisen rearrangements are known in which no allylic inversion occurs such as in the thermal analog, and furthermore p- as well as o-rearrangement is found. The intermediacy of free radical fragments has been suggested [75].

13-7. THE ACID STRENGTH OF EXCITED STATES

Excited states are subject not only to relaxation and chemical conversion, they also have equilibrium properties - which are, unfortunately, usually not amenable to measurement. An exception is their acid strength, which can be determined by means of the so-called Förster cycle [76]. The electronic absorption spectrum of the acid of interest is measured over a wide range of pH. The acid and its anion will have quite different λ_{max}. The frequencies are related to the energy differences ΔE_{HA} and ΔE_{A^-}; the pKa of the ground state corresponds to a free energy difference ΔG_{HA}, and the pKa of the excited state can then be calculated by completing the cycle as shown in Fig. 13-4. The combination of ΔE and ΔG values can be avoided by measuring the temperature dependence of pK_{HA} as well [77].

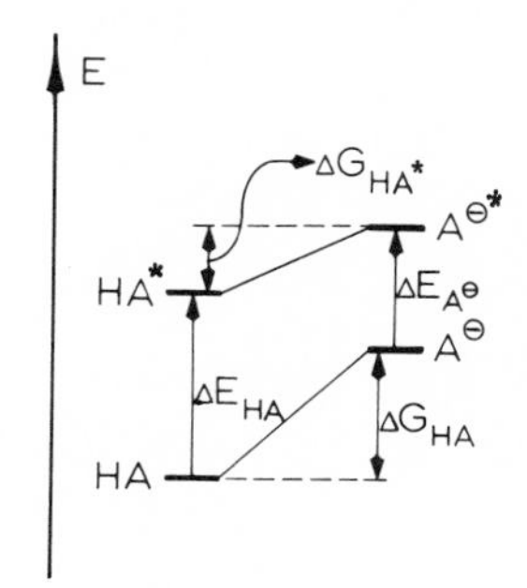

Fig. 13-4. The Forster cycle.

75. D. P. Kelly, J. T. Pinhey, and R. D. G. Rigby, Tetrahedron Lett., 5953 (1966).
76. T. Förster, Z. Elektrochem., 54, 42, 531 (1950).
77. A. Weller, Z. Elektrochem., 56, 662 (1952).

There is another uncertainty in that the absorption maxima may not be indicative of the energies of the excited electronic state, since one may actually have raised one or both species to excited vibrational states as well. For this reason a second measurement should be based on the fluorescence spectrum and its pH-dependence, and since fluorescence may underestimate the ΔE values (see the Jablonski diagram), perhaps the wisest course to follow it to take an average of the two results if they are widely different. A second reason for doing this is that because of Franck-Condon considerations, the excited state of the anion will not be solvated in the normal manner in the absorption experiment, and the ground state will not see the solvent in an equilibrium configuration in the fluorescence measurement [78]. Alternatively, pK_{HA^*} may be calculated from fluorescence intensities [79].

While the methods may yield answers that differ by as much as three pKa units, they agree that for all phenols examined, the excited states are far more acidic than the ground states, generally by 6-12 pKa units [79]. The reason for this is of course that the oxygen atom of phenols donates one of its nonbonding electrons to a π^* orbital spread out over the ring, and hence becomes far more electrophilic than is the case in the ground state. When $\pi^* \leftarrow \pi$ transitions are involved, the change in pKa is just opposite; thus, although the conjugate acid of a ketone is generally a very strong acid, it is only weakly so in the excited state.

The volume change upon ionization can also be determined for the excited state. This is done by measuring the pressure dependence of the pKa*, and the use of the equation

$$\Delta V^* = 2.303 \ RT \ \partial pKa^*/\partial p.$$

ΔV for ground state acid ionization is generally -10 to -20 cm^3/mole; the contraction may be ascribed to the electrostriction of the solvent surrounding the ions. For excited phenols, the contraction is generally only about half that value; this suggests - as do the pKa*-values themselves - that in excited phenoxides the charge is much more highly delocalized into the ring than in the ground state anions. With p-nitrophenols, ΔV^* is even positive, which hints at the highly dipolar character (126) of the excited phenol [80].

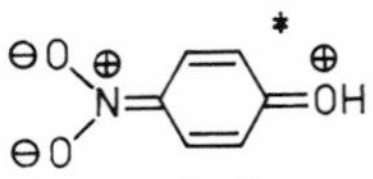

(126)

78. E. L. Wehry and L. B. Rogers, Spectrochim. Acta, 21, 1976 (1965).
79. For a discussion of the merits of these methods, see H. H. Jaffé and H. L. Jones, J. Org. Chem., 30, 964 (1965).
80. S. D. Hamann, J. Phys. Chem., 70, 2418 (1966).

13-8. PHOTOCHEMISTRY WITHOUT LIGHT

Excited states are not necessarily molecules that have just absorbed a quantum of light. Conceivably a lot of chemical energy may be built into a highly strained molecule, and if this is suddenly set free in the formation of a very stable molecule, this excess energy may - if the symmetry-based selection rules permit it - appear in the form of electronic excitation rather than thermal energy. Dioxetanes (127) have been found to decompose to two carbonyl fragments of which one is electronically excited. This species may sensitize others, and cause them to undergo photochemical transformations in the dark.

(127)

Thus, if trimethyldioxetane (128) is allowed to decompose in the presence of *trans*-stilbene, the products are acetone, acetaldehyde and *cis*-stilbene.

(128)

Similarly, if the substrate is 4,4-diphenylcyclohexadienone (57), lumiproduct (58) is obtained [81].

(128) (57) (58)

If acenaphthylene (129) is present, dimerization is the result, and the *cis*- and *trans*-dimers (130) and (131) form in the same ratio as in sensitized irradiation experiments [82].

---

81. E.H. White, J. Wiecko, and D.F. Roswell, *J. Amer. Chem. Soc.*, *91*, 5194 (1969).
82. E.H. White, J. Wiecko, and C.C. Wei, *J. Amer. Chem. Soc.*, *92*, 2167 (1970).

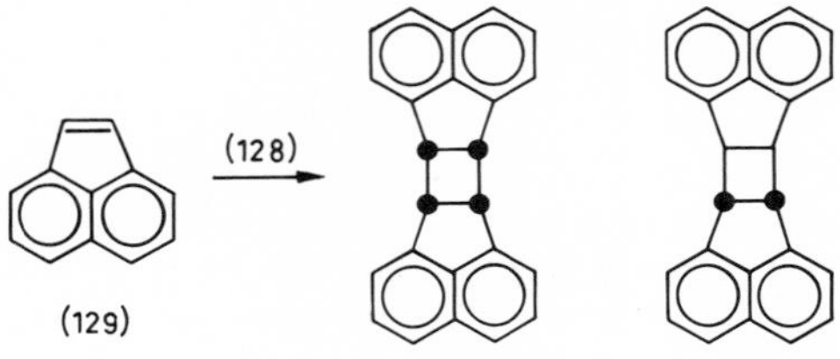

(128)

(129) (130) (131)

Since the successful sensitizers have an E_T of about 70 kcal/mole, at least that much energy must be released in the decomposition of (128). Recently, 3,3-dimethyl-1,2-dioxetane (132) has also been generated and shown to decompose to excited products [83] (cf. also p. 883).

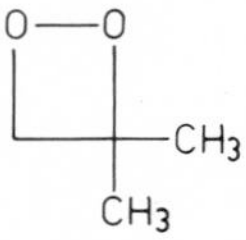

(132)

As detailed further in the next section, certain oxalate esters (133) upon oxidation by hydrogen peroxide yields transient 1,2-dioxetanedione (134); this decomposes to two molecules carbon dioxide of which one is electronically excited.

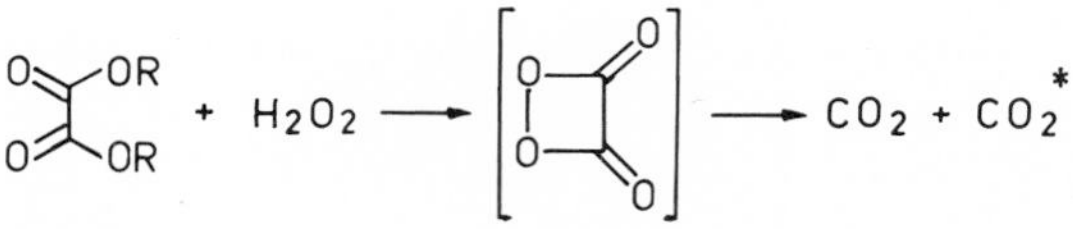

$$(133) + H_2O_2 \longrightarrow [(134)] \longrightarrow CO_2 + CO_2^*$$

This molecule may then bring about certain chemical transformations; among those that were demonstrated were the cyclization of o-tolylpropane-1,2-dione (69) and the cycloaddition of ethyl vinyl ether to phenanthraquinone (71) [84] (cf. p. 444).

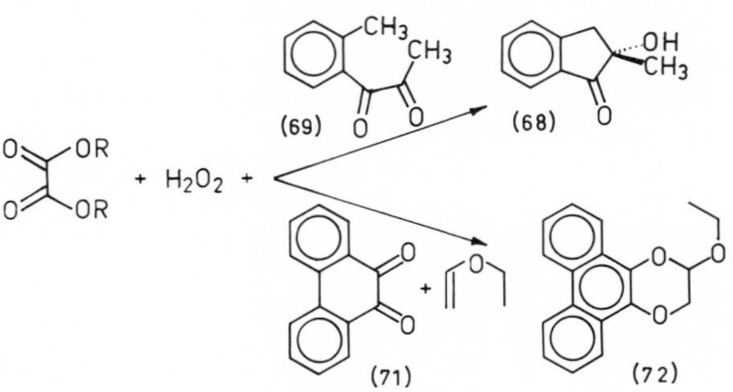

(69) (68)

(71) (72)

83. W.H. Richardson and V.F. Hodge, J. Amer. Chem. Soc., 93, 3996 (1971).
84. H. Güsten and E.F. Ullman, Chem. Commun., 28 (1970).

13-9. CHEMILUMINESCENCE

The emission of light from a chemical system is a fascinating thing to see, and many chemists have concerned themselves with it in recent years. If dioxetane (128) is decomposed without a substrate to accept and utilize the chemical energy thus available, light emission occurs centered at 435 nm, and hence this then is an example of chemiluminescence [82].

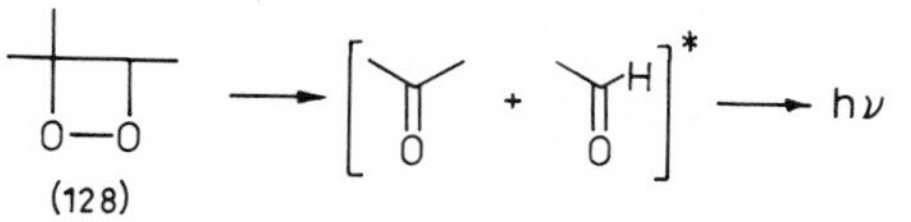

(128) → hν

Alternatively, if a suitable fluorescer such as anthracene is present, the excited product may induce its fluorescence, and induced phosphorescence (for instance with biacetyl) has likewise been achieved [81]. Relatively pure, monochromatic light can be obtained that way. If the dioxetane is decomposed in the presence of europium tris(thenoyltrifluoroacetonate)-1, 10-phenanthroline (135), 80% of the emission occurs at 613 nm with a quantum yield $\phi = 0.015$ [85].

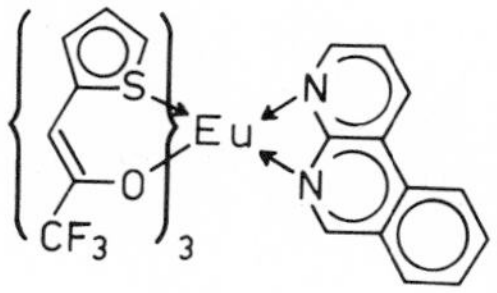

(135)

The mechanism of the dioxetane decomposition is of much interest at present. There are two main possibilities. The first of these, the concerted decomposition, is a reaction allowed by the Woodward-Hoffmann rules (see next Chapter, Section 14-3), and it should yield a singlet excited state.

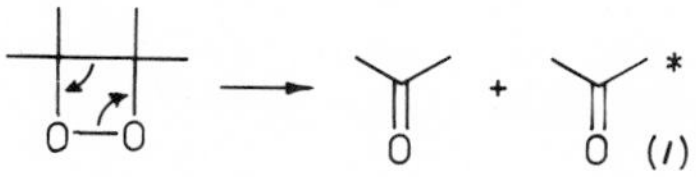

(I)

In the second possibility, a singlet diradical is first formed, which may or may not undergo intersystem crossing to a triplet before further decomposition.

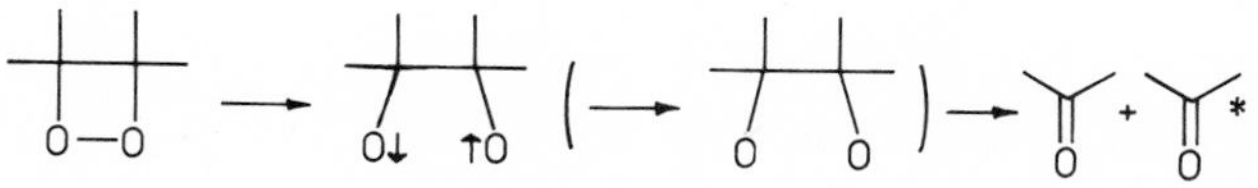

---

85. P. D. Wildes and E. H. White, J. Amer. Chem. Soc., 93, 6286 (1971).

Since induced fluorescence had been observed in several cases (see above), it was generally assumed that the first mechanism was operating. However, the evidence recently developing seems to favor the latter possibility. Thus, it has been found that 1,1-dimethyl- and 1-methyl-1-phenyldioxetanes (136) and (137), respectively, decompose at virtually identical rates; the authors expected that the transition state of the concerted reaction would be of lower energy for (124) because of the developing C=O double bond [86].

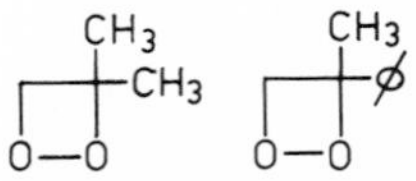

(136) (137)

Further strong evidence for the second mechanism is the fact that the excited carbonyl compound is apparently a triplet. Thus, it was already known that singlet acetone reacts with trans-1,2-dicyanoethylene (138) to give oxetane (139), and that the triplet instead isomerizes this substrate to the cis-isomer (140); when tetramethyldioxetane is decomposed in the presence of (138), the result is 50% (140) and less than 1% (139) [87].

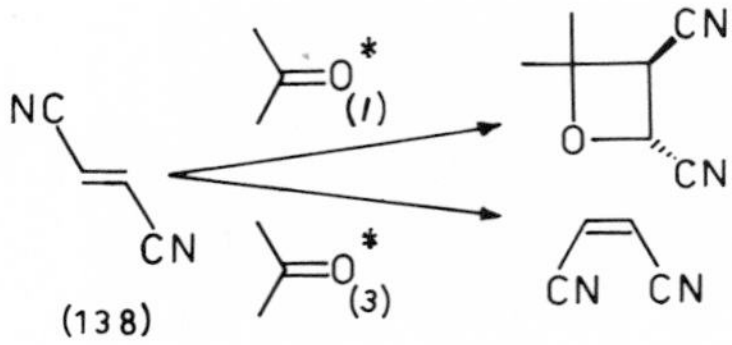

(138) (139) (140)

It should be remembered that fluorescence yields are generally low, so that the observation of weak emissions alone is not always conclusive.

The intermediacy of α-peroxylactones (141) in certain bioluminescing systems had been suspected for some time, but it had never been possible to isolate a simple member of this class of compounds. However, recently such a substance has been prepared. Use was made of the ability of t-butyl groups to stabilize a variety of small rings (see p. 204): t-butyl α-peroxylactone (142) proved to be obtainable in carbon tetrachloride solution at -10° [88].

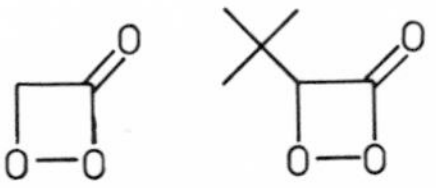

(141) (142)

---

86. W.H. Richardson, M.B. Yelvington, and H.E. O'Neal, J. Amer. Chem. Soc., 94, 1619 (1972).
87. N.J. Turro and P. Lechtken, J. Amer. Chem. Soc., 94, 2886 (1972) and Pure Appl. Chem., 33, 363 (1973).
88. W. Adam and J.C. Liu, J. Amer. Chem. Soc., 94, 2894 (1972).

The solution showed 9:1 nmr singlet signals at 8.90 and 4.52 τ, and had $\nu_{C=O}$ at 1875 cm^{-1} (compare α-lactones, $\nu_{C=O}$ = 1900 cm^{-1}). At room temperature the solution decomposes with light emission to give carbon dioxide and pivalaldehyde (143).

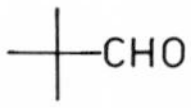

$(CH_3)_3C{-}CHO$

(143)

1,2-Dioxetanedione (carbon dioxide dimer), (134) has been much discussed as a source of chemiluminescence. It had been discovered by Chandross [89] that the oxidation of oxalyl chloride (144) with hydrogen peroxide in the presence of fluorescers led to light emission, the spectrum of which matched the fluorescence spectrum.

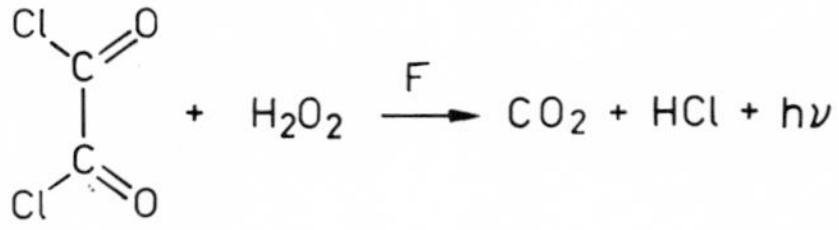

$$ClC(=O){-}C(=O)Cl + H_2O_2 \xrightarrow{F} CO_2 + HCl + h\nu$$

(144)

The mechanism of this reaction was elucidated by Rauhut [90] and shown to involve (134).

Thus, this material - although not isolable in any way - is volatile and long-lived enough to be swept out of the reaction mixture by means of an inert gas stream; it may then be used to induce fluorescence in another vessel. An m/e = 88 peak has been observed if such a gas stream is led directly into a mass spectrometer [91]. However, it was recently shown that if the ion residence time in the mass spectrometer is reduced to nearly zero (10^{-8} sec), the intensity of the 88 peak virtually vanishes [92]; this last piece of evidence if therefore at best uncertain at the moment.

---

89. E.A. Chandross, Tetrahedron Lett., 761 (1963).
90. (a) M.M. Rauhut, B.G. Roberts, and A.M. Semsel, J. Amer. Chem. Soc., 88, 3604 (1966); (b) M.M. Rauhut, Accounts Chem. Res., 2, 80 (1969).
91. H.F. Cordes, H.P. Richter, and C.A. Heller, J. Amer. Chem. Soc., 91, 7209 (1969).
92. J.J. DeCorpo, A. Baronavski, M.V. McDowell, and F.E. Saalfeld, J. Amer. Chem. Soc., 94, 2879 (1972).

A different type of chemical system capable of chemiluminescence is exemplified by luminol, (145), which upon basic oxidation gives rise to singlet excited 3-aminophthalate ion (146); this will fluorescence ($\phi = 0.01$), or can be used to produce induced fluorescence [93].

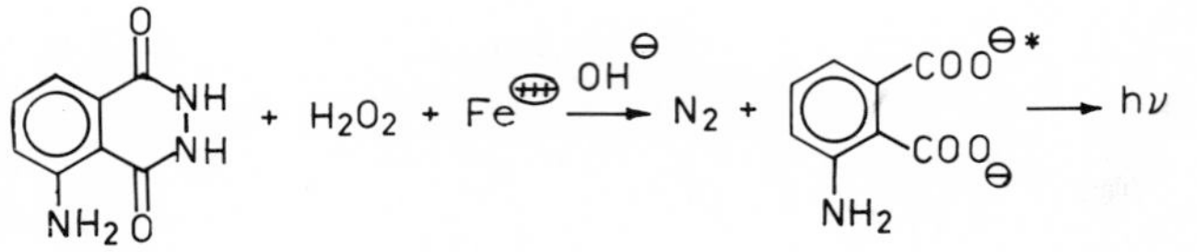

$$(145) + H_2O_2 + Fe^{3+} \xrightarrow{OH^-} N_2 + (146)^* \longrightarrow h\nu$$

(145) (146)

A great deal of work has already been done to improve the efficiency of this process by modification of the hydrazide. Some success has been achieved by incorporating a more efficient fluorescer as a substituent in the molecule [94]; thus, the benzo(ghi)perylene-1,2-dicarboxylic acid hydrazide (147) reaches a quantum yield of nearly 0.09. This is of course still a far cry from firefly luciferin, which has a quantum yield of 88% [95].

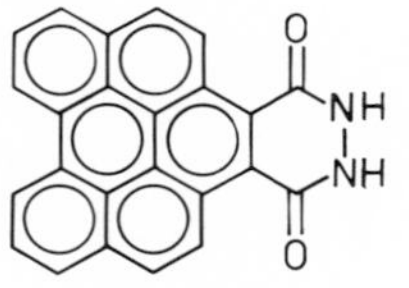

(147)

A vital requirement in efficient chemiluminescent reactions - beside those of sufficient energy and molecular orbital symmetry conservation - is that in the critical step the energy release must be sudden. That is to say, in the reaction profile there must be very little atomic motion during that step (see Fig. 13-5), since otherwise the energy is likely to get distributed in the various bonds as excess vibrational and hence thermal energy. This is the case especially in electron transfer reactions. Thus, if sodium naphthalenide is oxidized with a variety of

---

93. E.H. White and D.F. Roswell, Accounts Chem. Res., 3, 54 (1970).
94. (a) D.F. Roswell, V. Paul, and E.H. White, J. Amer. Chem. Soc., 92, 4855 (1970); (b) C.C. Wei and E.H. White, Tetrahedron Lett., 3559 (1971); (c) E. Rapaport, M.W. Cass, and E.H. White, J. Amer. Chem. Soc., 94, 3153 (1972).
95. H.H. Seliger and W.D. McElroy, Arch. Biochem. Biophys., 88, 136 (1960).

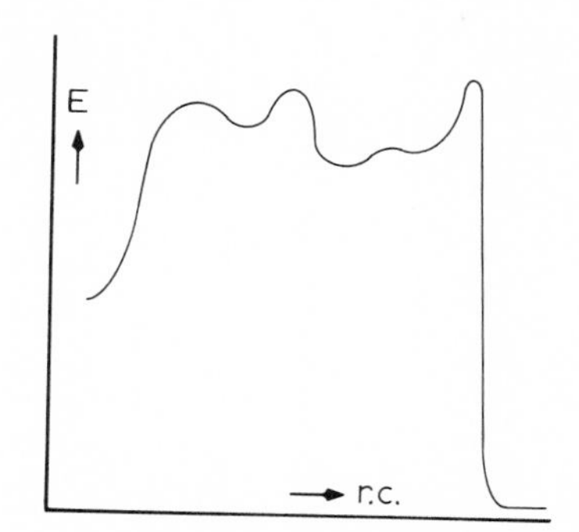

Fig. 13-5. A hypothetical multistep reaction in which the last step is chemiluminescent.

oxidizing agents, naphthalene fluorescence can be observed [96]. An especially elegant system is the electron transfer from radical anions to radical cations. Thus, naphthalene fluorescence is observed when this hydrocarbon, embedded in a low temperature matrix and γ-irradiated, is allowed to slowly warm up [97], or when a rapidly alternating current is used to electrolyze fluorescent hydrocarbons [98].

---

96. (a) E.A. Chandross and F.I. Sonntag, J. Amer. Chem. Soc., 88, 1089 (1966) and 86, 3179 (1964); (b) J.W. Haas and J.E. Baird, Nature, 214, 1006 (1967).
97. B. Brocklehurst, G. Porter, and J.M. Yates, J. Phys. Chem., 68, 203 (1964).
98. (a) D.M. Hercules, Science, 145, 808 (1964); (b) J.T. Maloy, K.B. Prater, and A.J. Bard, J. Amer. Chem. Soc., 93, 5959 (1971); (c) J.T. Maloy and A.J. Bard, J. Amer. Chem. Soc., 93, 5968 (1971).

Chapter 14

THE WOODWARD-HOFFMANN RULES

14-1. INTRODUCTION

An interesting item appeared in the Annalen der Chemie in 1958. Vogel [1] reported having observed that *cis*-3, 4-dicarbomethoxycyclobutene (1) is subject to ring opening to give the butadiene (2), and that it does so in a cleanly stereospecific manner. Only the *cis*-*trans*- isomer is obtained, and by the principle of microscopic reversibility the same stereochemistry should apply if it were possible to conduct the reaction in the opposite direction.

COOCH$_3$ COOCH$_3$ (1) $\xrightarrow{\Delta}$ COOCH$_3$ COOCH$_3$ (2)

---

1. E. Vogel, *Ann.*, *615*, 14 (1958).

Three years later Havinga [2] described some of the results of his investigations in vitamin D chemistry. He had already noted that lumisterol (3) undergoes photochemical ring cleavage to pre-calciferol (4) and subsequent reversible reclosure to ergosterol (5); now he found that (4) would also undergo thermal ring closure, but to pyrocalciferol (6) and isopyrocalciferol (7) - hence with stereochemistry opposite to that in the irradiation experiments.

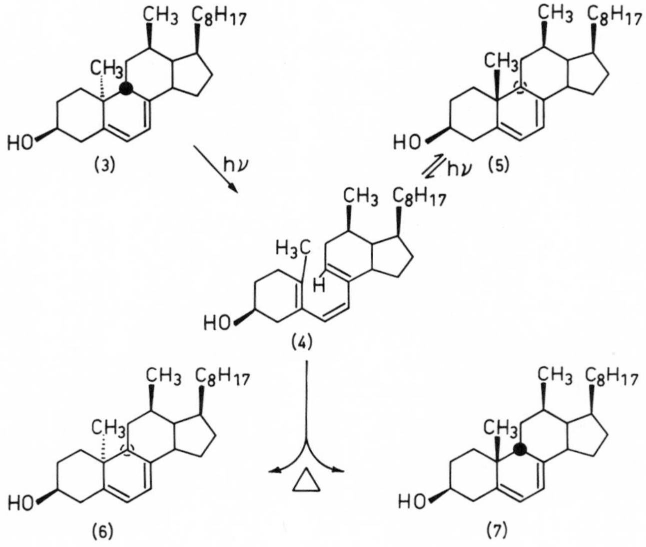

After a discussion of these facts in terms of steric control, he concluded: "As Prof. Oosterhoff pointed out, another factor that possibly contributed to the stereochemical difference between the thermal and the photo induced ring closure may be found in the symmetry characteristics of the highest occupied π-orbital of the conjugated hexatriene system. In the photo excited state this highest occupied orbital is antisymmetric with regard to the plane that is perpendicular to the bond 6,7 - making 'syn' approach less favorable."

Elsewhere in the same paper, after taking note of the fact that of these various ring-closed isomers only (7) upon irradiation undergoes further reaction to photoisopyrocalciferol (8), he concludes his argument of steric control with: "....., a process that is furthered by the symmetry characteristics of the orbital of the promoted electron in the butadiene system. Although it would seem rash to consider this qualitative reasoning as sufficient ...., it may indicate the line along which a more rigorous treatment could clarify the preference for formation of the bicyclohexene systems in the case of these <u>syn</u> substituted cyclohexadiene derivatives."

2. E. Havinga and J. L. M. A. Schlatmann, Tetrahedron, 16, 146 (1961).

These prophetic words were not widely noted, and several years had passed when Woodward ran into an unexpected problem in his vitamin B12 synthesis. In the course of this synthesis he obtained (9) which upon heating cyclized as hoped, but with stereochemistry (10) opposite to that expected (11) on the basis of steric and electronic arguments. It was learned on further investigation that if (10) were irradiated it would reopen to give the stereoisomer (12), and that this triene upon heating gave the desired compound (11).

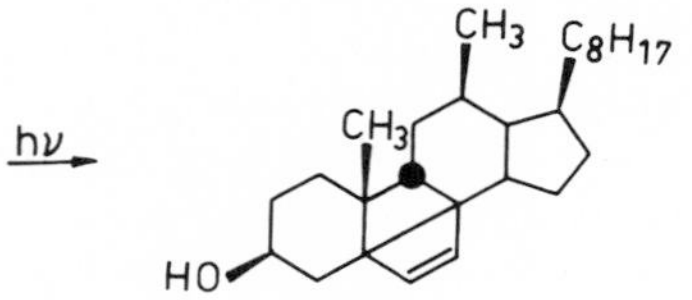

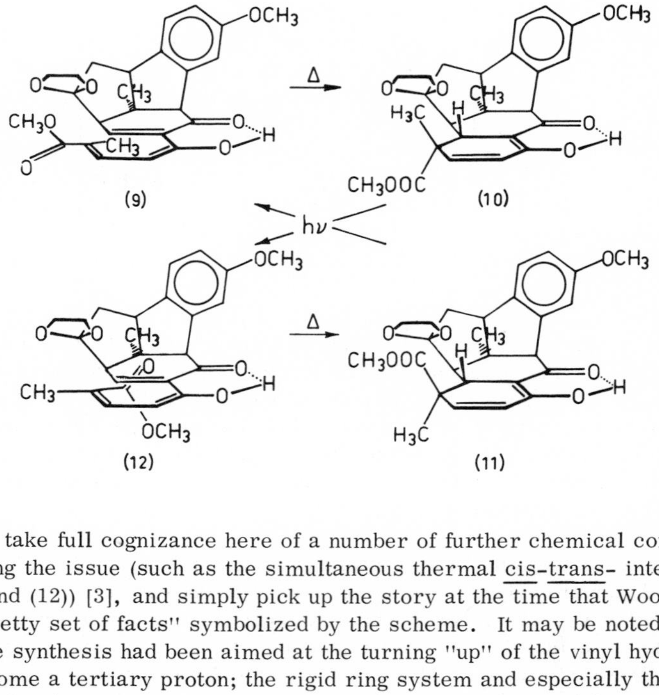

We need not take full cognizance here of a number of further chemical complications clouding the issue (such as the simultaneous thermal cis-trans- interconversion of (9) and (12)) [3], and simply pick up the story at the time that Woodward had the "very pretty set of facts" symbolized by the scheme. It may be noted that the design of the synthesis had been aimed at the turning "up" of the vinyl hydrogen about to become a tertiary proton; the rigid ring system and especially the ketal group make the alternative geometry impossible. The unexpected feature was the epimeric configuration of the CMeCOOMe group. Woodward realized that the basis

3. (a) R. B. Woodward, "Aromaticity", Special Publication No. 21 of the Chemical Society, Burlington House, London, 1967, p. 217; (b) Chem. Eng. News, 43, 38, (Dec. 6, 1965).

of these facts was provided by a new principle (known to the chemical theoreticians for a long time [4] but new to organic chemists), namely that of the conservation of orbital symmetry. In collaboration with Hoffmann he published this principle early in 1965, and during the remainder of that year these authors wrote a classic series of papers showing how virtually all concerted organic reactions - many of them known, many others predicted or implied - were subject to it. For each group of reactions the principle has its own specific consequences; these are now known as the Woodward-Hoffmann rules. In the next several years the Woodward-Hoffmann contributions were followed by a veritable torrent of papers by others, verifications in many cases, the scrutiny of apparent exceptions, new ways to derive or describe the rules, new reactions, further elaborations, reviews, and so on; rarely in the history of science had a new idea found such fertile territory in which to sprout.

Oosterhoff later recalled: "After Havinga and Schlattman had described their results, the idea crossed my mind that the stereochemistry might be controlled by orbital symmetry, or rather by the number of nodes in the upper occupied molecular orbital. After they left I tried to find a corresponding formulation in the Valence Bond theory, but a cursory application of this method did not reveal any difference to be expected from butadienes and hexatrienes. I realized that a more thorough analysis was needed but I forgot about it. It was therefore much to my surprise when several years later several people seemed to know about my rather casual suggestion and were referring to it. After reading the papers by Havinga and Schlattman, and by Woodward and Hoffmann I felt that I should again study the Valence Bond formulation. I suggested the problem to some of my students. It was found by van der Lugt that ionic structures make important contributions when orthogonal atomic orbitals are used, and by Mulder that cyclic permutations are essential. When these precautions are observed, the Molecular Orbital results are confirmed. Soon afterwards van der Hart and Mulder began using an even more powerful, generalized Valence Bond method, which led to new insights, and which covered the previous results as special cases." [5]

In each of the following sections we discuss one of these groups of reactions; while these discussions follow largely historical lines, we shall be bound by them only where it seems convenient from a didactic point of view [6].

---

4. While there are no references or predictions in the older literature pointing to the myriad of possibilities for this principle in organic chemistry, it lies at the heart of the correlation diagrams which had been in use for so long by quantum mechanicians and spectroscopists.
5. L. J. Oosterhoff, private communication. The results referred to are described in (a) J. J. C. Mulder and L. J. Oosterhoff, Chem. Commun., 305, 308, (1970); (b) W. T. A. M. van der Lugt and L. J. Oosterhoff, Chem. Commun., 1235 (1968) and J. Amer. Chem. Soc., 91, 6042 (1969); (c) W. J. van der Hart, J. J. C. Mulder, and L. J. Oosterhoff, J. Amer. Chem. Soc., 94, 5724 (1972).
6. For general reviews, see (a) R. Hoffmann and R. B. Woodward, Science, 167, 825 (1970); (b) R. Hoffmann and R. B. Woodward, Accounts Chem. Res., 1, 17 (1968); (c) R. B. Woodward and R. Hoffmann, Angew. Chem., Int. Ed. Engl., 8, 781 (1969); (d) G. B. Gill, Quart. Rev. (London), 22, 338 (1968); (e) R. E. Lehr and A. P. Marchand, "Orbital Symmetry: A Problem-Solving Approach", Academic Press, New York, 1972. Others of more restricted coverage are quoted below; furthermore, virtually every text book on organic chemistry since 1965 has at least a section devoted to it.

14-2. ELECTROCYCLIC REACTIONS [7]

Before we write the rule for these reactions, some terms need definition. We consider only concerted reactions. These are single step or "no-mechanism" reactions that involve no other intermediates; in multistep reactions the rules will be obeyed in each single, concerted step but the overall results are then not so simply predictable (see Fig. 14-1). Electrocyclic reactions are the cyclizations of completely conjugated terminal polyolefins (13) and the reverse ring opening reactions of the isomeric cycloalkapolyenes $C_nH_{(n+2)}$ (14) and their derivatives; the rules in general are not concerned with whether these reactions are forward or reverse, and whether the products or reactants are thermodynamically more stable - only with the problem of how the one is converted to the other.

$\{H-C=C-H\}_n$ with terminal $CH=CH_2$ groups (13) $\rightleftharpoons$ $\{H-C=C-H\}_n$ with terminal $CH-CH_2$ groups joined in a ring (14)

(13) (14)

The numbers and types of electrons involved are often indicated; thus Vogel's reaction of (1) would now be symbolized by 2π, $2\sigma \rightarrow 4\pi$. Since molecules (13) are or can be completely planar whereas (14) contains two methylene groups perpendicular to that plane, these terminal atoms undergo $\pi/2$-rotations during the reaction. If these two rotations occur in the same direction, the ring closure or opening is said to have conrotatory characteristics; conversely, if they move in the opposite sense, we speak of disrotatory ring formation or cleavage. Thus,

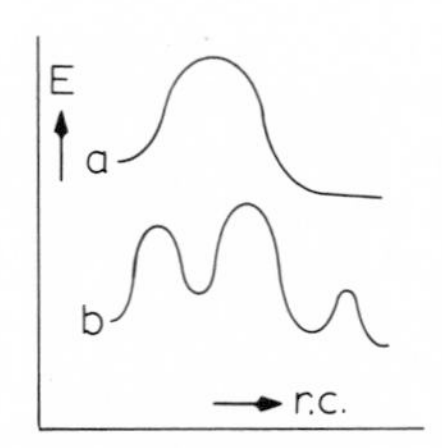

Fig. 14-1. Examples of a concerted reaction (a) and a stepwise reaction (b).

---

7. For review, see J. J. Vollmer and K. L. Servis, J. Chem. Educ., 45, 214 (1968).

Vogel's reaction of (1) is an example of conrotatory ring opening;

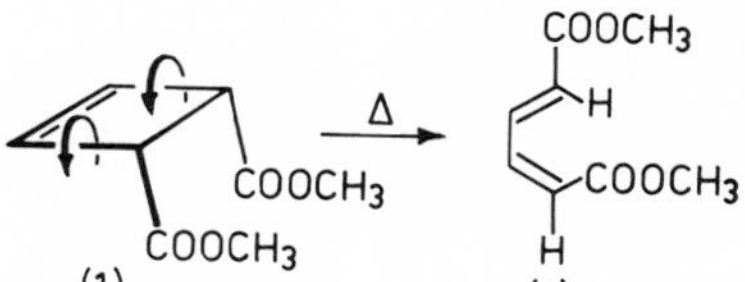

COOCH3
H
Δ
COOCH3
COOCH3
COOCH3
H
(1)
(2)

Havinga's thermal ring closures of pre-calciferol (4) exemplify the disrotatory process.

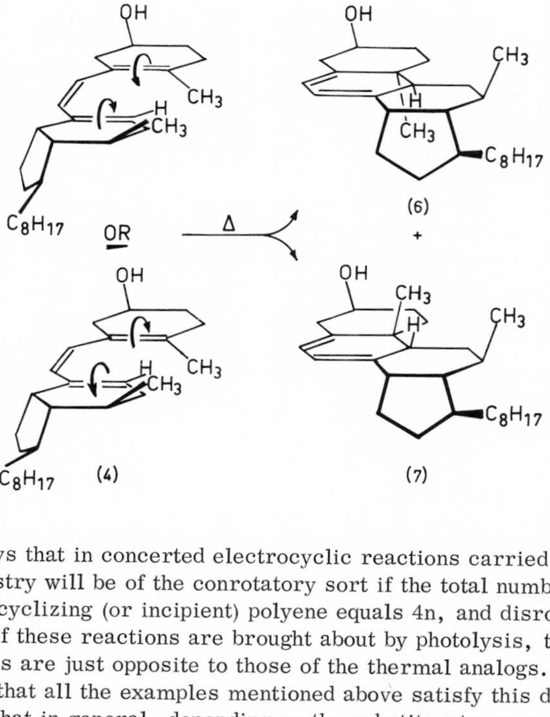

OH
OH
CH3
CH3
H
H
CH3
CH3
C8H17
(6)
C8H17
OR
Δ
+
OH
OH
CH3
H
CH3
H
CH3
CH3
C8H17
C8H17
(4)
(7)

The rule now says that in concerted electrocyclic reactions carried out thermally, the stereochemistry will be of the conrotatory sort if the total number of π-electrons in the cyclizing (or incipient) polyene equals 4n, and disrotatory if it equals (4n + 2); if these reactions are brought about by photolysis, the stereochemical consequences are just opposite to those of the thermal analogs. It should be readily verified that all the examples mentioned above satisfy this description. It may be noted that in general, depending on the substituents present, each substance of the sort described here will generate two stereoisomeric products since both the conrotations and the disrotations can be carried out in either one sense or the other; these isomers may be of the geometric, diastereomeric or optical type depending on the individual case. It happens frequently that strain and hindrance lead to an excess - or even the exclusive appearance - of one of the products; only if the reactant is inactive and the products are optical isomers of one another may equal amounts of both be expected.

The origin of this phenomenon [8] is easily traced on the basis of the molecular π orbitals describing completely conjugated polyenes. Figure 14-2 traces the general contour and modes of the orbitals of several linear conjugated systems (see also Section 1-11). According to the simplest rationale available, the upper occupied molecular orbital of the polyene decides the nature of the rotations; thus, in 1,3-butadiene, in order to provide for overlap between the lobes of the upper occupied π molecular orbital at atoms C_1 and C_4, conrotation is necessary in the thermal reaction. Disrotation would lead to a σ^* antibonding situation. Indeed, the whole purpose of the rules is to identify the most highly bound and hence most stable or lowest energy states and pathways, so that exceptions are scarcely imaginable. In the excited state, the highest occupied orbital has an extra node reversing the signs at C_4, so that disrotation is then the mode of the reaction (see Fig. 14-3). This extremely simple procedure should not need further elaboration.

It may be noted at this point that if the photochemical conversions are triplet sensitized, the reaction involves an intermediate of considerable life-time and can no longer be regarded as concerted. Thus, if we consider Z,Z-2,4-hexadiene, we see that the excited state has an electron in an orbital anti-bonding between atoms 1 and 2, and between 3 and 4 as well. The diradical might therefore very well isomerize to some degree - or even completely equilibrate - before closing.

One might of course raise the same point about the singlet excited state; however, this state has a very much shorter life time and hence - even though strictly speaking it is an intermediate - it has no such chance for isomerization. The distinction may seem a bit arbitrary, but in practice this has not given rise to difficulties.

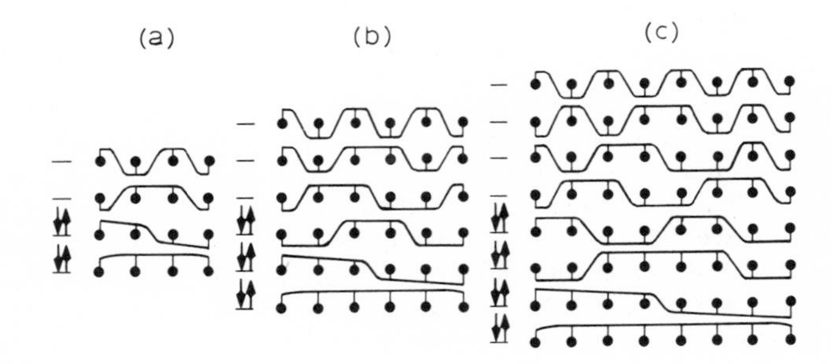

Fig. 14-2. The π molecular orbitals of 1,3-butadiene, 1,3,5-hexatriene and 1,3,5,7-octatetraene.

8. R.B. Woodward and R. Hoffmann, J. Amer. Chem. Soc., 87, 395 (1965).

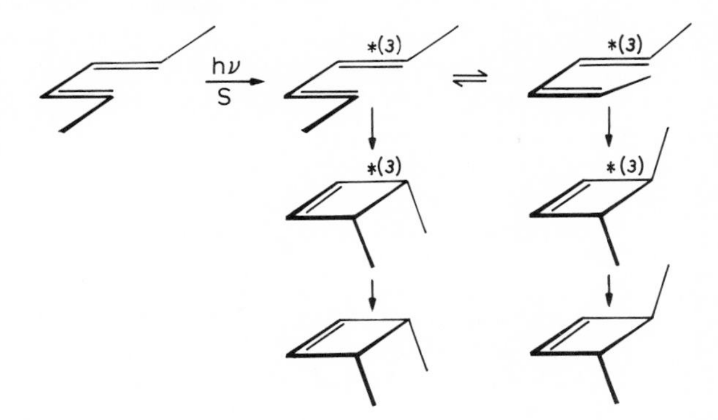

A further point worth reviewing is the question why only the highest occupied orbital of the polyene is considered. Since the answer to this is by no means obvious, a somewhat different rationale has been devised by Longuet-Higgins for these reactions [9]: A correlation diagram is constructed for all the orbitals involved in the reaction, symmetry being used explicitly for the correlations. If no great energy expenditures obviously occur, the reaction is allowed; otherwise it is forbidden. We will develop this point of view further in the next group of concerted reactions; however, it is worthwhile at this point to discuss just what is meant by the terms allowed and forbidden. Even if all correlations are level or downhill in energy from reactant to product, the reaction - which is then obviously allowed - is not demanded; since any steric questions are completely ignored, the activation energy may still be high. Conversely, a forbidden reaction may lead to such release of strain, for example, that it will take place even though forbidden (it may also find an alternative non-concerted pathway to the low energy product). Allowed and forbidden are therefore relative terms; an allowed reaction is more likely to be observed than a forbidden one - everything else being equal.

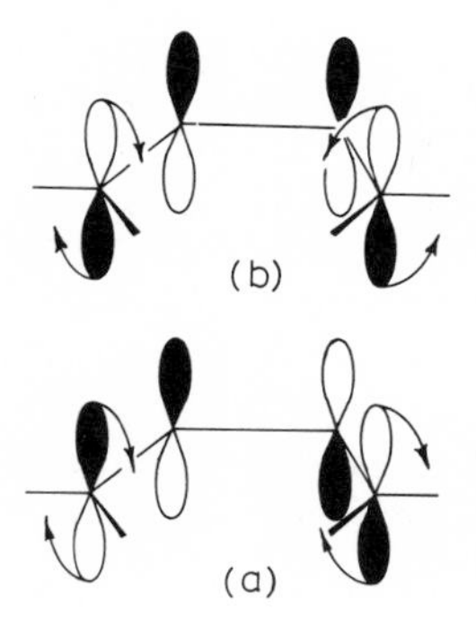

Fig. 14-3. The allowed rotations in butadiene cyclization, (a) in the ground state; (b) in the excited state. In every case, an allowed reaction also results if both termini rotate in the opposite direction.

---

9. H. C. Longuet-Higgins and E. W. Abrahamson, J. Amer. Chem. Soc., 87, 2045 (1965).

Still another question arises here: That of the odd-numbered chains. Examples such as allyl and 2,4-pentadienyl chains are just as subject to the particle-in-a-box analysis as their even-numbered cousins, and though experimentally these systems are less easily studied because we must then necessarily deal with ions or radicals, they are amenable to prediction on the basis of the rule. It is of interest in this connection that the rule predicts opposite behavior of the cations and corresponding radicals and anions (see Fig. 14-4). The ring opening of the cyclopropyl cation ($2\sigma \rightarrow 2\pi$) is especially interesting. It is expected to be disrotatory; however, a second feature causes the ring opening to occur in a unique and specific way. Carbonium cations are as a rule flat, with sp^2 hybrid bonds at 120°; since in the cyclopropyl ion one bond angle must necessarily be close to 60°, the angle strain is severe and hence the ion is even less stable than open carbonium ions (these questions are discussed in some detail in Chapter 21, Sec. 3). The reactions in which such ions normally form (such as the solvolysis of halides) are generally abnormally slow in the case of cyclopropyl. By contrast, the allyl cation which has two equivalent resonance structures is quite stable and readily formed. For this reason it might be supposed that in the solvolysis of cyclopropyl derivatives the cyclic ion will be by-passed altogether, and that the departure of the anion and the opening of the ring will occur in one concerted operation. This assumption is implicit in the prediction by Woodward and Hoffmann [8] that in such reactions the allylic isomer will be in effect formed in such a way that the newly formed p orbitals will begin to overlap with the back lobe at the carbonium ion site at the earliest moment (see Fig. 14-5). In other words, this electrocyclic ring opening is not merely governed by the general rule; of the two possible products only one obtains because of control by the leaving group. These predictions have been borne out in many experiments. Thus, the acetolysis of the trans,trans-2,3-dimethylcyclopropyl brosylate (15) is about 10,000 times faster than that of the all cis-isomer (16) [10]; the ring opening in the former case leads to the extra relief of strain caused by the eclipsed methyl groups in contrast to the reaction of (16).

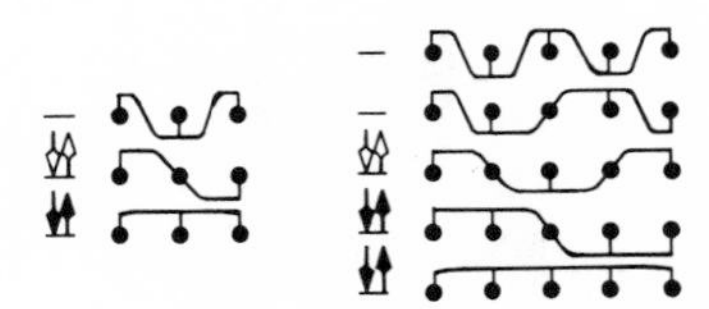

Fig. 14-4. The molecular orbitals of the allylic and 2,4-pentadienyl systems. Note that occupation of the non-bonding orbitals effectively reverses the symmetry of the highest filled level.

10. P. von R. Schleyer, G.W. Van Dine, U. Schöllkopf, and J. Paust, J. Amer. Chem. Soc., 88, 2868 (1966).

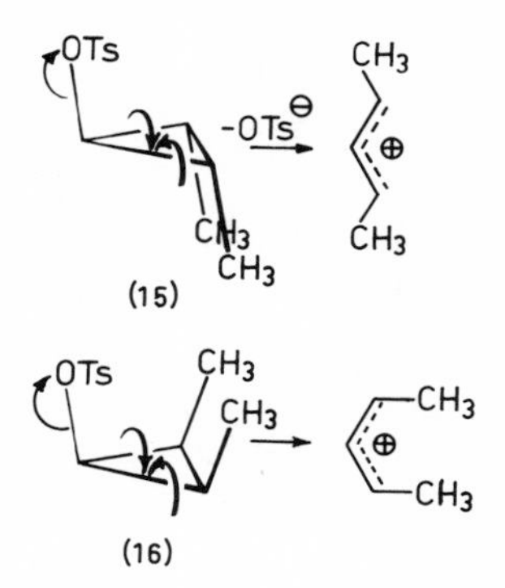

On the other hand, in bicyclic molecules this mode of ring opening can lead to an increase in strain. Thus, in (17) and (18) the relative rates are exactly reversed, with the latter now 10,000 times faster than the former [10].

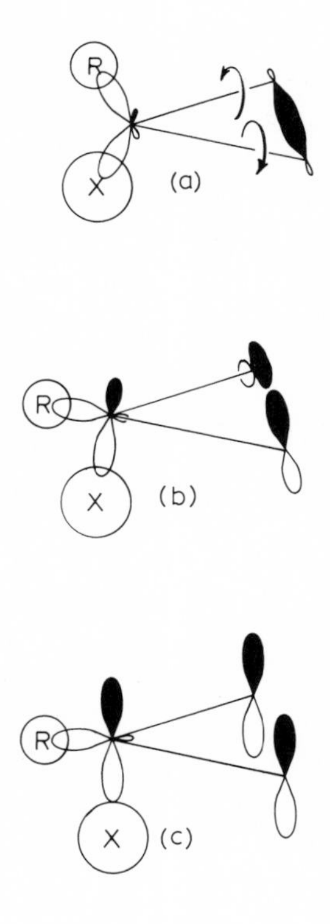

Fig. 14-5. Concerted cyclopropyl halide solvolysis, (a) at the start; (b) in progress; (c) when completed.

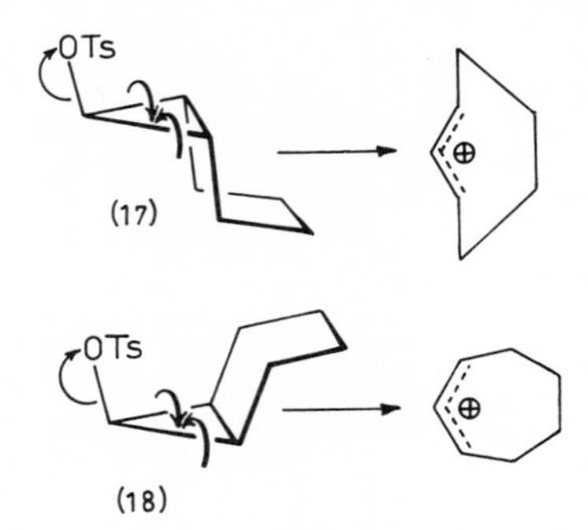

Similar observations have been made by others. Thus, under conditions in which anti halide (19) is stable, the epimer (20) rearranges to (21) - presumably via the allylic carbonium ion (22);

Cl Cl Cl $Cl^{\ominus}$

(19) (20) (21) (22)

in the case of the epimeric pair of dihalides (23) and (24), both rearrange, to give only (25) and (26), respectively, and without any cross-over [11].

(23) (25)

(24) (26)

11. (a) L. Ghosez, P. Laroche, and G. Slinckx, *Tetrahedron Lett.*, 2767 (1967); (b) L. Ghosez, G. Slinckx, M. Glineur, P. Hoet, and P. Laroche, *Tetrahedron Lett.*, 2773 (1967); (c) C.W. Jefford and R. Medary, *Tetrahedron Lett.*, 2069 (1966); (d) C.W. Jefford, E.H. Yen, and R. Medary, *Tetrahedron Lett.*, 6317 (1966).

The size of the ring is obviously important in these comparisons. Thus, in the series (27), (28) and (29) the relative solvolysis rates decrease in order $1{:}10^{-4}{:}10^{-8}$. This is in accord with expectations guided by the anticipated ring strains; however, in (30) the ring is smaller still, but it solvolyzes faster even than (27).

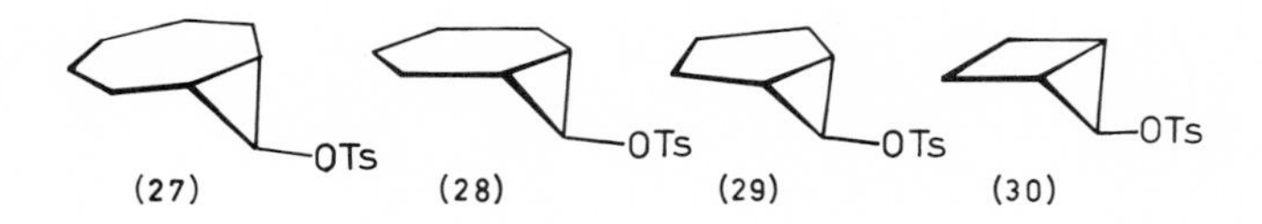

(27) (28) (29) (30)

Here we seem to have an exception then; but this is only apparent. The common edge is a long, weak bond, and the compound probably epimerizes to (31) with ease; the isomer then reacts rapidly, with relief of strain and in accord with the Woodward-Hoffmann rule. It may be noted that under the same conditions the ethers (32) and (33) interconvert readily; the diradical (34) may be a short-lived intermediate [12].

(30) → OTs → TsO (31)

OCH₃ CH₃O OCH₃

(32) (33) (34)

The ring strain is no longer important if the ring is sufficiently large. Thus, (35) is actually somewhat faster in methanolysis than the epimeric substance (36); both give stereospecific products [13].

---

12. (a) K. Fellenberger, U. Schollköpf, C.A. Bahn, and P.v.R. Schleyer, Tetrahedron Lett., 359 (1972); (b) J.J. Tufariello, A.C. Bayer, and J.J. Spadaro, Tetrahedron Lett., 363 (1972).
13. (a) C.B. Reese and A. Shaw, J. Amer. Chem. Soc., 92, 2566 (1970); (b) C.B. Reese and M.R.D. Stebles, Tetrahedron Lett., 4427 (1972).

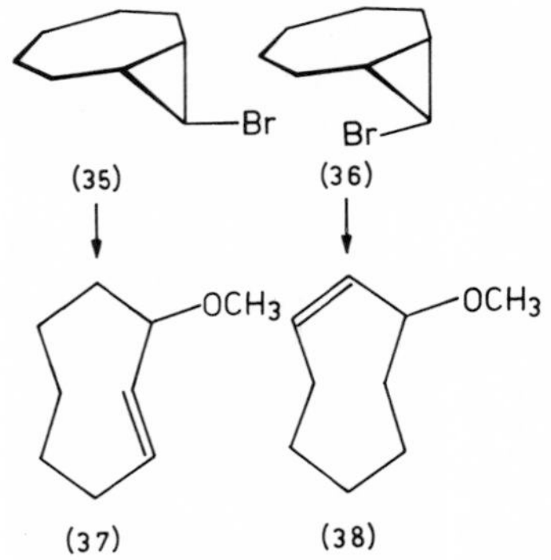

In those relatively rare cases in which the cyclopropyl moiety survives the ionization long enough to warrant claims of a cyclopropyl carbonium ion intermediate, the stereoelectronic control of the leaving group is lost and either product may then result from the electrocyclic ring opening. Thus, when the nitrosamines (39) and (40) are treated with base, both give the same product (41). These reactions involve the diazonium ions (42) and (43), and the cyclopropyl carbonium ions (44) and (45), respectively. The latter pair either equilibrate rapidly or, if they are planar, they are simply identical.

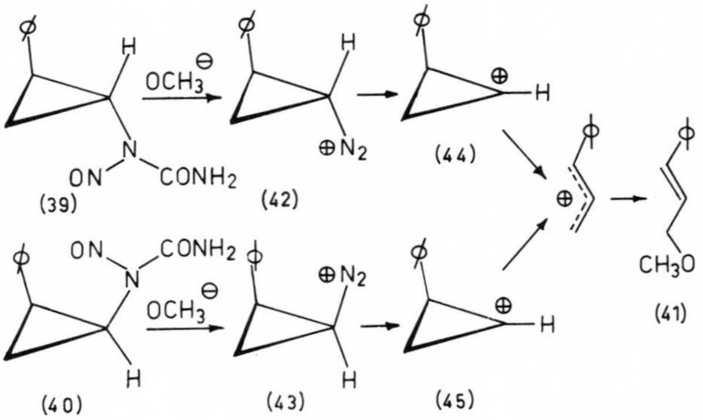

Evidence that the equilibration did not involve the prior stages (42) and (43) via the diazocyclopropane (46) is furnished by the fact that no deuterium incorporation occurs if methanol-O-d is used; evidence for the actual presence of ions such as (44) is furnished by the fact that small amounts of unopened solvolysis products such as (47) can be isolated [14].

14. W. Kirmse and H. Schütte, J. Amer. Chem. Soc., 89, 1284 (1967).

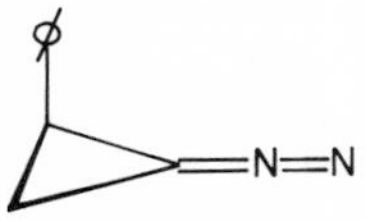

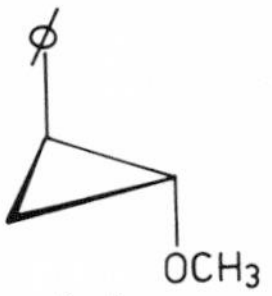

(46) (47)

N-chloroaziridines are known to invert slowly enough to permit the isolation of stereoisomers (cf. p. 180), and hence the opening of these rings is also open to test. The result is that in analogy to the cyclopropyl case *trans, trans-* dimethyl substitution (48) leads to an enormous solvolysis rate increase (10^5) compared to the parent compound (49). The products in this case are ammonia and acetaldehyde [15].

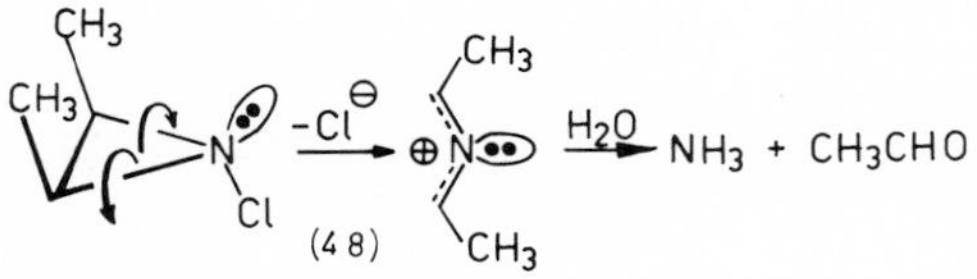

$\xrightarrow{-Cl^{\ominus}}$ $\xrightarrow{H_2O}$ $NH_3 + CH_3CHO$

(48) 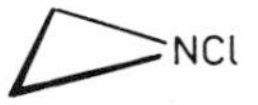 (49)

There are as yet no data on the stereochemical course of the opening of cyclopropyl radicals (see p. 637). Probably the most beautiful demonstration of the conrotatory nature of cyclopropyl anion opening and reclosure involves aziridines such as (50) and (51) ($2n, 2\sigma \rightleftharpoons 4\pi^{c}$);

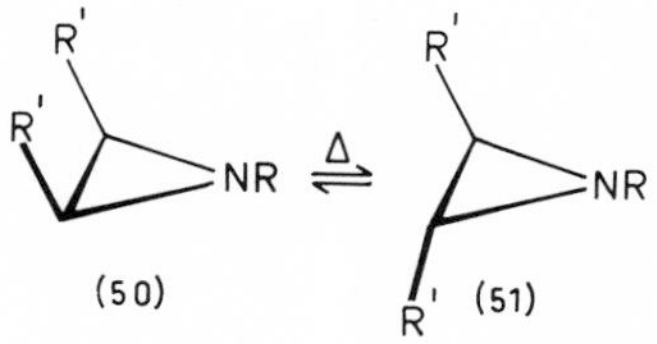

(50) $\overset{\Delta}{\rightleftharpoons}$ (51)

upon heating these species interconvert, and it is clear from the substituent effects of R and R' on the rates of these equilibrations that they involve zwitterions such as (52) as intermediates [16].

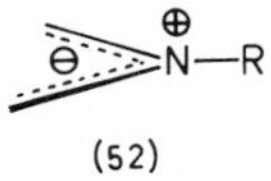

(52)

---

15. P. G. Gassman and D. K. Dygos, *J. Amer. Chem. Soc.*, *91*, 1543 (1969).
16. R. Huisgen, W. Scheer, G. Szeimies, and H. Huber, *Tetrahedron Lett.*, 397 (1966).

Huisgen has shown that although the equilibration of these zwitterions may be rapid, each ring opening or reclosure, whether thermally achieved or photochemically is stereospecific. The proof is that reactive 1, 3-dipolarophiles such as methyl acetylenedicarboxylate captures them (in a Diels-Alder type of cycloaddition) before they have a chance to isomerize; the argument is presented below in schematic form [17]. It assumes that the cycloaddition is also stereospecific and *cis*; this question is further considered in the next section.

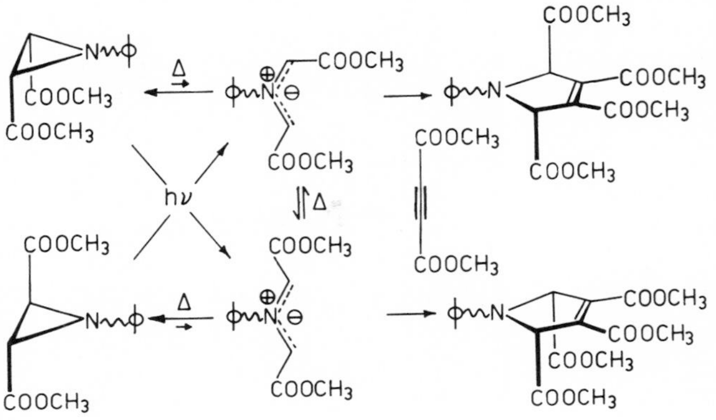

The opening of bicyclo[n.1.0] anions should be conrotatory, and hence strain-impeded. Some such reactions are known to occur under very vigorous conditions, though it is not known by which mechanisms [18]. The conrotatory nature of thermal, concerted $2\sigma, 2\pi \rightleftharpoons 4\pi$ reactions has now been demonstrated many times. Thus, warming of *E,Z*-1,3-cycloocta- and nonadiene (53) and (54) leads to *cis*-bicyclic isomers (55) and (56); the same compounds result from the photolysis of the *cis,cis*-analogs (57) and (58), respectively. The latter pair of dienes is thermally stable; they would have to produce the strained *trans*- bicyclic compounds (59) and (60) [19].

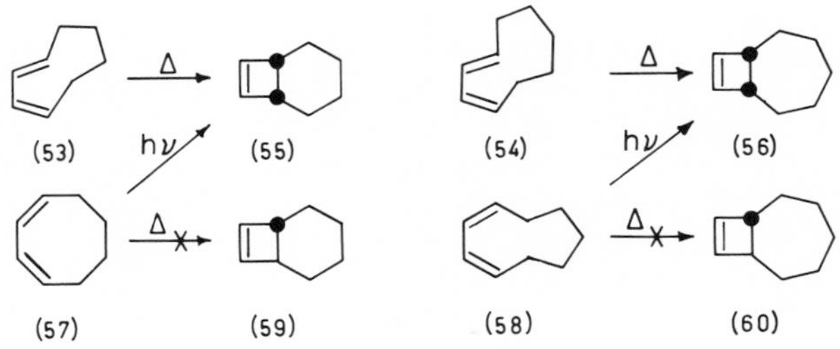

17. (a) R. Huisgen, W. Scheer, and H. Huber, *J. Amer. Chem. Soc.*, *89*, 1753 (1967); (b) R. Huisgen and P. Eberhard, *J. Amer. Chem. Soc.*, *94*, 1346 (1972), and work quoted there.
18. M.E. Londrigan and J.E. Mulvaney, *J. Org. Chem.*, *37*, 2823 (1972).
19. K.M. Shumate, P.N. Neuman, and G.J. Fonken, *J. Amer. Chem. Soc.*, *87*, 3996 (1965).

However, this agreement with theory is to some degree fortuitous. It was later found that the photolysis of (57) (which was carried out at rather high temperature) caused isomerization to (53) which then thermally goes on to (55); the intermediacy of (53) in this case could be proved by trapping it as the silver nitrate complex [20]. Compound (60) can be obtained by the irradiation of (54), but this observation is also obscured somewhat by the simultaneous isomerization of (54) to (58) and the appearance of more severely rearranged isomers [21].

Unlike the cyclopropyl cation ring opening, the $2\pi, 2\sigma \rightleftarrows 4\pi$ systems do not provide us at present with a theoretical basis for predicting which of the two possible conrotatory thermal (or disrotatory photochemical) modes will be preferred. Nevertheless, nature does have its prejudices in this regard. The ring opening of the cyclobutenone (61) in methanol for instance - which presumably involves ketene (62) as an intermediate - gives exclusively (63) if carried out thermally, whereas the same operation carried out photochemically gives isomer (64) [22].

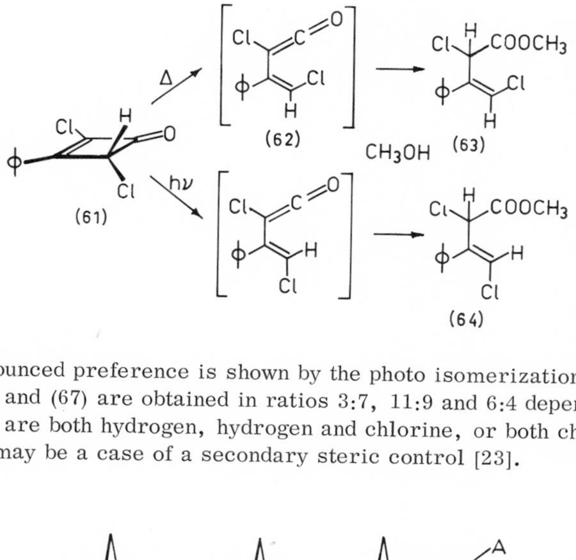

(61) (62) (63) (64) CH_3OH

A less pronounced preference is shown by the photo isomerizations of compounds (65); (66) and (67) are obtained in ratios 3:7, 11:9 and 6:4 depending on whether A and B are both hydrogen, hydrogen and chlorine, or both chlorine, respectively; this may be a case of a secondary steric control [23].

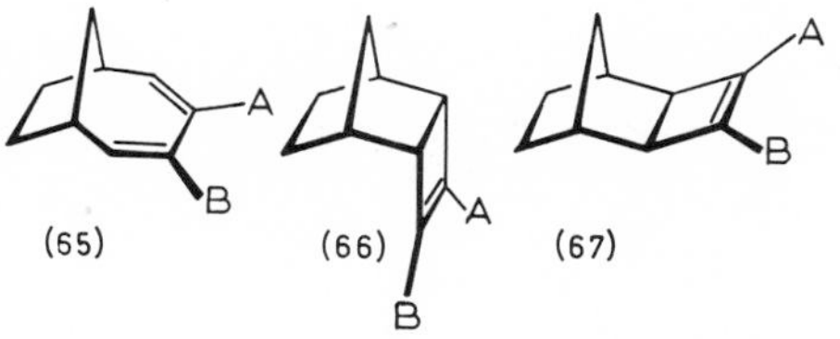

(65) (66) (67)

20. R.S.H. Liu, J. Amer. Chem. Soc., 89, 112 (1967).
21. K.M. Shumate and G.J. Fonken, J. Amer. Chem. Soc., 88, 1073 (1966).
22. J.E. Baldwin and M.C. McDaniel, J. Amer. Chem. Soc., 90, 6118 (1968).
23. C.W. Jefford and F. Delay, J. Amer. Chem. Soc., 94, 4794 (1972).

It has been possible in one instance to learn the difference in rate between allowed and forbidden reactions. The cyclobutene (68) at 280° gives both (69) and (70), but the latter in a yield of only 0.005%, and the difference in activation energies for the two reactions is approximately 11 kcal/mole. The high energy reaction may in fact not be the concerted and forbidden reaction, but the stepwise reaction involving the intermediate (71) [24].

(68) (69) (70) (71)

The thermal closure of pentadienyl cations has been reported to take a conrotatory course as required; thus, the dissolution of pentamethylpentadienyl alcohols (72) and (73) in strong acids which presumably gives carbonium ions (74) and (75), respectively, leads to the closed allylic ion (76) whereas alcohol (77) via cation (78) gives (79). All these reactions are very rapid, and only the cyclic ions can be directly observed even at low temperatures [25].

OH
H⊕
(72) (74)
OH
H⊕
(76)
(73) (75)
OH
H⊕
(77) (78) (79)

24. J.I. Brauman and W.C. Archie, *J. Amer. Chem. Soc.*, *94*, 4262 (1972).
25. P.H. Campbell, N.W.K. Chiu, K. Deugau, I.J. Miller, and T.S. Sorensen, *J. Amer. Chem. Soc.*, *91*, 6404 (1969).

These ring closures offer an interesting contrast with the thermal base promoted reaction of the benzamide derivative (80). This $6\pi \rightarrow 4\pi$, 2σ reaction is disrotatory and yields exclusively (81), the most hindered of the two stereoisomeric amarines. On prolonged treatment with base (81) is converted into an equilibrium of mixture containing 96% of (82); this latter product can also be obtained by irradiation of (80) [26]. This is but one of the numerous known cases of triene - or rather 6π electron - ring closures; a number of others were cited in Sec. 14-1.

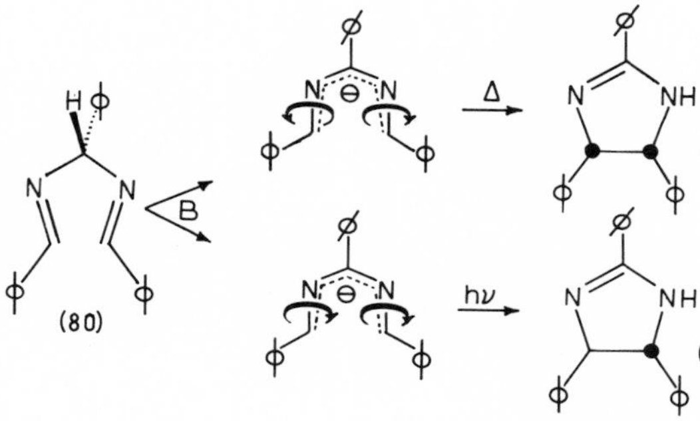

A slightly more complex case is that of the bicyclic Z, Z-diene (83). On heating this undergoes an allowed, conrotatory opening to the cyclic triene (84); on further heating this recloses by an allowed disrotatory path to (85) [27].

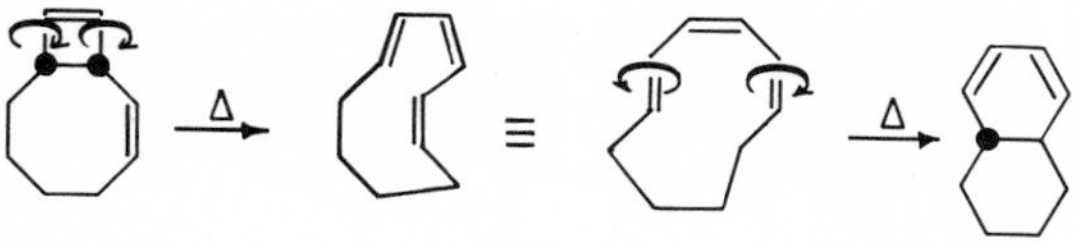

It is at present not yet clear whether there is some length for these conjugated chains at which the stereospecificity for ring closure or the reverse is lost. If the chains are long enough, it may be difficult to tell which conformation brings the terminal atoms sufficiently close together; furthermore, such a conformation may force other parts of the molecule in close proximity, thereby inducing side reactions which render the answer to this question difficult to trace.

The $8\pi \rightarrow 6\pi$, 2σ case is still beautifully clear-cut; in thermal ring closures only the conrotatory mode is observed. The resulting cyclooctatrienes undergo further disrotatory closure at somewhat higher temperatures; the entire system has been unraveled by Huisgen [28], who found that the equilibria I-IV have the constants 6 (at 16°), 15 (at 55°), 9 (at 35°) and 4 (at 60°), respectively.

---

26. D. H. Hunter and S. K. Sim, J. Amer. Chem. Soc., 91, 6202 (1969).
27. P. Radlick and W. Fenical, Tetrahedron Lett., 4901 (1967).
28. R. Huisgen, A. Dahmen, and H. Huber, J. Amer. Chem. Soc., 89, 7130 (1967), and Tetrahedron Lett., 1461 (1969).

In this case therefore all possible components are present in the equilibrium mixtures. The equilibrations are extremely rapid at 170°, and either system can be heated to that temperature for some time before contamination by components from the other becomes detectible. Evidently each of these molecules is isomerized and reformed in errorless ways thousands of times.

Extension to still longer conjugated chains has as yet not been reported. The photolysis of (86) leads to a mixture of (87) and (88); these two products interconvert thermally very rapidly and nmr coalescence of the two bridgehead protons is readily observable [29]. These are allowed reactions; the further conversion to decapentaenes was not reported, however.

(86) (87) (88)

Some 8 π electron anionic systems have been examined, but little is known as yet about stereochemistry. Thus, triene (89) reacts at -50° with butyllithium to give anion (90) which cyclizes at -30° to (91) - in a stereochemically unknown way [30].

(89) (90) (91)

29. S.W. Staley and T.J. Henry, J. Amer. Chem. Soc., 92, 7612 (1970).
30. (a) R.B. Bates, W.H. Deines, D.A. McCombs, and D.E. Potter, J. Amer. Chem. Soc., 91, 4608 (1969); (b) H. Kloosterziel and J.A.A. van Drunen, Rec. Trav. Chim. Pays-Bas, 88, 1084 (1969).

Conversely, the bicyclic diene (92) opens at -70° when the anion is irradiated [31]. This was thought to be an instance of disrotatory ring opening of the 7-membered ring; however since cyclopropyl hydrogen atoms tend to be acidic it seems at least as likely that disrotatory opening of the cyclopropyl ring occurred.

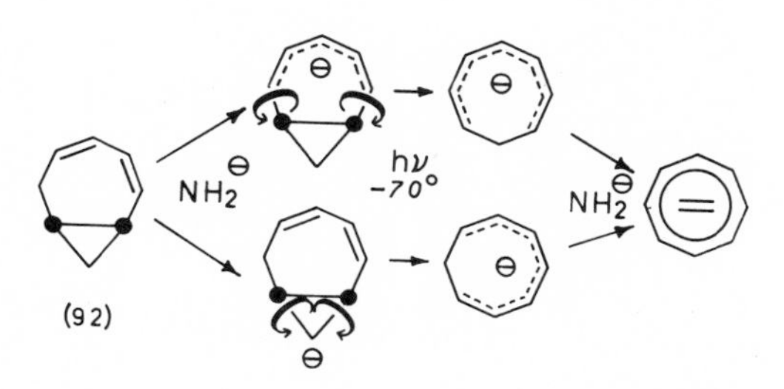

As hinted above (p. 477), in the more extensively conjugated systems electrocyclic reactions may occur between atoms that are not at the ends of the π chain. For the purpose of such reactions we may regard the molecule as a shorter conjugated molecule, with perhaps an unsaturated side chain. Thus, it may be noted for instance that the highest filled orbital of 1,3,5-hexatriene is the same as a combination of the highest filled orbitals of ethylene and of 1,3-butadiene. Therefore, if 3-vinylcyclobutene formation is observed, the stereochemistry should be the same as in any other butadiene reaction. Even completely conjugated cyclic systems may be so treated.

As an example we may consider the following reactions of [16]annulene, (93). Warming it results in a double disrotatory ring closure to give (94), while photolysis produces (95) *via* a double conrotatory closure [32].

(93) Δ → (94)

hν → (95)

31. H. Kloosterziel and G. M. Gorter-la Roy, Chem. Commun., 352 (1972).
32. G. Schröder, W. Martin, and J. F. M. Oth, Angew. Chem., Int. Ed. Engl., 6, 870 (1967).

In the photochemical interconversion of the dihydropyrene derivative (96) and corresponding metacyclophane (97), as is clear by inspection, the rearranging "cyclohexadiene part" of the molecule opens in conrotatory fashion as it should [33].

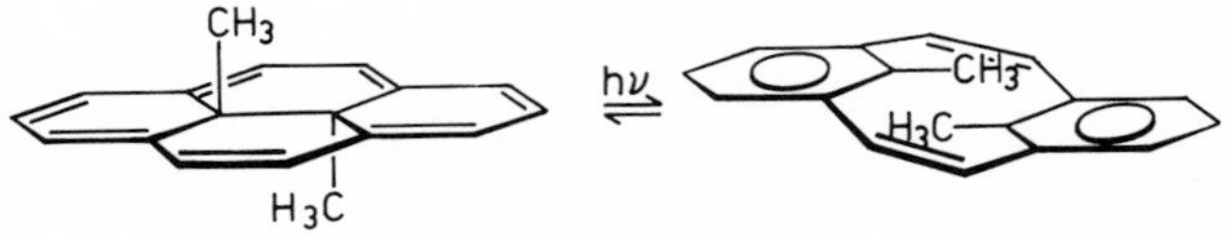

CH_3 hν H_3C CH_3 H_3C

(96) (97)

The combination of strain in (97) and aromaticity of (96) is such however, that thermal reversion also occurs. Unsaturated 7-membered rings such as azepine (98) undergo allowed photochemical bridging to give valence isomer (99) [34].

N $COOCH_3$ hν N $COOCH_3$

(98) (99)

The behavior of oxepin (100) is complicated by the extremely rapid (thermally allowed) valence tautomerism and by the fact that both tautomers readily undergo triplet sensitized conversion to phenol. In the absence of sensitizers, both benzene formation (by photoreduction) and isomerization to (101) occur; the use of light of long wavelength (> 310 nm) - which presumably only the triene can absorb - results in isomerization alone [35].

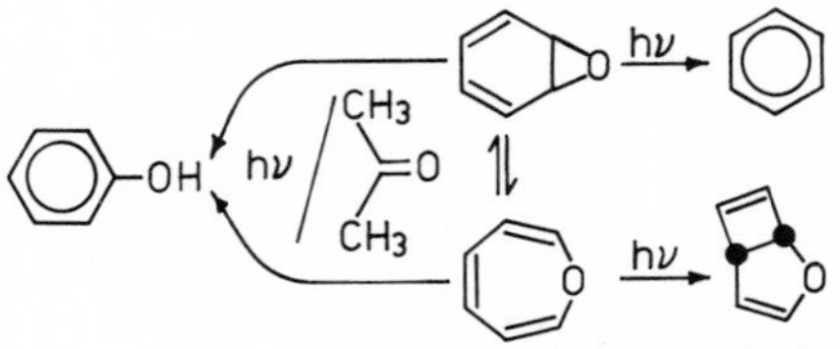

OH hν CH_3 O CH_3 O hν O hν O

(100) (101)

33. W. Schmidt, Helv. Chim. Acta, 54, 862 (1971).
34. L.A. Paquette and J.H. Barrett, J. Amer. Chem. Soc., 88, 1718 (1966).
35. J.M. Holovka and P.D. Gardner, J. Amer. Chem. Soc., 89, 6390 (1967).

Azonine (102) undergoes both photochemical and thermal bridging processes as shown below.

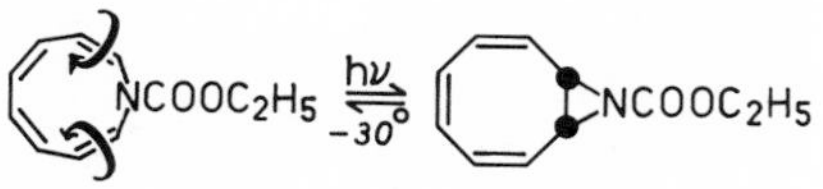

(102)

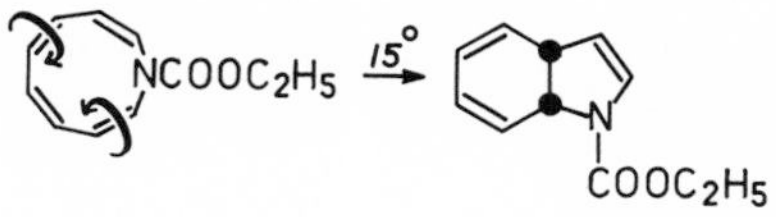

Both are disrotatory as required by the Woodward-Hoffmann rules [36]. Similarly, in the case of oxonin (103) the bicyclic isomer (104) obtains thermally; the geometric isomer (105) leads to (106) instead [37].

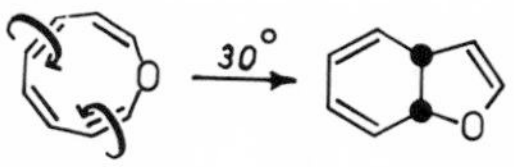

(103) (104)

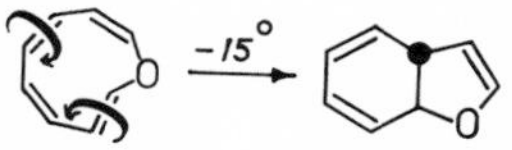

(105) (106)

With this series of successes and others to be described below, it is little wonder that organic chemists have developed such confidence in the reliability of the rules that these are now called upon to shore up proofs of structure or configuration of products and intermediates. For example, when it was observed that the 1,5-bisdibromocarbene adduct of cyclooctatetraene (107) upon treatment with methyllithium at -78° gave naphthalene, the conclusion was drawn that the bis-allene produced must have been the meso stereoisomer (108) rather than the dl-pair (109), since the former can undergo the necessary disrotatory ring closure; the latter would instead give rise to the rather strained looking affair (110) [38].

(107) 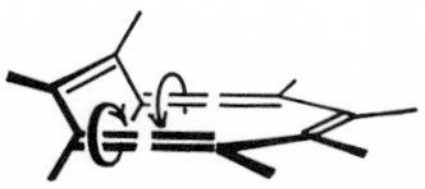 (108) 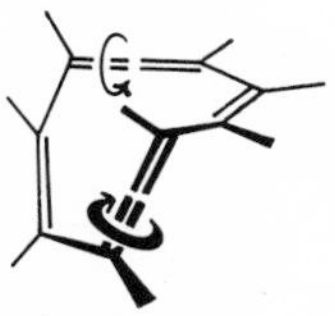 (109) 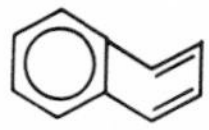 (110)

36. A.G. Anastassiou and J.H. Gebrian, Tetrahedron Lett., 5239 (1969).
37. S. Masamune, S. Takada, and R.T. Seidner, J. Amer. Chem. Soc., 91, 7769 (1969).
38. E.V. Dehmlow and G.C. Ezimora, Tetrahedron Lett., 4047 (1970).

In another instance, when the two isomeric cyclobutenes (111) and (112) were obtained, an assignment of configuration could be made on the basis of the fact that one of them formed a simple Diels-Alder adduct with dimethyl acetylenedicarboxylate whereas the other formed a diadduct.

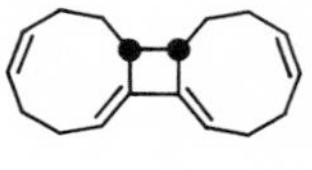

(111) 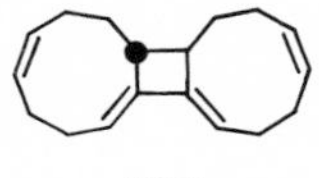 (112)

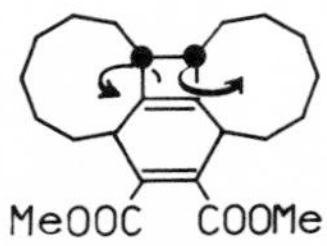

MeOOC COOMe MeOOC COOMe

(113) 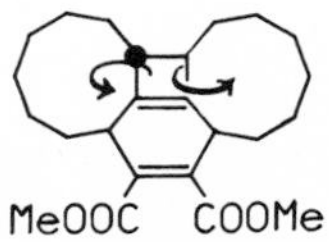 (114)

Since of the two mono-adducts (113) and (114) only the latter undergo a conrotatory ring opening to a strainless butadiene (116), the corresponding precursor must have been (112), with its two cyclobutane hydrogen atoms in *trans*-positions [39].

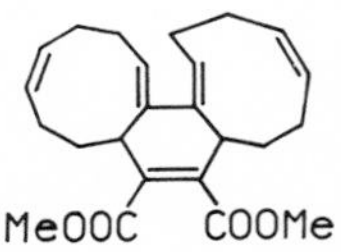

MeOOC COOMe

MeOOC COOMe MeOOC COOMe MeOOC COOMe

(115) 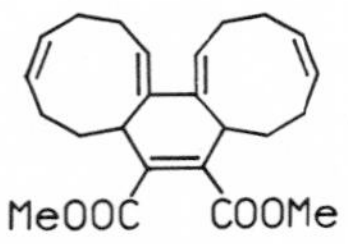 (116) 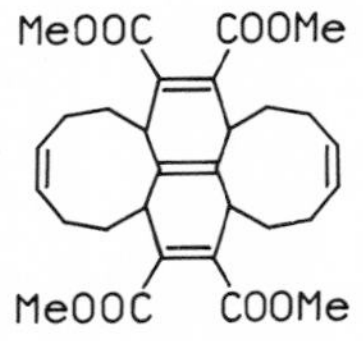 (117)

While assignments on this basis are not without hazards (one must be able to rule out multistep conversions, for instance), some isomerization schemes become so complex that it is almost impossible to sort them out without applying the Woodward-Hoffmann rules. The valence isomerizations of the $C_{10}H_{10}$ hydrocarbons provide one example and perhaps even more ambitious systems can be examined.

14-3. CYCLOADDITIONS [40]

Cycloadditions are somewhat like valence isomerizations in that they are difficult to define precisely. Originally the term was meant for Diels-Alder reactions and its analogs, and indicated the formation of single bonds between the ends

---

39. K. G. Untch and D. J. Martin, *J. Amer. Chem. Soc.*, *87*, 4501 (1965).
40. For review, see J. J. Vollmer and K. L. Servis, *J. Chem. Educ.*, *47*, 491 (1970), and the survey papers quoted in Sec. 1.

of two separate, completely conjugated, terminal polyenes. In the wake of the second Woodward-Hoffmann paper [41] however, it was soon realized that many related cases were subject to the selection rules for these reactions. Thus, additions to single bonds, independently connected polyene systems, fragments numbering more than two - all these features are possible and would have to be included in a precise description. We shall therefore forego a definition and make do with the original one, and note special cases as we encounter them. The words concerted and stepwise, and thermal and photochemical have the same meaning as before, and only the following additional terms should now be described. An addition is suprafacial if it occurs at the two termini on the same side of the molecule; it is antarafacial if it happens at opposite sides.

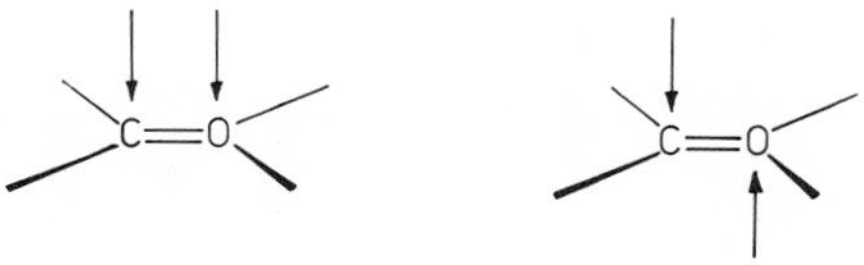

Thus, these terms are very similar to the words *cis*- and *trans*-, *syn*- and *anti*-, *exo*- and *endo*- and so on. A symbolic notation has become popular recently; an adequate example is the Diels-Alder reaction which is a $[{}_{\pi}4_{s} + {}_{\pi}2_{s}]$ cycloaddition. If σ electrons or antarafacial stereochemistry is involved, that can obviously be indicated in the symbol. It is also used for cycloreversions, which is the same reaction in the opposite direction. If we now assume for a moment that all fragments are operating suprafacially, the selection rule says that regardless of π- or σ-nature of the participating electrons, their total number must be $(4n + 2)$ if the reaction is to be allowed thermally, and $(4n)$ if it is to be carried out photochemically. Each time that the symbol *a* is encountered, these conclusions are reversed.

A simple though not rigorous proof of the sort presented in the preceding section would involve the drawing of the highest occupied molecule orbital (HOMO) of the one fragment and of the lowest unfilled molecular orbital (LUMO) of the other to show that bonding can occur at both ends simultaneously (see Fig. 14-6).

A more rigorous and satisfying procedure is available, however, which furthermore makes clear why the rules are considered to be based on the principle of conservation of orbital symmetry. We might imagine that two ethylene molecules approach one another in suprafacial manner to give cyclobutane, as in Fig. 14-7. We note that two principle symmetry elements are present: Planes P_1 and P_2. There are others, such as P_3, and the axes which are the intersections of these planes; however, these are not of interest since either they are redundant with respect to P_1 and P_2, or all orbitals are symmetric with respect to them (P_3).

---

41. R. Hoffmann and R. B. Woodward, *J. Amer. Chem. Soc.*, *87*, 2046 (1965).

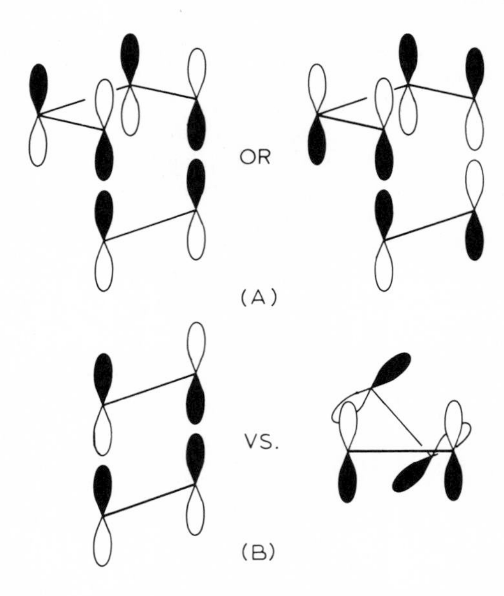

Fig. 14-6. (a) Butadiene and its LUMO and ethylene with its HOMO, and vice versa, in an allowed, thermal suprafacial cycloaddition; (b) the forbidden, thermal, suprafacial dimerization of ethylene and the allowed antarafacial analog.

The reactant orbitals involved in the reaction (the only ones therefore whose symmetry concern us) are the π and π^* orbitals. The former are combined to give SS and SA orbitals, i.e., new orbitals both of which are symmetric with respect to P_1, but only one of which is also symmetric about P_2. The other is antisymmetric about this plane (Fig. 14-8).

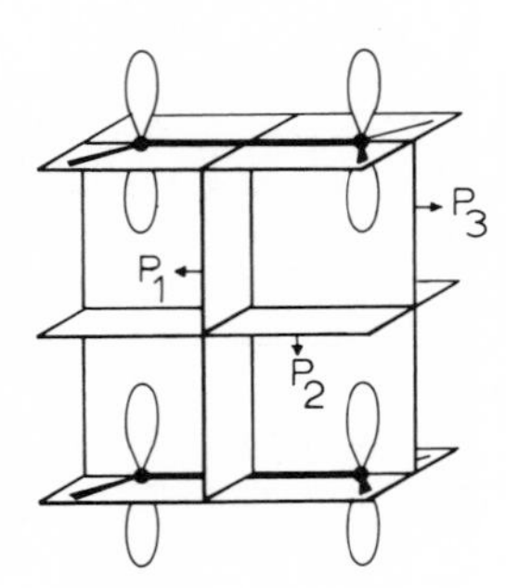

Fig. 14-7. Suprafacial approach of two ethylene molecules in a hypothetical reaction to give cyclobutane.

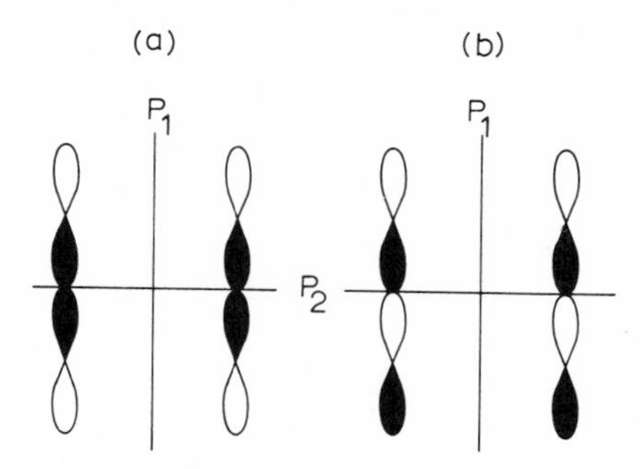

Fig. 14-8. (a) The SS combination of π orbitals; (b) the SA combination.

In the same way, the two π^* orbitals are combined to yield AS and AA orbitals. So long as the ethylene molecules are too far apart to interact, the SS and SA combinations are bonding and hence of low energy; the AS and AA combinations are still antibonding at that point.

On the product side, the newly formed orbitals are the two σ and σ^* orbitals. The former are combined to the two bonding orbitals SS and AS (see Fig. 14-9), and the latter to the antibonding SA and AA orbitals. Finally the results are given in the correlation diagram of Fig. 14-10. The correlations between orbitals are drawn on the basis of conservation of symmetry; it is at once obvious that if the reaction occurs, the two originally bonding π-SA electrons will become antibonding σ^*-SA electrons. That makes the reaction prohibitively expensive in energy, and one concludes that it is symmetry forbidden. On the other hand, raising one electron to the π^*-AS level has the effect of trading the energy deficit of one electron for the excess of another, and hence the photochemical reaction is allowed.

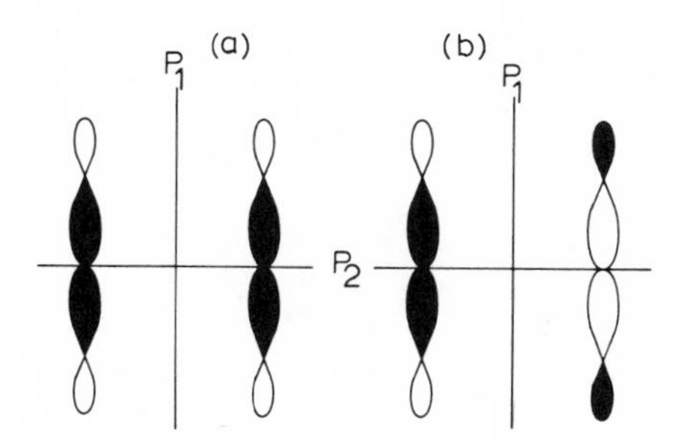

Fig. 14-9. (a) The SS combination of π orbitals; (b) the AS combination.

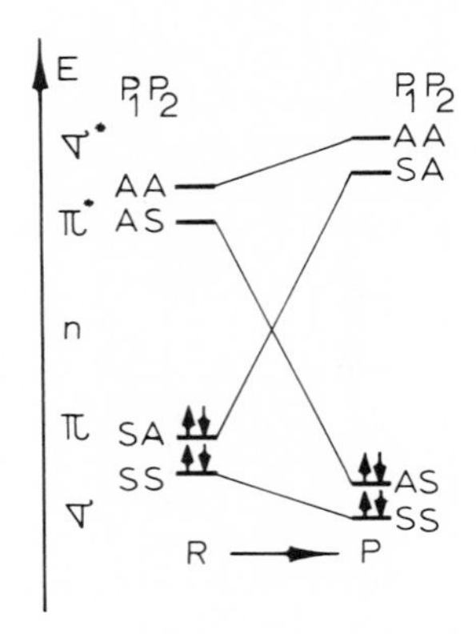

Fig. 14-10. The correlation diagram for the dimerization of ethylene to cyclobutane.

The Diels-Alder reaction can be analyzed in similar vein; only the P_1 plane need now be considered. The correlation (Fig. 14-11) shown is then readily derived; it shows that the thermal reaction is allowed since none of the orbitals occupied in the reactants correlates with an anti-bonding orbital in the product. Similar reasoning can be applied to any type of cycloaddition or cycloreversion, and the results are generalized in the Woodward-Hoffmann rule as phrased above.

The experimental evidence is in agreement with these predictions in the most astounding ways. With some exceptions scrutinized further below, thermal cyclobutane formation from ethylene cycloadditions does not occur. On the other hand, the photochemical reaction is quite common; a large number of examples were mentioned in Chapters 11 and 13, and no further comment seems necessary.

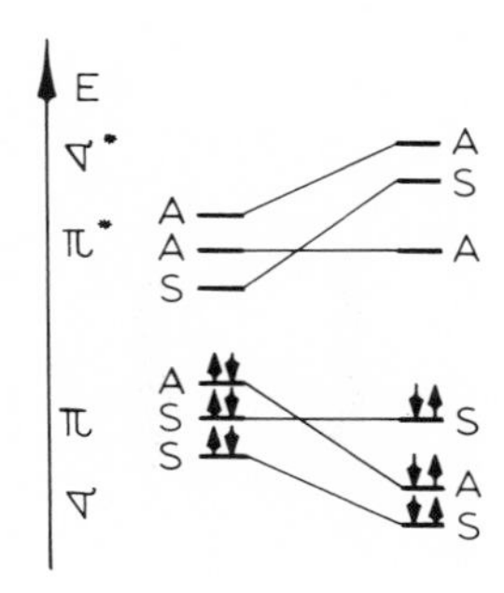

Fig. 14-11. The correlation diagram for the Diels-Alder reaction of ethylene with butadiene.

Thermal [2 + 2] cycloreversions are also forbidden. It is to this fact that so many highly strained small ring polycyclic compounds owe their isolability. The strain energy of a compound like prismane is enormous; if the resonance energy of benzene is added it seems a miracle that any prismane derivatives are known. But it turns out that any such isomerizations involve [$_\pi 2_s$ + $_\pi 2_s$]cycloreversions, and these compounds are therefore saved by extremely high activation energies; in fact, rather high temperatures are often necessary to accomplish aromatization. If the molecule is able to absorb a photon, cycloreversion can be done photochemically. The annelated cyclobutenes (118) and (119) provide an interesting example. The _trans_-epimer (118) yields the _E_-dodecaenyne (120) on photolysis in a [$_\pi 2_s$ + $_\pi 2_s$] cycloreversion. Electrocyclic ring opening would have yielded the highly strained _trans_-cyclohexene (121) and hence is not observed. The opposite is true of the _cis_-epimer (119); photolysis yields 1,1'-bicyclohexenyl (122), and only on prolonged exposure does (123) begin to appear [42].

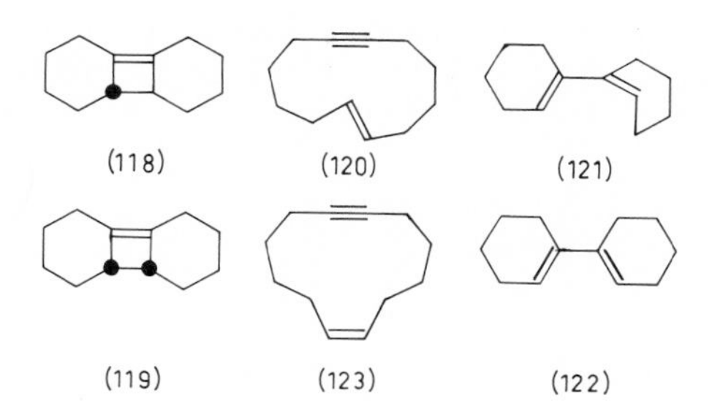

There are a number of known cases of thermal cyclobutane formation from olefins. Offhand there would seem to be three possible descriptions for such cases, and we discuss them here in order: These reactions may be non-concerted so that the symmetry rules are not applicable; they may involve the antarafacial mode for one of the olefins so that the rule is obeyed; they may be concerted and suprafacial so that the rule is violated. We begin with a discussion of the problem how to distinguish concerted and stepwise reactions.

Several criteria are available to judge whether a cycloaddition is concerted or not. One of these is the isotope effect. Bond formation and cleavage processes are characterized by small but measurable isotope effects (i.e., rate changes due to isotopic substitution), and hence in concerted reactions this effect should operate at both ends. In a stepwise reaction in which the first step is rate determining, the isotope effect will only contribute at one end. In practice this approach is a difficult one for experimental reasons (isotopic synthesis and degradation, analysis of small effects) as well as reasons of interpretation. Thus, one may not know for a non-concerted reaction whether the first or second step is rate determining; there is also the problem that when one end is labelled, either end may react, and a detailed appraisal shows that the net effect to be expected is even smaller than one might suppose. Nevertheless, Seltzer has made an analysis of

---

42. J. Saltiel and L.S.N. Lim, _J. Amer. Chem. Soc._, _91_, 5404 (1969).

the Diels-Alder reaction on this basis and concluded that it is a concerted [$_{\pi}2_{s}$ + $_{\pi}4_{s}$] cycloaddition; there is general agreement now that this conclusion is correct [43].

Another example of the use of isotope effects concerns the cycloaddition of vinyl ethers (124) to azo compounds such as (125) to give diazetidines (126).

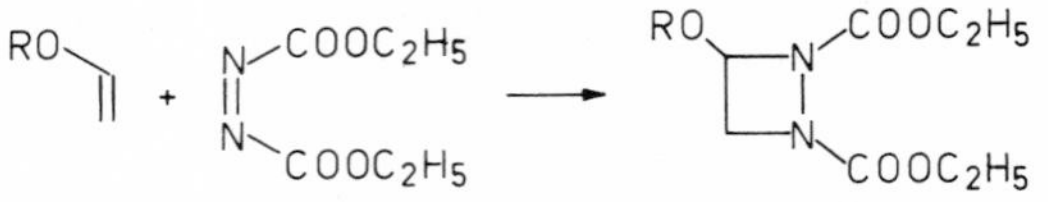

(124) (125) (126)

An isotope effect was observed at both carbon atoms, but in opposite direction. Deuterium substitution at the α-carbon atom led to a 10% decrease in rate, whereas at the β-carbon atom, each deuterium atom causes a 20% increase. On the basis of other known effects of deuterium substitution in various model reactions, this was interpreted to mean that a zwitterionic intermediate (127) was present.

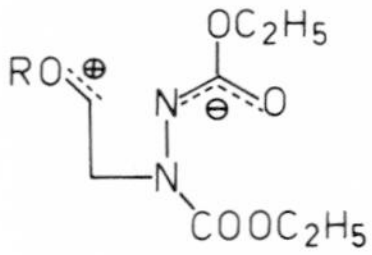

(127)

Support for this contention was found in the observation that if traces of water are present, the intermediate can be trapped and diverted to give a product such as (128) [44].

(128)

43. (a) S. Seltzer, J. Amer. Chem. Soc., 87, 1534 (1965); (b) M. Taagepera and E.R. Thornton, J. Amer. Chem. Soc., 94, 1168 (1972).
44. E. Koerner von Gustorf, D. V. White, J. Leitich, and D. Henneberg, Tetrahedron Lett., 3113 (1969).

A second means of determining concertedness is in the effect of pressure on the rate constant. In analogy to the temperature effect from which the energy and entropy of activation can be calculated, the pressure effect yields the volume of activation (see further Chapter 14). It is found that the formation of a bond invariably reduces the volume of the transition state, and of course two such bonds will do this in double measure. Eckert has shown for instance that the activation volume in the Diels-Alder reaction of dimethyl acetylenedicarboxylate (129) with cyclopentadiene is approximately equal to the overall volume decrease and hence both bonds were nearly formed when the transition state was reached (cf. also p. 505) [45].

$COOCH_3$ — $COOCH_3$ (129)

$\Delta V^{\ddagger}$ = -30 cc/mole → $COOCH_3$, $COOCH_3$

ΔV = -34 cc/mole → $COOCH_3$, $COOCH_3$

The catch here, of course, is again the assumption that the formation of the intermediate, if it occurs, will be rate limiting; if it is rapid and reversible, the activation parameters apply only to the final transition state (see Fig. 14-12).

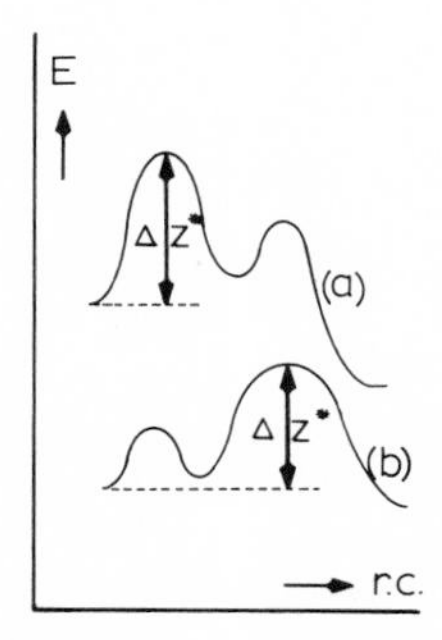

Fig. 14-12. The reaction profile and the activation parameter Z (a) if the formation of the intermediate is rate controlling, and (b) if it is reversible.

45. (a) R.A. Grieger and C.A. Eckert, J. Amer. Chem. Soc., 92, 2918, 7149 (1970), and Trans. Faraday Soc., 66, 2579 (1970); (b) C. Brun and G. Jenner, Tetrahedron, 28, 3113 (1972); (c) C. Brun, G. Jenner, and A. Deluzarche, Bull. Soc. Chim. France, 2332 (1972).

Perhaps more convincing on this score is an experiment by Stewart, who studied the dimerization of chloroprene (130). Various products obtain, some of which (131) result from [2 + 2] cycloaddition and others (132) which are due to Diels-Alder addition. Under pressure, the latter products are favored; thus, while at room pressure, they account for only 33% of the product, at 10,000 atmospheres this has increased to 90% [46].

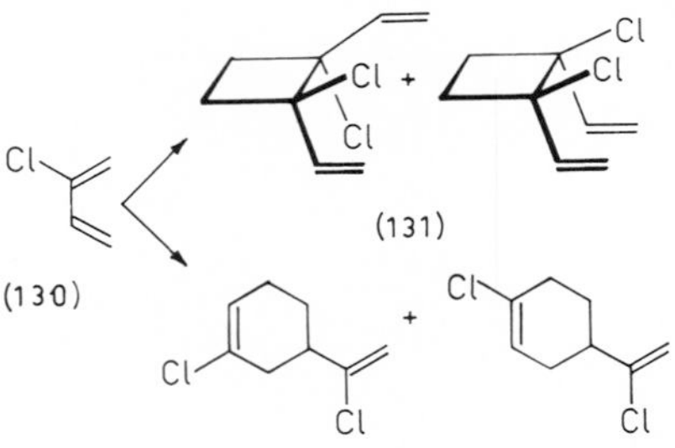

(130) (131) (132)

This criterion seems to be applicable also to concerted decomposition reactions; thus, the decomposition of β-bromoangelate ion (133) to give bromide, 2-butyne and carbon dioxide - which has been characterized independently as a concerted process [46a] - has an activation volume of +18 cm^3/mole [46b], a value twice as high as those of simple decarboxylations [49c].

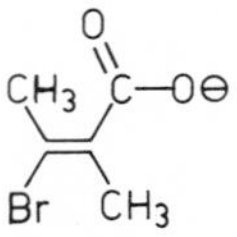

(133)

The most important of the criteria for concertedness is stereochemistry. In a concerted suprafacial cycloaddition, *cis*-substituents must remain *cis*; in a stepwise reaction, in which only one bond is formed at a time, the intermediate may have time to exercise its freedom of internal rotations, and hence stereospecificity may be lost. On the other hand, the intermediate may have been too short-lived, or the activation energy for the rotation may have been too high, or perhaps only one product is possible for thermodynamic reasons; all these questions have to be considered if stereospecificity is observed. In other words, this criterion is not absolutely definitive in all cases. It might be mentioned for example that although deuterium isotope effects on the rate led Koerner von Gustorf to his conclusion of a stepwise diazetidine formation, these same D atoms remain stereospecifically in place during the addition [42].

46. C.A. Stewart, *J. Amer. Chem. Soc.*, **93**, 4815 (1971) and **94**, 635 (1972).
46a. C.A. Grob and P.W. Schiess, *Angew. Chem., Int. Ed. Engl.*, **6**, 1 (1967).
46b. W.J. le Noble, R. Goitien, and A. Shurpik, *Tetrahedron Lett.*, 895 (1969).
46c. K.R. Brower, B. Gay, and T.L. Konkol, *J. Amer. Chem. Soc.*, **88**, 1681 (1966).

Tetrahaloethylenes tend to undergo [2 + 2] cycloadditions. The stereochemistry of this reaction has been carefully studied by Bartlett [47], and a number of fascinating details have come to light. Thus, the additions of tetrafluoroethylene (134) to cis- and trans-dideuterioethylene give identical mixtures; hence this reaction is completely non-stereospecific and hence non-concerted [48].

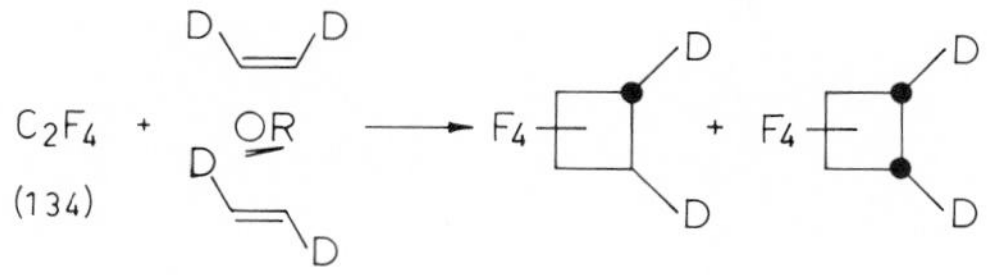

If (134) is added to cis-2-butene or trans-2-butene, stereospecificity is again lost but not completely; a small bias toward retention can be discerned in the product distribution. It was also noted that some of the unreacted butene had isomerized, and that this did not happen in the absence of (134), so that we do have a case here of reversible intermediate formation; thus, the overall reaction scheme is as shown below [49].

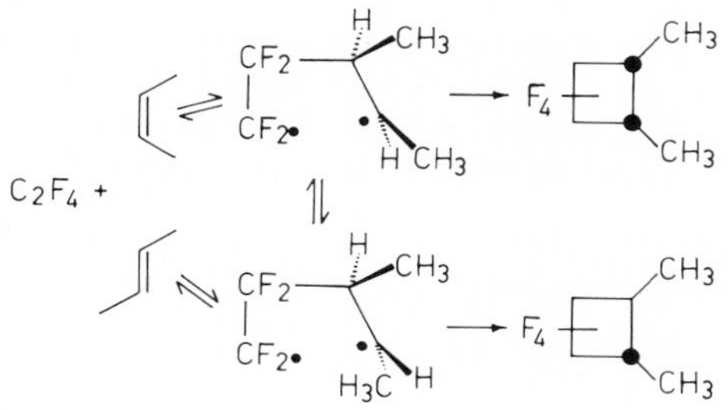

Competing processes occur if dienes are used. These can usually be recognized by the products, with the concerted process giving six membered rings and the diradical process producing cyclobutanes. It turns out that if Z- and E-1,2-difluoroethylene, (135) and (136) are compared in their reactions with cyclopentadiene, stereospecificity is largely lost in the [2 + 2] products but not in the [2 + 4] adducts [50].

---

47. P.D. Bartlett, Science, 159, 833 (1968).
48. P.D. Bartlett, G.M. Cohen, S.P. Elliott, K. Hummel, R.A. Minns, C.M. Sharts, and J.Y. Fukunaga, J. Amer. Chem. Soc., 94, 2899 (1972).
49. P.D. Bartlett, K. Hummel, S.P. Elliott, and R.A. Minns, J. Amer. Chem. Soc., 94, 2898 (1972).
50. R. Wheland and P.D. Bartlett, J. Amer. Chem. Soc., 92, 3822 (1970).

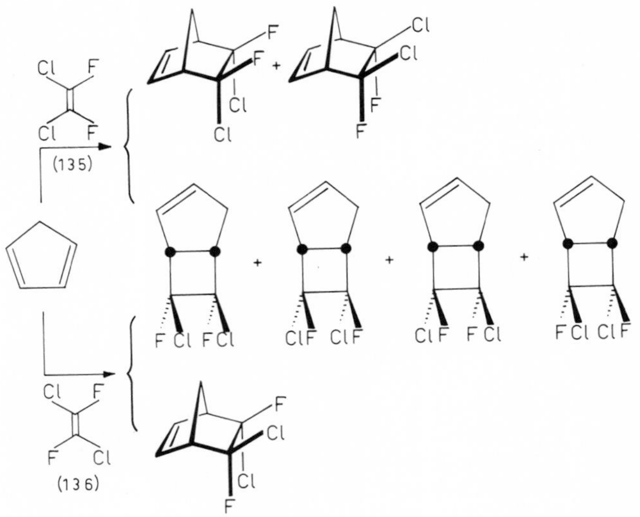

(135)

(136)

The initial diradical formation is not reversible in these reactions. This can be shown indirectly on the basis of the reasoning that if it is not, then the first step in the cycloreversion must be. This was shown to be so with the dichloro adducts (137) and (138). Brief heating to 400° led to isomerization mixtures in which (139) was the only common product; clearly reversion had occurred up to the diradical stage (but not beyond) [51].

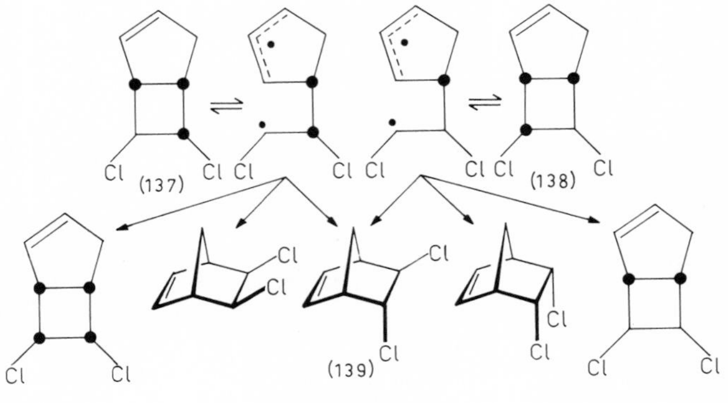

(137) (138) (139)

Many of these [2 + 2] cycloadditions are possible because chlorine atoms tend to stabilize radicals (Chr. 18). E,Z-1,3-cyclooctadiene (140) undergoes the reaction because the normal Diels-Alder product (141) would be too highly strained; it is non-stereospecific [52].

---

51. P.D. Bartlett, L.M. Stephenson, and R. Wheland, J. Amer. Chem. Soc., 93, 6518 (1971).
52. P.G. Gassman, H.P. Benecke, and T.J. Murphy, Tetrahedron Lett., 1649 (1969).

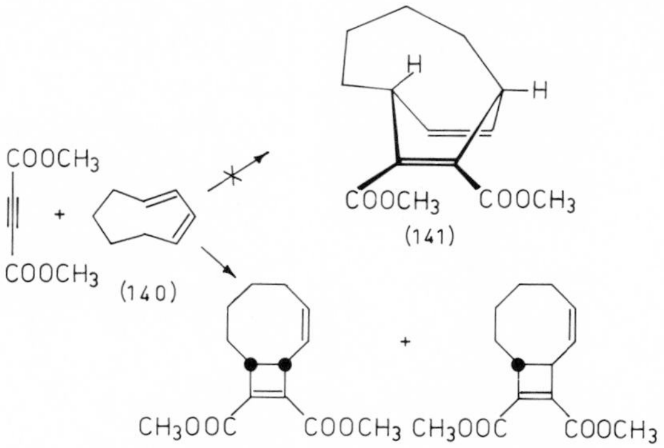

Steric hindrance also induces a [2 + 2] reaction in 1,8-divinylnaphthalene (142) at 425° to give (143) and (144). The reaction proceeds via diradicals; thus the E,E-dideuterated sample (145) gives a mixture of stereoisomers (146) [53].

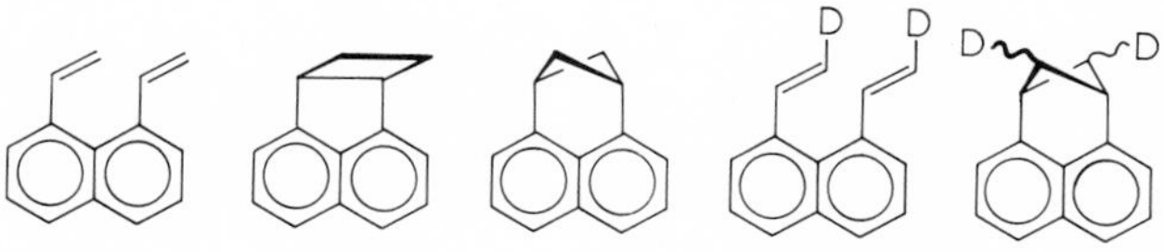

(142) (143) (144) (145) (146)

Bicyclo[4.2.0]octane undergoes reversion at 400°; if the labelled species (147) is studied, both cis- and trans-1,2-dideuterioethylene are obtained [54].

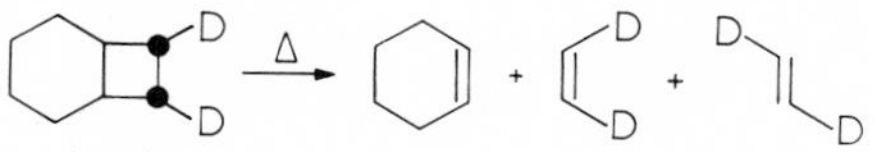

(147)

A number of [σ2 + π2] cycloadditions and reversions have been observed, especially in combinations of strained single bonds and ethylenes or acetylenes bearing electron withdrawing groups [49]. Two examples may be mentioned: The thermal isomerization of tricyclo[3.2.1.0^{2,4}]oct-6-ene, and the addition of methyl fumarate to [2.2]paracyclophane (148). The appearance of two stereoisomeric products in the latter reaction is indicative of a diradical intermediate [55].

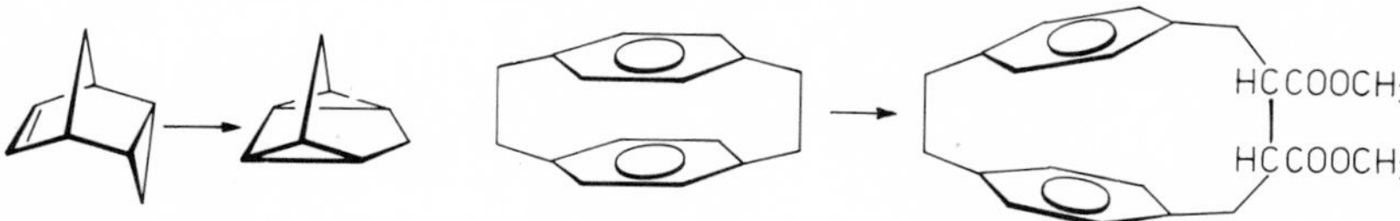

(148)

---

53. S.F. Nelsen and J.P. Gillespie, J. Amer. Chem. Soc., 94, 6238 (1972).
54. J.E. Baldwin and P.W. Ford, J. Amer. Chem. Soc., 91, 7192 (1969).
55. Reviewed by P.G. Gassman, Accounts Chem. Res., 4, 128 (1971).

An odd type of inversion process has been noted in some of these reactions, especially those of bicyclo[2.1.0]pentane, (149). They are of the free radical type, as is obvious from the fact that both fumaro- and maleonitrile ((150) and (151), respectively), give a mixture of adducts (152)-(154).

It has been found that the π-partner approaches from the "inside of the flap". Thus, the D-labeled compound (155) with maleic anhydride gives several products of which one, (156), clearly indicates endo-attack [56].

By the same token, in the cycloreversions of (157) [57], (158) and (159) [58] the five membered rings invert as shown.

---

56. P.G. Gassman, K.T. Mansfield, and T.J. Murphy, J. Amer. Chem. Soc., 91, 1684 (1969).
57. W.R. Roth and M. Martin, Ann., 702, 1 (1967).
58. E.L. Allred and R.L. Smith, J. Amer. Chem. Soc., 89, 7133 (1967).

It seems certain that the approach - or departure - of the π-fragment cannot be symmetrical and hence concerted, however; thus, even in the spirocompound (160) such inversion occurs.

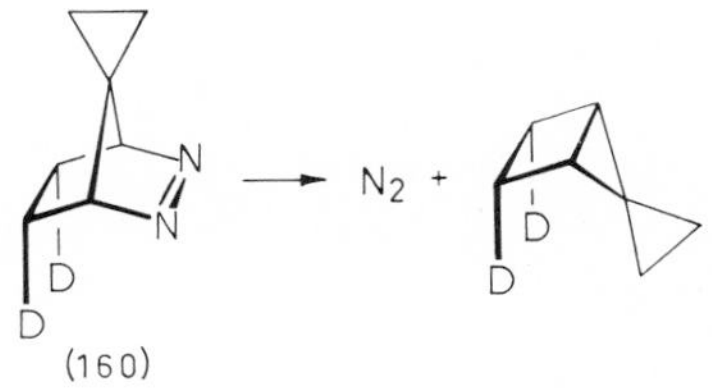

When the symmetrical inside approach is made completely impossible such as in the tricyclene (161), a different type of reaction occurs, but this is still best explained by a local, initial endo-attack on the strained sigma bond [59].

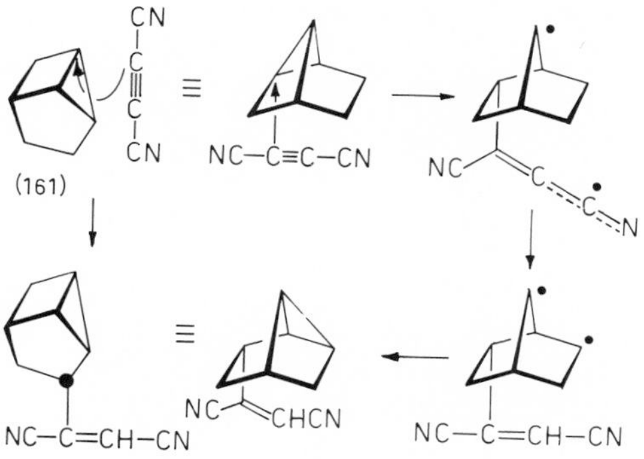

Cyclopropenones such as (162) provide us with a likely example of a thermal $[\pi 2 + \sigma 2]$ cycloaddition with a zwitterionic intermediate. While the mechanism has not been investigated, the easy electrocyclic ring opening of (162) to the oxyallylic isomer (163) is well-known (see p. 865) [60].

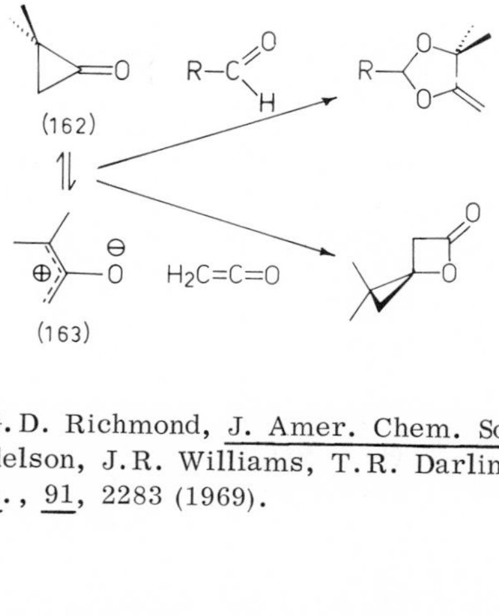

---

59. P. G. Gassman and G. D. Richmond, J. Amer. Chem. Soc., 92, 2090 (1970).
60. N. J. Turro, S. S. Edelson, J. R. Williams, T. R. Darling, and W. B. Hammond, J. Amer. Chem. Soc., 91, 2283 (1969).

In the reaction of norbornadiene (164) with (165), [2 + 2] and [2 + 4] cycloadditions compete. The use of polar solvents greatly favor the formation of (167) and (168), suggesting that (169) is intermediate [61].

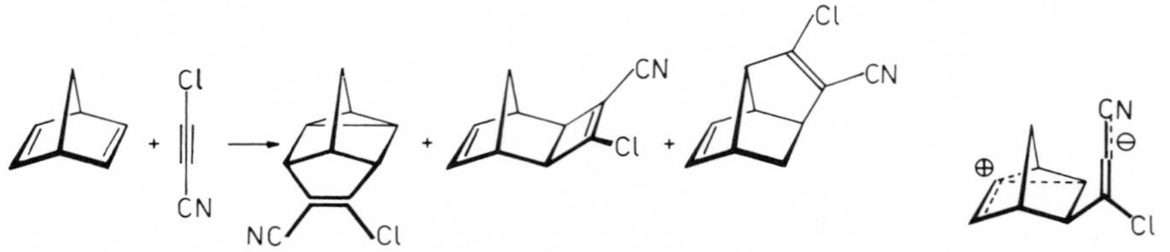

(164) (165) (166) (167) (168) (169)

The ground rules are completely different when one or both of the reacting molecules of the [2 + 2] cycloaddition have cumulenic double bonds; these reactions are invariably highly stereospecific. Thus, when β-lactams (170) are heated, CO_2 is eliminated and olefins obtain with complete retention [62].

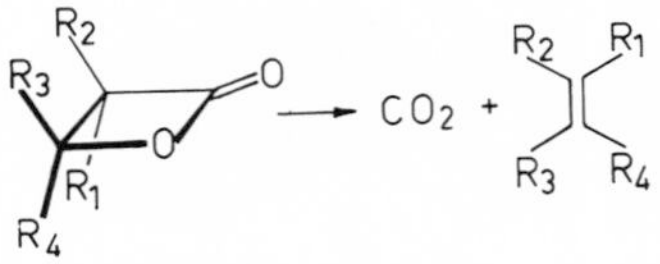

(170)

Conversely, when ethoxyketene (171) is generated in the presence of various olefins, cyclobutanones are obtained also with retention [63].

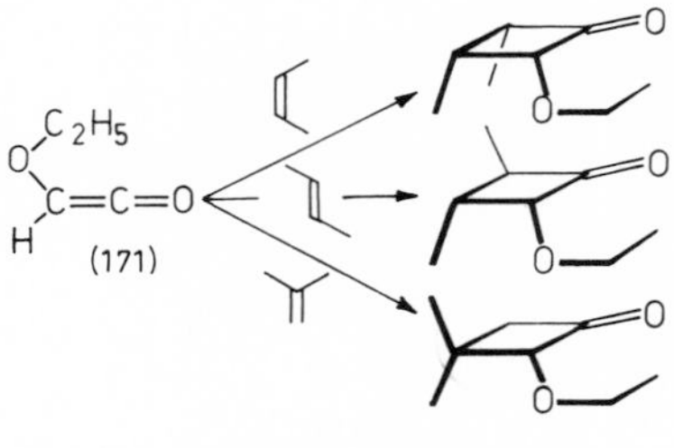

(171)

An interesting aspect of such ketene additions is that the most hindered products tend to form. For instance, ketenes (172) (X = Cl, Br or Me) all add to cyclopentadiene in such a way that the phenyl group is in the endo-position [64].

---

61. T. Sasaki, S. Eguchi, M. Sugimoto, and F. Hibi, J. Org. Chem., 37, 2317 (1972).
62. W. Adam, J. Baeza, and J.-C. Liu, J. Amer. Chem. Soc., 94, 2000 (1972).
63. T. DoMinh and O. P Strausz, J. Amer. Chem. Soc., 92, 1766 (1970).
64. W.T. Brady, F.H. Parry, R. Roe, and E.F. Hoff, Tetrahedron Lett., 819 (1970).

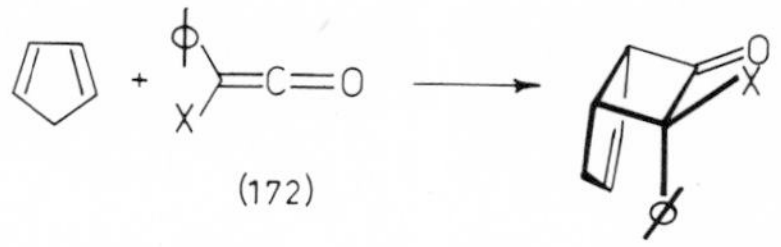

(172)

Likewise, when ketenes (173) are added to cyclopentadiene, the ratio (174) to (175) increases from 0.7 when R is methyl to 1.6 when R is ethyl (X = Cl), and from 3 when R is i-propyl to 100 when R is t-butyl (X = Br) [65].

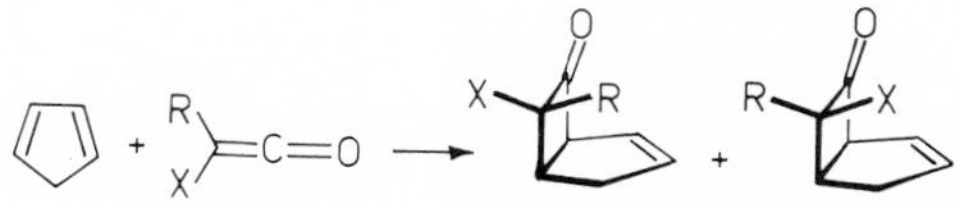

(173) (174) (175)

In addition, it is generally found that cis-disubstituted olefins add to ketenes much faster than trans- [66]; with the azobenzenes, addition to ketenes occurs with the cis-isomers only [67]. The explanation for these phenomena is that the ketene is added in antarafacial manner, so that we are dealing here with an allowed, thermal [$_\pi 2_s$ + $_\pi 2_a$] cycloaddition. In this type of addition, the least hindered approach to the transition state eventually involves actually the most hindered products, since the business end of the ketene must be twisted around before the energy minimum is reached (see Fig. 14-13). Since the ketenes are planar their stereochemical fate cannot be determined, but this is possible in the dimerization of allenes. It was found that both d- and l-1,2-cyclononadiene (176) give the meso-1,2-dimethylenecyclobutane adduct (177), whereas a d,l-mixture of (176) produces both d,l- and

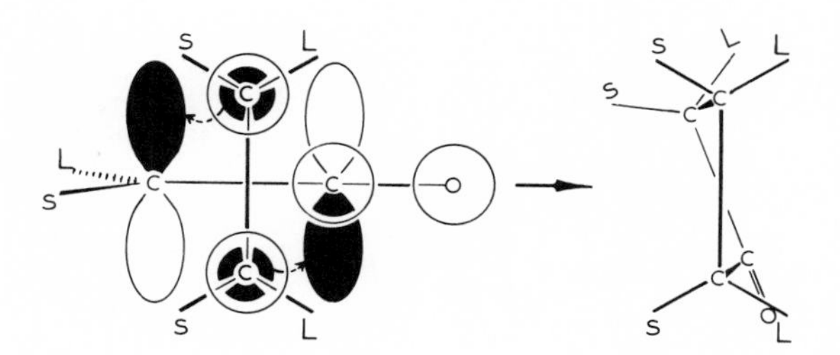

Fig. 14-13. The least hindered approach to the transition state in a [$_\pi 2_s$ + $_\pi 2_a$] cycloaddition. L is a large substituent, and S a small one.

65. W.T. Brady and R. Roe, J. Amer. Chem. Soc., 92, 4618 (1970).
66. N.S. Isaacs and P.F. Stanbury, Chem. Commun., 1061 (1970).
67. R.C. Kerber and T.J. Ryan, Tetrahedron Lett., 703 (1970).

meso- (177). This finding is compatible only with the tetrahedral approach characteristic of [$_{\pi}2_s$ + $_{\pi}2_a$] cycloaddition [68] (see Fig. 14-14). If the ketene cycloadditions are of the same type, we will have to conclude that it is the ketene that suffers inversion in these reactions.

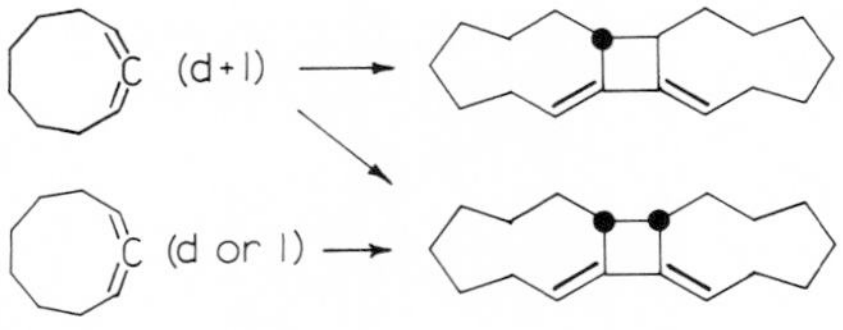

C (d+l) →

C (d or l) →

(176) (177)

Several measurements of isotope effects have been reported for the addition reactions of ketenes. In most cases concertedness and retention of the olefin were concluded to characterize the reaction, but little more can be said since they cannot inform us about the stereochemistry of the antarafacial part [69].

Longer cumulenes are also subject to [2 + 2] cycloadditions. The hexapentaene (178) dimerizes at the melting point of 185° to give (179) [70]. The 72 protons in this molecule give a sharp singlet in the nmr at 8.80 τ, which must be something of a record.

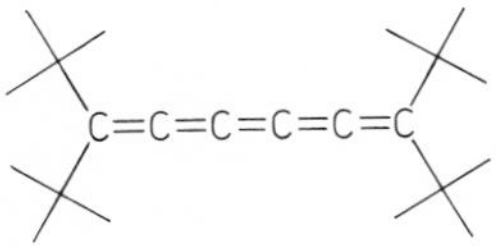

C=C=C=C=C=C

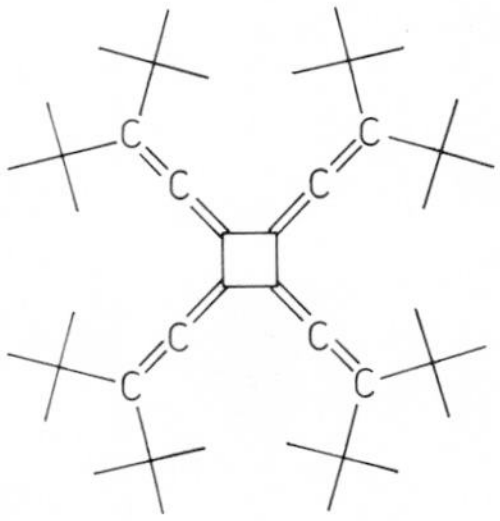

(178) (179)

68. (a) W.R. Moore, R.D. Bach, and T.M. Ozretich, J. Amer. Chem. Soc., 91, 5918 (1969). (b) The dimerization of many other allenes proceeds via diradicals; see e.g. T.L. Jacobs and R.C. Kammerer, J. Amer. Chem. Soc., 94, 7190 (1972).
69. (a) For instance, J.E. Baldwin and J.A. Kapecki, J. Amer. Chem. Soc., 92, 4874 (1970). However, see also, (b) W.R. Dolbier and S.H. Dai, J. Amer. Chem. Soc., 90, 5028 (1968) and 92, 1774 (1970).
70. H.D. Hartzler, J. Amer. Chem. Soc., 88, 3155 (1966).

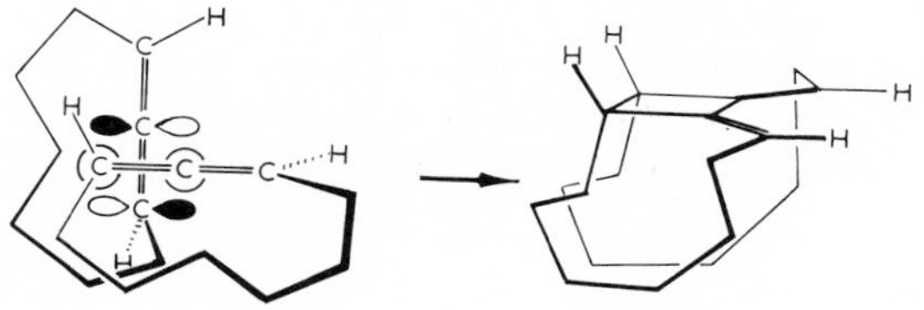

Fig. 14-14. Two identical 1,2-cyclononadiene molecules in their $[\pi 2_s + \pi 2_a]$ approach to meso-dimer (177).

The only cycloadditions known so far which truly seemed to go counter to the Woodward-Hoffmann predictions are those of the electron-rich olefins such as tetramethoxyethylene (180) to their highly electrophilic cousins such as azo compounds, nitroso compounds, tetracyanoethylene, and so on [71]. These reactions are stereospecific and characterized by retention. Thus, the cis- and trans-isomers of 1,2-bistrifluoromethyl-1,2-dicyanoethylene, (181) and (182), add with retention to tetramethoxyethylene.

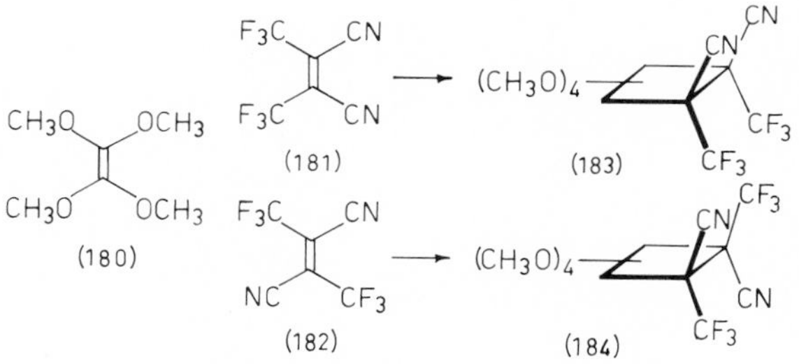

Since the electrophilic olefin most likely would be the inverting partner, retention probably occurred with both olefins. In spite of this stereospecificity, these reactions probably are not concerted. To begin with, charge transfer complexes (see Chr. 23) are almost certainly formed, and from there it is only a small step to a dipolar intermediate. This could almost certainly close before any rotations occurred (see also p. 488).

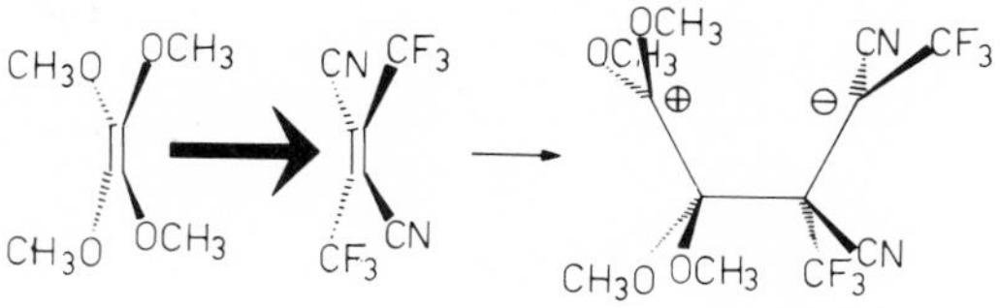

71. R.W. Hoffmann, U. Bressel, J. Gehlhaus, and H. Häuser, Chem. Ber., 104, 873 (1971).

Thermal [2 + 2] cycloreversions - if concerted and suprafacial - should produce an excited state fragment. The chemiluminescent decomposition of dioxetanes such as (185) may be an example [72]. It is in at least some cases likely that the reaction does not completely merit the description just given, however (see further p. 456).

OC_2H_5 OC_2H_5 ⟶ $HCOOC_2H_5$ + $HCOOC_2H_5^{*}$

(185)

Diels-Alder reactions ($[\pi 4_s + \pi 2_s]$) and the reverse are well known in synthesis, involving dienes and properly substituted olefins referred to as dienophiles. If the substituents of the latter are sufficiently electronegative, even benzene becomes subject to their addition. Thus, dicyanoacetylene adds to benzene on prolonged and severe heating to give low yields of adduct (186); in the presence of aluminum bromide - which forms an isolable complex with the acetylene - the reaction goes rapidly even at room temperature [73].

CN, CN, Δ, CN, CN

(186)

$AlBr_3$

$N{\equiv}C{-}C{\equiv}C{-}C{\equiv}N{-}AlBr_3$ ⊕ ⊖

An interesting recent application of the Diels-Alder reaction, the addition of biallenyl (187) to acetylenic dieophiles, is employed to prepare [2.2]paracyclophanes [74].

$COOCH_3$, $COOCH_3$, $COOCH_3$, $COOCH_3$, $(COOCH_3)_2$, $(COOCH_3)_2$

(187)

---

72. T. Wilson and A. P. Schaap, J. Amer. Chem. Soc., 93, 4126 (1971).
73. E. Ciganek, Tetrahedron Lett., 3321 (1967).
74. H. Hopf, Angew. Chem., Int. Ed. Engl., 11, 419 (1972).

Retro Diels-Alder reactions are also important in synthesis. They are particularly facile when the incipient double bond helps create an aromatic system. An elegant recent example is the synthesis of fulvenes (188), isoindoles (189), isobenzofurans (190) and other labile molecules by a general technique involving a Diels-Alder reaction with 3, 6-diphenyl-s-tetrazine (191), followed by an immediate retro reaction in which nitrogen is expelled; the product can be isolated, but alternatively a second retro reaction can be affected to give the desired product [75].

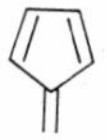

(188) 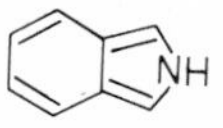 (189) 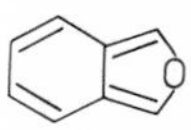 (190) 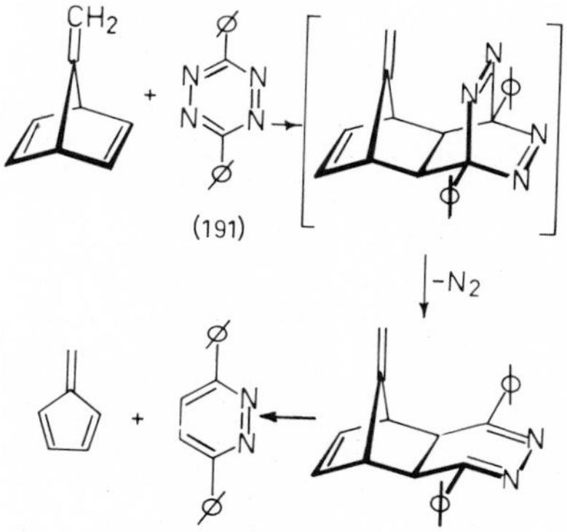 (191)

Nitrogen expulsion also occurs in the retro reaction of pyrazolinyl anions (192) [76]. Another case of such anionic cycloreversion is that of tetrahydrofuran [77]; the stereochemistry is not known in either instance.

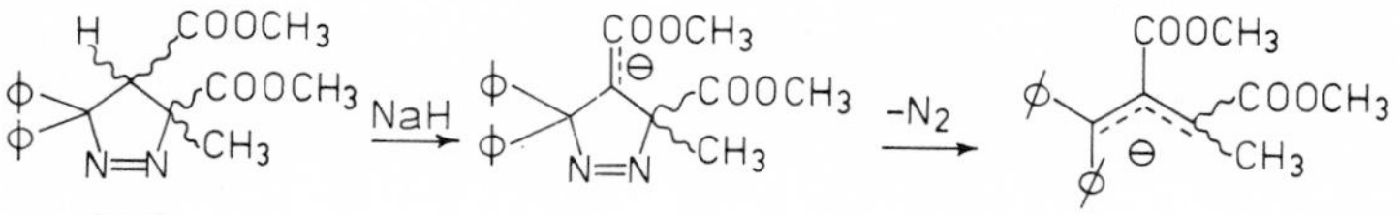

(192)

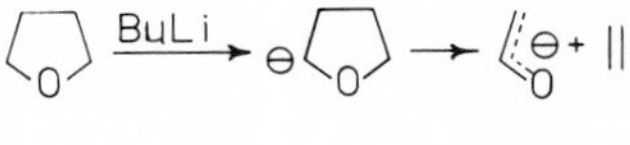

Much is known about the mechanism of the Diels-Alder reaction. Every indication is that the reaction is concerted (see p. 488 and further). An interesting experiment has been devised by Krantz to show that the Diels-Alder reaction is favored over [2 + 2] or biradical pathways even when all these pathways lead to the same product [78]. The substrate was the allylcyclohexadiene (193); one of

---

75. W. S. Wilson and R. N. Warrener, Chem. Commun., 211 (1972).
76. P. Eberhard and R. Huisgen, J. Amer. Chem. Soc., 94, 1345 (1972).
77. R. B. Bates, L. M. Kroposki, and D. E. Potter, J. Org. Chem., 37, 560 (1972).
78. A. Krantz, J. Amer. Chem. Soc., 94, 4020 (1972).

products obtained is (194). A deuterium label exclusively winds up in the position demanded by a [2 + 4] cycloaddition.

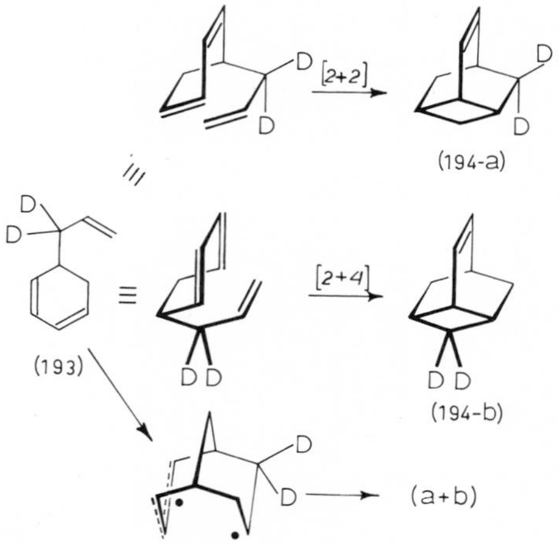

Solvent effects are often quoted in support of zwitterionic intermediates; however, a recent example clearly shows that this is not necessarily always true [79]. The reaction of (195) with (191) produces (196) 500 times faster in dimethyl sulfoxide then in chloroform. The reason for this effect is that (195) in chloroform has an internal hydrogen bond to the double bond, whereas in dimethyl sulfoxide this bond can involve the external base. Thus, (195) has only a single bond in the OH region in the ir and it is independent of concentration, but in basic solvents it has two. The rate differential disappears when the substituent is methoxy rather than OH.

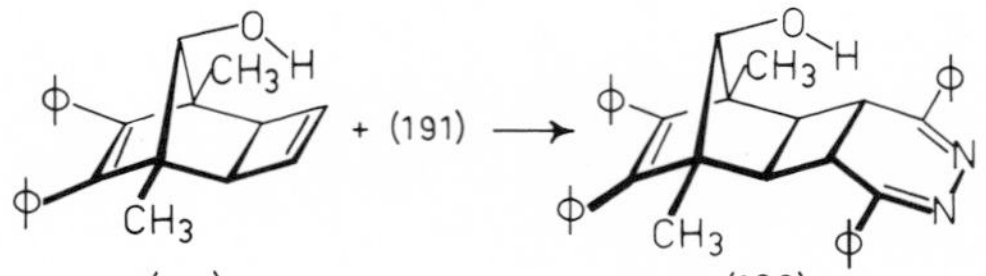

Photochemical [$_{\pi}4$ + $_{\pi}2$] reactions have been reported [80]. Thus, 9-cyanoanthracene (197) reacts with certain dienes such as E,E-2,4-hexadiene to give Diels-Alder products such as (198).

---

79. I.W. McCay, M.N. Paddon-Row, and R.N. Warrener, Tetrahedron Lett., 1401 (1972).
80. N.C. Yang, J. Libman, L. Barrett, M.H. Hui, and R.L. Loeschen, J. Amer. Chem. Soc., 94, 1406 (1972).

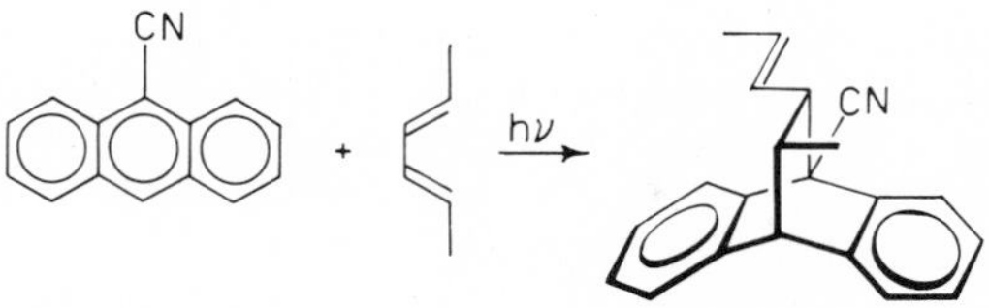

(197) (198)

The stereochemistry of the diene is preserved, although the Woodward-Hoffmann rules require either antarafacial addition or a non-concerted one. It is of course possible again that a diradical is involved which is too-short-lived to undergo rotations.

Certain substituted 1,3,5-hexatrienes are subject to internal photo Diels-Alder reactions; thus, (199) gives (200) - a reaction involving an excited state such as (201) and hence presumably an allowed [$\pi 4_s + \pi 2_a$] cycloaddition.

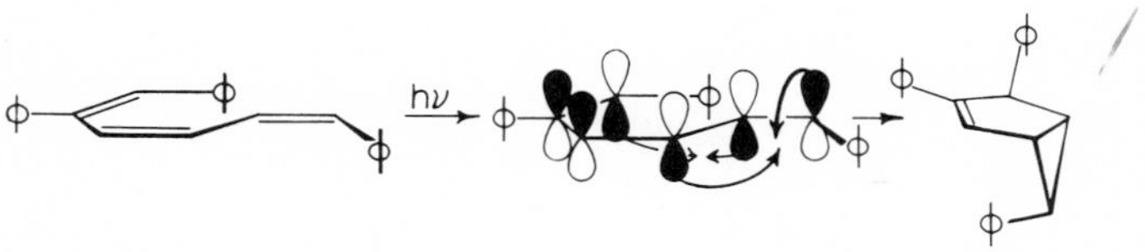

(199) (201) (200)

The problem here was that the isomer (202) gives the same product. However, a careful investigation showed the latter reaction to be a two-photon process. The first of these isomerizes (202) to (199), which is then a detectible intermediate [81].

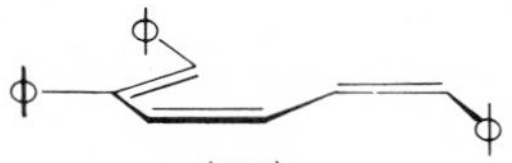

(202)

It may be noted that the doubly suprafacial requirement in thermal Diels-Alder reactions still leaves open many stereochemical choices. Thus, since both olefin and diene have two faces and since one end of the olefin may become attached to either end of the diene, eight isomers may form if both molecules are sufficiently asymmetrical (e.g. the cyclopentadiene (203) and olefin (204).

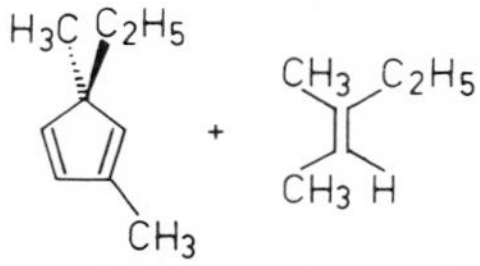

(203) (204)

81. A. Padwa, L. Brodsky, and S. Clough, J. Amer. Chem. Soc., 94, 6767 (1972).

Steric requirements will decide which isomers will predominate [82]; however, there is a remarkable feature about those reactions in which the dienophile is itself a conjugated diene. In such cases the reaction involves further specificity such that the double bonds wind up in the closest possible proximity. This rule has been known for a long time as Alder's endo-rule. It has intrigued organic chemists for a long time, partly because the products so obtained are ideal starting materials for photochemical conversion to various polycyclic structures and also because it often gives rise to highly hindered isomers when *exo*-addition could have proceeded with little or no strain at all. Thus, when cyclopentene adds to the cyclopentadienone (205), a mixture obtains in which *exo*- (206) predominates; but cyclopentadiene gives a mixture in which 97% has the *endo*-configuration (207) [83].

Many theories have been offered to account for this phenomenon [83]. It is now generally agreed that the driving force for it is the so-called secondary orbital interaction [84], as suggested in Fig. 14-15. The interaction is strong enough to lower the energy of the transition state significantly.

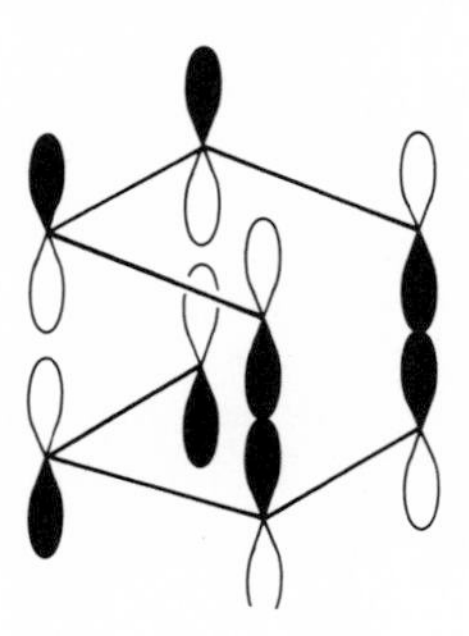

Fig. 14-15. Secondary orbital interaction during the approach of two dienes; a similar interaction is still possible at one pair of carbon atoms even if the *s-trans*-conformation is considered for the dienophile.

82. K. L. Williamson and Y.-F. L. Hsu, *J. Amer. Chem. Soc.*, *92*, 7385 (1970).
83. K. N. Houk, *Tetrahedron Lett.*, 2621 (1970).
84. R. Hoffmann and R. B. Woodward, *J. Amer. Chem. Soc.*, *87*, 4388 (1965).

The experiment by Eckert quoted earlier provides a direct confirmation of this interaction [45a]; thus, while it should evoke no surprise that the Diels-Alder products have a smaller volume than the starting materials, if the dienophile is itself a diene, then the volume of the transition state in smaller than either.

A phenomenon related to the [2 + 4] cycloaddition is the rearrangement of certain Diels-Alder products. For instance, the resolved adduct (208) upon heating rearranges smoothly to optically active (209).

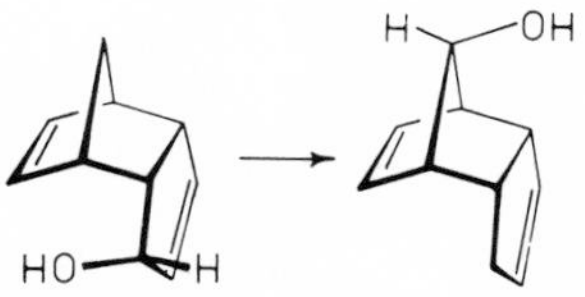

(208) (209)

Evidently a retro reaction begins, but then the molecule finds itself on the track of another forward Diels-Alder reaction and rearranges [85]. Alternatively, one may regard the reaction simply as a Cope rearrangement.

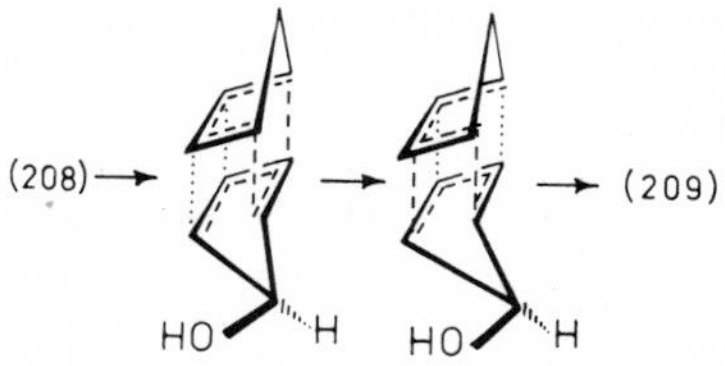

(208) → → → (209)

Another reaction closely related to the Diels-Alder reaction is the "ene synthesis" [86] and its reverse [87], symbolized by $[\sigma 2_s + \pi 2_s + \pi 2_s]$.

Endo-addition is the rule here also, as demonstrated in a beautiful experiment by Berson [88].

85. R.B. Woodward and T.J. Katz, Tetrahedron, 5, 70 (1959).
86. H.M.R. Hoffmann, Angew. Chem., Int. Ed. Engl., 8, 556 (1969).
87. W.R. Roth, Chimia, 20, 229 (1966).
88. J.A. Berson, R.G. Wall, and H.D. Perlmutter, J. Amer. Chem. Soc., 88, 187 (1966).

He finds that the ene reaction of maleic anhydride of cis-2-butene followed by complete hydrogenation of the product gives racemic 3,4-dimethylhexane (210), whereas the same sequence with trans-2-butene eventually affords the meso-product (211); exo-addition would obviously have reversed these results.

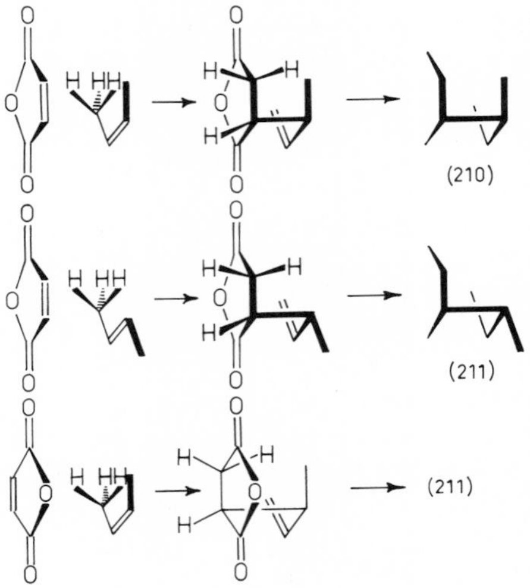

(210)

(211)

(211)

Under the stimulus of the symmetry based selection rules many new reactions have been found which are analogous to the Diels-Alder reaction but involve σ electrons. While it cannot be said that the detailed mechanisms are known in every case, all of them are permitted by the Woodward-Hoffmann rules. Thus, cyclopropanones add to furan to give adduct (212) ($\pi 4_S + \sigma 2_S$]) [60],

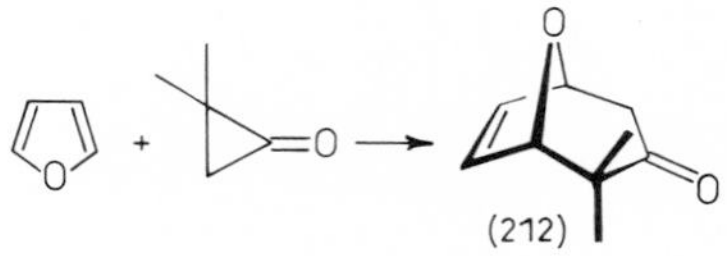

(212)

quadricyclane (213) reacts in [$\sigma 4_S + \pi 2_S$] fashion with acetylenes to give (214) [89], norbornadiene (215) in a [$\pi 2_S + \pi 2_S + \pi 2_S$] reaction adds tetracyanoethylene to give (216) [90], bicyclopentene (217) with the same reagents gives (218) [91] in a [$\pi 2_S + \sigma 2_S + \pi 2_S$] reaction, diademane (219) reverts to triquinacene (220) [92] and

89. C.D. Smith, J. Amer. Chem. Soc., 88, 4273 (1966).
90. A.T. Blomquist and Y.C. Meinwald, J. Amer. Chem. Soc., 81, 667 (1959).
91. J.E. Baldwin and R.K. Pinschmidt, Tetrahedron Lett., 935 (1971).
92. A. de Meijere, D. Kaufmann, and O. Schallner, Angew. Chem., Int. Ed. Engl., 10, 417 (1971).

7-oxatetracyclane (221) rearranges to oxepin (222) [93]. Both of these last two reactants were themselves prepared by photocycloadditions, and both relax to open species by [$\sigma 2_s + \sigma 2_s + \sigma 2_s$] processes.

COOCH3 + COOCH3 → COOCH3 COOCH3

(213) (214)

(215) + NC NC CN CN → (216) (CN)4

NC NC CN CN + (217) → (CN)4 (218)

hν Δ

(219) (220)

O COOCH3 COOCH3 hν O COOCH3 COOCH3 Δ O CH3OOC COOCH3

(221) (222)

The decomposition of many cyclic azo compounds is of interest in this connection, since the data now available permit a direct comparison with the diradical reaction of (223) discussed earlier (p. 494).

N N → N_2 +

(223)

---

93. E. Druckrey, M. Arguëlles, and H. Prinzbach, Chimia, 20, 432 (1966).

The thermolysis of (224) is 10^{11} times faster, and it yields 1,4-cyclohexadiene even though tricyclo[3.1.0.0^{2,4}]hexane (225) is stable under comparable conditions [94].

$\xrightarrow{\Delta}$ N_2 +

(224) (225)

Thus, the cyclopropyl group participates in the formation of the transition state, which is more stable than that of (223) by about RT ln 10^{11}, or 20 kcal/mole. The effectiveness of the cyclopropyl group depends on its precise orientation in a most remarkable way; thus, compared to (223), the thermolyses of (226), (227) and (228) are faster by factors of 10^{11}, 5 x 10^8 and 10, respectively.

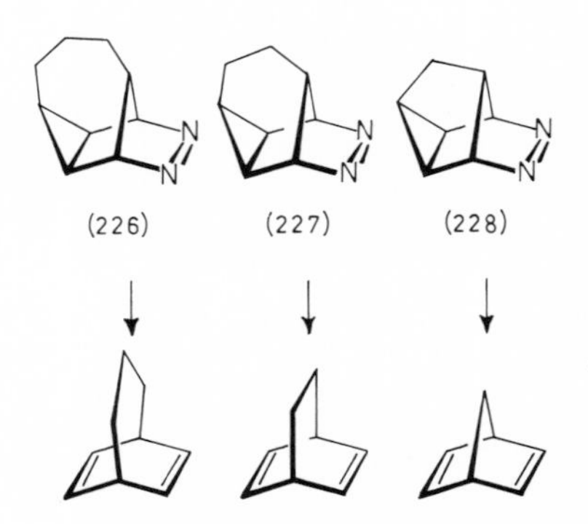

All give the corresponding dienes [95]. This remarkable finickiness of cyclopropyl groups for the precisely right orientation to participate has also been noted in carbonium ion reactions (p. 751f).

Cyclobutyl participation does not seem to occur in these reactions. If we compare (223), (229) and (230), it is found [96] that the relative rate constants are 1, 10^{17} and 10^4.

(229) (230)

---

94. E.L. Allred, J.C. Hinshaw, and A.L. Johnson, J. Amer. Chem. Soc., 91, 3382 (1969).
95. E.L. Allred and A.L. Johnson, J. Amer. Chem. Soc., 93, 1300 (1971).
96. E.L. Allred and J.C. Hinshaw, Chem. Commun., 1021 (1969) and Tetrahedron Lett., 387 (1972).

Thus, it would superficially seem that cyclobutyl participation, while not as powerful as that of the cyclopropyl group, does nevertheless take place. However, the rate enhancement is probably due only to the greater relief of strain of the four _syn_-oriented hydrogen atoms. Thus, the entropies of activation for these reactions are +10.5, -21 and +11 eu, respectively; moreover, while the reaction of (230) does produce 1,5-cyclooctadiene (231), in the early stages, the reaction mixture contains substantial amounts of (232) - which evidently then goes on to (231). Here we have a case of an allowed concerted reaction which prefers a stepwise pathway.

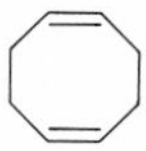

(231)  (232)

One new development in these reactions is the finding that the reversions of (233)-(235) are stereospecific and hence probably concerted whether carried out thermally or photochemically [97]. A possible cause of this result might be antarafacial expulsion of the nitrogen as in (236).

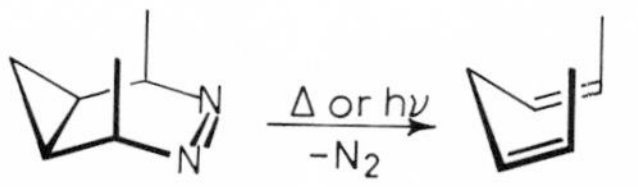

Δ or hν
$-N_2$

(233)

(234)

(235)

(236)

Many cycloadditions involving more than six electrons are now known. Anthracene undergoes a photochemical $[_\pi 4_s + _\pi 4_s]$ cycloaddition to dienes to give an adduct (237) [98].

97. J.A. Berson and S.S. Olin, _J. Amer. Chem. Soc._, _92_, 1086 (1970).
98. N.C. Yang and J. Libman, _J. Amer. Chem. Soc._, _94_, 1405 (1972).

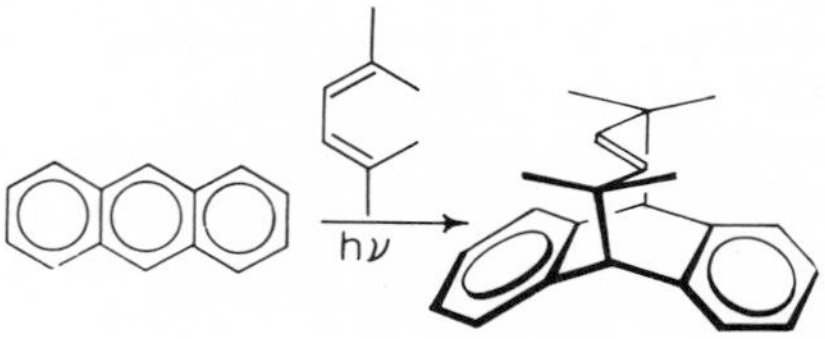

(237)

The pyrone (238) in the solid state gives photodimer (239), although [2 + 2] cycloaddition occurs if a dilute solution is photolyzed [99].

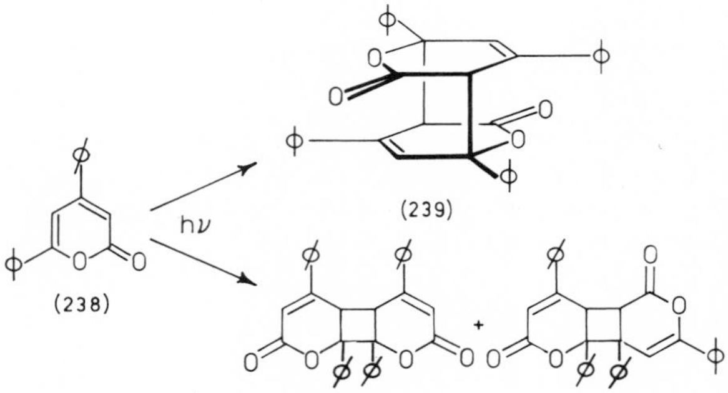

(238) (239)

The photochemical conversion of cyclopentadienenaphthoquinone endo-adduct (240) to (241) is a $[_{\pi}6_S + _{\pi}2_S]$ reaction [100],

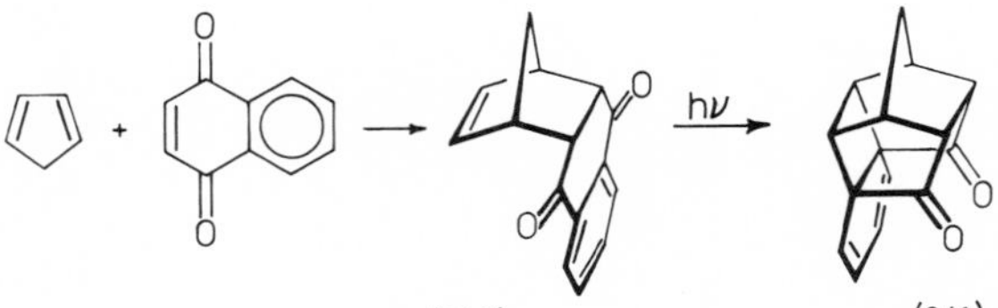

(240) (241)

and the photo addition of butadiene to the steroidal enone (242) is formally a [2 + 2 + 2 + 2] reaction [101].

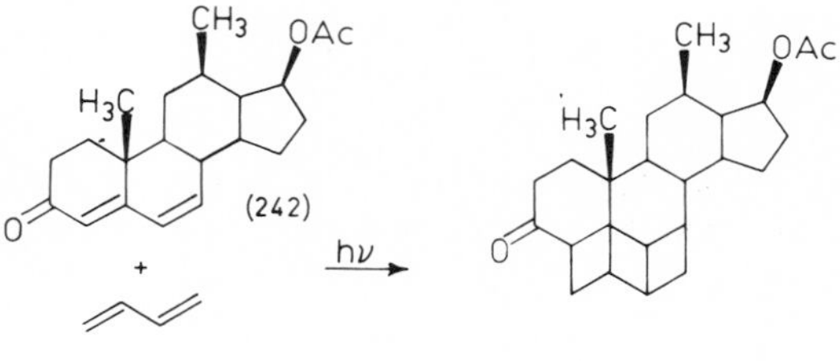

(242)

99. R.D. Rieke and R.A. Copenhafer, Tetrahedron Lett., 879 (1971).
100. A.S. Kushner, Tetrahedron Lett., 3275 (1971).
101. G.R. Lenz, Tetrahedron Lett., 3027 (1972).

Thermal cycloadditions involving ten electrons are of interest not only from a synthetic point of view but also because secondary orbital interactions now tend to be repulsive in nature so that *exo*-addition will be preferred - just opposite to what is found in the Diels-Alder reactions.

N-carbethoxyazepine (243) dimerizes when heated to 125°; the structure of the dimer is (244) [102]. Note that structure (245) is not obtained; unlike the actual product, its formation would involve several unfavorable secondary orbital interactions [103] (Fig. 14-16).

$COOC_2H_5$ (243) ⟶ H_5C_2OOC–N (244) N–$COOC_2H_5$ + H_5C_2OOC–N (245) N–$COOC_2H_5$

Diradicals are not involved; they *can* be generated at higher temperature but they rearrange to still another dimer (cf. p. 520).

Perhaps even more convincing are the several examples encountered by Houk. Cyclopentadienone (246) adds thermally to tropilidene (247) to give [2 + 4] product (248) and [4 + 6] product (249); note that (248) has the *endo*-configuration whereas (249) is *exo*.

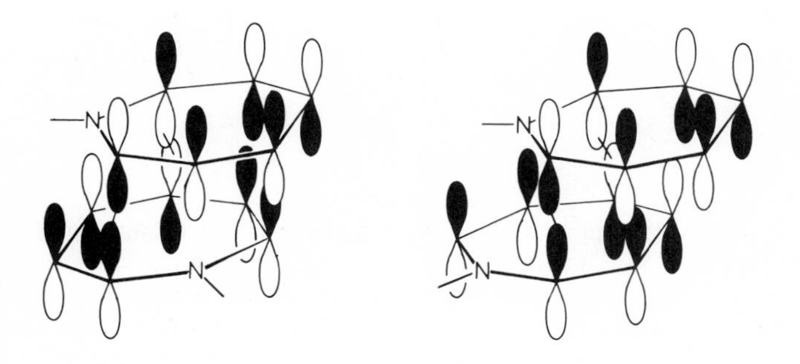

Fig. 14-16. Bonding and secondary orbital interactions in the formation of (244) and (245).

---

102. I.C. Paul, S.M. Johnson, J.H. Barrett, and L.A. Paquette, *Chem. Commun.*, 6 (1969).
103. L.A. Paquette, J.H. Barrett, and D.E. Kuhla, *J. Amer. Chem. Soc.*, *91*, 3616 (1969).

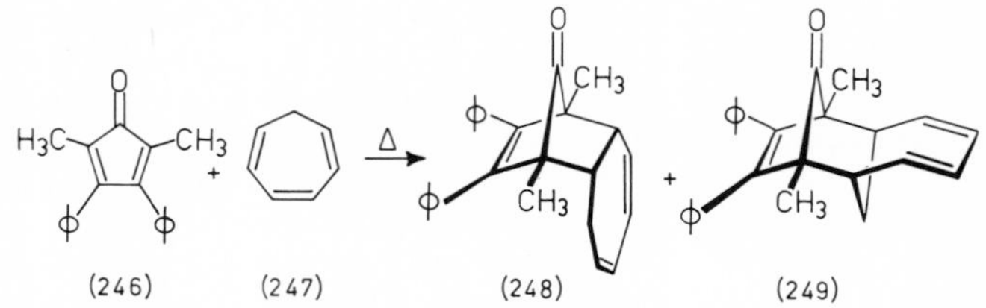

(246) (247) (248) (249)

The reaction is concerted; its rate is solvent independent - which is not expected if a zwitterionic intermediate were involved - and the diradical again gives rise to another product described in the following section (p. 520) [104]. Several additional features of interest may be observed if tropone (250) is used [105]. An endo-[2 + 4] product (251), an exo-[2 + 8] product (252) and an exo-[4 + 6] product (253) obtain in as many concerted reactions. The last of these products upon reduction gives an alcohol which reverts in [4 + 6] fashion (254).

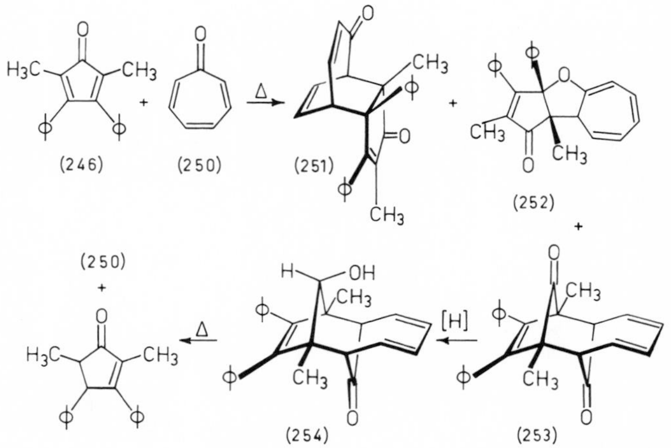

(246) (250) (251) (252) (253) (254)

The study of these systems is sometimes complicated by various side reactions and isomerizations. For instance, the addition of dimethylfulvene (255) to tropone results in endo-[2 + 4] product (256) (exact configuration unknown) and (257). The latter compound actually results from a 1,5-hydrogen shift (cf. next section) of [4 + 6] product (258) which can be spectroscopically observed. Longer reaction times result in another [4 + 6] cycloaddition to give diadduct (259) [106].

---

104. K. N. Houk and R. B. Woodward, J. Amer. Chem. Soc., 92, 4143 (1970).
105. K. N. Houk and R. B. Woodward, J. Amer. Chem. Soc., 92, 4145 (1970).
106. K. N. Houk, L. J. Luskus, and N. S. Bhacca, J. Amer. Chem. Soc., 92, 6392 (1970).

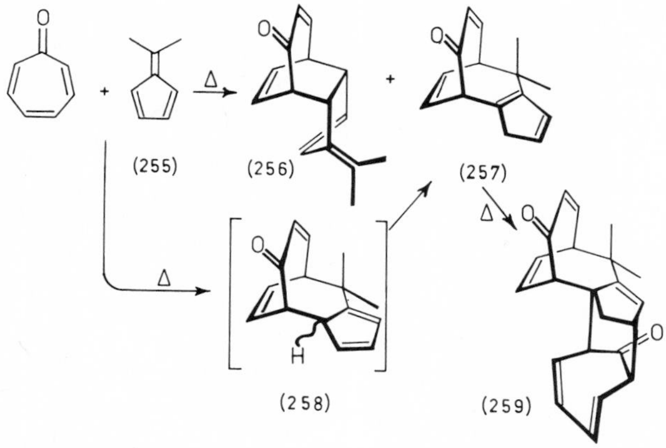

(255) (256) (257) (258) (259)

Additional examples are the reaction dipolar [$_{\pi}6_s$ + $_{\pi}4_s$] cycloaddition of dimethylfulvene to diazomethane to give (260) which tautomerizes to (261) [107],

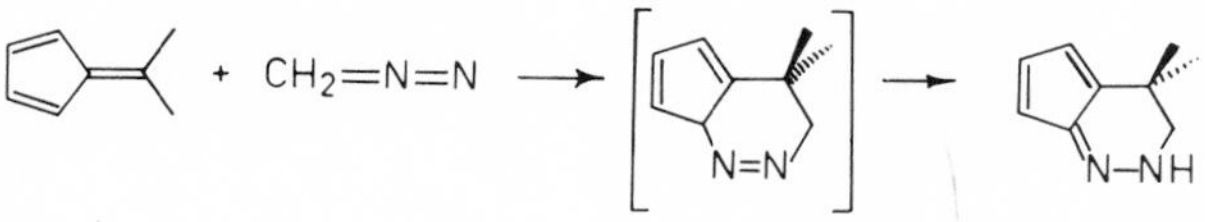

(260) (261)

the [$_{\pi}6_s$ + $_{\sigma}4_s$] addition to tropone to quadricyclane [108],

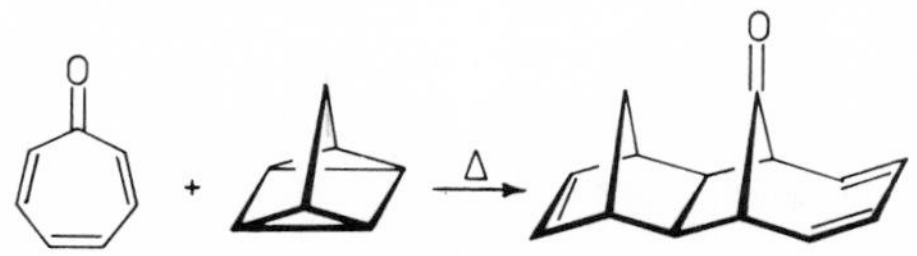

the [$_{\pi}8_s$ + $_{\pi}2_s$] additions to (262) [109],

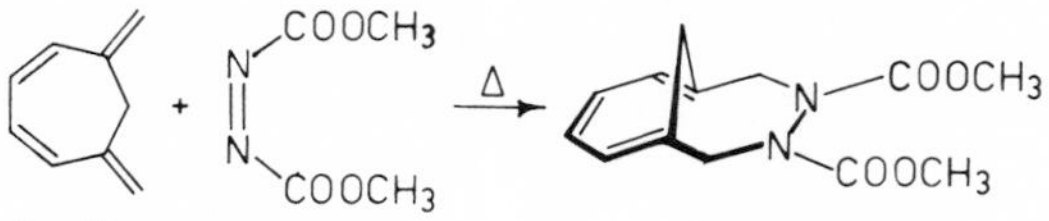

(262)

---

107. (a) K. N. Houk and L. J. Luskus, Tetrahedron Lett., 4029 (1970); (b) K. N. Houk and C. R. Watts, Tetrahedron Lett., 4025 (1970).
108. (a) H. Tanida and T. Tsushima, Tetrahedron Lett., 395 (1972); (b) H. Tanida, T. Tsushima, and Y. Terui, Tetrahedron Lett., 399 (1972).
109. G. C. Farrant and R. Feldmann, Tetrahedron Lett., 4979 (1970).

and the reaction of oxonin (263) with the triazolinone (264) to give (265); this product is either an example of $[_\pi 2_S + _\pi 2_S + _\pi 2_S]$ addition or $[_\pi 8_S + _\pi 2_S]$ addition to give (266), followed by disrotatory ring closure [110].

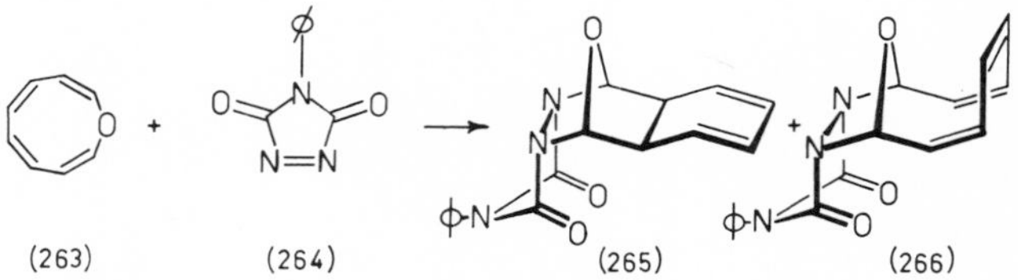

(263) (264) (265) (266)

The photodimerization of tropone in acetonitrile gives several adducts. Some of these of the forbidden type [111]; however, it was later found that the reaction involves diradicals.

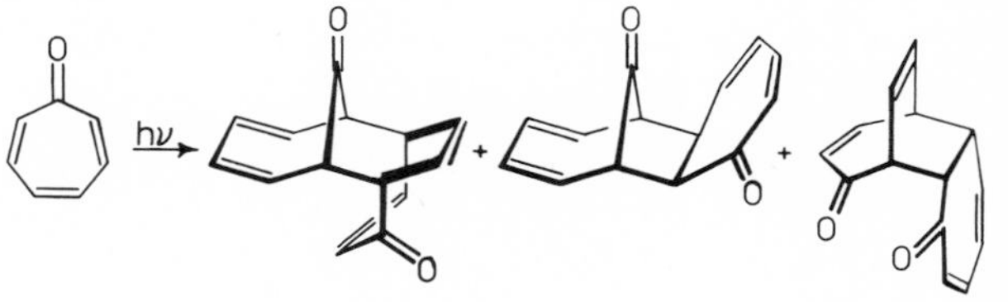

Thus, fluorenone - which has a triplet energy of 53 kcal/mole - sensitizes the reaction, and anthracene (E_t = 41 kcal/mole) quenches it [112]. In sulfuric acid the allowed $[_\pi 6_S + _\pi 6_S]$ photodimer (267) is found [113].

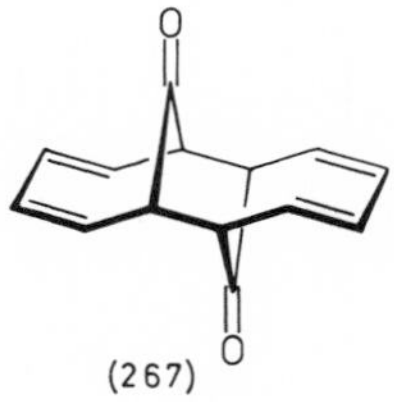

(267)

The current record seems to be the thermal reaction of fidecene (268) with tetracyanoethylene to give $[_\pi 2_S + _\pi 16_S]$ adduct (269), which immediately undergoes a 1,9-hydrogen shift to furnish (270) as the final product [114].

---

110. A.G. Anastassiou and R.P. Cellura, Chem. Commun., 484 (1970).
111. A.S. Kende, J. Amer. Chem. Soc., 88, 5026 (1966).
112. Chem. Eng. News, 45, 46, Sept. 18, 1967.
113. T. Mukai, T. Tezuka, and Y. Akasaki, J. Amer. Chem. Soc., 88, 5025 (1966).
114. H. Prinzbach, Pure Appl. Chem., 28, 281 (1971).

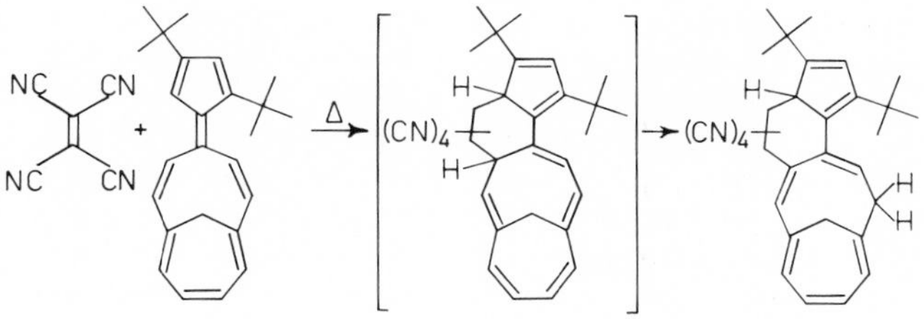

(268) (269) (270)

14-4. SIGMATROPIC REACTIONS

Sigmatropic shifts are concerted rearrangements in which a singly bound atom or group moves from one terminus of a conjugated system to another. The order [i, j] of such a reaction is given by the number of atoms between the forming and the breaking bond. Thus, allylic rearrangements are formally of the [1, 3] type (actually often not concerted);

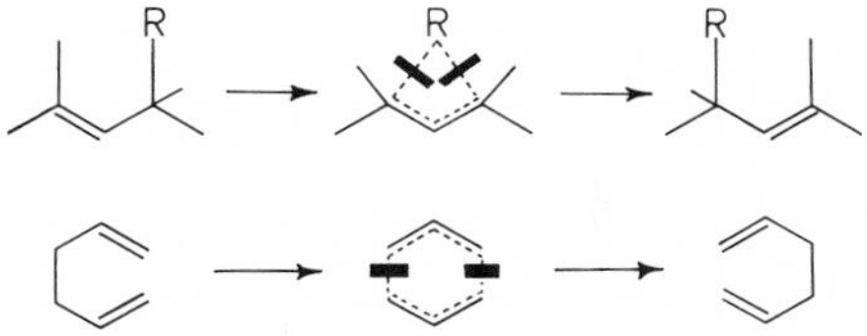

Cope rearrangements are of order [3, 3], and so on. In the process all the single bonds in the conjugated chain or chains become double, and vice versa. This should be remembered because confusion sometimes arises in the assignment of the order of sigmatropic shifts in cyclic systems; thus, the rearrangement of 5, 5-dimethylcyclopentadiene to the 1, 5-isomer is a 1, 5- and not a 1, 2-shift.

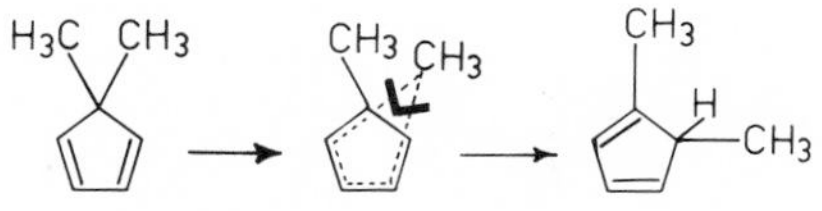

The selection rule here is that such concerted shifts are allowed thermally if (i + j) = 4n + 2, and photochemically if (i + j) = 4n, provided that both groups migrate suprafacially; if one of them is involved in the antarafacial manner, the requirement is just opposite. It might be noted parenthetically that when i = 1, antarafacial for the migrating group means: With inversion [115].

---

115. R. B. Woodward and R. Hoffmann, J. Amer. Chem. Soc., 87, 2511 (1965).

The origin of this phenomenon is the same as that of the groups of reactions discussed in earlier sections, and readily understood on the basis of the highest filled π orbital of the conjugated chain. Concerted allylic migration can occur in concerted fashion only in the excited state, or antarafacially, or with inversion (see Fig. 14-17). Antarafacial is of course hard to imagine with such a short chain (sterically it is virtually impossible); furthermore, inversion is not possible if hydrogen is the migrating group. It might be noted that many allylic derivatives such as allyl chloride readily undergo thermal 1, 3-shifts; however, these are known to be of an ionic nature, and hence as stepwise reactions do not fall within the boundaries of this chapter.

One of the simplest and best-known sigmatropic shifts is the Wagner-Meerwein rearrangement in carbonium ions. This is an allowed 1, 2-shift, occurring suprafacially and with retention of the migrating group, with the highest occupied molecular orbital of the π system holding just two electrons.

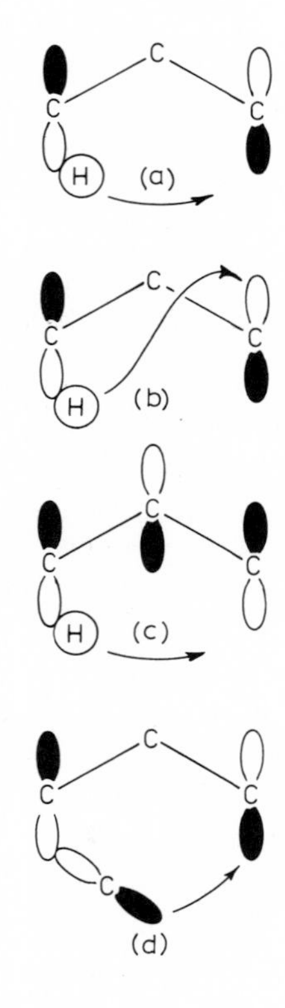

Fig. 14-17. (a) Forbidden 1, 3-hydrogen shift; (b) Allowed but sterically hindered antarafacial shift; (c) Allowed photochemical 1, 3-shift; (d) Allowed 1, 3-shift with inversion.

The overall reaction involves three stages: The generation of the carbonium ion, the 1,2-shift, and a final reaction of the ion to return to the neutral state.

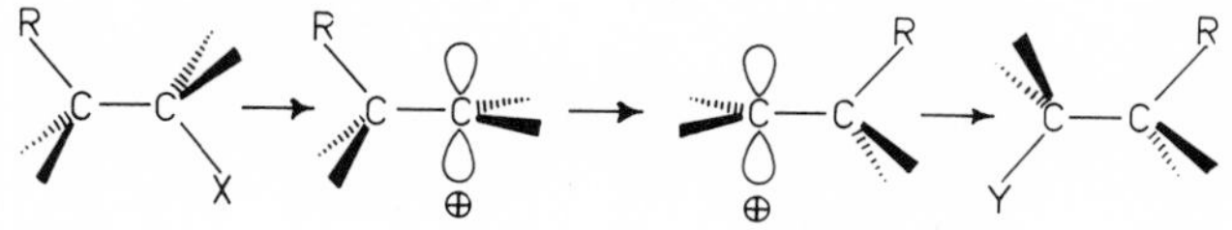

These three steps may occur in concert and often they do; in those cases both carbon atoms 1 and 2 are inverted. A large degree of racemization at C_1 and/or C_2 would indicate long-lived ions. Examples of all these possibilities are known (see Chr. 20). So far as the migrating group is concerned, its configuration is almost always retained, and the migration is always suprafacial. The following scheme shows how these stereochemical questions can in principle be studied in the laboratory.

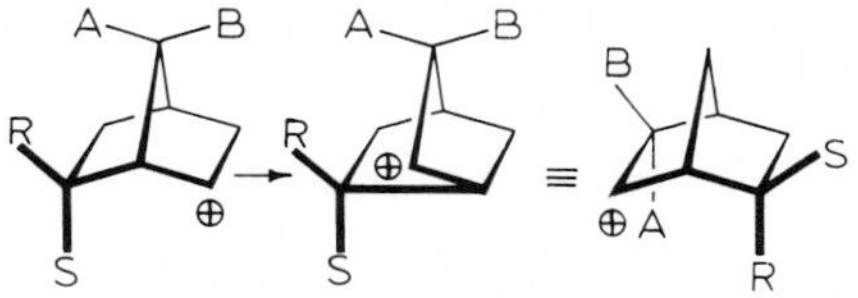

A large number of variations of this rearrangement have been described. They include the pinacol, Demyanov, Hofmann, Curtius, Wolff and Beckmann rearrangements; examples of these reactions are given below (whether the products shown are the final products may depend on the solvent). In each case, migration occurs to an electron deficient center. Which of the several available groups will migrate in open-chain compounds depends on the so-called migratory aptitude of these groups [116].

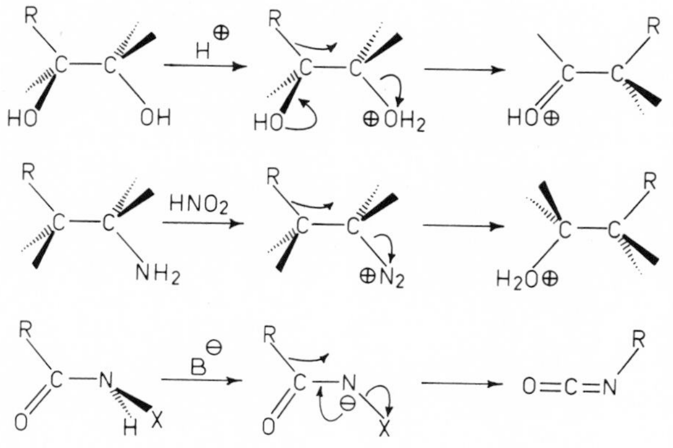

116. An excellent and comprehensive discussion of all aspects of the 1,2-shifts has been given by G.W. Wheland, "Advanced Organic Chemistry", 3rd Ed., Wiley, New York, 1960, Chr. XII.

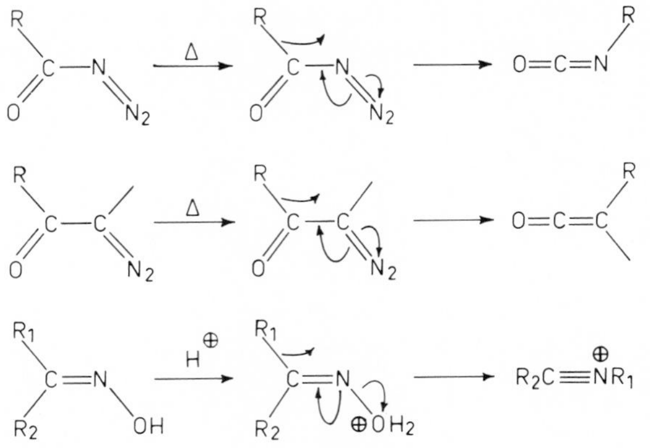

A number of anionic 1, 2-shifts are known also. For a time these were considered to be simple analogs of the cationic kind, with the same basic mechanism. As an example we might mention the Favorsky, Stevens, Wittig rearrangements.

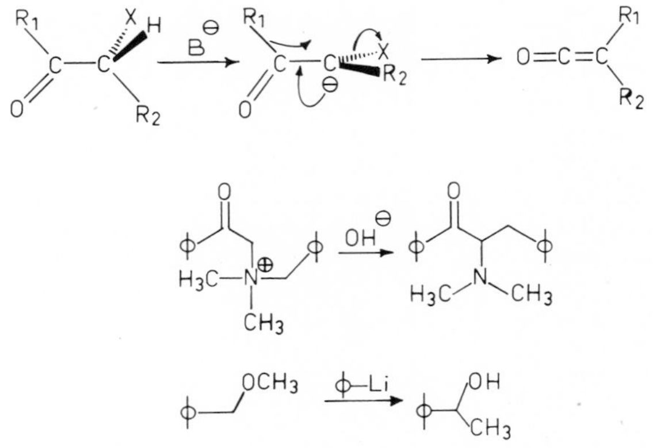

With the advent of the Woodward-Hoffmann rules it became clear that these reactions are forbidden [117]. In the Stevens rearrangement it was subsequently shown by Lepley [118] that the reaction is accompanied by the CIDNP phenomenon (see p. 72); radical pairs are therefore involved. Lansbury [119] drew a similar conclusion for the Wittig rearrangement on the basis of substituent effects in that reaction.

---

117. S.H. Pine, J. Chem. Educ., 48, 99 (1971).
118. A.R. Lepley, J. Amer. Chem. Soc., 91, 1237 (1969).
119. P.T. Lansbury, V.A. Pattison, J.D. Sidler, and J.B. Bieber, J. Amer. Chem. Soc., 88, 78 (1966).

A few thermal, suprafacial 1, 3-shifts have been observed. Those that are concerted should involve inversion of the migrating atom, and this appears to be the case. Thus, when *exo*-5-methylbicyclo[2. 1. 1]hex-2-ene (271) is heated to 120°, a 99% yield of the isomeric [3. 1. 0] product (272) is formed in which the methyl group is still *exo*; this means that inversion is occurring as implied by (273) rather than retention as suggested by (271) [120].

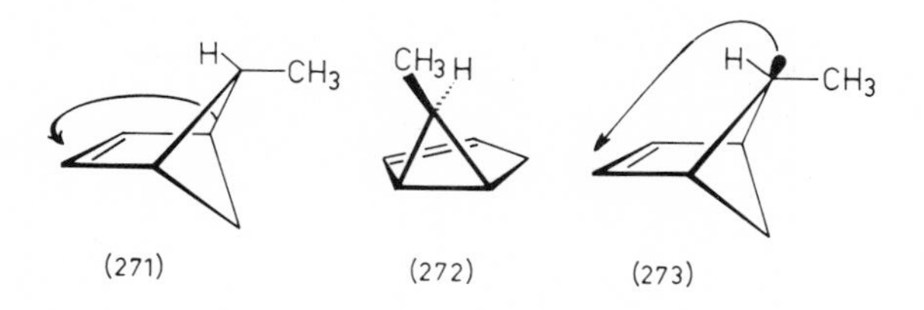

(271) (272) (273)

The argument loses some of its strength when it is realized that the *endo*-isomer (274) at 150° rearranges much less cleanly; the ratio of inversion (275) to retention (276) is only about 2:1. It is of course possible that the retention in this reaction results from a diradical intermediate; it might be noted also that retention here leads to the less hindered product (275).

(274) (275) (276)

A similar study has been done by Berson [121], who found that (277) upon heating to 300° gave rise exclusively to (278).

(277) (278)

120. W.R. Roth and A. Friedrich, *Tetrahedron Lett.*, 2607 (1969).
121. J.A. Berson, *Accounts Chem. Res.*, *1*, 152 (1968).

Similarly, if (279) is briefly heated, a small amount of norbornenyl isomer (280) is produced - exclusively with inversion, but when (281) is similarly treated, only (280) is obtained - with retention.

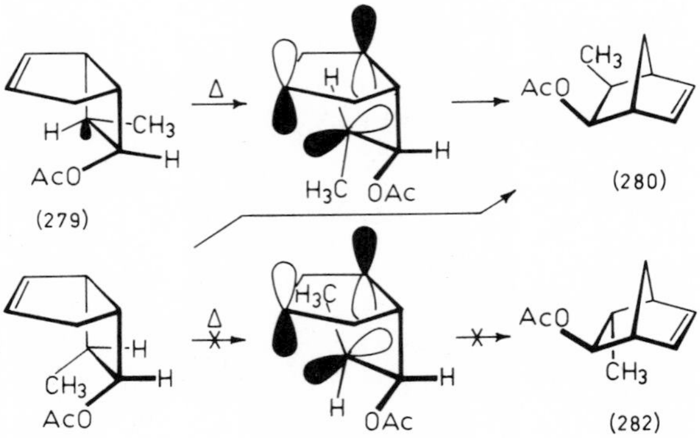

(279) (280) (281) (282)

The allowed pathway in the latter case would have required the methyl group to pass directly beneath the five-membered ring. This highly hindered situation is evidently avoided by the adoption of a diradical pathway. Support for this contention derives from the fact that upon more prolonged heating (281) slowly converts to (279), presumably _via_ the same diradical [122]. The general conclusion therefore is that allowed, thermal 1,3-shifts can occur under favorable circumstances, but the transition state is highly strained because of the necessary inversion and it seems that even a little extra hindrance may lead to the diradical alternative. Similarly, the latter pathway may be facilitated by delocalization of the single electrons: Thus, as noted earlier (p. 512), (249) and (244) readily rearrange to (283) and (284), respectively (the former involves in addition a Diels-Alder reaction).

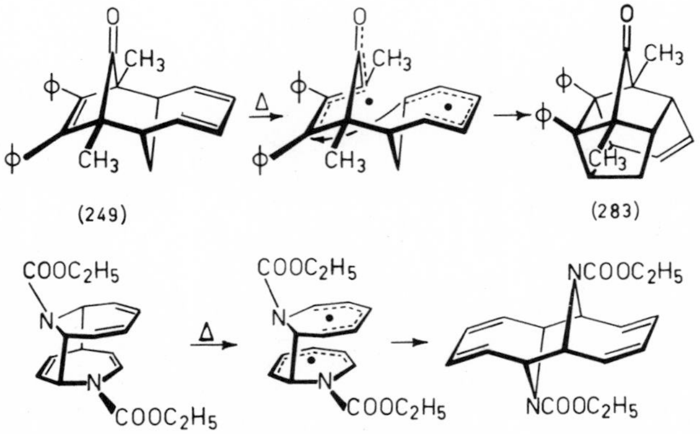

(249) (283)

(244) (284)

122. J.A. Berson and G.L. Nelson, J. Amer. Chem. Soc., 92, 1096 (1970).

1,3-Photoshifts have been reported to involve migration with retention, as required by the Woodward-Hoffmann rules. Thus, (285) upon irradiation gives (286) and (287) gives (288) [123].

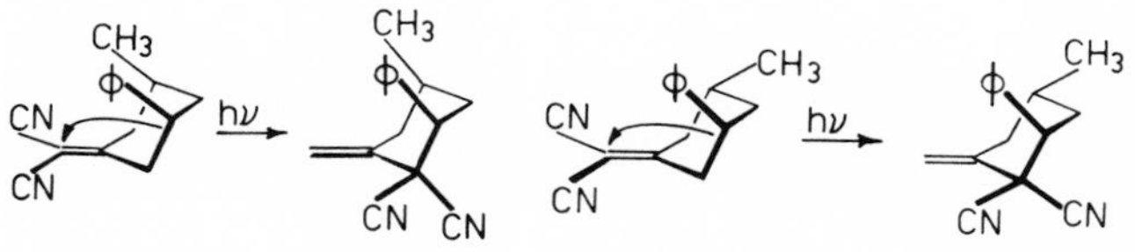

(285) (286) (287) (288)

In these reactions however, there is a distinct possibility of the formation of intermediate zwitterions or ion-pairs, which close before epimerization can occur:

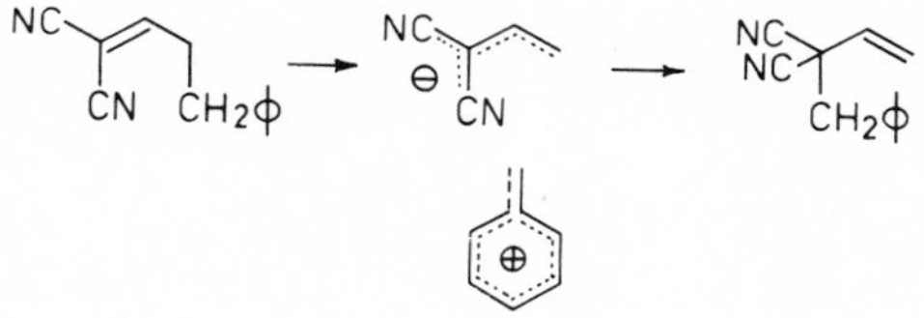

For example, while photolysis of optically active (289) gives (290) of retained configuration, thermolysis does so also [124].

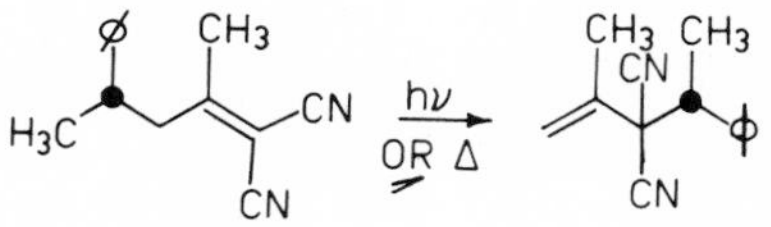

(289) (290)

Similarly, when the labelled bicyclo[3.2.0]hept-2-en-4-one (291) is irradiated, a 7-norbornenone (292) is obtained in which one of the CHD groups is completely scrambled; no such scrambling occurs in either the starting material or product, and an intermediate diradical (293) seems certain [125].

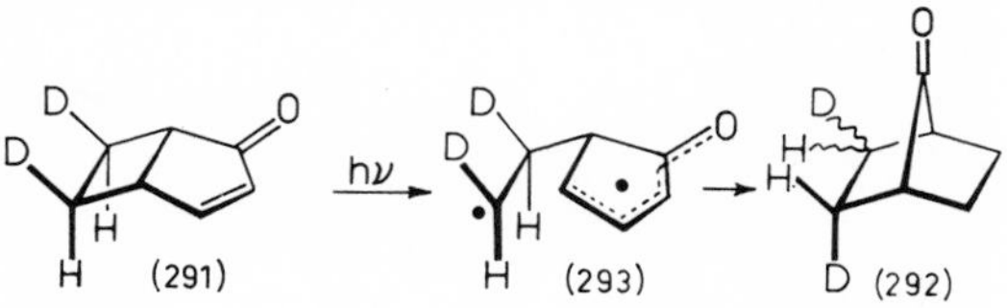

(291) (293) (292)

---

123. R.C. Cookson, J. Hudec, and M. Sharma, Chem. Commun., 107, 108 (1971).
124. R.C. Cookson and J.E. Kemp, Chem. Commun., 385 (1971).
125. R.L. Cargill, B.M. Gimarc, D.M. Pond, T.Y. King, A.B. Sears, and M.R. Willcott, J. Amer. Chem. Soc., 92, 3809 (1970).

Cationic 1,4-shifts are well-known. Thus, when (294) is dissolved in strong acid medium presumably to give the ion (295), the D-atom scrambles among the 5-membered ring positions even at -90°. That the 1,4-shifts occur with inversion is clear from the fact that if (296) is used, the 6-D-atom remains endo even at -20°, at which temperature the shifts must be very rapid [126].

(294) (295) (296)

Another example is the zwitterion (297), which can be prepared in several ways and which rearranges clearly with inversion of the migrating group to give (298) [127].

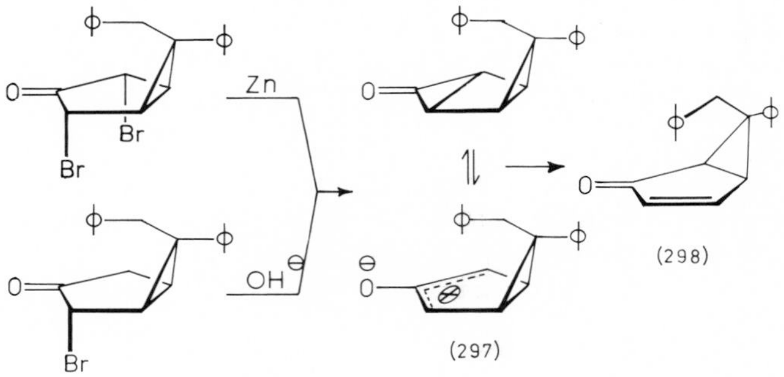

(297) (298)

In an anionic 1,4-shift a suprafacial reaction with retention is indicated. An example has been described by Schöllkopf [128], who found that the α-alkoxypyridine-N-oxide (299) rearranges to N-alkoxy-α-pyridone (300) with retention in the migrating group, with a very large negative activation entropy (-50 eu) and without observable CIDNP; by contrast, if the migrating group is benzhydryl as in (301), CIDNP is observed and $\Delta S^{\ddagger} \approx 0$. The reason for the fact that the radical pathway is competitive is probably the fact that nitroxide radicals such as (302) tend to be quite stable (see p. 629).

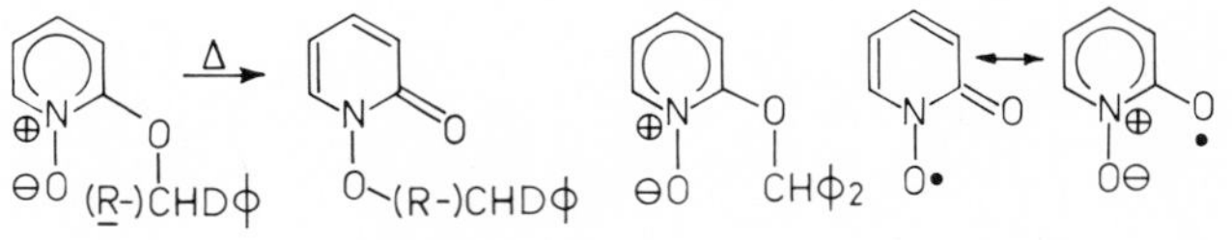

(299) (300) (301) (302)

---

126. P. Vogel, M. Saunders, N.M. Hasty, and J.A. Berson, J. Amer. Chem. Soc., 93, 1551 (1971).
127. T.M. Brennan and R.K. Hill, J. Amer. Chem. Soc., 90, 5614 (1968).
128. U. Schöllkopf and I. Hoppe, Tetrahedron Lett., 4527 (1970).

Suprafacial 1, 5-shifts are thermally very facile, and they have on occasion clouded searches for possible 1, 3-shifts. Thus, when 1-deuterioindene (303) is heated, one finds not only 3- but also 2-deuterioindene products (304) and (305), and hence these reactions are probably better explained by a series of 1, 5-shifts then a 1, 3-shift [129] - this in spite of the fact that the benzene resonance energy is lost in the process.

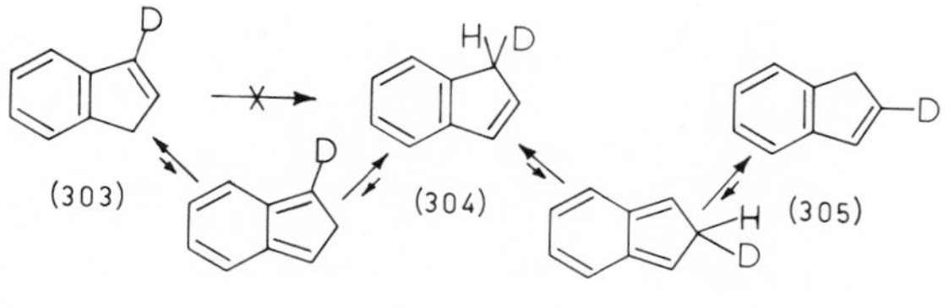

(303) (304) (305)

In the case of 5-trimethylsilylcyclopentadiene (306) the silicon moves around the ring via 1, 5-shifts so readily that all five ring protons become nmr equivalent at 80°. That 1, 3-shifts play no role in this case is supported by the observation that the 1- and 3-protons of indene (307) do not become nmr equivalent even at 180°. Yet even in this case 1, 5-migration does occur: The isoindene (308) can be trapped in high yield within a day at room temperature by means of the dienophile tetracyanoethylene to give [2 + 8] adduct (309) [130].

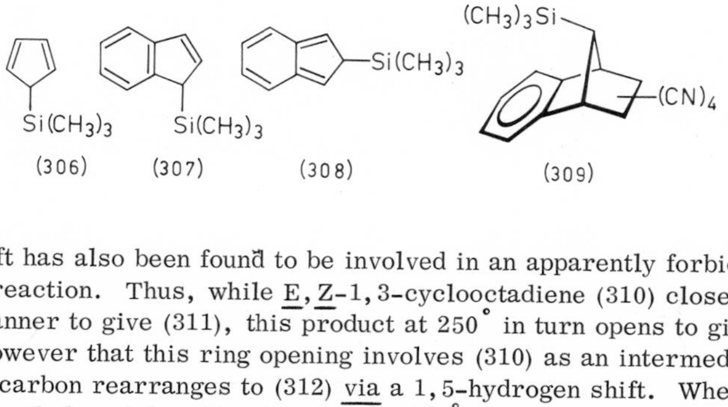

$Si(CH_3)_3$ $Si(CH_3)_3$ $-Si(CH_3)_3$ $(CH_3)_3Si$ $(CN)_4$

(306) (307) (308) (309)

A 1, 5-shift has also been found to be involved in an apparently forbidden electrocyclic reaction. Thus, while E, Z-1, 3-cyclooctadiene (310) closes at 80° in the allowed manner to give (311), this product at 250° in turn opens to give (312). It was found however that this ring opening involves (310) as an intermediate, and that this hydrocarbon rearranges to (312) via a 1, 5-hydrogen shift. When (311) is heated with dimethyl acetylenedicarboxylate to 110°, it gives the same Diels-Alder adduct as does (310) [131].

(310) (311) (312)

---

129. W. R. Roth, Tetrahedron Lett., 1009 (1964). Note that the 1, 5-shift could with equal right be called a 1, 9-shift in this case.
130. A. J. Ashe, Tetrahedron Lett., 2105 (1970).
131. (a) J. S. McConaghy and J. J. Bloomfield, Tetrahedron Lett., 3719 (1969); (b) J. J. Bloomfield, J. S. McConaghy, and A. G. Hortmann, Tetrahedron Lett., 3723 (1969).

The stereochemistry is now known to involve retention. Thus, the concerted 1,5-shift in the rearrangement of the tricyclododecatetraene (313) to (314) cannot take place any other way [132].

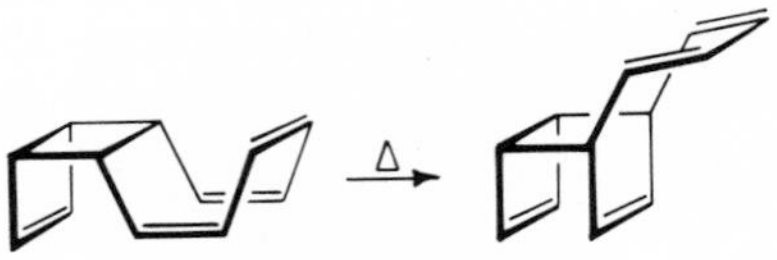

(313) (314)

Similarly, the dimethylspirodienes (315) and (316) rearrange thermally exclusively to (317) and (318), respectively (the five-membered ring of the product is subject to subsequent 1,5-proton shifts) [133].

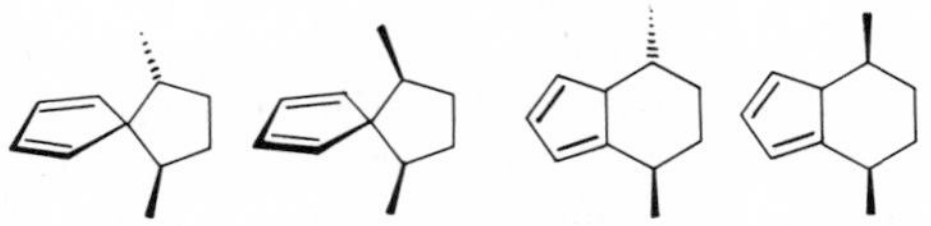

(315) (316) (317) (318)

A photochemical 1,5-shift has been observed in one instance: That of the tetramethylallene dimer (319) to give the product (320). The shift is almost certainly antarafacial; it should be relatively easy to shift a proton from the bottom of one dimethylethylene group to the top-face of the other in this case since the gem-dimethyl groups are likely to force the ring to non-planarity (321) [134].

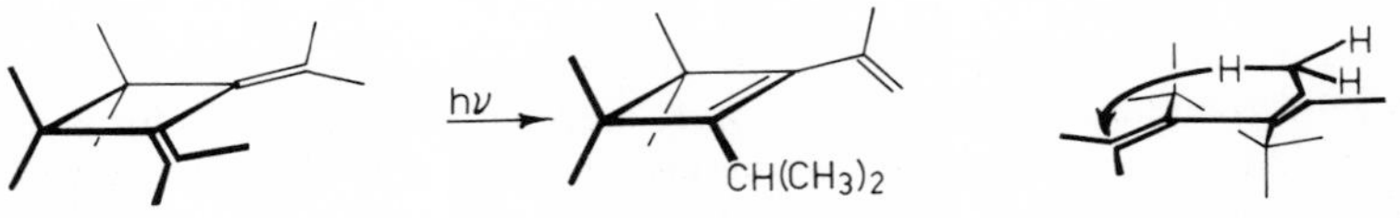

hν

$CH(CH_3)_2$

(319) (320) (321)

Anionic 1,6-shifts have been observed with pentadienyl anions such as (322).

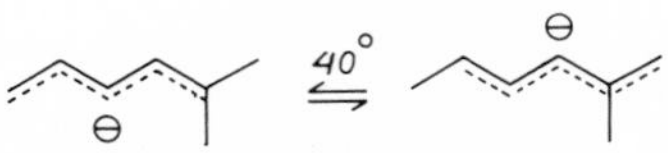

40°

(322)

---

132. L.A. Paquette and J.C. Stowell, Tetrahedron Lett., 4159 (1969).
133. M.A.M. Boersma, J.W. de Haan, H. Kloosterziel, and L.J.M. van de Ven, Chem. Commun., 1168 (1970).
134. E.F. Kiefer and J.Y. Fukunaga, Tetrahedron Lett., 993 (1969).

These shifts are intramolecular: If such ions and perdeuterated ones are mixed, no D-exchange occurs between them. The shift is evidently antarafacial as it should be since it stops if a cyclic species such as (323) is used. Photochemically it can still be done since the suprafacial transfer is then permitted [135].

hν
Δ

(323)

Photochemical 1,7-shifts are possible suprafacially and several are known; thus, the 1,4-bis(cycloheptatrienyl)benzenes (324) and (325) rearrange upon irradiation to (326) and (327), respectively [136], and 1-vinylcyclopentadiene (328) similarly gives 6-methylfulvene (329) [137].

hν

(324) (326)

hν

(325) (326)

hν

(328) (329)

Of the many thermal 1,7-shifts known [138], one of the earliest described in all stereochemical detail was the thermal equilibration of precalciferol (previtamin D) (4) and calciferol (vitamin D) (330); in that case only antarafacial transfer is possible for steric reasons [139].

---

135. R. B. Bates, S. Brenner, W. H. Deines, D. A. McCombs, and D. E. Potter, J. Amer. Chem. Soc., 92, 6345 (1970).
136. R. W. Murray and M. L. Kaplan, J. Amer. Chem. Soc., 88, 3527 (1966).
137. L. J. M. van de Ven, J. L. M. Keulemans-Lebbink, J. W. de Haan, and H. Kloosterziel, Chem. Commun., 1509 (1970).
138. See for instance (a) J. A. Berson and M. R. Willcott, J. Amer. Chem. Soc., 88, 2494 (1966) and references quoted there; (b) H. Heimgartner, J. Zsindely, H. J. Hansen, and H. Schmid, Helv. Chim. Acta, 53, 1212 (1970).
139. A. Verloop, A. L. Koevoet, and E. Havinga, Rec. Trav. Chim. Pays-Bas, 76, 689 (1957).

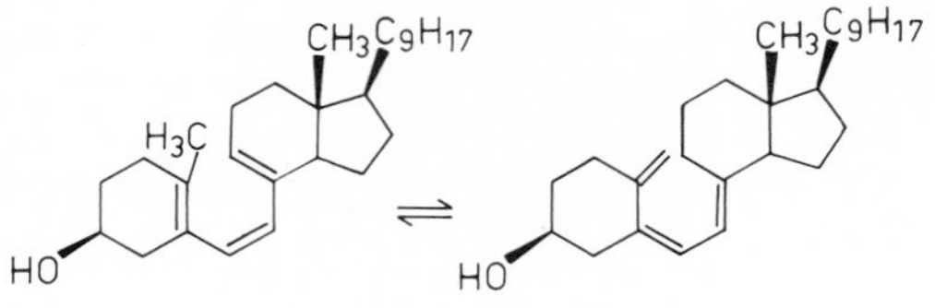

(4) (330)

An interesting 1,9-shift has been found by Rees, who observed that the nmr spectrum of (331) exhibits coalescence at 170° so as to indicate twofold symmetry. Since the nitrogen atom seems likely to migrate with inversion in this process, the migration must be antarafacial (332) - but this is only speculation at present [140].

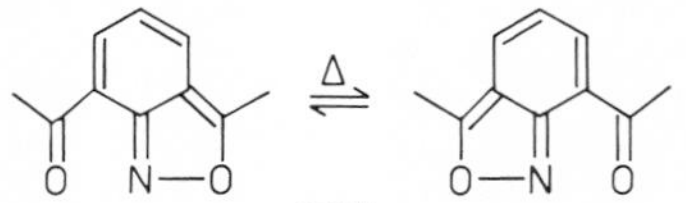

(331)  (332)

An extreme case of symmetry control in sigmatropic reactions is the photochemical, antarafacial 1,16-hydrogen shift in the corrin derivative (333) (note that the N-atom contributes two electrons); the reaction is followed by electrocyclic ring closure to give (334) [141].

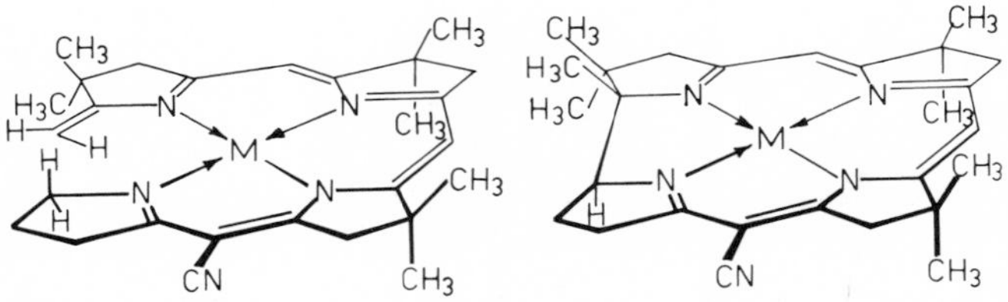

(333) (334)

When [3,3] or higher order sigmatropic reactions are considered, one additional feature arises, as follows. The transition state may be boat - or chair shaped - depending on whether the central atoms in the allylic fragments can engage in a bonding secondary orbital interaction or not. This problem was investigated by Woodward and Hoffmann also [142].

---

140. K. P. Parry and C. W. Rees, Chem. Commun., 833 (1971).
141. Y. Yamada, D. Miljkovic, P. Wehrli, B. Golding, P. Löliger, R. Keese, K. Müller, and A. Eschenmoser, Angew. Chem., Int. Ed. Engl., 8, 343 (1969).
142. R. Hoffmann and R. B. Woodward, J. Amer. Chem. Soc., 87, 4389 (1965).

Correlation diagrams are drawn up for the chair - and boat approaches of two allyl radicals; the symmetry elements relevant in the latter case are planes P_1 and P_2, and P_1 and axis C_2 in the former (see Fig. 14-18). The diagrams show at once that the chair approach is better than the boat-transition state, the main reason being that a non-bonding pair in the boat approach correlates with an anti-bonding pair in the product.

This development provided a rationale for the facts - already known - that both the Cope and Claisen rearrangements clearly prefer chair-shaped transition states [143]. Doering's experiments with 3,4-dimethylhexa-1,5-diene provide a beautifully simple insight into the problem [126]. Thus, the meso-isomer (335) upon heating to 225° gives exclusively E,Z-octa-2,6-diene (336), whereas the d,l-mixture (337) gives both Z,Z- and E,E-products (338) and (339) exactly as a chair-approach would demand [144].

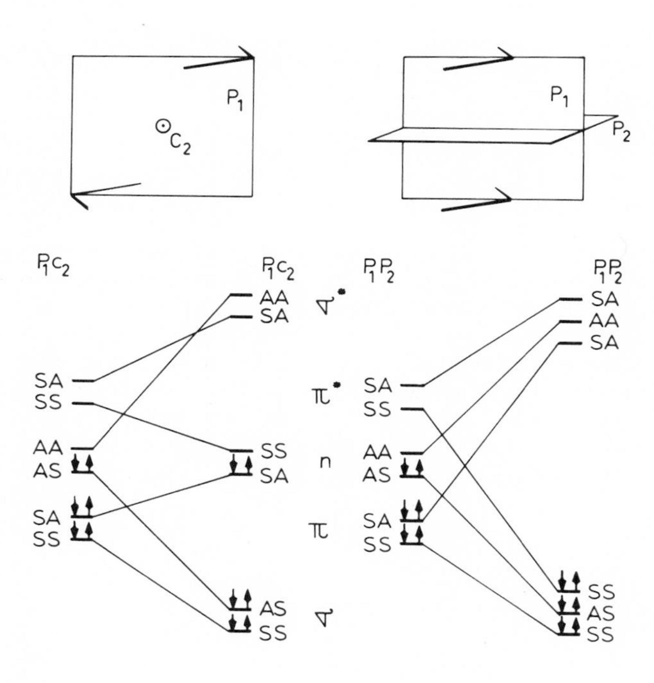

Fig. 14-18. Two possible approaches of a pair of allyl radicals, and the corresponding correlation diagrams.

---

143. For references, see R.P. Lutz, S. Bernal, R.J. Boggio, R.O. Harris, and M.W. McNicholas, J. Amer. Chem. Soc., 93, 3985 (1971).
144. W. von E. Doering and W.R. Roth, Tetrahedron, 18, 67 (1962).

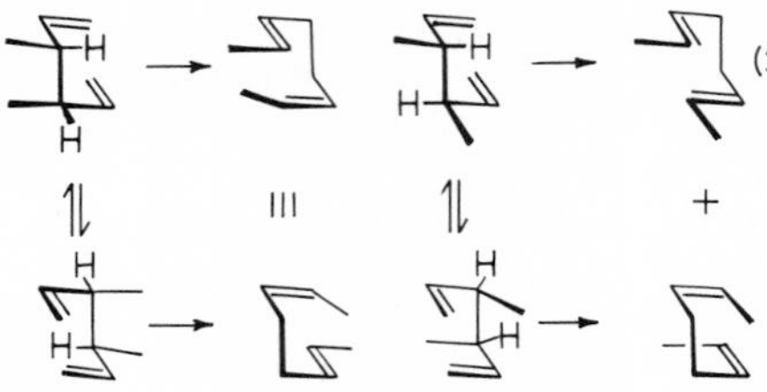

(339)

(335) (336) (337) (338)

The difference in energy between boat and chair transition states in the Cope rearrangement of 1, 5-hexadiene is now known to be about 11 kcal/mole. The resolved tetradeuteriohexadienes (340) and (341), with stereochemical descriptions *ERSZ* and *ESRZ*, respectively, racemize readily at 233°; this is possible only *via* a chair transition state.

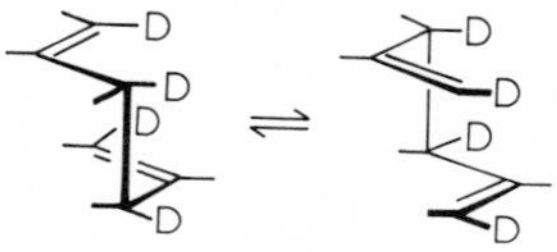

(340) (341)

Upon suitable oxidation these dienes produce *meso*-2, 3-dideuteriosuccinic acid. This racemization process can be continued for twenty half lives without producing isomers obtainable only via boat transition states, such as the *ERSE*- or *ESSZ*-configurations (these would be recognizable eventually as the *dl*-succinic acids). Only at much higher temperature does such contamination set in [145]. It has also been noted that if the allylic groups cannot approach closely, a diradical mechanism may intervene. Thus, while *cis*-1, 2-divinylcyclobutane gives *Z*, *Z*-1, 5-cyclooctadiene at 120°, the resolved *trans*-isomer (342) at 175° gives mostly racemic 4-vinylcyclohexene (343) [146].

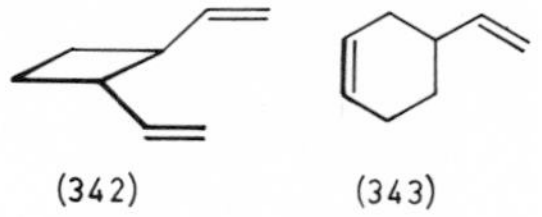

(342) (343)

Formal six π electron analogs to the Stevens rearrangement have been described. Thus, the salt (345) upon treatment with amide anion rearranges to (346), and gives very little (347); hence the allowed, concerted pathway apparently applies [147].

---

145. M. J. Goldstein and M. S. Benzon, *J. Amer. Chem. Soc.*, *94*, 7147, 7149 (1972).
146. G. S. Hammond and C. D. DeBoer, *J. Amer. Chem. Soc.*, *86*, 899 (1964).
147. V. Rautenstrauch, *Helv. Chim. Acta*, *55*, 2233 (1972).

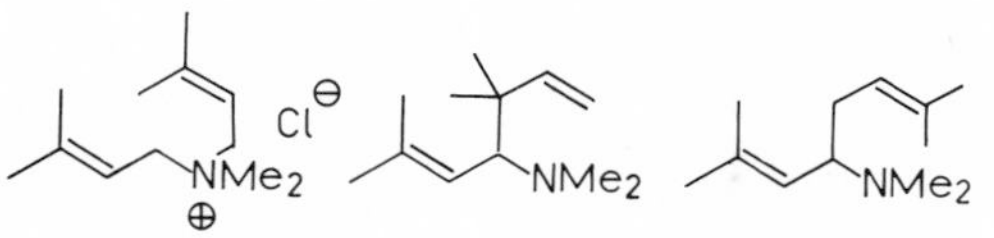

(345) (346) (347)

Photolytic suprafacial 3,3-sigmatropic shifts are forbidden, and no cases have been reported. The photo-retro-Claisen rearrangement of (253) to (348) does occur, but this reaction depends on sensitization and hence is likely not to be a concerted one [148].

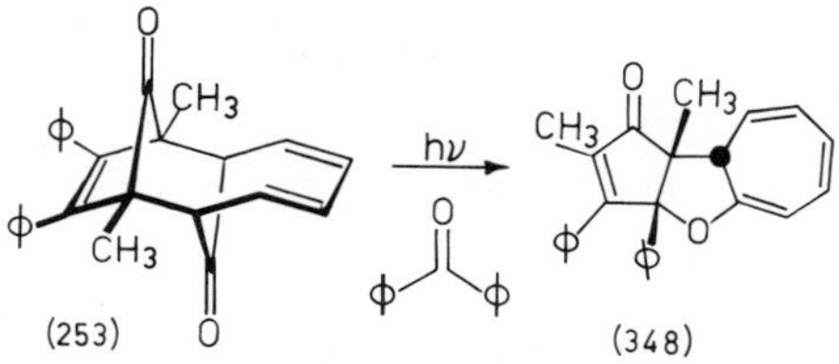

(253) (348)

A thermal 5,5-shift is known: The ether (349) rearranges at 185° to give phenol (350) [149], presumably *via* dienone (351) as an intermediate.

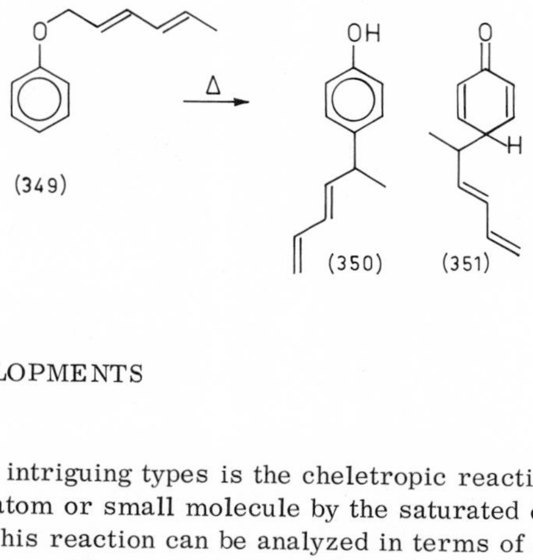

(349) (350) (351)

14-5. RELATED DEVELOPMENTS

One of the additional intriguing types is the cheletropic reaction, by which is meant the ejection of an atom or small molecule by the saturated ends of an otherwise conjugated chain: This reaction can be analyzed in terms of orbital symmetry, but these are not open to unambiguous tests because the fragment X is not amenable to stereochemical investigation [6,150].

---

148. K. N. Houk and D. J. Northington, *J. Amer. Chem. Soc.*, *93*, 6693 (1971).
149. G. Frater and H. Schmid, *Helv. Chim. Acta*, *51*, 190 (1968).
150. J. P. Snyder, R. J. Boyd, and M. A. Whitehead, *Tetrahedron Lett.*, 4347 (1972).

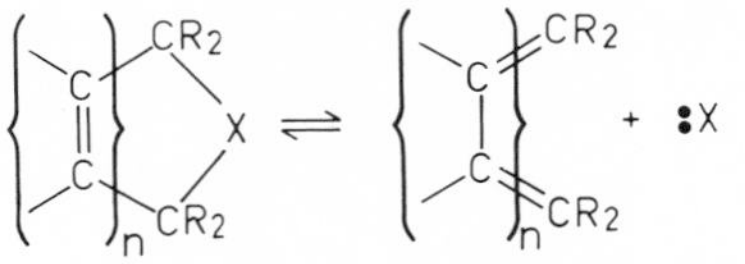

(352) (353) (354)

As an example, we may discuss the loss of sulfur dioxide from sulfone (355). The SO_2 molecule is originally bound by 4 σ electrons; two of these are going to wind up in the newly formed π system and two will become unshared sulfur electrons.

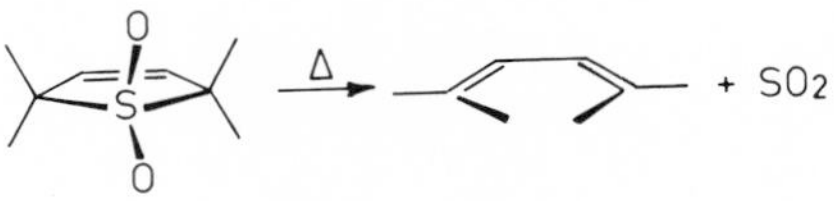

(355)

There are two ways for the hydrocarbon fragment to achieve the final geometry: Conrotatory and disrotatory. By the same token, there are two ways for the SO_2 molecule to recede: Linearly and non-linearly, as suggested by the equations below.

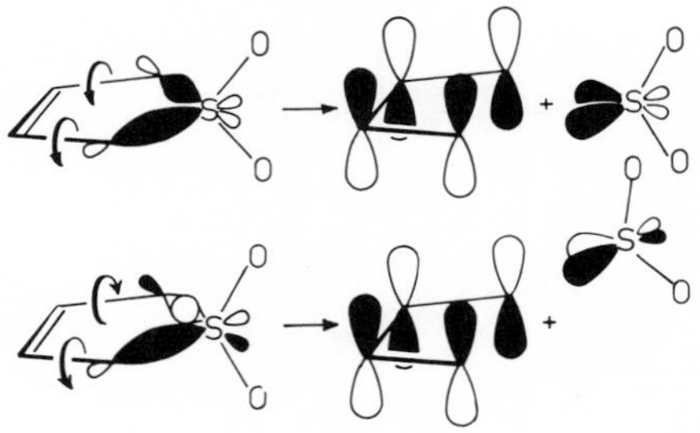

In this reaction, the ring opening occurs in a disrotatory way (and hence the sulfur dioxide departs in non-linear fashion). Thus, the sulfolene (356) produces only the Z,Z-hexadiene (357), whereas (358) gives the E,Z-isomer (359) [151].

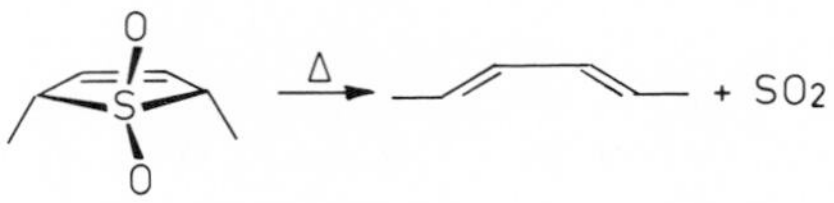

(356) (357)

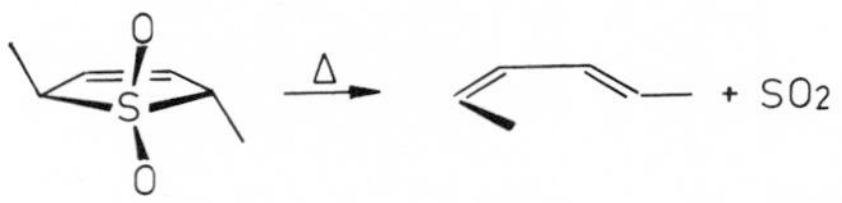

(358) (359)

151. (a) S. D. McGregor and D. M. Lemal, J. Amer. Chem. Soc., 88, 2858 (1966); (b) W. L. Mock, J. Amer. Chem. Soc., 88, 2857 (1966).

For similar reasons, the cyclic sulfone (360) requires a much higher temperature than the isomer (361) to produce 1, 3, 5-cyclooctatriene [152].

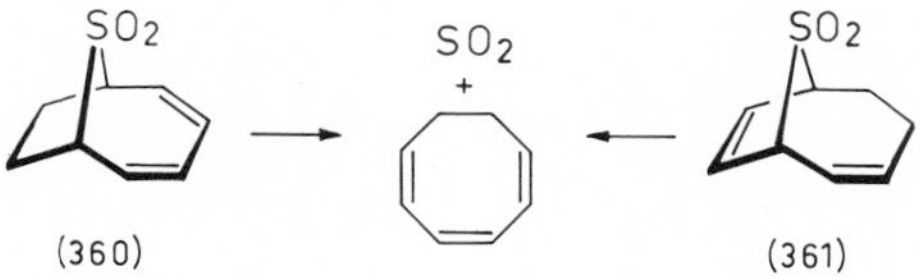

SO_2 SO_2 + SO_2

(360) (361)

The base-promoted formation of olefins from α-halosulfones and of acetylenes from α,α'-dihalosulfones via episulfones (362) and thiirene dioxides (363), respectively, have important synthetic applications [153]. The episulfoxide (364) is known to be able to eject the unstable fragment SO [154].

SO_2 SO_2 SO

(362) (363) (364)

The fragment that breaks off may be anyone of a number of small molecules such as N_2, CO, N_2O, S and so on; conversely, these considerations apply equally in the additions of oxygen atoms, carbenes, nitrenes, etc. to olefins and polyenes. Some of the reactions will be mentioned again later on (Chapters 16 and 17).

The so-called dyotropic reactions may well be amenable to the Woodward-Hoffmann analysis [155]. In these reactions, the two saturated ends of an otherwise conjugated system in concert exchange one pair of groups:

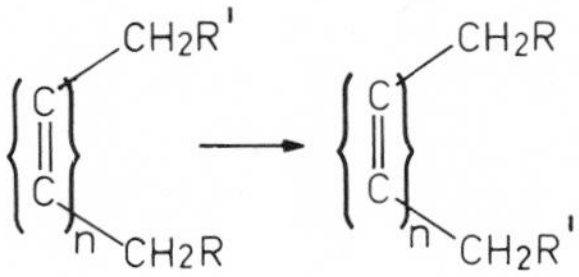

CH_2R' C=C n CH_2R → CH_2R C=C n CH_2R'

---

152. (a) W. L. Mock, J. Amer. Chem. Soc., 92, 3807 (1970). For further discussions, see (b) W. L. Mock, J. Amer. Chem. Soc., 92, 6918 (1970), and (c) F. G. Bordwell, J. M. Williams, E. B. Hoyt, and B. B. Jarvis, J. Amer. Chem. Soc., 90, 429 (1968).
153. (a) N. P. Neureiter, J. Amer. Chem. Soc., 88, 558 (1966); (b) L. A. Paquette, Accounts Chem. Res., 1, 209 (1968).
154. S. Saito, Tetrahedron Lett., 4961 (1968).
155. M. T. Reetz, Angew. Chem., Int. Ed. Engl., 11, 129, 130 (1972).

There may be one example: Thus, the two bromine atoms in (365) seem to change places easily [156], a reaction unaffected by changes in solvent, additives and so on.

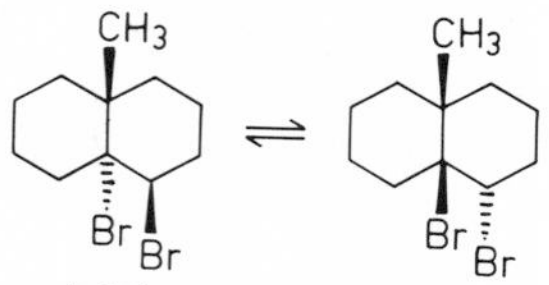

(365)

Group transfer reactions of the general type

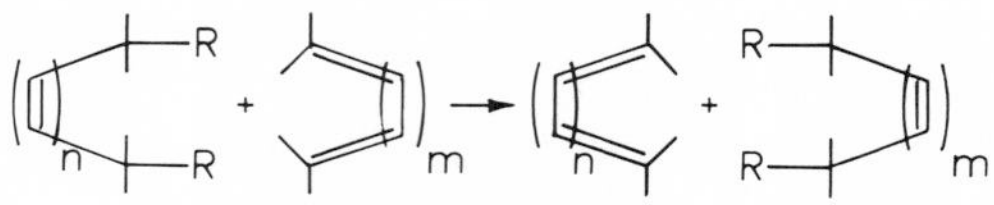

have been treated also: the cis-diimide reduction is the best known example.

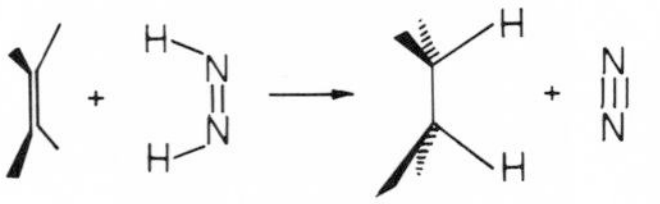

These and several other types of reactions can be viewed in alternative theoretical ways. Thus, Dewar [157] has described the "allowedness" or "forbiddenness" in terms of the aromaticity of transition states of all these pericyclic reactions; Zimmermann [158] has made a distinction between the so-called Hückel (or flat) molecules and the Möbius (or twisted) analogs; Epiotis [159] has called attention to systems in which highly dissimilar substituents greatly perturb the symmetry of the reacting molecules; Fukui based an early approach on his "frontier-orbitals" [160]. All these formulations lie outside the scope of this writing, however.

---

156. C.A. Grob and S. Winstein, Helv. Chim. Acta, 35, 782 (1952).
157. (a) M.J.S. Dewar, Angew. Chem., Int. Ed. Engl., 10, 761 (1971); (b) C.L. Perrin, Chem. Brit., 8, 163 (1972).
158. H.E. Zimmerman, J. Amer. Chem. Soc., 88, 1564, 1566 (1966), and Accounts Chem. Res., 4, 272 (1971).
159. N.D. Epiotis, J. Amer. Chem. Soc., 94, 1924, 1935, 1941, 1946 (1972).
160. (a) K. Fukui, Tetrahedron Lett., 2009 (1965); and later papers; (b) however, see also N.L. Bauld, C.S. Chang, and F.R. Farr, J. Amer. Chem. Soc., 94, 7164 (1972).

It is of interest here, however briefly, to mention one alternative effort to understand the stereochemistry peculiar to a number of reactions. This effort centered on the so-called least-motion principle, according to which the nuclei sought out their new positions on the basis of a minimized travelling distance. A number of successes have been achieved that way [161], but it is now clear that the atoms of a rearranging molecule occasionally choose complex trajectories to get to their final positions, and hence that electronic energies are far more important than the nuclear mass displacement in determining the stereochemical details.

A surge of interest has recently developed in the catalysis of forbidden reactions. It turns out that many reactions which are forbidden in the Woodward-Hoffmann sense can be brought about in the presence of certain transition metal catalysts [162], and conversely, some that are allowed and normally observed seem to be obstructed by metal complexation. Thus, chromium complexation in methyltropilidene (366) does not stop 1,5-shifts, (all of the isomers (367)-(369) are observable in succession) but it is clear that the endo-proton only moves around, and the epimeric compound (370) does not undergo these shifts at all [163].

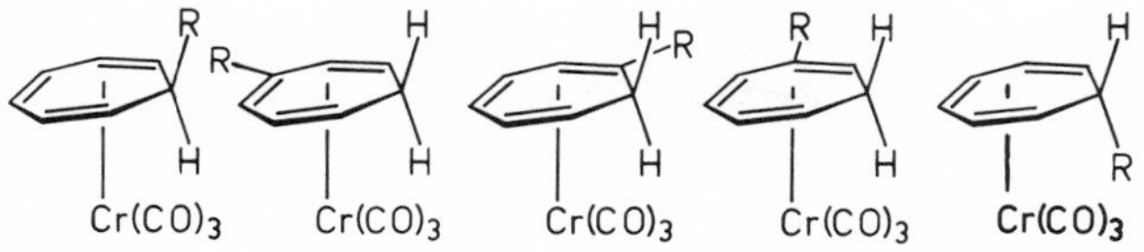

(366) (367) (368) (369) (370)

When catalysis occurs, it is, in most cases, not clear whether the metal-promoted reaction is one that normally in stepwise and slow, or whether the orbitals of the metal atom provide a new symmetry-allowed pathway; however, it is often striking to see how such catalysis can circumvent the road-block of forbiddenness. Thus, when norbornadiene is treated with cobalt derivatives, the $[\pi 2s + \pi 2s]$ dimer (371) is obtained [164].

(371)

161. For some examples, see O.S. Tee and K. Yates, J. Amer. Chem. Soc., 94, 3074 (1972).
162. For review, see (a) L.A. Paquette, Accounts Chem. Res., 4, 280 (1971); (b) R. Pettit, H. Sugahara, J. Wristers, and W. Merk, Disc. Faraday Soc., 47, 71 (1969).
163. M.I. Foreman, G.R. Knox, P.L. Pauson, K.H. Todd, and W.E. Watts, J. Chem. Soc., Perkin II, 1141 (1972).
164. D.R. Arnold, D.J. Trecker, and E.B. Whipple, J. Amer. Chem. Soc., 87, 2596 (1965).

Simple zeolite converts cyclopropene into dimer (372) [165].

(372)

When the alcohol (373) is exposed to ironpentacarbonyl, a suprafacial 1,3-shift occurs that is not observed with the epimer (374) [166].

(373) HO $Fe(CO)_5$ (374) OH O

The $[_{\pi}2_s + _{\pi}2_s]$ cycloreversion of (375) is readily brought about by silver ion [167].

(375) $Ag^{\oplus}$

Disrotatory ring opening of cyclobutene in hexamethyl Dewar-benzene (376) is easily achieved with rhodium complexes [168].

(376)

---

165. A. J. Schipperijn and J. Lukas, Tetrahedron Lett., 231 (1972).
166. F. G. Cowherd and J. L. von Rosenberg, J. Amer. Chem. Soc., 91, 2157 (1969).
167. W. Merk and R. Pettit, J. Amer. Chem. Soc., 89, 4787 (1967).
168. H. Hogeveen and H. C. Volger, Chem. Commun., 1133 (1967).

The conversion of quadricyclane (377) to norbornadiene which thermally requires 15 hours at 140° with palladium chloride takes one hour at -26° [169] - and so on.

(377)

In the ensuing investigations some remarkable discoveries were made. Silver ion in general seems to be able to promote isomerizations of the sort:

$$C_1-C_2-C_3-C_4 \xrightarrow{Ag^{\oplus}} C_1-C_3-C_2-C_4$$

particularly in caged systems. Thus, in the presence of that ion basketane (378) is converted into snoutane, (379) [170] and cubane (380) into cuneane (381) [143].

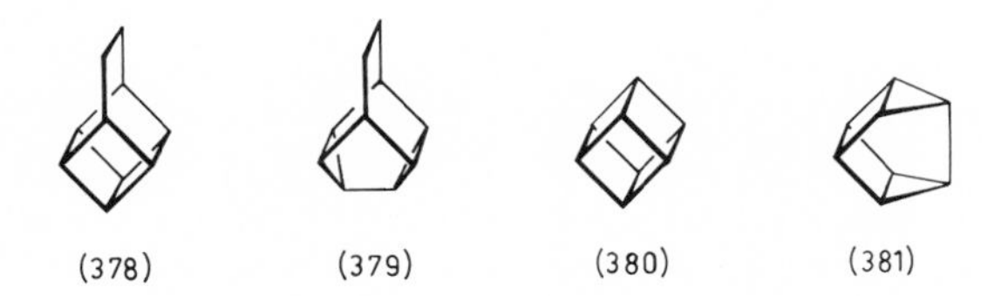

(378) (379) (380) (381)

Another discovery was that different metals often had completely different effects. For instance, when (380) is treated with rhodium compounds (382) is obtained [171].

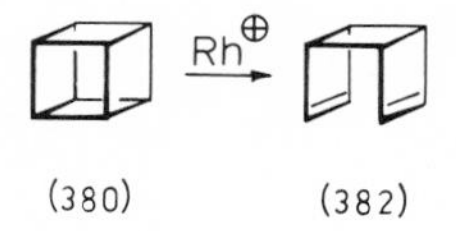

(380) (382)

169. H. Hogeveen and H.C. Volger, J. Amer. Chem. Soc., 89, 2486 (1967).
170. L.A. Paquette and J.C. Stowell, J. Amer. Chem. Soc., 93, 2459 (1971).
171. L. Cassar, P.E. Eaton, and J. Halpern, J. Amer. Chem. Soc., 92, 6366 (1970).

When tricyclo[4.1.0.$0^{2,7}$]heptane (383) is heated to 300°, (384) is formed; low temperature treatment with Ag^+ ion affords (385), treatment with $[Rh(CO)_2Cl]_2$ gives (386), and if PtO_2 is used, (387) is obtained beside (386) [172, 173].

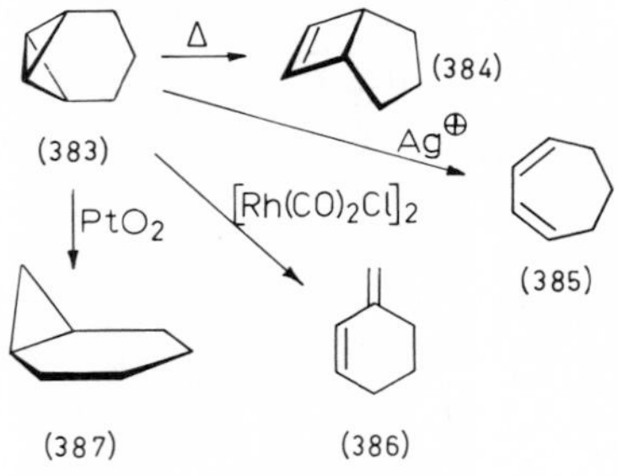

(383) (384) (385) (386) (387)

Some mechanistic features of these reactions are beginning to emerge. The Rh-catalyzed reaction of cubane involves complex (388) as an intermediate [171].

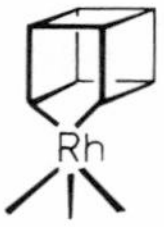

(388)

The catalyzed opening of bicyclobutanes such as (389) has been considered to go via metal complexes of carbene-carbonium ion hybrids [174].

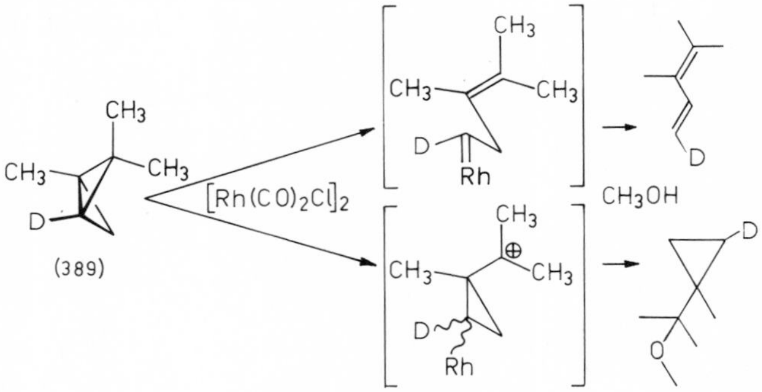

(389)

172. Chem. Eng. News, 48, 42, June 15, 1970.
173. P.G. Gassman and T.J. Atkins, J. Amer. Chem. Soc., 93, 1042 (1971).
174. P.G. Gassman and F.J. Williams, J. Amer. Chem. Soc., 92, 7631 (1970) and Chem. Commun., 80 (1972).

The metal-catalyzed rearomatization of certain benzene isomers such as prismanes seems to depend mostly on the Lewis acidity of the catalysts; thus, even simple 1, 3, 5-trinitrobenzene is effective in this capacity. However, the clearest conclusion that can as yet be drawn from these studies is that as yet no single known line of logic is capable of explaining all of the observed facts, and we shall have to wait for further results [175].

---

175. See e.g., (a) P.G. Gassman and F.J. Williams, J. Amer. Chem. Soc., 94, 7733 (1972), and papers immediately following; (b) L.A. Paquette, S.E. Wilson, R.P. Henzel, and G.R. Allen, J. Amer. Chem. Soc., 94, 7761 (1972), and papers immediately following.

Chapter 15

REACTIVE INTERMEDIATES; GENERAL CONSIDERATIONS

15-1. INTRODUCTION [1]

An intermediate is a minimum in the reaction profile somewhere between reactant and product (see Fig. 15-1). Intermediates by definition occur in nonconcerted reactions; these are characterized by the occurrence of at least two maxima. Of these maxima, each of which is "a" transition state, the highest is called "the" transition state.

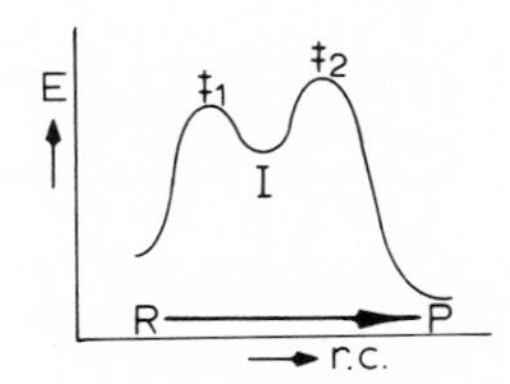

Fig. 15-1. A reaction profile showing a reactive intermediate.

1. For a brief survey and review, see N. S. Isaacs, Chem. Brit., 6, 206 (1970).

Intermediates may occur both before and after "the" transition state (Fig. 15-2).

Intermediates may differ widely in stability. The minimum must be deep enough to have at least one vibrational level; otherwise its life time will be so short - 10^{-14} sec or so - that it makes no sense to speak of an intermediate at all. On the other hand, if an intermediate is so stable as to be isolable at room temperature it is not considered an intermediate by the kineticist. Thus, if benzene is treated with an excess of bromine under sufficiently vigorous conditions, the product is p-dibromobenzene. Clearly, bromobenzene was an intermediate in the reaction, and if less bromine and more gentle conditions are used that compound can be isolated; it is then found to be a perfectly stable material. The synthetic chemist often refers to such substances as intermediates - even if he isolates and purifies them before going to the next stage in his synthesis. The question is further complicated by the possibility of using low temperatures to increase life times, or flash techniques to increase concentrations. For the purpose of this discussion, a substance is a reactive intermediate if it has at least one vibrational level but is not isolable at room temperature.

In this chapter we briefly discuss some of the techniques by means of which such intermediates can be hunted, the types of intermediates that have interested organic chemists thus far, and the properties of those species that are amenable to investigation.

15-2. GENERAL TECHNIQUES IN IDENTIFYING INTERMEDIATES

Kinetics is obviously one of the most important of the general techniques used in identifying intermediates [2]. In principle one can always write the rate law for any reaction profile, from the simplest to the most complex, at least in differential form. The approach is as follows. If it is suspected that a given reaction is

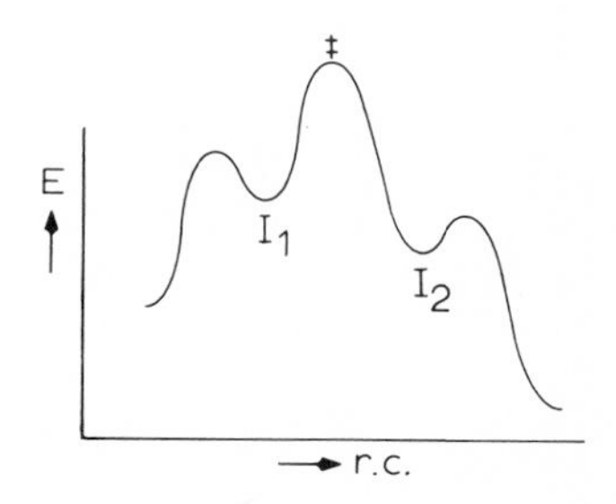

Fig. 15-2. Multiple intermediates.

---

2. For references, see R. H. DeWolfe, J. Chem. Educ., 40, 95 (1963).

non-concerted and that one or more intermediates are involved, the kinetics are written out, and the rate law is drawn up and compared with experiment. If agreement is obtained, one has evidence for the intermediate. This evidence is not conclusive; other mechanisms are normally possible that require the same rate law.

By way of example, let us consider the reaction:

$$HAX + OH^- \rightarrow HAOH + X^-$$

where A is some atom or group not further specified. Suppose furthermore that the following pathway is suspected:

Step 1: $$HAX + OH^- \underset{k_{-1}}{\overset{k_1}{\rightleftharpoons}} AX^- + H_2O$$

Step 2: $$AX^- \xrightarrow[\text{(slow)}]{k_s} A + X^-$$

Step 3: $$A + H_2O \xrightarrow[\text{(rapid)}]{k_r} HAOH$$

The energy profile will accordingly be that of Fig. 15-3. A complete treatment would include the reverse of step 2 and of step 3; however, in order to keep the discussion reasonably simple, we assume that the forward reaction of A is so much more favored than the return to AX^- that the latter may be neglected, and likewise that the product forming step is irreversible.

First, we write the differential rate laws for all species in which A is present:

$$\frac{d(HAX)}{dt} = -k_1(OH^-)\,(HAX) + k_{-1}(AX^-) \tag{1}$$

$$\frac{d(AX^-)}{dt} = k_1(OH^-)\,(HAX) - k_{-1}(AX^-) - k_s(AX^-) \tag{2}$$

$$\frac{d(A)}{dt} = k_s(AX^-) - k_r(A) \tag{3}$$

$$\frac{d(HAOH)}{dt} = k_r(A) \tag{4}$$

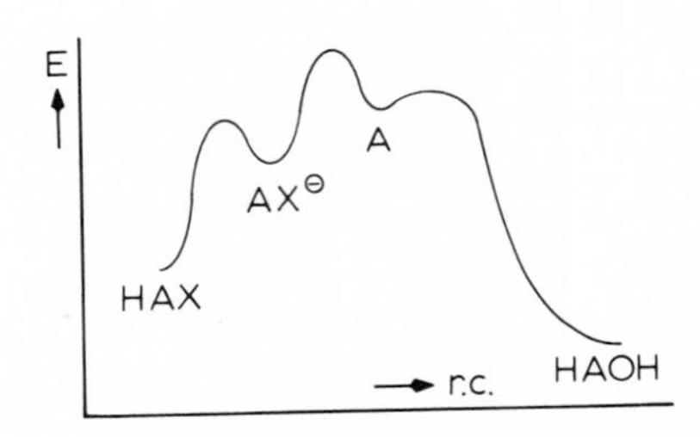

Fig. 15-3. The reaction profile suspected for the base promoted hydrolysis of HAX.

At all times the material balance applies:

$$[HAX]_o = [HAX]_t + [AX^-]_t + [A]_t + [HAOH]_t \tag{5}$$

We note that the (AX^-) and (A) vs t curves are very flat (see Fig. 15-4); i.e.,

$$\frac{d(AX^-)}{dt} \approx 0$$

and

$$\frac{d(A)}{dt} \approx 0$$

We simplify matters by assuming that these derivatives are exactly zero (the so-called steady state approximations). The concentrations of (A) and (AX^-) in the steady state then can be obtained.

$$[AX^-] = \frac{k_1}{k_{-1} + k_s} (OH^-) (HAX)$$

$$[A] = \frac{k_s}{k_r} (AX^-)$$

$$= \frac{k_1 k_s}{k_r(k_{-1} + k_s)} (OH^-) (HAX)$$

If we substitute these findings into eq 1, we obtain:

$$\frac{d(HAOH)}{dt} = \frac{k_1 k_s}{k_{-1} + k_s} (OH^-) (HAX)$$

The observed rate law is:

$$\frac{d(HAOH)}{dt} = k_{obs} (OH^-) (HAX)$$

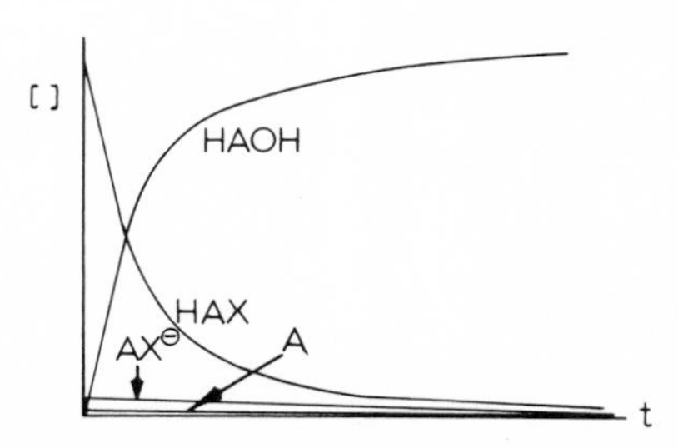

Fig. 15-4. The concentration vs. time curves of the various species containing the group A.

and the pathway assumed is therefore consistent with experiment. But another mechanism is possible: The S_N2 displacement reaction, a concerted bimolecular reaction characterized by a single energy maximum and the equation:

$$HAX + OH^- \xrightarrow{k2} HAOH + X^-$$

so that:

$$\frac{d(HAOH)}{dt} = k_2\ (OH^-)\ (HAX)$$

One cannot tell whether:

$$k_{obs} = k_2$$

or

$$k_{obs} = \frac{k_1 k_s}{k_{-1} + k_s}$$

Hence, kinetics alone cannot be used to determine the mechanism, and further experiments must be devised.

For instance, one possible extension might be to change the concentrations in such a way that one of the steady state approximations no longer holds. In the present case, if (OH^-) is made very high, the multistep mechanism may lead to first order kinetics. At very high pH, if HAX is not too weak an acid, one essentially converts all of it instantly into AX^-. The reaction now becomes first order in HAX alone (see Fig. 15-5) as a further increase in base concentration cannot increase the rate further. The reaction at such a high pH is of course very fast, and a special technique (the stopped flow apparatus) is required to measure it. One depends on very rapid mixing and flow through the observation chamber (Fig. 15-6), and the use of an oscilloscope and camera to follow the reaction in the chamber after

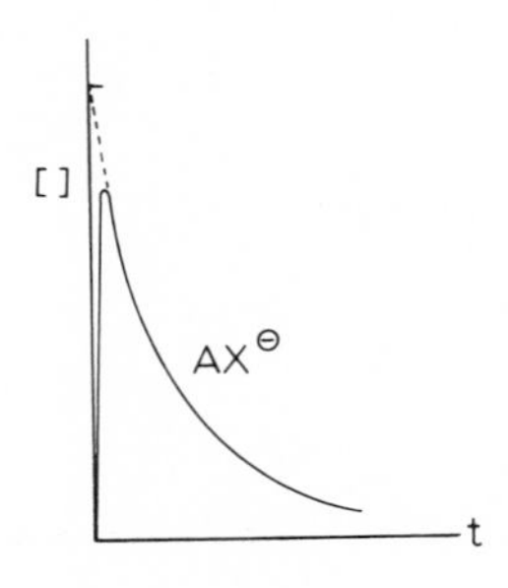

Fig. 15-5. The multistep reaction at high pH.

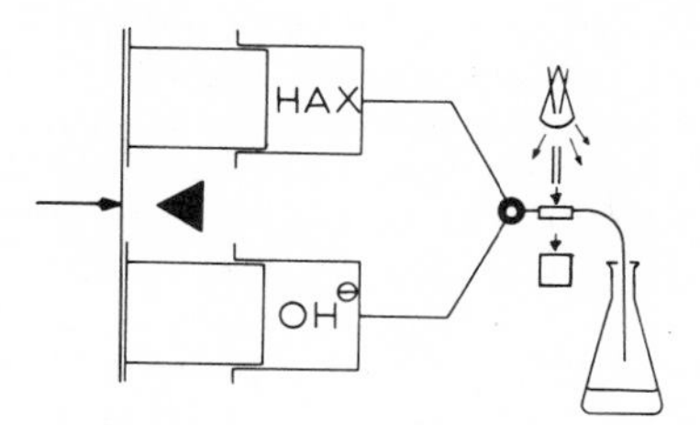

Fig. 15-6. Schematic illustration of the stopped flow device.

the flow has stopped. Reactions with a half life of a millisecond or so can be followed in this way; for still faster reactions, the various jump techniques can be used [3]. If this pH effect is indeed observed, one has stronger evidence than before; if not, one still does not know whether the acid is too weak to be completely ionized even at high pH, or whether the S_N2 alternative applies, and further experiments must again be devised. This situation is quite typical, and it illustrates the conclusion: Kinetics is a powerful tool in the hunt for intermediates, but it is not conclusive.

Closely related with kinetics is the use of activation parameters. According to the theory of absolute rates:

$$k = \frac{RT}{Nh}\, e^{-\Delta G^{\ddagger}/RT}$$

so that

$$\ln \frac{kNH}{RT} = -\frac{\Delta H^{\ddagger}}{RT} + \frac{\Delta S^{\ddagger}}{R}$$

and

$$\frac{\ln (k/T)}{d\,(1/T)} = -\frac{\Delta H^{\ddagger}}{R}$$

The slope and intercept of a plot such as in Fig. 15-7 allow one to calculate $\Delta H^{\ddagger}$ and $\Delta S^{\ddagger}$.

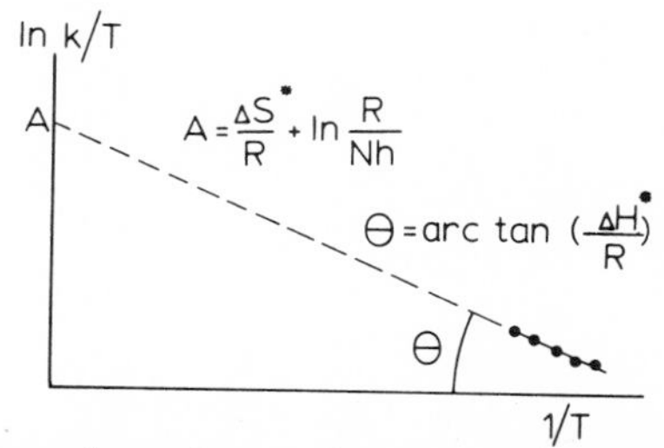

Fig. 15-7. The temperature dependence of a rate constant, and its relation to the activation parameters.

---

3. G.G. Hammes and L.E. Erickson, J. Chem. Educ., 35, 611 (1958).

Temperature changes of 10° generally cause changes in k of 2 - 4; since k can easily be measured to 1 - 2% precision and T can easily be held constant to ±0.01°, $\Delta H^{\ddagger}$, $\Delta S^{\ddagger}$ and $\Delta G^{\ddagger}$ can all be measured with reasonable precision. However, only a very small 1/T range is normally accessible, and since the activation parameters are only approximately independent of temperature one should be cautious when comparing rate data of reactions studied in widely different temperature ranges [4].

One finds in general that $\Delta S^{\ddagger}$ varies from perhaps -30 to +20 eu (entropy units, or cal/mole °K), and that $\Delta H^{\ddagger}$, $\Delta E^{\ddagger}$ and $\Delta G^{\ddagger}$ all range from 5 - 40 kcal/mole. The lower limit of 5 kcal/mole implies reactions so fast that special techniques (such as nmr coalescence at low temperature) are required to study them.

Since $\partial G/\partial p = V$, the theory of absolute rate also leads to a volume of activation which can be obtained from the pressure dependence of a rate constant [5]. One should be careful not to confuse this with the effect of relatively low pressures on the rates (rather than the rate constants) of gas phase reactions; these are essentially concentration effects. A pressure increase of 1000 atm generally changes a rate constant by a factor of 0.3 - 3, and since pressures to a few thousand atm are easily accessible in the laboratory, $\Delta V^{\ddagger}$ can also be measured with considerable precision; it is generally found to lie between -30 and +20 cm^3/mole. This parameter is somewhat pressure dependent and tends to level off at high pressures; it is therefore customary to compare data valid at zero pressure. Pressure dependence studies have been much less common than the use of temperature variations; chemists like to look at their solutions, and shy away from experiments that hide them from view. Still other parameters (such as the specific heat of activation [6]) have also been used.

As yet, it has not been possible to calculate (i.e., predict) any activation parameters with precision, and hence their use is still empirical. One of the difficulties is that even simple reactions usually involve more than one bond. Only very weak bonds can be thermally cleaved without any concurrent energy compensation (see p. 29); an example might be an oxygen-oxygen single bond, $\Delta E^{\ddagger}$ would then be about 30 kcal/mole. In such a process $\Delta S^{\ddagger}$ may be 20 eu or more; if the reaction is done at room temperature, $T\Delta S^{\ddagger}$ helps provide about 5 kcal/mole. $\Delta V^{\ddagger}$ might be expected to be about 10 cm^3/mole, and $p\Delta V^{\ddagger}$ is therefore only of the order of a calorie/mole - a negligible contribution to the overall $\Delta G^{\ddagger}$ of perhaps 25 kcal/mole. These numbers are more or less illustrative of the relative magnitudes of $\Delta E^{\ddagger}$, $T\Delta S^{\ddagger}$ and $p\Delta V^{\ddagger}$ in many chemical reactions of more complicated types as well.

---

4. R.C. Petersen, J. Org. Chem., 29, 3133 (1964).
5. W.J. le Noble, Progr. Phys. Org. Chem., 5, 207 (1967) and J. Chem. Educ., 44, 729 (1967).
6. K.T. Leffek, R.E. Robertson, and S. Sugamori, J. Amer. Chem. Soc., 87, 2097 (1965).

The simple breaking of bonds - from the strongest covalent bond to the weakest electrostatic interactions - is always characterized by an increase in energy, an increase in entropy and an increase in volume; the formation of bonds is usually characterized by an increase in energy in the initial stages (when the repulsive potentials of non-bonded systems must be overcome), by a decrease in entropy and a decrease in volume. The difficulty in applying such a generalization is that compensating processes almost always occur. Thus, although an ionization is formally a bond breaking process, concurrent bonding of several solvent molecules by the incipient ions takes place, and both $\Delta S^{\ddagger}$ and $\Delta V^{\ddagger}$ take on large negative values.

In the example quoted in our kinetics discussion, $\Delta V^{\ddagger}$ would probably be the most useful. A large number of data is available suggesting that in displacement or abstraction reactions this parameter usually has a value of about -5 to -10 cm^3/mole; if the experimental value in the hypothetical reaction just discussed is +15 cm^3/mole, this would be a strong indication of the bond breaking aspect of the rate controlling second step. This use of an activation parameter (comparison with known cases) is quite typical.

These parameters exemplify the belief that the transition state preceding a given intermediate already reflects the properties of that intermediate to some degree. Thus, as noted earlier (p. 250), a relatively stable intermediate is believed to be preceded by an early and low-lying transition state, (the Hammond postulate is usually considered to state the first of these adjectives). The Woodward-Hoffmann rules have taught us to be cautious with assertions that fast reactions are indicative of stable products and vice versa, but if due care is taken, the generalization is still very valuable.

Hammett sigma-rho studies represent another measure often taken of a reaction; a measure again in which transition state properties foreshadow those of the following intermediates [7]. It was noted by Hammett that reactions and equilibria of the side chains of m- and p-substituted benzene derivatives, which are presumably free from steric effects of the substituents, obey the relations

$$\log k/k_o = \sigma\rho$$

and

$$\log K/K_o = \sigma\rho$$

Here k and K represent the rate or equilibrium constant of the substituted benzene derivative and k_o (K_o) that of the parent or unsubstituted member; σ is the substituent constant, a measure of the nature of the substituent only and independent of the reaction, and ρ is the reaction constant - dependent of the nature of the reaction only and independent of the substituents. The reaction constant for the ionization of benzoic acids in water at 25° is defined as unity, so that the substituent constant for a group p-X

$$\sigma_X = pKa_{(\text{benzoic acid})} - pKa_{(\text{p-X-benzoic acid})}$$

7. For a brief review and references, see J. Shorter, Chem. Brit., 5, 269 (1969).

As experience grew. it developed that the substituent constants, which gave excellent linear correlations in many reactions, led to scatter in others. The cause of this is that while all substituents may be presumed to interact with nearby reaction centers in a direct electrostatic manner, in certain cases resonance may also play an important role. Thus, the effect of a *m*-methyl group on the ionization of benzoic acid must surely be a purely polar one; however, the effect of a *p*-dimethylamino group on the transition state in cumyl chloride solvolysis obviously has a large resonance contribution.

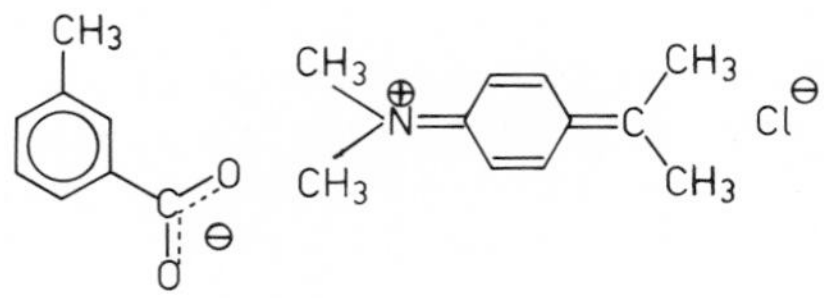

For this reason several groups of σ-constants are now in use. The most important of these are those based on benzoic acid and the σ^+-constants based on the rates of solvolysis of cumyl chlorides at 25° in a mixture of acetone and water containing 90% of the former (see also p. 695) [7a]. The reaction constant for the latter reaction (-4.54) is chosen so as to give the best possible agreement with σ-constants.

The substituent constants vary from +0.8 for *p*-NO_2 to -0.8 for *p*-OMe, and the reaction constants vary from perhaps -10 in reactions in which a positive ion attacks the *p*-position to small positive values in reactions of anions, such as the alkaline hydrolysis of benzoate esters (see Table I and II). Several important conclusions can be drawn in a given $\sigma\rho$-study. For example, by introducing substituted phenyl groups in various positions at and near the site of a reaction, one may be able to learn the sign and magnitude of partial charges at such sites during the reaction. Furthermore, by studying the question whether σ- or σ^+-constants give a better (more nearly linear) correlation with the observed rate constants, one may be able to learn something of the structure of the transition state. Finally, if a given correlation is not linear but curved or even shows an extremum, the operation of two competing mechanisms is clearly indicated [8].

The Hammett equations were the first of the so-called linear free energy relationships. Another well-known example is the Swain-Scott equation, which provides the basis of the so-called nucleophilicity constants of anions, on the basis of the rates of their displacement of bromide ion from methyl bromide (p. 691) [8a]. Still another is the Grunwald-Winstein equation, which deals with solvent effects. This equation is of importance primarily in solvolysis reactions and hence is further discussed in Chapter 20 (p. 696). However, it may be noted here that reactions in which charges develop will obviously go faster if polar solvents are used,

7a. H.C. Brown and Y. Okamoto, *J. Amer. Chem. Soc.*, *79*, 1913 (1957).
8. J.O. Schreck, *J. Chem. Educ.*, *48*, 103 (1971).
8a. C.G. Swain and C.B. Scott, *J. Amer. Chem. Soc.*, *75*, 141 (1953).

TABLE 15-1. Some Substituent Constants

Substituent	σ	σ^+
m-Me	-0.069	-0.066
m-F	+0.337	+0.352
m-Cl	+0.373	+0.399
m-Br	+0.391	+0.405
m-I	+0.352	+0.359
m-NO_2	+0.710	+0.674
m-OMe	+0.115	+0.047
p-Me	-0.170	-0.311
p-F	+0.062	-0.073
p-Cl	+0.227	+0.114
p-Br	+0.232	+0.150
p-I	+0.18	+0.135
p-NO_2	+0.778	+0.790
p-OMe	-0.268	-0.778

TABLE 15-2. Some Reaction Constants

Reaction	ρ
$H_2O + \phi\text{-OH} \rightleftharpoons \phi\text{-O}^- + H_3O^+$	+2.21
$H_2O + \phi\text{-NH}_3^+ \rightleftharpoons \phi\text{-NH}_2 + H_3O^+$	+2.94
$\phi\text{-SO}_3\text{Et} \xrightarrow{\text{EtOH}} \phi\text{SO}_3\text{H} + \text{Et}_2\text{O}$	+1.30
$\phi\text{-COOMe} + H_2O \xrightarrow{OH^-} \phi\text{-COO}^- + \text{MeOH}$	+2.37
$\phi\text{-COOH} + \text{MeOH} \xrightarrow{H^+} \phi\text{COOMe} + H_2O$	-0.52
$\phi_3\text{COH} + H^+ \underset{}{\overset{H_2O}{\rightleftharpoons}} \phi_3C^+ + H_2O$	-3.60

whereas reactions in which only neutral species play a role will be more or less indifferent to the solvent. The protic or aprotic nature of the solvent is also of great importance [9]; thus, small anions will be strongly hydrogen bonded in protic solvents such as water or alcohols. On the other hand, cations are well solvated in liquids such as dimethyl sulfoxide or hexamethyl phosphoramide.

Kinetic isotope effects are another important means of studying transition states. If, for example, in a given reaction a hydrogen atom gets abstracted, it is found that if the same reaction is done with deuterium the rate constant is considerably less. This is described as a primary isotope effect. The ratio is related to the difference in zero point energies (p. 31), and to the degree of bond breakage; a theoretical maximum of k_H/k_D of about 30 has been calculated [10], and in one instance a ratio of 24 has been found (in the 2, 6-lutidine catalyzed iodination of 2-nitropropane) [11]. Smaller (secondary) effects operate if isotopic substitution is studied in the vicinity of the reaction site.

The element effect is used now and then; thus, if in one hypothetical two-step example it makes little or no difference in the rate constant whether X is chlorine or iodine, one would suspect that the first transition state must be the highest - since A-X bond cleavage has evidently not occurred as yet in the rate controlling step. Solvolysis reactions can be classified to some degree on the basis of leaving group effects (p. 692). Structural effects have been used to great advantage. They cover so many aspects of mechanistic chemistry that they do not lend themselves to review here; but as a single example, it may be mentioned that ring size would be of obvious importance in reactions at the site of a carbon atom undergoing a hybridization change during a reaction.

The "capture" of intermediates is an idea that easily fires the imagination; unfortunately, the term is greatly abused. It refers to situations in which a substance is added to the reaction mixture and is converted to a substance in which the identity of the intermediate is apparently clearly recognizable. For instance, going back to our kinetic example, suppose that the addition of cyclohexene leads to the quantitative formation of the adduct shown below.

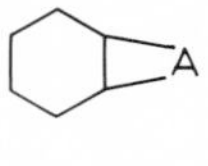

A

This product might then be claimed to prove that A was present and that it has been captured, but this is deceiving. If we look back at the reaction profile, we see that what is being assumed is that curve b is followed now instead of a (Fig. 15-8).

9. (a) M. R. J. Dack, Chem. Technol., 1, 108 (1971); (b) J. Miller and A. J. Parker, J. Amer. Chem. Soc., 83, 117 (1961).
10. R. R. Hentz, J. Chem. Educ., 35, 625 (1958).
11. L. Funderburk and E. S. Lewis, J. Amer. Chem. Soc., 86, 2531 (1964).

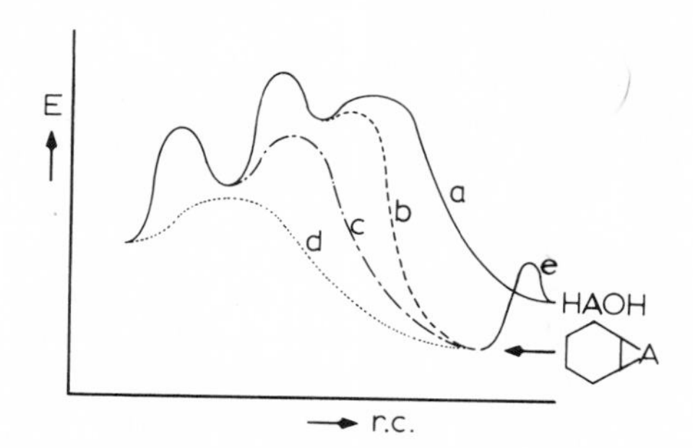

Fig. 15-8. The hydrolysis of HAX in the presence of a trap; various ways in which a new product may then arise.

Actually, the real path may be any of c, d, or a followed by e; cyclohexene may be reacting with HAX itself, or with AX^-, or with HAOH. The mere appearance of a suggestive product is therefore not nearly enough, and additional experiments are definitely required. The following further evidence would shore up the conclusion that A is an intermediate.

The rate law and the rate constant must not be affected by the presence of the trap, or in other words, the trap must not react with either HAX or AX^-. Thus, cyclohexene must not form the adduct observed with HAX in the absence of base. If this is the case, c and d can probably be ruled out. Furthermore, the trap must not form the adduct if it is added to a mixture in which the hydrolysis had already been allowed to go to completion; this would rule out curve b-e. The "traps" need not lead to suggestive products; for example, if the products are H_2A_2 and 3,3'-bicyclohexenyl, one might say that the intermediate was present and had been diverted from its normal course (so long as the basic kinetic criteria are met). The cyclohexene might be better described as a "lure" in that case. But one more word of caution is necessary. While one may thus be able to prove that an intermediate has been captured or diverted, one cannot thereby prove that it is A. For example, it may be A-----HOH or $A\text{-}OH_2$ - even if A is actually present also!

The acidity functions are useful in acid and base promoted reactions - especially in strong acid or base. They were devised by Hammett [12]. The problem was that in solutions more strongly acidic than, say, 0.1 N H^+, (H^+) becomes unreliable as a measure of the ability of the solution to transfer a proton. Thus (H^+) in pure sulfuric acid is quite small, yet this liquid is an exceedingly acidic medium in which few other molecules remain unprotonated. Hammett studied the behavior of a number of increasingly weak bases with spectra different from those of their conjugate acids. Thus, among the bases shown below aniline is already half protonated at a pH of 3, but 2,4,6-trinitroaniline is not measurably protonated even in H_2SO_4.

12. L. P. Hammett, "Physical Organic Chemistry", 2nd Edit., McGraw-Hill, New York, 1970.

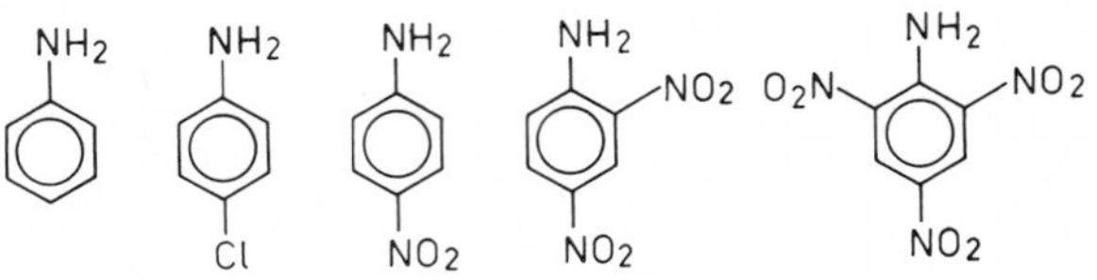

On this basis a function h_o was derived ($H_o = \log h_o$) which expresses the change of medium in highly acidic solvents much better than pH. Without going into precise details, one can in principle divide acid catalyzed reactions into two categories - one of which has rate constants proportional to (H^+), and the other has rate constants proportional to h_o. The former is said to involve a solvent molecule in the transition state, the latter is not. Thus, one could in principle distinguish between, say, A and A—OH_2. For highly basic solutions, h_- and H_- functions have similarly been devised. In recent years however, some difficulties have arisen with this technique that have to some extent discredited it; thus, it suggests the occurrence of primary carbonium ions in some cases. In these instances it has especially been in conflict with $\Delta V^{\ddagger}$ measurements, which lead to much more reasonable conclusions (Chapter 20, p. 698).

Product distributions represent a particularly fruitful way of studying intermediates, especially in reactions in which the intermediate can be generated in two or more ways, and in which it leads to more than one product. Suppose for instance that if HAX is allowed to react with base in the presence of cyclohexene and cyclopentene in equimolar amounts, and that then one obtains the corresponding adducts in a ratio of 5.30 to 1. Let us further assume that we succeed in synthesizing the diazocompound AN_2, and that this substance easily generates nitrogen, possibly *via*:

$$AN_2 \longrightarrow N_2 + (A) \xrightarrow{H_2O} HAOH$$

Now if that reaction is carried out in the presence of the two olefins in a 1:1 ratio, and if the same products should obtain and in the same ratio - then the evidence for an intermediate common to the two reactions is very strong. But it does not solve the problem completely.

First, once again we demonstrate the occurrence of a common intermediate, but not necessarily what intermediate. Second, the conditions must be the same in both experiments. This may not be so easy. For instance, AN_2 may not be soluble in water. If we do it in alcohol instead, and the product ratio is now 3.22, we have not proven anything. Since different reactions demand different conditions, this method is not as simple as it may appear, however powerful it may seem in principle.

A further requirement in this approach is that the selectivity must be reasonably high. A product ratio of 1 or close to 1 may merely mean that two different intermediates are involved, both hot, reactive, and reacting indiscriminately with both substrates.

Stereochemistry is a powerful tool, but not always applicable and rarely simple to use. As an example, if HAX is a resolvable molecule and if A is symmetrical, the product HAOH must be racemic. If this is indeed the case, one must still show that there are no other features that could lead to this observation. Thus, AX^- may rapidly racemize (perhaps demonstrable by observing the stereochemical stability of HAX in weakly basic medium). If HAOH is not racemic, a symmetrical intermediate A can still not be ruled out: If A is sufficiently short-lived it might capture water before the departing halide ion is far enough away to render the conclusion unassailable.

Scrambling of isotopic labels has very often been applied, and with great success; but again this method is not a simple one and much thought and preparation should precede its use. Suppose for example that HAX is really

$$P-Q\equiv Q-Q(P)(H)(X)$$

If one of the P atoms can be labeled for instance so as to give

$$\overset{**}{P}-Q\equiv Q-Q(P)(H)(X)$$

then the label should be scrambled in the product

$$\overset{*}{P}-Q\equiv Q-Q(P^*)(H)(OH)$$

There may be small isotope effects which may make the distribution less than exactly 50-50. Much work is usually necessary: The synthesis and degradation must be done in a demonstrated specific manner, and other ways in which symmetrization can occur must be ruled out; for example, a rapid equilibration

$$\overset{*}{P}-Q\equiv Q-Q^{\ominus}(P)(X) \rightleftharpoons (P^*)(X)\,{}^{\ominus}Q-Q\equiv Q-P$$

- which does not involve dissociation - would obviously lead to a wrong conclusion.

In those instances in which A is moderately stable and not so ephemeral that its existence can only be inferred, spectral observation may be important. The uv-visible and esr spectra are especially important because they are applicable to species in low concentration. Uv spectra can be employed to 10^{-5}M concentrations; there are many instances of the transient appearance of a color or uv absorption that first signalled the presence of an intermediate. Historically, it is probably the first evidence to be used for this purpose. Esr is useful for intermediates with one and sometimes more than one unpaired electron, at concentrations as low as 10^{-8}M. Nmr is important simply because it is so powerful in structural determinations, and CIDNP is becoming very important in reactions involving radical pairs. Infrared and Raman have played lesser roles; all of these techniques will be referred to at one point or another in subsequent chapters (see also Chapter 2).

If the concentration of an intermediate is apparently always so low that it cannot be observed directly, it is sometimes possible to raise its concentration by producing them all at once rather than piecemeal. Flash photolysis [13] and pulse radiolysis [14] can be applied to this end. For example, if our compound A=N=N undergoes the photolytic decomposition:

$$AN_2 \xrightarrow{h\nu} N_2 + A \longrightarrow P$$

one may hope to generate a high concentration of A by passing through a powerful flash of brief duration (10^{-7} sec or so) and then looking at the spectrum as suggested in Fig. 15-9. This has been done in many cases, but often the lifetime of A is so short that only secondary intermediates are observed.

Another way of preparing solutions of A in concentrations high enough to allow the observation of spectra is to use a solvent so unreactive toward A that its concentration builds up for lack of other things to do. A good example of this is the so-called super acid medium - usually mixtures of such compounds as liquid sulfur dioxide, fluorosulfonic acid, antimony pentafluoride, and so on. The mixture is preferably of such composition that its freezing point is very low.

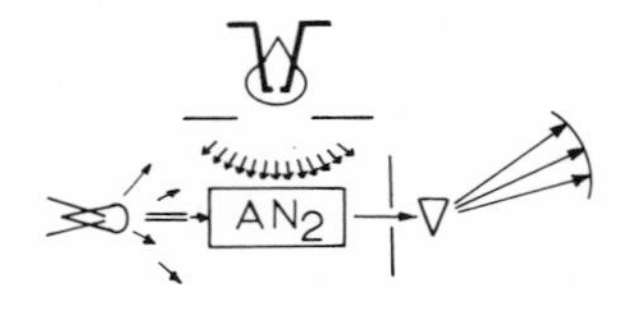

Fig. 15-9. Schematic flash photolysis.

---

13. L. S. Nelson, Science, 136, 296 (1962).
14. T. J. Kemp, Chem. Brit., 5, 255 (1969).

An extreme case of a special medium is something like solid argon, at 20°K or lower [15]. A highly dilute mixture of AN_2 vapor and argon is allowed to condense and freeze on a window held at liquid hydrogen temperature. Photolysis is then applied, and one obtains a matrix in which A is trapped and stabilized indefinitely (matrix isolation technique). It may be possible to study its chemical reactivity as well, by depositing alternate layers of AN_2 and a substrate and - after photolysis - by allowing A to diffuse to the substrate (by raising the temperature somewhat).

It should be remembered in all special medium and matrix work that the properties of the intermediate (because of the low temperature and unusual surroundings) may be different - even drastically different - from those in the more usual chemical solvents.

This list is still not complete. Additional approaches, such as the use of "tailor-made" substrates, will occasionally provide a deep insight in special cases. However, the overall conclusion is that none of the common methods can give us absolute assurance of one mechanism or another. The best we can do is to search for the intermediate by every experimental approach imaginable, and if none of the methods fail, a strong though still not absolute case can be made for it.

15-3. CLASSIFICATION OF INTERMEDIATES

There are various features that may make a molecule unstable enough that it can only be identified as a transient intermediate. The principal one among them is an abnormal valence state. Carbon atoms seek to surround themselves with eight electrons, and a deviation from that number can only lead to a molecule anxious to end the deviation. However, large deviations have been achieved. Thus, free, neutral carbon atoms with only four electrons can be generated and their properties can be studied. Univalent carbon species (carbynes) and divalent ones (carbenes) can likewise be generated and studied. Several analogs are known with atoms other then carbon short two electrons; these include the silylenes, nitrenes, and oxygen and sulfur atoms. Trivalent carbon atom intermediates are very common - carbonium ions, free radicals and carbanions are all of this sort. The carbanions do have eight electrons, and many are known as perfectly stable and isolable salts. Pentacovalent carbon atoms (with a two-electron, three-center bond) have been involved in the non-classical ions, and as noted earlier (p. 16), even hexacovalent carbon atoms are known in the stable carboranes.

Intolerable strains have led to other important intermediates; benzyne and small ring cycloalkynes, tetrahedrane, and the "anti-Bredt" compounds are the best known examples. The availability of a lower electronic state leaves a molecule in an unstable situation: Singlet oxygen is a good example of that. Antiaromaticity has already been encountered as a source of chemical instability, with cyclobutadiene and cyclopentadienone as examples. There are many molecules which are probably intrinsically not so unstable, but which are subject to isomerizations

---

15. J.S. Ogden and J.J. Turner, Chem. Brit., 7, 186 (1971).

to more stable substances via low activation energies (enols, norcaradiene), or which disproportionate readily (diimide) or which have symmetry allowed pathways to more stable fragments (cyclopropanone).

15-4. PROPERTIES OF INTERMEDIATES

Intermediates are molecules like any other - only more short-lived. We can therefore try to measure all the properties that interest us about more stable species, but of course the special conditions required greatly limit the studies that are possible. We will probably never know the refractive index of benzyne, for example.

In many cases it has been possible to learn something about the intermediate's reactivity and selectivity. These properties can be learned from the reactions that the intermediate undergoes, and especially from the care with which it selects from among the molecules in the mixture, that species with which it will form the final product. An intermediate that indiscriminately attacks any neighboring molecule is obviously not very stable. The opposite is true for an intermediate that selects very carefully; a relatively long life time is then implied.

Another property that can often be determined is the multiplicity. Among the more interesting reactive intermediates, some have the characteristics that they have an atom electron deficient by at least 2 in its valence shell, and that this atom is bound to at most two other atoms. In such cases the multiplicity can have either one of two values. The definition of the multiplicity was given on p. 7 as:

$$A = 2S + 1$$

where S is half the number of unpaired electrons; thus, methane - with all its electrons paired - is a singlet, the methyl radical is a doublet, and methylene (CH_2) may be either a singlet or a triplet, depending on whether the 2 unshared electrons are paired or not. If the two orbitals available are similar in energy, the triplet is likely to be the ground state (Hund's rule); if they differ widely in energy, the singlet will be the ground state. In many cases one can tell which species is initially generated on the grounds of spin conservation. For example, in the reaction

$$A_2\ddot{X}Y \longrightarrow A_2\ddot{X}^{\oplus} + Y^{\ominus}$$

A_2X and Y^- are either both singlets or they are both triplets. Since Y^- is not electron deficient it should be a singlet, and then A_2X must be one also.

In many cases the singlet and triplet species are not very far apart in energy, so that experiments must decide which is the ground state, and what particular state the intermediate is in when it reacts with other molecules.

The reactions of such electron deficient molecules are almost always oxidations, or acid reactions. A triplet can only engage in one electron oxidations; a singlet can react with a pair at once. For example, a common reaction of singlets is the concerted addition to multiple bonds to form three-membered ring compounds;

triplets can also form such products, but would do so via still another intermediate, the diradical which must undergo spin inversion before it can close. The result is a difference in stereochemistry, with singlets attacking stereospecifically, and triplets leading to mixtures. The use of this rule (Skell's rule), of course, depends on the "fact" that spin inversion is usually slow enough to allow large numbers of internal rotations to occur, so that conformational equilibrium can be reached.

For example, the photochemical conversion of cis- and trans-triazolines to the corresponding aziridines is largely stereospecific:

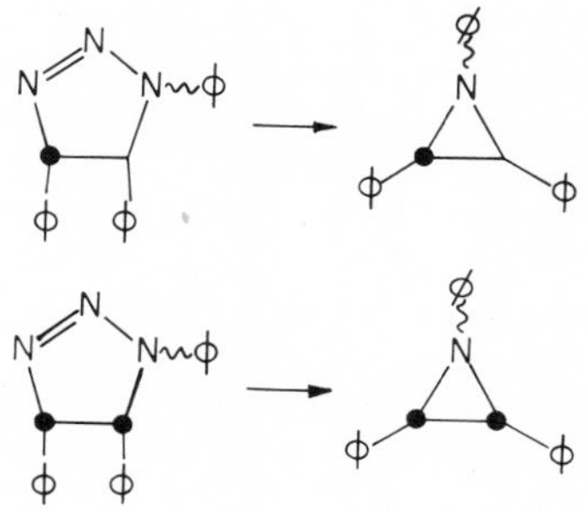

cis-triazoline gives a cis-trans product ratio of 4, and trans-triazoline gives a ratio of 1/3. The product ratios are unchanged if the solution is 5M in piperylene, an efficient triplet quencher [16]. Evidently, ring closure for the intermediate singlet diradical is much faster than single bond rotation. On the other hand, the specificity is largely lost if the reaction is sensitized by benzophenone: Both substrates produce the isomeric products in a 1.4 ratio. The triplet state is clearly much longer lived.

If the reactions of these two intermediates are different and if the one generated initially is the less stable of the two, it is in principle possible to catalyze the relaxation to the ground state and hence to bring about a change in products. One way is to conduct the reaction in the presence of some rare earth or transition metal ion having low level excited states of different multiplicity [17]; this is in fact a type of sensitization. Another way, is to use a solvent with one or more heavy atoms, such as iodobenzene or bromoform [18]. The heavy atoms have large orbits, and spin inversion is achieved by spin-orbit coupling.

The structure of the intermediate is of course of great interest, and the usual approach is to generate the intermediate by flash photolysis or in matrix isolation conditions so that spectroscopic methods can be used.

---

16. P. Scheiner, J. Amer. Chem. Soc., 88, 4759 (1966).
17. E. Gelles and K. S. Pitzer, J. Amer. Chem. Soc., 77, 1974 (1955).
18. M. Kasha, J. Chem. Phys., 20, 71 (1952).

Another direction of the research concerned with a given intermediate usually consists of its chemical stabilization. Within a given class of intermediates, some individuals are more stable than others, and it is desirable to identify what features this stability depends on. Once this is clear, then extreme examples with either stable or unstable character can be designed; in some cases the intermediate may become so unstable that it can no longer be generated; in others, so stable that it can be isolated. It may be noted that the area of stable molecules has likewise been the subject of studies in which various destabilizing features are magnified until the molecule is no longer capable of isolation - although it may still be capable of a transient existence as a reactive intermediate.

As we will see, the features that stabilize reactive intermediates are the same as those that stabilize already unreactive molecules. Thus, molecules with electron deficient centers are stabilized by electron donating substituents; cyclic molecules with a highly strained center are stabilized by relief of such strain, or by bulky substituents in the ring; substituents permitting delocalization are stabilizing here as they are in the realm of stable molecules. A favorite method has been the addition of certain metal compounds to which the intermediate may be bound as a ligand, and by which it may be stabilized in the process. Thus, o-quinodimethane is a highly unstable species, incapable of isolation and subject to immediate ring closure to benzocyclobutane; but, if it is generated in the presence of iron carbonyl compounds, the complex isolated is stable to 500°, at which temperature it begins to evolve benzocyclobutane [19] (see also Chapter 10).

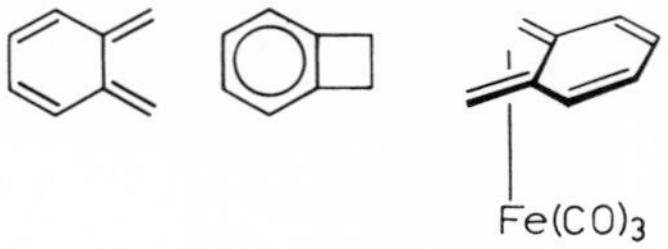

$Fe(CO)_3$

One should realize that the enormous stabilization claimed in some cases must be taken with a grain of salt: The properties of the stabilized intermediate may be so completely altered that it is really no longer proper to consider it a stabilized intermediate. Thus, the description of compounds such as $XYC{=}ML_n$ as carbene metal complexes is no more justified than the description of cyclopropanes as carbene ethylene complexes.

It was noted earlier (p. 538) that intermediates by definition occur in multistep reactions, to which the Woodward-Hoffmann rules do not apply. Nevertheless, one should realize that each single step is a concerted one in which MO symmetry is conserved, and that reactive intermediates are not likely to be captured or generated by pathways violating these rules.

19. W.R. Roth and J.D. Meier, Tetrahedron Lett., 2053 (1967).

Chapter 16

CARBENES AND RELATED INTERMEDIATES [1]

16-1. INTRODUCTION

In 1950, Hine startled the organic chemists by reporting [2] strong evidence that dichlorocarbene was involved in the reaction of chloroform with aqueous hydroxide:

$$HCCl_3 + OH^- \longrightarrow HCOO^- + CO + H_2O + Cl^-$$

---

1. Among the many excellent reviews that are available: (a) J. Hine, "Divalent Carbon", Ronald Press, New York, 1964; (b) W. Kirmse, "Carbene Chemistry", Academic Press, New York, 1964; (c) B. J. Herold and P. P. Gaspar, Forschr. Chem. Forsch., 5, 89 (1965); (d) W. Kirmse, Angew. Chem., Int. Ed. Engl., 4, 1 (1965); (e) J. O. Schreck, J. Chem. Educ., 42, 260 (1965); (f) G. G. Rozantsev, A. A. Fainzil'berg, and S. S. Novikov, Russ. Chem. Rev., English Transl., 34, 69 (1965); (g) G. Köbrich, Angew. Chem., Int. Ed. Engl., 5, 41 (1967); (h) R. A. Moss, Chem. Eng. News, 47, 60, June 16, 1969, and 50, June 30, 1969; (i) C. A. Buehle, J. Chem. Educ., 49, 239 (1972).
2. J. Hine, J. Amer. Chem. Soc., 72, 2438 (1950).

The reaction was found to be first order in chloroform and hydroxide, and hence the following mechanisms would fit the kinetics:

A. $$CHCl_3 + OH^- \xrightarrow{slow} CHCl_2OH + Cl^- \quad (S_N2 \text{ reaction})$$

$$CHCl_2OH \xrightarrow{rapid} \text{products}$$

B. $$CHCl_3 + OH^- \overset{rapid}{\rightleftharpoons} CCl_3^- + H_2O$$

$$CCl_3^- + H_2O \xrightarrow{slow} H_2\overset{+}{O}—\overset{-}{C}Cl_2 + Cl^- \quad (B_2 \text{ reaction})$$

$$H_2\overset{+}{O}—\overset{-}{C}Cl_2 \xrightarrow{rapid} \text{products}$$

C. $$CHCl_3 + OH^- \overset{rapid}{\rightleftharpoons} CCl_3^- + H_2O$$

$$CCl_3^- \xrightarrow{slow} CCl_2 + Cl^- \quad (B_1 \text{ reaction})$$

$$CCl_2 \xrightarrow{rapid} \text{products}$$

The considerations that point to the B_1 path are as follows. In D_2O, chloroform undergoes base catalyzed H-D exchange much faster than hydrolysis, as required by the B_1- and B_2 mechanisms; the S_N2 mechanism is of course not ruled out by this. Nucleophiles other than hydroxide displace chloride ion from the various chlorinated methanes in the following order of rates: $CH_3Cl > CH_2Cl_2 > CHCl_3 > CCl_4$, possibly because of increasing hindrance to backside displacement; however, with hydroxide ion the order is $CHCl_3 >> CH_3Cl > CH_2Cl_2 > CCl_4$. Thus, the reaction of chloroform with hydroxide appears to be of a special type. The activation volume is large and positive, which is difficult to reconcile with the S_N2 and B_2 mechanisms [3]. In methanolysis, the rate constant correlates well with the h- function, which provides some support for the B_1 mechanism [4]. Added chloride ion decreases the rate somewhat and this is not due to a simple change in ionic strength, since nitrate and perchlorate have no effect; iodide produces small amounts of dichloroiodomethane [1a]. The latter observations make sense if the second step in either a B_1- or B_2 mechanism is not completely irreversible. Chloroform gives phenyl orthothioformate with thiophenoxide ion in a reaction which is base catalyzed - a fact hard to understand in terms of an S_N2 reaction.

$$CHCl_3 + C_6H_5S^- \xrightarrow{-OH} CH(SC_6H_5)_3 + Cl^-$$

The Reimer-Tiemann reaction of chloroform and phenols in aqueous base to give o- and p-hydroxybenzaldehydes could similarly be explained in terms of dichlorocarbene [5]. When the base promoted solvolysis of chloroform is carried out in

---

3. W. J. le Noble and M. Duffy, J. Amer. Chem. Soc., 86, 4512 (1964).
4. R. A. M. O'Ferrall and J. H. Ridd, J. Chem. Soc., 5035 (1963).
5. J. Hine and J. M. van der Veen, J. Amer. Chem. Soc., 81, 6446 (1959) and J. Org. Chem., 26, 1406 (1961).

the presence of cyclohexene, the carbene is trapped and the bicyclic adduct (1) is obtained [6].

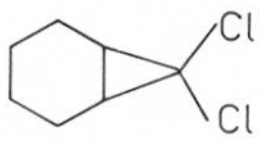

(1)

The papers by Hine and Doering have had a profound impact in organic chemistry. Although they were, of course, not the first chemists to uncover a mechanism and to base new preparations on their findings, the simplicity of Hine's reasoning, the novelty of a divalent carbon species and Doering's simple entry into functionalized cyclopropane derivatives have done much to catalyze the unprecedented development of organic chemistry in the past two decades. In this chapter we review some of the ways to generate carbenes and the reactions they generally undergo. This is followed by a discussion of various carbenes now known; as may become clear from that discussion, the structure, the chemical reactivity and the state of our knowledge all differ very widely from case to case. At the end of the chapter some analogs such as carbynes and silylenes are mentioned. Another group of carbene analogs, the nitrenes are considered in Chapter 17, and several free atoms with carbene-like properties (carbon, oxygen, sulfur) are discussed in Chapter 24. We may note here that the rapid growth of this field has produced some confusion in the nomenclature. Some chemists refer to carbenes as methylenes; others began to use the word carbene for singlets and methylene for triplets. Later it became fashionable to use the same word as for the corresponding radical followed by the ending idene.

16-2. THE GENERATION OF CARBENES

The generation of carbenes from haloforms by means of a base is quite general and virtually all conceivable combinations have been studied. Both the trihalomethide ion and carbene become more stable as the halogens become more electronegative (see below); on the other hand, fluoride is the worst leaving group. This combination of facts lead to an order of rates for carbene formation in which difluoroiodomethane is fastest, and fluoroform slowest. At the slow end of this spectrum the mechanism is somewhat different, the rates of deuterium exchange and hydrolysis are now similar [1a] and the activation volume, though still positive, is much smaller [7]. These facts are consistent with a concerted reaction in which the proton abstraction and loss of halide ion occur simultaneously (α-dehydrohalogenation).

---

6. W. von E. Doering and A.K. Hoffmann, J. Amer. Chem. Soc., 76, 6162 (1954).
7. W.J. le Noble, J. Amer. Chem. Soc., 87, 2434 (1965).

The trihalomethide ion can be obtained alternatively by the basic decarboxylation of trihaloacetic acids, and hence this is another pathway to dihalocarbenes [1a].

$$CCl_3COO^- \xrightarrow{\Delta} CO_2 + Cl^- + CCl_2$$

Still another method using chloroform, ethylene oxide and a trace of bromide depends on a series of equilibria, as follows:

$$\text{ethylene oxide} \underset{}{\overset{NEt_4^+\ Br^-}{\rightleftharpoons}} BrCH_2CH_2O^- \overset{CHCl_3}{\rightleftharpoons} HOCH_2CH_2Br + CCl_3^-$$

$$CCl_3^- \rightleftharpoons CCl_2 + Cl^- ; \quad CCl_2 \xrightarrow{\text{substrate}} \text{adduct}$$

The overall reaction occurs smoothly and reversibly at 150°; the epoxide and chloroform are not used up unless a substrate is present to remove the carbene [8].

Certain trihalomethylmercury derivatives [9] decompose under relatively gentle conditions (for example, in refluxing benzene), and hence they are useful precursors:

$$C_6H_5\text{-Hg-}CCl_2Br \overset{\Delta}{\rightleftharpoons} C_6H_5\text{-HgBr} + CCl_2$$

The rate of carbene release is independent of the concentration and nature of the substrate that finally catches it, hence free carbenes are involved. Since the starting material is prepared by essentially the reverse reaction, the phenylmercuric halide may be considered a way of storing these carbenes, permitting the experimenter to release them when it suits him. Alternatively, the carbene may be obtained from the mercury compound by setting the latter free with iodide ion:

$$C_6H_5\text{-HgCCl}_3 + I^- \longrightarrow Cl^- + C_6H_5\text{-HgI} + CCl_2$$

The cheletropic release or cycloelimination of the carbene with the simultaneous formation of a stable aromatic species is often effective [10]. Thus, some of the bridged [10]annulenes such as (2) readily decompose into naphthalene and a carbene [11], and adduct (3) upon photolysis ejects methylene to form phenanthrene [12].

---

8. (a) F. Nerdel and J. Buddrus, Tetrahedron Lett., 3585 (1965); (b) J. Buddrus, Angew. Chem., Int. Ed. Engl., 11, 1041 (1972).
9. D. Seyferth, Accounts Chem. Res., 5, 65 (1972).
10. R.W. Hoffmann, Angew. Chem., Int. Ed. Engl., 10, 529 (1971).
11. V. Rautenstrauch, H.J. Scholl, and E. Vogel, Angew. Chem., Int. Ed. Engl., 7, 288 (1968).
12. D.B. Richardson, L.R. Durrett, J.M. Martin, W.E. Putnam, S.C. Slaymaker, and I. Dvoretzky, J. Amer. Chem. Soc., 87, 2763 (1965).

Cl Cl

(2)

(3)

Dehalogenations with strong bases such as methyllithium are often effective even at temperatures well below 0°; thus bromotrichloromethane and butyllithium in the presence of olefins lead to dichlorocyclopropanes in 90% yield [13].

It is not always clear that all of these reactions produce the dichlorocarbene in a completely free state. A number of them have been found by Skell to lead to the same mixture of cyclopropanes when carried out in the same mixture of olefins [14]; however, in other cases it is equally clear that the carbene is being transferred without achieving a free state. This is especially believed to be so in the base promoted, low temperature dehalogenations, which are thought to occur via metal-halogen interchange [15]. The intermediate transfer agents are then referred to as carbenoids. An especially clear-cut case of such transfer is lithiotrichloromethane, which is stable in solution below -80°, but which reacts with cyclohexene at -100° to give 7,7-dichloronorcarane (1) [16].

The cycloeliminations likewise may similarly involve the transfer of an intermediate rather than its complete ejection - even in those cases in which the latter mode of reaction is "allowed". This difference in mechanism is often recognizable by means of a corresponding difference in selectivity; an intermediate which is transferred remains at least weakly bound, and hence will be less reactive and more discriminate in its choice of substrates than the free species. The loss of methylene by benzotropilidene (4) at 400° to give naphthalene and methylnaphthalenes is an example.

(4)

---

13. W.T. Miller and C.S.Y. Kim, J. Amer. Chem. Soc., 81, 5008 (1959).
14. P.S. Skell and M.S. Cholod, J. Amer. Chem. Soc., 91, 6035, 7131 (1969).
15. (a) G.L. Closs and L.E. Closs, J. Amer. Chem. Soc., 85, 99 (1963); (b) G.L. Closs and R.A. Moss, J. Amer. Chem. Soc., 86, 4042 (1964); (c) D.F. Hoeg, D.I. Lusk, and A.L. Crumbliss, J. Amer. Chem. Soc., 87, 4147 (1965).
16. (a) W.T. Miller and D.M. Whalen, J. Amer. Chem. Soc., 86, 2089 (1964). See also (b) G.L. Closs and J.J. Coyle, J. Amer. Chem. Soc., 84, 4350 (1962) and Ref. lg.

While methylene ordinarily indiscriminately attacks benzene and cyclohexane, in this reaction it appears to react only with benzene to give toluene, and a pathway involving species such as (5)-(8) seems likely [17].

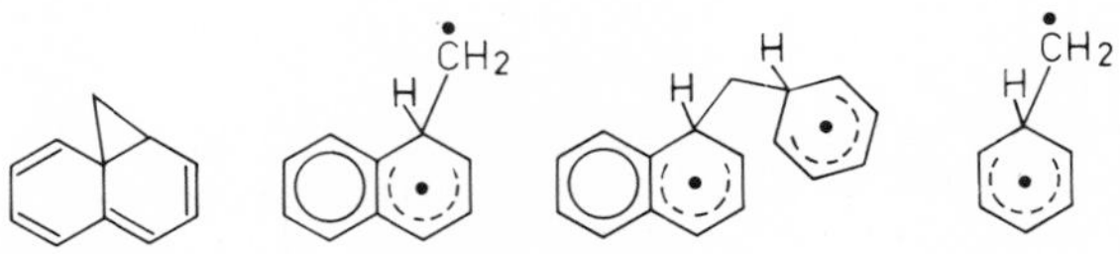

(5) (6) (7) (8)

Similar uncertainty exists in most of the following reactions, which are different from the foregoing in that they are not especially suited for dihalocarbenes. The well-known Smith-Simmons cyclopropanation of olefins with methylene iodide and zinc-copper couple is a good example in which a carbene is not free at any time, and is in effect transferred from one compound to another (see further p. 574). Carbene itself can be obtained by the photodecarbonylation of ketene in the gas phase. Diazomethane and diazirine have been similarly used; the former will also produce carbene in solution via the thermal route. Substituted carbenes can often be prepared from the carbonyl precursor by the Bamford-Stevens reactions. The carbonyl compound is first converted in a tosylhydrazone (9), and the product is heated in basic medium to give the tosylate anion, nitrogen and a carbene; the diazocompound is an intermediate in this sequence.

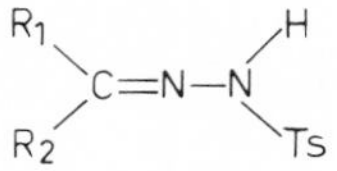

$$(R_1)(R_2)C{=}N{-}N(H)(Ts)$$

(9)

16-3. REACTIONS OF CARBENES

The chemistry of the various carbenes is characterized by variety, as will become obvious from the following sections; however, certain reactions seem quite common, and a brief discussion seems therefore useful.

Carbenes are electron deficient, and unless strong resonance interaction such as in (10) is possible, the reactions will be of the electrophilic type.

$$-\ddot{X}-C: \longleftrightarrow -\overset{\oplus}{X}=\overset{\ominus}{C}:$$

(10)

---

17. M. Pomerantz, A.S. Ross, and G.W. Gruber, J. Amer. Chem. Soc., 94, 1403 (1972).

A common reaction is that with anions. For example, the dihalocarbenes are subject to return of the halide ion, and many trihalomethanes can be synthesized in this way [1a]. Attack on neutral molecules (such as on water to give hydrolysis products) is of course also frequently observed.

A very important reaction is the addition to multiple bonds, not only because it affords a simple way (sometimes the only known way) to prepare functionalized cyclopropanes, but also because the stereochemical details often afford insights into the structure of the carbene and the mechanism of the reaction. First of all, one observes stereospecific addition, for example to cis- and trans-2-butene; this is usually considered an indication of a singlet multiplicity, and a concerted attack on the π-electron pair. As noted in Chapter 15 (p. 555), the appearance of a mixture of cis- and trans-products (especially that of identical mixtures) is considered evidence of a triplet in which spin inversion and subsequent ring closure is slower than single bond rotation. This was of course initially not at all certain, and there was much doubt about the use of this criterion when Skell started using it to distinguish singlet and triplet carbenes in the nineteen fifties (see further p. 858). Secondly, if cyclic olefins are used, only cis-bicyclo[n. 1. 0]alkanes are obtained as products, so that approach to the olefin is suprafacial and in the π-plane so far as the olefin is concerned. Thirdly, if bicyclic olefins are used, addition may occur on either side of the molecule (exo- or endo), and if the two substituents of the carbene are different, again two possible products may be obtained (syn- or anti-). If one approach is more favorable than another, the reaction is said to be regioselective. Of course, this is virtually always the case, and the use of the term makes sense only if one can identify the feature that favors the observed product. Unfortunately, the necessary experiments have as yet been carried out in only a few cases, and these have not yet yielded results of general utility.

Additions to CN double bonds have likewise been observed in many instances [18]. The addition of dichlorocarbene to diazocompounds (11) to give olefins (12) is of interest in that diazocompounds are often used to generate carbenes. The appearance of symmetrical olefins in such reactions should therefore not be interpreted as a dimerization of the carbene.

$$(R_1)(R_2)C{=}N{=}N \quad (11) \qquad (R_1)(R_2)C{=}CCl_2 \quad (12)$$

Attack on Wittig reagents (phosphorus ylids) (13) to give olefins is also known [19]. Addition to NN double bonds does not lead to a diaziridine; when azodicarboxylates (14) are used to trap dichlorocarbene, the product is the rearranged (15) [20].

18. H. Reimlinger, Angew. Chem., Int. Ed. Engl., 1, 156 (1962) and Chem. Ber., 97, 339 (1964).
19. Y. Ito, M. Okano, and R. Oda, Tetrahedron, 22, 2615 (1966).
20. D. Seyferth and H. M. Shih, J. Amer. Chem. Soc., 94, 2508 (1972).

$\phi_3P{=}CHCOOC_2H_5$ (13)

$H_5C_2OOC{-}N{=}N{-}COOC_2H_5$ (14)

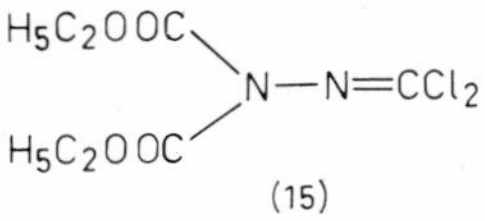

$(H_5C_2OOC)_2N{-}N{=}CCl_2$ (15)

Triple bond additions are also possible; they have been used (with dichlorocarbene) to produce cyclopropenones [21]. Addition to benzenes occurs less frequently, but at least one instance of addition even to hexafluorobenzene has been noted: Bis-trifluoromethylcarbene in that solvent gives the tropilidene (16) (or the corresponding norcaradiene (17)) [22]. This example is important because hexafluorobenzene has often been used as an "inert solvent" in carbene studies.

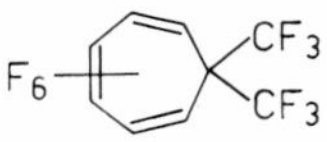

F_6–tropilidene–$C(CF_3)_2$ (16) F_6–norcaradiene–$C(CF_3)_2$ (17)

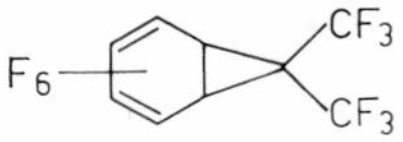

It may be noted that these additions constitute the reverse of the cycloelimination reactions, and by the principle of microscopic reversibility the motion of all atoms must be exactly the reverse; therefore, the same remarks apply here as in Chapter 14 (p. 530) [23]. The mechanism of the carbene attack on multiple bonds is difficult to study. Once the carbene is generated, the highest energy transition state in the overall reaction has passed and hence kinetics are of little or no help to us. The product distribution in competition experiments involving pairs of olefins (selectivity) and stereochemistry are now our principal sources of information. For example, the addition of dichlorocarbene to mixtures of ring-substituted styrenes gives rise to a Hammett plot in which a good correlation is obtained if σ^+ constants are used; $\rho = -0.62$: Evidently this carbene is an electrophilic reagent and the addition is a polar process, but not very much so [24].

Attacks on single bonds are common. In the absence of suitable substrates such attacks may be intramolecular; for example, when the hydrazone derivative (18) of pinacolone is heated in alcoholic base, the cyclopropane (19) is formed in 50% yield [25].

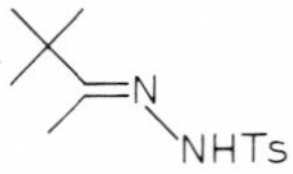

=N–NHTs (18)    (19)

21. E.V. Dehmlow, Tetrahedron Lett., 2317 (1965).
22. D.M. Gale, J. Org. Chem., 33, 2536 (1968).
23. R. Hoffmann, D.M. Hayes, and P.S. Skell, J. Phys. Chem., 76, 664 (1972).
24. D. Seyferth, J.Y. Mui, and R. Damrauer, J. Amer. Chem. Soc., 90, 6182 (1968).
25. L. Friedman and H. Shechter, J. Amer. Chem. Soc., 81, 5512 (1959).

The attack on carbon hydrogen bonds may be of two types. Hydrogen abstractions leading to radical pairs are usually ascribed to triplet state carbenes, and concerted insertion is thought to characterize singlets. However, there is precious little evidence on these points. In one study for example [26], the surprising result obtained that dichlorocarbene (from whatever source) inserts into the benzylic carbon hydrogen bond of (+)-2-phenylbutane with complete racemization, ruling out concerted attack in that case. The authors ascribed this to a carbonium ion-carbanion pair; however, this is not the whole story since racemization from such a pair means that it was sufficiently long lived to permit one or both members to undergo many rotations. The same authors also found that insertion by methylene itself (photolytically generated from diazomethane) into the α-CH bonds of the diacetate of *trans*-1,2-cyclopentanediol (20) occurs with inversion, the two acetate groups in (21) winding up in *cis*-positions! Evidently, the back-lobe of the CH bond is involved as in (22).

Similarly, it was recently found that *o*-(2-*endo*-norbornyl)-phenylcarbene (23) inserts into the 3-position in such a way as to give a *trans*-junction (24) [27].

Dimerization of carbenes almost never occurs, at least not directly. Indirect dimer formation is possible in some cases, for instance in the decomposition of diazocompounds as noted above (p. 563), and in solvolysis reactions if the carbene inserts in the CH bond of the parent and the product suffers rapid elimination of HX.

---

26. V. Franzen and R. Edens, *Ann.*, *729*, 33 (1969).
27. C.D. Gutsche, G.L. Bachman, W. Udell, and S. Bäuerlein, *J. Amer. Chem. Soc.*, *93*, 5172 (1971).

The reason for the non-occurrence of the reaction is simply that the concentration of the carbene is so low that these species cannot find each other. The story is different under the conditions of flash photolysis, or of a warming, inert matrix, and then dimer products are indeed common.

16-4. HALOCARBENES

Compared to other methylenes, dihalocarbenes are quite stable, and the greatest stability is associated with fluorine substitution. Thus it is found that in the mixed trihalomethide ions, the largest halogen atom is released as the anion, and in competitive experiments with mixtures of olefins, the greatest selectivity is found in those carbenes carrying the lightest halogen atoms (see Table 16-1) [28]. It may be noted that the alkylated olefins are attacked most rapidly in spite of steric constraints, as would be expected for an attack by electrophilic reagents. There is similar evidence that chlorocarbene and fluorocarbene are quite selective compared to methylene [29].

When 1, 1-difluorotetrachlorocyclopropane is heated to 200° it releases CF_2, rather than CCl_2 [30]. This is one of the few instances in which a carbene addition to a non-aromatic double bond is reversed; it is undoubtedly related to the relative stability of difluorocarbene. In this case one might indeed describe the cyclopropane as an ethylene carbene complex (cf. p. 556). It may also be noted that tetrafluoroethylene thermally dissociates to the carbene, and that the constant for this equilibrium at 1600°K equals unity [31]. The bond energy is about 70 kcal/mole,

TABLE 16-1. Relative Rate Constants for Addition in the Competition of Isobutylene With Another Olefin for a Carbene

Olefin	CFCl	CCl_2
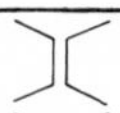	31	6.5
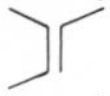	6.5	2.8
	1	1
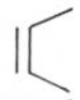	0.14	0.23
	0.097	0.15
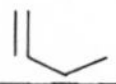	0.0087	0.011

28. R.A. Moss and R. Gerstl, J. Org. Chem., 32, 2268 (1967) and Tetrahedron, 23, 2549 (1967).
29. Y. Tang and F.S. Rowland, J. Amer. Chem. Soc., 89, 6420 (1967).
30. (a) J.M. Birchall, R.N. Haszeldine, and D.W. Roberts, Chem. Commun., 287 (1967); (b) N.C. Craig, T. Hu, and P.H. Martyn, J. Phys. Chem., 72, 2234 (1968).
31. G.A. Carlson, J. Phys. Chem., 75, 1625 (1971).

less than half the normal value for olefins [32]. The reason for the carbene stability can of course be debated; perhaps it is related to the fact that in the halogens below fluorine d orbitals are relatively accessible, which would make the electron deficient carbon atom relatively more reactive:

$$Cl-C-Cl \longleftrightarrow Cl-\overset{\oplus}{C}=\overset{\ominus}{Cl}$$

On the other hand, all halogens stabilize the carbene by overlap of their unshared electrons into the vacant orbital on carbon:

$$F-C-F \longleftrightarrow F-\overset{\ominus}{C}=\overset{\oplus}{F}$$

This picture fits well with the much greater stabilization observed when the carbon atom carries other first row atoms with unshared electrons (see Section 10 below).

Dibromocarbene was probably the first carbene to be investigated with a view to the specificity of the addition to the 2-butenes. The result was that this reaction is clearly stereospecific [33].

Since then all evidence on dihalocarbenes has tended to yield the same conclusion. This evidence includes spectral information. The stability of difluorocarbene has permitted the determination of its microwave spectrum. There is no Zeeman splitting, which is only compatible with a bent structure with all electrons paired; the bond angle was calculated to be 105° [34]. The infrared spectrum of this carbene generated in a matrix from 3,3-difluorodiazirine gives similar information [35], and so does matrix isolated dichlorocarbene [36]. Furthermore, oxygen does not scavenge difluorocarbene unless it is generated by means of the reaction [37]:

$$O(^3P) + C_2F_4 \longrightarrow CF_2O + CF_2$$

The addition of dihalocarbenes to double bonds is straightforward, but the use of substrates with strained double bonds has been observed to lead to rearranged products. For example, norbornene gives the bicyclo[3.2.1]octene derivative (25) [38].

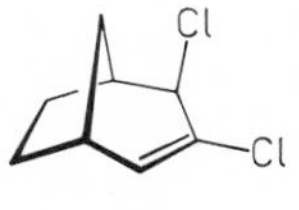

Cl

Cl

(25)

32. J.P. Simons, J. Chem. Soc., 5406 (1965).
33. P.S. Skell and A.Y. Garner, J. Amer. Chem. Soc., 78, 3409 (1956).
34. F.X. Powell and D.R. Lide, J. Chem. Phys., 45, 1067 (1966).
35. D.E. Milligan, D.E. Mann, M.E. Jacox, and R.A. Mitsch, J. Chem. Phys., 41, 1199 (1964).
36. A.K. Maltsev, R.G. Mikaelian, O.M. Nefedov, R.H. Hauge, and J.L. Margrave, Proc. Nat. Acad. U.S.A., 68, 3238 (1971).
37. J. Heicklen, N. Cohen and D. Saunders, J. Phys. Chem., 69, 1774 (1965).
38. C.W. Jefford, S. Mahajan, J. Waslyn, and B. Waegell, J. Amer. Chem. Soc., 87, 2183 (1965).

Addition to aromatic hydrocarbons occurs only if the latter are polynuclear; thus, ring expansion occurs when dichlorocarbene is generated in t-butyl alcohol in the presence of anthracene to give (26) [39].

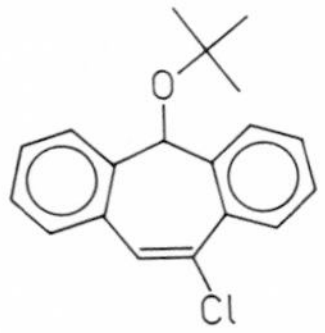

(26)

Other double bonds such as those in imines are susceptible to addition. For example, when a basic solution of chloroform in t-butyl alcohol is treated with imine (27), adduct (28) is formed. In this instance, it was shown that trichloromethide ion is not likely to be responsible: Only a minor amount of the product was formed from the anion (29) of N-phenyl-1,1,1-trichloro-2,2-diphenylethylamine [40].

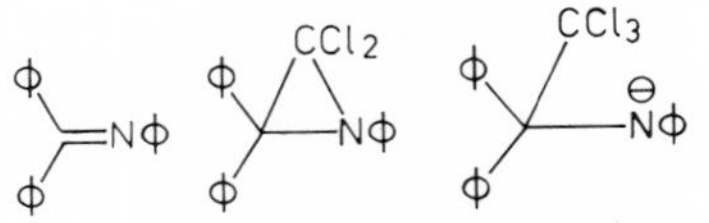

(27) (28) (29)

Several other cases of dihalocarbene additions to multiple bonds were noted earlier (p. 563). It appears that these carbenes prefer double bonds to triple linkages when given a choice, although steric factors may reverse this preference. For example, while enyne (30) gives an acetylenic cyclopropane (31), the analog (32) gives a cyclopropenone (33) [41].

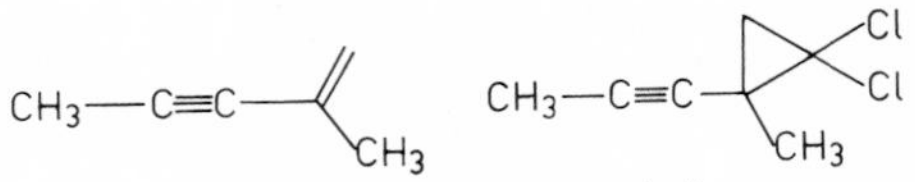

(30) (31)

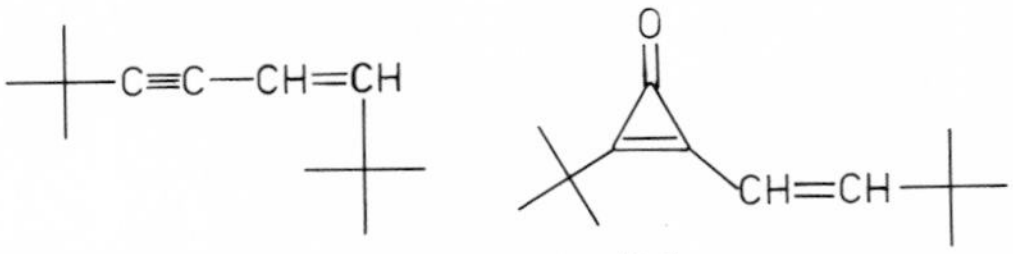

(32) (33)

---

39. F. Nerfel, J. Buddrus, W. Brodowski, J. Windhoff, and D. Klamann, Tetrahedron Lett., 1175 (1968).
40. J.A. Deyrup and R.B. Greenwald, Tetrahedron Lett., 321 (1965).
41. (a) E.V. Dehmlow, Tetrahedron Lett., 3763 (1966); (b) Ref. 1c.

Insertion in C-H bonds is fairly uncommon. Adamantane gives a high yield of 1-(dichloromethyl)adamantane (34) [42]; CH-bonds α to an ether linkage will also undergo this reaction to some extent, even in competition with double bonds. Thus, 1,4-dihydrofuran (35) gives both (36) and (37) [43].

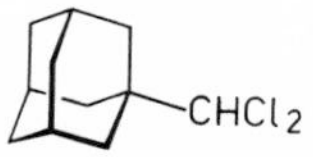

$CHCl_2$ (34)  (35) CCl_2 (36) $CHCl_2$ (37)

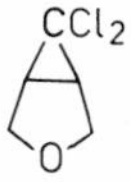

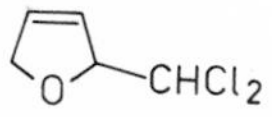

Insertion in C-C single bonds does not appear to occur at all. One apparent case is known: 1-phenyl-1-butyne (38) gives the cyclopropenone (39) [44].

$\phi—C\equiv C—C_2H_5$ (38)

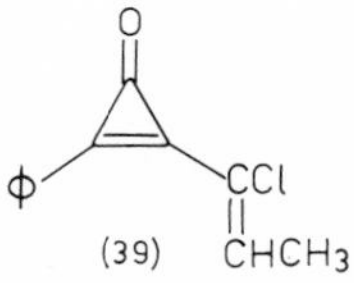

O, φ, CCl, $CHCH_3$ (39)

This is probably an indirect result of a more complex series of reactions involving a series of intermediates such as (40)-(42).

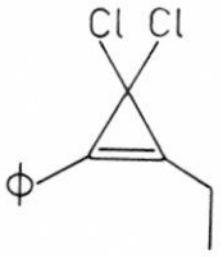

(40) 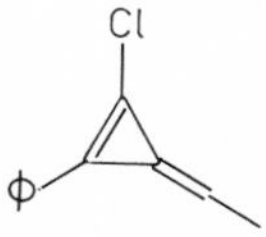 (41) 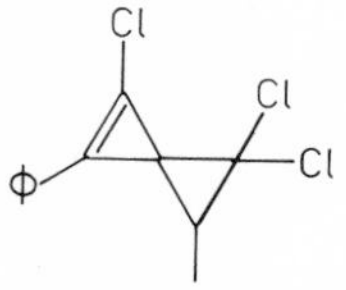 (42)

Thus, it was found that when the carbene is generated from trichloroacetate in the absence of strong base, the product is simply 1-phenyl-2-ethycyclopropenone. Weaker single bonds are generally more susceptible to dihalocarbene insertions. There are several examples that bear witness to this: The carbon-mercury bond in (43) is attacked by dichlorocarbene to give (44) [45], the silicon-carbon bond of

---

42. I. Tabushi, Z. Yoshida, and N. Takahashi, J. Amer. Chem. Soc., 92, 6670 (1970).
43. J.C. Anderson, D.G. Lindsay, and C.B. Reese, J. Chem. Soc., 4874 (1964).
44. E.V. Dehmlow, Tetrahedron Lett., 4003 (1965).
45. M.E. Gordon, K.V. Darragh, and D. Seyferth, J. Amer. Chem. Soc., 88, 1831 (1966).

(45) suffers insertion to yield (46) [46], and the tin–tin bond of hexamethyldistannane (47) is attacked to give (48) [47].

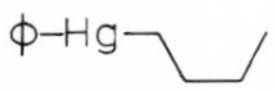

ϕ–Hg–CH₂CH₂CH₂CH₃ (43) ϕ–Hg–CCl_2–CH₂CH₂CH₂CH₃ (44)  (45) 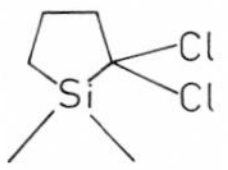 (46)

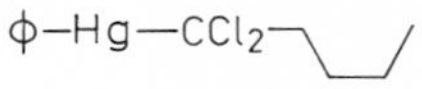

Silicon and germanium hydrides such as in (49) and (50) [48] offer easy targets, and yield products (51) and (52), respectively.

$Me_3SnSnMe_3$ (47) $Me_3SnCCl_2SnMe_3$ (48)

Et_3SiH (49) ϕ_3GeH (50) $Et_3SiCHCl_2$ (51) $\phi_3GeCHCl_2$ (52)

Unshared electrons are of course relatively easily attacked by electrophilic carbenes, as witness the capture of dihalocarbenes by added halide ions in solution [1a]. Another example is the carbylamine reaction, formerly used by organic chemists to demonstrate the presence of primary amines by means of the sickening odor of isocyanides (53) produced with chloroform and base.

$R-\overset{\oplus}{N}\equiv\overset{\ominus}{C}:$

(53)

Isocyanides themselves are not immune from further reaction; in methanol for example (54) is obtained from cyclohexyl isocyanide (55) [49]. Dichlorocarbene readily reduces dimethyl sulfoxide - a reaction that probably begins by carbene attack at the oxygen atom [50].

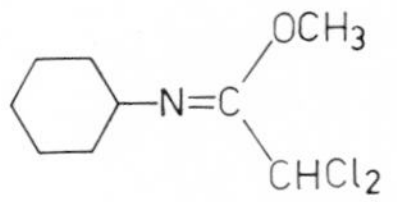

$C_6H_{11}-N=C(OCH_3)CHCl_2$ (54) $C_6H_{11}-\overset{\oplus}{N}\equiv\overset{\ominus}{C}$ (55)

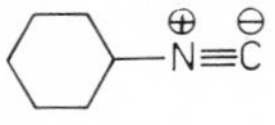

46. D. Seyferth, R. Damrauer, and S. S. Washburne, J. Amer. Chem. Soc., 89, 1538 (1967).
47. D. Seyferth and F. M. Armbrecht, J. Amer. Chem. Soc., 89, 2790 (1967).
48. D. Seyferth and J. M. Burlitch, J. Amer. Chem. Soc., 85, 2667 (1963).
49. A. Halleux, Angew. Chem., Int. Ed. Engl., 3, 752 (1964).
50. R. Oda, M. Mieno, and Y. Hayashi, Tetrahedron Lett., 2363 (1967).

Finally, dimerization - e.g. of difluorocarbene - may occur under low temperature conditions where the carbenes are sufficiently long-lived to survive until an encounter takes place [35].

Bistrifluoromethylcarbene furnishes an interesting contrast with difluorocarbene, since no resonance stabilization is now possible and only the inductive electron withdrawal is operating. While dihalocarbenes react only with polynuclear aromatic systems and insert only in activated CH bonds, bistrifluoromethylcarbene adds even to hexafluorobenzene as already noted (p. 564), and when it is generated in the presence of cyclohexene it does not only add to give (56), but it also indiscriminately inserts into the CH bonds to give products such as (57) and (58) [51].

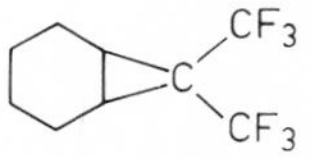

(56) 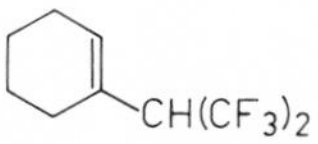 (57) 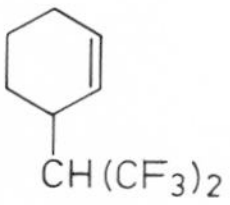 (58)

This carbene also demonstrates its reactivity in that it forms epoxides with some carbonyl compounds; thus, with carbonyl fluoride the species (59) is obtained [52].

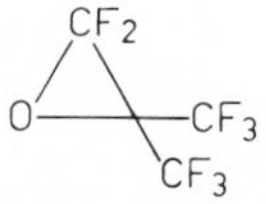

(59)

The ground states of this carbene and of trifluoromethylcarbene are triplets with 150° bond angles; this was deduced from esr spectra determined at 4.2°K in a matrix [53].

16-5. METHYLENE

This species was first observed and identified by Herzberg [54] in 1959. When diazomethane is subjected to flash photolysis, a transient new absorption occurs at about 1400 Å. This absorption is shifted a little when deuteriodiazomethane is

---

51. D.M. Gale, W.J. Middleton, and C.G. Krespan, J. Amer. Chem. Soc., 88, 3617 (1966).
52. E. Wasserman, L. Barash, and W.A. Yager, J. Amer. Chem. Soc., 87, 4974 (1965).
53. W. Mahler, J. Amer. Chem. Soc., 90, 523 (1968).
54. G. Herzberg and J. Shoosmith, Nature, 183, 1801 (1959).

used, and the shift is a little larger still for dideuteriodiazomethane, showing that the species contains two hydrogen atoms: The only reasonable conclusion is that it must be methylene.

Later work showed that the methylene so generated is an excited singlet state with a bond angle of 103° and bond lengths of 1.12 Å as in (60) [55].

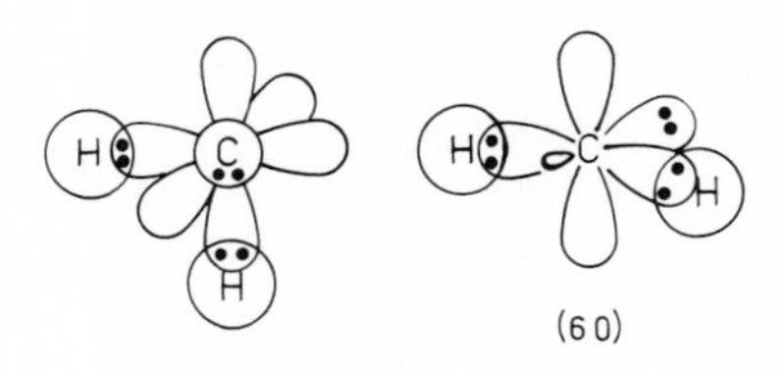

(60)

These authors also described triplet methylene as a linear or near-linear species such as (61).

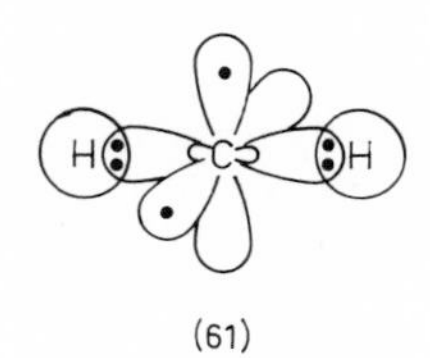

(61)

Later work confirmed many of these conclusions. Thus, the singlet was found to add to olefins in a stereospecific manner [56]. If diazomethane is photolyzed in a large excess of nitrogen, this stereospecificity is lost [57]; this suggests that the methylene degenerates to the triplet ground state if it undergoes a great many collisions. Unfortunately, this simple picture is made a little less certain by the observations that methylene formation from ketene and diazomethane is reversible; the parent carbene will react with carbon monoxide to form ketene [58] and with nitrogen to form diazomethane [59], so that the photolysis of these compounds are likely to be more complex processes than had been realized. It was also found that the stereospecificity of addition to cis-2-butene depends on the foreign gas pressure in a complex way [60]; the fraction of trans- product, which is low at low

---

55. G. Herzberg, Proc. Roy. Soc., A 262, 291 (1961).
56. (a) D.C. Blomstrom, K. Herbig, and H.E. Simmons, J. Org. Chem., 30, 959 (1965); (b) P.S. Skell and R.C. Woodworth, J. Amer. Chem. Soc., 78, 4496 (1956).
57. F.A.L. Anet, R.F.W. Bader, and A.M. Van der Auwera, J. Amer. Chem. Soc., 82, 3217 (1960).
58. T.B. Wilson and G.B. Kistiakowsky, J. Amer. Chem. Soc., 80, 2934 (1958).
59. (a) C.B. Moore and G.C. Pimentel, J. Chem. Phys., 41, 3504 (1964); (b) A.E. Shilov, A.A. Shteinman, and M.B. Tjabin, Tetrahedron Lett., 4177 (1968).
60. R.F.W. Bader and J.I. Generosa, Can. J. Chem., 43, 1631 (1965).

pressure and rises with pressure, is also high if the pressure is made extremely low. The authors felt that this loss of specificity at very low pressures was due to the lack of opportunity of "hot" cis- product to undergo deactivating collisions.

The following experiments may now be regarded as definitive. Methylene capture by the 2-butenes, which normally gives rise to considerable amounts of the isomerized product, becomes very stereospecific when a small amount of oxygen is present. The oxygen presumably rapidly removes the triplet. On the other hand, if a large amount of oxygen or if methyl iodide is present [61], or if the photolysis is sensitized by mercury atoms in the ^3P state [62], then all stereospecificity is lost; all this is in accord with the assumption that the triplet addition is non-stereospecific, and that it represents the ground state. The best recent evidence that methylene has a triplet ground state is the fact that an esr signal can be observed in a xenon matrix at 4.2°K, and that this signal is stable for many hours [63]. The esr results suggest [64] that the triplet is bent with a bond angle of 135°. Herzberg [65] subsequently was able to reinterpret his old spectra; they are now thought to agree with the esr results and with the results from quantum mechanical calculations [66].

Methylene is reactive and highly indiscriminate; for example, it inserts in any type of C-H bond and adds to benzene to form cycloheptatriene [67]. One of the odd features of methylene chemistry is that the singlet reactions cannot be completely suppressed. This has been interpreted by Frey as meaning that a tiny equilibrium amount of the singlet will ordinarily be present, and by studying product yields as a function of oxygen pressure he was able to deduce that the singlet state lies about 7 kcal/mole above the triplet [68].

Recently the new tool of CIDNP has been applied to help unravel the mechanistic details of methylene reactions. Thus, diazomethane photolysis in toluene produces ethylbenzene without the CIDNP effect, but if benzophenone is used to sensitize the reaction the emission is observed: Evidently the latter reaction goes via a radical pair and the former proceeds via direct insertion [69]. If the solvent is deuteriochloroform, the unsensitized reaction gives 1,1,2-trichlorodeuterioethane [1d] with the proton signal in emission; the sensitized reaction gives 1,1,1-trichlorodeuterioethane with the proton signal in enhanced absorption. Evidently the singlet is able to abstract chlorine to produce a radical pair [70]. Another

---

61. H.M. Frey, J. Amer. Chem. Soc., 82, 5947 (1960).
62. F.J. Duncan and R.J. Cvetanovic, J. Amer. Chem. Soc., 84, 3593 (1962).
63. R.A. Bernheim, H.W. Bernard, P.S. Wang, L.S. Wood, and P.S. Skell, J. Chem. Phys., 53, 1280 (1970).
64. E. Wasserman, V.J. Kuck, R.S. Hutton, and W.A. Yager, J. Amer. Chem. Soc., 92, 7491 (1970).
65. G. Herzberg and J.W.C. Jones, J. Chem. Phys., 54, 2276 (1971).
66. W.A. Yeranos, Z. Naturforsch., A26, 1245 (1971).
67. W. von E. Doering, R.G. Buttery, R.G. Laughlin, and N. Chaudhuri, J. Amer. Chem. Soc., 78, 3224 (1956).
68. H.M. Frey, Chem. Commun., 1024 (1972).
69. H.D. Roth, J. Amer. Chem. Soc., 94, 1761 (1972).
70. H.D. Roth, J. Amer. Chem. Soc., 93, 4935 (1971).

CIDNP experiment concerns the photolysis of diazomethane in carbon tetrachloride, which produces pentaerythrityl tetrachloride (66) via such intermediate radicals as chloromethyl, trichloromethyl and the species (62)-(65).

$\dot{C}H_2-CCl_3$ (62) $CH_2Cl-\dot{C}Cl_2$ (63) $CH_2Cl-CCl_2-\dot{C}H_2$ (64) $(CH_2Cl)_2\dot{C}Cl$ (65) $C(CH_2Cl)_4$ (66)

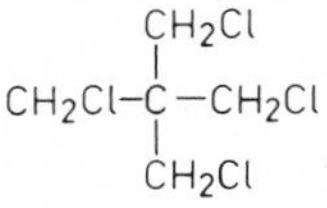

The chain nature of the reaction is clear from a very high quantum yield [71]. A small amount of 1,1,1,2-tetrachloroethane is also formed; its protons normally appear in strong emission, but this changes to enhanced absorption when benzophenone is added. Again the conclusion is that the singlet must be able to abstract chlorine [72].

Net methylene transfers can also often be effected by the thermal reaction with diazomethane; however, in many cases the pathway is known to involve a pyrazoline intermediate. For instance, duroquinone (67) reacts with diazomethane to give such adducts as (68) and (69), which under the action of either acid catalysts, heat or light loose nitrogen to give (70) and (71), respectively [73].

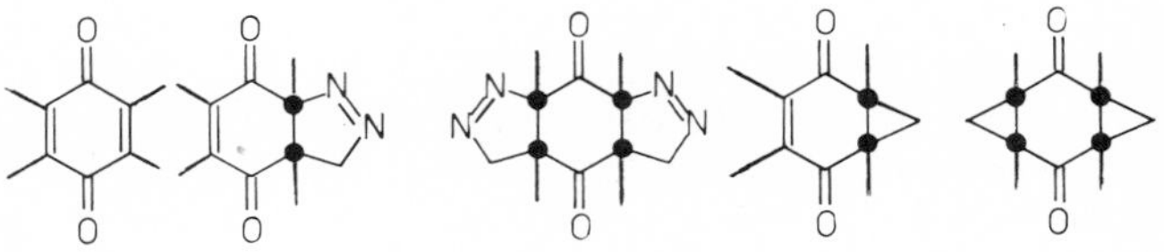

(67) (68) (69) (70) (71)

For synthetic purposes, organic chemists often employ the Smith-Simmons reaction of methylene iodide and a zinc-copper alloy for the cyclopropanation of olefins [74]. The reaction is stereospecific [75], and in fact, strongly regioselective; thus, it has been found that Δ^3-cyclopentenol (72) gives only the endo-bicyclic alcohol (73) [76].

---

71. W.H. Urry, J.R. Eiszner, and J.W. Wilt, J. Amer. Chem. Soc., 79, 918 (1957).
72. H.D. Roth, J. Amer. Chem. Soc., 93, 1527 (1971).
73. W.C. Howell, M. Ktenas, and J.M. MacDonald, Tetrahedron Lett., 1719 (1964).
74. H.E. Simmons and R.D. Smith, J. Amer. Chem. Soc., 81, 4256 (1959).
75. H.E. Simmons, E.P. Blanchard, and R.D. Smith, J. Amer. Chem. Soc., 86, 1347 (1964).
76. (a) S. Winstein, J. Sonnenberg, and L. de Vries, J. Amer. Chem. Soc., 81, 6523 (1959); (b) E.J. Corey and R.L. Dawson, J. Amer. Chem. Soc., 85, 1782 (1963).

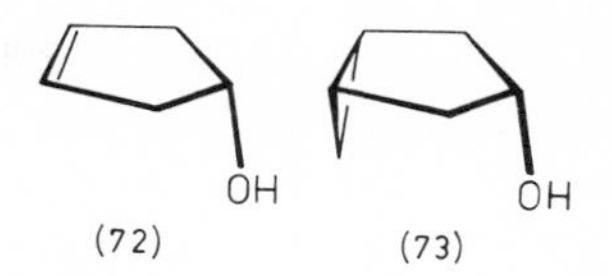

(72) (73)

Apparently the incipient methylene is complexed with the zinc, which in turn becomes coordinated with the oxygen. Solutions from which the metallic components have been filtered retain their activity for hours, although the additions to olefins are very rapid. This clearly indicates that these solutions do not contain free methylene, but a species at least loosely bound in some way, perhaps as $IZnCH_2I$. Methylene can be set free from methylene iodide, but this requires photolysis [77].

The chemistry of alkylcarbenes is not very different from that of methylene itself (aside perhaps in such obvious features as the possibility of internal hydrogen abstractions and insertions); however, new types of reactions become important if strain relief is present as a driving force. Cyclopropylidenes for instance undergo ring opening to the isomeric allenes [78]. These carbenes are readily obtained by the reaction of methyllithium with 1,1-dihalocyclopropanes, and since these compounds themselves are accessible via a carbene addition, this sequence amounts to a carbon atom insertion into olefins to give allenes. Fairly strained cyclic allenes such as (74) have been prepared this way [79], and a one-step two-carbon extension from triene (75) via (76) to pentaene (77) has been reported [80]. The reverse of the ring opening has also been observed in one instance: Irradiation of 1,2-cyclononadiene leads to tricyclononane (78) [81].

(74) (75) (76) (77) (78)

These reactions are very stereospecific and appear to involve a conrotation: Thus, the resolved precursors (79) and (80) both produce (81) in a state of very high optical purity [82].

(79) (80) (81)

77. D.C. Blomstrom, K. Herbig, and H.E. Simmons, J. Org. Chem., 30, 959 (1965).
78. W.R. Moore and H.R. Ward, J. Org. Chem., 27, 4179 (1962).
79. K.G. Untch, D.J. Martin, and N.T. Castellucci, J. Org. Chem., 30, 3572 (1965).
80. L. Skattebol, Tetrahedron Lett., 2175 (1965).
81. H.R. Ward and E. Karafiath, J. Amer. Chem. Soc., 90, 2193 (1968).
82. W.R. Moore and R.D. Bach, J. Amer. Chem. Soc., 94, 3148 (1972).

The opening of the ring and the departure of the leaving group are not concerted and the carbene is truly an intermediate; thus, it has been trapped in stereospecific fashion by added olefins [83]. In one instance an activated and suitably located internal CH bond leads to capture: (82) upon treatment with butyllithium leads to (83) (note: With retention), and the carbene ignores added olefins [84].

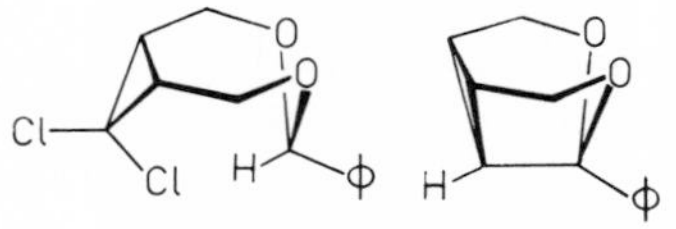

(82) (83)

Strain relief remains the principal reaction even when the carbene site is not in the ring itself. Cyclopropylcarbene itself gives, beside some bicyclobutane, the rearrangement products cyclobutene and 1,3-butadiene, and the fragmentation produces ethylene and acetylene [85]. The (n+3)-bicyclo[n.1.0]alkylcarbenes yield up to 50% acetylene [86], and on heating the tricyclic carbene precursor (84), Δ^3-cyclopentenylacetylene (85) is obtained as the exclusive product [87].

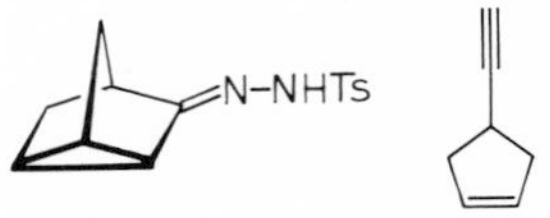

=N–NHTs

(84) (85)

An interesting extension of this led Bergman to investigate the carbene (86) which was obtained from two different precursors. The hope was that this carbene would be stabilized by homoaromatic character; however, fragmentation to give (87) is the only mode of reaction, and it is not yet clear whether some stabilization was achieved [88].

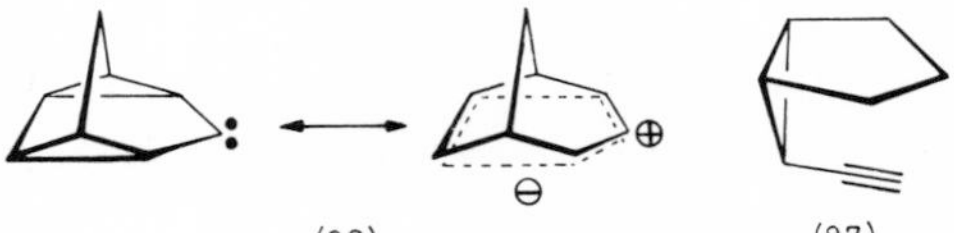

(86) (87)

83. W. M. Jones, M. H. Grasley, and D. G. Baarda, J. Amer. Chem. Soc., 86, 912 (1964).
84. B. Fraser-Reid, J. T. Brewer, and R. L. Sun, Tetrahedron Lett., 2779 (1969).
85. D. M. Lemal, F. Menger and G. W. Clark, J. Amer. Chem. Soc., 85, 2529 (1963).
86. W. Kirmse and K. H. Pook, Chem. Ber., 98, 4022 (1965).
87. S. J. Cristol and J. K. Harrington, J. Org. Chem., 28, 1413 (1963).
88. R. G. Bergman and V. J. Rajadhyaksha, J. Amer. Chem. Soc., 92, 2163 (1970).

16-6. ALKENYL- AND ALKYNYLCARBENES

Vinylcarbenes are interesting in that allylic resonance should stabilize them, and in that internal addition may lead to cyclopropenes.

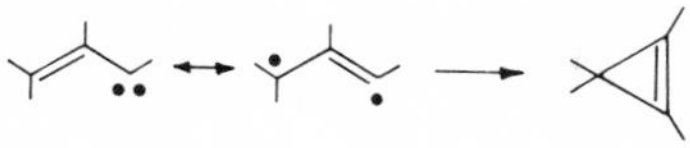

It is not obvious whether the conjugation should stabilize the singlet or the triplet state more, nor is it clear from experimental evidence what multiplicity is preferred. The reactions appear to be those of singlets, but this may be due to a lack of opportunity of the carbene to cool off before undergoing internal reaction, and to the unavoidable geometric constraints that will contribute to stereospecificity even if a diradical intermediate is involved. In one instance olefin and even benzene additions can be observed: If the diazodiene (88) is photolyzed in benzene, norcaradiene (89) is obtained.

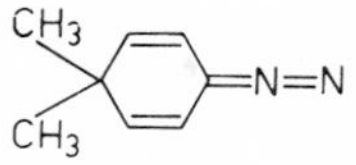

CH_3 CH_3 =N=N

(88)

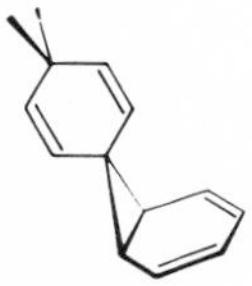

(89)

The additions are stereospecific, and they remain that way even if the substrate is present in low concentration in inert solvents (giving the carbene time to relax to its ground state configuration) and even if photosensitizers are used. Yet the ground state is a triplet, as is shown by the fact that at low temperatures an esr spectrum is observed [89]; evidently this carbene does not fit the Skell criterion of stereospecificity, with the triplet diradical intermediates undergoing spin inversion and closure more rapidly than internal rotations.

It is clear at any rate that the internal reactions do usually readily take place, and in fact, cyclopropenes are now most easily prepared that way [90]. For example, (90) in hot basic methanol solution forms 3,3-dimethylcyclopropene (91) [91], and the tetrachloride (92) readily reacts in basic solution to form cyclopropenone (93) [92].

---

89. M. Jones, A.M. Harrison, and K.R. Rettig, J. Amer. Chem. Soc., 91, 7462 (1969).
90. G.L. Closs and K.D. Krantz, J. Org. Chem., 31, 638 (1966).
91. G.L. Closs, L.E. Closs, and W.A. Böll, J. Amer. Chem. Soc., 85, 3796 (1963).
92. D.G. Farnum and P.E. Thurston, J. Amer. Chem. Soc., 86, 4206 (1964).

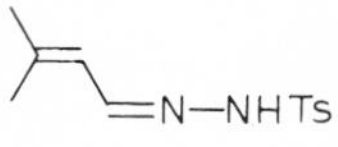

=N—NHTs Cl Cl Cl ϕ O ϕ OH

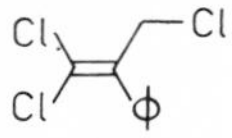

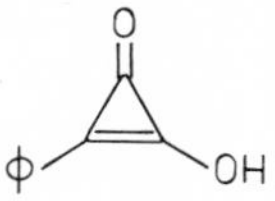

(90) (91) (92) (93)

These internal additions are sometimes reversible under certain conditions; thus, optically active 1,3-diethylcyclopropene (94) rapidly racemizes at 160-190° [93].

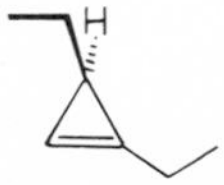

H

(94)

Highly strained small ring systems are in principle also possible with homo-allylic carbenes; thus, the diazobutene (95) upon photolysis gives bicyclobutane (96) beside much 1,3-butadiene [94].

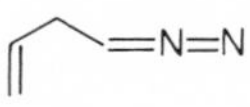

=N=N

(95) (96)

These findings inspired a great many searches for tetrahedrane by this route. Thus, the cyclopropenylcarbene precursor (97) was treated with butyllithium at -20° to give several dimers of the general structure (98); evidently an intermediate cyclobutadiene rather than a tetrahedrane resulted [95].

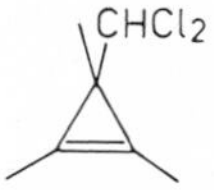

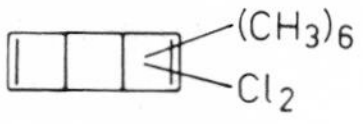

$CHCl_2$ $(CH_3)_6$ Cl_2

(97) (98)

Likewise, base catalyzed decomposition of (99) gives only methylacetylene, tolane and nitrogen as fragmentation products [96].

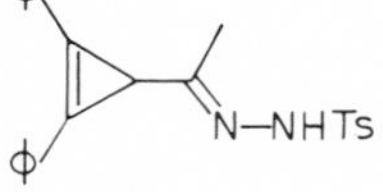

ϕ ϕ N—NHTs

(99)

93. E.J. York, W. Dittmar, J.R. Stevenson, and R.G. Bergman, J. Amer. Chem. Soc., 94, 2882 (1972).
94. D.M. Lemal, F. Menger, and G.W. Clark, J. Amer. Chem. Soc., 85, 2529 (1963).
95. G.L. Closs and V.N.M. Rao, J. Amer. Chem. Soc., 88, 4116 (1966).
96. N. Obata and I. Moritani, Bull. Chem. Soc. Japan, 39, 2250 (1966).

In view of what is now known of tetrahedrane, this substance is itself an unstable intermediate that decomposes by fragmentation, so that this reaction may have proceeded that way (cf. p. 861).

Another homoallylic carbene of great interest is the species (100) which might be stabilized by homoaromaticity. As in Bergman's case, (p. 576), it has been possible to generate this carbene but no information is available about its stability; the product is (101) [97]. Norbornadienylidene (102) is also known as an intermediate; its principal reaction is the formation of benzene and a free carbon atom (see Chapter 24).

(100) (101) (102)

Still another related species is cycloheptatrienylmethylene (103), one of the C_8H_8 valence isomers. It decomposes to benzene and acetylene, heptafulvene (104) and cyclooctatetraene [98].

(103) (104)

In the precursor (105), the double bond is still further away from the potential carbene site. Neither internal addition nor ring opening are possible because of the strain that would accompany these reactions, and the product obtained in greatest yield, (106), is due to intramolecular insertion [99].

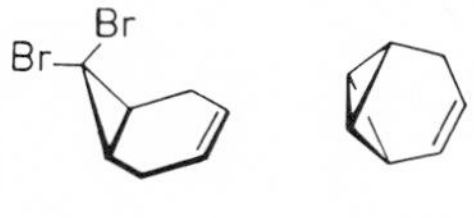

(105) (106)

Finally, propargylene (107), (derived from the corresponding diazo compound) is of interest in that it is a symmetrical resonance hybrid involving two identical structures (it is not known whether the two hydrogen atoms are colinear). It adds to olefins without stereospecificity, and hence it presumably is a triplet [100].

---

97. R.A. Moss, U.H. Dolling, and J.R. Whittle, Tetrahedron Lett., 931 (1971).
98. H.E. Zimmerman and L.R. Sousa, J. Amer. Chem. Soc., 94, 834 (1972).
99. G.W. Klumpp and J.J. Vrielink, Tetrahedron Lett., 539 (1972).
100. P.S. Skell and J. Klebe, J. Amer. Chem. Soc., 82, 247 (1960).

$$HC{\equiv}C{-}CHN_2 \longrightarrow HC{\equiv}C{-}\ddot{C}H \longleftrightarrow H\ddot{C}{-}C{\equiv}CH \longleftrightarrow H\dot{C}{=}C{=}\dot{C}H$$

(107)

16-7. ARYLCARBENES AND RELATED SPECIES

Arylcarbenes are stabilized by resonance like the allylic analogs. Diphenylcarbene has been generated by photolysis of diazodiphenylmethane in solid matrices and observed by esr and uv; at 77°K it is stable even in methanol and 2-methyl-2-pentene [101]. The triplet nature of this carbene has already been deduced [102] from its non-stereospecific additions and from the facts that it can abstract hydrogen and react with molecular oxygen to give benzophenone [103]. The spectra have been interpreted on the basis of a non-linear carbene; the bent nature of these species is perhaps best demonstrated by the fact that the low temperature matrix spectra of both α- and β-naphthylcarbenes reveal the presence of isomers such as (108) and (109) [104].

(108) (109)

The relative stability of these species has made it possible to study a number of dicarbenes. The irradiation of dilute solid solutions of bisdiazocompounds (110) gives rise to dicarbenes such as (111).

(110) (111)

101. (a) R.W. Murray, A.M. Trozzolo, E. Wasserman, and W.A. Yager, J. Amer. Chem. Soc., 84, 3213 (1962); (b) R.W. Brandon, G.L. Closs, and C.A. Hutchison, J. Chem. Phys., 37, 1878 (1962); (c) W.A. Gibbons and A.M. Trozzolo, J. Amer. Chem. Soc., 88, 172 (1966).
102. R.M. Etter, H.S. Skovronek, and P.S. Skell, J. Amer. Chem. Soc., 81, 1008 (1959).
103. W. Kirmse, L. Horner, and H. Hoffmann, Ann., 614, 19 (1958).
104. A.M. Trozzolo, E. Wasserman, and W.A. Yager, J. Amer. Chem. Soc., 87, 129 (1965).

Singlet, triplet or quintet configurations are now possible; the epr spectrum reveals the presence of two unpaired electrons so that the best structure is (112) [105].

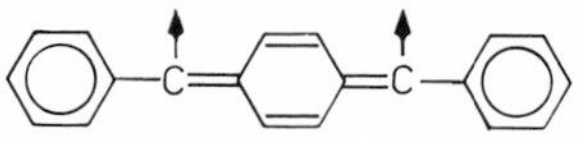

(112)

If the central ring is meta-substituted, a quintet state is indicated by esr [106] and by chemical evidence. Thus, it was found that when bisdiazocompound (113) is photolyzed with cis-2-butene, the products are only 2% cis,trans- and 98% cis,cis- (one obtains anti-anti- configuration (114), anti-syn- (115), and even the syn-syn- isomer); if a very dilute solution in cyclohexane was photolyzed, cis,trans- products added up to 25% and even 7% of trans,trans- adducts were formed [107].

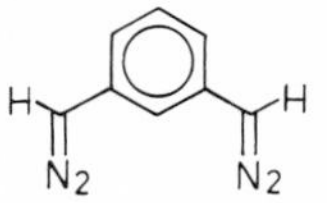

(113) 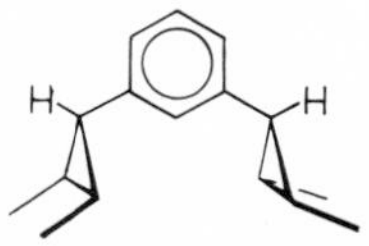 (114) 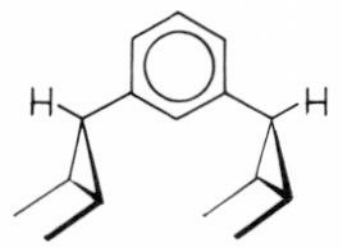 (115)

Dimesityldiazomethane (116) has been studied in the hope that steric inaccessibility would stabilize the carbene. Photolysis at -75° in cyclopentane gave tetramesitylethylene; if oxygen is present, dimesitylketone is obtained instead. No ketone is obtained if the oxygen is admitted an hour later; evidently the carbene does not survive that long at -75° [108].

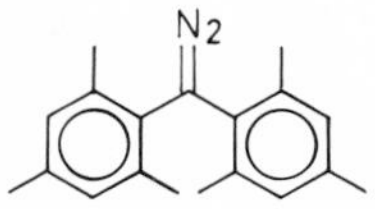

(116)

A very interesting development in phenylcarbene chemistry has occurred in the last few years: If they are generated at high temperature by thermolysis of diazocompounds, the carbenes produced are capable of degenerate rearrangements

105. R.W. Murray and A.M. Trozzolo, J. Org. Chem., 26, 3109 (1961).
106. (a) K. Itoh, Chem. Phys. Lett., 1, 235 (1967); (b) E. Wasserman, R.W. Murray, W.A. Yager, A.M. Trozzolo, and G. Smolinsky, J. Amer. Chem. Soc., 89, 5076 (1967).
107. S.I. Murahashi, Y. Yoshimura, Y. Yamamoto, and I. Moritani, Tetrahedron, 28, 1485 (1972).
108. H.E. Zimmerman and D.H. Paskovich, J. Amer. Chem. Soc., 86, 2149 (1964).

before they deactivate themselves by product formation. Thus, when o-, m- or p-tolylcarbenes are generated by thermolysis of the corresponding diazocompounds at 420°, all produce similar mixtures of benzocyclobutane and styrene [109]. In the main, two mechanisms for this reaction have been considered. Mechanism (a) is readily ruled out by labelling experiments.

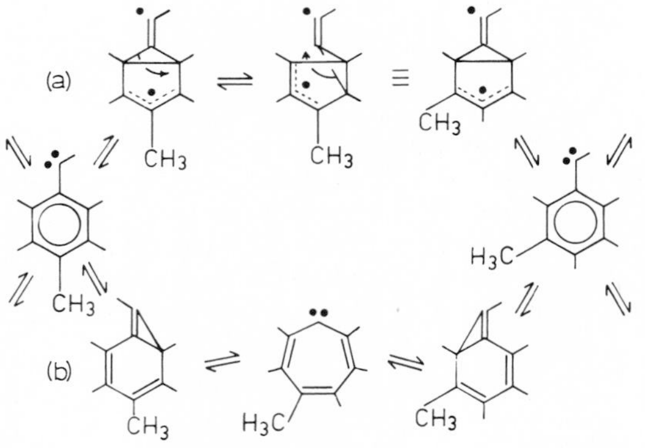

(a)

(b)

If the carbene carbon atom of the p-tolylcarbene is C^{13} as in (117), the cyclobutane obtained has the label exclusively in the 4-position (118) [110]; only at much higher temperatures (~600°) does scrambling occur [111].

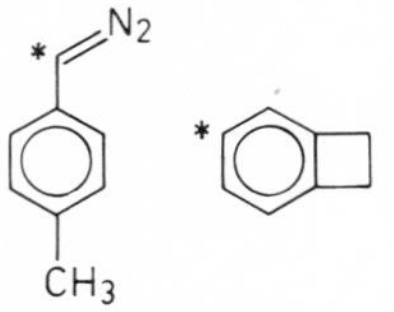

(117) (118)

In mechanism b, the bicyclic triene intermediates may have zwitterionic or diradical character such as (119) or (120), or they may merely represent transition states [112].

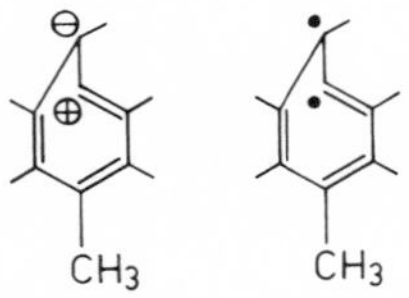

(119) (120)

109. W.J. Baron, M. Jones, and P.P. Gaspar, J. Amer. Chem. Soc., 92, 4739 (1970).
110. E. Hedaya and M.E. Kent, J. Amer. Chem. Soc., 93, 3283 (1971).
111. W.D. Crow and M.N. Paddon-Row, J. Amer. Chem. Soc., 94, 4746 (1972).
112. T. Mitsuhashi and W.M. Jones, J. Amer. Chem. Soc., 94, 677 (1972).

Whatever the details are in this series of events, the only atom moving around is the carbon atom to which the original carbene site was bound. Direct evidence for the tropylidene intermediate has been found in several experiments; for example the diazocompounds (121) and (122) give very similar mixtures of *cis*- and *trans*-stilbene and heptafulvalene (123) at 300° [113].

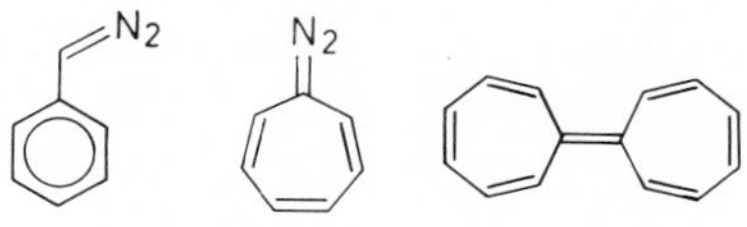

N_2 N_2

(121) (122) (123)

These isomerizations do not occur at lower temperatures although one exception is known: The diazocompound (124) forms product (125) when heated to only 80° in cyclohexane [114].

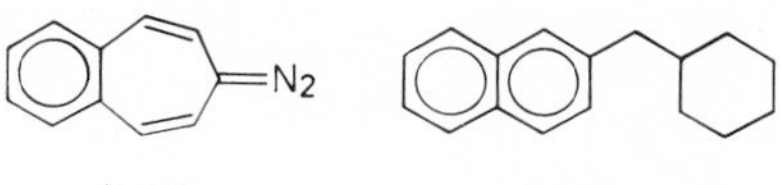

$=N_2$

(124) (125)

When tropylidene itself is generated at lower temperatures, heptafulvalene is the only product [115]. It is not captured by simple alkenes, but it does add stereospecifically to electron-poor olefins such as maleonitrile and fumaronitrile, (126) and (127), respectively; no change is observed if high dilution or heavy atom solvents are used [116]. This carbene furthermore adds to mixtures of *p*-substituted styrenes in such a way that ρ can be deduced to be +1.05 [117]. These findings suggest that tropylidene is a nucleophilic singlet, the electron-donating character being imparted by a resonance contribution from (128) [118]. Similar observations have been made in the decomposition of the cyclopropenylidene precursor (129) in basic medium [119].

---

113. (a) C. Wentrup and K. Wilczek, *Helv. Chim. Acta*, *53*, 1459 (1970); (b) J.A. Myers, R.C. Joines, and W.M. Jones, *J. Amer. Chem. Soc.*, *92*, 4740 (1970).
114. K.E. Krajca, T. Mitsuhashi, and W.M. Jones, *J. Amer. Chem. Soc.*, *94*, 3661 (1972).
115. W.M. Jones and C.L. Ennis, *J. Amer. Chem. Soc.*, *89*, 3069 (1967).
116. W.M. Jones, B.N. Hamon, R.C. Joines, and C.L. Ennis, *Tetrahedron Lett.*, 3909 (1969).
117. L.W. Christensen, E.E. Waali, and W.M. Jones, *J. Amer. Chem. Soc.*, *94*, 2118 (1972).
118. W.M. Jones and C.L. Ennis, *J. Amer. Chem. Soc.*, *91*, 6391 (1969).
119. W.M. Jones, M.E. Stowe, E.E. Wells, and E.W. Lester, *J. Amer. Chem. Soc.*, *90*, 1849 (1968).

(126) (127) (128) (129)

Cyclopentadienylidene (130) behaves quite differently. Its groundstate is a triplet, as has been deduced from an esr spectrum obtained at 77°K [120], but the singlet may have a contribution from structure (131).

(130) (131)

The reactions of diazocyclopentadiene upon photolysis show that the carbene addition is highly unselective but stereospecific, and remains so even on dilution with an inert solvent [121]. In the presence of pyridine, ylids such as (132) are readily formed [122].

(132)

Benzo annelation should stabilize this carbene somewhat. Indenylidene (133) has been generated but not yet studied in great detail [123]; but fluorenylidene (134) has been observed [124] to change its reaction with 1,1-dicyclopropylethylene upon dilution with decalin.

(133) (134)

---

120. E. Wasserman, L. Barash, A.M. Trozzolo, R.W. Murray, and W.A. Yager, J. Amer. Chem. Soc., 86, 2304 (1964).
121. R.A. Moss and J.R. Przybyla, J. Org. Chem., 33, 3816 (1968).
122. I.B.M. Band, D. Lloyd, M.I.C. Singer, and F.I. Wasson, Chem. Commun., 544 (1966).
123. D. Rewicki and C. Tuchscherer, Angew. Chem., Int. Ed. Engl., 11, 44 (1972).
124. N. Shimizu and S. Nishida, Chem. Commun., 389 (1972).

The product mixture of 10:1 (135) and (136) changes to a 1:3 ratio upon 30-fold dilution; product (136) is of course traceable as a radical product.

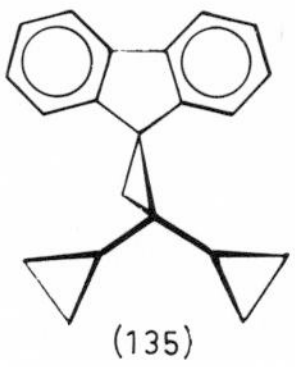

CHC_2H_5

(135) 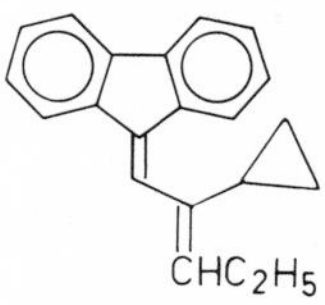 (136)

The stereospecificity in the additions to the 2-butenes decreases upon dilution with hexafluorobenzene, and it increases if 1,3-butadiene or oxygen are added (they presumably remove the triplet) [125]. Maleate and fumarate esters give identical mixtures [126] - all of these observations agree with the ground state triplet found by low temperature esr [127]. The ambivalent character of phenalene makes the carbene (137) an interesting objective. The generation of this intermediate has been reported, but too little is known as yet about its chemistry to warrant any conclusions regarding its relative stability or multiplicity [128].

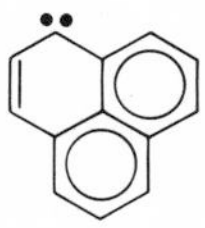

(137)

Pyran-2-ylidene (138) may well be a greatly stabilized carbene; it has not yet been generated in free form, however [129].

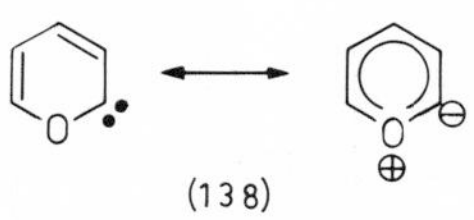

(138)

Another interesting approach to stabilization is the use of one electron donating and one withdrawing substituent, such as in (139). This allene is known, but its

125. M. Jones and K. R. Rettig, J. Amer. Chem. Soc., 87, 4013, 4015 (1965).
126. E. Funakubo, I. Moritani, T. Nagai, S. Nishida, and S. Murahashi, Tetrahedron Lett., 1069 (1963).
127. A. M. Trozzolo, R. W. Murray, and E. Wasserman, J. Amer. Chem. Soc., 84, 4990 (1962).
128. I. Murata, T. Nakazawa, and S. Yamamoto, Tetrahedron Lett., 2749 (1972).
129. (a) R. Gleiter and R. Hoffmann, J. Amer. Chem. Soc., 90, 5457 (1968); (b) C. W. Rees and E. von Angerer, Chem. Commun., 420 (1972).

carbenic character is nil; the vinyl protons do not appear at low field in the nmr, and solvent polarity does not affect the uv spectrum [130]. It is of course possible that the opposite behavior may characterize less highly annelated analogs.

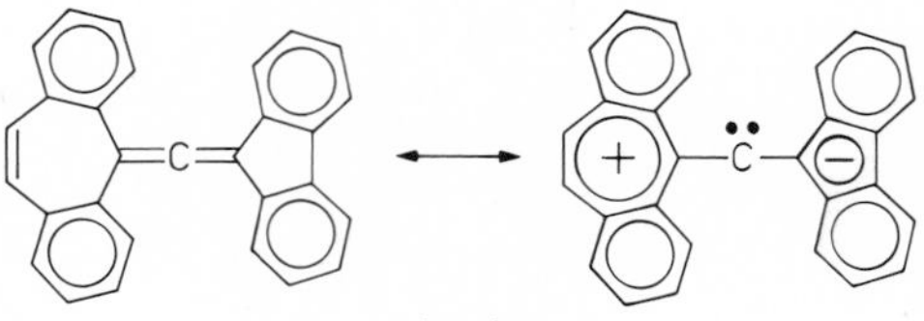

(139)

It might be pointed out that virtually all of the arylcarbene studies have made use of diazocompounds or tosylhydrazones as precursors. It turns out that the base catalyzed hydrolysis of halides does produce the corresponding anions, but these usually displace halide ion from another substrate molecule on the road to stilbene products [131] (cf. also p. 658). The kinetics are accordingly third order. In some instances certain substituents in the aryl groups lead to second order kinetics and hence carbenes may then well be involved, but trapping by means of olefins has not succeeded in these cases [132].

16-8. KETOCARBENES AND RELATED SPECIES

Ketocarbenes represent an interesting member in the series. They are often pictured as intermediates (140) in the Wolff rearrangement of α-diazoketones (141); upon warming or photolysis, nitrogen evolves and products form which can be traced to an intermediate ketene (142). For example, if the reaction is done in an alcohol, the ester (143) will obtain.

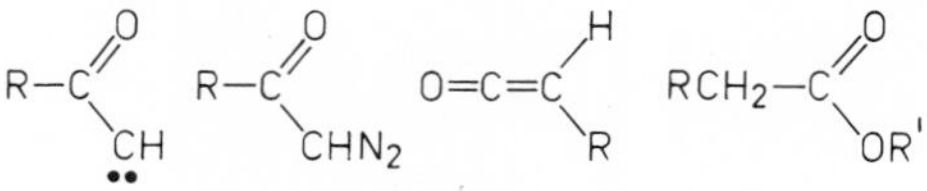

$R-C(=O)-\ddot{C}H$ (140) $R-C(=O)-CHN_2$ (141) $O=C=C(H)R$ (142) $RCH_2-C(=O)OR'$ (143)

Such products are not reminescent of carbene reactions, however, and it is generally believed that the loss of nitrogen and the 1,2-shift are concerted, so that the ketocarbene is not a discreet intermediate. But there are exceptions. For example, the diazo derivative of cyclopropenylglyoxal (144) upon irradiation gives the bicyclic ketone (145) [133].

---

130. L. Salisbury, J. Org. Chem., 37, 4075 (1972).
131. A. Ledwith and Y. S. Lin, Chem. Ind. (London), 1867 (1964).
132. D. Bethell and A. F. Cockerill, Proc. Chem. Soc., 283 (1964).
133. S. Masamune, J. Amer. Chem. Soc., 86, 735 (1964).

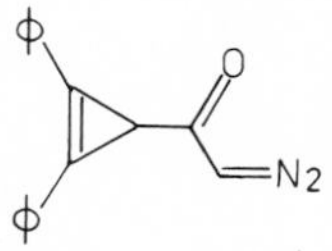

(144) 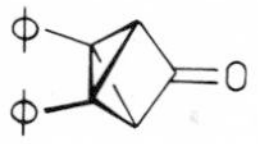 (145)

In some Wolff reactions small amounts of external carbene addition products are observed; thus, the diazo derivative (146) of phenylglyoxal, if warmed in the presence of 5-decyne gives a 20% yield of the adduct (147) [134].

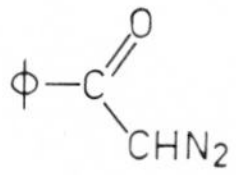

(146) 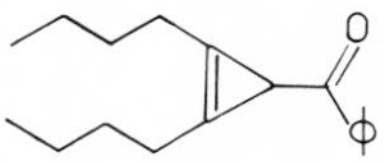 (147)

A particularly interesting example of ketocarbenes is the species (148), derivable from the zwitterionic diazonium oxide (149). In aqueous solution irradiation of (149) gives the Wolff-product (150); however, the carbene can be diverted.

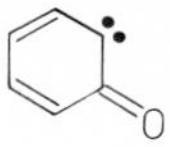

(148) 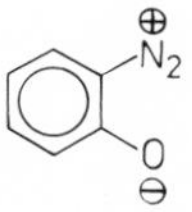 (149) 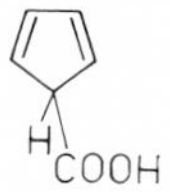 (150)

For instance, when tolane is present, 1,3-cycloaddition leads to the benzofuran (151) [135].

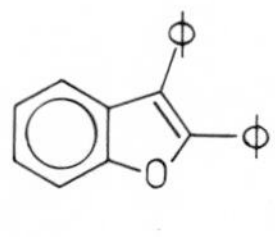

(151)

Because of the obvious analogy with allylic carbenes which rearrange to cyclopropenes, one might wonder whether ketocarbenes even rearrange to oxirenes. No such structures have as yet been detected, but the search has been stimulated by the fact that the photolysis of 1,2,3-thiadiazole (152) produces thiirene (153) which was noted to have a half life of several seconds [136].

---

134. N. Obata and I. Moritani, Bull. Chem. Soc. Japan, 39, 1975 (1966).
135. R. Huisgen, G. Binsch, and H. König, Chem. Ber., 97, 2868 (1964).
136. O.P. Strausz, J. Font, E.L. Dedio, P. Kebarle, and H.E. Gunning, J. Amer. Chem. Soc., 89, 4805 (1967).

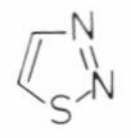

(152)  (153)

There is good evidence that oxirenes are intermediates in at least some photo-Wolff rearrangements. Thus, the C^{13}-labeled diazoketone (154) upon photolysis fragments to nitrogen, carbon monoxide and propylene; half of the label is found in the propylene and half in the carbon monoxide.

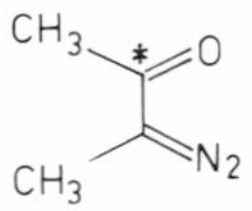

(154)

This means that the ketocarbene equilibrates with dimethyloxirene (155) before going on to dimethylketene (156) and fragmentation [137].

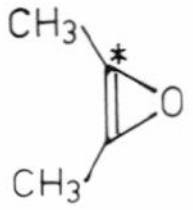

(155) 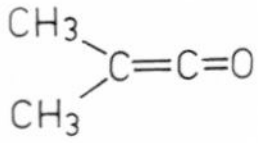 (156)

However, the same experiment with the C^{14}-labeled α-diazohomoadamantanone (157) yielded only the acid (158): No radioactivity could be found in the carbon dioxide upon decarboxylation, and hence the oxirene (159) did not intervene in any way in this reaction [138].

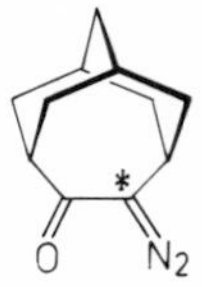

(157) 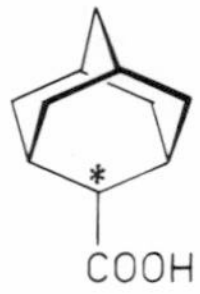 (158) 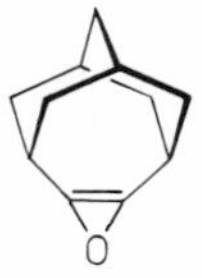 (159)

---

137. (a) I. G. Csizmadia, J. Font, and O. P. Strausz, J. Amer. Chem. Soc., 90, 7360 (1968). Oxirenes may be produced by ^3P oxygen atoms and acetylenes as intermediates toward ketenes; see (b) I. Haller and G. C. Pimentel, J. Amer. Chem. Soc., 84, 2855 (1962).
138. Z. Majerski and C. S. Redvanly, Chem. Commun., 694 (1972).

In several additional experiments of this sort the results have generally been that partial scrambling occurred [139]. The activation energies for scrambling and for rearrangement thus seem to be similar. It was observed furthermore that only the singlet carbene rearranges at all; benzophenone sensitization stopped scrambling altogether [140].

There are also some peculiar features in the decomposition of α-diazoesters. Wolff rearrangement occurs only in the photodecomposition; the only exception is methyl diazomalonate (160) - and then only if the decomposition is carried out at 500°. At that temperature methyl pyruvate (161) becomes an important product, presumably via steps (162)-(164) [141].

$COOCH_3$–C(=N_2)–$COOCH_3$ (160) $COOCH_3$–CO–CH_3 (161) $COOCH_3$–C:–$COOCH_3$ (162) $COOCH_3$–$COCH_3$=CO (163) $COOCH_3$–•$COCH_3$• (164)

There is little doubt that carbenes are intermediates in the photodecompositions; they are readily trapped by olefins, and they insert in CH-bonds. Dilution and sensitization greatly diminish the stereospecificity, so that the ground states are triplets [142]. A curious aspect of these photogenerated carbenes is that they seem relatively reluctant to undergo internal additions or insertions. For example, when the allyl diazoester (165) is photolyzed in cyclohexane, (166) is the only product [143];

N_2=CH–C(=O)–O–$CH_2CH=CH_2$ (165) C_6H_{11}–CH_2–C(=O)–O–$CH_2CH=CH_2$ (166)

---

139. (a) D.E. Thornton, R.K. Gosavi, and O.P. Strausz, J. Amer. Chem. Soc., 92, 1768 (1970); (b) G. Frater and O.P. Strausz, J. Amer. Chem. Soc., 92, 6654 (1970); (c) R.L. Russell and F.S. Rowland, J. Amer. Chem. Soc., 92, 7508 (1970); (d) S.A. Matlin and P.G. Sammes, J. Chem. Soc., Perkin I, 2623 (1972); (e) K.P. Zeller, H. Meier, H. Kolshorn, and E. Müller, Chem. Ber., 105, 1875 (1972).
140. S.A. Matlin and P.G. Sammes, Chem. Commun., 11 (1972).
141. D.C. Richardson, M.E. Hendrick, and M. Jones, J. Amer. Chem. Soc., 93, 3790 (1971).
142. M. Jones, A. Kulczycki, and K.F. Hummel, Tetrahedron Lett., 183 (1967).
143. W. Kirmse and H. Dietrich, Chem. Ber., 98, 4027 (1965).

similarly, when the t-amyl ester (167) is photolyzed in this solvent, one obtains a 95% yield of (168) and only 5% γ-lactones such as (169) [144].

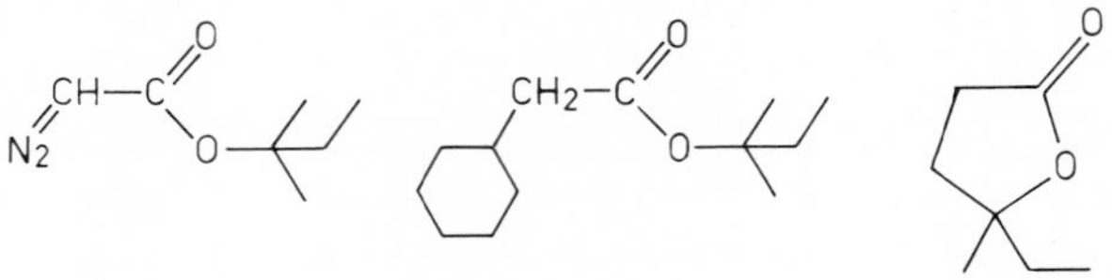

(167) (168) (169)

These facts are probably the result of a proximity effect: The short-lived carbene is not in the necessary cyclic conformation often enough. Thus, in N, N-disubstituted α-diazoamides one of the alkyl groups must, because of the amide resonance, lie fairly close to the carbene center; if (170) is photolyzed in dioxane, about equal amounts of (171) and (172) are obtained as the only products [145].

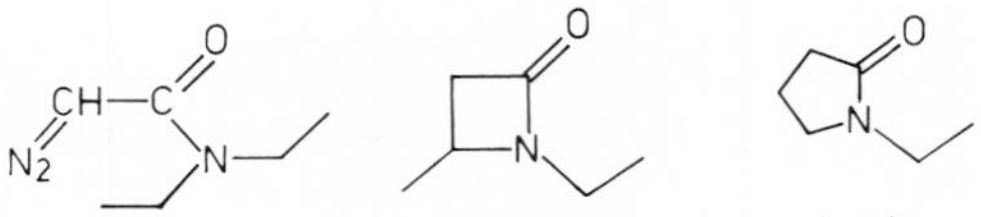

(170) (171) (172)

Likewise, the sulfonylcarbene derived from (173) gives fair yields of internal products such as (174) and (175) [146].

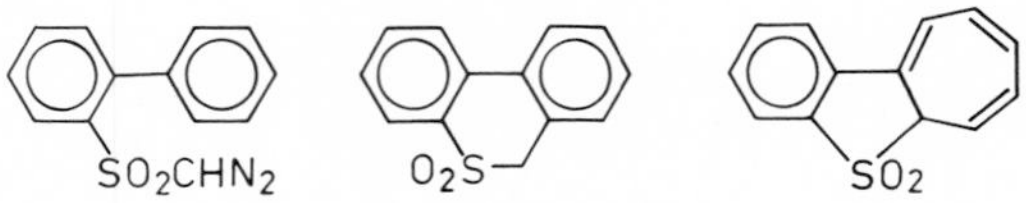

(173) (174) (175)

Dicyanocarbene has been generated by the thermolysis and photolysis of diazomalononitrile (176) [147], as well as by the reactions of triethylamine with bromomalononitrile (177) [148].

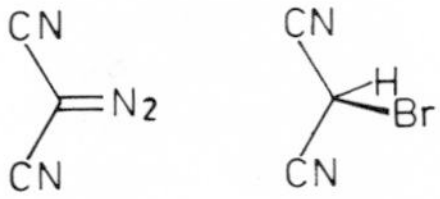

(176) (177)

---

144. W. Kirmse, H. Dietrich, and H.W. Bücking, Tetrahedron Lett., 1833 (1967).
145. R.R. Rando, J. Amer. Chem. Soc., 92, 6706 (1970).
146. R.A. Abramovitch, V. Alexanian, and E.M. Smith, Chem. Commun., 893 (1972).
147. E. Ciganek, J. Amer. Chem. Soc., 88, 1979 (1966).
148. J.S. Swenson and D.J. Renaud, J. Amer. Chem. Soc., 87, 1394 (1965).

This carbene is a ground state triplet as is obvious from its addition to double bonds, which becomes less stereospecific upon dilution of the reaction mixtures, and from its low temperature spectrum [149]. This carbene provides us with one of the rare examples of a 1,4-addition of a carbene: With cyclooctatetraene, a 20% yield of adduct (178) is obtained [150].

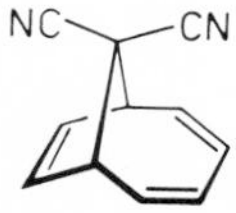

NC CN

(178)

16-9. VINYLIDENES AND THEIR ANALOGS

1,1-Dihaloolefins react with methyllithium in the presence of unsaturated substrates to give adducts such as (179) [151];

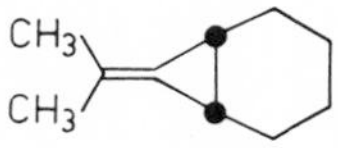

CH_3 CH_3

(179)

thus, the [3]-radialene (180) has been prepared that way from butatriene (181) [152].

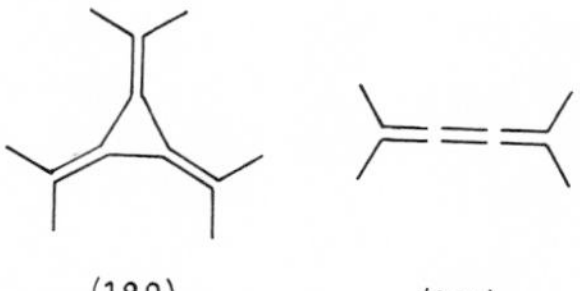

(180) (181)

The addition to styrenes has a ρ-value of -3.4 and hence this carbene is a much stronger electrophile than dichlorocarbene for example [153].

---

149. E. Wasserman, L. Barash, and W.A. Yager, J. Amer. Chem. Soc., 87, 2075 (1965).
150. A.G. Anastassiou, R.P. Cellura, and E. Ciganek, Tetrahedron Lett., 5267 (1970).
151. (a) H.D. Hartzler, J. Amer. Chem. Soc., 86, 526 (1964); (b) M.S. Newman and A.O.M. Okorodudu, J. Amer. Chem. Soc., 90, 4189 (1968).
152. G. Köbrich, H. Heinemann, and W. Zündorf, Tetrahedron, 23, 565 (1967).
153. M.S. Newman and T.B. Patrick, J. Amer. Chem. Soc., 91, 6461 (1969).

Isocyanides and carbon monoxide formally also fall in this group, but they are stabilized to such an extent by an unshared pair at the neighboring atom that they are usually not considered among the carbenes. Still, if a sufficiently tempting target is available, the carbenic nature of the isocyanide asserts itself: Thus, phenyl isocyanide adds to the electron-rich and highly strained cycloheptyne (182) to give (183) [154].

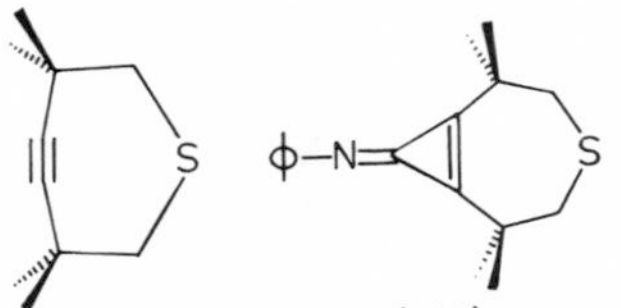

S ϕ—N= S

(182) (183)

Cumulenic carbenes are known. For example, 1-chloro-3-methylbuta-1,2-diene (184) and 3-chloro-3-methylbut-1-yne (185) produce carbene (186) when treated with t-butoxide; this carbene can be trapped to give allenic products such as (187).

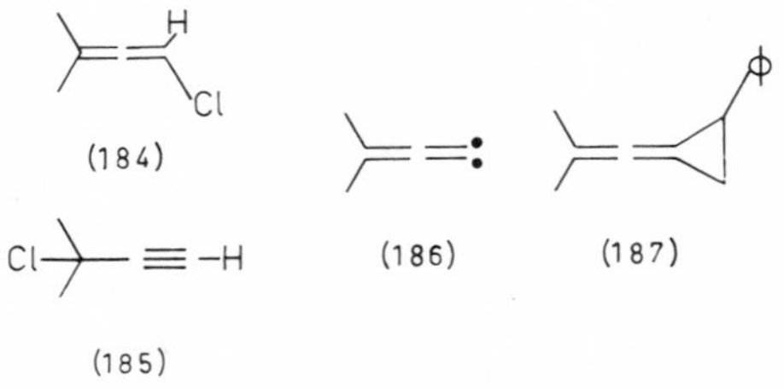

H Cl

(184)

Cl ≡-H

(185)

(186) (187)

If a mixture of substrates is used, both reagents give the same product mix, and all products form stereospecifically. The reactions of (184) and (185) with aqueous alcoholic base has been studied with great care [155]. Both reactions in D_2O are accompanied by rapid D-exchange. The principal products are the acetylenic alcohol (188) and ether (189). Small amounts (~1%) of the allenic isomer form when the acetylenic substrates solvolyze. The ambident nature of the carbene is indicated by the fact that with methyl malonate, both (190) and (191) are obtained.

154. A. Krebs and H. Kimling, Angew. Chem., Int. Ed. Engl., 10, 409 (1971).
155. (a) G. F. Hennion and D. E. Maloney, J. Amer. Chem. Soc., 73, 4735 (1951); (b) V. J. Shiner and J. W. Wilson, J. Amer. Chem. Soc., 84, 2402 (1962); (c) V. J. Shiner and J. S. Humphrey, J. Amer. Chem. Soc., 89, 622 (1967); (d) H. D. Hartzler, J. Amer. Chem. Soc., 83, 4990, 4997 (1961) and J. Org. Chem., 29, 1311 (1964); (e) W. J. le Noble, J. Amer. Chem. Soc., 87, 2434 (1965); (f) W. J. le Noble, Y. Tatsukami, and H. F. Morris, J. Amer. Chem. Soc., 92, 5681 (1970); (g) A. F. Bramwell, L. Crombie, and M. H. Knight, Chem. Ind. (London), 1265 (1965); (h) G. F. Hennion and J. F. Motier, J. Org. Chem., 34, 1319 (1969).

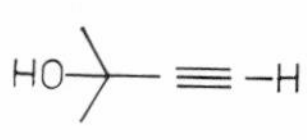

(188) 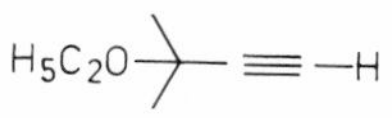 (189) 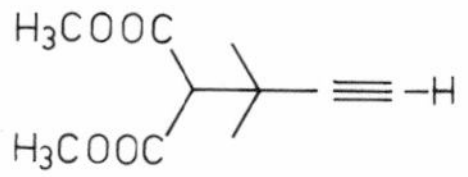 (190) 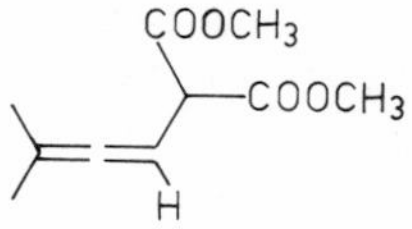 (191)

The activation volume for these reactions is only barely larger than zero, and a considerable zwitterionic character (192) for the intermediate carbene is thereby again indicated.

(192)

It was also learned that during the hydrolysis of (185), the addition of chloride ion has much less than the expected effect on the yield of isomer (184), so that the latter is apparently formed not by capture of free chloride ion but from a carbene-anion pair intermediate (193) which preceded the free carbene.

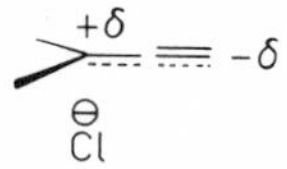

(193)

The di-*t*-butylpropargyl acetate (194) similarly reacts with *t*-butoxide to give a carbene (195) that can be captured with olefins. In the absence of olefins, it reacts with the precursor anion to give the cumulene (196). On melting at 185°, the latter product undergoes a remarkable cycloaddition to form (197) [156].

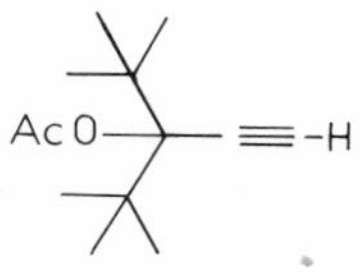

(194) 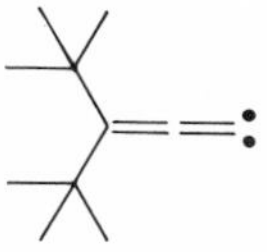 (195) 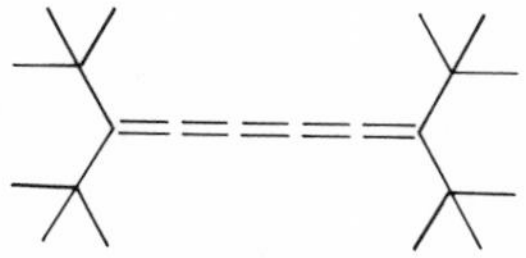 (196)

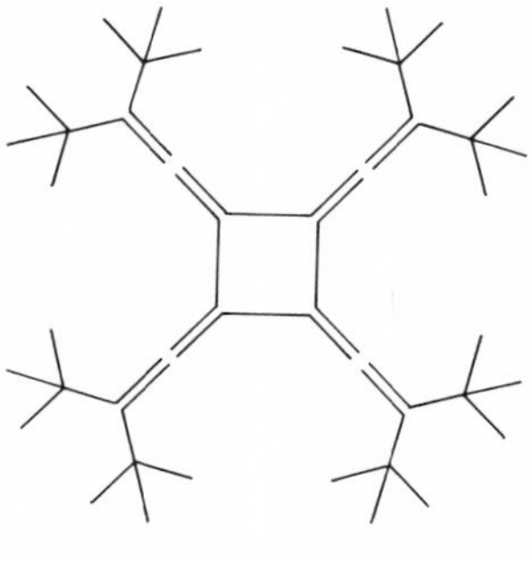

(197)

---

156. H. D. Hartzler, *J. Amer. Chem. Soc.*, **88**, 3155 (1966) and **93**, 4527 (1971).

A related species is diazomethylene, C=N=N, formed by irradiation of cyanogen azide. The esr spectrum measured at 4°K shows this triplet species to be linear [157].

16-10. AZA- AND OXAMETHYLENES

Since electron deficiency and the accessibility of the unfilled or half filled orbitals are the features that make carbenes reactive, it seems natural that chemists should attempt to stabilize them by electron donation via resonance. The synthesis of stable divalent carbon compounds would be an important result of such research; another useful result would be the likelihood that such carbenes, even if not isolable, could at least be generated under unusually gentle conditions.

It had of course been noticed early that dihalocarbenes are much less reactive than would be expected from a purely inductive substituent effect, and it may be expected that nitrogen and oxygen substitution will further stabilize carbenes - as it does in carbon monoxide and isocyanides. Such resonance stabilization can be readily demonstrated. For example, when (dimethoxymethyl)trimethoxysilane (198) is heated with cyclohexene at 125°, tetramethoxysilane (199) and 7-methoxynorcarane (200) are formed. The rate of the reaction is independent of the nature of the added olefin, and all olefins add stereospecifically [158]. Methoxymethyltrimethoxysilane (201) does not release CH_2 even at 300°.

$(CH_3O)_2CH—Si(OCH_3)_3$ (198) $Si(OCH_3)_4$ (199) [bicyclic]—OCH_3 (200) $CH_3OCH_2—Si(OCH_3)_3$ (201)

Carbenes such as dimethoxy- and bisdimethylaminocarbene are readily released by the thermolysis of the appropriately 7,7-substituted norbornadienes (202) [159], quadricyclanes (203) [160] and cycloheptatrienes (204) [161]; with substituents other than amino- or methoxy, these reactions proceed either very slowly or not at all.

---

157. E. Wasserman, L. Barash, and W.A. Yager, J. Amer. Chem. Soc., 87, 2075 (1965).
158. W.H. Atwell, D.R. Weyenberg and J.G. Uhlmann, J. Amer. Chem. Soc., 91, 2025 (1969).
159. (a) R.W. Hoffmann and H. Häuser, Tetrahedron, 21, 891 (1965); (b) D.M. Lemal, R.A. Lovald, and R.W. Harrington, Tetrahedron Lett., 2779 (1965).
160. D.M. Lemal, E.P. Gosselink, and S.D. McGregor, J. Amer. Chem. Soc., 88, 582 (1966).
161. (a) A.P. ter Borg, E. Razenberg, and H. Kloosterziel, Rec. Trav. Chim. Pays-Bas, 84, 1230 (1965); (b) R.W. Hoffmann and J. Schneider, Tetrahedron Lett., 4347 (1967).

(202) (203) (204)

In view of this stability, it would seem reasonable to assume that tetraaminoethylenes might readily dissociate, and it has indeed been found that the normally colorless species (205) sometimes forms colored solutions, and that it reacts with an electrophilic olefin such as tetracyanoethylene to give the adduct (206) [162].

(205) (206)

Alas, it was learned later that a mixture of (205) and the tetra-p-tolyl analog even under drastic conditions does not give the di-p-tolyl cross-over product expected from such facile dissociation [163]. In fact, the compound thought to be (206) later turned out to be a type of charge transfer salt (207) [164].

(207)

Sulfur is also effective as a carbene stabilizer; thus, the tosylhydrazones (208) readily decompose in warm basic solution to give the formal carbene dimers (209) [165].

$(CH_3S)_2C{=}N{-}NHTs$ (208) $(CH_3S)_2C{=}C(SCH_3)_2$ (209)

---

162. H.W. Wanzlick, Angew. Chem., Int. Ed. Engl., 1, 75 (1962).
163. D.M. Lemal, R.A. Lovald, and K.I. Kawano, J. Amer. Chem. Soc., 86, 2518 (1964).
164. H.W. Wanzlick and B. Lachmann, Z. Naturforsch., B24, 574 (1969).
165. U. Schöllkopf and E. Wiskott, Angew. Chem., Int. Ed. Engl., 2, 485 (1963).

The reaction is not diverted by cyclohexene, but 1,1-diethoxyethylene gives a low yield of the adduct (210), and triphenylphosphine gives a quantitative yield of the ylid (211) [166].

$C(OC_2H_5)_2$ / $C(SCH_3)_2$ (cyclopropane) $(CH_3S)_2C{=}P\phi_3$

(210) (211)

The tris(phenylmercapto)methyllithium salt (212) in fact seems to be an example of a substance in reversible equilibrium with a carbene (213) at room temperature. At 20°, it slowly decomposes to dimer (214); this reaction is evidently self-inhibited; it can be slowed down also by added lithium thiophenoxide and not by other lithium salts. A mixture of (212) and (215) rapidly leads to the formation of all four possible cross-product salts, and if an olefin is added, cyclopropanes such as (216) form [167].

$(\phi\text{-S})_3CLi$ $(\phi\text{-S})_2C{:}$ $(\phi\text{-S})_2C{=}C(\text{S-}\phi)_2$ $(CH_3\text{-}\phi\text{-S})_3CLi$ $(\phi\text{-S})_2C$(cyclopropane)

(212) (213) (214) (215) (216)

Another dithiacarbene, the species (217) which is generated from tosylhydrazone (218) is of interest in that it undergoes a sigmatropic shift to form (219) [168].

TsNH-N

(217) (218) (219)

Silicon substitution appears to be much less effective: Trimethylsilyldiazomethane is thermally quite stable and catalysis by cuprous chloride is required for its decomposition. The carbene so generated adds readily to cyclohexene [169].

16-11. CARBENE COMPLEXES [170]

A rather different kind of stabilization is that by certain metals. Since so many metalcarbonyls are known, one might perhaps expect that metals might bond

---

166. D. M. Lemal and E. H. Banitt, Tetrahedron Lett., 245 (1964).
167. D. Seebach, Chem. Ber., 105, 487 (1972).
168. J. E. Baldwin and J. A. Walker, Chem. Commun., 354 (1972).
169. D. Seyferth, A. W. Dow, H. Menzel, and T. C. Flood, J. Amer. Chem. Soc., 90, 1080 (1968).
170. For review, see D. J. Cardin, B. Cetinkaya, and M. F. Lappert, Chem. Rev., 72, 545 (1972).

with carbenes other than CO. Fischer and Maasböl [171] found that successive treatments of tungsten hexacarbonyl with methyllithium, acid and diazomethane gave a compound $W(CO)_5(MeCOMe)$ that has no carbonyl absorption in the infrared; they considered it a pentacarbonyltungsten complex of methoxymethylcarbene. X-ray diffraction measurement [172] has shown the structure to be as shown below.

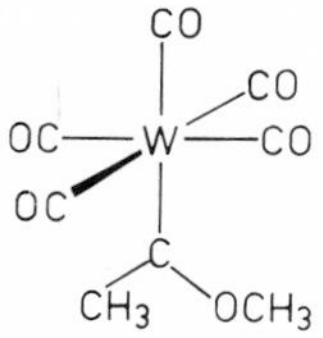

Even compounds clearly involving tetravalent carbon atoms only such as (220) are referred to as carbene complexes [173].

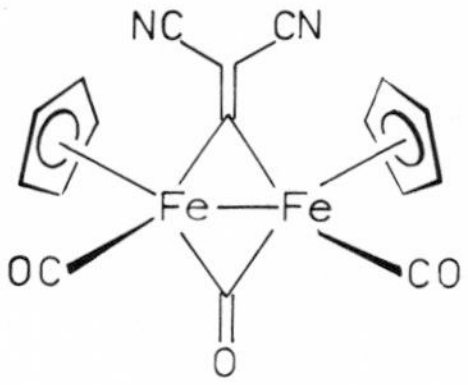

(220)

In some ways the complexes do mimic the reactions of free carbenes. For example, treatment of pentahaptocyclopentadienyldicarbonylcarbenyliron with tetrafluoroboric acid in the presence of 2-butenes gives the dimethylcyclopropanes in a stereospecific manner [174], and the chromium carbonyl complex (221) reacts with methyl *E*-crotonate (222) to give cyclopropane derivatives (223) [175].

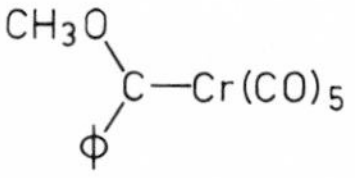

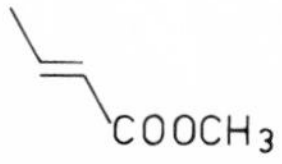

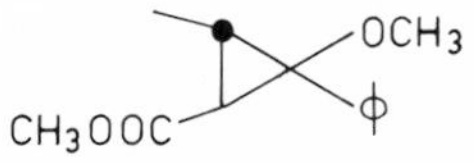

(221) (222) (223)

The corresponding tungsten complex reacts with the Wittig reagent (224) to give vinyl ether (225) [176].

---

171. E.O. Fischer and A. Maasböl, *Angew. Chem., Int. Ed. Engl.*, **3**, 580 (1964).
172. O.S. Mills and A.D. Redhouse, *Angew. Chem., Int. Ed. Engl.*, **4**, 1082 (1965).
173. R.B. King and M.S. Saran, *J. Amer. Chem. Soc.*, **94**, 1784 (1972).
174. P.W. Jolly and R. Pettit, *J. Amer. Chem. Soc.*, **88**, 5044 (1966).
175. E.O. Fischer and K.H. Dötz, *Chem. Ber.*, **103**, 1273 (1970).
176. C.P. Casey and T.J. Burkhardt, *J. Amer. Chem. Soc.*, **94**, 6543 (1972).

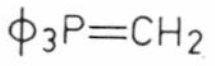

$\phi_3P{=}CH_2$ (224)

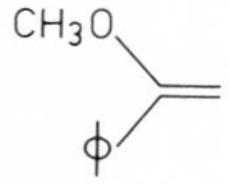

$CH_3O(\phi)C{=}CH_2$ (225)

The dismutation of Wanzlick's olefins, unsuccessfully attempted in thermal fashion by Lemal (see p. 595), succeeds readily in the presence of rhodium complex (226) [177].

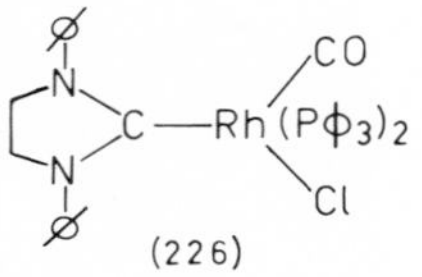

(226)

However, it seems likely that no free carbenes are involved in any of these reactions. Thus, while the complexes (227) with chromium, molybdenum and tungsten all stereospecifically donate the carbene ligand to various olefins, if mixtures of olefins are employed different product ratios are obtained with each of the metals.

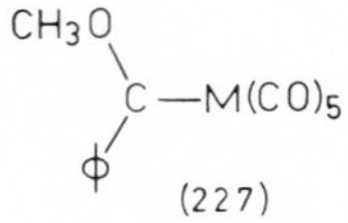

$CH_3O(\phi)C{-}M(CO)_5$ (227)

Clearly the carbene moiety is not set free, but merely transferred [178]. We encountered a similar situation earlier with the Smith-Simmons reagent (p. 574). As a result, it is somewhat problematical whether these compounds should be referred to as complexes (see also p. 556 and 841); probably only Seyferth's mercury carbene complexes - which are capable of reversible, thermal dissociation - deserve the name (p. 560).

16-12. CARBYNES

Carbyne itself (or methyne, CH) has been generated by flashing of methane. Its rate of disappearance can be studied by following one of several absorptions, for example at 3143 Å.

---

177. D. J. Cardin, M. J. Doyle, and M. F. Lappert, Chem. Commun., 927 (1972).
178. K. H. Dötz and E. O. Fischer, Chem. Ber., 105, 1356 (1972).

Acetylene is the principal product immediately after the flash; note that ethylene would be expected from methylene, and ethane from methyl radicals. In the later stages the remaining carbyne reacts with the methane to form ethylene and a hydrogen atom, and the latter finally winds up as molecular hydrogen. Evidence for these reactions derives from the fact that a mixture of methane and tetradeuteriomethane produces ethylene and mono-, tri- and tetradeuterioethylene but no dideuterioethylene [179].

A more conventional route has been followed by Strausz [180], who succeeded in generating carbethoxycarbyne by the photolysis of the mercury bisdiazoacetate (228). This reaction in an inert matrix at 77° K produced an esr signal only interpretable as due to the doublet carbyne.

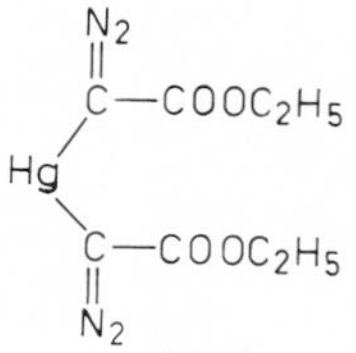

(228)

Its reactions consist of an addition (or insertion) and abstraction - in that order. This conclusion is supported by the fact that with cyclohexene the addition product is predominantly of *endo*- configuration (229); the carbene is known under these conditions to give primarily the *exo*- epimer (230).

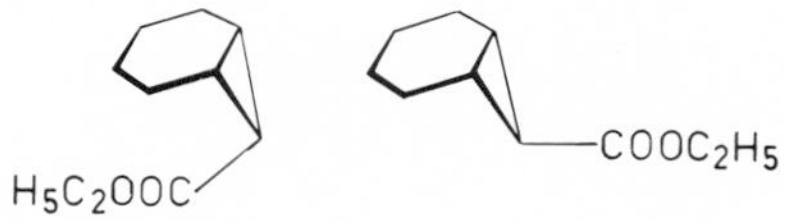

(229) (230)

The addition is highly stereospecific; *trans*-2-butene gives only (231), and *cis*-2-butene only (232) and (233) [181].

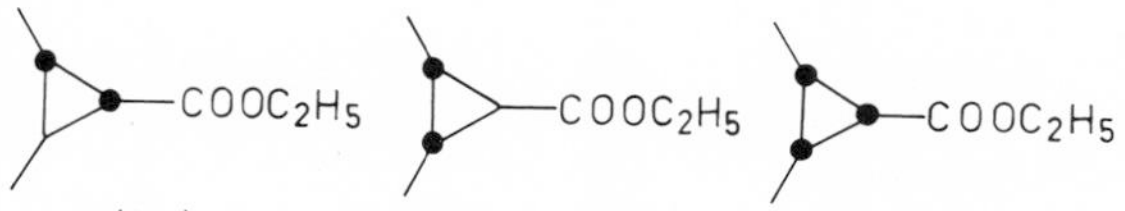

(231) (232) (233)

---

179. W. Braun, J.R. McNesby, and A.M. Bass, Abstr. 152nd ACS meeting, 1966.
180. T. DoMinh, H.E. Gunning, and O.P. Strausz, *J. Amer. Chem. Soc.*, *89*, 6785 (1967).
181. O.P. Strausz, T. DoMinh, and J. Font, *J. Amer. Chem. Soc.*, *90*, 1930 (1968).

16-13. SILYLENES AND RELATED INTERMEDIATES [182]

If one compares the stability of dichlorocarbene with that of lead chloride, a stability sequence for divalent species of group IV suggests itself in which dihalosilylenes would occupy an intermediate position [183]. Difluorosilylene [184] is indeed much less reactive than difluorocarbene. It can be generated by passing tetrafluorosilane over elemental silicon at 1150° [185].

It was found to have a half life of 150 seconds at room temperature and at 0.1 torr pressure. Such stability easily permits the measurement of its microwave spectrum; the bond angle was deduced to be 101°, which compares with 120° for the isoelectronic sulfur dioxide molecule. Thus the silicon orbitals in this species clearly have a great deal of p character [186]. Dichlorosilylene has a bond angle of 120° - a result of a low temperature matrix study by means of infrared [187]. The same is true for dichlorogermylene, stannous chloride and lead chloride [188].

Difluorosilylene will insert into carbon fluorine bonds, for example of hexafluorobenzene to give (234),

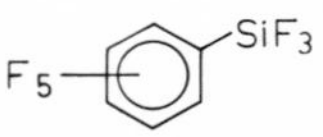

F_5–C_6–SiF_3

(234)

and it will add in 1,4-fashion to 1,3-butadiene and to benzene, to give products such as (235), where n = 2-8; the C-Si-bond is evidently prone to further insertion [189].

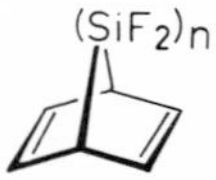

$(SiF_2)_n$

(235)

---

182. For review, see W.H. Atwell and D.R. Weyenberg, Angew. Chem., Int. Ed. Engl., 8, 469 (1969).
183. J.L. Margrave, K.G. Sharp, and P.W. Wilson, Fortschr. Chem. Forsch., 26, 1 (1972).
184. (a) J.C. Thompson and J.L. Margrave, Science, 155, 669 (1967); (b) J.L. Margrave and P.W. Wilson, Accounts Chem. Res., 4, 145 (1971).
185. P.L. Timms, R.A. Kent, T.C. Ehlert, and J.L. Margrave, J. Amer. Chem. Soc., 87, 2824 (1965).
186. V.M. Rao, R.F. Curl, P.L. Timms, and J.L. Margrave, J. Chem. Phys., 43, 2557 (1965).
187. D.E. Milligan and M.E. Jacox, J. Chem. Phys., 49, 1938 (1968).
188. L. Andrews and D.L. Frederick, J. Amer. Chem. Soc., 92, 775 (1970).
189. (a) J.C. Thompson and J.L. Margrave, Inorg. Chem., 11, 913 (1972); (b) P.L. Timms, D.D. Stump, R.A. Kent, and J.L. Margrave, J. Amer. Chem. Soc., 88, 940 (1966).

Another example is the addition of ethylene to give (236) [190].

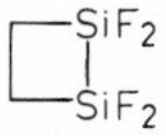

$$\text{cyclo-}CH_2CH_2SiF_2SiF_2$$

(236)

If the silylene is generated in only trace quantities in the presence of 1,3-butadiene, products such as (237) can be isolated. Since oxygen does not affect the yield, the intermediate is probably a singlet [191].

$$\text{1,1-difluoro-1-silacyclopent-3-ene } (SiF_2)$$

(237)

Silylene itself is presumably involved in disproportionations of disilane; thus, a mixture of perdeuteriodisilane and methylsilane at 375° gives tetradeuteriosilane and (238) [192].

$$SiHD_2{-}SiH_2{-}CH_3$$

(238)

Similar reactions were observed with germanium hydrides, implicating germylene as well. Low temperature matrix photolysis of silane yields some silylene (as well as some SiH); the infrared spectrum reveals its structure as that of a bent singlet [193].

Dimethylsilylene was actually the first of this group to be discovered [194]. Thus, dichlorodimethylsilane reacts with sodium vapor and trimethylsilane clearly to give pentamethyldisilane only. 1,4,5,6-Tetraphenyl-7,7-dimethylsilabenzonorbornadiene (239) reacts with tolane to give products (240) and (241) suggesting this intermediate also;

---

190. J.C. Thompson, J.L. Margrave, and P.L. Timms, Chem. Commun., 566 (1966).
191. Y.N. Tang, G.P. Gennaro, and Y.Y. Su, J. Amer. Chem. Soc., 94, 4355 (1972).
192. P. Estacio, M.D. Sefcik, E.K. Chan, and M.A. Ring, Inorg. Chem., 9, 1068 (1970).
193. D.E. Milligan and M.E. Jacox, J. Chem. Phys., 52, 2594 (1970).
194. (a) P.S. Skell and E.J. Goldstein, J. Amer. Chem. Soc., 86, 1442 (1964); (b) O.M. Nefedov and M.N. Manakov, Angew. Chem., Int. Ed. Engl., 3, 226 (1964); (c) H. Gilman, S.G. Cottis and W.H. Atwell, J. Amer. Chem. Soc., 86, 1596 (1964).

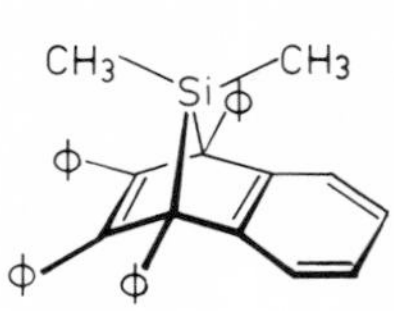

(239) 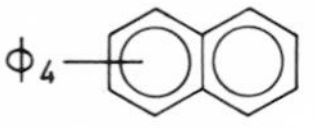 (240) 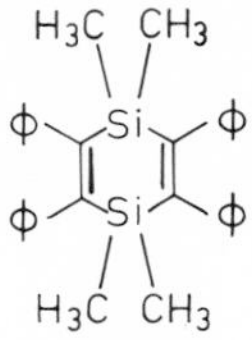 (241)

the silacyclopropene (242) may be an intermediate in this reaction (see also p. 338).

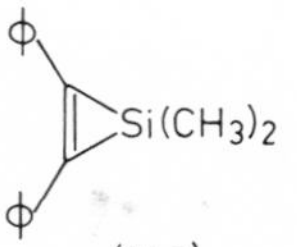

$Si(CH_3)_2$

(242)

Dimethylsilylene is clearly also involved in the reaction of (243) to give (244) and (245) [195].

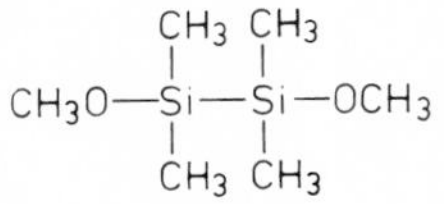

$CH_3O{-}Si(CH_3)_2{-}Si(CH_3)_2{-}OCH_3$ (243)

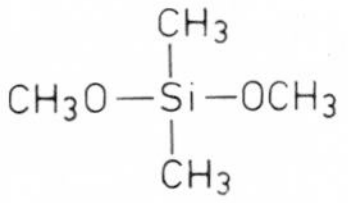

$CH_3O{-}Si(CH_3)_2{-}OCH_3$ (244)

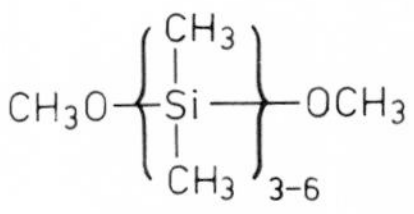

$CH_3O{-}[Si(CH_3)_2]_{3-6}{-}OCH_3$ (245)

The reaction is first order in (243); the intermediate can be observed in the mass spectrum of the reaction mixture, and acetylenes will trap it, i.e., divert the reaction from its normal course without affecting the rate law or the rate constant, and give products such as (246) [196].

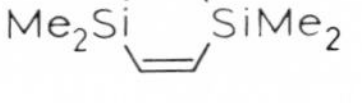

Me_2Si $SiMe_2$

(246)

A general preparation of dialkylsilylenes is now available; it is based on a general synthesis of 2,2-dialkyltrisilanes, and photolytic ejection of the 2-silicon atom as the silylene [197].

---

195. W.H. Atwell and D.R. Weyenberg, J. Amer. Chem. Soc., 90, 3438 (1968).
196. D.R. Weyenberg and W.H. Atwell, Pure Appl. Chem., 19, 343 (1969).
197. H. Sakurai, Y. Kobayashi, and Y. Nakadaira, J. Amer. Chem. Soc., 93, 5272 (1971).

There are as yet no unequivocal instances of silacyclopropanes resulting from additions of silylenes to double bonds. The reaction of diphenyldichlorosilane with lithium metal in tetrahydrofuran in the presence of pure cyclohexene gives octaphenyltetrasilacyclobutane, and no silacyclopropane.

There are a few examples of stable metal-silylene complexes such as (247), which reacts with tolane in a way suggesting electron deficiency on the silicon atoms [198].

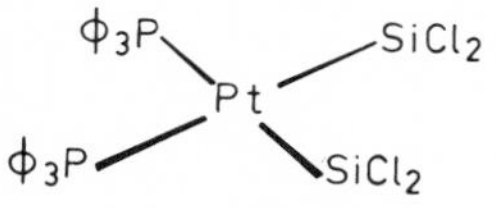

(247) 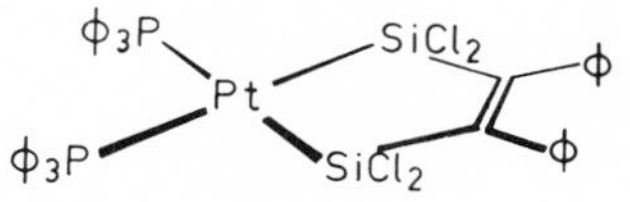 (248)

Some unstable pentacarbonylgermylene- and stannylenechromium complexes are also known [199].

Little is known about the carbenic behavior of other Group IV elements in their divalent state. Trichlorogermane is ether solution reacts with 1,3-butadiene to give (249) [200]. Germanium difluoride and diiodide are stable, at least in the gas phase; the latter will insert in C-I bonds under certain conditions and react with triple bonds and phosphines to give the adducts (250) and (251) (cf. also p. 338).

(249) 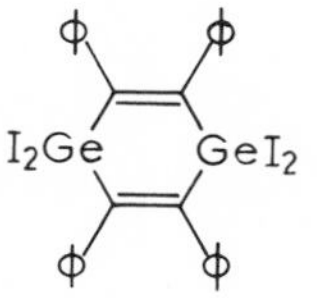 (250) (251)

Perhaps the most outlandish carbene analog yet studied is boron monofluoride, made by high temperature reactions of boron trifluoride and elemental boron. It inserts in B-F and SiF bonds, but further than that, little is known as yet about this species [201].

---

198. G. Schmid and H. J. Balk, Chem. Ber., 103, 2240 (1970).
199. T. J. Marks, J. Amer. Chem. Soc., 93, 7090 (1971).
200. O. M. Nefedov and M. N. Manakov, Angew. Chem., Int. Ed. Engl., 5, 1021 (1966).
201. Chem. Eng. News, 44, 50 (Sept. 19, 1966).

Chapter 17

NITRENES

17-1. INTRODUCTION

It is of course more difficult to prepare and study species with an open sextet on nitrogen than on carbon. It should therefore not be surprising that nitrenes are not as well-known as carbenes. Only a few series are relatively well characterized; the carbonylnitrenes, the arylnitrenes and the aza- and halonitrenes. We will discuss them in that order [1].

17-2. CARBONYLNITRENES

Among the more common examples are the carbalkoxynitrenes, which can be generated by heating or photolyzing the corresponding azidoformates (1) or by treatment of N-nosylurethanes (2) with a base.

1. Reviews: (a) L. Horner and A. Christmann, Angew. Chem., Int. Ed. Engl., 2, 599 (1963); (b) W. Lwowski, Angew. Chem., Int. Ed. Engl., 6, 897 (1967); (c) W. Lwowski, "Nitrenes", Interscience, New York, 1970; (d) R.A. Abramovitch and B.A. Davis, Chem. Rev., 64, 149 (1964); (e) S. Hünig, Helv. Chim. Acta, 54, 1721 (1971).

In early experiments it was shown that this type of nitrene is capable of insertion into C-H bonds and of addition to double bonds; thus, benzene is transformed into azepine (3) and into N-phenylurethane (4), and cyclohexene into aziridine (5) and N-cyclohexenylurethanes (6) [2].

N=N=N—COOR (1)　　$NCOOCH_3$ (3)　　ϕ—$NHCOOCH_3$ (4)

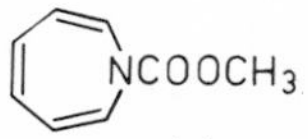

NsNH—COOR (2)　　$NCOOCH_3$ (5)　　$NHCOOCH_3$ (6)

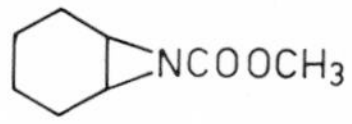

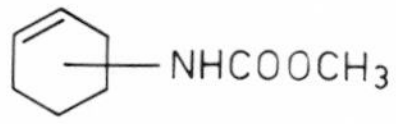

Several cases of internal insertion have been noted; for example, t-butyl azidoformate upon heating gives the cyclic product (7) [3].

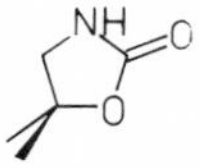

NH O O

(7)

There is strong evidence that a free intermediate is involved. For example, in competition experiments with pairs of olefins, the product ratios are the same whether (1) or (2) is used as precursor [4]. Furthermore, nitrogen is developed at the same rate in cyclohexane and cyclohexene, so that the thermolysis reactions presumably do not involve a triazoline such as (8) [5].

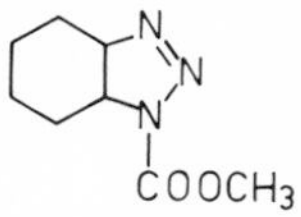

N N N $COOCH_3$

(8)

It is also clear however that these nitrenes are short-lived. When they are generated under flash photolysis conditions only the secondary decomposition product NCO can be observed [2].

---

2. (a) R.S. Berry, D. Cornell, and W. Lwowski, J. Amer. Chem. Soc., 85, 1199 (1963); (b) D.W. Cornell, R.S. Berry, and W. Lwowski, J. Amer. Chem. Soc., 87, 3626 (1965).
3. (a) R. Puttner and K. Hafner, Tetrahedron Lett., 3119 (1964); (b) T.J. Prosser, A.F. Marcantonio, C.A. Genge, and D.S. Breslow, Tetrahedron Lett., 2483 (1964); (c) R. Kreher and D. Kühling, Angew. Chem., Int. Ed. Engl., 4, 69 (1965).
4. (a) W. Lwowski, T.J. Maricich, and T.W. Mattingly, J. Amer. Chem. Soc., 85, 1200 (1963); (b) W. Lwowski and T.J. Maricich, J. Amer. Chem. Soc., 87, 3630 (1965).
5. W. Lwowski and T.W. Mattingly, J. Amer. Chem. Soc., 87, 1947 (1965).

In a matrix isolation experiment designed to gain knowledge about the multiplicity of ground state carbethoxynitrene, no esr signal could be found at 77°K [5]. Although some nitrenes such as *p*-fluorobenzenesulfonylnitrene have since been recognized that way [6], it should be noted that coupling with the nitrogen quadrupole may broaden the signals to the point where they are unobservable.

In fact, it is now considered that the nitrene initially formed is a singlet which then undergoes a relatively slow intersystem crossing. The singlet can add to benzene, add to olefins in a stereospecific manner, and insert into various carbon hydrogen bonds with high selectivity. The triplet adds to olefins also, but nonstereospecifically; it abstracts hydrogen atoms and is selectively removed from a mixture by styrene [7]. The evidence for these generalizations is as follows.

When carbomethoxynitrene is generated by either the azide or the solvolytic route in the presence of *iso*-pentane (9), various insertion products such as (10)-(13) form in such a ratio as to suggest that primary, secondary and tertiary carbon hydrogen bonds are attacked in the ratio of 1:10:30 [8].

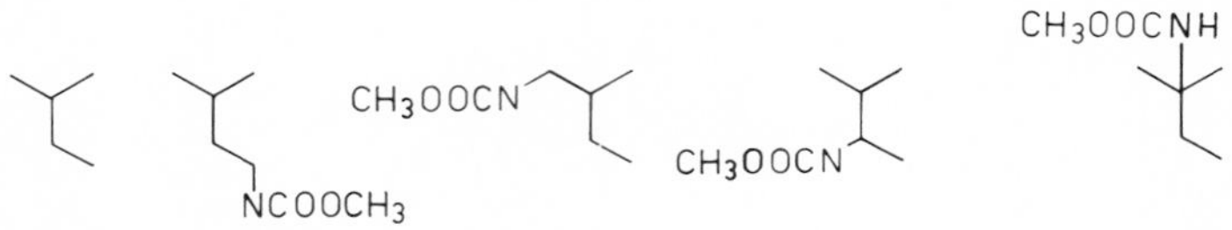

$NCOOCH_3$ CH_3OOCN CH_3OOCN CH_3OOCNH

(9) (10) (11) (12) (13)

It is noteworthy that this spread in selectivity of 30 is much greater than that of carbomethoxycarbene; in that case the corresponding number is only 3 [9]. This selectivity of the nitrene is not affected by lowering the concentration of the hydrocarbon substrate, (although the yield of insertion products then decreases) or by introducing styrene. This generally selective insertion process can be used to functionalize hydrocarbons - which is often a difficult task [10].

If optically active 3-methylhexane (14) is employed as the substrate, the tertiary insertion product (15) is formed with complete retention [7].

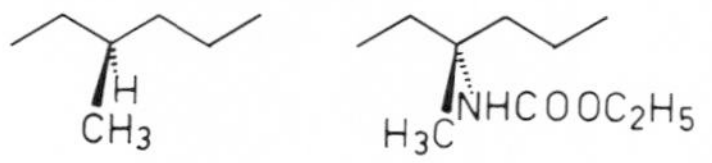

H CH_3 H_3C $NHCOOC_2H_5$

(14) (15)

6. R. M. Moriarty, M. Rahman, and G. J. King, *J. Amer. Chem. Soc.*, **88**, 842 (1966).
7. J. M. Simson and W. Lwowski, *J. Amer. Chem. Soc.*, **91**, 5107 (1969).
8. W. Lwowski and T. J. Maricich, *J. Amer. Chem. Soc.*, **86**, 3164 (1964).
9. W. von E. Doering and L. H. Knox, *J. Amer. Chem. Soc.*, **83**, 1989 (1961).
10. (a) D. S. Breslow, E. I. Edwards, R. Leone, and P. von R. Schleyer, *J. Amer. Chem. Soc.*, **90**, 7097 (1968); (b) J. Meinwald and D. H. Aue, *Tetrahedron Lett.*, 2317 (1967).

This is not expected of a nitrene with triplet multiplicity; if a triplet would insert in aliphatic carbon hydrogen bonds at all, it would be via a radical pair resulting from H-abstraction. Such a pair presumably would racemize to some degree before collapsing. It has similarly been shown that the nitrene thermally derived from optically active 2-methylbutyl azidoformate (16) inserts with retention into the tertiary C-H bond to give (17) [11].

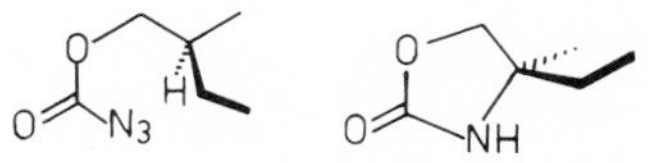

(16) (17)

On the other hand, hydrogen abstraction by triplet carbalkoxynitrenes occurs, especially when the incipient radical is fairly stable. Thus, carbethoxynitrene attacks the 9- and 1- positions of anthracene to give (18) and (19), respectively.

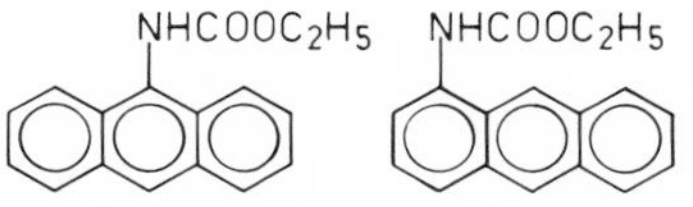

$NHCOOC_2H_5$ $NHCOOC_2H_5$

(18) (19)

The ratio depends on the concentration; it may be only 2 when the anthracene concentration is high, and rise to a value of 9 as the concentration is reduced to a low value. Low values are obtained again if cumene is present. These experiments suggest that a singlet formed initially is to some extent relaxing to a triplet ground state [12].

It was further noted that the competition between addition and insertion processes in the reaction of ethyl azidoformate with cyclohexene was dependent on the concentration of the latter substance: As it decreased, so did the proportion of insertion products [13]. If one of the 2-butenes was used as a substrate, a lowering of the olefin concentration also resulted in a lowering of the stereospecificity of the addition process [14]. The addition of styrene resulted in an increase in this stereospecificity [15]. In another study supporting these contentions, a mixture of benzene and cyclohexane was employed to trap carbethoxynitrene. An azepine - N-cyclohexylurethane mixture is obtained, the yield of which decreases if increasing amounts of methylene chloride are also present. However, the ratio of these

---

11. G. Smolinsky and B.I. Feuer, J. Amer. Chem. Soc., 86, 3085 (1964).
12. A.J.L. Beckwith and J.W. Redmond, Chem. Commun., 165 (1967).
13. (a) W. Lwowski and F.P. Woerner, J. Amer. Chem. Soc., 87, 5491 (1965); (b) D.S. Breslow, T.J. Prosser, A.F. Marcantonio, and C.A. Genge, J. Amer. Chem. Soc., 89, 2384 (1967).
14. W. Lwowski and J.S. McConaghy, J. Amer. Chem. Soc., 87, 5490 (1965).
15. J.S. McConaghy and W. Lwowski, J. Amer. Chem. Soc., 89, 2357 (1967).

products remains constant. By contrast, if benzene and cyclohexene are present, dilution again lowers the yield of azepine but does not greatly affect the yield of aziridine [16]. Finally, it has been noted that if the photochemical generation of such nitrenes is sensitized by acetophenone, then ethyl urethane (20) and 3,3'-bicyclohexenyl (21) become important products.

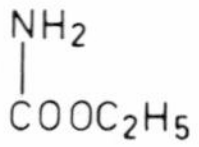

NH_2–$COOC_2H_5$

(20) 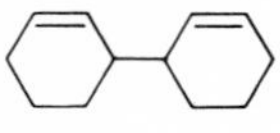 (21)

1,3-Dipolar addition of these nitrenes has been observed in several cases; for example, with nitriles, 1,3,4-oxadiazoles (22) are formed [17] and with acetylenes, 1,3-oxazoles (23) can be made [18].

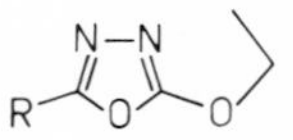

(22) 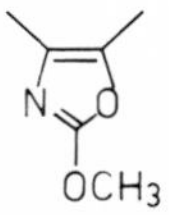 (23)

Sulfonylnitrenes have been reported [19]; they are obtainable by thermolysis of the corresponding azides such as (24).

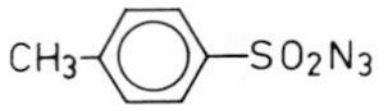

CH_3–C_6H_4–SO_2N_3

(24)

These intermediates have been observed to add to trivalent phosphorus [20]; they will undergo insertion and abstraction reactions but with less selectivity than the carbalkoxynitrenes [21]. Methanesulfonylnitrene will insert in benzene to give a 54% yield of N-methanesulfonylanilide (25); however, if tetracyanoethylene is present, the main product is the bicyclic compound (26) suggesting that N-methanesulfonylazepine (27) is an intermediate [22].

---

16. W. Lwowski and R. L. Johnson, Tetrahedron Lett., 891 (1967).
17. (a) W. Lwowski, A. Hartenstein, C. deVita, and R. L. Smick, Tetrahedron Lett., 2497 (1964); (b) R. Huisgen and H. Blaschke, Ann., 686, 145 (1965); (c) Ref. 3a.
18. J. Meinwald and D.H. Aue, J. Amer. Chem. Soc., 88, 2849 (1966).
19. D.S. Breslow and M.F. Sloan, Tetrahedron Lett., 5349 (1968).
20. I.T. Kay and B.K. Smell, Tetrahedron Lett., 2251 (1967).
21. (a) R.A. Abramovitch, C.I. Azogu, and I.T. McMaster, J. Amer. Chem. Soc., 91, 1219 (1969); (b) M.F. Sloan, T.J. Prosser, N.R. Newburg, and D.S. Breslow, Tetrahedron Lett., 2945 (1964).
22. R.A. Abramovitch and V. Uma, Chem. Commun., 797 (1968).

ϕ—$NHSO_2CH_3$

(25) (26) 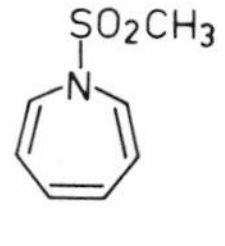 (27)

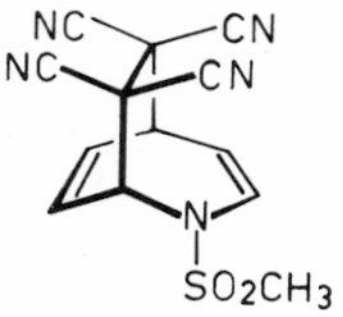

Simple ketonitrenes are thought to be involved in the Curtius rearrangement of acid azides (28) to give isocyanates (29) [23].

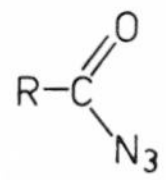

R—N=C=O

(28) (29)

As in the Wolff rearrangement (see p. 586), it is not always clear whether the reaction is concerted or whether the nitrene is a discrete intermediate in this reaction. Some observations have been made that support the latter possibility. Thus, when 1,1-dimethyldecalin-10-carboxylic acid azide (30) is photolyzed, the product consists of about equal amounts of compounds resulting from Curtius rearrangement (31) and from internal nitrene insertion (32) [24].

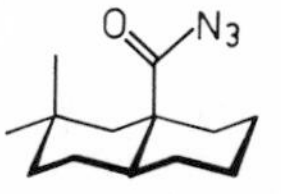

(30) 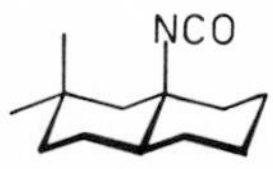 (31) 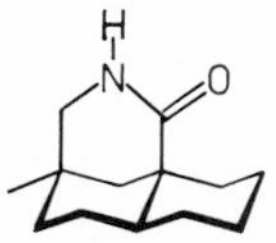 (32)

Furthermore, when pivaloyl azide (33) is irradiated in cyclohexane, one again obtains (34) and (35), respectively [25],

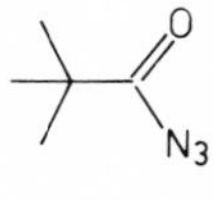

(33) 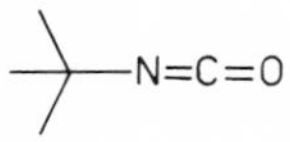 (34) 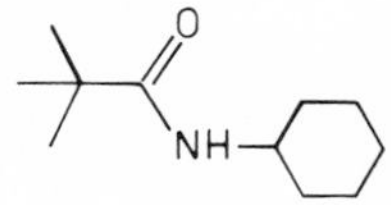 (35)

and ethoxalyl azide (36) under these conditions gives rise to insertion (37) only [26].

---

23. G.W. Wheland, "Advanced Organic Chemistry", 3rd Ed., Wiley, New York, 1960; Chr. 12.
24. R.F.C. Brown, Austral. J. Chem., 17, 47 (1964).
25. W. Lwowski and T.G. Tisue, J. Amer. Chem. Soc., 87, 4022 (1965).
26. T. Shingaki, M. Inagaki, M. Takebayashi, R. Lebkücher, and W. Lwowski, Bull. Chem. Soc. Japan, 43, 1912 (1970).

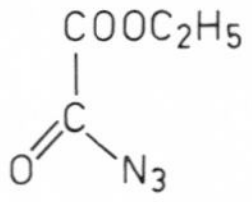

$$\underset{(36)}{C_2H_5OOC-C(=O)-N_3} \qquad \underset{(37)}{C_2H_5OOC-C(=O)-HN-C_6H_{11}}$$

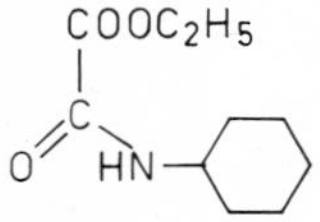

There are two factors that have an important effect on the direction of this reaction. It appears to be vital that the nitrene be generated photochemically; the thermal decomposition gives rise exclusively to isocyanates or their derived products. Thus, the pivaloylnitrene obtained in the thermolysis of (33) in any medium gives t-butyl isocyanate only, by what must be a concerted pathway [25]. It is, of course, not too surprising that photolysis should give rise to a higher energy intermediate.

More puzzling is the observation that alcohols seem to reinforce the tendency to give Curtius products. When ethoxalylnitrene is obtained by photolysis in alcohols, for example, a mixture of Curtius and nitrene products is obtained ((38), and (39) and (40), respectively) [26].

$$\underset{(38)}{C_2H_5OOC-NH-COOR} \qquad \underset{(39)}{C_2H_5OOC-CONH_2} \qquad \underset{(40)}{C_2H_5OOC-CONHOR}$$

In thermolyses in alcohols, the former product is the only one. The alcohol effect is observed, in fact, with the carbalkoxynitrenes: When carbethoxynitrene is generated in methanol [27] or t-butanol [3a], Curtius products such as (41) obtain, albeit in low yield.

$$\underset{(41)}{ROOC-NHOC_2H_5}$$

The same is true of at least one sulfonylnitrene: When benzenesulfonyl azide is heated in methanol, the product is phenyl sulfamate (42) [28].

$$\underset{(42)}{\phi-NHSO_3CH_3}$$

27. W. Lwowski, R. DeMauriac, T.W. Mattingly, and E. Scheiffele, Tetrahedron Lett., 3285 (1964).
28. W. Lwowski and E. Scheiffele, J. Amer. Chem. Soc., 87, 4359 (1965).

It seems conceivable that the alcohol effect is due to hydrogen bonding to the carbonyl group. In support, it has been noted that boron trifluoride forms adduct (43) with acyl azides; these complexes are unstable and rearrange even below 0° to Curtius products [29].

(43)

An interesting member of this series is the species NCN, usually obtained by the thermolysis or photolysis of cyanogen azide (44). It adds readily to aromatic compounds [30] to form azepines (45); cyclooctatetraene gives both 1,2- and 1,4-addition products (46) and (47) [31].

$NC-N_3$

(44) (45) (46) (47)

The thermal reactions appear to be preceded by the formation of triazolines (48) [32].

(48)

The cyanonitrene products suggest that two spin states are at work. Thus, when cyanonitrene is generated in either cis- or trans-1,2-dimethylcyclohexane, stereospecific insertion occurs into the tertiary C-H bonds; if the solvent is methylene chloride, much of this specificity is lost, and in methylene bromide both epimers give an identical mixture [33]. The methylene bromide undoubtedly

29. E. Fahr and L. Neumann, Angew. Chem., Int. Ed. Engl., 4, 593 (1965).
30. F.D. Marsh and H.E. Simmons, J. Amer. Chem. Soc., 87, 3529 (1965).
31. A.G. Anastassiou, J. Amer. Chem. Soc., 87, 5512 (1965).
32. (a) Chem. Eng. News, 43, 29, Dec. 27 (1965); (b) F.D. Marsh, J. Org. Chem., 37, 2966 (1972); (c) M.E. Hermes and F.D. Marsh, J. Org. Chem., 37, 2969 (1972).
33. (a) A.G. Anastassiou and J.N. Shepelavy, J. Amer. Chem. Soc., 90, 492 (1968); (b) A.G. Anastassiou and H.E. Simmons, J. Amer. Chem. Soc., 89, 3177 (1967); (c) A.G. Anastassiou, J. Amer. Chem. Soc., 88, 2322 (1966).

acts by providing the heavy atoms that promote spin inversion; the result is hydrogen abstraction rather than concerted insertion, and finally, collapse of the radical pair (the technique of CIDNP has not yet been applied to this problem). This interpretation was confirmed in a flash photolysis experiment in which both the initially formed singlet and the subsequently formed triplet could be observed in the uv; the esr [34] and ir spectra [35] of matrix isolated cyanonitrene have likewise confirmed these assignments.

17-3. ARYLNITRENES

These nitrenes can be generated by the thermolysis or photolysis of aryl azides. In one instance, the analog N-sulfinylaniline (49) was used successfully; the sulfur monoxide fragment could be detected by microwave spectroscopy [36].

$$\phi\text{—N=S=O}$$

(49)

Ferrocenylnitrene is a rare example of the thermal generation of such a species from an isocyanate [37].

The ground state is a triplet as shown for example by a study of the effect of added diluants and sensitizers on the product composition [38]. The singlet formed initially has somewhat nucleophilic character because of resonance (50).

$$C_6H_5\text{—}\ddot{N} \longleftrightarrow \overset{\oplus}{C_6H_5}\text{=}\overset{\ominus}{N}$$

(50)

Thus, phenylnitrene is selective in its insertion reactions; a study of its reactions with iso-pentane shows the ratio of attacks on tertiary, secondary and primary carbon hydrogen bonds to be approximately 200:10:1 [39]. Attack on aromatic CH bonds does not appear to occur, but an exception is known with a nitrene bearing a

---

34. E. Wasserman, L. Barash, and W.A. Yager, J. Amer. Chem. Soc., 87, 2075 (1965).
35. D.E. Milligan, M.E. Jacox, J.J. Comeford, and D.E. Mann, J. Chem. Phys., 43, 756 (1965).
36. S. Saito and C. Wentrup, Helv. Chim. Acta, 54, 273 (1971).
37. R.A. Abramovitch, R.G. Sutherland, and A.K.V. Unni, Tetrahedron Lett., 1065 (1972).
38. A. Reiser and L.J. Leyshon, J. Amer. Chem. Soc., 93, 4051 (1971).
39. J.H. Hall, J.W. Hill, and H. Tsai, Tetrahedron Lett., 2211 (1965).

deactivating substituent in the ring: *p*-cyanophenylnitrene has been observed to insert into the aromatic C-H bonds of *N*,*N*-dimethylaniline to give (51) and (52) [40].

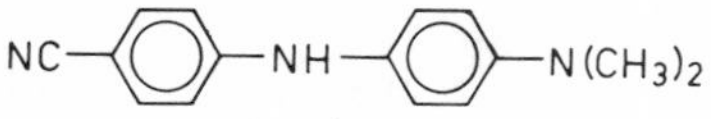

(51) 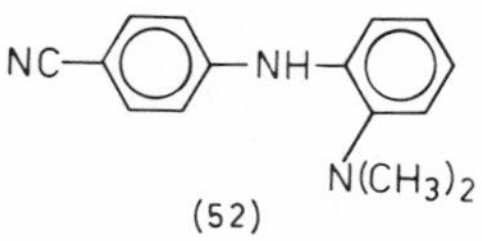 (52)

In some instances when nitrenes or nitrenoid intermediates have been generated in alcohols, the products of nucleophilic attack on the ring (such as (53)) have been observed [41].

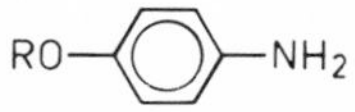

(53)

If an *o*-substituent is present, insertion may occur there; for instance, *o*-*n*-butylphenyl azide (54) gives products (55) and (56) [42], and *o*-azidobiphenyl leads to the formation of carbazole (57) (sensitized reaction gives azo derivative (58) instead) [43].

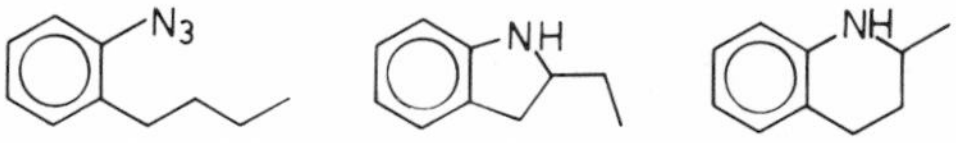

(54) (55) (56)

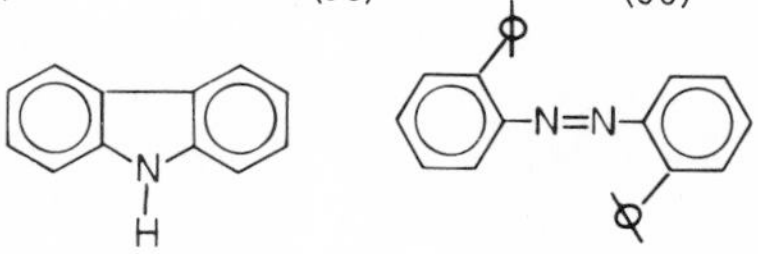

(57) (58)

It should be pointed out however that when phenyl azides bearing *o*-substituents with unshared electrons are thermolyzed, it is in fact not sure whether free nitrenes are involved at all [44]. Thus, although substituents in the *p*-position

---

40. (a) R.A. Abramovitch and E.F.V. Scriven, *Chem. Commun.*, 787 (1970); (b) R.A. Abramovitch, S.R. Challand, and E.F.V. Scriven, *J. Org. Chem.*, *37*, 2705 (1972).
41. R.J. Sundberg and R.H. Smith, *Tetrahedron Lett.*, 267 (1971).
42. G. Smolinsky and B.I. Feuer, *J. Org. Chem.*, *29*, 3097 (1964).
43. (a) J.S. Swenton, T.J. Ikeler, and B.H. Williams, *J. Amer. Chem. Soc.*, *92*, 3103 (1970); (b) B. Coffin and R.F. Robbins, *J. Chem. Soc.*, 1252 (1965).
44. L.K. Dyall and J.E. Kemp, *J. Chem. Soc. (B)*, 976 (1968).

generally have little effect on the rates, in the o-position the same substituents affect the rate very greatly. It was suggested by the authors that the loss of nitrogen is assisted by unshared electrons on neighboring groups, and that free nitrenes do not intervene.

The same question has arisen with respect to the intermediates involved in the reduction of nitro- and nitroso compounds by means of phosphites and phosphines. Since the products of these reductions often resemble those of the azide decompositions, nitrene involvement seems likely; thus o-nitrobiphenyl reduction gives carbazole [45]. In the presence of secondary amines, 3H-azepines such as (59) are formed; these had earlier been observed in azide decomposition and they could arise via the rearranged intermediate (60) [45].

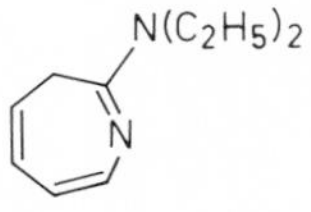

$N(C_2H_5)_2$

(59)  (60)

Adducts such as (61)-(62) are also found and easily explained in terms of nitrenes [46].

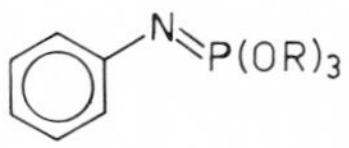

$N{=}P(OR)_3$ O OR P—OR N

(61) 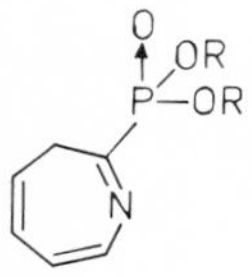 (62)

Recently, evidence has convincingly demonstrated that at least not all of these reactions can involve the free nitrenes, however. For example, when the azide (63) is heated, (64) and (65) are among the products. Analysis of the mixture shows that the migratory ratio phenyl/methyl equals 1.8. The reduction of nitrosocompound (66) gives the same products, but now the ratio equals 160.

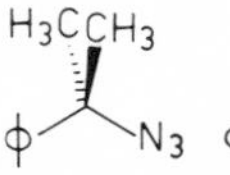

$H_3C\,CH_3$ ϕ N_3 — CH_3 CH_3 N ϕ — CH_3 N CH_3 ϕ — $H_3C\,CH_3$ ϕ NO

(63) 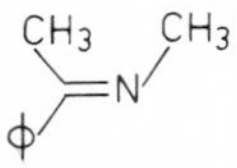 (64) 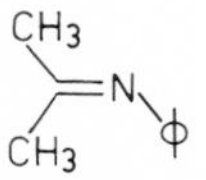 (65) 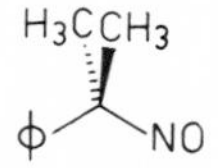 (66)

---

45. R.A. Odum and M. Brenner, J. Amer. Chem. Soc., 88, 2074 (1966).
46. (a) R.J. Sundberg, J. Amer. Chem. Soc., 88, 3781 (1966); (b) R.J. Sundberg, B.P. Das, and R.H. Smith, J. Amer. Chem. Soc., 91, 658 (1969).

Furthermore, in this reaction the disappearance of the starting materials is somewhat faster than the appearance of the products. Both this time lag and the increased selectivity require that the reduction involves a different intermediate - possibly (67) [47].

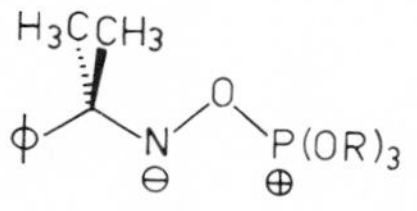

(67)

We ignore this question of nitrene vs nitrenoid intermediacy for the moment, and instead emphasize the synthetic value of these species. An o-nitro group leads to the interesting group of benzofuroxans. It was shown by means of low temperature nmr that the isomers (68) and (69) rapidly equilibrate [48], presumably via o-dinitrosobenzenes.

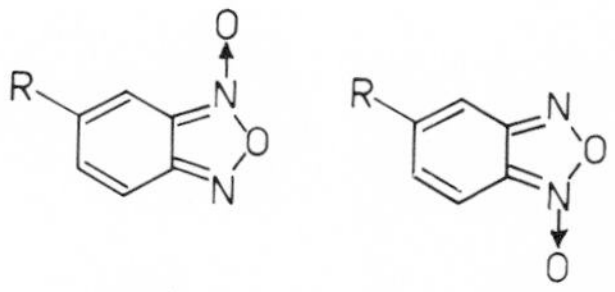

(68) (69)

The presence of an o-thiophenoxy group results in rearrangement; for example, the phosphite reduction of (70) gives phenothiazine (71), clearly via (72) [49].

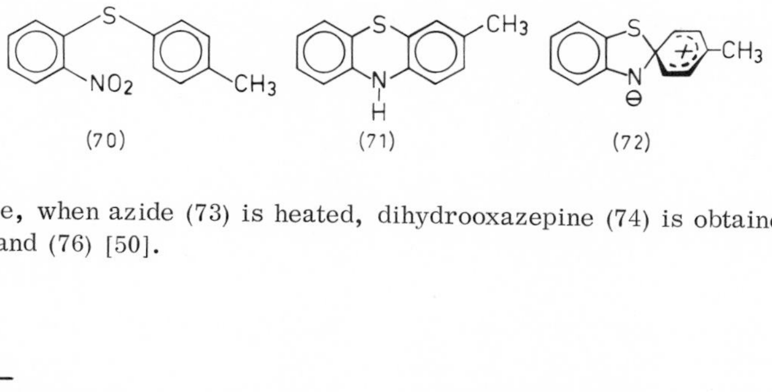

(70) (71) (72)

Likewise, when azide (73) is heated, dihydrooxazepine (74) is obtained, possibly via (75) and (76) [50].

47. R.A. Abramovitch, J. Court, and E.P. Kyba, Tetrahedron Lett., 4059 (1972).
48. A.R. Katritzky, S. Oksne, and R.K. Harris, Chem. Ind. (London), 990 (1961).
49. (a) J.I.G. Cadogan, R.K. Mackie, and M.J. Todd, Chem. Commun., 736 (1968); (b) J.I.G. Cadogan and S. Kulik, J. Chem. Soc. (C), 2621 (1971); (c) J.I.G. Cadogan, Accounts Chem. Res., 5, 303 (1972).
50. J.I.G. Cadogan and P.K.K. Lim, Chem. Commun., 1431 (1971).

(73) (74)

(75) (76)

It may be noted here in passing that the reverse of nitro- and nitroso group reductions also occurs; the generation of phenylnitrene or -nitrenoid species in the presence of oxygen gives the nitro compound. An esr study of the matrix reactions revealed (77) as an intermediate [51].

$\phi-\dot{N}-O-\dot{O}$

(77)

One of the interesting things about arylnitrenes is that they are prone to rearrangements if generated at high temperatures. Thus, phenylnitrene itself at 600-1000°K gives 1-cyanocyclopentadiene [52]; 3- and 6-methyl-2-azidopyridines (78) and (79) at 380°C both give 5-methyl-2-cyanopyrrole (80), and 4- and 5-methyl-2-azidopyridines similarly lead to an identical mixture of 3- and 4-methyl-2-cyanopyrroles [53]. These observations are reminiscent of the phenylcarbene rearrangements, (see p. 581); if the mechanism is the same, a nitrene to carbene rearrangement must be involved, proceding via such species as (81) and (82).

(78) (79) (80) (81) (82)

N^{15}-labeling experiments show that the two nitrogen atoms in α-pyridylnitrene do indeed become equivalent [54].

51. (a) R.A. Abramovitch and S.R. Challand, Chem. Commun., 964 (1972); (b) J.S. Brinen and B. Singh, J. Amer. Chem. Soc., 93, 6623 (1971); (c) R.A. Abramovitch, C.I. Azogu, and R.G. Sutherland, Chem. Commun., 134 (1971).
52. W.D. Crow and C. Wentrup, Tetrahedron Lett., 4379 (1967).
53. W.D. Crow and C. Wentrup, Chem. Commun., 1082 (1968).
54. (a) W.D. Crow and C. Wentrup, Chem. Commun., 1387 (1969); (b) C. Wentrup, Chem. Commun., 1386 (1969).

It has furthermore been suggested that the appearance of azepines when phenylnitrenes are generated in the presence of secondary amines (see p. 614) similarly indicates a nitrene-carbene rearrangement [55].

Several polynitrenobenzenes are known. o-Dinitrenobenzene (product from the diazide) decomposes to give Z, Z-muconitrile (83) [56]; the same product is obtained by the lead tetraacetate oxidation of 2-aminobenzotriazole (84) [57].

CN CN N N–NH$_2$ N

(83) (84)

The irradiation of matrix-isolated p-diazidobenzene at 77°K gives rise to a triplet esr signal that was at first attributed to p-dinitrenobenzene [58]; however, it was later shown that the azocompound (85) gives the same signal, and hence the species observed is probably (86) [59].

N$_3$ N=N N$_3$ N N–N N

(85) (86)

17-4. HALO- AND AZANITRENES

Relatively little is known about halonitrenes, although all of them have been observed by ir in low temperature matrices [60]. Several emissions resulting from a microwave discharge through nitrogen trifluoride [61] and through nitrogen-chlorine mixtures [62] have been attributed to singlet halonitrenes returning to the triplet ground state. It has furthermore been found that the reaction of difluoramine with hydroxide ion to give difluorodiazene is much faster than with any other

---

55. R. J. Sundberg, M. Brenner, S. R. Suter, and B. P. Das, Tetrahedron Lett., 2715 (1970).
56. (a) J. H. Hall, J. Amer. Chem. Soc., 87, 1147 (1965); (b) J. H. Hall and E. Patterson, J. Amer. Chem. Soc., 89, 5856 (1967).
57. C. D. Campbell and C. W. Rees, Chem. Commun., 192 (1965).
58. (a) A. M. Trozzolo, R. W. Murray, G. Smolinsky, W. A. Yager and E. Wasserman, J. Amer. Chem. Soc., 85, 2526 (1963); (b) A. Reiser, H. M. Wagner, R. Marley, and G. Bowes, Trans. Faraday Soc., 63, 2403 (1967).
59. B. Singh and J. S. Brinen, J. Amer. Chem. Soc., 93, 540 (1971).
60. (a) D. E. Milligan and M. E. Jacox, J. Chem. Phys., 40, 2461 (1964); (b) J. J. Comeford and D. E. Mann, Spectrochim. Acta, 21, 197 (1965).
61. (a) A. E. Douglas and W. E. Jones, Can. J. Phys., 44, 2251 (1966); (b) W. E. Jones, Can. J. Phys., 45, 21 (1967).
62. R. Colin and W. E. Jones, Can. J. Phys., 45, 301 (1967).

anion [63] and that this reaction has a positive volume of activation (+7 cm^3/mole); the formation of fluoride in the bimolecular reaction with acetate, by contrast, has an activation volume of -18 cm^3/mole [64]. The second order rate law for the hydrolysis - first order in hydroxide and in difluoramine - is unaffected by the presence of certain lures such as thiophenol or hydroxylamines; these materials do not react with difluorodiazene or with difluoramine itself, but they divert the reaction from its normal course, and are oxidized to diphenyl disulfide and to the corresponding nitroxides, respectively. The latter products are easily detected by esr. When difluoramine is passed over solid potassium hydroxide, a weak visible light emission can be seen which has approximately the same wavelength as that observed in the microwave discharge experiments [65]. All of these observations are consistent with a B_1 scheme, involving in succession difluoramide anion, fluoronitrene singlet and fluoronitrene triplet; the latter species then abstracts hydrogen from difluoramine or another available hydrogen source on the way to final products. It may be noted that the multiplicity of ground state fluoronitrene is suggested by its ability to abstract hydrogen; a wide energy gap between singlet and ground state triplet seems reasonable if it is realized that fluoronitrene is isoelectronic with oxygen.

Another series that should be stabilized by the presence of a neighboring atom with unshared electrons is that of the aminonitrenes [66].

A variety of methods have been employed to generate them; common among these are the oxidation of hydrazines with lead tetraacetate [67], the decomposition of N-azidoamines [68], the photolysis of N-aminoaziridines [69], the reaction of secondary amines with difluoramine [70] and the treatment of secondary amines with Angeli's salt (Na_2ONNO_2) [71]. The nucleophilic nature of these intermediates is clear from their additions to double bonds. Thus, the oxidation of (87) and various analogs in the presence of the 2-butenes leads to very stereospecific addition even in high dilution [72], and the yields of aziridines are highest when electron-poor olefins such as tetrachloroethylene are used [73].

---

63. (a) G.A. Ward and C.M. Wright, J. Amer. Chem. Soc., 86, 4333 (1964); (b) A.D. Craig and G.A. Ward, J. Amer. Chem. Soc., 88, 4526 (1966); (c) W.T. Yap, A.D. Craig, and G.A. Ward, J. Amer. Chem. Soc., 89, 3442 (1967).
64. W.J. le Noble and D.N. Skulnik, Tetrahedron Lett., 5217 (1967).
65. W.J. le Noble, E.M. Schulman, and D.N. Skulnik, J. Amer. Chem. Soc., 93, 4710 (1971).
66. B.V. Ioffe and M.A. Kuznetsov, Russ. Chem. Rev., English Transl., 41, 131 (1972).
67. R.S. Atkinson and C.W. Rees, Chem. Commun., 1230 (1967).
68. R. Ahmed and J.P. Anselme, Can. J. Chem., 50, 1778 (1972).
69. (a) D.W. Jones, Chem. Commun., 884 (1972); (b) T.L. Gilchrist, C.W. Rees, and E. Stanton, J. Chem. Soc. (C), 988 (1971).
70. Y. Hata and M. Watanabe, Tetrahedron Lett., 3827 (1972).
71. D.M. Lemal and T.W. Rave, J. Amer. Chem. Soc., 87, 393 (1965).
72. (a) R.S. Atkinson and C.W. Rees, J. Chem. Soc. (C), 772 (1969); (b) D.J. Anderson, T.L. Gilchrist, D.C. Horwell, and C.W. Rees, J. Chem. Soc. (C), 576 (1970).
73. D.J. Anderson, T.L. Gilchrist, D.C. Horwell, and C.W. Rees, Chem. Commun., 146 (1969).

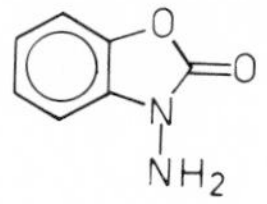

(87)

Other reactions that are observed now and then are fragmentation, a 1,2-shift to produce an azo linkage, attack on azocompounds to give azimines, and addition to dimethyl sulfoxide. For example, the treatment of aziridines with difluoramine and the oxidation of N-aminoaziridine both produce nitrogen and ethylene [70]. The oxidation of (88) gives benzonitrile and nitrogen, but adduct (89) is obtained if dimethyl sulfoxide is present; similarly, oxidation of (90) produces the diacetylene (91) [74].

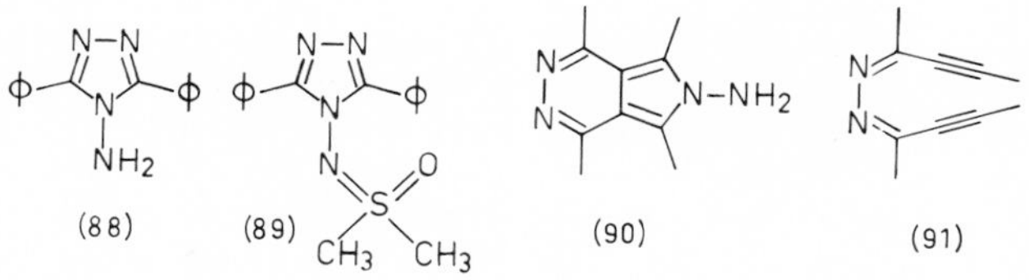

(88) (89) (90) (91)

When diallylamine is treated with difluoramine, azocompound (92) is obtained [75].

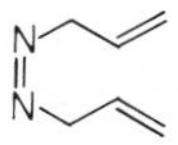

(92)

The oxidation of (93) produces 3-cinnolinol (94) [72a].

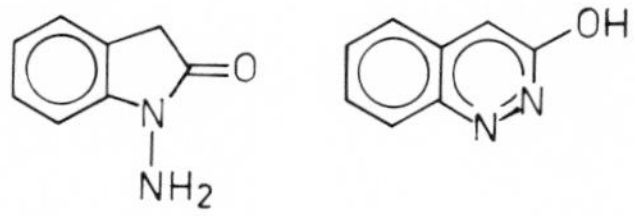

(93) (94)

The decomposition of azide (95) in (96) leads to azimine (97) [66].

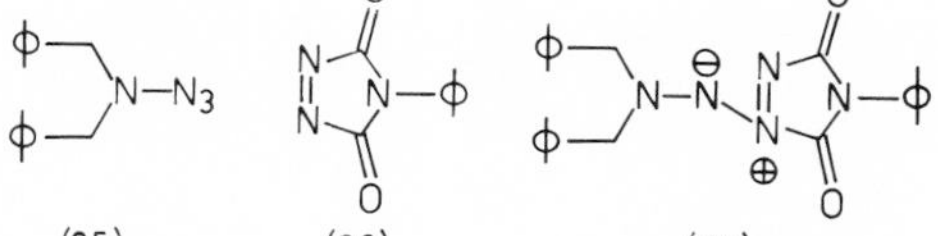

(95) (96) (97)

74. (a) K.K. Mayer, F. Schröppel, and J. Sauer, Tetrahedron Lett., 2899 (1972); (b) K. Sakai and J.P. Anselme, Bull. Chem. Soc. Japan, 45, 307 (1972).
75. J.E. Baldwin, J.E. Brown, and G. Höfle, J. Amer. Chem. Soc., 93, 788 (1971).

Finally if (98) is oxidized in the presence of acetylenes such as (99), the rearranged adduct (100) is obtained; it might be noted that the unrearranged adduct (101) would be antiaromatic [76].

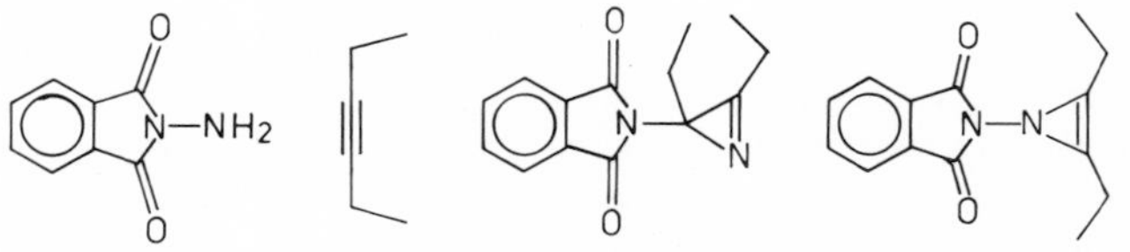

(98) (99) (100) (101)

A remarkable case of a stabilized nitrene has been reported by Smith [77]. Mild warming of 1,4-diphenyl-5-azido-1,2,3-triazole (102) gives nitrogen and a deep red, isolable substance that can be reduced with hydrogen to the 5-amino analog (103), and which has phenyl nmr signals are all shifted downfield compared to the parent substance. The absence of any esr signals, a study of the ir spectrum and elemental analysis led the authors to propose that the product was the free, resonance stabilized nitrene (104) in equilibrium with a small amount of a ring opened isomer (105).

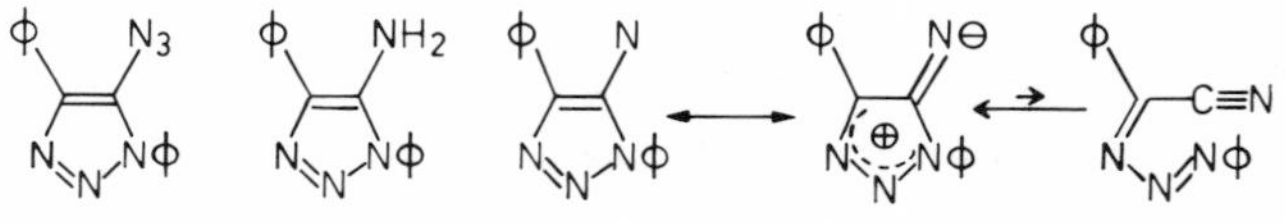

(102) (103) (104) (105)

Perhaps similarly, photolysis of single crystal cyanuric triazide (106) at -160° gives an esr spectrum which persists more or less indefinitely even at room temperature [78].

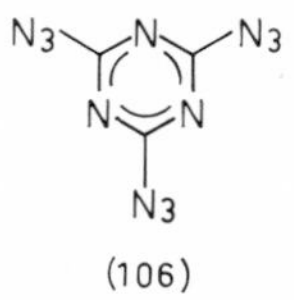

(106)

The oxidation of methoxyamine (107) gives the nitrene NOMe [79]. In the presence of olefins this intermediate gives N-methoxyaziridines (108), which do not invert (nmr time scale) even at 130° (see p. 178).

---

76. D.J. Anderson, T.L. Gilchrist, and C.W. Rees, Chem. Commun., 147 (1969).
77. P.A.S. Smith, L.O. Krbechek, and W. Resemann, J. Amer. Chem. Soc., 86, 2025 (1964).
78. R.M. Moriarty, M. Rahman, and G.J. King, J. Amer. Chem. Soc., 88, 842 (1966).
79. S.J. Brois, J. Amer. Chem. Soc., 92, 1079 (1970).

H_2NOCH_3 (107) $\quad$ $CH_3O{-}N{<}$ $\begin{matrix}C(CH_3)_2\\ |\\ C(CH_3)_2\end{matrix}$ (108)

17-5. RELATED NITRENES

As far as chemical reactivity is concerned, one of the least known examples in this chapter is the parent nitrene NH (sometimes called imidogen). It can be generated by photolysis of hydrazoic acid in aqueous solution [80], in organic substrates [81] and in inert solid matrices at low temperature [82], or by flash thermolysis of ammonia [83]. These studies all have suggested that the ground state is a triplet. There is a substantial body of literature suggesting [84] that chloramine readily loses hydrochloric acid to give this nitrene and that this intermediate plays a vital role in the Raschig synthesis of hydrazine from chlorine and ammonia. The treatment of chloramine at -75° with phenyllithium in tetramethylethylene indeed produces traces of tetramethylaziridine [85]. Similarly, the base promoted hydrolysis of chloramine has been claimed to proceed via NH [86]; however, chloramine, methylchloramine and dimethylchloramine all solvolyze in aqueous base at similar rates [87], and all have similar, negative activation volumes [88]. These reactions are clearly displacements.

A few alkylnitrenes have been described as triplets in inert matrices at 4°K, on the strength of their esr spectra [89]. Trityl azides (109) eliminate nitrogen and give imines (110) when photolyzed or thermolyzed [90], and it is thought that the reaction proceeds through a discrete nitrene stage in the photochemical reaction, but not in the thermal one.

$\phi_3C{-}N_3$ (109) $\quad$ $\phi_2C{=}N\phi$ (110)

---

80. I. Burak and A. Treinin, J. Amer. Chem. Soc., 87, 4031 (1965).
81. D.W. Cornell, R.S. Berry, and W. Lwowski, J. Amer. Chem. Soc., 88, 544 (1966).
82. (a) D.E. Milligan and M.E. Jacox, J. Chem. Phys., 41, 2838 (1964); (b) K. Rosengren and G.C. Pimentel, J. Chem. Phys., 43, 507 (1965).
83. (a) M.W.P. Cann and S.W. Kash, J. Chem. Phys., 41, 3055 (1964); (b) C.E. Melton and P.H. Emmett, J. Phys. Chem., 68, 3318 (1964).
84. J. Fischer and J. Jander, Z. Anorg. Allgem. Chem., 313, 14, 37 (1961).
85. J. Jander, Angew. Chem., Int. Ed. Engl., 6, 1087 (1967).
86. L.F. Audrieth and R.A. Rowe, J. Amer. Chem. Soc., 77, 4726 (1955).
87. M. Anbar and G. Yagil, J. Amer. Chem. Soc., 84, 1790 (1962).
88. W.J. le Noble, Tetrahedron Lett., 727 (1966).
89. E. Wasserman, G. Smolinsky and W.A. Yager, J. Amer. Chem. Soc., 86, 3166 (1964).
90. F.D. Lewis and W.H. Saunders, J. Amer. Chem. Soc., 89, 645 (1967) and 90, 3828, 7031 (1968).

The support for this contention is that if one of the aromatic rings bears an electron withdrawing substituent, this tends to lower the migratory aptitude of that ring in the thermal reactions, but in the photochemical analogs the rings compete on a statistical basis regardless of substituent (see also p. 614). It was also found that in the presence of compounds with a weakly bound hydrogen atom (for example, tri-n-butyltin hydride), the latter reaction gives much trityl amine; this is not true of the thermal reaction.

There are as yet not very many instances of nitrenes stably coordinated in metal complexes. The parent nitrene is believed to be a ligand in (111), a substance which is itself an intermediate in the decomposition of azidopentaamine-ruthenium-III complexes (112) [91].

$(H_3N)_5RuNH^{\oplus\oplus}$ $(H_3N)_5RuN_3^{\oplus\oplus}$

(111) (112)

Fluoroalkylnitrenes are known in complexed form with several transition metals such as the iridium complex (113) [92], and simple alkylnitrenes are known as ligands in some rhenium complexes such as (114) [93]; in no case has the release of this ligand been accomplished, however.

$(\Phi_3P)_2ClIrNCH(CF_3)_2$

(113)

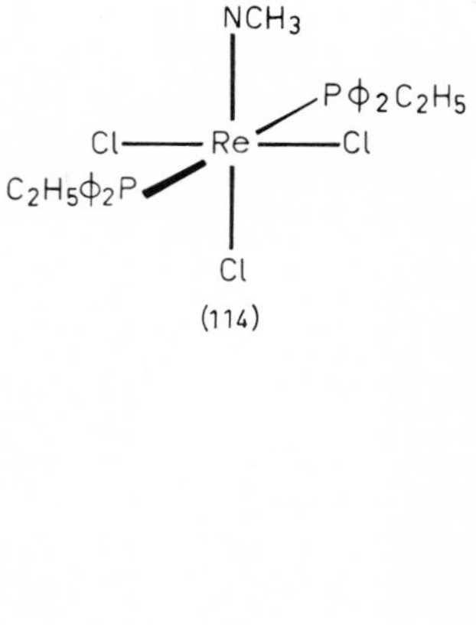

NCH_3, $P\Phi_2C_2H_5$, Cl—Re—Cl, $C_2H_5\Phi_2P$, Cl

(114)

91. B.C. Lane, J.W. McDonald, F. Basolo, and R.G. Pearson, J. Amer. Chem. Soc., 94, 3786 (1972).
92. M.J. McGlinchey and F.G.A. Stone, Chem. Commun., 1265 (1970).
93. J. Chatt, J.R. Dilworth, and G.J. Leigh, J. Chem. Soc. (A), 2239 (1970).

Chapter 18

FREE RADICALS

18-1. INTRODUCTION

Of all the intermediates in organic chemistry, the free radicals have the longest history. It is however a history full of early claims and denials, of experiments and interpretations proven wrong, and finally, of disbelief and skepticism. The principal reason for all this early turmoil was that there were no reliable and generally accepted ways to measure molecular weights, and hence no experiments to distinguish between radicals and dimers; indeed, the very concepts necessary to support the notion of free radicals were shaky or missing altogether. Thus, little distinction was made between radicals as functional groups - transferable from one molecule to another and surviving as units in many reactions - and free radicals as substances, isolable as pure materials. Even the first enduring claim (from Gomberg in 1900) met with years of doubt and denial.

What Gomberg found [1] was that a reaction which rationally should have produced hexaphenylethane - a compound expected to be colorless and relatively unreactive - instead gave rise to a yellow solution, the color of which reversibly

---

1. M. Gomberg, Chem. Ber., 33, 3150 (1900).

deepened on heating, and which extremely readily reacted with oxygen to give a peroxide $C_2O_2\phi_6$ and with halogens to give $CX\phi_3$. Molecular weight determinations gave "low" results. If the color was discharged with oxygen, it soon reappeared again. The removal of the solvent gave a white solid which absorbed oxygen extremely readily. All these data suggested an equilibrium:

$$\underset{(1)}{C_2\phi_6} \rightleftharpoons \underset{(2)}{2\,\dot{C}\phi_3}$$

Gradually Gomber's contemporaries were convinced that he was right, and the idea was elaborated and expanded: New free radicals were prepared, based on still more highly crowded groups, and on aromatic groups permitting still greater delocalization.

A second milestone in free radical chemistry was Paneth's experiment [2]. He found that when tetraethyllead vapor was passed through a glass tube and a portion of the tube was heated, a lead mirror began to form some distance downstream, and butane was present in the effluent gas. Paneth found that if another section of the tube further upstream were heated not only did another mirror begin to form but the one already there disappeared, and no butane was then obtained. If the two sections are very far apart, so that it takes an appreciable length of time for the decomposition products to reach the original mirror, the disappearing act fails. Paneth did not find it hard to understand these results, and at that point free radical chemistry was no longer confined to such laboratory curiosities as the perphenylated alkanes. As the techniques for hunting, finding and studying radicals became more sophisticated and reliable, ever more fleeting and unstable radicals joined the list, which now includes both free hydrogen atoms [3] and free electrons (Chapter 24 - Sec. 9) in solution. Almost any type of radical can now be generated and studied virtually at will [4].

To mention a few random examples, the thermal decompositions of alkyl peroxides (3), acyl peroxides (4), azo compounds (5) and *t*-butyl peresters (6) are frequently used [4a]; the photolysis of *t*-butyl hypochlorites (7) is a reliable way [5]; the Kolbe electrolysis of carboxylates can be used [6];

---

2. F. Paneth and W. Hofeditz, *Chem. Ber.*, *62*, 1335 (1929).
3. (a) W.A. Pryor and R.W. Henderson, *J. Amer. Chem. Soc.*, *92*, 7234 (1970); (b) P. Neta, *Chem. Rev.*, *72*, 533 (1972).
4. For review, see (a) C. Walling, "Free Radicals in Solution", Wiley, New York, 1957; (b) G.W. Wheland, "Advanced Organic Chemistry", 3rd Edit., Wiley, New York, 1960; (c) W.A. Pryor, "Free Radicals", McGraw-Hill, New York, 1966; (d) *Pure Appl. Chem.*, *15*, No. 1 (1967) (a condensed version of this reference is given in *Chem. Eng. News*, *44*, 90, Oct. 3, 1966); (e) E. Müller, A. Rieker, K. Scheffler, and A. Moosmayer, *Angew. Chem., Int. Ed. Engl.*, *5*, 6 (1966).
5. C. Walling and B.B. Jacknow, *J. Amer. Chem. Soc.*, *82*, 6108 (1960).
6. W.B. Smith and H.G. Gilde, *J. Amer. Chem. Soc.*, *81*, 5325 (1959).

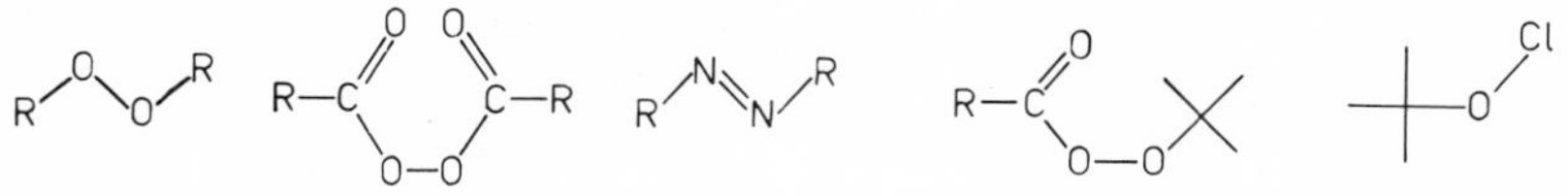

(3) (4) (5) (6) (7)

a convenient and very specific way is the photolysis of a mixture of t-butyl peroxide, triethylsilane and a bromide at -40°, presumably via the sequence [7]:

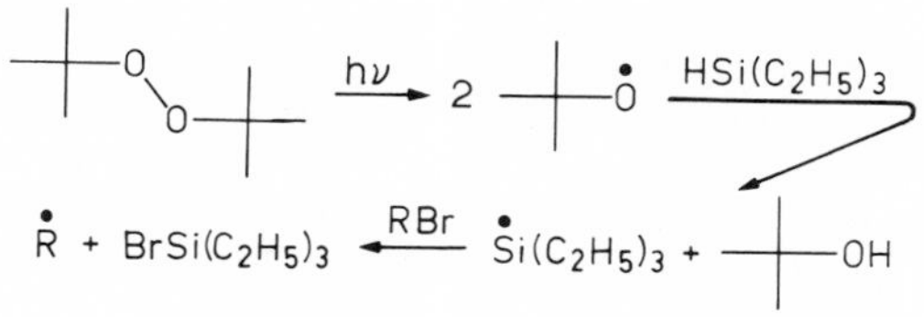

$$(CH_3)_3C{-}O{-}O{-}C(CH_3)_3 \xrightarrow{h\nu} 2\ (CH_3)_3C{-}\dot{O} \xrightarrow{HSi(C_2H_5)_3} \dot{S}i(C_2H_5)_3 + (CH_3)_3C{-}OH \xrightarrow{RBr} \dot{R} + BrSi(C_2H_5)_3$$

In fact, the technique can be as simple as the passing of a solution of the species of interest in an inert solvent through the cone of a flame; thus, aqueous solutions of sodium pivalate (8) produce the dimer product (9) that way [8].

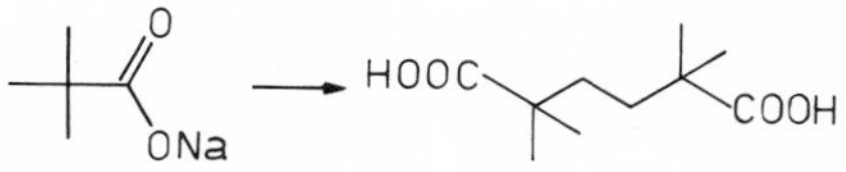

(8) (9)

The Hunsdiecker reaction of bromine with silver carboxylates such as (10) and (11) to give bromides (12) and (13) clearly involves radicals [9], and so on.

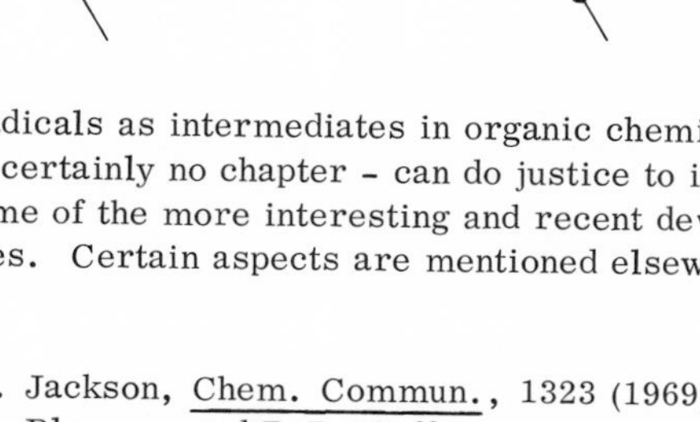

(10) (11) (12) (13)

The role of free radicals as intermediates in organic chemistry is so pervasive that no book - and certainly no chapter - can do justice to it. This writing is therefore limited to some of the more interesting and recent developments, and to a few historical snatches. Certain aspects are mentioned elsewhere in this book;

---

7. A. Hudson and R.A. Jackson, Chem. Commun., 1323 (1969).
8. C.S. Cleaver, L.G. Blosser, and D.D. Coffman, J. Amer. Chem. Soc., 81, 1120 (1959).
9. D.E. Applequist and A.H. Peterson, J. Amer. Chem. Soc., 82, 2372 (1960).

these include the role of radicals in photochemistry (Chr. 13), in photo-oxidation (Chr. 24, Sec. 8) and in some non-concerted reactions (Chr. 14). The topics of polymerization and autoxidation are perhaps the principal ones omitted here [4a].

18-2. RADICAL STABILITY

In this section, which is meant only as a survey, we ignore the question of thermodynamic vs. kinetic stability, and do not concern ourselves with such distinctions as stability to dimerization vs. stability to abstraction, to oxygen and so on.

As it turns out, hindrance, delocalization, chelation and - in some as yet unclear way - the nature of the adjacent atoms all play a role. In the triarylmethyls the stability of the radical - in this case thermodynamic, compared to dimer - is governed much more by the question of hindrance than that of delocalization. Thus, tris-(2,6-dimethoxyphenyl)methyl (14) is completely unassociated even in the solid state; on the other hand the sesquixanthydryl dimer (15) gives no evidence of dissociation even in solution [10].

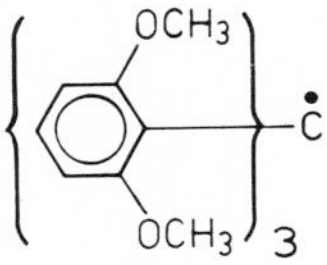

OCH3 OCH3 3 C

(14) 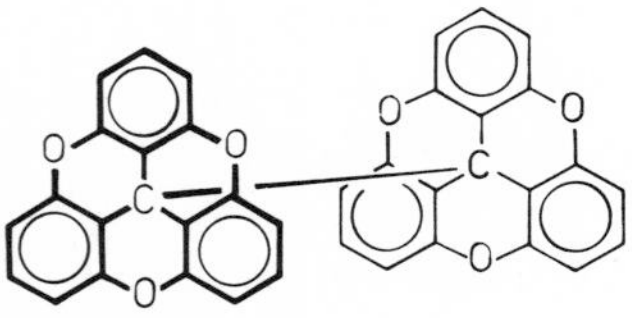 (15)

This is all the more remarkable since the potential xanthydryl radical is quite capable of planarity as models show, and since furthermore the three rings in (14) are far from coplanar - this is clear from the fact that the coupling constants of the ring protons in the esr spectrum are substantially smaller than those of triphenylmethyl itself.

An odd fact about the radical-dimer equilibrium of triarylmethyls is that p-substituents occasionally affect it very strongly. Thus, while triphenylmethyl itself (2) is completely associated in the pure solid, the tris-p-t-butyl substituted analog (16) remains free under those circumstances.

10. M. J. Sabacky, C. S. Johnson, R. G. Smith, H. S. Gutowsky, and J. C. Martin, J. Amer. Chem. Soc., 89, 2054 (1967).

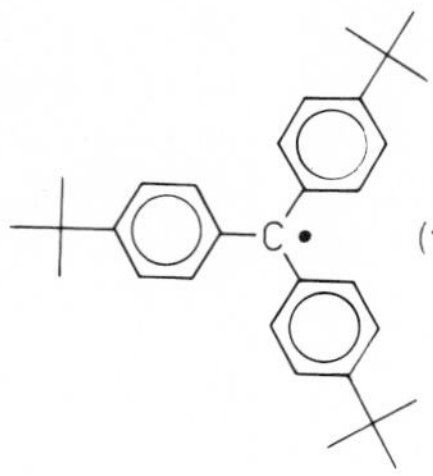

(16)

The reason for this came to light in 1968: Almost 70 years after Gomberg's initial discovery, a routine nmr investigation revealed the fact that the structure of (1) is not that of hexaphenylethane, but that of the quinoid species (17) [11].

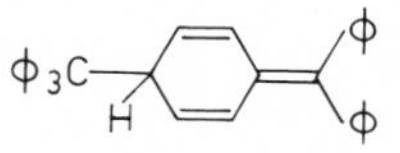

(17)

While several other such dimers were uncovered as well, at least some hexaphenylethanes do seem capable of existence; thus, the C^{13} nmr spectrum of the 9-phenylfluorenyl dimer has revealed the structure in that case to be (18) [12].

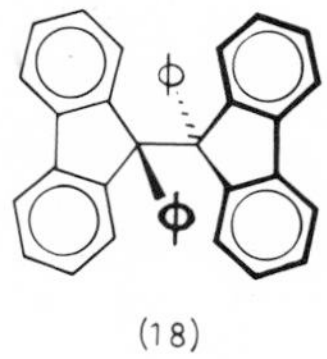

(18)

If the steric requirements are great enough, even diphenylmethyl radicals can be isolated. Thus, perchlorodiphenylmethyl (19) is a paramagnetic solid stable even to oxygen [13].

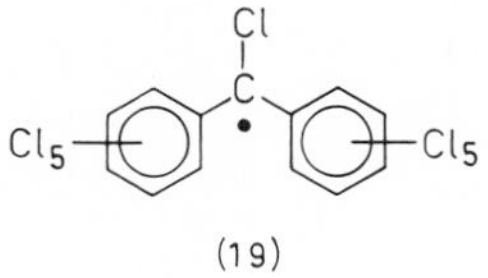

(19)

---

11. (a) H. Lankamp, W. T. Nauta, and C. Maclean, Tetrahedron Lett., 249 (1968); (b) W. B. Smith, J. Chem. Educ., 47, 535 (1970).
12. H. A. Staab, K. S. Rao, and H. Brunner, Chem. Ber., 104, 2634 (1971).
13. (a) M. Ballester and J. Riera, J. Amer. Chem. Soc., 86, 4505 (1964); (b) M. Ballester, J. Riera, J. Castafier, C. Badia, and J. M. Monso, J. Amer. Chem. Soc., 93, 2215 (1971).

It should not be assumed that resonance has no effect, however. Delocalization of the unpaired electron into the chlorine d orbitals is probably important in (19). The radical (20), often called Koelsch radical, is not so severely hindered and yet it can not only be isolated as a pure solid but it is completely inert to oxygen as well [14].

(20)

Both hindrance and delocalization also play a role in other types of radicals. Thus, the oxidation of phenols - which usually produces quinoid structures - leads to isolable radicals (21) if the hydroxy group is flanked by t-butyl groups [15].

(21)

In this case, the importance of delocalization into the ring is shown by the fact that (21) has a carbonyl absorption in the ir at 1660 cm^{-1}. An extreme example of this group is the radical galvinoxyl (22) [16], which is isolable and stable in air.

(22)

14. C.F. Koelsch, J. Amer. Chem. Soc., 79, 4439 (1957).
15. (a) V.D. Pokhodenko, V.A. Khizhnyi, and V.A. Bidzilya, Russ. Chem. Rev. (English Transl.), 37, 435 (1968); (b) L.M. Strigun, L.S. Vartanyan, and N.M. Emanuel, Russ. Chem. Rev. (English Transl.), 37, 421 (1968).
16. G.M. Coppinger, Tetrahedron, 18, 61 (1962).

Chelation may in some instances have a stabilizing effect on radicals. Taube has reported the preparation of the deeply colored solids (23) and (24) derived from lithium and 2,2'-bipyridyl [17] (from these two compounds he was then able to prepare many others in which atoms such as silicon (25) are stabilized in odd valence states). The symmetrical nature of these species is exemplified by boron compound (26); since by coincidence $a_{B^{11}}$ and a_N are equal at 2.7 gauss, the esr spectrum of this compound shows a beautifully simple pattern with signals equally spaced in the ratio 1:5:15:31:49:61:61:49:31:15:5:1 [18].

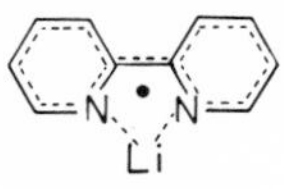

(23) 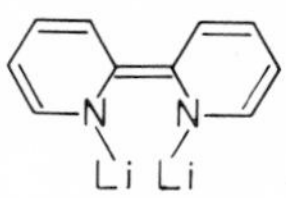 (24) 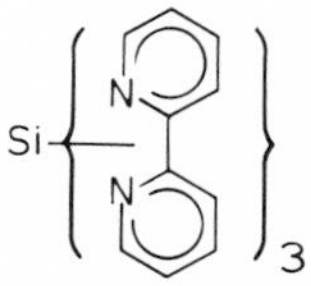 (25) 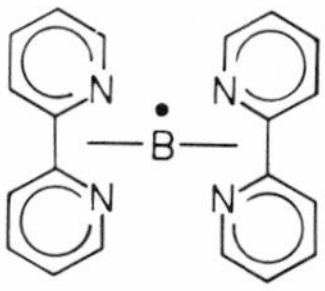 (26)

The nitroxides (27) are a very stable group of radicals [19], especially if bulky or electronegative groups flank the N-O group.

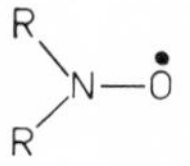

(27)

The stability reminds one of the inorganic radicals NO and NO_2. While dimerization to a peroxide with its weak O-O bond would not lower the energy of the system very much, these radicals do not seem to be able to stabilize themselves by hydrogen abstraction like most carbon radicals - this in spite of the large O-H bond energy. This inertness has been used in many ways. Thus, one of the interesting examples of nitroxide radicals is t-butylferrocenylnitroxide, (28), which is stable in the solid state.

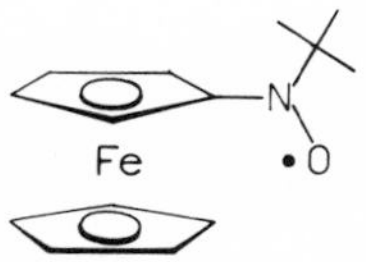

(28)

17. S. Herzog and R. Taube, Z. Chem., 2, 208 (1962).
18. M.A. Kuck and G. Urry, J. Amer. Chem. Soc., 88, 426 (1966).
19. E.G. Rozantsev and V.D. Sholle, Synthesis, 3, 190, 401 (1971).

Its X-ray diffraction pattern has been determined, and it shows that no distortion occurs so as to permit Fe-O bonding [20] (this question has also been raised with respect to the ferrocenylcarbonium ion; see p. 816). The liquid radical di-*t*-butylnitroxide (29) can be used as a solvent for other radicals; the feature that makes such a solvent especially attractive is that because of its rapid spin exchange with the solute radical it is able to sharpen the nmr signals of radicals to the point where they can be measured [21].

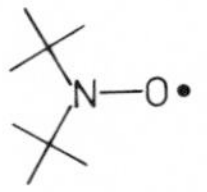

(29)

Hydroxylamines such as (30) may be considered precursors of such radicals, and these compounds are sometimes used to indicate the presence of radicals themselves too short-lived for direct observation. The radical (31) may be used that way; it is stable in solution, and its esr spectrum (a 1:1:1 triplet) is readily detected [22] (see further Sec. 18-3).

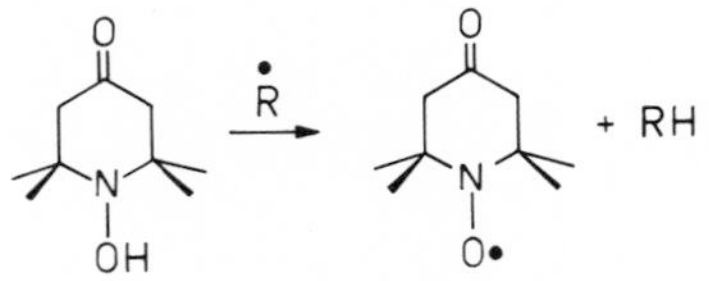

(30) (31)

It has been postulated that the electronegativity of the substituents on the nitrogen atom may be important in determining the radical stability. Thus, bistrifluoromethylnitroxide (32) is a stable, purple gas at room temperature. In order to discuss this point we momentarily digress to mention the McConnell equation [23] (cf. p. 71)

$$a = Q\rho$$

where a is a coupling constant for a given atom, ρ is the spin density on that atom and Q is a constant depending especially on the type of atom; for instance, it has the approximate value of -25 for hydrogen. Now in most nitroxides a_N is about 15-20 gauss, but in the case of (32) it was found that $a_N = 8.2$ and $a_F = 9.3$ gauss [24].

---

20. A.R. Forrester, S.P. Hepburn, R.S. Dunlop, and H.H. Mills, *Chem. Commun.*, 698 (1969).
21. R.W. Kreilick, *J. Amer. Chem. Soc.*, *90*, 2711 (1968).
22. For an example, see W.J. le Noble, E.M. Schulman, and D.N. Skulnik, *J. Amer. Chem. Soc.*, *93*, 4710 (1971).
23. H.M. McConnell and D.B. Chestnut, *J. Chem. Phys.*, *28*, 107 (1958), and earlier papers.
24. W.D. Blackley and R.R. Reinhard, *J. Amer. Chem. Soc.*, *87*, 802 (1965).

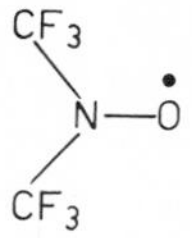

(32)

Since these coupling constants are temperature dependent in an opposite sense (presumably the N-C bond rotation is slower at lower temperature), many authors have recently preferred delocalization as the important feature here [25] rather than the electronegativity.

If α-H is present, this atom is readily abstracted. One can try to prevent the abstraction reactions by making use of Bredt's rule such as in (33); however, slow dimerization still occurs to give product (34) [26].

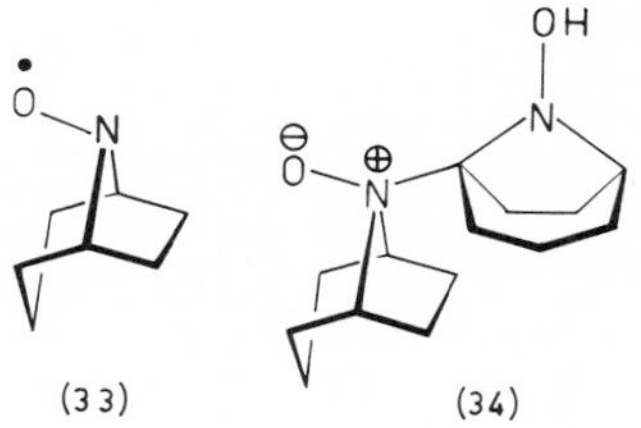

(33) (34)

Kosower and Poziomek have prepared an interesting analog of the nitroxyl radicals: 1-alkyl-4-carbomethoxypyridinyl (35) is isolable as a green liquid, moderately stable in pure form at room temperature [27].

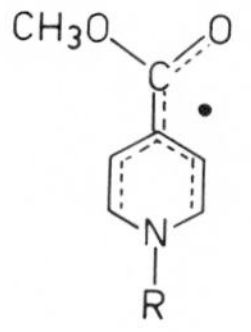

(35)

A different type of analog is provided by certain hydrazyls such as 2,2-diphenylpicrylhydrazyl (36). There appears to be no evidence for any dimer in this case. Pure crystals of (36) have been compressed to 100,000 atmospheres or more, but although the dimer should have a smaller volume than the free radical, no dimerization could be affected. The only observable change was a tremendous,

25. P.J. Scheidler and J.R. Bolton, J. Amer. Chem. Soc., 88, 371 (1966).
26. G.D. Mendenhall and K.U. Ingold, J. Amer. Chem. Soc., 94, 7166 (1972).
27. E.M. Kosower and E.J. Poziomek, J. Amer. Chem. Soc., 85, 2035 (1963).

reversible drop in resistivity, from about 10^{13} ohm cm at room pressure to 10^4 at the highest pressures; evidently the unpaired electron under these conditions is not localized to one free radical alone, but it winds up in a conductance band of the whole crystal, and a semi-conductor results [28].

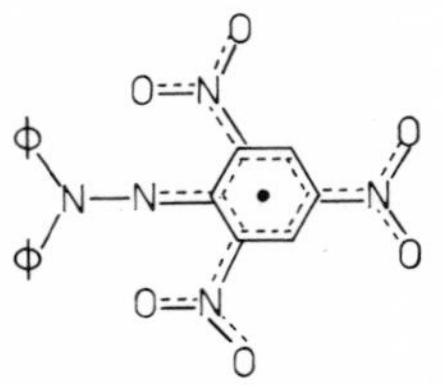

(36)

The verdazyls are further examples of this type; compounds such as (37) are isolable and stable to air [29].

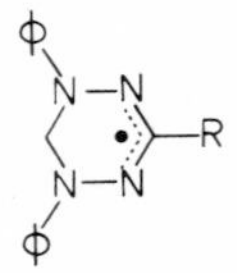

(37)

Such a symmetrical structure (verifiable by means of esr) is also a feature of the α-nitronylnitroxides, (38); they are isolable and stable to oxygen [30].

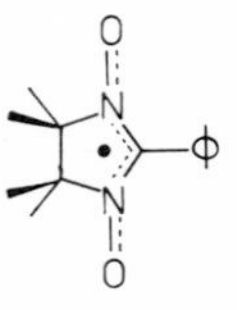

(38)

Since most radicals are neutral species, the matrix isolation technique is ideally suited to study the less stable species. Thus both benzoyl bromide and benzaldehyde can be irradiated in dilute films at 77°K to give the same orange colored radical ϕ-C=O, the esr spectrum of which can then readily be measured.

---

28. H. Inokuchi, I. Shirotani, and S. Minomura, Bull. Chem. Soc. Japan, 37, 1234 (1964).
29. See for example R. Kuhn, F. A. Neugebauer, and H. Trischmann, Monatsh. Chem., 97, 525 (1966) and earlier references quoted there.
30. (a) J. H. Osiecki and E. F. Ullman, J. Amer. Chem. Soc., 90, 1078 (1968); (b) E. F. Ullman, J. H. Osiecki, D. G. B. Boocock, and R. Darcy, J. Amer. Chem. Soc., 94, 7049 (1972).

If the matrix in the former experiment is allowed to warm up, recombination occurs with the bromine atoms, and both color and esr signal disappear at 140°K [31]. A similar experiment based on a brief (1 μsec) but powerful laser flash has revealed that the esr spectrum in the first few microseconds after the flash appears in emission - a phenomenon resembling CIDNP [32].

Often quite simple experiments suffice to study the properties of radicals. The entire theoretical 10-line spectrum of t-butyl can be observed at 130°K upon the irradiation of pivalic acid (39) [33].

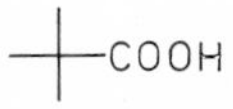

$(CH_3)_3C-COOH$

(39)

In an even more remarkable experiment, small traces of methyl iodide and D- and C^{13} labeled methyl iodides were absorbed on glass surfaces and then photolyzed at room temperature; the glass subsequently exhibited the typical 4-, 7- and 8-line esr spectra of methyl, trideuteriomethyl, and C^{13}-labelled methyl for several days afterwards [34]. Thus, the sensitivity of esr (to perhaps 10^{17} spins or so) enables one to study virtually any radical desired, stable or not.

Finally, it may be mentioned that radicals may be stabilized by transition metal bonding; thus, species such as (40) have been isolated and characterized [35].

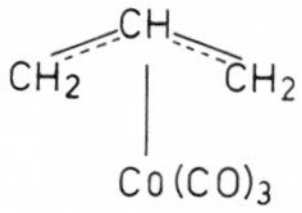

$(CH_2{\cdots}CH{\cdots}CH_2)Co(CO)_3$

(40)

18-3. SPIN TRAPPING

Reactive free radicals play a role in many reactions. Although it is always possible to study any radical at low temperature, it is often desirable to learn something about the nature of the reaction under the conditions at which it takes place; this often means conditions under which the radicals are too short-lived to

31. U. Schmidt, K.H. Kabitzke, and K. Markau, Angew. Chem., Int. Ed. Engl., 4, 355 (1965).
32. P.W. Atkins, I.C. Buchanan, R.C. Gurd, K.A. McLauchlan, and A.F. Simpson, Chem. Commun., 513 (1970).
33. H. Shields and P. Hamrick, J. Chem. Phys., 42, 443 (1965).
34. J. Turkevich and Y. Fujita, Science, 152, 1619 (1966).
35. R. Seip, Acta Chem. Scand., 26, 1966 (1972).

be detected directly. One of the early solutions to this problem was the use of radical traps. The reaction experiment would be carried out as before, but in the presence of some material from which a hydrogen atom is readily abstracted so as to leave a much more stable radical, or a material which is itself a stable radical that will scavenge its more reactive cousins [36]. These approaches usually give information only about the fact that radicals are intervening, and not about the identity of these radicals.

In order to solve that problem the technique of spin trapping has been devised, evidently more or less simultaneously by several workers [37].

The general idea is that when radicals are being generated in a given medium, if a suitable nitroso compound such as 2-nitroso-2-methylpropane (41) is present, addition will occur to give a long-lived nitroxide, and the coupling constants (both a_N and a_H) will then reveal the identity of the radical.

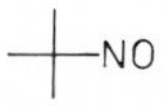

—NO

(41)

In many cases, the nitroxide derivative is already known, and only comparison is required, but in others identification depends on more general experience. Thus, a_N is now known to be 10-15 gauss for alkyl radicals, nearly 30 gauss for alkoxy radicals and only 6-9 gauss for acyl radicals. Nitrones such as (42) have a similar use. In this case $a_{\alpha\text{-H}}$ can also be used.

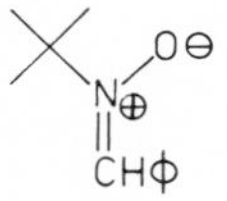

O⊖
N⊕
CHϕ

(42)

It is usually 2-3 gauss, but now the hydrogen atoms near the original radical site are further away, making it more difficult to identify the radical if its splittings are not already known.

As an example, the succinimidyl radical might be mentioned. This species had often been mentioned as a likely intermediate in the allylic brominations with N-bromosuccinimide, (43) [38].

---

36. See for instance, H. J. Shine, J. A. Waters, and D. M. Hoffman, J. Amer. Chem. Soc., 85, 3613 (1963).
37. (a) A. Mackor, T. A. J. W. Wajer, T. J. de Boer, and J. D. W. van Voorst, Tetrahedron Lett., 2115 (1966); (b) C. Lagercrantz, J. Phys. Chem., 75, 3466 (1971) and earlier work quoted there; (c) E. G. Janzen, Accounts Chem. Res., 4, 31 (1971) and earlier work quoted there.
38. J. H. Incremona and J. C. Martin, J. Amer. Chem. Soc., 92, 627 (1970).

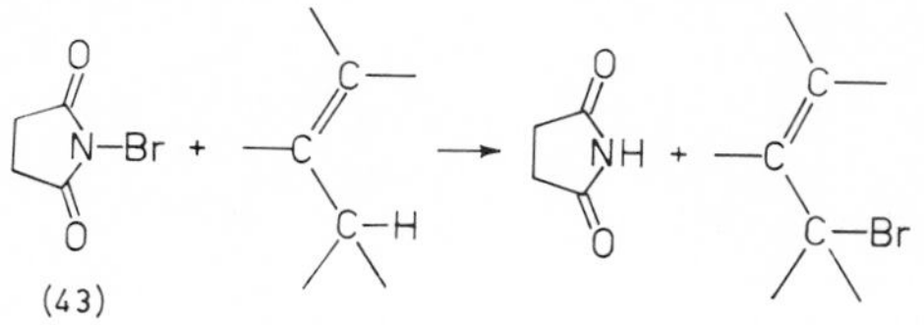

(43)

In the presence of (41), a species (44) forms with 9 lines of equal intensity - indicative of two non-equivalent nitrogen atoms (a_{N1} = 16.4 gauss, a_{N2} = 1.81 gauss) [39].

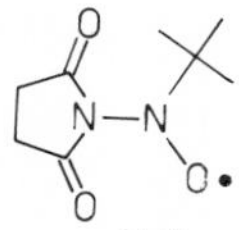

(44)

Similarly, if (41) is added to a solution in which acetyl peroxide is decomposing, an esr signal consisting of a 1:1:1 triplet of 1:3:3:1 quartets (a_N = 15.2 gauss, a_H = 11.3 g) is observed [40] showing that methyl radicals have been trapped to give (45).

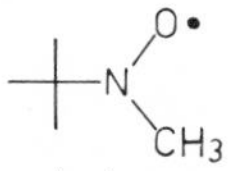

(45)

The photoreduction of benzophenone by alcohols also clearly involves radicals, as was shown by the trapping of these species by means of (41) [41].

These experiments require a certain caution in that anions may add to the spin traps as well, and if subsequent oxidation is possible, for instance by adventitious oxygen, then a signal may be obtained even though no radicals were involved. Thus, the anionic polymerization of styrene initiated by butyllithium can produce nitroxide signals this way [42]. Conversely, if an easily oxidized species such as iodide or thiophenol is present, some nitroxides may be reduced to the hydroxylamine and thus escape detection [43]. Other nitroxides such as the species (46) readily disproportionate to give (41) and (47) [44].

---

39. C. Lagercrantz and S. Forshult, *Acta Chem. Scand.*, *23*, 708 (1969).
40. M.J. Perkins, P. Ward, and A. Horsfield, *J. Chem. Soc., B*, 395 (1970).
41. I.H. Leaver and G.C. Ramsay, *Tetrahedron*, *25*, 5669 (1969).
42. A.R. Forrester and S.P. Hepburn, *J. Chem. Soc., C*, 701 (1971).
43. (a) H.J. Emeleus, P.M. Spaziante, and S.M. Williamson, *Chem. Commun.*, 768 (1969); (b) K. Murayama and T. Yoshioka, *Bull. Chem. Soc. Japan*, *42*, 1942 (1969).
44. R.L. Craig and J.S. Roberts, *Chem. Commun.*, 1142 (1972).

(46) (47)

For these and other reasons [45] various alternative spin traps have recently been proposed, among them 1-nitrosoadamantane (48) [46], perdeuterio-t-nitrosobutane (49) [47] and benzonitrile-N-oxide (50) [48]; a discussion of the merits and demerits of each one is beyond the scope of this book, however.

NO $(CD_3)_3C-NO$ $\phi-C\equiv\overset{\oplus}{N}-\overset{\ominus}{O}$

(48) (49) (50)

18-4. THE SHAPE OF RADICALS [49]

The methyl radical has been examined in the vacuum uv by Herzberg who described it as planar or nearly planar [50]. This same conclusion has also been reached on the basis of other studies, but it is also clear that the difference in energy between planar and pyramidal radicals is not large.

The rates of thermal decomposition of the t-butyl peresters (51)-(54) are in the relative order 1:0.4:0.05:0.0004 [51]. This clearly suggests that radicals prefer to be planar, though by not merely as wide a margin as carbonium ions (see Chr. 20).

(51) (52) (53) (54)

45. C.M. Camaggi and M.J. Perkins, J. Chem. Soc., Perkin II, 507 (1972).
46. J.W. Hartgerink, J.B.F.N. Engberts, T.A.J.W. Wajer, and T.J. de Boer, Rec. Trav. Chim., Pays-Bas, 88, 481 (1969).
47. R.J. Holman and M.J. Perkins, J. Chem. Soc., C, 2324 (1971).
48. B.C. Gilbert, V. Malatesta, and R.O.C. Norman, J. Amer. Chem. Soc., 93, 3290 (1971).
49. C. Rüchardt, Angew. Chem., Int. Ed. Engl., 9, 830 (1970).
50. G. Herzberg, Proc. Roy. Soc. (London), A262, 291 (1961).
51. (a) J.P. Lorand, S.D. Chodroff, and R.W. Wallace, J. Amer. Chem. Soc., 90, 5266 (1968); (b) R.C. Fort and R.E. Franklin, J. Amer. Chem. Soc., 90, 5267 (1968).

The esr spectra of the 1-adamantyl and 1-bicyclo[2.2.2]octyl radicals (generated by photolysis of (52) and (53) at -120°) have been measured, and the coupling constants deviate considerably from those normally found for tertiary radicals [52].

If one of the bond angles must be much smaller than 120°, the radical evidently prefers to be pyramidal. Thus, the 7-oxa-2-norbornyl radical (55) has been generated in solution, and its esr spectrum shows that $a_{3H\text{-endo}}$ equals about 20 gauss and $a_{3H\text{-exo}}$ is about 40 gauss.

(55)

If the radical were planar, these values should have been approximately equal; as it is, we must conclude that the 2-hydrogen atom is bent in the *endo*-direction [53]. Similarly, an abnormal value for the α-hydrogen atom coupling in the cyclopropyl radical has indicated that this radical is non-planar also; in this case the four β-couplings are all equal so that rapid inversion must be occurring [54]. Studies of this radical are made difficult by the fact that they are very unstable to ring opening and conversion to the allyl radical. Unlike the corresponding carbonium ion (p. 811) however, the formation and opening of the ring are not concerted, and in the presence of a good hydrogen atom donor small quantities of cyclopropanes can be obtained [55]. A slight extrapolation brings us to the vinyl radical - which is clearly bent. Thermolysis of the *cis*- and *trans*- isomeric perester pair (56) and (57) in the presence of cumene gives exclusively *cis*- and *trans*-1-propenyl methyl ether, respectively; thus the vinyl radicals are able to retain their original geometries until they disappear by hydrogen abstraction [56].

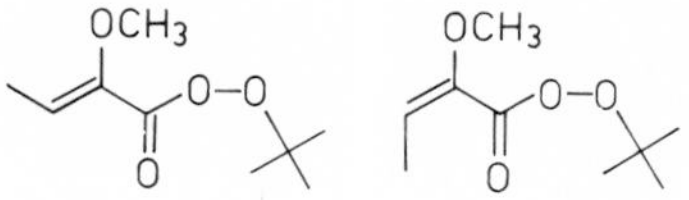

(56) (57)

Substitution generally tends to stabilize radicals: Tertiary radicals are formed more easily than the secondary ones, and so on. This is true especially when planarity is possible, and the conclusion is that alkyl-substitution reinforces

---

52. P.J. Krusic, T.A. Rettig, and P. von R. Schleyer, *J. Amer. Chem. Soc.*, *94*, 995 (1972).
53. T. Kawamura, T. Koyama, and T. Yonezawa, *J. Amer. Chem. Soc.*, *92*, 7222 (1970).
54. R.W. Fessenden and R.H. Schuler, *J. Chem. Phys.*, *39*, 2147 (1963).
55. (a) S. Sustmann, C. Rüchardt, A. Bieberbach, and G. Boche, *Tetrahedron Lett.*, 4759 (1972); (b) S. Sustmann and C. Rüchardt, *Tetrahedron Lett.*, 4765 (1972).
56. M.S. Liu, S. Soloway, D.K. Wedegaertner, and J.A. Kampmeier, *J. Amer. Chem. Soc.*, *93*, 3809 (1971).

the slight preference of radicals to be planar. The reason for this is steric hindrance; when aryl substituents are considered, delocalization may also help [49].

The resonance interaction of an unpaired electron with neighboring unsaturation can be studied by means of the conformational integrity of allylic radicals, since such resonance imparts a degree of double bond character to a normally single bond preventing its normal freedom of rotation.

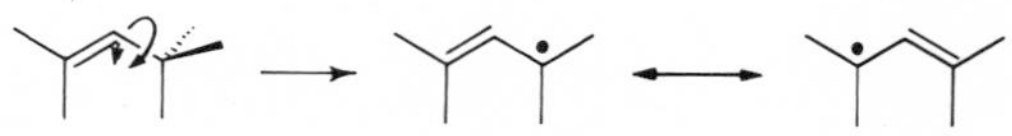

In the gas phase where such radicals have a relatively long life time, conformational integrity is not maintained; thus, the three azo compounds (58)-(60) all give an identical mixture of the six octadienes (61)-(66) when heated in the gas phase [57].

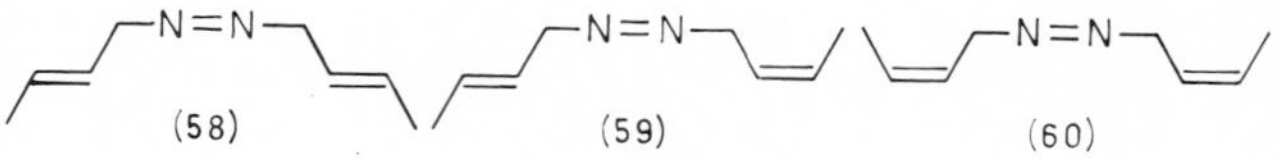

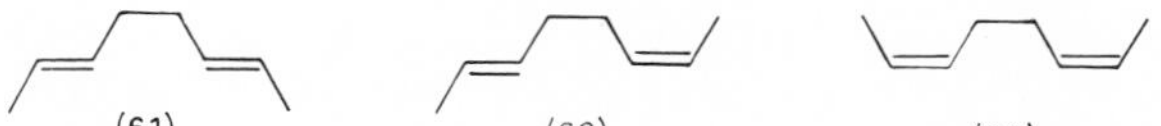

In solution the freedom of internal and external motions is much more severly impeded, and in that phase allylic radicals do indeed have conformational stability. Thus, when a mixture of *trans*-2-butene and *t*-butyl hypochlorite is irradiated, a chain reaction occurs in which the principal product, 1-chloro-2-butene, if purely *trans*-; likewise, *cis*-2-butene gives only the *cis*-isomer [58].

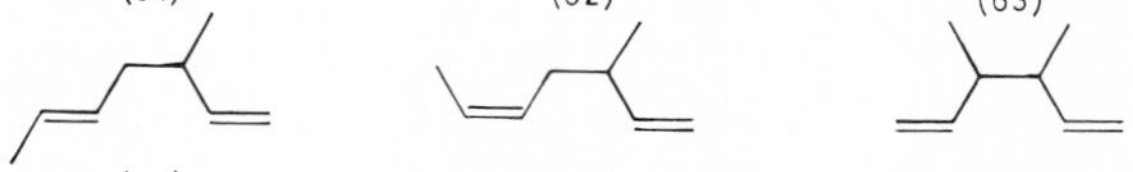

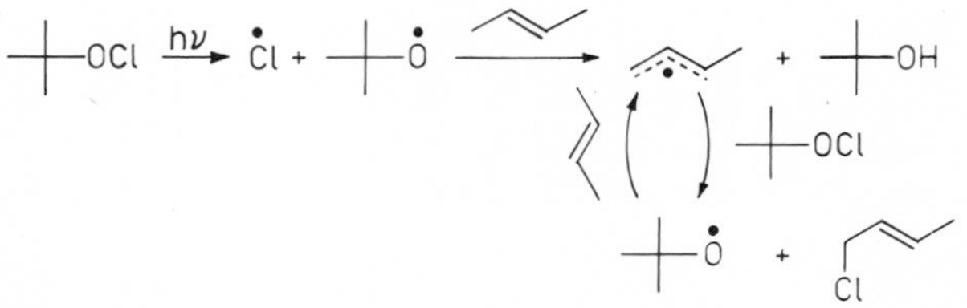

Among the carbon radicals, there is one other type that seems to be pyramidal - namely the fluorine substituted species. This has again been concluded from esr spectra [59].

---

57. R.J. Crawford, J. Hamelin, and B. Strehlke, *J. Amer. Chem. Soc.*, **93**, 3810 (1971).
58. C. Walling and W. Thaler, *J. Amer. Chem. Soc.*, **83**, 3877 (1961).
59. R.W. Fessenden and R.H. Schuler, *J. Chem. Phys.*, **43**, 2704 (1965).

The other radicals of group IV all seem to be pyramidal. Silicon radicals have been generated by hydrogen abstraction with the t-butoxy radical and their esr splittings are known (silicon contains a few percent Si^{29} of spin 1/2). The SiH_3 radical has a silicon coupling constant of 266 gauss, and this declines only slightly in the series to $SiMe_3$; all of these radicals are thought to be pyramidal since these high values of a indicate a high s character of the unpaired electron [60]. These conclusions are born out by ir studies of matrix isolated SiF_3 at $4^{\circ}K$; the spectrum can be interpreted only in terms of a pyramidal radical [61].

Still another approach is the examination of radicals generated from optically active precursors. If the resolved germanium compound (67) is treated with benzoyl peroxide in refluxing carbon tetrachloride, the chloride (68) is obtained with retained configuration; the intermediate germyl radicals must therefore be pyramidal and have a fair activation energy to inversion [62].

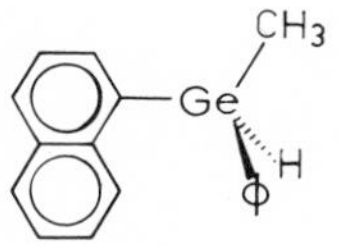

(67) 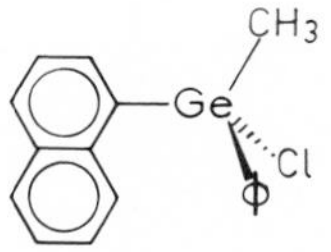 (68)

The controversy about the nature of certain carbonium ions (Chr. 20) induced similar arguments in free radical chemistry. "Participation" seems well established in some cases. Thus, t-butyl 8-phenylthio-1-pernaphthoate (69) decomposes 4000 times faster than t-butyl 1-pernaphthoate (70), the principal product is unrearranged, (71), and hence the rate enhancement is not due to the formation of an unusually stable product [63], and a "bridged" or "non-classical" intermediate radical (72) seems likely.

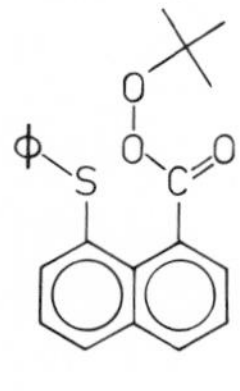

(69) 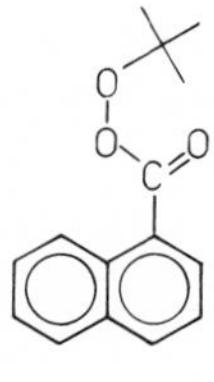 (70) 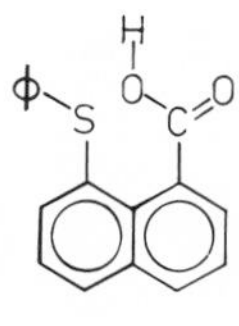 (71) 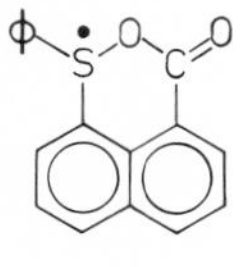 (72)

Bromine has similarly been considered able to participate. Thus, it has been noted that in photobrominations of alkanes, the dibromo- products are usually

60. P.J. Krusic and J.K. Kochi, J. Amer. Chem. Soc., 91, 3938 (1969).
61. D.E. Milligan, M.E. Jacox, and W.A. Guillory, J. Chem. Phys., 49, 5330 (1968).
62. H. Sakurai and K. Mochida, Chem. Commun., 1581 (1971).
63. T.H. Fisher and J.C. Martin, J. Amer. Chem. Soc., 88, 3382 (1966).

vicinal; furthermore, when resolved 1-bromo-2-methylbutane (73) is photobrominated, the vicinal product (74) is optically active [64].

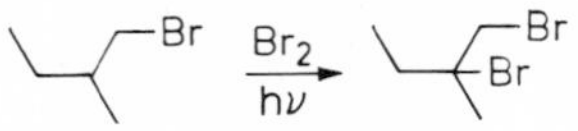

(73) $\xrightarrow[h\nu]{Br_2}$ (74)

A similar finding had been reported by Kharasch [65] as early as 1940. Similarly, in the *cis*- and *trans*-4-bromo-1-*t*-butylcyclohexanes (75) and (76) the former gives exclusively the trans vicinal dibromide (77); (76) reacts more slowly, and is much less selective [66]. Participation such as in (78) is clearly indicated.

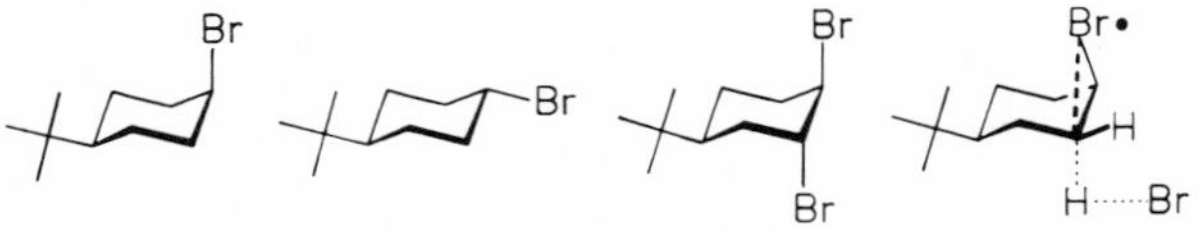

(75) (76) (77) (78)

However, these results were disputed by Tanner, who claimed that vicinal dibromides are not formed to any great degree early in the reaction, although in the later stages those products do build up. He suggested [67] that bromine atoms abstract a β-hydrogen atom and that the resulting radical sheds a bromine atom to yield an olefin which then undergoes immediate ionic *trans*-addition.

This alternative has now been ruled out. Ronneau has found that if radiobromine labelled substrate is used and if HBr and Br_2 are trapped and removed at any stage, very little inorganic radioactivity is encountered. On the other hand, if the vicinal dibromide is isolated it turns out that nearly all activity is still in the original position [68]. Shortly afterwards Skell [69] and Traynham [70] demonstrated that Tanner's experiments could not be repeated; at 1% reaction the product ratio was the same as that obtained when the reaction is over.

64. P.S. Skell, D.L. Tuleen, and P.D. Readio, *J. Amer. Chem. Soc.*, *85*, 2849 (1963).
65. H.C. Brown, M.S. Kharasch, and T.H. Chao, *J. Amer. Chem. Soc.*, *62*, 3435 (1940).
66. P.S. Skell and P.D. Readio, *J. Amer. Chem. Soc.*, *86*, 3334 (1964).
67. D.D. Tanner, D. Darwish, M.W. Mosher, and N.J. Bunce, *J. Amer. Chem. Soc.*, *91*, 7398 (1969).
68. C. Ronneau, J.P. Soumillion, P. Dejaifve, and A. Bruylants, *Tetrahedron Lett.*, 317 (1972).
69. P.S. Skell and K.J. Shea, *J. Amer. Chem. Soc.*, *94*, 6550 (1972).
70. J.G. Traynham, E.E. Green, Y.S. Lee, F. Schweinsberg, and C.E. Low, *J. Amer. Chem. Soc.*, *94*, 6552 (1972).

Carbon participation has also been claimed and doubted, and it seems likely that it occurs only as a prelude to rearrangement. An early report [71] that anti-7-norbornenyl halides (79) are reduced by tri-n-butyltin deuteride (80) to give exclusively the retained product (81) - an observation nicely accounted for by a bridged radical such as (82) - appeared in 1968.

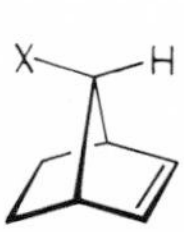

(79) 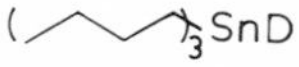 (80) 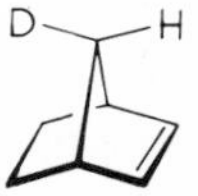 (81)  (82)

There is a possibility of experimental error in this case, however; a very similar experiment by Russell [72] drew the conclusion that the reduction was not stereospecific. Cristol [73] found that both syn- and anti-benzonorbornenyl bromides (83) give 50-50 mixtures of product.

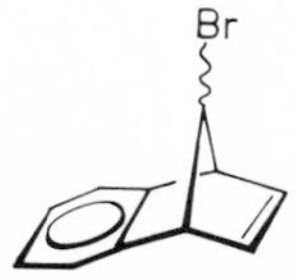

(83)

Monoreduction of the dichloro analog (84) is fairly specific to give (85) but this is opposite to what would be expected if participation were an important factor [74].

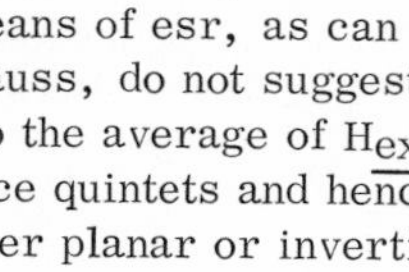

(84) 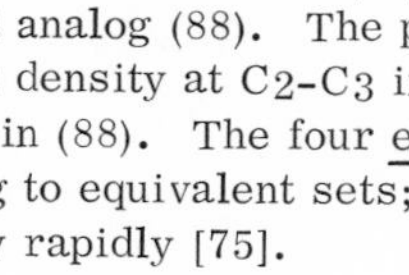 (85)

During photolysis of the perester (86), the norbornenyl radical (87) can be studied directly by means of esr, as can the saturated analog (88). The proton splittings, indicated in gauss, do not suggest a large spin density at C_2-C_3 in (87); it is in fact similar to the average of H_{exo} and H_{endo} in (88). The four exo- and endo-protons produce quintets and hence both belong to equivalent sets; hence in (88) the radical is either planar or inverting extremely rapidly [75].

---

71. J. Warkentin and E. Sanford, J. Amer. Chem. Soc., 90, 1667 (1968).
72. G.A. Russell and D.W. Lamson, J. Amer. Chem. Soc., 91, 3967 (1969).
73. S.J. Cristol and A.L. Noreen, J. Amer. Chem. Soc., 91, 3969 (1969).
74. B.B. Jarvis and J.B. Yount, Chem. Commun., 1405 (1969).
75. P. Bakuzis, J.K. Kochi, and P.J. Krusic, J. Amer. Chem. Soc., 92, 1434 (1970).

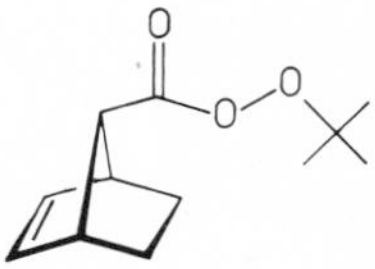

(86) 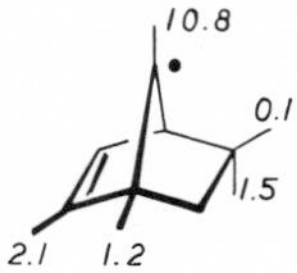 (87) 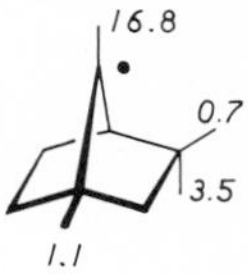 (88)

In other cases carbon participation fares just as badly. Benzoyl peroxide initiated addition of bromotrichloromethane to benzonorbornadiene produces only adduct (89) and no (90) or (91) such as might have been expected from bridged species such as (92) or (93) [76].

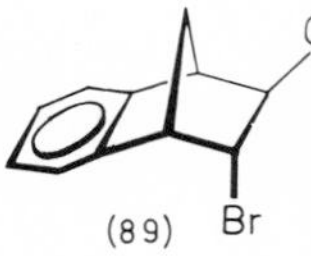

(89) 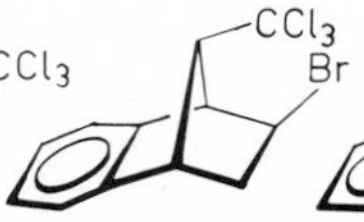 (90)  (91)

(92)  (93)

Cholesteryl and cholestanyl chlorides (94) and (95) can be dechlorinated by means of sodium biphenyl via a reaction known to involve radicals - and each gives unrearranged hydrocarbon. This result clearly rules out the common hybrid intermediate (96) [77].

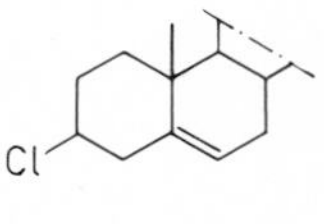

(94) 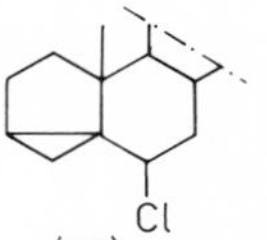 (95) 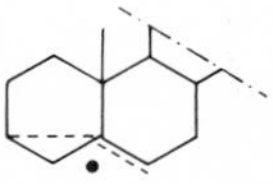 (96)

Phenyl bridging in open chain radicals is conceivable in the following case found by Urry [78]: free radical hydrogen abstraction from aldehyde (97) leads to products (98)-(100).

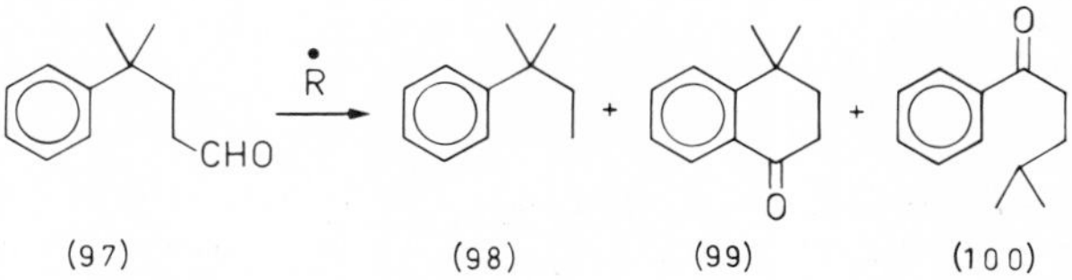

(97) (98) (99) (100)

76. L.E. Barstow and G.A. Wiley, Tetrahedron Lett., 865 (1968).
77. S.J. Cristol and R.V. Barbour, J. Amer. Chem. Soc., 88, 4262 (1966).

However, such bridging is not concomitant with radical formation. Krusic and Kochi [79] studied the photolysis of certain phenyl substituted acyl peroxides by means of esr; the 2-phenyltetramethylethyl radical (101) showed a septet and 2-phenylethyl itself (102) gave rise to a triplet of triplets, so that structures such as (103) are ruled out.

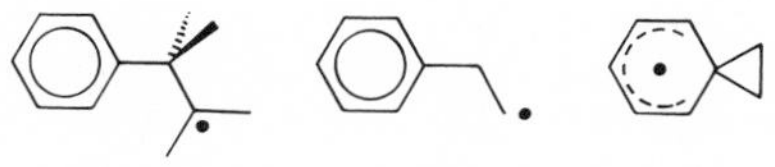

(101) (102) (103)

Bartlett has found that t-butyl per-2-exo-norbornylcarboxylate (104) is only three times faster in decomposition than its epimer (105), so that there is no evidence for a bridged 2-norbornyl radical (106) either [80].

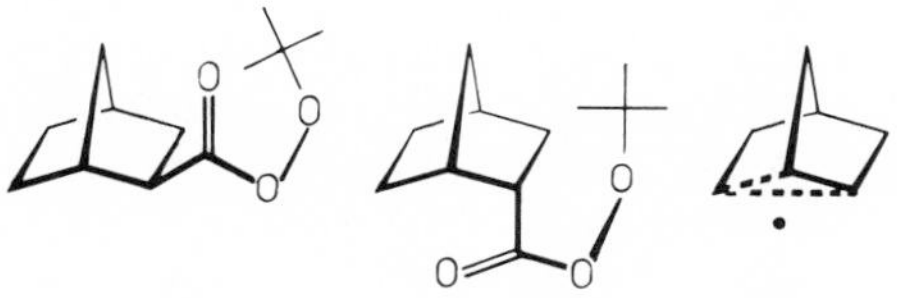

(104) (105) (106)

Finally, a fairly strong claim of cyclopropyl interaction with neighboring radical centers can be made. Martin [81] has found that the thermolysis of certain symmetrical azo compounds to give radicals does seem to be accelerated by neighboring cyclopropyl groups; thus, in the series (107)-(109) the relative rates are 1, 26 and 240, and in the series (110)-(114) they are 1, 27, 360, 285 and 2500, respectively.

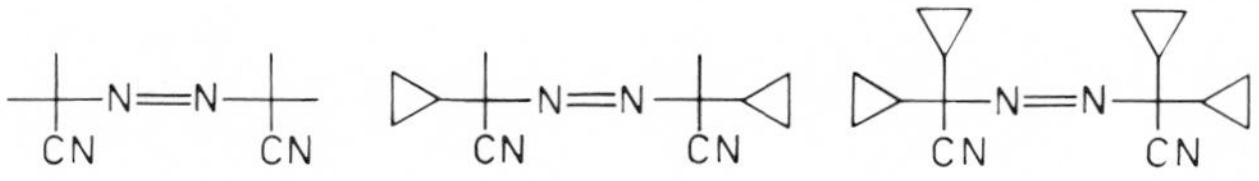

(107) (108) (109)

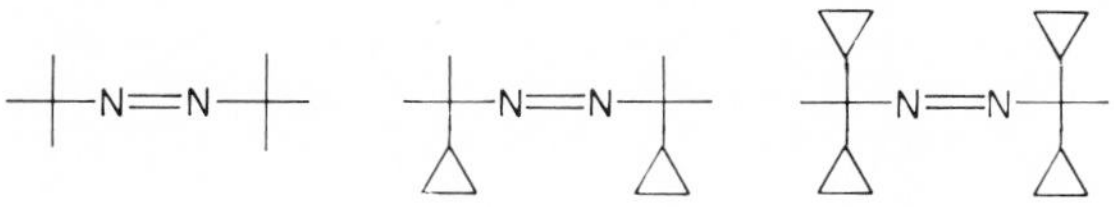

(110) (111) (112)

78. W.H. Urry, D.J. Trecker, and H.D. Hartzler, J. Org. Chem., 29, 1663 (1964).
79. (a) J.K. Kochi and P.J. Krusic, J. Amer. Chem. Soc., 91, 3940 (1969); (b) D.J. Edge and J.K. Kochi, J. Amer. Chem. Soc., 94, 7695 (1972).
80. P.D. Bartlett and J.M. McBride, J. Amer. Chem. Soc., 87, 1727 (1965).
81. J.C. Martin, J.E. Schultz, and J.W. Timberlake, Tetrahedron Lett., 4629 (1967).

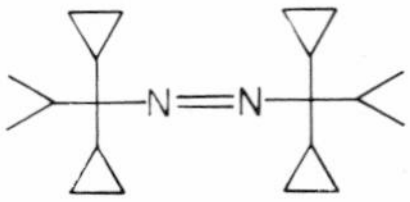

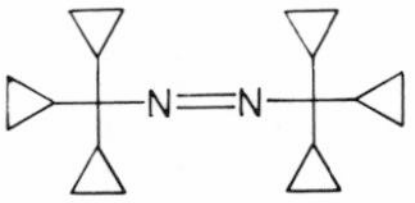

(113) (114)

The reason for these rate enhancements cannot be relief of strain via concerted ring opening of the cyclopropyl group. While ring-opened products such as (115) do form (from (108)), unrearranged products predominate, or can be made predominant by the addition of triphenylmethyl dimer (to give (116) e.g.).

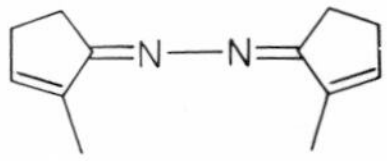

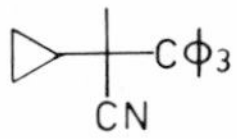

(115) (116)

Relief of ground state crowding is also unlikely to be responsible, since (113) and (114) should then have had similar rates. In recent years, the concertedness of the two C-N bond cleavages has been questioned (cf. next section), and if diazenyl radicals are involved - however fleetingly - it would seem likely that the transition state has a dipolar nature (117) - which would be reduced in energy by cyclopropyl accommodation of positive charge (p. 751).

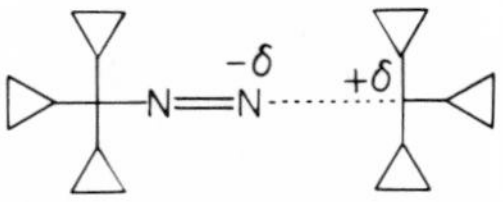

(117)

In this connection it is of interest that esr studies have shown that cyclopropyl groups bound to aromatic radicals prefer the bisected conformation as they do in carbonium ions; this preference is enhanced by positive charge and attenuated by negative charge [82]. The most direct evidence against delocalization of the impaired electron into the cyclopropyl moiety comes from Kochi [83]; who found that the cyclopropylcarbinyl and homoallyl radicals (118) and (119) are both stable at -140°, and hence they are not contributing structures to a common hybrid such as (120).

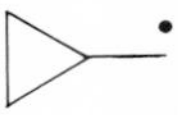

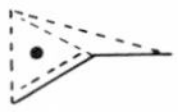

(118) (119) (120)

---

82. (a) N. L. Bauld, J. D. McDermed, C. E. Hudson, Y. S. Rim, J. Zoeller, R. D. Gordon, and J. S. Hyde, J. Amer. Chem. Soc., 91, 6666 (1969); (b) N. L. Bauld, R. Gordon, and J. Zoeller, J. Amer. Chem. Soc., 89, 3948 (1967).
83. J. K. Kochi, P. J. Krusic, and D. R. Eaton, J. Amer. Chem. Soc., 91, 1877 (1969).

18-5. CAGED RADICALS

When a mixture of azomethane (121) and perdeuterioazomethane is heated in the gas phase, a statistical mixture of ethane, ethane-d_3 and ethane-d_6 is obtained; however, if the same experiment is repeated in i-octane solution, the ethane-d_3 is absent [84].

$$CH_3N{=}NCH_3$$

(121)

This result is one of the prettier demonstrations of a phenomenon known as the cage effect. Radicals are necessarily always generated from their precursor molecules in neighboring pairs, and in the liquid phase such pairs are caged for a brief time, i.e., some time elapses before they diffuse apart. During that time they may combine; once apart, they become subject to a new set of reactions. The combination reactions must obviously be fast, to compete successfully with diffusion. Caged radicals may in fact occur in the gas phase if the density is reasonably high. Thus, azomethane in compressed propane upon photolysis gives ethane and methane; if nitrous oxide is present, the yield of ethane is unaffected, but that of methane is severely depressed [85]. The oxide is said to be a scavenger - removing the separate radicals, but not the caged ones. By means of scavengers, it can be determined which products are formed in the cage and which are not; the yields of the latter depend on the scavenger concentration, and those of the former do not. Thus, the yield of recombination product (122) from azo compound (123) has been measured in the presence of iodine, bromine and 2,2-diphenylpicrylhydrazyl; it was the same in all cases and independent of the additive concentration unless this was made extremely high [86].

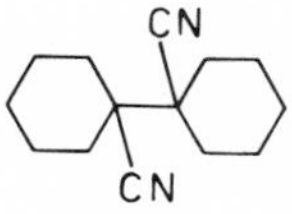

CN CN CN CN N=N

(122) 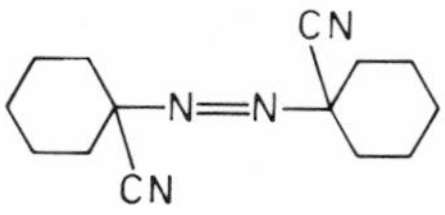 (123)

The product distribution is also very often affected strongly by such factors as solvent viscosity and pressure. Thus, when phenylazotriphenylmethane (124) is decomposed in the presence of iodine at one atmosphere, 3% of cage product (125) is obtained; if the same experiment is carried out at 2500 atmospheres, the yield goes up to 70% [87].

84. R.K. Lyon and D.H. Levy, J. Amer. Chem. Soc., 83, 4290 (1961).
85. R.K. Lyon, J. Amer. Chem. Soc., 86, 1907 (1964).
86. H.P. Waits and G.S. Hammond, J. Amer. Chem. Soc., 86, 1911 (1964).
87. (a) R.C. Lamb and J.G. Pacifici, J. Phys. Chem., 70, 314 (1966). Cf. also (b) R.C. Neuman and J.V. Behar, Tetrahedron Lett., 3281 (1968); (c) R.C. Neuman, Accounts Chem. Res., 5, 381 (1972).

ϕ—N=N—Cϕ_3 Cϕ_4

(124) (125)

The behavior of radicals in the cage has been illuminated by some beautiful experiments by Bartlett [88]. He finds that both meso- and dl-azo-bis-3-methyl-2-phenyl-2-butane, (126) and (127), respectively, give an identical 1:1 mixture of meso- and dl-hydrocarbons (128) and (129), as well as disproportionation products (130) and (131).

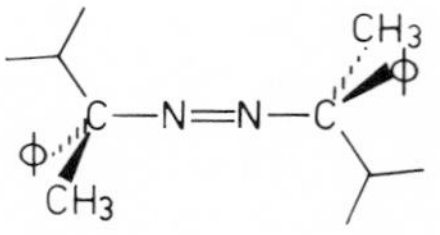

(126) 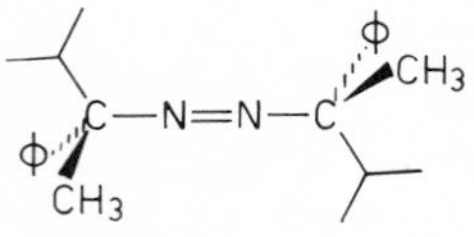 (127)

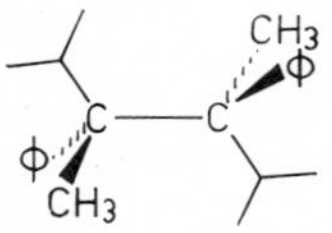

(128) 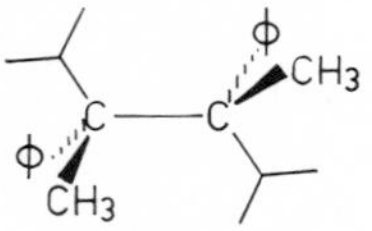 (129) 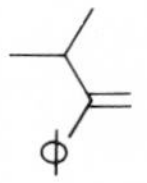 (130) 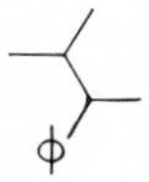 (131)

This means that the caged radicals are able to rotate freely in the cage, and do so before combining to product. When the same reaction is carried out by photolysis in a rigid glass at -196° and subsequent warming, (126) gives only (128) and (127) gives only (129); the rotational freedom is gone then. Furthermore, at the low temperature, the pair can be observed directly by esr; approximately half of them have their distant spins still paired, and in half of them they are unpaired (triplet). Similar experiments and results have been reported by Greene [89].

It seems likely that the radical pairs that produce these final compounds are not always the first stage of the reaction: The two C-N bonds may not cleave simultaneously, and hence there may be a prior pair such as (132).

R—N=N• R•

(132)

This has been shown by Seltzer, who concluded from isotope effects on the rate that the two bonds must be breaking in succession in α-phenylethylazomethane, (133) [90].

---

88. P.D. Bartlett and J.M. McBride, Pure Appl. Chem., 15, 89 (1967).
89. F.D. Greene, M.A. Berwick, and J.C. Stowell, J. Amer. Chem. Soc., 92, 867 (1970).
90. S. Seltzer and F.T. Dunne, J. Amer. Chem. Soc., 87, 2628 (1965).

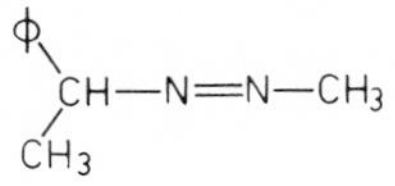

φ—CH(CH$_3$)—N=N—CH$_3$

(133)

In some instances nitrogen coupling can be seen directly by means of esr [91].

In another type of experiment it was learned that the rate of decomposition of (134) is approximately twice that of (135) - as expected on statistical grounds [92].

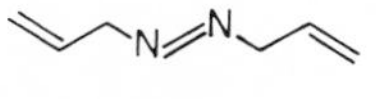

(134) 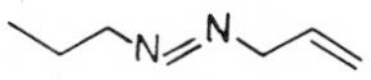 (135)

One of the more surprising developments has been the observation that the cage effects are not significantly reduced if both radicals bear like charges; this conclusion has been reached for the azocompound (136) and its conjugate diacid (137) [93].

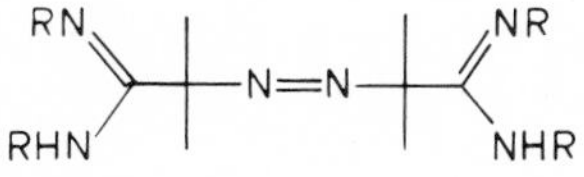

(136) 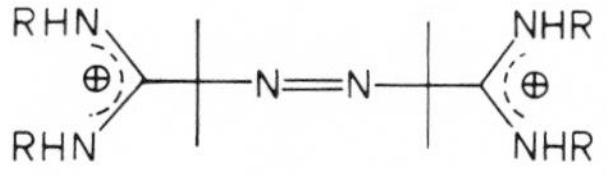 (137)

There are other types of compounds that give rise to caged radicals as well. The peroxides are of interest in that three bonds are candidates for cleavage. Thus, acetyl peroxide (138) forms ethane and methyl acetate, suggesting that both (139) and (140) are present as paired radicals. Furthermore, Martin has learned that if carbonyl O^{18} - labelled (141) is used, the unused portion of the starting material contains significant amounts of (142) - thus, the geminate pair actually consists of a pair of acetyloxy radicals (143) [94].

---

91. (a) P.B. Ayscough, B.R. Brooks, and H.E. Evans, J. Phys. Chem., 68, 3889 (1964); (b) P. Stilbs, G. Ahlgren, and B. Åkermark, Tetrahedron Lett., 2387 (1972).
92. (a) K. Takagi and R.J. Crawford, J. Amer. Chem. Soc., 93, 5910 (1971). For further evidence, see (b) N.A. Porter, M.E. Landis, and L.J. Marnett, J. Amer. Chem. Soc., 93, 795 (1971); (c) R.J. Crawford and K. Takagi, J. Amer. Chem. Soc., 94, 7406 (1972).
93. G.S. Hammond and R.C. Neuman, J. Amer. Chem. Soc., 85, 1501 (1963).
94. J.W. Taylor and J.C. Martin, J. Amer. Chem. Soc., 88, 3650 (1966) and 89, 6904 (1967).

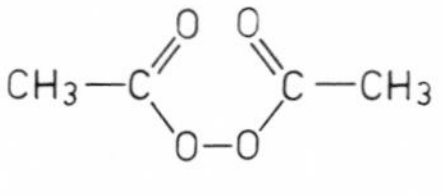

$CH_3-C(=O)-O-O-C(=O)-CH_3$ (138) $(\dot{C}H_3\ \ \dot{C}H_3)$ (139) $(\dot{C}H_3\ \ \cdot CO_2-CH_3)$ (140)

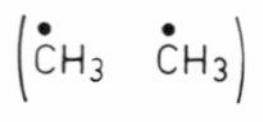

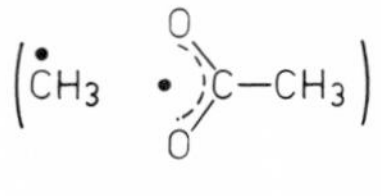

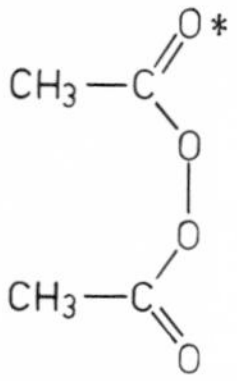

(141) 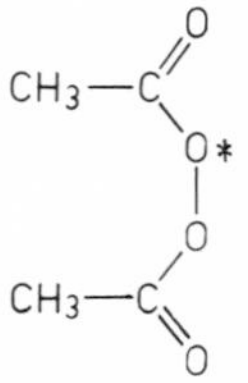   (142) 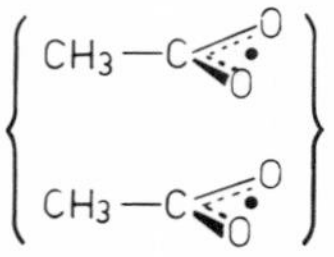   (143)

Radical pairs are of course most conveniently obtained by decomposition, but they may also form by coincidental encounter during the diffusion process. Pairs of this sort can of course not be produced in a high concentration, but CIDNP provides a very sensitive method of hunting for them. The interaction of the two spins in a cage causes nuclear polarization which in turn can be observed in the nmr of the products formed by the radicals. It has been found that CIDNP also occurs in certain chain processes in which the radicals propagate by displacement (abstraction). Thus, when phenylacetyl peroxide (144) is decomposed in the presence of bromotrichloromethane, the emissions observed in the nmr could be explained only in terms of encounters of radicals not present in the original cage [95].

$$(\phi-CH_2-COO)_2$$

(144)

18-6. POLYRADICALS

An interesting situation derives when the two radicals in a cage are bound together and unable to diffuse apart. This happens not only when a bond breaks in a cyclic molecule, but probably also when a forbidden cycloaddition occurs. Since the spins may interact not only directly but also through the bonds, much interest has centered on such diradicals recently [96].

The cyclophanes provide an example of such diradicals in which the spins are many atoms apart. The conclusion that such diradicals are formed rests on the following evidence. The resolved [2.2]paracyclophane (145) racemizes at 200° at a rate which is independent of solvent polarity. If heated in the presence of _p_-diisopropylbenzene, the parent cyclophane is converted in part into the open chain

95. C. Walling and A.R. Lepley, _J. Amer. Chem. Soc._, _93_, 546 (1971).
96. (a) L. Salem and C. Rowland, _Angew. Chem., Int. Ed. Engl._, _11_, 92 (1972); (b) L. Salem, _Pure Appl. Chem._, _33_, 317 (1973).

compound (146), and if it is done in the presence of either methyl maleate or fumarate, the same mixture of addition products (147) is obtained [97].

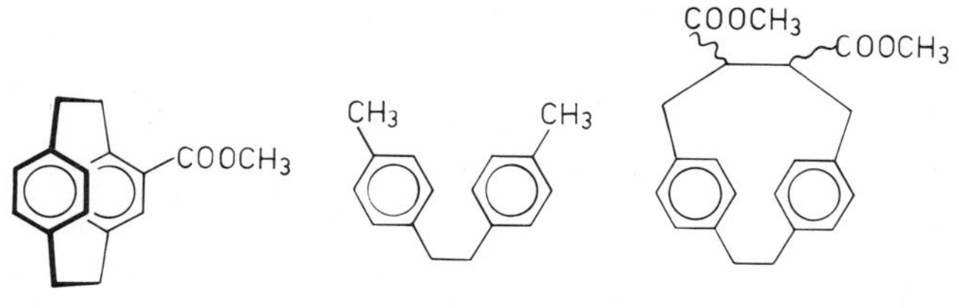

(145) (146) (147)

These facts are all most easily accounted for in terms of initial bond cleavage of one of the bridges followed by rotations of the rings and reclosure or radical reactions.

In forbidden [2 + 2] cycloadditions the radical centers are much closer. Bartlett has made a study of the relative rates of closure and internal rotation by measuring the amounts of product in the [2 + 2] cycloaddition of 1,1-dichlorodifluoroethylene to the three isomeric 2,4-hexadienes. The mixtures that are obtained make clear that there is time for perhaps 10 rotations of the newly formed single bonds before closure, but the allylic part of the radical retains its conformation [98].

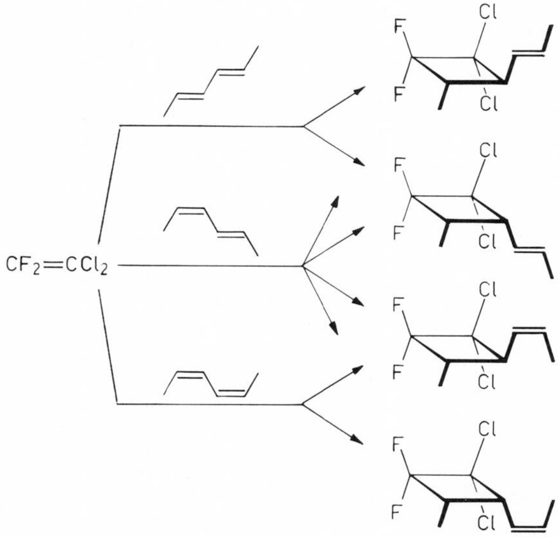

$CF_2{=}CCl_2$

97. (a) H. J. Reich and D. J. Cram, J. Amer. Chem. Soc., 91, 3517 (1969); (b) M. H. Delton and D. J. Cram, J. Amer. Chem. Soc., 94, 1669 (1972); (c) D. J. Cram and J. M. Cram, Accounts Chem. Res., 4, 204 (1971).
98. L. K. Montgomery, K. Schueller, and P. D. Bartlett, J. Amer. Chem. Soc., 86, 622 (1964).

Trimethylene (148) has been claimed to be an intermediate in the thermolysis of 1-pyrazoline (149).

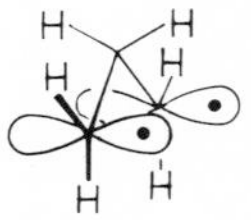

(148)  (149)

The products of this reaction are cyclopropane and propylene. The intermediacy of (148) was deduced from the fact that both (150) and (151) give an identical mixture of products (cis- and trans-2-butene, 1-butene, and methylcyclopropane with 50% of the deuterium cis- and 50% trans- to the methyl group) [99].

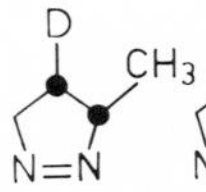

(150) 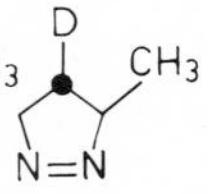 (151)

However, Bergman has suggested that these reactions involve a diazenium diradical [100]. The azocompound (152) upon thermolysis gives mostly (153) rather than (154), which is better explained via stages (155) and (156).

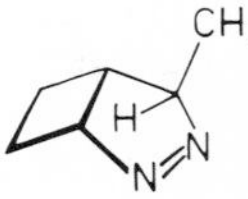

(152) 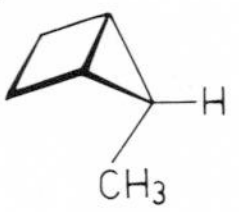 (153) 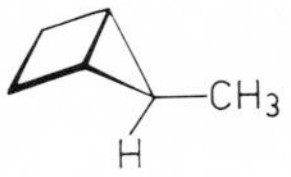 (154)

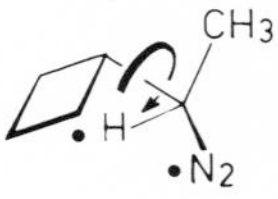

(155) 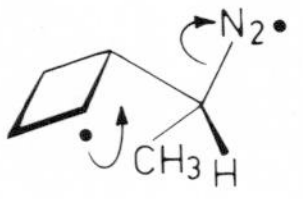 (156)

Trimethylenemethane is an interesting variant of diradical (148). If 3-methylenecyclobutanone (157) or 4-methylene-1-pyrazoline (158) is photolyzed at -196°, an esr spectrum is obtained which is clearly attributable to the triplet diradical (159) [101]: 7 lines are seen, and both $\Delta m = 1$ and $\Delta m = 2$ transitions

99. (a) R.J. Crawford and A. Mishra, J. Amer. Chem. Soc., 88, 3963 (1966); (b) R.J. Crawford and G.L. Erickson, J. Amer. Chem. Soc., 89, 3907 (1967); (c) R.J. Crawford and L.H. Ali, J. Amer. Chem. Soc., 89, 3908 (1967).
100. D.H. White, P.B. Condit, and R.G. Bergman, J. Amer. Chem. Soc., 94, 7931 (1972).
101. P. Dowd and K. Sachdev, J. Amer. Chem. Soc., 89, 715 (1967).

can be demonstrated (cf. also p. 384) [102]. If (158) is heated, methylenecyclopropane (160) is obtained, but labelling studies show that the three positions are scrambled before the product is formed [103] (cf. also p. 211 and 217).

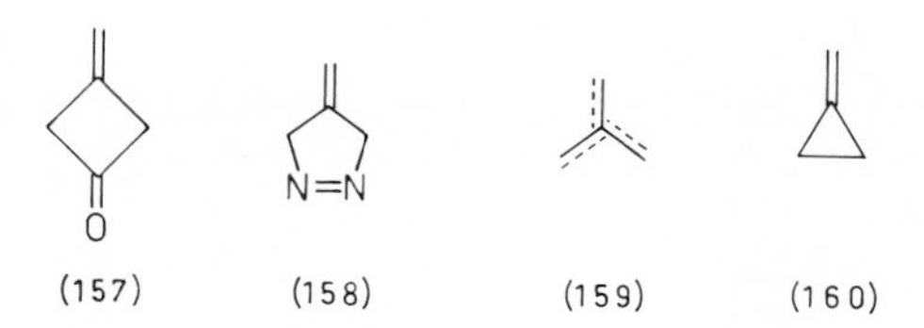

(157) (158) (159) (160)

The increased stability of (159) as compared to (148) is obvious in that trimethylenemethane can give rise to dimeric products such as 1,4-dimethylenecyclohexane, and it can be caught by other intermediates that are themselves not very stable; thus, if (161) and (162) are heated together with potasium in the vapor phase, cyclobutadiene adduct (163) can be isolated [104].

Cl Cl
I I

(161) (162) (163)

It is still an open question whether the high temperature intermediate is the same triplet diradical as the low temperature species. Both (164) and (165) upon heating give the same equilibrium mixture containing also (166) and (167); but when resolved (164) is used, the (166) and (167) appearing initially are optically active, so that a non-planar intermediate must play some role [105].

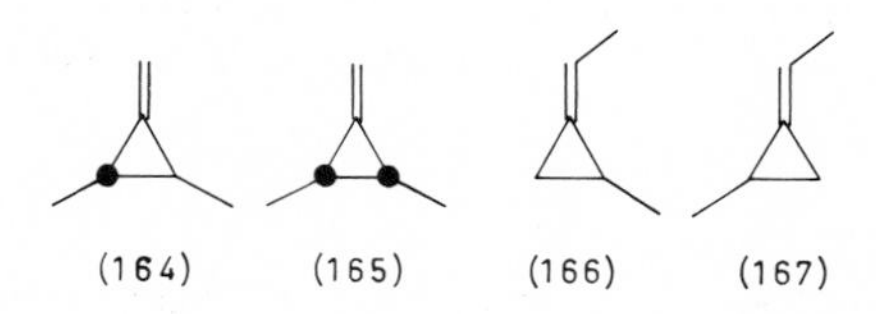
(164) (165) (166) (167)

The planar species is stabilized by bonding with irontricarbonyl, and this complex (168) has been isolated under ordinary laboratory conditions by Emerson [106].

---

102. (a) P. Dowd, A. Gold and K. Sachdev, J. Amer. Chem. Soc., 90, 2715 (1968); (b) P. Dowd, Accounts Chem. Res., 5, 242 (1972).
103. R.J. Crawford and D.M. Cameron, J. Amer. Chem. Soc., 88, 2589 (1966).
104. R.G. Doerr and P.S. Skell, J. Amer. Chem. Soc., 89, 3062 (1967).
105. (a) J.J. Gajewski, J. Amer. Chem. Soc., 93, 4450 (1971); see also (b) J.A. Berson, R.J. Bushby, J.M. McBride, and M. Tremelling, J. Amer. Chem. Soc., 93, 1544 (1971).
106. G.F. Emerson, K. Ehrlich, W.P. Giering, and P.C. Lauterbur, J. Amer. Chem. Soc., 88, 3172 (1966).

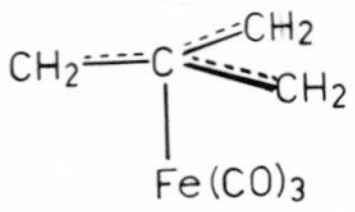

CH_2—C(CH_2)(CH_2)—Fe(CO)$_3$

(168)

More complex diradicals of this sort can be devised. The tetradeuterio-1,2-dimethylenecyclobutane (169) equilibrates at 275° with (170) and (171), evidently via tetramethylenethane, (172) [107].

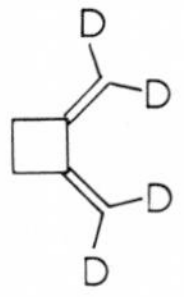

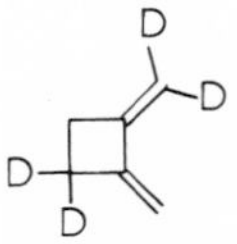

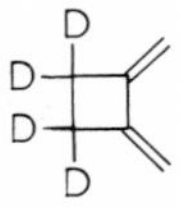

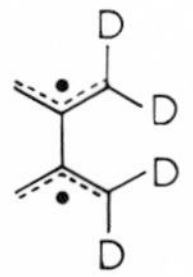

(169) (170) (171) (172)

If unlabeled (169) is added, no dideuterio products are obtained, hence no reversion to allene occurs [108]. Photolysis of ketone (173) at -196° gives rise to a triplet whose esr spectrum can be recorded [109].

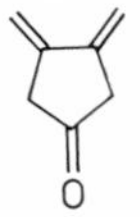

(173)

Making use of strain relief, Bauld designed an analog that opens thermally even at 50°: (174) at that temperature dimerizes to (175) and (176). If E,E-2,4-hexadiene is present, a mixture of (177) and (178) is obtained, the hence a triplet species (179) is indicated [110].

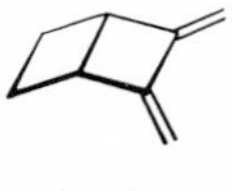

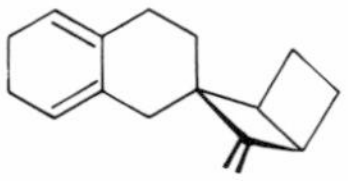

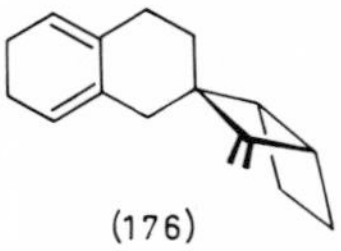

(174) (175) (176)

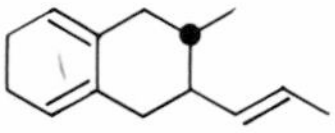

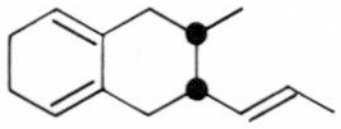

(177) (178) (179)

---

107. (a) J. J. Gajewski and C. N. Shih, J. Amer. Chem. Soc., 89, 4532 (1967); (b) W. von E. Doering and W. R. Dolbier, J. Amer. Chem. Soc., 89, 4534 (1967).
108. J. J. Gajewski and C. N. Shih, J. Amer. Chem. Soc., 94, 1675 (1972).
109. P. Dowd, J. Amer. Chem. Soc., 92, 1066 (1970).
110. (a) C. S. Chang and N. L. Bauld, J. Amer. Chem. Soc., 94, 7593 (1972); (b) N. L. Bauld and C. S Chang, J. Amer. Chem. Soc., 94, 7594 (1972).

Diradicals of a rather different type are responsible for at least some cases of thermochromism - the reversible color changes that occur on heating or grinding certain solids. Bianthrone (180) is an example: The normally yellow compound reversibly becomes green on heating. Since the green form is paramagnetic, its structure has been postulated to be (181) [111].

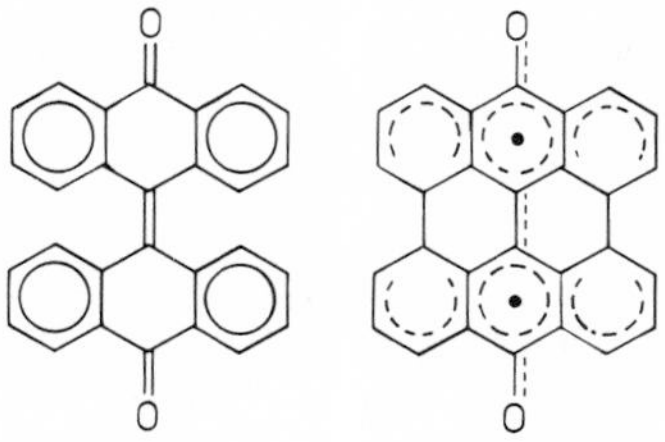

(180) (181)

A related case involving the hexaphenylbiimidazolyl (182) is thought to involve free monoradicals (183) [112].

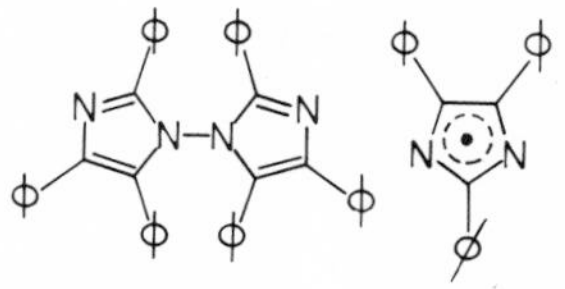

(182) (183)

A number of rather more stable diradicals are known, most of which are relatively simple extensions of stable monoradicals. One of these is known as the Chichibabin hydrocarbon (184), which gives rise to a weak esr signal possibly attributable to a small equilibrium quantity of diradical (185) [113].

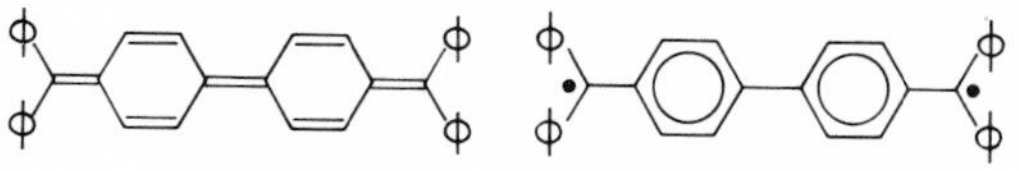

(184) (185)

By building up hindrance so as to prevent the planar quinoid isomer from forming, pure diradicals can be obtained which are easily demonstrated to be paramagnetic; (186) is an example [4e].

---

111. R. B. Woodward and E. Wasserman, J. Amer. Chem. Soc., 81, 5007 (1959).
112. T. Hayashi and K. Maeda, Bull. Chem. Soc. Japan, 38, 685 (1965).
113. C. A. Hutchison, A. Kowalsky, R. C. Pastor, and G. W. Wheland, J. Chem. Phys., 20, 1485 (1952).

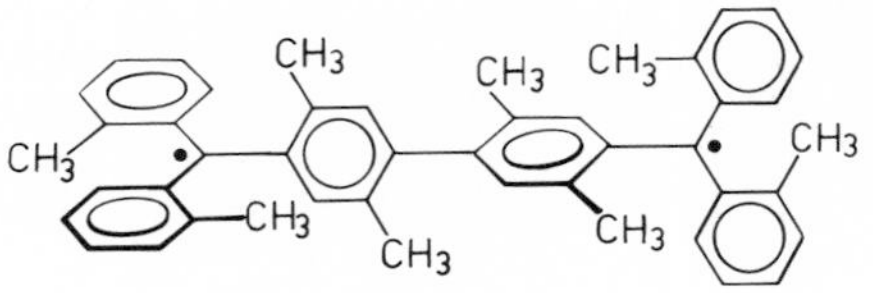

(186)

An attempt to prepare (187), the diradical analog of 2,2-diphenylpicrylhydrazyl, led to the interesting observation that the material was diamagnetic; it is not clear as yet whether this compound is a spin-paired singlet diradical or a valence isomer such as (188) [114].

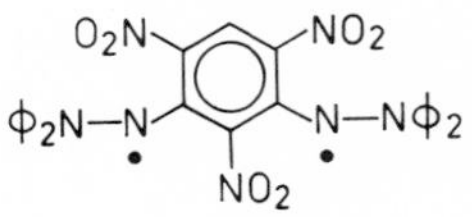

(187) 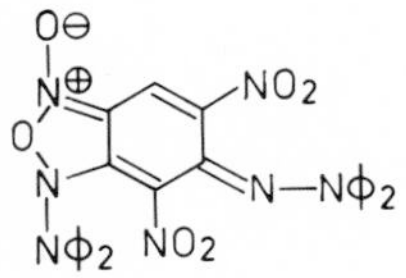 (188)

In this connection it is of interest that solid nitroso compounds which are colorless and presumed to be dimeric (189) often exhibit weak esr signals, suggesting [115] a small equilibrium amount of dinitroxides (190).

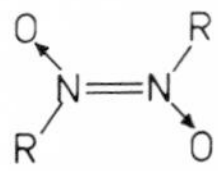

(189) 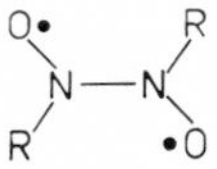 (190)

The diradical 1,1'-ethylenebis(4-carbomethoxypyridinyl) (191) has been isolated [116]. It is paramagnetic and its esr spectrum can be determined in dilute solution; association of an undetermined sort hinders the measurement at higher concentrations.

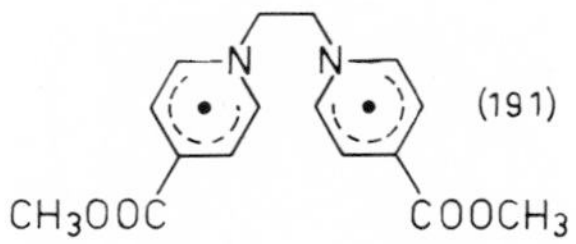

(191)

114. J. Heidberg and J.A. Weil, J. Amer. Chem. Soc., 86, 5173 (1964).
115. W. Theilacker, A. Knop, and H. Uffmann, Angew. Chem., Int. Ed. Engl., 4, 688 (1965).
116. (a) E.M. Kosower and Y. Ikegami, J. Amer. Chem. Soc., 89, 461 (1967); See also (b) Y. Ikegami, H. Watanabe, and S. Seto, J. Amer. Chem. Soc., 94, 3274 (1972).

Perhaps the most clear-cut and stable diradicals yet known are the galvinoxyl analogs (192) [117] and (193) [118]; both are ground state triplets.

(192) (193)

Several 1,3,5-trisubstituted phenyl triradicals have been reported - among them (194) [119], which is paramagnetic, very sensitive to oxygen and partly associated; verdazyl analog (195) [120], whose magnetic susceptibility reveals a true quartet with three unpaired and uncoupled electrons, and trisnitroxide (196) [121].

(194) (195)

(196)

---

117. N.C. Yang and A.J. Castro, J. Amer. Chem. Soc., 82, 6208 (1960).
118. E.A. Chandross, J. Amer. Chem. Soc., 86, 1263 (1964).
119. G. Schmauss, H. Baumgärtel, and H. Zimmermann, Angew. Chem., Int. Ed. Engl., 4, 596 (1965).
120. R. Kuhn, F.A. Neugebauer, and H. Frischmann, Angew. Chem., Int. Ed. Engl., 4, 72 (1965).
121. M.B. Neiman, E.G. Rozantsev, and V.A. Golubev, Bull. Acad. Sci. U.S.S.R. (English Transl.), 529 (1965).

The latter compound, a red stable solid, shows a 1:1:1 esr triplet at room temperature, suggesting that the three spins are then effectively insulated, but at 130° this has changed to the 7 line spectrum due to three equivalent nitrogen atoms. The phosphite triradical (197) likewise has a 7 line esr spectrum even at room temperature [121].

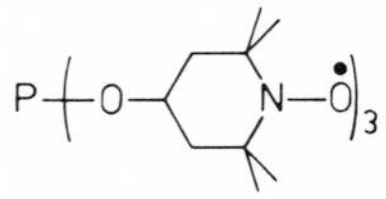

(197)

Similar behavior is shown by the few tetraradicals known; (198) has a 9 line esr spectrum, and (199) has a 5 line spectrum at room temperature changing to 9 lines at 140° [122].

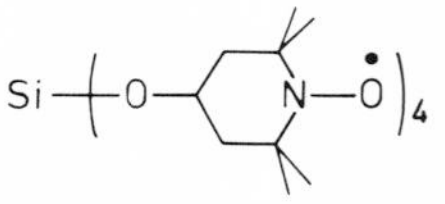

(198) 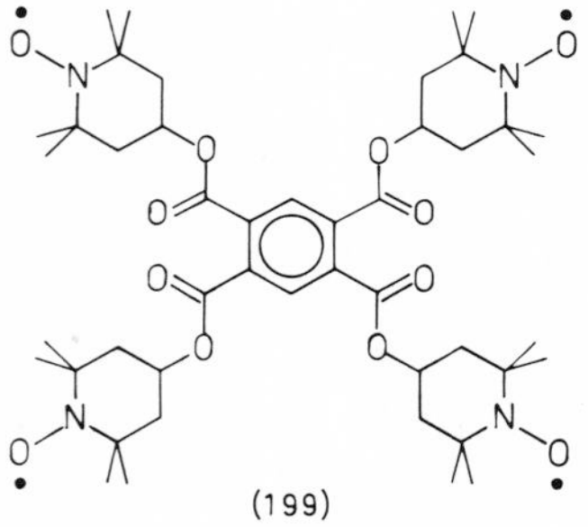 (199)

18-7. ION RADICALS

A remarkable example of the intervention of radical anion intermediates has been described by Kornblum. If the ambident anion (200) (see Chr. 22, Sec. 2), derived from 2-nitropropane is treated with a wide variety of alkylating agents, the displacement occurs at the site of the oxygen atom to give product (201). There is one exception: p-nitrobenzyl derivatives such as (202) produce the carbon alkylate (203) in addition.

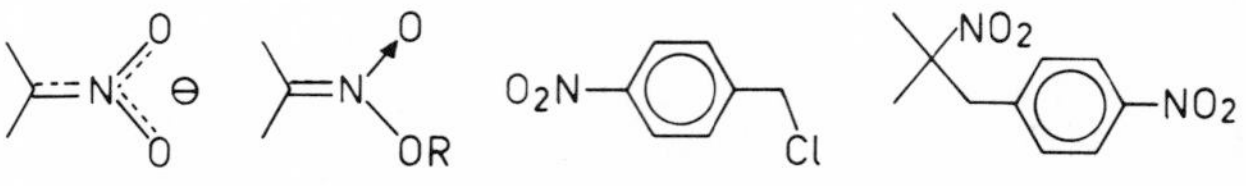

(200) (201) (202) (203)

122. E.G. Rozantsev and V.A. Golubev, Bull. Acad. Sci. U.S.S.R. (English Transl.), 695 (1965).

Another peculiar circumstance is that while in ambident anion reactions one normally finds an increase in the "C/O ratio" as the leaving group in (201) is varied from chloride to iodide ion, in the present case the reverse (92/6 to 7/86) is observed [123].

The chain reaction mechanism devised to account for these facts is as follows [124].

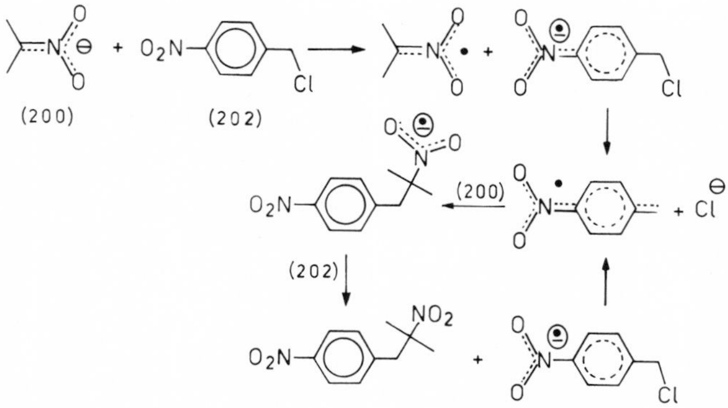

(200) (202)

As the reaction proceeds an esr signal could indeed be observed. If other nitro compounds capable of forming radical anions are added, the chain is broken and stopped; thus, in the presence of p-dinitrobenzene the C/O ratio for the chloride can be reduced to 2/72. Cupric chloride likewise has a dramatic effect on this ratio - even in concentrations as low as 10^{-6}. The overall net result is the simple product expected of S_N2 displacement reaction; but a further interesting distinction is that unlike any other substrate, the p-nitrobenzyl derivatives are subject to this reaction even when they are tertiary. Thus, p-nitrocumyl chloride (204) readily undergoes displacement by thiophenoxide to give (205) [125]; the reaction is depressed by the admission of oxygen or the addition of p-dinitrobenzene.

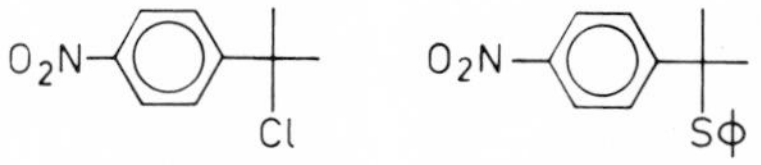

(204) (205)

A further elegant development was the insight that anions which are not capable of initiating the chain may nevertheless be able to keep it going ("entrainment"). This indeed turned out to be the case. While azide ion for example does not react

123. R.C. Kerber, G.W. Urry, and N. Kornblum, J. Amer. Chem. Soc., 86, 3904 (1964) and 87, 4520 (1965).
124. (a) N. Kornblum, R.E. Michel, and R.C. Kerber, J. Amer. Chem. Soc., 88, 5660, 5662 (1966); (b) G.A. Russell and W.C. Danen, J. Amer. Chem. Soc., 88, 5663 (1966).
125. N. Kornblum, T.M. Davies, G.W. Earl, N.L. Holy, R.C. Kerber, M.T. Musser, and D.H. Snow, J. Amer. Chem. Soc., 89, 725 (1967).

with α, p-dinitrocumene (206), in the presence of a small amount of (200) one can obtain a 97% of the azido derivative (207) [126].

$$O_2N-C_6H_4-C(CH_3)_2-NO_2 + N_3^{\ominus} \xrightarrow{(200)} O_2N-C_6H_4-C(CH_3)_2-N_3$$

(206) (207)

These reactions furnish an example of the process of electron transfer in organic reactions. Another instance of this is the apparently simple reaction of the trityl cation with t-butoxide anion to give the corresponding ether. A strong esr signal, identifiable as that of the trityl radical, can be observed immediately after the addition. This signal intensifies, and after reaching a maximum eventually dies away; thus, the suggestion is strong that the reaction begins by electron transfer [127]. Still further examples are the base catalyzed formation of stilbene from p-nitrobenzyl chloride [128] (until recently a mystery: cf. p. 586) and the reactions of benzenediazonium ions with iodide [129], phenoxide [130] and amide ions [131].

Radical anions are generally stable species if generated by exposure to alkali metals or by electrolytic reduction [132]. The very great sensitivity of the esr technique is very helpful in this area of study, but as always with sensitive probes, great care must be exercised concerning the purity of the compound under study lest some contaminant lead one astray. Thus, the exposure of adamantane to alkali metals at -150° leads to a five line spectrum of approximately the right intensities for four equivalent protons; this led to the claim that the unpaired electron must have been trapped inside the cavity surrounded by the adamantane nucleus (208) [133].

(208)

126. N. Kornblum, R.T. Swiger, G.W. Earl, H.W. Pinnick, and F.W. Stuchal, J. Amer. Chem. Soc., 92, 5513 (1970).
127. K.A. Bilevitch, N.N. Bubnov and O.Y. Okhlobystin, Tetrahedron Lett., 3465 (1968).
128. G.L. Closs and S.H. Goh, J. Chem. Soc., Perkin II, 1473 (1972).
129. P.R. Singh and R. Kumar, Austr. J. Chem., 25, 2133 (1972).
130. N.N. Bubnov, K.A. Bilevitch, L.A. Poljakova, and O.Y. Okhlobystin, Chem. Commun., 1058 (1972).
131. J.K. Kim and J.F. Bunnett, J. Amer. Chem. Soc., 92, 7463, 7464 (1970).
132. For review, see (a) L.L. Miller, J. Chem. Educ., 48, 168 (1971); (b) F. Gerson, Pure Appl. Chem., 28, 131 (1971); (c) F.D. Popp and H.P. Schultz, Chem. Rev., 62, 19 (1962); (d) P. Zuman, Chem. Eng. News, 46, 94, March 18, 1968; (e) G.A. Russell, Science, 161, 423 (1968).
133. K.W. Bowers, G.J. Nolfi, and F.D. Greene, J. Amer. Chem. Soc., 85, 3707 (1963).

However, later it turned out that the spectrum observed was actually the inner portion of the seven-line spectrum of benzene radical anion - a contamination of 10^{-7} sufficed to bring this situation about [134].

Radical anions have also provided a number of examples of how a molecule can seem more symmetric than it is because of accidental equality of coupling constants. Thus, a nine line spectrum with $a_H = 7.65$ gauss obtains when 1,2-dimethylenecyclobutane (209) is electrolytically reduced at $-90°$, leading to the claim of tetramethyleneethane radical anion (210) [135]. Later, however, this was withdrawn when it turned out that 1,2-dimethylenecyclopentane (211) also had a nine line spectrum ($a = 6.70$ gauss); the esr equivalence of the two sets of four hydrogen atoms is now believed to be coincidental [136].

(209) (210) (211)

Another likely pitfall to be considered is that of the asymmetry that can be induced by solvation. Thus, the *m*-dinitrobenzene radical anion has been studied in ethanol solvent, and it was observed that the two nitrogen atoms are not equivalent. In dimethylformamide the expected 1:2:3:2:1 spectrum is observed. It seems probable that solvation in ethanol restricts the extra electron to one nitrogen atom [137]. Rapid exchange must surely occur, but it should further be remembered that esr involves much higher energy than nmr, and averaging occurs only at reaction rates several orders of magnitude faster than those important in nmr.

Exposure to potassium in ethereal solvents converts [2.2] paracyclophane into its radical anion; a complex spectrum obtains showing that the electron is located in one ring and split not only by the protons of that ring but also by the potassium atom; the radical anion is thus part of an ion-pair. The removal of the cation *via* the addition of a little hexamethylphosphoramide causes the collapse into the expected nine line pattern. Under these conditions, the [4.4] homolog has only 5 lines: Thus, if the rings are far enough apart the extra electron is restricted to one of them. Similarly, the radical anion of 1,2-diphenylethane shows only one ring interaction, though that of diphenylmethane has two [138]. Conversely, complete symmetry (or extremely rapid equilibration) has been shown to exist this

---

134. M.T. Jones, *J. Amer. Chem. Soc.*, *88*, 174 (1966).
135. N.L. Bauld and G.R. Stevenson, *J. Amer. Chem. Soc.*, *91*, 3675 (1969).
136. N.L. Bauld, F. Farr, and G.R. Stevenson, *Tetrahedron Lett.*, 625 (1970).
137. W.E. Griffiths, C.J.W. Gutch, G.F. Longster, J. Myatt, and P.F. Todd, *J. Chem. Soc.*, *B*, 785 (1968).
138. (a) F. Gerson and W.B. Martin, *J. Amer. Chem. Soc.*, *91*, 1883 (1969); see also (b) D.J. Williams, A.O. Goedde, and J.M. Pearson, *J. Amer. Chem. Soc.*, *94*, 7580 (1972).

way in the radical anions of hexamethyl [3] radialene (212) [139], norbornadiene (213) [140] and the propellane (214) [141].

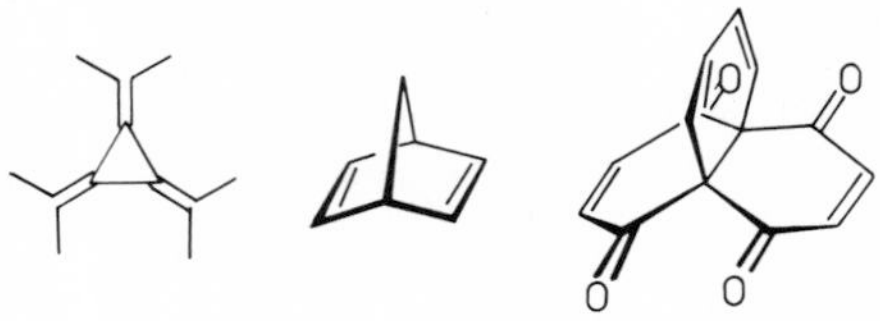

(212) (213) (214)

Reasonably stable radical anions can generally be expected only if the unoccupied orbital accepting the electron is not at a very high energy level. In practice that means that saturated molecules having only σ^* levels empty cannot accept an extra electron. At the other extreme, a molecule such as 6,6-diphenylfulvene (215) is very easily converted into the radical anion (216) [142].

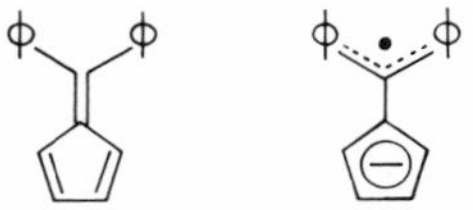

(215) (216)

Likewise, the esr spectrum of nitronylnitroxide (217) which consists of five (1:2:3:2:1) doublets in dimethylsulfoxide solution, collapses to five singlets upon addition of t-butoxide ions [143].

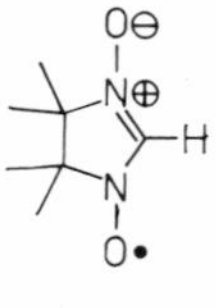

(217)

Electronegative substituents may similarly facilitate electron acceptance; for instance, nitrosotrifluoromethane dimer (218) can be electrolytically reduced at low voltage to a radical anion with a 35 line spectrum (a_N = 18 gauss, a_F = 7.5 gauss) [144].

---

139. F. Gerson, E. Heilbronner, and G. Köbrich, Helv. Chim. Acta, 48, 1525 (1965).
140. H. Hogeveen and E. de Boer, Rec. Trav. Chim. Pays-Bas, 85, 1163 (1966).
141. R. Bar-Adon, S. Schlick, B. L. Silver, and D. Ginsburg, Tetrahedron Lett., 325 (1972).
142. C. M. Camaggi, M. J. Perkins, and P. Ward, J. Chem. Soc., B, 2416 (1971).
143. D. G. B. Boocock, R. Darcy, and E. F. Ullman, J. Amer. Chem. Soc., 90, 5945 (1968).
144. J. L. Gerlock and E. G. Janzen, J. Amer. Chem. Soc., 90, 1652 (1968).

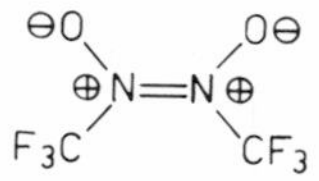

(218)

The semidiones (219) are of importance in this connection; these are stable and easily handled radical anions that can usually be prepared readily from ketones having an α-methylene group [132e].

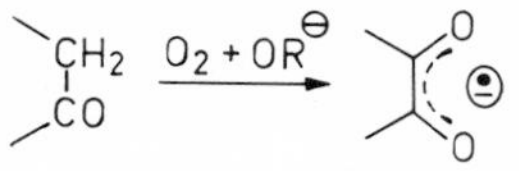

(219)

Many aliphatic structures have been examined by esr in this way; for instance, the basketane semidione has primarily a quintet structure with $a_H = 0.53$ gauss showing that the major part of the spin density outside the dione group is on the hydrogen atoms shown in (220). On this basis and in view of the fact that a_H for a free hydrogen atom is 508 gauss the spin density is about 0.001 at each of these atoms; these were visualized by Russell as indicating a contribution from structures such as (221) [145].

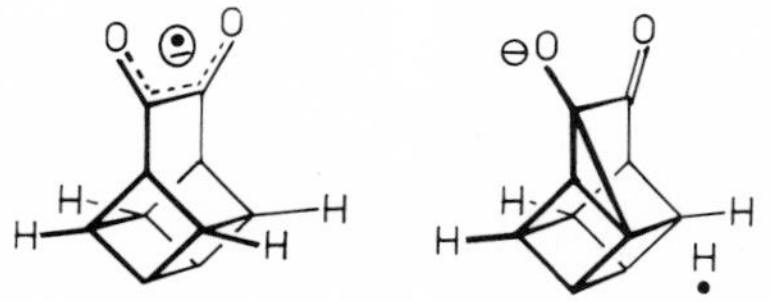

(220) (221)

In the cycloalkylsemidiones (222), a_{H_α} is found to be several times smaller for cyclopropyl than for any other member of the series, suggesting that cyclopropyl adopts the bisected conformation (223) in these species [146].

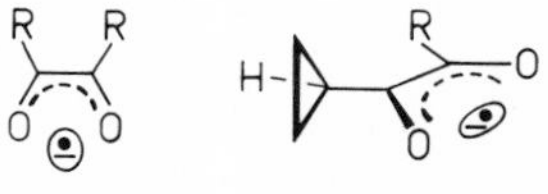

(222) (223)

Radical cations also play a role as reaction intermediates in electron transfer reactions [147]; thus, the oxidation of aliphatic amines to aldehydes by chlorine dioxide and other one-electron oxidizing agents begins this way [148].

---

145. G.A. Russell, G.W. Holland, and K.Y. Chang, J. Amer. Chem. Soc., 89, 6629 (1967).
146. G.A. Russell and H. Malkus, J. Amer. Chem. Soc., 89, 160 (1967).
147. A. Ledwith, Accounts Chem. Res., 5, 133 (1972).
148. L.A. Hull, G.T. Davis, D.H. Rosenblatt, H.K.R. Williams, and R.C. Weglein, J. Amer. Chem. Soc., 89, 1163 (1967).

$$R'\text{-}CH_2NR_2 + ClO_2 \rightarrow ClO_2^- + R'\text{-}CH_2NR_2^{\bullet+} \rightarrow R'\text{-}CHO + HNR_2$$

These species can often be preserved for study if properly generated by electro-oxidation, or by chemical means such as dissolution in sulfuric acid (aluminum chloride-nitromethane mixtures have also worked in certain cases [149]). For example, tetrakisdimethylaminoethylene radical cation (224) has been obtained by polarographic oxidation. It shows a complex signal, with coupling to four equivalent nitrogen atoms (a_N = 4.85 gauss) and two sets of twelve protons each (a_{H1} = 3.28 and a_{H2} = 2.84 gauss) [150].

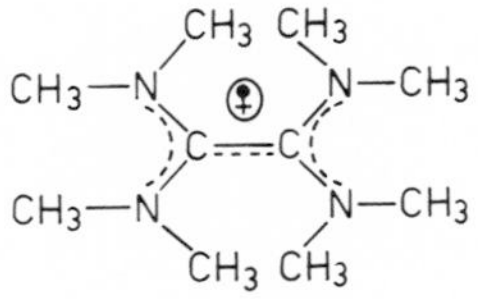

(224)

In some instances radical cation salts can be isolated. Chlorine oxidation of (225) gives the stable, purple salt (226), which has a simple quintet esr spectrum. Further oxidation produces the aromatic dicationic dithioleum salt (227), while exposure to sodium bisulfite regenerates (226) [150a, b].

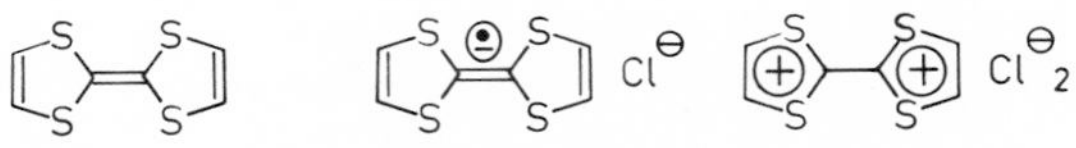

(225) (226) (227)

As with the anion counterparts, one of the important uses of these cation radicals is the information the esr spectra of aromatic species furnish about the Hückel coefficients of the orbital the unpaired electron occupies: The coupling constants are essentially proportional to these coefficients squared [132b]. In many cases both anion and cation radicals have been scrutinized, so that experimental values can then be assigned to the Hückel coefficients of both the highest occupied and lowest empty orbitals of the neutral molecule, and satisfying agreement is obtained in most instances with the calculated values. Thus, in perylene (228), these coefficients are predicted by theory to be the same; the cation and anion radical spectra are virtually superimposable. Additional simple examples

149. W.F. Forbes and P.D. Sullivan, J. Amer. Chem. Soc., 88, 2862 (1966).
150. K. Kuwata and D.H. Geske, J. Amer. Chem. Soc., 86, 2101 (1964).
150a. F. Wudl, G.M. Smith and E.J. Hufnagel, Chem. Commun., 1453 (1970); (b) H.N. Blount and T. Kuwana, J. Amer. Chem. Soc., 92, 5773 (1970).

of the use of such esr spectra are those of the radical ions of (229) [151], (230) [152] and (231) [153].

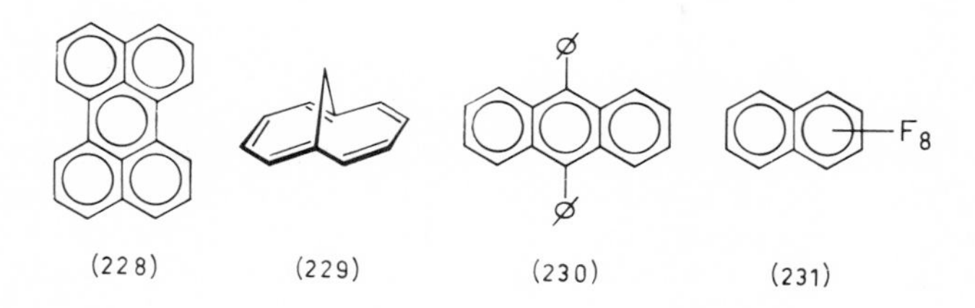

(228) (229) (230) (231)

Since solutions of many radical ions are so easily accessible, they have a future as simple laboratory reagents. The naphthalene radical anion in tetrahydrofuran reacts with 1,4-dichloropentane to give the bicyclic epimeric pair (232) [154], and with water to yield the Birch reduction product (233) [155]. The well-known Birch reduction itself obviously also involves radical anions [156].

CH_3

(232) (233)

In principle radical polyions should give information about still higher (or lower) levels. In practice such species become rapidly less stable as the charge is multiplied, and since ion-multiplets are virtually unavoidable, the esr spectra are made more complex because of association equilibria and counterion coupling. Nevertheless, in several cases success has been achieved. Thus, 1,1,3,3-tetraphenylallyl radical dianion (234) is stable in tetrahydrofuran at -50°; a doublet is observed by esr both for this species and for the neutral radical. The absence of further couplings suggests that the rings cannot be coplanar with the allyl skeleton [157]. The tropyl dianion (235) has been prepared; its esr spectrum is easily observable with $a_H = 3.48$ and $a_{Na} = 1.74$ gauss (the fact that $a_H = 2a_{Na}$ leads to a deceptively simple spectrum) [158].

---

151. (a) F. Gerson, E. Heilbronner, W.A. Böll, and E. Vogel, Helv. Chim. Acta, 48, 1494 (1965); (b) F. Gerson, K. Müllen, and E. Vogel, J. Amer. Chem. Soc., 94, 2924 (1972).
152. L.O. Wheeler, K.S.V. Santhanam, and A.J. Bard, J. Phys. Chem., 70, 404 (1966).
153. N.M. Bazhin, N.E. Akhmetova, L.V. Orlova, V.D. Shteingarts, L.N. Shchegoleva, and G.G. Yakobson, Tetrahedron Lett., 4449 (1968).
154. J.F. Garst, Accounts Chem. Res., 4, 400 (1971).
155. S. Bank and W.D. Closson, Tetrahedron Lett., 1349 (1965).
156. J.K. Brown, D.R. Burnham, and N.A.J. Rogers, Tetrahedron Lett., 2621 (1966).
157. P. Dowd, J. Amer. Chem. Soc., 87, 4968 (1965).
158. N.L. Bauld and M.S. Brown, J. Amer. Chem. Soc., 87, 4390 (1965).

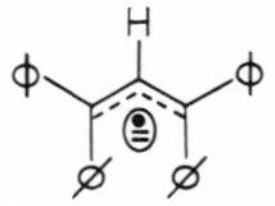

(234)  (235)

The yellow tetrahydrofuran solution of Koelsch radical (20) provides us with an interesting example. Its esr spectrum gives evidence of coupling to eight equivalent protons; thus, the phenyl ring is presumably twisted far out of the plane of the rest of the molecule. Exposure to one equivalent of potassium changes the color to blue, and leads to the disappearance of the esr spectrum. A second equivalent of metal changes the color to green, and a new esr spectrum appears different from that of the neutral species; interaction now occurs with only three pairs of protons, so that the unpaired electron must be localized in one fluorenyl ring. Still more potassium gives the orange trianion; again the esr spectrum disappears. The admission of small amounts of oxygen allows one to trace all these changes in reverse [159]. Similar experiments can be done with phenanthraquinone (236); now the purple trianion has an unpaired electron that signals its presence in the esr quintuplet signal. The purple trianion of oxalate (237) has also been observed as a singlet by esr [160].

(236) 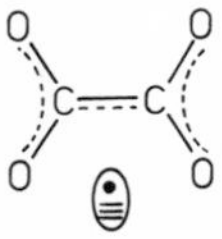 (237)

159. E.G. Janzen and J.G. Pacifici, J. Amer. Chem. Soc., 87, 5504 (1965).
160. N.L. Bauld, J. Amer. Chem. Soc., 86, 3894 (1964).

Chapter 19

BENZYNE [1]

19-1. INTRODUCTION

Benzyne (dehydrobenzene) has turned out to be the culprit in a number of highly puzzling observations in certain nucleophilic aromatic substitutions. For example, it had long been known that the treatment of many aromatic halides with strong bases resulted not only in the direct replacement of the halogen, but - in certain cases - also in the formation of isomeric products in which the new substituent occupies a position ortho to that formally held by the halogen. Wittig [2] found in 1942 that fluorobenzene and phenyllithium form a reaction mixture which apparently contains o-lithiobiphenyl: If water is added, biphenyl is obtained and the addition of benzophenone results in the formation of trityl alcohol (1).

---

1. For review, see (a) J. F. Bunnett, J. Chem. Educ., 38, 278 (1961); (b) G. Wittig, Angew. Chem., Int. Ed. Engl., 4, 731 (1965); (c) T. Kauffmann, Angew. Chem., Int. Ed. Engl., 4, 543 (1965); (d) R.W. Hoffmann, "Dehydrobenzene and Cycloalkynes", Academic Press, New York, 1967.
2. G. Wittig, Naturwiss., 30, 696 (1942).

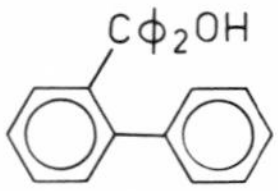

$C\phi_2OH$

(1)

Wittig postulated that the reagents first form *o*-fluorophenyllithium which then eliminates lithium fluoride to give zwitterion (*2*).

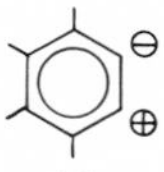

(2)

He also looked for evidence for a neutral, triply bonded benzyne (3); however, he reported that *o*- and *m*-fluoroanisole gave only *o*- and *m*-methoxybiphenyl, respectively, apparently ruling out a common intermediate (4).

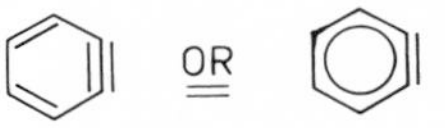

OCH_3

OR

(3) 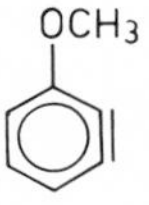 (4)

Roberts studied the base-catalyzed formation of aniline from chlorobenzene in liquid ammonia. He found that when one of the *o*-carbon atoms is ^{14}C, the radiolabel becomes equally distributed between the 1-position on the one hand, and the two *o*-positions on the other. This requires a symmetrical benzyne. The same result obtained if 1-labelled iodobenzene was used [3]. Huisgen then checked Wittig's earlier work and found that it had been partly in error. Not only *did* both methoxyfluorobenzenes lead to common products, but it was found that this case was not unique; thus, if α- and β-fluoronaphthalenes are treated successively with phenyllithium and carbon dioxide, both reactions give the same product distribution of (5) and (6) (in the latter reaction, (7) was also obtained) [4].

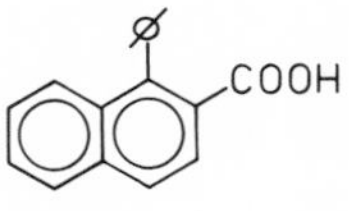

ϕ COOH COOH ϕ COOH ϕ

(5) 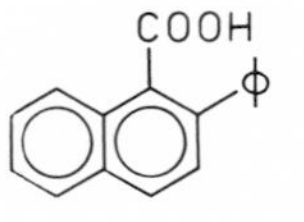 (6) 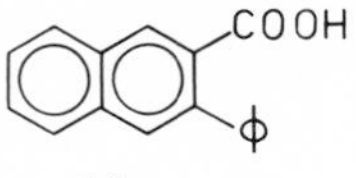 (7)

Further evidence is as follows.

3. J.D. Roberts, H.E. Simmons, L.A. Carlsmith, and C.W. Vaughan, *J. Amer. Chem. Soc.*, *75*, 3290 (1953).
4. R. Huisgen and H. Rist, *Naturwiss.*, *41*, 358 (1954).

CH_3 Br OCH_3

(8)

Di-o-substituted halobenzenes such as (8) do not react with amide anion [5]. A competition experiment involving a mixture of bromobenzene and 2,6-dideuterio-bromobenzene has yielded an isotope effect of 5.5 [6]. Potassium triphenylmethide does not react with chlorobenzene in liquid ammonia until amide ion is added in trace amounts [7]; this ion also catalyzes the formation of diphenylsulfide from bromobenzene and thiophenoxide ion [8]. All six α- and β-chloro-, bromo- and iodonaphthalenes give the same mixture of α- and β-naphthylpiperidines when treated with base in piperidine [9] - and so on.

All this evidence tends to prove that benzyne is symmetrical; however, it is still possible that benzyne is weakly complexed with solvent, an alkali halide ion-pair, or other species. It is now thought, however, that benzyne is indeed present in free form. Thus, Brower [10] has found that the formation of benzyne from an aryl halide is characterized by a positive activation volume, so that the transition state of the benzyne formation is apparently one characterizing decomposition without compensating complexation.

Huisgen has based an argument for free benzyne on the competition method [11]. Thus, benzyne adds to furan and cyclohexadiene in Diels-Alder fashion to give (9) and (10), respectively; if both substrates are present, the product ratio is the same (21:1) by whichever one of several routes the benzyne is generated (cf. next section; see also p. 674).

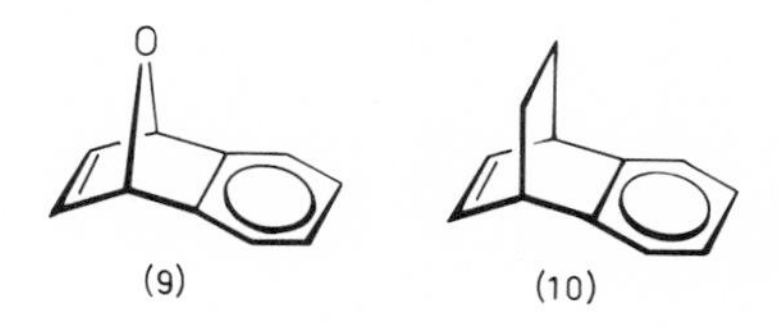

(9) (10)

In one of the first demonstrations of the finite, if short lifetime of free benzynes, Hoffmann found that if the intermediate is generated by passing argon

5. R.A. Benkeser and W.E. Buting, J. Amer. Chem. Soc., 74, 3011 (1952).
6. J.D. Roberts, D.A. Semenow, H.E. Simmons, and L.A. Carlsmith, J. Amer. Chem. Soc., 78, 601 (1956).
7. (a) R.E. Wright and F.W. Bergstrom, J. Org. Chem., 1, 179 (1936); (b) R.A. Seibert and F.W. Bergstrom, J. Org. Chem., 10, 544 (1945).
8. J.F. Bunnett and T.K. Brotherton, J. Org. Chem., 23, 904 (1958).
9. J.F. Bunnett and T.K. Brotherton, J. Amer. Chem. Soc., 78, 155, 6265 (1956).
10. K.R. Brower, personal communication.
11. (a) R. Huisgen and R. Knorr, Tetrahedron Lett., 1017 (1963); (b) M. Jones and M.R. DeCamp, J. Org. Chem., 36, 1536 (1971).

saturated with di-(*o*-iodophenyl)mercury through a hot zone at 750°C, the emerging gas maintains its ability to add to furan some distance downstream from where it was formed [12].

The mass spectrum of benzyne can be measured by the use of various flash techniques; thus, Berry [13] found that soon after a flash, the m/e = 76 peak disappeared while an m/e = 152 peak simultaneously grew in - the entire process requiring only 300 microseconds under the experimental conditions employed. By a similar technique the authors were also able to measure the ultraviolet spectrum of this species, and to calculate its rate of dimerization to biphenylene (11): At least one out of every 100 collisions appears to be effective [14].

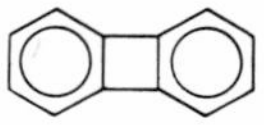

(11)

Nevertheless, it should be remembered that benzyne may have one or more short-lived precursors, and that their responsibility must be ruled out in experiments designed to study the behavior of benzyne. For example, *o*-benzenediazonium carboxylate (12) may proceed to benzyne *via* the zwitterion (13) or its isomer (14): Thus, aniline reacts with the starting material to give *p*-aminoazobenzene (15), conceivably *via* one of these species [15].

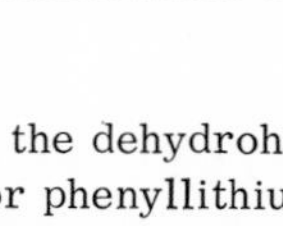

$COO^{\ominus}$, $N_2^{\oplus}$ (12) $\ominus$, $N_2^{\oplus}$ (13) N=N (14) N=N, NH_2 (15)

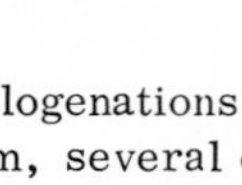

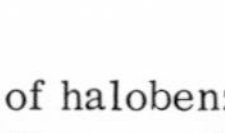

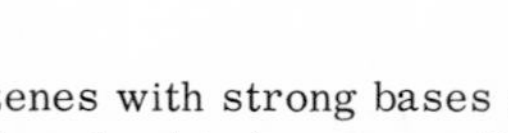

19-2. PREPARATION OF BENZYNE

Beside the dehydrohalogenations of halobenzenes with strong bases such as amide ion or phenyllithium, several other reactions lead to benzyne. If Grignard reactions are attempted with *o*-dihalobenzenes such as (16), benzyne products obtain instead [16].

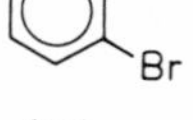

F, Br (16)

12. R.W. Hoffmann and H.F. Ebel, *Angew. Chem., Int. Ed. Engl.*, *3*, 145 (1964).
13. R.S. Berry, J. Clardy, and M.E. Schafer, *J. Amer. Chem. Soc.*, *86*, 2738 (1964).
14. M.E. Schafer and R.S. Berry, *J. Amer. Chem. Soc.*, *87*, 4497 (1965).
15. S. Yaroslavsky, *Chem. and Ind. (London)*, 765 (1965).
16. G. Wittig and L. Pohmer, *Chem. Ber.*, *89*, 1334 (1956).

The action of heat or light upon the zwitterionic salt (12) leads to benzyne [17]. A modification of this method allows the generation of benzyne under very gentle and non-basic conditions: The in situ diazotation of anthranilic acid (17) with i-amyl nitrite in methylene chloride [18].

NH_2
COOH
(17)

The pyrolysis of o-halobenzoate salts has been claimed to produce benzyne [19]; however, since some benzyne additions fail under these conditions, this claim has also been disputed [20].

The photolyses of o-diiodobenzene [21] and of the zwitterionic iodonium salt (18) [22] produce benzyne, as do the thermolysis of phthalic anhydride [23] and of the benzothiadiazole (19) [24].

$\oplus$ Iϕ, $COO^{\ominus}$ (18) SO_2, N, N (19)

Benzyne can be generated by the lead tetraacetate oxidation of 1-aminotriazole (20) [25], and by the ethyl diphenylphosphinite (21) reduction of nitroso compounds such as (22) [26]; an especially gentle method is to cover the salt (23) (stable when wet) with tetrahydrofuran at 0° and to permit the mixture to warm up to room temperature; fragmentation then produces benzyne [27]. Still further methods are known, but these are used more rarely and need not be mentioned here.

N, N, N, NH_2 (20) $\phi_2POC_2H_5$ (21) N, N, N, NO (22) N, N, N, $\overset{\oplus}{Li}$ $\overset{\ominus}{N}$—N=N—NHTs (23)

17. (a) R.S. Berry, G.N. Spokes, and R.M. Stiles, J. Amer. Chem. Soc., 82, 5240 (1960), and 84, 3570 (1962); (b) M. Stiles, R.G. Miller, and U. Burckhardt, J. Amer. Chem. Soc., 85, 1792 (1963).
18. L. Friedman and F.M. Logullo, J. Amer. Chem. Soc., 85, 1549 (1963).
19. H.E. Simmons, J. Org. Chem., 25, 691 (1960).
20. J.K. Kochi, J. Org. Chem., 26, 932 (1961).
21. J.A. Kampmeier and E. Hoffmeister, J. Amer. Chem. Soc., 84, 3787 (1962).
22. E. Le Goff, J. Amer. Chem. Soc., 84, 3786 (1962).
23. E.K. Fields and S. Meyerson, Chem. Commun., 474 (1965).
24. G. Wittig and R.W. Hoffmann, Chem. Ber., 95, 2718 (1962).
25. C.D. Campbell and C.W. Rees, Proc. Chem. Soc., 296 (1964) and J. Chem. Soc. (C), 742 (1969).
26. J.I.G. Cadogan and J.B. Thomson, Chem. Commun., 770 (1969).
27. M. Keating, M.E. Peek, C.W. Rees, and R.C. Storr, J. Chem. Soc., Perkin I, 1315 (1972).

19-3. THE STRUCTURE OF BENZYNE

There are two reasonably likely structures for benzyne. In the first of these (24), there is a true triple bond; two carbon atoms have sp hybrid orbitals.

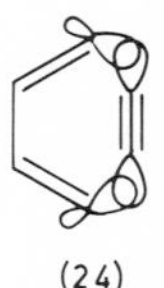

(24)

Severe distortions are necessary to obtain appreciable overlap between the sp^2- and sp hybrid orbitals, which then accounts for the reactivity of benzyne. In resonance language, we would write this structure as (25).

(25)

In the alternative structure (26), benzene would be essentially undisturbed except for the removal of two neighboring hydrogen atoms. There would be some overlap between adjacent sp^2 orbitals; this would be a very weak bond, consistent with observed reactivity. Wittig's zwitterion (2) would now be considered a resonance structure of (26).

(26)

If the electrons have parallel spins, the ground state would be a triplet state; hydrogen abstraction is rarely if ever observed, however (cf. p. 677), so that this triplet state is not a likely description [28].

A tentative choice has been made between the two structures [29]. It was argued that the shorter the C_1C_2 distance could be, the more stable would be the structure (26), since a shorter distance would provide better overlap between the weakly interacting sp^2-orbitals. In analogy to the corresponding bond lengths in the parent molecules, one would expect bond lengths to increase in the series (27), (28) and (29).

---

28. T. Yonezawa, H. Konishi, and H. Kato, Bull. Chem. Soc. Japan, 42, 933 (1969).
29. R. Huisgen, W. Mack, and L. Möbius, Tetrahedron, 9, 29 (1960).

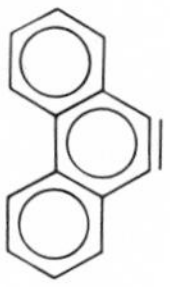

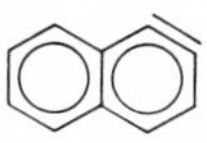

(27) (28) (29)

Correspondingly, one should expect selectivity to decrease in the same order i.e., benzyne should be less selective than 9-phenanthryne. When each of these arynes is allowed to choose between phenyllithium and lithium piperidide, the ratios were 0.08, 0.2 and 0.3, respectively. Thus, the effect predicted on the basis of structure (26) is indeed found. The argument is not strong, however, since one cannot with equal confidence predict what sequence of selectivity should apply on the basis of structure (25).

19-4. REACTIONS OF BENZYNE

Dimerization is a reaction requiring two short-lived individuals to come together, and hence biphenylene formation is primarily a gas phase product. The trimeric product triphenylene (30) is also observed in some instances (see p. 679).

(30)

Benzyne is of course a reactive species; two of its carbon atoms tend to be electron deficient, and so we may expect to see generally electrophilic behavior. If one of the carbon atoms adjacent to the triple bond carries an atom or group that readily provides electrons, cyclization may occur. Bunnett has shown that a wide variety of heterocyclic compounds may be prepared this way [30]; for example, the benzoannelated species (31)-(34) have been prepared by means of a strong base treatment of (35)-(38), respectively.

---

30. (a) B. F. Hrutfiord and J. F. Bunnett, J. Amer. Chem. Soc., 80, 2021 (1958); (b) J. F. Bunnett and B. F. Hrutfiord, J. Amer. Chem. Soc., 83, 1691 (1961).

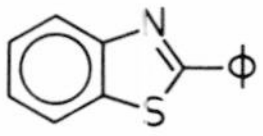

(31) 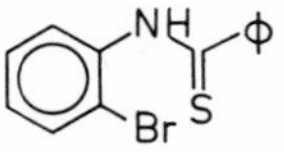 (35)

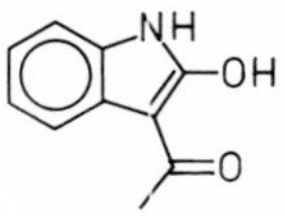

(32) 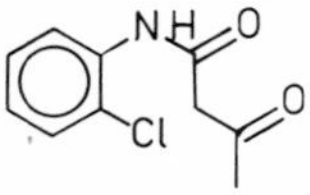 (36)

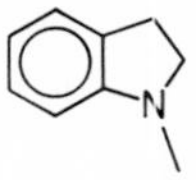

(33) 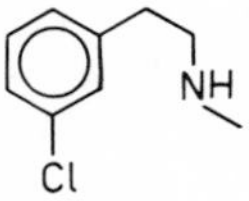 (37)

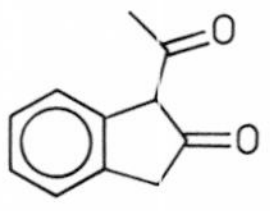

(34) 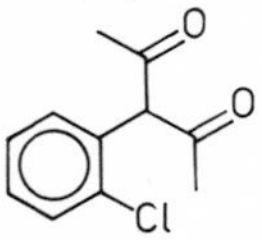 (38)

Equally important are the Diels-Alder reactions of benzyne with conjugated systems. Examples involving cyclohexadiene and furan were already encountered. The [c]benzofuran (39) is especially effective in trapping benzynes; 7-oxanorbornadienes such as (40) are then obtained.

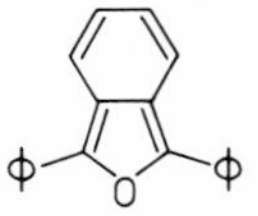

(39) 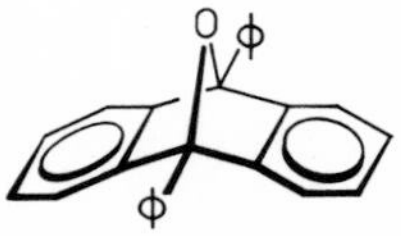 (40)

Pyrroles (41) are also attacked in this fashion, although the product first formed then undergoes further reaction to yield β-naphthylamines (42) [31].

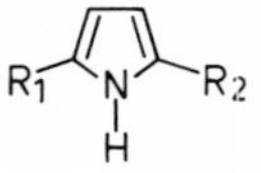

(41) 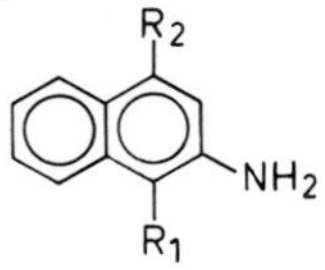 (42)

---

31. E. Wolthuis, D. Vander Jagt, S. Mels, and A. De Boer, J. Org. Chem., 30, 190 (1965).

Tetracyclone (43), one of the few stable cyclopentadienones, is often used to trap benzynes. The benzonorbornadienone adducts (44) usually split off carbon monoxide rapidly to give the corresponding 1, 2, 3, 4-tetraphenylnaphthalenes (45).

(43) (44) (45)

Benzyne in fact can be added to cyclopentadienide anion; if o-bromofluorobenzene is added to a mixture of magnesium and cyclopentadienylmagnesium bromide, benzonorbornadiene (46) is obtained after neutralization [32].

(46)

Even benzene [33] and naphthalene [34] are readily (46) attacked by benzyne in this fashion to give benzo- and dibenzobarrelene, (47) and (48), respectively.

(47) (48)

The formation of trypticenes (49) and their isomers (50) from benzyne and anthracene is a standard reaction to show the presence of the former.

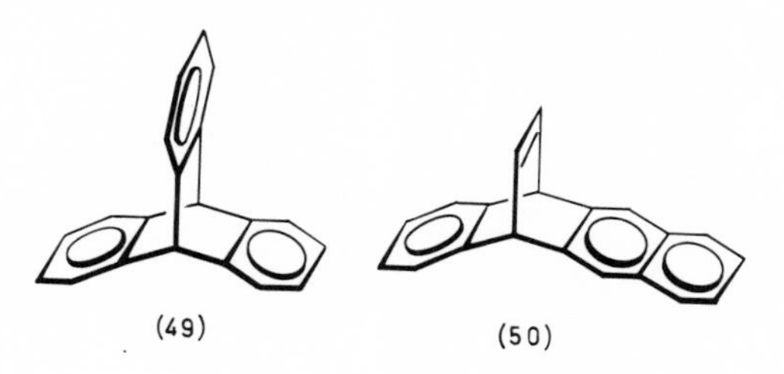

(49) (50)

---

32. W.T. Ford, R. Radue, and J.A. Walker, Chem. Commun., 966 (1970).
33. J.P.N. Brewer and H. Heaney, Tetrahedron Lett., 4709 (1965).
34. R.G. Miller and M. Stiles, J. Amer. Chem. Soc., 85, 1798 (1963).

Of the two products formed, the first predominates with anthracene itself, but the presence of electron-withdrawing substituents such as cyano on the central ring diverts the benzyne to one of the terminal rings - an observation attesting to the electrophilic nature of benzyne [35]. Attack on a terminal ring can also be induced by 1,4-dimethoxy substitution. Interestingly, it was found that the ratio of products resulting from attack on the central and substituted terminal rings was independent of the method of benzyne generation, showing that all these reactions involve a common intermediate. However, there was one exception [36]: When the precursor was diphenyliodonium-2-carboxylate, the preference for the terminal ring was less pronounced. This was attributed to the deactivation of the substituted ring by the iodobenzene formed in the reaction, presumably because of the formation of the charge transfer complex (51) (Chr. 23). The same effect was noted with other benzyne precursors if iodobenzene was independently added.

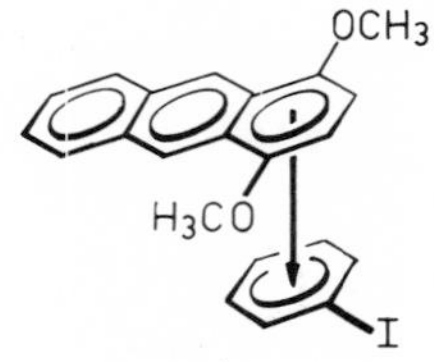

OCH_3 H_3CO I

(51)

The addition of an electrophilic benzyne to alkoxylated rings may be followed by a retro Diels-Alder reaction; this is a useful preparation of dialkoxyacetylenes. These compounds, otherwise not readily available, have been prepared from 2,3-dialkoxynaphthalenes (52) and tetrachloroanthranilic acid (53), presumably *via* adduct (54), and giving tetrachloroanthracene (55) as by-product [37].

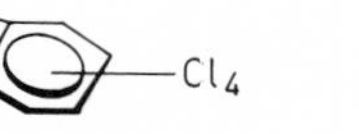

OCH_3 OCH_3 Cl_4 NH_2 COOH

(52) 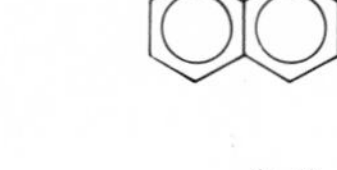 (53)

OCH_3 CH_3O Cl_4 Cl_4

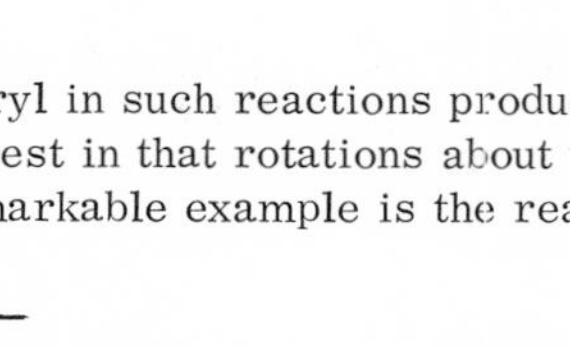

(54) 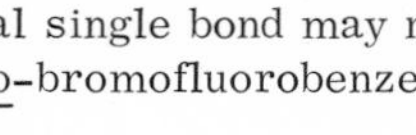 (55)

9,9'-Bianthryl in such reactions produces 9,9'-bitrypticyl, a molecule of theoretical interest in that rotations about the central single bond may not be possible [38]. A remarkable example is the reaction of *o*-bromofluorobenzene and 3,4-

---

35. B.H. Klanderman, *J. Amer. Chem. Soc.*, *87*, 4649 (1965).
36. B.H. Klanderman and T.R. Criswell, *J. Amer. Chem. Soc.*, *91*, 510 (1969).
37. J. Font, F. Serratosa, and L. Vilarrasa, *Tetrahedron Lett.*, 4743 (1969).
38. C. Koukotas, S.P. Mehlman, and L.H. Schwartz, *J. Org. Chem.*, *31*, 1970 (1966).

dichlorotetramethylcyclobutene (56) with lithium amalgam in ether, which leads to benzotetramethylbicyclo[2.2.0]hexa-2,5-diene (a "hemi-Dewar naphthalene"). This molecule apparently results from the Diels-Alder addition of benzyne to tetramethylcyclobutadiene [39]; if both species were indeed free, it would represent a rare instance of one transient intermediate reacting with another in solution.

The reaction of benzyne with phenyl isocyanate [40] may be mentioned as an example of a Diels-Alder addition of benzyne to give a heterocyclic product (58).

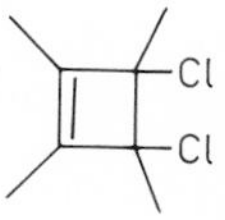

(56) 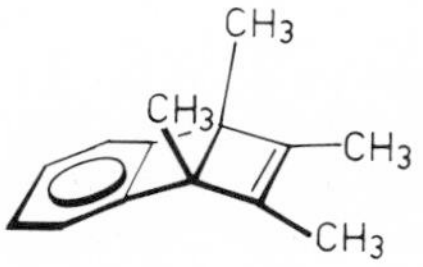 (57) 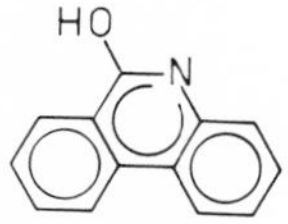 (58)

Finally, additions to more widely separated centers have been observed as well. Tetrachlorobenzyne reacts with norbornadiene to give a product (59) formally resulting from a homo-Diels-Alder addition [41].

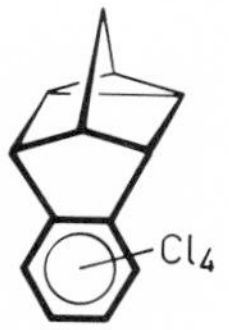

(59)

This same mixture also gives an example of a [2 + 2] cycloaddition of benzyne; some adduct (60) is obtained. Similarly, tetrachlorobenzyne undergoes a cycloaddition to norbornene to give exclusively _exo_-addition product (61) [41].

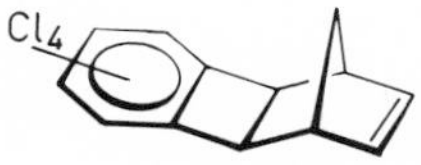

(60)  (61)

Berson and Pomerantz found that _o_-benzenediazoniumcarboxylate reacts with 3,3-diphenylcyclopropene (62) to give dibenzonorbornadiene (63), presumably _via_ an

---

39. D.T. Carty, Tetrahedron Lett., 4753 (1969).
40. J.C. Sheehan and G.D. Daves, J. Org. Chem., 30, 3247 (1965).
41. (a) H. Heaney and J.M. Jablonski, Tetrahedron Lett., 2733 (1967); (b) H. Heaney, J.M. Joblonski, K.G. Mason, and J.M. Sketchley, J. Chem. Soc., (C), 3129 (1971).

initial cycloaddition to give (64) [42]. Vinyl ethers and esters often add to benzyne to give benzocyclobutene derivatives [43].

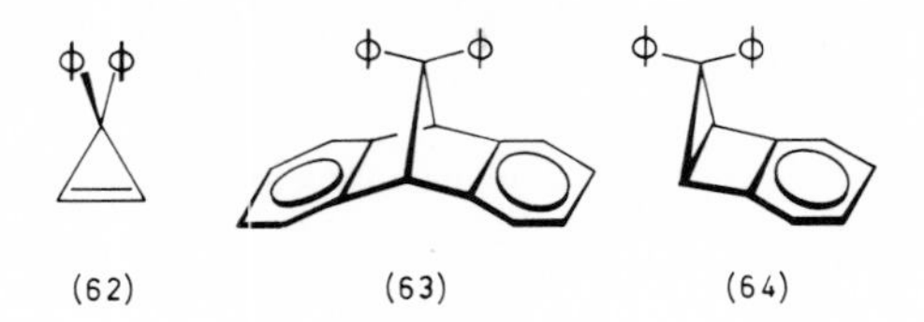

(62) (63) (64)

The mechanism of these cycloadditions has come under study in recent years. According to an analysis of such reactions by Hoffmann [44], the antarafacial approach involves such distortions that it is unlikely. If the addition is suprafacial for both systems, it cannot be concerted, but of course the [2 + 4] cycloadditions would be. Evidence supporting these suppositions has come forth from several laboratories.

E- and *Z*-1-methoxypropene add to benzyne to give a mixture of *cis*- and *trans*-cyclobutenes (65) and (66) [45].

CH_3 OCH_3 CH_3 OCH_3

(65) (66)

Other vinyl ethers [45] and ethylene chloride [46] gave similar results, all suggesting a 2-step mechanism. In each case, some bias toward retention is observed, so that the bond rotation is not much faster than ring closure. By contrast however, the [2 + 4] cycloaddition to *E*,*E*-2,4-hexadiene gives exclusively the stereospecific product (67) [47].

(67)

42. J.A. Berson and M. Pomerantz, *J. Amer. Chem. Soc.*, *86*, 3896 (1964).
43. H.H. Wasserman and J. Solodar, *J. Amer. Chem. Soc.*, *87*, 4002 (1965).
44. D.M. Hayes and R. Hoffmann, *J. Phys. Chem.*, *76*, 656 (1972).
45. (a) I. Tabushi, R. Oda, and K. Okazaki, *Tetrahedron Lett.*, 3743 (1968); (b) L. Friedman, R.J. Osiewicz, and P.W. Rabideau, *Tetrahedron Lett.*, 5735 (1968); (c) H.H. Wasserman, A.J. Solodar, and L.S. Keller, *Tetrahedron Lett.*, 5597 (1968).
46. M. Jones and R.H. Levin, *Tetrahedron Lett.*, 5593 (1968).
47. (a) M. Jones and R.H. Levin, *J. Amer. Chem. Soc.*, *91*, 6411 (1969); (b) R.W. Atkin and C.W. Rees, *Chem. Commun.*, 152 (1969).

There is some evidence that the intermediate which is permitting the bond rotations is a diradical rather than a zwitterion. Thus, in the [2 + 2] cycloadditions to E-cyclooctene, both (68) and (69) are observed; but rearranged products typical of carbonium ion reactions, arising for example from 1,2-hydride shifts, transannular reactions and so on are not observed. The solvent dependence so typical of carbonium ion reactions is also absent in this reaction [48].

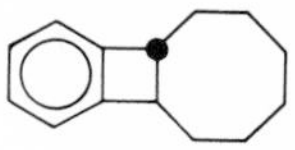

(68) 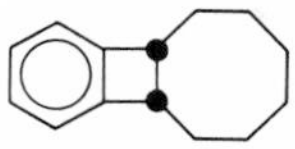 (69)

It may have been noted that the examples of benzyne attacks on double bonds involve either strained or electron rich double bonds. Simple double bonds usually do not lead to [2 + 2] cycloadditions, unless highly electrophilic benzynes such as tetrafluorobenzyne are used. Olefins often do react with benzynes, but by different pathways. Cyclohexene for instance gives 3-phenylcyclohexene by a mechanism that points to an allowed ene reaction. For example, the use of cyclohexene-1,2-d_2 gives only the 2,3-d_2 product (70) and none of the 1,2-d_2-isomer (71); this rules out hydrogen abstraction (72) as an intermediate step and strongly suggests an ene transition state (73) [49].

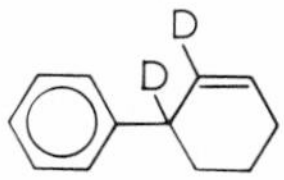

(70) 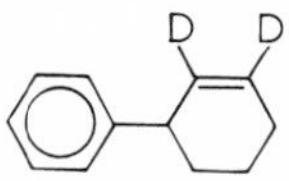 (71) 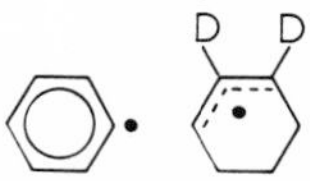 (72) 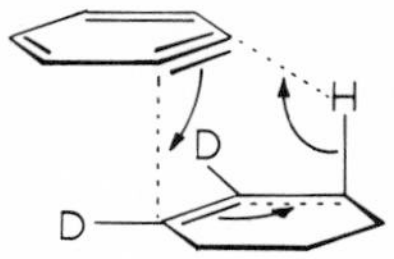 (73)

Another complex case arises with styrene. In this reaction 9-phenyl-9,10-dihydrophenanthrene (74) is obtained; the author [50] considered this a case of a [2 + 4] addition to give intermediate (75) which then reacts with a second molecule of benzyne to give the product.

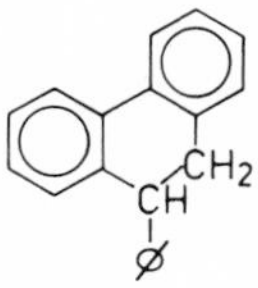

(74) 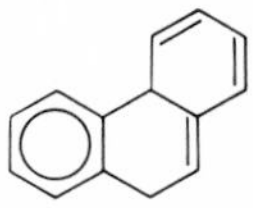 (75)

48. P.G. Gassman and H.P. Benecke, Tetrahedron Lett., 1089 (1969).
49. G. Ahlgren and B. Åkermark, Tetrahedron Lett., 3047 (1970).
50. W.L. Dilling, Tetrahedron Lett., 939 (1966).

1-Vinylnaphthalene (76) can be used to give a modest yield of chrysene (77), and 3-vinylbenzo(b)thiophene (78) similarly produces some 9-thiabenzo(a)fluorene (79) [51].

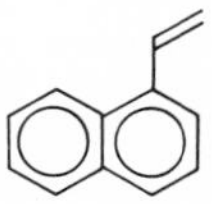

(76) 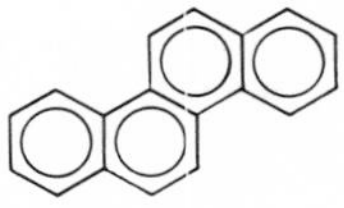 (77) 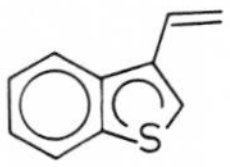 (78) 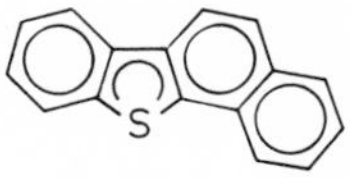 (79)

3,4-Dimethoxyphenylacetylene (80) reacts similarly with benzyne to give phenanthrene derivative (81) [52].

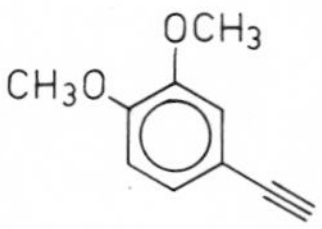

(80) 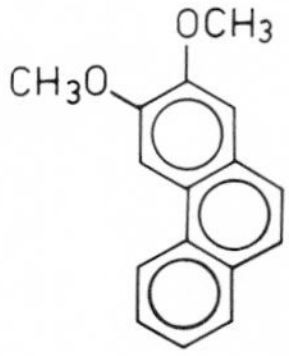 (81)

Activated triple bonds such as in 1-N-dimethylaminopropyne react with benzyne in a complex manner, and any pathway detailing how the products (82) and (83) arise must be regarded as speculative at present [53].

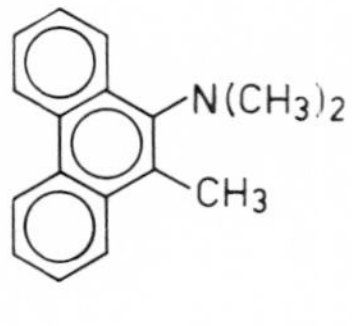

(82) 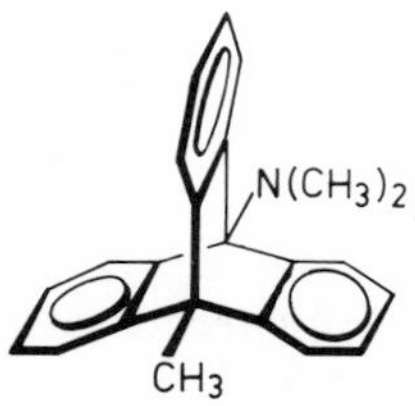 (83)

An additional aspect of benzyne additions to multiple bonds is of interest here: The incredible ability of silver ion to alter the course of the reaction [54]. If benzyne is generated in benzene, one ordinarily obtains mostly benzobarrelene, some biphenylene, and traces of biphenyl and benzocyclooctatetraene. The trace products become virtually the sole products when a small amount of silver ion is present even though all products are stable to this ion. Even if the metal ion

51. T.G. Corbett and Q.N. Porter, Austr. J. Chem., 18, 1781 (1965).
52. S.F. Dyke, A.R. Marshall, and J.P. Watson, Tetrahedron, 22, 2515 (1966).
53. J. Ficini and A. Krief, Tetrahedron Lett., 4143 (1968).
54. L. Friedman, J. Amer. Chem. Soc., 89, 3071 (1967).

concentration is only 10^{-9}, the effect on the product distribution is already easily measurable. Evidently the silver ion complexes with benzyne and thereby drastically alters its chemical nature. Silver ion is known to have a catalytic effect in other cycloadditions as well (cf. p. 534).

A few instances are known in which benzyne adds to longer conjugated systems; thus, the thermal decomposition of o-benzenediazonium carboxylate in the absence of other trapping compounds gives some [2 + 6] cycloaddition product (84) [55], and the [2 + 8] reaction with 8-cyanoheptafulvene (85) gives the azulene derivative (86) [56].

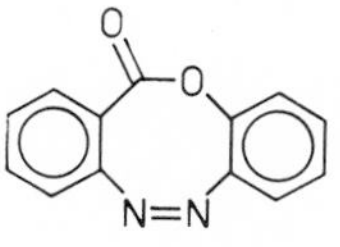

(84) 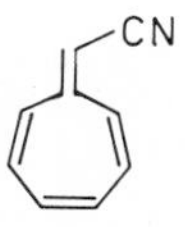 (85) 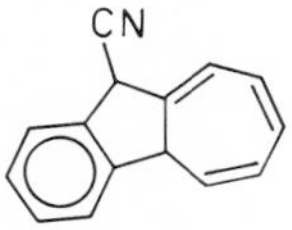 (86)

The trimerization of benzyne is often an important side reaction. Thus, when o-fluorophenylmagnesium bromide is prepared in tetrahydrofuran, an 85% yield of triphenylene is isolated. Since the same conditions usually produce some biphenylene, it was initially thought that triphenylene simply results from the incorporation of a third benzyne molecyle into biphenylene. However, it was found by Heaney that when a substituted biphenylene is added to the reaction mixture, this does not get incorporated into the triphenylene [57]. On the other hand, when benzyne is generated in the presence of the known and stable Grignard derivative of 4,4'-dimethyl-2-fluoro-2'-iodobiphenyl (87), 2,7-dimethyltriphenylene (88) is obtained. This suggests that the intermediate is not biphenylene, but the 2-halo-2'-metallo-derivative of biphenylene. When biphenylene is present in a mixture in which a highly electrophilic benzyne is generated, it does get captured - but the product is a barrelene such as (89) rather than triphenylene [58].

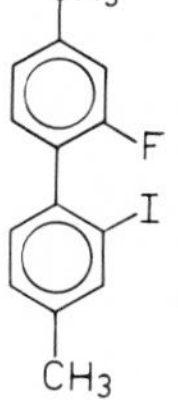

(87)  (88)  (89)

55. T. Miwa, M. Kato, and T. Tamano, Tetrahedron Lett., 2743 (1968).
56. M. Oda and Y. Kitahara, Bull. Chem. Soc. Japan, 43, 1920 (1970).
57. H. Heaney and P. Lees, Tetrahedron Lett., 3049 (1964).
58. H. Heaney, K. G. Mason, and J. M. Sketchley, Tetrahedron Lett., 485 (1970).

The reaction of benzyne with various Lewis bases has already been mentioned. In some instances the direction of attack of the base is so different from the site preferentially approached by cations in the parent hydrocarbon that the benzyne route is the preferred one for preparing derivatives. An example is pyrene, which is attacked by electrophilic reagents in the 1-position to the virtual exclusion of the C-2 atom. 1-Halopyrene in the reaction with potassium amide in liquid ammonia gives primarily 2-aminopyrene (90) [59]; which in turn is readily converted into other 2-pyrene derivatives.

(90)

Nucleophiles much weaker than amide will attack benzyne. Halide ions [60], alcohols and carboxylic acids [61] all will readily add. Thioethers are cleaved by benzynes [62], and even ethers can be fragmented that way if the benzyne is electrophilic enough [63]; for example, tetrachlorobenzyne and ethyl ether give ethylene and the tetrachlorophenetole (91).

(91)

Benzyne itself attacks tetrahydrofuran if 2-benzenediazonium carboxylate is heated in this solvent (if moist) to give (92) and (93) [64].

(92) (93)

59. A. Jensen and A. Berg, Acta Chem. Scand., 19, 520 (1965).
60. R.W. Hoffmann, G.E. Vargas-Nunez, G. Guhn, and W. Sieber, Chem. Ber., 98, 2074 (1965).
61. M. Stiles, R.G. Miller, and U. Burckhardt, J. Amer. Chem. Soc., 85, 1792 (1963).
62. V. Franzen, H.I. Joschek, and C. Mertz, Ann., 654, 82 (1962).
63. J.P.N. Brewer, H. Heaney, and J.M. Jablonski, Tetrahedron Lett., 4455 (1968).
64. E. Wolthuis, B. Bouma, J. Modderman, and L. Sytsma, Tetrahedron Lett., 407 (1970).

A number of studies have been carried out to learn the directive effect of substituents in benzyne, and hence the mode of reaction of benzyne with nucleophiles [65]. In such studies the starting compounds are _o_- or _p_-substituted aryl halides; either way one can produce only one of the two possible monosubstituted benzynes. By measuring the ratio of products, one can then calculate the _m_/_p_- or _m_/_o_-ratio, which in turn can be used to evaluate the importance of inductive and resonance effects. Alternatively, one can let the benzyne select between two nucleophiles.

The results may be described in general terms with a few examples. We may expect that electron-withdrawing groups (whether operating inductively or by resonance) will direct nucleophiles to the _p_-position, and that the reverse resonance effect will apply to donating groups. This is exactly as observed. The _p_/_m_-ratio in the capture of amide ion and/or ammonia is 30 for the cyano group, 3 for fluoro, 1/2 for dimethylamino, and 0 for oxide anion [65b].

The electrophilic nature of benzyne is clear from the following experiment. If the benzyne is generated in an ethyl alcohol solution of a lithium halide, the ratio of halobenzenes to ethyl phenyl ether varies in the series chloride, bromide and iodide from 2.4, to 25, to 120, respectively [60]. Finally, the selectivity of arynes seem to become less with all substituents. Evidently the polarization of the aryne triple bond that results from the introduction of a substituent tends to make it more reactive.

1,3-Dipolar additions to benzyne are known to occur. Nitrosobenzene is believed to add to benzyne in this fashion in the first of several steps eventually leading to N-phenylcarbazoles (94) [66];

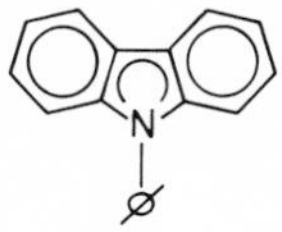

N
|
ø

(94)

azides react with benzyne to give triazoles (95) [67].

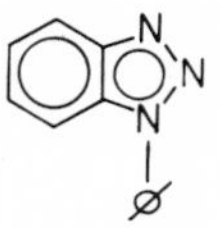

N
N
N
|
ø

(95)

65. (a) Refs. 60 and 61; (b) G. B. R. de Graaff, H. J. den Hertog, and W. C. Melger, _Tetrahedron Lett._, 963 (1965); (c) E. R. Biehl, E. Nieh, and K. C. Hsu, _J. Org. Chem._, _34_, 3595 (1969).
66. G.W. Steinhoff and M. C. Henry, _J. Org. Chem._, _29_, 2808 (1964).
67. G.A. Reynolds, _J. Org. Chem._, _29_, 3733 (1964).

19-5. ANALOGS OF BENZYNE

In the preceding paragraphs mention has already been made of other benzynes such as the naphthalynes, 9-phenanthryne, and so on. Benzyne analogs are also encountered among non-benzenoid and heterocyclic aromatic species.

When 4-bromo-2,3:6,7-dibenzotropone is treated with potassium *t*-butoxide, the intermediate resulting from the loss of hydrobromic acid is readily intercepted with furan, tetracyclone, and so on, to give products such as (96) [68].

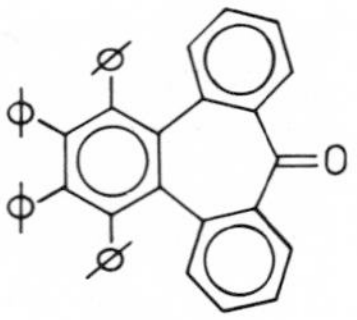

(96)

Similarly, all three of the bromotropolones can be converted into dehydrotropolones capable of being trapped by azide ion to give (97), and so on [69]. When 2-bromo-1,6-methano[10]annulene is treated with base in the presence of tetracyclone, the dehydro-product (98) is obtained [70].

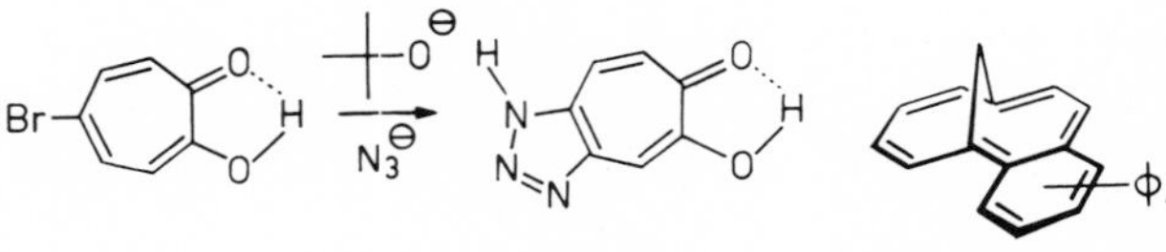

(97) (98)

Benzynequinone has been trapped by Rees in the oxidation of the corresponding aminotriazole (99) with lead tetraacetate; tetracyclone was used as the trapping agent to prepare (100) [71].

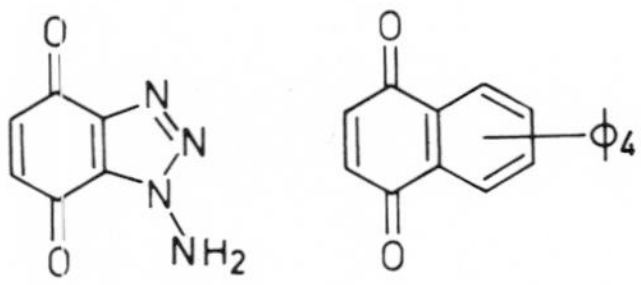

(99) (100)

---

68. W. Tochtermann, K. Oppenländer, and U. Walter, *Chem. Ber.*, *97*, 1318, 1329 (1964).
69. T. Yamatani, M. Yasunami, and K. Takase, *Tetrahedron Lett.*, 1725 (1970).
70. W.A. Böll, *Angew. Chem., Int. Ed. Engl.*, *5*, 733 (1966).
71. C.W. Rees and D.E. West, *Chem. Commun.*, 647 (1969) and *J. Chem. Soc. (C)*, 583 (1970).

Even cyclopentadienyl anion is apparently capable of producing a benzyne analog. Thus, when a solution of potassium diazocyclopentadiene-2-carboxylate (101) is heated in benzene containing dicyclohexyl-18-crown-6 ether (102) and tetracyclone, some of the tetraphenylindene (103) can be isolated after protonation [72].

$N_2^{\oplus}$ $CO_2^{\ominus}$ $K^{\oplus}$ ϕ_4

(101) (102) (103)

It is necessary to have the trapping agent present in rather high concentration - testimony to the highly reactive nature of the intermediate. Likewise, ferrocyne may be involved in the reaction of chloroferrocene with butyllithium to give several products such as butylferrocene and biferrocenyl [73]. If 2-chloromethylferrocene is taken as the substrate, comparable amounts of 2- and 3-_n_-butylmethylferrocene are obtained. Quenching of the reaction with carbon dioxide leads to the acid (104), so that surely the lithioderivative (105) must be present during the reaction [74].

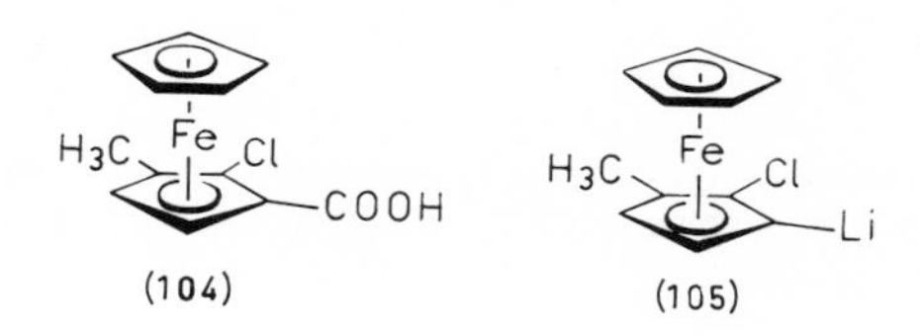

(104) (105)

As an example of the many hetarynes [75], the pyridynes may be mentioned. Treatment of 3-bromopyridine with sodamide in liquid ammonia containing acetophenone gives rise to products such as 4-phenacylpyridine (106) [76].

$CH_2CO\phi$

(106)

Pyridine-3-diazonium-4-carboxylate (107) has been flash-photolyzed to give fragmentation products, but also dimers (108) most likely derived from 3-pyridyne [77];

---

72. J.C. Martin and D.R. Bloch, _J. Amer. Chem. Soc._, _93_, 451 (1971).
73. J.W. Huffman, L.H. Keith, and R.L. Asbury, _J. Org. Chem._, _30_, 1600 (1965).
74. J.W. Huffmann and J.F. Cope, _J. Org. Chem._, _36_, 4068 (1971).
75. For review, see (a) Ref. 1c; (b) T. Kauffmann and R. Wirthwein, _Angew. Chem., Int. Ed. Engl._, _10_, 20 (1971).
76. R. Levine and W.W. Leake, _Science_, _121_, 780 (1955).
77. J.M. Kramer and R.S. Berry, _J. Amer. Chem. Soc._, _93_, 1303 (1971).

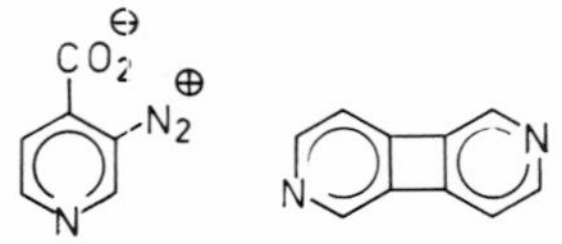

(107) (108)

when 2-chloropyridine-N-oxide is treated with potassium amide in liquid ammonia, both the 2- and 3-aminopyridine-N-oxides (109) and (110) are obtained [78].

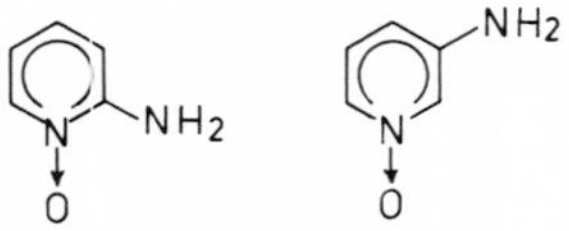

(109) (110)

Another off-shoot in the benzyne field is that of intermediates resulting from the elimination of non-adjacent substituents. Both m- and p-benzenediazonium carboxylates upon flash photolysis produce transient products which in a mass spectrometric analysis show an m/e = 76 peak. p-Benzyne is apparently as intermediate or transition state in the isomerization of Z-1,2-diethynylethylene. The labelled species (111) and (112) interconvert at 200° with a 30 minute half life without giving any isomers such as (113) or (114). If hydrogen donors are present, benzene is formed; carbon tetrachloride gives p-dichlorobenzene [79]. Oddly enough, a different species seems to be involved in the base promoted ammonolysis of (115) in the presence of diphenylisobenzofuran: (116) is then obtained [80].

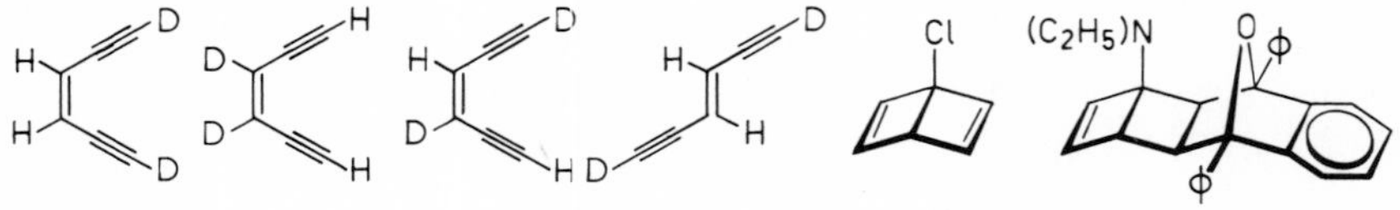

(111) (112) (113) (114) (115) (116)

Lead tetraacetate oxidation of 1-aminoaphtho(1,8-de)triazine (117) produces an intermediate (peri-naphthalyne) that adds stereospecifically to 1,2-dichloroethylene to give (118), to ethyl acetylenedicarboxylate to give (119), and to benzene to produce fluoranthene (120) [81].

---

78. T. Kauffmann and R. Wirthwein, Angew. Chem., Int. Ed. Engl., 3, 806 (1964).
79. R.R. Jones and R.G. Bergman, J. Amer. Chem. Soc., 94, 660 (1972).
80. R. Breslow and J. Napierski, quoted in Ref. 79 (Footnote 9).
81. (a) C.W. Rees and R.C. Storr, Chem. Commun., 193 (1965) and J. Chem. Soc. (C), 760, 765 (1969); (b) R.W. Hoffmann, G. Guhn, M. Preiss, and B. Dittrich, J. Chem. Soc. (C), 769 (1969); (c) J. Meinwald and G.W. Gruber, J. Amer. Chem. Soc., 93, 3802 (1971).

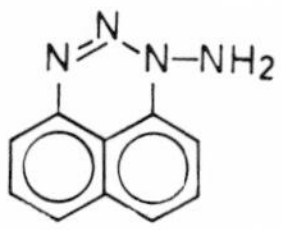

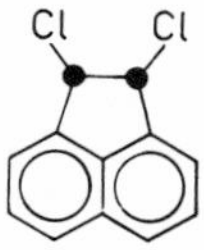

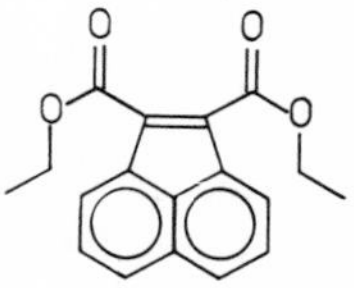

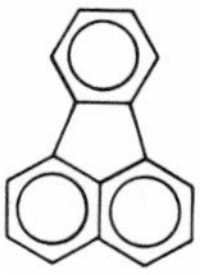

(117) (118) (119) (120)

Likewise, when 2-bromo-6-ethoxypyridine (121) is treated with potassium amide in liquid ammonia, a mixture of 2- and 4-amino-6-ethoxypyridine, (122) and (123), is obtained, suggesting a 2,4-dehydropyridine intermediate (124) [82].

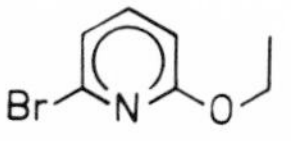

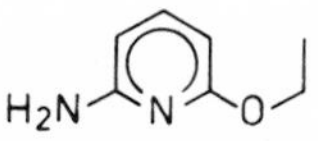

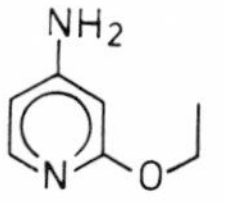

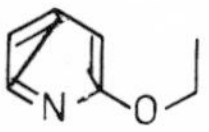

(121) (122), (123) (124)

In fact, the species in which the strained triple bond is present does not have to be aromatic. Cyclooctatrienyne is evidently the initial product when bromocyclooctatetraene is treated with t-butoxide ion, because if phenyl azide or tetracyclone are present the usual type of products, (125) and (126) are obtained [83].

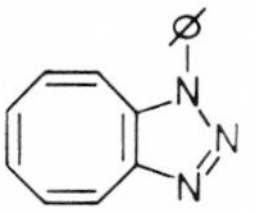

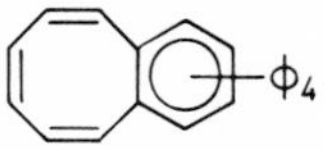

(125) (126)

Bromobullvalene when treated with the same base leads to dimer (127), and trimer (128); the intermediate bullvalyne (129) can also be trapped by the usual reagents [84].

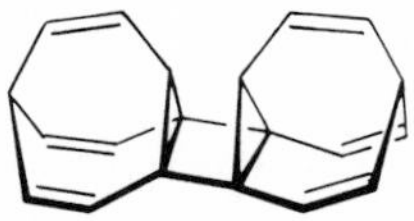

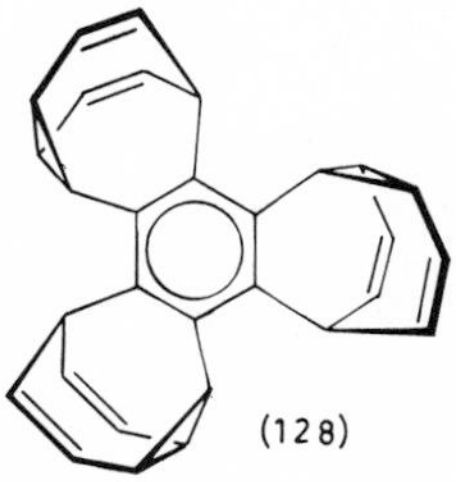

(127) (128) (129)

---

82. H. Boer and H. J. den Hertog, Tetrahedron Lett., 1943 (1969).
83. (a) A. Krebs, Angew. Chem., Int. Ed. Engl., 4, 953 (1965); (b) A. Krebs and D. Byrd, Ann., 707, 66 (1967); (c) A. S. Lankey and M. A. Ogliaruso, J. Org. Chem., 36, 3339 (1971).
84. (a) G. Schröder and J. F. M. Oth, Angew. Chem., Int. Ed. Engl., 6, 414 (1967); (b) R. J. Böttcher, H. Röttele, G. Schröder, and J. F. M. Oth, Tetrahedron Lett., 3935 (1968).

Both hydrocarbons have temperature dependent nmr spectra reminiscent of that of the parent bullvalene: The dimer spectrum at -30° consists of two peaks at 3:1 intensities in the vinyl and aliphatic regions; at 120° this reversibly becomes a singlet. The trimer at -60° shows three peaks at 4.15, 6.5 and 7.7 τ in the ratio of 4:2:2; at 60° there are only two peaks in a 6:2 ratio.

The simple cycloalkynes provide an excellent background against which to judge the instability of various arynes; in their case it is possible to vary the ring size and study the effect of this handle on reactivity.

Cyclobutyne has been sought very carefully by Wittig, and to date it can only be said that no proof of its existence has been produced; nevertheless, it is instructive to review the search because it illustrates how easy it is to draw wrong conclusions from superficial evidence [85]. When 1,2-dibromocyclobutene is treated with magnesium in the presence of diphenylisobenzofuran, the adduct (130) is produced in 8% yield.

(130)

This result of course suggests that cyclobutyne forms and that it undergoes Diels-Alder addition to the isobenzofuran. However, if the metal is left out of the reaction, adduct (131) forms, and the first-mentioned product (130) finally obtains if the magnesium is added later.

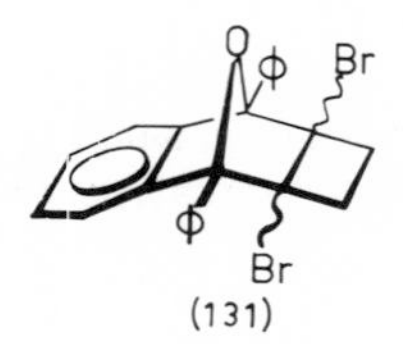

(131)

In the case of 1,2-dibromocyclopentene, Wittig [86] found that the successive additions of diphenylisobenzofuran and magnesium gave a simple Diels-Alder adduct (132) in no more than 0.08% yield under a certain set of conditions; this product could be dehalogenated with magnesium, and the material (133) so obtained would react with a second molecule of isobenzofuran to give a bis-adduct (134).

(132) (133) (134)

85. G. Wittig and E.R. Wilson, Chem. Ber., 98, 451 (1965).
86. G. Wittig, J. Weinlich, and E.R. Wilson, Chem. Ber., 98, 458 (1965).

When magnesium and the trap are simultaneously present under otherwise identical conditions, the _bis_-adduct is again obtained, but now in slightly more than 1% yield. This difference is the main support for Wittig's claim that cyclopentyne must have been formed.

Much stronger evidence arises from labelling studies [87]. It is now known that both 1-chlorocyclopentene and 1-chlorocyclohexene can be dehydrohalogenated with phenyllithium. If a carbon labelled cyclohexene is employed, with labelling equally at the 2- and 6-positions, one quarter of the activity product shows up at the 1-position in the 1-phenylcyclohexene. This result is _demanded_ by a cyclohexyne intermediate; however, it is also _permitted_ by the isomeric intermediate 1,2-cyclohexadiene (135); the only additional assumption required is that phenyl group attack on the central and on the two terminal atoms is equally probable.

(135)

While this may seem a bit far-fetched, it should be pointed out that 1,2-cycloalkadienes become progressively more stable relative to isomeric cycloalkynes in medium-sized rings as these are decreased in size [88]. It was further found however, that 1-chloro-2-methylcyclohexene does not give a coupling product with phenyllithium, whereas 1-chloro-6-methylcyclohexene reacts with this base to give both 3- and 6-methyl-1-phenylcyclohexene. The intermediate (135) _can_ play a role under different conditions, as was shown in another labelling study by Bottini [89].

At the other end, the smallest stable cycloalkyne is the eight membered cyclooctyne (cf. p. 221); though reactive, it is isolable, and its structure has been measured by means of electron diffraction [90]. 1,2-Cyclooctadiene can be preserved long enough in cold solutions for ir measurements [91]. Even cycloheptyne can be prepared in solution and briefly stored and manipulated at -80° [92]; these semi-stable cycloalkynes can hence be subjected to a selectivity study. The numbers in Table 19-1 represent competition ratios for phenyllithium and lithium piperidide in refluxing ether [93]. It seems obvious that the smallest ratios are those in which the angle strain is at a maximum.

87. L.K. Montgomery, F. Scardiglia, and J.D. Roberts, _J. Amer. Chem. Soc._, _87_, 1917 (1965).
88. W.R. Moore and H.R. Ward, _J. Amer. Chem. Soc._, _85_, 86 (1963).
89. (a) A.T. Bottini, F.P. Corson, R. Fitzgerald, and K.A. Frost, _Tetrahedron Lett._, 4753, 4757 (1970). Cf. also (b) P. Mohanakrishnan, S.R. Tayal, R. Vaidyanathaswamy, and D. Devaprabhakara, _Tetrahedron Lett._, 2871 (1972).
90. J. Haase and A. Krebs, _Z. Naturforsch._, _A26_, 1190 (1971).
91. E.T. Marquis and P.D. Gardner, _Tetrahedron Lett._, 2793 (1966).
92. A. Krebs and H. Kimling, _Angew. Chem., Int. Ed. Engl._, _10_, 509 (1971).
93. L.K. Montgomery and L.E. Applegate, _J. Amer. Chem. Soc._, _89_, 5305 (1967).

TABLE 19-1. Selectivity of Various Cycloalkynes and Benzynes

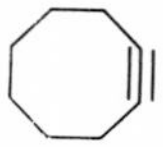	21
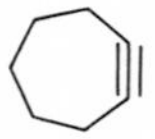	21
	5.2
	2.6
	4.4
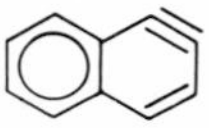	5.8
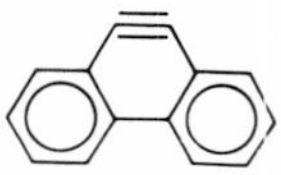	12.8

A few cases of benzyne stabilization by a transition metal have been reported. Thus, the species (136) has been prepared by heating nickel tetracarbonyl with o-diiodobenzene [94]; to date this assignment has not yet been supported by X-ray diffraction measurements.

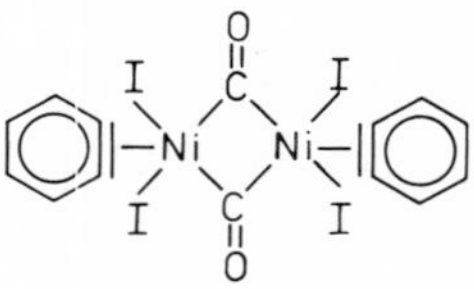

(136)

94. E.W. Gowling, S.F.A. Kettle, and G.M. Sharples, Chem. Commun., 21 (1968).
95. C.W. Bradford, R.S. Nyholm, G.J. Gainsford, J.M. Guss, P.R. Ireland, and R. Mason, Chem. Commun., 87 (1972).
96. G.B. Robertson and P.O. Whimp, J. Organometal. Chem., 32, C69 (1971).

Such data do support the assignment (137) for an osmium complex encountered by Bradford and Nyholm [95]. There is also a cyclohexyne complex (138); the CC triple bond length is 1.29 Å as determined by X-ray diffraction [96].

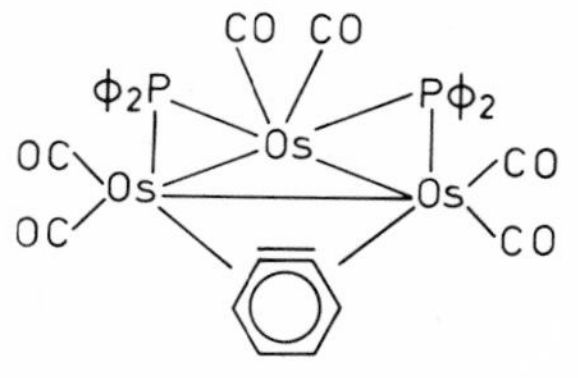

(137)

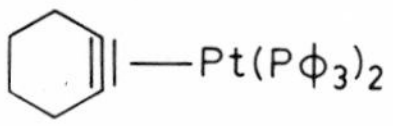

(138)

Chapter 20

CARBONIUM IONS [1]

20-1. INTRODUCTION

It has been learned in the past two decades that the ionization of organic halides and esters is a complex process, involving several intermediates in succession. The reaction may symbolically be written as follows:

$$\begin{array}{ccccccc} RX & \rightleftharpoons & (R^{\oplus}X^{\ominus}) & \rightleftharpoons & R^{\oplus}\|X^{\ominus} & \rightleftharpoons & R^{\oplus} + X^{\ominus} \\ \downarrow & & \downarrow & & \downarrow & & \downarrow \\ P & & P & & P & & P \\ S_N2 & & \longrightarrow & & \longrightarrow & & S_N1 \end{array}$$

1. For review, see (a) A. Streitwieser, "Solvolytic Displacement Reactions", McGraw-Hill, New York, 1962; (b) C.K. Ingold, "Structure and Mechanism in Organic Chemistry", 2nd Ed., Cornell University Press, Ithaca, New York, 1969; (c) P. von R. Schleyer and G. Olah, "Carbonium Ions", Interscience, New York, 1968; (d) N.C. Deno, Chem. Eng. News, 42, 88 (Oct. 5, 1964); (e) G.D. Sargent, Quart. Rev. (London), 20, 301 (1966); (f) D. Bethell and V. Gold, "Carbonium Ions, an Introduction", Academic Press, New York, 1967.

At each of the four major stages, further progress toward ionization competes with return, solvent competes with added nucleophiles, Wagner-Meerwein rearrangements compete with hydride shifts, and so on.

In order to keep the discussion manageable, we shall in this chapter begin by making the simplifying assumption that S_N2 reactions proceed directly from RX itself, and we shall ignore the differences between the three ionization stages. The major emphasis will be on the process of participation in the formation of the carbonium ion, and on the nature of the ion so produced. In the following chapter we will discuss ion pairs, dicarbonium ions, and various analogs of carbonium ions.

20-2. THE DISPLACEMENT REACTION [1a]

In order to understand the topic of carbonium ions, it is first desirable to consider the closely related S_N2 displacement reaction, which is believed by most chemists to be a concerted, one-step reaction (however, see Chr. 21):

$$\text{PQRCX} + \text{Y:} \longrightarrow \text{PQRCY} + \text{X:}$$

In the transition state the central carbon atom is essentially pentacovalent (bound to 5 different groups, though not sharing 10 electrons):

Experimentally, these reactions are characterized by several important features. The kinetics are second order, first in each of the reactants. Inversion accompanies the displacement process; the displacement occurs from the rear. The rate of the reaction depends on the effectiveness of the attacking species (its nucleophilicity), on that of the leaving group, on the substituents P, Q and R of the carbon atom and on the solvent.

The nucleophilicity of a group can be quantitatively expressed by the Swain-Scott constants: The logarithm of the rate constant characterizing the reaction of a given nucleophile with methyl bromide in water relative to the reaction of methyl bromide with water itself [2]. In general, this constant is larger for anions than for neutral species (e.g., hydroxide ion: 4.2, water: 0); it is larger for the larger atoms within a group (e.g., hydrosulfide ion: 5.1), and larger for atoms of lower atomic numbers within a row (e.g., chloride ion: 3.0). While these numbers are useful in qualitative discussions, they should not be taken too seriously since other values will obviously obtain with other substrates and solvents.

---

2. C.G. Swain and C.B. Scott, J. Amer. Chem. Soc., 75, 141 (1953).

There is also a degree of constancy in the ratio of rate constants of reactions involving various leaving groups [3]. Generally, iodide and arylsulfonate are very good and fluoride is very poor, with the ratio of these two about 10^5.

The carbon atom undergoing displacement must be accessible, and hence large groups bound to it will inhibit displacement. The reaction may be accelerated by unsaturated groups such as phenyl or vinyl, which are able to delocalize such partial charges as may develop on the central carbon atom during the displacement. The following relative rates (averages for a number of S_N2-reactions) are given by Streitwieser [4]: methyl: 1; ethyl: 0.03; i-propyl: 0.003; neopentyl: 10^{-6}; allyl: 1; benzyl: 3.

The solvent effect is determined principally by the net change in charge occurring during the reaction: If charges are formed, a polar solvent will accelerate the reaction, and if charges are neutralized, a non-polar solvent will be better. In reactions involving no net change in charge, reasonable guesses may be made on the basis of the solvation required by the incoming and leaving groups; solvation of small ions will be tighter than that of large and/or delocalized ions. Further insight may be gained by the realization that certain solvents are especially well constructed for solvating negative charges (hydroxylic solvents), and others (aprotic solvents) are unusually well suited for solvating positive charge (dimethylsulfoxide, hexamethylphosphoramide).

20-3. CARBONIUM IONS IN SOLVOLYSIS

When a comparison is made of the rates of displacement of various substrates with alkoxide ions in the corresponding alcohols, one finds that the product formation in many cases is first order only in substrate, and independent of the concentration of lyate ion. This tends to be the case especially with tertiary substrates such as t-butyl halides, and with substrates able to delocalize charge at the site of the central carbon atom, such as allylic and benzylic halides. Such reactions are usually described as S_N1 reactions. They are considered to involve prior ionization to carbonium ions, which are subsequently rapidly scavenged by the solvent to give the product in protonated form.

These reactions are not always easily distinguished from reactions involving solvent molecules as the displacing agent; such reactions are of course also first order, so that the rate law cannot help us. The difference between these two pathways is small and difficult to assess. What we might do to learn about such differences is to study some stable carbonium ions, which in solvolysis reactions would surely form via the S_N1 pathway. In Chapter 9 we encountered some highly stable positive carbon ions such as the tropylium system; however, the corresponding precursors are already ionized even in the pure form and hence their behavior in solvolytic media is not typical.

---

3. Ref. 1a, p. 30.
4. Ref. 1a, p. 13.

A more suitable group is that of triarylmethyl derivatives: The combination of hindrance offered by the aryl groups and of their ability to delocalize charge (for example, structure (1)) makes them ideal candidates.

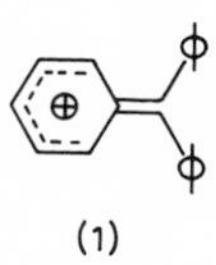

(1)

As a rule, rates and rate changes are difficult to measure in the triarylmethyl substrates since they usually solvolyze about as rapidly as they can be put into solution; however, ionization can be demonstrated by means of conductance measurements to occur in even moderately polar, aprotic solvents such as sulfur dioxide [5]. In fact, in the presence of a stable and non-nucleophilic counterion, the triarylmethyl carbonium ion salt becomes a stable and isolable species; thus, hexachloroantinonates, perchlorates and tetrafluoroborates of triarylmethyl carbonium ions are isolable salts [6] that give conducting solutions, and differ in their spectral properties from covalent derivatives such as the alcohols [7].

Such carbonium ions can often be generated even in aqueous acid _via_ the process:

$$Ar_3COH + H^{\oplus} \overset{K_R}{\rightleftharpoons} Ar_3C^{\oplus} + H_2O$$

and their stability can be expressed in terms of pK_R, the pH at which the alcohol and the carbonium ion have equal concentrations. Thus trityl alcohol has a pK_R of -6, tri-_p_-anisylcarbinol of 0.8, and the alcohol (2) has one of 9.1; the concentrations of the ions and hence the pK_R values can be measured either by uv or nmr [8].

(2)

---

5. P. Walden, _Chem. Ber._, _35_, 2018 (1902).
6. H.J. Dauben, L.R. Honnen, and K.M. Harmon, _J. Org. Chem._, _25_, 1442 (1960).
7. A. Hantzsch, _Chem. Ber._, _54_, 2573 (1921).
8. (a) V. Gold and B.W.V. Hawes, _J. Chem. Soc._, 2102 (1951); (b) N.C. Deno, J.J. Jaruzelski, and A. Schriesheim, _J. Amer. Chem. Soc._, _77_, 3044 (1955); (c) J.C. Martin and R.G. Smith, _J. Amer. Chem. Soc._, _86_, 2252 (1964); (d) A.E. Young, V.R. Sandel, and H.H. Freedman, _J. Amer. Chem. Soc._, _88_, 4532 (1966); (e) S.V. McKinley, J.W. Rakshys, A.E. Young, and H.H. Freedman, _J. Amer. Chem. Soc._, _93_, 4715 (1971).

It is found that in such systems substituents able to donate or withdraw electrons have considerable effects on the solvolysis rate. For instance, the rate of formation of triarylmethyl ethyl ethers from the corresponding chlorides in a mixture of ether and alcohol is increased by a factor of 100 upon the introduction of a *p*-methoxy group, and retarded by about as much when a *p*-nitro group is used [9].

A most revealing experiment has been reported by Breslow [10]. When the triarylcarbinol (3) is dissolved in cold sulfuric acid, the nmr spectrum of the solution has three different methyl signals indicative of the sulfonium salt (4).

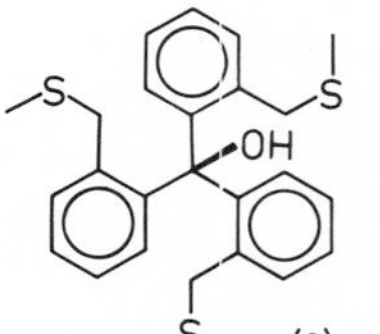

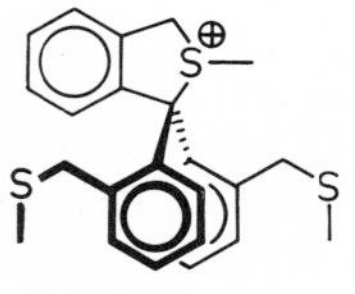

In this salt, one sulfur atom is bound to the carbonium ion center, and the two others are bound to rings that rotate sufficiently slowly so that the methyl groups find themselves in magnetically different environments. When the temperature is raised, exchange of the methyl groups sets in. If the interchange occurred by the S_N2-displacement of one sulfur atom by another, two of the methyl signals should coalesce and the third should remain different; however, all three coalesce simultaneously, which suggests that an S_N1 process is leading to the true carbonium ion (5).

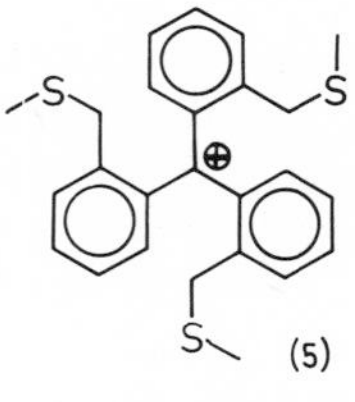

We may conclude that at least in the case of the trityl substrates, carbonium ions can and do exist, and that they may be intermediates in displacement reactions even when the ultimate displacing agent is forcibly held nearby. The ready solvolysis of these compounds makes them unsuited for general comparative use, and other models are desirable.

Bridgehead substituted caged structures are good models in that the rear-side displacement so characteristic of S_N2 reactions is impossible. On the other hand,

---

9. A.C. Nixon and G.E.K. Branch, *J. Amer. Chem. Soc.*, *58*, 492 (1936).
10. (a) R. Breslow, L. Kaplan, and D. LaFollette, *J. Amer. Chem. Soc.*, *90*, 4056 (1968); (b) R. Breslow, S. Garratt, L. Kaplan, and D. LaFollette, *J. Amer. Chem. Soc.*, *90*, 4051 (1968).

other effects now operate to lead to certain complications that make them unsuitable for comparison. Thus, it is generally accepted that carbonium ions are planar, because ions such as 1-norbornyl (6) - which cannot assume such geometry - form extremely slowly even under the most favorable circumstances.

(6)

Thus, 1-chlorobicyclo[2.2.1]heptane ethanolyzes some 10^{13} times more slowly than t-butyl chloride [11]. In somewhat less rigid bridgehead-substituted structures such as the 1-adamantyl derivatives, solvolysis is more rapid; carbonium ion stabilization by overlap with the tertiary centers in the rear has been invoked to explain this (7) [12].

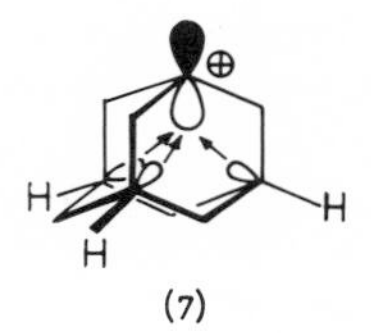

(7)

Three substances deserve special mention in any discussion of model solvolysis substrates. One group consists of cumyl chloride (8) and its ring-substituted cousins.

ϕ—$CX(CH_3)_2$

(8)

The σ^+-constants are based on them; as noted earlier (p. 546), these constants are derived from the relative hydrolysis rates of these compounds at 25° in 90% aqueous acetone [13]. (In passing, it is regretted that this confusing term (percent aqueous solvent) is in such general use, since neither the standard (water or the organic solvent) nor the measure (% by volume, weight or number of moles) is defined, and in fact some chemists use the term with a different meaning than others. One should consult the original paper in each case to make sure of the author's meaning of the term. In the present case it means that 10 volume units of water were mixed with 90 volume units of acetone.)

---

11. R.C. Fort and P. von R. Schleyer, Adv. in Alicyclic Chem., 1, 283 (1966).
12. R.C. Fort and P. von R. Schleyer, J. Amer. Chem. Soc., 86, 4194 (1964).
13. H.C. Brown and Y. Okamoto, J. Amer. Chem. Soc., 79, 1913 (1957).

t-Butyl halides are also important models. The intervention of a carbonium ion is supported by the observation that the solvolysis of t-butyl bromide in ethanol is about 1000 times faster than that of i-propyl bromide [14], and in formic acid [15] this ratio is about 10^7. It would be logical to expect a reaction involving ionization to be accelerated greatly by the use of a more polar solvent. This solvent effect can be described quantitatively to some extent by the Grunwald-Winstein relation [16]:

$$\log k/k_o = mY$$

Here k is the solvolysis rate constant for a given substrate in a certain solvent, k_o is that in the standard solvent (80% "aqueous ethanol"), Y is termed the ionizing power of the solvent (by definition: Zero for 80% aqueous ethanol), and m is a measure of the sensitivity of the reaction to solvent change (by definition: Unity for t-butyl chloride). Y has a value of -2 for pure ethanol, +2 for formic acid, and +3.6 for water; trifluoroacetic acid and fluorosulfonic acid [17] are much more ionizing still (for another measure of solvent polarity, see p. 852). The factor m has a value of approximately 1 for tertiary substrates, but only about 0.3 for most secondary substrates [18]; this difference is used to support the contention that secondary esters solvolyze with a measure of solvent assistance in the rear, and the tertiary analogs without it.

Since there is a great deal of interest in the possibility of S_N1 solvolysis with secondary substrates, the 2-adamantyl system (9) is still another model that has attracted much attention [19]: The axial hydrogen atoms in the rear hinder displacements.

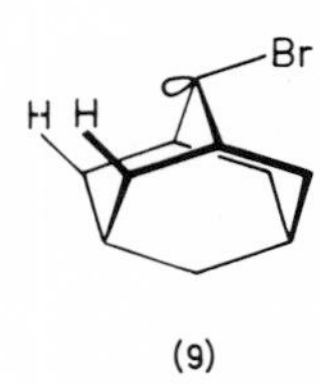

(9)

Schleyer has indeed found that this compound's behavior in solvolysis is "limiting"; i.e., by all criteria that can be applied to it, it behaves as though it leads to a carbonium ion which is free from covalent bonds to the solvent, although it is of course solvated. Thus, its m-value is 1.03, compared to 0.43 for isopropyl bromide [20].

---

14. S. Winstein, E. Grunwald, and H.W. Jones, J. Amer. Chem. Soc., 73, 2700 (1951).
15. L.C. Bateman and E.D. Hughes, J. Chem. Soc., 945 (1940).
16. E. Grunwald and S. Winstein, J. Amer. Chem. Soc., 70, 846 (1948).
17. (a) I.L. Reich, A. Diaz, and S. Winstein, J. Amer. Chem. Soc., 91, 5635 (1969); (b) A. Diaz, I.L. Reich, and S. Winstein, J. Amer. Chem. Soc., 91, 5637 (1969).
18. S. Winstein and H. Marshall, J. Amer. Chem. Soc., 74, 1120 (1952).
19. J.L. Fry, C.J. Lancelot, L.K.M. Lam, J.M. Harris, R.C. Bingham, D.J. Raber, R.E. Hall, and P. von R. Schleyer, J. Amer. Chem. Soc., 92, 2538 (1970).
20. D.J. Raber, R.C. Bingham, J.M. Harris, J.L. Fry, and P. von R. Schleyer, J. Amer. Chem. Soc., 92, 5977 (1970).

There are additional criteria by which the question of a carbonium ion intermediate *vs.* concerted solvent displacement can be approached. First, it is found that in limiting cases, α-methyl groups cause a large rate acceleration. This factor is about 10^8 for (9); for *i*-propyl bromide it is only 10^4 [21]. Secondly, a good nucleophile such as azide ion can often be used as a trap; i.e., it can capture the intermediate without affecting the rate law [22]. The borohydride ion has also been much used for this purpose; its capture leads to the formation of a hydrocarbon [23]. In those reactions in which solvolysis competes with elimination, the ratio of products (e.g., olefins to alcohols) should be independent of the leaving group if a free carbonium ion is an intermediate [24]. Another criterion is the sensitivity of the rate to the leaving group. Thus, the ratio of rates ROTs/RBr may be as large as 10^4 in extreme S_N1 solvolysis, and it may be less than 1 for extreme S_N2 reactions [25].

The stereochemistry provides an important criterion. If the substrate is a resolved compound and the center undergoing solvolysis is also the center of asymmetry, the picture of a simple, planar carbonium ion calls for complete racemization [26]. It should be remembered however, that planarity of the carbonium ion center does not necessarily mean planarity - and thus optical inactivity - for the whole ion. For example, X-ray diffraction has shown that while the central carbon atom of tris-*p*-aminotrityl perchlorate (10) is planar, the three rings lie out of this plane by 30° [27].

(10)

Solvolysis of the resolved triaryl derivative (11) gives largely retained alcohol; evidently the carbonium ion, though long-lived, is not stable anough to have time for racemization via rotation of the bond indicated ((12)).

---

21. J.L. Fry, J.M. Harris, R.C. Bingham, and P. von R. Schleyer, *J. Amer. Chem. Soc.*, *92*, 2540 (1970).
22. J.M. Harris, D.J. Raber, R.E. Hall, and P. von R. Schleyer, *J. Amer. Chem. Soc.*, *92*, 5729 (1970).
23. H.M. Bell and H.C. Brown, *J. Amer. Chem. Soc.*, *88*, 1473 (1966).
24. J.H. Bayless, F.D. Mendicino, and L. Friedman, *J. Amer. Chem. Soc.*, *87*, 5790 (1965).
25. H.M.R. Hoffmann, *J. Chem. Soc.*, 6753 (1965).
26. For other methods of determining the stereochemistry in solvolysis reactions, see C.A. Kingsbury and W.B. Thornton, *J. Amer. Chem. Soc.*, *88*, 3159 (1966).
27. L.L. Koh and K. Eriks, *Acta. Cryst.*, *B27*, 1405 (1971).

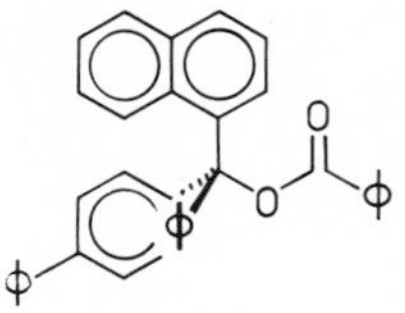

(11) 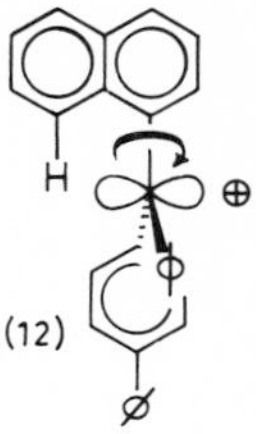 (12)

Even sulfuric acid solutions of such ions are optically active, and the solutes can be hydrolyzed back to the original resolved carbinols [28]. Thus, caution is in order in the application of these criteria, as several of them have failed to give very clear-cut answers in some instances. However, when all those that are applicable to a given substrate point in the direction of limiting solvolysis, most organic chemists seem confident that a reasonably free carbonium ion is indeed involved.

An entirely different approach has been developed for carbonium ions generated via strong acid protonation of alcohols, ethers, acetals, esters and orthoesters. In principle the same question arises here: Whether the solvent assists in the departure of the leaving water or alcohol molecule from the protonated precursor (13), or whether a free carbonium ion intermediate intervenes.

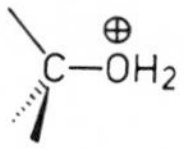

$C-\overset{\oplus}{O}H_2$

(13)

In analogy to the S_N2 and S_N1 pathways, these reactions are referred to as the A2- and A1 mechanisms. The first experimental distinction between them was based on the Hammett H_o function (see p. 549); however, this has led to certain discrepancies that have undermined its credibility. For example, oxetane, which by the A1 mechanism would give rise to a primary carbonium ion, has nevertheless been assigned to hydrolyze that way on the basis of a correlation of the rate constant with the H_o function [29].

Results that are at least intuitively more satisfactory are obtained by means of studies of the effect of hydrostatic pressure on the rate. In those cases where the A1 mechanism is operative, one expects a positive volume of activation, since a simple bond cleavage characterizes the transition state; in the A2 pathway, displacement should lead to a transition state more compact than the initial state.

---

28. (a) B. L. Murr and C. Santiago, J. Amer. Chem. Soc., 88, 1826 (1966); (b) B. L. Murr and C. Santiago, J. Amer. Chem. Soc., 90, 2964 (1968); (c) B. L. Murr and L.W. Feller, J. Amer. Chem. Soc., 90, 2966 (1968).
29. J.G. Pritchard and F.A. Long, J. Amer. Chem. Soc., 78, 2667 (1956) and 80, 4162 (1958).

Thus, methyl orthoformate in acid-catalyzed hydrolysis gives an activation volume of +2.4 [30], diethyl acetal of 0.0 [30], and oxetane of about -5.5 cm^3/mole [31]. This spectrum of values - from positive for incipient tertiary or alkoxy-stabilized ions to negative for simple oxygen-methylene linkages - appears to be generally applicable, and to date no exceptions have been discovered.

Further important insights have been gained by means of studies of salt effects [1a], isotope effects [32], and the dissection of activation parameters of solvolysis reactions in mixed solvents [33]. Some references to these phenomena will be made in passing below; a detailed discussion is not within the scope of this book, however.

20-4. THE GENERATION OF CARBONIUM IONS

Solvolysis is usually suitable for tertiary systems, or for carbonium ions stabilized in some way so that little or no solvent assistance is required. In more difficult cases, the use of excellent leaving groups such as the triflate (trifluoromethanesulfonate) anion and/or the use of highly non-nucleophilic (superacid) media may produce the desired result; as we shall see in the following sections, almost any halide, ester or alcohol can be made to generate stable carbonium ions in such media as sulfuric acid, fluorosulfonic acid, antimony pentafluoride, etc. Thus, even such an unlikely species as the trisazidomethyl cation has been isolated as the hexachloroantimonate salt from the highly unreactive carbon tetrachloride and tetrachloroantimonium azide [34].

If the substrate is an olefin, only protonation is necessary and no leaving group needs to be removed. The additions to olefins of a number of reagents such as the halogens are likewise thought to proceed *via* carbonium ions.

Nitrous acid deamination is another way to generate carbonium ions. They are preceded as unstable intermediates by the corresponding diazonium ions which then lose nitrogen. Since this method sometimes leads to different products than solvolysis, many chemists in fact suspect that both of these ions may be involved in the product-forming steps [35].

Skell [36] has pointed out that the carbene reaction of trihalomethanes in basic alcohols - which produces much carbon monoxide - is a closely analogous reaction which may involve the species R-O═C-X as a precursor. It may have some use as the only known way to generate carbonium ions in basic medium.

---

30. J. Koskikallio and E. Whalley, *Trans. Faraday Soc.*, *55*, 809 (1959).
31. W.J. le Noble and M. Duffy, *J. Phys. Chem.*, *68*, 619 (1964).
32. S.E. Scheppele, *Chem. Rev.*, *72*, 511 (1972).
33. E.M. Arnett and D.R. McKelvey, *Record Chem. Progr.*, *26*, 185 (1965).
34. U. Müller and K. Dehnicke, *Angew. Chem., Int. Ed. Engl.*, *5*, 841 (1966).
35. C.J. Collins, *Accounts Chem. Res.*, *4*, 315 (1971).
36. P.S. Skell and I. Starer, *J. Amer. Chem. Soc.*, *81*, 4117 (1959).

20-5. PARTICIPATION AND NON-CLASSICAL IONS

As should be evident from the foregoing discussions in the last several chapters, molecules with electron deficient carbon atoms are generally highly reactive, and they rapidly react in a variety of ways so as to fill their vacant or half vacant orbitals. Carbonium ions are no exception to this rule. The presence of a positive charge is an additional complication that makes their study a difficult task. Basically there are two ways in which a carbonium ion center can end its electron shortage: By reaction with either external electron sources (i.e., other molecules in possession of loosely held or unshared electrons), or by drawing on internal sources. In both cases, if such sharing begins even as ionization gets under way, we say that the ionization process is being assisted, or that participation takes place. External assistance is the mark of the S_N2 process; the internal process is referred to as the neighboring group effect, or anchimeric assistance. In either case, electrons are supplied to the rear of the carbon atom from which the leaving group is departing.

In the decade of the forties a long and systematic study began in order to learn on what internal resources an incipient carbonium ion might be able to draw. As the possibilities considered became more and more marginal, an intense and vigorous controversy arose over what conclusions should be drawn, and what the criteria should be. Although a great many organic chemists applied their talents to this problem, the two principals in the dispute were Professors H. C. Brown of Purdue University, and S. Winstein of U. C. L. A. It would seem that the careers of these two men have been marked by the competition between them, starting perhaps with their applications for the position of instructor at the Illinois Institute of Technology in 1940 [37]. Winstein landed the job; during a seminar much later at UCLA Brown [38] took much pleasure in revealing the contents of a consolation letter he had received at that time saying that he had been considered greatly overqualified!

In this controversy, a great deal of ingenuity was brought to bear, and basic questions were raised about the validity of the tools used by the physical organic chemist; for these reasons it seems worthwhile to review this type of reaction in detail and to point out some of the fascinating results that were obtained.

Basically there are three types of anchimeric assistance: Assistance by nonbonding electrons, by π electrons, and by σ electrons. Covalently bound atoms with unshared electrons may clearly lower the energy of the transition state and hence enable or facilitate carbonium ion formation by sharing such electrons with the incipient carbonium in center as in process II.

---

37. P. Bartlett, J. Amer. Chem. Soc., 94, 2161 (1972).
38. H.C. Brown, "Boranes in Organic Chemistry", Cornell University Press, Ithaca, New York, 1972.

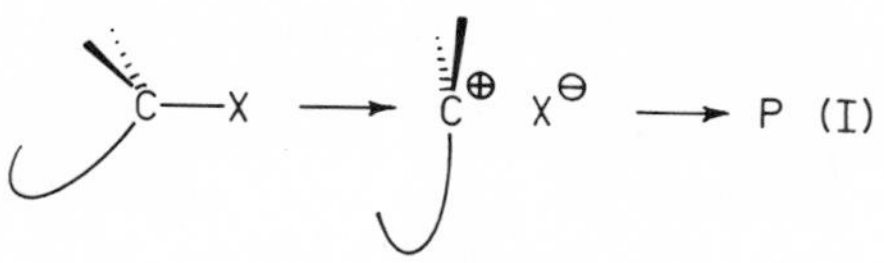

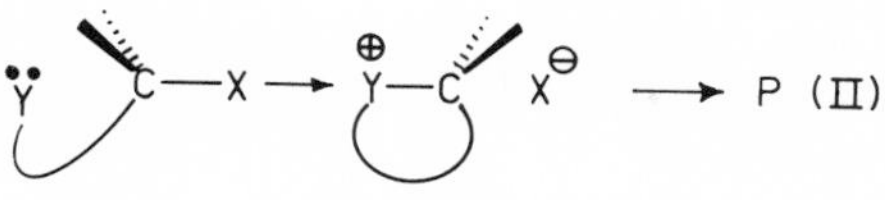

In the less controversial cases, the group Y: may be one already known for its highly nucleophilic character. In such cases it is usually found that the rate of the reaction is many orders of magnitude faster than in the absence of the group Y:, and in the final product the atom Y may still be bound to the carbon atom from which X left. We may refer to this general phenomenon as n-participation (n for non-bonding), but one finds frequent reference to more detailed terminology such as N-, O- and S-participation.

In the intermediate, the positive charge centers on Y rather than on the carbon atom. The greater stability of this intermediate as compared to the simple, open carbonium ion is the reason for the acceleration. As remarked earlier on p. 249, it is important to realize that chemists who study kinetics and mechanisms believe that lowering the energy of the product will lower the energy of the transition state leading to that product. This principle does not exactly have the status of a fundamental law of nature, but it correlates so many chemical observations that its value at least as an empirical rule is beyond any doubt. Now the realities of what can conveniently be measured in the laboratory force us to turn around and use the rule in the converse sense: The activation energies ΔEa can readily be measured, and large differences in rate between otherwise similar reactions are interpreted as involving initial products (intermediates) of greatly differing stability. The controversy arose in part because each chemist is free to decide which $\Delta\Delta Ea$ value is large enough to signify a stabilized intermediate. These difficulties plague us especially when we consider C-participation, whether it be π-, phenyl-, or σ-participation.

In order to create some kind of order in the maze of relative rates, Schleyer [39] derived an equation now known as the Foote-Schleyer relation, designed to show how fast the solvolysis rate should be for a given structure. It should be realized that such equations - which arbitrarily divide the problem into unrelated pieces - are necessarily empirical; still this is a remarkable improvement over mere verbal argument. The equation is:

$$\log k_{rel.} = 1/8\,(1715 - \nu_{CO}) + 1/2\,E_r\,\Sigma(1 + \mathrm{Cos}\,3\theta) + I + \frac{IS - TS}{2.3RT}$$

The rate constant thus obtained is relative to the cyclohexyl derivative with the same leaving group and under the same conditions. The first of the four terms is related to angle strain. The carbonium ion has an sp^2-hybridized carbon atom,

39. P. von R. Schleyer, J. Amer. Chem. Soc., 86, 1854 (1964).

which should be subject to the same considerations of strain as is the carbon atom of the carbonyl group. Foote [40] had found that solvolysis rates fit a linear correlation with carbonyl stretching frequencies. Both the bond holding the leaving group and the carbonyl double bond get stronger (more s character) as the C-C-C angle becomes smaller; the first term is a reflection of that fact.

The second term is related to torsional strain - the change in energy as the carbon-carbon bonds rotate. The planar carbonium ion center and the neighboring tetrahedral carbon atoms interact in a way almost independent of conformation; hence we need to compare only the initial states. E_r is the barrier to single bond rotation, taken to be 3.6 kcal as in ethyl chloride. θ is the smaller of the two dihedral angles between CX and one of the neighboring carbon atoms (hence: 60° at most, as in (14)). Thus, if both neighboring carbon atoms are completely eclipsed, the rate will be accelerated by a factor of $10^{5.4}$.

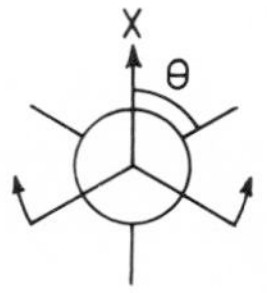

(14)

The third term allows a correction to be made for substitution by electronegative groups (inductive effects). Such effects may be estimated on the basis of known inductive effects on acid strength (p. 414). The difficulties inherent in such estimates are perhaps best illustrated by an observation by Peterson, who found that the ketone (15) formolyzes nearly three times more slowly than (16) [41]. Evidently such effects are able to make themselves strongly felt as far as ten bonds away!

(15)  (16)

The fourth term calls for an estimate of the relief or increment in steric interactions between the initial and transition states. There is unfortunately little agreement between chemists about the magnitude - and sometimes even sign - of this term; therefore we will discuss it in a little more detail.

Instances of steric acceleration are easy to find. Thus, Bartlett found that trimethyl-, trineopentyl-, and tri-t-butylcarbinyl p-nitrobenzoates hydrolyze at relative rates of 1, 560 and 13,000 [42]. Krapcho found the acetolysis rate

---

40. C.S. Foote, J. Amer. Chem. Soc., 86, 1853 (1964).
41. P.E. Peterson and D.M. Chevli, experiments quoted in Chem. Eng. News, 45, 47 (Sept. 18, 1967).
42. P.D. Bartlett and T.T. Tidwell, J. Amer. Chem. Soc., 90, 4421 (1968).

constants of (17)-(19) to be in the order 1:72:120 [43], and Battiste found that (20) acetolyzes 5000 times faster than (21) [44].

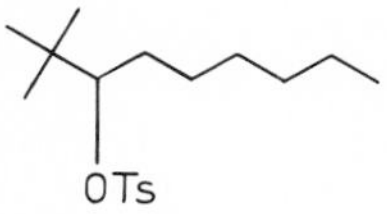

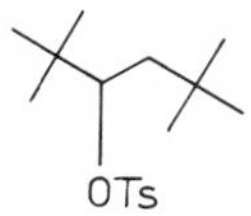

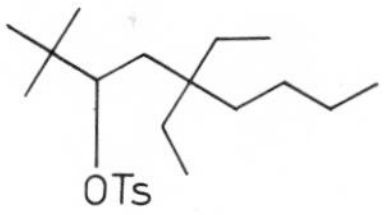

OTs OTs OTs

(17) (18) (19)

BsO BsO

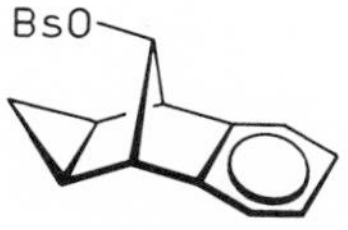

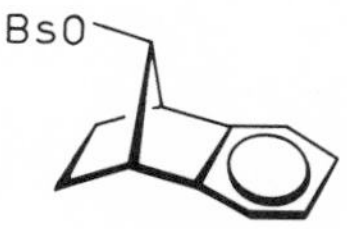

(20) (21)

An extreme case was reported by Brown, who found that the *cis,cis,cis*-perhydro-9-phenalenyl *p*-nitrobenzoate (22) hydrolyzes some 10^7 times faster than the *cis,cis,trans*-stereoisomer (23) [45].

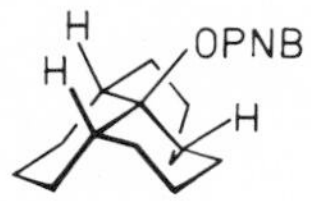

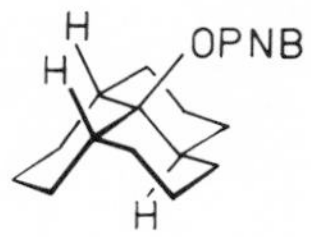

H H OPNB H H H OPNB H

(22) (23)

Examples of steric retardation are less in number. Brown has pointed out that the borohydride reduction of ketones - which has a transition state (24) similar to that of carbonium ion generation by solvolysis (25) - is indeed subject to hindrance.

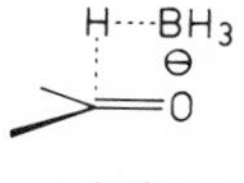

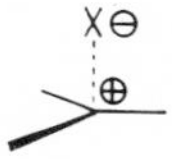

H····BH$_3$ ⊖ O X⊖ ⊕

(24) (25)

This is clear from the reduction of ketones (26)-(29), whose rate constants are in the ratio 1:40:320:207.

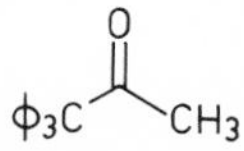

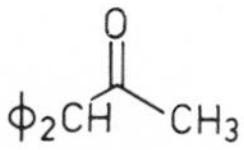

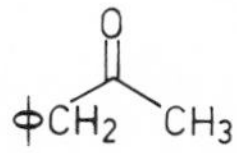

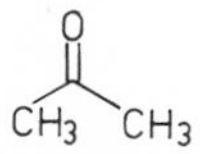

O O O O

ϕ$_3$C CH$_3$ ϕ$_2$CH CH$_3$ ϕCH$_2$ CH$_3$ CH$_3$ CH$_3$

(26) (27) (28) (29)

---

43. A. P. Krapcho, B. S. Bak, R. G. Johanson, and N. Rabjohn, *J. Org. Chem.*, *35*, 3722 (1970).

44. M. A. Battiste, P. F. Ranken, and R. Edelman, *J. Amer. Chem. Soc.*, *93*, 6276 (1971).

45. H. C. Brown and W. C. Dickason, *J. Amer. Chem. Soc.*, *91*, 1226 (1969).

In actual carbonium ion reactions this factor is rarely very large, and somewhat contradictory; thus, while Brown [46] found that the tricyclic tertiary substrate (30) hydrolyzes some 4000 times faster than its epimer (31), Rothberg found that the acetolysis rate of the 2-endo-norbornyl tosylate (32) is 20 times slower than the apparently highly hindered tetracyclic tosylate (33) [47].

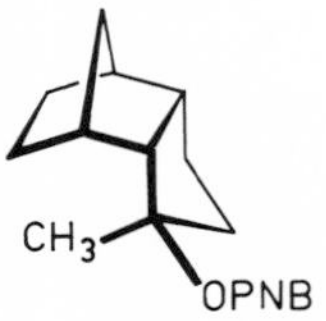

CH_3 OPNB PNBO CH_3 OTs OTs

(30) 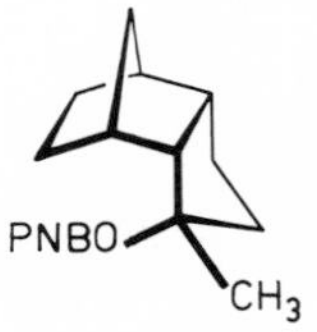 (31)  (32)  (33)

The prediction that the solvolysis of substrates such as (33) must be hindered is based on the assumption that the anion's position in the transition states must be the same ideal one, perpendicularly above the plane of the developing carbonium ion as in t-butyl chloride (34), but this assumption is unproven. In fact, Rothberg's result suggests that neither the route of departure of the anion nor its final destination in the ion pair (Chr. 21) are of great energetic importance, and in view of such uncertainty we cannot even always be sure of whether the reaction will feature an increase or relief of steric strain.

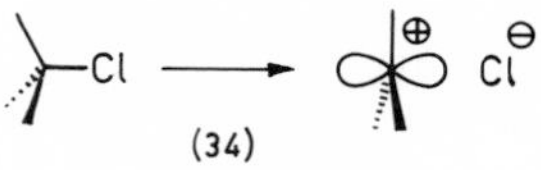

Cl ⟶ ⊕ $Cl^{\ominus}$

(34)

Although this point is destined to remain a thorny one for some time, the net result of the Foote-Schleyer equation is that it predicts the relative rates of a wide variety of substrates fairly well, i.e., within a factor of 1-10. It never overestimates the rates very much, but in a number of cases it underestimates rate constants by a very wide margin, and these are practically all cases in which anchimeric assistance has been claimed [48]. On the other hand, assistance has also been claimed in cases in which the rates had been predicted correctly without it by the Foote-Schleyer equation. The 7-norbornyl esters (35) furnish an example.

X

(35)

---

46. H. C. Brown, I. Rothberg, and D. L. Vander Jagt, J. Amer. Chem. Soc., 89, 6378 (1967).
47. I. Rothberg, Chem. Commun., 268 (1968).
48. P. von R. Schleyer, J. Amer. Chem. Soc., 86, 1856 (1964).

They solvolyze very slowly, probably because of angle strain and eclipsing in the intermediate cation; yet there seems to be σ-participation by one of the neighboring carbon-carbon bonds. The reaction gives rise to the rearranged exo-2-bicyclo-[3.2.0]heptyl acetate (36) and unrearranged 7-norbornyl acetate; both can be shown [49] (by means of 2,3-deuterium labelling) to be largely of retained configuration. This information suggests that there was internal assistance (37), that the incipient ion was stabilized by resonance (38), and that it was not simply a symmetrical 7-norbornyl ion.

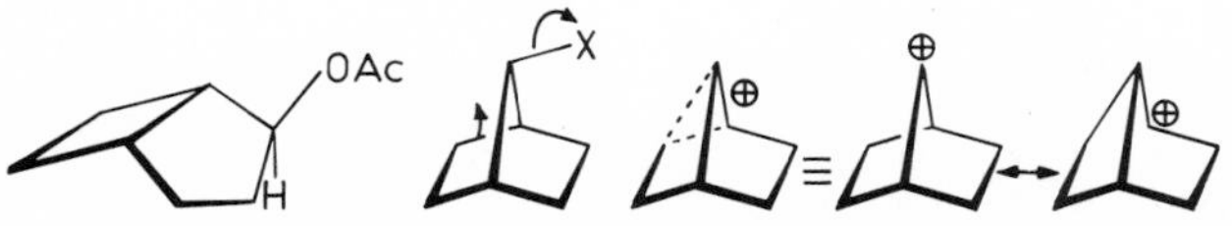

(36) (37) (38)

Because of these uncertainties, many chemists prefer not to try to estimate what the rate should be on some absolute basis, but to choose a model reaction for comparison. The choice of the model is obviously important in controversial cases.

Among the other criteria, the nature of the product is often suggestive. A product in which Y is bound to the former carbonium ion center suggests - but does not prove - that Y was already bound to some degree in the intermediate, and hence in the transition state preceding the intermediate. Likewise, the absence of cyclic products suggests but does not require that participation and a stabilized intermediate are not involved.

It is sometimes possible to generate the intermediate itself under extreme conditions such as very low temperatures, in the absence of any nucleophile, and so on, and hence to preserve it long enough and in high enough concentration for study. Nmr is the favorite tool in such cases; Raman and photo electron spectroscopy are also becoming popular. The results so obtained show us the nature of these intermediates under somewhat atypical conditions, and not all chemists have accepted them as decisive.

The stereochemistry is an important criterion. If we obtain a non-cyclized product as the result of two displacements, retention would be expected. This is in fact frequently observed.

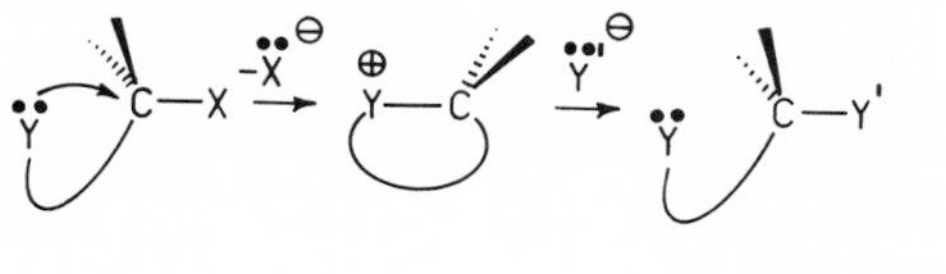

49. (a) P.G. Gassman and J.M. Hornback, J. Amer. Chem. Soc., 89, 2488 (1967); (b) F.B. Miles, J. Amer. Chem. Soc., 89, 2488 (1967) and 90, 1265 (1968).

Needless to say, it is in general hard to prove that the retention is the result of an initial displacement by Y: rather than by a solvent molecule; but if the group Y' is also a solvent molecule it is hard to see why the solvent should assist in the ionization process on the one side, and then displace its own kind in a final attack to give product on the other.

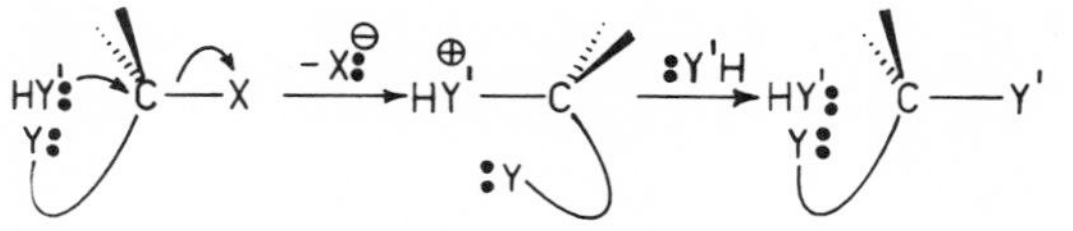

Substituents have great value as probes in carbonium ion reactions. Thus, methyl groups at C-X generally tend to increase the rate of solvolysis by approximately 10^{5-8} in limiting cases (the hydrolysis rates of methyl and ethyl perchlorate are about equal!) [50]; if in a given case the effect is much smaller than usual, it appears likely that the ion is stabilized in the transition state. Phenyl groups normally have even larger effects ($\sim 10^8$) and hence they have been used. It must be kept in mind, however, that such substituents may alter the site of the reaction greatly, and hence have an effect on the course of the reaction, so drastic that comparisons are no longer valid. Substituents at Y alter the molecules in a less deep-seated way, but in this connection there is little background information for comparison, and there is no general agreement as to what is to be expected. The substituent effect most commonly employed is that of p-groups in a phenyl group already present at either C-X or Y; a good correlation with σ^+-constants is indicative of a carbonium ion reaction, and a large negative ρ-value (e.g., -5) means that a large part of the charge is located at the atom to which the phenyl group is bound. A more subtle probe is the secondary isotope effect. Thus, the substitution of deuterium for hydrogen at either groups C-X or Y may have small effects on the reaction rate. The smallness of the effect makes it more difficult to measure (e.g., $k_H/k_D \approx 1.025$ for 2-adamantyl benzenesulfonate); the variation from case to case is even smaller, and our theoretical insight is probably not great enough to understand the precise meaning of these small (though real) effects.

Another criterion is the response to the nucleophilicity of the solvent. The externally (solvent) assisted reaction and internally assisted reactions compete, and often either mode can be made dominant by the choice of a proper solvent. In ethanol for example, which is a relatively nucleophilic solvent, the solvolysis of methyl brosylate is 4000 times faster than that of neopentyl brosylate - presumably because of hindrance to solvent displacement; but in fluorosulfonic acid the reaction is a million times slower - because of methyl participation [51].

Charge delocalization tends to have the effect that nearby solvent molecules are less rigidly oriented and compressed, and hence both $\Delta S^{\ddagger}$ and $\Delta V^{\ddagger}$ tend to have less negative values in such cases; however, this is of course most useful when the transfer of charge from C to Y in the transition state is only partial, and

50. D. N. Kevill and B. Shen, Chem. Ind. (London), 1466 (1971).
51. A. Diaz, I. L. Reich, and S. Winstein, J. Amer. Chem. Soc., 91, 5637 (1969).

not complete or nearly complete. The most clear-cut cases arise when the charge is equally divided between two or more centers because of symmetry. The picture is further complicated by the fact that the formation of an additional bond may lead to a more negative value of $\Delta V^{\ddagger}$.

Intermediates dependent on participation by Y for their stability may have symmetries they would not have acquired otherwise, and in such cases isotopic labelling may be informative; but this and other special techniques such as the construction of "tailor-made" substrate molecules are best discussed when we encounter the appropriate examples.

Perhaps the most interesting ions that are produced by anchimeric assistance are those thought to result from carbon participation. In many instances one of the carbon atoms in these intermediates is pentacovalent; i.e., it is bound to five other atoms - to two of these by means of a two-electron, three-center bond. Such ions are referred to as non-classical, and it is the proof of their existence and structure that has provided the focus of the carbonium ion controversy. The 7-norbornyl ion (38) described above would be an example; the 2-carbon atom which undergoes the Wagner-Meerwein shift would be the pentacovalent one in the intermediate ion.

It is perhaps also appropriate at this point to discuss what nomenclature should be used in describing such ions. While the term carbonium ion has been in almost universal use, the term is not very appropriate since onium ions have traditionally meant ions with an atom in an expanded valence state (for example, ammonium, oxonium and sulfonium ions). Olah has proposed the use of the terms carbenium ions for classical species $^{+}CR_3$, carbonium ions for the non-classical ions, and carbocations as a generic term for all [52]. While his system is a perfectly consistent and logical one, the problem with it is that it introduces elements of confusion and prejudice into a controversial subject. Thus, if a writer referred to the 7-norbornyl carbonium ion it would not even be clear whether he was using conventional nomenclature or the new system. It seems better therefore to use the conventional names until all organic chemists agree on the structures of all these ions; perhaps the International Union of Pure and Applied Chemistry may be expected to judge when the time is appropriate.

20-6. n-PARTICIPATION

There is no unsolved problem concerning the ability of a nitrogen atom to participate. Tetraalkylammonium ions are well-known, stable materials, and the appearance of cyclic products in solvolysis reactions of substrates with nitrogen atoms at some distance from the reacting center, coupled with enormously enhanced rates has left no doubt in the minds of most investigators about the details of this reaction. For example, the cyclization of 4-bromobutylamine to give pyrrolidine is approximately a million times faster than the hydrolysis of 1-bromobutane under

52. G.A. Olah, J. Amer. Chem. Soc., 94, 808 (1972).

comparable conditions [53]. Oxonium salts are less stable, but they are known to exist, and for reasons similar to those described above, singly bound oxygen atoms are generally accepted as able to participate effectively in ionization reactions. For example, *trans*-2-acetoxycyclohexyl brosylate (39) reacts with acetic acid 600 times faster than the *cis*-isomer (40) [54]; it is converted into *trans*-1,2-diacetoxycyclohexane (41) and if the initial ester is resolved, racemic product is obtained. These facts are consistent with acetyl participation leading a symmetrical intermediate (42).

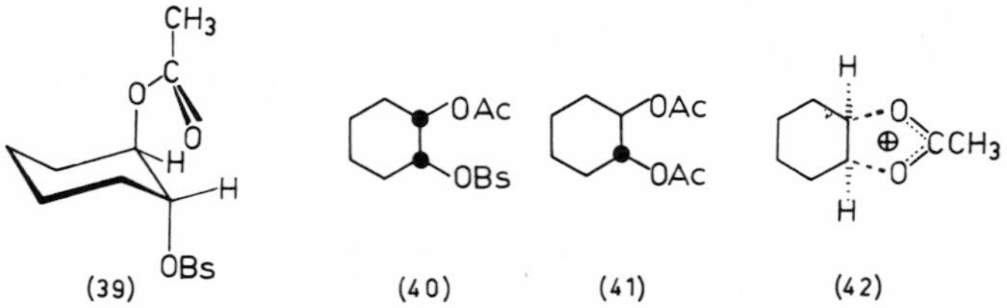

(39) (40) (41) (42)

An instructive case of oxygen participation has been described by Spurlock [55]. When *endo*-2-norbornyl brosylate (43) is compared with the oxa analog (44), the former acetolyzes about 100,000 times faster. This fact must be ascribed to an inductive effect; the electronegative oxygen atom interferes with the necessary polarization of the C-OBs bond. However, for the *exo*-epimers (45) and (46), the rates are about equal. The major product in each case *exo*-acetate (47).

(43) (44) (45) (46) (47)

A 1-D label leads to a 50/50 mixture of (48) and (49). The most reasonable interpretation is that the *exo*-substrate solvolyzes with O-participation to give (50) directly, and that (44) solvolyzes without such assistance to an open carbonium ion which subsequently rearranges to the oxonium ion.

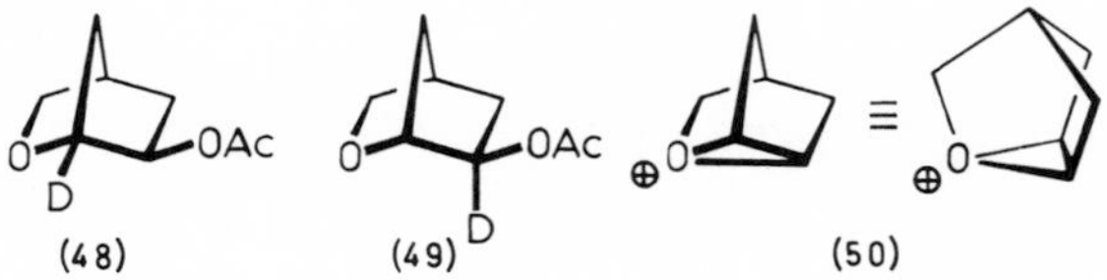

(48) (49) (50)

---

53. G. Salomon, *Helv. Chim. Acta*, *19*, 743 (1936).
54. S. Winstein, E. Grunwald, and L. L. Ingraham, *J. Amer. Chem. Soc.*, *70*, 821 (1948).
55. L. A. Spurlock and R. G. Fayter, *J. Amer. Chem. Soc.*, *94*, 2707 (1972).

The problem becomes more difficult when the electronegativity is increased still further. A classic experiment was carried out by Winstein and Lucas [56] to show that bromine participation can occur [57]. Erythro-3-bromo-2-butanol (51) was found to react with hydrobromic acid to give meso-2,3-dibromobutane (52), and the threo-bromoalcohol gave a d,l-mixture of isomeric dibromides: Evidently these reactions occur with complete retention.

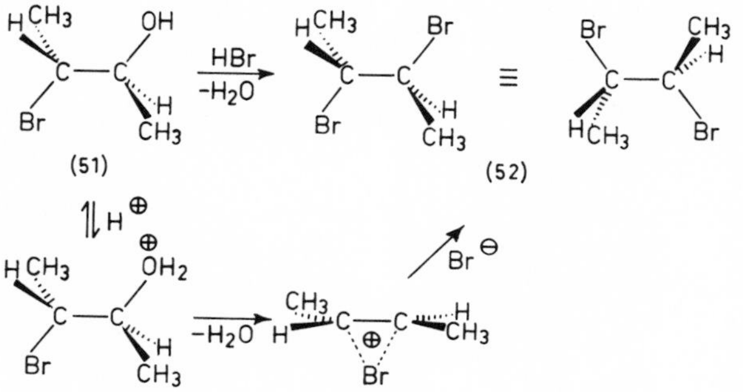

Obviously, if participation had not occurred and an open carbonium ion had formed, free access by bromide ion to either face of the ion would have led to stereochemical cross-over. The bromonium ion in which the halogen atom is bound to adjacent carbon atoms is presumably the same as the one postulated to account for the trans-bromination of olefins:

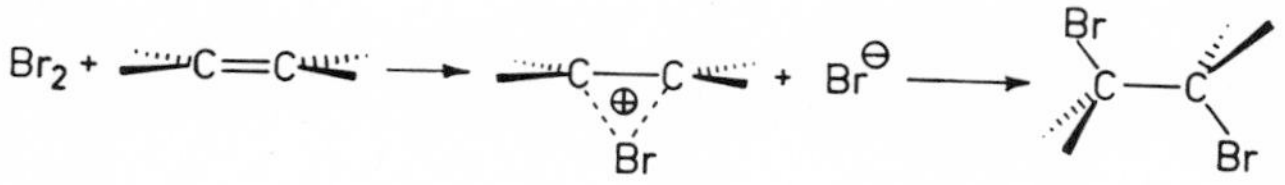

The existence of such ions has recently found strong support in the bromination of adamantylideneadamantane (53), which leads to the isolation of a salt claimed to be (54) [58].

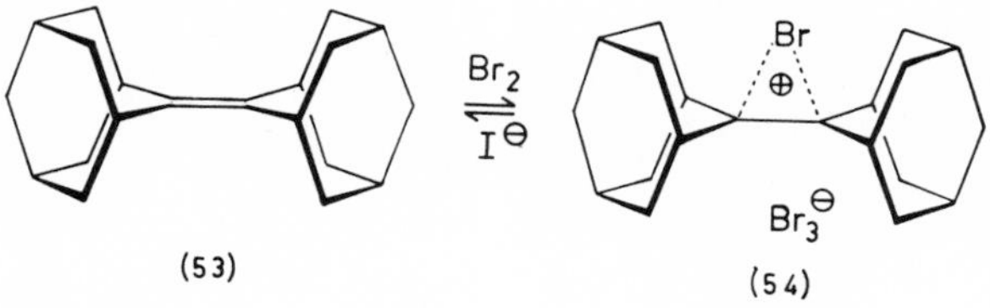

Access to the positively charged site is apparently made difficult by the crowded environment: Reversal of the reaction occurs with iodide ion. No X-ray diffraction has as yet been reported.

---

56. S. Winstein and H. J. Lucas, J. Amer. Chem. Soc., 61, 1576 (1939).
57. J. G. Traynham, J. Chem. Educ., 40, 392 (1963).
58. J. Strating, J. H. Wieringa, and H. Wynberg, Chem. Commun., 907 (1969).

There are examples of more distant bromine participation as well. Thus, it has been found that bromine reacts with the norbornene derivative (55) to give two products (56) and (57) in about a 7:1 ratio [59].

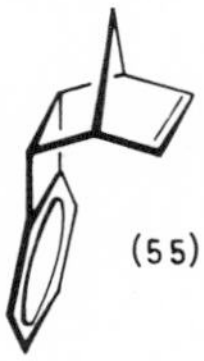

(55) 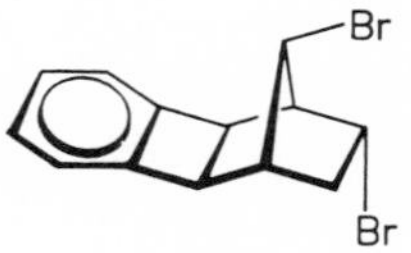 (56) 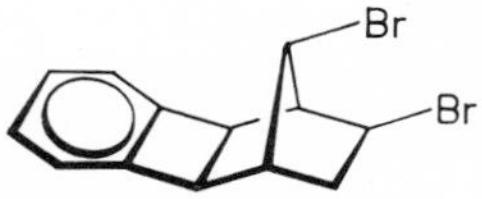 (57)

The pathway giving rise to (57) would normally be expected to be the only one operating (see Sec. 20-11), via bromonium ion (58), a Wagner-Meerwein shift, and finally exo-attack on the ion (59).

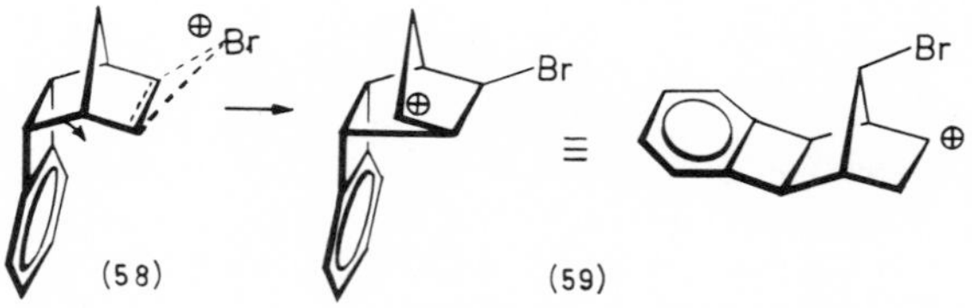

(58) (59)

The fact that the bromide ion preferentially adds from the endo-direction was attributed to a bromonium ion of the type (60).

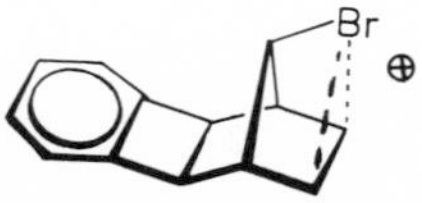

(60)

4-Bromobutyl nosylate reacts with acetic acid to give the corresponding bromobutyl acetate. If the starting material is deuterated as in (61), 22% of the product turns out to be rearranged (62) [60].

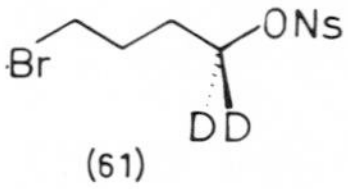

(61) 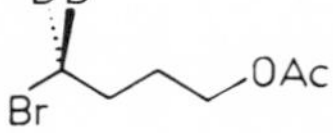 (62)

---

59. M. Avram, I. Pogany, F. Badea, I.G. Dinulescu, and C.D. Nenitzescu, Tetrahedron Lett., 3851 (1969).
60. W.S. Trahanovsky, G.L. Smyser, and M.P. Doyle, Tetrahedron Lett., 3127 (1968).

The dilemma that sometimes accompanies such evidence can be well illustrated with this example. Since the amounts of the two products are not equal, a long-lived cyclic intermediate such as (63) is ruled out (no isotope effect could conceivably give rise to a ratio of products of about 7:2 as observed); and the conclusion would be that 44% of the substrate solvolyzed with bromine participation and the rest without it.

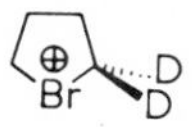

(63)

Another way to interpret the data is the formation of an open ion without participation, which rearranges at a rate several fold smaller than the rate of solvent capture. The stage intermediate between the equilibrating ions (64) and (63) would be a transition state rather than an intermediate.

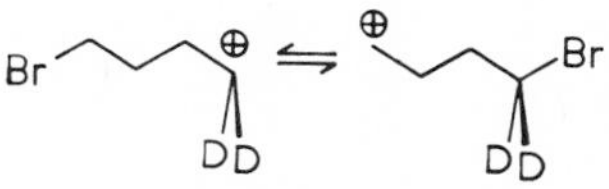

(64)

Furthermore, these two possibilities may be merely the extremes between which reality lies; both may be operating simultaneously to different extents, or a bromonium ion may be formed which is not quite symmetrical, possibly because of the proximity of the leaving group, and so on. These questions become more and more difficult to answer as the participation becomes more and more marginal.

The search for chlorine participation also furnishes us with instructive examples. 4-Chloro-1-pentyl nosylate (65) acetolyzes to give 73% (66) and 27% rearranged product (67).

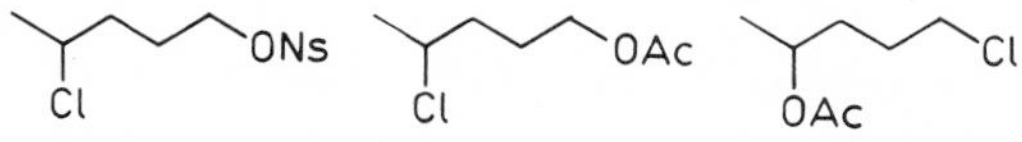

(65) (66) (67)

Thus, it would appear as though halogen participation had once again provided a pathway for the reaction. First of all, even if the participation stands up to further tests in this case, the evidence just quoted does not necessarily support the formation of a chloronium ion. In the present case rearrangement from the primary ion to the secondary ion provides a driving force for participation, and it is quite likely that concerted ionization and rearrangement (*i.e.*, participation) occurred, not to form a chloronium ion but a secondary carbonium ion. In other words, rearrangement can be the driving force for participation if it is not degenerate. Secondly, it is found that if the medium is buffered (with sodium acetate), virtually no rearrangement occurs at all during solvolysis. Obviously, the strong acid formed during the reaction (*p*-nitrobenzenesulfonic acid) can regenerate the ion from the acetate product again and again, and so enhance the overall probability of rearranged product. This demonstrates the need for buffered media; however, in the early studies this was often not done, and the conclusions are thereby weakened. Thirdly, if the medium is buffered trifluoroacetic acid, rearrangement is virtually the exclusive reaction mode. Participation by an internal group clearly becomes more important if solvent assistance is removed as a possibility; however this does not necessarily strengthen the argument for a chloronium ion, since any ion is expected to live longer in such media and hence to have more time to scramble [61].

Low temperature nmr supports the possible existence of true chloronium ions. When 1,4-dichlorobutane is dissolved at -60° in a superacid medium such as fluorosulfonic acid - antimony pentafluoride mixtures, the nmr spectrum is indicative of a symmetrical species (68) [62]. A very rapidly equilibrating mixture of ions (69) would of course also account for the spectrum.

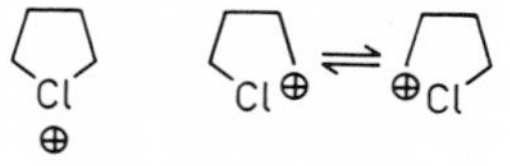

(68) (69)

However, recently the isolation of pure, solid dimethylhalonium hexafluoroantimonates has left no room for doubt at this point. Furthermore, Raman spectral measurements support symmetrical structures for these substances [63]: Chemical shift differences in the infrared region are of the order of 10^{13} Hz, rather than a few hundred as in the nmr, and hence the argument of possibly rapid equilibration no longer provides an escape route (see Chr. 2, p. 54). A chloronium ion has also been observed in the chlorination of adamantylideneadamantane [64], but it is much less stable than the bromine analog. However, one should not conclude that chloronium ions intervene in all cases where they conceivably could. A superacid medium is very different from such relatively nucleophilic media as acetic acid or aqueous acetone. Olah's evidence only shows that such ions *can* exist, and extrapolation back to solvolysis remains a hazardous proposition.

Even fluorine is now claimed to be able to participate in cation formation; thus, the addition of trifluoroacetic acid to (70) gives rise, in part, to rearranged products such as (71) [65].

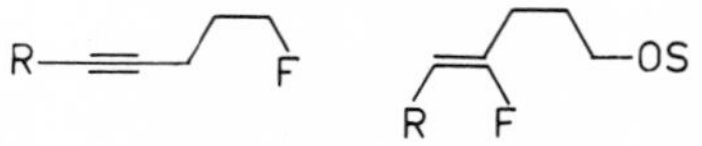

(70) (71)

---

61. (a) P.E. Peterson and J.F. Coffey, J. Amer. Chem. Soc., 93, 5208 (1971); (b) P.E. Peterson, Accounts Chem. Res., 4, 407 (1971).
62. G.A. Olah, J.M. Bollinger, and J. Brinich, J. Amer. Chem. Soc., 90, 2587 (1968).
63. G.A. Olah and J.R. DeMember, J. Amer. Chem. Soc., 92, 718 (1970).
64. J.H. Wieringa, J. Strating, and H. Wynberg, Tetrahedron Lett., 4579 (1970).
65. P.E. Peterson, R.J. Bopp, and M.M. Ajo, J. Amer. Chem. Soc., 92, 2834 (1970).

Further studies are of course necessary to substantiate the mechanistic details. A xenonium ion has been detected in the gas phase by ion cyclotron resonance spectroscopy: A mixture of methyl fluoride, xenon and hydrogen produces up to 35% methylxenonium ions (72) [66]. It has also been shown that the Me-Kr bond sometimes remains intact during the β-decay of $Me^{82}Br$, and hence the existence - however precarious - of the methylkryptonium ion (73) may be considered proved [67].

$CH_3Xe^{\oplus}$ (72) $CH_3Kr^{\oplus}$ (73)

An entirely different approach toward measuring participation in general is to interfere with it in some way. Thus, Gassman [68] determined that the methanolysis rate for (74) is only barely ten times slower than that of (75); on the basis of the inductive effect, a factor of 10^4 times slower or so would have been expected. The product is unrearranged. Thus, nitrogen participation is occurring despite the fact that this violates Bredt's rule ((76)).

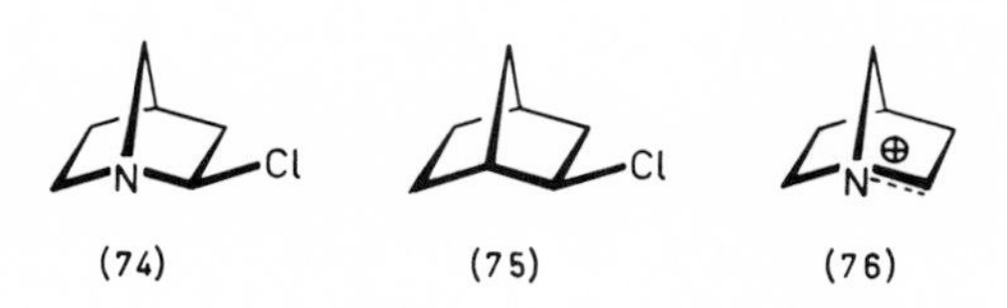

(74) (75) (76)

20-7. Π-PARTICIPATION [69]

The many examples of assistance by distant non-bonding electrons in ionization reactions soon led to the question whether remote bonding electrons could do the same. This would be easiest for π electrons, since these are energetically less strongly held and geometrically more accessible. For a demonstration, it would be preferable to use a rigid structure relatively free from the various conformational motions that characterize open chain molecules.

One of the best demonstrations that - given such favorable conditions - π electrons can participate was provided by Winstein [70], who found the relative

---

66. D. Holtz and J. L. Beauchamp, Science, 173, 1237 (1971).
67. T.A. Carlson and R.M. White, J. Chem. Phys., 39, 1748 (1963).
68. P.G. Gassman, R.L. Cryberg, and K. Shudo, J. Amer. Chem. Soc., 94, 7600 (1972).
69. For review, see (a) H. Tanida, Accounts Chem. Res., 1, 239 (1968); (b) S. Winstein, Quart. Rev. (London), 23, 141 (1969).
70. (a) S. Winstein, M. Shatavsky, C. Norton and R.B. Woodward, J. Amer. Chem. Soc., 77, 4183 (1955); (b) S. Winstein and C. Ordronneau, J. Amer. Chem. Soc., 82, 2084 (1960).

acetolysis rate constants of (77)-(81) to be in the ratio 10^7:1:10^4:10^{11}:10^{14}. These ratios are so large that they cannot be measured directly, but must be deduced by extrapolation of data in widely different temperature ranges. These extrapolations are not without hazard (see p. 544); however, precise information is obviously not necessary here.

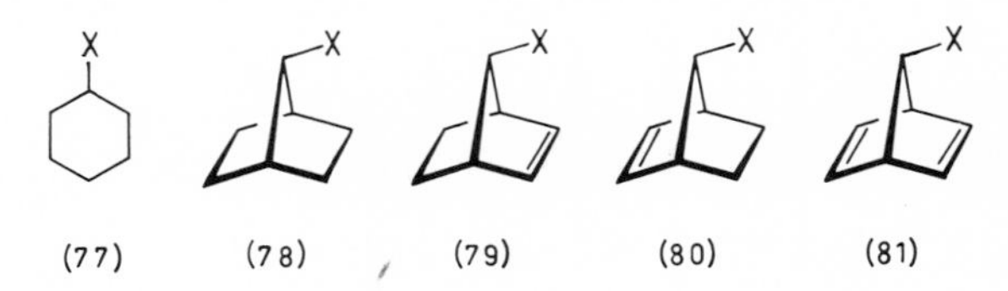

(77) (78) (79) (80) (81)

The unusually slow reaction of 7-norbornyl tosylate was noted earlier (p. 704); the factors responsible for it however are more than compensated for by a double bond. With a double bond in the anti-position the solvolytic reactivity becomes extremely high, and this is attributed to π-participation. The carbonium ion intermediate was described as a non-classical one (82). Note that the 7-carbon atom is bound to five other atoms, to two of these via a two-electron, three center bond. This ion may be considered an ethano-bridged bishomocyclopropenium ion - conceivably stabilized by homoaromatic character. This ion is eventually attacked from the rear by the solvent to give the product (83) with overall retention.

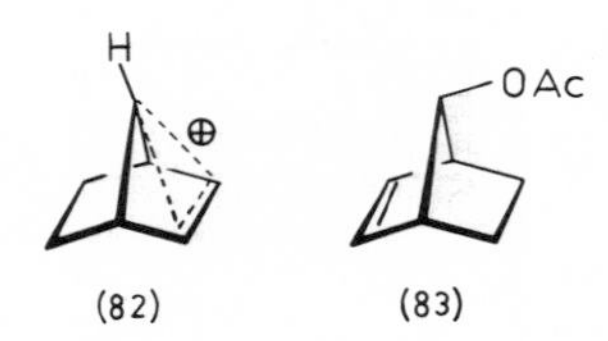

(82) (83)

The syn isomer is accelerated also, though to a lesser degree; in this case a rearranged product (84) is formed, and the enhanced rate is undoubtedly due to concerted ionization and rearrangement ((85)) to give a classical, allylic ion (86).

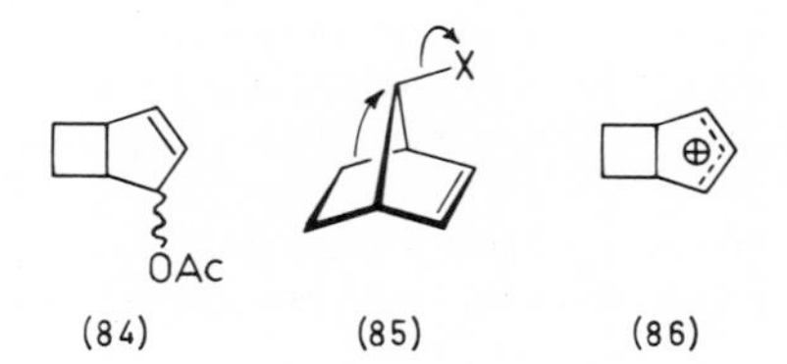

(84) (85) (86)

In the 7-norbornadiene system an extreme is reached which is pictured as involving extended participation of the type (87). Originally it was considered possible that the non-classical ion might be not merely (88), but that it might have the six-fold symmetry of (89). The hydrolysis product is unrearranged norbornadienol (90).

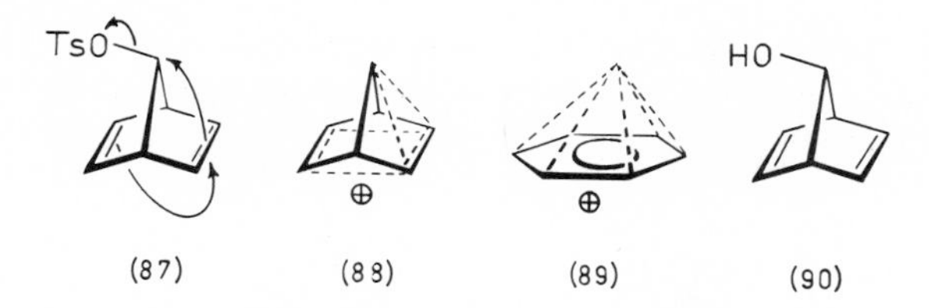

(87) (88) (89) (90)

Another observation suggestive of π-participation is that solvolysis reactions of this type have larger volume contractions as the transition states are reached. In the present case for instance, 7-anti-norbornenyl brosylate has an activation volume of -21 cm^3/mole as compared to -17 for cyclopentyl tosylate. The greater contraction is ascribed to the additional bonding resulting from participation [71]. π-Participation of this type is so effective that the ketone (91) will enolize to make it possible: Acetolysis is 10^4 times faster than that of the epimer, exclusively gives retained ketoacetate, and in AcOD leads to the deuterated product (92) [72].

(91) (92)

The enormous rate accelerations have left no doubt about the reality of π-participation in this case; however, it has not been universally accepted that such participation necessarily leads to non-classical ions. It may be noted at this point that there are two possible reasons why a given non-classical ion might be more stable than the classical one. One of these is resonance, which should be especially effective if the contributing structures are equal, as is the case here. This alone is of course not enough, as was noted on p. 242. Another way to describe the situation is provided by the molecular orbital method. We noted in Chr. 9 (p. 360) the stability due to homoaromaticity: The 7-norbornenyl ion is a bishomocyclopropenium ion. Whichever way one looks at it, the energy should be lowered if the π pair can be used for additional bonding. A second possible reason why a non-classical ion might be more stable than its classical counterpart is that its structure is sometimes less highly strained (this is probably not the case here, however). On the other hand, the classical ion does have one energy advantage: Since the solvation energy is proportional to the square of the charge, the concentrated charge of the classical ion will be solvated more tightly than the delocalized one of the non-classical species.

Now if participation must be accepted, and if the intermediate must reflect somehow the energy of the transition state preceding it, what is the alternative to the non-classical ion? Brown in particular has advocated the possibility that the

---

71. W.J. le Noble, B.L. Yates, and A.W. Scaplehorn, J. Amer. Chem. Soc., 89, 3751 (1967).
72. P.G. Gassman, J.L. Marshall, and J.M. Hornback, J. Amer. Chem. Soc., 91, 5811 (1969).

driving force for the reaction is not the formation of a non-classical ion but rather that of the tricyclic ion, which he presumes undergoes rapid degenerate Wagner-Meerwein rearrangements (93).

(93)

In support of his view he was able to show that if hydrolysis is carried out in the presence of borohydride ion, (94) is one of the products (15%) in addition to norbornene (95) (70%) and 7-*anti*-norbornenol (96) (6%) [73]. If no borohydride is present, (96) is the only product and Winstein [74] particularly noted the absence of any tricyclic alcohol (97); but Brown countered with the statement that this compound was unknown and that it might readily rearrange to the *anti*-7-norbornenol normally observed.

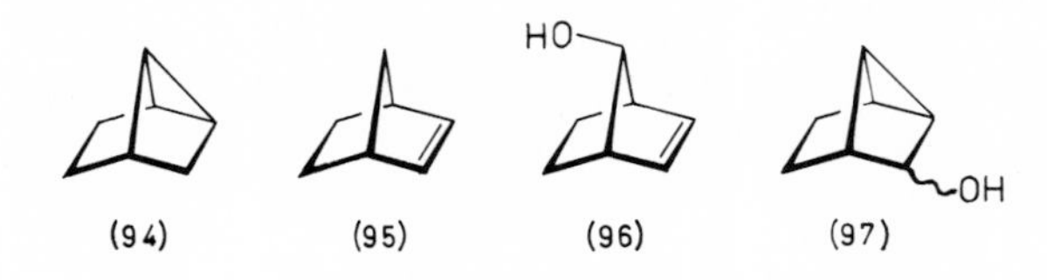

(94) (95) (96) (97)

This argument received support from Tanida [75] who showed that 7-norbornadienyl chloride (98), which normally gives the corresponding methyl ether during methanolysis, produces ether (99) instead if the methanol is made strongly basic. Evidently this tricyclic ether rearranges to the diene under neutral conditions; the product obtained is only the thermodynamically controlled one.

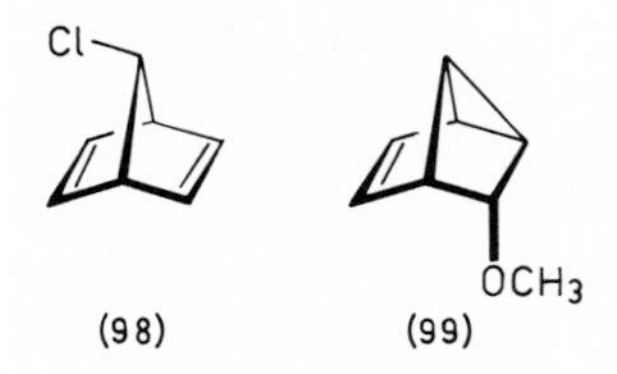

(98) (99)

Significantly however, it may be noted that even when the tricyclic ether forms, this substance is exclusively in the *endo*-form as would be expected if the *exo* side were protected by the three-center bond. If the ion were classical, another reason has to be invoked for this fact. The reason Brown has given in such cases is that the classical Wagner-Meerwein pair of ions have the new bond shifting back and

---

73. H.C. Brown and H.M. Bell, J. Amer. Chem. Soc., 85, 2324 (1963).
74. S. Winstein, A.H. Lewin, and K.C. Pande, J. Amer. Chem. Soc., 85, 2324 (1963).
75. H. Tanida, T. Tsuji, and T. Irie, J. Amer. Chem. Soc., 88, 864 (1966).

forth so rapidly that the solvent cannot accumulate at the exo-side. This notion, humorously christened the "windshield wiper effect" by the Winstein group, is reminiscent of Kekulé's oscillating double bonds in benzene, or - to make a comparison of more recent vintage - to the bond shifts in the degenerate valence isomerizations (Chr. 21).

The problem of the nature of the product in the solvolysis of 7-norbornenyl chloride was further examined by Winstein [76]. It was found that in buffered methanol in which endo-6-methoxytricyclo[3.2.0.0^{2,7}]heptane is fully stable,

OCH_3

only 0.3% of this material is formed; however, in 4M methoxide ion as much as 50% of this compound obtains. Methanol and methoxide ion clearly differ very greatly in the position they prefer to approach, and the nature of the product is not very informative regarding the structure of the ion except in that both products are formed by attack on the side opposite to that of the three center bond (or the oscillating two center bonds). Both the bicyclic and tricyclic ethers give the same ion upon the addition of acids; these two reactions are referred to as the π- and σ-routes to the carbonium ion - terms that one encounters continually in the carbonium ion literature. The tricyclic alcohol (100) was later synthesized by Tufariello [77]; while it is rearranged to 7-norbornenol by acids, it is stable under neutral conditions, and hence it is not an intermediate in the buffered hydrolysis of 7-norbornenyl derivatives.

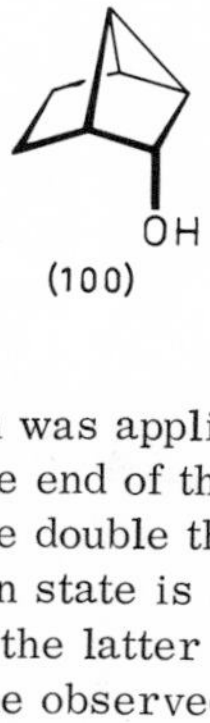

OH

(100)

Another approach to the problem was applied by Gassman, who reasoned that if the participation stemmed from one end of the double bond, then the rate effect of 2,3-dimethyl substitution should be double that of a 2-methyl group (statistical effect); alternatively, if the transition state is symmetrical, then the former rate effect should be the square of that of the latter (the free energy of activation should be lowered by twice the amount). The observed relative rates are 1:13:150. Had the statistical effect been observed instead, one could have made a powerful case for the tricyclic ion intermediate (see further p. 726) [77a].

---

76. A. Diaz, M. Brookhart, and S. Winstein, J. Amer. Chem. Soc., 88, 3133 (1966).
77. J.J. Tufariello, T.F. Mich, and R.J. Lorence, Chem. Commun., 1202 (1967).
77a. P.G. Gassman and D.S. Patton, J. Amer. Chem. Soc., 91, 2160 (1969).

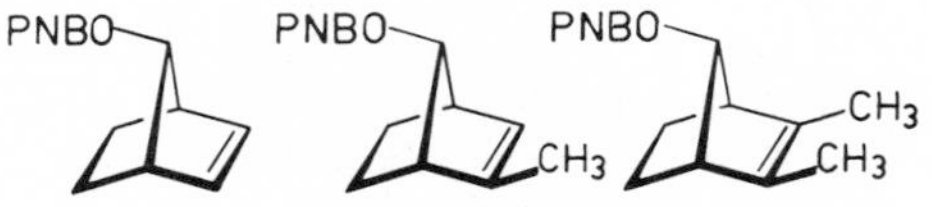

In further studies into the nature of these carbonium ions, much use has been made of nmr and superacid media. Thus, 7-norbornadienyl chloride dissolves in liquid sulfur dioxide containing one equivalent of silver fluoroborate to give silver chloride and an ionic solution, the nmr spectrum of which is informative in the sense that certain structures can be ruled out: Four signals are observed at 2.4, 3.7, 4.7 and 6.5 τ in the ratio of 2:2:2:1. Thus, the gross features of the spectrum are in agreement with either the non-classical structure or a Wagner-Meerwein classical pair of tricyclyl ions equilibrating rapidly enough to give rise to an average spectrum; the more symmetrical structures as (89) or static, classical ions such as the 7-norbornadienyl ion or a single tricyclyl ion are obviously ruled out [78].

In a search for any rapid degenerate rearrangements that might go on in the ion, Winstein studied the temperature dependence of the nmr spectrum. The ion begins to decompose at 45°C, in part to tropylium ion. Nevertheless, at 77°C another process can still be observed; while the 2.4 τ signal remains sharp, the others broaden showing that some sort of exchange is taking place. The use of deuterated substrates such as (101) shows that five of the seven carbon atoms exchange positions in the process, since the signal ratio then becomes 2:1.6:1.6:0.8.

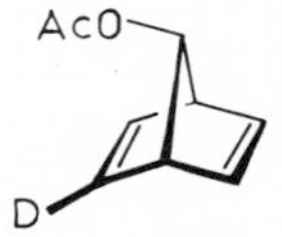

(101)

Evidently the C_2-C_3 bridge wanders around the periphery of ("circumambulates") the five-membered ring in processes such as depicted below [79].

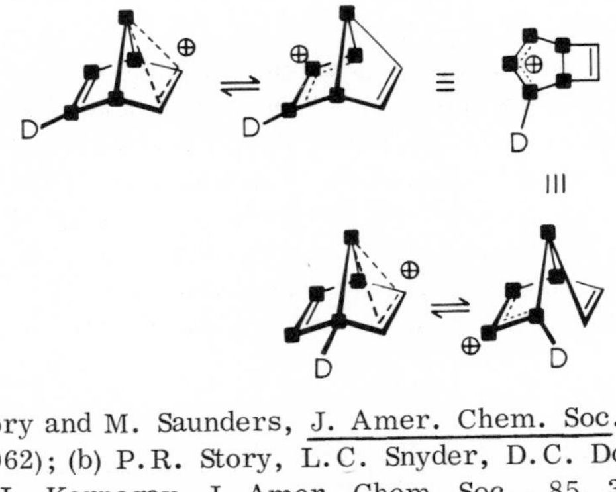

78. (a) P.R. Story and M. Saunders, J. Amer. Chem. Soc., 82, 6199 (1960) and 84, 4876 (1962); (b) P.R. Story, L.C. Snyder, D.C. Douglass, E.W. Anderson, and R.L. Kornegay, J. Amer. Chem. Soc., 85, 3630 (1963).

79. R.K. Lustgarten, M. Brookhart, and S. Winstein, J. Amer. Chem. Soc., 89, 6350 (1967) and 94, 2347 (1972).

At about 0° or above, the C_2-C_3 bridge also slowly incorporates deuterium; this can be most simply visualized as "bridge flipping".

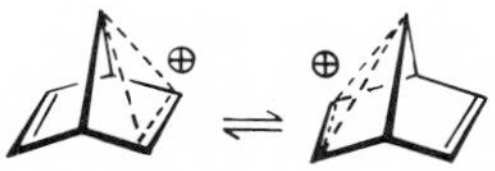

The rate of flipping rapidly increases as 7-methyl- [80], 7-phenyl- or 7-methoxy-substituents are introduced [81]; this is reasonable since the transition states for flipping must be symmetrical, classical 7-norbornadienyl ion structures, which are presumably lowered in energy by these substituents.

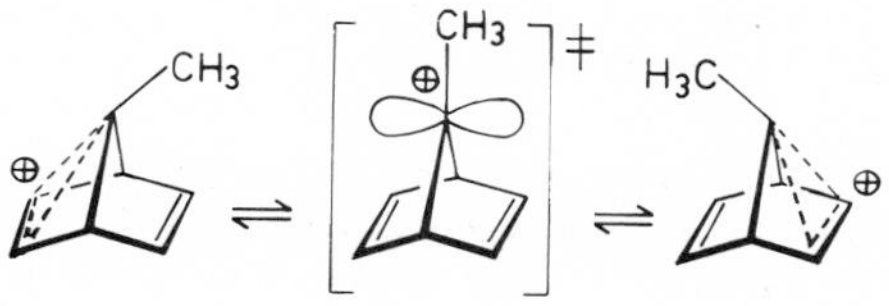

With a 7-methoxy group, hindered rotation (or inversion) is furthermore observed; at low temperature, the two bridgehead protons are non-equivalent but at 0° their signals coalesce.

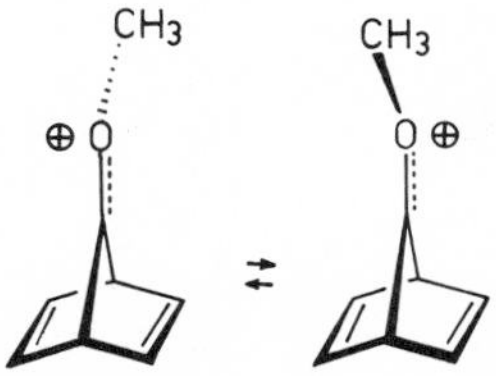

Since no vinyl proton coalescence is observed it appears likely that the 7-(7-methoxynorbornadienyl) cation is a classical symmetrical ion [82]. Clearly, the 7-norbornadienyl carbonium ion is a fascinating species capable of many degenerate changes that could easily remain hidden from view. For all these processes, the rates and activation free energies can be deduced from the coalescence temperatures or the rate of intensity changes. However, none of these observations allow us to distinguish between the non-classical ion and the Wagner-Meerwein pair with absolute certainty.

---

80. M. Brookhart, R.K. Lustgarten, and S. Winstein, J. Amer. Chem. Soc., 89, 6352 (1967).
81. M. Brookhart, R K. Lustgarten, and S. Winstein, J. Amer. Chem. Soc., 89, 6354 (1967).
82. R.K. Lustgarten, M. Brookhart, and S. Winstein, Tetrahedron Lett., 141 (1971).

The nmr spectrum of the 7-norbornenyl ion has been reported by Winstein and Richey [83]. Extraction of a methylene chloride solution of 7-norbornenol into a superacid medium at -50° gives a stable solution of the ion; the chemical shifts are given below.

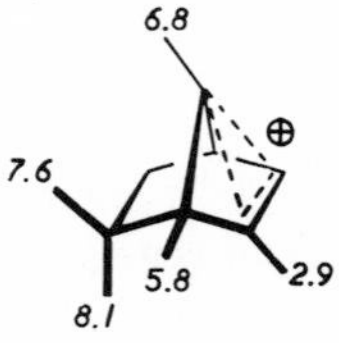

6.8
7.6
5.8
2.9
8.1

The symmetrical nature of the spectrum again points to either a non-classical ion or a pair of equilibrating classical ones.

Another insight into the kinetic behavior of 7-norbornenyl derivatives was obtained by adding a 7-substituent with an enormous capacity to stabilize positive charge. It was found [84] that a 7-anisyl group virtually completely swamps any π-participation; the relative hydrolysis rates for (102)-(104) are 1, 2.4 and 0.4, respectively.

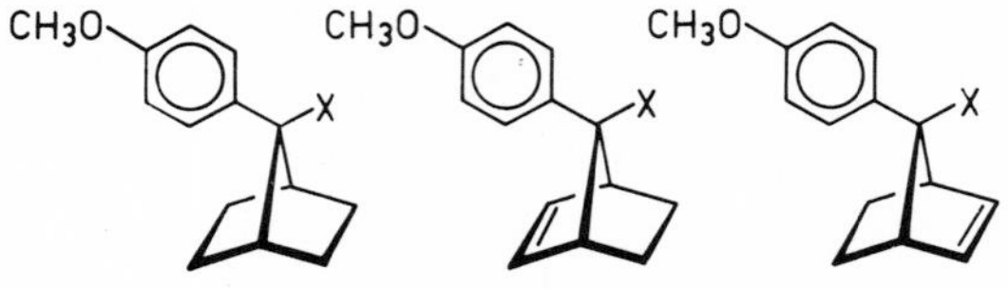

CH_3O X (102) CH_3O X (103) CH_3O X (104)

Equally interesting, the reaction loses its stereospecificity at the same time; both _syn_- and _anti_- substrates in methanol give an identical mixture of _syn_- and _anti_-methyl ethers. If the 7-substituent is a phenyl group with a less stabilizing _p_-substituent, some π-participation still occurs, and retention of configuration is then observed [85]. Gassman's kinetic data can be summarized as follows. The hydrolysis of 7-(7-phenylnorbornyl) _p_-nitrobenzoates is characterized by a ρ-value of -5.3; that of the corresponding _anti_-norbornenyl esters also has this ρ-value so long as the σ^+-value of the _p_-substituent is -0.8 or more, but ρ becomes -2.3 for less stabilizing substituents as π-participation sets in [86].

Since a 7-anisyl group evidently removes π-participation altogether, the resulting ion must be a classical one, and hence its nmr spectrum can be expected to

83. (a) M. Brookhart, A. Diaz and S. Winstein, _J. Amer. Chem. Soc._, 88, 3135 (1966); (b) H.G. Richey and R.K. Lustgarten, _J. Amer. Chem. Soc._, 88, 3136 (1966).
84. P.G. Gassman, J. Zeller, and J.T. Lumb, _Chem. Commun._, 69 (1968).
85. P.G. Gassman and A.F. Fentiman, _J. Amer. Chem. Soc._, 91, 1545 (1969).
86. P.G. Gassman and A.F. Fentiman, _J. Amer. Chem. Soc._, 92, 2549 (1970).

be revealing. The chemical shifts are indeed indicative [87] that changes have been brought about by the 7-p-anisyl group in the norbornenyl ion: The vinyl protons are shifted upfield by about 1 ppm, and the aryl hydrogen atoms have chemical shifts normal for a p-methoxybenzyl cation. By contrast, in 7-phenylnorbornenyl ions with less stabilizing substituents these atoms are shifted upfield relative to where they are found in the corresponding benzyl cations.

A further argument for the non-classical nature of the 7-norbornenyl ion is based on the fact that 2-, 2,3- and 7-methyl substituents have very little effect on the nmr spectrum. One would expect [88] that a 2-methyl group would shift the equilibrium of a Wagner-Meerwein pair in the direction of the tertiary ion, and that the cyclopropyl hydrogens would then appear at 7-8 τ as they do in cyclopropylcarbinyl ions [89]. The C^{13}-nmr spectra and the $J_{C^{13}H}$ constants have been measured for the 7-norbornenyl and 7-norbornadienyl ions at -150° and these studies likewise support their non-classical nature; for the details of the argument the original paper should be consulted [90].

If the 7-norbornadienyl system provides perhaps the best demonstration for π-participation, it also is a good starting point to explore the question how unfavorable conditions have to be made to stop it. Several questions may be raised: The distance of the π electrons from the carbonium ion center, the number of intervening bonds, the effect of substituents at both the sites of ionization and the π bond, the rigidity of the system, the strain of the double bond, the symmetry of the resulting ion, the orientation of the p and π orbitals, and so on. That at least some of these aspects of the reaction will matter may be illustrated by means of the familiar allylic ion; ionizations leading to such ions are generally very rapid. But what happens when the π orbital and the developing p orbital cannot be parallel? This is actually the case with 1(2-methyleneadamantyl) tosylate (105), which acetolyzes about 10,000 times more slowly than 1-adamantyl tosylate (106) and 2,2-dimethyladamantyl tosylate (107) [91]! All these compounds lead to unrearranged acetates.

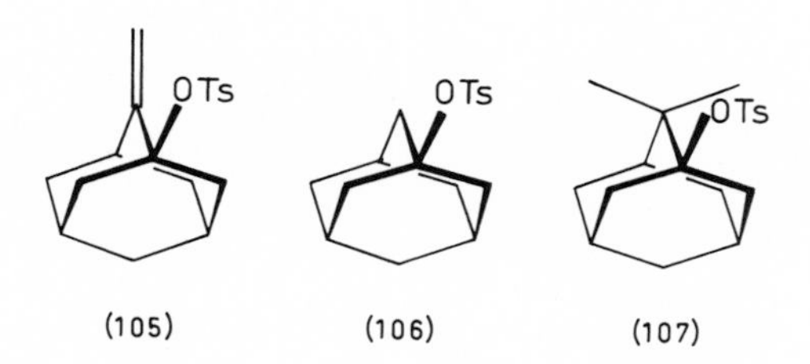

---

87. H.G. Richey, J.D. Nichols, P.G. Gassman, A.F. Fentiman, S. Winstein, M. Brookhart, and R.K. Lustgarten, J. Amer. Chem. Soc., 92, 3783 (1970).
88. R.K. Lustgarten, P.G. Gassman, D.S. Patton, M. Brookhart, S. Winstein, H.G. Richey, and J.D. Nichols, Tetrahedron Lett., 1699 (1970).
89. H.G. Richey, "Carbonium Ions", Vol. III, G.A. Olah and P. von R. Schleyer, Ed., Interscience, New York, 1968.
90. G.A. Olah and A.M. White, J. Amer. Chem. Soc., 91, 6883 (1969).
91. B.R. Ree and J.C. Martin, J. Amer. Chem. Soc., 92, 1660 (1970).

The number of intervening carbon atoms can be made greater with little or no sacrifice in effectiveness, so long as a well-oriented, rigid structure is considered.

Thus, the tetracyclic system (108) acetolyzes at a rate of about 10^{10} relative to that of 7-norbornyl

(108)

to give the saturated acetate (109) [92], and (110) gives the "birdcage hydrocarbon" (111) and the related alcohol (112) during hydrolysis at a relative rate of 2×10^{13} [93].

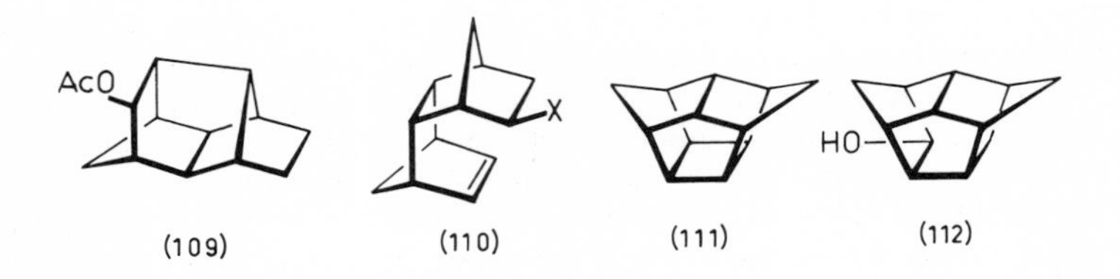

(109) (110) (111) (112)

The diene (113) apparently gives rise to the type of extended participation reminiscent of norbornadienyl derivatives; the relative rate is 10^{14} and only unrearranged alcohol obtains [94].

(113)

Acetolysis of the unsaturated brosylate (114) evidently occurs without participation; thus, the reaction is only twice as fast as that of the epimer (115), and six times slower than that of (116).

---

92. S. Winstein and R. L. Hansen, Tetrahedron Lett., No. 25, 4 (1960).
93. P. Bruck, D. Thompson, and S. Winstein, Chem. Ind. (London), 590 (1960).
94. E. L. Allred and J. C. Hinshaw, Tetrahedron Lett., 1293 (1968).

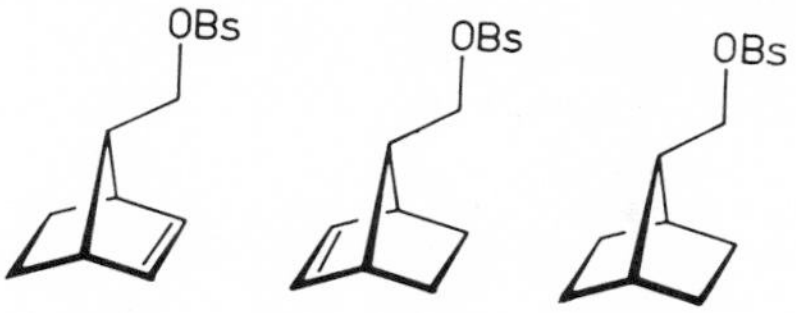

(114) (115) (116)

Evidently the distance between the reacting center and the π orbital has become slightly too large (the fact that we are dealing with a primary rather than a secondary center should make the electron demand greater if anything) [95]. The products are largely unrearranged acetates. By contrast, the corresponding ethyl brosylate (117) solvolyzes at the very fast rate of 2×10^5 compared to either (118) or (119).

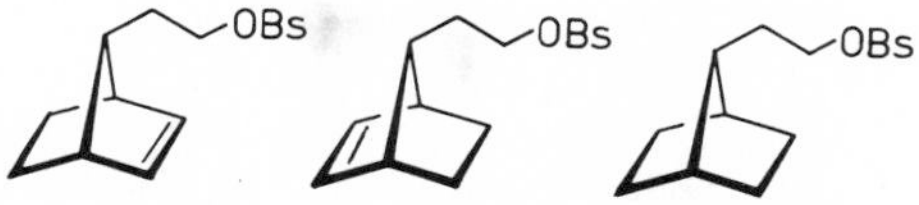

(117) (118) (119)

The carbonium ion center is now within range again, and in spite of the increased rotational flexibility π-participation is again dominant. The products are the tetracyclic hydrocarbon (120) and the tricyclic acetates (121) and (122); the last of these two depends on an initial hydride shift [96].

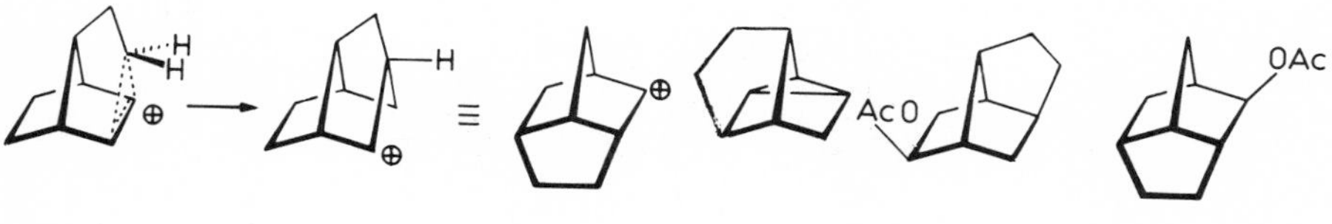

(120) (121) (122)

The effect of small changes in strain is suggested by the substrates (123)-(126), for which the relative rate constants are 1, 2×10^8, 10^{11}, and 5×10^{14}, respectively [97].

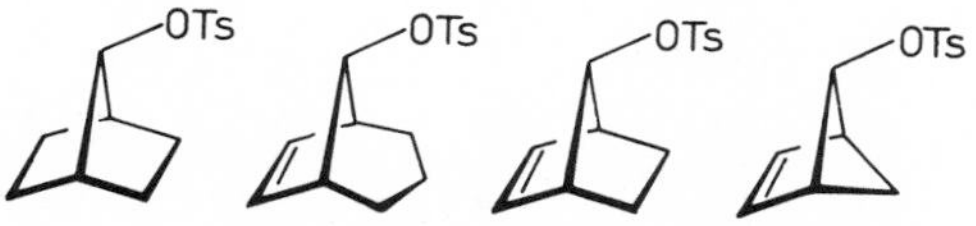

(123) (124) (125) (126)

95. R.K. Bly and R.S. Bly, J. Org. Chem., 31, 1577 (1966).
96. R.S. Bly, R.K. Bly, A.O. Bedenbaugh, and O.R. Vail, J. Amer. Chem. Soc., 89, 880 (1967).
97. (a) B.A. Hess, J. Amer. Chem. Soc., 91, 5657 (1969); (b) S. Masamune, S. Takada, N. Nakatsuka, R. Vukov, and E.N. Cain, J. Amer. Chem. Soc., 91, 4322 (1969).

The variation in chain length affects the strain at both the double bond and the carbonium ion site, in both the intial and transition states; while the results are probably not predictable, they do suggest that strain relief may characterize these reactions.

Another interesting situation arises with an extension of the participating group to a conjugated system. Tanida found that (127) gives retained acetate as well as (128) and (129) at a rate only 10^4 times greater than 7-norbornyl brosylate.

BsO— OAc OAc

(127) (128) (129)

Thus, participation occurs, but it is not as effective as a simple double bond, probably because the upper occupied π orbital in (127) has a node right behind the leaving group [98].

Another revealing result was obtained with the allylic substrates (130) and (131). The latter p-nitrobenzoate hydrolyzes in acetone some 250 times more slowly than the former. This rate retardation is judged to be more severe than can be explained on the basis of an inductive effect of the vinyl group, and it was concluded that the cation was subject to destabilization by its bishomoantiaromatic character [99].

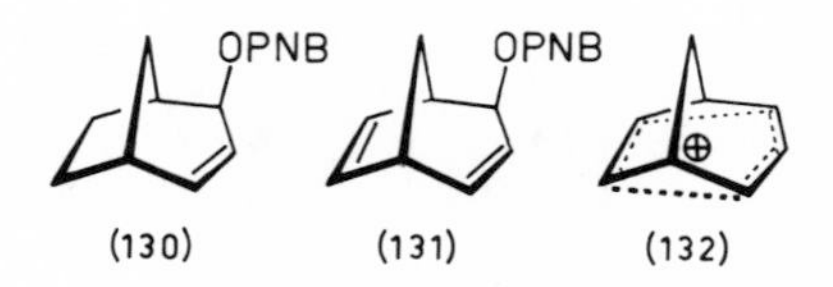

(130) (131) (132)

The 3,5-dinitrobenzoates (133) and (134) are both slower in solvolysis than (135), and these reactions are considered not to be enhanced by π-participation - that in spite of the fact that the bishomotropylium ion (136) can readily be generated in super acid medium [100]. This is an important example showing that conclusions applicable to these solutions are not necessarily valid in more nucleophilic media.

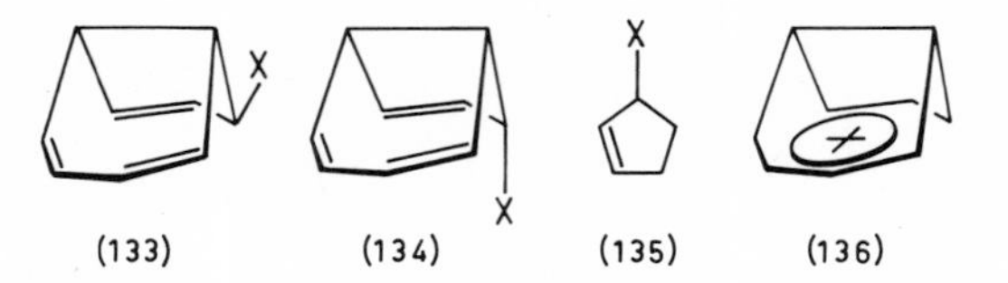

(133) (134) (135) (136)

---

98. T. Tsuji, H. Ishitobi, and H. Tanida, Tetrahedron Lett., 3083 (1972).
99. A. F. Diaz, M. Sakai, and S. Winstein, J. Amer. Chem. Soc., 92, 7477 (1970).
100. D. Cook, A. Diaz, J. P. Dirlam, D. L. Harris, M. Sakai, S. Winstein, J. C. Barborak, and P. von R. Schleyer, Tetrahedron Lett., 1405 (1971).

The lowering of anchimeric assistance by electron donating substituents such as *p*-anisyl at the carbonium ion center was already mentioned (p. 720). Electron withdrawing substituents at the site of the double bond should have the same effect. That this is indeed so was nicely shown by Farnum [101] in a study of the addition of bromine to olefins (137)-(141).

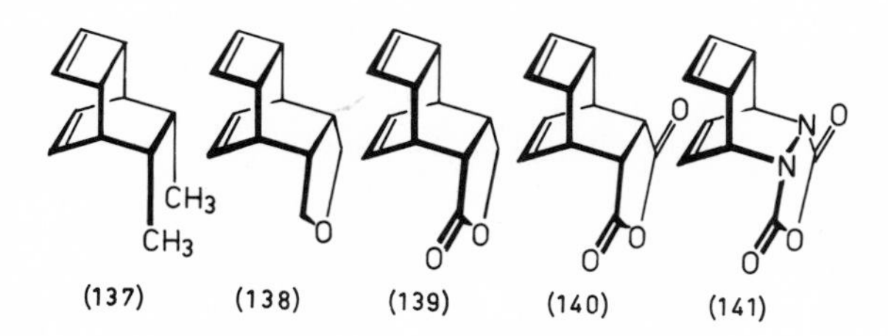

(137) (138) (139) (140) (141)

In these substrates bromine participation is expected to compete with π-participation, and these modes of reaction should give products such as (142) and (143), respectively.

Br
Br
CH3
CH3
Br
Br
CH3
CH3

(142) (143)

When this series of compounds is allowed to react with one mole of bromine per mole of diene, the reaction is retarded as more and more electronegative substituents are encountered, and the amount of dibromoolefin increases at the expense of the tetracyclic product.

π-Assistance generally becomes less effective when the double bond is merely *able* to assume a favorable position, than when it is *forced* to be there by a perfectly rigid structure, as the following examples show.

An interesting example was found by Bartlett and Lawton. 2-Δ^3-Cyclopentenylethyl nosylate (144) [102] solvolyzes faster than the saturated and Δ^2-analogs [103], by 1-3 orders of magnitude depending on the solvent, and much larger

---

101. D. G. Farnum and J. P. Snyder, *Tetrahedron Lett.*, 3861 (1965).
102. (a) R. G. Lawton, *J. Amer. Chem. Soc.*, *83*, 2399 (1961); (b) P. D. Bartlett, S. Bank, R. J. Crawford, and G. H. Schmid, *J. Amer. Chem. Soc.*, *87*, 1288 (1965); (c) P. D. Bartlett and G. D. Sargent, *Science*, *140*, 379 (1963).
103. W. D. Closson and G. T. Kwiatkowski, *Tetrahedron*, *21*, 2779 (1965).

—ONs

(144)

deuterium isotope effects obtain [104]. The reaction produces exclusively exo-2-norbornyl derivatives, presumably via the symmetrical 2-norbornyl ion:

The symmetrical nature of the transition state is strongly suggested by the fact that 3-methyl- and 3,4-dimethyl substitution cause rate increases of 7 and 38, respectively. Participation also occurs in certain cyclohexenylmethyl derivatives; thus, methyl or methoxy substituents at the vinylic carbon atoms as in (145) bring about rate increases of 10-100, and norbornyl derivatives such as (146) are obtained [105].

OBs

CH_3O CH_3 O CH_3

(145) (146)

On the other hand, an attempt to arrive at the same result via the acetolysis of trans-3-vinylcyclopentyl bromide failed; only unrearranged acetate formed at a rate not larger than that of cyclopentyl bromide itself [106].

These results generally show that π-participation is not so overwhelming as it was in the rigid, symmetric 7-norbornenyl esters. Stereospecifically formed polycyclic products and small rate enhancements by the double bond have also been observed in the case of (147) to give (148) [107],

104. C.C. Lee and E.W.C. Wong, Tetrahedron, 21, 539 (1965).
105. H. Felkin and C. Lion, Chem. Commun., 60 (1968).
106. J.P. Schaefer and J. Higgins, J. Org. Chem., 32, 553 (1967).
107. (a) G. Le Ny, Compt. rend., 251 1526 (1960); (b) C. Chuit, F. Colard, and H. Felkin, Chem. Commun., 118 (1966).

ONs OAc

(147) (148)

of (149) to give (150) [108],

OTs OAc

(149) (150)

of (151) to give (152) [109],

OTs OH

(151) (152)

of (153) to yield (154) [110],

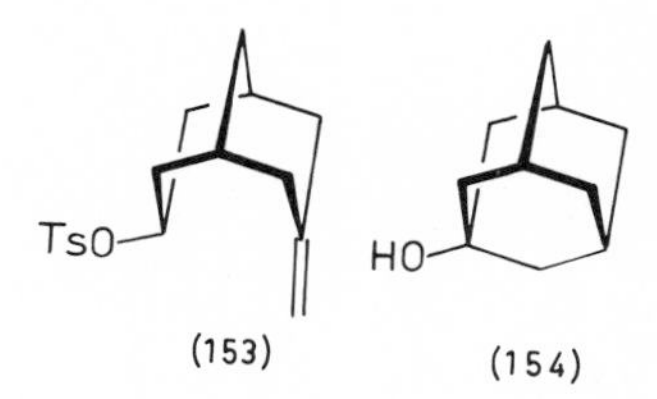

(153) (154)

---

108. (a) W. Kraus, W. Rothenwöhrer, W. Kaiser, and M. Hanack, Tetrahedron Lett., 1705 (1966); (b) H. Felkin, G. Le Ny, C. Lion, W.K.D. Macrosson, J. Martin, and W. Parker, Tetrahedron Lett., 157 (1966); (c) A.C. Cope, D.L. Nealy, P. Scheiner, and G. Wood, J. Amer. Chem. Soc., 87, 3130 (1965).
109. (a) T.L. Westman and R.D. Stevens, Chem. Commun., 459 (1965); see also (b) H.L. Goering, H.H. Espy, and W.D. Closson, J. Amer. Chem. Soc., 81, 329 (1959).
110. (a) M. Eakin, J. Martin, and W. Parker, Chem. Commun., 206 (1965); (b) H. Stetter, J. Gärtner, and P. Tacke, Angew. Chem., Int. Ed. Engl., 4, 153 (1965).

of (155) to product (156) [111],

(155) (156)

and of (157) [112]. Where these reactions were noted in the medium size rings, they were sometimes referred to as trans-annular effects.

(157)

The simplest system of all, at least conceptually, is the allylcarbinyl skeleton; the tosylate (158) formolyzes about four times faster than *n*-butyl tosylate, and the ion undergoes scrambling of the three methylene groups before solvent capture to give a mixture of unrearranged products and rearranged ones (159) and (160).

(158) (159) (160)

In each of the products, scrambling is extensive; however, it is not so complete so as to suggest a tetrahedral tricyclobutonium ion (161). The results can be accounted for on the basis of equilibrating bicyclobutonium ions (162) [113].

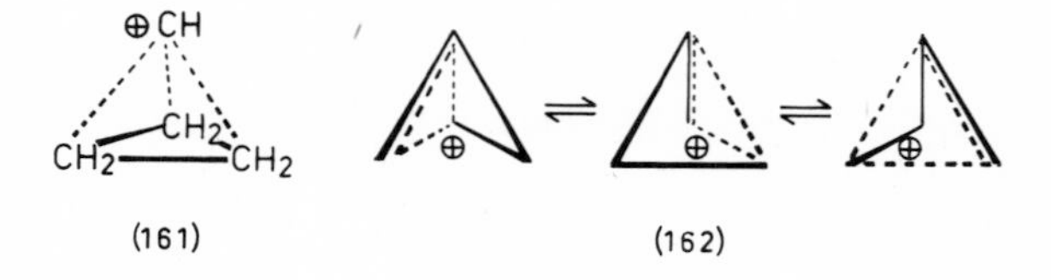

(161) (162)

111. (a) S. Winstein and E. M. Kosower, *J. Amer. Chem. Soc.*, *81*, 4399 (1959); (b) R. Sneen, *J. Amer. Chem. Soc.*, *80*, 3977, 3982 (1958).
112. (a) S. J. Cristol, T. C. Morrill, and R. A. Sanchez, *J. Amer. Chem. Soc.*, *88*, 3087 (1966); however, see also (b) E. N. Peters and H. C. Brown, *J. Amer. Chem. Soc.*, *94*, 7920 (1972).
113. (a) M. S. Silver, M. C. Caserio, H. E. Rice, and J. D. Roberts, *J. Amer. Chem. Soc.*, *83*, 3671 (1961); (b) K. L. Servis and J. D. Roberts, *J. Amer. Chem. Soc.*, *86*, 3773 (1964).

Methyl substitution at the site of the double bond greatly increases the rate; for example, (γ,γ-dimethylallyl)carbinyl tosylate formolyzes 4500 times faster than the parent compound [114]. The same ion can also be prepared by the sigma route from either cyclopropylcarbinyl and cyclobutyl esters, and we shall defer its further discussion until we consider σ-participation (see p. 755).

What conclusions can we draw at this point? There appears to be universal agreement that π participation occurs at least in those cases where enormous rate increases are observed. It is equally clear that it is sometimes absent, and that this absence is most easily demonstrated if no cyclization occurs and if there is a small rate retardation because of induction by the double bond. Perhaps the only question is precisely where to draw the line.

As far as the nature of the intermediate is concerned, there is no concensus. The problem of the non-classical vs. the equilibrating ions is essentially that depicted on p. 241: Do the curves have a low maximum or a shallow minimum? In superacid media the evidence is little short of overwhelming that the symmetrical ions can exist. In spite of great efforts in many laboratories, no nmr evidence has ever revealed the coalescence expected at some low temperature if equilibration indeed occurs, which means that in every case the activation energies must be 5-10 kcal/mole or less. While Raman analysis could in principle circumvent this problem, this technique cannot yet be applied at the fundamental level to the relatively complex ions encountered here (although it does have empirical value). Perhaps only X-ray diffraction or photo-electron spectroscopy can be relied upon with certainty, and these require a solid phase. Even if such an experiment succeeded in a given case (p. 786), that would not end all doubt; regardless which school of thought would see its predictions blessed by the results, the other would no doubt insist that this was just a special case not invalidating its general thesis in the slightest. We have in fact already encountered one instance of an ion capable of generation in super acid which has declined to intervene in simple solvolysis (p. 724), and several instances of stable species involving hexacovalent carbon (see p. 16).

Perhaps the weakest point in the armor of the non-believers is their failure to provide a viable alternative explanation of the stereochemical results. Inaccessibility of part of a molecule by a permanent bond is acceptable to the chemist, and there is ample precedent for it; but the proposition that the same result can be affected by a shifting bond has not become popular. Inevitably, one wonders what powers the windshield wiper. But no other attempt to explain retention has been put forward. We might also say a word about the connection between the concepts of participation and non-classical ions. Not everyone agrees that these are separate and unrelated questions [115]. Thus, if one keeps in mind that a simple one-step ionization reaction must surely occur at the time level of a molecular vibration with all atoms moving in concert from their initial to final positions, it is hard to visualize how a symmetrical species such as 2-(Δ^3-cyclopentenyl)ethyl brosylate

114. (a) K. L. Servis and J. D. Roberts, J. Amer. Chem. Soc., 87, 1331 (1965); (b) M. Hanack, S. Kang, J. Häffner, and K. Görler, Ann., 690, 98 (1965).
115. P. D. Bartlett and G. D. Sargent, J. Amer. Chem. Soc., 87, 1297 (1965).

can be subject to symmetrical participation while reaching the transition state only to diverge to an unsymmetrical classical norbornyl ion on the down slope (the initial symmetrical ion could of course subsequently rearrange to an unsymmetrical species). No fundamental law of nature is violated however - at least not in the complex liquid state where solvent molecules are part of the overall picture - and this argument is not decisive.

Finally, if the case for non-classical ions is so strong, why does doubt continue to be expressed?

Many chemists find it difficult to swallow the proposal that in a solvent containing atoms with unshared electrons, a developing or fully developed carbonium ion center should prefer to be "solvated" by bonding carbon electrons even if these are necessarily adjacent to the site. "You might as well ask me to accept that a sodium ion prefers gasoline to water", one of the non-believers remarked once. Add to this intuition the fact that a great deal of the support for non-classical ions derives from new and relatively untried techniques such as nmr, computer calculations and so on, and one has some idea why this proposal has been so stoutly resisted.

20-8. PHENYL PARTICIPATION

π-Participation must decrease as the π electrons are held in orbitals of lower energy. Since aromatic π electrons are obviously much lower in energy, phenyl participation is likely to be of marginal importance, and hence the controversy that characterized carbon participation in general was not at its least lively here.

Again, we may begin with a symmetrical, rigid system which has the best possible orientation of the appropriate orbitals. The 7-benzonorbornenyl system fits that description, and a convincing case for phenyl participation has been made by Tanida and Winstein [116].

First of all, substituents on the ring have large effects; thus, the relative rate constants of acetolysis for (163) with R is MeO, Me, H, Cl and NO_2 are 55, 6, 1, 0.05 and 1.4×10^{-4}, respectively; the reaction constant, $\rho \approx -5$ [117].

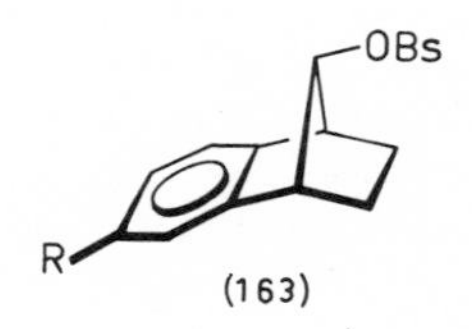

(163)

116. For review, see H. Tanida, Accounts Chem. Res., 1, 239 (1968).
117. (a) H. Tanida, J. Amer. Chem. Soc., 85, 1703 (1963); (b) H. Tanida, T. Tsuji, and H. Ishitobi, J. Amer. Chem. Soc., 86, 4904 (1964).

On the same scale, anti-7-norbornenyl has a relative rate constant of 3×10^6, and 7-norbornyl of 3×10^{-5}. Clearly phenyl participation is less effective than simple π-participation, but equally clearly, it does exist.

In the symmetrically disubstituted systems (164) the effect of the two groups R on the rate constant is precisely the square of that of a single group; for example, the relative rate is 36 when R = Me, and 3000 when R = OMe.

(164)

Evidently the free energy of activation is altered precisely twice as much by two groups as by one; and the conclusion is that the transition state is symmetrical [118].

7-Methyl substitution does not greatly affect the phenyl participation; in (165), the anti-syn rate ratio is still nearly 500 (1200 in the parent compounds).

(165)

In contrast to the syn-product, the anti-product purely has the retained configuration [119]. Further support comes from the following observation. While in the series (164) the OMe/NO_2 rate ratio is about 4×10^5, in the syn-epimers it is less than 50. In another comparison, with R = OMe, the anti/syn rate ratio is 40,000; with R = NO_2, it is 4.

The products are likewise revealing; whereas the anti-esters give unrearranged products, the syn-epimers lead to a different bicyclic system (166). The driving force in this instance is provided by the benzylic resonance gained by rearrangement to ion (167).

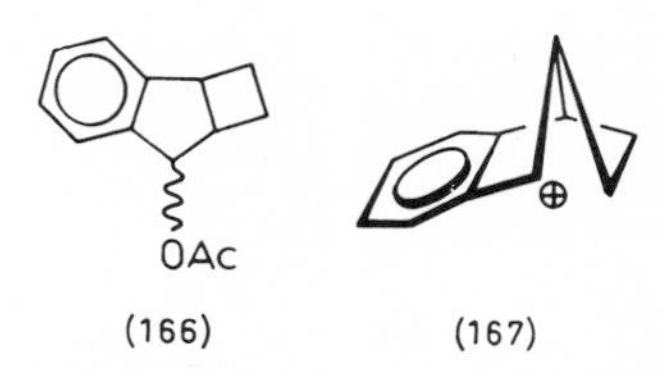

(166) (167)

118. H. Tanida and H. Ishitobi, J. Amer. Chem. Soc., 88, 3663 (1966).
119. H. Tanida, Y. Hata, S. Ikegami, and H. Ishitobi, J. Amer. Chem. Soc., 89, 2928 (1967).

Complete retention was observed in the acetolysis of both the syn- and anti-nitrodibenzonorbornadienyl brosylates (168) and (169), though the former solvolyzes many orders of magnitude faster than the latter [120].

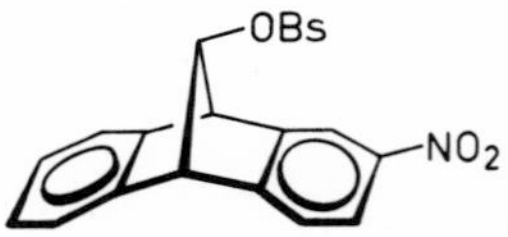

(168) 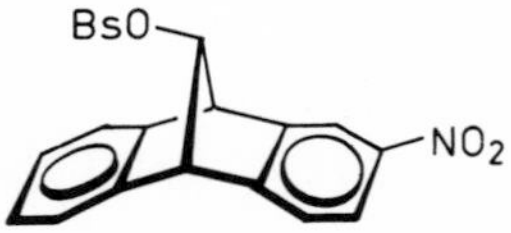 (169)

As remarked above, phenyl participation is less effective than π-participation; just how much less can further be shown by the rate ratio of the 7-bromobenzonorbornadienes (170) and (171). The former acetolyzes about 44000 times faster than the latter. Both give products with complete retention [121].

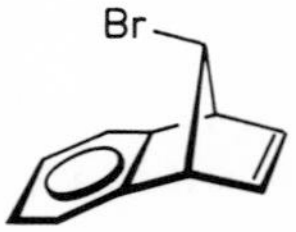

(170) 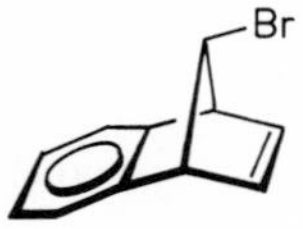 (171)

With the 7-benzonorbornenyl derivatives as our starting point, we may again try various extensions to see where phenyl participation ends. In the [2.2]paracyclophane system (172), participation is expected to be impossible since in the ion the vacant p orbital is parallel to both rings.

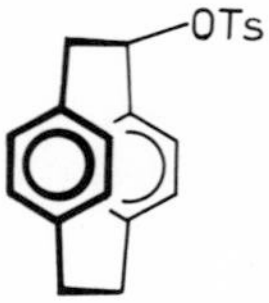

(172)

Yet assistance does occur by all the usual standards: In spite of the angle strain, the rate is a hundred times greater than that of i-propyl tosylate, and the products are of completely retained configuration. Evidently, this already highly strained molecule manages to distort further in such a way as to take advantage of the ring electrons (it is not yet known which ring) [122].

---

120. H. Tanida, T. Tsushima, and T. Irie, Tetrahedron Lett., 4331 (1970).
121. (a) J.W. Wilt and P.J. Chenier, J. Amer. Chem. Soc., 90, 7366 (1968); (b) S.J. Cristol and G.W. Nachtigall, J. Amer. Chem. Soc., 90, 7132 (1968).
122. R.E. Singler and D.J. Cram, J. Amer. Chem. Soc., 93, 4443 (1971).

In the somewhat less rigid systems (173), (174) and (175), the relative acetolysis rates are 1, 20, and 1, respectively. In the case of (174), ρ is found to be -2.9, and rearranged products (176) obtain which are easily traceable to ion (177); phenyl participation seems to run out as the flexibility and number of intervening bonds increase [123].

OBs BsO BsO OAc HOAc

(173) (174) (175) (176) (177)

A subtle change in the picture is introduced with the consideration of the 5-benzonorbornenyl system (178).

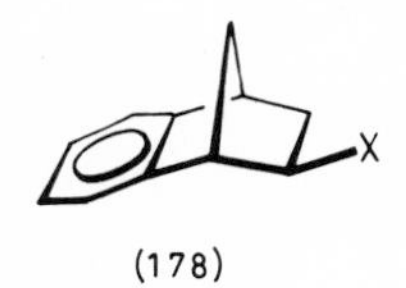

(178)

In the 7-substituted case the equal rate effects of symmetrically placed substituents showed that the nucleophilic action of the π electrons arises from the edge of the phenyl ring as in (179), or possibly even the face as in (180) or (181).

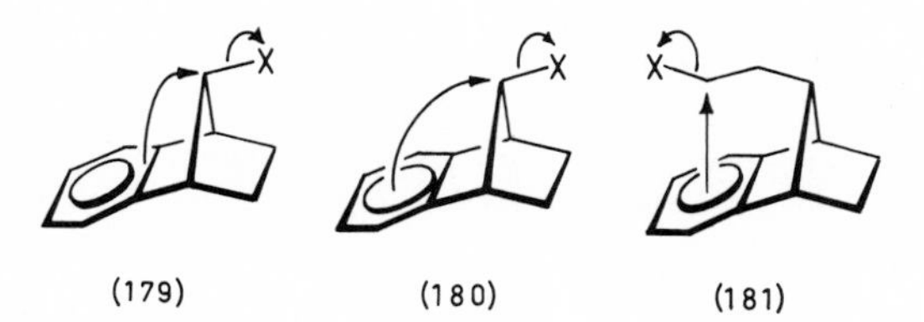

(179) (180) (181)

With 5-substitution, the phenyl contribution must come from one of its corners, and it is more appropriate to consider the intermediate ion a σ-complex than a π-complex (Chr. 23). Nevertheless, the assisting electrons are ultimately supplied by the π-system, and we may continue to consider these cases instances of π-participation (for a discussion of the nature of the intermediate, see p. 746). At any rate, indications are that phenyl participation is more effective when the electrons can be supplied by the face or edge of the ring than when they must flow through a corner. Thus, the [9]phane tosylates (182) have been studied; the numbers shown are relative rates when the anion departs from that position [124].

---

123. R. Muneyuki and H. Tanida, J. Amer. Chem. Soc., 90, 656 (1968).
124. D. J. Cram and M. Goldstein, J. Amer. Chem. Soc., 85, 1063 (1963).

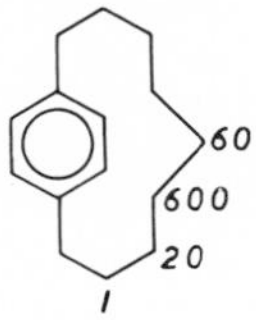

60
600
20
1

(182)

In the 5-substituted case Tanida has again made important contributions. An important criterion in these studies is the exo-/endo acetolysis rate ratio, which for the 5-benzonorbornenyl brosylates (183) equals 15,000:1, presumably because phenyl can participate when the brosylate anion is leaving in the exo-direction, but not when it is departing on the endo- side (see also p. 775) [125].

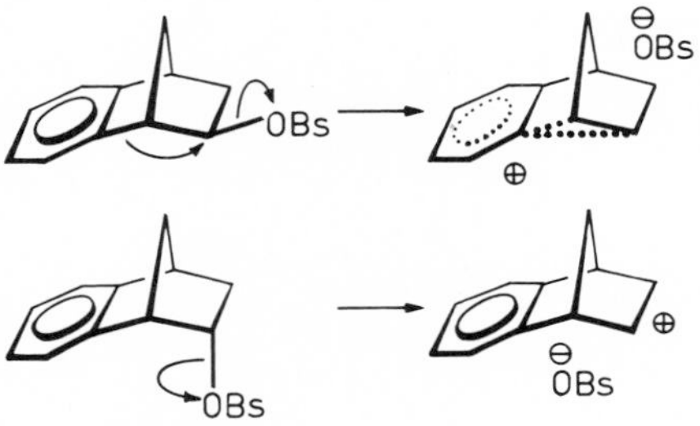

OBs
OBs
OBs
OBs

(183)

Winstein has shown that the rate ratio is about 62,000 if polarimetric rates are used instead of titrimetric rates. The argument is as follows. Ionization should be measured by the rate at which ions are produced rather than by the rate at which titratable acid is formed. As described in the next chapter, the first intermediates in the ionization process are not free carbonium ions but ion-pairs. These are subject to return, and hence they may escape detection by titration; however, since the cation in the pair now being considered has a plane of symmetry, its formation may be detected by measuring the rate at which optical activity is lost when resolved substrate is used. Both the product and substrate formed by return are racemic [126]. The two rate constants are the same in the endo- case, but for the exo-substrate they differ by a factor of 4.1. It is unfortunate that this important information is almost always missing in our discussions of carbonium ions, albeit for good reasons. The ion-pair does not always have the necessary symmetry characteristics, and horrendous amounts of time and effort are usually required to prepare a reasonable quantity of resolved material; in addition, there is no guarantee that even more sensitive probes may not turn up even more extensive return (see our discussion of ion-pairs).

---

125. H. Tanida, T. Tsuji, and S. Teratake, J. Org. Chem., 32, 4121 (1967).
126. J.P. Dirlam, A. Diaz, S. Winstein, W.P. Giddings, and G.C. Hanson, Tetrahedron Lett., 3133 (1969).

Originally Tanida concluded that phenyl participation was ineffective in this case: Dimethoxy substitution in (184) increased the *exo*-rate by only 16 and the *endo*-rate by 1.6, hence the *exo-/endo* ratio by 10. However, with one methoxy-group *ortho*- and one *meta*- to the site upon which the electron demand is made, it is not clear whether these facts are informative [127].

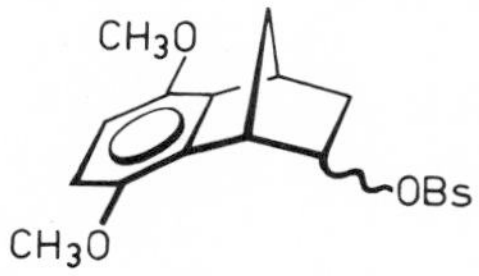

(184)

Indeed, with a single *m*-methoxy group as in (185) the *exo*-rate declines by 0.65; but the *p*-substituent in (186) increases the rate by a factor of 150, and the *exo-/endo* ratio by 60 [128].

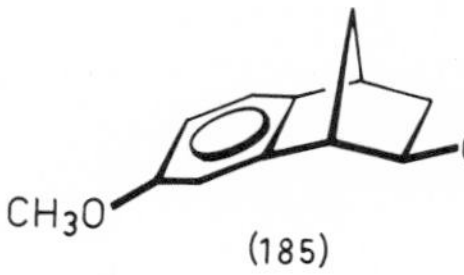

(185) 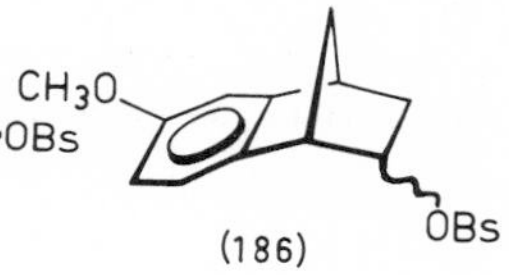 (186)

All *exo*-substrates give only retained (i.e. *exo*-) products, so that phenyl participation may be assumed to have intervened with varying degrees of effectiveness in each case. But the story is different when deactivating substituents are introduced. Thus, (187) and (188) acetolyze 10^5 and 10^3 more slowly than the unsubstituted parent and give rise to about equal amounts of *exo*- and *endo*- products.

(187)  (188)

Similarly, the *exo/endo* rate ratio in the dinitro-substituted molecule is only 3.7, and if the resolved dinitro substrate is used, the product is not completely racemic. Substitution of this kind can evidently decrease phenyl participation, or perhaps suppress it altogether [129].

---

127. H. Tanida, T. Tsuji and S. Teratake, *J. Org. Chem.*, *32*, 4121 (1967).
128. D.V. Braddon, G.A. Wiley, J. Dirlam, and S. Winstein, *J. Amer. Chem. Soc.*, *90*, 1901 (1968).
129. (a) H. Tanida, H. Ishitobi, T. Irie, and T. Tsushima, *J. Amer. Chem. Soc.*, *91*, 4512 (1969); (b) H. Tanida, T. Irie, and T. Tsushima, *J. Amer. Chem. Soc.*, *92*, 3404 (1970).

The *p*-substituent in the benzo ring provides a sensitive probe for studies of the effect of substituents at the carbonium ion center, by means of the *exo*/*endo* rate ratio criterion. This probe is unhampered here by the steric encumbrances which make its use so much more doubtful in the parent 2-norbornyl ion (see p. 768).

For instance, a 2-methyl substituent (189) reduces the rate acceleration by a *p*-methoxy substituent from 150 to 16; evidently, the presence of the methyl group does not reduce the demand on and operation of phenyl participation to zero [130].

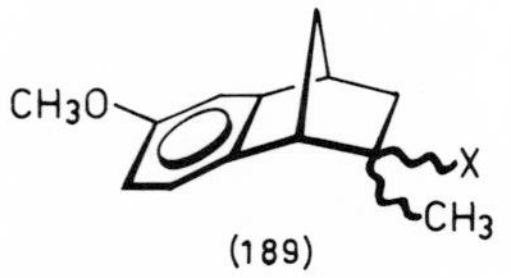

(189)

A *p*-methoxy substituent in the benzo ring of 2-(5-benzonorbornenyl-2-phenyl) brosylate (190) increases the *exo*/*endo* ratio from 3000 to 9000, indicating that there is still some demand when the carbonium ion center is benzylic *and* when the *p*-substituent in the benzo ring is methoxy [131].

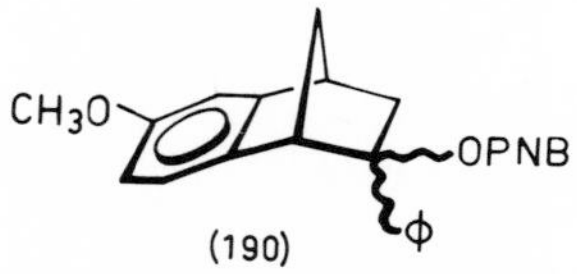

(190)

With an unsubstituted benzo ring, the *exo*/*endo* ratio remains virtually constant at about 3000 as the *p*-position of the 2-phenyl ring is varied from trifluoromethyl to methoxy, so that the *p*-methoxy substituent in the benzo ring is indeed necessary to provide assistance [132]. But when it is there, even a 2-*p*-anisyl group does not appear to reduce the benzo ring assistance to zero; the *exo*/*endo* rate ratio is then about 7000 [130].

A major stumbling block early in these studies was Brown's observation that 2-methyl- and 2-phenyl substitution, as in (191) and (192) respectively, have apparently no effect on the *exo*/*endo* rate ratio [133, 134].

---

130. J. P. Dirlam and S. Winstein, *J. Amer. Chem. Soc.*, *91*, 5905 (1969).
131. J. P. Dirlam and S. Winstein, *J. Amer. Chem. Soc.*, *91*, 5907 (1969).
132. H. C. Brown, S. Ikegami, and K. T. Liu, *J. Amer. Chem. Soc.*, *91*, 5911 (1969).
133. H. C. Brown and G. L. Tritle, *J. Amer. Chem. Soc.*, *88*, 1320 (1966).
134. H. C. Brown, *Chem. Brit.*, *2*, 199 (1966) and *Chem. Eng. News*, *45*, 87, June 5, 1967.

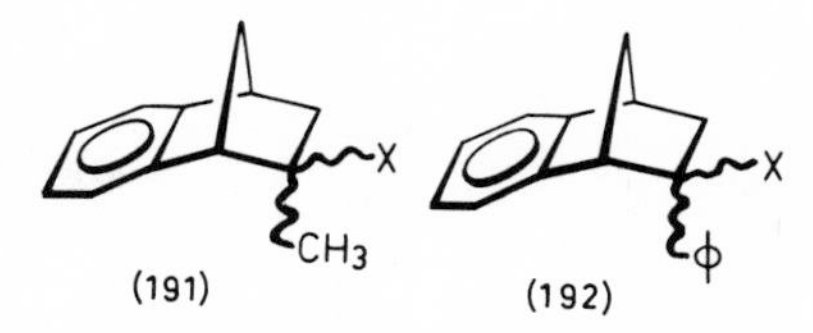

(191) (192)

He argued forcefully that if this ratio did not change when the need for participation was drastically reduced, then it has no value as an argument that there was participation in any other case.

It appears in retrospect that this ratio does have such value, but that it may be subject to unforeseen secondary influences if 2-substituents are used [1e]. Hence, such comparisons are valid only if more remote substituents are employed. For instance, a 3-spirocyclopropyl group has been used as a probe in (193). As we shall see below, the cyclopropyl group is also very able to supply electrons to a neighboring cationic center (see p. 751); the need for participation by the benzo ring should therefore be greatly diminished. In this case the _exo/endo_ rate ratio is indeed reduced to 12 [135].

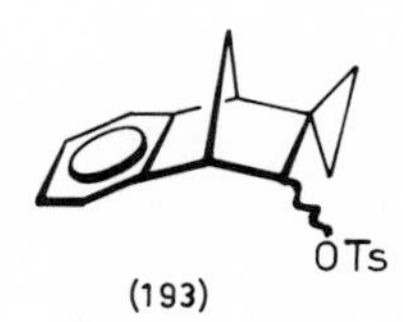

(193)

All this leaves one puzzling question: Why should _p_-anisyl substitution at the carbonium ion center be able to completely swamp phenyl participation in 7-norbornenyl-, but not in the 5-benzo-2-norbornenyl derivatives, although in the former it is obviously much more effective than in the latter.

A similar controversy about phenyl participation has centered around open chain systems, in which the principal early contributions were made by Cram, who reported an exhaustive study of the 3-phenyl-2-butyl and related systems [136]. To the drawbacks already described in the preceding case, we must add here the circumstance that an unstrained starting material is now expected to provide assistance although it must assume a strained structure to do it. It is not appropriate to fully describe all the circumstances and complications that surrounded this work (ion-pair return, elimination, hydrogen migrations and direct displacement by solvent are all competing reactions), and we shall restrict ourselves to the main stereochemical and radiochemical results.

When (+)-_threo_-3-phenyl-2-butyl tosylate (194) is solvolyzed, one obtains completely racemic _threo_- solvolysis products (195). Furthermore, if one of the

---

135. D. Lenoir, P. von R. Schleyer, and J. Ipaktschi, _Ann._, _750_, 28 (1971).
136. For reviews and listings of the early work, see (a) D. J. Cram, _J. Amer. Chem. Soc._, _86_, 3767 (1964); (b) W. B. Smith and M. Showalter, _J. Amer. Chem. Soc._, _86_, 4136 (1964).

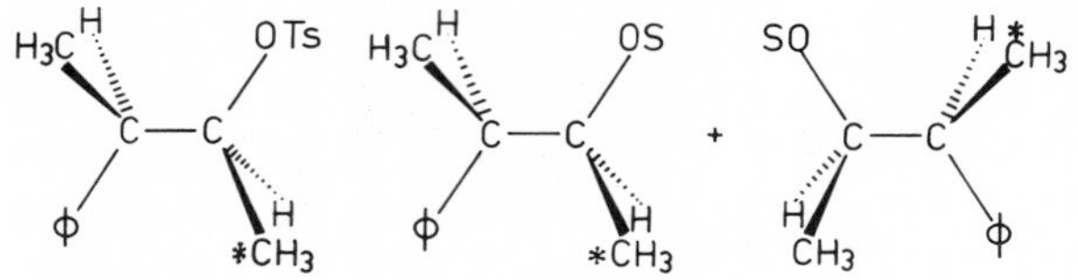

(194) (195)

methyl groups is labeled with C^{14} and the product is reresolved, the label is found exclusively in the positions indicated. When the (+) erythro-diastereomer (196) is employed, the product has the completely retained erythro-configuration (197).

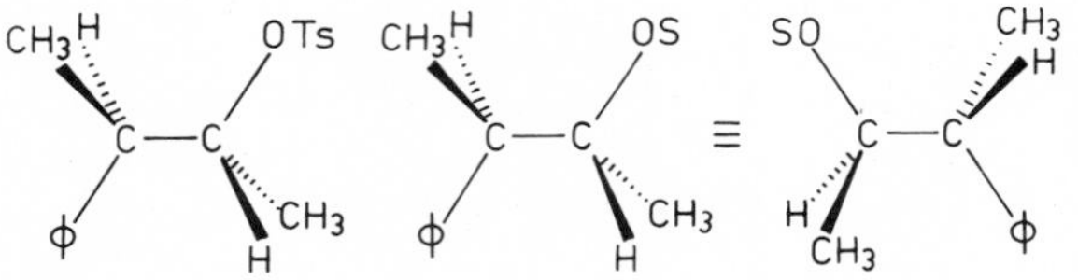

(196) (197)

If a mechanism is assumed involving either static, open ions such as (198), free to undergo internal rotations and to undergo collapse with solvent at either side,

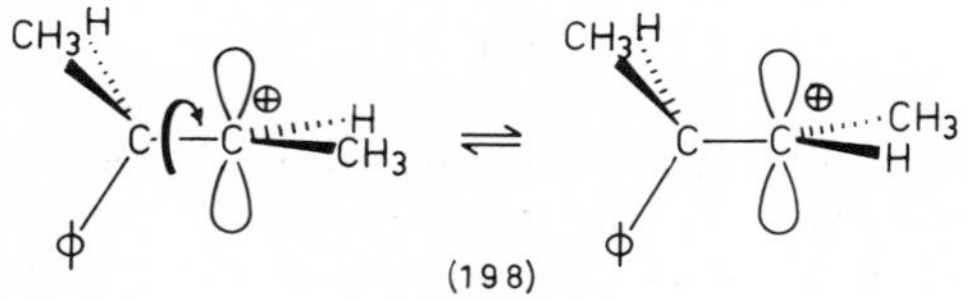

(198)

or involving rapidly equilibrating ions such as (199) with the same properties of rotation and solvent capture, these results cannot be explained.

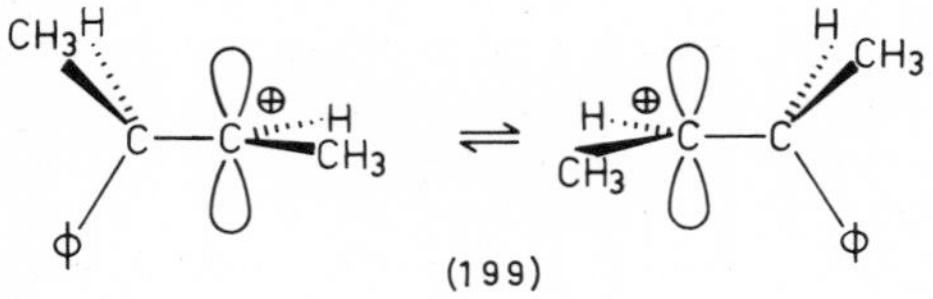

(199)

The results require either a symmetrical phenyl bridged ion which can react with solvent at either C_2 or C_3, but only on the side opposite the phenyl group, or a pair of open ions equilibrating much faster than they are reacting with solvent but which are nevertheless not free to undergo internal rotation. Cram postulated the former intermediate, and that the ionization leading to these species must be concerted, i.e., characterized by phenyl participation.

On the other hand, Brown has emphasized the fact that these reactions are not noted for unusually fast rates as are many others in which participation is operating, and he therefore preferred the view that no phenyl participation is occurring, and that the ions are rapidly equilibrating, classical ions incapable of internal rotation [137]. In this paper Brown eloquently set forth his concern over the virtual absence of equilibrating ions in the literature, the inadequacy of the arguments used to claim participation and non-classical ions, the imprecise definitions of the latter, and so on. It should be read by all who would like to savor the spirit of this controversy.

The question then became: What effect would a β-phenyl group have on the rate if it did not participate? The phenyl group will have a steric and inductive influence on neighboring atoms, and it may be deceiving to say that β-phenyl does not assist a given reaction merely because the rate constants of the substituted and parent molecules are equal. What is needed is a model in which a β-phenyl group cannot participate, and these are hard to imagine. One possible example has been published: The benzonorbornyl derivative (200) [138], for which models were said to indicate that overlap of the empty p orbital with the aromatic π cloud is precluded. It solvolyzes more slowly than the analogs (201)-(204) by factors of 12, 47, 9, and 800, respectively; the difference between (200) and the neophyl derivative (204) is especially revealing.

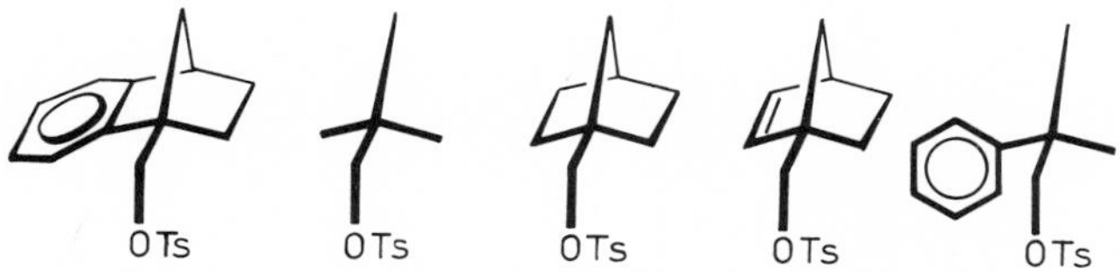

OTs OTs OTs OTs OTs

(200) (201) (202) (203) (204)

Another approach, tailor-made for rate studies of phenyl participation, has been the introduction of p-substituents. Thus, Brown compared the rates of formolysis of various β-phenylethyl tosylates (205) with those of borohydride reduction of the corresponding ketones and found that p-methoxy substitution increased the former more relative to the latter. He conceded that this was an indication of participation, but considered it very weak, π-type and likely to lead to equilibrating π complexes (206) [139].

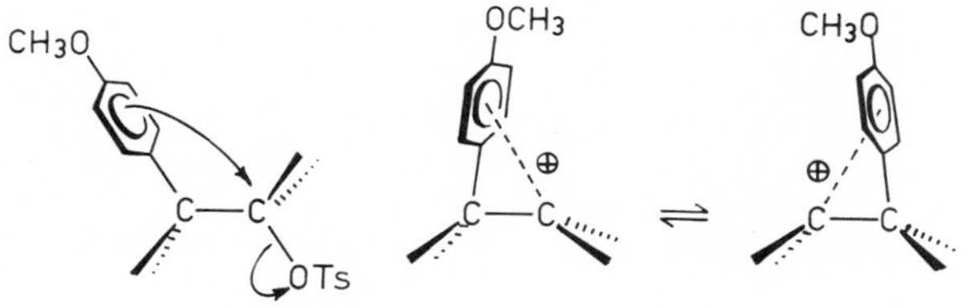

CH_3O C—C OTs OCH_3 C—C ⊕ ⇌ CH_3O ⊕ C—C

(205) (206)

137. H.C. Brown, K.J. Morgan, and F.J. Chloupek, J. Amer. Chem. Soc., 87, 2137 (1965).
138. J.W. Wilt, C.A. Schneider, J.P. Berliner, and H.F. Dabek, Tetrahedron Lett., 4073 (1966).
139. H.C. Brown, R. Bernheimer, C.J. Kim, and S.E. Scheppele, J. Amer. Chem. Soc., 89, 370 (1967).

Cram went in the opposite direction by studying the effect of a *p*-nitro group on the stereochemistry of the 3-phenyl-2-butyl system, and found that retention was no longer observed, and inversion (from *threo*-substrate into *erythro*-products and vice versa) occurred instead [140]. In a comprehensive study Brown then made use of a cyclic structure to reduce the amount of work necessary to study the stereochemical aspects of the reaction as a function of *p*-substituents [141].

X
retention
OS
Y
inversion
OS
Y
Y

He found that the *cis*-substrates give rise predominately to elimination and to small amounts of inverted solvolysis products. On the other hand, the *trans*-substrates gave rise to solvolysis products varying from almost complete retention (98-2 for *p*-methoxy) to complete inversion (100-0 for *p*-nitro). Yet, remarkably, both reactions had approximately equal and small ρ-values and *p*-methoxy/*p*-nitro rate ratios; the contradiction between rates and stereochemical results persisted (much larger *p*-substituent effects on the rate were reported for the 3-phenyl-2-butyl system, however: $\rho = -2.4$) [142]. It is of interest to note that the activation volume for solvolysis of β-phenyl ethyl derivatives has been found to change from -7 cm^3/mole for *p*-methoxy to -13 cm^3/mole for the *p*-nitro substrate [142a].

Meanwhile, the belief was taking hold that in the β-phenyl derivatives (and perhaps elsewhere), one feature that needed to be considered was the simultaneity of assisted and unassisted (or rather: Solvent-assisted) reactions. The idea was first proposed by Winstein [143] to allow the dissection of rate constants in terms of k_S (solvent-assisted ionization or displacement by solvent, and k_Δ (internally assisted ionization or participation). In much of the work that had been reported on β-phenyl group effects there were gradual changes in the deviations from Hammett plots, stereochemistry and type of product as the phenyl substituent was

140. (a) D. J. Cram and J. A. Thompson, *J. Amer. Chem. Soc.*, *89*, 6766 (1967); (b) J. A. Thompson and D. J. Cram, *J. Amer. Chem. Soc.*, *91*, 1778 (1969).
141. C. J. Kim and H. C. Brown, *J. Amer. Chem. Soc.*, *91*, 4286, 4287 (1969).
142. C. J Kim and H. C. Brown, *J. Amer. Chem. Soc.*, *91*, 4289 (1969).
142a. C. Yamagami and A. Sera, *Chem. Lett.*, 741 (1972).
143. S. Winstein, *Bull. Soc. Chim. France*, C 55, (1951).

varied. While it was not a priori necessary to exclude from consideration the possibility of a single reaction changing from weak or no assistance to strong participation as the phenyl substituents were varied, the concept of simultaneous and competing reactions is simpler to visualize and it fits all the known facts so neatly that it has won general acceptance - at least in the case of phenyl-assisted reactions.

Perhaps the arguments have been best summarized by Schleyer in four papers [144].

Thus, on the basis of a rate extrapolation from deactivating substituents, one can estimate what k_S for a given activating substituent will be, and hence from the observed rate how large k_Δ must be. Similarly, one can estimate these two rate constants on the basis of products (the k_S path giving inverted and the k_Δ path giving retained products in the first approximation). On the basis of such methods one may consistently obtain the same fractions for the solvent- and anchimerically assisted pathways, and thus learn how these vary with solvent and p-substituent.

Meanwhile the extremely non-nucleophilic and ionizing solvent trifluoroacetic acid had become popular [145], and its use now makes it possible to study phenyl participation in cases where it otherwise would have been marginal or absent. For instance, 1-phenyl-2-propyl tosylate solvolyzes 17 times faster than i-propyl tosylate in this solvent, even though it is 6 times slower in alcohol. By the same token, resolved substrate leads to 100% retention, although in alcohol 93% inversion is observed [146].

Even in 2-phenylethyl tosylate, phenyl participation can be important; it solvolyzes 4 times more slowly than ethyl tosylate in alcohol, but 3000 times more rapidly in trifluoroacetic acid. 2,2-Dideuterio- substrate then gives completely scrambled product [147]. On the basis of a comparison of the solvolysis rates of ethyl and neophyl tosylates in various solvents, it was estimated that k_S accounts for more than 99% of the reaction in alcohol, that k_Δ accounts for more than 99.99% in trifluoroacetic acid, and that the two are comparable in acetic acid [148]. In fact, not only has the degree of C^{14}-scrambling in the acetolysis of 2-phenylethyl-1-C^{14} tosylate been utilized to calculate the values of k_S and k_Δ, but $\Delta S^{\ddagger}$ has been determined separately for each path from the temperature dependence of this scrambling [149].

---

144. (a) C.J. Lancelot and P. von R. Schleyer, J. Amer. Chem. Soc., 91, 4291 (1969); (b) C.J. Lancelot, J.J. Harper, and P. von R. Schleyer, J. Amer. Chem. Soc., 91, 4294 (1969); (c) C.J. Lancelot and P. von R. Schleyer, J. Amer. Chem. Soc., 91, 4296 (1969); (d) P. von R. Schleyer and C.J. Lancelot, J. Amer. Chem. Soc., 91, 4297 (1969).
145. See e.g., P.E. Peterson and J.F. Coffey, Tetrahedron Lett., 3131 (1968).
146. J.E. Nordlander and W.J. Kelly, J. Amer. Chem. Soc., 91, 996 (1969).
147. (a) J.E. Nordlander and W.G. Deadman, Tetrahedron Lett., 4409 (1967); See also (b) W.H. Saunders, S. Asperger, and D.H. Edison, J. Amer. Chem. Soc., 80, 2421 (1958).
148. A. Diaz, I. Lazdins, and S. Winstein, J. Amer. Chem. Soc., 90, 6546 (1968).
149. J.L. Coke, F.E. McFarlane, M.C. Mourning, and M.G. Jones, J. Amer. Chem. Soc., 91, 1154 (1969).

As might be expected, $\Delta S^{\ddagger}$ for the assisted reaction is less negative; the more highly dispersed positive charge is less effective in orienting the local solvent molecules.

^{14}Carbon has also been used to trace the solvent effect on these competing reactions. The labelled species (207) acetolyzes 3% more slowly than the unlabelled compound if R is methoxy, and equally rapidly if R is H; but the isotope effect is also observable in the latter compound in formolysis and trifluoroacetolysis [150].

(207)

An ingenious approach was then used by Snyder to determine the stereochemistry for the solvolysis of 2-phenylethyl derivatives [151]. Threo- and erythro- ϕ-CHD-CHDOTs have distinctly different AB nmr patterns, and on that basis it was learned that these compounds trifluoroacetolyze with complete retention. Furthermore, it was learned from a comparison of the solvent effect on the degree of retention and that on the degree of carbon scrambling that the percentage loss of α-label is in each case half that of the percentage product with retained configuration - a neat piece of evidence for the validity of the assumption of the simultaneous and competing solvolytic processes described in the preceding pages (Table 20-1).

Other aromatic systems have been known to participate. For instance, α- and β-naphthylethyl tosylates solvolyze somewhat faster than the phenylethyl derivative,

TABLE 20-1. Comparison of Scrambling and Stereochemistry in β-Phenylethyl Solvolysis.

Solvent	% loss of α-label	% product with retained configuration
CF_3COOH	50	98
HCOOH	43.5	81
CH_3COOH	6.5	13

150. Y. Yukawa, T. Ando, M. Kawada, K. Token, and S. G. Kim, Tetrahedron Lett., 847 (1971).
151. R. J. Jablonski and E. I. Snyder, Tetrahedron Lett., 1103 (1968) and J. Amer. Chem. Soc., 91, 4445 (1969).

and extensive scrambling of the two methylene groups occurs [152]; on the other hand, it was found that in norbornadienes (208) and (209), (208) acetolyzes 6 times rapidly than its epimer.

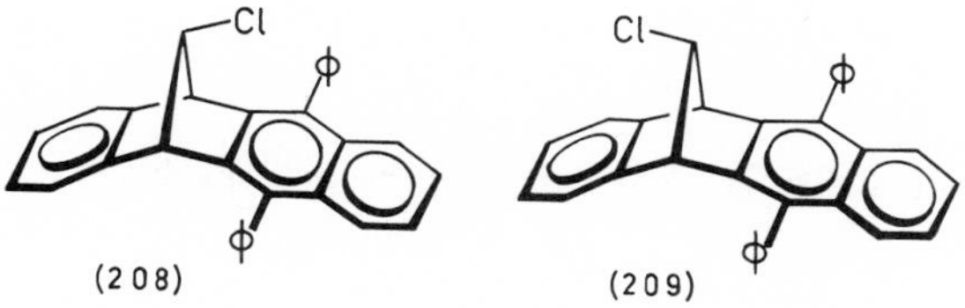

(208) (209)

Both give completely retained products, so that bridge flipping does not occur [153]. 9-Fluorenylmethyl tosylates (210) formolyze rapidly to give phenanthrenes, with benzylic ion resonance (211) and aromatization undoubtedly promoting the ionization and elimination [154].

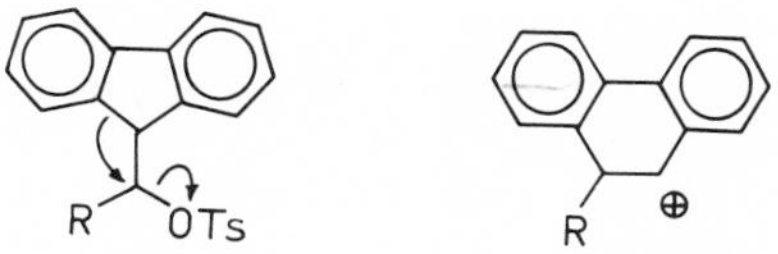

(210) (211)

A highly interesting result was obtained with 2-(-9-anthryl)-ethyl-1, 1-d_2 tosylate (212) which gives completely scrambled product in acetolysis (phenylethyl tosylate under the same conditions solvolyzes for only 10% by the k_Δ path). Moreover, in buffered aqueous dioxane one observes beside 15% scrambled 2-(9-anthryl) ethanol an 85% yield of spiro-compound (213), which is rapidly converted into the alcohol under acidic conditions [155].

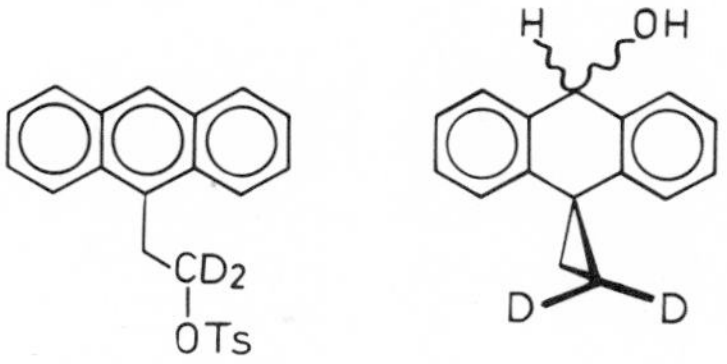

(212) (213)

A pyrrole ring is an extremely effective "participant"; thus, 2-(3-tryptophyl)-ethyl-1, 1-d_2 tosylate (214) acetolyzes 10^4 times faster than the corresponding indenyl analog to give unrearranged but completely scrambled acetate [156].

---

152. C.C. Lee and A.G. Forman, Can. J. Chem., 43, 3387 (1965) and 44, 841 (1966).
153. S.J. Cristol and G.W. Nachtigall, J. Amer. Chem. Soc., 90, 7132 (1968).
154. F.A.L. Anet and P.M.G. Bavin, Can. J. Chem., 43, 2465 (1965).
155. L. Eberson, J.P. Petrovich, R. Baird, D. Dyckes, and S. Winstein, J. Amer. Chem. Soc., 87, 3504 (1965).
156. W.D. Closson, S.A. Roman, G.T. Kwiatkowski, and D.A. Corwin, Tetrahedron Lett., 2271 (1966).

D D OTs N H

(214)

Similarly impressive evidence was found for the β-(1-azulyl)ethyl tosylate (215), which presumably reacts via the spirotropylium ion (216) [157].

OTs +

(215) (216)

A special case is the reaction of p-hydroxyphenylethyl esters in t-butanol containing one equivalent of base so as to ionize the phenol. The p-oxide anion (217) now makes phenyl participation (referred to as Ar_1^--3 participation) overwhelming; the rate is increased by 10^6 or so, and the product is the unstable but isolable spiro compound (218) [158].

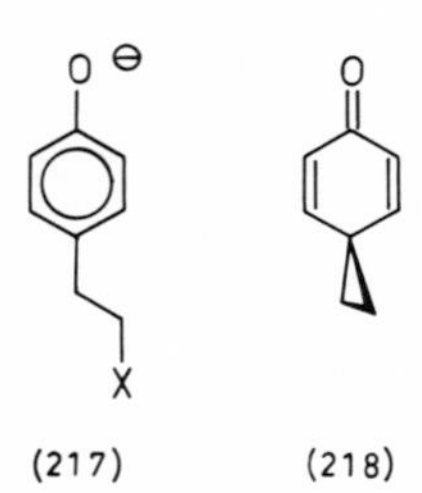

(217) (218)

Even Ar_1^--5 participation is effective, as indicated by a rate acceleration of at least 50 in the reaction of (219) to give (220) [159]. Another example is the reaction of (221) to give (222) [160].

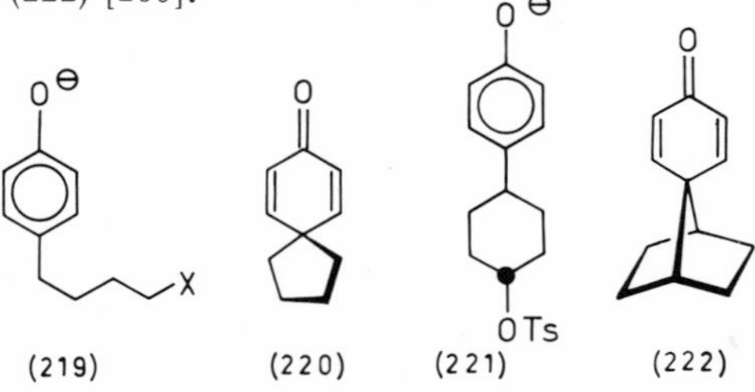

(219) (220) (221) (222)

157. R.N. McDonald and J.R. Curtis, J. Amer. Chem. Soc., 93, 2530 (1971).
158. R. Baird and S. Winstein, J. Amer. Chem. Soc., 85, 567 (1963).
159. R. Baird and S. Winstein, J. Amer. Chem. Soc., 84, 788 (1962).
160. R. Barner, A.S. Dreiding, and H. Schmid, Chem. Ind. (London), 1437 (1958).

Such reactions have the small, negative activation volumes of about 0 to -5 cm^3/mole characteristic of S_N2 displacement reactions rather than of about -20 cm^3/mole as is typical of S_N1-solvolyses such as of ω-p-methoxyphenylbutyl derivatives [161]; clearly the transition state has no carbonium ion characteristics at all. Ar_1-5 participation also occurs in the tricyclic system (223), which hydrolyzes to (224);

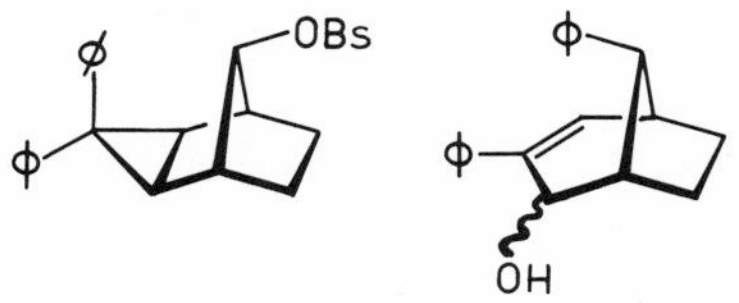

(223) (224)

ionization with phenyl participation gives rise to a cyclopropyl ion which undergoes Woodward-Hoffmann allowed ring opening to the isomeric allylic ion [162]. 3-(p-Hydroxyphenyl)propyl esters do not exhibit Ar_1-4 participation. Only second order displacements occur to give polymeric products; 1,1-d_2-labelling does not result in scrambling. Evidently the blend of strain and distance which governs the occurrence of participation renders it ineffective if four-membered rings must be produced [163].

Ar_2-participation is also known; usually indanes or tetralins [164] are the result. Actually, it is not always clear whether these compounds result from initial Ar_1-participation followed by ring expansion, or from direct Ar_2 attack. Jackman found that the labelled brosylate (225) gives tetralin (226) with the deuterium pair in the positions indicated; evidently 74% (2 x 37) of the reaction therefore occurred via the expansion route, and 26% directly [165].

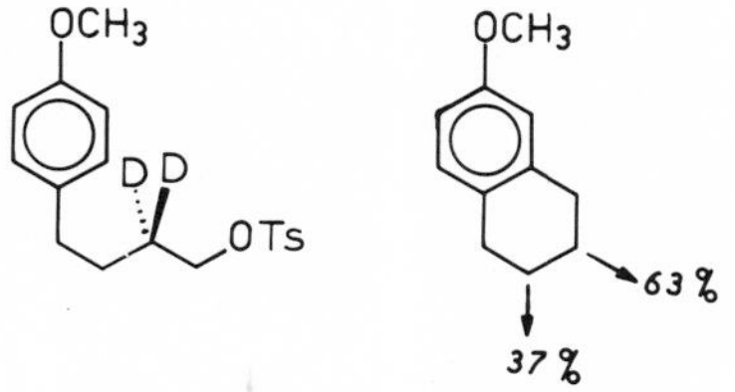

(225) (226)

As far as the nature of the ions is concerned, it will be recognized that truly non-classical ions with a carbon atom bound to five centers can be involved in only a few cases, such as 7-norbenzonorbornenyl ions (227), and in those cases rapidly equilibrating ions are an alternative description.

---

161. W.J. le Noble and B. Gabrielsen, Tetrahedron Lett., 45 (1970).
162. J.W. Wilt and T.P. Malloy, J. Amer. Chem. Soc., 92, 4747 (1970).
163. W.J. le Noble and B. Gabrielsen, Tetrahedron Lett., 3417 (1971).
164. P.G. Duggan and W.S. Murphy, Chem. Commun., 770 (1972).
165. V.R. Haddon and L.M. Jackman, J. Amer. Chem. Soc., 93, 3832 (1971).

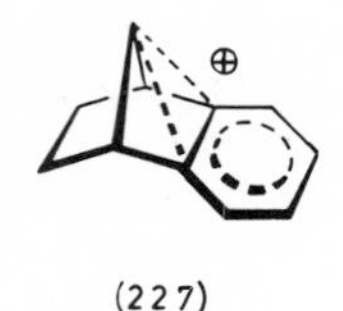

(227)

When corner - rather than edge - assistance is provided, the ion involves no such carbon atoms but its description is still a matter of considerable disagreement. Cram, Winstein and others prefer the structure (228), which implies considerable delocalization of charge in the three membered ring [166];

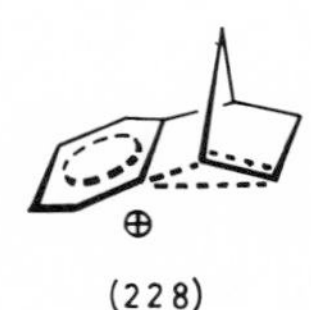

(228)

Brown and Deno [167] consider that such an ion is a simple σ-complex with the charge located in the five-carbon chain and delocalized as in a pentadienyl cation (229), and the most extreme position is taken by W. Hückel ("Alle Grenzformeln für vermeintliche Phenoniumionen schweben in der Luft") [168].

(229)

As an example of a simple cyclohexadienylium ion we may mention the heptamethylbenzenonium ion (230), given here with its nmr chemical shifts in τ units [169].

7.0
7.5
7.2
8.3

(230)

166. L. Eberson and S. Winstein, J. Amer. Chem. Soc., 87, 3506 (1965).
167. (a) H.C. Brown, K.J. Morgan and F.J. Chloupek, J. Amer. Chem. Soc., 87, 2137 (1965); (b) N.C. Deno, Progr. Phys. Org. Chem., 2, 129 (1964).
168. W. Hückel, J. Prakt. Chem., Series 4, 28, 27 (1965).
169. W. von E. Doering, M. Saunders, H.G. Boyton, H.W. Earhart, E.F. Wadley, W.R. Edwards, and G. Laber, Tetrahedron, 4, 178 (1958).

The ion can be isolated as the tetrachloroaluminate salt, and its 1:2:2:2 nmr spectrum indicates that it is not a π complex with one methyl group above the center of the ring. The signals merge between 45 and 75°; if hexamethylbenzene is added, its signal remains distinct so that the methyl exchange is intramolecular [170]. Detailed multiple resonance experiments further revealed that the methyl group rotates *via* 1, 2-shifts, and not by random migrations across the ring (*via* a face-centered π-complex) [171]. The 2-methyl group at 7.2 τ represents a decrease of about 1 ppm from its position in toluene, presumably due to deshielding by the nearby positive charge.

When 2-(mesityl)ethyl chloride (231) is dissolved in a mixture of antimony pentachloride and sulfur dioxide at -60°, the resulting nmr spectrum exhibits methyl signals in a 1:2 ratio at 7.4 and 7.6 τ, about half a ppm upfield from those in heptamethylbenzenonium ion.

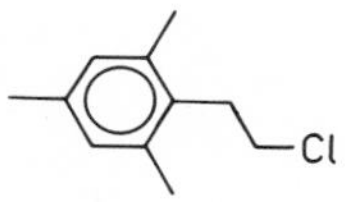

Cl

(231)

At the same time, a four proton signal appears as a singlet at 6.5 τ, sharply downfield from the normal cyclopropyl position of 9.5-10 τ. Clearly considerable charge has been siphoned into the cyclopropyl ring. A similar result has been obtained for the phenonium ion derived from 2-*p*-anisylethyl chloride [172], but an early claim [173] to the generation of a stable phenonium ion from 3-phenyl-2-butanol turned out to be in error [174].

The very first case of nmr observation of a phenonium ion is the anthrylethyl ion [175].

Generation of this ion by a σ route, i.e., the dissolution of the spiro alcohol (232) in an antimony pentafluoride-sulfur dioxide mixture at -60°, gave a solution (233), the nmr spectrum of which showed that H_a had moved from 4.6 to 0.4 τ, the aromatic protons from the 2.0-3.4 τ region to 1.8-2.3 τ, and the A_2B_2 pattern at 8.8 τ had become a singlet at 6.5 τ.

---

170. V.A. Koptyug, V.G. Shubin, and A.I. Rezvukhin, Bull. Acad. Sci. USSR, (English Transl.), 192 (1965).
171. B.G. Derendyaev, V.I. Mamatyuk, and V.A. Koptyug, Tetrahedron Lett., 5 (1969).
172. (a) G.A. Olah, E. Namanworth, M.B. Comisarov, and B. Ramsey, J. Amer. Chem. Soc., 89, 711 (1967); (b) G.A. Olah and R.D. Porter, J. Amer. Chem. Soc., 93, 6877 (1971).
173. G.A. Olah and C.U. Pittman, J. Amer. Chem. Soc., 87, 3509 (1965).
174. (a) M. Brookhart, F.A.L. Anet, and S. Winstein, J. Amer. Chem. Soc., 88, 5657 (1966); (b) M. Brookhart, F.A.L. Anet, D.J. Cram, and S. Winstein, J. Amer. Chem. Soc., 88, 5659 (1966).
175. L. Eberson and S. Winstein, J. Amer. Chem. Soc., 87, 3506 (1965).

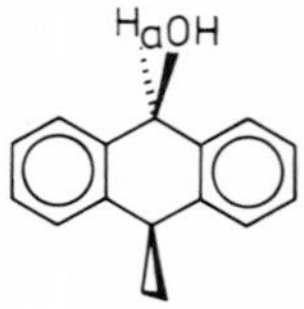

(232) 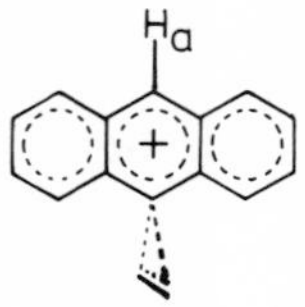 (233)

The authors argued that the large shift of the cyclopropyl protons showed the presence of much charge in the three membered ring, in spite of the fact that the charge could be fully located in a benzhydryl position. If on the other hand one were observing equilibrating ions, the shift would be much larger, and that of H_A much less. Similar observations were made by the protonation of spiro anthrone (234);

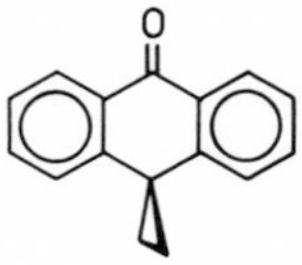

(234)

Zollinger [176] in fact has argued on the basis of uv and nmr evidence that the spiroketones themselves have dipolar character of the sort suggested by (235).

(235)

Further experiments have been done to rule out equilibrating ions as an alternative [177]. Thus, the cis- and trans-dimethylspiroketones (236) and (237) can be dissolved in fluorosulfonic acid at -65° to give indefinitely stable solutions the nmr spectra of which are different, e.g., in that the cis-substrate unlike the trans- leads to different vinyl protons. Upon warming to -40°, both rearrange via a hydride shift to the more stable benzyl ion (238).

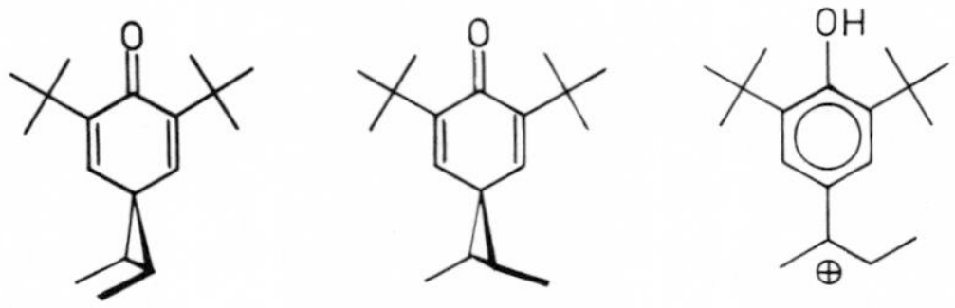

(236) (237) (238)

176. P. Rys, P. Skrabal, and H. Zollinger, Tetrahedron Lett., 1797 (1971).
177. D. Chamot and W. H. Pirkle, J. Amer. Chem. Soc., 91, 1569 (1969).

Clearly we do not have equilibrating ions: It is one thing to say that rapid equilibration can preserve configuration of an ion during solvolysis, in the brief interval between formation and collapse, but it is quite another to say that it can prevent single bond rotations indefinitely. Similar conclusions have been reached in a study of the protonation of the diastereomeric anthrones (239) [178].

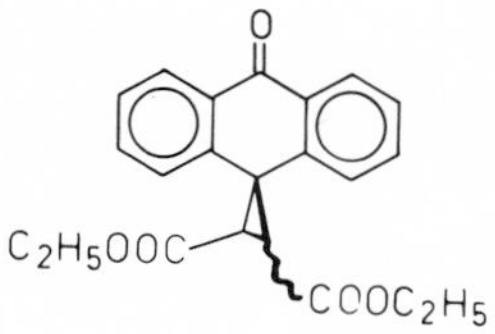

(239)

20-9. HOMOPROPARGYL AND HOMOALLENIC SYSTEMS

Triple bonds have π electrons of relatively high energy, and hence participation should occur in favorable cases in the formation of electron deficient centers. On the other hand, the linear shape of the acetylenic linkage may make it sterically difficult or impossible, and the first product formed - the vinyl cation - is not very stable as in known from the slow solvolysis rates of various vinyl derivatives (see Sec. 21-3).

Several examples have been reported [179]. Thus, 6-phenyl-5-hexynylbrosylate (240) acetolyzes to give a substantial amount (36%) of the rearranged product (241) [180],

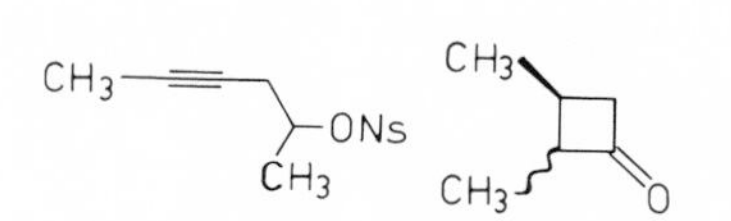

(240) (241)

and 5-hex-2-ynyl tosylate (242) solvolyzes to give *cis*- and *trans*-2,3-dimethylcyclobutanone (243) [181].

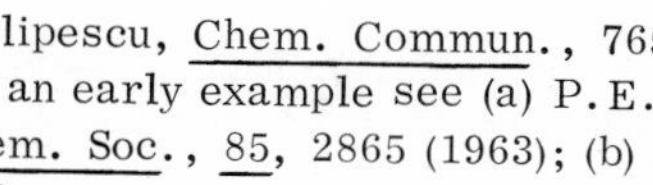

(242) (243)

178. J.W. Pavlik and N. Filipescu, *Chem. Commun.*, 765 (1970).
179. For a listing including an early example see (a) P.E. Peterson and J.E. Duddey, *J. Amer. Chem. Soc.*, *85*, 2865 (1963); (b) J.W. Wilson, *Tetrahedron Lett.*, 2561 (1968).
180. W.D. Closson and S.A. Roman, *Tetrahedron Lett.*, 6015 (1966).
181. M. Hanack, S. Bocher, K. Hummel, and V. Vött, *Tetrahedron Lett.*, 4613 (1968).

An instructive publication by Ward and Sherman may be mentioned at this point [182]. These authors found that 4-phenyl-3-butyn-1-yl brosylate (244) formolyzes to give a 30% yield of cyclopropyl phenyl ketone (245); however, this product also forms at a very fast rate from the addition product (246) of formic acid to the acetylene, and the addition is a side reaction which can be shown to compete with solvolysis to give (247). It was shown that the rearranged product is not a result of triple bond participation, and it is therefore a good example of the fact that rearrangement alone is not a sufficient criterion for participation.

(244) 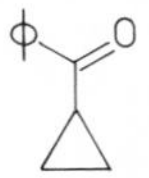 (245) 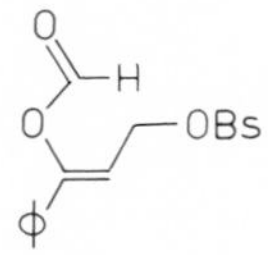 (246) 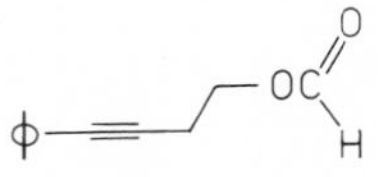 (247)

Participation by allenic double bonds is less handicapped by an unfavorable geometry. For example, Bly found that the allenic brosylate (248) undergoes acetolysis several orders of magnitude faster than the saturated analog and in fact, faster than the corresponding olefin. The product is mostly a rearranged acetate (249) [183]

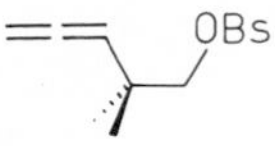

(248) 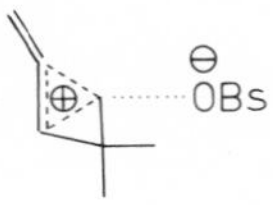 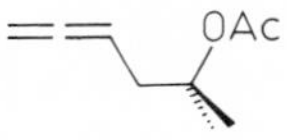 (249)

(if the resolved substrate (250) is used, the product is still optically active) [184].

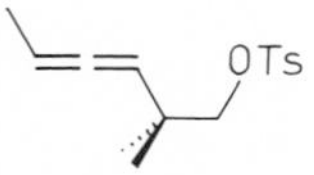

(250)

Methyl substitution at the α-position of (251) causes a rate acceleration of only 40--another indication of participation [185].

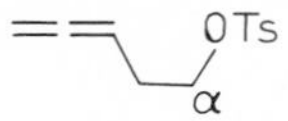

(251)

182. H.R. Ward and P.D. Sherman, J. Amer. Chem. Soc., 89, 1962 (1967).
183. R.S. Bly, A.R. Ballentine, and S.U. Koock, J. Amer. Chem. Soc., 89, 6993 (1967).
184. T.L. Jacobs and R. Macomber, Tetrahedron Lett., 4877 (1967).
185. M. Santelli and M. Bertrand, Tetrahedron Lett., 3699 (1969).

20-10. Σ-PARTICIPATION: THE CYCLOPROPYL GROUP

In pursuit of the question just how tightly held bonding electrons must be before they can no longer be tempted to participate in the stabilization of incipient electron-deficient carbon, the next logical step is the cyclopropyl group, with its bonding electrons in single but strained bonds (see p. 202).

As in other instances of participation, the effect is most obvious in rigid, symmetrical substrates in which the orientation is favorable compared to epimeric molecules. Two groups have reported [186] on the four isomers (252)-(255), whose relative acetolysis rates are in 10^4, 10, 1, and 10^{12}, respectively, while that of (256) is 10^9.

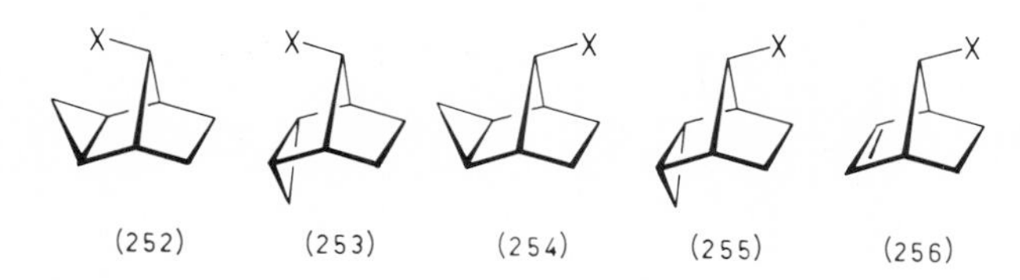

The syn compounds give highly complex product mixtures. Incipient rearrangements account for their enhanced rates; in the case of (252) relief of steric hindrance could also be responsible.

The anti-endo-isomer is extremely reactive in solvolysis and gives rise exclusively to (257), consistent with participation leading to the intermediate trishomocyclopropenium ion (258). Interestingly, (254) is the slowest of the whole group. Evidently the cyclopropyl group can provide electrons with an edge but not with its face. This is exactly what the Walsh model (p. 202) predicts.

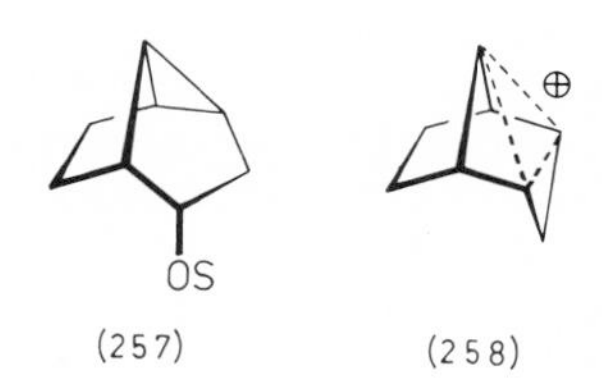

An ingenious modification of this study was carried out by Coates to answer the question of whether the rearrangement provides the driving force for participation [187]. He used (259); the relative solvolysis rate is about 10^{10}, and the product is unrearranged. Actually, the rearrangement of (260) is now degenerate, and hence it cannot provide the driving force.

---

186. (a) H. Tanida, T. Tsuji, and T. Irie, J. Amer. Chem. Soc., 89, 1953 (1967); (b) M. A. Battiste, C. L. Deyrup, R. E. Pincock, and J. Haywood-Farmer, J. Amer. Chem. Soc., 89, 1954 (1967); (c) J. S. Haywood-Farmer and R. E. Pincock, J. Amer. Chem. Soc., 91, 3020 (1969).
187. R. M. Coates and J. L. Kirkpatrick, J. Amer. Chem. Soc., 92, 4883 (1970).

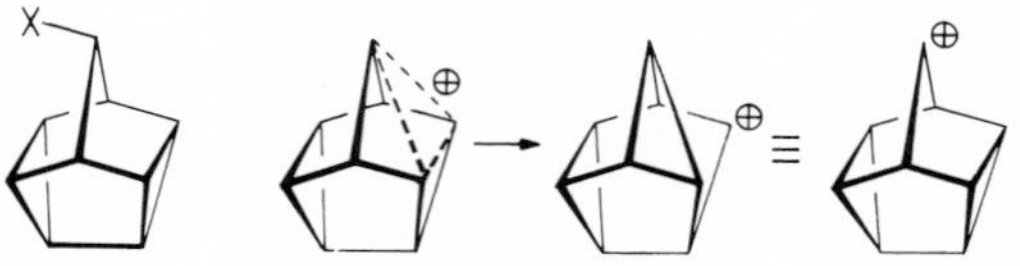

(259) (260)

The deuterio substrates (261) and (262) give rise to alcohols in which the deuterium is equally distributed among three positions as shown.

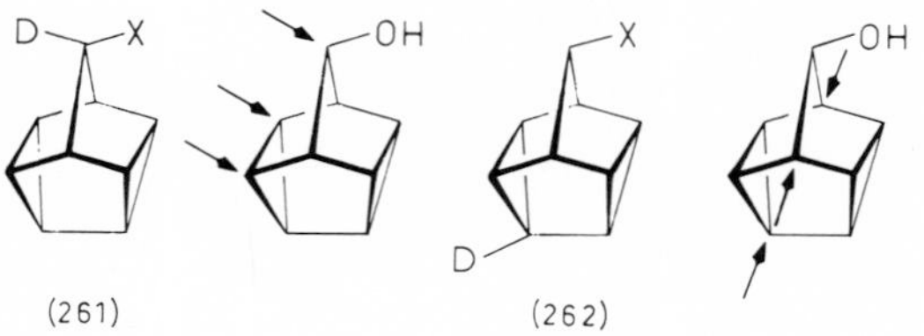

(261) (262)

These findings are suggestive of the symmetrical ion (263).

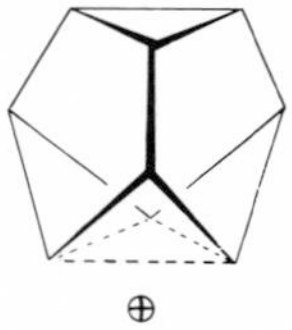

(263)

The rigidity of these molecules appears to be a factor in their rapid solvolysis rates. Thus, the epimeric 5-bicyclo[3.1.0]hexyl tosylates themselves have a much more modest rate ratio. The scrambling and stereochemistry in that experiment led Winstein to propose the concept of homoaromaticity (see p. 364).

The rate data suggest that the cyclopropyl group is more effective than a double bond in providing assistance. This conclusion is supported by a study of Gassman, who found that an α-p-anisyl group, which is enough to completely swamp π-participation in the norbornenes, is not enough to eliminate assistance by the cyclopropyl group: (264) hydrolyzes 4000 times faster than (265) to give (266) [188].

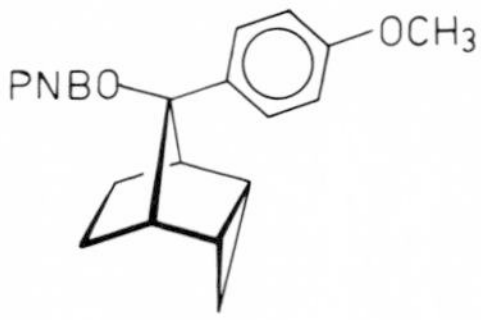

(264) 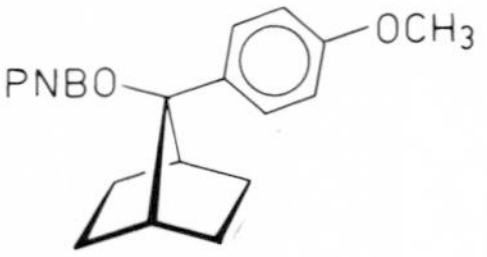 (265) 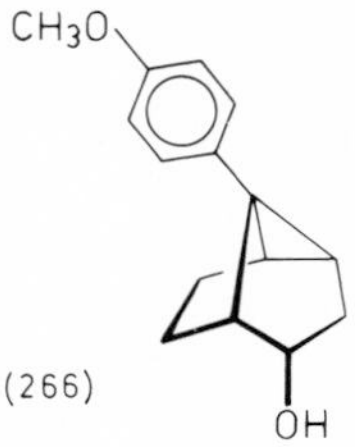 (266)

188. P.G. Gassman and A.F. Fentiman, J. Amer. Chem. Soc., 92, 2551 (1970).

The relative effectiveness of π bonds and cyclopropyl groups does not always come out this way, however. The acetolysis rates of the syn- and anti-epimers (267) and (268) are identical, for example [189].

(267) (268)

Sargent found a similar lack of assistance in (269) as compared to (270); he considered this a result of interference by the cyclopropyl bridgehead hydrogen atoms [190].

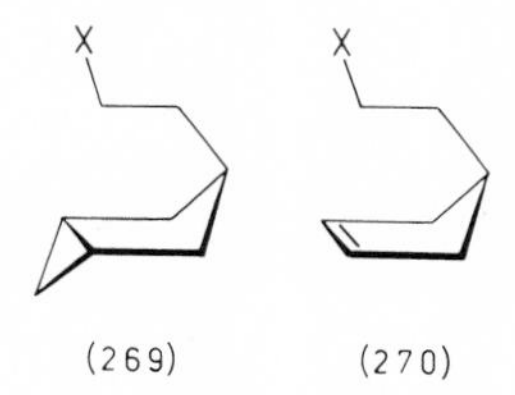

(269) (270)

While the number of intervening bonds is large in these cases, that alone does not seem to be responsible. The number is equally large in the tosylate (271), which acetolyzes 5000 times faster than the cyclohexyl tosylate and gives the ring expanded product (272) (the olefin (273) is twice as fast again, and gives rise to 1-adamantyl acetate (274)) [191].

(271) (272) (273) (274)

A similar conclusion also follows from the series (275)-(277), whose relative rates are $10^{4.5}$, 1, and 10^7, respectively [192].

---

189. R. Muneyuki, T. Yano, and H. Tanida, J. Amer. Chem. Soc., 91, 2408 (1969).
190. G.D. Sargent, R.L. Taylor, and W.H. Demisch, Tetrahedron Lett., 2275 (1968).
191. M.A. Eakin, J. Martin, and W. Parker, Chem. Commun., 955 (1967).
192. M.A. Battiste, J. Haywood-Farmer, H. Malkus, P. Seidl, and S. Winstein, J. Amer. Chem. Soc., 92, 2144 (1970).

(275) (276) (277)

Another striking series of rate data was obtained by Dewar for formolysis [193]; neither β-cyclopropyl nor methyl-substituted cyclopropyl groups enhanced the rate very greatly: The relative rates for the series (278)-(283) is 1, 0.9, 1.33, 3.24, 2.56 and 3.64, respectively.

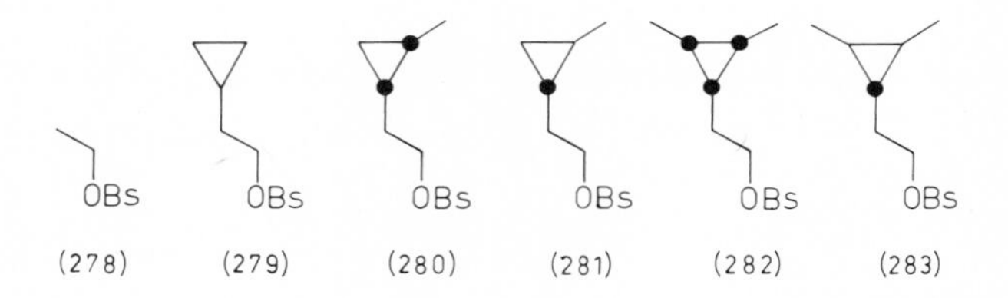

(278) (279) (280) (281) (282) (283)

Similarly, 2-n-alkyl bromides all have nearly identical solvolysis rates; by contrast, one finds the relative rates of the series (284)-(287) to be 1, 3.7, 770 and 165, respectively.

OTs OTs OTs OTs

(284) (285) (286) (287)

The orientation of the cyclopropyl group and the rigidity of the system are clearly important. If the ring is in just the right position and rigidly held there, it is very effective; otherwise its assistance is marginal or absent altogether. An extreme example of non-assistance is the 1-nortricyclyl ester (288) [194]; it requires a triflate leaving group, 50-50 aqueous alcohol and a temperature of 255° for a half-life of about an hour!

OTf

(288)

---

193. (a) M. J. S. Dewar and J. M. Harris, J. Amer. Chem. Soc., 92, 6557 (1970); (b) R. R. Sauers and R. W. Ubersax, J. Org. Chem., 88, 495 (1966).
194. (a) R. C. Bingham, W. F. Sliwinski, and P. von R. Schleyer, J. Amer. Chem. Soc., 92, 3469 (1970); (b) S. A. Sherrod, R. G. Bergman, G. J. Gleicher, and D. G. Morris, J. Amer. Chem. Soc., 92, 3469 (1970) and 94, 4615 (1972).

This surely must be a record, and it is due to the strain that would be present in the planar bridgehead carbonium ion. Evidently the nearby face of the cyclopropyl group does not stabilize it. Corner participation has been demonstrated in some special cases such as (289); these will be considered later (p. 775) along with other norbornyl derivatives.

(289)

It may be said at this point, however, that such participation is not extremely effective, and thus we are left with the conclusion that evidently the edge of the cyclopropyl is most efficient by far in the stabilization of nearly positive charges; it is furthermore necessary that the developing lobe of the incipient empty p orbital point to this edge. Thus, no evidence for participation was found in the trans-fused cyclopropane derivative (290), in which this orbital and the edge are in close proximity but parallel [195]. Such participation (291) would at any rate be forbidden.

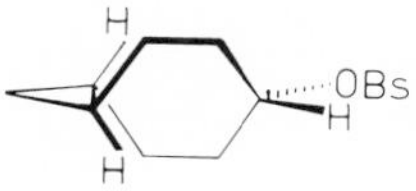

(290) 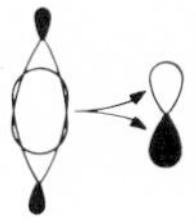 (291)

A very interesting series is that of the acetates (292)-(295). The first three of these hydrolyze at identical rates, but (295) is a thousand times faster. Some positive charge is clearly being delocalized into the three-membered ring [196] (see also p. 748).

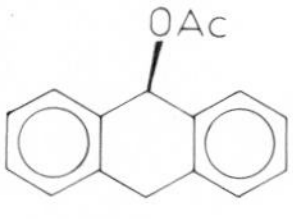

(292) 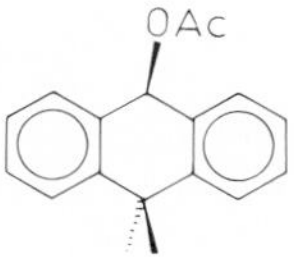 (293) 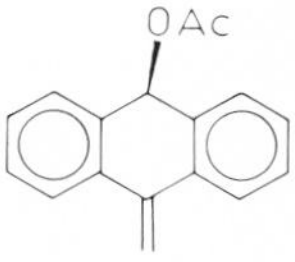 (294) 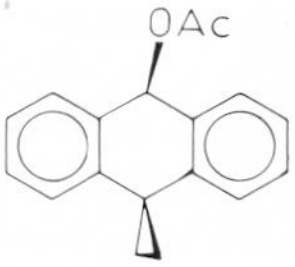 (295)

So far we have considered cyclopropyl derivatives which have two or more carbon atoms between the carbonium ion site and the participating group. At this point we shall consider some of the most interesting literature on the cyclopropylcarbinyl ion (296), which is the analog of the homoallylic cation.

---

195. P.G. Gassman, J. Seter, and F.J. Williams, J. Amer. Chem. Soc., 93, 1673 (1971).
196. R. Leute and S. Winstein, Tetrahedron Lett., 2475 (1967).

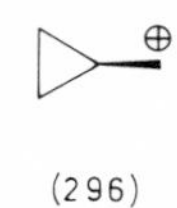

(296)

The cyclopropyl ring has been noted by many authors to behave in a way suggesting an ability to conjugate with unsaturated groups [197]. This is obvious even in such simple properties as the vapor pressure; thus, quadricyclanone (297) is less volatile than quadricyclanol (298) [198].

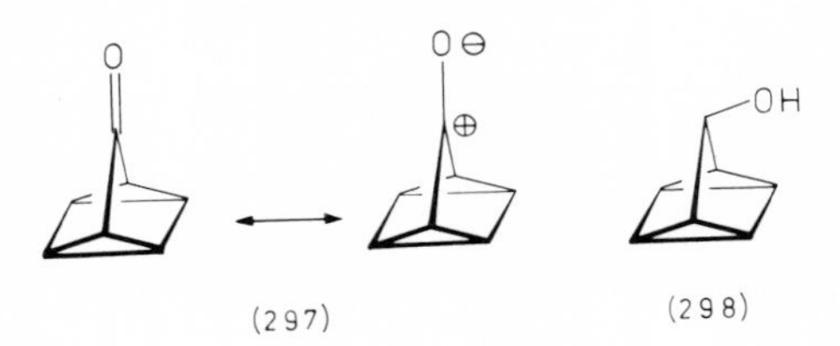

(297) (298)

However, the most dramatic evidence for this type of conjugation is the interaction of a cyclopropyl group bound to a carbonium ion center. Again, we shall begin our discussion by considering a rigid system.

The tricyclic *p*-methoxybenzoate (299) is about 10^{22} times faster than the 7-norbornyl analog; it provides a σ-route to the 7-norbornenyl ion [199].

OPMB

(299)

Surely (299) is one of the most reactive substrates known; it stands in remarkable contrast to the isomeric nortricyclyl derivative (288).

Another notable comparison is that of the epimeric substrates (300) and (301). The former reacts about 10^6 times more slowly than the latter; without the cyclopropyl bond, the *exo*/*endo* ratio is much *greater* than 1. Evidently the α-cyclopropyl group is more useful in providing assistance than β-phenyl - so long as it is oriented properly.

---

197. M.Y. Lukina, *Russ. Chem. Rev. (English Transl.)*, *31*, 419 (1962).
198. P.R. Story and S.R. Fahrenholtz, *J. Amer. Chem. Soc.*, *86*, 1270 (1964).
199. (a) J.J. Tufariello and R.J. Lorence, *J. Amer. Chem. Soc.*, *91*, 1546 (1969); (b) J. Lhomme, A. Diaz, and S. Winstein, *J. Amer. Chem. Soc.*, *91*, 1548 (1969).

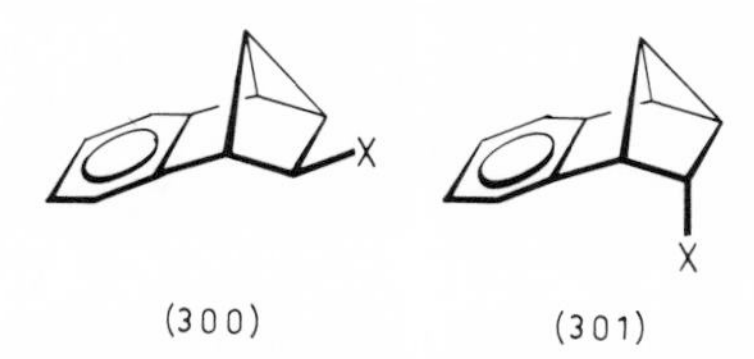

In the dibenzosemibullvalenylcarbinyl system (302), where the cyclopropyl and the phenyl groups compete from equal distance, it is the latter that participate; the product is then (303) [200].

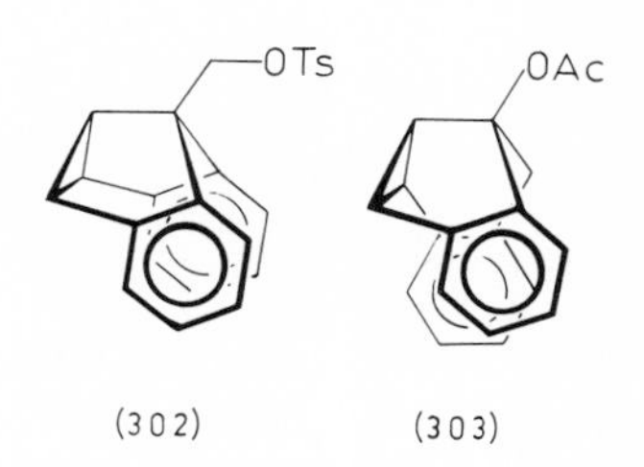

Another case of much interest involves the nortricyclyl and quadricyclyl substrates (304) and (305) [201], which solvolyze 10^9 and 10^{10} times faster than the 7-norbornyl analog, respectively.

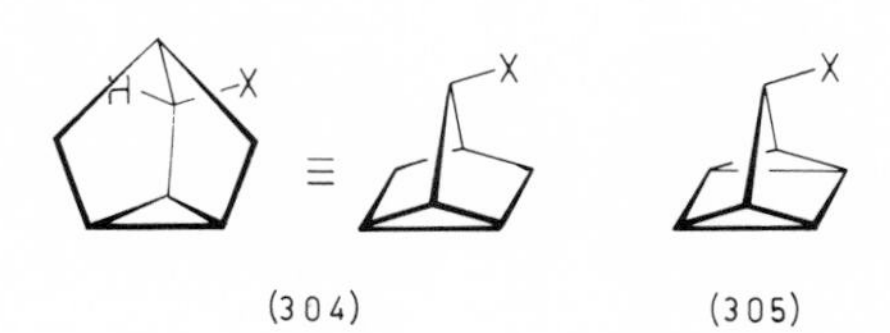

Clearly the carbonium ion center interacts with only one of the cyclopropyl groups. The ion is of course closely related to the norbornadienyl ion but it is not identical with it; acetolysis gives 70% norbornadienyl acetate and 30% quadricyclyl acetate, and alcoholysis in the presence of sodium borohydride give only quadricyclane [202]. Precisely the opposite conclusion must be drawn from the series (306)-(312), in which the solvolysis rates are in the ratio of 4×10^7, 3×10^7, 2×10^7, 10^6, 10^6, 40, and 1, respectively [203].

---

200. L.A. Paquette and G.H. Birnberg, J. Amer. Chem. Soc., 94, 164 (1972).
201. (a) S. Winstein, H.M. Walborsky, and K. Schreiber, J. Amer. Chem. Soc., 72, 5795 (1950); (b) H.G. Richey and N.C. Buckley, J. Amer. Chem. Soc., 85, 3057 (1963).
202. (a) P.R. Story and S.R. Fahrenholtz, J. Amer. Chem. Soc., 86, 527 (1964); (b) M. Brookhart, R.K. Lustgarten, D.L. Harris, and S. Winstein, Tetrahedron Lett., 943 (1971).
203. L. Birladeanu, T. Hanafusa, B. Johnson, and S. Winstein, J. Amer. Chem. Soc., 88, 2316 (1966).

X X X

(306) 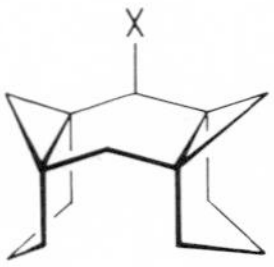 (307) 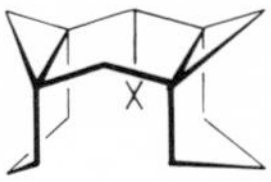 (308)

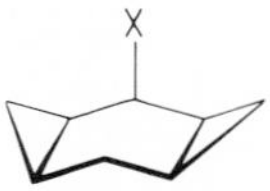

X X X X

(309) 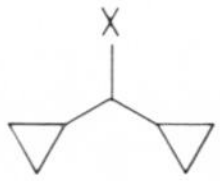 (310) 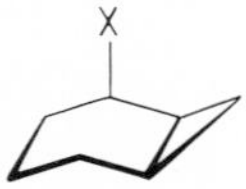 (311) 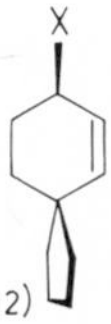 (312)

Several features about this series are interesting to note since they appear contradictory to what we have seen in examples given earlier. Thus, the epimers (307) and (308) react about equally fast. The open and more rigid substrates (309) and (310) which differ greatly in rigidity are also equal in rate. Finally, the tricyclic system (309) is very much faster than (311). Thus, here is at least one case where more than one cyclopropyl group is involved at the same time.

Another series of interest in this connection has been published by Hart. The relative hydrolysis rates of (313)-(316) are 1, 250, 25,000 and 25,000,000 [204].

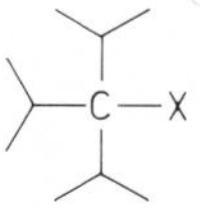

C—X C—X C—X C—X

(313) 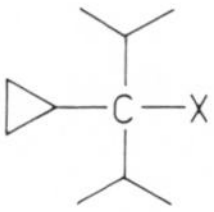 (314) 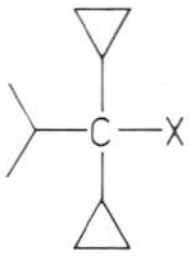 (315) 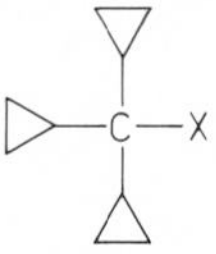 (316)

Similarly, the relative hydrolysis rates of (317)-(321) are 1, 1000, 0.3, 300, and 300, respectively [205].

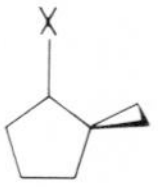

X X X X X

(317) 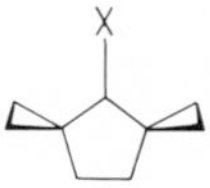 (318) 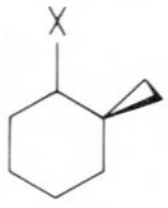 (319) 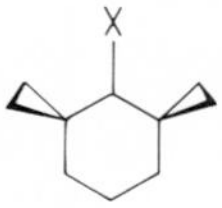 (320) 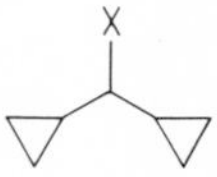 (321)

---

204. (a) H. Hart and P.A. Law, J. Amer. Chem. Soc., 86, 1957 (1964); (b) H. Hart and J.M. Sandri, J. Amer. Chem. Soc., 81, 320 (1959).
205. (a) A.P. Krapcho, R.C.H. Peters, and J.M. Conia, Tetrahedron Lett., 4827 (1968). However, see also (b) T. Tsuji, I. Moritani, S. Nishida, and G. Tadokoro, Tetrahedron Lett., 1207 (1967); (c) T. Tsuji, I. Moritani, and S. Nishida, Bull. Chem. Soc. Japan, 40, 2338 (1967).

Exhaustive studies have been carried out with the unsubstituted cyclopropylcarbinyl esters themselves by Roberts [206]. A striking observation is that such esters and the isomeric cyclobutyl and allylcarbinyl esters give rise to identical mixtures of the three solvolysis products, so that a common intermediate is strongly suggested. The rates are somewhat faster than would be expected from comparison with various model compounds [207]. As noted earlier, equilibrating bicyclobutonium ions account better for the results of labelling studies than does the highly symmetrical tricyclobutonium ion [208].

8, 9-Dehydro-2-adamantyl 3, 5-dinitrobenzoate (322) solvolyzes in aqueous acetone 10^{8-9} times faster than the 2-adamantyl analog, and 10^3 times faster than the rate calculated on the basis of the Foote-Schleyer relation [209]; the only product is unrearranged alcohol.

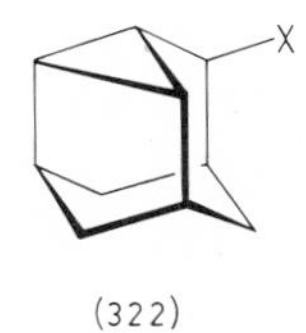

(322)

If 2-deuterium labelled substrate is used, the label is extensively though not completely scrambled among the 2-, 8- and 9- positions. The incompleteness of the scrambling rules out a three-fold symmetrical "tricyclobutonium ion" (323) also in this case, and the authors considered that a rapid equilibration between three bicyclobutonium ions (324) was in agreement with the facts.

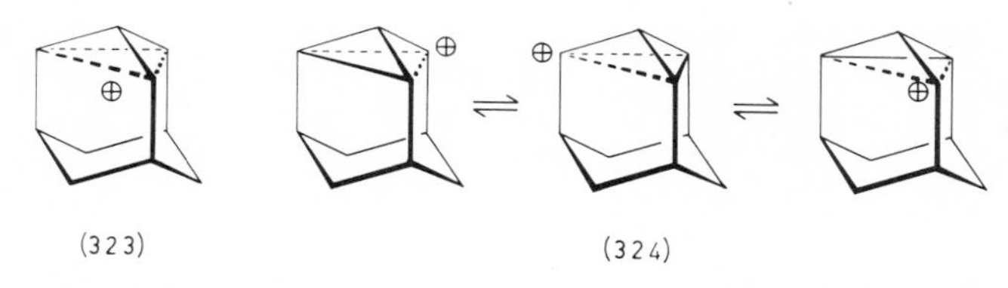

(323) (324)

As far as the charge distribution in the incipient cyclopropylcarbinyl ion is concerned, we note first of all that methyl groups cause the rate enhancements shown in (325) [210].

---

206. For references to the early work, see for example; K. L. Servis and J. D. Roberts, J. Amer. Chem. Soc., 87, 1331 (1965).
207. J. D. Roberts and R. H. Mazur, J. Amer. Chem. Soc., 73, 2509 (1951).
208. (a) R. H. Mazur, W. N. White, D. A. Semenow, C. C. Lee, M. S. Silver, and J. D. Roberts, J. Amer. Chem. Soc., 81, 4390 (1959); (b) M. C. Caserio, W. H. Graham, and J. D. Roberts, Tetrahedron, 11, 171 (1960).
209. J. E. Baldwin and W. D. Foglesong, J. Amer. Chem. Soc., 90, 4303 (1968).
210. (a) M. Nikoletic, S. Borcic, and D. E. Sunko, Tetrahedron, 23, 649 (1967); (b) D. D. Roberts, J. Org. Chem., 31, 2000 (1966).

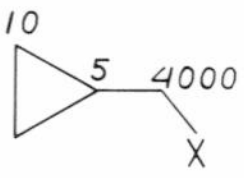

(325)

This suggests that the more distant β-carbon atoms in the cyclopropyl group are called upon for a greater contribution than the α-atom.

The acid catalyzed racemization of resolved cyclopropylethanol has about the same rate as that of oxygen exchange; this suggests - but it does not require - that the cyclopropylcarbinyl ion has the symmetrical "bisected" conformation (326) rather than the "parallel" conformation (327) [211].

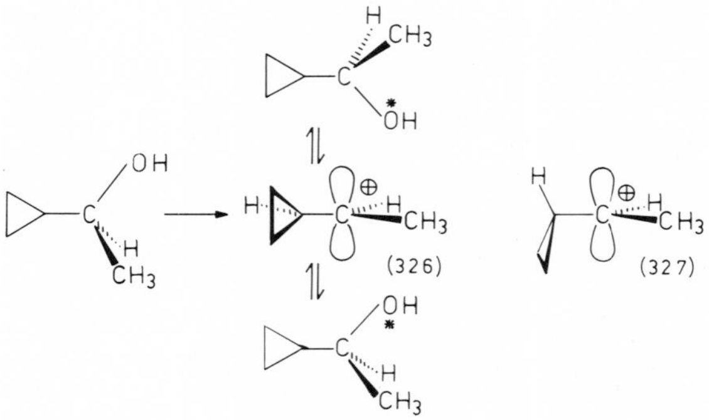

(326) (327)

Further support for this symmetrical structure was found by Schleyer, who found that 2-trans-methyl- and 2, 3-trans-trans-dimethyl substitution as in (328) and (329) caused rate enhancements of 11 and 124, respectively [212].

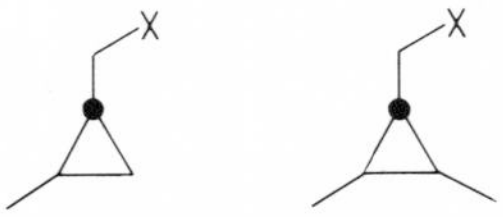

(328) (329)

The rigidity of this conformation is indicated by the stereochemistry of the reaction: If the deuterium labeled mesylate (330) is hydrolyzed, the hydrogen atom is scrambled in the three products (331)-(333) as might be expected from equilibrating bicyclobutonium ions; but perhaps more interesting is the observation that it winds up exclusively in the cis- positions. In spite of the complexity of this reaction, it is completely stereospecific [213].

---

211. H.G. Richey and J.M. Richey, J. Amer. Chem. Soc., 88, 4971 (1966).
212. P. von R. Schleyer and G.W. Van Dine, J. Amer. Chem. Soc., 88, 2321 (1966).
213. Z. Majerski and P. von R. Schleyer, J. Amer. Chem. Soc., 93, 665 (1971).

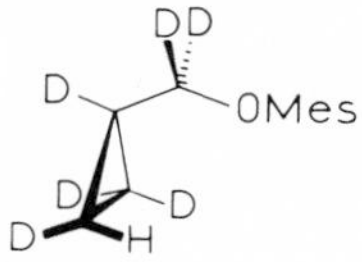

DD D OMes D D D H 22% OH 78% OH 60% 40% 26% OH 0%

(330) 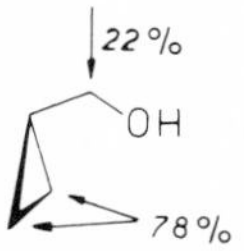 (331) 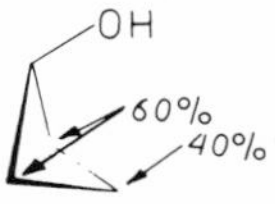 (332) 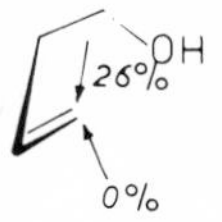 (333)

The bisected structure applies evidently even when the cyclopropyl group is at a considerable distance from the carbonium ion center (but not conjugated with it). Thus, Brown found the following interesting series of relative hydrolysis rates for the cumyl chlorides (334)-(340) [214]: They are in the ratio 1:2:4:18:157:172:37.

Cl Cl Cl Cl Cl Cl Cl

(334) 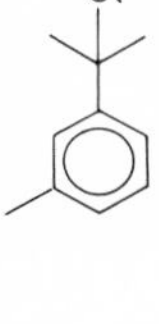 (335) 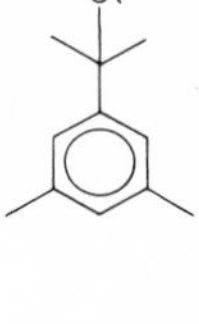 (336)  (337)  (338) 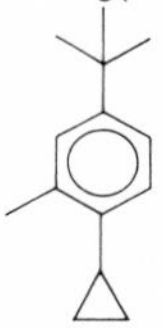 (339) 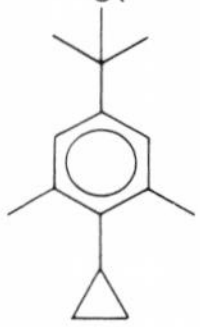 (340)

Clearly, the advantage of a cyclopropyl over an *i*-propyl group depends on the ability of the former to assume the bisected conformation as in (341) and (342); if this is prevented as in (343), this ability disappears.

(341) 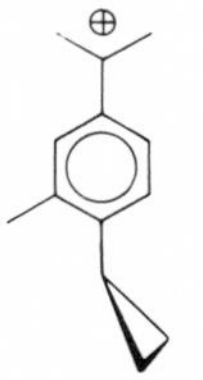 (342) 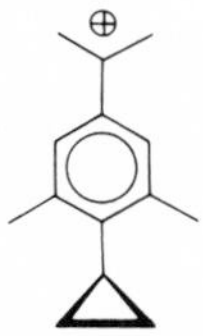 (343)

From an analysis of the nitration rates in various substituted cyclopropylbenzenes, Stock has deduced that parallel *p*-cyclopropyl has a σ^+-value of -0.1, and the bisected conformation has a σ^+-value of -0.4 [215]. The value for *i*-propyl is -0.28 [216].

If the parallel conformation is rigidly enforced, the inductive effect of the cyclopropyl group in fact retards ionization. Evidence was found by Martin [217] in

---

214. H.C. Brown and J.D. Cleveland, *J. Amer. Chem. Soc.*, *88*, 2051 (1966).
215. L.M. Stock and P.E. Young, *J. Amer. Chem. Soc.*, *94*, 4247 (1972).
216. L.B. Jones and V.K. Jones, *Tetrahedron Lett.*, 1493 (1966).
217. (a) B.R. Ree and J.C. Martin, *J. Amer. Chem. Soc.*, *92*, 1660 (1970); (b) V. Buss, R. Gleiter, and P. von R. Schleyer, *J. Amer. Chem. Soc.*, *93*, 3927 (1971).

the acetolyses of (344)-(347), whose rates are in the ratio of 1:2.3:10^{-4}:6.5 x 10^{-3}.

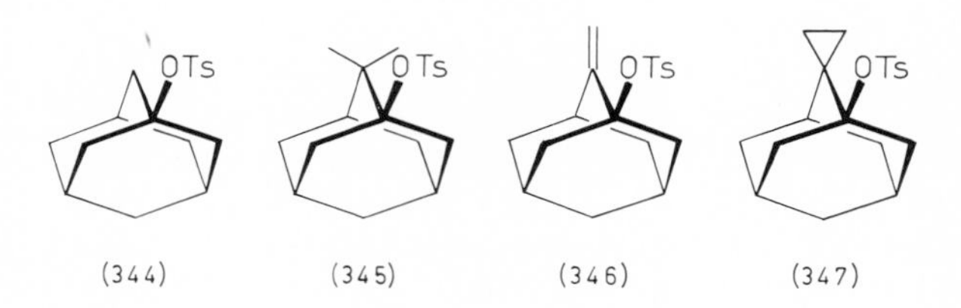

(344) (345) (346) (347)

More quantitative information on this point is now available [218]. If we compare the rates of acetolysis at 25° of the tosylate pairs (348)-(349), (350)-(351), (352)-(353) and (354)-(355), we find that the ratios decrease from 2.5 x 10^8, to 4 x 10^5, to 4 x 10^3 to 5 x 10^{-3}; at the same time the angles indicated in (356)-(359) increase from 0 to 30, to 60, to 90°, respectively.

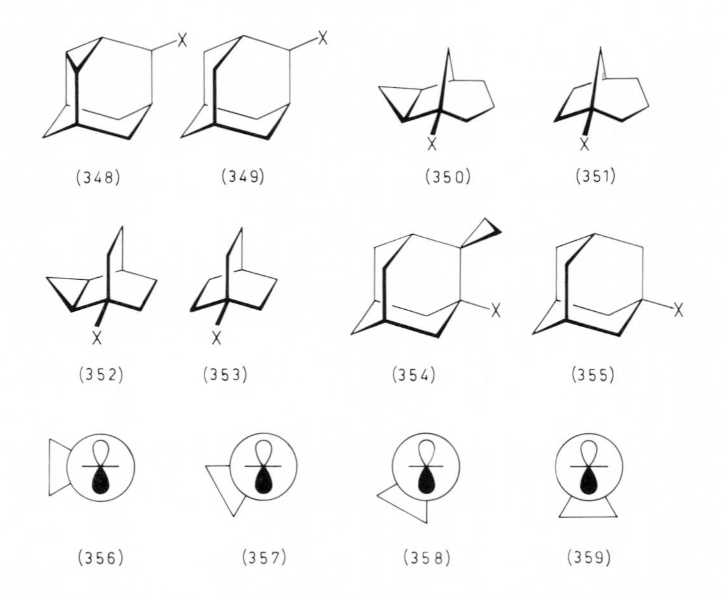

(348) (349) (350) (351)

(352) (353) (354) (355)

(356) (357) (358) (359)

In the trishomobarrelene chloride (360) and the trishomobullvalene chloride (361) the angles are about 60°; the interaction with three cyclopropyl rings however leads to rate increases of 10^{10} or more when compared to (362) [219].

---

218. Y.E. Rhodes and V.G. DiFate, J. Amer. Chem. Soc., 94, 7582 (1972).
219. (a) A. de Meijere, O. Schallner, and C. Weitemeyer, Angew. Chem., Int. Ed. Engl., 11, 56 (1972). For additional studies of this sort, see (b) K.B. Wiberg and T. Nakahira, J. Amer. Chem. Soc., 93, 5193 (1971); (c) P.G. Gassman, E.A. Williams, and F.J. Williams, J. Amer. Chem. Soc., 93, 5199 (1971); (d) A.S. Kende, J.K. Jenkins, and L.E. Friedrich, Chem. Commun., 1215 (1971).

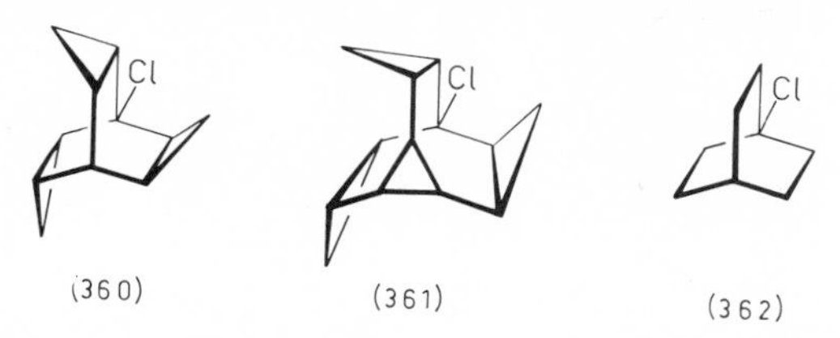

(360) (361) (362)

Once again however, one should be careful not to extrapolate such data to less rigid carbon skeletons. Thus, in the series (363)-(365) the data take on a different hue, the relative rates being 1, 6×10^2 and 2×10^5, respectively [205c].

(363) (364) (365)

The rate effect of a 90° twist in the cyclopropyl conformation is therefore about 10^7 or so, which corresponds to roughly 10 kcal/mole. Interestingly, cyclopropylcarboxaldehyde also has the bisected conformation as shown by nmr [220].

Beside the orientation effect, there is another factor to be considered in evaluating the possibility of cyclopropyl assistance. If the ions to be expected are antihomoaromatic, no stabilization will result; or, to put in another way, the interaction will then be forbidden by the requirement of orbital symmetry conservation. To mention an example, tosylate (366) acetolyzes without cyclopropyl assistance to a mixture of (367) and (368).

(366) (367) (368)

Such participation in the present case would have required the symmetry forbidden [2 + 2] interaction, and the trishomocyclopentadienyl cation (369) [221].

(369)

In the deuterium labelled (370), scrambling indicates the formation of (371) rather than (372). In the tosylate (373), the heptahomotropylium ion (374) would be allowed, but deuterium labelling shows that it does not intervene (see p. 363) [222].

220. T. Yonezawa, H. Nakatsuji, and H. Kato, Bull. Chem. Soc. Japan, 39, 2788 (1966).

221. J.B. Lambert, F.R. Koeng, and A.P. Jovanovich, J. Org. Chem., 37, 374 (1972).

222. R.W. Thies, M. Sakai, D. Whalen, and S. Winstein, J. Amer. Chem. Soc., 94, 2270 (1972).

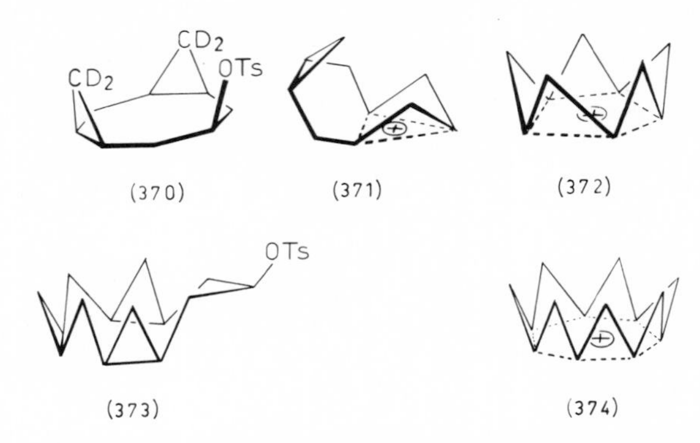

The physical properties of the ions can be studied at leisure in super acid solutions. While alkyl- and cycloalkyl ions have no uv absorption above 220 nm, cyclopropylcarbinyl ions have a band at 290 nm with $\epsilon_{max} \approx 10^4$ [223].

The nmr is extremely informative. Thus, the t-butyl ion has a methyl resonance at 6.2 τ; protons β to a carbonium ion center are shifted down by about 2.8 ppm. The protons in cyclopropane itself normally appear at about 10 τ. The cyclopropyldimethylcarbonium ion first of all shows two methyl resonances, at 7.4 and 6.8 τ; evidently at the temperature of the solution (-38°) the rotation of the cyclopropyl group is slow on the nmr time scale, and one of the methyl groups is cis and the other is trans to it. This is incontrovertable evidence for the bisected structure of the cyclopropylcarbinyl ion; the barrier to rotation is 13.7 kcal/mole [224].

The relatively high field for the methyl resonances shows that the carbinyl carbon atom is bearing less than a full electronic positive charge. Both the α- and β-protons on the ring are shifted down by about 3 ppm - evidence for the presence of considerable charge in the ring [225].

Further evidence to the effect that the charge is delocalized to the β-atoms at least as much as to C_α was found by Martin [226]. When triarylcarbinols are

---

223. G.A. Olah, C.U. Pittman, R. Waack, and M. Doran, J. Amer. Chem. Soc., 88, 1488 (1966).
224. D.S. Kabakoff and E. Namanworth, J. Amer. Chem. Soc., 92, 3234 (1970).
225. (a) G.A. Olah, M.B. Comisarow, C.A. Cupas, and C.U. Pittman, J. Amer. Chem. Soc., 87, 2997 (1965); (b) C.U. Pittman and G.A. Olah, J. Amer. Chem. Soc., 87, 2998 (1965); (c) N.C. Deno, J.S. Liu, J.O. Turner, D.N. Lincoln, and R.E. Fruit, J. Amer. Chem. Soc., 87, 3000 (1965); (d) N.C. Deno, H.G. Richey, J.S. Liu, D.N. Lincoln, and J.O. Turner, J. Amer. Chem. Soc., 87, 4533 (1965); (e) C.U. Pittman and G.A. Olah, J. Amer. Chem. Soc., 87, 5123 (1965).
226. T. Sharpe and J.C. Martin, J. Amer. Chem. Soc., 88, 1815 (1966).

dissolved in trifluoroacetic acid, the corresponding carbonium ions are generated; in the following structures the numbers indicate downfield shifts in Hz as a result of the ionization.

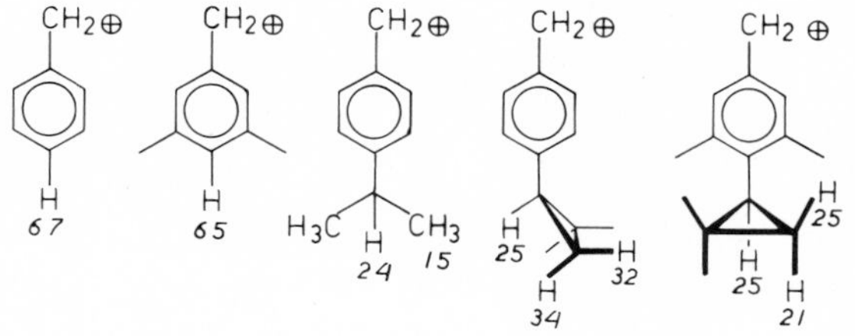

The parent cyclopropylcarbinyl ion was studied in detail by Olah [227]. In a mixture of antimony pentafluoride and fluorosulfuryl chloride at -80°, one finds a one-proton signal at 3.5 τ consisting of a pair of quartets with J values of 6.5 and 8.0 Hz, and a pair of three-proton doublets at 5.4 and 5.9 τ. This is clearly in agreement with either a tricyclobutonium (375) ion or equilibrating bicyclobutonium ions such as (376).

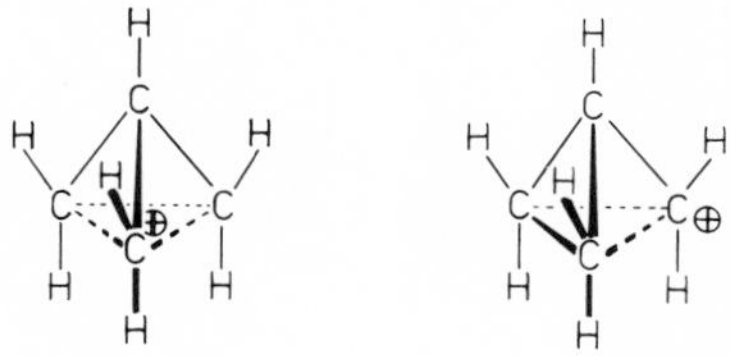

(375)

(376)

The methylene group ^{13}C shifts observed and those calculated on the basis of equilibrating classical ions disagree by a wide margin, and on that basis Olah ruled out the latter, but our inability to make a prediction for non-classical ions makes this conclusion somewhat less than satisfying.

An alternative approach has been published by Winstein [228], who compared the nmr spectra of various hydroxy substituted ions. He found that at low temperature hydroxyl protons resulting from the protonation of carbonyl groups exchange sufficiently slowly with the super acid medium so that their signals can be observed separately; thus, acetaldehyde gives rise to a pair of doublets at -4.8 and -5.1 τ in about a 5:1 ratio with coupling constants of 8.5 and 18.5 Hz, respectively. These are ascribed to syn- and anti-conformations, (377) and (378), respectively, on the basis of the Karplus rule.

---

227. G.A. Olah, D.P. Kelly, C.L. Jeuell, and R.D. Porter, J. Amer. Chem. Soc., 92, 2544 (1970).
228. M. Brookhart, G.C. Levy, and S. Winstein, J. Amer. Chem. Soc., 89, 1735 (1967).

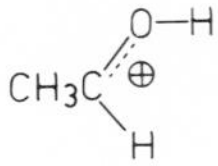

(377) 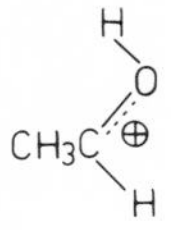 (378)

These chemical shifts are typical of simple aldehydes and ketones; however, when charge can be delocalized away from the hydroxyl group, upfield shifts occur. A number of these have been reproduced below; the data (in τ values) speak for themselves.

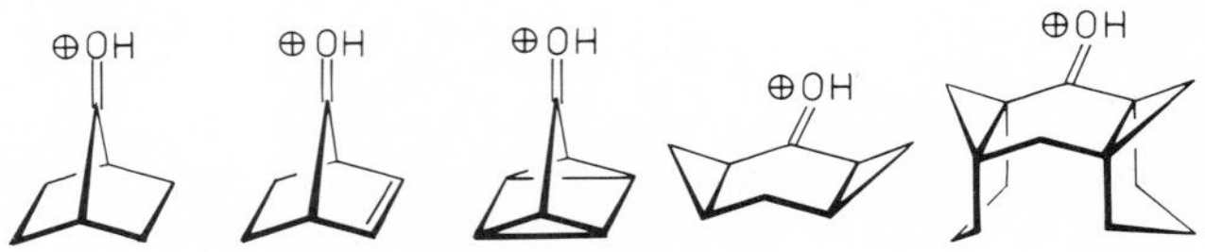

-4.3 -2.8 -2.2 -1.8 -1.0

-1.5 -0.8

Sigma participation by cyclobutyl groups is less dramatic than that by cyclopropyl rings, but it can still be quite significant as is clear from the following series of relative acetolysis rates: Those of (379)-(383) are in the ratio 0.06:3.6: 1.1:72,000:1.

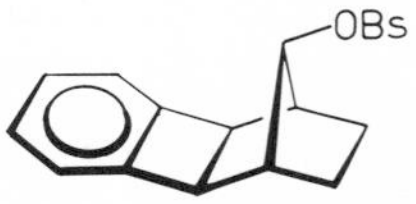

(379) 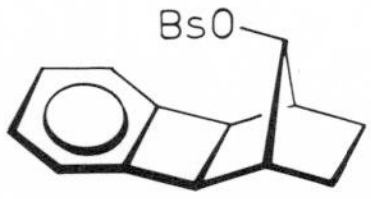 (380)

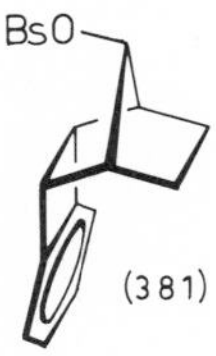

(381) 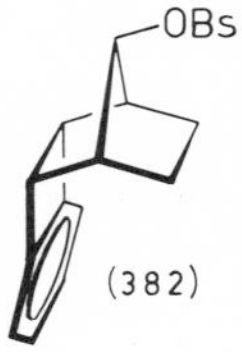 (382)  (383)

Rearrangement is extensive to exclusive in these reactions; thus (380) gives 80% (384) [229].

229. M.A. Battiste and J.W. Nebzydoski, J. Amer. Chem. Soc., 91, 6887 (1969) and 92, 4450 (1970).

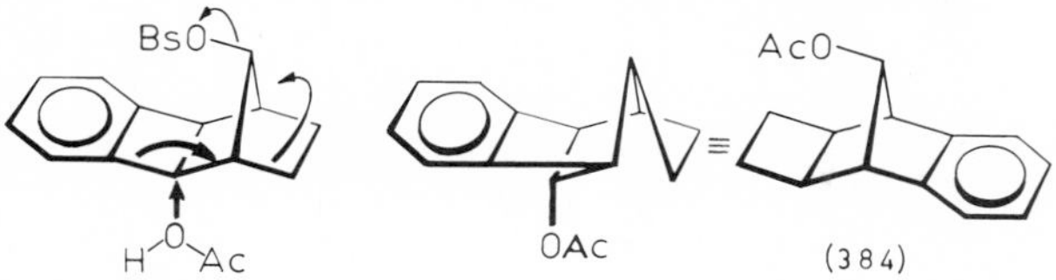

BsO AcO OAc H O Ac

(384)

Similarly, Winstein reported the acetolysis rates of (385)-(388) [230] to be in the ratio of 1:1.2:20,000:600,000;

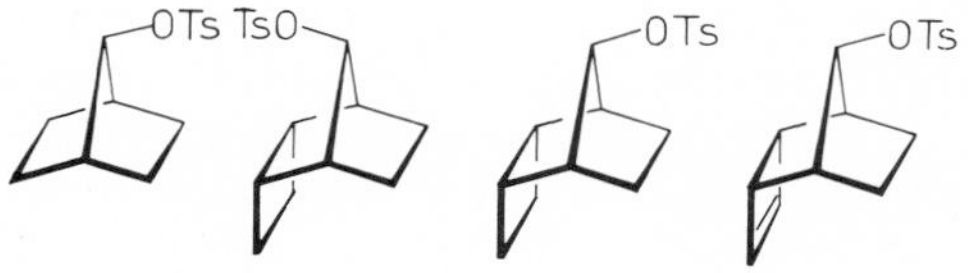

OTs TsO OTs OTs

(385) (386) (387) (388)

the reaction of (387) gives about equal amounts of (389) and (390).

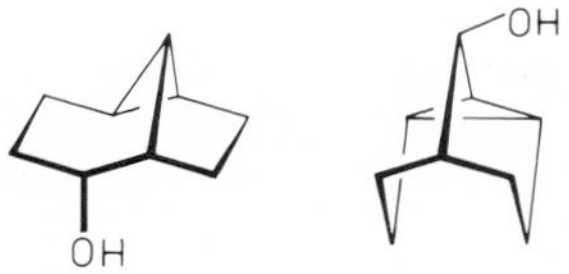

OH OH

(389) (390)

Beside the rearrangements already mentioned, small rings are a vital part of many carbonium ions subject to degenerate rearrangements [231]. As examples one might mention (391), which undergoes a five-fold degenerate rearrangement such that the inside substituent always remains inside - the Woodward-Hoffmann rules indeed require such 1,4-shifts to go with inversion.

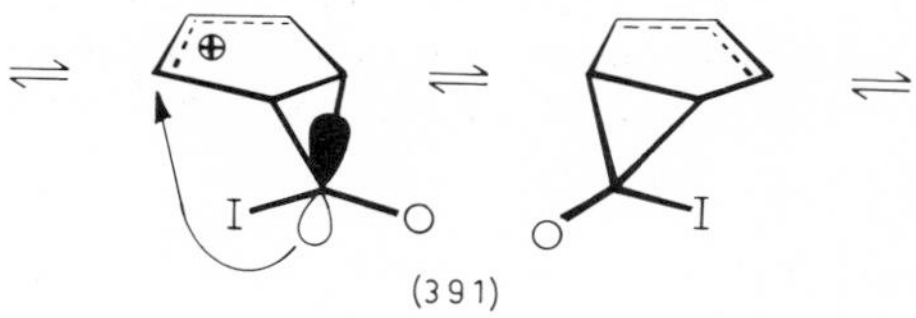

⊕ I O O I

(391)

Another is the barbaralyl ion (392) which is completely degenerate to the nmr even at -135° [232].

---

230. M. Sakai, A. Diaz, and S. Winstein, J. Amer. Chem. Soc., 92, 4452 (1970).
231. R.E. Leone and P. von R. Schleyer, Angew. Chem., Int. Ed. Engl., 9, 860 (1970).
232. See for example, (a) S. Yoneda, S. Winstein, and Z. Yoshida, Bull. Chem. Soc. Japan, 45, 2510 (1972); (b) R. Hoffmann, W.D. Stohrer, and M.J. Goldstein, Bull. Chem. Soc. Japan, 45, 2513 (1972).

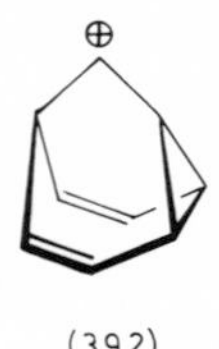

(392)

The cyclopropylcarbinyl ion is one of the few more controversial carbonium ions for which the isoelectronic borane is known. Many organic chemists had hopes at one time that the preparation of species such as (393) would prove very informative about the nature of the corresponding ions; apparently such compounds have proved elusive, however. Cyclopropyldimethylborane (394) is now known, but it has turned out a bit disappointing: Even at -100° there is no evidence of any restricted rotations, and the six methyl protons are still equivalent in a sharp singlet [233].

CH_3
B

B CH_3 CH_3

(393) (394)

20-11. Σ-PARTICIPATION: THE NORBORNYL ION [234]

We may now consider the question whether σ-participation can be involved in less highly strained systems. Although numerous studies could be cited, we shall limit our attention here chiefly to the 2-exo-norbornyl derivatives (395). This case is simplified by the fact that skeletal rearrangements cannot be a driving force for the reaction - any of the fast bond shifts that characterize carbonium ions merely regenerate the ion. The reason that this structure has attracted so much attention is that it is constructed more or less ideally for σ-participation; hence the case for this phenomenon more or less stands or falls with the evidence for it in the norbornyl derivatives.

Winstein's proposal is that the two σ electrons of the 1,6-bond assist in the ionization process and that the symmetrical ion (396) is formed as a result. The solvent finally attacks either the 1- or 2-position - again with inversion - to give an enantiomeric pair of products (397) [235].

---

233. A.H. Cowley and T.A. Furtsch, J. Amer. Chem. Soc., 91, 39 (1969).
234. For review, see: (a) G.E. Gream, Rev. Pure Appl. Chem., 16, 25 (1966); (b) Ref. 1e; (c) Ref. 134.
235. S. Winstein and D. Trifan, J. Amer. Chem. Soc., 74, 1147, 1154 (1952).

(395)

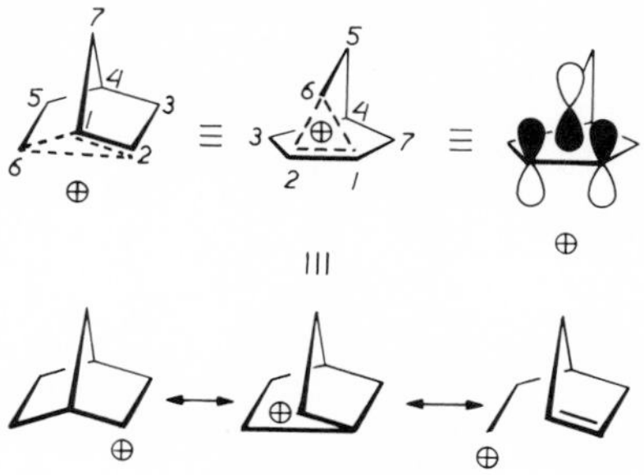

(396)

(397)

It may be noted here that several recent reviewers have ascribed the proposal to Wilson [236]. In his 1939 paper on the rearrangement of camphene hydrochloride (398) to isobornyl chloride (399) via the carbonium ion (400), Wilson casually mentioned that "it is possible that it is mesomeric between this and the corresponding isobornyl structure". No preference was expressed, no evidence was cited, and, in fact, it is presently believed that this ion is a classical one.

(398) 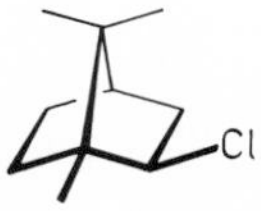 (399)  (400)

The following points - and counterpoints - are usually made in discussions of this proposal. As far as the stereochemistry is concerned, when resolved 2-*exo*-norbornyl esters are acetolyzed, the *exo*-acetate obtained is completely racemic [235].

This of course would be neatly accounted for in terms of a symmetrical, non-classical intermediate such as (396). The *endo*-substrate gives about 93% racemic *exo*-product, and so the classical 2-norbornyl ion formed without participation must partially rearrange. It should be realized, however, that there are other pathways by which racemization can be accomplished. Thus, rapid Wagner-Meerwein rearrangements would lead to racemization as well,

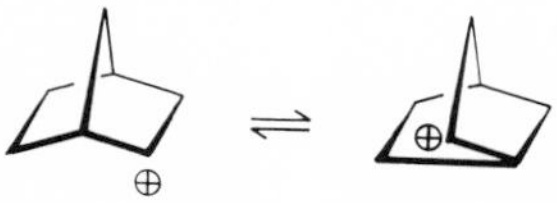

236. T.P. Nevell, E. de Salas, and C.L. Wilson, *J. Chem. Soc.*, 1188 (1939).

and so would 3 → 2 and 6 → 1 hydride shifts - all of these reactions are known to occur (see p. 780).

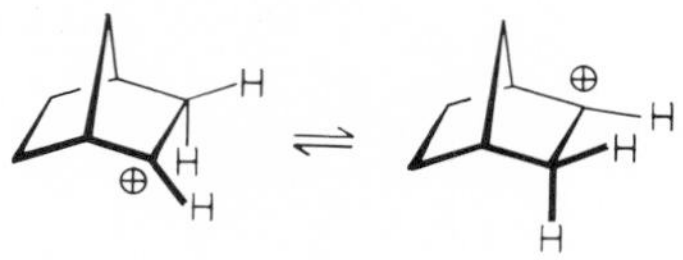

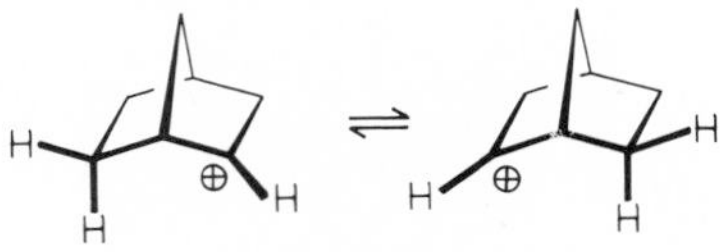

A second stereochemical feature is that the 2-norbornyl ion captures solvent virtually exclusively at the exo-side; only a few hundredths of a per cent is endo-product. Furthermore, the small amount (4%) of product due to elimination is about 98% tricyclane (401) and 2% norbornene (402). These observations are readily interpreted on the basis of a non-classical ion [237].

(401)  (402)

The product forming step is in a sense the microscopic reverse of the solvolysis step [238], and preferred solvent capture from the exo- side therefore goes hand in hand with an exo- solvolysis rate considerably greater than that of endo-solvolysis. The evidence on that score is reviewed further below (p. 774).

Isotope scrambling has been used also in these studies. A symmetrical ion requires that C_1 and C_2 become equivalent, as do C_3 and C_7. The solvolyses of 2,3-C^{14} - and of 2-H^3-norbornyl brosylates have been studied in great detail. It was learned that scrambling goes far beyond what would be required by a single symmetrical ion. The hydride shifts just mentioned must indeed be occurring; thus, if a C^{14}-label is originally present equally at the C_2- and C_3-positions, there is no way in which a simple non-classical ion can give rise to a product in which C_5 and C_6 are labelled. Yet this is the case. Roberts [239] found the data conveyed by the following scheme and noted the simplest way to explain these observations is to assume that the solvolysis goes 55% by way of a non-classical norbornyl ion

---

237. S. Winstein, E. Clippinger, R. Rowe, and E. Vogelfanger, J. Amer. Chem. Soc., 87, 376 (1965).
238. H. L. Goering and C. B. Schewene, J. Amer. Chem. Soc., 87, 3516 (1965).
239. J. D. Roberts, C. C. Lee, and W. H. Saunders, J. Amer. Chem. Soc., 76, 4501 (1954).

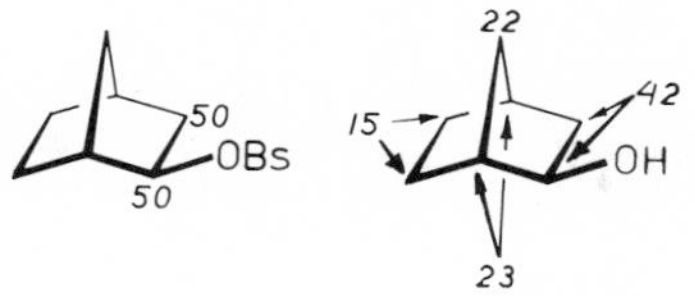

and 45% by way of a still more highly symmetrical "nortricyclonium ion" (403).

(403)

Lee's tritium experiment can be summarized by the following percentages [240]:

Process	2-exo-2H^3 (at 25°)	2-endo-2H^3 (at 115°)
S_N2	0	5
All carbons equivalent	10	5
C_1 and C_2 equivalent	45	43
C_1, C_2 and C_6 equivalent	45	47

The "all carbons equivalent" reaction clearly requires 3 → 2 as well as 6 → 2 hydride shifts.

An independent generation of the carbonium ion has led to contradicting results. Thus, hydrochloric acid addition to 2,3-dideuterionorbornene in pentane at -78° gives 38% of a mixture of (404) and (405), 50% of (406) and 12% of (407).

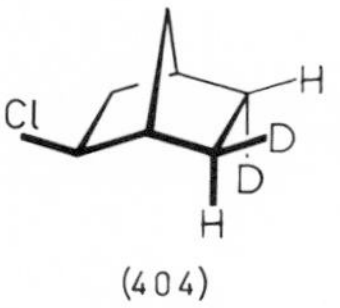

(404) 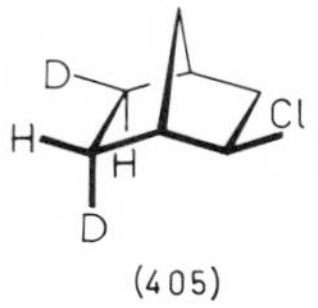 (405) 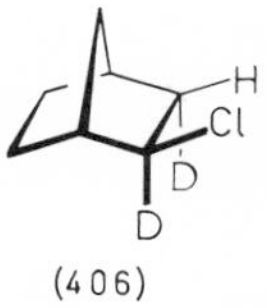 (406) 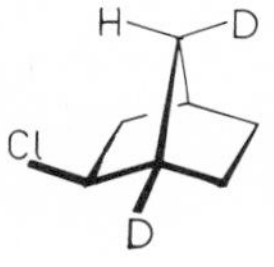 (407)

The first two products must arise from a 6 → 2 hydride shift; the last two should be formed in equal amounts if protonation to give a non-classical ion is the exclusive first step. This is evidently not so [241]. Brown has uncovered several such discrepancies. Thus, the addition of hydrochloric acid to the olefin (408) in ether at 0° leads to a mixture of (409) and (410) "corresponding to only 52-56% scrambling" if only a 1-2 minute exposure is allowed; 100% scrambling occurs only upon 2 hour exposure [242].

240. C.C. Lee and L.K.M. Lam, J. Amer. Chem. Soc., 88, 2831 (1966).
241. J.K. Stille, F.M. Sonnenberg, and T.H. Kinstle, J. Amer. Chem. Soc., 88, 4922 (1966).
242. H.C. Brown and K.T. Liu, J. Amer. Chem. Soc., 89, 466 (1967).

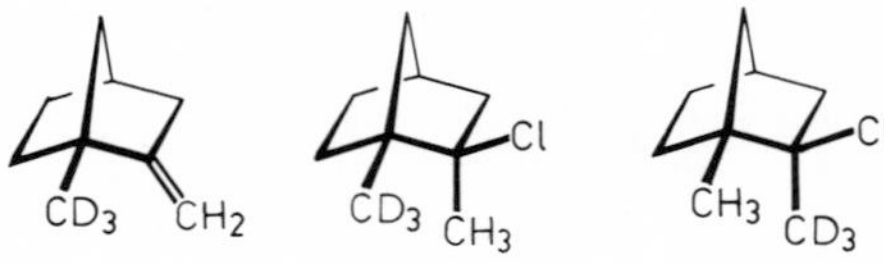

(408) (409) (410)

If deuterochloric acid is added to norbornene in methylene chloride at -78°, one obtains a mixture of (411)-(413) in the ratio of 60:34:6 [243].

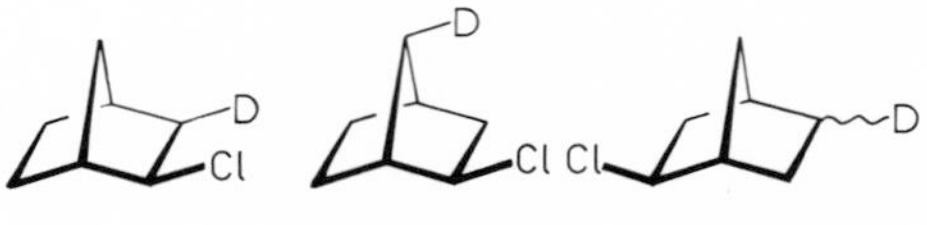

(411) (412) (413)

Even the trifluoroacetic acid addition does not always lead to a completely scrambled mixture [244]; (414)-(416) are obtained in the ratio 37:26:37.

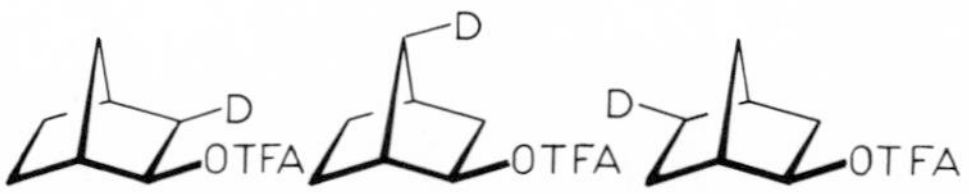

(414) (415) (416)

Clearly one cannot invoke a single pathway involving a symmetrical ion in any of these cases. In the last example mentioned, Brown pointed out that the addition of trifluoroacetic acid was 99.98% *exo* and 0.02% *endo*: This casts suspicion on the argument of high stereospecificity as well. One might counter that there could be a concerted (i.e., non-ionic) component in the reaction. Brown has argued that there should not be; thus, the addition of acetic acid to apobornene (417) under these conditions leads to a mixture of (418)-(421) in the ratio of 70:0.06:7:24 [245]. Thus, one obtains 70% unrearranged acetate in which an *exo*/*endo* product ratio of about 1000 is observed.

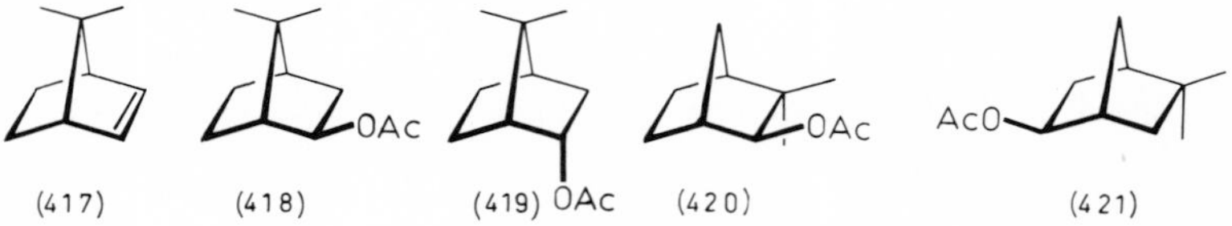

(417) (418) (419) (420) (421)

---

243. H.C. Brown and K.T. Liu, *J. Amer. Chem. Soc.*, *89*, 3900 (1967).
244. H.C. Brown, J.H. Kawakami, and K.T. Liu, *J. Amer. Chem. Soc.*, *92*, 5536 (1970).
245. H.C. Brown, J.H. Kawakami, and K.T. Liu, *J. Amer. Chem. Soc.*, *92*, 3816 (1970).

If molecular AcOH were added in a single, concerted step, one should expect that the syn-7-methyl group would hinder the exo-approach and in fact, reagents that do approach double bonds in concerted fashion (such as diimide) are hindered by the presence of a 7-syn-methyl group. Obviously, these experiments cast serious doubt on the bridged norbornyl ion [246].

Recently Cristol [247] has scrutinized the addition process of benzonorbornadiene. He noted that the addition of deuteriochloric acid in deuteroacetic acid gave a ratio of chloride to acetate product of 94:6, and that this ratio did not vary as the concentration of DCl was varied over a hundredfold range. The chloride product was completely scrambled; the acetate had only 40% 7-d. Cristol's explanation is that the acetic acid adds to some degree in a concerted fashion. While this result does not automatically explain all of Brown's results, it does show that cycloaddition is a possibility that should carefully be ruled out in each case.

We now turn to the question of rates. From our earlier discussion it will be clear that a carbonium ion center is stabilized by various groups; among these, the methyl group stabilizes the center inductively, and a phenyl group lowers its energy by charge delocalization. One should expect that a center that is already stabilized in some way cannot take the same degree of advantage of such substitution, and hence the effect of methyl or phenyl substitution has often been used to detect such stabilization. Thus, methyl and phenyl substitution increase the solvolysis rates of i-propyl derivatives by about 5×10^4 and 3×10^8, respectively, but of benzhydryl substrates by only about 350 and by 10^4. In exo-2-norbornyl chloride, these numbers are 6×10^4 and 3×10^8 [248].

Three points need to be considered to strengthen this argument. First, it would seem possible that whereas the secondary norbornyl ions are non-classical in nature, the tertiary ions obtained by substitution are not and therefore the two should not be compared. For example, Goering has shown that unlike its parent, resolved 1,2-dimethyl-2-norbornyl esters (422) hydrolyze to give only partially racemic products ((423)-(425)), so that the intermediate ion cannot have been the symmetrical (426) [249].

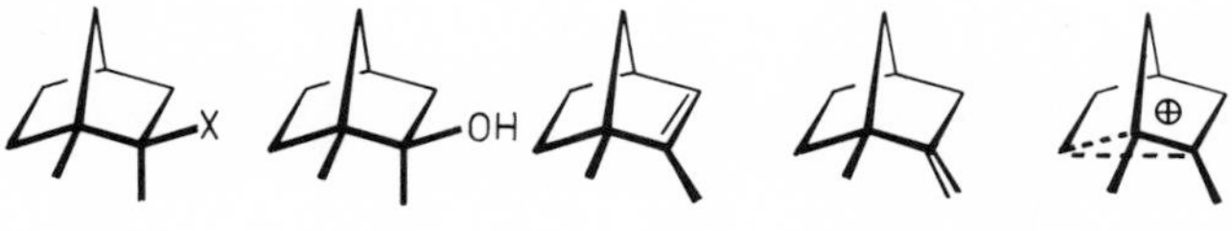

X OH ⊕

(422) (423) (424) (425) (426)

---

246. (a) H.C. Brown and K.T. Liu, J. Amer. Chem. Soc., 92, 200 (1970) and 93, 7335 (1971); (b) H.C. Brown and J.H. Kawakami, J. Amer. Chem. Soc., 92, 201 (1970). See also (c) W.C. Baird and J.H. Surridge, J. Org. Chem., 37, 1182 (1972).
247. S.J. Cristol and J.M. Sullivan, J. Amer. Chem. Soc., 93, 1967 (1971).
248. H.C. Brown and M.H. Rei, J. Amer. Chem. Soc., 86, 5008 (1964).
249. (a) H.L. Goering and K. Humski, J. Amer. Chem. Soc., 90, 6213 (1968); (b) H.L. Goering and J.V. Clevenger, J. Amer. Chem. Soc., 94, 1010 (1972).

However, one can then extrapolate this argument by saying that if the tertiary ions are classical and hence form without sigma participation, then the various other criteria that have been used to argue the non-classical nature of the secondary ions should fail here. But Brown observed that the solvolysis of the dimethyl system and its parent have amazingly similar energy profiles [250].

Secondly, it is argued that i-propyl esters do not solvolyze by a limiting pathway, and that substrates such as 2-adamantyl would be more suitable. The effects of 2-methyl and 2-phenyl substituents in that case are indeed several orders of magnitude larger (see p. 697).

The third objection is that the argument itself is not a good one, that the introduction of a substituent may affect not only the charge density at the carbonium ion center but that it may have additional influences (such as steric and conformational effects) that make the numbers of 5×10^4 and 3×10^8 unreliable guide lines [1e].

More distant groups can be expected to have much smaller effects. Thus, 1-substituents have been studied and found to lead to rate increases of 50 for methyl, 4 for phenyl, 46 for amino and 4 for carboxylate [251]; however, while these rate changes suggest that participation is occurring, this argument is open to the objection that the participation may not be of the sigma type (for instance, n-participation by carboxylate), or that its driving force is rearrangement to an ion stabilized by direct substitution (427) rather than non-classical delocalization.

(427)

If bridging imparts a favorable state of low energy on the norbornyl ion, reactions leading to this ion should be unusually rapid when compared to a suitable standard. Since anyone is free to choose his own model as suitable, one might expect much controversy here. The facts do not disappoint us in this regard.

Winstein chose the endo-epimer as his standard. When the rates of production of benzenesulfonic acids from both isomeric esters in acetic acid are compared, one finds that exo-benzenesulfonates have rates about 350 times faster than their epimers [235]. If racemization rates are compared, one observes that internal return occurs in the exo-substrates and that correction for this leads to an exo/endo rate ratio of about 1600 [237]. A high ratio is indicative of either a low energy of the exo-transition state, or of a high energy for the endo-activated complex. Thus, Winstein chose the former. Brown then observed that the titrimetric

---

250. H.C. Brown and M.H. Rei, J. Amer. Chem. Soc., 90, 6216 (1968).
251. (a) H.C. Brown, F.J. Chloupek, and M.H. Rei, J. Amer. Chem. Soc., 86, 1246 (1964); (b) J.W. Wilt and W.J. Wagner, J. Amer. Chem. Soc., 90, 6135 (1968).

exo/endo rate ratios for the 2-methyl- and 2-phenyl substituted norbornyl substrates are 83 and 143, respectively - values not that much less than that of the parent [252]. Similarly, the ratios in the series (428) with Y is methoxy, hydrogen, trifluoromethyl, and nitro are 280, 143, 188 and 114, respectively [253].

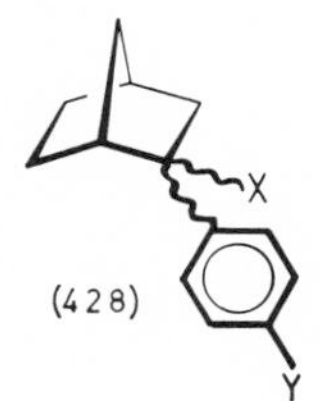

(428)

When certain stabilizing groups are added in the 5, 6 position the exo/endo ratios increase; thus, the ratios for the brosylates (429)-(432) are 350, 1400, 6000, and 8000, respectively [254].

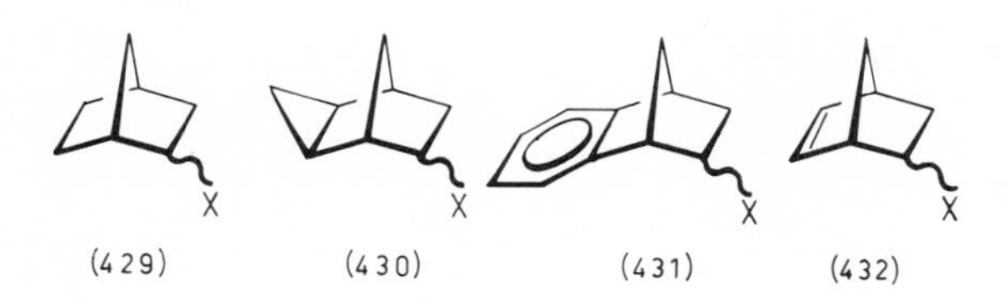

(429) (430) (431) (432)

Brown then showed that the ratio of the benzonorbornenyl substrate does not change significantly upon the further addition of a 2-methyl or 2-phenyl substituent [255]. Only when the effect of substituents in the phenyl ring of the benzonorbornenyl substrate on the exo/endo rate ratio was then measured did it become clear that participation was occurring and that the mere use of 2-substituents could be misleading; in the compounds (433)-(435), the ratio is 3.2×10^5, 1.6×10^3 and 200, respectively [256].

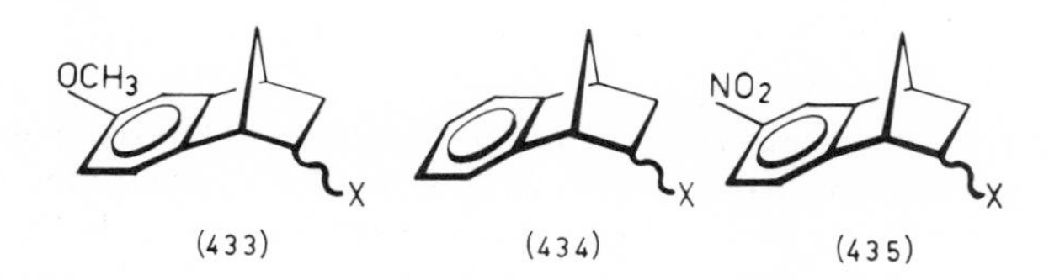

(433) (434) (435)

However, with the introduction of the benzo group we have changed the nature of the participation essentially from σ to π.

---

252. H. C. Brown, F. J. Chloupek, and M. H. Rei, J. Amer. Chem. Soc., 86, 1248 (1964).
253. K. Takeuchi and H. C. Brown, J. Amer. Chem. Soc., 90, 2693 (1968).
254. A. K. Colter and R. C. Musso, J. Org. Chem., 30, 2462 (1965).
255. H. C. Brown and G. L. Tritle, J. Amer. Chem. Soc., 88, 1320 (1966).
256. (a) H. Tanida, H. Ishitobi, and T. Irie, J. Amer. Chem. Soc., 90, 2688 (1968); (b) H. C. Brown and G. L. Tritle, J. Amer. Chem. Soc., 90, 2689 (1968).

A number of chemists have tried to introduce substituents in various remote parts of the norbornyl molecules with the objective of interfering with sigma participation; a reduced exo/endo rate ratio should result. The following examples should be mentioned. The introduction of a trimethylene bridge in the exo-exo-5, 6-positions as in (436) should make it more difficult for the 5, 6-carbon bridge to assume a symmetrical position between the 1- and 2- carbon atoms as in (437). Indeed the ratio is now reduced to 11 [257].

(436) 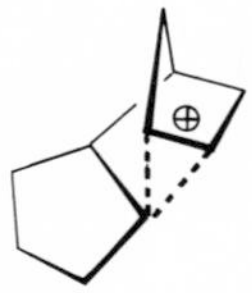 (437)

Exactly the same argument applies to an exo-4, 5-trimethylene bridge as in (438); in that case the ratio is reduced to 8 [258].

(438)

Still another case is (439), which has a rate ratio of 20 [259].

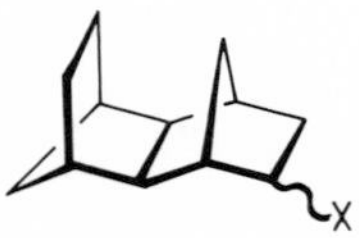

(439)

If a substituent is present in the 3-position which can stabilize positive charge, the reduced need for sigma participation should likewise decrease the exo/endo ratio. Wilcox and Jesaitis have uncovered two examples [260]. In the cyclopropyl substituted molecule (440), the acetolysis rate ratio is down to 3, and in the

257. (a) K. Takeuchi, T. Oshika, and Y. Koga, Bull. Chem. Soc. Japan, 38, 1318 (1965); see also (b) R. Baker and T. J. Mason, J. Chem. Soc., Perkin II, 18 (1972) and J. Chem. Soc. (B), 1144 (1971).
258. E. J. Corey and R. S. Glass, J. Amer. Chem. Soc., 89, 2600 (1967).
259. (a) L. de Vries and S. Winstein, J. Amer. Chem. Soc., 82, 5363 (1960); see also (b) R. K. Howe, P. Carter, and S. Winstein, J. Org. Chem., 37, 1473 (1972).
260. (a) C. F. Wilcox and R. G. Jesaitis, Tetrahedron Lett., 2567 (1967); (b) C. F. Wilcox and R. G. Jesaitis, Chem. Commun., 1046 (1967).

allylic system (441) it is down to 4. Finally, in the 3-psinortricyclyl ion (442) it is down to 0.64 [261].

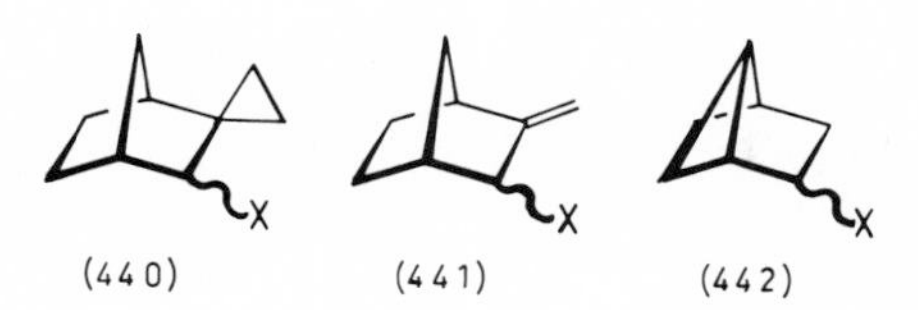

(440) (441) (442)

By the same token, carbonium ion destabilizing groups should lower the exo/endo rate ratio if bound to carbon atoms C_5 and C_7. For (443), the ratio has been found to be 40 [262]; for (444), it has been found to be 0.4 [263]; and for (445), it was down to 0.16 [264].

O OBs CH3OOC OBs O OTs

(443) (444) (445)

The exo-tosylate give much endo-product in this case, and the elimination (8%) proceeds to give tricyclone (446) and 7-norbornenone (447) in about equal amounts.

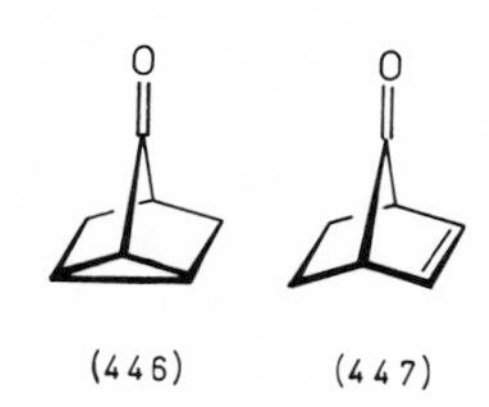

(446) (447)

It may be recalled that the parent substrate gave tricyclane and norbornene in a 50:1 ratio. This reaction seems to be somewhat different from the other 2-norbornyl solvolyses - perhaps the deactivation by the 7-oxo group has produced non-limiting behavior. In contrast to these data, it has been found that the exo/endo ratio is virtually unaffected by syn- and anti-chloro substituents [265].

---

261. R.K. Lustgarten, J. Amer. Chem. Soc., 93, 1275 (1971).
262. J.C. Greever and D.E. Gwynn, Tetrahedron Lett., 813 (1969).
263. G.W. Oxer and D. Wege, Tetrahedron Lett., 457 (1971).
264. (a) P.G. Gassman and J.L. Marshall, Tetrahedron Lett., 4073 (1965); (b) P.G. Gassman and J.L. Marshall, J. Amer. Chem. Soc., 87, 4648 (1965) and 88, 2822 (1966).
265. P.G. Gassman and J.M. Hornback, J. Amer. Chem. Soc., 94, 7010 (1972).

There is one result involving a 2-substituent that stands in stark contrast with Brown's data. Traylor [266] studied the exchange of methanol-d_4 with norcamphor dimethyl ketal (448) under the catalytic effect of a trace of acid to give initially (449) and (450).

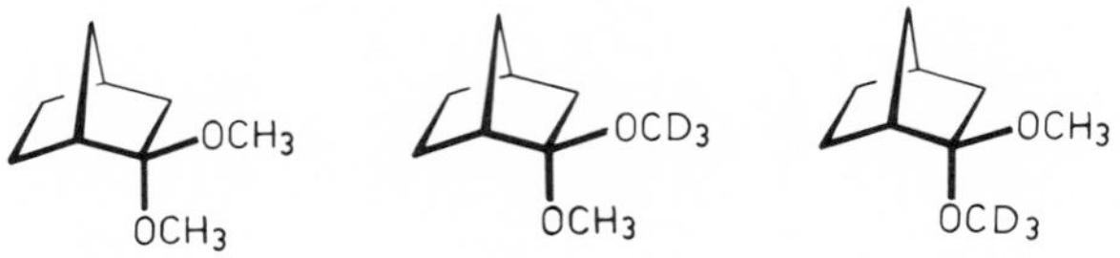

(448) (449) (450)

Since the capture of methanol by the 2-methoxynorbornyl ion is simply the reverse of acid catalyzed ionization, the ionization rate ratio is equal to the exchange ratio (*exo*/*endo*), which was found to be equal to 16. In the presence of borohydride ion the ketal forms *exo*- and *endo*-methyl norbornyl ethers in a 1:16 ratio; *exo* capture of this anion is also 16 times gaster than *endo*. While those facts would seem to support the non-classical ion, they are subject to the same objections that are often made to explain why 2-methyl or 2-phenyl substituents have no major effect on *exo*/*endo* rate ratios.

Isotope effects have uniformly been claimed to support participation; however, this support is somewhat weakened by the fact that here our experience and theoretical insight are not so great that we know with certainty exactly what to expect. We often have to rely on the reasoning that if unusual isotope effects on the rate constant are observed, this indicates that the reaction is unusual in some way - perhaps because of participation.

One result of extraordinary interest has been obtained however, namely that of 6-D substitution. Ordinarily γ-isotope effects on rates are far too small to measure, but in the series of brosylates (451)-(454) the k_H/k_D ratio was found to be 0.99, 0.98, 1.11, and 1.09, respectively, experimental limits in all cases being 0.01 [267].

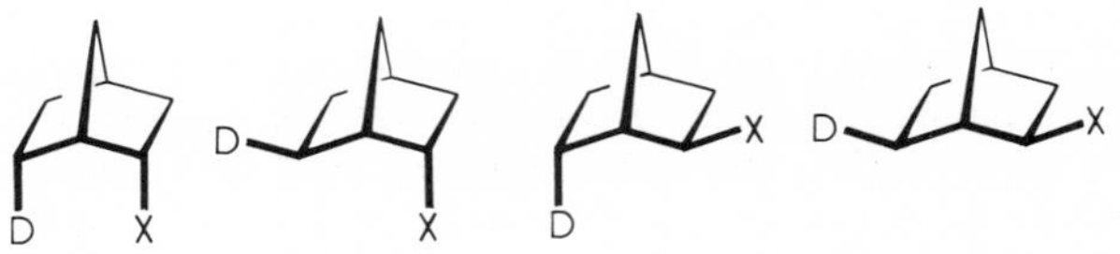

(451) (452) (453) (454)

The large k_H/k_D ratio is observed regardless whether the 6-D atom is *exo* or *endo*, but only when the leaving group is *exo*! Regardless of one's theoretical understanding of isotope effects, the most obvious explanation for the effect is

266. T.G. Traylor and C.L. Perrin, *J. Amer. Chem. Soc.*, *88*, 4934 (1966).
267. (a) B.L. Murr, A. Nickon, T.D. Swartz, and N.H. Werstiuk, *J. Amer. Chem. Soc.*, *89*, 1730 (1967); (b) J.M. Jerkunica, S. Borcic, and D.E. Sunko, *J. Amer. Chem. Soc.*, *89*, 1732 (1967).

bonding between C6 and C2, so that we are dealing with a β- rather than a γ-isotope effect. Goering [268] later showed that 6,6-dideuteration had no such effect on 1,2-dimethylnorbornyl esters (455); this fact supports his stereochemical results according to which the 1,2-dimethylnorbornyl ion is classical.

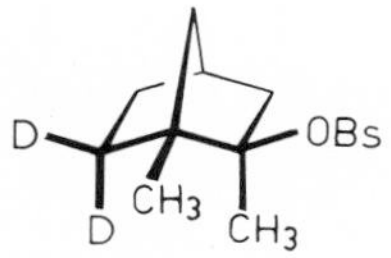

(455)

A great variety of less routine kinetic experiments have been applied to the problem, and we review some of them here.

Charge delocalization leads to a loosened solvent shell, and the volume of activation may consequently be expected to be less negative than is ordinarily the case in unassisted ionization reactions. This effect has indeed been found in several instances; thus, the activation volumes for 2-exo- and endo-norbornyl brosylate and for cyclopentyl brosylate are -14.3, -17.8 and -17.7 cm^3/mole, respectively [269].

A remarkable series of relative rate constants was reported by Tanida, who measured the nitration rates at the α- and β-positions in the benzocycloalkane systems (456)-(461). The ratio k_β/k_α in the first five of these vary from 0.72 to 1.50, but for (461) it equals 3.63, suggesting a transition state leading toward (462) in that case. Similarly, when the solvolysis rates of (463)-(466) are compared, the ratio between the latter two is 3.5 times greater than that of the former pair [270].

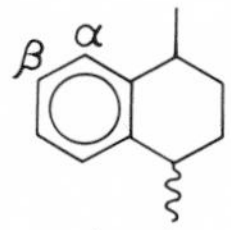

(456) 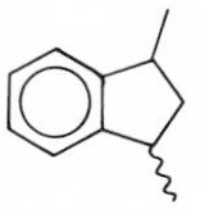 (457) 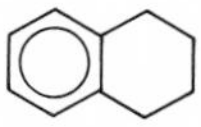 (458) 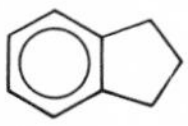 (459)  (460) 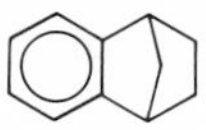 (461) 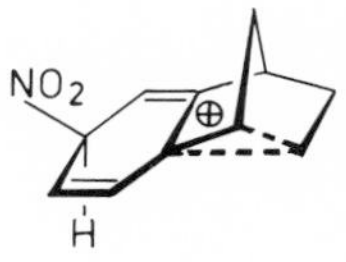 (462)

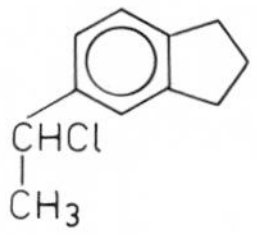

(463) 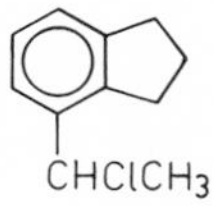 (464) 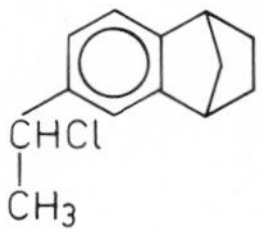 (465) 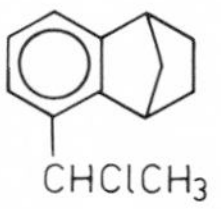 (466)

268. H. L. Goering and K. Humski, J. Amer. Chem. Soc., 91, 4594 (1969).
269. (a) W. J. le Noble and B. L. Yates, J. Amer. Chem. Soc., 87, 3515 (1965); (b) Ref. 71.
270. H. Tanida and R. Muneyuki, J. Amer. Chem. Soc., 87, 4794 (1965).

The hydride shifts already referred to provide a fascinating insight into the carbonium ion. Collins [271] found that the diol (467) undergoes the pinacol rearrangement in sulfuric acid at 0° to give only the endo-phenyl ketone (468). This presumably indicates a pathway involving, in succession, ions (469)-(472).

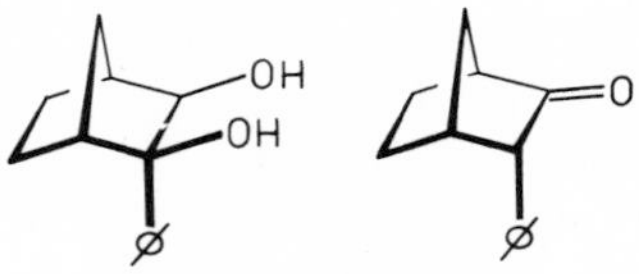

(467) (468)

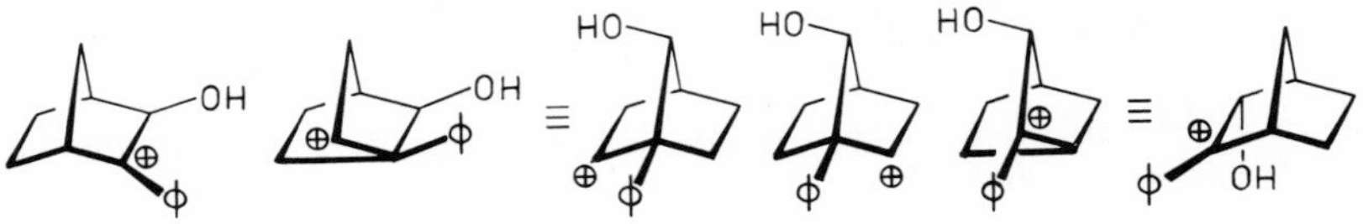

(469) (470) (471) (472)

While the relative configurations of the diol and ketone are not known and though other paths are conceivable, the 3 → 2 shift that must occur evidently only takes place on the exo- side.

Similarly, Berson observed that the acetolysis of optically active brosylate (473) gave (474) as its product, and none of its enantiomer (475) [272].

(473) (474) (475)

A simple endo-3,2-shift would have sufficed to produce (475); instead, the ion goes through two Wagner-Meerwein shifts, a 6,2-shift and an exo-3,2-shift to reach the product stage. Evidently, if needs be the secondary norbornyl ion will choose a highly circuitous route to rearrange to the tertiary isomer.

The 6,2-hydride shifts occur exclusively at the endo-side of the molecule, as becomes clear, for instance, from the fact that dideuteriodiol (476) upon dehydration gives product (477) [273].

---

271. C.J. Collins, Z.K. Cheema, R.G. Werth, and B.M. Benjamin, J. Amer. Chem. Soc., 86, 4913 (1964).
272. (a) J.A. Berson, J.H. Hammons, A.W. McRowe, R.G. Bergman, A. Remanick, and D. Houston, J. Amer. Chem. Soc., 87, 3248 (1965). For an exception, see (b) P. Wilder and W.C. Hsieh, J. Org. Chem., 36, 2552 (1971).
273. B.M. Benjamin and C.J. Collins, J. Amer. Chem. Soc., 88, 1556 (1966).

(476) (477)

These stereospecific shifts as in (478) are often quoted as evidence for a non-classical norbornyl ion.

(478)

There is independent evidence that hydride shifts can occur only if the C-H bond and the p orbital are lined up reasonably neatly [274]. For example, 1,2- and 2,1-hydride shifts on the adamantane nucleus have to occur across 90° and 60° dihedral angles, as in (479) and (480), respectively; they are strongly inhibited. For example, heating chlorotrideuterioadamantane (481) for an hour at 100° in the presence of antimony pentachloride led to no scrambling at all [275]!

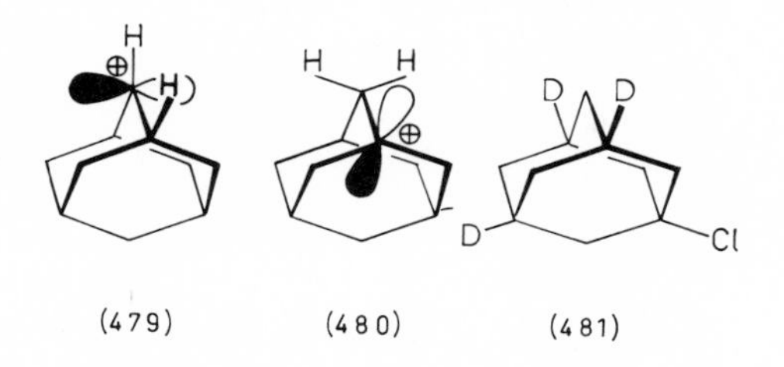

(479) (480) (481)

Shifts involving the adamantyl nucleus do occur, but only in intermolecular fashion. Methyl shifts can occur intramolecularly, but only via Wagner-Meerwein shifts; the C-Me bond never breaks as labelling experiments show [276]. The reaction is conceivable via ions (482)-(485).

---

274. P. von R. Schleyer, Angew. Chem., Int. Ed. Engl., 8, 529 (1969).
275. P. Vogel, M. Saunders, W. Thielecke and P. von R. Schleyer, Tetrahedron Lett., 1429 (1971).
276. Z. Majerski, P. von R. Schleyer, and A. P. Wolf, J. Amer. Chem. Soc., 92, 5731 (1970).

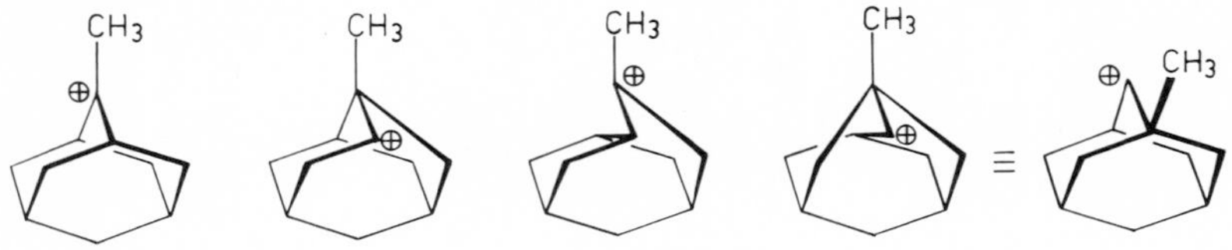

(482) (483) (484) (485)

An interesting test was devised by Bartlett [277]. Since carbonium ions and radicals are ordinarily both planar, if exo/endo rate ratios are to be ascribed to steric effects, they should also be observed in radical fission reactions. The fragmentation rates of (486) and (487) are in the ratio of only 4, however.

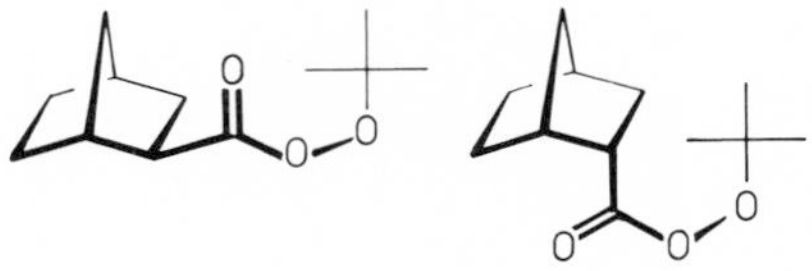

(486) (487)

Another approach to the norbornyl ion has been to reduce the time available to the norbornyl ion to undergo rearrangements by the use of a counterion which returns with a highly nucleophilic atom in such a way that the ionization can be demonstrated to have taken place. The isomerization of thiocyanates (488) to isothiocyanates (489) via ionization is an example.

R−S−C≡N R−N=C=S

(488) (489)

When 2-exo-norbornyl thiocyanate is heated to 150° in sulfolane, such isomerization takes place. Very little exchange with radiosulfur labelled isothiocyanate ion occurs, so that the counterion does not move very far away (only "intimate ion-pairs" are involved). If the 2-deuteriocompound is used, the label appears in the product only at C_1 and C_2; none appears at C_3 or C_6. Thus, although no hydride shifts have had time to take place, the norbornyl ion has fully equilibrated in the Wagner-Meerwein sense. By the same token, resolved thiocyanate gives only completely racemic isomer; this also rules out concerted, non-ionic isomerization (490) [278].

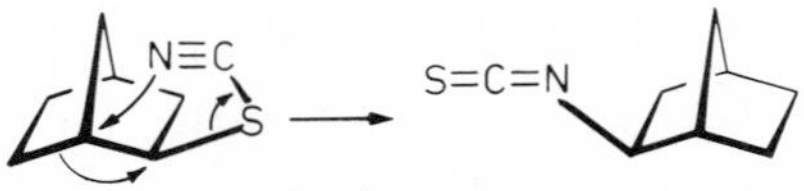

(490)

277. P.D. Bartlett and J.M. McBride, J. Amer. Chem. Soc., 87, 1727 (1965).
278. L.A. Spurlock and T.E. Parks, J. Amer. Chem. Soc., 92, 1279 (1970).

Oddly enough, 2-Δ^3-cyclopentenylethyl thiocyanate (491) isomerizes without skeletal reorganization to the isothiocyanate (492) [279].

SCN NCS

(491) (492)

Evidently, even though bonding to the sp^2-carbon atoms has already begun, it is not complete in the ion-pair stage, and a suitably located thiocyanate ion can extricate the original framework.

The results of another such study have been presented by Corey [280], who carried out the isomerization and the racemization of optically active 2-exo-norbornyl-m-carboxylphenyl benzenesulfonate (493) in basic medium.

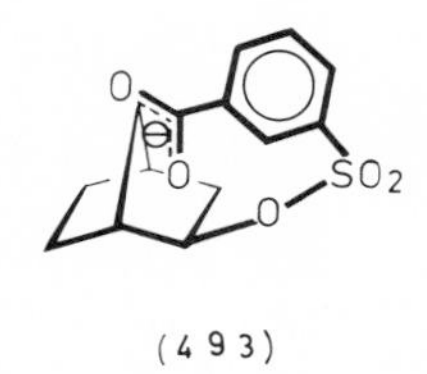

(493)

The nucleophilicity and proximity of the carboxylate group should give the 2-norbornyl ion precious little opportunity to equilibrate; nevertheless, racemization is complete in the isomerization process.

A similar study of the racemization of optically active 2-norbornyl chloride catalyzed by radioactive mercuric chloride revealed that exchange occurred with only one of the two chlorine atoms: The exchange rate is half as great as the racemization rate. Thus the intermediate is (494) rather than (495) for which the ratio would be 0.67, or (496) for which it would be zero [281].

(494) (495) (496)

279. L.A. Spurlock and W.G. Cox, J. Amer. Chem. Soc., 91, 2961 (1969).

280. E.J. Corey, J. Casanova, P.A. Vatakencherry, and R. Winter, J. Amer. Chem. Soc., 85, 169 (1963).

281. (a) J.P. Hardy, A.F. Diaz, and S. Winstein, J. Amer. Chem. Soc., 94, 2363 (1972); (b) J.P. Hardy, A. Ceccon, A.F. Diaz, and S. Winstein, J. Amer. Chem. Soc., 94, 1356 (1972).

The possibility of alternative generation of the norbornyl ion may be considered here. The π route has already been considered (see p. 726). We are left with a puzzle here: Goering's experiments with optically active 1,2-dimethylnorbornyl esters show that the corresponding ion is classical, Bartlett's methyl substitution experiments in the π approach to the ion seem to say that it is symmetrical. While these results are not absolutely irreconcilable, they are a good example of the difficulty of finding a simple solution that adequately explains all known data in this field.

Another σ-route is available via 2-norpinyl derivatives (497); it was found that unrearranged, retained (498) was formed as well as endo- and exo-norbornyl ethers. It was proposed that the first two products arose from non-classical ion (499), and that the rearrangement to the 2-norbornyl ion was responsible for the exo- product.

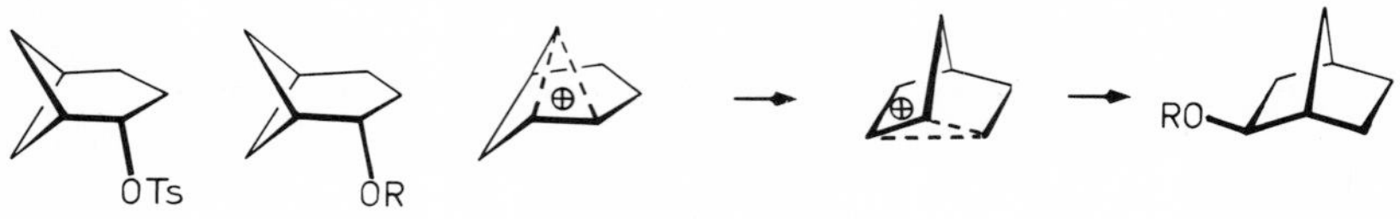

OTs OR RO

(497) (498) (499)

This material becomes the main product in solvents of low nucleophilicity such as 2,2,2-trifluoroethanol [282]. The opposite of this type of reaction has also been observed; brosylate (500) acetolyzes to give (501)-(503), whereas (504) gives only (501) [283].

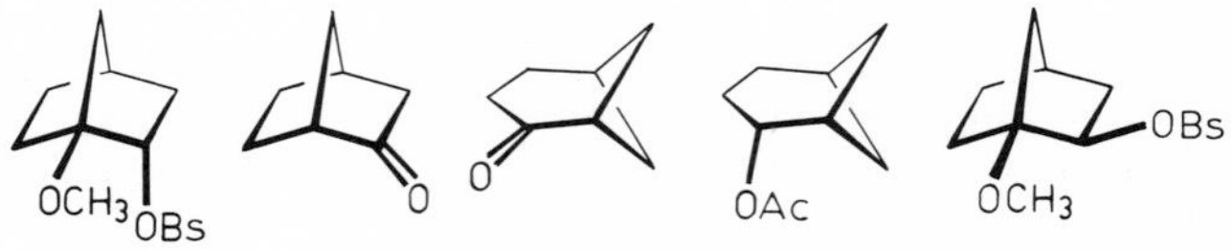

OCH_3 OBs O O OAc OBs OCH_3

(500) (501) (502) (503) (504)

The leaving group would ordinarily not be expected to make much difference on the nature of the ion, but the deamination of amines - which finally leads to nitrogen as the leaving group - may be exceptional. Thus, optically active 2-exo-norbornylamine upon deamination gives a substantial amount of endo-products and it does not lead to complete racemization [284]. Thus, the products must arise from a different intermediate, and a classical 2-norbornyl ion is the simplest possibility. Several other possibilities have been considered; the "hot", classical carbonium ion which is different from normal ions because of the exothermic formation step (nitrogen being an excellent leaving group), and a diazonium ion subject

282. W. Kirmse and R. Siegfried, J. Amer. Chem. Soc., 90, 6564 (1968).
283. Y. Lin and A. Nickon, J. Amer. Chem. Soc., 92, 3496 (1970).
284. E.J. Corey, J. Casanova, P.A. Vatakencherry, and R. Winter, J. Amer. Chem. Soc., 85, 169 (1963).

to partial intervention of a S_N2 displacement reaction may be mentioned among them [285]. Collins recently interpreted the results of a labelling study (much too complex for a full presentation here) to mean that the 2-norbornyl ions derived from amines retained their classical characteristics even after they had suffered a hydride shift and Wagner-Meerwein rearrangements [286]. Needless to say, these observations have contributed their share to obscure an already murky subject.

Many 2-norbornyl derivatives can be dissolved in super acid media to give stable solutions of the norbornyl ion. Unlike the precursor neutral molecules which have highly complex nmr spectra, the spectrum of the ion [287] at room temperature shows only a single sharp peak at 6.5 τ. Some process or processes must therefore be occurring by means of which all hydrogen atoms become equivalent. The 3,2- and 6,1-hydride shifts and the Wagner-Meerwein rearrangement or non-classical resonance provide the necessary pathways; it is a simple exercise to show that all eleven hydrogen atoms can reach any position in the ion.

When such solutions are cooled down to -60°, the spectrum resolves into three bands at 4.6, 6.9 and 7.8 τ with relative areas of 4, 1 and 6, respectively. Evidently the 3,2 hydride shifts have become slow at this temperature, but the other processes are still too rapid [288] to lend to further resolution of the signals. The 6 → 2 shift can be "frozen out" at -154°; the signal of area 4 splits into a pair of two each (at 3 and 7 τ), and broadening simultaneously occurs in the band of area 6. Thus, either the Wagner-Meerwein rearrangement is still rapid even at this temperature (which means that it has an activation energy of no more than 5 kcal/mole), or the ion is non-classical (no activation energy at all) [289]. Upon further cooling one is thwarted by the inevitable freezing of the solution.

Winstein has pointed out the similarities shown below between the 2-norbornyl and 7-norbornenyl ions, both in structure and in nmr spectra. Thus, if there is a good case for the non-classical nature of the one, the same conclusion is likely to apply to the other [290].

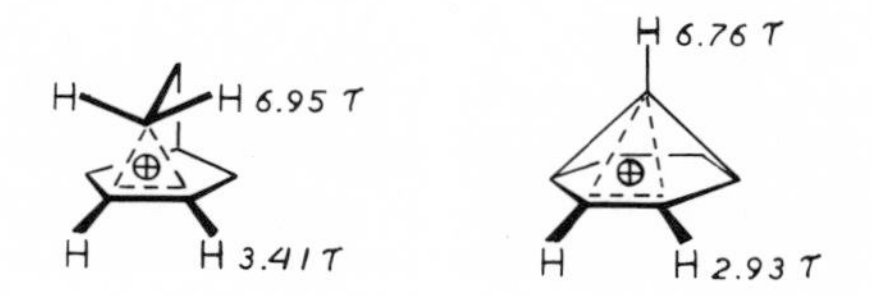

285. C.J. Collins and B.M. Benjamin, J. Amer. Chem. Soc., 92, 3182 (1970).
286. V.F. Raaen, B.M. Benjamin, and C.J. Collins, Tetrahedron Lett., 2613 (1971).
287. P. von R. Schleyer, W.E. Watts, R.C. Fort, M.B. Comisarov, and G.A. Olah, J. Amer. Chem. Soc., 86, 5679 (1964).
288. (a) M. Saunders, P. von R. Schleyer, and G.A. Olah, J. Amer. Chem. Soc., 86, 5680 (1964); (b) F.R. Jensen and B.H. Beck, Tetrahedron Lett., 4287 (1966).
289. G.A. Olah and A.M. White, J. Amer. Chem. Soc., 91, 3956 (1969).
290. R.K. Lustgarten, M. Brookhart, S. Winstein, P.G. Gassman, D.S. Patton, H.G. Richey, and J.D. Nichols, Tetrahedron Lett., 1699 (1970).

Stabilizing substituents reduce the need for bridging and are likely to lead to equilibrating ions: Examples of this have been proved by nmr. Thus, Schleyer showed in 1963 [291] that the 1,2-di-p-anisyl-norbornyl ion below -80° has two aryl signals in the nmr spectrum. The phenanthrene derivative (505) dissolves in fluorosulfonic acid to give the equilibrating ions (506), recognizable by a uv spectrum very similar to that of 9-protonated phenanthrene and by the broadening of all proton signals in the nmr save that of the bridgehead proton at -110° [292]. The 1,2-dimethoxynorbornyl ion has two methoxy signals (at 5.1 and 6.2 τ) that do not coalesce until the temperature goes up to 7° [293].

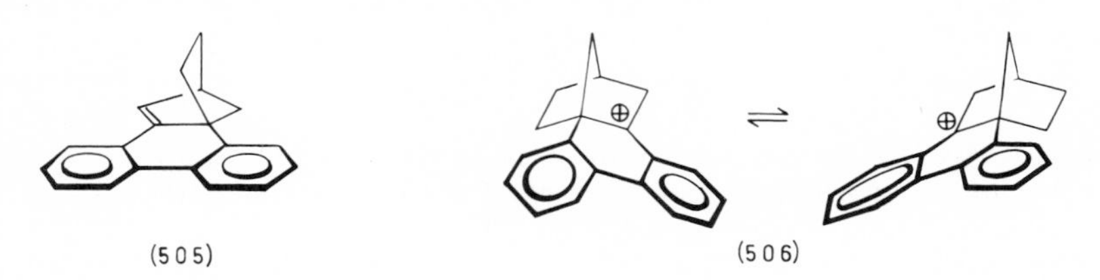

The Raman spectrum of the norbornyl ion solution is remarkably similar to that of nortricyclane, and on this basis Olah [294] prefers the "corner-protonated cyclopropane" structure for this ion - not significantly different from that postulated by Winstein. Some of these features (Raman and C^{13} nmr) do not show up in the 2-methyl- and 2-phenyl-2-norbornyl ions, and hence these are better described as classical ions [295]. Powerful support for these contentions derives from electron spectroscopy.

The carbon 1s spectra of the t-butyl ion shows 2 peaks, of relative intensities 1:3, with the smaller one of these at 3.5 eV higher energy than the larger. Isobutene shows only a single peak at still slightly lower energy, and the tropylium ion shows up as a single peak at the same energy as the methyl carbons in the t-butyl ion [296]. All these observations are just as expected on the basis of the binding energies of 1s electrons likely to exist in charged, slightly charged, and uncharged carbon atoms. The norbornyl ion shows two signals, differing by only 1.7 eV; the high and low energy peaks are in the ratio of two to five. The 2-methylnorbornyl ion has two signals differing by 3.7 eV in the ratio of one to seven. Because of the very high energy of the transitions and correspondingly short time scale (10^{-16} sec), equilibrations are irrelevant; static classical ions would appear no different from equilibrating ones [297].

---

291. P. von R. Schleyer, D.C. Kleinfelter, and H.G. Richey, J. Amer. Chem. Soc., 85, 479 (1963).
292. R.M. Cooper, M.C. Grossel, and M.J. Perkins, J. Chem. Soc., Perkin II, 594 (1972).
293. A. Nickon and Y. Lin, J. Amer. Chem. Soc., 91, 6861 (1969).
294. G.A. Olah, A.M. White, J.R. DeMember, A. Commeyras, and C.Y. Lui, J. Amer. Chem. Soc., 92, 4627 (1970).
295. G.A. Olah, J.R. DeMember, C.Y. Lui, and A.M. White, J. Amer. Chem. Soc., 91, 3958 (1969).
296. G.A. Olah, G.D. Mateescu, L.A. Wilson, and M.H. Gross, J. Amer. Chem. Soc., 92, 7231 (1970).
297. G.A. Olah, G.D. Mateescu, and J.L. Riemenschneider, J. Amer. Chem. Soc., 94, 2529 (1972).

It may be noted here that the two peaks in the norbornyl spectrum are not clearly separated. The high energy peak appears as a shoulder on the main signal, and a curve resolver was required to determine the 2:5 ratio. This result is therefore at variance with suggestions that the charge is largely concentrated at the bridging atom. In fact, had such suggestions indeed been correct, the carbon 1s spectrum might well have been very similar to that of a classical ion! The 2-methyl substituted ion is a case in point: Is it a classical ion with all of the charge at C_2, or a non-classical one with most of the charge at C_6? Whatever our prejudices are, the spectrum does not tell us which carbon atoms are represented by the peaks. At any rate, the lack of two clearly separated peaks in the spectrum of the 2-norbornyl ion - regardless of their intensities - is highly convincing proof that the 2-norbornyl ion is non-classical, and that the 2-methyl norbornyl ion is classical - under the conditions of this experiment. That is to say, in frozen arsenic trifluoride. What this means in terms of the more conventional solvolytic displacements in the nucleophilic solvents still remains to be decided.

20-12. ALTERNATIVES AND EXTENSIONS

The dilemma posed by arguments that convincingly if not compellingly point to opposite conclusions, has led to a search for alternative interpretations. The first of these was proposed by Brown, who argued that perhaps the exo/endo rate ratios were due not so much to unusually rapid rates for the exo- derivatives but to unusually slow endo-solvolysis. The reason given for this is that the leaving group must depart in the direction of the 6-endo-hydrogen atom and thus be hindered as in (507) [298].

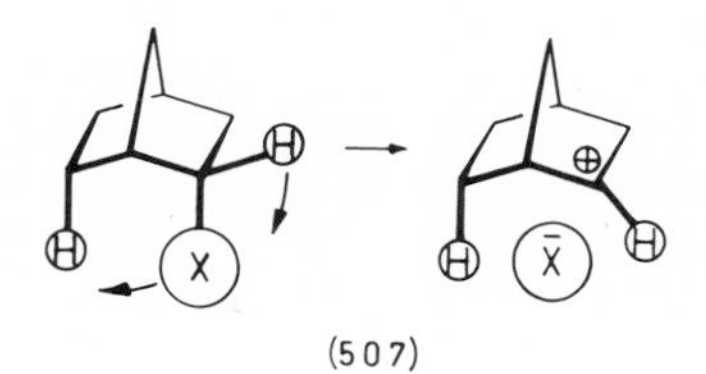

(507)

This point of view at first sight appears very reasonable, but whether it can account for the rate ratio is comething else again. If Brown's picture is to be used as a guide, one should expect enormous rate retardations for such compounds as (508) and (509).

(508) (509)

298. H.C. Brown, Spec. Publ. #16, Chem. Soc., 140 (1962).

But in fact, the rate retardations for both of these species are only a factor of 3 [299]. If we compare the brosylates (510) and (511), the former acetolyzes more than 300 times faster than the latter [300].

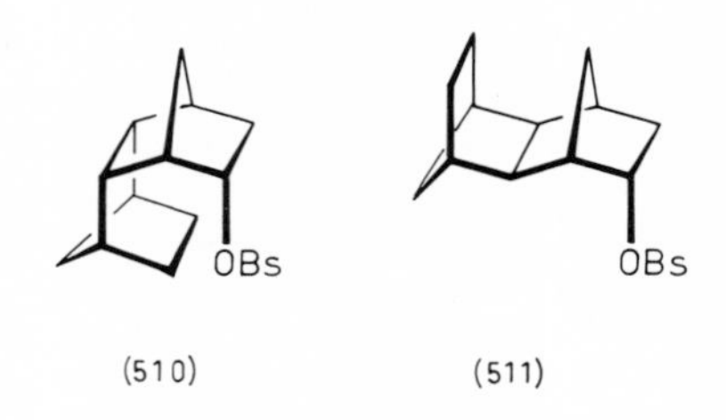

(510) (511)

Furthermore, it is difficult to see why the exo/endo ratio for the substrates (512) should be much less than 1 as observed [301], nor can this type of hindrance account for the reduction of the exo/endo rate ratio of the norbornyl substrate by so many remote substituents, for the difference in exo- and endo- activation volumes [302], and so on (cf. also p. 704).

(512)

An ingenious experiment has been devised by Nickon to answer the question of whether exo- or endo-2-norbornyl ion reactions should be singled out as exceptional: The brexyl derivative (513) is both at the same time [303], and hence it ought to show the same exceptional behavior as does the exceptional 2-norbornyl epimer.

---

299. (a) P. von R. Schleyer, as quoted by K. B. Wiberg and G. R. Wenzinger, J. Org. Chem., 30, 2278 (1965) (Table V, footnote d); (b) I. Rothberg, J. C. King, S. Kirsch and H. Skidanow, J. Amer. Chem. Soc., 92, 2570 (1970).
300. P. Carter and S. Winstein, J. Amer. Chem. Soc., 94, 2171 (1972).
301. (a) C.W. Jefford, J. Gunsher, and B. Waegell, Tetrahedron Lett., 3405 (1965); (b) C.A. Grob and A. Weiss, Helv. Chim. Acta, 49, 2605 (1966).
302. W. J. le Noble and A. Shurpik, J. Org. Chem., 35, 3588 (1970).
303. A. Nickon, H. Kwasnik, T. Swartz, R.O. Williams, and J. B. DiGiorgio, J. Amer. Chem. Soc., 87, 1613 (1965).

X X X X
X X X
(513)

The results of this study have not yet been published, but exactly the same arguments can be made with Freeman's highly strained tricyclic molecule (514) [304]; the species solvolyzes at the same rate as 2-exo-norbornyl.

X X X X
X X X (514)

Freeman has also reported the interesting epimeric pair of deltacyclyl brosylates (515), related to Nickon's system. The exo/endo rate ratio is 65; either stereoisomer gives only exo-acetate, and if the labelled substrate (516) is used, a 50-50 mixture of (517) and (518) results.

OBs D OBs OAc D OAc D

(515) (516) (517) (518)

Finally, dissolution of the alcohol in superacid at -80° permits the determination of the nmr spectrum of the ion. All results support a symmetrical structure (519) [305].

⊕

(519)

---

304. P.K. Freeman, R.B. Kinnel, and T.D. Ziebarth, Tetrahedron Lett., 1059 (1970).
305. P.K. Freeman and D.M. Balls, Tetrahedron Lett., 437 (1967).

Another alternative, proposed by Schleyer, is known as the torsional effect [306]. The 2-norbornyl ion is not completely rigid, but the C_2-C_3 bridge may be twisted a bit as shown below:

When the trigonal center is "up", the substituents at C_1 and C_2 and those at C_3 and C_4 are eclipsed; when the carbonium ion site is down, these groups are skewed. For steric reasons an endo-leaving group must lead to the less favorable conformation, and hence be slower in ionization. The 3,2-exo-hydride shift has a similar explanation. However plausible this sounds, apparently few other organic chemists have been able to explain their results by means of this effect.

A similar comment applies to still another point of view that was offered by Jensen [307]. It has long been known that the ability of p-alkyl groups to promote Friedel-Crafts reactions depends on the number of hydrogen atoms at the carbon bound to the p-carbon atom. This has been attributed to greater hyperconjugation of the C-H bond than of C-C bonds [308]. If for some reason the C_1-C_6 bond in the norbornyl group had an unusual ability to hyperconjugate, this might help explain the problem of the norbornyl ion. The classical, bridged and hyperconjugated ions are shown below [308a].

The following relative rates were reported for the Friedel-Crafts benzoylation (H = 1):

306. P. von R. Schleyer, J. Amer. Chem. Soc., 89, 699, 701 (1967).
307. F.R. Jensen and B.E. Smart, J. Amer. Chem. Soc., 91, 5686, 5688 (1969).
308. J.W. Baker and W.S. Nathan, J. Chem. Soc., 1844 (1935).
308a. It will be noted of course that - as is customary in the literature - the latter of these three structures has been used throughout this chapter for the symmetrically bridged ion.

t-butyl = 398, i-propyl = 519, ethyl = 563, methyl = 633, 7-norbornyl = 822, 2-endo-norbornyl = 1040, 2-exo-norbornyl = 1630, 1-norbornyl = 1790. The difference between exo and endo is clearly there; whether it is enough to account for the solvolysis ratios cannot be said. It may be emphasized that it cannot explain all the observations made with the norbornyl ion, because this type of assistance is distinguished from bridging in that the molecular geometry is simply that of the classical ion. Thus, the experiments in which exo-endo rate ratios were affected by distant groups that resist bridging by introducing strain cannot be explained this way.

The concept of carbon participation in solvolysis may well be extended still further. Thus, the fact that neopentyl tosylate solvolysis, ordinarily so slow in solvents such as ethanol, becomes so rapid in trifluoroacetic acid (p. 706) is considered indicative of methyl participation. The use of trideuteromethyl groups does not affect the rate, but CH_3 does migrate in preference to CD_3 (by a factor of 1.3). This means that the participation is of a special type in this case: Absent in the rate controlling ionization [309].

The concept of non-classical ions likewise may be capable of use in other areas as well. Protonated methane (520) has been found in the mass spectrum of methane, and simple alkanes can be converted into similar species in superacid solutions [310]. Pentahydroborane (521) is apparently an intermediate in the hydrolysis or borohydride ion; at high pH, borohydride undergoes exchange with D_2O and the hydrolysis rate slows down, suggesting that hydroxide is returning some of the intermediate to borohydride after the deuteron introduced has become chemically equivalent with the four hydrogen atoms [311].

BH_5

(520) (521)

309. W.M. Schubert and W.L. Henson, J. Amer. Chem. Soc., 93, 6299 (1971).
310. G.A. Olah, G. Klopman, and R.H. Schlosberg, J. Amer. Chem. Soc., 91, 3261 (1969).
311. (a) G.A. Olah, P.W. Westerman, Y.K. Mo, and G. Klopman, J. Amer. Chem. Soc., 94, 7859 (1972); (b) M.M. Kreevoy and J.E.C. Hutchins, J. Amer. Chem. Soc., 94, 6371 (1972).

Chapter 21

INTERMEDIATES RELATED TO CARBONIUM IONS

21-1. ION-PAIRS

The neutral substrates that give rise to carbonium ions do so in stages. In the first of these, a so-called tight ion-pair forms in which the carbonium ion and the leaving group are in neighboring positions, held together by electrostatic attraction, and solvated and hemmed in by surrounding solvent molecules. These formations have a finite lifetime, and hence they are intermediates.

Several lines of evidence point to their existence [1]. When α,α-dimethylallyl chloride (1) is acetolyzed, both solvolysis to (2) and rearrangement to the γ,γ-isomer (3) occur.

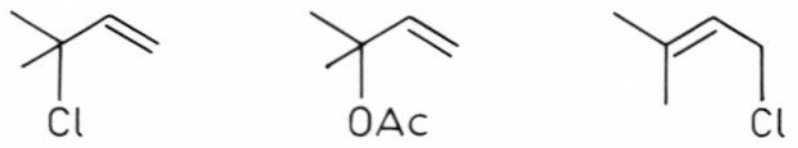

Cl OAc Cl

(1) (2) (3)

1. For a review of the early work, see H. L. Goering, Record Chem. Progr., 21, 109 (1960).

Since the isomer solvolyzes much more slowly, the first order rate plot consists of 2 branches (Fig. 21-1). This could be due to *external return* of the free chloride ion in solution to the resonance stabilized allylic ion; however, the product distribution is not affected by adding chloride ion, and hence the return is *internal* [2].

In other instances, solvolysis rates show a discrepancy when the production of acid is followed as compared to experiments in which racemization is studied. Thus, optically active, *threo*-3-phenyl-2-butyl tosylate (4) acetolyzes about five times more slowly than the solutions racemize.

H CH3 OTs

ϕ H CH3

(4)

The latter reaction is not a concerted isomerization [3], for the polarimetric rate constant responds to solvent changes in about the same way as does the titrimetric one. Hence the two reactions are thought to involve a common intermediate, the ion-pair, which then partitions between collapse to covalency (either the (+) or (-) enantiomer) and separation and solvolysis [4].

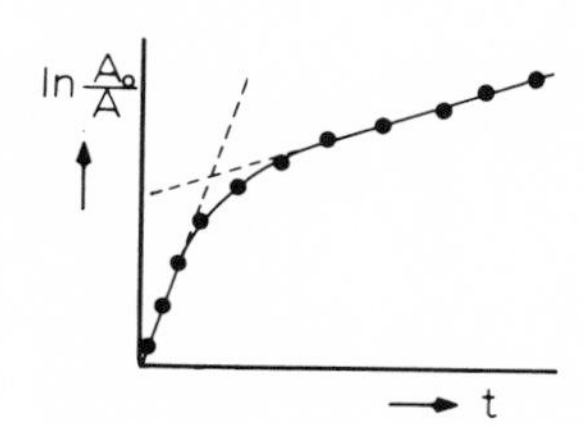

Fig. 21-1. Solvolysis marked by simultaneous isomerization.

---

2. W.G. Young, S. Winstein, and H.L. Goering, *J. Amer. Chem. Soc.*, *73*, 1958 (1951).
3. Such isomerizations are known; *e.g.*, allylic azides are in mobile equilibrium, and the rates are virtually independent of solvent ((a) A. Gagneux, S. Winstein, and W.G. Young, *J. Amer. Chem. Soc.*, *82*, 5956 (1960)), and even phase ((b) W.J. le Noble, *Progr. Phys. Org. Chem.*, *5*, 207 (1967); see p. 231). The rearrangement of thion- to thiolbenzoates has similar characteristics ((c) S.G. Smith, *J. Amer. Chem. Soc.*, *83*, 4285 (1961)); the isomerization of thiocyanates to isothiocyates is a reaction of intermediate type ((d) P.A.S. Smith and D.W. Emerson, *J. Amer. Chem. Soc.*, *82*, 3076 (1960)).
4. S. Winstein and K.C. Schreiber, *J. Amer. Chem. Soc.*, *74*, 2165 (1952).

It may be noted that in most cases the ion-pair can only return to give the original molecule. Its presence is then not directly detectible; its intervention is obvious only if return is accompanied by rearrangement or racemization.

Very detailed studies have been published by Goering concerning the ion-pairs of allylic benzoates. Thus, if we consider the system (5), both cation and anion are in principle symmetrical. It is possible in this case to measure the rates of solvolysis, cis-trans isomerization, racemization, radiocarbon scrambling and oxygen exchange. In aqueous acetone one finds that racemization is five times faster than solvolysis; neither k_t or k_p are affected by added common-ion salts, and if labelled p-nitrobenzoate ion is present the label does not become incorporated. Thus, the pair (6) is indicated.

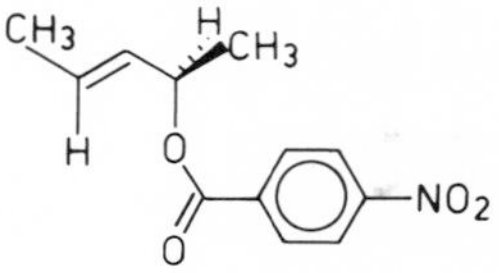

(5)

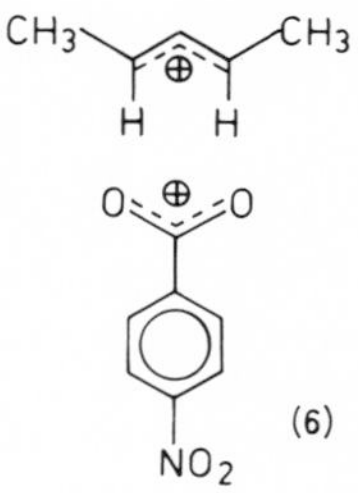

(6)

Oxygen exchange (between carbonyl and either positions) occurs at the same rate as racemization; this rules out a 1,3 sigmatropic shift such as (7).

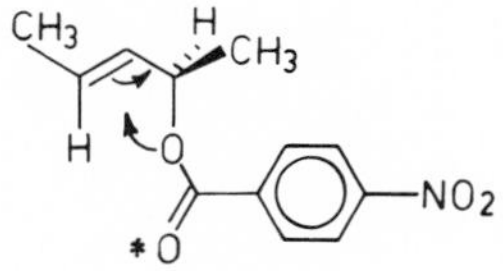

(7)

If the racemized ester is reresolved and the oxygen label is identified in the inverted substrate, one finds that the oxygen atoms are scrambled there; this rules out a 3,3 shift such as (8).

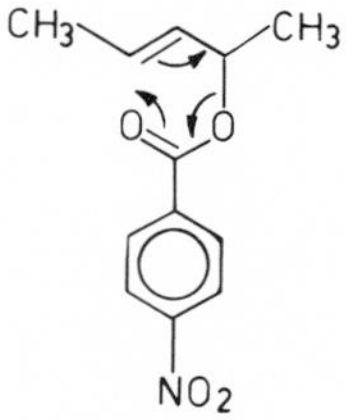

(8)

The scrambling is less then complete though, and so although the isomerization must involve the ion-pair, the anion is not completely free to rotate relative to the cation [5]. Similar experiments have been done with benzenesulfonate esters [6].

By means of such studies the importance of a number of simple processes within the ion-pair can be evaluated; but interestingly, *cis-trans* isomerization of the allylic cation is not observed. The activation energy for this process is clearly much higher than that of the others. The tetramethylallyl ion (9) has been observed in sulfuric acid, and coalescence of the methyl groups does not set in until the temperature is 55° [7].

(9)

We note parenthetically at this point that Sneen [8] has recently suggested the intervention of two isomeric ion-pairs (10) in the reactions of allyl chlorides [8]; however, the evidence - a bit too complex for recounting here - is as yet not unassailable.

(10)

Further evidence for the intervention of ion-pair intermediates in solvolysis has been found by Brower. Since ionization is characterized by a volume decrease (electrostriction), it might be supposed that solvolysis will be favored over return by the application of pressure (see also p. 706). Such an effect has indeed been found by Brower [9].

The independent-generation method has been applied [10] successfully to the ion-pair intermediate. From measurements of the ethanolysis and oxygen scrambling rates of benzhydryl benzoate (11) we know that the ion-pairs (12) partition about equally between return and dissociation; the reaction of diphenyldiazomethane (13) with benzoic acid, which in principal should begin *via* the same pair, indeed also gives about equal amounts of benzhydryl benzoate and the ethyl ether (14).

---

5. H.L. Goering and M.M. Pombo, *J. Amer. Chem. Soc.*, **82**, 2515 (1960).
6. H.L. Goering and R.W. Thies, *J. Amer. Chem. Soc.*, **90**, 2967, 2968 (1968).
7. N.C. Deno, R.C. Haddon, and E.N. Nowak, *J. Amer. Chem. Soc.*, **92**, 6691 (1970).
8. (a) R.A. Sneen and W.A. Bradley, *J. Amer. Chem. Soc.*, **94**, 6975 (1972); (b) R.A. Sneen and P.S. Kay, *J. Amer. Chem. Soc.*, **94**, 6983 (1972); (c) R.A. Sneen and J.V. Carter, *J. Amer. Chem. Soc.*, **94**, 6990 (1972).
9. K.R. Brower, *J. Amer. Chem. Soc.*, **94**, 5747 (1972).
10. A.F. Diaz and S. Winstein, *J. Amer. Chem. Soc.*, **88**, 1318 (1966).

$$\underset{(11)}{\phi_2CH{-}OCO\phi} \qquad \underset{(12)}{\phi_2\overset{\oplus}{C}H\ \overset{\ominus}{O_2}C\phi} \qquad \underset{(13)}{\phi_2C{=}N_2} \qquad \underset{(14)}{\phi_2CHOCH_2CH_3}$$

Important information has also been obtained from studies of the salt effect [11]. It had been known that added non-nucleophilic salts such as lithium perchlorate may accelerate solvolysis reactions, and that this effect is well described by the equation:

$$k = k_o\ (1 + b\ [LiClO_4])$$

The constant b has values typically close to 10 (depending on the system) [12].

This salt effect evidently operates through the exchange of the ion-pair with a less nucleophilic gegenion which is less likely to collapse. In one case this was demonstrated when such return did occur. Thus, i-cholesteryl acetate (15) isomerizes in acetic acid to cholesteryl acetate (16) if a catalytic quantity of perchloric acid is present.

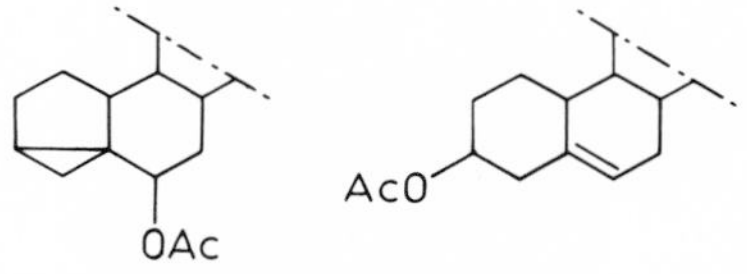

OAc

(15)

AcO

(16)

The rate-law is: $R = k\ [HClO_4]$

rather than: $R = k\ [i\text{-}ROAc]\ [HClO_4]$

The following scheme explains the observed rate law; the perchlorate ion pair separation is the slow step..

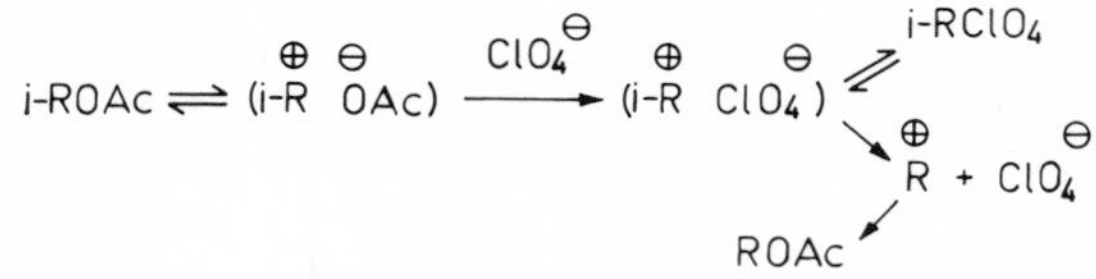

$$i\text{-}ROAc \rightleftharpoons (i\text{-}\overset{\oplus}{R}\ \overset{\ominus}{OAc}) \xrightarrow{ClO_4^{\ominus}} (i\text{-}\overset{\oplus}{R}\ \overset{\ominus}{ClO_4}) \rightleftharpoons i\text{-}RClO_4$$

$$(i\text{-}\overset{\oplus}{R}\ \overset{\ominus}{ClO_4}) \rightarrow \overset{\oplus}{R} + \overset{\ominus}{ClO_4}; \quad \overset{\oplus}{R} \rightarrow ROAc$$

The perchloric acid is temporarily diverted to some neutral species (presumably the perchlorate ester resulting from collapse) during the isomerization as can be shown by indicators [13]. Such esters are not unheard of; in fact, even *gem*-diperchlorates can be prepared [14].

In certain cases very much larger accelerations by such salts are observed, though only at very low concentrations; this observation is now called the special

11. S. Winstein, E. Clippinger, A.H. Fainberg, and G.C. Robinson, J. Amer. Chem. Soc., 76, 2597 (1954).
12. S. Winstein and K.C. Schreiber, J. Amer. Chem. Soc., 74, 2171 (1952).
13. A. Ehret and S. Winstein, J. Amer. Chem. Soc., 88, 2048 (1966).
14. K. Baum, J. Amer. Chem. Soc., 92, 2927 (1970).

salt effect [15]. This is interpreted as follows. Ionization traverses in succession the tight or intimate ion-pair and then still another intermediate stage: The loose or solvent-separated ion-pair, before free ions finally obtain. The normal salt effect, as stated earlier, operates by exchange of the tight ion-pair with perchlorate to a new ion-pair which is much less likely to return; the special salt effect operates similarly through the solvent separated ion-pair.

Further evidence was obtained when it became known that bromide ion is able to trap the loose ion-pair and convert it into covalent bromide, but not the tight pair - which still returns [16].

There is also a convincing experiment by Goering bearing on this point. Optically active and carbonyl-oxygen-labelled p-chlorobenzhydryl p-nitrobenzoate (17) undergoes return unaffected by added p-nitrobenzoate; in the ester, scrambling is faster than racemization which in turn is faster than solvolysis.

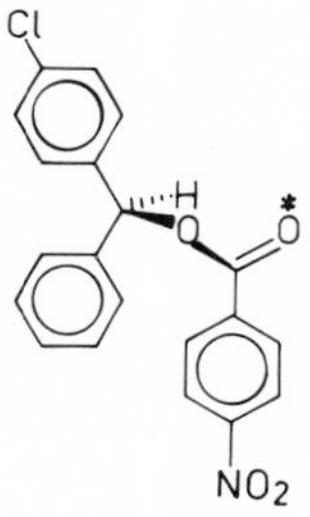

Cl

H

O

O*

NO_2

(17)

If azide ion is added, return still occurs because scrambling continues; however, the ester racemization stops. Evidently the racemization in this case (more difficult than in the allylic cases referred to above because it requires the turn-around of the cation) is associated with the loose pair, but anion rotations already occur in the tight pair [17].

Return is sometimes a feature that introduces undesired complications, and several means are available to reduce its importance; among these are added salts as discussed above, more polar solvents [18], non-nucleophilic leaving groups

---

15. S. Winstein, E. Clippinger, A.H. Fainberg, and G.C. Robinson, J. Amer. Chem. Soc., 76, 2597 (1954) and Chem. Ind. (London), 664 (1954).
16. (a) S. Winstein, P.E. Klinedienst, and E. Clippinger, J. Amer. Chem. Soc., 83, 4986 (1961). For some recent cases involving the special salt effect, see (b) S.J. Cristol, A.L. Noreen, and G.W. Nachtigall, J. Amer. Chem. Soc., 94, 2187 (1972); (c) I.L. Reich, A.F. Diaz, and S. Winstein, J. Amer. Chem. Soc., 94, 2256 (1972).
17. H.L. Goering and J.F. Levy, J. Amer. Chem. Soc., 86, 120 (1964).
18. S. Winstein, E.C. Friedrich, and S. Smith, J. Amer. Chem. Soc., 86, 305 (1964).

such as the triflate ion [19] and the use of fragmentation reactions to generate the ions [20]. Thus, chlorosulfates (18) ionize via fragmentation, with a sulfur trioxide molecule separating the ions [21].

$$ROSO_2Cl$$

(18)

Deno has similarly observed [22] the fragmentation of protonated β-xanthylethyl substrates (19) to isobutylene, water and ion (20).

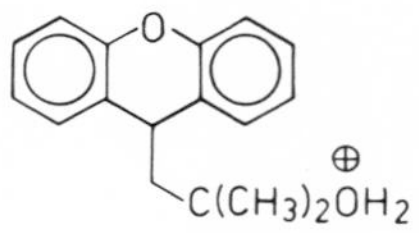

$C(CH_3)_2\overset{\oplus}{O}H_2$

(19) 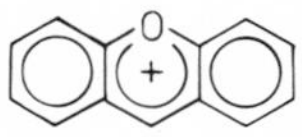 (20)

Likewise, Roberts [23] has used such a reaction to avoid return in the solvolysis of (21) to give N-methyl-4-pyridone (22) and a carbonium ion.

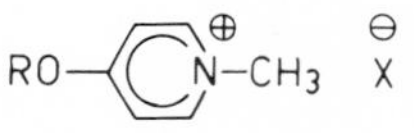

RO–$\overset{\oplus}{N}$–CH_3 $\overset{\ominus}{X}$ O= N–CH_3

(21) 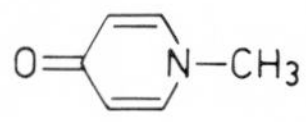 (22)

We conclude at this point that solvolysis is much more complex than mere ionization, and that in those cases which are amenable to analysis the following general scheme applies.

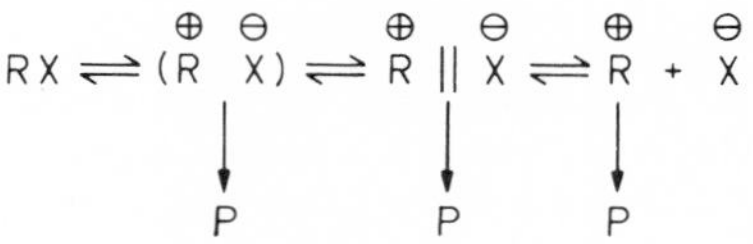

$$RX \rightleftharpoons (\overset{\oplus}{R}\ \overset{\ominus}{X}) \rightleftharpoons \overset{\oplus}{R} \| \overset{\ominus}{X} \rightleftharpoons \overset{\oplus}{R} + \overset{\ominus}{X}$$

$$\downarrow P \qquad \downarrow P \qquad \downarrow P$$

19. (a) T.M. Su, W.F. Sliwinski, and P. von R. Schleyer, J. Amer. Chem. Soc., 91, 5386 (1969); (b) F. Effenberger and K.E. Mack, Tetrahedron Lett., 3947 (1970).
20. For reviews of this important reaction, see (a) C.A. Grob and P.W. Schiess, Angew. Chem., Int. Ed. Engl., 6, 1 (1967) and (b) C.A. Grob, Angew. Chem., Int. Ed. Engl., 8, 535 (1969).
21. E. Buncel and J.P. Millington, Proc. Chem. Soc., 406 (1964).
22. N.C. Deno and E. Sacher, J. Amer. Chem. Soc., 87, 5120 (1965).
23. M. Vogel and J.D. Roberts, J. Amer. Chem. Soc., 88, 2262 (1966).

Until recently it was assumed that the S_N2 substitution reaction occurs with neutral RX, but Sneen has pointed out that there is no kinetic evidence to rule out the possibility that all or most of S_N2 reactions actually involve the tight ion-pair [24]. As yet this pathway cannot be regarded as proved in any instance, however.

Zwitterions are ion-pairs in which the ions are held together not only by electrostatic abstraction but also by a covalent bond or a chain of such bonds. Among the zwitterions the so-called ylids are especially important [25]; they contain the system (23).

$$\rangle \overset{\ominus}{C}—\overset{\oplus}{X}$$

(23)

X may be boron [26], nitrogen [27], phosphorus [28], sulfur [29]; several of these are isolable compounds. In those cases in which X is a second row element the dipole moment may be less than expected on the basis of a full electronic charge: Empty d orbitals then permit some charge neutralization. In a few rare cases both charges may be on carbon atoms [30], as is clear from the following example. The cyclopropane (24) reacts with methanol at 150° to give the ring-opened product (25). If resolved material is used, racemization turns out to be 100 times faster than methanolysis. Both reactions are first order and respond to solvent changes in the manner expected if the zwitterion (26) were an intermediate.

---

24. (a) H. Weiner and R.A. Sneen, J. Amer. Chem. Soc., 87, 287, 292 (1965); (b) R.A. Sneen and J.W. Larsen, J. Amer. Chem. Soc., 91, 6031 (1969); (c) R.A. Sneen and H.M. Robbins, J. Amer. Chem. Soc., 94, 7868 (1972); (d) J.M.W. Scott, Can. J. Chem., 48, 3807 (1970); (e) D.J. McLennan, Tetrahedron Lett., 2317 (1971); (f) D.J. Raber, J.M. Harris, R.E. Hall, and P. von R. Schleyer, J. Amer. Chem. Soc., 93, 4821 (1971); (g) F.G. Bordwell and T.G. Mecca, J. Amer. Chem. Soc., 94, 2119 (1972).
25. For review, see (a) P.A. Lowe, Chem. Ind. (London), 1070 (1970); (b) R.F. Hudson, Chem. Brit., 7, 287 (1971).
26. H. Jäger and G. Hesse, Chem. Ber., 95, 345 (1962).
27. (a) I.B.M. Band, D. Lloyd, M.I.C. Singer, and F.I. Wasson, Chem. Commun., 544 (1966); (b) R. Appel, H. Heinen, and R. Schöllhorn, Chem. Ber., 99, 3118 (1966); (c) J. Streith and J.M. Cassal, Tetrahedron Lett., 4541 (1968).
28. Chem. Eng. News, 45, 47 (April 17, 1967).
29. (a) H. Behringer and F. Scheidl, Tetrahedron Lett., 1757 (1965); (b) A. Hochreiner, Monatsh. Chem., 97, 823 (1966); (c) H. Nozaki, K. Kondo, and M. Takaku, Tetrahedron Lett., 251 (1965).
30. (a) E.W. Yankee and D.J. Cram, J. Amer. Chem. Soc., 92, 6328, 6329, 6331 (1970); (b) D.J. Cram and A. Ratajczak, J. Amer. Chem. Soc., 90, 2198 (1968).

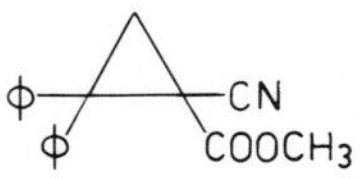

(24) 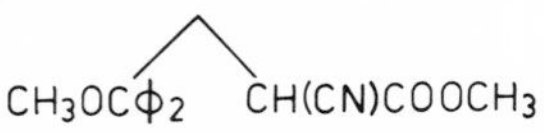 (25) 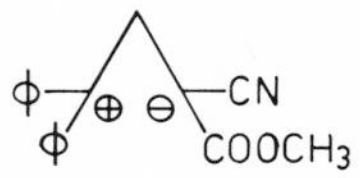 (26)

The abnormal Cope rearrangement of (27) to form (28) rather than (29) seems to be an ion-pair reaction; thus, in the presence of borohydride some (30) is obtained [31].

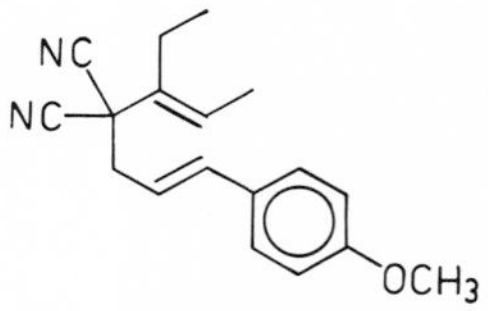

(27) 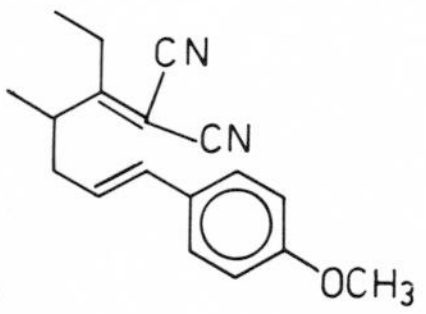 (28)

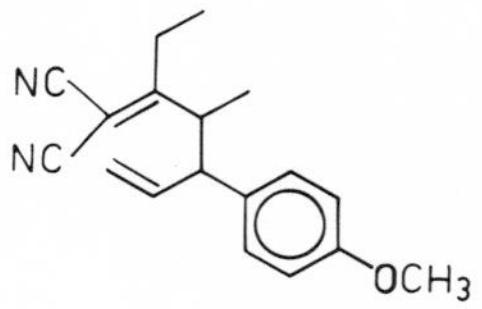

(29) 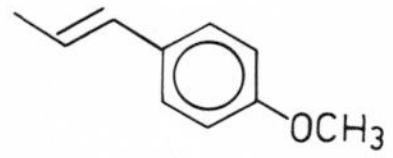 (30)

Ion-pairs of higher charge are more stable to dissociation, and many of these have long been known to the inorganic chemists to exist in water. They can readily be detected by means of conductance methods, and their temperature [32] and pressure [33] sensitivities have thus been determined. In organic media, a number of stable ion-pairs have been characterized by means of spectral measurements. Thus, anilinium acetate pairs have been found in acetic acid by means of uv [34]. Perhaps the best known instances are the salts such as lithium fluorenide. Such salts have uv spectra that are sensitive to solvent and concentration changes, and these are readily interpreted in terms of both tight and loose pairs (31) and (32) [35]. Thus, except under conditions of extreme dilution, conductivity measurements show that very little of the salt is dissociated (33);

31. D.C. Wigfield, S. Feiner, and K. Taymaz, Tetrahedron Lett., 891, 895 (1972).
32. J.H.B. George, J. Amer. Chem. Soc., 81, 5530 (1959).
33. (a) F.H. Fisher and D.F. Davis, J. Phys. Chem., 69, 2595 (1965); (b) S.D. Hamann, P.J. Pearce, and W. Strauss, J. Phys. Chem., 68, 375 (1964); (c) M. Nakahara, K. Shimizu, and J. Osugi, Rev. Phys. Chem. Japan, 40, 12 (1970).
34. G.W. Ceska and E. Grunwald, J. Amer. Chem. Soc., 89, 1371 (1967).
35. T.E. Hogen-Esch and J. Smid, J. Amer. Chem. Soc., 87, 669 (1965) and 88, 307 (1966).

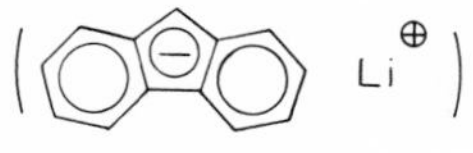

(31) 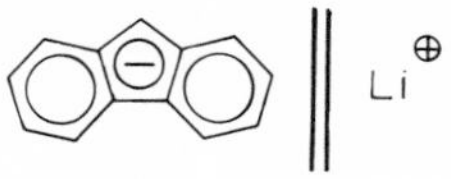 (32) 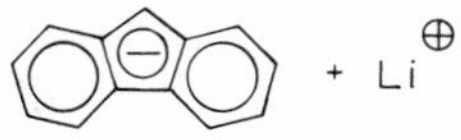 (33)

but either pair can be made dominant by the proper choice of solvent [36], cation, addends (such as DMSO) [37], temperature and pressure [38]. Esr spectra have similarly been employed by Hirota to detect several types of ion-pairs involving radical anions such as sodium naphthalene (34) (also see p. 656) [39].

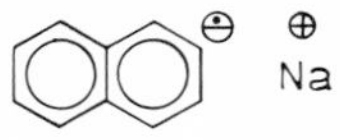

(34)

Finally, ion-pairs have been found to play an important role in anionic polymerization [40] (the so-called living polymers), but a discussion of this phenomenon would lead us too far afield. Instead, we conclude at this point that the evidence for ion-pairs is nothing short of overwhelming, and turn to several topics of the carbonium ion field for which some understanding of ion-pairs is desirable. Among these is the "memory effect" [41], one example of which may be described here [42]. 7-Norbornylcarbinyl-, 2-bicyclo[2.2.2]octyl- and 2-(Δ^3-cyclohexenyl)ethyl brosylates (35)-(37) all acetolyze to give the identical mixture (38)-(42);

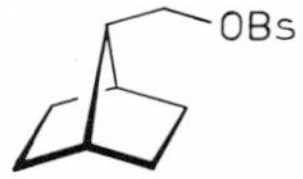

(35) 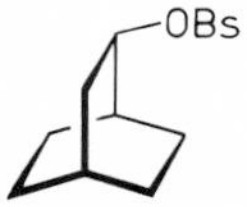 (36) 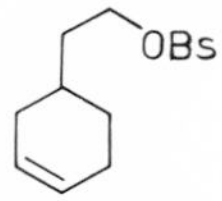 (37)

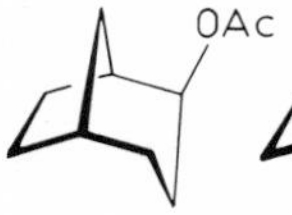

(38)  (39)  (40)  (41) 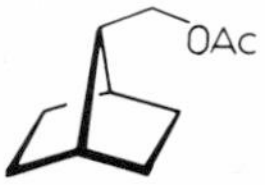 (42)

36. K.H. Wong, G. Konizer, and J. Smid, J. Amer. Chem. Soc., 92, 666 (1970).
37. J. Smid, Angew. Chem., Int. Ed. Engl., 11, 112 (1972).
38. (a) W.J. le Noble and A.R. Das, J. Phys. Chem., 74, 3429 (1970); (b) S. Claesson, B. Lundgren, and M. Szwarc, Trans. Faraday Soc., 66, 3053 (1970).
39. (a) N. Hirota and R. Kreilick, J. Amer. Chem. Soc., 88, 614 (1966); (b) N. Hirota, J. Amer. Chem. Soc., 90, 3603 (1968); (c) N. Hirota, R. Carraway and W. Schook, J. Amer. Chem. Soc., 90, 3611 (1968); (d) E. de Boer, Rec. Trav. Chim. Pays-Bas, 84, 609 (1965).
40. (a) M. Szwarc, Science, 170, 23 (1970); (b) L.L. Chan and J. Smid, J. Phys. Chem., 76, 695 (1972); (c) P.E.M. Allen, D.O. Jordan, and M.A. Naim, Trans. Faraday Soc., 63, 234 (1967).
41. Reviewed by J.A. Berson, Angew. Chem., Int. Ed. Engl., 7, 779 (1968).
42. J.A. Berson and M.S. Poonian, J. Amer. Chem. Soc., 88, 170 (1966).

each of these products can be traced to the 2-bicyclo[2.2.2]octyl ion (43).

(43)

If the deuterated substrates (44) and (45) are hydrolyzed, the products include (46) and (47) in unequal amounts: (44) gives mostly (46) and (45) gives primarily (47).

(44) (45) (46) (47)

Since the classical bicyclooctyl ion is symmetrical, these two products should have been formed in a one to one ratio; hence it was concluded that bridging occurs - preferably to the rear of the developing cationic sites as in (48)-(50).

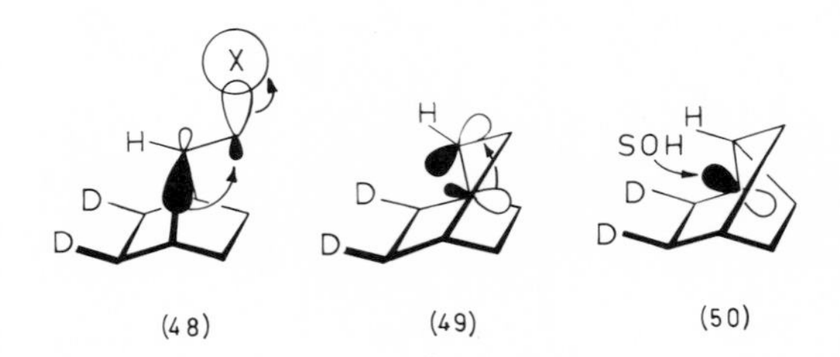

(48) (49) (50)

Thus, although the bicyclo[2.2.2]octyl ion (49) is involved, this does not lead to symmetrization. This type of observation is sometimes referred to as the memory effect. We should note however that the rearrangements may well occur within the ion-pair, and that the gegenion in this pair decides from which side bridging shall occur.

Another and perhaps simpler example has been described by Fort [43]. He found that the cis- and trans- substrates (51) and (52), though both have identical heats of combustion and though both should give the same (flat) carbonium ion, hydrolyze with rate constants in the ratio of 4:1.

43. R.C. Fort, R.E. Hornish, and G.A. Liang, J. Amer. Chem. Soc., 92, 7558 (1970).

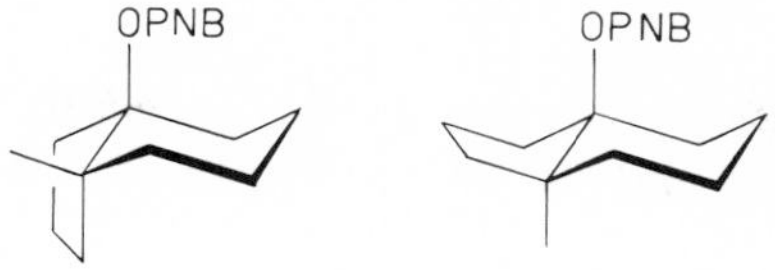

OPNB OPNB

(51) (52)

Fort described the rate ratio to conformationally different ions, but again the location of the counterion may be responsible.

Much attention has been drawn by multiple rearrangements [44], understood to be a succession of Wagner-Meerwein shifts. That multiple shifts can occur is clear from the preparations of such compounds as adamantane, diamantane and triamantane from precursor isomers (53) [45], (54) [46], and (55) [47], respectively (hydride shifts must also occur in these examples).

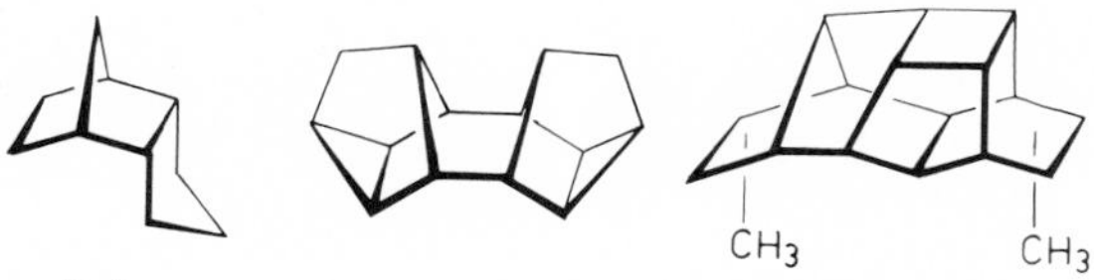

CH_3 CH_3

(53) (54) (55)

Some of the systems subject to multiple rearrangements are degenerate. An example of a degenerate cation is the barbaralyl ion (56); only a single signal is observed in the nmr spectrum at -125° in superacid solvents (see also Chr. 20) [48].

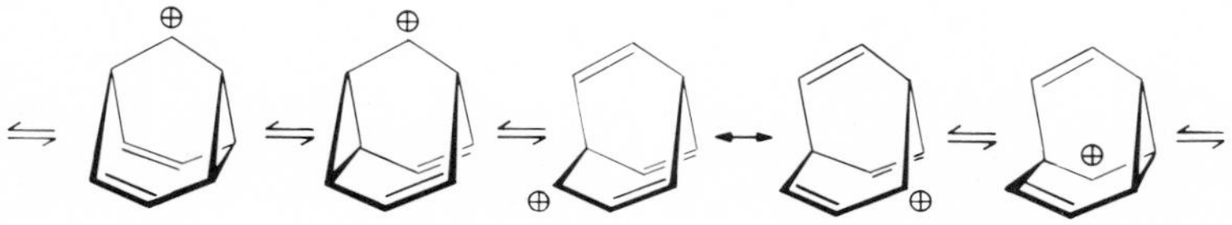

(56)

44. Reviewed by R.E. Leone and P. von R. Schleyer, Angew. Chem., Int. Ed. Engl., 9, 860 (1970).
45. R.C. Fort and P. von R. Schleyer, Chem. Rev., 64, 277 (1964).
46. T.M. Gund, V.Z. Williams, E. Osawa, and P. von R. Schleyer, Tetrahedron Lett., 3877 (1970).
47. R.C. Bingham and P. von R. Schleyer, Fortschr. Chem. Forsch., 18, 1 (1971).
48. P. Ahlberg, D.L. Harris, and S. Winstein, J. Amer. Chem. Soc., 92, 4454 (1970).

Homocubyl tosylate (57) is characterized by rapid formolysis, and the use of a deuterium label indicates complete scrambling; thus it is clear that the ion (58) is subject to an endless succession of Wagner-Meerwein shifts before solvent capture [49].

(57) (58)

However, the tetradeuterated tosylate (59) during acetolysis gives a product which has no deuterium in the bridge, in contrast to the epimer (60). In this less ionizing medium, a free ion is evidently not formed, and the presence of the gegenion restricts the Wagner-Meerwein shifts to those that can occur at the rear of the carbon atom facing it.

(59) (60)

The net result for (59) is as though the $C_5H_5^+$ ring were rapidly rotating while sandwiched between the anion and the C_4D_4 ring [50]. Similar reactions also occur in the acetolysis of the bishomocubyl mesylate (61) to give acetates (62)-(65) [51].

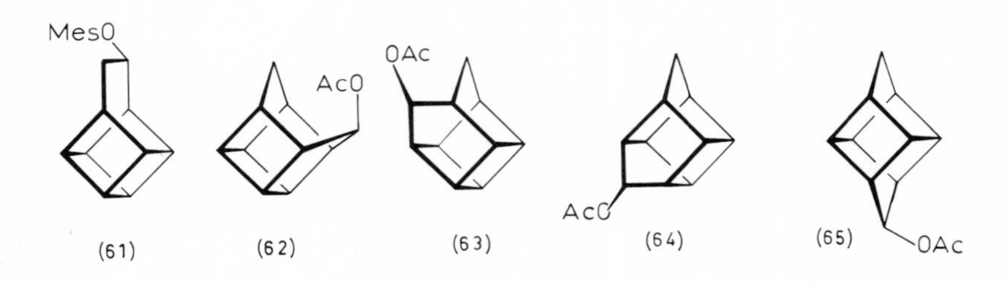

(61) (62) (63) (64) (65)

---

49. P. von R. Schleyer, J.J. Harper, G.L. Dunn, V.J. DiPasquo, and J.R.E. Hoover, J. Amer. Chem. Soc., 89, 698 (1967).
50. J.C. Barborak and R. Pettit, J. Amer. Chem. Soc., 89, 3080 (1967).
51. W.G. Dauben and D.L. Whalen, J. Amer. Chem. Soc., 88, 4739 (1966) and 93, 7244 (1971).

Ion-pairs also play a remarkable role as intermediates in reactions in which the negative ion is a carbanion (cf. Chr. 22). Thus, Cram has studied the D-labeled, resolved 9-methylfluorenes (66) by measuring the rates of exchange (k_e) and of racemization (k_α), and the effect of substituents and medium on these two. Examples were found in which k_e/k_α equals 0.5 (exchange with inversion), 1 (exchange with racemization) or greater than 1 (exchange with retention); however, the most interesting examples are those in which the ratio is less than 0.5 (racemization without exchange). This happens for instance in tri-*n*-propylamine-*t*-butanol with the substrate (67).

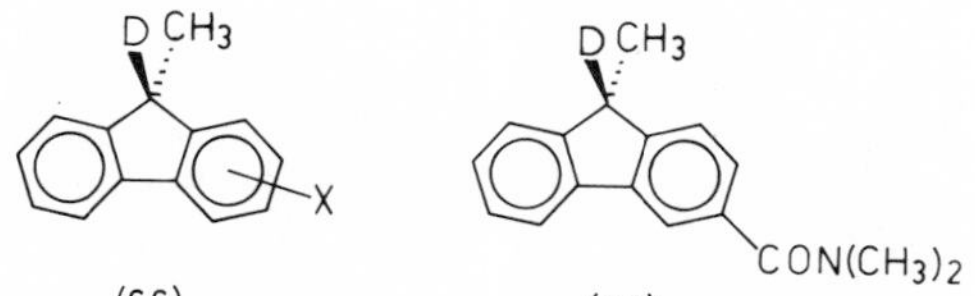

(66) (67)

Cram reasons that the deuteron is taken off by the amine, that the ammonium ion remains paired with the carbanion *via* a deuterium bond. The cation need not remain at the original site: Because of resonance, the negative charge is delocalized. If the cation makes it all the way to the oxygen atom, it may slide to the other side of the anion, and return all the way to the 9-carbon atom without ever exchanging the deuteron with the proton pool (the "conducted-tour mechanism").

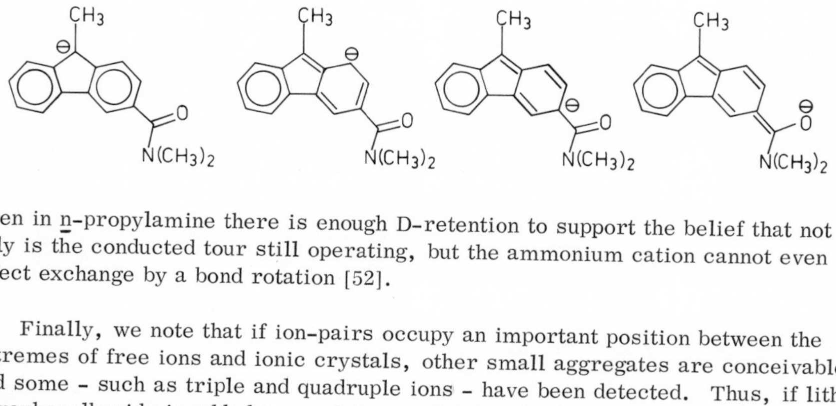

Even in *n*-propylamine there is enough D-retention to support the belief that not only is the conducted tour still operating, but the ammonium cation cannot even affect exchange by a bond rotation [52].

Finally, we note that if ion-pairs occupy an important position between the extremes of free ions and ionic crystals, other small aggregates are conceivable and some - such as triple and quadruple ions - have been detected. Thus, if lithium tetraphenylboride is added to a solution containing the lithium-anthraquinone ion-pairs, the triple ion (68) is clearly recognizable by its 1:2:3:4:3:2:1 signal [53]. Another development of interest is the evidence for the existence of pairs of ions of like charge, but a detailed consideration of such species is outside the scope of this writing [54].

---

52. D. J. Cram, W. T. Ford, and L. Gosser, J. Amer. Chem. Soc., 90, 2598 (1968).
53. K. S. Chen and N. Hirota, J. Amer. Chem. Soc., 94, 5550 (1972).
54. P. Hemmes, J. Amer. Chem. Soc., 94, 75 (1972).

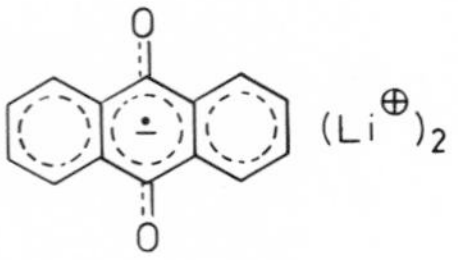

$(Li^{\oplus})_2$

(68)

21-2. DICARBONIUM IONS

Dicarbonium ions comprise an area that has been plagued by misinterpretations more than any other. Thus, pentamethylbenzoyl trichloride has been observed to dissolve in pure sulfuric acid with an i-factor of 5; this was interpreted on the basis of a dicarbonium ion - with the reasonable assumption that it would relieve crowding [55].

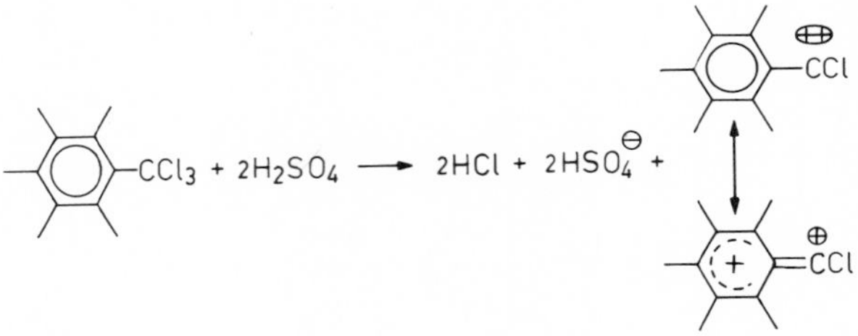

$$-CCl_3 + 2H_2SO_4 \longrightarrow 2HCl + 2HSO_4^{\ominus} + \quad -\overset{\oplus}{C}Cl \longleftrightarrow (+)=\overset{\oplus}{C}Cl$$

This interpretation requires that hydrochloric acid does not react with sulfuric acid; however, Gillespie found that an equilibrium reaction virtually completely converts this substance into chlorosulfuric acid:

$$HCl + H_2SO_4 \longrightarrow ClSO_3H + H_3O^{\oplus} + HSO_4^{\ominus}$$

and that the explanation of the i-factor is [56]:

$$ArCCl_3 + H_2SO_4 \longrightarrow ArCCl_2^{\oplus} + ClSO_3H + H_3O^{\oplus} + 2HSO_4^{\ominus}$$

Such trichlorides furthermore react with antimony pentachloride to give red salts which in turn react with tropilidene to give the covalent dichloride (69) [57].

---

55. H. Hart and R.W. Fish, J. Amer. Chem. Soc., 80, 5894 (1958), 82, 5419 (1960), and 83, 4460 (1961).
56. (a) R.J. Gillespie and E.A. Robinson, J. Amer. Chem. Soc., 87, 2428 (1965) and 86, 5676 (1964); (b) N.C. Deno, N. Friedman and J. Mockus, J. Amer. Chem. Soc., 86, 5676 (1964).
57. H. Volz and M.J. Volz de Lecca, Tetrahedron Lett., 3413 (1965).

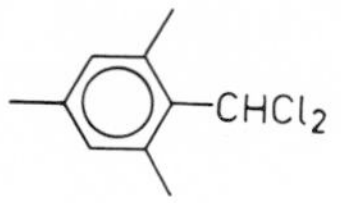

(69)

Another possible approach is the generation of dicarbonium ions stabilized by aromatic ring structure. Thus, when the cyclobutene derivative (70) is treated with stannic chloride, a salt is obtained that was described as a dicarbonium ion salt (71) [58].

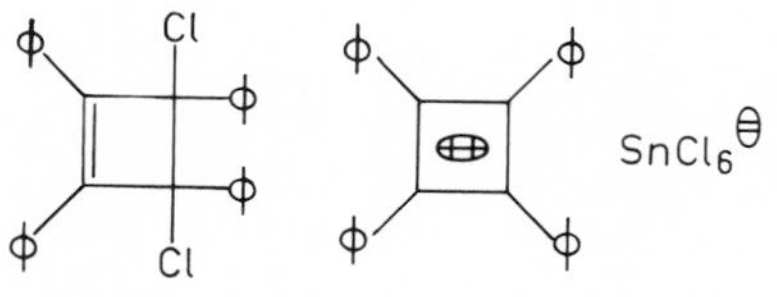

(70) (71)

However, X-ray diffraction studies later showed it to be a monocarbonium ion salt (72) [59].

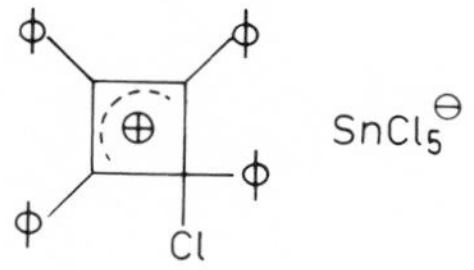

(72)

It has also been reported that the hydroxyketone (73) dissolves in sulfuric acid to give a red solution the nmr of which suggests the structure (74) [60]; however even if this structure is correct it is likely to be better described as the more trivial (75).

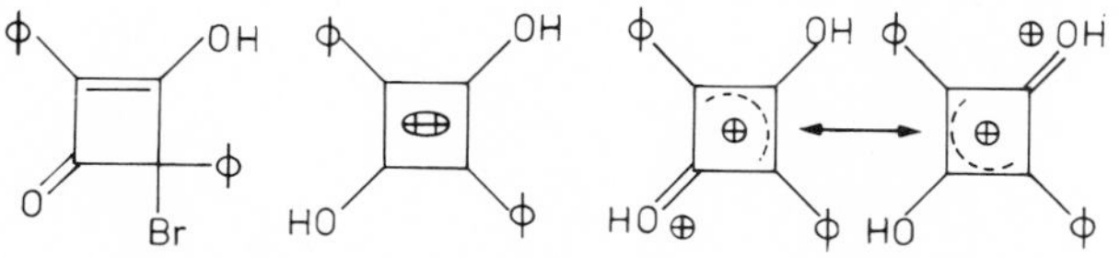

(73) (74) (75)

---

58. H. H. Freedman and A. M. Frantz, J. Amer. Chem. Soc., 84, 4165 (1962).
59. (a) R. F. Bryan, J. Amer. Chem. Soc., 86, 733 (1964); (b) H. H. Freedman and A. E. Young, J. Amer. Chem. Soc., 86, 734 (1964).
60. D. G. Farnum and B. Webster, J. Amer. Chem. Soc., 85, 3502 (1963).

On the other hand, if the cyclobutene (76) is dissolved in a mixture of antimony pentafluoride and sulfur dioxide, the nmr spectrum of the solution initially shows three singlets between 7 and 8 τ in the ratio of 1:2:1 ((77)), but on standing a new spectrum evolves with a single peak at 6.3 τ. On quenching in methanol the diether (78) is formed. It is hard to take issue with Olah's contention that the 6.3 peak is due to the tetramethylcyclobutadiene dication (79) [61] (see further Chr. 9).

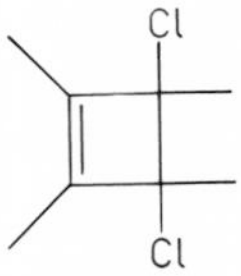

(76) 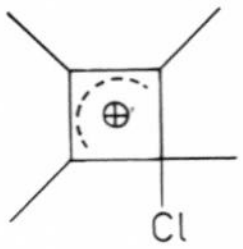 (77) 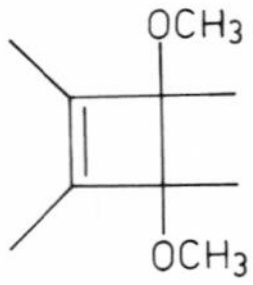 (78) 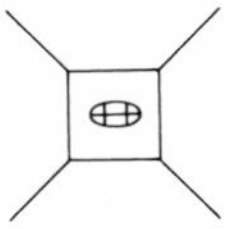 (79)

A related approach has been based on the homoaromaticity concept. A possible bishomocyclobutanienyl dication has been suggested on the basis of the relative acetolysis constants, which are 1.5 x 10^{-3}, 0.3, 3 x 10^{-6}, and 1, respectively [62].

The rate ratio in the last two compounds is rather large due to the inductive effect of the second tosyloxy group; by contrast this group retards much less in the unsaturated species. If the two groups leave simultaneously, they might leave the bishomodication (80) behind.

(80)

But in the related molecule (81), Tanida observed that the initial acetolysis product was (82), and that it could be intercepted; clearly the reaction to form (83) is stepwise [63].

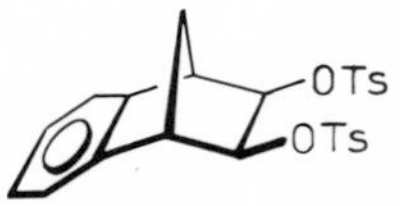

(81) 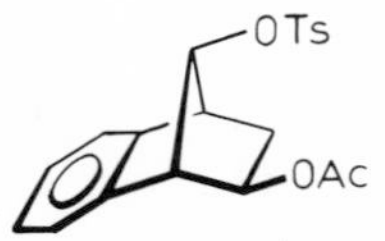 (82) 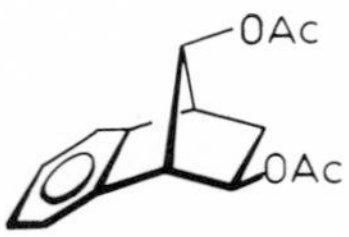 (83)

61. G.A. Olah, J.M. Bollinger, and A.M. White, J. Amer. Chem. Soc., 91, 3667 (1969).
62. J.B. Lambert and A.G. Holcomb, J. Amer. Chem. Soc., 91, 1572 (1969).
63. (a) H. Tanida and T. Tsushima, Tetrahedron Lett., 3647 (1969); see also (b) J.B. Lambert, H.G. Smith, and A.P. Jovanovich, J. Org. Chem., 35, 3619 (1970).

Lambert then found that (84) does not solvolyze significantly more slowly than the exo, exo- isomer; his interpretation of these relatively rapid rates now is that the presence of the inductively retarding second ester group reinforces the need for π-participation [64].

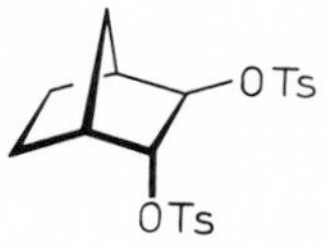

OTs
OTs

(84)

Greater success has been obtained in the more simple extensions of the highly stable triarylcarbonium ions and aromatic cations. Thus, Hart has found that the molecules (85) and (85) have i-factors in sulfuric acid of 5 and 7, respectively, clearly indicative of dicarbonium ions [65].

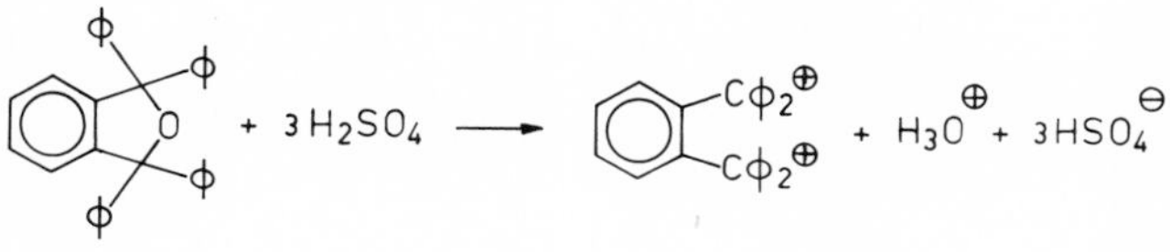

$$(85) + 3\,H_2SO_4 \longrightarrow C_6H_4(C\phi_2^{\oplus})_2 + H_3O^{\oplus} + 3\,HSO_4^{\ominus}$$

(85)

$$\phi_2COH-C_6H_4-COH\phi_2 + 4\,H_2SO_4 \longrightarrow \phi_2C^{\oplus}-C_6H_4-C^{\oplus}\phi_2 + 2\,H_3O^{\oplus} + 4\,HSO_4^{\ominus}$$

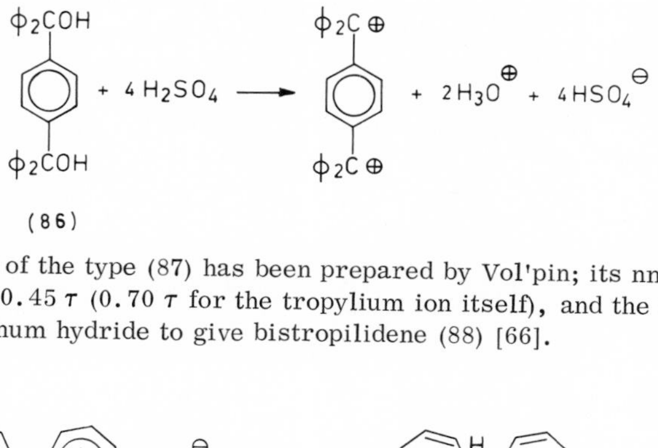

(86)

A ditropylium salt of the type (87) has been prepared by Vol'pin; its nmr spectrum shows a singlet at 0.45 τ (0.70 τ for the tropylium ion itself), and the salt reacts with lithium aluminum hydride to give bistropilidene (88) [66].

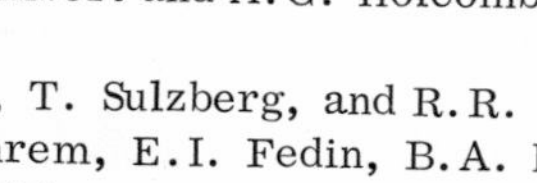

$(PCl_6^{\ominus})_2$

(87)

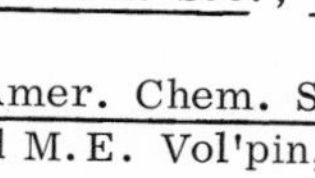

H
H

(88)

64. J.B. Lambert and A.G. Holcomb, J. Amer. Chem. Soc., 93, 2994, 3952 (1971).
65. H. Hart, T. Sulzberg, and R.R. Rafos, J. Amer. Chem. Soc., 85, 1800 (1963).
66. I.S. Akhrem, E.I. Fedin, B.A. Kvasov, and M.E. Vol'pin, Tetrahedron Lett., 5265 (1967).

Nmr spectra of solutions of acyl fluorides in antimony pentafluoride show the presence of acylium ions, and the extension to the difunctional analogs simply leads to diacylium ions of the type (89). The salts have proved to be isolable [67].

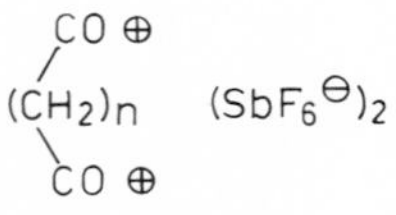

$CO^{\oplus}$
$(CH_2)_n$ $(SbF_6^{\ominus})_2$
$CO^{\oplus}$

(89)

When n is 1, the positive centers are evidently too close and a complex is formed, the nmr spectrum of which suggests the structure (90). However, when oxalyl difluoride is used, it is claimed that the salt (91) can be formed by means of a somewhat mysterious sequence of temperature changes.

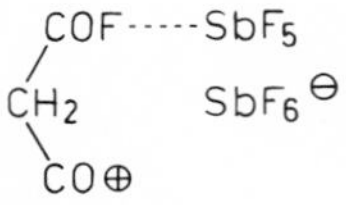

$COF \cdots SbF_5$
CH_2 $SbF_6^{\ominus}$
$CO^{\oplus}$

(90)

$CO^{\oplus}$
$CO^{\oplus}$ $(SbF_6^{\ominus})_2$

(91)

The infrared spectrum showed the presence of C≡O bonds at 2182 and 2235 cm^{-1}, and the normal band at about 1700 cm^{-1} was absent [68]. We may conclude this section with the description of another experiment by Olah [69] who found that upon dissolution in super acid media, each of the permethylated hydrocarbons (92), (93) and (94) is oxidized - and (95) is protonated - to a common ion likely to be (96).

$-(CH_3)_8$ $-(CH_3)_8$ $-(CH_3)_8$

(92) 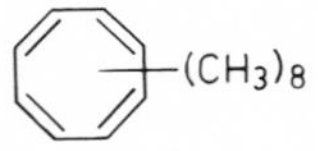 (93) 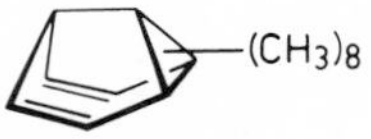 (94)

$-(CH_3)_8$

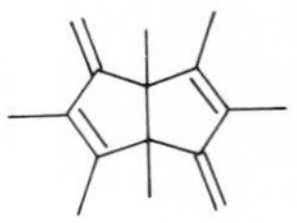

(95)  (96)

The nmr spectrum has 3 singlet peaks at 6.7, 7.7 and 7.9 τ in a 2:1:1 ratio; this rules out the possibly aromatic 6 π electron system (97), though not the structurally less likely species (98).

---

67. (a) G.A. Olah, S.J. Kuhn, W.S. Tolgyesi, and E.B. Baker, J. Amer. Chem. Soc., 84, 2733 (1962); (b) G.A. Olah and M.B. Comisarow, J. Amer. Chem. Soc., 88, 3313 (1966).
68. M. Adelhelm, Angew. Chem., Int. Ed. Engl., 8, 516 (1969).
69. J.M. Bollinger and G.A. Olah, J. Amer. Chem. Soc., 91, 3380 (1969).

(CH$_3$)$_8$ (CH$_3$)$_8$

(97) (98)

21-3. UNSATURATED CARBONIUM IONS

We have already observed that certain ions are greatly destabilized. There are for example ions which for geometric reasons cannot become planar (1-norbornyl), ions derived from primary carbon compounds (methyl), ions with inductively withdrawing substituents nearby and yet unable to reach the carbonium ion site and hence to stabilize it by electron sharing (1-cyano-2-norbornyl), ions derived from substrates subject to angle strain (7-norbornyl) and ions that are too crowded to be solvated very well (neopentyl). Another category of such species are those in which the carbonium ion site is part of an unsaturated or pseudosaturated system such as phenyl, vinyl and cyclopropyl ions. The use of highly reactive leaving groups such as triflate are making studies of these systems possible now in reasonable temperature ranges [70].

The solvolysis of cyclopropyl esters appears to be concerted with ring opening. This conclusion is supported by the stereospecificity (see p. 468), by the fact that α-methyl substitution increases the rate only moderately (by 10^3 or so), and that the rates are much faster (by 10^4 or so) than that of 7-norbornyl even though the angle strain is greater [71]. It is possible however to preserve the cyclopropyl moiety by the proper use of substituents; thus, (99) and (100) methanolyze to give (101) with the cyclopropyl ring still intact [72].

SΦ Cl Cl SΦ SΦ OCH$_3$

(99) (100) (101)

Evidently the sulfur atom stabilizes the α-carbonium ion site to such a degree that ring opening to the allylic isomer is not worthwhile. A variation of this theme is the solvolysis of 1-cyclopropylcyclopropyl esters (102), which occurs some 10,000 times faster than that of 1-i-propylcyclopropyl analogs and leads to largely unrearranged product [73]. In principle one could also hope to preserve the ring by

70. (a) T.M. Su, W.F. Sliwinski, and P. von R. Schleyer, J. Amer. Chem. Soc., 91, 5386 (1969); (b) F. Effenberger and K.E. Mack, Tetrahedron Lett., 3947 (1970).
71. P. von R. Schleyer, W.F. Sliwinski, G.W. Van Dine, U. Schöllkopf, J. Paust, and K. Fellenberger, J. Amer. Chem. Soc., 94, 125 (1972).
72. U. Schöllkopf, E. Ruben, P. Tonne, and K. Riedel, Tetrahedron Lett., 5077 (1970).
73. (a) J.A. Landgrebe and L.W. Becker, J. Amer. Chem. Soc., 89, 2505 (1967) and 90, 395 (1968); (b) B.A. Howell and J.G. Jewett, J. Amer. Chem. Soc., 93, 798 (1971).

making the possible allylic ion highly strained; for example, in propellane (103) solvolysis with ring opening could generate ion (104) which violates Bredt's rule. In fact the dichloride dimerizes even at room temperature to give (105); the strain in (103) is evidently even greater than in (104) [74].

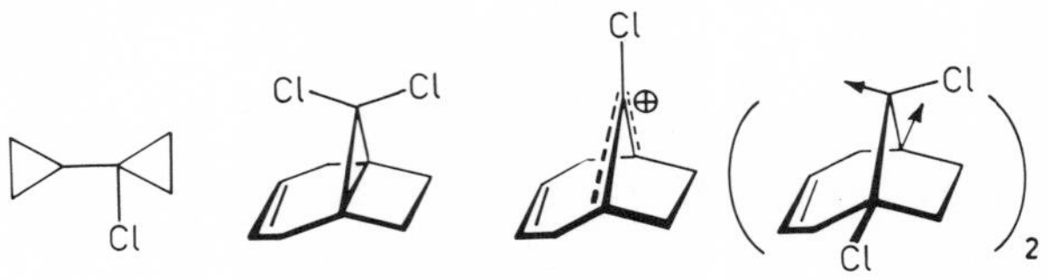

(102) (103) (104) (105)

It was discovered by Grob that vinyl bromides could solvolyze by the S_N1 mechanism if incipient ions were simultaneously of the benzylic or allylic type [75]; indicative of this mechanism are insensitivity to added lyate ions, sensitivity to solvent polarity, a very large ρ-value and a β-deuterium isotope effect similar to that observed for saturated cationic substrates [76].

The use of the triflate leaving group permits the routine study of the solvolysis of vinyl derivatives [77]. The questions that have been raised about vinyl ions are similar to those already encountered in our discussion of the more stable ions, and the methods used toward their solution are likewise similar.

The shape of such vinyl ions may obviously be either linear or bent. In several solvents and with several nucleophiles as trapping agents, cis- and trans-products were obtained in 1:1 ratio from (106) and (107); this suggests either a linear ion or a pair of readily equilibrated bent ions [78].

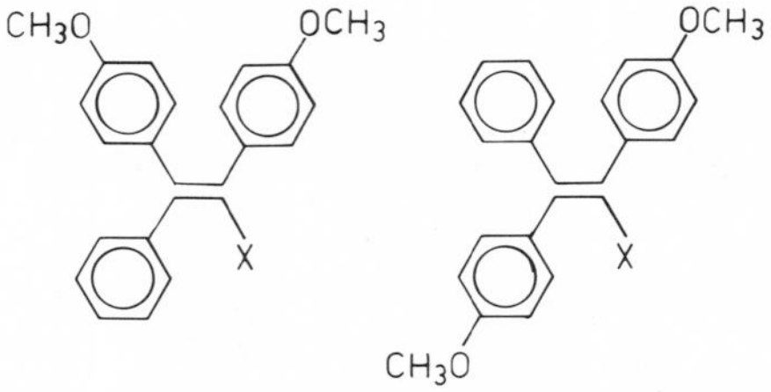

(106) (107)

74. P. Warner, R. LaRose, C. Lee, and J.C. Clardy, J. Amer. Chem. Soc., 94, 7607 (1972).
75. (a) C.A. Grob and G. Cseh, Helv. Chim. Acta, 47, 194 (1964); (b) C.A. Grob and R. Spaar, Tetrahedron Lett., 1439 (1969); (c) C.A. Grob and H.R. Pfaendler, Helv. Chim. Acta, 54, 2060 (1971).
76. W.M. Jones and D.D. Maness, J. Amer. Chem. Soc., 92, 5457 (1970).
77. (a) P.J. Stang and R. Summerville, J. Amer. Chem. Soc., 91, 4600 (1969); (b) review by M. Hanack, Accounts Chem. Res., 3, 209 (1970).
78. Z. Rappoport and Y. Apeloig, J. Amer. Chem. Soc., 91, 6734 (1969).

Perhaps more informative is the fact that vinyl ions which are necessarily bent are generated much more slowly than those that can assume a linear shape, as is obvious from the relative rate constants of the following series of triflates [79].

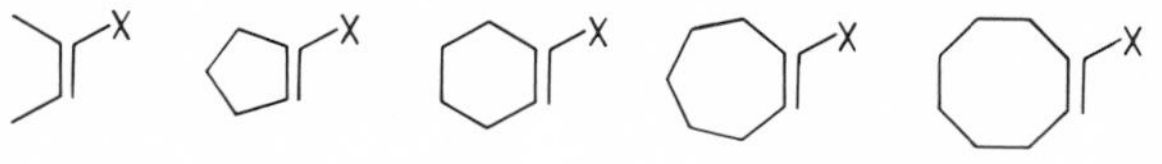

1 0.00003 0.01 0.3 2.7

Methyl substitution at the β-vinyl position seems to have little effect (solvolysis with α-hydrogen does not yet seem to have been accomplished); however, methyl in allylic positions seems to increase solvolytic rates very greatly, as can be seen from the following relative rate data [80].

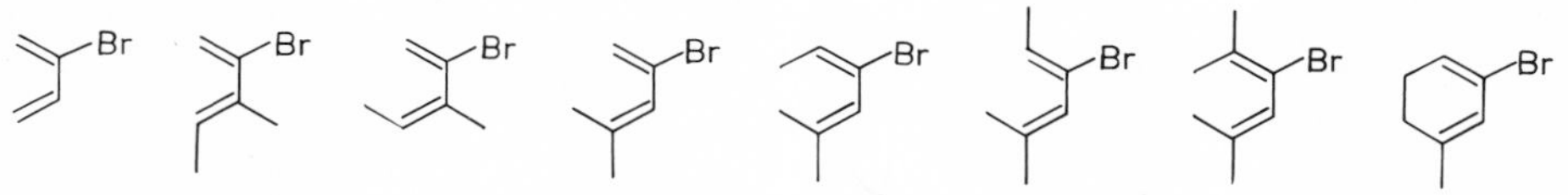

1 75 120 32 000 43 000 17 000 44 000 ~0

Neighboring group assistance has been found to occur in the case of sulfur [81]. If the α-ring in (108) is C^{14}-labelled, the activity winds up to an equal degree in both aryl rings in the products. If the solvent is methanol, the methyl vinyl ether is obtained with the aryl groups still trans; this means that the solvolysis took place with retention (in nitromethane, thiophenes (109) are obtained).

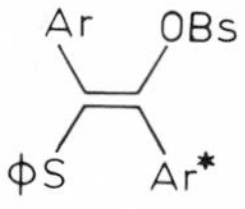

(108) 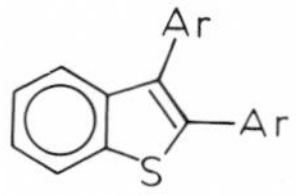 (109)

Finally, if the two aryl groups are in fact different, identical product mixtures are invariably obtained. All this argues for the symmetrical intermediate thiirenium ion (110), which indeed may have some aromatic stability.

79. W.D. Pfeifer, C.A. Bahn, P. von R. Schleyer, S. Bocher, C.E. Harding, K. Hummel, M. Hanack, and P.J. Stang, J. Amer. Chem. Soc., 93, 1513 (1971).

80. (a) C.A. Grob and R. Spaar, Helv. Chim. Acta, 53, 2119 (1970); (b) C.A. Grob and R. Nussbaumer, Helv. Chim. Acta, 54, 2528 (1971).

81. (a) G. Capozzi, G. Melloni, G. Modena and U. Tonellato, Chem. Commun., 1520 (1969); (b) A. Burighel, G. Modena, and U. Tonellato, Chem. Commun., 1325 (1971); (c) G. Modena and U. Tonellato, J. Chem. Soc. (B), 374 (1971).

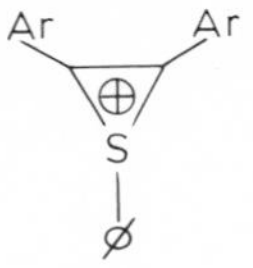

(110)

On the other hand, phenyl participation does not seem to occur in these systems; thus (111) gives a _50-50 cis-trans_ product mixture corresponding to the complete absence of stereospecificity [82].

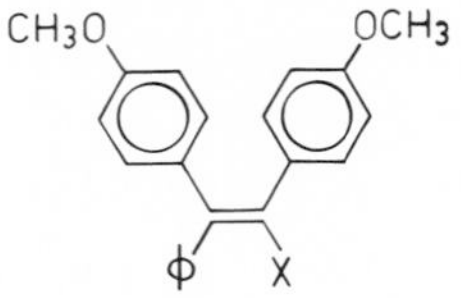

(111)

Rappoport discovered that in such systems _cis-trans_ isomerization of the substrate is sometimes observed concurrent with solvolysis; ion-pair return is clearly a factor in these reactions [83]. Hydride shifts have not yet been observed here, but a methyl shift has been claimed by Griesbaum in the addition of hydrochloric acid to the acetylene (112) on the basis of the products obtained ((113)-(116)).

(112) (113) (114) (115) (116)

Thus, (115) cannot arise from (113), and (116) cannot be obtained from 1-chloro-3, 3-dimethylbut-1-ene under the same conditions; hence these products must have arisen from a methyl shift in the vinyl cation [84].

The isomeric ions (117) should be obtainable from either allenic or propargylic precursors. Both pathways have been realized [85], but otherwise little is known as yet about these ions.

---

82. (a) G. F. P. Kernaghan and H. M. R. Hoffmann, _J. Amer. Chem. Soc._, _92_, 6988 (1970); (b) Z. Rappoport and Y. Apeloig, _Tetrahedron Lett._, 1817 (1970).
83. Z. Rappoport and Y. Apeloig, _Tetrahedron Lett._, 1845 (1970).
84. K. Griesbaum and Z. Rehman, _J. Amer. Chem. Soc._, _92_, 1416 (1970).
85. (a) H. G. Richey, J. C. Philips, and L. E. Rennick, _J. Amer. Chem. Soc._, _87_, 1381 (1965); (b) M. D. Schiavelli, S. C. Hixon, H. W. Moran, and C. J. Boswell, _J. Amer. Chem. Soc._, _93_, 6989 (1971).

(117)

21-4. METAL COMPLEXES OF CARBONIUM IONS

It is not possible to do justice to all the work that has been reported in this field, but the following observations are of interest. The complexed benzyl chloride (118) solvolyzes some 10^5 times faster than the uncomplexed molecule, so that the chromium atom is evidently able to satisfy an electron demand [86].

CH_2Cl

$Cr(CO)_3$

(118)

When the two 2-endo-benzonorbornenyl substrates (119) and (120) are compared, the latter solvolyzes about ten times more slowly than the former: The inductive effect is electron withdrawing.

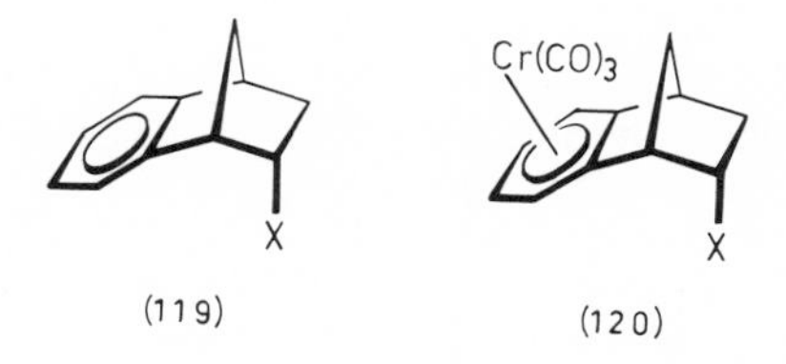

(119) (120)

This ratio increases to several hundred when the corresponding exo-epimers are compared; thus, the effect is magnified when the inductive effect can interfere with phenyl participation. The fast exo-rate is fully restored however if the chromium atom is in the endo-position as in (121); thus, in spite of this inductive effect, chromium can enhance solvolysis if it is in a position to participate directly [87].

X

$Cr(CO)_3$

(121)

86. W. S. Trahanovsky and D. K. Wells, J. Amer. Chem. Soc., 91, 5870 (1969).
87. (a) R. S. Bly and R. C. Strickland, J. Amer. Chem. Soc., 92, 7459 (1970); (b) D. K. Wells and W. S. Trahanovsky, J. Amer. Chem. Soc., 92, 7459 (1970).

This same question has been much debated in the case of ferrocenylmethyl derivatives (122), which solvolyze at very rapid rates [88]; it is not clear whether this should be explained as an example of benzylic resonance or of direct iron participation. A related question is whether the ion stabilized in trifluoroacetic acid has structure (123) or (124) (a metal-stabilized benzylic cation or a fulvene complex); this problem is as yet unresolved [89].

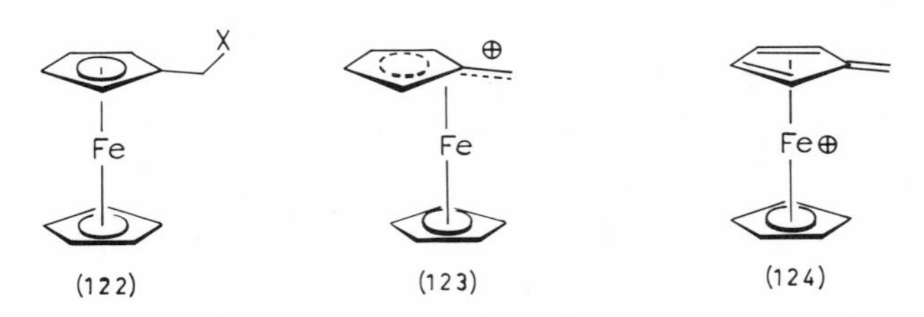

(122) (123) (124)

21-5. POSITIVE CHARGE AT OTHER ATOMS

There is evidence that trivalent cations are difficult to form if the charge must be centered on other atoms of Group IV. Thus, fluorotrimethylsilane does not ionize in cold antimony pentafluoride [90], and efforts to detect triphenylsilonium and -germonium ions have not been successful [91].

It is of course also more difficult to generate the equivalent nitrogen cations [92]. It has long been known that the Schmidt reaction of ketones with hydrazoic acid in sulfuric acid and the Beckmann rearrangement of oximes - both reactions proceeding to amides - involve an incipient nitrenium ion; however, in view of the specificity of the latter reaction it was supposed that the formation of the ion and the shift are concerted, and that the free ion stage is bypassed [93].

$$RR'C{=}O \xrightarrow{HN_3} RR'C(OH)N_3 \xrightarrow{H^{\oplus}} RR'C{=}N{-}N_2^{\oplus} \longrightarrow RC(=O){-}NHR' \xleftarrow{H^{\oplus}} RR'C{=}N{-}OH$$

Lansbury has discovered evidence that the charge at nitrogen does get developed sufficiently fully to permit cationic side reactions if the conditions are favorable. Thus, with the oxime (125) in which the electron-deficient nitrogen atom is

---

88. J.H. Richards and E.A. Hill, J. Amer. Chem. Soc., 81, 3484 (1959).
89. For a recent discussion, see M. Hisatome and K. Yamakawa, Tetrahedron, 27, 2101 (1971).
90. G.A. Olah and Y.K. Mo, J. Amer. Chem. Soc., 93, 4942 (1971).
91. G.P. Sollott and W.R. Peterson, J. Amer. Chem. Soc., 89, 6783 (1967).
92. S. Hünig, Helv. Chim. Acta, 54, 1721 (1971).
93. P.A.S. Smith, "Molecular Rearrangements", P. de Mayo, Ed., Interscience, New York, 1963; Chr. 8.

necessarily near the *t*-butyl group, the principal product in phosphoric acid is (126). Both of the lactams (127) and (128) also form: Evidently the stereospecificity is absent in this case. The corresponding indanone (129) also gives these products in the Schmidt reaction, though not in the same ratio; the intermediate is not so free that it gives identical mixtures regardless of generation. Finally, the orientation of the reacting groups appears to be critical: (130) gives only Beckmann products [94].

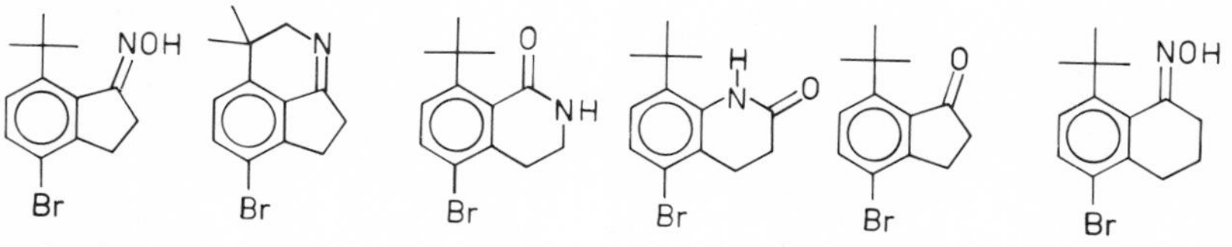

(125) (126) (127) (128) (129) (130)

Possibly an important feature of this reaction is that in non-cyclic oximes, nitrilium ions (131) are sometimes detectible or even isolable as intermediates [95]; this stage is clearly not possible in the cyclic analogs.

$$R_2-C\equiv \overset{\oplus}{N}-R_1$$

(131)

If one considers the difference between vinyl and alkyl cations, one might suppose that simple nitrenium ions stand a much better chance of being detected than the imonium ions just mentioned. An important factor in the field of nitrenium ions was the discovery by Haberfield in 1960 that N-chloro-N-methylaniline was an intermediate in the ring chlorinations of N-methylaniline [96]. This fact had long been suspected, but Haberfield [96] put an end to these speculations by simply isolating the material.

At about the same time it was learned that treatment of N-chloroazacyclononane (132) with silver ion gives rise to indolizidine (133) [97]. This transannular reaction had every appearance of an intermediate nitrenium ion, and with that news the rapid development of nitrenium ion chemistry began, principally in Gassman's laboratory [98].

---

94. (a) P.T. Lansbury, J.G. Colson, and N.R. Mancuso, J. Amer. Chem. Soc., 86, 5225 (1964); (b) P.T. Lansbury and N.R. Mancuso, Tetrahedron Lett., 2445 (1965) and J. Amer. Chem. Soc., 88, 1205 (1966).
95. (a) Y. Yukawa and T. Ando, Chem. Commun., 1601 (1971); (b) R.N. Loeppky and M. Rotman, J. Org. Chem., 32, 4010 (1967); for a detailed study of the Beckmann rearrangement, see (c) C.A. Grob, H.P. Fischer, W. Raudenbusch, and J. Zergenyi, Helv. Chim. Acta, 47, 1003 (1964).
96. P. Haberfield and D. Paul, J. Amer. Chem. Soc., 87, 5502 (1965).
97. O.E. Edwards, D. Vocelle, J.W. ApSimon, and F. Haque, J. Amer. Chem. Soc., 87, 678 (1965).
98. P.G. Gassman, Accounts Chem. Res., 3, 26 (1970).

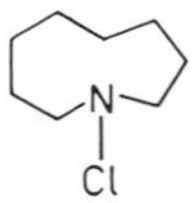

(132) 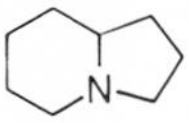 (133)

Nitrenium ions resemble carbonium ions in many ways. Thus, rearrangements are common, as the nitrogen atom tries to pass on the hot potato of positive charge to its neighbors. N-chloroisoquinuclidine (134) methanolyzes to give the rearranged product (135) [99].

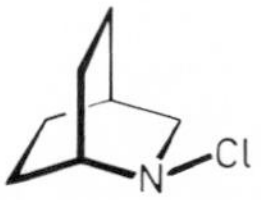

(134) 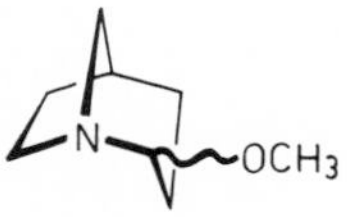 (135)

Internal return occurs; for example, (136) gives a good yield of (137).

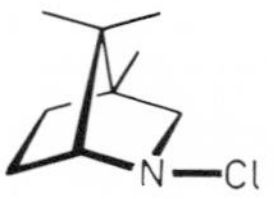

(136)  (137)

Interestingly, this reaction is strongly catalyzed by silver ion - this is a rare case of a carbonium ion getting chloride back from silver chloride [100]. The ions can be generated by the π route; for instance, when (138) is heated in methanol, products such as (139)-(141) are obtained [101].

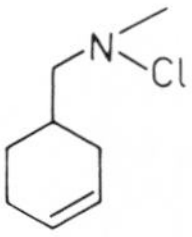

(138) 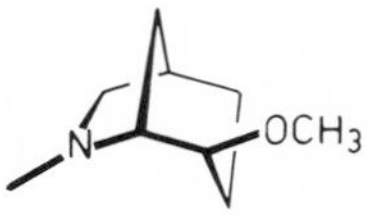 (139) 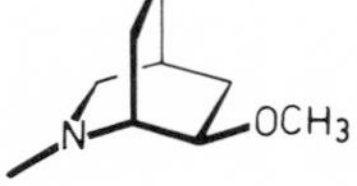 (140) 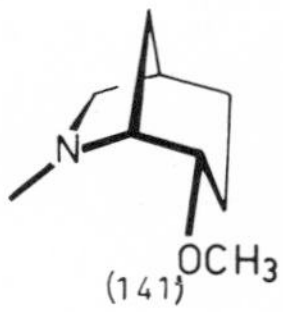 (141)

Furthermore, these reactions are not limited to N-chlorocompounds: A variety of other leaving groups has now been found to produce similar results [102]. The

99. P.G. Gassman and B.L. Fox, J. Amer. Chem. Soc., 89, 338 (1967).
100. P.G. Gassman and R.L. Cryberg, J. Amer. Chem. Soc., 90, 1355 (1968) and 91, 2047 (1969).
101. P.G. Gassman and J.H. Dygos, Tetrahedron Lett., 4745 (1970).
102. (a) P.G. Gassman and G.D. Hartman, Chem. Commun., 853 (1972); (b) P.G. Gassman, K. Shudo, R.L. Cryberg, and A. Battisti, Tetrahedron Lett., 875 (1972); (c) P.G. Gassman and K. Shudo, J. Amer. Chem. Soc., 93, 5899 (1971); (d) J.P. Fleury, J.M. Biehler, and M. Desbois, Tetrahedron Lett., 4091 (1969).

formation rate of these ions is nicely correlated by means of σ^+-constants; the rearrangement of N-chloroanilines to ring-chlorinated products has a ρ-value of -6.35 [103]. N-chloroaziridines solvolyze in a reaction concerted with ring opening; thus, solvolysis of (142) is some 10^5 times faster than that of the parent aziridine (since the initial product (143) hydrolyzes further the stereospecificity or mode of the ring opening is as yet not established) [104].

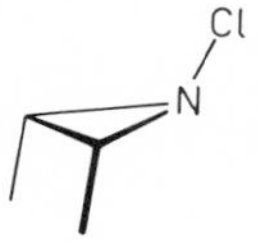

Cl

N

$CH_3CH{=}\overset{\oplus}{N}{=}CHCH_3$

(142) 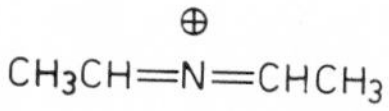 (143)

One important difference between carbonium and nitrenium ions is that the latter can at least on paper exist in two spin states: Singlet and triplet [95b, 105]. Gassman had noted that in the solvolysis of (136) a small amount of (144) forms, and that this might be due to hydrogen abstraction by a triplet nitrenium ion.

NH

(144)

He then found that if heavy atom solvents such as bromoform are added, this product becomes virtually the only one. Evidently the first intermediate is a singlet nitrenium ion (as spin conservation requires), and this partitions between rapid return or collapse and relaxation to the triplet ground state; another important conclusion is that these ions must have a reasonable life time ("have a discrete existence"). As yet there is little knowledge about the behavior of nitrenium ion precursors in magic acid, but there are some examples; for instance azoxybenzenes (145) can be converted in dications (146) [106].

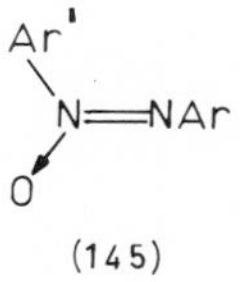

Ar'

N=NAr

O

$Ar'{-}N{=}N{-}Ar$ (⊕⊕)

(145) 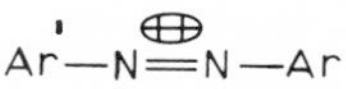 (146)

---

103. P.G. Gassman and G.A. Campbell, J. Amer. Chem. Soc., 93, 2567 (1971) and 94, 3891 (1972).
104. P.G. Gassman and J.H. Dygos, J. Amer. Chem. Soc., 91, 1543 (1969).
105. P.G. Gassman and R. L. Cryberg, J. Amer. Chem. Soc., 91, 5176 (1969).
106. G.A. Olah, K. Dunne, D.P. Kelly, and Y.K. Mo, J. Amer. Chem. Soc., 94, 7438 (1972).

The onium ions in nitrogen chemistry are well-known, stable, and of no special interest; however, one somewhat unexpected item that might be mentioned here is the fact that aryldiazonium ions scramble the two nitrogen atoms slowly as they hydrolyze to phenols [107]. The analogous reaction of aryl isocyanides to cyanides is well-known, but phenylacetylide ion does not rearrange even under very vigorous conditions [108].

Incipient oxenium ions have also been mentioned as intermediates, especially in perester decompositions; thus, hydroperoxide (147) reacts with nosyl chloride in pyridine to give some (148), and hypobromite (149) gives only this product [109].

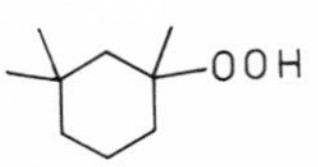

(147)  (148) 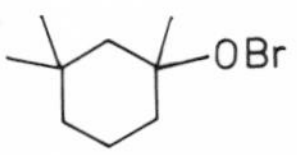 (149)

The solvolysis of perbenzoate esters (150) in methanol shows features (solvent dependence, migratory aptitudes) suggesting that neighboring carbon participation is extremely strong and that the free oxenium ion is bypassed in favor of the α-alkoxycarbonium ion (151) [110].

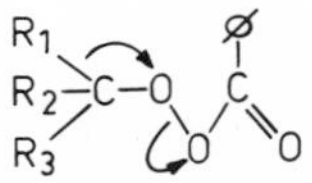

(150) 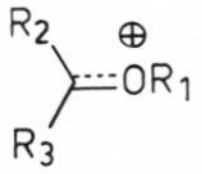 (151)

107. (a) J.M. Insole and E.S. Lewis, *J. Amer. Chem. Soc.*, **85**, 122 (1963) and **86**, 32 (1964); (b) E.S. Lewis and R.E. Holliday, *J. Amer. Chem. Soc.*, **88**, 5043 (1966) and **91**, 426 (1969).
108. J. Casanova, M. Geisel, and R.N. Morris, *J. Amer. Chem. Soc.*, **91**, 2156 (1969).
109. R.A. Sneen and N.P. Matheny, *J. Amer. Chem. Soc.*, **86**, 3905 (1964).
110. E. Hedaya and S. Winstein, *J. Amer. Chem. Soc.*, **89**, 1661 (1967).

Chapter 22

CARBANIONS

22-1. INTRODUCTION

Carbanions have been encountered repeatedly elsewhere in this book (Chrs. 9, 12, 14 and 20). Even more than free radicals, their presence in organic chemistry is so pervasive that their complete description cuts across virtually every other area in that field. This chapter is therefore limited to a few special topics that include the reactions of ambident anions, the preferred shape of these ions, and counterion chelation [1].

22-2. AMBIDENT ANIONS [2]

It had been known for a long time that the isomer obtained in the preparations of nitriles and isonitriles (isocyanides) depends on whether sodium or silver cyanide

1. For a much more complete description, see D. J. Cram, "Fundamentals of Carbanion Chemistry", Academic Press, New York, 1965, and Chem. Eng. News, 41, 92 (Aug. 19, 1963).
2. Reviews: (a) S.A. Shevelev, Russ. Chem. Rev. (English Transl.), 39, 844 (1970); (b) R. Gompper, Angew. Chem., Int. Ed. Engl., 3, 560 (1964); (c) A. L. Kurts and I. P. Beletskaya, Bull. Acad. Sci. USSR, Chem. Sci. (English Transl.), 781 (1970).

is used, and that a similar difference exists in the reactions of sodium and silver nitrite.

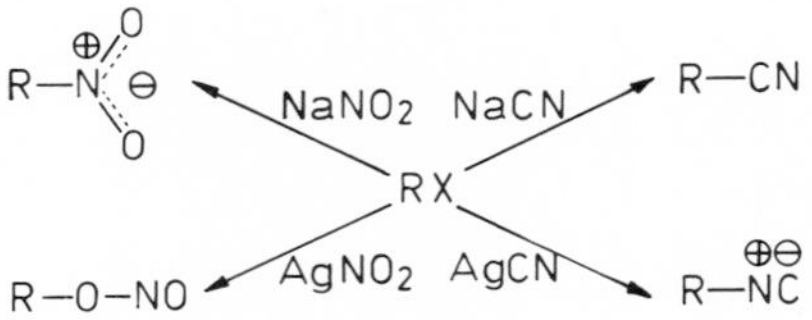

In 1954, Brändström in a review [3] attempted to rationalize the alkylations of salts of keto-enol tautomeric compounds by a mechanism involving initial chelation and solvation, then exchange of solvent molecule and alkylating agent in the solvation sphere, and finally reaction.

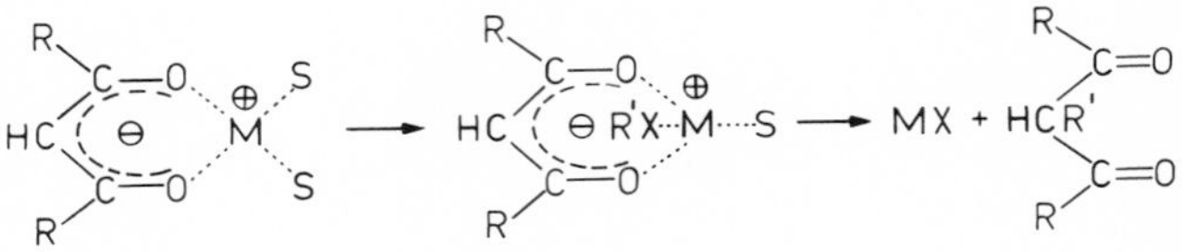

The following year Kornblum in a classic paper drew all these reactions together [4]. He termed the anions involved in these reactions ambident anions; these are anions therefore for which resonance structures can be written such that the negative charge must in part be located at several different atoms. Cyanide ion (1), nitrite anion (2), the anions derived from ethyl acetoacetate (3), acetylacetone (4), phenol (5), fluorenone oxime (6) and even simple ketones (7) and imines (8) all fall in this category [5].

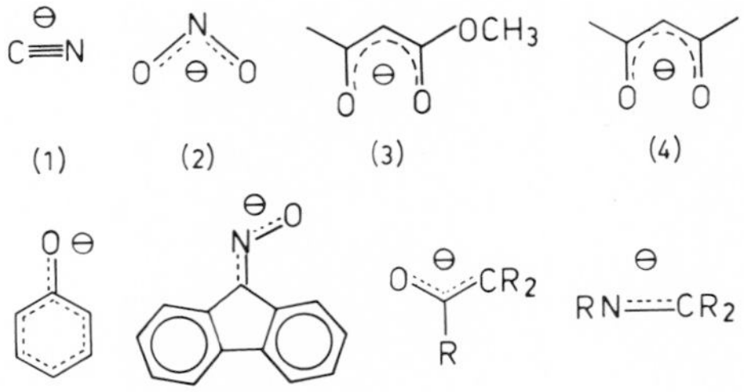

Kornblum's paper has a resounding impact, and because of the importance of many of these species in synthesis, the simple tool available for study (product analysis), and the variety of choices possible in conditions and structure, investigations were

3. A. Brändström, Arkiv Kemi, 6, 155 (1953) and 7, 81 (1954).
4. N. Kornblum, R.A. Smiley, R.K. Blackwood, and D.C. Iffland, J. Amer. Chem. Soc., 77, 6269 (1955).
5. For a fairly complete listing and review, see ref. 2b.

touched off in many laboratories in search of generalizations that would lead to a comprehensive understanding of the field.

It should be said at the outset that these generalizations seek to explain the kinetic product ratios. It is therefore important in each study that experiments be done that guarantee the stability of the products; for if these can interconvert by any pathway under the reaction conditions, then the results describe the thermodynamic ratios. This point is perhaps obvious, but such experiments are not always done.

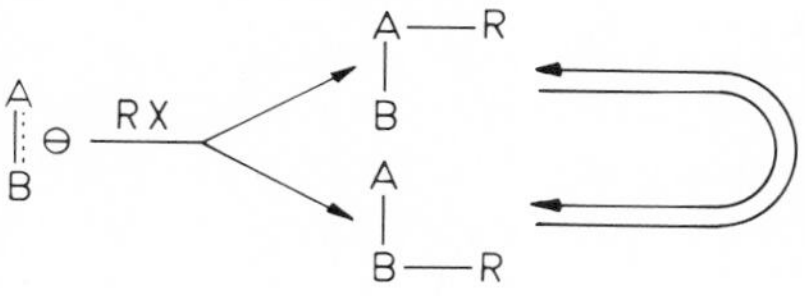

kinetic control thermodynamic control

The need for them was recently demonstrated in a study of the reaction of benzoylacetone anion (9) with acetyl chloride [6]. Under gentle conditions the enol ester (10) is the main product; this rearranges under somewhat more vigorous conditions to (11), and if treated to still higher temperatures, the triketone (12) finally obtains.

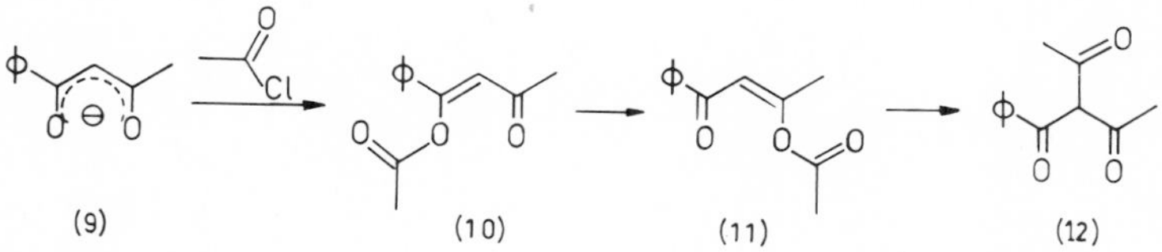

(9) (10) (11) (12)

In this case as in many others, the kinetically and thermodynamically favored products are not the same. The initial attack often occurs at the site of the most electronegative atom, since the charge is concentrated principally there; on the other hand, if the products are considered, resonance structures suggest that this site in the end would have been able to satisfy its electron demand better if the alternative reaction had occurred.

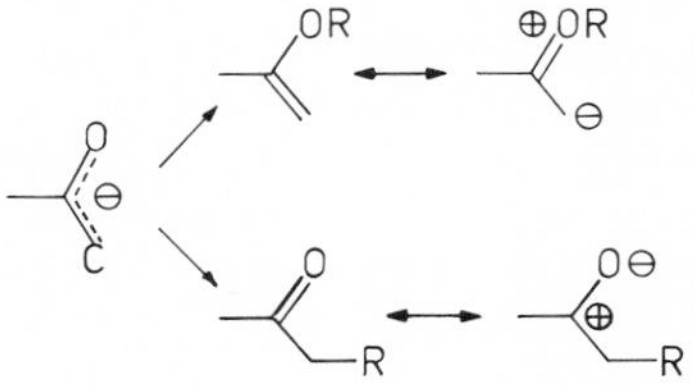

---

6. M. Suama, Y. Nakao, and K. Ichikawa, Bull. Chem. Soc. Japan, 44, 2811 (1971).

The picture is similar with ambident anion protonation, where enols are generally less stable than keto forms unless resonance changes the relative stability; but one important factor that plays a role in the tautomeric ratios - internal hydrogen bonding - is absent here.

A second factor that should be kept in this discussion is that the actual product distribution - however important from a synthetic or technological point of view - gives a somewhat exaggerated picture of these site preferences. Thus, if in a given case 90% of the reaction is O-alkylation and 10% occurs at carbon, then $\Delta\Delta G^{\ddagger} = -RT \ln O/C \approx 1.3$ kcal/mole at room temperature - while the free energies of activation themselves might be 25 kcal/mole or more. It should be added to this that activation free energies are temperature dependent, and that small differences in the entropy and enthalpy terms can easily change a result such as principal C-alkylation at one temperature to chiefly O-alkylation at another.

While it is clear that the resonance structures such as (13) and (14) are not equal in weight, it would be desirable to know what the actual charge distribution is. To this end Kloosterziel has compared the nmr spectrum of the anion of crotonaldehyde (15) with that of the anion of 1,4-pentadiene, (16). The chemical shifts (in τ) clearly indicate the difference, and show that the charge is located principally on the oxygen atom [7] - at least in the medium of liquid ammonia. The coupling constants between protons on adjacent atoms are all in the range of 10-15 Hz, indicating that both anions have the W-conformations indicated.

O, $CH_2^{\ominus}$ ⟷ $O^{\ominus}$, CH_2 — 5.89, 5.18, 6.22, 3.65, 2.21 — 6.99, 5.65, 7.42, 3.87

(13) (14) 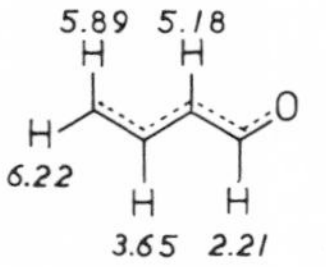 (15) 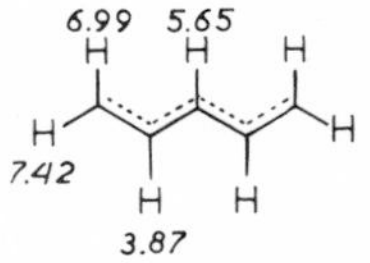 (16)

In the remainder of this section we discuss the relative reactivities in terms of the reaction medium, the nature of the anion, and that of the alkylating agent.

One of the most important features in these alkylation and acylation reactions is the medium. At one extreme, they can be carried out heterogeneously with only the alkylating agent and the salt of the ambident anion present; at the other, they can be done in media in which the salt is not only in solution but present in dissociated form. In this spectrum of media it is obvious that one is changing the equilibrium constants of the following processes. In this scheme we begin with an ionic solid in which each ion is surrounded by as many counterions as steric factors will allow; reactions are then possible only at the surface of the crystallites. If a solvent is added, competition between ionic and ion-dipole interactions may lead to dissolution. In organic media the salt is then likely to be present in the form of ion aggregates of undetermined size. Dilution causes dissociation to smaller

---

7. G. J. Heiszwolf and H. Kloosterziel, Rec. Trav. Chim. Pays-Bas, 86, 807 (1967).

clusters, then to ion-pairs, and finally to individual ions. These ionic solutions cannot exist unless at least one of the two ions can interact strongly with the solvent; if the latter is able to engage in hydrogen bonding (such as alcohols), the anion is especially solvated; if it is a good acceptor of hydrogen bonds (*i.e.*, it is strongly basic), then the metal ion is primarily responsible for the dissociation.

There is an eminently sensible generalization which explains a variety of effects observed in ambident anion chemistry, namely: The freer the anions, the greater the tendency for alkylation at the most electronegative sites [8]. As we shall see below, virtually all known effects of solvents, concentrations, additives, counterions and even structure are explainable in terms of this generalization.

Both counterions and solvent molecules (if of the hydrogen bonding variety) are likely to be near and around the site of highest electron density, which means that alkylation at the alternative positions requires the least initial reorganization. If on the other hand the anion is relatively unencumbered, the alkylating agent is more likely to be able to penetrate to the more electronegative atom.

It is not always easy to say by which branch or branches of the manifold of ionic species the products appear. In principal the anion may be present free and in all the types of associated forms, and since the equilibrations between them are likely to be rapid, the products may entirely derive from a form present in only minute amounts - if it reacts much more rapidly than the others. In spite of this problem, many results can be readily rationalized. Thus, when potassium *p-t*-octylphenoxide (17) is treated with allyl halides in toluene, a medium in which the salt is insoluble, the product is initially *o*-allyl-*p-t*-octylphenol (18). As the reaction progresses and the properties of the medium change accordingly, the remaining salt of (17) goes into solution; simultaneously, the product changes to the ether (19). In the final stages (19) is in fact the only product [9].

t-C_8H_{17}–C_6H_4–$O^{\ominus}$ (17) *t*-C_8H_{17}–C_6H_3(allyl)–OH (18) *t*-C_8H_{17}–C_6H_4–O–allyl (19)

Common ion addition should promote C-alkylation of β-ketoesters. Several such sodium salts have been examined for this feature, and C-alkylation was indeed raised by the addition of sodium perchlorate [10].

The effect to be expected of dilution has been elegantly demonstrated by a group of Russian chemists [11]. They reasoned that with salts of β-ketoesters dilution should produce more free ions, and furthermore that while the associated

---

8. W. J. le Noble, *Synthesis*, 2, 1 (1970), and references quoted there.
9. N. Kornblum and A. P. Lurie, *J. Amer. Chem. Soc.*, 81, 2705 (1959).
10. G. Bram, F. Guibé, and M. F. Mollet, *Tetrahedron Lett.*, 2951 (1970).
11. A. L. Kurts, A. Macias, I. P. Beletskaya, and O. A. Reutov, *Tetrahedron Lett.*, 3037 (1971).

anions probably have the U-conformation (20), electrostatic repulsion should favor the W-conformation (21) for the free species. The results are in excellent agreement. When the potassium salt, dissolved in low concentration (0.01 M) in dimethylformamide (22) (DMF; one of the dipolar solvents), is treated with ethyl tosylate, the O/C ratio (products (23)/(24)) is about 8; furthermore, almost all of the O-product is in the Z-form. When the concentration is gradually raised, finally to 7.5 M, the product (24) becomes predominant; at the same time, the oxygen alkylated product begins to contain a significant amount of the E-isomer (25).

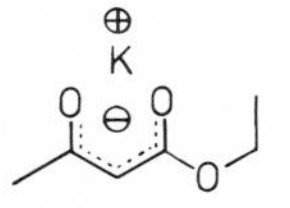

(20) 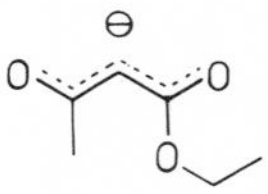 (21) 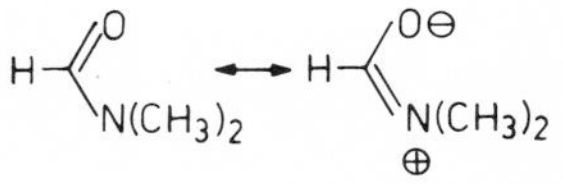 (22)

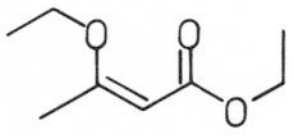

(23) 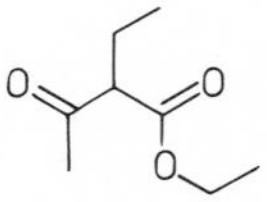 (24) 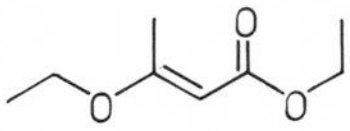 (25)

A similar finding was reported by Miller [12]. With salts of ethyl sodioacetoacetate only the Z- and E O-alkylated products (26) and (27) are obtained if (28) is the alkylating agent.

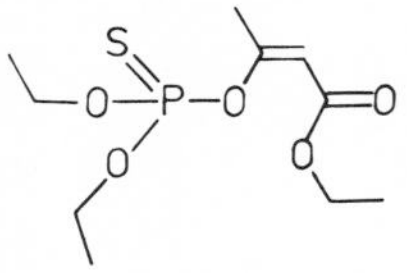

(26) 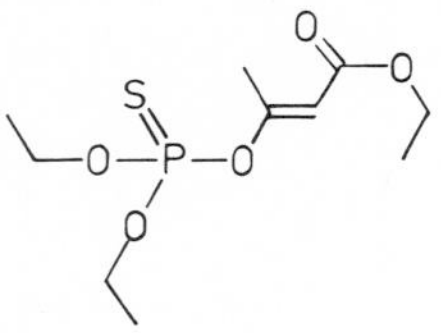 (27) 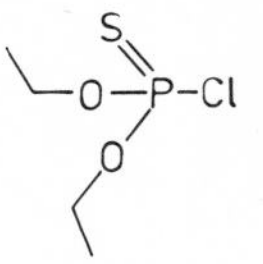 (28)

In highly polar solvents (27) is nearly the exclusive product, but in benzene the opposite is true. Furthermore, the ratio trans/cis declines as the concentration is raised [13].

These experiments tie in nicely with some done by Zaugg. He found [14] that the addition of certain "additives" such as (29), to a reaction mixture consisting of

---

12. B. Miller, H. Margulies, T. Drabb, and R. Wayne, Tetrahedron Lett., 3801 (1970).
13. B. Miller, H. Margulies, T. Drabb, and R. Wayne, Tetrahedron Lett., 3805 (1970).
14. (a) H.E. Zaugg, B.W. Horrom, and S. Borgwardt, J. Amer. Chem. Soc., 82, 2895 (1960); (b) H.E. Zaugg, J. Amer. Chem. Soc., 82, 2903 (1960).

ethyl sodio-*n*-butylmalonate (30) and *n*-butyl halides in benzene, greatly increased the alkylation rate, presumably by breaking up large clusters of the salt. Later he was able to show [15] that these same salts have uv spectra which are cation dependent in non-polar media such as glyme (29) but not in the dipolar medium (22).

(29) 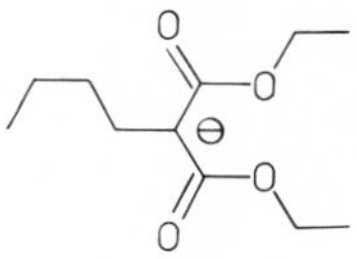 (30)

A quantitative dissection was carried through by Smith [16]. He studied the reaction of sodium 9-fluorenone oximate (31) with methyl iodide in a mixture of *t*-butyl alcohol and acetonitrile, which gives both *N*- and *O*-methylated products (32) and (33).

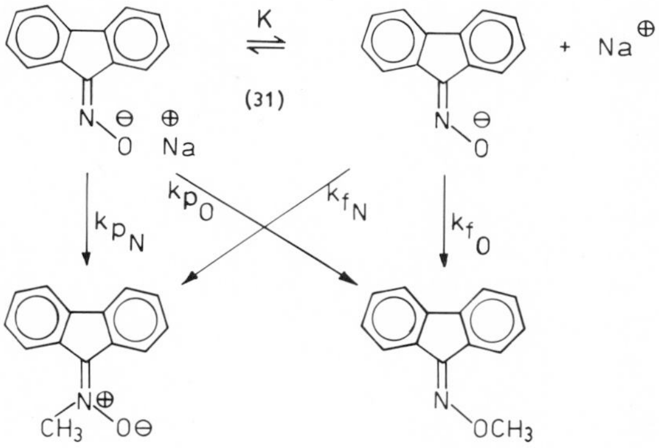

(31)

(32) (33)

Conductance measurements permitted the determination of the equilibrium constant K for ion-pair dissociation, and a combination of rate and product measurements allowed the evaluation of the four rate constants. This analysis revealed that the free anions have an N/O ratio of about 0.5, whereas that of the ion-pair is more than 2. Both of these values depended on extrapolations from observations in the manageable part of the concentration range; but they could be measured directly in later experiments done in the presence of reagents that powerfully solvate sodium ion (crown ethers, see p. 836 and others that generate free sodium ions and thereby encourage association of the oximate (e.g., sodium tetraphenylboride)) [17].

---

15. H.E. Zaugg and A.D. Schaefer, *J. Amer. Chem. Soc.*, *87*, 1857 (1965).
16. S.G. Smith and D.V. Milligan, *J. Amer. Chem. Soc.*, *90*, 2393 (1968).
17. (a) S.G. Smith and M.P. Hanson, *J. Org. Chem.*, *36*, 1931 (1971). A similar effect is now known in β-ketoesters: See (b) G. Bram, F. Guibé, and P. Sarthou, *Tetrahedron Lett.*, 4903 (1972).

If ionizing, basic solvents such as dimethylformamide (22) are used as alkylation solvents, a powerful tendency to give oxygen-alkylated products develops. Thus, benzyl bromide reacts with sodium β-naphthoxide in tetrahydrofuran to give 60% (34) and 36% (35); in (22) the former product only is formed (97%) [18]. An even more extreme case is the acetylation of benzoylacetone anion (9); whereas 90% of (12) forms in benzene, none is obtained at all in (22) [6].

(34) (35)

In 1964, Brower used the even more ionizing and basic solvent dimethyl sulfoxide (36) (DMSO, [19]) to achieve about 20% of oxygen alkylation (with ethyl bromide) of ethyl acetoacetate - a reaction which had until that time only rarely been observed [20] (it was found some time afterwards that small amounts of O-alkylates sometimes obtain also in alcoholic solvents [21]).

Eventually the most basic solvent of all, hexamethylphosphoramide, (37) [22] (HMPA [22]) came into use.

(36) (37)

The chemical shift of the proton of chloroform is often used to measure this basicity. In dioxane, this proton is shifted downfield by 0.6 ppm, in acetone by 0.9 ppm, in dimethylformamide (22) by 1.3 ppm, in (36) by 1.3 ppm, and in (37) by over 2 ppm. It has also been noted that benzyl magnesium chloride is colored red in this solvent: One of the rare cases of dissociation of a Grignard reagent. In (37) still further increases in O-alkylation were observed (see also p. 829) [23].

---

18. N. Kornblum, R. Seltzer, and P. Haberfield, J. Amer. Chem. Soc., 85, 1148 (1963). These authors explained their data on an alternative basis, however.
19. (a) For review of the properties of this solvent, see D. Martin, A. Weise, and H.J. Niclas, Angew. Chem., Int. Ed. Engl., 6, 318 (1967); see also (b) E.J. Corey and M. Chaykovsky, J. Amer. Chem. Soc., 87, 1345 (1965).
20. K.R. Brower, R.L. Ernst, and J.S. Chen, J. Phys. Chem., 68, 3814 (1964).
21. S.T. Yoffe, K.V. Vatsuro, E.E. Kugutcheva, and M.I. Kabachnik, Tetrahedron Lett., 593 (1965).
22. For review, see H. Normant, Angew. Chem., Int. Ed. Engl., 6, 1046 (1967) or Russ. Chem. Rev. (English Transl.), 39, 457 (1970).
23. (a) W.J. le Noble and H.F. Morris, J. Org. Chem., 34, 1969 (1969); (b) A.L. Kurts, A. Macias, I.P. Beletskeya, and O.A. Reutov, Russ. J. Org. Chem. (English Transl.), 7, 2323 (1971).

The degree of dissociation depends not only on the solvent, but also on the cation. One would intuitively expect that the larger cations should interact more weakly with the anion; it has indeed been found that the potassium salts of β-diketones are dissociated more than those of sodium, and these in turn more than the lithium salts [24]. Several studies have been carried out concerning the effect of the metal ion on alkylation ratios. Thus, when ethyl acetoacetate is allowed to react with *n*-propyl chloride in (36), the percentage oxygen alkylated product declines from 47 to 23 from potassium to lithium [25]; similar observations apply to acetylacetone [26], 2-carbethoxycyclopentanone (38) [27], the octalin derivative (39) [28] and butyrophenone (40) [29].

(38) (39) (40)

It was furthermore noted that if ethyl tosylate was used to alkylate ethyl acetoacetate in DMF, not only did the O/C ratio steeply decline from cesium to lithium, but so did the trans/cis ratio of the enol ether [11]. Miller found a similar metal effect in his ratios of (26) and (27) [12]. Divalent metals may be expected to be especially tightly bound, and not surprisingly it was found that sodium salts of β-ketoesters give much more O-alkylated products than the corresponding magnesium salts [30].

In the most basic solvents, dissociation may be far enough advanced that only the free anions are alkylated and no cation effect should be observable. Zook [31] has noted that the alkylations of (40) in DMSO were much faster than in glyme, gave much more O-alkylation and were more nearly independent of cation. The cation effect of ethyl acetoacetate, easily demonstrable in DMSO [25], has disappeared in HMPA [23]. The free ion is then not virtually the only species present, as can be shown by conductance measurements, but it is the only one reacting [32].

---

24. D.C. Luehrs, R.T. Iwamoto, and J. Kleinberg, Inorganic Chemistry, 4, 1739 (1965).
25. (a) W.J. le Noble and J.E. Puerta, Tetrahedron Lett., 1087 (1966); (b) G. Brieger and W.M. Pelletier, Tetrahedron Lett., 3555 (1965).
26. A.L. Kurts, N.K. Genkina, I.P. Beletskaya, and O.A. Reutov, Bull. Acad. Sci. USSR, Div. Chem. Sci. (English Transl.), 1237 (1971), and Doklady Akad. Nauk USSR (English Transl.), 188, 775 (1969).
27. D.M. Pond and R.L. Cargill, J. Org. Chem., 32, 4064 (1967).
28. F.H. Bottom and F.J. McQuillin, Tetrahedron Lett., 1975 (1967).
29. (a) H.D. Zook and T.J. Russo, J. Amer. Chem. Soc., 82, 1258 (1960); (b) H.D. Zook and W.L. Gumby, J. Amer. Chem. Soc., 82, 1386 (1960).
30. J.P. Ferris, B.G. Wright, and C.C. Crawford, J. Org. Chem., 30, 2367 (1965).
31. H.D. Zook and J.A. Miller, J. Org. Chem., 36, 1112 (1971).
32. A.L. Kurts, A. Macias, I.P. Beletskaya, and O.A. Reutov, Tetrahedron, 27, 4759 (1971).

O-alkylation can be promoted not only by solvents able to solvate cations for polar reasons, but also by those that can do it because of a favorable geometry. The alkylation of (40) for example is faster by about 10^4 in diglyme (41) than in ether, and the O/C ratio is also much greater. Conductance studies furthermore show that the salt is dissociated to a much greater extent as well [33].

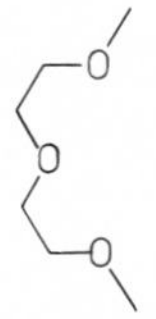

(41)

Solvents able to engage the ambident anion with hydrogen bonds promote attack at the less electronegative site. Sodium phenoxide, for example, which with allyl chloride produces only allyl phenyl ether (42) at room temperature in solvents such as DMF or tetrahydrofuran (THF), gives about 50% of *o*- and *p*-allylphenols (43) and (44) in water and fluorinated alcohols, and up to 75% in phenol itself [34].

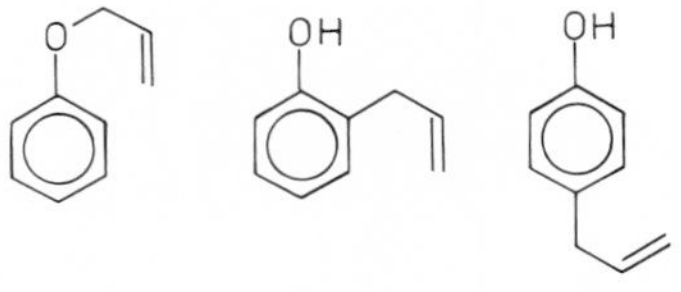

(42) (43) (44)

The proposition that freeness of the anion promotes attack at its most electronegative site is nicely borne out by the pressure effects that have been observed. If it is realized that pressure promotes dissociation into ions (see p. 801) as well as the solvation of these ions, it should be anticipated that O/C ratios should increase in those cases in which ion-pairs are largely responsible for the reaction, but should decrease in others in which the separate anions might become better solvated. This is exactly as found. The allylation of phenol in water for example which normally produces (42), (43), and (44) in the ratio of 60:20:20, at 7000 atmospheres gives 40:20:40. Thus, it appears that the increased solvation sphere interferes strongly with the oxygen reaction, less so with the *o*-positions and not at all with the *p*-position. Similarly, in methanol in this pressure range (43) increased from 0.5 to 5%, and (44) from less than 0.1 to 3% [35].

---

33. H.D. Zook, T.J. Russo, E.F. Ferrand, and D.S. Stotz, *J. Org. Chem.*, 33, 2222 (1968).
34. N. Kornblum, P.J. Berrigan, and W.J. le Noble, *J. Amer. Chem. Soc.*, 82, 1257 (1960) and 85, 1141 (1963).
35. (a) W.J. le Noble, *J. Amer. Chem. Soc.*, 85, 1470 (1963); (b) Ref. 20.

The methylation of (31) on the other hand gives more O-alkylate under pressure if Smith's conditions are employed, but the O/N ratio decreases at high dilution. In the latter case the pairs are presumably largely dissociated already and pressure then promotes H-bonding by the t-butyl alcohol [36]. In still another case the benzylation of 1-methyl-2-naphthoxide ion to give (45) and (46) leads to an increase of (46) under pressure if methanol is the solvent, but the same reaction is unaffected by pressure in glyme [37].

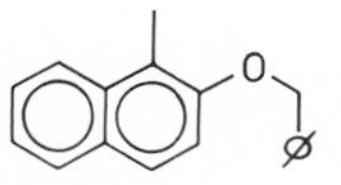

(45) 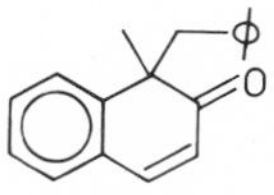 (46)

An ingenious new type of alkylation procedure has recently been described that may open up new possibilities in directing this reaction ("ion-pair extraction reactions"). Use is made of a rapidly stirred two-phase system, for instance chloroform and water. The former contains the organic reagents, the latter the base and some tetraalkylammonium salt. The ambident anion is transported into the organic layer as an ion-pair with the large ammonium cation as escort, and alkylated there; the cation then returns to the water to start the cycle again. One of the advantages is that one avoids the many secondary reactions that often occur in the harsh medium of aqueous base [38]. The technique can be extended to other reactions as well, for instance to the use of inorganic oxidizing agents [39].

Subtle structural effects in the ambident anion may lead to large changes in relative reactivity. When cyclic and acyclic β-ketoesters such as (47) and (48) are compared, the former reacts much more slowly and gives rise to less O-alkylate.

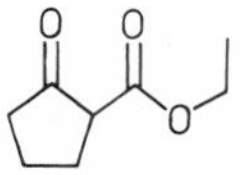

(47) 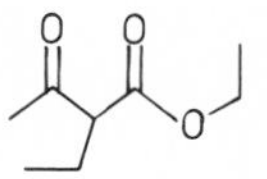 (48)

The reason for this is its inability to assume a W-conformation, and hence its reduced tendency to form free anions; this is also clear from conductance measurements [40]. As the ring is enlarged, the O-alkylation reaction becomes more important; thus, the C/O-ratio varies from 11 for (49) to 0.3 for (50) (with ethyl α-bromoacetate in a mixed solvent) [41].

36. W. J. le Noble and S. K. Palit, Tetrahedron Lett., 493 (1972).
37. W. J. le Noble, J. Phys. Chem., 68, 2361 (1964).
38. A. Brändström and U. Junggren, Acta Chem. Scand., 23, 2536, 3585 (1969), and Tetrahedron Lett., 473 (1972).
39. W. P. Weber and J. P. Shepherd, Tetrahedron Lett., 4907 (1972).
40. S. J. Rhoads and R. W. Holder, Tetrahedron, 25, 5443 (1969).
41. A. Chatterjee, D. Banerjee, and S. Banerjee, Tetrahedron Lett., 3851 (1965).

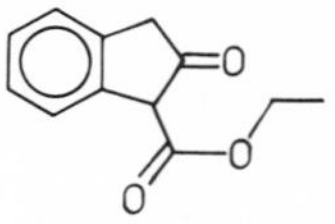

(49) 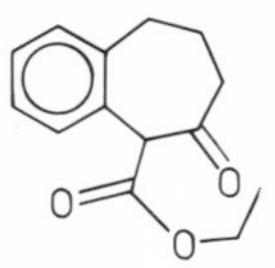 (50)

The larger rings are much less rigid; non-planarity would probably favor an increase in the charge density at oxygen.

Steric effects may become very large. Thus, under conditions that lead to exclusive O-methylation of phenoxide ion, the i-propylation of 2, 6-di-t-butylphenoxide gives only C-alkylation [42]. Substituents that exert large electronic effects can also play an important role (see for example p. 833).

Much has been written about the effect of the alkylating agent. Kornblum, noting that the ambident anion reactions involving silver ion tend to favor alkylation at the most electronegative site, put forward a suggestion that became known as the S_N1-rule: The greater the degree of carbonium ion character at the site of the carbon atom being substituted, the greater is its tendency to react with the anionic site of greatest electron density [4]. Although the rule nicely rationalized the results available at that time, further predictions made on that basis have not always fit. One of the more serious difficulties is that allyl and benzyl halides, which are capable of S_N1-solvolysis when other primary halides are not, tend to give C alkylation with phenols and β-ketoesters. Recently the cyclopropylcarbinyl group was found to react this way also [43]. Another problem is that with p-substituted benzyl halides, the nature of the substituent has no effect on the C/O ratio in ethyl acetoacetate [23]. The leaving group effect (see further p. 833) is such that among the halogens, the iodide is consistently best in promoting C-alkylation. Finally, stable carbonium ion salts such as tropylium bromide give only C-alkylation with 2-nitro-2-propane [44] and ethyl acetoacetate [45] (however, these findings could be the result of thermodynamic control; see also p. 834). The rule has been defended [2a] with the statement that these are S_N2 reactions, and hence that the ability of these reagents to solvolyze via an S_N1 path is immaterial; however, it may be said that all ambident anion reactions are S_N2.

As an alternative, an S_N2 rule has been proposed [25a]; the greater the S_N2 reactivity, the greater the tendency toward substitution at the least electronegative site. This alternative generalization is not as squarely in contradiction to the original one as might appear at first sight; it should be recalled that S_N1 and S_N2 reactivities often vary in the same direction in a given series of substrates.

It is also quite possible that such rules are simply the apparent results of other factors. For one, the steric environment of the alkylating agent is important

---

42. N. Kornblum and R. Seltzer, J. Amer. Chem. Soc., 83, 3668 (1961).
43. B. Miller and K.H. Lai, Chem. Commun., 334 (1971).
44. M. Bersohn, J. Amer. Chem. Soc., 83, 2136 (1961).
45. M.E. Volpin, I.S. Akhrem, and D.N. Kursanov, J. Gen. Chem. USSR, (English Transl.), 30, 1207 (1960).

as already noted. In free β-ketoester anions, the singly bound oxygen atom is much more accessible than the thrice-bonded carbon atom, and the overwhelming tendency of neopentyl halides to react at oxygen may simply have a steric origin [23]. Another factor that is capable of rationalizing many results [2b] is the so-called principle of hard and soft acids and bases [46]. Pearson noted that throughout chemistry, equilibria and kinetics involving the congregation of two groups at one site are favored if these groups are of comparable polarizability. Polarizability (or softness) is associated with unsaturation, high atomic number, large size and neutral or negative oxidation state, and hardness with high electronegativity, small size and positive oxidation state. Although the physical basis for the principle is somewhat obscure, it does readily accommodate the tendency of allylic and benzylic substrates to alkylate at carbon. Ethyl trifluoroacetoacetate (51) in HMPA gives exclusively O-alkylation when the parent compound does not [47]; in this case the trifluoromethyl group of course hardens the base, which consequently prefers to react with the O-site.

$$CF_3COCH_2COOC_2H_5$$

(51)

Perhaps the strongest support is to be found in the leaving group effect; if O-alkylation is desired, chlorides are better than bromides and much better than iodides; sulfates, sulfonates, onium salts and so on are superior by far [48].

An entirely unexpected result that is not readily understood on any basis yet described is the fact that β-dicarbonyl derivatives in DMSO attack optically active 2-bromobutane with inversion on the way to O-alkylate but with retention in the formation of the carbon derivative. If this result stands up it may well lead to new insight into the details of these reactions [49].

The radical anion mechanism operative when the alkylating agent is a p-nitrobenzyl halide is another good example of the unexpected complications sometimes encountered in this field (see p. 656); still another is the very large leaving

46. (a) J.O. Edwards and R.G. Pearson, J. Amer. Chem. Soc., 84, 16 (1962); (b) B. Saville, Angew. Chem., Int. Ed. Engl., 6, 928 (1967); (c) R.G. Pearson, J. Amer. Chem. Soc., 85, 3533 (1963); (d) R.G. Pearson, Science, 151, 172 (1966); (e) R.G. Pearson and J. Songstad, J. Amer. Chem. Soc., 89, 1827 (1967); (f) Chem. Eng. News, 43, 90 (May 31, 1965). However, see also (g) R.S. Drago and R.A. Kabler, Inorg. Chem., 11, 3144 (1972); (h) R.G. Pearson, Inorg. Chem., 11, 3146 (1972).
47. (a) A.L. Kurts, A. Macias, P.I. Dem'yanov, I.P. Beletskaya, and O.A. Reutov, Proc. Acad. Sci., Chem. Ser., USSR (English transl.), 195, 920 (1970); (b) A.L. Kurts, N.K. Genkina, A. Macias, I.P. Beletskaya, and O.A. Reutov, Tetrahedron, 27, 4759, 4769, 4777 (1971).
48. (a) Refs. 2, 8, 11, 23, 25, 26 and 27; (b) G.J. Heiszwolf and H. Kloosterziel, Rec. Trav. Chim. Pays-Bas, 89, 1153 1217 (1970); (c) W.S. Murphy and D.J. Buckley, Tetrahedron Lett., 2975 (1969); (d) R. Chong and P.S. Clezy, Tetrahedron Lett., 741 (1966).
49. M. Suama, T. Sugita, and K. Ichikawa, Bull. Chem. Soc. Japan, 44, 1999 (1971).

group effect of (9-anthryl)methyl derivatives in the reaction with 2-nitro-2-propane: The O-alkylation product 9-anthraldehyde (52) forms in 85% yield when the leaving group is chloride, and the C-methylation product (53) obtains in 90% yield when it is trimethylamine.

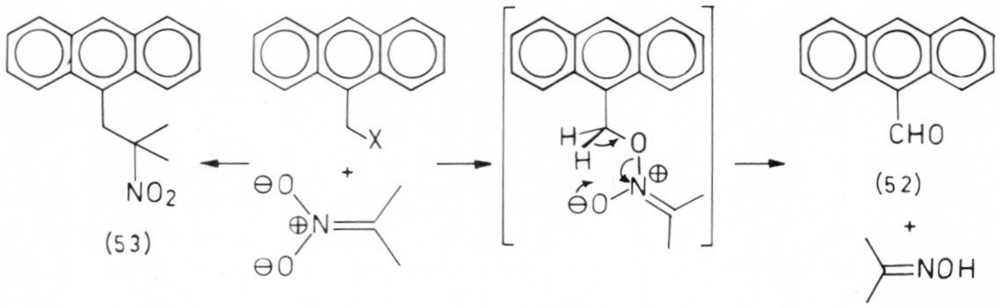

It was found that the latter reaction was actually an indirect one; the intermediate (54) could be isolated, and treatment of this species with a second ambident anion (not necessarily the same) gives the apparently normal C-alkylation product [50].

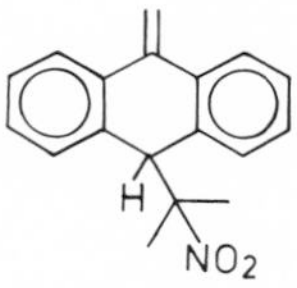

The reaction of sodiodipivaloylmethane (55) with trityl chloride is of interest in that the electrophile is also ambident. Of the four products that seem a priori possible, two are obtained: (56) and (57), in a 10:1 ratio (see also p. 627) [51].

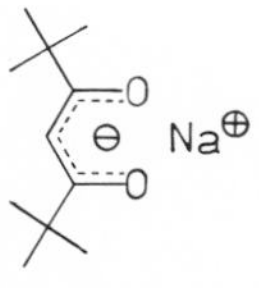

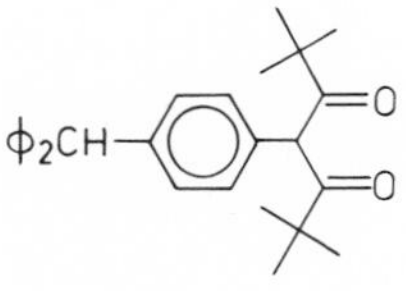

Hauser discovered that an excess base can often generate a dianion, and that such dianions are frequently alkylated at the "second" and more reactive site [52]. Thus, when ethyl acetoacetate is treated with two equivalents of sodium hydride,

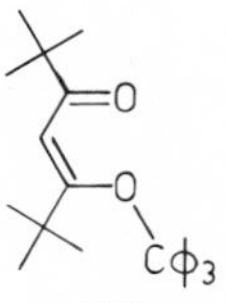

---

50. C.W. Jaeger and N. Kornblum, J. Amer. Chem. Soc., 94, 2545 (1972).
51. H.E. Zaugg, R.J. Michaels, and E.J. Baker, J. Amer. Chem. Soc., 90, 3800 (1968).
52. (a) K.G. Hampton and C.R. Hauser, J. Org. Chem., 30, 2934 (1965) and earlier papers. For another interesting dianion salt, see (b) D.O. DePree, J. Amer. Chem. Soc., 82, 721 (1960).

dianion (58) forms which reacts with various halides to give (59) [53]. Another dividend of these studies is the finding that highly hindered anions such as mesitoate, formerly considered unesterifiable, are readily converted by methyl iodide into methyl esters in HMPA [54].

(58) (59)

22-3. COUNTERION CHELATION

The alkylation rates of phenoxides in etheral solvents turn out to be affected by the nature of the solvent in a remarkable way. With n-butyl chloride at 25° for example, the rates in THF, glyme, tetraglyme and hexaglyme are in the ratio 1:8:1200:3600 [55]. The authors considered this to be a consequence of the fact that the polyglymes are geometrically especially well constructed for cation solvation - leaving the anion more inclined to attack.

This effect was demonstrated very clearly by Smid [56]. When the uv spectra of 9-fluorenyl salts in tetrahydrofuran are examined, evidence is found for ion-pair equilibria as described earlier (see p. 800); the nmr spectrum of THF is shifted upfield somewhat because of the ring current in the anion.

$$, M^{\oplus} \rightleftharpoons // M^{\oplus}$$

When a glyme is added, even in stoichiometric quantities, several drastic changes occur. The nmr spectrum of the THF returns to normal, but that of the glyme is now shifted upfield by amounts for exceeding those that applied to the THF. At the same time, the uv spectrum is shifted toward that of the loose pair. A new equilibrium should now be written (g = glyme):

$$F^{\ominus}, M^{\oplus} + ng \rightleftharpoons F^{\ominus} // M^{\oplus} \cdot g_n$$

53. L. Weiler, J. Amer. Chem. Soc., 92, 6702 (1970).
54. P.E. Pfeffer, T.A. Foglia, P.A. Barr, I. Schmeltz, and L.S. Silbert, Tetrahedron Lett., 4063 (1972).
55. (a) J. Ugelstad, A. Berge, and H. Listou, Acta Chem. Scand., 19, 208 (1965). For review of solvent effects in general, see (b) A.J. Parker, Quart. Rev. (London), 16, 163 (1962).
56. L.L. Chan and J. Smid, J. Amer. Chem. Soc., 89, 4547 (1967).

A plot of log $[F^-//M^+]/[F^-, M^+]$ vs log [glyme] gives n; for glyme itself, n = 2.5, and for the higher glymes it equals 1. The equilibrium constants can readily be measured and from their temperature dependence one finds that ΔH is typically -7 kcal/mole.

A big step forward was taken with the synthesis [57] of cyclic polyethers [58] such as (60) which can envelop the cation completely. The stability constants for a process such as depicted below can be measured by a variety of means [59], and the results neatly fit what might be expected on a geometric basis.

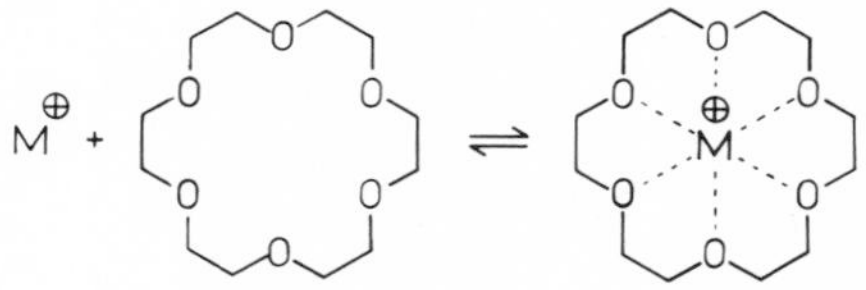

(60)

Thus, sodium ion fits best in the cavity of crown-18 ethers, but cesium reaches a maximum in the equilibrium constant with crown-21. With softer metal ions, K increases furthermore as the oxygen atoms are replaced by sulfur or nitrogen. Many of these complexes are now known in the pure crystalline state, and the X-ray diffraction patterns of a number of them have been determined [60].

Applications of these remarkable facts have not been long in coming. The cyclic polyethers are even more effective than their linear cousins in promoting the loosening of ion-pairs; in some cases stability constants larger than 10^8 were observed [61]. Several more complex analogs such as nonactin (61) have been discovered in biological systems, where they probably play a role in metal ion transport [59b]. As a result of such complexation solubilities can often be increased enormously; thus, barium sulfate is quite soluble in water in the presence of the macrocyclic aminoether (62), a member of the "cryptates" [62].

---

57. (a) C.J. Pedersen, J. Amer. Chem. Soc., 89, 7017 (1967); (b) J. Dale and P.O. Kristiansen, Chem. Commun., 670 (1971); (c) R.N. Greene, Tetrahedron Lett., 1793 (1972).
58. For review, see (a) Chem. Ind. (London), 1402 (1970); (b) J.J. Christensen, J.O. Hill, and R.M. Izatt, Science, 174, 459 (1971); (c) C.J. Pedersen and H.K. Frensdorff, Angew. Chem., Int. Ed. Engl., 11, 16 (1972).
59. (a) H.K. Frensdorff, J. Amer. Chem. Soc., 93, 600, 4684 (1971); (b) R.M. Izatt, D.P. Nelson, J.H. Rytting, B.L. Haymore, and J.J. Christensen, J. Amer. Chem. Soc., 93, 1619 (1971).
60. (a) M.A. Bush and M.R. Truter, Chem. Commun., 1439 (1970); (b) B. Metz, D. Moras, and R. Weiss, J. Amer. Chem. Soc., 93, 1806 (1971).
61. (a) U. Takaki, T.E. Hogen-Esch, and J. Smid, J. Amer. Chem. Soc., 93, 6760 (1971); (b) K.H. Wong, G. Konizer, and J. Smid, J. Amer. Chem. Soc., 92, 666 (1970).
62. (a) J.M. Lehn, Angew. Chem., Int. Ed. Engl., 9, 175 (1970); (b) J.M. Lehn, J.P. Sauvage, and B. Dietrich, J. Amer. Chem. Soc., 92, 2916 (1970).

(61) (62)

More important, many inorganic anions can be put into solution in organic media this way where they appear greatly more reactive than in water. For instance, potassium hydroxide, thus dissolved in benzene, will hydrolyze methyl mesitoate - a reaction not possible in water [58a]; olefins can be oxidized in high yield by potassium permanganate similarly dissolved in benzene [63], and so on. Even the alkali metals themselves can be dissolved in some media. Potassium has been dissolved in THF; the blue solutions which are indefinitely stable at -78° and decompose only slowly at room temperature, show an esr spectrum best explainable in terms of free (though solvated) electrons [64]. Further applications of these compounds are likely to be reported in profusion.

22-4. THE SHAPE OF CARBANIONS

A great many experiments could be quoted all of which suggest that simple carbanions are pyramidal, like amines, and that they invert fairly easily. A few of these will be mentioned.

The reaction of 1-iodoadamantane (63) with excess *t*-butyllithium followed by benzaldehyde quenching gives both *t*-butylphenyl- and 1-adamantyl-phenylcarbinol, (64) and (65), respectively [65]; this suggests that 1-adamantyllithium was present and that the non-planar bridgehead site is not an insuperable problem for a carbanionic species.

---

63. D. J. Sam and H. E. Simmons, *J. Amer. Chem. Soc.*, *94*, 4024 (1972).
64. (a) J. L. Dye, M. G. DeBacker, and V. A. Nicely, *J. Amer. Chem. Soc.*, *92*, 5226 (1970); (b) J. L. Dye, M. T. Lok, F. J. Tehan, R. B. Coolen, N. Papadakis, J. M. Ceraso, and M. G. DeBacker, *Ber. Bunsen Ges.*, *75*, 659 (1971).
65. P. T. Lansbury and J. D. Sidler, *Chem. Commun.*, 373 (1965).

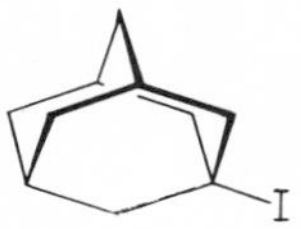

(63) 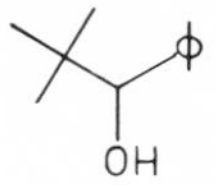 (64) 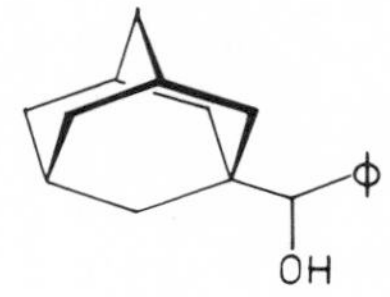 (65)

In a related experiment Streitwieser [66] found that 1-H-undecafluorobicyclo[2. 2. 1] heptane (66) exchanges its lone proton with tritium (in tritiated methanol-methoxide) even more readily then 2-H-heptafluoropropane, so that in this case again the carbanion is apparently not uncomfortable in the bridgehead position.

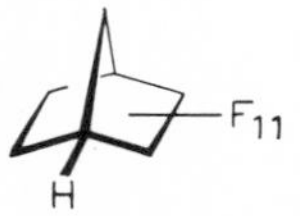

(66)

Similarly, it has been observed [67] that whereas 9-fluorofluorene (67) is ten times weaker in acidity than fluorene itself, benzal fluoride (68) is more than ten thousand times stronger than toluene; the interpretation is that in planar anions such as the resonance stabilized "dibenzocyclopentadienyl" anion, fluorine substitution has an effect quite different from that in other carbanions - which are therefore presumably non-planar.

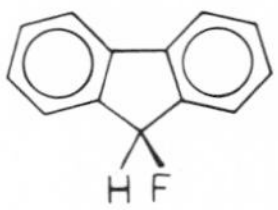

(67) (68)

X-ray diffraction studies of solid carbanion structures such as (69) have also shown that the carbanion is non-planar [68].

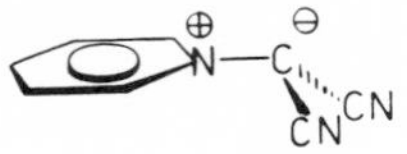

(69)

66. A. Streitwieser and D. Holtz, *J. Amer. Chem. Soc.*, 89, 692 (1967).
67. A. Streitwieser and F. Mares, *J. Amer. Chem. Soc.*, 90, 2444 (1968).
68. C. Bugg, R. Desiderato, and R. L. Sass, *J. Amer. Chem. Soc.*, 86, 3157 (1964).

As in the case of radicals, the cyclopropyl structure inhibits planarity still more, and cyclopropyllithium and sodium are not merely non-planar but preserve configurational stability in reactions such as carboxylation [69].

By the same token, vinyl carbanions are non-linear. When resolved (70) is treated with deuteromethanol-methoxide, exchange occurs at a rate 140 times greater than racemization [70]. Cyclopropene has been found to undergo hydrogen exchange readily in basic media - and only at the vinyl positions (see p. 417) [71].

(70)

These studies are of course always complicated by the presence of ion-pairs, as the conducted-tour reactions of Cram (p. 805) demonstrate. If crown-ethers are added to the solutions in which these reactions occur, exchange and racemization are greatly increased in rate and "iso-inversion" is stopped [72].

Homoenolate anions occupy a special niche in carbanion chemistry. Nickon found that optically active camphenilone (71) racemizes at high temperature in the presence of *t*-butoxide ion; furthermore, the compound incorporates in the limit three deuterium atoms if *d*-labeled *t*-butanol is used. These observations are compatible with the following scheme [73]. An alternative involving reversible *t*-butoxide anion addition can be ruled out because esters such as (72) were found *not* to be convertible into (71).

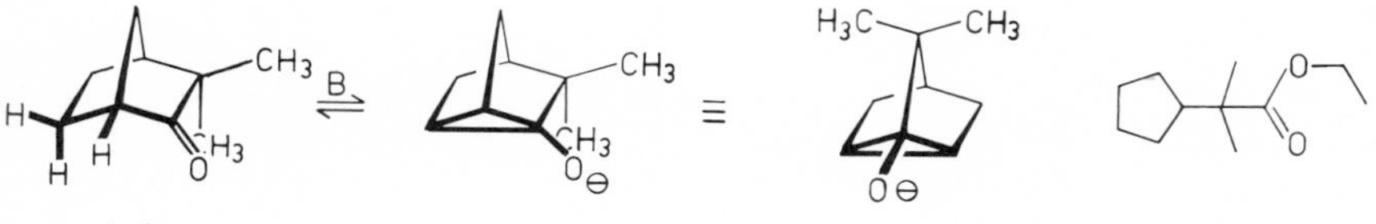

(71) (72)

69. (a) D.E. Applequist and A.H. Peterson, *J. Amer. Chem. Soc.*, *83*, 862 (1961); (b) J.B. Pierce and H.M. Walborsky, *J. Org. Chem.*, *33*, 1962 (1968).
70. H.M. Walborsky and I.M. Turner, *J. Amer. Chem. Soc.*, *94*, 2273 (1972).
71. E.A. Dorko and R.W. Mitchell, *Tetrahedron Lett.*, 341 (1968).
72. (a) J.N. Roitman and D.J. Cram, *J. Amer. Chem. Soc.*, *93*, 2225, 2231 (1971); (b) S.M. Wong, H.P. Fischer, and D.J. Cram, *J. Amer. Chem. Soc.*, *93*, 2235 (1971).
73. (a) A. Nickon and J.L. Lambert, *J. Amer. Chem. Soc.*, *84*, 4604 (1964) and *88*, 1905 (1966); (b) A. Nickon, J.H. Hammons, J.L. Lambert, and R.O. Williams, *J. Amer. Chem. Soc.*, *85*, 3713 (1965); (c) A. Nickon, J.L. Lambert, and J.E. Oliver, *J. Amer. Chem. Soc.*, *88*, 2787 (1966).

Because of its easy return to ketone, it was considered possible that the homoenolate was not merely the anion of a cyclopropanol but that it had true enolate character (73). Later it proved possible to prepare 1-hydroxynortricyclene (74), and either acid, base, or mild heating converts it into 2-norbornanone (75) [74].

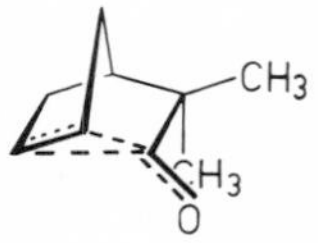

CH_3 CH_3 O OH O

(73)  (74)  (75)

A related phenomenon is the exchange behavior of bridgehead ketols such as (59) in basic media [75]. Prolonged exposure to warm D_2O-OD^- led to the introduction of four deuterium atoms into the molecule (in addition to OH-exchange), suggesting that equilibration must be occurring between (77) and (78), with a species such as (79) either as intermediate or transition state.

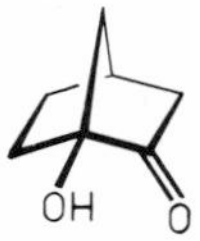

OH O O⊖ O O ⊖O O ⊖ O ≡ O ⊖ O

(76) 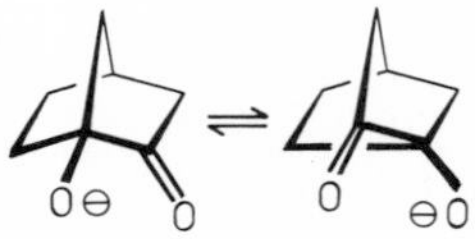 (77) (78) 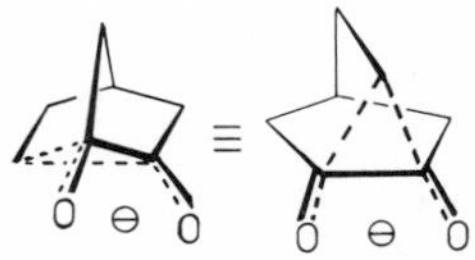 (79)

A homoenolization reaction involving (80) can be ruled out since it would lead to the exchange of six protons, and the same is true for another alternative involving 1,2-shifts and the species (81).

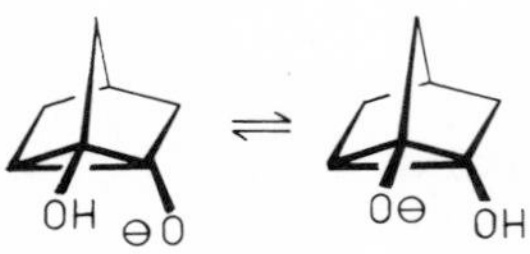

OH ⊖O O⊖ OH O⊖ O O ⊖O ⊖O O

(80) 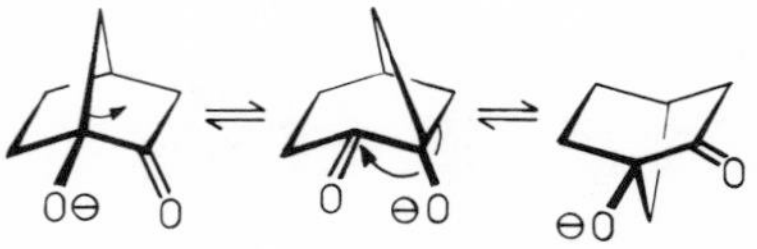 (81)

The sesquiketol (82) was mentioned by these authors as an intriguing variant; it should exchange all but one of its protons via the anion (83); but at the moment this is still in the realm of speculation.

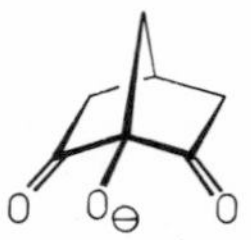

O O⊖ O O ⊖ O O

(82) 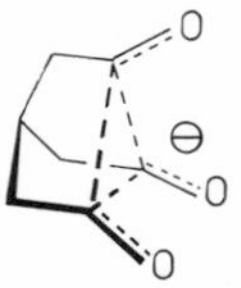 (83)

74. A. Nickon, J. L. Lambert, R. O. Williams, and N. H. Werstiuk, J. Amer. Chem. Soc., 88, 3354 (1966).
75. A Nickon, T. Nishida, and Y. Lin, J. Amer. Chem. Soc., 91, 6860 (1969).

Chapter 23

COMPLEXES [1]

23-1. INTRODUCTION

In 1971, Geluk [2] reported that the treatment of adamantanone with nitric acid resulted in the formation of a stable, solid adduct that could be recrystallized from n-hexane, that had a melting point of 60°, a composition of $C_{10}H_{14}O.HNO_3$ a carbonyl frequency shifted down from 1720 to 1670 cm^{-1} and an nmr spectrum in $CDCl_3$ not noticeably different from the parent compound except for a general downfield shift of a few tenths of one ppm. The author referred to this and several additional such materials formed by adamantane derivatives as complexes, and thereby extended an old chemical tradition, of calling complexes all these compounds which are in some way odd or unusual, and which do not readily fit the line-for-every-electron-pair representations that are the hallmark of our language. Very often the term complex persists even after the nature of the species has been elucidated - thus we have the complex inorganic ions, the π-complexes, σ-complexes, charge-transfer complexes, organometallic complexes, clathrate

---

1. For review, see L.J. Andrews and R.M. Keefer, "Molecular Complexes in Organic Chemistry", Holden-Day, San Francisco, Calif., 1964.
2. H.W. Geluk, Tetrahedron Lett., 4473 (1971).

complexes [3], hydrogen-bonded complexes, and so on. The organometallic complexes were discussed in Chr. 10; in the following sections we briefly review the π- and σ complexes in their various forms.

23-2. SIGMA COMPLEXES

It was found by Melander [4] that in many aromatic substitution reactions no significant hydrogen isotope effects are operating. This means that the C-H bond being substituted is not yet breaking in the transition state, and that some addition process must be occurring leading to an intermediate ArHX. Hughes and Ingold had shown that the fragment X is an electrophilic, positive ion [5], and hence many authors favored a structure such as (1) for this species. It was called a sigma complex [6]; because of Wheland's successful quantum mechanical description of it, it is also often called the Wheland intermediate. It is in fact a cyclohexadienium ion. Not everyone agreed that the intermediate in fact had structure (1); (2) was proposed as an alternative [7].

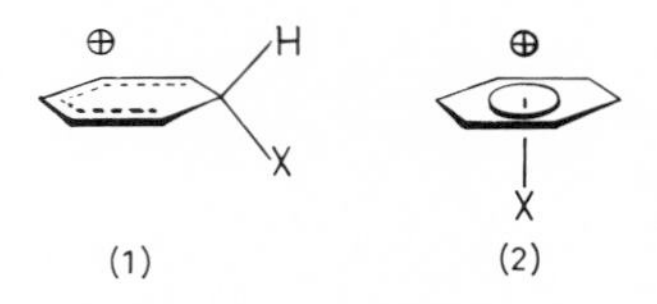

In 1958, Olah's use of cold superacid solutions enabled him to prepare and isolate such complexes. Thus, while toluene is miscible with neither boron trifluoride or with hydrofluoric acid, a 1:1:1 mixture of all three at -50° becomes homogeneous, colored, conducting, and insoluble in most organic media. If deuterofluoric acid is used, decomposition of the product led to o- and p-deuterotoluenes [8]. Bromomethylcyclohexadienes such as (3) give the same material when treated with silver tetrafluoroborate [9], and if reagents such as nitrosyl- or alkyl fluorides were used, decomposition of the complexes leads to substitution products such as o- and p-nitro- or alkyltoluenes [10]; hence Olah favored the σ-complex structure.

---

3. (a) M.M. Hagan, J. Chem. Educ., 40, 643 (1963); (b) K.A. Kobe and L.R. Reinhart, J. Chem. Educ., 36, 300 (1959).
4. L. Melander, Arkiv Kemi, 2, 211 (1950).
5. C.K. Ingold, "Structure and Mechanism in Organic Chemistry", 2nd Ed., Cornell University Press, Ithaca, N.Y., 1969.
6. G.W. Wheland, J. Amer. Chem. Soc., 64, 900 (1942).
7. M.J.S. Dewar, "The Electronic Theory of Organic Chemistry", Oxford Univ. Press, London, 1949.
8. G.A. Olah and S.J. Kuhn, J. Amer. Chem. Soc., 80, 6535 (1958).
9. G.A. Olah, A.E. Pavlath, and J.A. Olah, J. Amer. Chem. Soc., 80, 6540 (1958).
10. G.A. Olah and S.J. Kuhn, J. Amer. Chem. Soc., 80, 6541 (1958).

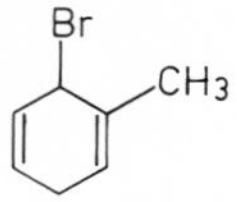

(3)

Proof of this structure came with the application of nmr. Thus, when aluminum chloride, methyl chloride and hexamethylbenzene were mixed, four signals in a 1:2:2:2 ratio are observed; this rules out a structure such as (2) and establishes (4) as the structure of this species [11].

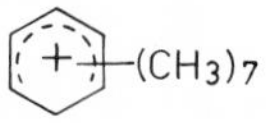

(4)

All four of the signals merge at 50-70°; evidently rapid Wagner-Meerwein shifts occur within the ion [12]. The nature of these species was further clarified when it was found that the ion (5) has a singlet at 1.17 τ with $J_{C^{13}H}$ = 163 Hz and one at double the intensity at 5.95 τ with $J_{C^{13}H}$ = 128 Hz; clearly the site at which the cation becomes bound assumes sp^3 - hybridization in the process [13].

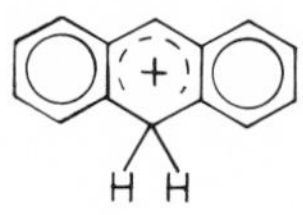

(5)

It may seem surprising that σ-complex formation apparently goes on even when basic substituents are present. Thus, 1,3,5-triaminobenzene at room temperature is protonated at the ring to produce (6) several times more readily than at the nitrogen atoms to give (7) [14];

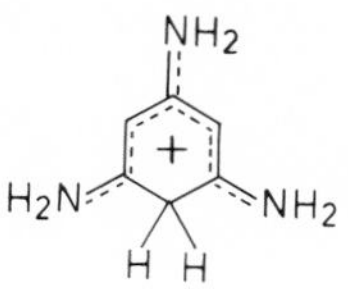

(6) 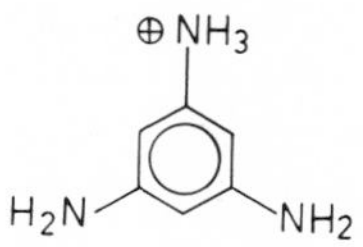 (7)

---

11. (a) W. von E. Doering, M. Saunders, H.G. Boyton, H.W. Earhart, E.F. Wadley, W.R. Edwards, and G. Laber, Tetrahedron, 4, 178 (1958); (b) G.A. Olah, J. Amer. Chem. Soc., 87, 1103 (1965).
12. V.A. Koptyug, V.G. Shubin, and A.I. Rezvukhin, Bull. Acad. Sci. USSR (English transl.), 192 (1965).
13. V.A. Koptyug, I.S. Isaev and A.I. Rezvukhin, Tetrahedron Lett., 823 (1967).
14. T. Yamaoka, H. Hosoya, and S. Nagakura, Tetrahedron, 26, 4125 (1970).

similarly, the α,α-dimethylbenzyl carbanion upon neutralization gives both cumene (8) and the methylenecyclohexadienes (9) and (10) [15].

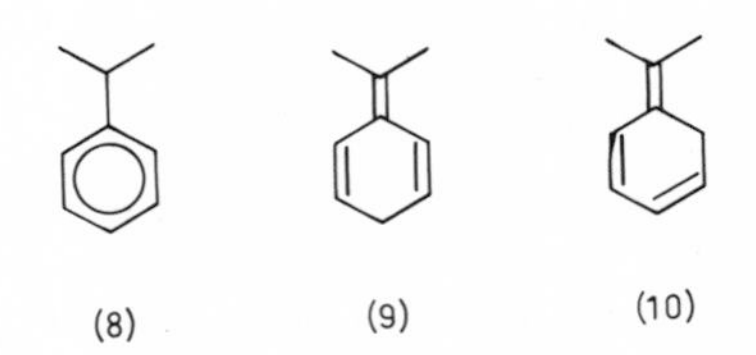

(8) (9) (10)

A measure of the stabilities of the σ-complexes has been provided by Mackor [16] on the basis of the distribution coefficients of various aromatic hydrocarbons in the two phase system consisting of heptane and solutions of alkali fluorides and boron trifluoride in hydrofluoric acid. An extremely wide range of constants was found; for example, those of benzene, toluene, m-xylene and mesitylene are in the ratio of $1:10^3:10^6:10^9$.

23-3. MEISENHEIMER COMPLEXES [17]

In aromatic compounds bearing electron-withdrawing groups, nucleophilic substitution at the ring by an anion is often observed; for example:

$$\text{F-C}_6\text{H}_3(\text{NO}_2)_2 \xrightarrow[-F^{\ominus}]{CH_3O^{\ominus}} \text{CH}_3\text{O-C}_6\text{H}_3(\text{NO}_2)_2$$

Some of these reactions can be catalyzed by Lewis bases; thus, in the nucleophilic equivalent of the Friedel-Crafts reaction, potassium fluoride and perfluoropropene react with perfluoropyridine to give perfluoro-4-isopropylpyridine [18]. Meisenheimer had noted early [19] that color changes sometimes indicated the presence of an intermediate in these reactions, and these intermediates, which now bear his name, are the anionic equivalent of σ-complexes [20].

---

15. G.A. Russell, J. Amer. Chem. Soc., 81, 2017 (1959).
16. E.L. Mackor, A. Hofstra, and J.H. van der Waals, Trans. Faraday Soc., 54, 66, 186 (1958).
17. For review, see M.J. Strauss, Chem. Rev., 70, 667 (1970).
18. R.D. Chambers, R.A. Storey, and W.K.R. Musgrave, Chem. Commun., 384 (1966).
19. J. Meisenheimer, Ann., 323, 205 (1902).
20. J. Miller, J. Amer. Chem. Soc., 85, 1628 (1963).

The sp^3-hybrid nature of these anions has been supported by nmr. Thus, solutions containing the anion (11) have a single sharp methoxy and a single sharp m-hydrogen peak [21]; the salt (12) can be isolated in pure form, and its symmetrical shape is likewise supported by nmr [22].

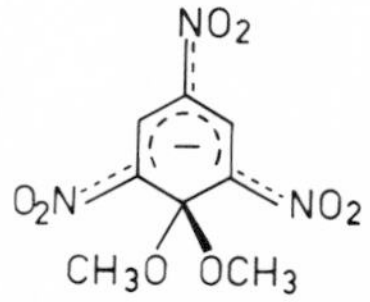

(11) 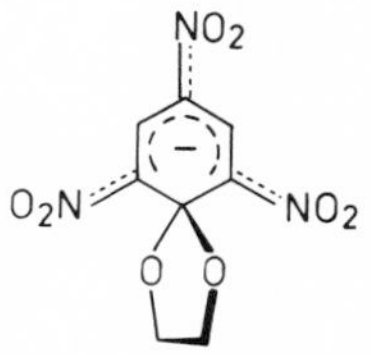 (12)

23-4. Pi-COMPLEXES

From the solubilities of hydrogen chloride and hydrogen bromide gases in heptane containing various aromatic hydrocarbons, Brown concluded that these acids form weak complexes with the rings [23]. In accord with ir studies that show the stretching frequencies of the halides to move to lower values in these media [24], it was proposed that these hydrogen bonding interactions resulted in π-complexes in which the proton was weakly bound directly to the π-cloud. The equilibrium constants for complex formation of hydrogen chloride for benzene, toluene, m-xylene and mesitylene at -78° are 1:1.4:1.8:2.3 - a very much lower sensitivity to methyl groups than in the σ-complexes. The π-complexes may play a role in some aromatic substitution reactions. Thus, the brominations of benzene and m-xylene in acetic acid take place at relative rates of $1:10^6$; since the same ratio occurs in their σ-complex formation constants, σ-complexes are logically regarded as the intermediates whose formation is rate controlling in this reaction. On the other hand, the nitration rates of these same substrates with nitronium fluoroborate in sulfolane are in the ratio of 1:1.8, so that these reactions are best regarded as proceding through rate controlling π complex formation [25]. These relative rates are determined by means of competition experiments, and since reactions such as those with nitronium ions are essentially instantaneous, the question has been raised whether imperfect mixing of the mixture of aromatic hydrocarbons with the nitrating medium might be responsible for the low ratio [26]. It was observed for

---

21. P. Caveng, P. B. Fischer, E. Heilbronner, A. L. Miller, and H. Zollinger, Helv. Chim. Acta, 50, 848 (1967).
22. R. Foster, C. A. Fyfe, and J. W. Morris, Rec. Trav. Chim. Pays-Bas, 84, 516 (1965).
23. H. C. Brown and J. J. Melchiore, J. Amer. Chem. Soc., 87, 5269 (1965).
24. W. Gordy, J. Chem. Phys., 9, 215 (1941).
25. G. A. Olah, S. J. Kuhn, and S. H. Flood, J. Amer. Chem. Soc., 83, 4571, 4581 (1961).
26. J. H. Ridd, Accounts Chem. Res., 4, 248 (1971).

example that bibenzyl (13) under these conditions - if mixed in a 1:1 mole ratio with the nitrating agent gave as much as 80% of dinitro products (14) even though only 25% should have been obtained on a statistical basis.

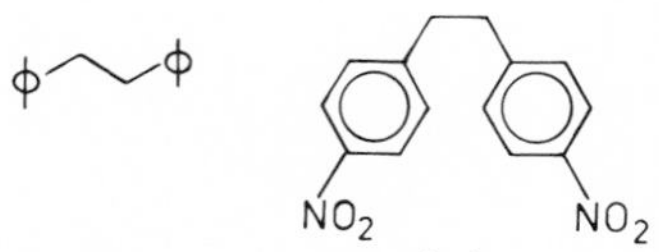

(13) (14)

23-5. CHARGE TRANSFER COMPLEXES

The π-complexes described in the preceding section are but a few examples of a very large number of the so-called charge transfer complexes. These are compounds consisting of a donor and an acceptor molecule in which an electron is partially transferred from the former to the latter. The donor molecule may be an electron-rich aromatic compound, olefin or molecule with unshared pairs (these could be said to give rise to n-complexes); the acceptor may be a proton or Lewis acid, a halogen, electron-poor aromatic ring or olefin, or a transition metal ion. Many of these complexes are characterized by the appearance of a new absorption in the visible or uv part of the spectrum at a wavelength longer than any of the component spectra; this is the so-called charge transfer band. Such a band is usually observable, but its absence should not be taken to mean that no complex is formed. This was nicely demonstrated by Dewar [27], who showed that substances such as 2-(1-pyrenyl)ethyl tosylate (15) form intramolecular complexes although no charge transfer band is observable. The evidence for such complexes in this case is found in the nmr spectrum; thus, the ring protons of the tosyloxy group lie at higher field than in simple aliphatic tosylates, and they return to their normal positions when a stronger π-acceptor such as the quinone (16) is added also.

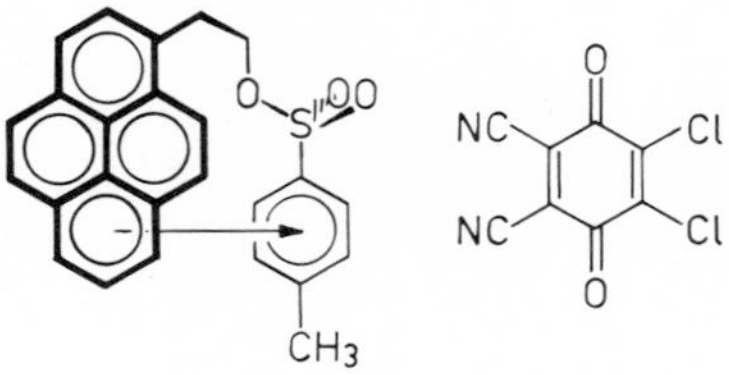

(15) (16)

In a related case, a remarkable solvent dependence was found in the nmr spectrum of formamidines such as (17). The C-N bond has some double bond character because of resonance of the type indicated, and consequently the methyl groups demonstrate a coalescence phenomenon as the temperature is varied.

---

27. M. D. Bentley and M. J. S. Dewar, Tetrahedron Lett., 5043 (1967).

(17)

It turns out that the coalescence temperature is much higher in benzene than in chloroform: Evidently in the former the positively charged nitrogen atom complexes with the benzene - which enhances the importance of the dipolar structure in that solvent [28].

The structure of these complexes has been studied by many means. Thus, the X-ray diffraction pattern has been determined [29] for the weak, solid complex for bromine with benzene, which is isolable at low temperatures; the structure is shown as in (18).

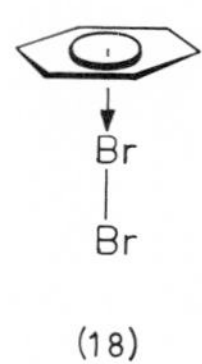

(18)

The same conclusion can be drawn from infrared and far infrared studies; thus, a benzene solution of iodine shows the fundamental I_2 stretching frequency which normally would be infrared-inactive. This frequency is furthermore somewhat dependent on the nature of the donor: Electron-rich donors shift it to lower values compared to donors that are not so well-off [30]. Infrared similarly shows that the carbonyl stretching frequency undergoes a red shift upon the mixing of aldehydes with iodine in carbon disulfide. The shift is the largest for the complexes (19) formed with α,β-unsaturated aldehydes such as crotonaldehyde (20), and less so for aromatic aldehydes, aliphatic aldehydes and chloral (21), in that order [31].

(19) (20) (21)

28. J.P. Marsh and L. Goodman, Tetrahedron Lett., 683 (1967).
29. O. Hassel and K.O. Stromme, Acta Chem. Scand., 13, 1781 (1959).
30. E.K. Plyler and R.S. Mullikan, J. Amer. Chem. Soc., 81, 823 (1959).
31. (a) E. Augdahl and P. Klaboe, Acta Chem. Scand., 16, 1637, 1647, 1655 (1962); (b) P. Klaboe, J. Amer. Chem. Soc., 85, 871 (1963).

Dipole moments can be measured also. In some cases such as with triethylamine and iodine in dioxane the moment is so large (12.5 D) that the complex must be regarded as essentially an ion-pair (it equals 12 D for the amine picrate salt) [32]; others however have preferred a virtually covalent bond since the N-I distance by X-ray diffraction was found to be only 2.3 Å (the sum of the van der Waals radii is 3.7 Å; the sum of the covalent radii is 2.0 Å([33].

In most instances now known, the charge transfer band is the most prominent feature signalling the presence of the complex. Both the wavelength and the intensity increase as the highest occupied orbital of the donor lies higher in energy, and as the lowest empty orbital of the acceptor is lowered [34]; thus, while salts such as tropylium perchlorate (22) are colorless, the iodide (23) is brightly yellow [35].

(22) (23)

Stable, colored charge transfer salts such as (24) result when sodium pentamethoxycyclopentadienide is mixed with the appropriate tetrafluoroborate [36].

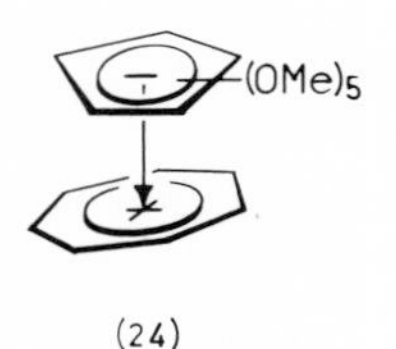

(24)

When one of the components is a neutral molecule, paramagnetism may result from the transfer [37]. Thus, when tropylium salts are mixed with tetracene in methylene chloride, an esr spectrum is observable then can be identified as that of the radical cation (25).

(25)

---

32. K. Toyoda and W.B. Person, J. Amer. Chem. Soc., 88, 1629 (1966).
33. S. Kobinata and S. Nagakura, J. Amer. Chem. Soc., 88, 3905 (1966).
34. (a) M.J.S. Dewar and A.R. Lepley, J. Amer. Chem. Soc., 83, 4560 (1961); (b) A.R. Lepley, J. Amer. Chem. Soc., 84, 3577 (1962).
35. K.M. Harmon, F.E. Cummings, D.A. Davis, and D.J. Diestler, J. Amer. Chem. Soc., 84, 3349 (1962).
36. E. Le Goff and R.B. LaCount, J. Amer. Chem. Soc., 85, 1354 (1963).
37. H.J. Dauben and J.D. Wilson, Chem. Commun., 1629 (1968).

A remarkable example was described recently, in which p-phenylenediamine and 2,3-dichloro-5,6-dicyanoquinone were mixed in a stopped flow apparatus. Esr spectra of both (26) and (27) could be identified, although that of the former disappeared immediately after the flow was stopped [38].

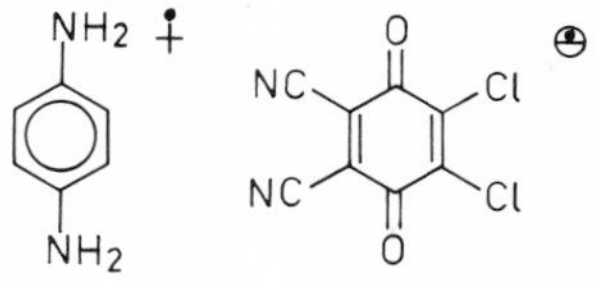

(26) (27)

The precise orientation of the two partners is presumably not too important since the interaction is weak. This was demonstrated by White, who studied the shape and location of the charge transfer bands of (28), (29) and (30), and found that all were nearly identical [39].

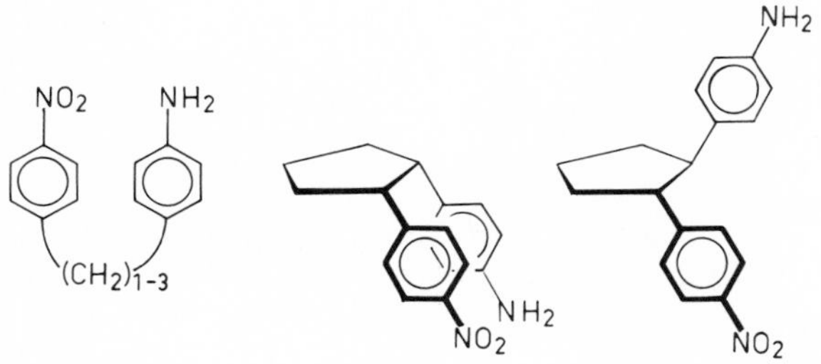

(28) (29) (30)

On the other hand, the distance between the partners is quite critical when they are close; even modest pressures of a few kbar result in dramatic increases in intensity which can only partly be ascribed to increases in the degree of association [40], and are at least partly due to increased extinction coefficients [41].

The electronic nature of these complexes has been elucidated principally by Mullikan [42], who describes them in terms of contributions from non-bonded and dative forms. In the ground states of all but the most extreme cases, the non-bonded form is predominant, and the opposite is true in the excited state.

---

38. N.H. Kolodny and K.W. Bowers, J. Amer. Chem. Soc., 94, 1113 (1972).
39. W.N. White, J. Amer. Chem. Soc., 81, 2912 (1959).
40. (a) A.H. Ewald and J.A. Scudder, J. Phys. Chem., 76, 249 (1972); (b) A. H. Ewald, Trans. Faraday Soc., 64, 733 (1968).
41. (a) J.R. Gott and W.G. Maisch, J. Chem. Phys., 39, 2229 (1963); (b) W.H. Bentley and H.G. Drickamer, J. Chem. Phys., 42, 1573 (1965).
42. R.S. Mullikan and W.B. Person, "Molecular Complexes: A Lecture and Reprint Volume", Wiley, New York, 1969.

$$D \rightarrow A \equiv D \quad A \longleftrightarrow \overset{\oplus}{D} \quad \overset{\ominus}{A}$$

Because of the reverse polarity of many excited state molecules, it happens frequently that excited molecules will complex with their ground state congenitors; such complexes are known as excimers (or exciplexes if the ground state partner is a different molecule) [43]. This phenomenon can be recognized by changes in the fluorescence spectrum, and by the concentration dependence of these changes.

Charge transfer complexes have been used in many areas of organic chemistry. The impregnation of certain absorbants with acceptors such as picric acid (31) improves their ability to separate aromatic mixtures [44].

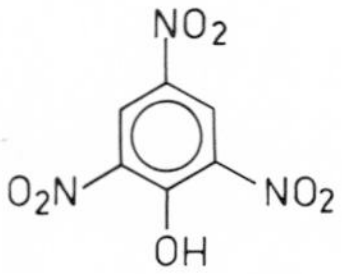

NO_2 O_2N NO_2 OH

(31)

Enantiomorphic pairs such as hexahelicene (32) have been separated via the formation of diastereomeric complexes with (-) (33) [45].

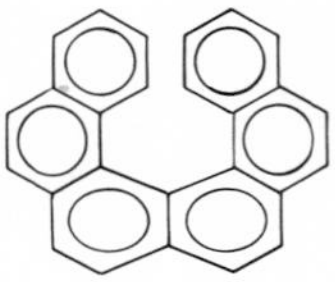

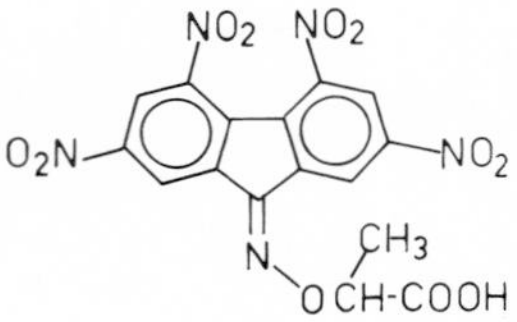

NO_2 NO_2 O_2N NO_2 N CH_3 O CH-COOH

(32) (33)

The optical purity of a given enantiomer can sometimes be gauged through complexation. Thus, the partially resolved alcohol (34) in pure, resolved α-(1-naphthyl)ethylamine (35) shows two F^{19} doublets and two unequal proton quartets - due of course to the two diastereomeric complexes of which one is shown (36).

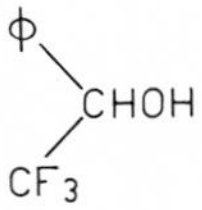

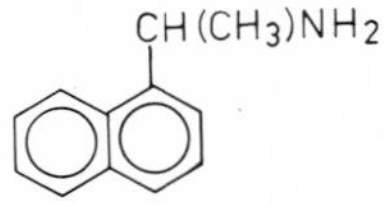

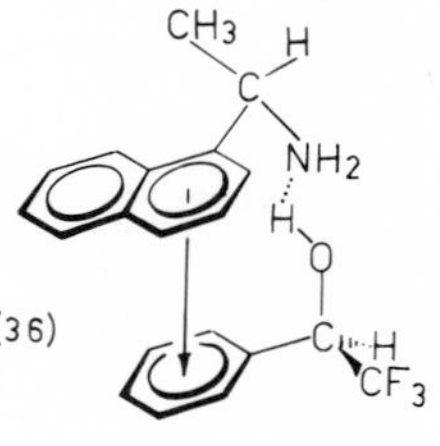

ϕ CHOH CF_3 CH(CH_3)NH_2 CH_3 H C NH_2 H O C H CF_3

(34) (35) (36)

43. (a) J. B. Birks, Nature, 214, 1187 (1967); (b) N. Mataga and K. Ezumi, Bull. Chem. Soc. Japan, 40, 1355 (1967).
44. L. H. Klemm, D. Reed, and C. D. Lind, J. Org. Chem., 22, 739 (1957).
45. M. S. Newman and D. Lednicer, J. Amer. Chem. Soc., 78, 4765 (1956).

In each case the carbinyl proton is shifted downfield compared to the alcohol alone, since it is presumably located between the rings in a region where it is deshielded by both ring currents; the fluorines are shifted upfield because of the charge transfer. These shifts are at a maximum if the alcohol has a 2,4-dinitrophenyl group, and the uv spectrum reveals a charge transfer band in that case [46].

In another case a charge transfer complex was used to show that "through-space coupling" is possible if the conditions are favorable. A comparison of the uv spectrum of cyclophanes (37)-(39) and of (40) shows that only (38) has a charge transfer band (at 302 nm), and in that compound the protons show up as a quintet - demonstrating H-F coupling which does not occur in (40) [47].

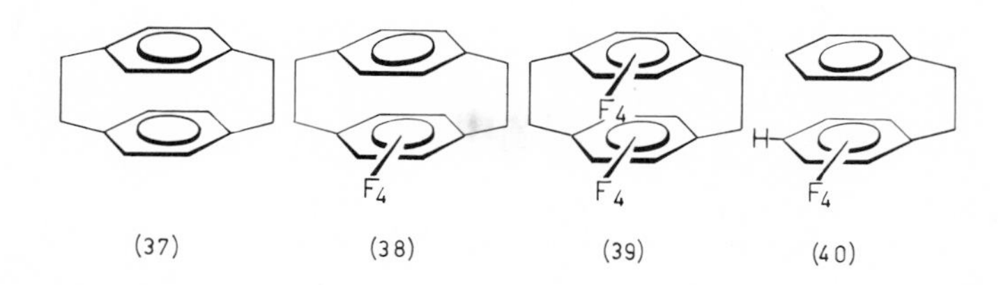

(37) (38) (39) (40)

Catalysis by reversibly formed charge transfer complexes is undoubtedly very common in biological reactions, and the possibility even in simple organic chemistry has already been demonstrated by Colter [48]. Thus, the acetolysis of 9-(2,4,7-trinitro)fluorenyl tosylate (41) can be much increased in rate by the presence of donors such as phenanthrene and dimethoxynaphthalenes; obviously, the formation of the carbonium ion is greatly facilitated in the presence of a donor.

NO_2

O_2N NO_2

OTs

(41)

The involvement of a complex is clear from the fact that the activation volume for the solvolysis is much more negative if a donor is present [49]. Brown has attributed a case of phenyl participation to the intermediacy of a π complex in which the phenyl group stabilizes the carbonium ion by π electron donation (see p. 739). Even acceptors may function as catalysts in suitable cases. Thus, biphenyl

46. W.H. Pirkle and S.D. Beare, J. Amer. Chem. Soc., 89, 5485 (1967).
47. R. Filler and E.W. Choe, J. Amer. Chem. Soc., 91, 1862 (1969).
48. (a) A.K. Colter, S.S. Wang, G.H. Megerle, and P.S. Ossip, J. Amer. Chem. Soc., 86, 3106 (1964); (b) A.K. Colter, F.F. Guzik, and S.H. Hui, J. Amer. Chem. Soc., 88, 5754 (1966); (c) A.K. Colter and S.H. Hui, J. Org. Chem., 33, 1935 (1968).
49. R.K. Williams, J.J. Loveday, and A.K. Colter, Can. J. Chem., 50, 1303 (1972).

racemization passes through a planar transition state, and since planarity is promoted by complexation, it is no surprise that 2,4,7-trinitrofluorenone catalyzes such racemizations [50].

Still another elegant application is the use of complexes in the evaluation of solvent polarity [51]. Kosower found that the position of the charge transfer band of 1-ethyl-4-carbomethoxypyridinium iodide (42) is remarkably dependent on solvent, with polar solvents such as water producing the greatest blue shifts.

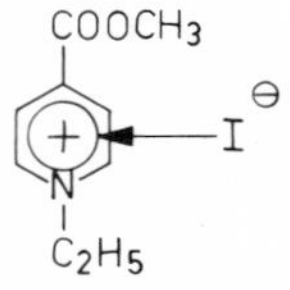

$COOCH_3$

$I^{\ominus}$

N

C_2H_5

(42)

This is considered due to the fact that the ground state is essentially an ion-pair and the excited state has much covalent character. It was noted that if the transition energies (in kcal/mole) were plotted against the composition in aqueous organic solvents, all these gave rise to linear correlations easily extrapolated to a common value for water. Other solvents similarly have characteristic transition energies; these are called Z-values. These numbers characterize the polar nature of the solvent. They can much more readily be measured then the kinetic Y-values (see p. 696), and the range covers solvents in which solvolysis does not occur; however, for those media in which both Y- and Z-values can be determined, a linear relation exists between the two.

---

50. A.K. Colter and L.M. Clemens, J. Amer. Chem. Soc., 87, 847 (1965).
51. (a) E.M. Kosower, J. Amer. Chem. Soc., 80, 3253, 3261, 3267 (1958); (b) A. Ray, J. Amer. Chem. Soc., 93, 7146 (1971).

Chapter 24

MISCELLANEOUS INTERMEDIATES

24-1. INTRODUCTION

The intermediates described so far have been in the lime light a great deal, but there are a number of others which are perhaps a little less important but nonetheless interesting. Each mirrors in its own way the histories of their better-known cousins. In this chapter we describe the stories of some of them.

24-2. CARBON ATOMS [1]

Carbon atoms can be generated in several ways. When malonic acid is dehydrated with phosphorus pentoxide, the resulting carbon suboxide (1) can be isolated, and photolyzed to give carbon atoms, which can be trapped in situ. It is not certain however that the atoms are free at any time; it is possible that the carbene (2) is trapped and that the product subsequently loses a second carbon monoxide fragment [1].

---

1. For review, see G. Czichocki, Z. Chem., 10, 423 (1970).

O=C=C=C=O (1)   C=C=O (2)

Photolysis of cyanogen azide (3) in a low temperature solid matrix likewise produces carbon atoms; here definitely two stages and an intermediate cyanonitrene (4) are involved (see p. 611) [2].

N=N=N—C≡N (3)   N—C≡N (4)

The thermal decomposition of the sodium salt of quadricyclanone tosylhydrazone (5) presumably yields the carbene (6) which subsequently eliminates a carbon atom [3].

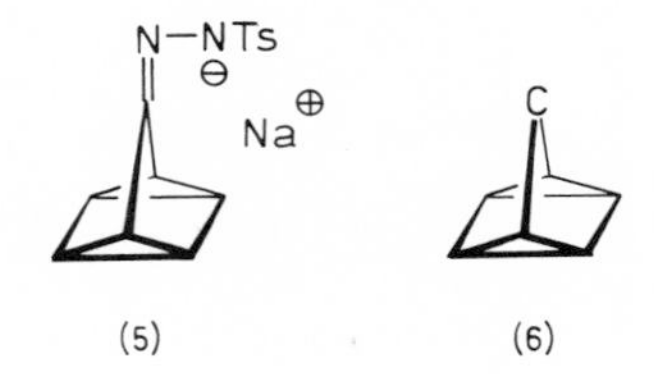

(5)   (6)

5-Tetrazolyldiazonium chloride (7) decomposes into three molecules of nitrogen and a carbon atom [4].

C–N≡N.HCl (tetrazolyl, ⊕)

(7)

The nuclear (γ, n) reaction of C^{12} converts this atom into C^{11}; if an organic compound is irradiated, the recoil energy of the reaction liberates the C^{11} atoms. They emerge with great velocity, and an inert gas moderator is required to study the reaction of the thermalized atoms with organic substrates. This technique has the built-in advantage that the products are radioactive and hence they can be monitored although the reaction can obviously be done only on a small scale (with perhaps 10^8 atoms or so); the half life of C^{11} is only 20 minutes so that rapid work-up is required [5]. Finally, if elemental carbon is heated to 2200°, some

---

2. (a) D. E. Milligan, M. E. Jacox, and A. M. Bass, J. Chem. Phys., 43, 3149 (1965); (b) D. E. Milligan and M. E. Jacox, J. Chem. Phys., 44, 2850 (1966).
3. P. B. Shevlin and A. P. Wolf, Tetrahedron Lett., 3987 (1970).
4. P. B. Shevlin, J. Amer. Chem. Soc., 94, 1379 (1972).
5. (a) C. MacKay and R. Wolfgang, Science, 148, 899 (1965); (b) Chem. Eng. News, 43, 40 (July 19, 1965).

carbon vapor is produced and this consists largely of carbon atoms (54%); however, C_2 and C_3 fragments are also present (11 and 35% respectively). A carbon arc produces a carbon vapor under more extreme conditions, and C_1-C_4 fragments are then present in a 68:24:7:1 ratio. A further difference is that the latter method, unlike the thermal technique, produces a significant fraction of these species in various excited states [6]. The electronic states of carbon are the ^3P ground state, a ^1D state (1.3 eV above the ground state) and a ^1S state (2.7 eV above the ground level); the two excited singlets have been described as metastable [7]. The carbon vapor must be condensed in an inert matrix if the primary reactions of carbon atoms are to be studied; a further problem may then develop since even at 30°K diffusion may set in and produce further fragments C_4-C_9 [8]. On the other hand, the vaporization of carbon can be carried out indefinitely and on a fairly large scale; it seems to be the method of choice in spite of the fact that mixtures are produced.

The reactivity of individual carbon atoms can be gauged from the fact that they will readily attack carbon monoxide at 14°K to give (2); they react with fluorine to give difluorocarbene and with chlorine to give dichlorocarbene; these species can then be studied spectroscopically [9]. With hydrogen chloride, chlorocarbene is produced, and above 15°K this carbene goes on to insert a second time and give methylene chloride [10]. Carbon atoms can deoxygenate acetone to give carbon monoxide and propylene [11]; a mixture of phosgene (8) and 2-butene is stereospecifically converted into carbon monoxide and the adduct (9) [12].

$COCl_2$

(8)

CH_3 Cl H_3C Cl

(9)

The deoxygenations of ethylene oxide and of THF to give ethylene have been found to occur with carbon atoms from several sources [3, 4, 11a]; those of the _cis_- and _trans_-but-2-ene oxides (10) and (11) give mixtures of the two 2-butenes, _and_ hence a diradical intermediate such as (12) is thought to play a role [13]. Desulfurization has likewise been observed [14].

---

6. P.S. Skell and P.W. Owen, _J. Amer. Chem. Soc._, _94_, 1578 (1972).
7. G. Herzberg, "Atomic Spectra and Atomic Structure", Dover Publications, New York, 1944; p. 142.
8. K.R. Thompson, R.L. DeKock, and W. Weltner, _J. Amer. Chem. Soc._, _93_, 4688 (1971).
9. D.E. Milligan and M.E. Jacox, _J. Chem. Phys._, _47_, 703 (1967).
10. M.E. Jacox and D.E. Milligan, _J. Chem. Phys._, _47_, 1626 (1967).
11. (a) P.S. Skell, J.H. Plonka, and R.R. Engel, _J. Amer. Chem. Soc._, _89_, 1748 (1967); (b) P.S. Skell and J.H. Plonka, _J. Amer. Chem. Soc._, _92_, 836 (1970).
12. P.S. Skell and J.H. Plonka, _J. Amer. Chem. Soc._, _92_, 2160 (1970).
13. J.H. Plonka and P.S. Skell, _Chem. Commun._, 1108 (1970).
14. K.J. Klabunde and P.S. Skell, _J. Amer. Chem. Soc._, _93_, 3807 (1971).

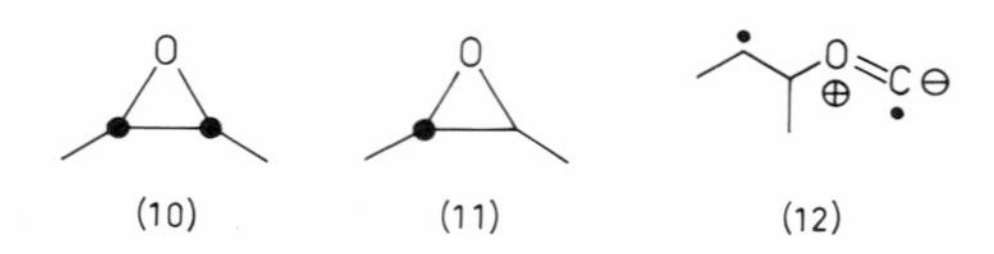

(10) (11) (12)

Insertion products are obtained if the substrate is trimethylsilane, (13). The product distribution suggests that thermally generated carbon vapor, containing evidently mostly ^{3}P carbon atoms, selectively attacks the SiH bond; but the arc produced vapor, which has a substantial fraction of carbon atoms in the ^{1}S and ^{1}D states, attacks the CH and SiH bonds indiscriminately [6].

$(CH_3)_3SiH$

(13)

A puzzling situation developed when the reaction with olefins was examined. When C^{11} atoms are used, the reaction with ethylene gives rise mostly to acetylene and some allene [15]; moderation with neon and deposition on a cold surface reverses the relative yields, but no cyclopropanes are found [16]. This is reasonable since cyclopropenylidenes invariably collapse to form allenes [17]. Similar observations were made with chemically produced carbon atoms [3,4].

Carbon atoms from an arc however were claimed to add to two olefin molecules to give spiropentanes; if *cis*-2-butene was admitted to a matrix in which the carbon atoms had been frozen two seconds or less, the reaction was largely stereospecific to produce (14), but if the matrix was aged longer than that, products (14) and (15) formed in equal amounts, and *trans*-2-butene gave (15), (16) and (17) in the ratio of 3:3:2.

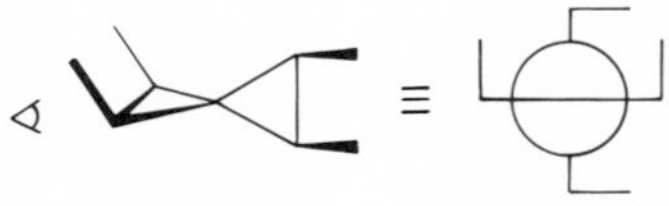

(14) 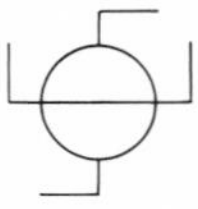 (15) 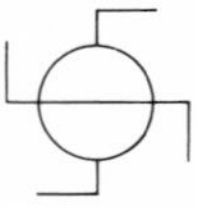 (16) 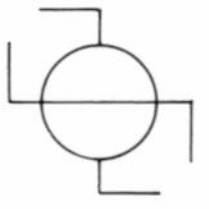 (17)

These observations were in accord with a ^{3}P state of the carbon atom which should add the first olefin molecule stereospecifically and the second non-stereospecifically [18]. A reinvestigation later revealed, however, that spiropentanes do not result from arc-carbon atom plus olefin reactions at all: The products with the 2-butenes are *cis*- and *trans*-1,3-pentadiene (18) and (19), and 2,3-pentadiene (20) [19]. The first two of these products are formed with considerable stereospecificity.

15. J. Nicholas, C. MacKay, and R. Wolfgang, *J. Amer. Chem. Soc.*, **88**, 1610 (1966).
16. J.E. Nicholas, C. MacKay, and R. Wolfgang, *J. Amer. Chem. Soc.*, **87**, 3008 (1965).
17. L. Skattebol, *Acta Chem. Scand.*, **17**, 1683 (1963).
18. P.S. Skell and R.R. Engel, *J. Amer. Chem. Soc.*, **87**, 1135, 1135, 2493 (1965) and **88**, 3749 (1966).
19. P.S. Skell, J.E. Villaume, J.H. Plonka, and F.A. Fagone, *J. Amer. Chem. Soc.*, **93**, 2699 (1971).

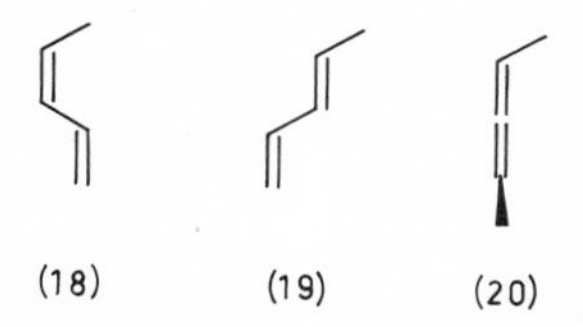

(18) (19) (20)

The principal products of carbon atoms with benzene are toluene and tropilidene [3, 20], cyclopentadiene gives benzene, fulvene and ring-opened products [21], and cyclopropane gives mostly methylenecyclopropane [22]. Most of these reactions can be understood in terms of initial CH insertion. Similar insertions occur with carbon-chlorine bonds [23].

The acetylene formation which always accompanies the reactions of carbon atoms with various organic substrates is also due to insertion. It has been investigated by means of D-labelling. The labelled substrates (21)-(24) all give C_2H_2 and C_2D_2 only and very little or no C_2HD: Evidently insertion occurs first, followed by a specific break-up of the carbene (25) [24].

CD_3—CH_3 (21)

CD_3—CH_2—CH_3 (22)

CH_3—CD_2—CH_3 (23)

CD_3—CH_2—CD_3 (24)

CD_3 ∿ HC—CH ∿ CD_3 (25)

The C_2 present in carbon vapor apparently engages mostly in hydrogen abstraction reactions. Thus, when methanol and carbon vapor are codeposited, dimethoxymethane, acetylene and ethylene form. If the vapor is obtained from radiolabelled graphite, the specific activity of the latter two products is double that of the former, but if the methanol is labelled, only the dimethoxymethane is radioactive. The conclusion is that C_2 is indeed present in the vapor [25]. It was further learned that if a mixture of acetone and perdeuteroacetone is used as substate, some C_2HD is obtained but it is completely suppressed by toluene. The C_2H_2 and C_2D_2 products are uneffected. This was interpreted as meaning that both singlet and triplet states are present; the singlet abstracts both hydrogens at once, and the triplet does it in two steps [26].

---

20. (a) J. L. Sprung, S. Winstein, and W. F. Libby, J. Amer. Chem. Soc., 87, 1812 (1965); (b) T. Rose, C. MacKay, and R. Wolfgang, J. Amer. Chem. Soc., 89, 1529 (1967).
21. T. Rose, C. MacKay, and R. Wolfgang, J. Amer. Chem. Soc., 88, 1064 (1966).
22. R. R. Engel and P. S. Skell, J. Amer. Chem. Soc., 87, 4663 (1965).
23. P. S. Skell and R. F. Harris, J. Amer. Chem. Soc., 87, 5807 (1965).
24. H. J. Ache and A. P. Wolf, J. Amer. Chem. Soc., 88, 888 (1966).
25. P. S. Skell and R. F. Harris, J. Amer. Chem. Soc., 88, 5933 (1966).
26. (a) P. S. Skell and J. H. Plonka, J. Amer. Chem. Soc., 92, 5620 (1970). For recent further work, see (b) P. S. Skell, F. A. Fagone, and K. J. Klabunde, J. Amer. Chem. Soc., 94, 7862 (1972); (c) P. S. Skell, J. E. Villaume, and F. A. Fagone, J. Amer. Chem. Soc., 94, 7866 (1972).

C_3 is present in carbon vapor to the extent of 12-20% [27]. They add to olefins to give bisethanoallenes (26). If cis- or trans-2-butene is admitted to the matrix after depositing of the vapor, the former gives only (27) whereas the trans- substrate gives an equimolar mixture of (28) and (29): Thus the C_3 ground state is a singlet (note that the formulas (26)-(29) should not be confused with (14)-(17)). If the depositions of the vapor and the substrates are simultaneous, both 2-butenes produced some (30), so that an excited triplet state is implicated in the vapor [28].

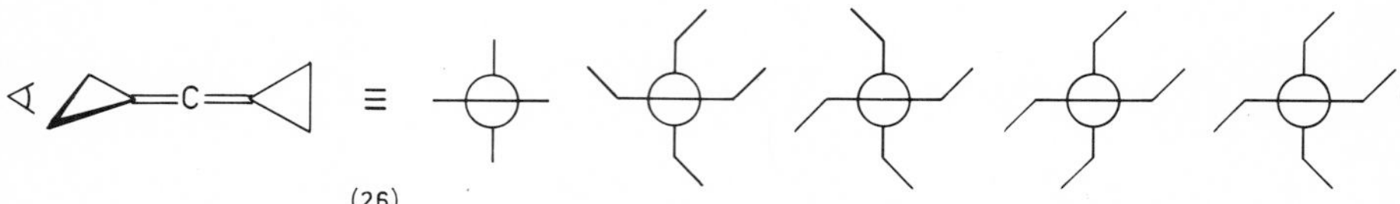

(26) (27) (28) (29) (30)

24-3. REACTIONS OF ATOMS OTHER THAN CARBON [29]

Neutral atoms with 6 valence electrons provide a further analog to carbene chemistry, and some of their properties have been studied.

Oxygen atoms can be generated in either the triplet ^3P ground state by the irradiation of nitrous oxide at 245 nm wavelength, or in the ^1D singlet state by the irradiation of nitrogen dioxide at 229 nm. The triplet atoms will insert into acetylenic C-H bonds to form ketenes [30] and in aromatic C-H bonds to form phenols [31]. They add to double bonds in non-stereospecific fashion. The singlet atoms add to olefins in a way which is believed initially to be stereospecific, but this result is clouded by the fact that the hot product formed initially undergoes some isomerization [32]; nevertheless these early studies were vital in the evolution of Skell's rule (see p. 555).

Sulfur atoms provide one of the very few clear-cut examples of a violation of this rule. ^1D sulfur atoms can be generated by photolysis of carbonyl sulfide, COS, with radiation at 230-255 nm. These atoms insert into C-H bonds with very little selectivity to give mercaptans; thus, with propane as substrate, n-propyl- and i-propylmercaptan obtain in a 3:1 ratio, and with i-butane, i-butyl- and t-butylmercaptans form in a 9:1 ratio [33]. Inert gas lowers this insertion ability to zero,

---

27. P. S. Skell and R. F. Harris, J. Amer. Chem. Soc., 91, 699 (1969).
28. P. S. Skell, L. D. Wescott, J. P. Golstein, and R. R. Engel, J. Amer. Chem. Soc., 87, 2829 (1965).
29. For review, see O. M. Nefedov and M. N. Manakov, Angew. Chem., Int. Ed. Engl., 5, 1021 (1966).
30. I. Haller and G. C. Pimentel, J. Amer. Chem. Soc., 84, 2855 (1962).
31. E. Grovenstein and A. J. Mosher, J. Amer. Chem. Soc., 92, 3810 (1970).
32. S. Sato and R. J. Cvetanovic, Can. J. Chem., 36, 1668 (1958).
33. (a) A. R. Knight, O. P. Strausz, and H. E. Gunning, J. Amer. Chem. Soc., 85, 2349 (1963); (b) K. S. Sidhu, I. G. Csizmadia, O. P. Strausz, and H. E. Gunning, J. Amer. Chem. Soc., 88, 2412 (1966).

and consequently it is assumed that the deactivation process reduces the ^1D sulfur atoms to ground state ^3P atoms which react with COS and with one another to form sulfur. The ^3P atoms can also be generated directly in a mercury-sensitized photolysis [33]. It was further found that both the ^1D and ^3P sulfur atoms add to the 2-butenes to form episulfides with complete stereospecificity [34]. It is not known whether this violation of the Skell rule should be attributed to rapid spin inversion and ring closure of the diradical, or to an unusually slow rotation of the carbon-carbon bond. It is indeed ironic that ^3P oxygen and sulfur atoms should differ in this way. In the late fifties, when the validity of Skell's rule was at best uncertain, the behavior of the ground state oxygen atom was widely quoted in support of it. One can only surmise how this story would have evolved if Cvetanovich had started his work with sulfur atoms instead of oxygen.

Nitrogen atoms can be obtained from molecular nitrogen via microwave discharges. They usually react with organic substrates to give HCN and sometimes nitriles - possible via steps such as indicated in (31)-(34) [35].

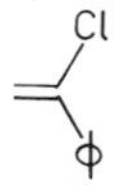

(31) 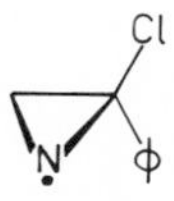 (32) $CH_2Cl—C(\phi)=N\cdot$ (33) $\phi—CN$ (34)

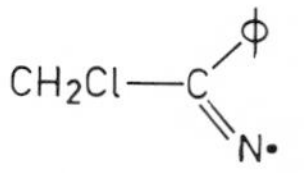

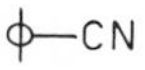

Silicon atoms are present in silicon vapor. They insert readily into SiH bonds; thus, trimethylsilane (35) is converted into the trisilane (36) [36].

$(CH_3)_3SiH$ (35)

$(CH_3)_3SiSiH_2Si(CH_3)_3$ (36)

Si_2 has been recognized by ir in a matrix in which silane had been photolyzed - as was SiH, the singlet SiH_2, and pyramidal SiH_3 [37].

We may note at this point that there now is some evidence for the existence of doubly bound silicon. Thus, if (37) is heated to 350°, naphthalene and a polymer - $(SiMe_2)_{2n}$ - form; the reaction is first order in (37) alone. If anthracene is present, the disilene is captured, and (38) is obtained [38]. If (39) is heated to 600°

---

34. (a) K.S. Sidhu, E.M. Lown, O.P. Strausz, and H.E. Gunning, J. Amer. Chem. Soc., 88, 254 (1966); (b) E.M. Lown, H.S. Sandhu, H.E. Gunning, and O.P. Strausz, J. Amer. Chem. Soc., 90, 7164 (1968).
35. J.J. Havel and P.S. Skell, J. Amer. Chem. Soc., 94, 1792 (1972).
36. (a) P.S. Skell and P.W. Owen, J. Amer. Chem. Soc., 89, 3933 (1967); (b) P.W. Owen and P.S. Skell, Tetrahedron Lett., 1807 (1972).
37. D.E. Milligan and M.E. Jacox, J. Chem. Phys., 52, 2594 (1970).
38. (a) G.J.D. Peddle, D.N. Roark, A.M. Good, and S.G. McGeachin, J. Amer. Chem. Soc., 91, 2807 (1969); (b) D.N. Roark and G.J.D. Peddle, J. Amer. Chem. Soc., 94, 5837 (1972).

and the products are frozen out at -196°, the matrix shows a sharp ir peak at 1400 cm^{-1}. This band has been assigned as the double bond stretching frequency of (40); at -120° it disappears irreversibly, and the product of the reaction is (41) [39].

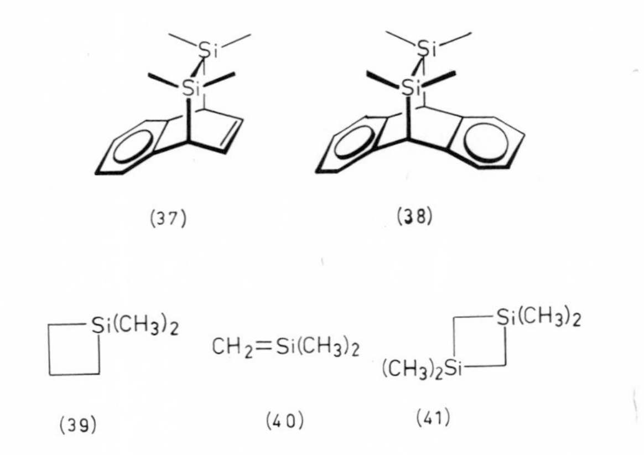

24-4. STRAINED HYDROCARBONS

Once the limit of strain possible in a given class of isolable compounds has been reached, it is still possible to search for the next more highly strained member in the realm of unstable intermediates.

Cyclobutadiene is a good example. The lack of success in its attempted synthesis is due to a combination of angle strain and electronic factors (see p. 278). Nevertheless, it is now clear that this hydrocarbon was precent as a transient intermediate in many of the reactions that were hopefully employed in the synthetic work. Thus, the dimeric (42) is often observed. Pettit found that this material obtains in the ceric-IV oxidation of cyclobutadieneirontricarbonyl (43), and that an adduct (44) formed instead if the oxidation were carried out in the presence of methyl propriolate (45).

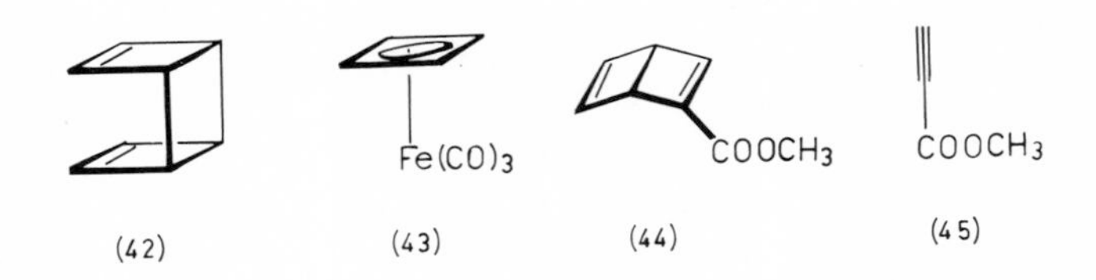

Furthermore, a volatile product could be swept out of the oxidation mixture with an inert gas, and if the condensate at 77°K were allowed to warm up with (45), (44) formed in the trap [40]. The extremely facile Diels-Alder reactions are now used

39. T. J. Barton and C. L. McIntosh, Chem. Commun., 861 (1972).
40. (a) J. D. Fitzpatrick, L. Watts, G. F. Emerson, and R. Pettit, J. Amer. Chem. Soc., 87, 3254 (1965); (b) Chem. Eng. News, 43, 38 (Aug. 23, 1965); (c) R. Pettit, Pure Appl. Chem., 17, 253 (1968).

to recognize intermediate cyclobutadiene; products such as (42) and (44) obtain in the pyrolysis of photo-α-pyrone (46) [41], in the photolysis of the tricyclodecatriene (47) [42], in the zinc dehalogenation of trans-3, 4-diiodocyclobutene (48) [43] and in the lithium-amalgam dechlorination of perchlorocyclobutene (49) [44]; one instance [44] even a mixed dimer (50) could be prepared via the simultaneous generation of two cyclobutadienes.

(46) (47) (48) (49) (50)

Tetrahedrane (51) is another good example. The various vain attempts to prepare it and its derivatives (see p. 206) were capped with a publication [45] suggesting that this molecule - if it has normal CC bond lengths - is less stable than its isomer cyclobutadiene by no less than 70 kcal/mole, and that it should decompose to acetylene without activation energy; in other words, that it should not have a finite lifetime under any circumstances.

(51)

Wolf investigated the reaction of carbon atoms generated by photolysis of carbon suboxide, C_3O_2 (hence possibly the species :C=C=O) with cyclopropene (53) [46]. This reaction produces substantial amounts of acetylene, and the possibility exists that this product has a tetrahedrane precursor (54)-(56).

(53) (54) (55) (56)

---

41. E. Hedaya, R.D. Miller, D.W. McNeil, P.F. D'Angelo, and P. Schissel, J. Amer. Chem. Soc., 91, 1875 (1969).
42. R.D. Miller and E. Hedaya, J. Amer. Chem. Soc., 91, 5401 (1969).
43. E.K.G. Schmidt, L. Brener, and R. Pettit, J. Amer. Chem. Soc., 92, 3240 (1970).
44. R. Criegee and R. Huber, Chem. Ber., 103, 1862 (1970).
45. R.J. Buenker and S.D. Peyerimhoff, J. Amer. Chem. Soc., 91, 4342 (1969).
46. H.W. Chang, A. Lautzenheiser, and A.P. Wolf, Tetrahedron Lett., 6295 (1966).

Use was made of the labelled olefin (57). This should produce dideuterotetrahedrane, and in turn C_2H_2, C_2HD and C_2D_2 should then form in the ratio 1:4:1. Mass spectrometric analysis was used to verify that this ratio did indeed obtain, and that the acetylenes and the cyclopropene undergo no exchange under the conditions used. In a double labelling experiment involving both (57) and (58) it was shown that the C_2H_2 was inactive, that one third of the total radioactivity was present in the C_2D_2 and two thirds in the C_2HD. In the actual event, the ratios were not so simple because of incomplete labelling of (57); however, if an allowance is made for that, the results can be explained only with intermediate (51) or - what seems much less likely - a set of rapidly equilibrating open, diradical forms (59) of tetrahedrane [47].

O=C=C=C=O

(57) (58) (59)

It seems conceivable that the decomposition to acetylene might be reversible in a favorable case. Staab [48] has prepared the interesting hydrocarbon (60), in which two acetylenic linkages are rigidly held in crossed positions. If the resolved species should become available, its racemization might be indicative of the reverse reaction <u>via</u> the symmetrical (61) and (62).

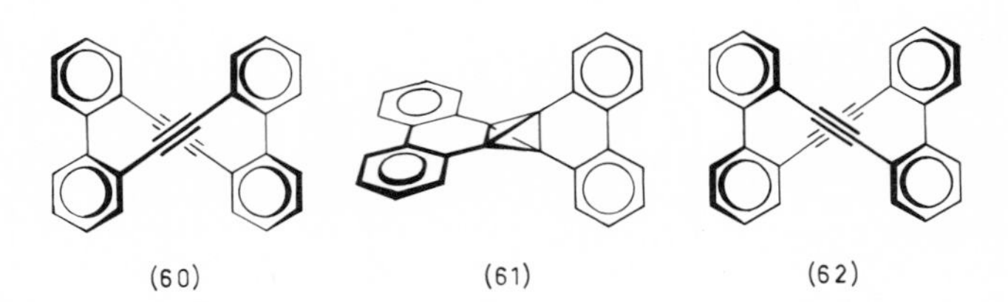

(60) (61) (62)

Among the propellanes the smallest isolable members are the tricyclic species (63) and (64) discussed on p. 208; in these compounds all four valences of each of the bridgehead carbons lie in one hemisphere with the carbon nucleus at the center.

(63) (64)

47. (a) P. B. Shevlin and A. P. Wolf, <u>J. Amer. Chem. Soc.</u>, <u>92</u>, 406 (1970); see also (b) R. F. Peterson, R. T. K. Baker, and R. L. Wolfgang, <u>Tetrahedron Lett.</u>, 4749 (1969).
48. (a) H. A. Staab and E. Wehinger, <u>Angew. Chem., Int. Ed. Engl.</u>, <u>7</u>, 225 (1968). See also (b) E. H. White and A. A. F. Sieber, <u>Tetrahedron Lett.</u>, 2713 (1967).

It is possible that the next smaller member (65) may be capable of transient existence [49]. Thus, if 1,4-dichloronorbornane (66) is treated in succession with lithium-sodium alloy and carbon dioxide, (67) is obtained but (68) is not; perhaps (69) is formed, which could then eliminate LiCl and give (65); this surely would rapidly be converted to (70) and thence to (67).

(65) (66) (67) (68) (69) (70)

The smallest molecule known to violate Bredt's rule, 1,2-norbornene (71), is evidently an intermediate in the dehalogenation of exo-1,2-diiodonorbornane (72) with methyllithium; in the presence of furan the two Diels-Alder adducts (73) and (74) are obtained [50].

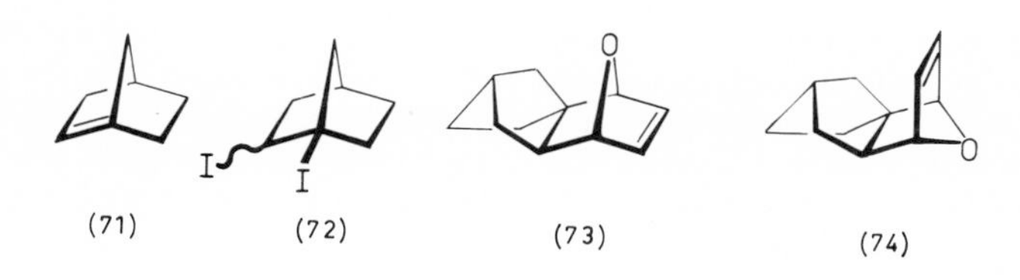

(71) (72) (73) (74)

The product ratio was found to be independent of the base and of the type of 1-iodo-2-halonorbornane. Similar evidence has been adduced for the existence of 1-adamantene (75); 1,2-diiodoadamantane (76) with strong bases gives (77) and or related dimers. Since the two p orbitals in (75) are perpendicular, it seems possible that the "olefin" is really a triplet diradical [51].

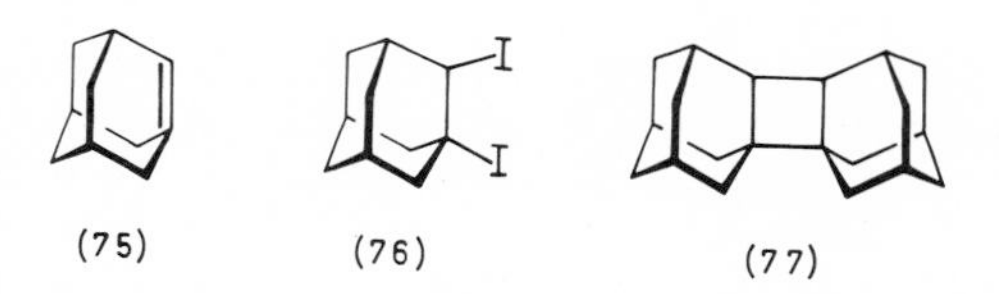

(75) (76) (77)

Angle strain at olefinic positions is at a maximum in cyclopropene and in $\Delta^{1,5}$-bicyclo[3.2.0]heptene (78); recently however, this series was extended to the corresponding [2.2.0]hexene (79). Thus, pyrolysis of (80) leaves a deposit on a

---

49. (a) C. F. Wilcox and C. Leung, J. Org. Chem., 33, 877 (1968), quoted by (b) K. Wiberg, E. C. Lupton, and G. J. Burgmaier, J. Amer. Chem. Soc., 91, 3372 (1969); (c) P. G. Gassman, A. Topp, and J. W. Keller, Tetrahedron Lett., 1093 (1969).
50. (a) R. Keese and E. P. Krebs, Angew. Chem., Int. Ed. Engl., 10, 262 (1971) and 11, 518 (1972); see also (b) G. L. Buchanan, N. B. Kean, and R. Taylor, Chem. Commun., 201 (1972).
51. (a) D. Lenoir, Tetrahedron Lett., 4049 (1972); (b) D. Grant, M. A. McKervey, J. J. Rooney, N. G. Samman, and G. Step, Chem. Commun., 1186 (1972).

nearby cold finger which upon warming yields the dimer (81). If cyclopentadiene is present during the warming-up period, (82) is obtained; if the deposit is distilled into methylene chloride at -50°, a solution is obtained which shows a singlet at 6.76 τ. Warming this solution then gives rise to the dimer (81); this points to the olefin (79) as the origin of the 6.76 τ signal [52].

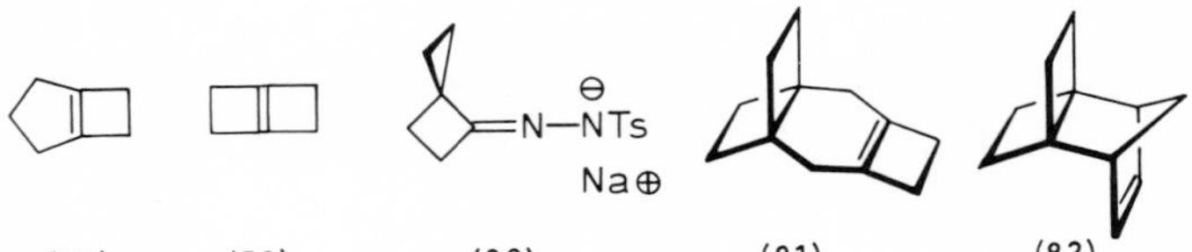

=N—NTs
Na⊕

(78) (79) (80) (81) (82)

24-5. OXYGENATED SMALL RINGS

Cyclopropanone and its derivatives were mentioned on p. 217, and here we only discuss the evidence that such species are intermediates in the Favorsky rearrangement, in which α-haloketones upon treatment with base give rearranged carboxylic acids. Thus, α-chlorocyclohexanone (83) gives cyclopentylcarboxylic acid (84), and the intermediates would be - in succession - (85) and (86) [53].

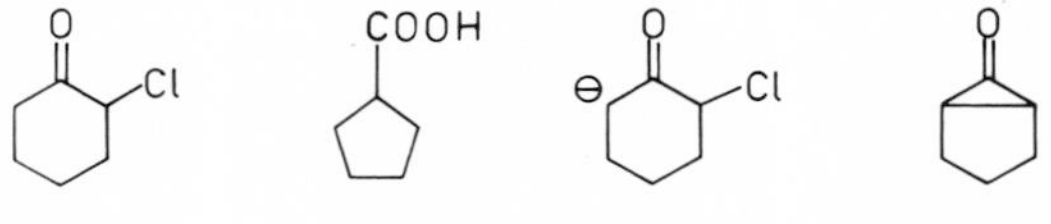

O Cl COOH O Cl O

(83) (84) (85) (86)

If α-radiocarbon labelled (83) is used, the activity in the product is found to be distributed equally among the positions 1 and 2 [54]. If phenoxide ion is present, α-phenoxycyclohexanone is obtained; that this product results from the interception of (86) and not from simple S_N2 displacement or S_N2' attack on the enol is clear from the fact that if the 1,2-labelled precursor (87) is used, the product has half the activity at C_1 and half at C_2 and C_6. S_N2 reaction would have given (88) and S_N2' would have produced (89) [55].

---

52. K.B. Wiberg, G.J. Burgmaier, and P. Warner, J. Amer. Chem. Soc., 93, 246 (1971).
53. For references and review, see (a) N.J. Turro, R.B. Gagosian, C. Rappe, and L. Knutsson, Chem. Commun., 270 (1969); (b) A.A. Akhrem, T.K. Ustynyuk, and Y.A. Titov, Russ. Chem. Rev. (English transl.), 39, 732 (1970).
54. R.B. Loftfield, J. Amer. Chem. Soc., 73, 4707 (1951).
55. W.B. Smith and C. Gonzalez, Tetrahedron Lett., 5751 (1966).

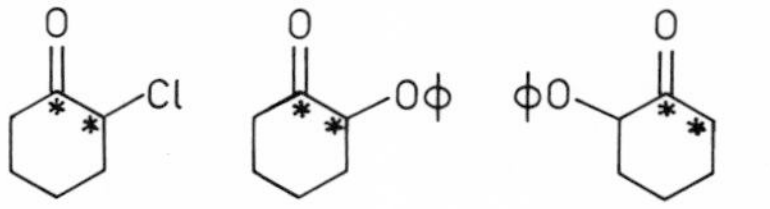

(87) (88) (89)

The fact that the tetrahaloketone (90) gives acid (91) rules out the possibility that both proton and halogen leave from the same carbon atom to give an intermediate carbene such as (92) - at least in that case [56].

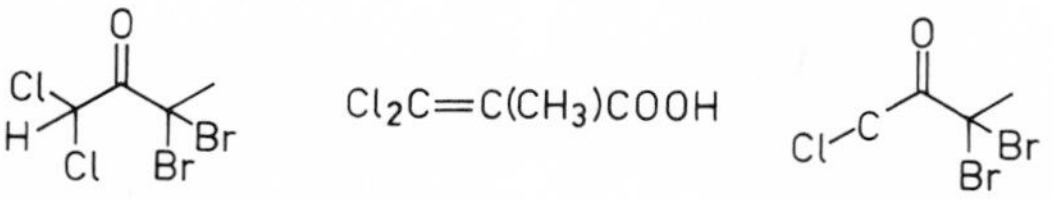

$Cl_2C{=}C(CH_3)COOH$

(90) (91) (92)

The capture of intermediate cyclopropanones has been reported in some cases. Thus, when the ketone (93) is treated with base in the presence of furan, (94) is obtained [57].

(93) (94)

It is of course possible in all these cases that the intermediate is in the open dipolar form; some evidence that this is actually so is provided by the fact that the ketone (95) gives, beside the normal product (96), a ketone (97) possibly derived from the zwitterion (98) [58].

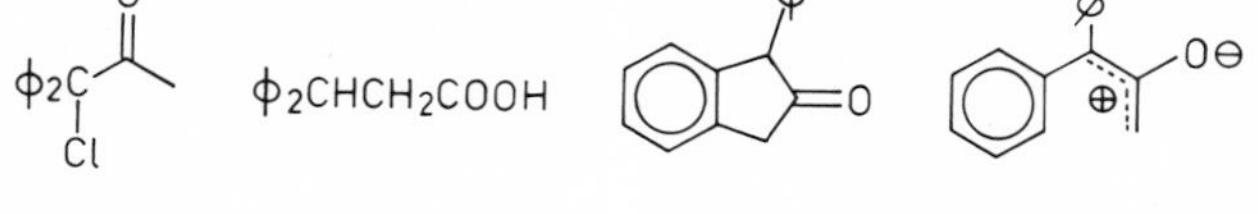

ϕ_2CHCH_2COOH

(95) (96) (97) (98)

Bordwell also found that loss of chloride is rate limiting in the reaction of ketone (99) since k_H/k_D is 1.0 if the 2, 6, 6-trideuterioanalog is used; but with the bromide

56. C. Rappe, Acta Chem. Scand., 20, 862 (1966).
57. A.W. Fort, J. Amer. Chem. Soc., 84, 4979 (1962).
58. (a) F.G. Bordwell and R.G. Scamehorn, J. Amer. Chem. Soc., 93, 3410 (1971). See also (b) A.J. Fry and R. Scoggins, Tetrahedron Lett., 4079 (1972).

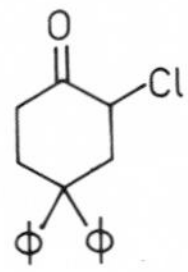

(99)

$k_H/k_D = 4.1$ and furthermore $k_{Br}/k_{Cl} \approx 100$, so that anion formation may be becoming concerted with halide loss if the halogen is bromine [59]. Finally, if the leaving halogen is further away as for instance in (100), a homo-Favorsky rearrangement occurs with the cyclobutanone (101) as stable product [60].

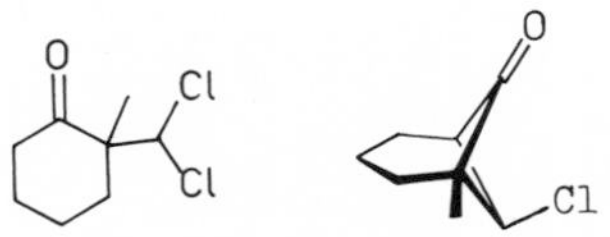

(100) (101)

The stability of α-lactones (102) is even less than that of the cyclopropanones. Their role as intermediates has been postulated in several reactions; thus, the fact that α-bromopropionate ion solvolyzes much faster than i-propyl bromide is explained that way [61].

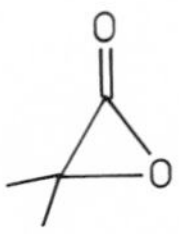

(102)

Persuasive evidence for the intermediacy of an α-lactone was found by Wheland and Bartlett [62] in the ozonization of diphenylketene (103) at -78° to polybenzilic ester (104). This polymer instantly precipitates out of the solution even at -100°. It does not react with methanol under these conditions after it has been formed; however, if methanol is present from the start, the polymer does not appear and (105) is formed instead. Thus, the methanol is trapping an intermediate which most likely is the α-lactone (106). Hexafluoroacetone can be used in the same role to give (107).

---

59. F.G. Bordwell, R.R. Frame, R.G. Scamehorn, J.G. Strong, and S. Meyerson, J. Amer. Chem. Soc., 89, 6704 (1967).
60. E. Wenkert, P. Bakuzis, R.J. Baumgarten, C.L. Leicht, and H.P. Schenk, J. Amer. Chem. Soc., 93, 3208 (1971).
61. E. Grunwald and S. Winstein, J. Amer. Chem. Soc., 70, 841 (1948); see also Refs. 62 and 63.
62. R. Wheland and P.D. Bartlett, J. Amer. Chem. Soc., 92, 6057 (1970).

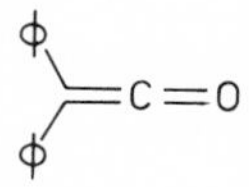

(103) 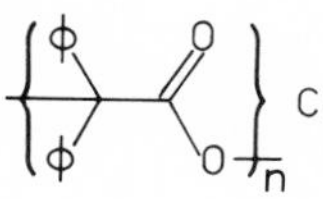 (104) 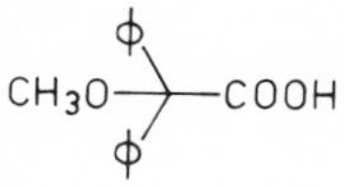 (105) 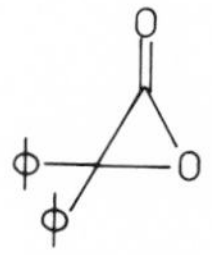 (106) 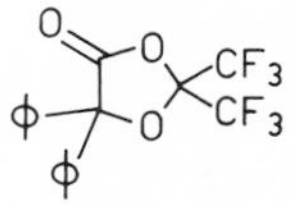 (107)

The same polymer is also formed when diphenyldiazomethane is photolyzed at -78° in carbon dioxide; evidently the carbene adds to the C=O double bond to give (106). Di-t-butylketene can likewise be ozonized at -78°. As in many other instances, the t-butyl groups seem to stabilize the small ring; thus, the polyester does not begin to precipitate out of the solution till it has been warmed to -20°. It seems possible that the first product is a true α-lactone, and that the diphenyl analog is an open zwitterion (108). The dioxolanedione (109) upon photolysis at -80° also affords a polybenzilic ester. Once again the appearance can be prevented by the presence of methanol; the product is then the acid (110), so that an α-lactone was present as an intermediate in that case also [63].

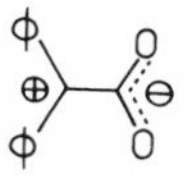

(108) 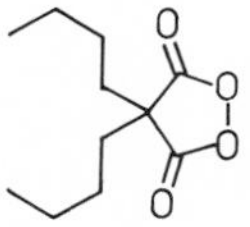 (109) 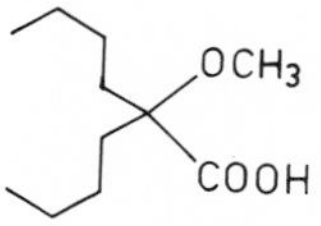 (110)

Even the next extrapolation to carbon trioxide (111) has been realized. It was first postulated by Taube as an intermediate in the reaction of ozone and carbon dioxide [64]; later it was recognized by ir in a low temperature matrix [65]. This study also revealed that it did not have the open structure (112).

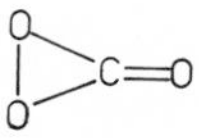

(111) 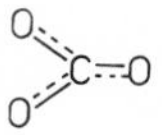 (112)

The possible role of carbon dioxide dimer (or 1,2-dioxetanedione) (113) in chemiluminescence has already been discussed (p. 457); its existence has been verified by mass spectroscopy [66].

---

63. W. Adam and R. Ruckтäschel, J. Amer. Chem. Soc., 93, 557 (1971).
64. (a) D. Katakis and H. Taube, J. Chem. Phys., 36, 416 (1962). See also (b) W.B. DeMore and C.W. Jacobsen, J. Phys. Chem., 73, 2935 (1969).
65. N.G. Moll, D.R. Clutter, and W.E. Thompson, J. Chem. Phys., 45, 4469 (1966).
66. H.F. Cordes, H.P. Richter, and C.A. Heller, J. Amer. Chem. Soc., 91, 7209 (1969).

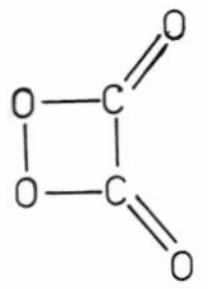

(113)

An almost bewildering variety of highly oxygenated small ring intermediates has been suggested to account for the ozonization of olefins; thus, the reaction of olefin (114) leads to the corresponding epoxide, acetaldehyde and cyclohexanone via the intermediates shown below. In this scheme, (115) is usually referred to as the Criegee intermediate, and (116) as the Staudinger molozonide. The latter intermediate can be diverted by ketones readily subject to Bayer-Villiger oxidation (of ketones to esters by peracids) to the dioxetanes (117); the presence of the latter is demonstrated by the fact that rapid heating of the mixture then gives rise to chemiluminescence. A description of the proof of the complete scheme is beyond the scope of this book, however [67].

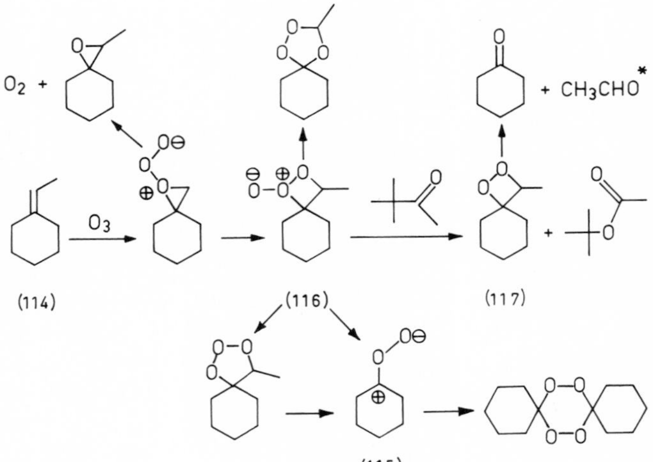

O_2 + ... + CH_3CHO^*

O_3

(114) (116) (117)

(115)

We close this section with a reference [68] to the fact that three-membered rings (118) often differ very little in energy from dipolar open forms such as (119). Thus, cyclopropane is a known, stable closed molecule, while ozone is open; in one case ((120)) both forms have been isolated [69]. Further studies of this phenomenon will surely turn up much of interest.

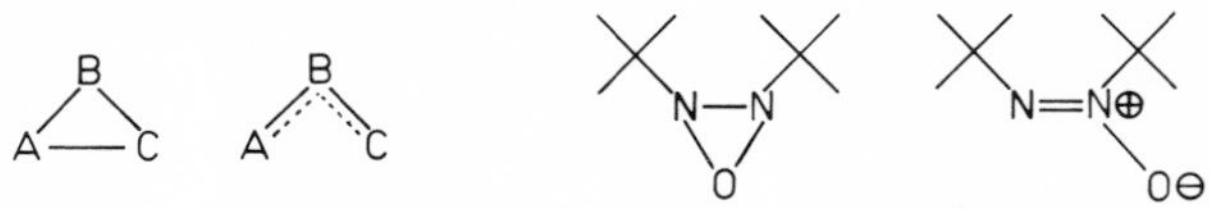

(118) (119) (120)

67. For references, see P.R. Story, E.A. Whited, and J.A. Alford, J. Amer. Chem. Soc., 94, 2143 (1972).
68. F.D. Greene and S.S. Hecht, J. Org. Chem., 35, 2482 (1970).
69. S.S. Hecht and F.D. Greene, J. Amer. Chem. Soc., 89, 6761 (1967).

24-6. HALOKETENES, SULFENES AND SULFINES

The substances (121), (122) and (123) may be thought of as organic derivatives of carbon dioxide, sulfur trioxide and sulfur dioxide, respectively.

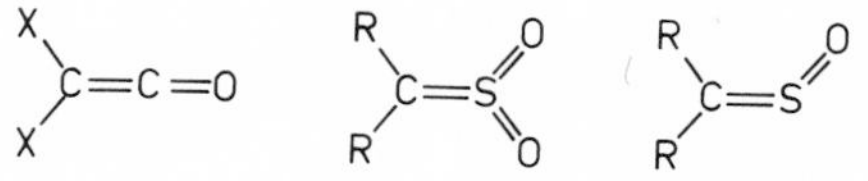

$X_2C{=}C{=}O$ (121) $R_2C{=}SO_2$ (122) $R_2C{=}S{=}O$ (123)

Ketenes - including the parent compound - have been known for many years. Most are stable in the absence of protic acids, which add to them to give acetic acid derivatives. Exceptions are the halogenated ketenes, which resist any preparation; only in recent years has it proved possible to generate them as intermediates and trap them - sometimes in ways useful in synthesis [70].

If the acid chloride (124) is treated with zinc at temperatures below -5°, the ir shows no ketene bands during the reaction which gives tetrafluoroethylene and carbon monoxide; but if acetone is present, the ketene is trapped and (125) is formed [71].

$BrCF_2{-}COCl$ (124) (125)

When (126) is treated with triethylamine, a polymer $(C_2Cl_2O)_n$ is formed but the monomer can be intercepted with cyclopentadiene to give the adduct (127); the latter substance reacts with base to give tropolone [72].

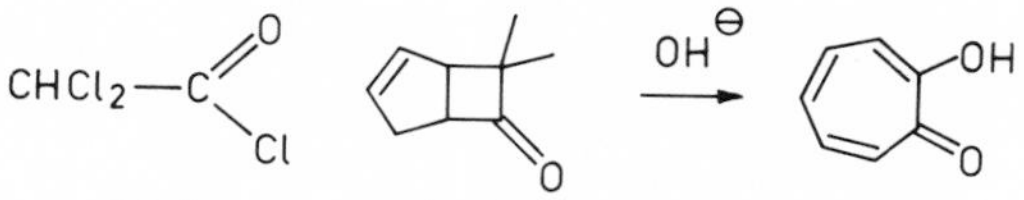

$CHCl_2{-}COCl$ (126) (127) $\xrightarrow{OH^{\ominus}}$ tropolone

Monohaloketenes can be generated in the same way, and this has been found to be a useful synthesis for 2-alkyltropones [73].

---

70. For review, see W.T. Brady, Synthesis, 415 (1971).
71. D.C. England and C.G. Krespan, J. Org. Chem., 33, 816 (1968).
72. (a) H.C. Stevens, D.A. Reich, D.R. Brandt, K.R. Fountain, and E.J. Gaughan, J. Amer. Chem. Soc., 87, 5257 (1965); (b) G. Ghosez, R. Montaigne, and P. Mollet, Tetrahedron Lett., 135 (1966).
73. W.T. Brady and J.P. Hieble, Tetrahedron Lett., 3205 (1970).

An interesting report concerns the photolysis of the cyclobutenedione (128). Apparently nothing happens; however, if the reaction is carried out in the presence of methanol, the succinate ester (129) is formed, and with phenyl isocyanide the adduct (130) is obtained. Evidently the photolysis produces the bisketene (131) which reverts to (128) in the absence of a trap; at -78° however, the ketene is stable and observable by ir [74].

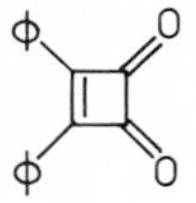

(128) 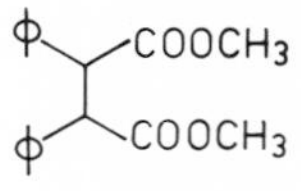 (129) 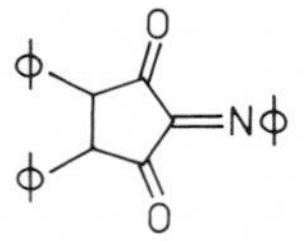 (130) 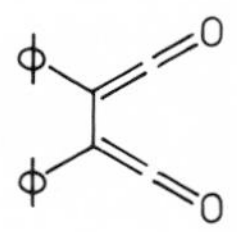 (131)

Sulfenes (132) resemble ketenes in both their preparation and reactions [75]. They can be prepared from sulfonyl chlorides with an α-hydrogen atom (133), and are subject to [2 + 2] cycloaddition to activated double bonds such as those found in vinyl ethers and enamines to give products such as (134).

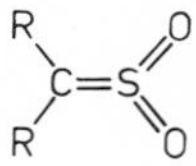

(132) 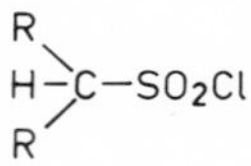 (133) 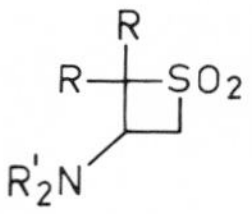 (134)

Even a case of addition to pyridine is known: The parent sulfene reacts with it to give the pyrido[1,2-d]dithiazine derivative (135) [76].

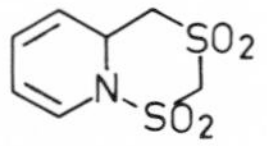

(135)

A case of [2 + 4] cycloaddition is also known; the parent sulfene adds to enamine (136) to give sulfonate (137) [77].

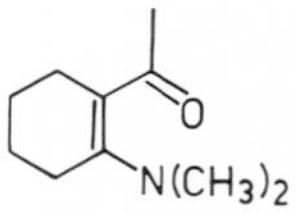

(136) 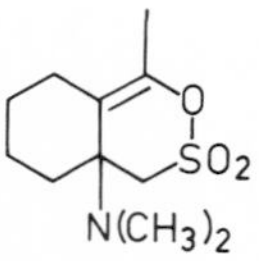 (137)

---

74. N. Obata and T. Takizawa, Chem. Commun., 587 (1971).
75. For review, see (a) T. J. Wallace, Quart. Rev. (London), 20, 67 (1966); (b) G. Opitz, Angew. Chem., Int. Ed. Engl., 6, 107 (1967).
76. J. S. Grossert, Chem. Commun., 305 (1970).
77. G. Opitz and E. Tempel, Angew. Chem., Int. Ed. Engl., 3, 754 (1964).

If the sulfene is generated in the presence of methanol-o-d, the deuterosulfonate (138) is obtained, showing that the direct reaction did not occur [78].

$$DCR_2{-}SO_2CH_3$$

(138)

The reality of sulfene intermediacy is supported by the fact that there are several alternative reactions by means of which they can be generated. One of these involves a Wolff-type of rearrangement from a carbene: Irradiation of α-diazosulfone (139) in methanol gives (140), presumably via sulfonylcarbene (141) and sulfene (142) [79].

$\phi{-}SO_2CHN_2$ (139) $\phi{-}CH_2SO_3CH_3$ (140) $\phi{-}SO_2\ddot{C}H$ (141) $\phi{-}CH{=}S(=O)_2$ (142)

Simple carbenes may also form sulfenes by capture of sulfur dioxide [75b]. If no trapping reagent is present, dimerization to such products as (143) may occur [80]. Retro Diels-Alder reaction of (144) at 220° also gives a sulfene [81].

(143) 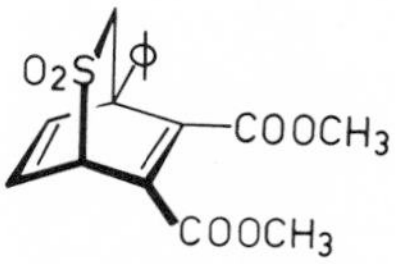 (144)

In a case such as the sulfonyl chloride (145), a sulfene (146) is produced for which dimerization is not feasible, and a polymer (147) is obtained [82].

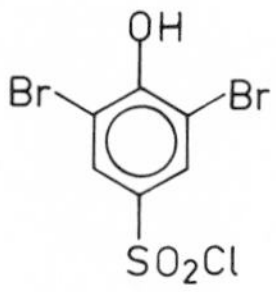

(145) 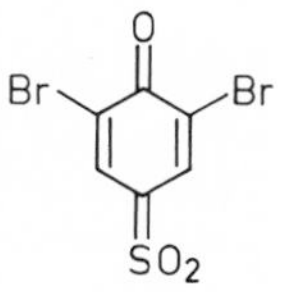 (146) 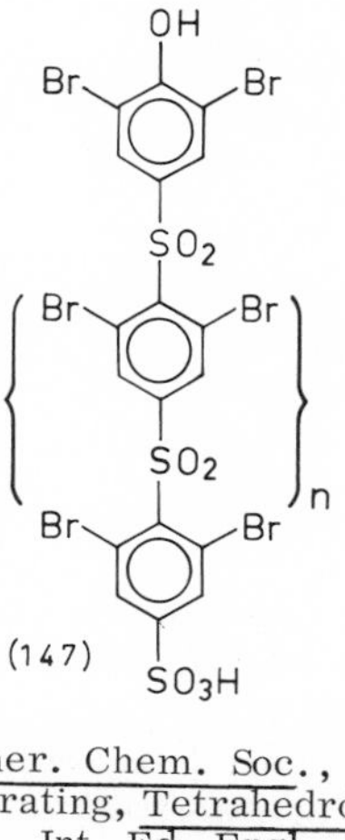 (147)

78. W.E. Truce and R.W. Campbell, J. Amer. Chem. Soc., 88, 3599 (1966).
79. R.J. Mulder, A.M. van Leusen, and J. Strating, Tetrahedron Lett., 3057 (1967).
80. G. Opitz and H.R. Mohl, Angew. Chem., Int. Ed. Engl., 8, 73 (1969).
81. J.F. King and E.G. Lewars, Chem. Commun., 700 (1972).
82. S. Oae and R. Kiritani, Bull. Chem. Soc. Japan, 38, 1543 (1965).

Thermolysis followed by low temperature quenching has led to the direct detection of sulfenes. Thus, the pyrolysis of (148) at 600° gives phtalimide (149), presumably *via* (150), and sulfene; the latter fragment can be chemically intercepted with ethylamine to give sulfonamide (151) [83].

NSO_2CH_3 (148) NH (149) OH, N (150) $CH_3SO_2NHC_2H_5$ (151)

Generation of sulfene by thermolysis of either chlorosulfonylacetic acid (152) or methanesulfonic anhydride (153) at 650° followed by cold finger condensation leads to a deposit that can be studied by ir; both vinyl CH and C=S stretching frequencies are recognizable. Subsequent exposure to methanol, HCl or DCl then gives the normal sulfene products [84].

$ClSO_2CH_2COOH$ (152) $(CH_3SO_2)_2O$ (153)

In solution at somewhat higher temperatures sulfene apparently gives the dimeric sulfene (154), and this material is stable for at least several days at -40°.

$CH_3SO_2CH{=}SO_2$

(154)

Thus, if methanesulfonyl chloride is treated with triethylamine in acetonitrile at -40° and the resulting ammonium salt is removed by filtration, the subsequent addition of deuteroalcohols leads to derivatives (155) [85].

$CH_3SO_2CHDSO_2OR$

(155)

Most of these reactions have a counterpart in the sulfinyl halides; sulfines are then obtained. Oxidation of thioketones and thioaldehydes by means of peroxyacids

---

83. W. J. Mijs, J. B. Reesink, and U. E. Wiersum, Chem. Commun., 412 (1972).
84. J. F. King, R. A. Marty, P. de Mayo, and D. L. Verdun, J. Amer. Chem. Soc., 93, 6304 (1971).
85. G. Opitz, M. Kleeman, D. Bücher, G. Walz, and K. Rieth, Angew. Chem., Int. Ed. Engl., 5, 594 (1966).

also leads to these products. The sulfines are more stable than the corresponding sulfenes; several such as (156) have been isolated [86]. The sulfine group is nonlinear as is indicated by the nmr non-equivalence of the two methyl groups of (157) [86].

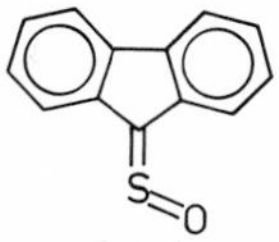

(156) 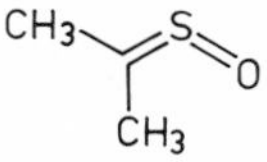 (157)

The sulfines will add to activated double bonds; thus, the dehydrochlorination of (158) leads to the sulfine (159), which can be added to diphenylketene to give (160) [87].

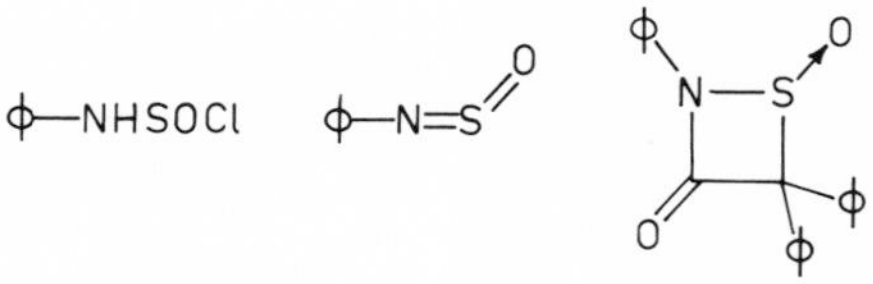

ϕ—NHSOCl ϕ—N=S=O

(158) (159) (160)

24-7. DIAZENE AND ITS DERIVATIVES [88]

The first attempt to prepare this compound N_2H_2, which is also frequently referred to as diimide, was reported by Thiele in 1892 [89]. He found that the decarboxylation of dipotassium azodicarboxylate (161) led only to nitrogen and hydrazine.

$$^{\ominus}OOC—N{=}N—COO^{\ominus}$$

(161)

---

86. W.A. Sheppard and J. Diekmann, J. Amer. Chem. Soc., 86, 1891 (1964).
87. G.R. Collins, J. Org. Chem., 29, 1688 (1964).
88. For review, see (a) S. Hünig, H.R. Müller, and W. Thier, Angew. Chem., Int. Ed. Engl., 4, 271 (1965); (b) C.E. Miller, J. Chem. Educ., 42, 254 (1965); (c) S. Hünig, Helv. Chim. Acta, 54, 1721 (1971).
89. J. Thiele, Ann., 271, 127 (1892).

In 1910, Raschig [90] attempted two alternative preparations: The pyrolysis of salts of benzenesulfonyl hydrazide (162) and the sodium chlorate oxidation of hydrazine. Both reactions led only to nitrogen and hydrogen.

$\phi{-}SO_2NHNH_2$

(162)

He commented that the hydrazine in Thiele's reaction must be due to the reduction of the azodicarboxylate by the hydrogen. Apparently both Thiele and Raschig felt that diazene might well be an intermediate too unstable to permit its isolation.

Several nearly simultaneous publications in 1961 made it clear that this compound would play an important role in organic chemistry. Hünig [91] and Corey [92] found that olefins could be reduced in the presence of hydrazine and an oxidizing agent, a remarkable reaction in which an intermediate product in the hydrazine oxidation reduces the double bond. Oxygen, ferricyanide, and mercuric oxide were all effective in bringing this about [93]. They assumed that the intermediate was diazene and that in the reducing step two hydrogen atoms were transferred simultaneously; for example, phenylpropiolic acid (163) gave only *cis*-cinnamic acid (164). Furthermore they found that benzenesulfonylhydrazide and azodicarboxylate ion gave rise to similar reductions [94].

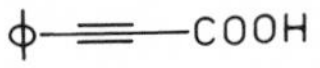

$\phi{-}C{\equiv}C{-}COOH$

(163)

H H
φ COOH

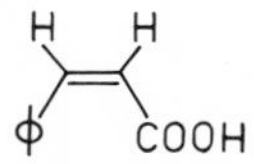

(164)

Corey also reported that maleic acid is reduced by N_2D_2 to the *meso*-d^2-succinic acid, whereas fumaric acid produces the *d,l*-mixture [95]. Similar observations were published slightly later by van Tamelen [96], who also found that in contrast to non-polar multiple bonds such as C=C, C≡C and N=N, polar ones such as C=N, C≡N, C=O, S=O, N=O, etc. are either not reduced at all or only with difficulty [97].

---

90. F. Raschig, *Angew. Chem.*, *23*, 972 (1910).
91. S. Hünig, H.-R. Müller, and W. Thier, *Tetrahedron Lett.*, 353 (1961).
92. E.J. Corey, W.L. Mock, and D.J. Pasto, *Tetrahedron Lett.*, 347 (1961).
93. J.M. Hoffman and R.H. Schlessinger, *Chem. Commun.*, 1245 (1971).
94. See also J.W. Hamersma and E.I. Snyder, *J. Org. Chem.*, *30*, 3985 (1965).
95. E.J. Corey, D.J. Pasto, and W.L. Mock, *J. Amer. Chem. Soc.*, *83*, 2957 (1961).
96. (a) E.E. van Tamelen, R.S. Dewey, and R.J. Timmons, *J. Amer. Chem. Soc.*, *83*, 3725 (1961); (b) R.S. Dewey and E.E. van Tamelen, *J. Amer. Chem. Soc.*, *83*, 3729 (1961).
97. (a) E.E. van Tamelen, R.S. Dewey, M.F. Lease, and W.H. Pirkle, *J. Amer. Chem. Soc.*, *83*, 4302 (1961). See also (b) E.W. Garbisch, S.M. Schildcrout, D.B. Patterson, and C.M. Sprecher, *J. Amer. Chem. Soc.*, *87*, 2932 (1965) and (c) E.E. van Tamelen, M. Davis, and M.F. Deem, *Chem. Commun.*, 71 (1965).

Since these discoveries much has been learned about both the generation and the subsequent reactions of diazene. Among the more elegant preparations one now finds Corey's thermal elimination of anthracene from (165) [98] and Wolinsky's reaction of hydrazine and carbon tetrachloride [99].

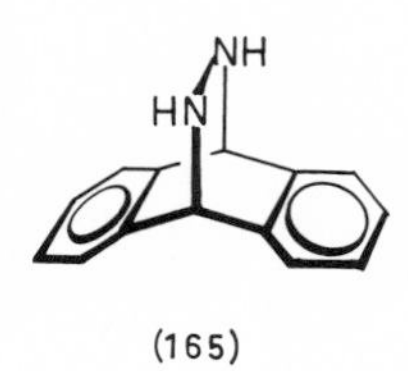

(165)

In the latter reaction copper-II catalysis and an induction period occur; these observations as well as the formation of chloroform point to radical intermediates.

Diazene has been found to be a key intermediate in certain chloramine (NH_2Cl) reactions. Chloramine is itself an intermediate in the industrial Raschig process by which ammonia is oxidized with chlorine to hydrazine [100]. It had long been known that solvolyzing solutions of chloramine have reducing porperties, and it was sometimes assumed that nitrene (NH) might be involved, especially since an insertion product (hydroxylamine) can be demonstrated to be present as yet another intermediate. Later it was found that the reaction of hydroxide with chloramine has a negative activation volume, and that the same is true of dimethylchloramine (which cannot form a nitrene); thus these are simple displacement reactions [101]. The hydroxylamine was demonstrated then to react with dichloramine to give diazene [102].

Microwave discharges through hydrazine vapor produce diazene under conditions in which it can be deposited in a low temperature matrix. These deposits are yellow in color; the color disappears irreversibly at -135° as disproportionation to nitrogen and hydrogen occurs. Ir and uv studies indicate that the compound is (166) rather than (167) or (168); this information is a bit difficult to square with the rationalizations formerly offered for the facile reduction of homopolar bonds (these explanations involved (167)) [103].

---

98. E.J. Corey and W.L. Mock, J. Amer. Chem. Soc., 84, 685 (1962).
99. J. Wolinsky and T. Schultz, J. Org. Chem., 30, 3980 (1965).
100. J. Fischer and J. Jander, Z. Anorg. Allg. Chem., 313, 14 (1961).
101. (a) W.J. le Noble, Tetrahedron Lett., 727 (1966); see also (b) M. Anbar and G. Yagil, J. Amer. Chem. Soc., 84, 1790 (1962).
102. E Schmitz, R. Ohme, and G. Kozakiewicz, Z. Anorg. Allg. Chem., 339, 44 (1965).
103. (a) S.N. Foner and R.L. Hudson, J. Chem. Phys., 28, 719 (1958); (b) E.J. Blau and B.F. Hochheimer, J. Chem. Phys., 41, 1174 (1964); (c) K. Rosengren and G.C. Pimentel, J. Chem. Phys., 43, 507 (1965); (d) A. Trombetti, J. Chem. Soc. (A), 1086 (1971) and Can. J. Phys., 46, 1005 (1968).

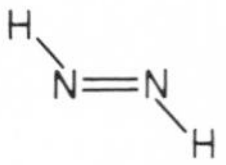

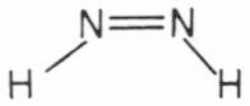

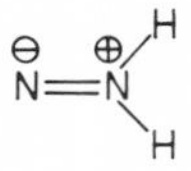

(166) (167) (168)

Monosubstituted diazenes are also unstable substances. Their involvement in the deamination by difluoramine (HNF_2) was first postulated in 1963 by Baumgardner [104]. Some time after that Cohen and Nicholson reported that the acid-catalyzed methanolysis of low concentrations of N-phenyl-N'-benzoyldiazene (169) leads to methyl benzoate, nitrogen and benzene; in the presence of various radical traps, evidence for phenyl radicals was found.

ϕ—N=N—COϕ

(169)

The authors postulated that phenyldiazene was formed in the initial step [105]. Simultaneously, Kosower and Huang discovered that if phenylazoformic acid (170) is decarboxylated in dilute solution at 25°, at moderate pH, and under rigorous exclusion of oxygen, phenyldiazene can be preserved and observed by uv for several hours [106]. Traylor later found that borohydride reduction of benzenediazonium salts also give such solutions [107]. Subsequently, many simple alkyldiazenes have been prepared in solution as well [108].

ϕ—N=N—COOH

(170)

The free radical nature of the reactions of these species have been explained as possibly due to a low-lying triplet state of these diazenes; this argument is reasonable if it is remembered that oxygen has a triplet ground state, and the excited triplet state for ethylene is not extremely high in energy [108].

Several of the intermediates mentioned above are involved in the base promoted hydrolysis of chloroacetylhydrazide, (171), to give acetate and chloride ions, nitrogen, hydrazine and some acetylhydrazide. On the basis of these products the

---

104. (a) C.L. Bumgardner, K.J. Martin, and J.P. Freeman, J. Amer. Chem. Soc., 85, 97 (1963); (b) C.L. Bumgardner and J.P. Freeman, J. Amer. Chem. Soc., 86, 2233 (1964); (c) R.J. Baumgarten, J. Chem. Educ., 43, 398 (1966).
105. (a) S.G. Cohen and J. Nicholson, J. Org. Chem., 30, 1162 (1965); (b) J. Nicholson and S.G. Cohen, J. Amer. Chem. Soc., 88, 2247 (1966).
106. E.M. Kosower and P.C. Huang, J. Amer. Chem. Soc., 87, 4645 (1965).
107. C.E. McKenna and T.G. Traylor, J. Amer. Chem. Soc., 93, 2313 (1971).
108. E.M. Kosower, Accounts Chem. Res., 4, 193 (1971).

reaction was postulated to involve anion (172) and subsequently - *via* a concerted fragmentation reaction - ketene and diazene; these intermediates would then go on to give the products observed [109].

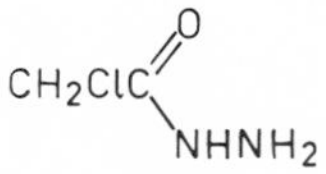

$CH_2ClC(=O)NHNH_2$

(171)

$CH_2ClC(=O)NH\overset{\ominus}{N}H$

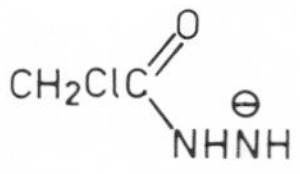

(172)

By traversing the same sequence more than once, even trichloroacetylhydrazide can be reduced all the way to acetic acid. However, it was found that the activation volume of the reaction is negative, so that fragmentation cannot be part of the rate controlling step. It was learned furthermore that N, N-dimethyl substitution does not significantly lower the rate of chloride ion formation, whereas N, N'-dimethyl substitution lowers it by 10^8. The reaction of (171) becomes base independent at a pH of 11, so that the acid producing the anion intermediate must be fairly strong. These various data suggest that the rate controlling step is the formation of the α-lactam (173). A transient, yellow intermediate can be seen at high pH; its spectrum and its rate of disappearance can be determined by means of stopped flow experiments. The spectrum does not match that of ketene, and the rate law for its disappearance (first order) rules out diazene disproportionation. Thus, the only reasonable alternative is acetyldiazene (174) or its corresponding anion. This then goes on to acetate and diazene, and thence to nitrogen and hydrazine, in rapid steps [110]. It may be noted that a few stable α-lactams such as (175) are known, but all are colorless [111].

N-NH2 (173)

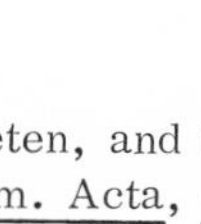

$CH_3C(=O)N{=}NH$ (174)

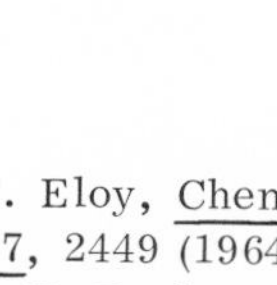

(175)

109. (a) R. Buyle, A. Van Overstraeten, and F. Eloy, Chem. Ind. (London), 839 (1964); (b) R. Buyle, Helv. Chim. Acta, 47, 2449 (1964); (c) H. Paulsen and D. Stoye, Chem. Ber., 99, 908 (1966); (d) H. Paulsen and D. Stoye, "The Chemistry of Amides", J. Zabicky, Ed., Wiley, New York, 1970; Chr. X.
110. (a) W. J. le Noble and Y.-S. Chang, J. Amer. Chem. Soc., 94, 5402 (1972) and J. Chem. Educ., 50, 418 (1973); see also (b) R. Ahmed and J. P. Anselme, Tetrahedron, 28, 4939 (1972).
111. I. Lengyel and J. C. Sheehan, Angew. Chem., Int. Ed. Engl., 7, 25 (1968).

24-8. SINGLET OXYGEN [112]

Oxygen enters many organic reactions, and it usually does so in one of two ways. In the so-called autoxidations an initiator molecule is converted into radicals which start a chain reaction involving oxygen in its triplet ground state. Thus, the commerical oxidation of cumene to phenol and acetone is such a process. The other main group of reactions of oxygen are the sensitized photooxidations; in these reactions a sensitizer such as eosin or methylene blue somehow promotes the oxidation of the organic substrate upon irradiation. In the Schönberg-Schenck mechanism, the sensitizer triplet reacts with the oxygen ground state to give a peroxide of unspecified structure; this "complex" then transfers oxygen to the substrate [113]. The Kautsky mechanism [114] differs in that the excited triplet sensitizes the oxygen; that is, converts it to excited singlet oxygen but does not become bound to it. The singlet oxygen then enters into reaction with the substrate. Kautsky did an experiment which virtually proved his point of view: He absorbed the substrate and the sensitizer on separate silica particles, allowed oxygen to flow through a mixture of these particles and upon irradiation still obtained oxidation products. In spite of this finding the former mechanism was generally adopted until 1964; but before proceding, we should briefly digress to discuss the oxygen molecule [115]. If we combine all the appropriate atomic orbitals to molecular orbitals, a correlation diagram for the approach of two oxygen atoms obtains which is shown in Fig. 24-1. This diagram shows that the ground state has its two

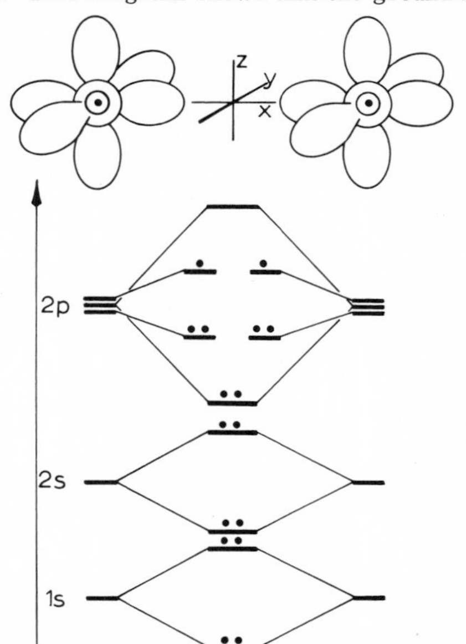

Fig. 24-1. The molecular orbital diagram of oxygen.

112. For review, see C.S. Foote, Accounts Chem. Res., 1, 104 (1968).
113. (a) A. Schönberg, Ann., 518, 299 (1935); (b) G.O. Schenck, Angew. Chem., 69, 579 (1957).
114. (a) H. Kautsky and H. de Bruijn, Naturwissenschaffen, 19, 1043 (1931); see also (b) E.F.J. Duynstee, Chem. Weekblad, 67, 21 (Sept. 10, 1071).
115. For review, see D.R. Kearns, Chem. Rev., 71, 395 (1971).

highest energy electrons in two degenerate orbitals, and hence by Hund's rules the molecule is a ground state triplet ($^3\Sigma$). The diagram also shows that two relatively low lying excited singlet states are available; the $^1\Delta$ state in which the two electrons are in the same two orbitals but with spins paired, and the $^1\Sigma$ state, in which they are paired but one of them is now in the next higher level. These levels are 22 and 37 kcal/mole above the ground state, respectively. The return of the molecule from either state to the ground level is spin forbidden and in the absence of quenchers, the states are indeed long lived; 45 minutes for the $^1\Delta$ state [116] and 7 sec for the $^1\Sigma$ excited molecule [117]. In the presence of other molecules and especially in solution, these life times are much shorter and must be expressed in terms of small fractions of a second [118].

The switch to the Kautsky mechanism began with the discovery that the red chemiluminescence which accompanies the reaction of hydrogen peroxide with hypochlorite was due to a transition in which a pair of $^1\Delta$ oxygen molecules return to the ground state [119]. Indeed, spin conservation requires that the oxygen must be in a singlet state when it is first generated in this reaction; but here was proof that the excited oxygen so produced has a sufficient life time in solutions to undergo a bimolecular relaxation.

Very soon thereafter, Foote showed that the peroxides formed in sensitized photooxidations could equally well be obtained by mixing the substrate, hydrogen peroxide and hypochlorite in a common solvent [120]. Corey generated singlet oxygen by means of a radiofrequency discharge unit and found that the emerging gas was capable - unlike ordinary oxygen - of oxidizing anthracene to the same peroxide (176) that can be obtained by the sensitized photooxidations [121].

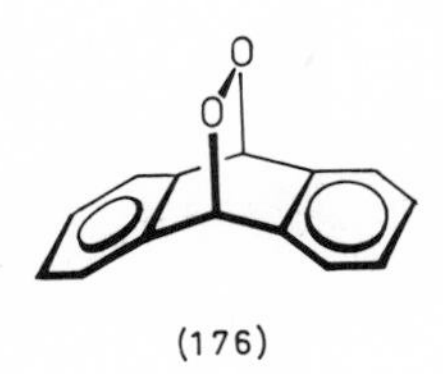

(176)

116. W.H.J. Childs and R. Mecke, Z. Physik, 68, 344 (1931).
117. R.M. Badger, A.C. Wright, and R.F. Whitlock, J. Chem. Phys., 43, 4345 (1965).
118. (a) R.P. Steer, R.A. Ackerman, and J.N. Pitts, J. Chem. Phys., 51, 843 (1969); (b) I.D. Clark and R.P. Wayne, Proc. Roy. Soc. (London), A314, 111 (1969); (c) D.R. Adams and F. Wilkinson, Trans. Faraday Soc. II, 586 (1972).
119. (a) A.U. Khan and M. Kasha, J. Chem. Phys., 40, 605 (1964); (b) S.J. Arnold, E.A. Ogryzlo, and H. Witzke, J. Chem. Phys., 40, 1769 (1964).
120. (a) C.S. Foote and S. Wexler, J. Amer. Chem. Soc., 86, 3880 (1964); (b) C.S. Foote, S. Wexler, W. Ando, and R. Higgins, J. Amer. Chem. Soc., 90, 975 (1968).
121. E.J. Corey and W.C. Taylor, J. Amer. Chem. Soc., 86, 3881 (1964).

It was furthermore shown by Wilson that when two substrates compete for oxygen in a photo-oxidation, the product ratio is the same as when singlet oxygen is used in a dark reaction. She also ruled out another possible alternative - that of oxidation of excited triplet substrates by showing that tetramethylethylene is readily oxidized by singlet oxygen; this olefin has a triplet energy far above that of any of the sensitizers normally used [122].

There also happens to be a substrate which leads to more than one product: Limonene (177) forms no fewer than six hydroperoxides ((178)-(183)). They are obtained in the same ratio whether photooxidation or hydrogen peroxide-hypochlorite exposure is used [112].

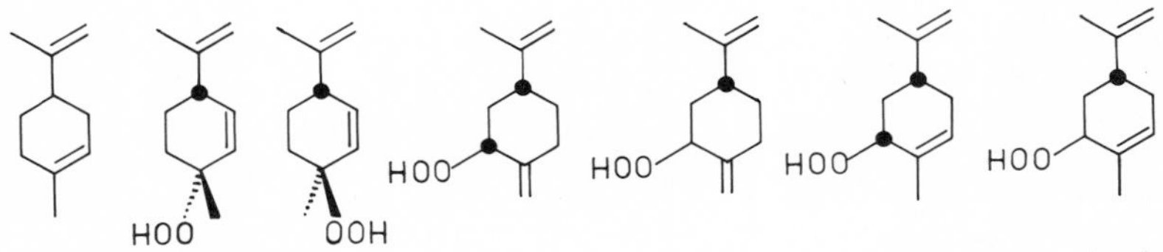

HOO OOH HOO HOO HOO HOO

(177) (178) (179) (180) (181) (182) (183)

After these insights became widely known, an alternative way to prepare singlet oxygen - more suitable perhaps to the needs of organic chemists - was found. It was learned that low temperature ozonization of triphenylphosphite gave a moloxide (184) which decomposes upon warming to -35° into the phosphine oxide and singlet oxygen; in the presence of substrates sensitive to singlet oxygen the appropriate oxidation products were found [123].

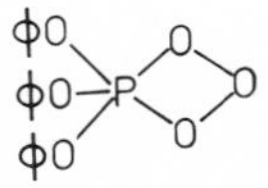

$(\phi O)_3P$ (O–O–O ring)

(184)

Another such reaction is that of peroxyacetylnitrate (185) with hydroxide to give acetate, water, nitrite and singlet oxygen.

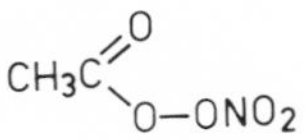

$CH_3C(=O)O{-}ONO_2$

(185)

The importance of this fact is that (185) has been recognized as a constituent in polluted air, and thus singlet oxygen is thereby linked to that problem also [124].

---

122. T. Wilson, J. Amer. Chem. Soc., 88, 2898 (1966).
123. R.W. Murray and M.L. Kaplan, J. Amer. Chem. Soc., 90, 537 (1968) and 91, 5358 (1969).
124. R.P. Steer, K.R. Darnall, and J.N. Pitts, Tetrahedron Lett., 3765 (1969).

Potassium perchromate (K_3CrO_8) requires only lukewarm water to decompose to singlet oxygen and chromate [125]. The detection of singlet oxygen is now also a simple affair. One of the easiest ways rests on the fact that the excited oxygen can bleach the red dye rubrene (186) in the dark to a colorless endo-peroxide (187). Alternatively, esr can be used [126].

(186)

(187)

Singlet oxygen can be efficiently quenched by means of tertiary amines such as 1,4-diazabicyclo[2.2.2] octane, (188) [127]; thus, rubrene is not bleached when such amines are present. The amines themselves do not have a triplet state at 22 kcal/mole or below, nor are they chemically attacked, and the mechanism of the quenching process is not known [127].

(188)

The reactions of singlet oxygen are likely to be those of the $^1\Delta$ state; that is clearly the substance formed in at least some of the experiments quoted above, and it has a much greater life time. However, the question is not yet completely settled. Beside the red luminescence at 44 kcal/mole, 60 kcal emissions (480 nm) have been observed, presumably resulting from $^1\Delta$ plus $^1\Sigma$ deactivations [128], and there are proposals that these emissions do not originate simply in bimolecular collisions, but from an intermediate singlet O_4 ("bimol"). For example, although methylene blue has triplet and singlet energies of 34 and 39 kcal/mole respectively, $^1\Delta$ singlet oxygen is able to excite it, and since ternary collisions are extremely unlikely it would seem that the process must involve O_4 [129]. The problem is further complicated by the fact that certain products may be generated in excited states as well (see below).

---

125. J.W. Peters, J.N. Pitts, I. Rosenthal, and H. Fuhr, J. Amer. Chem. Soc., 94, 4348 (1972).
126. E. Wasserman, R.W. Murray, M.L. Kaplan, and W.A. Yager, J. Amer. Chem. Soc., 90, 4160 (1968).
127. C. Ouannes and T. Wilson, J. Amer. Chem. Soc., 90, 6527 (1968). One often sees this compound referred to as DABCO in the literature.
128. C. Balny, J. Canva, P. Douzou, and J. Bourdon, Photochem. Photobiol., 10, 375 (1969).
129. J. Stauff and H. Fuhr, Ber. Bunsenges. Physik. Chem., 73, 245 (1969).

The reactions of singlet oxygen are numerous. Many olefins such as (189) give rise to allylic hydroperoxides such as (190); these products can usually also be obtained by radical chains involving triplet oxygen [112].

(189) 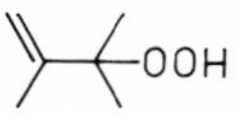 (190)

Dienes and polyaromatic species may give peroxides resulting from [2 + 4] cycloaddition. Anthracene and rubrene were already mentioned, but there are others. Thus, the furan (191) in methanol gives (192), probably *via* an *endo*-peroxide, and tetracyclone (193) gives the diketone (194) presumably in the same way [112].

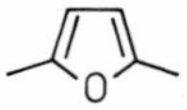

(191) 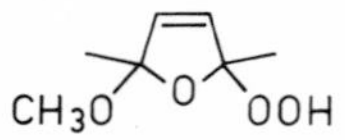 (192) 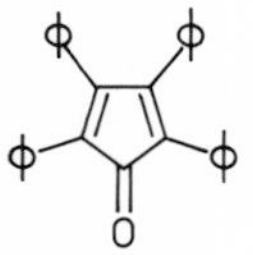 (193) 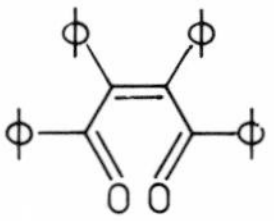 (194)

In a number of cases, molecules with activated double bonds but without allylic hydrogen atoms are broken altogether. Thus, enamines lead to amides plus ketones and vinyl ethers to esters and ketones. It was found that these reactions begin with a [2 + 2] cycloaddition followed by a like reversion. For instance, 10,10'-dimethyl-9,9'-biacridylidene (195) upon treatment with singlet oxygen at -78° gives the corresponding acridone (196) in its singlet excited state, and fluorescence is observed; this is as would be expected from the dioxetane intermediate (197) [130].

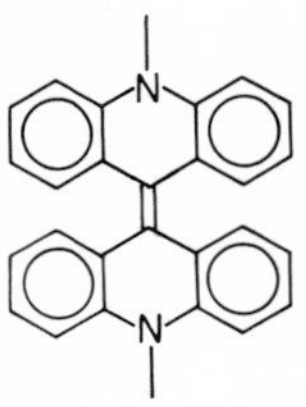

(195) 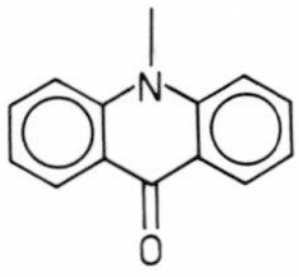 (196) 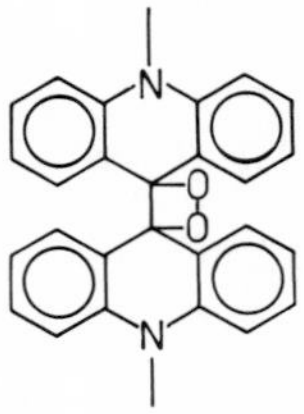 (197)

Similar observations were reported for bifluorenylidene (198) [131].

---

130. F. McCapra and R.A. Hann, *Chem. Commun.*, 442 (1969).
131. W.H. Richardson and V. Hodge, *J. Org. Chem.*, *35*, 1216 (1970).

(198)

In some cases the dioxetanes have proved to be isolable. Cis- and trans-diethoxyethylene (199) and (200) at -78° give dioxetanes (201) and (202), respectively, the nmr spectra of which are clearly different; but both give ethyl formate (203) upon warming [132].

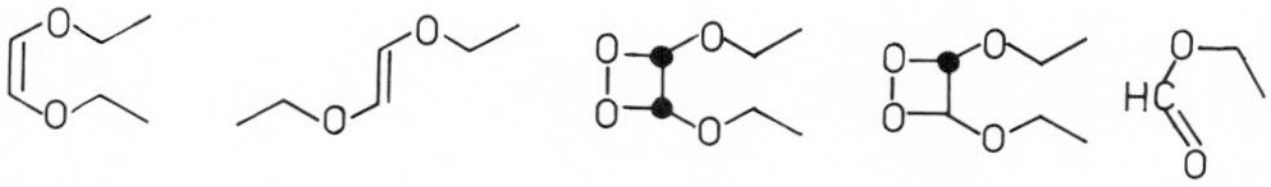

(199) (200) (201) (202) (203)

p-Dioxene (204) and 1,3-dioxole (205) give the dioxetanes (206) and (207), respectively; these decompose at reasonable rates only at 60°. These decompositions to diformate esters are chemiluminescent, and suitable additives can be made to fluoresce [133].

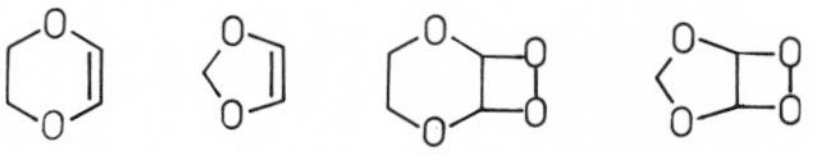

(204) (205) (206) (207)

Tetramethoxyethylene (208) likewise gives a stable dioxetane, (209) which decomposes only slowly to dimethyl carbonate (210) at 60° [134]; (209) is a ketal of Rauhut's carbon dioxide dimer (211) (see p. 457).

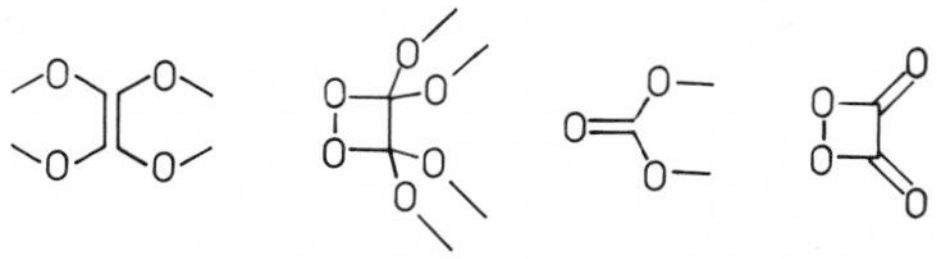

(208) (209) (210) (211)

Another close relative of this compound is the intermediate (212), which is involved in the singlet oxygen oxidation of diphenylketene [135].

---

132. P. D. Bartlett and A. P. Schaap, J. Amer. Chem. Soc., 92, 3223 (1970).
133. A. P. Schaap, Tetrahedron Lett., 1757 (1971).
134. S. Mazur and C. S. Foote, J. Amer. Chem. Soc., 92, 3225 (1970).
135. L. J. Bollyky, J. Amer. Chem. Soc., 92, 3230 (1970).

(212)

Tetrathioethylenes such as (213) are oxidized to give dithiooxalates and disulfides (214). A dioxetane may still be involved, although no chemiluminescence is observed and no fluorescence can be induced [136].

(213) (214)

An exceptional case is that of the non-activated olefin adamantylideneadamantane (215) (see also p. 709). It has four allylic hydrogen atoms, but the allylic hydroperoxides violate Bredt's rule and hence their formation is clearly out of the question here. A dioxetane (216) is formed which can be isolated and which has a melting point of 163°. It explodes at 240°. A refluxing glycol solution is strongly luminescent; the product is adamantanone [137].

(215) (216)

Beside the [2 + 2] and [2 + 4] cycloadditions of singlet oxygen already discussed above, there is some evidence that a [2 + 6] analog may occur as well. Tropilidene can be photo-oxidized to a mixture which upon reduction yields (217) as one of the products, suggesting that the bicyclic peroxide (218) was one of the initial components of the oxidation mixture [138].

(217) (218)

---

136. W. Adam and J.C. Liu, J. Amer. Chem. Soc., 94, 1206 (1972).
137. J.H. Wieringa, J. Strating, H. Wynberg, and W. Adam, Tetrahedron Lett., 169 (1972).
138. A.S. Kende and J.Y.C. Chu, Tetrahedron Lett., 4837 (1970).

Among other singlet oxygen reactions are the oxidation of nitrogen to nitrate, [139] of sulfides to sulfoxides [140] and of certain methyl groups such as those of tropinone (219) and pseudopelletierine (220) to aldehyde groups [141]; this latter reaction, although achieved by sensitized photooxidation and inhibited by known singlet oxygen quenchers, does not occur with hydrogen peroxide hypochlorite mixtures, however.

(219)  (220)

Finally, it is beginning to appear that singlet oxygen may even have a role in free radical autoxidations. In these and other chain processes the mechanistically oriented organic chemist often glibly assumes that there are always some unknown radicals R around that will initiate the chain, but the true nature of this adventitious radical is usually a mystery. Rawls and van Santen [142] found that singlet oxygen is an excellent initiator for the autoxidation of fatty acid esters such as methyl linoleate (221). The conjugated peroxide (222) was formed; this then thermally decomposes to provide the initiating radicals.

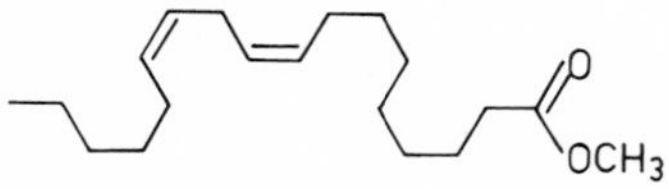

(221) 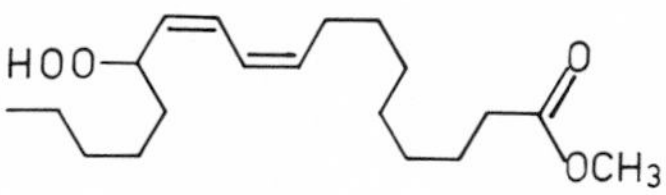 (222)

Thus, it turns out that singlet oxygen scavengers such as tetramethylethylene and tetracyclone inhibit autoxidation. The authors pointed out that in most natural products low concentration of pigments are present that may serve as the sensitizers in the photo-oxidative production of singlet oxygen.

24-9. SOLVATED ELECTRONS [143]

It is well known that alkali metals can be reversibly dissolved in liquid ammonia to give blue (dilute) to copper colored (concentrated) solutions, and that such

139. M. Anbar, J. Amer. Chem. Soc., 88, 5924 (1966).
140. C.S. Foote and J.W. Peters, J. Amer. Chem. Soc., 93, 3795 (1971).
141. M.H. Fisch, J.C. Gramain, and J.A. Oleson, Chem. Commun., 13 (1970).
142. H.R. Rawls and P.J. van Santen, Tetrahedron Lett., 1675 (1968).
143. For review, see (a) S.R. Logan, J. Chem. Educ., 44, 344 (1967); (b) K. Eiben, Angew. Chem., Int. Ed. Engl., 9, 619 (1970).

solutions serve as reducing agents, for instance, in the Birch reduction process (see also p. 663). The metal atoms in these solutions are dissociated to ions and free electrons; the esr spectrum of the latter can be determined. The additions of certain catalysts such as ferric ion promotes the reaction of these electrons to give colorless amide ions and hydrogen:

$$e^{\ominus} + NH_3 \rightleftharpoons NH_2^{\ominus} + 1/2\, H_2$$

The reaction is reversible [144]; thus, the same color and esr signal can be produced by bubbling hydrogen into a sodamide solution in liquid ammonia; the equilibrium constant for the above process is about 10^5. By measuring the temperature and pressure dependences of the esr signal intensity, one can measure the thermodynamic properties characteristic of the reaction; thus, it has been learned for instance from the large, negative volume change that the electron in ammonia is located in a large cavity in that liquid with a 3Å radius [145].

Hydrated electrons [146] can best be generated by the radiolysis of water. Proving the presence of such a species has proved difficult [147] since these solutions are very unstable, especially if they are acidic; the electrons rapidly combine with protons to give hydrogen atoms which can to some degree mimic their reducing properties (it is now known that hydrated electrons displace chloride ion from chloroacetic acid whereas hydrogen atoms abstract hydrogen from it) [148]. However, it could be shown from the effect of the ionic strength on the rate of reduction that one of the reducing species had a charge of -1 [147]. Furthermore, by means of a flash technique the visible - uv spectrum has become known; it is similar to that in ammonia. With this knowledge, certain scavengers have been identified such as H^+, O_2, CO_2, H_2O_2, N_2O and so on. These electrons have been used to partially reduce such metal ions as Zn^{++} and Mn^{++} to the monopositive species, the spectra of which could subsequently be observed. One remarkable difference between hydrated and ammoniated electrons is that the former are located in a much smaller cavity (perhaps 1 Å [149]. Irradiation of various solid salts such as sodium chloride also produces free trapped electrons called F-centers [150]).

---

144. E.J. Kirschke and W.L. Jolly, Science, 147, 45 (1965).
145. (a) U. Schindewolf, Angew. Chem., Int. Ed. Engl., 6, 575 (1967); (b) U. Schindewolf, R. Vogelsgesang, and K.W. Böddeker, Angew. Chem., Int. Ed. Engl., 6, 1076 (1967); (c) K.W. Böddeker, G. Lang, and U. Schindewolf, Angew. Chem., Int. Ed. Engl., 8, 138 (1969).
146. G. Lepoutre and J. Jortner, J. Phys. Chem., 76, 683 (1972).
147. E.J. Hart, Science, 146, 19 (1964).
148. E. Hayon and A.O. Allen, J. Phys. Chem., 65, 2181 (1961).
149. R.R. Hentz and D.W. Brazier, J. Chem. Phys., 54, 2777 (1971).
150. (a) R.B. Gordon, Amer. Scientist, 47, 361 (1959); (b) M.J. Blandamer, L. Shields, and M.C.R. Symons, J. Chem. Soc., 3759 (1965).

It is now known that the photolysis of certain aqueous anions such as phenoxide [151] and iodide [152] produces hydrated electrons and that these can be used in chemical synthesis; thus, benzonitriles have been reduced to benzaldehydes that way [153], and a general phenol to aniline conversion depends on them [154]. For those experiments that must be done in organic media, it is possible to prepare stable, blue solutions of electrons at preparative concentrations in hexamethylphosphoramide. Such solutions have esr signals split into 35 lines, presumably due to interactions with four nearby solvent molecules and six more distant ones (phosphorus splittings) [155]; however, others have claimed that this signal is due to a naphthalene impurity [156] (see also p. 658).

Even dielectrons ($e_2^=$) are now discussed as a possibility. Thus, concentrated solutions of alkali metals are less magnetic than expected. Concentrated aqueous sodium hydroxide glasses irradiated at 77°K and warmed to 120°K to remove trapped electrons as judged by color and esr, redevelop these signals when quickly recooled to 77°; the interpretation is that electron pairs trapped in a single cavity slowly diffuse apart [157]. Efforts to find the expected spectrum of such dielectrons to date have failed, however [158].

---

151. J. Jortner, M. Ottolenghi, and G. Stein, J. Amer. Chem. Soc., 85, 2712 (1963).
152. J.G. Calvert and J.N. Pitts, "Photochemistry", Wiley, New York, 1966; p. 271.
153. J.P. Ferris and F.R. Antonucci, Chem. Commun., 1294 (1971).
154. R.A. Rossi and J.F. Bunnett, J. Org. Chem., 37, 3570 (1972).
155. H.L.J. Chen and M. Bersohn, J. Amer. Chem. Soc., 88, 2663 (1966).
156. R. Catterall, L.P. Stodulski, and M.C.R. Symons, J. Chem. Soc. A, 437 (1968).
157. J. Zimbrick and L. Kevan, J. Amer. Chem. Soc., 89, 2483 (1967).
158. G. Czapski and E. Peled, Chem. Commun., 1303 (1970).

AUTHOR INDEX*

*Reference numbers are in parentheses.

E

F

G

H

K

L

M

N

Q

R

T

W

Y

Z

SUBJECT INDEX

C

D

E

I

J

K

Q

R

S

T

U

V